Sustainable Horticulture
Today and Tomorrow

Raymond P. Poincelot
Fairfield University, Connecticut

Prentice Hall

Upper Saddle River, New Jersey 07458

Library of Congress Cataloging-in-Publication Data

Poincelot, Raymond P., 1944-
    Sustainable horticulture : today and tomorrow / Raymond P. Poincelot.
        p. cm.
    ISBN 0-13-618554-1
    Incudes bibliographical references and index.
        1. Sustainable horticulture.   I. Title.

SB319.95 .P65 2003
333.76'16—dc21   2002034598

Editor-in-Chief: Stephen Helba
Executive Editor: Debbie Yarnell
Associate Editor: Kimberly Yehle
Production Editor: Lori Dalberg, Carlisle Publishers Services
Production Liaison: Janice Stangel
Director of Manufacturing and Production: Bruce Johnson
Managing Editor: Mary Carnis
Marketing Manager: Jimmy Stephens
Manufacturing Buyer: Cathleen Petersen
Design Director: Cheryl Asherman
Senior Design Coordinator: Miguel Ortiz
Cover Designer: Amy Rosen
Cover Images: Raymond P. Poincelot
Composition and Interior Design: Carlisle Communications, Ltd.
Printing and Binding: Courier Westford

Pearson Education LTD.
Pearson Education Australia PTY, Limited
Pearson Education Singapore, Pte, Ltd.
Pearson Education North Asia Ltd.
Pearson Education Canada, Ltd.
Pearson Educacíaon de Mexico, S.A. de C.V.
Pearson Education—Japan
Pearson Education Malaysia, Pte. Ltd.

10 9 8 7 6 5 4 3 2 1
ISBN 0-13-618554-1

Contents

Chapter 3   *Plant Parts and Functions*   63

Chapter 4   *Plant Processes*   98

Chapter 5 *Plant Development*   125

Chapter 6   *Plants and Their Environment*   159

SECTION  *Three*

INTRODUCTION TO HORTICULTURAL PRACTICES   201

Chapter 7   *Plant Propagation*   203

Chapter 8   *Conventional Horticultural Crop Breeding*   250

Chapter 9   *Biotechnology*   280

Chapter 12   *Home Horticulture Basics*   418

Chapter 13   *Managing Enclosed Plant Environments*   465

Chapter 14   Controlling Growth by Physical, Biological, and Chemical Means   497

Chapter 15   Plant Protection   529

Chapter 18   *Fruits and Nuts*   761

Preface

This book is a comprehensive introduction to the emerging discipline of sustainable horticulture. Students with minimal background in the plant sciences or those students with horticultural experience will both find this book a useful introduction to all aspects of sustainable horticulture. This textbook can be used for a one- or two-semester course in sustainable horticulture for either nonscience students or students majoring in horticulture or a related area. Education, extension, and research professionals interested in sustainable horticulture will find this book to be a good introduction and resource handbook for their professional activities.

The first section, meant to be an introduction to sustainable horticulture, consists of Chapter 1. This chapter helps the reader to understand the driving force behind sustainable horticulture and its *raison d'être*. All students, regardless of their background, should read this chapter, which sets the tone for the rest of the book.

In the second section (Chapters 2 through 6), the student is introduced to the foundations of horticultural science that underlie all forms of horticulture, from conventional through sustainable to organic. Little or no background in science is needed to understand this section. Students with botanical, horticultural, or plant science backgrounds can bypass this first section, except for Chapter 6. Chapter 6 emphasizes environmental interactions and their effects on plant growth. It also offers a thorough coverage of soil properties and the importance of organic matter. It is recommended that students with plant science backgrounds look over the summaries for Chapters 2 through 5 and read Chapter 6 before moving to the third section.

The science behind horticulture, once mastered, enables the student to comprehend the factual basis for applied horticultural science, that is, the horticultural practices in the third section. The emphasis in Section Three is on sustainable approaches to practicing horticulture. Sustainable practices are environmentally sound, resource conserving, economically reasonable, and socially acceptable ways to practice horticulture today and many tomorrows from now. In logical sequence this leads the student to the fourth section, horticultural production, with its special emphasis on horticultural crops. The sections on horticultural practices and horticultural production are sufficiently detailed to provide a stimulating challenge to those students who already possess some background in horticultural science.

My belief is that sustainable horticulture can only be understood and properly practiced when it is seen as the sum of the foregoing four parts. However, a holistic appreciation of horticulture depends on the final weaving of one more element into the horticultural fabric: artistry. Landscaping and ornamental horticulture are not complete without the artistic touch.

It is my hope that the student will achieve both an understanding and appreciation of sustainable horticulture through the use of this text. The following text features will help to achieve the foregoing purpose and set it apart from other horticultural texts.

General Features

1. Designed for Learning. I have attempted a logical progression of informative chapters that moves from basic horticultural science to sustainable horticultural practices and finally to crops and their cultivation. Connections are made between science and practice to help explain the "why" of plants or sustainable activities. For example, the two types of photosynthesis (C_3 and C_4) are made more understandable when the competition between Kentucky bluegrass and crabgrass is shown to be a C_3 and C_4 photosynthetic battle. My goal is to maintain readability throughout, even in the most complex and difficult areas of science such as photosynthesis or biotechnology. Difficult concepts are presented in a concise, simplified manner without sacrificing accuracy. The flow is often initially general followed by specific detail and examples. I have endeavored to provide definitions (italicized entries) and examples throughout to enhance comprehension and to maintain reading flow. Many figures, photographs, and tables are included to illustrate, explain, and expand on subjects. Coverage is especially heavy for practices, equipment, and horticultural crops.

My goal is to accelerate learning and to improve retention. Each chapter provides several features to facilitate the learning process. All chapters first present the reader with a list of key topics to be covered and a list of learning objectives for the chapter. To reinforce the learning process for each chapter, I have summarized key materials at the chapter's end. Immediately after the summary is a list of exercises (key questions) that test your new knowledge and refer you back for further study, if needed.

Additional learning aids and a means to explore chapter topics in greater depth are also found at the end of the chapter. A list of Internet resource sites helps you learn and explore. Each site was checked for content and reliability at the time of writing this book. Sites were not always found for all key topics or were sometimes deemed unacceptable. If you experience difficulty with any site over time, please let me know by e-mail (see end of preface). Even the best of sites can move to a new URL, or not be kept up, or vanish altogether. An extensive bibliography is given at the conclusion of each chapter for an in-depth exploration of the chapter's topics. The listings are complete and recent, except for certain older books that have become "horticultural classics." The range is such as to include both simple and highly technical references to satisfy the needs of all students.

2. Contemporary Nature. The latest published research and research in progress on sustainable ways to produce horticultural crops are both included. For example, the environmental impact of conventional horticulture in the United States, the events that led to the newly emerging field of sustainable horticulture, the emergence of low-input and sustainable ways to grow crops, improved ways to utilize natural resources, natural control of plant pests, the advent of genetically modified organisms, and increased energy conservation are just some of the recent developments cited here. This text will take you on a contemporary, ground-floor tour through horticulture as it undergoes the transformation from nonsustainability to sustainability.

3. Comprehensive, but Flexible. Much subject matter is presented overall in order to give a detailed and comprehensive picture of sustainable horticulture. While some might view the basic sciences in the second section as unnecessary, horticultural practices are better understood and managed when their scientific underpinnings are known. The four sections provide sufficient material for a two-semester course in sustainable horticulture. It is also possible to conduct a one-semester course by concentrating on Sections One, Three, and Four.

4. Scientifically Correct, but Pragmatic. Plants are cited by common and scientific names. Plant nomenclature is based on the authority established by *Hortus Third: A Concise Dictionary of Plants Cultivated in the United States and Canada* (the staff of the Liberty Hyde Bailey Hortorium, Macmillan Publishing Co., Inc., 1976). In certain cases, more recent specialized publications were utilized. Scientific names were used for insects and diseases, as well as for plants, whenever possible. Pesticides and growth regulators are cited by chemical and trade or common names. Although Chapters 2 through 6 contain basic science, Chapters 7 through 18 contain pragmatic sustainable practices and plant cultural information that follow from, and are directly related to, the earlier science, leading to a better understanding of these applied practices.

5. **Balanced Coverage.**   This book provides a national approach to practices and horticultural crops, unlike the somewhat regionalized coverage provided in some other books. Coverage of both home gardening and commercial horticulture is presented to satisfy the needs of both types of horticulturists.

Some division exists between horticulturists favoring the "organic" or natural approach versus those favoring the "chemical" approach. Both sides are presented here. The author believes both sides have pros and cons, and the best path is the blending of the good from both into an integrated, sustainable horticultural practice. Technology from the simplest to the most complex is included. For example, conventional plant breeding and biotechnology receive equal treatment.

Crops are covered uniformly and completely: fruits, herbs, nuts, ornamentals, and vegetables. All are important to horticulture. City, suburban, and country horticulture are discussed. All exist and the well-trained horticulturist must have a knowledge of all of them.

Specific Features

1. **Section One:**   (Introduction to Sustainable Horticulture.) Chapter 1 explores the historical foundations of sustainable agriculture, horticultural implications, the problems of current conventional horticulture, and what steps are needed to approach sustainability. This chapter also provides the compelling case for why sustainable horticulture has emerged in the professional and public domains. It also answers the question of what exactly is sustainable horticulture.

2. **Section Two:**   (Introduction to Horticultural Science.) This section consists of Chapters 2 through 6, which contain the basic plant science behind horticulture and its practices. First is plant classification (Chapter 2) with special emphasis on scientific classification and taxonomy as it applies to horticulture. The naming and classification of plants is presented as simply as possible to make it easier to understand. Horticultural classification, a valuable communication tool, is also examined. Plant structures (Chapter 3) are next with special attention being placed on macrostructures rather than microstructures. Plant metabolic processes (Chapter 4), particularly those associated with good productivity of plants, come next, followed by plant development (Chapter 5). Emphasis is placed on macroaspects: vegetative and reproductive cycles. The section concludes with the relationships between environmental factors and how the plant responds to them (Chapter 6). Soils, organic matter, and erosion receive special attention. Completion of this section enables the student to understand why certain practices are used and how they influence sustainability.

3. **Section Three:**   (Introduction to Horticultural Practices.) This section contains the applied aspects and is divided into nine chapters. We begin with comprehensive coverage of asexual and sexual propagation and interim care (Chapter 7), a basic and important horticultural practice. Propagation leads into conventional plant breeding (Chapter 8), the source of new horticultural cultivars. Biotechnology and its application to the production of genetically modified plants (Chapter 9) is covered next because of its current potential and its critical role in tomorrow's horticulture. Next energy conservation (Chapter 10), now a critical factor in horticultural management, is given special attention. Chapter 11 provides an in-depth analysis of sustainable environmental and resource management practices from a commercial perspective. Chapter 12 does the same, but from a home garden viewpoint. Chapter 13 looks at indoor environments, greenhouses, and homes and their sustainable management. Chapter 14 concentrates on the control of plant development, with a focus on pruning practices. Plant protection (Chapter 15) comes next and provides an examination of vectors causing plant problems and how to treat them. Treatments stress IPM and organic approaches.

4. **Section Four:** (Introduction to Horticultural Production.) This section deals largely with horticultural crops, which are grouped by ornamentals (Chapter 16), vegetables and herbs (Chapter 17), and fruits and nuts (Chapter 18). Chapter 17 includes a discussion of herbs and plants grown for natural products, which are sometimes ignored or minimized in other texts. Career opportunities in the horticultural industries associated with these crops are explored. The economic significance of these fields and their basic thrust is examined. A large amount of plant cultural information is concisely expressed in this last section.

Comments are welcome. If possible I will incorporate responses in future editions of this text. For those who wish to communicate by e-mail, I am available at rpoincelot@mail. fairfield.edu. My closing thought is, hopefully, your inspiration. The fields of horticulture are a joy to the eye and mind, but a deeper appreciation is yours only through questioning, study, and involvement.

Acknowledgments

My special thanks go to Betsy, who with her love and encouragement, helped me through the long, busy preparation of this text. Thank you to my grown children, Ray, Daniel, and Wendy; Daniel's wife, Tina; and Betsy's son, Josh, for adding to my life. To my first grandson, Jacob Scott: Your future is why I wrote this book.

I thank all those professionals, friends, and students who discussed horticulture with me. You are too numerous to be cited here, but your thoughts can be found here. Thank you to Laura and the Biology Department for their patience.

I thank the following professors who critically reviewed parts or all of the entire manuscript: April Hill, Fairfield University; Richard Durham, University of Kentucky; Charles Francis, University of Nebraska; Christopher Gunter, Purdue University; John Preece, Southern Illinois University; Judith Hough-Goldstein, University of Delaware; Mary Ann Gowdy, University of Missouri-Columbia; Lewis Jett, University of Missouri; Richard Evans, U.C. Davis; Steve Gustafson, University of Tennessee at Martin; Edward W. Bush, Louisiana State University; Robert R. Dockery, Western Piedmont Community College; Brian Maynard, University of Rhode Island; and Vern Grubinger, University of Vermont-Extension.

Many thanks to those individuals, organizations, institutions, and industries that supplied photographs or permission to quote from their publications.

Finally, my thanks to all those people at Prentice-Hall, Inc., who helped make this book possible: Sandy Hakanson, Janice Stangel, Charles Stewart, Jr., Debbie Yarnell, and Kimberly Yehle. Thanks also to Lori Dalberg at Carlisle Publishers Services.

Raymond P. Poincelot

SECTION One

INTRODUCTION TO SUSTAINABLE HORTICULTURE

This section consists of one chapter that provides an introduction to sustainable horticulture. As a discipline, sustainable horticulture is very new. Although the sustainability concept for agriculture originated sometime in the late 1970s, the activity in sustainable agriculture started in the 1980s. Sustainable horticulture appeared somewhat later, when the emphasis shifted from grains, forestry, and animal husbandry to horticultural crops. Given the newness of the discipline, an introduction is appropriate. One needs to understand the problems that brought about the shift to sustainability, the history of the transition process, and the actual changes that occurred.

This chapter starts with an exploration of the roots of agriculture and horticulture and moves on to the recent history of the period that gave rise to sustainable agriculture/horticulture. Next the chapter examines what sustainable horticulture really involves. Sustainability ultimately applies to resources, the environment, profitability, and sociological needs. The question of why the need exists is examined in the context of what is wrong with conventional horticulture. The chapter culminates with a plan to bring sustainability to horticulture. Regardless of what is covered in the textbook for instructional activity, this chapter is a must. It sets the stage for the chapters that follow.

Chapter I

Moving to Sustainability

OVERVIEW

Horticulture is old, tracing its beginnings back some 10,000 years ago to the dawn of agriculture. Horticulture was reasonably sustainable until the American Civil War. Pesticides and fertilizers were natural products and little energy other than human and horse or mule sweat was used. Sustainability decreased after the Civil War with the advent of chemical fertilizers and even more in the 1940s with the advent of synthetic pesticides and the introduction of ever more diesel- and gasoline-dependent machinery to replace human and animal labor. Monocultures and less replacement of organic matter dealt sustainability a final blow. Although down, sustainability was not out, in that it survived on the fringe as organic horticulture. Today we are seeing a return to sustainability, driven in part by the loss of environmental quality and diminishing resources resulting from unsustainable practices.

CHAPTER CONTENTS

Roots of Agriculture and Horticulture
History of Sustainable Agriculture/Horticulture
Defining Sustainable Agriculture/Horticulture

Threats to Environmental and Resource
Sustainability
Toward a More Sustainable Horticulture

OBJECTIVES

- ❧ To understand the historical events that occurred prior to and during the movement from conventional to sustainable horticulture
- ❧ To define the term *sustainable horticulture* in a reality-based manner so that farmers, productivity, profits, the environment, and society all benefit

- ❧ To understand how the current practices work against environmental and resource sustainability
- ❧ To learn how we need to modify horticultural practices, devise new policies, and bring about sociological changes in order to achieve horticultural sustainability

ROOTS OF AGRICULTURE AND HORTICULTURE

In terms of historical agriculture, U.S. agriculture is but a blip on the timescale. Agriculture's roots possibly extend as far back as 10,000 years ago, when the hunter-gatherers switched to crop production. The origins of well-organized systems of crop production can be traced back to the Middle East around 7,000 years ago. Native Americans probably had agricultural operations as early as 7,000 years ago.

In our earliest settlements, American farmers worked hard to achieve self-sufficiency and to expand the acreage of quality cropland. Later colonists/farmers started to produce food for sale to others that helped improve the quality of rural life and helped bring about population centers that became cities. George Washington and Thomas Jefferson, seeking to better farm practices, became advocates for new farming practices. Washington promoted crop rotations, composted livestock wastes, and better planting practices. Jefferson constantly looked for new crops that better served the farmer and consumer. Farming began to spread westward from the original colonies as better soils were discovered.

The Golden Age

Most of the accelerated development in U.S. agriculture has occurred since the Civil War. Improved plows and mechanical planters and reapers appeared after the Civil War and helped raise production. Chemical fertilizers, pesticides, and big fuel-driven, labor-saving machines become available after World War II and pushed our productivity and technology into a great leap forward. Plant breeding efforts and government policies from the 1950s on drove the productivity level even higher. All of this activity was characterized first as the Agricultural Revolution and then the Green Revolution. Farms increased in size and, as the corporate farm became the norm, small family farms no longer dominated agriculture. Lately, biotechnology promises to push the technology envelope even further.

Crop and livestock production has more than doubled in the last 50 years. One farmer supplies the food and fiber needs for 78 people. One hour of today's farm labor is equivalent to 14 hours of 1920s farm work. We lead the world in food production, both for our own citizens and for export. Our agricultural development during its short existence has brought about greater productivity and more advanced technology than all the world's agricultural efforts during the last 6,600 years!

We owe much of our success and greatness to an agricultural juggernaut second to none. No one wants to return to an agricultural system that required many hours of back-breaking labor by much greater numbers of farmers in order to feed a greatly reduced population. From a societal perspective, these changes were beneficial.

Seeds of Discontent

In the last few decades, uneasiness with horticultural practices began to surface. Some people started to question what this rapid-developmental success cost us in terms of resources, environmental wellness, and societal values: topsoil depletion, contamination of water by farm chemicals, energy concerns, the disappearance of family farms, poor working conditions and living standards for farm laborers, increasing costs of production, and the deplorable state of economic and social conditions in rural communities. Actually, those who advocated and practiced organic farming had been raising questions about resources and environmental wellness since the 1940s. The difference was that now the concerns were coming from more mainstream sources and even friends of conventional agriculture. Sustainable agriculture traces its roots to those concerns and questions.

HISTORY OF SUSTAINABLE AGRICULTURE/HORTICULTURE

The first seeds of dissent appeared in a 1911 book, *Farmers of Forty Centuries*. Its author, Franklin King, was opposed to the agricultural dogma that was gaining ground. The popular view was that mined and manufactured fertilizers would bring about great increases in agricultural productivity. King felt that the older practices were the key to retaining soil health and permanence. Sound familiar? One could imagine just changing the date to our times and the material would be current. King, of course, lost.

The biodynamic farming and organic agriculture movements influenced the roots of sustainable agriculture/horticulture. Biodynamic farming, organic farming, and other forms of farming are defined on the Web (see the Internet Resources section at the end of the chapter for definitions). Biodynamic farming was publicly started by Rudolf Steiner in 1924. Ehrenfried Pfeiffer first popularized it in the United States in 1938 through his book, *Bio-Dynamic Farming and Gardening*. Organic agriculture got its start in England. The first reference to organic agriculture is attributed to Lord (Baron) Walter Ernest Christopher James Northbourne in a 1940 publication entitled *Look to the Land*. Ironically, Lord Northbourne was urging his English countrymen to switch from organic methods to the use of newly emerging farm practices, chemical fertilizers, and pesticides. The motivation was to increase food production because of World War II. A second person in England also influenced the organic agriculture movement in the United States, Lady Evelyn Barbara Balfour. She authored a 1943 book called *The Living Soil*.

The organic farming movement in the United States was likely influenced also by an American publication, *Five Acres and Independence*, written by M. G. Kains in 1935. Kains favored the small farm and the utilization of practices that became the mainstay of organic farming: composting and, the use of green manure, cover crops, and farmyard manure. Jerome I. Rodale founded the organic farming movement in the United States. His first publication in 1945 was *Pay Dirt: Farming and Gardening with Composts*. Sir Albert Howard, who is credited with the development of the Indore composting process, heavily influenced Rodale. In fact, Sir Howard wrote the foreword in this particular Rodale book and also wrote an earlier 1940 seminal book on soil health, *An Agricultural Testament*. A later Rodale book, *The Organic Front,* appeared in 1948. In this work, Rodale refers to the movement as *organiculture*. Rodale went on to found the magazine called *Organic Gardening and Farming,* and is remembered as the father of organic farming in the United States. His son, Robert Rodale, continued his work until his untimely death in Russia in 1990. The well-known Rodale Institute, a place for organic farming research, helped to place and legitimize organic farming in public and scientific circles.

During the 1940s other individuals were unhappy with the changes going on in agriculture, such as Edward Hubert Faulkner, who in 1943 published *Plowman's Folly*. Faulkner found fault with plowing, and especially the bad effects of the moldboard plow on soil health. Conservation tillage likely got its start with Faulkner. Louis Bromfield, author of the 1945 and 1948 publications, *Pleasant Valley* and *Malabar Farm,* disliked the direction farming was going with its fight against nature, rather than using nature as an ally. Bromfield favored traditional agricultural practices that were being abandoned in favor of the new, science-driven agriculture. He advocated the value of working with nature through the use of practices to restore soils and to maintain soil productivity: crop rotations and animal manners. He also bemoaned the changing attitudes of farmers as stewards that cared for their land and animals and avoided abuses of natural resources. Commercialization and economic forces were changing the perspective of the farmer. While the title "father of sustainable agriculture" has not been assigned to anyone, Louis Bromfield is a likely candidate. Another voice was that of Aldo Leopold in 1949, author of *A Sand County Almanac and Sketches Here and There*. Leopold was an advocate for caring and respecting the land; he felt that a land and conservation ethic should be a product of social evolution.

Voices on the Rise

Other voices were heard in the 1950s, but the nosiest battle against mainstream agriculture was that waged by J. I. Rodale on behalf of organic farming. Voices of dissent were often ignored or disparaged during the 1950s and 1960s, given the fantastic productivity and the entrenchment that came with great success of the agricultural establishment. The next landmark to move us closer to sustainable agriculture was a book called *Silent Spring,* written by Rachel Carson in 1962. Carson developed a chilling case against the dangers of pesticides. In 1964, Beatrice Hunter, became a proponent of alternate methods to pesticides with her book *Gardening without Poisons.* Many scientists were also researching alternative methods for insect control, primarily biological control. Some of them of note include Harry S. Smith, Robert Van den Bosch, and C. B. Huffaker. Numerous publications appeared and helped influence the movement toward integrated pest management (IPM). One that I feel was influential was *Biological Control* by Robert Van den Bosch and P. S. Messenger, published in 1973. Another early book that likely influenced the sustainable agriculture movement was the Meadows, Randers, and Behrens book *The Limits to Growth,* which pointed out the limitations of resources, including agricultural inputs. Agricultural policy also came under criticism in 1977 with the publication of Wendell Berry's classic book, *The Unsettling of America: Culture and Agriculture.* Berry believed that existing policy did not favor responsible stewardship of the land.

The likely first use of the term *sustainable agriculture* in print appeared in 1978 in the publication *Toward a Sustainable Agriculture.* This book is the proceedings of the First conference of the International Federation of Organic Agriculture Movements (IFOAM). The late Robert Rodale also attributes the use of the term *sustainable agriculture* in conversations during the late 1970s to Lady Balfour. Organic agriculture also received a boost in respectability with the appearance of R.C. Oelhaf's *Organic Agriculture: Economic and Ecological Comparisons with Conventional Methods.* This 1978 publication made an interesting case for farming organically. In 1979 Charles Walters, Jr., and C. J. Fenzau introduced a primer on ecoagriculture. Their book, *An ACRES U.S.A. Primer,* borrowed heavily from the organic perspective and based its practices on good science. In that same year, Jeavons advocated a system of biodynamic French intensive farming. Organic agriculture finally achieved mainstream legitimacy in 1980 when the U.S. Department of Agriculture (USDA) in cooperation with the Rodale Institute released the famous report, *Report and Recommendations on Organic Farming.* This report offered an overview of organic farming, including status, methods, implications for environmental and food quality, economic assessment, and research recommendations. Also in 1980, Wes Jackson's book *New Roots for Agriculture* offered new directions for agriculture.

The Battle Wages

The 1980s were a period of frenzied activity for sustainable agriculture. It was characterized by a flurry of protesting voices and maneuvers to seize the high ground with new forms of agriculture. Feelings were sometimes bruised as attacks and counterattacks were mounted. Things were somewhat confused as each party sought to define the new agriculture from their perspective. As a result, many new forms of agriculture were named, defined, and advocated (Table 1–1). Sustainable agriculture competed with agroecology, biodynamics, biological agriculture, ecological agriculture, environmentally sound agriculture, low-input sustainable agriculture, organic farming, permaculture, and regenerative agriculture.

It is beyond the scope of this book to provide a detailed history and complete list of publications and players during this tumultuous period. The reader is encouraged to use the references and Web sites listed at the chapter's end to find more details. My only intent is to give some highlights.

First, let's deal with the bruised feelings. Established agriculture does have some remarkable achievements to its credit. These pluses were sometimes lost in the discussion about what was wrong in conventional agriculture. Clearly, the United States had achieved dramatic levels of agricultural productivity, including surpluses that helped offset partially our chronic trade imbalances. U.S. consumers have an abundance of food choices at reasonable prices. The dominance of the United States in terms of global power results from the combined productivity and efforts of the combined agricultural/industrial/military complex.

Agriculture was used to "pats on the back" and much praise from the government and consumers. It came as somewhat of a shock then when the collective voices of complaint became loud enough to garner some political and social attention. Let's face it. When you

TABLE 1–1   Agricultural Systems, Approaches, and Tools

Type	Origin	Philosophy
Agroecology	Michael Altieri/U.S.	A system that promotes ecological and social sensitivity in agriculture and places high emphasis on ecological sustainability of the production system; pays close attention to ecological phenomena in the field
Alternative agriculture	Garth Youngberg/U.S.	Nontraditional approaches and unconventional systems of agriculture
Biodynamic farming	Rudolf Steiner/Austria	Shares some elements with organic farming, but also has a spiritual component and herbal additives to guide composting
Biological/ecological farming	Europe/Far East	Involves minimal to no chemical pesticides and has elements of biodynamic, natural, organic, and sustainable approaches
Conventional	U.S.	Reliance on chemical-intensive approaches to pests and fertilization, heavy monoculture, highly mechanized and industrialized, often large farms
Integrated pest management (IPM)	U.S.	Practice that relies heavily on natural, biological, and cultural practices with minimal emphasis on pesticides and a goal of restoring natural ecosystems where natural controls keep pests in check
Kyusei nature farming	Japan	Similar to organic farming, but also employs microbial inoculates to enhance microbial diversity of the soil
Low-input sustainable agriculture	USDA	Similar to conventional, but with moderate to high reductions of chemical inputs; some IPM usage
Organic farming	J. I. Rodale/U.S. IFOAM/Europe	Use of no manufactured chemicals, reliance on natural materials for fertilization, use of natural and biological controls for pests, mechanical cultivation for weeds, special attention to soil organic matter maintenance, crop rotations, cover crops, green manure, compost
Permaculture	Bill Mollison/Australia	A carefully planned landscape from garden to farm level that utilizes mixtures of plants, especially herbaceous and woody perennials (for permanence), animals, humans, and structures for low-maintenance, efficient crop production
Regenerative agriculture	Robert Rodale/U.S.	Emphasis on regeneration of renewable resources and a relationship to economic and social concerns
Sustainable agriculture	Many in U.S. and Europe	Varies from being essentially organic farming to systems that use some to all of the practices common to organic farming and some chemical inputs and IPM

are steeped in a system that appears to work well and is accepted by most, it is hard to deal with attacks on your life's work. The natural response is to think that the attackers don't know what they are talking about. You first try to discredit them or steam-roll them. When that tactic doesn't work, you tend to get nasty. Of course, this only heats up the dialogue and tends to make the other side hit harder. Civility and constructive debate tend to suffer. It becomes hard to acknowledge problems and to deal with them in an efficient manner. At some point, enough pressure is applied and the system begrudgingly starts to deal constructively with the need for some change.

On the other hand, the bearers of problems lose sight of the fact that criticism is hard to take, regardless of the manner in which it is offered. Sometimes the attacks are constructive, sometimes destructive. Many voices are speaking, not one. Power for change is diluted. The players for change jockey for power, each believing that his or her solution is the best. The opposition does not have a united front, making it easier for the established group to discredit, disparage, and ignore. In time, the players compromise and offer a united position with a common solution. The sticking point is usually the agreement on the wording of a solution, as each side tries to protect its turf. The final solution incorporates the main points of each side, so that each player can point to his or her part of the product. Only then can enough pressure be brought to bear such that constructive change becomes possible.

This painful period for sustainable and established agriculture lasted roughly one decade, from 1980 to 1990. One key event was the 1982 conference, Agricultural Sustainability in a Changing World Order, that was cosponsored by the Food, Land and Power Program at Pomona College and the W. K. Kellogg Foundation. This conference brought together several key players. The conference addressed the questions of why current agricultural practices were not sustainable and what steps were needed to achieve a sustainable system. Many different viewpoints were evident at this meeting, as can be gleaned from the 1984 proceedings (see references).

A second key conference, also sponsored by the W. K. Kellogg Foundation, was held at the University of Florida in 1982. The conference, Agriculture, Change and Human Values, dealt with the issues and social trends that have formed both the U.S. and world agricultural communities. Other areas included human/social aspects of farming, sustainability, environmental impact, and nutrition. This conference also published proceedings in 1984.

Here to Stay

Ultimately, the Agriculture, Change and Human Values conference led to an academic journal, *Agriculture and Human Values* (first issue published in 1984), and the Agriculture, Food and Human Values Society. A third development spearheaded by Garth Youngberg was the establishment of the Institute of Alternative Agriculture (later called the Henry A. Wallace Institute for Alternative Agriculture and now the Henry A. Wallace Center for Agricultural Policy at Winrock International) in 1983, which published *Alternative Agriculture* news. In turn, this event led to the 1986 publication of their academic journal, the *American Journal of Alternative Agriculture*. A third academic journal, the *Journal of Sustainable Agriculture*, founded by me in 1989, appeared in 1990. The appearance of academic journals signified a key change in the debate about sustainable agriculture. Such events generally mean that the area now has academic credibility and that interest is high enough to support a subscription base. The above journals are full of materials relating to agriculture during the transition period to the present. Some existing academic journals also contained articles of relevance to sustainable agriculture, such as *Advances in Agronomy, Rural Sociology, Journal of Soil and Water Conservation,* and *Biological Agriculture and Horticulture.*

A number of books also appeared during this time period. Again, only highlights are used here to illustrate the "flavor" of the times. In 1981 Neil Sampson's book *Farmland or Wasteland: A Time to Choose* opened the dialogue. Two early books in 1983 were *Sustainable Food Systems* (Dietrich Knorr, editor) and *Environmentally Sound Agriculture* (William Lockeretz, editor). The former was more concerned with sustainability after the farm gate and the latter with sustainable farm production. Also in 1983 Pye, Patrick, and Quarles in *Groundwater Contamination in the United States* covered the contribution of agriculture to groundwater pollution. While it first appeared in 1983 as an extension publication in Vermont, the more noted version of *The Soul of Soil: A Guide to Ecological Soil Management* (Gershuny and Smillie) appeared in 1986.

In 1984, Jackson, Berry, and Colman edited *Meeting the Expectations of the Land.* Also in 1984 Bezdicek and Powers edited *Organic Farming: Current Technology and Its Role in a Sustainable Agriculture.* Japan's Fukuoka brought his natural farming philosophy to our attention with an American publication in 1985. In 1986 my own thoughts on the problems of agriculture and where we should go appeared as *Toward a More Sustainable Agriculture.* Also in 1986, *New Directions for Agriculture and Agricultural Research* and *State of the World: A Worldwatch Institute Report on Progress Toward a Sustainable Society* appeared (edited by K. A. Dahlberg and written by Brown et al., respectively). In 1987 M. A. Altieri carefully presented ecological agriculture in *Agroecology: The Scientific Basis of Alternative Agriculture.* The case for conservation tillage was well presented by Little in 1987 with *Green Fields Forever.* Postel made a case against agricultural pesticides in 1987 with *Defusing the Toxic Threat.*

A large number of books have appeared in the 1990s that promote either sustainable agriculture or some subset of it, such as agroecology, biodynamic farming, or organic farming. The case for alternative agriculture was put forth in 1989 by the National Research Council's publication titled *Alternative Agriculture.* In 1990, the Soil and Water Conservation Society published *Sustainable Agricultural Systems,* and Francis, Flora, and King edited *Sustainable Agriculture in the Temperate Zones.* The USDA connected the environment to agriculture in its 1991 yearbook, *Agriculture and the Environment.* Hudson and Harsch of the USDA also presented the basic principles of sustainable agriculture in 1991. Callaway and Francis proposed the breeding of crops for sustainable agriculture in 1993 with *Crop Improvement for Sustainable Agriculture. Sustainable Agriculture Systems* by Hatfield and Karlen appeared in 1994. These five publications brought sustainable agriculture into the arena as a serious contender for guiding the future of agriculture. Its acceptance in academic sectors was ensured, leading to a guide for teaching sustainable agriculture in 1991 from the Center for Integrated Agricultural Systems located in Madison, Wisconsin.

The case for permaculture was summarized best in Mollison's 1990 book, *Permaculture: A Practical Guide for a Sustainable Future,* and for biodynamic agriculture by Sattler and Wistinghausen's 1992 book, *Bio-Dynamic Farming Practice.* Organic farming was promoted by Schwenke (*Successful Small-Scale Farming: An Organic Approach,* 1991) and Coleman (*The New Organic Grower: A Master's Manual of Tools and Techniques for the Home and Market Gardener,* 1995). Smith et al. (1994 and 1998) covered both organic and low-input farming with *The Real Dirt: Farmers Tell about Organic and Low-Input Practices in the Northeast.* Ecological approaches to farming were covered by Bender (1992), Soule and Piper (1992) and Gliessman (1998) with *Future Harvest: Pesticide-Free Farming, Farming in Nature's Image: An Ecological Approach to Agriculture,* and *Agroecology: Ecological Processes in Sustainable Agriculture,* respectively. A natural tie-in to sustainable and organic farming is community-supported agriculture (CSA), where local residents pay farmers for a guaranteed share of the harvest. A good guide to CSA was published in 1998 by the Center for Sustainable Living.

During this same time frame, a number of institutions helped shape and guide dialogue, policy, and research concerning sustainable agriculture. Some of these places have been around for a number of years, such as the Land Institute in Salina, Kansas, the Rodale Institute and the Rodale Research Center in Kutztown, Pennsylvania, and the World Resources Institute in Washington, D.C. Others are newer, such as the Henry A. Wallace Center for Agricultural Policy at Winrock International (established 1983) and the Leopold Center for Sustainable Agriculture (established 1987). Each of these institutes has published numerous materials pertaining to some aspect of sustainable agriculture. The Land Institute focuses on the marriage of ecology and agriculture, the Rodale Institute and Rodale Research Center are heavily involved with research on organic and regenerative agriculture, and the World Resources Institute is involved with resource monitoring and conservation. The Henry A. Wallace Center for Agricultural Policy at Winrock International is concerned with research, evaluation and policy analysis to promote the development of and to shape favorable policies on sustainable agriculture. The Leopold Center for Sustainable Agriculture is involved with research and information on responsible stewardship of farm resources and farm profitability.

The USDA and Congress were also active during this period. Faced with resource and environmental problems associated with agriculture, Congress responded to the challenge and pressure. In 1985, Congress enacted the National Agricultural Research, Extension and Teaching Policy Act Amendments. Under this act, the USDA established the Low Input Sustainable Agriculture program. In 1990 this program was reauthorized under Title XVI (Subtitle B) of the Food, Agriculture, Conservation and Trade Act (FACTA). The program's name was changed to the Sustainable Agriculture Research and Education Program (SARE). SARE was very active as of the late 1990s, and currently with competitive research and education grants and likely will remain so, unless severe budget cuts curtail its activity. The United States is divided into four SARE regions: Northeast, North Central, Southern, and Western. Two information centers on sustainable agriculture were also established: the Alternative Farming Systems Information Center (AFSIC) and the Appropriate Technology Transfer for Rural Areas (ATTRA). The former is located within the National Agricultural Library in Beltsville, Maryland, and is an excellent information resource on all things associated with alternative and sustainable agriculture. ATTRA is in Fayetteville, Arkansas, and is a great resource for farmers and others interested in enhancing the rural community through sustainable agricultural development. The various states' Cooperative Extension Services are now also offering workshops and other training to those farmers interested in sustainable agriculture.

Also by the end of 1990, a number of state (land grant) and private universities offered courses, options, programs, and even a bachelor degree in sustainable agriculture. A few also have master's and doctorate programs in sustainable agriculture. Currently, educational and internship opportunities exist in large numbers throughout the United States (see the Internet Resources at the end of this chapter).

Defining Sustainable Agriculture/Horticulture

What is sustainable agriculture/horticulture? This question is not an easy one to answer. The problem has its roots in issues of "turf protection," different perceptions, and degree of exclusivity. Established or conventional agriculture, because of problems associated with resource overuse and abuse, was in need of change. The question became one of how much change. Would the alteration of existing practices suffice? Is a radical change needed such that the entire system is rejected and replaced with a new paradigm? Most in the existing system would say some changes would suffice. Some outside the system would say a new order is needed.

The Foundation

Very early in the game, three sets of philosophy emerged: economics, ecology, and sociology. These sets have also been placed in the context of schools of thought: productivity, stewardship, and community. Many of those operating within the conventional agricultural system were economically driven. Any changes would have to work in the economic sense. Can my costs for changing practices be recovered within the system such that the system remains profitable? In all fairness, this concept cannot be ignored. The farmer has to make a living just like anyone else. Farmers need to feel that the return on their work and efforts is fair and equitable. From the sociological standpoint, these questions arose: Do these changes increase the amount of income that is spent on food? Do these changes reduce my choice and variety of food items? From the ecological perspective, the question is do these choices decrease the detrimental effects on resources and the environment?

Clearly, we could create agricultural systems that embody one of these concepts. For example, we can design an agricultural system that preserves environmental quality, conserves resources, and uses as many renewable resources as technology allows. But one has to take reality into consideration. Is this system profitable enough to be adopted on a widespread level? Will food prices rise too high such that sociological/political pressures become a serious consideration? One might say that the current system, conventional agriculture, was designed for profitability at the expense of the environment, but did work reasonably well in the sociological/political context for many years. On the other hand, we can design systems that produce cheap food only to the detriment of the agricultural workers who become an underclass in society. The hard part is to design a system that satisfies all realities—economic, ecological, and sociological/political. Compromise becomes necessary and, while all parties might not be happy, they can live with the system.

Contenders for the High Ground

Some systems did exist alongside conventional agriculture, such as organic agriculture, permaculture, and biodynamic agriculture. These systems used practices that had less impact on resources and the environment. All had their advocates, but could these systems be "sold" to all parties as the new order? Organic farming clearly has advocates in the United States and Canada and is profitable, because it has some consumer acceptance. In some countries, like New Zealand and the Netherlands, organic agriculture has a solid economic base. Permaculture has a following in Australia and somewhat in the United States, as does biodynamic farming. Still, many in conventional agriculture resist adoption of the practices that constitute organic farming or other alternative systems. Whether rightly or wrongly, many farmers and others in the system do not believe that an all-organic agriculture can work in an economic and sociological/political context. Similarly, biodynamic agriculture and permaculture have strong roots outside the United States, but are unlikely to be accepted as the new paradigm for U.S. and Canadian agriculture.

Given this situation, other paradigms were proposed: biological agriculture, ecological agriculture, alternative agriculture, regenerative agriculture, and low-input sustainable agriculture. Advocates of biological and ecological agriculture embrace many of the concepts found in organic farming, but have moved beyond just the adoption of a set of practices. While somewhat oversimplified, biological agriculture is an offshoot of the organic farming movement in Europe, and ecological farming is an American movement.

Rather than looking at practices as a way of maximizing a single crop or a system component, instead practitioners of biological/ecological agriculture attempt to deal with biotic interactions and the ultimate optimization of the entire ecosystem. Proponents have also sought to expand the scientific bases behind their system. By moving away from the term *organic* and grounding their system in science, these agricultural forms have distanced themselves from the "emotional baggage" and accusations of "not enough good science" associated with the term *organic agriculture*. While this direction helped to avoid the negative connotations that resulted from years of acrimonious debate between proponents of organic and conventional agriculture, it failed to capture the hearts and minds of conventional agriculturists. Essentially, these systems are a fresh start, a redefinition and repackaging, an attempt to expand the acceptability of organic practices with more consideration given to scientific and social aspects.

These systems also met the fate of organic agriculture in the United States. Wholesale adoption by conventional agriculture is unlikely. Another offshoot of the organic movement in the United States resulted in the term *regenerative agriculture*. This approach marked a move toward the regeneration of renewable resources that was believed to be the most critical component for sustaining agriculture. It was also felt that the idea of a regenerated agriculture would connote the idea of renewal and reconstruction that would have economic and social appeal, making it acceptable to farmers and urban dwellers. Regenerative agriculture did not capture mainstream attention. If its proponent, Robert Rodale, had not died prematurely, one wonders whether the term would have ultimately come to dominate the agricultural scene.

Alternative agriculture involved the concept of bringing together a number of players under one umbrella. The idea was that farmers would have a choice of systems, that is, any alternative set of practices to conventional agriculture. The problem here became one of perception. While it brought together organic farming, ecological agriculture, biodynamic agriculture, and permaculture and gave strength in numbers to these systems, it failed to become acceptable to conventional agricultural. Rightly or wrongly, conventional farmers and their supporters viewed this umbrella as a "slap in the face." Their form of farming was so unacceptable that an "alternative" was needed. Conventional agriculture was disenfranchised by this approach.

The conventional system offered its own attempt at resolving the problem: low-input sustainable agriculture (LISA). It was meant to be an umbrella term that could embrace organic farming and some forms of alternative agricultural, while allowing conventional farmers to make some changes and be included under the umbrella. For example, if a conventional farmer decreased the amount of fertilizer used and added some integrated pest management practices, the farmer could qualify as a LISA farmer. While well intentioned, the term ultimately failed to become completely accepted. Part of the problem is that the umbrella term was somewhat confusing. Organic farming is clearly a sustainable form of agriculture and does use low inputs (no chemicals), but does sometimes require more labor (higher input) to achieve suitable productivity. Essentially, organic farming does not quite fit under LISA.

The Compromise

The ultimate compromise was to accept the concept of sustainable agriculture. The term captures hearts and minds. Who would be against sustaining agriculture? It is as American as apple pie. Actually, it has global acceptability. The term is a very big umbrella. Essentially, it allows all players to reside within the term, from organic farmers to alternative agriculturists to conventional farmers changing practices from conventional tillage to low- or no-till practices, from pesticides to IPM, and from heavy to low applications of fertilizers. While no one is totally happy, it does allow each group to lay claim to sustainable agriculture, to protect its "turf," and to proselytize. Essentially, it is a win–win situation. Each group is free to pursue its goals and attempt to dominate the national system.

What Is It?

So what is sustainable agriculture? It depends on whom you ask. A number of definitions have been proposed, some by individuals and others by groups. A few examples will suffice. In 1988, Edwards proposed that sustainable agriculture use "integrated systems of agricultural production less dependent on high inputs of energy and synthetic chemicals and more management-intensive than conventional monocultural systems. These systems maintain, or very slightly decrease productivity, maintain or increase net income for the farmer, are ecologically desirable, and protect the environment."

Several definitions were put forth in 1989–1990 (see Internet Resources at the end of this chapter for thorough coverage of definitions and references). The American Society of Agronomy defined sustainable agriculture as "one that, over the long-term, enhances environmental quality and the resource base on which agriculture depends, provides for basic human food and fiber needs, is economically viable, and enhances the quality of life for farmers and society as a whole." Francis and Youngberg in 1990 defined it as follows: "Sustainable agriculture is a philosophy based on human goals and on understanding the long-term impact of our activities on the environment and on other species. Use of this philosophy guides our application of prior experience and the latest scientific advances to create integrated, resource-conserving, equitable farming systems. These systems reduce environmental degradation, maintain agricultural productivity, promote economic viability in both the short and long term, and maintain stable rural communities and quality of life." The *Congressional Record* defines sustainable agriculture as "an integrated system of plant and animal production practices having a site specific application that will, over the long-term: satisfy human food and fiber needs, enhance environmental quality and the natural resources base on which the agricultural economy depends, make the most efficient use of nonrenewable resources and on-farm resources and integrate, where appropriate, natural biological cycles and controls, sustain the economic viability of farm operations and enhance the quality of life for farmers and society as a whole." The *Journal of Sustainable Agriculture* in its statement of purpose in the first issue in 1990 defined the term as follows: "Sustainable agriculture is a system in which: (1) resources are kept in balance with their use through conservation, recycling and/or renewal, (2) practices preserve agricultural resources and prevent environmental damage to the farm and offsite land, water and air, (3) production, profits, and incentives remain at acceptable levels, and (4) the system works in concert with socioeconomic realities."

All definitions incorporate the themes of maintaining productivity, protecting environmental quality and ecological soundness, and socioeconomic soundness. All definitions accommodate the schools of thought that lead to the formation of sustainable agriculture: productivity, stewardship, and community. While we may not agree on what the perfect definition is, we do agree with goals that must be met if we are to have a sustainable agriculture. Given the great diversity of agriculture, both in terms of crops, practices, and personal philosophies and the variability of climates, soils, and geographic regions, it is perhaps easy to understand why we cannot arrive at one definition fits all! In a sense, we are moving toward a sustainable agriculture (Figure 1–1), but we may never arrive at a totally sustainable agriculture in the United States, or indeed, worldwide.

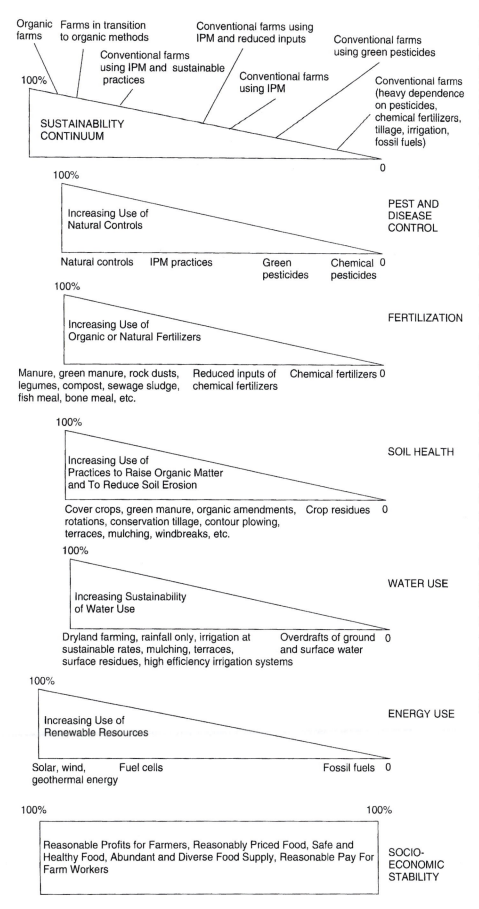

FIGURE 1-1 The sustainability continuum and horticultural crop production. A number of parameters are involved in sustainable horticulture. Some have the potential to be highly unsustainable, such as pest and disease control, fertilization, soil health, water use and energy use. Ideally, we would like to approach 100% sustainability for these parameters. For all, except for energy use, some existing forms of sustainable agriculture are closing in on this goal. Organic farming is closest, but even this form has a dependence on fossil fuels, hence none have reached the end of the sustainability continuum. One parameter, socio-economic stability, essentially is and must be close to 100% for any existing form of horticulture. Social, political, and economic realities make it mandatory. Given the advances in fuel cell technologies, we are likely to move ahead in future years to a higher point on the continuum. As energy prices rise and alternative renewable forms of energy then become competitively priced, we will close the final gap.

THREATS TO ENVIRONMENTAL AND RESOURCE SUSTAINABILITY

The field of sustainable agriculture/horticulture is still in its infancy. Best estimates are that about 5% of U.S. farmers use mostly sustainable practices. As such, many current agricultural practices continue to threaten the sustainability of the energy, soil, and water resource bases and do cause off-site pollution. Horticultural resources are diminished by overuse and environmental contamination and sometimes by forces external to agricultural systems. Plants can be damaged by environmental pollution. Let us examine these challenges to the resource base.

Soil Problems

Losses of agricultural soil are disturbing. The average annual rates of soil erosion by wind and water on U.S. cropland approach 1 and 2 billion metric tons, respectively. Wind erosion prevails in the Great Plains states, whereas water erosion is spread throughout the United States, affecting some 167, 168, and 54 million hectares (413, 415, and 133 million acres, respectively) of cropland, rangeland, and pastureland, respectively. The average yearly rate of water erosion for cropland is 12.3 metric tons/hectare (4.7 tons/acre), but in some areas the loss is far worse. Estimates indicate that U.S. soil is eroding at an average rate of 10 to 16 times that of soil formation. Estimates are that 44% of our cropland is eroding faster than soil loss tolerances.

These average numbers are for all forms of agriculture. Horticulture does contribute, but how much? Most U.S. soil erosion arises from cattle grazing or the production of crops to feed cattle. Still, some 14% of soil erosion results from the horticultural production of crops grown for human consumption. In California, the horticultural producer of about 50% of our vegetables and fruits, soil is eroding at some 80 times the rate of soil formation. Every year of horticultural production means the loss of enough topsoil to fill a line of dump trucks 178,850 miles (286,160 kilometers) long! The line of dump trucks would be long enough in 15 months to stretch from the earth to our moon at its closest point! Looked at in this perspective, horticulturists might feel that they are better stewards of the land than most other agriculturists, but we clearly can't rest on our laurels!

Another soil complication resulting from wind and water erosion is desertification. This process occurs in arid to semiarid regions and results in the conversion of productive cropland to unproductive desert. The U.S. Southwest is experiencing severe to very severe desertification, and the northcentral area, moderate desertification. Again, much of this process results from cattle grazing, but some arises from the production of irrigated vegetable and crop production.

Even in horticulture, unabated erosion and desertification can bring about lower crop yields. Current estimates are that erosion reduces crop productivity by 15% to 30%. Soil erosion causes loss of organic matter, silt, and clay, resulting in decreased capacity to retain water and nutrients. Plant nutrients are lost, soil structure deteriorates, and rooting depth is reduced. When topsoils are deep, little short-term change in crop productivity is noted. However, some studies demonstrate an eventual fall in crop yields on fertilized, eroded soils; decreases become larger as soils go from slightly to severely eroded. Yield losses at present erosion rates are projected at 3% to 4% one hundred years from now.

The price of erosion goes beyond actual and future food loss. Crop yields are currently maintained, but only with additional nutrients and water at an annual cost of $1.1 to $8.0 billion. Costs of $135 million are attributed to the water pollution resulting from eroded sediment. The total direct and indirect costs for losses and damage from soil erosion and runoff have been estimated to be as high as $25 billion annually.

Another factor to keep in mind is the historical context of erosion. Unabated erosion and desertification are associated with the downfall of some past civilizations. The Fertile Crescent, Mesopotamia, was a thriving civilization that supported large numbers of people in a high-quality area of cropland. Today, this area of the Middle East supports fewer people on a largely desert-like area. Similar soil declines also took place in Pre-Columbian southwestern United States and Central America. A similar process is occurring in parts of Africa today, such as Chad, Ethiopia, Somalia, and Sudan, leading to lower crop production, famine, and the failing of these countries.

Competition for farmland also threatens our access to agricultural soils. From 1958 to 1977, 8% of our cropland and 18% of our forestland were lost. Only pastureland and rangeland increased, showing a 12% gain. Prime farmland is especially jeopardized. Annually, 400,000 hectares (988,000 acres) were lost to nonfarm use during this same period. The level terrain and good drainage so crucial for growing crops are equally attractive to developers and builders. Farmland loss slowed during the 1990s. A 3% reduction in farmland was seen from 1987 to 1997. Urbanization of farmland appears to have slowed to a less threatening level.

Another problem is soil compaction. Heavy farm machinery with multiple passes in the same area over time can cause soil compaction. This creates a densely packed soil layer (traffic pan) that cause the loss of micropores. In turn, this hinders water movement, resulting in standing water, and increases the potential for erosion losses by water. Compacted soils warm slower, resulting in delayed, poorer germination. Organic matter also decomposes slower in colder soils and nitrogen losses increase. Root development and water/nutrient availability are also reduced. Ultimately, all of this reduces crop yields. Sometimes yields can be maintained, but additional irrigation and fertilizer are needed. Production costs must increase to compensate for soil compaction.

Water Consumption

The sustainability of water resources is especially critical. Agriculture uses 41% of the water in the United States. Water use or withdrawal and water consumption are two different measures. *Water use* refers to taking water and applying it for a purpose, such as irrigation or drinking. When water after use is not returned to its source, such as a surface water body or groundwater, the correct term is *water consumption*. Agricultural activities account for some 80% to 85% of all water consumption, or roughly 400 billion liters daily (106 billion gallons). About 94% of this total is used for irrigation and 4% for livestock. Irrigation occurs on about 18.2 million hectares (45 million acres) of our cropland and requires 8.3% of the farm energy budget. Crops produced by irrigation account for about one-fourth of the total dollar value of U.S. crops. In horticulture, vegetable and fruit producers are the main users of irrigation.

Most irrigated cropland is found in 17 western states that have limited surface water. In the West, irrigation accounts for 85% of all water use. Nationally, 60% of our irrigation is derived from surface water and 40% from groundwater, but in some western states groundwater provides up to 94% of the water budget. Sustainability is threatened in that the withdrawal of groundwater exceeds its replenishment by 25% in the United States as a whole. Groundwater withdrawal occurs in California, a regional of heavy fruit and vegetable production. Arizona also has overdraft problems associated with some horticultural crops. In some parts of the Gulf states, the overdraft is as high as 77%. Near Lubbock, Texas, withdrawal is about 50 times the recharge rate! Roughly half the acreage irrigated with groundwater is in areas where aquifers are declining at least 1 foot per year.

Future demands for water also threaten horticultural sustainability. The need for higher crop production for increased population might raise irrigation needs. In addition, modest increases in water needs can be expected from the general public and industry. Competition for water and its cost may also increase.

Energy Consumption

Horticulture is heavily dependent on energy usage. High yield levels with few farm workers are made possible by the use of heavy machinery. The machinery aids in the manipulation of soil and water as well as crop cultivation and harvesting. The use of this machinery does have a downside. Soil compaction can result and energy supplies are not bottomless, at least when fossil fuels are considered. Soil compaction has been covered under soil problems. Let's turn to energy.

A profile of U.S. agriculture reveals some energy-intensive practices (see Chapter 10). Although energy expenditures on fertilizers, farm machinery, tillage, irrigation, crop drying, and pesticides are substantial, these uses represent a small fraction of the entire energy expended by this country. Even within the food system, more than 80% of the energy is used for processing and distribution. Although after-farm energy consumption is beyond the scope of horticulture, such data help to put energy consumption for crop production into proper perspective.

Data on energy-intensive practices combined with energy audits on crops lead to strategies for saving energy. However, efficient policy development for energy conservation, applied research, and improved technologies also requires careful analyses of comparative energy expenditures for crops. Policy determinations must also be based on current data for types of products, geographical regions, acreage, and agricultural methods. For horticulturists, the big surprise is that vegetable and fruit production is the third most energy-intensive form of crop production. Much of the energy required is a result of heavy inputs of fertilizers, pesticides, fuels for machinery, and irrigation.

Hardship situations on a regional basis can be readily identified. At first glance, the national energy expenditure for greenhouse heating (see Chapter 10) is minimal. But on closer examination, the total data set reveals that the major expenditure in the north for the greenhouse industry is for heat.

Sometimes total energy information masks other problems. For example, fertilizers are the single most energy-intensive input (see Chapter 10). Most of the energy is needed to chemically process nitrogen. Energy conservation or using the biological process to fix nitrogen into forms usable by plants such as legumes, would appear to provide part of the solution. But consider phosphate fertilizer, which requires much less energy than nitrogen for processing. Worldwide usable supplies are limited to less than a century at current use. To make fertilizer supply sustainable is both an energy and a natural resource problem.

These kinds of information provide guidelines for determining what horticultural areas should receive high priority for energy conservation or for changes in crop production. Such choices must be balanced against socioeconomic, political, and dietary constraints, and the resulting solutions will not be easy ones to implement.

Water Pollution

Paradoxically, agriculture itself is the most widespread cause of non-point-source water pollution. A recent report from the U.S. Environmental Protection Agency (EPA) cited runoff from agricultural lands as the biggest source of pollution in our water bodies. The major agricultural pollutant is sediment from erosion of cropland, but poor management of fertilizers, irrigation runoff, pesticides, and animal wastes contribute substantially. About two-thirds of the pollutants of streams arises from agricultural activities and slightly less for lakes. The specific contribution from horticultural activities as part of agriculture is difficult to quantify, but is certainly a contributing factor.

Erosion from croplands contributes roughly one-half of the sediment in our waterways. On a daily basis, the Mississippi River carries 1.5 million tons of sediment to the ocean. Much of the soil comes from farms raising grains. The other half arises from timber cutting, construction, and mining. Sediment threatens our water ecosystems and their usefulness. Off-site costs are difficult to assess, but the cropland contribution is estimated at $2.2 billion annually.

Runoff from cropland and recycled irrigation water contains various salts from fertilizers, wastes, and even soil. This salinity causes crop damage, reduced yields, metal corrosion, soils unsuited for crops, and reduced water quality for drinking and agricultural uses. Nitrate is of special concern because of infant nitrate sensitivity and the link to carcinogenic nitrosamines. Excessive phosphorus and nitrogen lead to the eutrophication of surface waters, a decline in fish populations, and reduced recreational quality. A prominent example is the Kesterson National Wildlife Refuge in California, where the influx of various irrigation contaminants severely affected wildlife and the aquatic environment. Another example is the area in the Gulf of Mexico near the entry of the Mississippi River, where nutrients from upstream agriculture have created a dead zone resulting from low oxygen (hypoxia) levels. This hypoxic area is currently about the size of New Jersey each spring when runoff peaks. While it shrinks at other times and used to disappear in the fall, it now persists year-round.

Pathogens in runoff from feedlots, from land treated with sewage sludge or manure, and from areas where manure is stored can pose a hazard to water supplies. A survey of 10 states indicated that human and animal wastes were one of the top three contaminants of groundwater.

Pesticides from runoff and leaching also appear in surface and groundwater. Stream and groundwater contamination from pesticides has been found in most states. From 1991 to 1995 the U.S. Geological Survey tested for 84 pesticides in 5,000 samples collected from 20 major hydrologic basins. At least one pesticide was found in every stream and in one-half of the tested wells. Each stream sample had an average of seven to eight pesticides. Pesticides were from agricultural, surburban, and urban sources. Many were chemicals used in horticultural applications on the commercial and home levels. In California, pesticides are found in the drinking water of some 1 million people.

Saltwater contamination of freshwater has been produced by overdrafts of surface or groundwater for agriculture. Such contamination has been observed in California, Florida, Georgia, Louisiana, New Mexico, and Texas. These overdrafts have also produced a related problem: settling of the earth. Subsidence has damaged agricultural facilities in California and Texas.

Agricultural water usage also impacts riparian habitats through conversion of wild habitat to farmland, water diversion, and the pollution of existing riparian habitats with sediment from erosion and contamination by pesticides and fertilizers. Riparian habitats are important for the preservation of wildlife, genetic resources, preservation of natural enemies for biological control, buffer zones for farms, and maintenance of water quality.

Pesticides

Pesticide usage in crop production is considerable. Usage has grown heavily from the 1940s until now. As of the early 2000s, some 1–2 billion pounds of pesticides were used in agriculture each year. In terms of easier math, that is roughly up to eight pounds per person per year. Farmers have some 25,000 different products to choose from and spend about $4 billion annually on such products.

Four major crops drive the numbers. About 70% and 80% of insecticide and herbicide usage, respectively, is for corn, cotton, soybeans, and wheat. Horticultural crop applications are less, but hard numbers are difficult to come by. Collectively, about 30% of pesticide use in the United States is used for pests that inhabit houses, gardens, lawns, parks, and other recreational areas. A good guess is that about a third to a quarter of pesticides are used for farm and garden applications on vegetables, fruits, and ornamentals.

When first used in the 1940s, pesticides were quite effective. Crop losses fell from about 30% to essentially negligible numbers. Today, we are losing about 37% of our crops to pests and diseases. If we were to not use pesticides now, one wonders what the damage level would be. What went wrong? How did pests stage a comeback and actually become worse than they were? Several events have contributed to the decreasing effectiveness of pesticides. Some pests have developed resistance against pesticides over time. Often their populations have exploded when this occurs, leading to substantial losses before the next generation of pesticide can be rushed into use. About 500 plus major insect pests currently exhibit resistance to one or more pesticides. Usually, the next generation of pesticides costs more money and is more toxic. In 2001, the cost for research, development, and government approval of a new pesticide was from $80 to $120 million. New pesticides are only developed for major crops where major use will justify the investment. Many horticultural crops are not considered for new pesticide development because of insufficient acreage. Alternatives to pesticides become the only answer over the long term.

Another problem is that pesticides kill beneficial insects as well as harmful ones, leading to the emergence of secondary pests as primary pests. Secondary pests were kept low in numbers by predators and parasites. While some damage results from the activity of secondary pests, their low numbers produce tolerable damage. Loss of their natural enemies from pesticides allows secondary pest populations to increase. Plant damage rises from their increased activity to the point that these pests are now considered primary pests. Loss of bees from pesticides also reduces pollination and yields that have to be offset by the importation of beehives into fields.

Pesticides also pollute water supplies. Residues have turned up in water supplies in essentially every state. Concern over pesticide exposure in our water and as residues on our food or in internal tissues of meat continues to mount. Several pesticides are now found in fish, waterfowl, and humans. In Connecticut pesticides not used for many years have been found in birds during pathology studies involving West Nile virus.

Some pesticides are carcinogens; exposure to them over long time periods, especially in childhood, increases the risk of developing cancer. Other pesticides are toxic; some 300,000 farm workers are poisoned each year. Other pesticides can lead to spontaneous abortions and birth defects. The manufacture of pesticides consumes energy and petroleum feedstocks that could be used for other societal needs. Note, however, that not all pesticides are involved in these problems. Pesticides, like any chemicals, vary in terms of chemical properties, usefulness, and environmental implications.

Another problem is the indirect costs of pesticides, which include those for bird losses, crop losses, decreases in pollination, farm animal losses, fish losses, government regulations, honeybee losses, natural enemies losses, public health impacts, and water contamination. The total environmental and societal costs for these items are about $8 billion annually.

Erosion of Genetic Resources

Germplasm is a key agricultural resource. Of serious concern is the erosion of genetic resources, a process that threatens the sources of raw material for the introduction and breeding of new crops, the continued introduction of pest resistance, and the genetic engineering of traits, such as nitrogen fixation. Genetic diversity is being lost as unique environments, such as the South American rain forests, are destroyed. Estimates place yearly plant extinctions at several hundred species; in addition, thousands of indigenous crop varieties have already been lost. Once lost, germplasm is irreplaceable.

Plant breeders use wild relatives to improve plants and to supply new crop cultivars. For example, one valuable trait is resistance to insects and pathogens. As resistance breaks down, the germplasm pool is used as a source of new resistance. In addition, changes in methods of cultivation or market demands may require new cultivars; genetic diversity is fundamental to these innovations.

Genetic vulnerability is another threat. A high degree of narrowing of the germplasm base for commercial use in response to the needs of farmers and consumers has occurred in our widely cultivated crops. The devastating corn blight of 1970 raised concern about the susceptibility of our crops to major outbreaks of disease. Again, should changes be required to decrease genetic uniformity, new germplasm is essential.

TOWARD A MORE SUSTAINABLE HORTICULTURE

Reaching the goal of horticultural sustainability is not easy. It is even questionable whether a completely sustainable horticulture is possible, given limitations on some technologies and resources. Still, it is desirable to work toward a more sustainable or almost completely sustainable system, both in terms of a better environment and for future generations, that they too may eat well and affordably.

Sustainable horticulture is also not just a set of practices to be adopted, but instead is a system that includes the individual farms, the local ecosystems, and the communities connected to these farms from the local to the global context. When viewed from this perspective, it becomes easier to consider the consequences of changing practices on the ecosystem and communities that connect to the farm. This vision also means that research and education needs to be conducted and disseminated in an interdisciplinary manner. Both research and education for sustainable horticulture become the product of joint efforts. Input is required not only by scientists from multiple disciplines, but also from farmers, consumers, and policy makers. Working with systems involves considerable complexity as one works out the ripple effects that one or more changes out of many possibilities might have on all of the interconnected parameters. Computer models must be designed for effective analyses and interpretations of sustainable horticulture systems.

Sustainable horticulture challenges farmers to think about the long-term implications of their practices and the associated environmental consequences and resource dynamics. It also means that consumers must become more involved. Consumers need to be more educated about horticulture and more involved in their food systems. A key goal for all parties is to understand horticulture from an ecological perspective. We must think about ecosystems and nutrient/energy dynamics. We must consider interactions among plants, animals, insects, and other organisms in the combined farm/local community context, or the agroecosystem. All of this focus must be weighed carefully against profit, community, and consumer needs.

Making the transition to sustainable horticulture is a process that will occur slowly and involve a number of small steps. Each step will be influenced by several factors: economics, feasibility, agricultural policy, consumer pressures, and personal beliefs. Each small step will bring us further along the continuum of sustainable horticulture. Finally, it is important to realize that the goal of sustainable horticulture is not just the farmer's responsibility, but everyone's responsibility. Changes might also have costs, such as slight increases in food prices or taxes, or even alterations of the food system. These costs are a small price to pay, given the long-term benefits of sustainable horticulture.

Some Desirable Practices

Many steps can move a farm closer to sustainability. Sustainable practices involve a number of choices. All steps might not work for every farm, because of regional and crop variation. Any choice depends in part on availability of input, climate, the farmer's personal goals, region, pests, soil characteristics, and topography. Despite the site-specific and individual nature of sustainable farming, some practices have general value for sustainable horticulture.

Management Decisions.   Management decisions need to be weighed carefully, and not reached on the basis of minimal consideration. Management decisions need to be crafted on the basis of several factors: profitability, environmental/social considerations, individual beliefs/goals, and lifestyle choices. For example, certain practices might be profitable and environmentally sound, but require so much technology and skill that the farmer becomes a slave to the process. Management decisions should promote sustainability and better the environment and community, but not at unacceptable expense to the farmer.

Input Reduction.   Conventional inputs do not need to be abandoned or completely replaced. One favored practice for the conventional to sustainable changeover is to reduce inputs that cost too much, consume energy, and damage the environment. Reduced usage of fertilizers and pesticides fall into this category.

Integrated pest management works well for many farmers seeking to become more sustainable. IPM is a highly sustainable practice in which pests are controlled with a combined package of biological, cultural, physical, and chemical tools such that economic, health, and environmental risks are minimized.

Conventional farmers do not have to switch from chemical to organic fertilizers in order to move toward a more sustainable horticulture. Organic farming does require this switch, but nonorganic sustainable farmers utilize conventional fertilizers in a more efficient manner. One way is to apply less chemical fertilizer through adoption of better soil testing techniques and precision applications. Even better is to use less chemical fertilizer supplemented with locally available nutrients, such as on-farm resources (compost and manure) or legume cover crops/rotations. The idea is to utilize integrated systems having both chemical and biological inputs. Such systems should be designed to give efficient nutrient loads, reduced pollution, and energy/input costs (or they stay at least the same), and acceptable profitability.

Soil Conservation.   A number of practices help in the control of soil erosion: conservation tillage (no-till and reduced tillage), cover crops, managing irrigation to minimize runoff, mulching, strip-cropping, and windbreaks. However, more than erosion control is involved in sustaining soils.

Soil health is a key component for sustainable production of horticultural crops. Soil is a dynamic material with a living fraction. Damage to the living part from chemicals and excessive tillage leads to reduced soil health. Another important aspect of soil health is organic matter. Maintaining and increasing organic matter can improve soil aggregation, soil tilth, and diversity of soil microbial life.

Poorly managed, less healthy soils require greater inputs of nutrients, pesticides, and water to maintain crop productivity. Better managed soils, with efficient nutrient and water management practices, produce better crops that have optimal vigor and are less susceptible to pests. Many pests cause worse damage when crops are stressed from poor water and nutrient management.

Water Conservation.   A number of strategies can help reduce water consumption. Among these approaches are more efficient, reduced volume irrigation systems, using lower water-requiring or drought-tolerant crops, leaving fields fallow on alternate years (adoption of dryland farming), and mulches.

Water Quality.   Crop production impacts water quality by causing contamination with pesticides and nutrients from fertilizers. Additional problems include salinization. Reduced inputs, IPM, not farming to the water's edge, and the use of organic fertilizers are important strategies for the reduction of water pollution. The creation or maintenance of riparian buffer zones and wetlands protection also play valuable roles. These areas remove nutrients and pesticides, thus reducing the amounts that enter surface waters. In addition, these areas serve to preserve wildlife and provide a habitat for beneficial insects.

Salinity is a problem resulting from irrigation in arid areas. Fields can become salinized and salt levels in surface water can become unacceptable. Improved drainage and flushing can remove field salts, but disposal of the contaminated water can pollute water bodies. Other solutions include reduced volume irrigation and increased use of salt-tolerant crops. More drastic solutions are to remove the irrigated horticultural cropland from production or conversion to production of drought-tolerant forages, forest crops, or wildlife habitat.

Air Quality.   Some agricultural activities impact air quality: dust from field machinery and wind erosion; emissions from farm trucks, cars, field machinery, and fertilizers (nitrous oxide); pesticide drift from spraying operations; and smoke from agricultural burning. Reduced chemical inputs offer some improvements in terms of less pollution from fertilizers, pesticides, and tillage. Adoption or increased use of IPM also helps offset air pollution by pesticides. The incorporation of crop residues into the soil, cover crops, and windbreaks can reduce dust from farm machinery and wind erosion. The incorporation of crop residues also leads to less agricultural burning and, hence, less smoke pollution.

Cover Crops.   Clover, rye, and vetch can be grown as cover crops following the harvest of vegetable or other horticultural crops. Depending on how cover crops are used, they can prevent soil erosion, increase organic matter, or add nitrogen to the soil. Primary benefits to the farmer include erosion control, improved soil quality and nutrient levels, and weed suppression. Benefits on the agroecosystem level are also noted. Cover crops can help to conserve soil moisture, retain soil and nutrients, and improve water infiltration and soil water storage. Cover crops in orchards and vineyards create a favorable environment for beneficial insects and spiders. This change can help reduce pest problems and pesticide use. Diversified usage of cover crops is preferred when creating favorable sites for beneficial insects and spiders. The mixture helps to attract wider ranges of helpful organisms. Mixtures of plants also guard against habitat failure, should one cultivar or species fail in a given year.

Crop Diversity.   This strategy can reduce the risk of failure from severe pest problems, extreme weather, and unexpected market conditions. Diversity of horticultural crops and other farm plantings is preferred over monoculture practices. A mix of crops guards against the loss of the entire crop and its detrimental effects on income. Diversification also reduces economic risks associated with extreme price fluctuations. The use of mixed crops also allows for crop rotations that can suppress weeds, diseases, and insects. Mixed crops and other farm plantings, such as underutilized trees and shrubs, increase populations of beneficial insects, improve soil conservation, and enhance wildlife habitats.

One way to diversify is to adopt agroforestry. This practices utilizes a range of trees on the farm. Some can be planted in pastures, such as walnut trees. This use yields nuts for sale or use and provides shade for cattle and sheep. Trees can also be incorporated into a well-managed wood lot operation to supply firewood and profit from sales for firewood and lumber. Trees and shrubs can be encouraged along streams as riparian buffer strips to reduce water pollution. Trees and shrubs in general improve the habitat for beneficial insects and other wildlife.

When seeking to diversify crops, strive to select those best adapted for your local soil and climatic conditions. Look for insect- and disease-resistant cultivars whenever possible. Consider specialty crops that have high value and also crops likely to sell well in local markets.

Also consider adding livestock to crop-only farms. Livestock offers food, profit, and fertilizer. Mixed crop and livestock farms also have other advantages. Restricting row crops to level land and utilizing slopes for pasture or forages can reduce soil erosion. You can also rotate pasture and forage crops to improve soil quality and to reduce erosion. Livestock can also eat crops unsuitable for market because of drought or insect/disease damage, allowing you to sell better forage and regroup some of your losses.

Crop Marketing.   One strategy for achieving economic sustainability is to improve marketing operations. Direct marketing of agricultural goods to consumers is a good add-on to existing markets. Farmers' markets, roadside stands, and community-supported agriculture (CSA) can raise your bottom line.

CSA is becoming popular where farms are in proximity to urban areas. The consumer contracts with the farmer for a share of the weekly harvest; an up-front fee is paid. The combined fees allow the farmer to purchase seeds, fertilizer, and other farm needs and also guarantee a salary for the season. The consumer gets fresh, quality food at a reasonable price. Some arrangements also involve the consumer helping out with weeding or harvest at certain times for a lower fee. The farmer's risk is reduced and profitability is improved. A plus is that farmers who end up with houses all around their farms end up with better relations in the community, thus reducing the risk of problems from encroaching urbanization.

Changes Needed Beyond the Farm Gate

The success of sustainable farming also depends on changes of attitudes in society and policy changes in government. A complex relationship exists between the farmer, the consumer, institutions allied with agriculture, and the government. The current system is often at odds with sustainable horticulture. Farmers who want to become sustainable farmers often face challenges and difficulties arising from the current system. Some areas need to be rethought, redesigned, and updated. Areas of special concern include the following.

Policy Reform.　Existing policies on the local, state, and federal levels often hinder the goals of sustainable horticulture. Many policies do not promote the tenets of sustainable farming on an equal basis. Policies are needed that simultaneously espouse and link environmental health, economic profitability, and social and economic equity. Alliances, coalitions, and professional lobbyists are needed to exert pressure on all levels to bring about these changes.

An obvious example is the restructuring of commodity and price support programs. These programs need changes that allow farmers to maximize the benefits of gains in productivity that occur with more sustainable practices. Tax and credit policies currently support corporate concentration of farms and absentee ownership. Sustainable farming needs policies that encourage a diverse and decentralized system of family farms. While some changes in government and land grant university research policies now emphasize the development of sustainable alternatives, we need bigger changes. When one looks at budgeted amounts for sustainable and conventional farming programs and research in these institutions, it is apparent that we have a ways to go. The consumer needs education to the effect that cosmetic defects in organic and sustainably produced fruits and vegetables are safer than picture-perfect products treated with numerous pesticides. Government, land grant universities, and food system institutions need to support and help educate the consumer.

Land Loss.　The loss of farmland to urban uses is a particular concern in many states. This problem is seen in many urban areas experiencing rapid growth and escalating land values. Expansion and high land values often encourage sellouts by farmers. For sustainable horticulture, this loss is troubling on two counts. Valuable cropland and production are lost, but so to is a window of opportunity that applies specifically to sustainable horticulture.

Farms close to urban areas that operate in a sustainable manner are likely to be patronized and preserved, given the public demand for safer foods produced in a sustainable manner. Opportunities also exist for these farms to enter the new, profitable areas of community-supported agriculture, which was discussed previously under crop marketing.

Comprehensive new policies are needed to build public support for the preservation and protection of prime farmland. Outreach programs to the public and the farmer are needed to encourage markets for sustainably produced foods and to educate both parties about the value of sustainability. Outreach programs are also needed to help farmers undergo the transition from conventional to sustainable farming systems. Professionals involved in decisions about farmland and land use also need to be taught the values of sustainable horticulture.

Labor Equity.　One problem that exists is that horticultural labor generally does not enjoy the accepted social standards and legal protections found in other fields of employment. If we are to accept the sustainable qualifier, sustainability is needed on all fronts, including labor conditions. Policies and programs are clearly needed if sustainable farming systems are to provide employment that not only offers decent wages and health benefits, but also safe working conditions and social justice. To meet the latter goal, we must provide year-round labor and reasonable housing for migrant farm laborers. These goals will not be reached without the support of government policy, corporate and private farmers, and land grant university professionals involved in assessing the impacts of new horticultural technologies.

Rural Community Economics.　Rural communities throughout the farm areas of the United States are frequently described in terms of economic depression and environmental deterioration. The reasons for this negative change are complex and involve many parts of the food system. The move to a sustainable horticulture gives us an opportunity to assess the role of family farms and rural communities in the long-term viability of the United States. Policy makers and those who influence policy need to consider what role sustainable agriculture can play in stimulating more diversified horticultural production on family and corporate farms and in improving the economics of rural communities.

Consumers and the Food Market. Consumers can influence the movement toward sustainability in the food system. Food purchases can send strong messages that tell producers and retailers what consumers consider priorities. In this area, sustainable horticulture has a strong sell in that the products are reasonably priced, safe, and easy on the environment. These factors can be used to sway consumers in terms of product selection. The key is to educate the consumer about sustainable horticulture and what sets our products apart from others. We need to raise consumer awareness so that environmental quality, resource use, and social justice issues shape shopping decisions. Ways to enhance the marketability of sustainable produce, such as a specific food label like organic producers have, are needed. Improved advertising to both educate and attract consumers is also required. Certain critical coalitions organized around improving the food system could play a key role in bringing this information to the consumer. Issues need to be clarified, new policies put into place, and an understanding gained of the implications of how food is produced and how it affects the environment, diet, and health of those who produce and consume it.

SUMMARY

Agriculture and its horticultural component began some 10,000 years ago. Our own horticultural heritage is but a blip on this scale. In Colonial times almost everyone farmed as a matter of self-sufficiency. Gradually we moved toward a model where many farmed, but produced enough so that others could buy their surplus. After the Civil War, improved plows and mechanical planters and reapers increased farm production even more.

After World War II, the development of new chemical technologies made possible the switch from manure to chemical fertilizers. New chemical pesticides also allowed farmers to switch from natural checks on insect pests and diseases and cultivation against weeds to control of pests, diseases, and weeds with pesticides, fungicides, and herbicides, respectively. Labor-saving machines driven by fuels instead of horses or mules also appeared. Plant breeders gave us crops that thrived with these new technologies. Government policy makers aided farmers in their embrace of technology. Collectively, these changes ended our agrarian society and ushered in an era of fewer, but highly productive farmers that fed the country with the help of technology.

While successful from the productivity viewpoint, some problems existed. The cost of success were the abuse of natural resources, the overuse of water, the pollution of the environment (especially water), and the ever higher dependence on nonrenewable fossil fuels. Another victim was the family farm, which was replaced by larger corporate farms. The organic farmers were the first group to publicly react to these problems. Such farmers advocated no chemicals, instead choosing to fertilize with manure or other natural nutrient sources and to control pests with natural controls and weeds by cultivation. Biodynamic farmers were also another splinter group. The book *Silent Spring* raised public consciousness about pesticide dangers. The concept of sustainable agriculture arose in the late 1970s. In 1980, organic farming achieved mainstream status, being recognized by the U.S. Department of Agriculture.

Many individuals and institutions sought to redefine agriculture in terms of sustainability during the 1980s and into the 1990s. Many conventional farmers, government, agricultural educators and researchers, and manufacturers of agricultural chemicals resisted the concept of sustainable agriculture. Some viewed sustainable agriculture as an attack against all that they stood for, others thought that productivity would be compromised, and others dismissed it as a fringe movement. The fact that sustainable agriculture didn't have one voice didn't help. Numerous groups in the United States and other countries jockeyed for position as the leaders of the movement. These groups advocated agroecology, biodynamic farming, biological agriculture, ecological agriculture, environmentally sound agriculture, low-input sustainable agriculture, organic farming, permaculture, and regenerative agriculture.

Gradually the movement coalesced, with the various parties coming together under one "umbrella," sustainable agriculture. Academic research journals appeared in support of this movement, as did numerous books. Prominent nongovernment institutions lent their support and eventually the U.S. Department of Agriculture recognized its legitimacy. Research funds appeared from government sources and regional programs for sustainable agriculture research and education (SARE) came into existence through the land grant university system. Government-sponsored information systems were also created (Alternative Farming Systems Information Center and Appropriate Transfer for Rural Areas).

During much of this period, sustainable agriculture also sought to define itself. Many definitions emerged, as several forms of alternative farming existed. The ultimate definition made it possible for these various systems to coexist under the label of sustainable agriculture. There was room for the organic farmer at one end of the sustainable agriculture continuum as there was for the conventional farmer at the other end who adopted conservation tillage, IPM, and lower inputs of fertilizer. To work, the definition needed to recognize four basic needs: (1) Resources must be maintained through conservation, recycling, and/or renewal. (2) Farming practices must not pollute the environment on or off the farm. (3) Levels of production, profitability, and the incentive to farm must not be compromised. (4) The system must operate in the realm of social and economic reality.

Sustainable horticulture is in its infancy, perhaps even more so than sustainable agriculture. Numerous problems exist. High losses of farm soil from water and wind erosion have occurred. Increased inputs of water and nutrients are needed to compensate for reduced soil quality. High losses of soil organic matter cause productive soils to become desert-like. Surface water and groundwater supplies are diminished by overdrafts and water is also polluted with nutrients, pesticides, and soil. Heavy use of fossil fuels, a nonrenewable resource, continues. Pesticides cause losses of beneficial insects and possible long-term damage to human health. Our plant genetic resources are jeopardized.

Much research, education, consumer awareness, and political effort will be needed to move the United States to a sustainable horticulture. Several practices on the farm can help bring about the transition. These involve more technological savvy on the part of the farmer; reduced inputs of water, fertilizer, and pesticide; wider adoption of integrated pest management; increased inputs of organic matter into soil; conservation tillage; and more use of cover crops, crop rotations, and crop diversity. Better marketing opportunities will need to be available, such as community-supported agriculture, a type of contractual, pick-your-own vegetables and fruits situation.

Changes will also be needed beyond the farm. Food and agriculture policies that hinder sustainable horticulture must be revised. Land use for farms will need to be protected. Better conditions for farm workers are needed. Rural communities will need help to halt their economic and population losses. Consumers will need education to raise their awareness and to encourage support for vegetables and fruits produced by sustainable horticultural farms.

EXERCISES

Mastery of the following questions shows that you have successfully understood the material in Chapter 1.

1. How does the roughly three and one-half centuries of American Colonial agricultural practices to the present time compare to the length of time humans have practiced agriculture?
2. Major agricultural changes occurred in the United States after the Civil War and World War II. What were they and how did they advance agricultural productions? What initiated another leap forward in productivity, starting in the 1950s?
3. What events brought about a questioning of the consequences of our very productive agriculture?
4. What role did organic agriculture play in the development of sustainable agriculture?
5. What role did Rachel Carson play in the development of sustainable agriculture?
6. What event brought about mainstream legitimacy for organic farming?
7. A number of alternative agricultural forms arose in the 1980s. What were they called?

8. What two conferences and three academic journals launched the beginnings of serious sustainable agriculture?
9. A number of institutions played a role in promoting sustainable agriculture from the 1980s onward. What were their names?
10. What do LISA and SARE have to do with sustainable agriculture? What were their origins?
11. What are the basic tenets of sustainable agriculture? What is the general working definition of sustainable agriculture?
12. What soil problems occur with conventional agriculture?
13. What water problems occur with conventional agriculture?
14. What energy problems occur with conventional agriculture?
15. What problems do pesticides cause?
16. What problems occur with genetic resources?
17. What do we need to do in terms of future directions to achieve sustainable agriculture? What specific changes in practices are needed on the farm? What changes are needed beyond the farm?

ACADEMIC JOURNALS

Agriculture and Human Values
American Journal of Alternative Agriculture
Journal of Sustainable Agriculture

INTERNET RESOURCES

Alternative Farming Systems Information Center
  http://www.nal.usda.gov/afsic/

Biodynamic Farming
  http://www.attra.org/attra-pub/biodynamic.html

Collection of papers by Ikerd
  http://www.ssu.missouri.edu/faculty/jikerd/papers/default.htm

Educational and training opportunities in sustainable agriculture
  http://www.nal.usda.gov/afsic/AFSIC_pubs/edtr12.htm

Links to many sustainable agriculture organizations
  http://dir.yahoo.com/Science/Agriculture/Sustainable_Agriculture/Organizations/

Organic farming
  http://www.ofrf.org http:www.attra.org/attra-pub/organiccrop.html

Permaculture
  http://www.attra.org/attra-pub/perma.html

Publications on sustainable agriculture
  http://www.nal.usda.gov/afsic/agnic/agnic.htm#print
  http://chla.library.cornell.edu/
  http://ianrwww.unl.edu/ianr/csas/extvol6.htm

Searchable sites and databases
  http://www.nal.usda.gov/afsic/agnic/agnic.htm#search

Searchable site for *Journal of Sustainable Agriculture*
  http://www.haworthpressinc.com/store/product.asp?sku=J064

Sustainable agriculture bibliography, definitions and terms (including agroecology, biodynamic farming, organic farming, permaculture, etc.)

http://www.nal.usda.gov/afsic/AFSIC_pubs/srb9902.htm

http://www.nal.usda.gov/afsic/AFSIC_pubs/tracing.htm

http://agroeco.org/principles_and_strategies.html

http://www.sarep.ucdavis.edu/concept.htm

Sustainable Agriculture Network; SAN is the communications and outreach arm of the Sustainable Agriculture Research and Education (SARE) program of the U.S. Department of Agriculture

http://www.sare.org/

Sustainable agricultural organizations and publications

http://www.attra.org/attra-rl/susagorg.html

Sustainable agricultural resources for teachers, K–12

http://www.nal.usda.gov/afsic/AFSIC_pubs/k-12.htm

Sustainable farming internships and apprenticeships

http://www.attra.org/attra-rl/intern.html

RELEVANT REFERENCES

Altieri, Miguel A. (ed.) (1987). *Agroecology: The Scientific Basis of Alternative Agriculture.* Westview Press, Boulder, CO. Second edition published 1995.

American Society of Agronomy (1989). Decision Reached on Sustainable Ag. *Agronomy News* (Jan.): 15. ASA, Madison, WI.

Balfour, Evelyn Barbara, Lady (1943). *The Living Soil.* Faber and Faber, London. 248 pp.

Bender, Jim (1992). *Future Harvest: Pesticide-Free Farming.* University of Nebraska Press, Lincoln. 159 pp.

Berry, Wendell (1977). *The Unsettling of America: Culture and Agriculture.* Sierra Club Books, San Francisco. 228 pp.

Besson, J. M., and H. Vogtmann (eds.) (1978). *Toward a Sustainable Agriculture.* International Federation of Organic Agriculture Movements, Oberwil, Switzerland. 243 pp.

Bezdicek, D. F., and J. F. Powers (1984). *Organic Farming: Current Technology and Its Role in a Sustainable Agriculture.* American Society of Agronomy, Crop Science Society of America, and Soil Science Society of America, Madison, WI. 192 pp.

Bromfield, Louis (1948). *Malabar Farm.* Harper, New York. 405 pp.

Bromfield, Louis (1945). *Pleasant Valley.* Harper, New York. 300 pp.

Brown, L. R., W. U. Chandler, C. Flavin, C. Pollock, S. Postel, L. Starke, and E. C. Wolf (1986). *New Directions for Agriculture and Agricultural Research and State of the World: A Worldwatch Institute Report on Progress Toward a Sustainable Society.* Worldwatch Institute, Washington, DC.

Callaway, M. Brett, and Charles A. Francis (1993). *Crop Improvement for Sustainable Agriculture.* University of Nebraska Press, Lincoln. 261 pp.

Carson, Rachel (1962). *Silent Spring.* Houghton-Mifflin, Boston.

The Center for Integrated Agricultural Systems (1991). *Toward a Sustainable Agriculture: A Teacher's Guide.* CIAS, Madison, WI. 151 pp.

Center for Sustainable Living (1998). *The Community Supported Agriculture Handbook: A Guide to Starting, Operating or Joining a Successful CSA.* CSL, Wilson College, Chambersburg, PA. 42 pp.

Coleman, Eliot (1995). *The New Organic Grower: A Master's Manual of Tools and Techniques for the Home and Market Gardener* (2nd ed.). Chelsea Green Publishing, White River Junction, VT. 340 pp.

Congress (1990). Farm Bill: Food, Agriculture, Conservation, and Trade Act of 1990. Public Law 101-624, Title XVI, Subtitle A, Section 1603. Government Printing Office, Washington, DC.

Edwards, Clive A. (1988). *The Concept of Components of Sustainable Agriculture.* In C. A. Francis and J. W. King (eds.), Sustainable Agriculture in the Midwest, Proc. North Central Regional Conference, Agr. Res. Div. and Coop. Ext. Service, Univ. Nebraska, Lincoln, NE.

Edwards, Clive A., R. Lal, P. Madden, R. Miller, and G. House (eds.) (1990). *Sustainable Agricultural Systems.* Soil and Water Conservation Society, Ankeny, IA. 696 pp.

Faulkner, Edward Hubert (1943). *Plowman's Folly.* Grossett and Dunlap, New York. 155 pp.

Francis, Charles A., Cornelia Butler Flora, and L. D. King (eds.) (1990). *Sustainable Agriculture in Temperate Zones,* Wiley, New York.

Francis, C. A., and G. Youngberg (1990). *Sustainable Agriculture: An Overview.* In: Sustainable Agriculture in Temperate Zones (C. A. Francis, C. B. Flora, and L. D. King, eds.), Wiley, New York.

Fukuoka, Masanobu (1985). *The Natural Way of Farming: The Theory and Practice of Green Philosophy.* Japan Publications, Tokyo. 280 pp.

Gershuny, G., and J. Smillie (1986). *The Soul of Soil: A Guide to Ecological Soil Management* (2nd ed.). AgAccess, Davis, CA. 174 pp. Third edition published 1995.

Gliessman, Stephen R. (1998). *Agroecology: Ecological Processes in Sustainable Agriculture.* Sleeping Bear Press, Chelsea, MI. 357 pp.

Hatfield, J. L., and D. L. Karlen (1994). *Sustainable Agriculture Systems.* Lewis Publishers, Boca Raton, FL. 316 pp.

Howard, Albert, Sir (1940). *An Agricultural Testament.* Oxford University Press, Oxford. 253 pp.

Horne, J. E., and M. McDermott (2001). *The Next Green Revolution: Essential Steps to a Healthy, Sustainable Agriculture.* Food Products Press, Binghamton, NY. 312 pp.

Hudson, William J., and Jonathan Harsch (1991). *The Basic Principles of Sustainable Agriculture: An Introduction for Farmers, Environmentalists, the Public, and Policy-makers.* U.S. Dept. of Agriculture, Cooperative Extension Service, Washington, DC. 32 pp.

Hunter, Beatrice T. (1964). *Gardening without Poisons.* Houghton-Mifflin, Boston. 318 pp.

Jackson, Wes (1980). *New Roots for Agriculture.* University of Nebraska Press, Lincoln. 151 pp. New edition published 1985.

Jackson, Wes, Wendell Berry, and Bruce Colman (eds.) (1984). *Meeting the Expectations of the Land: Essays in Sustainable Agriculture and Stewardship.* North Point Press, San Francisco.

Jeavons, J. (1979). *How to Grow More Vegetables Than You Ever Thought Possible on Less Land Than You Can Imagine.* Ten Speed Press, Palo Alto, CA. 116 pp.

Kains, M. G. (1935). *Five Acres and Independence: A Practical Guide to the Selection and Management of the Small Farm.* Long out of print, a revised reprinting is available from Dover Publications through Amazon.com.

King, Franklin Hiram (1911). *Farmers of Forty Centuries (or) Permanent Agriculture in China, Korea, and Japan.* Harcourt, Brace, New York. 379 pp.

Knorr, Dietrich (ed.) (1983). *Sustainable Food Systems.* AVI Publishing, Westport, CT.

Leopold, Aldo (1949). *A Sand County Almanac and Sketches Here and There.* Oxford University Press, New York. 226 pp.

Little, Charles E. (1987). *Green Fields Forever.* Island Press, Washington, DC. 192 pp.

Lockeretz, William (ed.) (1983). *Environmentally Sound Agriculture.* Praeger Publishers, New York. 426 pp.

Meadows, D. H., D. L. Meadows, J. Randers, and W. W. Behrens III (1972). *The Limits to Growth.* Universe Books, New York. 205 pp.

Mollison, Bill (1990). *Permaculture: A Practical Guide for a Sustainable Future.* Island Press, Washington, DC. 579 pp.

National Research Council (1989). *Alternative Agriculture.* National Academy Press, Washington, DC. 448 pp.

Northbourne, Walter Ernest Christopher James, Lord (Baron) (1940). *Look to the Land.* Dent, London. 206 pp.

Oelhaf, R. C. (1978). *Organic Agriculture: Economic and Ecological Comparisons with Conventional Methods.* Allanheld, Osmun, and Company, Montclair, NJ. 271 pp.

Pell, Katharine M. (1992). *Sowing Fields of Dreams: Science, Sustainable Agriculture, and Public Agricultural Research.* Honors Thesis, Wesleyan University, Middletown, CT. 183 pp.

Pfeiffer, Ehrenfried (1938). *Bio-Dynamic Farming and Gardening.* Anthroposophic Press, New York.

Poincelot, R. P. (1986). *Toward a More Sustainable Agriculture.* AVI Publishing, Westport, CT.

Postel, S. (1987). *Defusing the Toxic Threat: Controlling Pesticides and Industrial Waste.* Worldwatch Institute, Washington, DC.

Pye, V. I., R. Patrick, and J. Quarles (1983). *Groundwater Contamination in the United States.* University of Pennsylvania Press, Philadelphia. 315 pp.

Rodale, Jerome Irving (1948). *The Organic Front.* Rodale Press, Emmaus, PA. 198 pp.

Rodale, Jerome Irving (1945). *Pay Dirt: Farming and Gardening with Composts.* Devin-Adair Company, New York. 242 pp.

Sampson, R. Neil (1981). *Farmland or Wasteland: A Time to Choose—Overcoming the Threat to America's Farm and Food Future.* Rodale Press, Emmaus, PA.

Sattler, Friedrich, and Eckard V. Wistinghausen (1992). *Bio-Dynamic Farming Practice.* Bio-Dynamic Agricultural Association, Stourbridge, West Midlands, UK. 333 pp.

Schwenke, Karl (1991). *Successful Small-Scale Farming: An Organic Approach* (2nd ed.). Storey Communications, Pownal, VT. 134 pp.

Smith, Miranda, Northeast Organic Farming Association and Cooperative Extension (eds.) (1994). *The Real Dirt: Farmers Tell about Organic and Low-Input Practices in the Northeast.* Northeast Region Sustainable Agriculture Research and Education (SARE) Program and Northeast Organic Farming Association, Burlington, VT. 264 pp. Second edition published 1998.

Soule, Judith D., and Jon K. Piper (1992). *Farming in Nature's Image: An Ecological Approach to Agriculture.* Island Press, Washington, DC. 286 pp.

U.S. Department of Agriculture (1991). *Agriculture and the Environment: The 1991 Yearbook of Agriculture.* USDA, Washington, DC. 325 pp.

U.S. Department of Agriculture Study Team on Organic Farming (1980). *Report and Recommendations on Organic Farming.* USDA, Washington, DC. 94 pp.

Van den Bosch, Robert, and P. S. Messenger (1973). *Biological Control.* Intext Educational Publishers, New York. 180 pp.

Walters, Charles, Jr., and C. J. Fenzau (1979). *An ACRES U.S.A. Primer.* ACRES U.S.A., Raytown, MO. 465 pp. Second edition published 1996.

VIDEOS AND CD-ROMs*

The Agricultural Revolution (Video, 1985).
Family Owned Farms (Video, 1997).
The Farmer in Changing America (Video, 1973).
Houses in the Fields (Video, 1995).
Rural America (Video, 1995).
Sustainable Environments (Video, 1994).
Women of Rural America (Video, 1995).

*Available from Insight Media, NY (www.insight-media.com/IMhome.htm)

INTRODUCTION TO HORTICULTURAL SCIENCE

*H*orticulture has its etymological origins in the Latin words *hortus* and *cultura*, which mean "garden" and "culture," respectively. Today its meaning goes far beyond the culture of gardens. Horticulture is the branch of plant agriculture concerned with the intense cultivation of garden crops produced for food, medicine, enjoyment, recreation, and general environmental improvement. Horticulture is one of the tripartite branches of plant agriculture; the other two are agronomy and forestry. Agronomy pertains to field crops, such as cereals and fodders, and forestry is associated with nonfood tree crops and their products.

Crops grown in the horticultural sense in gardens and farms include vegetables, fruits, ornamentals, nuts, herbs, and plants used for medical purposes. These crops and their purposes are the basis for the three divisions of horticulture: olericulture or vegetable culture; pomology or fruit culture; and ornamental horticulture, which includes three general areas: (1) floriculture, or the culture of cut flowers, potted flowers, and foliage plants and greenery; (2) landscape horticulture, or the design and construction of sites along with the planting and maintenance of woody and herbaceous ornamentals, bulbs, and related crops; and (3) production nursery, or the production of landscape horticultural crops. Some consider propagation to be a fourth division, and, indeed, it can be considered as separate or as an integral part of the preceding three divisions. The area of propagation is associated with the reproduction of horticultural crops by seed and vegetative techniques, as well as the improvement of plant material through plant breeding.

Horticulture touches on many scientific disciplines. These areas include physiology, biochemistry, genetics, and pathology of plants, as well as botany, soil science, ecology, climatology, physics, taxonomy, entomology, economics, food technology, art, architecture, and landscape design. In a sense, horticulture is a blend of science, art, technology, and aesthetics that has stimulated interest over the centuries.

Nevertheless, it is the scientific part—the mixture of sciences providing information about plants and their relationship with the environment—that forms the foundation of horticulture. Without a thorough understanding of this foundation, we cannot hope to design practices that are sustainable. Horticultural science provides the means to understand the art of horticulture, to improve horticultural technology, to increase crop production in a sustainable manner, and to deepen our appreciation of horticulture's aesthetic value. It is with this part of horticulture, the foundation of horticultural science, that Section Two is concerned.

Chapter 2

Plant Names

OVERVIEW

Plant nomenclature, or the naming of plants, has two approaches: common names and scientific names. Common names are useful in everyday conversation. Fellow gardeners can talk about the merits of their tomatoes or the beauty of their roses. Homeowners can talk about their grass problems or decide between a rhododendron and a mountain laurel at the local nursery. The apartment dweller might wax eloquently about his or her African violets. At the supermarket, we try to decide which lettuce is best for our salad: leaf, romaine, or iceberg.

However, common names have their limitations in that they are not always ideal terms for purposes of identification or serious communication. Scientific names convey exact identification when used correctly, but are not likely to be part of a normal conversation involving plants. Both types of names serve useful roles in horticulture.

Horticulturists also like to group or classify plants based on their use, growth habits, or even how plants are related in the evolutionary sense. The functional approach based on use and form is called *horticultural classification*. Such terminology is very useful when the horticulturist or consumer selects plants for landscape and garden uses. Relationships based on evolution are central to plant (scientific) classification. As with common versus scientific names, horticultural classification is the "working language" and plant classification is the "intellectual language." The landscaper, gardener, and homeowner are more interested in the former, while the horticultural scientist or the botanist needs the latter. Both types of classification have their applications in horticulture.

CHAPTER CONTENTS

Limitations of Common Names
History of Plant Nomenclature and Classification
Classification and Taxonomy Made Easy

Horticultural Classification of Plants
Plant Identification Keys

OBJECTIVES

* Realizing why common names are not universally accepted
* Understanding the history of science-based systems for grouping and naming plants
* Learning how plants are grouped (classification) based on evolutionary relationships and how each plant has a unique name (taxonomy)

* Understanding the cultivar and its horticultural value
* Learning the horticultural jargon associated with plants
* Discovering how to identify plants whose name you don't know

LIMITATIONS OF COMMON NAMES

Let's get back to common names. Many plants have common names, especially if the plants are popular, useful, or attractive. Such names are helpful, but not always guaranteed to deliver exact identification of the plant in question. Many faults are associated with common names because they are derived illogically and inconsistently. For example, the same common name can be applied to different plants or a plant can have more than one common name. Other complications are that the plant's name can vary from one region to the next, such as from New England to the Midwest. And common names change by country, even if the plant doesn't.

How about simple common names like Indian paintbrush? Which plant do you mean and what plant does another person mean? Three plants have this common name: *Asclepias tuberosa, Castilleja californica,* and *Castilleja coccinea.* In New England, you might mean either *Asclepias tuberosa* or *Castilleja coccinea,* but in California, you are thinking of *Castilleja californica.* Someone else in New England might refer to *Asclepias tuberosa* as butterfly weed, pleurisy root, tuberroot, or chigger flower. All of these common names would be correct, because they, too, are names for this same plant. Other examples are *Agastache cana, Azolla caroliniana,* and *Cynanchum assyrifolium,* which are all commonly called mosquito plant. The common names Moses-in-a-boat, Moses-in-the-bulrushes, Moses-in-the-cradle, and Moses-on-a-raft all refer to *Rhoeo spathacea.* In a different vein, we call it eggplant, but the French call it aubergine.

Talk about confusion and lack of identity! You will note that the italicized names appear to be more exact. They are, because they are scientific names; more specifically, species names. This type of name is used by scientists to identify an organism and does not lead to the confusion found with common names. It took a long time to get to this state of affairs. So how did all this ultimately get resolved? Let's look next at the history of nomenclature and classification.

HISTORY OF PLANT NOMENCLATURE AND CLASSIFICATION

The first use of plant names is lost in unrecorded history. Early humans were confronted with numerous plants. By trial and error, these plants were likely divided into edible food plants and harmful plants. With time, plants were probably subdivided into groups used for food, flavorings, beverages, dyes, ornaments, medicines, hallucinogens, and poison. Plants, so necessary for survival, presumably became items of trade. As language evolved, these plants were likely given names. Names became useful to describe the plants that had become so essential to human activities. Other words were likely used to indicate the qualities that distinguished one plant group from another. Thus the naming and grouping of plants, or their *taxonomy* (science of naming) and *classification* (science of grouping based on some relationship), started.

Plants were probably cultivated for food, the origins of agriculture, some 10,000 years ago. Pictorial records of food plants survive on the walls of the ancient Egyptian pyramids and temples. Lists of plant names appeared in an ancient Assyrian compilation. A treatise on agriculture existed in ancient Carthage. The writings of ancient China refer to plants. Plants had become an essential part of the ancient world, but very few ancient writings about plants survive.

Ancient Greece and Rome

The first surviving documentation of a plant classification scheme goes back to ancient Greece. These writings came from Theophrastus of Eresus (370–285 B.C.), who studied under the famous philosophers Plato and Aristotle. Later he became an assistant to Aristotle. Much of his time was spent in the botanical gardens of Athens that had been established by Aristotle. Theophrastus is known as the father of botany because of his work with plants.

His publication, *Historia Plantarum,* or translated, *History of Plants,* marks the start of the science of botany. In his various writings, Theophrastus made numerous observations about plants, including classification attempts based on form and texture. He also examined seed germination, seedling development, internal structures, flower parts, leaves, roots, and much more. He grouped plants based on their life spans: annuals, biennials, and perennials (one, two, and many growing seasons, respectively).

Theophrastus classified plants into four groups: herbs, half-shrubs, shrubs, and trees. He observed and recorded families among the flowering plants. One of these was the parsley family, which today is the designated family Umbelliferae. He noted other observations among the birches, poplars, alders, conifers, cereals, and thistles. One of his names for a plant group, the Greek word *asparagus,* survives today as a genus name, *Asparagus.*

The Roman culture that followed wrote much about agriculture and horticulture. The richer Romans even had formal gardens. Their name for a garden, as opposed to the larger farm, was *hortus.* Their word for cultivation of the garden was *cultura,* which comes from the verb *colere,* "to cultivate." These two words are the origin for our modern term *horticulture.* However, Roman culture emphasized the practical aspects much more than the scientific ones favored by the Greeks.

Middle Ages

The Dark Ages and the medieval period that followed the fall of Rome gave birth to little progress in the classification and naming of plants. Indeed, some horticultural information even disappeared and was rediscovered later. The exception was large books devoted to the hand illustration and description of plants used for medicinal purposes. These great books, the herbals, were the works of German, English, and Italian herbalists. These books arose during the 15th and 16th centuries (approximately 1470–1670). Although impressive books, you would not likely survive the medical applications. The belief was that the plant part's shape dictated its use, the so-called "doctrine of signatures." If a leaf was shaped like a heart, it was used for the treatment of heart disorders.

The system of naming plants reached a critical impasse. The basic flaw was that plants of a similar group (closely related or alike plants, such as oaks) were named by using their basic name (the genus or plural, genera), followed by a string of descriptive adjectives to identify each member of the group. However, as more plants were discovered, the strings of adjectives got longer and longer. In some cases an entire paragraph was needed to name the plant. For example, suppose you had two kinds of oaks that had different types of acorns. These oaks (the genus) might be named the "oak with pointed acorns" and the "oak with rounded acorns." However, suppose a third oak was found that had pointed acorns, but its leaves differed. Now the names of those two oaks might become "the oak with pointed acorns and round-edged leaves" and "the oak with pointed acorns and sawtooth-edged leaves." Names became increasingly cumbersome with this approach. Latin was usually used in this system.

Linnaeus, the Father of Classification

The scientists of the late 1600s and the early 1700s became increasingly vocal about these problems. Questions were raised about the usefulness of the naming systems and classification systems that dated back to ancient Greece and Rome. Some suggestions for reform were proposed. The scientist that finally reformed naming and classification was Carolus Linnaeus (1707–1778), later knighted to Carl von Linné. For his efforts, Linnaeus is remembered today as the father of taxonomy. Linnaeus was a physician and also a professor of botany and medicine at the University of Uppsala in Sweden. All physicians had to be knowledgeable about plants (botany) because most medications were plant products.

Binomial Nomenclature.   Linnaeus is remembered for two accomplishments: his standardized method of naming plants and his classification scheme. Both were published in his book *Species Plantarum* (*Species of Plants,* Figure 2–1). Let's look first at his naming method, which is called *binomial nomenclature.* This method uses two words to name the plant, as we do with people. For example, we can think of George Smith as opposed to Susan

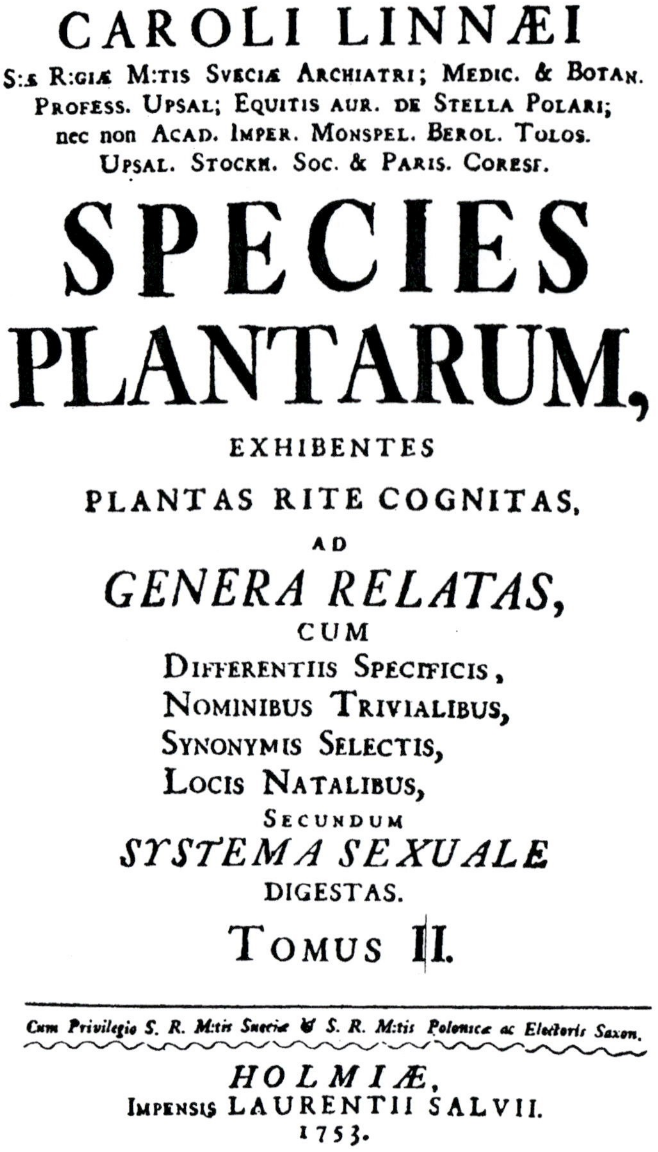

FIGURE 2–1   A reproduced title plate from *Species Plantarum* (1753) in which Linnaeus introduced the binomial nomenclature system of taxonomy and also a classification system based on sexual characteristics. Volume II is the section that covers botanical materials.

CAROLI LINNÆI

S:ᴇ R:ɢɪᴀ M:ᴛɪꜱ Sᴠᴇᴄɪᴀ Aʀᴄʜɪᴀᴛʀɪ; Mᴇᴅɪᴄ. & Bᴏᴛᴀɴ.
Pʀᴏꜰᴇꜱꜱ. Uᴘꜱᴀʟ; Eǫᴜɪᴛɪꜱ ᴀᴜʀ. ᴅᴇ Sᴛᴇʟʟᴀ Pᴏʟᴀʀɪ;
ɴᴇᴄ ɴᴏɴ Aᴄᴀᴅ. Iᴍᴘᴇʀ. Mᴏɴꜱᴘᴇʟ. Bᴇʀᴏʟ. Tᴏʟᴏꜱ.
Uᴘꜱᴀʟ. Sᴛᴏᴄᴋʜ. Sᴏᴄ. & Pᴀʀɪꜱ. Cᴏʀᴇꜱꜰ.

SPECIES PLANTARUM,

EXHIBENTES

PLANTAS RITE COGNITAS,

ᴀᴅ

GENERA RELATAS,

ᴄᴜᴍ
Dɪꜰꜰᴇʀᴇɴᴛɪɪꜱ Sᴘᴇᴄɪꜰɪᴄɪꜱ,
Nᴏᴍɪɴɪʙᴜꜱ Tʀɪᴠɪᴀʟɪʙᴜꜱ,
Sʏɴᴏɴʏᴍɪꜱ Sᴇʟᴇᴄᴛɪꜱ,
Lᴏᴄɪꜱ Nᴀᴛᴀʟɪʙᴜꜱ,
Sᴇᴄᴜɴᴅᴜᴍ
SYSTEMA SEXUALE
DIGESTAS.

Tᴏᴍᴜꜱ II.

Cum Privilegio S. R. M:tis Sueciæ & S. R. M:tis Polonicæ ac Electoris Saxon.

HOLMIÆ,
Iᴍᴘᴇɴꜱɪꜱ LAURENTII SALVII.
1753.

Jones. These two-part names, or binomials, work to name and identify people, so why not plants. While we might think Linnaeus clever for using this similar scheme, actually an earlier scientist (Caspar Bauhin) had suggested the two-term naming system for possible use. These binomials were called *species,* which, like the term *sheep,* is both singular and plural. Linnaeus gave species names to a great number of plants, as well as numerous animals. This concept of species and how they are named survives today and forms the basis for the taxonomy of all living organisms, not just plants.

Linnaeus also proposed that the language for naming plants be standardized worldwide and be Latin. This choice was probably based on the use of Latin as a scholarly language among the scientists of that time. It was also a good choice for other reasons. Being a dead language, once named, plant names would remain constant and not subject to change, as with a living language. It was also a good choice on political grounds. Since no country used Latin, no nationalistic pride would be hurt. It was also a universal language among scholars, regardless of what their mother language was.

So how does all of this work? Let's go back to oaks. The genus for oaks is *Quercus.* Prior to Linnaeus, oaks were named by starting off with *Quercus* and then a string of descriptive Latin terms followed. These terms talked about acorn shape, but since more than one oak had similar shaped acorns (the white oak, swamp white oak, and scarlet oak have egg-shaped acorns), other terms had to be added as more oaks were discovered. You could add something about the cup-like structure that attaches to one end of the acorn. The cups on the white oak acorn are shallow and cup-like in shape, covering only one-quarter of the acorn. Since the scarlet oak has a thick, deep top-shaped cup, that would allow us to name these two oaks. The swamp white oak has a deep cup also, but its shape is cup-like and it covers one-third of the acorn. So all three of these oaks can be distinguished with a few descriptive terms. However, as more oaks were named, this set of terms became too limited to distinguish each oak from one another. The solution was to add additional terms describing acorn colors and patterns, leaf shape and color, and bark appearance and color. Oak names got very long.

Enter Linnaeus and his binomial nomenclature. The white oak became *Quercus alba*, the swamp white oak, *Quercus bicolor*, and the scarlet oak, *Quercus rubra*. Names based on many lines simply became two words. The original genus was retained, but the string of descriptive terms was replaced by one adjective-like word. For the white oak, the adjective in Latin was *alba*, which means "white" and refers to the white undersides of the leaves. The swamp white oak was identified by the term *bicolor*, the Latin word for "two colors." It refers to the fact that the leaves were green on top and had white undersides. Because the term *alba* had already been used, some other descriptive term had to be applied. The red oak was identified with the term *rubra*, meaning red, which refers to the scarlet color of the leaves in the fall. In this manner, unique names for each oak species were formulated.

Artificial Classification. Linnaeus also classified many plants. His classification approach consisted of cataloging all known plants into 24 groups; each group was based on their sexual characteristics (he used the male part of the flower, the stamen). Group 24 consisted of plants with no known sexual characteristics, the nonflowering plants, such as algae, mosses, ferns, and fungi. This type of classification, the use of only one or very few characteristics, is known as *artificial classification*. This method does not survive today, with one exception. Guides to wildflowers for amateurs often group wildflowers by color. Once you have the color of the flower, red, for example, you go to the red-tabbed pages. There you find pictures and names of numerous wildflowers with red flowers.

Natural Classification. Although artificial classification continued for some time with modifications, it was eventually replaced by *natural classification*. Scientists became disenchanted with artificial classification, because it grouped some plants together that clearly were not related. For example, cacti and cherry flowers both have numerous stamens. Hence these two plants would be placed together, based on the artificial system using stamens. The early to middle 1800s saw the development of a natural classification system. This form of classification is based on overall resemblance of external morphology (form). Unlike artificial classification, as many morphological characteristics as possible are used. The natural system is dependent on degrees of similarity. The fundamental unit, the species, was retained. Similar species were placed into a common genus, similar genera were placed into a family, and so forth.

Still, even this system had its flaws. For example, based on overall similarity, cacti from our American Southwest would be grouped with some euphorbia from Africa. Their appearances are similar (Figure 2–2), both grow in deserts, and sharp spines protect both. Today we know that these plants are products of convergent evolution. These plants do not share a common ancestor, even though they look alike. Instead, these plants are the result of adaptation to a similar environment, the desert. Of course, the scientists of that time were unaware of this problem, given that the theory of evolution did not then exist.

Darwin, Evolution, and Phylogenetic Classification. With the appearance of Darwin's *Origin of Species* in 1853, the concept of evolution helped to explain some of the discrepancies inherent in the natural classification system. Taxonomists began to realize that only a classification system based on evolution could give a true picture of the relationship of the various living and extinct plants. Classification based on evolution is termed *phylogenetic*

FIGURE 2-2 The cactus of the southwest United States (left, *Lemaireocereus pruinosus*) and the euphorbia (right, *Euphorbia horrida nova*) from an African desert are examples of convergent evolution. (Photograph by Laura Dancho.)

classification. By the end of the 1800s, classification began to reflect the concept of evolution. With this approach, related plants are those that have genetic similarities as a consequence of shared descent, that is, they share a common ancestor. In this manner, cacti and euphorbia would not be related, since they do not share a common ancestor and are not genetically similar.

Classification Today. So where are we today? Somewhere in between two systems. Current classification is a mix of natural and phylogenetic approaches. The phylogenetic system is much preferred. Two problems prevent the total conversion to phylogenetics: the collection and interpretation of data. The sheer number of plants overwhelms the scientist who attempts to classify plants based on phylogenetics (this type of scientist is called a *systematist*; the science is *systematics*). This picture is further complicated by several problems. These problems include the loss of intermediate relatives from extinction, gaps in the fossil record that thwart our attempts to recreate the line of descent, the appearance of new species, the ability of some species to breed together and produce hybrids, and the appearance of polyploids (plants with major alterations in their chromosomal numbers). Even evolution doesn't cooperate. It is assumed that evolution causes a progression from simple to ever increasing levels of complexity. In some cases, yes; in others, no. Some plants show an evolutionary pattern toward greater simplicity (reduction). On top of all of this, the determination of genetic relations is also a tedious and complicated laboratory process (more about this later). The establishment of plant lines through evolution, the tracing to common ancestors, and the sheer volume of needed data pose a very difficult task indeed.

Given the above problems with data collection, it is easy to understand why systematists disagree on how things should be interpreted. For example, if three groups of related plants have a similar fruit, but discernible differences among their flowers, two interpretations are possible. You could say that the three groups belong in the same family, given the shared fruit, but that they can be further classified into three subunits (subfamilies) based on flower differences. Another view is that these three groups are actually three families, based on differences among flowers. Systematists who favor the first view are called *lumpers*, in that they tend to take a more conservative, simplified approach. The others are called *splitters* and take a more liberal view and prefer more complexity.

Even given these difficulties, the phylogenetic system is preferred. New relationships, previously unseen with the natural system, can emerge and be recognized. For example, two plants might be related, but the link is not apparent because of the extinction of an intermediate form. Using the evolutionary approach and a fossil, the relationship becomes apparent. Another possibility is that two plant groups might look unrelated, but be descendants of a common ancestor. They now appear considerably different because of reproductive isolation brought about by some ancient geological catastrophe. With this knowledge derived from evolution, the two groups can now be shown to be related.

To establish these evolutionary relationships, systematists must rely heavily on scientific evidence. These data are usually the result of extensive morphological (form), biochemical (amino acid sequences), DNA and RNA (nucleotide sequences; see Chapter 9, Biotechnology), cytological (cell science), physiological, and ecological analyses. Many tools of science have become important in establishing phylogenetic relationships. These include gel electrophoresis for examining proteins and DNA or RNA, automated nucleotide sequencers, scanning electron microscopy for seeing plant cell and pollen structures, and gas chromatographs for the study of the chemical composition of plant products. Given the complex and extensive amounts of data, computers have become the indispensable companions of the systematists. Much of today's systematics is known as *molecular systematics* because of the heavy dependence on molecular biology techniques.

The conversion from natural to phylogenetic classification is by no means complete. We do use the phylogenetic classification system developed by Adolph Engler and Karl Prantl. This system was developed from the late 1800s through the early 1970s and has been updated ever since. Major updates were accomplished by Arthur Cronquist, an American botanist and Armen Takhtajan, a Russian botanist. The most recent phylogenetic classification update is the work of the Angiosperm Phylogeny Group (APS) in 1998. Today, we know that parts of it are phylogenetically incorrect. We now know that some of their phylogenetic approaches about the nature of primitive flowers were wrong. Corrections, especially with flowering plants, are being made even as you read. The problem is that the complete revision of this 20-volume phylogenetic classification has not been justified to date by the changes. So it stands and is widely used. In time, given enough corrections, the needed revision will take place. Taxonomists will also have to reach some agreement about which of the proposed phylogenetic versions (splitters versus lumpers) should be adopted. Systematists around the world are committed to comprehensive phylogenetic classification of all organisms (not just plants) by the year 2020!

CLASSIFICATION AND TAXONOMY MADE EASY

Does the horticulturist need to know how to classify and name plants? If so, to what extent is this skill needed? The answer depends on what level of horticulture you practice. Certainly, the horticultural scientist should thoroughly understand the concepts of classification and taxonomy. Scientific names should be an everyday part of his or her professional life. What level of scientific nomenclature does the nursery professional that sells plants need? This professional needs some understanding of how nomenclature works, the general relationships of plants, and the ability to handle scientific names. Many plants, especially the more unusual ones, are sold based on their scientific name. The backyard horticulturist can get by with very little, if any, knowledge about classification or taxonomy. Yet many of the more expert gardeners are surprisingly knowledgeable about such things, given their high level of interest in getting the "right plant." And what do you, the student, need to know? You need to understand the mechanics of the system, so that you have a solid foundation for the things that follow.

Let's start first with phylogenetic classification. Classification involves the arrangement of organisms, plants in our case, into groups that get progressively smaller. As the group gets smaller, the number of shared characteristics resulting from evolution increases. The overall effect of classification is to show the evolutionary relationships among the various plants. Each group is called a *taxon,* or plural, *taxa.* The larger groups, with only a few shared characteristics, are major taxa, the smaller groups with many shared characteristics, minor taxa. The taxa have different names that reflect their size and the degree of that group's relationship. Each taxon is separated into smaller units by the taxon that follows it.

To understand the mechanics, let's relate classification of plants to something you already know: computer file hierarchical organization. On your computer, you likely have created a directory that you use for your work. Let's say that you create a directory on your hard disk that you call **DATA**. Then you break this into smaller, more manageable units, called subdirectories or folders. These in turn are divided into more folders, until you reach a specific document, or file. **DATA** is the sum totals of all your files and is the equivalent of the kingdom containing all plants. Each file is a recognizable unique entity and is comparable to a plant species. The folders in between these two are the equivalent of the classification groups between the kingdom and species.

Your computer file system might look like this (not all folders are shown to simplify the arrangement):

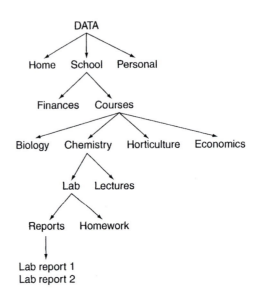

Classification is very similar to this file system. It follows a hierarchical arrangement:

MAJOR TAXA

Plant Kingdom (equivalent to data)
    Division (equivalent to home, school, or personal)
        Class (equivalent to finances and courses)
            Order (equivalent to biology, chemistry, horticulture, and economics)
                Family (equivalent to lab and lectures)

MINOR TAXA

Genus (equivalent to reports and homework)
    Species (equivalent to lab report 1, 2)
        Variety (equivalent to major revision of a lab report)
            Form (equivalent to minor revision of a lab report)
                Individual (equivalent to adding lab report 3 to list)

The Plant Kingdom

The plant kingdom (*Plantae* in Latin) is the largest taxon, collectively holding all known plants on earth, about 250,000 known species! The tropical rain forests hold many as yet unidentified plants that could greatly swell this number. While we are only interested in plants, other organisms also merit their own kingdoms. Others include the animal, fungal, moneran (bacteria), and protist (one- or few-celled organisms more complex than bacteria) kingdoms. Some divide the bacteria into two kingdoms, based on the fact that some bacteria are able to live in extreme environments (extremophiles). These bacteria are able to thrive in very hot or very cold environments, as well as in environments enriched in salt, methane, sulfur, and so forth.

A more recent approach, based in part on bacterial differences and evolutionary evidence (ribosomal RNA sequences), suggests that a higher level beyond the kingdom exists. These higher levels are called *domains*. Three domains are recognized: Bacteria, Archaea, and Eukarya. The Bacteria domain and the Archaea domain share a common ancestor and arose before the third domain. True bacteria are found in Bacteria and the extremophile bacteria in Archaea. Both of these groups have simple cells that are of the prokaryotic type. Prokaryotes lack membrane-bound nuclei and membrane-bound organelles. Eventually at some common ancestor many millennia later, Archaea split off the evolutionary line and this shared ancestor also gave rise to the Eukarya domain. This domain contains all the eukaryotes, whose cells have membrane-bound nuclei and other membrane-bound organelles (eukaryotic cells). This domain contains four eukaryotic kingdoms, plants, animals, fungi, and protists. The fifth kingdom, Monera, is split between the Bacteria and Archaea domains.

Whether there are five or six kingdoms, we, as horticulturists, are only interested in the plant kingdom. For our purposes, plants are defined as multicellular organisms containing green pigments called chlorophyll and cellulose in their cell walls. There are several other defining elements, but we will not worry about them.

Plant Divisions

Next the plant kingdom is divided into *divisions*. The division is used with plants. With animals, the term is changed to *phylum*, plural *phyla*. These different uses are a result of history. Many scientists feel today that plant divisions should be called phyla in the interest of consistency. How many divisions depend on whether you are a splitter or lumper. The lumper would argue that 2 divisions are needed, the splitter, 12. A further complication is that sometimes intermediate separations of a taxon are made and designated by the prefix *sub-*, for example, *subdivision*. To understand their differences of opinion, we need to look at what sets the various types of plants apart.

The more primitive plants are called nonvascular plants. These plants do not have specialized conducting tissues (analogous to pipelines) that move water and food around the plant body. Conducting tissues are called xylem, which conducts water and minerals from the root upward, and phloem, which moves food (sugar) from the leaves where it is made to where it is needed. Xylem and phloem are vascular tissues. Plants without these conducting tissues are called nonvascular plants. Surviving plants of this type include the mosses. Mosses and allied nonvascular plants (liverworts and hornworts) are called *bryophytes*. Plants with vascular systems (xylem and phloem) are more advanced products of evolution and are called *tracheophytes*. Unlike the bryophytes, this group has very large numbers of plants that span the evolutionary continuum from slightly more advanced than bryophytes to extremely advanced. An idea of when the various groups of plants first appeared in the evolutionary continuum is found in Table 2–1.

If you were a lumper, you could divide all the plants in the plant kingdom into two divisions that recognized their basic difference, the absence or presence of a vascular system. One would be called division Bryophyta (the bryophytes) and the other division Tracheophyta (the tracheophytes). Note here that the ending *-phyta* indicates the taxon or group is a division. If you were a splitter, you would say that neither division Bryophyta nor division Tracheophyta was acceptable. These two divisions would not give any indication of the magnitude of the evolutionary differences that existed in either bryophytes or tracheophytes. The lumper would acknowledge these differences on the next level of the classification scheme, the order. The "splitter" would insist that these differences were considerable and should be recognized as early as possible, that is, on the division level. I happen to agree with the latter view. So the classification scheme so far would look like this:

Kingdom Plantae (Plants)

1. Division Bryophyta (mosses)
2. Division Hepatophyta (liverworts)
3. Division Anthocerotophyta (hornworts)
4. Division Psilotophyta (whisk ferns)
5. Division Lycopodophyta (club mosses)

TABLE 2–1 *Geological Eras and Life Forms*

Era[2]	Period	Epoch	Comments
Precambrian: 4.5 bya[3]			Prokaryotic origins, 3.5 bya; eukaryotic origins, 1.5 bya. Soft-bodied invertebrates and first fungi, 700 mya.
Paleozoic: 570 mya[3]	Cambrian: 570 mya		Hard-shelled invertebrates, **sea weeds**. Mild climate, extensive seas over present continents.
	Ordovician: 510 mya		Mild climate. Begins with major extinction event. Crustaceans and mollusks dominate. Invasion of land by early fish and **land plants** likely. Seas shallow and cover much of U.S. Glaciers on Africa near end of period.
	Silurian: 439 mya		Mild climate. Begins with major extinction event. Continents flat. First **fossil plants**. First jawed fishes. Origins of **non-vascular bryophytes**, likely from a charophycean green **alga**.
	Devonian: 408 mya		Age of fishes. **Land plants** diversify: origin of **seedless vascular plants: ferns, lycophyte trees, giant horsetails**. Few **seed vascular plants** appear from line of **protogymnosperms (seedless gymnosperms): seed ferns** and **cordaites (primitive conifers)**. Extinction of primitive **vascular plants**. Insects appear. Seas cover most land. Some mountains.
	Carboniferous: 362 mya	Mississippian: 362 mya	Warm. Swampy. Amphibians appear and dominate. Origin of reptiles. **Forests** appear (Age of **ferns** and **lycophytes**, mostly tree types). Formation of coal.
		Pennsylvanian: 322 mya	
	Permian: 290 mya		Glaciers in southern hemisphere early. Appalachian Mountains formed. Some areas very dry. Origin of **conifers, cycads, ginkgoes**. Earlier **forest** types wane. Reptiles diversify. Largest mass extinction event occurs at end.
Mesozoic: 245 mya	Triassic: 245 mya		Continents mountainous and joined together. Large arid areas. **Forests of gymnosperms and ferns**. First dinosaurs and first mammals.
	Jurassic: 208 mya		Mild climate. Low areas, covered heavily by seas. **Gymnosperms**, esp. **cycads** dominate. Birds appear.
	Cretaceous: 145 mya		Climate uniform. Sea levels high. Africa and South America separate.
			Angiosperms appear. Insects diversify and dominate. Age of reptiles. Dinosaurs extinct at end.
Cenozoic: 65 mya	Tertiary: 65 mya	Paleocene: 65 mya	Mild to cool climate. Continental seas disappear. Early insect-eating mammals and primates appear.
		Eocene: 56.5 mya	Mild to very warm. Australia separates from Antarctica. India collides with Asia. Grasslands form. Major spread of mammals and birds. **Forests** shrink.
		Oligocene: 35.4 mya	Rise of Alps, Himalayas. South America separates from Antarctica. Volcanoes in Rocky Mts. Browsing mammals, monkey-like primates. Many **modern plant genera** evolve.
		Miocene: 23.2 mya	Moderate climate. Glaciers again in South America. **Forests** shrink further and **grasslands** spread. Grazing animals, apes.
		Pliocene: 5.2 mya	Cooler. Deserts form. First man-apes. Much mountain-building. Glaciers in northern hemisphere. Uplift at Panama joins North and South America.
	Quaternary: 1.6 mya	Pleistocene: 1.6 mya	Fluctuating cold to mild. Two-dozen glacial events. Many mountains form.
		Recent: 10,000 ya	Age of humans starts. **Agriculture** appears 10,000 years ago. Extinction of many large mammals and birds. (Wooly mammoth, saber-toothed tiger).

[1] Plant appearances are indicated in boldface type.

[2] Eras are major time units and are divisible into periods, which in turn are subdivided into epochs.

[3] Billion years ago and million years ago.

6. Division Equisetophyta (horsetails)
7. Division Pteridophyta (ferns)
8. Division Gingophyta (ginkgo tree)
9. Division Cycadophyta (cycads)
10. Division Coniferophyta (conifers)
11. Division Gnetophyta (gnetophytes)
12. Division Anthophyta (flowering plants)

Divisions of Horticultural Interest

Not all of these divisions are important from the horticultural view. Nonvascular plants, the first three divisions (mosses, liverworts, and hornworts), are of essentially no interest. In fact, some horticulturists take a dim view of mosses in the lawn. When utilized, mosses tend to be part of the naturalized landscape, such as in "woodsy" settings. These early plants in the evolutionary continuum not only have no vascular system, but they are also seedless. Spores are used for reproduction.

The next step in the evolutionary continuum was the appearance of vascular systems. These primitive plants are recognized in divisions 4 through 7. With the exception of the ferns (division Pteridophyta), horticulturists have little interest in these plants. None of these plants produce by seed, but continue the more primitive trait of reproduction by spores. A significant leap on the evolutionary continuum was the appearance of the first seed plants that produced seeds in cone-like structures (divisions 8 through 11). These plants are collectively called *gymnosperms*. Only the 10th division (division Coniferophyta) is of much interest here, because it contains the conifers, plants with very pronounced cones like the pine, spruce, fir, balsam, and cedar. Division 8 consists of only one plant, the gingko tree, which is the only surviving species in this division. Gingko trees are used sometimes in horticulture. Division 9 contains the cycads, primitive, ancient plants that are few in numbers and are endangered species. Some of these make very unusual house or greenhouse plants and are desired by collectors of the unusual. Division 11 contains some rather unusual plants that are seldom seen in horticulture and show some characteristics that make them appear to be a transition group between the preceding conifers and division 12. The most recent jump in plant evolution was the appearance of flowering plants (angiosperms) that produced seeds in fruits. These are represented by division 12 (division Anthophyta).

To summarize, only a few divisions are of interest to the horticulturist. The most primitive are the mosses. These nonvascular plants reproducing by spores were among some of the earliest plants in terms of evolution. As the evolution of plants continued, the next step was the vascular plants, but reproduction was still by spores. Ferns are the best horticultural example in this group. A later step in evolution was the appearance of the first vascular seed plants. The seed was borne unprotected in a cone-like structure. This group, the gymnosperms, is best represented in the horticultural sense by the conifers. Christmas trees and their relatives dominate this group and many of them form the mainstay of landscape plantings. The most recent step in evolution was the emergence of flowering plants, vascular seed plants with flowers and seeds protected inside fruits. These plants, the angiosperms, are essential for food production and landscaping.

Classes

All of these divisions can be divided into classes and so on. However, to keep it simple, we will look only at the flowering plants in detail. The division Anthophyta is broken into two classes, which represent a fundamental difference found in flowering plants. Evolution created two separate pathways or directions in this group. One direction gave us the monocots, and the other, the dicots (Figures 2–3 and 2–4). You can easily tell monocots from dicots by looking at the leaves. The veins on monocots are parallel, like railroad tracks. Some plants with this leaf pattern include tulips, lilies, palms, grasses, onions, and orchids. Dicots exhibit a nonparallel venation; examples are maples, oaks, tomatoes, apples, and

FIGURE 2–3   The lily is an example of a common monocot. Two distinguishing features of monocots are visible: parallel leaf venation and flower parts in threes or multiples thereof. Shown here is an Oriental hybrid lily cultivar, Stargazer. (Author photograph.)

FIGURE 2–4   An example of a dicot (dahlia). Dicots have netlike venation and flower parts in fours, fives or multiples thereof. This cultivar, 'Kevin Floodlight,' is a large type referred to as "dinnerplate" dahlias. (Author photograph.)

roses. Other differences exist. Monocots have one cotyledon or seed leaf in their seeds, hence, the name *monocot,* which is short for monocotyledon. *Dicots* (short for dicotyledons) have two. Other differences have to do with the anatomy of stems, roots, and flowers. Monocots are placed in the class Liliopsida (note the class ending, *-opsida*) and dicots in Magnoliopsida. There are about four times as many dicots as monocots.

Orders and Families

Things get complicated after this point. Given arguments over splitting and lumping, as well as differences of opinion on interpretation of evolutionary data, we have some competing schemes. What scheme you use depends on which prominent scientist you believe. You can break the classes into anywhere from 69 to 83 orders, which in turn can be divided into 387 to 440 families. Let's not worry about it. The order and family endings are *-ales* and *-aceae,* respectively. For more details on these different classification schemes, see the Internet listings at the end of this chapter. A few examples of orders are Polypodiales, Coniferales, Commelinales, and Rosales. These orders are associated with ferns, conifers, angiosperms (monocot type), and angiosperms (dicot type), respectively. While each contains several families, a single family of horticultural note from each, respectively, is the osmunda fern family (Osmundaceae), pine (Pinaceae), grass (Poaceae), and rose (Rosaceae) family.

Families are groups that can be well known to horticulturists. Families are named after the most prominent member of the group, such as the rose family, Rosaceae. The family ending is *-aceae*. A few exceptions to this ending exist, such as the sunflower family, Compositae. Newer names have been put forth for these exceptions. Compositae then becomes Asteraceae, and Gramineae becomes Poaceae (grass family). Given the large number of families, we cannot list them all here. A few families of horticultural note from the flowering plant division, Anthophyta, are offered in Table 2–2. Given that house plants alone come from some 118 families, we are just scratching the surface.

In a similar fashion, the division Coniferophyta (conifers and other gymnosperms) can be broken down into classes, then orders, and finally families. For example, the yew is found in the class Taxopsida. Its family name is Taxaceae, the yew family. The well-known pine is found in Pinaceae, the pine family.

TABLE 2–2 *Some Angiosperm Families of Horticultural Note*

Family Name	English Name	Dicot (D) or Monocot (M)
Arecaceae	Palm family	M
Asteraceae	Aster family	D
Begoniaceae	Begonia family	D
Brassicaceae	Mustard family	D
Bromelicaceae	Pineapple family	M
Cactaceae	Cactus family	D
Caryophyllaceae	Carnation family	D
Cornaceae	Dogwood family	D
Fabaceae	Pea family	D
Iridaceae	Iris family	M
Liliaceae	Lily family	M
Musaceae	Banana family	M
Oleaceae	Olive family	D
Orchidaceae	Orchid family	M
Poaceae	Grass family	M
Rosaceae	Rose family	D
Solanaceae	Nightshade family	D
Theaceae	Tea family	D

The Genus

Once we get to the taxon after the family, the *genus*, or plural *genera,* we get into territory of greater interest to the horticulturist. The collective units that comprise the family, the genera, are relatively small groups of closely related plants or other organisms (species) that share some very close phylogenetic characteristics. The number of plants in a genus lends itself well to studies of biochemical, cytological, ecological, and genetic relationships. Sometimes these studies, when completed, lead to the shifting of one or more plants from one genus to another genus, or even the elimination of a genus. Classsification is in continual flux, as new phylogenetic data arise.

The genus is a simple concept. All maples that we see in nature and in horticultural usage belong to the maple genus, *Acer.* In a similar fashion, all oaks are found in the oak genus, *Quercus.* The cultivated onion and its cultivated relatives, chive, garlic, leek, and shallot, along with their wild relatives, form the genus *Allium.* The pine and its relatives comprise the genus *Pinus,* and the garden pink and its relatives, both cultivated and wild, form the genus *Dianthus.*

The Species, Or How Plants Are Named

The collective units that comprise the traditional genus are called *species* (Figure 2–5). Each species is comprised of plants (or other organisms) possessing unique and essentially identical morphological and phylogenetic characteristics that are capable of interbreeding among themselves. The species may be assumed to represent a continuing succession of like plants from one generation to the next.

The species is written always in the context of the genus to which it belongs. Using the genus name as the base, or part of the binomial nomenclature, we can distinguish the red maple from the sugar maple by invoking the next taxon, the species. For example, the red maple species is *Acer rubrum* and the sugar maple is *Acer saccharum.* The species names of the various maples are formed from the genus (*Acer*) and a descriptive adjective called the specific epithet (*rubrum* or *saccharum*).

Acer rubrum, or the red maple, has reddish tiny flowers in the spring and brilliant red leaves in the fall. *Acer saccharum,* the sugar maple, is used to make maple syrup. By looking at them, we can recognize them as two different species and see that they share enough

FIGURE 2–5 *Verbascum thapsus.* This species is a common biennial weed naturalized throughout the United States. Some common names include flannel mullein and woolly mullein. Native Americans and colonists used it in the past for various medicinal purposes. (Author photograph.)

characteristics to be related; that is, both are maples. The shared genus, or the maple relationship, is evident from the shared *Acer.*

However, the different specific epithet, or the second part of the binomial nomenclature, tells us that they are distinct species of maples. The word *rubrum* is a Latin word, meaning "red." Our word *ruby* is derived from it. The word *saccharum* is Latin for "sweet" and the name for the artificial sweetener, *saccharin,* is derived from it. In this instance, the Latin is more or less obvious. More often than not, the meaning or descriptive term meant by the specific epithet isn't always that apparent. In a similar manner, the white oak would be *Quercus alba. Alba* is Latin for "white." Our word *albino* comes from it. The red oak would be the species *Quercus rubra.* Why the change from *rubrum* to *rubra?* Gender. The specific epithet has to have the same gender as the genus term. One is masculine, the other feminine. We have Linnaeus to thank for this useful means to name plants and to secure their place in the taxonomic scheme.

Your own name is somewhat analogous. Your family name is like the genus and your given name is the equivalent of the specific epithet. Your name appears on the professor's class list in the following fashion: Smith, Mary. Essentially, that is the species format! The only difference is that you are a unique individual, whereas many thousands of essentially genetically identical red maples are all collectively identified or named by the species term *Acer rubrum.*

It is worth noting at this time that you can always recognize the species format when you see it. The taxon species always consists of two words, the first of which is always capitalized and the second always lowercase. The two words are always written in italics and, of course, in Latin. On rare occasions, Greek, but in the form of Latinized Greek, is used. So you should have no trouble recognizing *Begonia semperflorens* as a species. This begonia comes originally from Brazil and is referred to as the wax begonia. Its cultivated descendents are widely used as bedding plants in the garden and as house plants. You should be aware that the species concept and the other taxonomy aren't just for plants. Other organisms use this same system. For example, *Homo sapiens* is the species that you belong to, thinking humans.

The Workings of Classification

Let's now look at how this all works. In Tables 2–3 and 2–4 you will find two species of plants with their placement in the taxonomic scheme.

Below the Species

Does that complete the picture? Well, not just yet. You may recall the original list of taxa seemed to contain some taxa below the species level. That might puzzle you. If the species

TABLE 2–3   *Classification of a Common Lawn Grass, Kentucky Bluegrass*

Taxon	Name	Description*
Kingdom	Plantae	Multicellular organisms rooted in soil or organic matter. Contain chlorophyll in cellular structures called chloroplasts that capture energy from the sun to make sugar by a process called photosynthesis.
Division	Anthophyta	Plants having a vascular system for food and water conduction. Reproduction involves the pollination of flowers by wind or certain creatures followed by double fertilization that produces a fruit containing seeds. Known as flowering plants or angiosperms.
Class	Liliopsida	Characterized by parallel venation of leaves, one cotyledon in the seed, flower parts often in threes, and randomly placed stem vascular bundles. Known as monocots.
Order	Commelinales	Leaves are fibrous. Flowers are relatively inconspicuous.
Family	Poaceae	Grass family. Reedy-like stems. Often spread by rhizomes underground. Green inconspicuous flowers.
Genus	*Poa*	Genus containing more than 100 species of temperate grasses, often tufted. A few are important lawn grasses.
Species	*Poa pratensis*	Kentucky bluegrass. Best lawn grass species. Widely used in central and northern United States. Tufted, spreads easily and has attractive blue green color.

*Not meant to be the complete detailed characteristic description of each taxon. Only meant as descriptive information to help you place each taxon in context.

TABLE 2–4   *Classification of a Common Landscape Plant, White Pine*

Taxon	Name	Description*
Kingdom	Plantae	Multicellular organisms rooted in soil or organic matter. Contain chlorophyll in cellular structures called chloroplasts that capture energy from the sun to make sugar by a process called photosynthesis.
Division	Coniferophyta	Plants having a vascular system for food and water conduction. Reproduction involves the pollination of cones by wind followed by a single fertilization that produces a seed borne unprotected on the scales of the cone. Known as conifers or one of the gymnosperms. Trees of large size. About 550 species.
Class	Coniferopsida	Characterized by needled or scale-like leaves.
Order	Coniferales	Order containing six families of temperate and tropical evergreens. A few species are deciduous.
Family	Pinaceae	Pine family. Others in this family include firs, cedar, larch, spruce, Douglas fir, and hemlock.
Genus	*Pinus*	Pine genus; containing many species of familiar temperate, needled evergreens highly valued for their lumber and landscape use. Found throughout northern temperate regions.
Species	*Pinus strobus*	Eastern white pine. Good landscape plant. Widely used in eastern United States. Has five needles in each sheath.

*Not meant to be the complete detailed characteristic description of each taxon. Only meant as descriptive information to help you place each taxon in context.

identifies the plant in an unequivocal manner, then why have other taxa below it? Good question. The answer is that evolution is not a stagnant process. As we read, species are changing as a result of evolutionary pressures. Categories below the species help us sort out those changes that have occurred in some species. That's why we need the taxa called variety, form, and individual. Of course, if the change is large enough to totally change some plants in the species, the new plants might merit a whole new species name! These taxa below the species are often of considerable interest to the horticulturist. Varieties and forms are frequently more pleasing, colorful, or more attractive than the species itself.

So what are these lesser changes? Certain degrees of morphological or phylogenetic variance from the essential identifying features of the species occur. If a group within the species shows minor but consistent differences in conjunction with specific geographical or ecological distribution, it may be classed as a subspecies. If the differences are not associated with distribution, it may only warrant the status of *variety* (Latin, *varietas*). The term *form* (Latin, *forma*) is for very minor variances, such as flower or fruit color. These differences are inheritable and should show predominately in the offspring. These fine distinctions may be lost in practice, but they do exist and serve to designate variances from the species norm. In fact, many of our interesting horticultural plants are a variety or form of an otherwise common species (Figure 2–6).

Plants of horticultural significance are often forms, varieties, subspecies, hybrids, and cultivars (see later section). As such it is often necessary to include them in the scientific name. Forms may be designated as *Astrophytum myriostigma* f. *nudum,* varieties as *Astrophytum myriostigma* var. *coahuilense* (or *Astrophytum myriostigma coahuilense*), and subspecies as *Pachypodium lealii* ssp. *Saundersii.* Hybrids, the result of crossbreeding two different plants, require special name forms. The names of hybrids are designated by a preceding X, such as *Clematis* X *jackmanii*, a hybrid between *C. ianuginosa* and *C. viticella.*

Sometimes a unique plant, an *individual,* (Latin, *individuum*) appears with characteristics that attract the eye of the horticulturist. This plant, an unusual result of chance mutation or genetic recombination, is known as a *sport.* The sport is propagated vegetatively, since it might not reproduce true from seed. The group of plants propagated by vegetative means is called a *clone.* The clonal progenitor is the *ortet,* and the individual offspring is designated a *ramet.*

These chance mutations have considerable economic value to the horticultural industry. The Red Delicious apple originated as a mutation in Iowa in 1895. This mutation mutated again in 1925, giving rise to the Red Delicious of commerce that we use today. Today, the Red Delicious apple is the number one apple in world commerce!

FIGURE 2–6 The variety is below the species level. Shown here is *Morus alba pendula* (weeping mulberry) which differs from the upright species, *Morus alba* (white mulberry). The weeping variety is often grafted onto the upright species to produce an umbrella-like tree form. (Author photograph.)

Fine-Tuning of Scientific Names

Is scientific nomenclature perfect? No, even though the scientific name is unique, some minor discrepancies can arise. Sanctioned name changes are sometimes accepted slowly, which leads to the use of two scientific names during the period of transition. The problems of naming some plants can be so complex as to cause disagreement among the experts. A world-level organization, the International Commission on Botanical Nomenclature, exists to mediate nomenclature disputes and questions. Nevertheless, professionals and amateurs now use scientific names as a matter of course. The need for proper identification far outweighs any resistance to their use.

Those of us who have studied Latin or have good memories may find the scientific names easy to learn. Many do not find it easy beyond the few cases where the Latin words suggest the English equivalent or where the scientific name is already used in the common sense. Those scientific names used in the common sense include such genera as *Chrysanthemum, Delphinium, Geranium, Calceolaria, Begonia, Philodendron, Caladium, Coleus, Cyclamen,* and *Dracaena.* For those who find it difficult, the only help is the association of the scientific name with exposure to plant material over a long period of time, reinforced by reference to the literature.

At times scientific names are written as *Agave parviflora* Torr. or *Sedum frutescens* Rose. The nonitalicized part refers to the scientist who has established the scientific name. Most of the names of the authorities, unless short, are abbreviated. The authority is not cited during popular writing or common speech, but only when a sense of botanical or historical accuracy is desired.

The Cultivar

While this completes our discussion of the taxa of scientific nomenclature and taxonomy, it is not quite the end. Many cultivated plants have a practical significance to humans and especially to the horticulturist. A special horticultural taxonomic category unique to plants, the cultivar, is used to designate these cultivated plants. The *cultivar,* a contraction of "cultivated variety," should be differentiated from the naturally occurring botanical variety discussed previously. Cultivars may be thought of as plants that have originated and persisted under cultivation; they are clearly distinguished by characters of a morphological, biochemical, cytological, or physiological nature. These characteristics are retained through asexual and often sexual reproduction. Different cultivars of a particular plant are often collectively termed a *group,* another horticultural category. For example, certain cultivars of rhododendrons can be placed into a specific, named group as a means to define their specific relationship.

Some cultivars are the result of unplanned, long-term, farmer selection, that is, the cumulative effect of saving the seeds from the best plants for next year's crop. This process occurred during cultivation for up to several thousands of years for wheat, rice and corn. The time span for corn is so long that we don't know the actual species that gave rise to our current cultivars of corn. We know that it was likely a wild grass species, but which one is still uncertain. The term *cultigen* is used to refer to plants that exist only in cultivation and whose origins remain unknown. The ancestry of wheat is known, going back some 8,000 plus years to a hybrid of two wild grasses.

Some cultivars are more modern, that is, created by plant breeders with deliberate crosses of plants during the last century or so. Modern cultivars exist for most of our garden and farm vegetables, such as for tomatoes and the newer cultivars of corn, wheat, and rice. Many ornamental cultivars also exist for flowers, shrubs, grasses, and trees. More recently, cultivars have been created in weeks to a few years, thanks to the modern technology of genetic engineering.

Modern cultivars are of special interest to the horticultural business world, because these cultivars can be patented/copyrighted. Legally protected cultivars can produce large profits for those who create them. Cultivars that attract a large share of the corn, wheat, or tomato market can produce considerable profit. For example, about 500,000 acres of fresh market and processed tomatoes are produced annually in the United States. Add on to this the fact that about 40 million gardeners grow tomatoes each year. If your new cultivar were to attract the lion's share of the tomato seed market, you stand to make a lot of money! Corn and wheat acreage makes these numbers look tiny by comparison when one considers that the United States produces one-half of the world's corn!

Cultivar Types.   Cultivars consist of five basic types. Asexually reproduced cultivars come from clones (*clonal cultivar,* Figure 2–7) or apomicts *(apomictic cultivar).* The Red Delicious apple is a clonal cultivar. Apomicts are unique plants produced from seed that did not arise through the usual sexual means, but from an asexual process. Certain cultivars of citrus and lawn grasses are apomictic cultivars. Sexually produced cultivars arise from self-fertilizing plants that breed true naturally (*line cultivars*), from cross-fertilizing plants maintained true through selection and isolation (*inbred line cultivars*), or from selectively crossbred or hybridized plants (*hybrid cultivars,* Figure 2–8). Many vegetables, grains, and flowers are line, inbred, and hybrid cultivars.

FIGURE 2–7   Most fruit trees are asexually propagated and hence are clonal cultivars. The apple cultivar, 'Red Delicious,' arose as a chance mutation and is undoubtedly today the most globally famous clonal cultivar. (Author photograph.)

FIGURE 2–8   This cultivar is sexually reproduced rather than asexually. Specifically, this popular petunia used widely in hanging baskets and window boxes is a hybrid cultivar designated 'Purple Wave.' (Author photograph.)

Cultivar Names.   A cultivar name can be written several ways. The cultivar name utilizes the language of the country where it originated, is not italicized, is set off by single quote marks and is limited to a maximum of three words. The cultivar can be expressed in several contexts. One might find *Sedum guatemalense* cv. Aurora (or *Sedum guatemalense* 'Aurora'). This form is most useful when one wishes to establish a complete identity for the cultivar, that is, link the cultivar to the species from which it came. Scientists and professional horticulturists use this format a lot. A less formal approach is corn 'Silver Queen.' Gardeners and trade personnel often use this form. Sometimes the reference is dropped entirely if the cultivar's reference plant is well known, for example, 'Big Boy,' 'Early Girl,' and 'Better Boy.' Most gardeners and farmers would recognize these names immediately as tomato cultivars. Sometimes this form is even shortened to simply Big Boy, Early Girl, and Better Boy, as in seed catalogs and seed packages where printing costs are lowered by dropping the single quote mark. This latter form is not technically correct, but "shortcuts" do happen.

HORTICULTURAL CLASSIFICATION OF PLANTS

Horticulturists have devised other schemes for categorizing the many plants that constitute the horticultural realm. These schemes collectively form the basis of a horticultural plant classification. This system of classifying plants is useful, since it is based on practical criteria such as usage, growth habits, and ability to tolerate environmental conditions. As such, it enables the horticulturist to select the proper plant for any given use, condition, or site and to predict the final appearance of the mature plant.

Growth Habits

Plants can be conveniently classified on the basis of growth habit or other significant physiological characteristics. *Nonwoody plants* with comparatively soft tissues, of which the aerial portion is relatively short lived, are called *herbaceous* or *succulent* plants. *Woody plants* have long-lived, dense, strong aerial tissues that increase in length and diameter each year. Herbaceous plants can be divided further on the basis of their stem growth habit. Seed-producing herbaceous plants with self-supporting stems are *herbs* (Figure 2–9). Herbaceous plants with a trailing or climbing habit are *vines* (Figure 2–10). Woody plants with self-supporting stems

FIGURE 2–9   Oregano is a good example of an herb in the horticultural classification sense in that it is a herbaceous plant that produces seeds and has upright stems. It also is a good example for confusion resulting from terms having double meanings, in that in the culinary sense it is also an herb used in the kitchen. (Author photograph.)

FIGURE 2–10   The climbing string bean in the garden seen here growing on netting is a typical vine. It has a twining vine habit. (Author photograph.)

FIGURE 2–11 *Clematis X jackmanii* hybrids are vines that become woody with age, so can technically be considered lianas. Other well-known lianas are climbing roses, trumpet vine (*Campsis radicans*), *Wisteria,* and certain species of *Allamanda.* (Author photograph.)

FIGURE 2–12 *Forsythia* is an attractive spring-flowering deciduous shrub. This one, *Forsythia suspensa* 'Sieboldii' is popular because of its graceful, arching branches. (Author photograph.)

FIGURE 2–13
Deciduous trees (center and left) add considerable beauty to the landscape during the growing season. (Author photograph.)

are either *trees* or *shrubs;* those that climb or trail are correctly designated as *lianas* (Figure 2–11). Often the term *liana* is overlooked and the term *vine* is substituted. *Shrubs* (Figure 2–12) are usually low with several shoots or trunks produced at the base, as opposed to the single trunk and distinct, elevated head of the *trees* (Figure 2–13). The distinction between trees and shrubs can be blurred by horticultural practices like pruning or training or by the effects of environmental conditions. Plants that grow on trees or other plants, but do not take moisture or nutrients from the host, are known as *epiphytes* (Figure 2–14). They obtain their essential nutrients from decaying organic matter that collects around their aerial roots.

FIGURE 2–14
Bromeliads are epiphytes grown outdoors in warmer areas of California and Florida, and indoors elsewhere. Shown here is a species of *Tillandsia* which in its native habitat would be found growing in tree canopies in the subtropics and tropics. Instead it is mounted on a board (barely visible at bottom) normally hung on a wall and sprayed occasionally with water and nutrients. (Author photograph.)

(A)   (B)

FIGURE 2–15   (A) The Colorado blue spruce is a common evergreen tree characterized by bluish needles. Shown here is a dwarf cultivar (*Picea pungens* 'Moerheim'). A red pine (*Pinus resinosa*) can be seen in the background that was deformed after infestation by European pine shoot moth and red pine scale. The tree died a few years later after this photo because these insects were not controlled. (B) The Japanese Andromeda is a common evergreen shrub. Shown here is *Pieris japonica* 'Valentine' noted for its reddish-pink flowers. (Author photographs.)

Foliage Habits

Foliage-retention habits form the basis of a major distinction between plants. Those that are always in leaf are called *evergreens* (Figure 2–15); those that lose all their leaves during a portion of the year (usually winter) are known as *deciduous*. Evergreens do shed their leaves, but at a gradual rate over one or more years and therefore are never without leaves. The persistence of leaves can vary with climate; some plants that are evergreen in the tropics become deciduous in cooler regions.

Size and Shape

Size is frequently an important factor in horticultural classification. Terms like *dwarf, small, medium,* and *tall* may be applied to shrubs or trees. These categories play an important part when a plant is chosen for landscape purposes. Another factor of importance to landscapers is the shape of the plants. Classifications for trees include *globular* (roundheaded), *fastigiate* (columnar), *pyramidal, picturesque* (irregular), *weeping* (drooping), and *horizontally branching.* Shrubs can be classed as *weeping, upright,* or *horizontally branching.* Vines can be described as *trailing, clinging,* or *twining.*

Environmental Tolerance

Plants are often classified on the basis of their tolerance toward environmental conditions such as shade, sun, drought, wet soils, hot or cold temperatures, infertile soils, air pollution, road or sea salts, alkalinity. This information is helpful when we select plants.

Some of these categories are especially detailed, such as the cold temperature or acid-to-alkaline tolerances of plants. In the cold temperature category, we find a division of large regions, such as the United States and Canada, into hardiness zones based on the limits of the average annual minimal temperatures (see Chapter 12). Plants would be called *hardy* or *tender,* depending on their ability to survive the minimal temperature of a zone. Additional refinements of this category are the distinctions made for woody plants of *wood hardiness* and *flower bud hardiness.* A woody plant might be capable of survival at certain low temperatures, but flower buds might not. *Root hardiness* is also important, especially in container-grown plants wintered over. Temperatures in the root zone could be lower in a container than would be encountered in the ground; failure to recognize this point could result in extensive winterkill. Such factors can restrict plants to certain areas, which explains, for instance, the centralization of the apricot industry in California. Horticultural crops are classified even further on the basis of temperature preference during the growing season. Thus, corn, tomatoes, peppers, eggplants, and melons are referred to as *warm season* crops, whereas lettuce, spinach, and cole crops are known as *cool season* crops.

Life Spans

Horticulturists are often concerned with classification of the life spans of plant material. Plants that finish their growth cycle from seed through maturity and death in one growing season are called *annuals* (Figure 2–16). Plants that require two seasons to complete their growth cycle are known as *biennials* (Figure 2–17). Vegetative growth occurs during the first growing season, and blooming, seed formation, and death occur in the second growing season. Plants that continue to grow for several or more growing seasons and usually do not die after seed formation are designated *perennials* (Figure 2–18). These life-span classifications are not absolute. Subtropical perennials grown in cooler areas cannot live through the winter; thus they become annuals in those areas. Certain root crops are biennials, but they are harvested after one season, so they are treated as annuals. Annuals and biennials are usually herbaceous, whereas perennials can be herbaceous or woody.

Usage

The classification of plants by use is probably the most important to the horticulturist. Essentially, three use categories exist: (1) edible plants, (2) ornamental plants, and (3) plants that are sources of natural products used as medicines, drugs, condiments, beverages, oils, insecticides, and gums or resins.

FIGURE 2–16 The zinnia is a common annual in gardens. (Author photograph.)

FIGURE 2–17 Parsley, a member of the carrot or parsley family (Apiaceae), is a biennial. (Author photograph.)

FIGURE 2–18 *Echinacea purpurea*, a herbaceous perennial, is a popular garden perennial and also the source of the herbal remedy, echinacea, used to build up the immune season during cold and flu seasons. (Author photograph.)

Edible Plants.   Each of these groups can be divided further. The *edible* plants consist of *fruits* and *vegetables*. Generally, fruits are considered to be the mature, ripened, seed containing ovary and sometimes accessory parts of a flowering plant (botanical fruit) that when edible tastes sweet and is served as a dessert (Figure 2–19). *Vegetables* are any edible part of a herbaceous plant (including actual botanical fruit) served raw or cooked with the main course of the meal (Figure 2–20). These definitions are not botanical ones and are not absolute, because certain exceptions arise. For example, freshly picked sweet corn fits the definition of a botanical fruit, but is eaten as a vegetable. Rhubarb fits more closely the definition of a botanical vegetable (plant parts other than botanical fruit), but its use as a dessert makes it a fruit to most of us. The concepts of fruits and vegetables can also vary with nationalities and geographical regions.

FIGURE 2–19   The orange is an example of a fruit, both in the common and horticultural sense. It is also a woody evergreen perennial tree. (Author photograph.)

FIGURE 2–20   Lettuces (cultivars of *Lactuca sativa*) are typical vegetables, both in the common and horticultural sense. They are herbaceous annuals. Shown here is a red looseleaf lettuce cultivar called 'Red Sails.' (Author photograph.)

Fruit plants are frequently woody perennials, with those of the temperate zone usually being deciduous and those of the tropics evergreen. Fruit subdivisions include *tree fruits, small fruits,* and *nuts.* Deciduous tree fruits include the *pome fruits* (apple and pear) and the *drupe fruits* (stone fruits such as peach, plum, apricot, and cherry). Nuts harvested from deciduous trees include the walnut, pecan, almond, and filbert. Deciduous small fruits encompass grapes, red and black raspberries, blackberries, loganberries, cranberries, strawberries, currants, gooseberries, and blueberries. Evergreen tree fruits consist of citrus fruits (lime, lemon, orange, and grapefruit) and miscellaneous fruits (fig, persimmon, date, avocado, mango, guava, and papaya). Tropical evergreen nut trees include the coconut, Brazil nut, cashew, and macadamia. Other evergreen fruits of the warmer areas are the banana and pineapple.

Vegetables can be divided into two categories; those harvested for (1) aerial parts or (2) underground portions. The category of aerial parts is divided further into stems and/or leaves plus fruits and/or seeds. Underground portions may be roots, tubers and rootstocks, or bulbs. Vegetables grown for their stems and/or leaves include the *cole* or *cabbage crops* (cabbage, Chinese cabbage, brussels sprouts, kohlrabi, kale, cauliflower, collards, mustard, and broccoli), *greens* or *pot herbs* (New Zealand spinach, Swiss chard, spinach, and turnip greens), *salad crops* (lettuce, endive, celery, cress, watercress, and parsley), and *miscellaneous crops* (asparagus, rhubarb, and artichoke). Vegetables grown for their fruits and/or seeds are the *legumes* or *pulse crops* (snap beans, lima beans, soybeans, and peas), the *solanaceous fruit crops* (tomato, pepper, and eggplant), the *vine crops* or *cucurbits* (squash, pumpkin, watermelon, cucumber, and melon), and *miscellaneous crops* (sweet corn and okra). Vegetables grown as *root crops* include the radish, carrot, turnip, rutabaga, salsify, sweet potato, cassava, and parsnip. Those harvested for their *tubers* and *rootstocks* are the potato, Jerusalem artichoke, and yam. Vegetables used for their *bulbs* or *corms* include the onion, garlic, shallot, leek, and taro.

Ornamentals.   Ornamental plants are usually separated into two groups: (1) *nursery plants* and (2) *flowering* plus *foliage plants.* Flowering and foliage plants are separated according to their life span: annuals, biennials, and perennials. Popular annuals include marigolds, petunias, impatiens, zinnias, and snapdragons. Some biennials of note are sweet william, Canterbury bells, and foxglove. A few perennials would be chrysanthemum, delphinium, hyacinth, columbine, and iris. Plants grown in flower gardens, perennial borders, and indoors (house plants, greenhouse plants) are grouped under flowering and foliage plants.

Nursery plants are an extensive division of the ornamentals. Subdivisions of this group are *lawn* or *turf plants,* such as bluegrass, bentgrass, Bermuda grass, fescues, and zoysia; *ground covers* like pachysandra, *hosta,* periwinkle *(Vinca minor),* and sedum; *vines* such as English ivy *(Hedera helix),* clematis, glorybower *(Clerodendrum),* and wisteria; *shrubs* (usually deciduous) like viburnum, flowering quince, lilac, and hydrangea; *evergreens* (trees and shrubs) such as blue spruce, rhododendron, andromeda, and Japanese holly *(Ilex crenata);* and *trees* (usually deciduous) like maple, oak, mountain ash, and dogwood.

Natural Products.   Plants are frequently sources of natural products. This use category has several divisions. Medicines and drugs such as witch hazel, digitalis, quinine, and cocaine are derived from *medicinal plants.* Condiments such as pepper, basil, oregano, thyme, ginger, and cinnamon come from *herb* and *spice plants.* Coffee, tea, and cocoa are derived from *beverage plants.* Cooking oils can come from *oil-bearing plants,* for example, from seeds of sunflower, soybeans, or canola. Insecticides such as pyrethrum, rotenone, ryania, sabadilla, hellebore, and nicotine were first isolated from plants, although most of them have been replaced by synthesized analogues. Plants are also sources of gums, resins, and turpentine.

PLANT IDENTIFICATION KEYS

At times the horticulturist encounters plants that are difficult to identify. A first recourse is to consult some of the excellent books of plant descriptions. If a careful search fails, the plant may not be in these plant manuals because the taxonomist or horticulturist felt it was of little value or too rare to be included. The possibility that it has not yet been given a name exists too. Under these circumstances the plant can be sent to a professional taxonomist at a well-known university, herbarium, or botanical garden. Once identified, the horticulturist can verify the identity by comparison with dried, pressed specimens found in herbarium collections. If indeed the plant has not been named previously, a specimen of it too will enter the herbarium collection.

Plants can be identified through some manuals by an analytical key arranged in dichotomous form. The *key* confronts the reader with series of contrasting, paired choices. As one choice of the pair is eliminated, the correct one determines which set of choices is consulted next. Again one choice is eliminated, and the next set of choices is chosen, and so on, until the plant is identified or "keyed out."

For example, a key to the vascular seed plants would start out with two choices. One would lead you to the gymnosperms (conifers) and the other to the angiosperms (flowering plants). Choice 1a would detail facts specific to conifers, but not flowering plants. The descriptive material would describe seeds borne on scales arranged to form a cone and other conifer-specific facts. Choice 1b would describe seeds contained within a fruit that appeared after a flower and other flowering plant details. If 1a matched, you would be directed to some other number, likely 2a and 2b. This choice set would start you with conifers and help to lead you to the appropriate category from which your plant-in-hand came, such as pines or yews. If 1b matched, you would have been directed to choices that dealt with flowering plants. Keys are currently available both in book and software formats. A simple example is shown in Table 2–5.

At times the key leads to confusion on the species level, since the diagnostic features of the key are morphological characteristics (leaf, twig, seed, flower, fruit, or others), unlike the sexual, phylogenetic, or other features used for plant classifications. On questions of identity with taxa below species, such as variety or form, it often becomes necessary to consult specialized monographs dealing with the plant group under study. Successful use of keys requires an understanding of the morphological terms (often complex), patience, practice, and a knowledgeable person to verify your answers on your initial attempts.

TABLE 2–5   *Dichotomous Key of Some New England Trees**

A.	Needle-like leaves	B or BB
AA.	Broad-leaved	G or GG
B.	Needles in bunches of 2, 3, 5	C or CC
BB.	Needles not in bundles, many found along branchlets	E or EE
C.	Needles in bunches of 5	Eastern white pine
CC.	Needles in bundles of 2 or 3	D or DD
D.	Needles in bundles of 2, cones usually curved	Jack pine
DD.	Needles in bundles of 3	Pitch pine
E.	Cones on end of branch, needles flat, not prickly	Eastern hemlock
EE.	Cones pendant (hanging down) or upright, not at end of branch	F or FF
F.	Cones pendant, needles often prickly and four-sided	Spruce (white, black, or red)
FF.	Cones upright/purplish, needles usually blunt or notched at the tip/light colored underneath, soft to the touch	Balsam fir
G.	Broad-leaved, simple	H or HH
GG.	Broad-leaved, compound	K or KK
H.	Maple tree group: only group with leaves opposite along twig and having 3 to 5 lobed leaves	I or II
HH.	Birch tree group: leaves alternate on twig and oval, heart shaped at base, or triangular	L or LL
I.	Leaf stem reddish, sharp angled sinuses	Red maple (swamp)
II.	Leaf stem not reddish, sinuses not sharp angled	J or JJ
J.	Leaf has finely toothed edges, green bark with whitish vertical stripe	Striped maple (moose wood)
JJ.	Leaf coarsely toothed, light colored underneath, bark dark or somewhat greenish, not striped	Mountain maple
K.	Long narrow, toothed leaflets, 3 times long as wide	Mountain ash
KK.	Other trees with compound leaves	Locust, sumac, ash, hickory, walnut
L.	White bark peeling in wide (1 inch plus) papery layers	Paper (white) birch
LL.	Bark yellow to silver-gray, peels into thin (less than 1 inch wide) curly strips, crushed twigs smell of wintergreen	Yellow birch

*Key courtesey of Leon Barkman, retired Professor of Biology at Housatonic Community College and currently an Appalachian Mountain Club Hut Naturalist. This key is used for tree identification around the AMCs mountain huts.

SUMMARY

Common names for plants are fine in normal conversation and in written form for most people. However, when exact identification is needed, as in botany, plant biochemistry, or use of medicinal plants, common names fail. The main reasons are that common names are not individually assigned or unique to each plant and names vary with geographical change or language. Names used by scientists, or scientific names, ensure reliable identification of a plant when it matters. Horticulturists need to be familiar with the latter approach (taxonomy). Humans also like to group things into categories (classification) so that relationships can be studied and knowledge advanced. It is human nature to wonder about and understand relationships among living things.

Both taxonomy and classification started before recorded history, as humans used plants for food, medicine, and other natural products. Some written documents survive from ancient Egypt and China regarding plants used by people, but the first documentation of classification and taxonomy appeared in ancient Greece. The word horticulture is derived from the Romans, but they were not responsible for much classification or taxonomy. By the 1600s, as the number of known plants increased and similarities between plants arose more often, plant names had become very cumbersome, because they consisted of strings of adjectives following a noun.

Carolus Linnaeus finally brought order to the plant name chaos in 1753 when he published *Species Plantarum,* which set forth both a taxonomic and classification process. His classification scheme did not survive, because improvements in understanding plant relationships led to better systems. His system of taxonomy, however, persists today and is used for all organisms, not just plants, earning him the title father of taxonomy. His naming system, termed *binomial nomenclature,* relied on two words to name plants, such as we do with names for people. These names were designated as species names and Latin was used because it was a common language across the globe for scientists. For example, the red maple became *Acer rubrum.* Linnaeus's system allows one to assign a unique name to each different plant. Names are always in Latin, italicized, and the first and second words are always uppercase and lowercase, respectively. The first word designates the genus, *Acer* = maple, and the second, the specific epithet, specifies an adjective (*rubrum* = red).

His classification scheme used sexual characteristics, essentially flowers, to classify plants. Classifications based on one characteristic are known as artificial classification. One form of artificial classification survives today in wildflower guides, which base identification on flower color and color photographs. It soon became apparent that more characteristics needed to be considered in the process of looking for relationships among plants. Systems that followed were based on many characteristics, including flower structures, many anatomical, morphological, physiological, and even biochemical ones. These systems were known as natural classification systems.

When Darwin proposed his theory of evolution in 1853, it set change in motion. Eventually, scientists realized that relationships were determined by evolution and that classification needed to reflect that reality. For example, some plants had very similar morphology and physiology, yet it was hard to believe they were related, given their geographic isolation. An example of this is the cacti of our Southwest and the euphorbia of African arid areas. Plants in separate geographic areas that are unrelated can develop similar features when subjected to the same unfavorable climatic features over long periods of time. We now understand this process and know this phenomenon as convergent evolution. With the introduction of evolution into classification, the new system became known as phylogenetic classification.

The tool we use to determine evolutionary relationships is the determination of the nucleotide sequences of DNA and RNA, which reflect the evolutionary heritage in today's plants. This approach is known as *molecular systematics.* Currently we have made great strides in converting natural classification to the phylogenetic system. However, the conversion is not complete. Much time, effort, and cost are involved with nucleotide sequence determinations and the subsequent analyses. In addition, a good fossil record is necessary to finalize relationships to extinct plants. The fossil record has gaps, especially with flowering plants. We have, nevertheless, made considerable progress. We know that the evolutionary line for plants started with simple, small spore-bearing (seedless) plants without vascular systems (internal structures for water and food movement). An example of that type of plant would be the mosses. The next step was the appearance of seedless plants with vascular systems, such as today's ferns. Ferns were followed by vascular plants that produced seeds, but no flowers, as represented by today's gymnosperms, such as the conifers. The final step was the appearance of seed vascular plants with flower and fruits, today's flowering plants, the angiosperms. The majority of the plants of interest to horticulturists in decreasing order are flowering plants, conifers, and ferns.

The mechanics of classification involve a hierarchical approach. Groups of related plants start off in large groups, and these are split into smaller and smaller groups where the relationship becomes closer and closer. All organisms are placed into three domains, based on the evolutionary features of their cells. Plants are put into the same domain as animals, fungi, and other complex cellular organisms. The next level down is the kingdom where plants have their own kingdom, the plant kingdom. Plants are then broken down further and further into smaller groups in the following sequence: division (also known as phyla), class, order, family, genus, species, variety, form, and individual. These groups or categories are generically known as taxa.

While there are several divisions, those of interest to horticulturists include Pteridophyta (ferns), Coniferophyta (conifers such as pine, spruce, yew, etc.), and Anthophyta (flowering plants). These are broken down gradually based on evolutionary relationships. By the time we get to the family level, we are talking about the rose family or the grass family, and at the genus level, the maple or oak genus (*Acer* and *Quercus,* respectively). When we arrive at the species level, we have specific plants, such as red maple (*Acer rubrum*). Taxa below this point reflect minor variations from the species norm, such as variegated leaves. These changes aren't large enough to merit a new species name, but often are of great horticultural interest. The presence of a third descriptive term on the species name indicates varieties and forms. If an X appears in the middle of a species name, it indicates a cross between two species, which is a hybrid. The individual taxon is reserved for a spontaneous mutation that causes great departure from the species norm, which, if given time and survival, might become a new species.

Another type of name is also found in horticulture, the cultivar. *Cultivar* is a combination of the two words "cultivated variety." A cultivar is any plant that has been modified by human action. For example, corn as we know it today bears no resemblance to the original wild grass cultivated thousands of years ago. Farmers, by saving seeds from the best plants year after year, gradually brought about modifications in corn. This process is called *artificial selection,* as opposed to natural selection by evolutionary forces. Cultivars can also be produced by selective crossing of two plants (plant breeding) or by gene transfer in the laboratory (genetic engineering). These two ways of creating cultivars can be patented, which provides an incentive for constant improvements of plants. Cultivar names are in the language of the country of origin and usually indicated by single quote marks around two or three words, such as 'Big Boy' tomato or 'Purple Wave' petunia. If crossing two different parent plants produces the cultivar, it is a hybrid cultivar. If the plant is produced by vegetative propagation (cloning), it is a clonal cultivar.

A form of horticultural jargon also exists; it is used to describe plants and has value for choosing plants for food and landscaping purposes. Some terms describe tissues, such as herbaceous (soft) or woody (hard). Other terms indicate foliage retention habits, such as deciduous or evergreen. Other terms describe life expectancy, such as annual (one growing season) or perennial (many growing seasons). Some terms describe growth habits, such as shrub, tree or vine; or size (dwarf) or form (weeping). Terms can also be used to describe environmental tolerance to pollution or how the plant is used, such as vegetables and fruits (edible plants) or ornamentals (landscape plants). These terms can be broken down further, such as *Cole* crops (cabbage, broccoli, cauliflower, etc.) or foliage plants (house plants grown primarily for their attractive foliage).

Plant identification is usually accomplished with what is known as a key. A key presents a series of two choices. When you answer one, you are directed to another set, which leads you to another set, until eventually the plant is identified. Types of questions can be broad at first, such as is the venation parallel? If yes, you are directed to questions involving monocot flowering plants; if no, dicot flowering plants. Questions at the end can be very detailed, such as does the plant have five needles or two needles per bundle? Keys are available in written form and also computerized format.

EXERCISES

Mastery of the following questions shows that you have successfully understood the material in Chapter 2.

1. Scientists do not use common names for plants. What are the disadvantages of common names from the perspective of the scientific community?
2. Problems with the early methods of naming plants scientifically resulted in a new approach to naming by Carolus Linnaeus. What were the problems?
3. Carolus Linnaeus developed a naming system for plants and other living organisms called *binomial nomenclature*. How does it work and what are its advantages?

4. What is the difference between taxonomy and classification? What is the purpose of classification?

5. Three approaches to classification have existed historically. Name these three approaches. Explain how each system works. Were there any weaknesses or problems with the first two systems? What system is currently used today and, what, if any, problems are present with that approach?

6. Explain the hierarchical nature of classification. What are the names and hierarchical arrangement of the taxa used in classification today? Why was the concept of domains introduced recently?

7. What plant divisions are of interest to horticulturists? What kinds of plants are found in those divisions?

8. Explain the concepts of genus and species. How is a species named? Are there any taxa below the species taxon? If yes, what are they and why are they needed?

9. Explain the concept of a cultivar. Why are cultivars important to horticulturists? How is a cultivar named? Is there more than one type of cultivar? If so, what are they and how do they differ?

10. Horticulturists use certain terms in order to group plants based on growth habits and usage. Give some examples of these terms and their definitions. Why are such terms useful in horticulture? Why are perennials more useful in horticulture than annuals?

11. Explain how a plant key works.

INTERNET RESOURCES

Comparison of four approaches to flowering plant classification currently in use
http://www.csdl.tamu.edu/FLORA/newgate/gateopen.htm

Index of all current and valid published suprageneric names
http://matrix.nal.usda.gov:8080/cgi-bin/starfinder/0?path=suprag.txt&id= anon&pass= &OK=OK

Integrated taxonomic information system with numerous and worldwide links
http://www.itis.usda.gov/itis/index.html

Plant taxonomy resources, includes links to information on plant taxonomy, identification, geographic distribution, ethnobotany, phytochemistry, and ecology
http://deal.unl.edu/CYT_agnic/siteForUser/index.html

Rules and codes for naming plants from the International Association for Plant Taxonomy
http://www.bgbm.fu-berlin.de/iapt/default.htm

Searchable database of high-quality images and horticultural descriptions for hundreds of unique species and cultivars
http://www.hcs.ohio-state.edu:90/pmi_4/FMPro?-db=PMI4&-lay=input&-format= plantsnew.html&-view

Taxonomy of landscape plants
http://members.tripod.com/~Hatch_L/tlan1.html

Tree of life, phylogenetic relationships of ferns, gymnosperms, and angiosperms
http://tolweb.org/tree?group=Embryophytes&contgroup=Green_Plants

Vascular plant images based on family names
http://www.csdl.tamu.edu/FLORA/gallery.htm

Vascular plant nomenclature at the family level
http://www.inform.umd.edu/PBIO/fam/revfam.html

RELEVANT REFERENCES

Anderson, Edward F. (2001). *The Cactus Family*. Timber Press, Portland, OR. 776 pp.

APS (1998). *An Ordinal Classification for the Families of Flowering Plants*. Annals of the Missouri Botanical Garden 85:531–553.

Baumgardt, John Philip (1982). *How to Identify Flowering Plant Families: A Practical Guide for Horticulturists and Plant Lovers*. Timber Press, Portland, OR. 269 pp.

Brako, Lois, Amy Y. Rossman, and David F. Farr (1995). *Scientific and Common Names of 7000 Vascular Plants in the United States*. American Phytopathological Society, St. Paul, MN. 295 pp.

Bryan, John E. (1989). *Bulbs* (two volumes). Timber Press, Portland, OR. 451 pp.

Chin, Wee Yeow (1998). *Ferns of the Tropics*. Timber Press, Portland, OR.

Coombes, Allen J. (1994). *Dictionary of Plant Names*. Timber Press, Portland, OR.

Cope, Edward A. (2001). *Muenscher's Keys to Woody Plants*. Cornell University Press, Ithaca, NY. 368 pp.

Cope, Edward A. (1986). *Native and Cultivated Conifers of Northeastern America*. Cornell University Press, Ithaca, NY. 224 pp.

Cronquist, A. (1988). *The Evolution and Classification of Flowering Plants* (2nd ed.). New York Botanical Garden. Bronx, NY. 555 pp.

Crow, Garrett E., and C. Barre Hellquist (2000). *Aquatic and Wetland Plants of Northeastern North America: Pteridophytes, Gymnosperms and Angiosperms: Dicotyledons* (Vol. 1). University of Wisconsin Press, Madison.

Crow, Garrett E., and C. Barre Hellquist (2000). *Aquatic and Wetland Plants of Northeastern North America: Monocotyledons* (Vol. 2). University of Wisconsin Press, Madison.

Cullen, James (1997). *The Identification of Flowering Plant Families: Including a Key to Those Native and Cultivated in North Temperate Regions* (4th ed.). Columbia University Press, New York. 192 pp.

Darke, Rick (ed.) (1995). *Manual of Grasses*. Timber Press, Portland, OR.

Dirr, Michael A., Bonnie Dirr (illustrator), Margaret Stephan (illustrator), Asta Sadauskas (illustrator), and Nancy Snyder (illustrator) (1998). *Manual of Woody Landscape Plants: Their Identification, Ornamental Characteristics, Culture, Propagation and Uses* (5th ed.). Stipes Publishing, Champaign, IL. 1187 pp.

Dressler, Robert L. (1993). *Phylogeny and Classification of the Orchid Family*. Timber Press, Portland, OR.

Eggli, U., and H. E. K. Hartmann (eds.) (2001). *Illustrated Handbook of Succulent Plants* (6 volumes). Springer-Verlag, New York.

Foote, Leonard E., and Samuel B. Jones, Jr. (1989). *Native Shrubs and Woody Vines of the Southeast: Landscaping Uses and Identification*. Timber Press, Portland, OR.

Foster, F. Gordon (1984). *Ferns to Know and Grow*. Timber Press, Portland, OR.

Graf, Alfred Byrd (1992). *Hortica: Color Cyclopedia of Garden Flora in All Climates—Worldwide—and Exotic Plants Indoors*. Roehrs Company, East Rutherford, NJ. 1216 pp.

Graf, Alfred Byrd (1992). *Tropica: Color Cyclopedia of Exotic Plants and Trees from the Tropics and Subtropics* (4th ed.). Roehrs Company, East Rutherford, NJ. 1152 pp.

Harris, James G., and Melinda Woolf Harris (2001). *Plant Identification Terminology: An Illustrated Glossary*. Spring Lake Publishing, Spring Lake, UT. 216 pp.

Henderson, Andrew, Gloria Galeano, and Rodrigo Bernal (1995). *Field Guide to the Palms of the Americas*. Princeton University Press, Princeton, NJ. 335 pp.

Hollingsworth, Peter M., Richard M. Bateman, and Richard Gornall (eds.) (1999). *Molecular Systematics and Plant Evolution*. Taylor & Francis, New York. 448 pp.

Jelitto, Leo, and Wilhelm Schacht (1990). *Hardy Herbaceous Perennials* (two volumes). Timber Press, Portland, OR.

Magee, Dennis W., Abigail Rorer (illustrator), and Harry E. Ahles (1999). *Flora of the Northeast: A Manual of the Vascular Flora of New England and Adjacent New York*. University of Massachusetts Press, Amherst. 1264 pp.

Malcolm, Bill, and Nancy Malcolm (2000). *Mosses and Other Bryophytes: An Illustrated Glossary*. Timber Press, Portland, OR.

Niklas, Karl J. (1997). *The Evolutionary Biology of Plants.* University of Chicago Press, Chicago, IL. 449 pp.

Petrides, George A. (1973). *A Field Guide to Trees and Shrubs: Northeastern and North-Central US and Southeastern and South-Central Canada.* Houghton-Mifflin, Boston.

Petrides, George A., and Roger Tory Peterson (illustrator) (1998). *A Field Guide to Eastern Trees.* Houghton-Mifflin, Boston.

Petrides, George A., Roger Tory Peterson (ed.), and Olivia Petrides (illustrator) (1998). *A Field Guide to Western Trees* (2nd ed.). Houghton-Mifflin, Boston.

Singh, Gurcharan (1999). *Plant Systematics.* Science Publishers, Enfield, NH. 258 pp.

Stuessy, Tod F. (1990). *Plant Taxonomy.* Columbia University Press, New York. 514 pp.

Symonds, George W. D., and A. W. Merwin (photographer) (1973). *The Shrub Identification Book.* William Morrow & Co., New York. 379 pp.

Symonds, George W. D., and Stephen V. Chelminsky (1973). *The Tree Identification Book.* William Morrow & Co., New York. 272 pp.

Takhtajan, A. (1980). *Outline of the Classification of Flowering Plants (Magnoliophyta).* Bot. Rev. 46:225–359.

Taylor, T. W., and E. L. Taylor (1993). *The Biology and Evolution of Fossil Plants.* Prentice-Hall, Upper Sadde River, NJ. 982 pp.

Uva, Richard H., Joseph C. Neal, and Joseph M. DiTomaso (1997). *Weeds of the Northeast.* Cornell University Press, Ithaca, NY.

van Gelderen, D. M., and J. R. P. van Hoey Smith (1996). *Conifers: The Illustrated Encyclopedia.* Timber Press, Portland, OR.

Walters, Dirk R., and David J. Keil (1996). *Vascular Plant Taxonomy* (4th ed.). Kendall/Hunt Publishing Company, Dubuque, IA. 622 pp.

Woodland, Dennis W. (1997). *Contemporary Plant Systematics.* (2nd ed.) Andrews University Press, Berrien Springs, MI. 619 pp.

CLASSIC REFERENCES

The following books may be out of print and available only through used book channels. However, many of these books still have some value in plant identification and taxonomy and clearly have historical value.

Backberg, C. (1977). *Cactus Lexicon* (English ed.). Blandford Press, Dorset, England. 828 pp.

Bailey Hortorium at Cornell University Staff (1976). *Hortus Third.* Macmillan, New York. 1290 pp.

Bailey, L. H. (1949). *Manual of Cultivated Plants Most Commonly Grown in the Continental United States and Canada* (rev. ed.). Macmillan, New York. 1116 pp.

Benson, Lyman (1992, updated in 1992). *The Cacti of the United States and Canada.* Stanford University Press, Palo Alto, CA. 1044 pp.

Benson, Lyman (1979). *Plant Classification* (2nd ed.). D.C. Heath and Company, Lexington, MA. 901 pp.

Blake, S. F., and A. C. Atwood (1942, 1961). *Geographical Guide to Floras of the World; Annotated List with Special Reference to Useful Plants and Common Plant Names. I. Africa, Australia, North America, South America, and Islands of the Atlantic, Pacific, and Indian Oceans. II. Western Europe: Finland, Sweden, Norway, Denmark, Iceland, Great Britain with Ireland, Netherlands, Belgium, Luxembourg, France, Spain, Portugal, Andorra, Monaco, Italy, San Marino, and Switzerland,* Miscellaneous Publications 401 and 797. U.S. Department of Agriculture, Washington, DC.

Borg, J. (1976). *Cacti: A Gardener's Handbook for Their Identification and Cultivation.* Blandford Press, Dorset, England. 512 pp.

Britton, Nathaniel Lord, and Addison Brown (1970). *An Illustrated Flora of the Northern United States and Canada* (three volumes; reprint of 1913 version). Dover Publications, New York. 2052 pp.

Elias, Thomas S. (1980). *The Complete Trees of North America: Field Guide and Natural History.* Times Mirror Magazines, New York. 948 pp.

Flint, Harrison L. (1983). *Landscape Plants for Eastern North America: Exclusive of Florida and the Immediate Gulf Coast.* John Wiley & Sons, New York. 677 pp.

Gilmour, J. S. L. et al. (1969). *International Code of Nomenclature for Cultivated Plants* (Vol. 64). Regnum Vegetabile, Utrecht, Netherlands. 32 pp.

Graf, A. B. (1975). *Exotica: Pictorial Cyclopedia of Exotic Plants* (8th ed.). Roehrs Company, East Rutherford, NJ.

Harrison, Charles R. (1975). *Ornamental Conifers.* Hafner Press, New York. 224 pp.

Jacobsen, H. (1976). *Handbook of Succulent Plants: Descriptions, Synonyms and Cultural Details for Succulents Other than Cactaceae* (three volumes; English ed.). Blandford Press, Dorset, England. 1445 pp.

Jaeger, Edmund C. (1978). *Desert Wild Flowers* (rev. ed.). Stanford University Press, Stanford, CA. 322 pp.

Little, Elbert L. (1980). *The Audubon Society Field Guide to North American Trees: Eastern Region.* Alfred A. Knopf, New York. 714 pp.

Little, Elbert L. (1980). *The Audubon Society Field Guide to North American Trees: Western Region.* Alfred A. Knopf, New York. 639 pp.

Mathias, Mildred E. (ed.) (1976). *Color for the Landscape: Flowering Plants for the Subtropical Climates.* California Arboretum Foundation, Arcadia. 210 pp.

Ouden, P. den, and B. K. Boom (1978). *Manual of Cultivated Conifers* (English ed.). Martinus Nijhoff, Boston. 526 pp.

Plowden, C. C. (1970). *A Manual of Plant Names* (2nd ed.). Philosophical Library, New York.

Rehder, A. (1940). *Manual of Cultivated Trees and Shrubs Hardy in North America Exclusive of the Subtropical and Warmer Temperate Regions* (2nd ed.). Macmillan, New York. 996 pp.

Rickett, Harold William (1961–1966). *Wildflowers of the United States. The Northeastern States, The Southeastern States, Texas, The Southwestern States, The Northwestern States, The Central Mountains and Plains* (six volumes). McGraw-Hill, New York.

Sargent, Charles Sprague (1965). *Manual of Trees in North America* (two volumes, reprint of 1922 ed.). Dover Publications, New York. 934 pp.

Stafleu, F. A., et al. (1972). *International Code of Botanical Nomenclature* (Vol. 82). Regnum Vegetabile, Utrecht, Netherlands. 426 pp.

Stearn, W. T. (1966). *Botanical Latin: History, Grammar, Syntax, Terminology, and Vocabulary.* Hafner Press, New York.

Thomas, Graham Stuart (1976). *Perennial Garden Plants or the Modern Florilegium.* David McKay Company, New York. 389 pp.

Tyron, Rolla M., and Alice F. Tryon (1982). *Ferns and Allied Plants with Special Reference to Tropical America.* Springer-Verlag, New York. 857 pp.

Williams, John G., and Andrew E. Williams (1983). *Field Guide to Orchids of North America.* Universe Books, New York. 143 pp.

VIDEOS AND CD-ROMs*

The Biology of Plants (Video, 1998).

Non Flowering Plants (CD-ROM, 1999).

Plant Kingdom (Video, 2001).

Seedless Plants (Video, 1991).

*Available from Insight Media, NY (www.insight-media.com/IMhome.htm)

Chapter 3

Plant Parts and Functions

OVERVIEW

At first glance there would appear to be few essential similarities among the plants that horticulturists favor. Variety and diversity would seem to be the obvious descriptive factor. But however diverse they may look, all of these plants carry out the same basic internal processes and possess great similarity in the construction of their seemingly diverse structures.

Generally, the plant body (Figure 3–1) has two basic parts. The part below the ground is called the root and the aboveground part, the shoot. Roots anchor the plant, absorb minerals and water from the soil, and conduct the absorbed materials to the stem base. Some roots also serve as food storage structures.

The shoot functions include support, food manufacture, reproduction, conduction, and sometimes food storage. Stems, leaves, buds, and sometimes reproductive structures collectively form the shoot. The joint where a leaf may be borne or is borne on the stem is called a *node*. The space between the nodes is known as the *internode*. *Lateral buds* are usually located at the leaf base between the leaf and stem angle. A *terminal bud* is present at the stem apex (tip). Buds are capable of growing into either branches that duplicate the existing shoot structures, or flowers, or both. Reproductive structures include flowers, fruits, and seeds on flowering plants and cones and seeds on conifers.

The whole plant can be broken down into smaller and smaller units, as we go from observing with the human eye to using a light microscope and finally to using an electron microscope. In order of decreasing size, these morphological structures and anatomical features are organs, tissue systems, tissues, and the cell with its cellular components. Roots, shoots, buds, leaves, flowers, fruits, cones, and seeds can be broken down into tissue systems, then tissues, and finally cells.

CHAPTER CONTENTS

OBJECTIVES

- To understand the structure and functions of roots and their relationship to horticultural practices
- To understand the structure and functions of nonreproductive parts: stems, buds, and leaves
- To understand the importance of modified stems, leaves, and flowers in terms of evolution and horticulture

- To understand the structure and functions of reproductive parts: flowers, fruits, cones, and seeds
- To understand the types of cells and tissues found in plants and what their functions are

FIGURE 3–1 A generalized diagram of the vascular plant body found with most horticultural plants. See text for explanation. The dormant terminal bud and accessory ("contingency") buds would be more characteristic of woody rather than herbaceous plants and fruits/seeds would not be found with the few seedless vascular plants of horticultural interest, such as ferns. Cones and seeds rather than fruit and seeds would be found with conifers. (From Raymond P. Poincelot, *Horticulture: Principles & Practical Applications,* (1980). Reprinted by permission of Prentice-Hall, Upper Saddle River, NJ.)

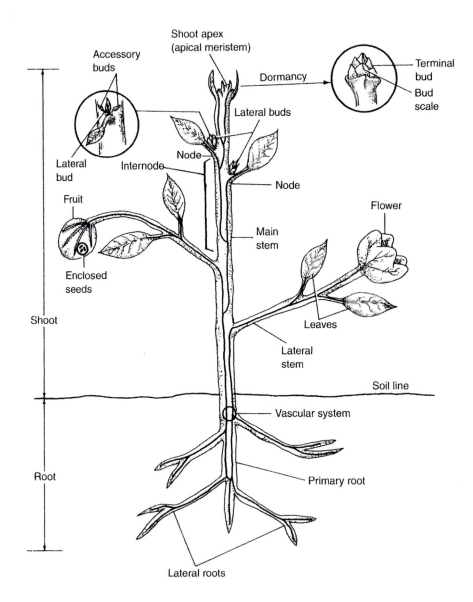

ROOT STRUCTURE

The general microscopic regions of a young root (found at tips of growing roots) are shown in Figure 3–2. The processes of cell division and elongation, initiated by the *root apex* (an *apical meristem,* or physiologically young cells that form new tissues and maintain themselves by cell division), occur primarily in the *zone of elongation.* The *root cap* protects the root apex from mechanical injury and acts as a lubricant as the root pushes through the soil particles. In the *zone of maturation* the cells have or almost have attained physiological ma-

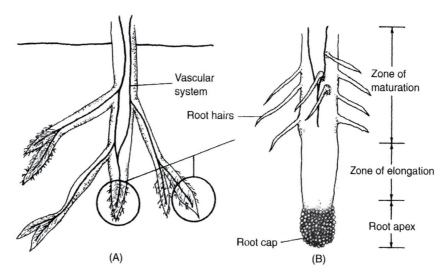

FIGURE 3–2 (A) Generalized diagram of vascular plant roots. See text for explanation. (B) Root tip. The zone of maturation contains the root hairs where much of the water and mineral absorption occurs prior to translocation upward in the xylem (part of vascular system). Many plant roots also have a symbiotic relationship with fungi (mycorrhizae), which enhances water/nutrient absorption. Food reaches the roots via phloem in the vascular system. (From Raymond P. Poincelot, *Horticulture: Principles & Practical Applications,* (1980). Reprinted by permission of Prentice-Hall, Upper Saddle River, NJ.)

turity. Root hairs are produced in this zone. The tissues of the zone of maturation are highly specialized for functions of absorption and translocation of water, minerals, and food (see the later discussion on xylem, phloem, and vascular systems and also Figures 3–3 and 3–4).

Taproots and Fibrous Roots

The root arises from the lower end or radicle of the embryo during germination of the seed. This early or primary root system grows downward into the soil and begins the root system of the plant. If the root system consists mostly of lateral roots near the surface of the soil, it is known as a *fibrous root system*. When the system consists predominately of the primary root growing downward with minimal branching, it is called a *taproot system*.

Taproots are found on plants such as the carrot, dandelion, and hickory nut, as opposed to fibrous-rooted plants like lawngrass. Plants with fibrous root systems respond quicker to fluctuations in the water and nutrient supply than do plants with taproots, since the fibrous root system tends to be shallow. Plants with taproots fare well during droughts, because their roots reach deeper down into the wetter parts of soil. Some plants have both a lower taproot and upper fibrous roots; the taproot increases access to the water supply, and the fibrous upper roots provide better nutrient uptake in the more fertile upper layers of soil.

Adventitious Roots

Another type of root, the *adventitious root,* is produced by many plants. The term *adventitious* in biology refers to appearing at an unusual place. Adventitious roots appear at locations other than the normal root system. These roots develop from aerial portions of the stem, such as the prop roots of corn, or from rhizomes, cuttings, stolons, and corms. Adventitious roots originate within or near the vascular tissues of the stem, which are involved in food and water conduction.

Adventitious roots can help some plants to spread or colonize adjacent areas. Adventitious roots are also one of the mainstays of commercial plant propagation (see Chapter 7). Adventitious root formation, such as on plant cuttings, can be enhanced with hormonal treatments for purposes of commercial propagation. Essentially, horticulturists take advantage of the ability of plants to form adventitious roots and apply this knowledge for commercial gain.

ROOT FUNCTION

Root systems can account for more than half the dry weight of the whole plant. The primary function of the roots is absorption of water and nutrients and transport of absorbed materials to the base of the stem. Secondary functions are anchorage and support of the shoot. In addition, roots serve as food-accumulating organs in some plants.

Some roots become rich in accumulated foods in the form of sugars and starch. Many plants with food storage roots provide a valuable addition to our diet. Food storage roots become enlarged and fleshy and form thick taproots in some plants. Plants with edible food storage taproots are the beet, carrot, jicama, parsnip, radish, rutabaga, salsify, and turnip. In other plants the branch roots can become swollen and function as storage organs. Roots of this type are known as *tuberous* (or storage) *roots* and occur on the dahlia, Jerusalem artichoke, and sweet potato.

Root systems are ideally suited to absorption, since the irregular arrangement of their many complex branches creates a very large surface area in contact with the soil. Much of this surface area derives from the numerous root hairs associated with areas of new growth. Root hairs slough away as they get older and farther back on the maturing root. Much of the absorption is through the root hairs, although some absorption occurs in young roots through their epidermal cells (the outermost single layer of cells).

ROOTS AND HORTICULTURAL IMPLICATIONS

The type of root system does have horticultural implications. A plant with a taproot system is difficult to *transplant* (removal of the plant by digging it up and placing it at a new site). The problem is that much of the deep taproot gets left behind in the soil. The plant is unable to survive this setback. Plants with taproots are frequently grown as container stock at nurseries to eliminate this problem. The whole taproot, usually in a spiral or corkscrew form in the soil ball of the container, is intact when you transplant the plant to your garden or landscape site. Plants with fibrous roots are much easier to transplant, because most of the root system can be removed during transplanting operations.

Root hairs also explain why, even with care, the transplanting of any plant causes a setback. No matter how careful you are when removing the plant, you do some damage to root hairs and lose some absorption capability. Depending on the extent, the plant may wilt or not. Some time is needed to restore this loss. With care, this time can be kept to a minimum. For this reason, plants grown in individual containers transplant better than field-dug plants. Nursery stock in containers is preferred over ball and burlap stock, and bedding plant plugs over flats.

The instant lawn from sod rolls is made possible because grasses have a fibrous root system. Grass is grown on huge sod farms, such as those in Ohio. When orders arrive for sod, a machine with a sharp blade slices through the soil just below the root zone of the grass. While slicing the sod from the soil, the machine also rolls the sod blanket. The shallow aspect of the grass roots is such that only a minimal amount of soil is lost. The sod rolls are transported by truck to various nurseries and sold either directly to consumers or to landscape contractors who install instant lawns.

One horticultural practice comes from understanding the function and placement of root hairs. Applying fertilizer in a circular band from the drip zone outward for woody plants is the most efficient way to fertilize trees or shrubs. The drip zone is the circular area below the outer edges of the ball of foliage. The reason is that the root ball in the soil generally exceeds the diameter of the foliage ball. The root hairs are found at the end of roots only and do most of the absorption of nutrients. Essentially, the greatest concentration of root hairs will be found beyond the drip zone.

Certain fungi in contact with roots, *mycorrhizae,* increase the absorption of water and nutrients by the roots of many important horticultural plants. The term *myco* means "fungus" in Latin, and *rhizae* means "root," hence *mycorrhizae* essentially means fungus-root. In exchange for their help on absorption, the fungus is given some nutrients from the plant. These nutrients increase the growth and reproduction of the fungus. These fungal partners can actually live separately from plants in the soil also. When they exist apart, the growth and reproduction of these isolated fungi are much slower.

Mycorrhizae can be especially important in those mature root areas that no longer have root hairs. It is estimated that some 90% of horticultural plants have mycorrhizal partners. This knowledge helps explain why certain plants propagated in nurseries and grown in home landscapes don't grow as well as their counterparts in natural environments. The nursery plants are missing their mycorrhizal partners. Sometimes nurseries will add some soil from natural sites to inoculate their container or field plants with the appropriate mycorrhizal fungi. More on mycorrhizae can be found in Chapter 6.

STEM STRUCTURE AND FUNCTION

Shoots are composed of stems, leaves, buds, and sometimes reproductive parts. The stem supports the leaves and reproductive structures; it also conducts the nutrients and water absorbed by the roots to the leaves. Some stems have limited food production abilities when young and green. Stems and stem modifications may also act as food storage organs. Plant forms are also determined by the structure and growth of the stems.

The stem has many specialized tissues (Figures 3–3 and 3–4), because of complex stem functions such as growth, support, and conduction. Herbaceous stems grow primarily in

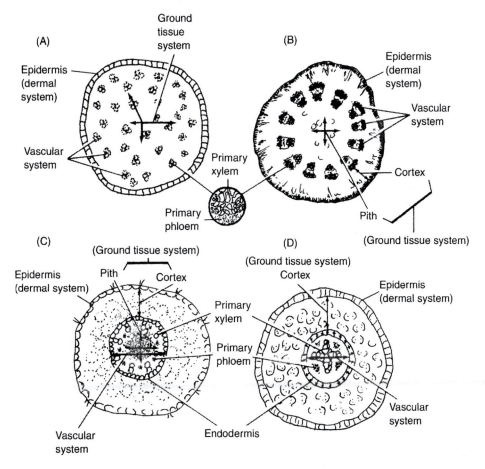

FIGURE 3–3   Cross sections through the monocot (herbaceous) stem (A), herbaceous dicot stem (B), monocot (herbaceous) root (C), and herbaceous dicot root (D). Monocots do not become woody (some become fibrous such as palms) because no vascular cambium is present. Dicots can be either herbaceous or woody. Woodiness requires vascular cambium activity (see woody dicots, Figure 3–4). Monocot stems are easily distinguished from herbaceous or woody dicot stems on the basis of vascular system symmetry (dicots have symmetry, monocots do not). Tissues shown here derived from the shoot apical meristem and are involved in lengthening of the stem. Tissues are discussed later in this chapter. (From Raymond P. Poincelot, *Horticulture: Principles & Practical Applications,* (1980). Reprinted by permission of Prentice-Hall, Upper Saddle River, NJ.)

FIGURE 3–4 Cross section through the woody dicot stem (upper) and woody dicot root (lower). Woodiness and increases in stem or root diameter result from activities of the two lateral meristems, the vascular cambium and the cork cambium. As the stem thickens, the epidermis and later the cortex and primary phloem are obliterated. Tissue destruction is similar in the root, except that the endodermis is also destroyed. While not shown here, cross sections through woody gymnosperm stems and roots look somewhat the same and have similar tissue losses. (From Raymond P. Poincelot, *Horticulture: Principles & Practical Applications,* (1980). Reprinted by permission of Prentice-Hall, Upper Saddle River, NJ.)

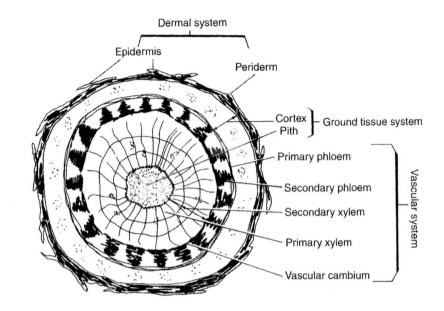

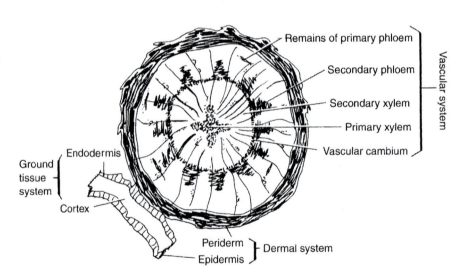

length and woody stems in length and diameter. Growth in length is initiated by the apical meristem (Figure 3–5), and growth in diameter by lateral meristems (*vascular cambium* and *cork cambium*). Upright growth of plants resulting from a rigid stem with one actively growing point is the norm. Multiple stems and several growing points produce shrubby or bushy growth.

BUDS

A bud (Figure 3–5) is an embryonic structure (either stem, flower, or both) that can grow actively or maintain the potential of growth while in a dormant stage. *Terminal buds* occur at the tips of stems and *lateral (axillary) buds* are found in the leaf axils (angle where leaf is attached to the stem). Some species have *accessory buds* that are located above or to the side of the axillary bud. Many buds develop into leafy stems; these are termed *leaf buds.* *Flower buds,* such as those found on elm or cherry, produce flowers, and *mixed buds,* such as those found on the apple, can give rise to leafy stems and flowers.

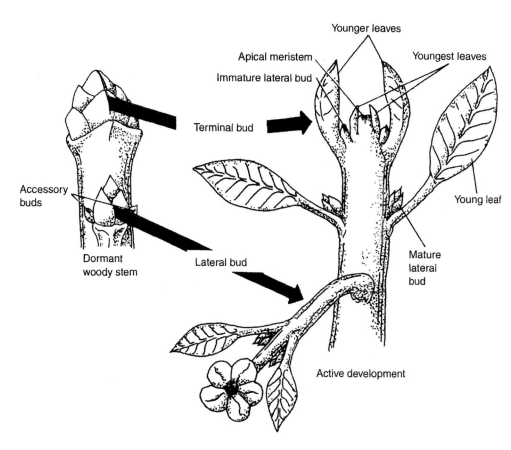

Younger leaves

Apical meristem

Immature lateral bud

Younger leaves

Youngest leaves

Terminal bud

Accessory buds

Young leaf

Dormant woody stem

Lateral bud

Mature lateral bud

Active development

FIGURE 3–5 Buds and shoot development. Terminal buds contain an apical meristem and are found on tips of growing shoots. These buds are involved in stem lengthening, production of leaves and lateral buds (leafy stems), and formation of flowers in some cases. Lateral buds give rise to leafy stems and/or flowers. Accessory buds can produce (when present) similar structures in event of lateral bud destruction. While buds are found both on herbaceous and woody plants, they are more pronounced in the latter case, especially dormant buds. (From Raymond P. Poincelot, *Horticulture: Principles & Practical Applications*, (1980). Reprinted by permission of Prentice-Hall, Upper Saddle River, NJ.)

Bud scales, often found in an overlapping pattern, cover the buds of most woody plants. Bud scales reduce water loss from the bud and protect it from mechanical injury. The buds of some woody plants have no bud scales, and such buds are called *naked buds*. In these buds the outer embryonic leaves are well developed and covered with hairs or scales; such leaves protect the younger, enclosed leaves. Nonwoody or herbaceous plants have less conspicuous buds than woody plants.

Adventitious shoot buds arise from areas other than expected, such as from a root, or at developmental stages other than expected, such as from mature cells no longer in the meristematic condition. Many cells, even mature ones, can return to the meristematic state and produce adventitious shoot buds (or roots). This makes vegetative propagation possible. Many roots, including food storage roots, can also be useful for propagation, if they have the capability of producing adventitious shoot buds.

In root cuttings, adventitious buds originate from the pericycle (outermost tissue of root vascular system). Some plants are preferentially propagated from their roots, such as raspberries that have annoying prickles on their stems. In leaf cuttings, such as the African violet, adventitious shoot buds and adventitious roots arise from mature cells at the base of the leaf blade or petiole. More information on propagation can be found in Chapter 7.

This knowledge helps us to understand the dandelion problem. No matter how hard you try to pull out dandelions in the lawn, they always seem to return. Dandelions have a deep taproot. When you pull the plant, or cut it with a hoe, you leave behind some of the taproot. These taproots are especially good at forming adventitious shoot buds, which produce a new dandelion from the damaged taproot. The only way to eradicate dandelions is with chemicals (herbicides) that kill the entire plant. You could instead just leave them and harvest their young greens for salad, while enjoying the beauty of their flowers and subsequent silky, blowing seeds.

STEM MODIFICATIONS

Several modifications of the normal, upright stem exist. These include both aboveground and belowground modifications. Such modifications arose over long periods of time as plants evolved under changing conditions. Stem alterations are of interest to the horticulturist, since many are useful as food sources or for propagation purposes.

Climbing Stems

Aboveground modifications of significance include climbing and succulent stems. The advantage enjoyed by the first climbing stems was their ability to literally grow over the competition. This probably led to them gaining a successful niche in the evolutionary stream of plants. *Climbing stems* are of several types: those that lean or clamber over supports *(Allamanda),* those that twine around supports *(Bougainvillea, Clematis, Wisteria),* and those that grasp supports with tendrils (grape) or holdfasts (English ivy).

Succulent Stems

Succulent stems are enlarged, with very tiny or absent leaves; this modification permits maximal food manufacture and water storage in the stem, and minimal water loss through leaves. Plants with succulent stems are highly adapted for survival in arid and desert areas. Such stem forms probably arose as areas of wetness suffered prolonged drought. Plants that evolved forms suitable for water storage (thick stems) and minimal loss of water (tiny or no leaves) would have survived the changing conditions. Examples are cacti and *Euphorbia.*

Offsets

Lesser aboveground modifications of the stem include offsets, runners, and spurs (Figure 3–6). The first two modifications arise from the *crown,* the portion of the stem near the ground. The crown may simply be a point of transition between the roots and stem, such as with trees and shrubs, or short, compact stems that give rise to new shoots, such as with herbaceous perennials. Crowns of many herbaceous perennials (asparagus, chrysanthemum, day lily, and rhubarb) can be divided to produce new plants. *Offsets* (sometimes called *offshoots*) are shortened, thickened stems of rosette-like appearance produced from the crown of some plants, such as many bromeliads and succulents. These are particularly useful for propagation.

Runners

Specialized stems that arise from leaf axils at the crown of a plant, such as is seen with *Ajuga* and strawberry, are called *runners.* A common house plant, the spider plant *(Chlorophytum comosum),* also has runners. When grown in a hanging basket, the runners dangle like spiders on a web, since there is no supporting soil. In their natural habitat, runners grow horizontally above the surface of the ground for some distance, giving the appearance of "running" along the surface. At the nodes of the runner, new plants arise and root.

Spurs

Short, lateral branches with closely spaced nodes sometimes found on woody branches are called *spurs.* Flowering and fruit production occur mostly on the spurs of some fruit trees, such as certain Delicious apple cultivars.

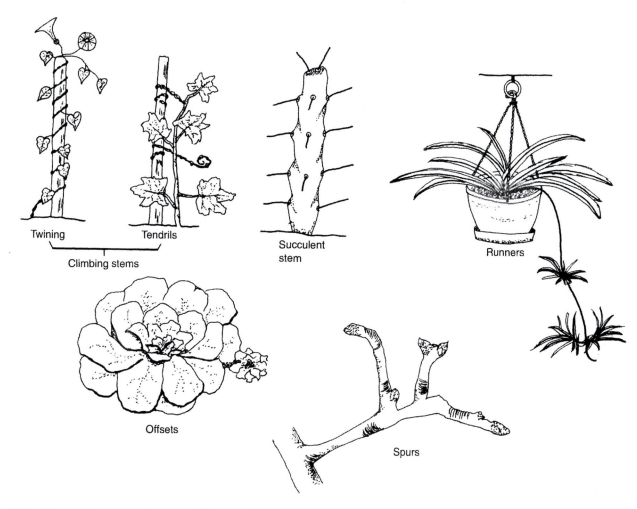

FIGURE 3–6 Aboveground stem modifications. Climbing stems are represented here by morning glory (left) and grape (right). Grape (*Vitis* cultivars) tendrils are also modified stems, but most climbing stem tendrils are modified leaves. Other aboveground stem modifications include succulent stems (*Euphorbia*), runners (Spider plant, *Chlorophytum*), offsets (*Echeveria*), and spurs (apple). See text for details. (From Raymond P. Poincelot, *Horticulture: Principles & Practical Applications*, (1980). Reprinted by permission of Prentice-Hall, Upper Saddle River, NJ.)

Offsets, Runners, and Propagation

The evolutionary advantage of offsets and runners is ease of natural propagation beyond seeds. Offsets, if broken off or left behind when the main plant dies, can produce adventitious roots and become a new plant. Runners can form adventitious roots at points that contact soil and become new plants. Both stem modifications enhance the colonizing ability and survival of the plant that produces them. Horticulturists can take advantage of these stem forms for commercial propagation, such as using offsets to propagate cacti and runners for strawberries.

Bulbs

Buried stem modifications include bulbs, corms, rhizomes, stolons, suckers, and tubers (Figure 3–7). A *bulb* consists of a short, fleshy stem axis (*basal plate*) containing a flower rudiment or growing point at its apex, plus an extensive overlayment of fleshy scales. Bulbs are produced by some monocots for food storage and reproduction; they are valued for purposes of propagation and some for our food. Their food storage ability and being deep in the soil give bulbs

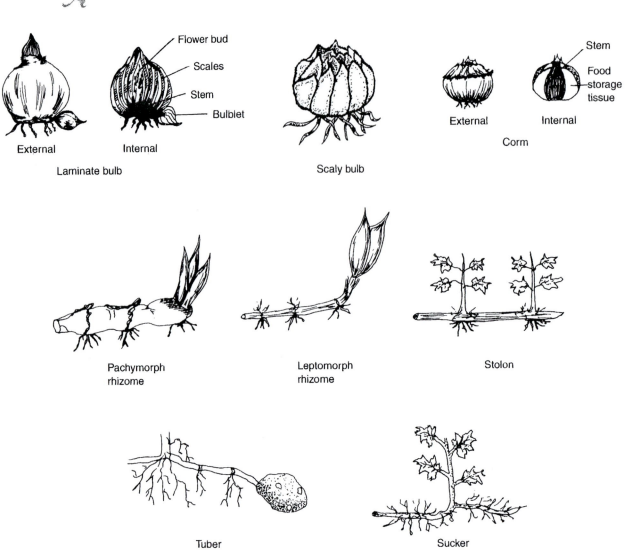

FIGURE 3–7   Belowground stem modifications. The laminate bulb is represented here by hyacinth (*Hyacinthus*), scaly bulb by lily (*Lilium*), corm by *Crocus*, pachymorph and leptomorph rhizomes by *Iris* and lily-of-the-valley, respectively, stolon by mint (*Mentha* sp.), tuber by white potato (cultivars of *Solanum tuberosum*), and sucker by red raspberry (cultivars of *Rubus*). (From Raymond P. Poincelot, *Horticulture: Principles & Practical Applications,* (1980). Reprinted by permission of Prentice-Hall, Upper Saddle River, NJ.)

(and other belowground stem modifications) the ability to survive hard winters or dry periods in a dormant stage. When conditions improve, the bulb sends leaves up through the soil.

Scales of bulbs may be present in continuous, concentric layers (*tunicate* or *laminate bulbs*) that are dry and papery outside and fleshy inside. Examples are the Amaryllis, daffodil, hyacinth, onion, and tulip. Scales are easily seen in the onion when you slice it. The concentric arrangement of scales also enables one to prepare deep-fried onion rings.

Bulbs without an outer dry covering and a less structured arrangement of the scales are known as *scaly* or *nontunicate bulbs*. An example is the lily (*Lilium* sp.). Miniature bulbs developed below ground from the parent bulb are referred to as *bulblets* and *offsets* when mature, or as *bulbils* when formed above ground on the stem.

Corms

A *corm* is the enlarged base of a stem axis enclosed by a few, dry scales. Corms are mostly stem structure with obvious nodes and internodes, in contrast to the bulb, which is mostly scales. When cut in half lengthwise, the bulb appears as layers of fleshy scales over a small, roughly triangular piece of tissue at the base. This smallish tissue is actually stem tissue. The corm by contrast appears solid and is predominately stem tissue. The corm, like the bulb, is valued for its use in propagation. Examples are the *Crocus, Colocasia, Gladiolus,* and *Tritonia.*

Fleshy scales are actually leaves modified for food storage. When plants evolved into bulbs, the buried leaves could evolve over time in two directions. The buried leaves would not photosynthesize, so they had the option of disappearing or changing functions. In bulbs, a functional change took place. Leaves became scales that stored food. In corms, they disappeared and the stem tissue both stored food and made leaves that appeared above ground.

Rhizomes

A *rhizome* is a specialized primary stem structure that grows horizontally at the ground surface or underground. Instead of compression, as with bulbs and corms, the rhizome is essentially equivalent to laying the whole stem on the ground and covering it with soil. Leaves and flowers will seem to appear directly from the ground. The rhizome, while root-like in appearance, differs from roots by the presence of nodes and internodes. Buds and scale-like leaves are sometimes found at the nodes. Most rhizomes are produced by monocots and the lower plant groups.

Two types of rhizomes are known: *pachymorph,* which is thick, fleshy, short, and found as a many-branched clump, and *leptomorph,* which is slender with long internodes. Examples are the *Iris* and lily-of-the-valley (*Convallaria*), respectively. Intermediate forms are called *mesomorphs.* The production of roots, shoots, and flowering stems from rhizomes makes them useful for propagation.

Stolons

Stolons are modified stems similar to the runner. Unlike runners, a shallow layer of soil usually (but not always) covers stolons (the basal or lowest branches). Stolons also tend to be more numerous than runners, often forming a crawling mass of surface-hugging stems. Because of their ability to root and form new plants, stolon-producing plants tend to be weedy and invasive. A good example of this type of plant is mint. In fact, runners and stolons are often confused. Stolons are useful for propagation, such as with Bermuda grass and mint (*Mentha* sp.).

Suckers

A *sucker* is a shoot that arises in the soil from an adventitious root bud. However, in practice, shoots arising from the crown or stem near the soil's surface are also referred to as suckers. Suckers can be used for propagation purposes, such as with the blackberry and red raspberry.

Tubers

A *tuber* is a belowground modified stem structure, essentially a terminal swelling of a stolon or rhizome. You might think of it as a "swollen stolon." Tubers are useful for propagation. White potatoes are propagated from tuber sections. A true tuber should be distinguished from a tuberous root. A true tuber has all the parts of a stem, that is, "eyes" which are actually nodes and contain buds, whereas a tuberous root has the external and internal structure of a root. The white potato and *Caladium* have tubers; the sweet potato and dahlia have tuberous roots. The so-called "eyes" on white potatoes are actually stem buds. If left long enough in your warm pantry, potato eyes elongate into spindly stems.

LEAF FORM AND STRUCTURE

Leaves may be thought of as a flattened or expanded portion of the stem. Many of the kinds of cells and tissues found in the stem are present in leaves. However, the growth of the leaf, unlike the stem, is limited. The design of the leaf serves to present a large surface area for the absorption of light energy needed for photosynthesis (see Chapter 4). The main two leaf functions are food manufacture (photosynthesis) and water loss by evaporation (transpiration).

FIGURE 3–8 Leaf form and venation examples. (A) Form is simple, venation is net-like pinnate. Examples: Elm, Birch (*Ulmus, Betula*). (B) Form is compound palmate, venation is net-like palmate. Example: *Clematis.* (C) Form is compound pinnate, venation is parallel. Example: Parlor palm (*Chamaedorea elegans*). The position of the lateral bud establishes (A) as a leaf and (B) and (C) as leaflets forming a compound leaf as the bud is located in the angle between the stem and leaf petiole. (From Raymond P. Poincelot, *Horticulture: Principles & Practical Applications,* (1980). Reprinted by permission of Prentice-Hall, Upper Saddle River, NJ.)

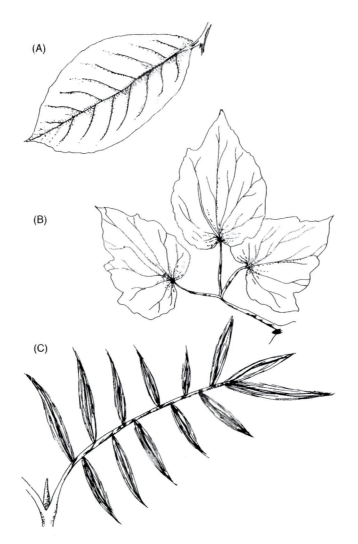

(A)

(B)

(C)

Sometimes leaves are needle-like or scaly, as seen on conifers such as pines and juniper, respectively. The leaf-like structure on ferns is referred to as a *frond.*

The typical leaf has two parts: the *blade,* or thin, flattened expanded portion, and the *petiole,* or stalk. Leaves without petioles are said to be *sessile.* This condition is found in the grasses, including corn. The base of the petiole sometimes has leaf-like or scale-like structures called *stipules.* This feature is especially prominent in the garden pea, *Pisum sativum,* where the stipules are actually larger than the leaf parts.

On the lower and often the upper surface of the leaf are lines or ridges called *veins* (Figure 3–8). These are a continuation of the vascular system of the stem. The vein arrangement (venation) is usually parallel in monocots and net-like in dicots. This characteristic makes it very easy to tell a monocot from a dicot just by examining the leaf. Net-like veins may assume either the form of a feather, a strong main vein with lateral smaller veins (*pinnately veined*), or the form of the fingers spreading from the palm of the hand (*palmately veined*). Veins are usually less conspicuous on conifers.

Leaves have either a simple or compound form (Figure 3–8). *A simple leaf* has an individual blade that may be indented. The *compound leaf* consists of a blade divided into several parts called *leaflets.* Compound leaves can be either *palmately* or *pinnately* compound (Figure 3–8). In the former, all the leaflets are attached at the petiole tip, in contrast to the latter, where the leaflets are attached to the sides of a central stalk like the barbs of a feather.

The leaf (Figure 3–9) is usually surrounded by an upper and lower single cell layer (*epidermis*) that is covered by a waxy coat (*cuticle*). A few plants have a multiple layer epidermis, such as *Nerium oleander.* The lower epidermis is rich in pores (*stomata* or *stomates*) through which gases and water vapor move. The upper layer has fewer or no stomata. Wa-

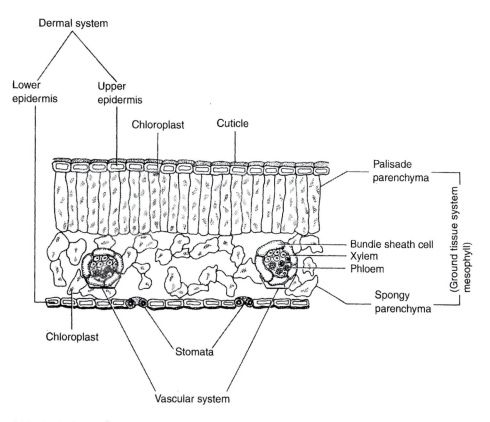

FIGURE 3–9   Leaf cross section. Many leaves of horticultural interest have a differentiated mesophyll with palisade and spongy parenchyma cell layers. The number of palisade parenchyma layers varies from one to three, depending on plant nutrition and species. Exceptions include certain desert and tropical plants that have an upper and lower palisade parenchyma layer with spongy parenchyma between, or some plants such as grasses and sweet corn where no clear cut distinction between palisade and spongy parenchyma cells is apparent. Depending on the type of photosynthesis, some plants will have a second type of chloroplast in their bundle sheath cells (bundle sheath chloroplast, see later in this chapter) as opposed to those in the mesophyll tissue (mesophyll chloroplast). (From Raymond P. Poincelot, *Horticulture: Principles & Practical Applications*, (1980). Reprinted by permission of Prentice-Hall, Upper Saddle River, NJ.)

ter lilies are an exception; their stomates are on the upper surface. Sandwiched between the epidermal layers is the mesophyll tissue, the site of photosynthesis.

LEAF MODIFICATIONS

Leaf modifications hold special interest for the horticulturist. These modifications include those functioning as food storage structures, such as thickened petioles, basal leaf portions that become bulb scales and thickened fleshy leaves. Others are modifications for climbing and defense, such as tendrils and spines, respectively. Some modifications may be edible, such as the *thickened petioles* of rhubarb or celery, the *bulb scales* of onions, or the *fleshy leaves* of cabbage. Others may be inedible, like the thick, fleshy leaves of succulents. *Tendrils* are found on vines and *spines* on cacti.

One needs to beware of imitations. Not every protective, sharp structure on plants is a spine. Spines, being leaf modifications, are found just below an axillary bud or shoot, and should be differentiated from thorns and prickles. *Thorns* are modified branches that arise from the leaf *axil* or just above it. *Prickles* are random outgrowths of the surface stem tissues (epidermis, periderm). Thorns are found on the hawthorn tree (*Crataegus* sp.), while prickles are found on roses and raspberries. Similarly, not all tendrils are leaf modifications. Pea tendrils are, but grape tendrils are actually branch modifications. Another misleading modification is the cladophyll. This misleading modification occurs when a stem looks like a leaf. An example of this structure can be found on asparagus. The asparagus spear has small scales; these scales are the true leaves. Unharvested asparagus produces lacy, graceful "leaves." These leaves are actually cladophylls. Similar structures are found on certain orchid cacti, such as *Epiphyllum*.

FLOWERS

The reproductive structures of angiosperms (flowers, Figure 3–10) vary greatly in appearance. However, they share fundamental similarities in their basic structural plan. The stalk of a solitary flower or a collective mass of flowers (inflorescence) is called a *peduncle*. The stalk of one of the flowers in an inflorescence is called a *pedicel*. The *receptacle* is a somewhat enlarged area at the apex of the flower stalk, to which the floral parts are attached. Four kinds of floral organs that may be thought of as modified leaves arise from the receptacle. These structures are the sepals, petals, stamens, and pistils (Figure 3–10).

Sepals

Sepals are found at the base of the flower and enclosed the blossom when it was in bud. Sepals protect the developing flower in the bud from dehydration. Sepals are usually small, green, and leaf-like (sometimes colored); collectively, the sepals are called the *calyx*. In some flowers the sepals can be mistaken for petals because of their similarity in color and size. In that case the individual parts are not called sepals and petals, but are referred to as *tepals* (examples include lily and tulip).

Petals

Above the calyx and toward the center are the *petals*, collectively known as the *corolla*. Petals are usually brightly colored and sometimes fragrant from perfume and sticky with a sugary substance (nectar) produced from glands. Together the calyx and corolla are called the *perianth*. Petals play a key role in those flowers that depend on bees or other pollinators for *pollination* (transfer of pollen).

FIGURE 3–10 A typical flower that shows all possible structures and their relative positions. See text for details. Flowers having sepals, petals, stamens, and pistils are termed complete flowers. (From Raymond P. Poincelot, *Horticulture: Principles & Practical Applications,* (1980). Reprinted by permission of Prentice-Hall, Upper Saddle River, NJ.)

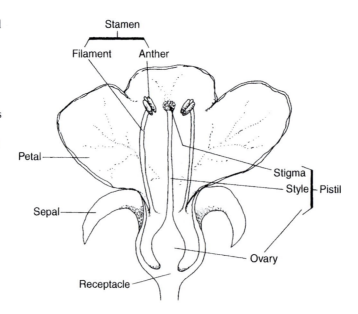

Color and Pollination. Color serves to attract honeybees, which, unlike most insects, have color vision. Bees are the dominant pollinators of plants used for horticultural purposes. Many flowers also have "honey guides," a series of lines that lead the bee along the petals to where the pollen and nectar are to be found. We cannot see these lines, but the bee can, because the bee has vision in the ultraviolet range. If we shine ultraviolet light on petals and photograph them with ultraviolet-sensitive film, we can see these lines on the photograph. Hummingbirds and other long-beaked birds are attracted to long tube-like red flowers. This flower shape works best with their elongated, narrow beaks. Long tube-like red flowers usually have no fragrance, because birds do not have a well-developed sense of smell. This arrangement is an excellent example of coevolution, where two organisms evolved together to the point of codependence.

Fragrance and Pollination. Some insects, such as moths and butterflies, have keenly developed senses of smell. They are attracted by the fragrance of flowers. Some fragrances are pleasant and we also enjoy them, such as with roses and hyacinths. Other flowers have unpleasant smells, such as the carrion flower, which smells like rotting meat. The flower also looks like rotting meat, with its mottled brown color. Carrion flowers are pollinated by flies, which are attracted by the stench. They are foiled in their attempts to lay eggs, but the flower is pollinated in the process.

Nectar and Pollination. Nectar and pollen (to a lesser extent) are used as food sources by the visiting insects and other pollinators. Bees convert the nectar into honey and hummingbirds use nectar to fuel their very high metabolism. Nectar is sugar and water and is a high-energy food. Nectar is produced at the base of petals in glands called *nectaries*. This reward reinforces the visits of insects and other pollinators.

Unusual Forms of Pollination. Some flowers have developed very unusual pollinating mechanisms. Some plants have flowers produced at ground level that are pollinated by snails. Other flowers are white, open at night, and are pollinated by either bats or night moths. Still others mimic the shape of the female version of the insect, as with certain orchids. The male arrives with ideas of copulation, but ends up pollinating the flower. Beetles are lured and then trapped for several hours by some flowers by rapid closure of petals. In their frenzy to escape, the beetles pollinate the flowers. Finally, small mammals called marsupials in Australia pollinate some flowers.

Stamens, Pistils, and Whorls

The *stamens* (male reproductive organ) are located above the petals. Typically, they consist of an elongated stalk, the *filament* that supports an enlarged, pollen-containing structure called the *anther*. In the center of the flower is the female reproductive organ(s) or the *pistil(s)*. This structure is somewhat flask shaped. The swollen, basal part (the *ovary*) is connected by a stalk-like portion (the *style*) to an enlarged, terminal portion (the *stigma*). The mature pollen grains are released from the disintegrating anther and carried to the stigma of the same or another flower by insects, wind, and rain (pollination). If the stigma is receptive, fertilization follows. The ovary, fertilized contents, and sometimes accessory parts then give rise to the fruit and seeds contained therein.

Pistils may be either simple or compound, depending on the numbers of carpels present in the structure. *Carpels* evolved from leaves that became modified for reproduction. These leaf-like reproductive structures eventually enfolded, bearing the ovaries (or ovules). A *simple pistil* has one carpel, and a *compound pistil* bears two or more fused carpels. Pistils having multiple carpels possess chambers or compartments in the ovary area. The arrangement of the ovules on the carpels is a useful characteristic from the viewpoint of classification.

The floral organs are usually arranged in whorls. In dicots (apple, cabbage, cacti, carnation, columbine, magnolia, maple, and poppy) the number of floral organs (sepals, petals, and so on) in each whorl is four or five or multiples thereof. In monocots (corn, lily, orchid, and palm) the factor is three or multiples thereof. Some families of both may have more than 10 stamens or pistils arranged in a spiral, rather than a whorl. This arrangement is considered to be more primitive in an evolutionary sense.

FLOWER MODIFICATIONS: FUSION AND REDUCTION

Modifications of the floral organs exist (Figure 3–11). In some flowers the petals are fused to form a lobed tube called a *corolla*. Examples are the petunia and morning glory. These lobes correspond to the number of petals. Fusion of the other floral organs is possible. Fusion of like parts and unlike parts occurs. Fusion of like parts, such as petals or sepals, is called *connation*. Fusion of unlike parts, such as a stamen to a petal, is termed *adnation*. The position of the ovary in relationship to the fused or unfused floral organs is also significant in plant classification. Fusion is considered a more recent and more advanced development on the evolutionary time frame.

Flowers that contain all four floral organs are called *complete flowers* (lily and cherry); *incomplete flowers* (grasses, maple, and willow) lack some or have only vestigial floral organs (Figure 3–12). Incomplete flowers came after complete flowers in terms of evolution. The lack of vestigial parts is termed *reduction* by evolutionists. Some incomplete flowers are wind pollinated, especially those with inconspicuous or absent petals.

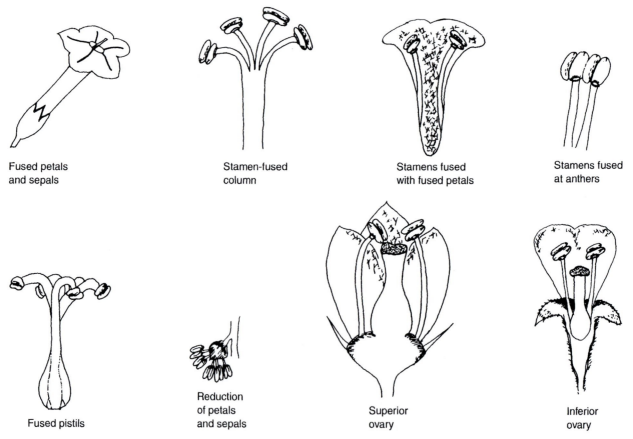

Fused petals and sepals

Stamen-fused column

Stamens fused with fused petals

Stamens fused at anthers

Fused pistils

Reduction of petals and sepals

Superior ovary

Inferior ovary

FIGURE 3–11   Floral organ modifications. Fusion, reduction, and the inferior ovary are more recent evolutionary developments and are considered more evolutionarily advanced relative to nonfusion, no reduction, and the superior ovary. (From Raymond P. Poincelot, *Horticulture: Principles & Practical Applications*, (1980). Reprinted by permission of Prentice-Hall, Upper Saddle River, NJ.)

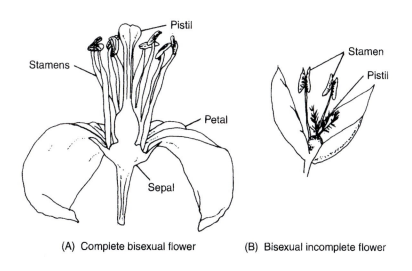

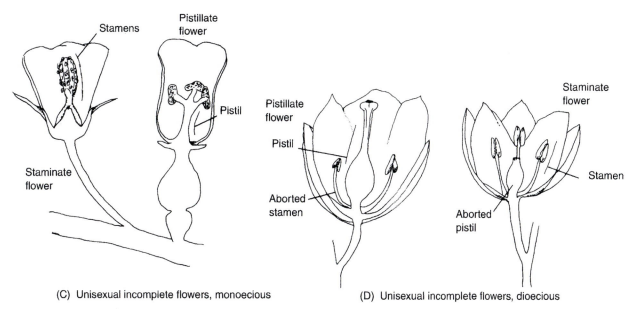

FIGURE 3–12   Flower sexuality. See text for details. Examples shown: (A) grapefruit (*Citrus* X *paradisi*). (B) Grasses. (C) Squash (cultivars of *Cucurbita*). (D) Asparagus (cultivars of *Asparagus officinalis*). Cross-pollination and self-pollination are both possible for types shown by (A), (B), and (C). Only cross-pollination is possible in (D). Self-pollination is not always assured, however, in (A), (B), and (C), since plants can have preventative mechanisms against it. These mechanisms include dichogamy (stamens and pistils mature at different times such that a receptive pistil occurs on another plant), heterostyly (differences in style length within a species favor pollen removal by insects in one flower and stigma contact in another), and incompatibility (pollen tube fails to reach ovules on same plant but does on a separate plant within the species). (From Raymond P. Poincelot, *Horticulture: Principles & Practical Applications,* (1980). Reprinted by permission of Prentice-Hall, Upper Saddle River, NJ.)

Monoecious and Dioecious Plants

Flowers that have both functional stamens and pistils, such as the tomato flower, are said to be *bisexual* (also *perfect* or *hermaphroditic*); flowers lacking either the stamens or pistils are termed *unisexual* (Figure 3–12). All complete flowers are bisexual in that they contain both stamens and pistils. The converse is not true; for example, some bisexual flowers might be missing petals. All unisexual flowers are incomplete, but not all incomplete flowers are unisexual.

FIGURE 3–13 Regular (Petunia) and irregular flowers (Pansy) have radial and bilateral symmetry, respectively. The latter is considered more advanced evolutionarily. (From Raymond P. Poincelot, *Horticulture: Principles & Practical Applications*, (1980). Reprinted by permission of Prentice-Hall, Upper Saddle River, NJ.)

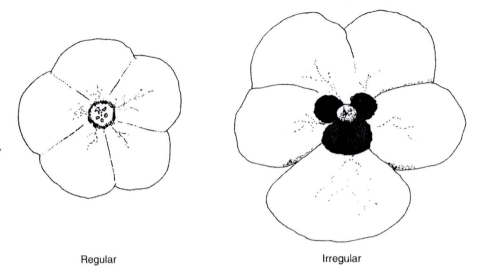

Regular   Irregular

The unisexual condition leads to two possibilities. Incomplete flowers without pistils are called *staminate flowers* (tassel of corn); those without stamens are called *pistillate flowers* (*silk* of corn). Plants with both pistillate and staminate flowers on the same plant are termed *monoecious* (begonia, corn, cucumber, oak, squash, and walnut); *dioecious* plants (holly and spinach) have pistillate flowers and staminate flowers on separate plants.

Understanding monoecious and dioecious plants is important to the horticulturist with regard to ease of pollination, productivity of fruits and vegetables, and production of seeds (especially hybrid seed). For example, if you have only one holly plant, you will not get the colorful red berries. You need to have a second plant of the opposite sex or you need to be fortunate enough to have a neighbor within pollinating distance who has the opposite-sexed holly cultivar.

Symmetry

Flower arrangement, both in terms of its individual structure and placement on the stem, varies (Figures 3–13 and 3–14). Flowers that can be divided into two similar parts by more than one longitudinal plane (radically symmetrical) are called *regular* or *actinomorphic* (petunia or rose). Flowers that can be divided into two similar parts by only one longitudinal plane (bilateral symmetry) are termed *irregular* or *zygomorphic* (orchid and pansy). Flowers with bilateral symmetry are a more recent evolutionary event. The advantage is that their irregular shape often presents a convenient "landing platform" for pollinators, such as the lower, longer, and more out-thrust petal of an orchid. This feature would enhance pollination and increase the number of progeny.

Flowers may be singular or borne in groups or clusters termed *inflorescences*. The arrangement of the flowers in the inflorescence varies, which provides the basis for several recognized forms: *catkin, corymb, dichasium, head, panicle, raceme, spike,* and *umbel.* Some of the more common forms with examples are shown in Figure 3–14. Because these may exhibit a degree of constancy for a genus or even family, the type of inflorescence is often useful in plant identification.

FRUITS

A *fruit* is the ripened ovary or group of ovaries and its contents. Some fruits also fuse with adjacent parts of the ovary. These fruits are more correctly termed *accessory fruits.* Seeds are usually present in the ovary. Fruits arise only from flowering plants, since floral organs are needed for fruit production. The above definition is for botanical fruits and should be

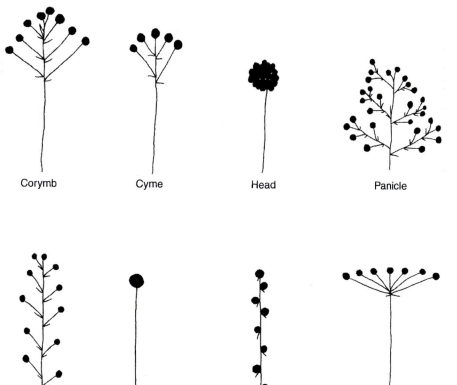

Corymb   Cyme   Head   Panicle

Raceme   Single   Spike   Umbel

FIGURE 3–14 Various types of inflorescences. The main stalk of an inflorescence or of a single flower (not in a group) is termed a peduncle, while the stalks of individual flowers in an inflorescence are called pedicels. Most inflorescences are easily distinguished, but some confusion arises with the corymb vs. the cyme. The outer flower opens first in the corymb, whereas the central flower opens first with the cyme. (From Raymond P. Poincelot, *Horticulture: Principles & Practical Applications*, (1980). Reprinted by permission of Prentice-Hall, Upper Saddle River, NJ.)

distinguished from the popular usage of fruits and vegetables. In the popular sense some botanical fruits are called vegetables (corn, cucumber, eggplant, squash, and tomato, for example).

Fruits may be classified as simple, aggregate, or multiple fruits. Simple fruits (Figure 3–15) in turn may be fleshy or dry. Dry fruits are either dehiscent (open at maturity) or indehiscent (not open at maturity).

Simple Fruits

A *simple fruit* develops from the ovary of one simple or compound pistil and can be either fleshy or dry. The grape, peach, and tomato are all fleshy, simple fruits. When these simple fruits have other floral parts fused to the ovary, they are termed *simple accessory fruits*. In simple fruits, the ripened, enlarged ovary wall is termed the *pericarp*; it may contain up to three distinct layers from the outside and inward: *exocarp* (skin), *mesocarp*, and *endocarp*. The soft or hard pericarp can be either dry or fleshy. Dry tissue consists of dead sclerenchyma cells with suberized or lignified walls. Fleshy tissue is composed of living parenchyma cells.

The three kinds of *simple fleshy fruits* are the berry, drupe, and pome. The *berry* (examples are eggplant, grape, and tomato) has a fleshy pericarp with a thin skin. The *pepo* is a berry fruit with a hard rind (cucumber, pumpkin, and squash), and the *hesperidium* is a berry fruit with a leathery skin and radial partition (citrus fruits). Note that the botanical berry is not the same as the edible fruits termed "berries" in the horticultural sense.

Drupes or *stone fruits* have a thin-skinned exocarp, a fleshy mesocarp, and a stony endocarp. Examples are the almond, apricot, cherry, coconut, olive, peach, and plum. *Pomes* have an endocarp that forms a dry, paper-like core, such as the apple and pear. Because pomes have other floral parts fused to the ovary, they are simple *accessory* fruits. In the apple, the white fleshy part came from flower petals that became fleshy. The core came from the pistil part. If you turn the apple over and see little, black triangular dry things in the

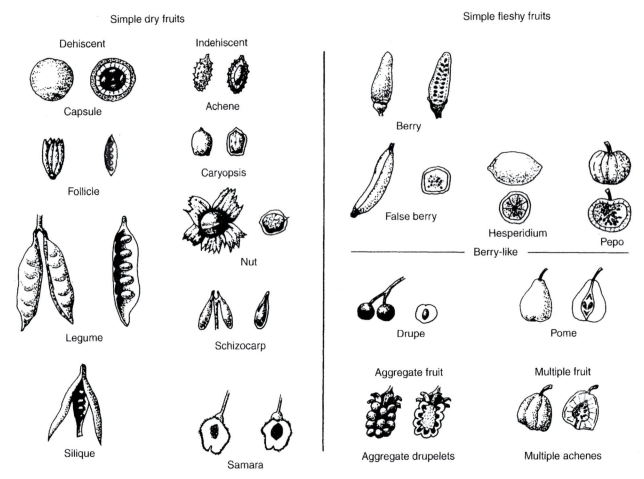

FIGURE 3–15 Various types of botanical fruits (seed portions are darkened). See text for explanation of each. Examples illustrated here are as follows: Brazil nut (*Bertholletia excelsa*, capsule), garden columbine (*Aquilegia vulgaris*, follicle), mimosa tree (*Albizia julibrissin*, legume), cole crops or cultivars of *Brassica* (silique), Transval daisy (*Gerbera jamosonii*, achene), sweet corn (cultivars of *Zea mays saccharata*, caryopsis), filbert (*Corylus avellana*, nut), chervil (*Anthriscus cerefolium*, schizocarp), American elm (*Ulmus americana*, samara), tabasco pepper (*Capsicum frutescens*, berry), banana (cultivars of *Musa acuminata* and *Musa x paradisiaca*, false berry), lime or other cultivars of *Citrus* (hesperidium), pumpkin or other cultivars of *Cucurbita* (pepo), cherry or other stone fruit cultivars of *Prunus* (drupe), quince (*Cydonia oblonga*, pome), blackberry and other cultivars of *Rubus* (aggregate), and fig (cultivars of *Ficus carica* (multiple). (From Raymond P. Poincelot, *Horticulture: Principles & Practical Applications*, (1980). Reprinted by permission of Prentice-Hall, Upper Saddle River, NJ.)

base indentation, you see the remains of the sepals. Other simple accessory fruits that are completely fleshy are termed *false berries* (banana and cranberry), since in the true berry the ovary is the complete and only fleshy part.

As mentioned earlier, simple dry fruits are either *dehiscent* or *indehiscent*. Dehiscent fruits open on maturity and release several to many seeds while still attached to the plant. Seeds are released later after fruits have fallen off the plant with indehiscent types. Dehiscent forms include the *legume* (found in the pea), which has one carpel and two dehiscing sutures (open at maturity); *follicle* (found in columbine), with one carpel and one dehiscing suture; *silique* (found in mustard), with two carpels and two dehiscing halves and a membranous portion remaining with attached seeds; and *capsule* (found in the Brazil nut), with two or more carpels and dehiscing pores, slits, or top.

Indehiscent forms include the *achene* (found in sunflower), which has one carpel and one loose seed; *samara* (found in maple), which is a winged achene with sometimes two carpels; *nut* (found in walnut), which is a large achene with a thick, hard pericarp; *caryopsis* or *grain* (found in corn), which has one carpel and one seed, with the seed and pericarp joined at all points; and *schizocarp* (found in parsley), which has two or more carpels, each usually containing one seed, which separate at maturity, but each retains its seed.

Aggregate and Multiple Fruits

Aggregate fruit consists of clusters of individual fruits that develop from the several pistils of a single flower (Figure 3–15). These pistils share a common receptacle. These fruits may consist of individual fruits that are drupes or achenes. Examples are the raspberry, blackberry, and strawberry. Aggregate fruits can also be accessory fruits. For example, the red fleshy part of the strawberry came from the receptacle of the flower, the hard, tiny crunchy parts came from the ovaries.

Multiple fruits (Figure 3–15) arise from the many flowers found on a compact inflorescence. The fleshy parts of these fruits are often fused floral organs (*accessory structures*). The pineapple is an example.

Parthenocarpic Fruits

Parthenocarpic fruits are seedless. These result from the phenomenon called *parthenocarpy,* or the production of fruit even though fertilization does not happen. Some seedless fruits are not naturally parthenocarpic, since they have resulted from the abortion of the young, fertilized ovaries, such as by spraying with growth regulators. Certain varieties of seedless grapes ('Thompson Seedless') are induced. Other varieties of seedless grapes and cultivated varieties of the pineapple, banana, clementines, the Washington navel orange, and some kinds of fig are parthenocarpic fruits. These types of fruits are highly valued in horticulture, because many people prefer their fruits and preserves free of seeds.

CONES

The conifers do not have flowers or fruits. Instead these plants have two types of cones. One type of cone produces pollen (microsporangiate cone) and is smaller and less scaly than the other type. The second cone produces ovules (ovulate cone) that, after a single fertilization, bear seeds on the cone scales. Pollen-bearing cones are usually on the lower branches and appear yellowish when the pollen is mature. Ovulate cones are found on the upper branches and are brown and scaly. This arrangement promotes wind-assisted cross-pollination. These cones are typical of pines.

Not all conifers have "pine cones." Some, such as juniper, have cones with fused, fleshy scales. These cones look more like blueberries. Juniper "berry-cones" are used to flavor gin. Yew ovulate cones have only one ovule surrounded by a red, fleshy cup-like structure (the *aril*). The seed of the yew cone is toxic and sometimes attractive to children because of the red aril.

SEEDS

A seed is an embryonic structure developed by flowering plants and conifers after fertilization. Seeds have the potential to develop into a plant. Seeds differ so much in shape, size, seed coats, and other characteristics that it is difficult to form any meaningful classification.

Seeds (Figure 3–16) are bounded by a *seed coat* that has developed from the integuments of the ovule. Seed coats vary from thin and papery, such as that found on peanuts after removal of the shell, to hard and thick, such as with the coconut. On some seed coats the *micropyle,* the opening in the integuments of the ovule through which the sperm enters, is still seen as a small pore. The *hilum,* or scar left by the stalk that attached the seed to the *placenta,* is usually visible. Both the micropylar and hilum scars are visible on bean seeds.

FIGURE 3–16
Structures of various seeds of horticultural crops. See text for details. Perisperm (beet) has stored food like the endosperm in the embryo sac, but it is found outside the embryo sac. Corn seeds are the most evolutionarily advanced seeds in that they have a well-developed plumule and radicle along with protective structures for each as they emerge (coleoptile and coleorhiza, respectively). Magnolia is representative of the more primitive end of the evolutionary scale. (From Hudson T. Hartmann and Dale E. Kester, *Plant Propagation: Principles and Practices*, 3rd edition, © 1975, p. 59 by permission of Prentice-Hall, Upper Saddle River, NJ.)

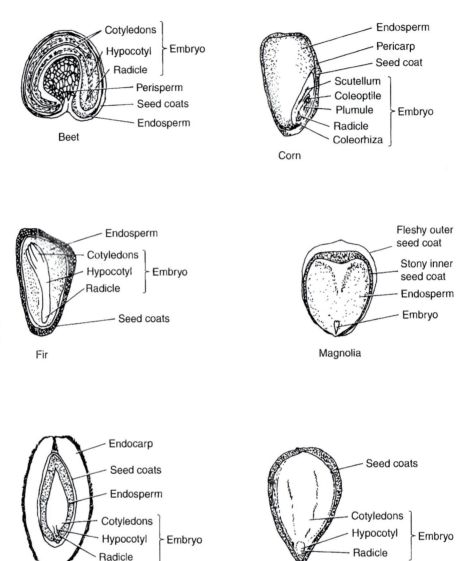

The *seed embryo* is usually fully developed when the seed is mature and ready for dispersal. A few plants, such as the orchid, have immature embryos that develop after dispersal. The structural features of the embryo vary somewhat. Conifer seeds have cotyledons, a simple apical meristem, and a hypocotyl. Embryos in the flowering plants have three basic parts: cotyledon(s), epicotyl, and hypocotyl.

Cotyledons are specialized seed leaves; one is found in seeds of monocots, two in seeds of dicots, and several in seeds of conifers. In some plants cotyledons function as a food source during germination. In others cotyledons serve as absorbing organs to transfer food from the endosperm to the developing embryo after germination. In some seeds, the cotyledons also become green and leaf-like and photosynthetic for a short time. Cotyledons, when green, bear no resemblance in shape to the true leaves that follow later.

The *epicotyl* is the region above the cotyledon(s). It varies in complexity from being only a basic apical meristem to a full-formed shoot bud, the *plumule*. In simpler seeds, the area below the cotyledon(s) is the hypocotyl–root axis. The end of this structure has an apical meristem that becomes the root. In more complex seeds the *hypocotyl* is a transition zone between the cotyledon(s) and radicle. The *radicle* is an embryonic root. The most advanced seeds in the evolutionary sense also have some additional structures: coleoptile, coleorhiza, and scutellum. The coleoptile and coleorhiza are protective sheaths that protect the plumule and radicle, respectively, during emergence from the seed. The *scutellum* is a massive, more specialized cotyledon that functions in food absorption. These structures are found in the corn seed.

A nutrient tissue, the *endosperm,* is present in many seeds. Endosperm is rich in oil and/or protein. In some seeds the embryo and seedling use the endosperm during germination. In other seeds the endosperm has been utilized during the development of the seed prior to germination. Seeds that have depleted the endosperm utilize the cotyledons as food storage organs. These seeds have fleshy cotyledons, such as is found in bean and pea seeds. Seeds with rich endosperms often have thin, papery cotyledons.

Endosperm-rich seeds form an important part of our diet. Seeds with oil-rich endosperms are the sources of cooking oils, such as olive, peanut, sunflower, and canola seeds. Seeds with protein-rich endosperms are often used as foods, such as rice and wheat. Some seeds are rich in both; seeds of corn, cotton, and soybean are used as both oil and protein sources.

MICROSTRUCTURES

The various plant structures are composed of cells. Cells in turn are organized into tissues, and these in turn into tissue systems. Each of these is examined, starting with the basic structure, the cell.

Cells

Cells are the structural units of plants and animals. This concept is universally accepted. Some algae are composed of one cell, but the more complex plants utilized in horticulture contain many billions of cells. These cells are fundamentally alike, but as they grow and differentiate, changes in size and form become commonplace. Sizes vary from 10 to 200 μm for cells of higher plants to 0.5 meters for the largest bark fiber cells. Cell forms are spherical, irregular, cylindrical, spindle shaped, and brick shaped. Cells vary considerably in their functions. Some of these functions include food manufacture, transportation of food and water, food storage, reproduction, prevention of water loss, and support. There are even unspecialized (meristematic) cells that give rise to other cells.

Given this diversity, it is apparent that there is no such thing as a typical plant cell. However, two groups can be defined: (1) cells involved in the metabolic activities of the plant and (2) cells responsible for mechanical support and the conduction of fluids. Cells involved in the latter functions developed from the first group through growth and differentiation. Therefore, a metabolically active plant cell may be considered a typical plant cell for purposes of this section. A diagrammatic illustration of such a plant cell is shown in Figure 3–17.

Plant cells, exclusive of their cell walls, are referred to as *protoplasts*. These may be produced easily by enzymic digestion of the cell wall. Protoplasts are essentially membrane-bound units that consist of two parts, the nucleus and cytoplasm. The outer, boundary membrane is the *plasma membrane (plasmalemma)*. Although less used now, the term *protoplasm* denotes the nucleus and the living cell material in which it is situated.

Cytoplasm.   The *cytoplasm* is 75% to 90% water, contains proteins, lipids (fatty materials), carbohydrates, and lesser amounts of other organic and inorganic compounds. The chemical compounds are present in dissolved, colloidal, and particulate states. This viscous fluid phase (*cytosol*) contains various specialized cytoplasmic bodies, called *organelles*, and several membrane systems. Much emphasis has been placed on the cytoplasmic protein present as enzymes, since these have very important functions in the vital processes of the cell. The organelles (chloroplasts, mitochondria, and nucleolus) have been studied extensively for the same reasons. These organelles can be seen to move around the cell, a process known as *cytoplasmic streaming*. This movement might help the exchange of cellular materials within the cell or between cells.

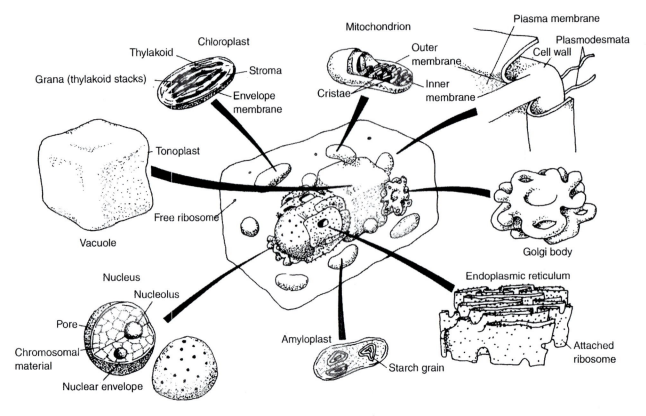

FIGURE 3–17   A typical metabolically active plant cell is shown, which demonstrates various structures and organelles. (From Raymond P. Poincelot, *Horticulture: Principles & Practical Applications,* (1980). Reprinted by permission of Prentice-Hall, Upper Saddle River, NJ.)

The cytoplasm also contains a complex network of protein filaments called actin filaments, intermediate filaments, and microtubules. This collection is referred to as a *cytoskeleton*. The collective cytoskeleton is likely involved in the structural support of the cell. Actin filaments probably play a role in several cellular processes, such as cytoplasmic streaming, cell wall construction, nuclear movement after cell division, and the growth of pollen tubes. Microtubules are involved in cell plate formation, chromosomal division, and orderly growth of the cell wall. The role of intermediate filaments in the cell is presently unknown.

Cell Walls.   The cell wall presence in plant cells helps to distinguish them from animal cells. The *cell wall* surrounds the plant cell and contains enzymes that facilitate the absorption, secretion, and transport of various biochemicals. In general, the cell wall is reasonably permeable to most solvents and solutes. The cell wall also plays a role in cellular defense against bacterial and fungal pathogens. Oligosaccharins in the cell wall may possibly act as hormones in regulating development. The cell wall is a relatively rigid structure that gives support and protection to the cell. Cell walls are produced by the cytoplasm of the cell.

Cells of actively dividing cells, immature tissues, and soft tissues have only a *primary cell wall* (Figure 3–18), as do mature cells involved in photosynthesis, respiration, and secretion. This wall type is composed of pectin, cellulose, and hemicellulose and also contains enzymes and glycoproteins. Some primary walls also contain a waxy material, *cutin,* such as those in the epidermis. Primary cell walls have thin areas called primary pit fields. Adjacent cells are often connected through these primary pit fields by cytoplasmic threads called *plasmodesmata.*

In fibrous and woody tissues, a *secondary wall* is formed inside the primary wall (Figure 3–18). The secondary wall contains cellulose, hemicellulose, and lignin. Some secondary walls, such as those in cork, also contain waxes and *suberin,* a fatty material. The

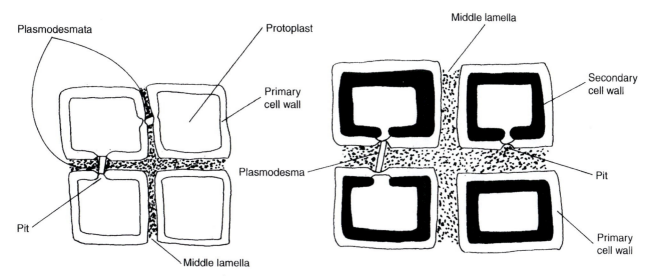

FIGURE 3–18 Primary and secondary cell walls. See text for details. Protoplasts are connected through cell wall pits by membrane/cytoplasm strands termed plasmodesmata. The middle lamella is rich in pectin-like materials and helps to "glue" cells together. In time the middle lamella can partially disappear, leading to intercellular spaces through which gases and water vapor can flow. (From Raymond P. Poincelot, *Horticulture: Principles & Practical Applications*, (1980). Reprinted by permission of Prentice-Hall, Upper Saddle River, NJ.)

secondary wall has large numbers of roughly circular holes called *pits*. Most pits are adjacent to primary pit fields found in the primary wall. Often pits in two separate adjacent cells are lined up to produce pit pairs. When a secondary wall is present, the primary wall will contain lignin, as will the intercellular space (*middle lamella*), a pectinaceous layer found between adjacent cell walls (Figure 3–18). Lignin increases cell wall rigidity and accounts for the greater stiffness of fibers such as wood, hemp, and jute.

The chemical constituents of the cell walls are polymers. *Cellulose* is a homopolymer that consists of linked simple sugar units (glucose) from which one oxygen and two hydrogen atoms have been removed. Hundreds or thousands of these linked sugars are present in a single cellulose molecule. *Hemicelluloses* are heteropolymers of galactose, mannose, xylose, and arabinose (simple sugars); *pectins* are heteropolymers of modified forms of glucuronic and galacturonic acids. *Lignin* is a very complex polyphenolic polymer and is not a carbohydrate.

These cell wall materials are useful horticultural products. Paper and its products are derived from cellulose. Pectins are used in jellies and medicines. Lignin has many uses in the synthetic rubber, food, drug, and ceramic industries. The manufacture of synthetic resin, adhesives, vanilla, pigments, and tanning agents depends on lignin.

Plasmalemma. The *plasma membrane (plasmalemma)* is found immediately adjacent to the cell wall (Figure 3–17). This membrane is the delineating feature of the protoplast. This membrane is easily exposed by enzymic digestion of the cell wall. The plasma membrane contains lipids, proteins, and smaller amounts of carbohydrates. Through this limiting, double membrane, the cell maintains selective permeability with the external environment, as well as the organization and maintenance of its internal environment.

Endoplasmic Reticulum. Another membrane system of the cell is known as the *endoplasmic reticulum* (Figure 3–17). It appears as a network of channels and vesicles bounded by membranes. The endoplasmic reticulum varies in amount, size, and shape and appears to extend from the plasma membrane to the nucleus, mitochondria, and Golgi bodies. It actually appears to be continuous with the Golgi bodies, nuclear membrane, and plasma membrane, but only surrounds the mitochondria. Rough and smooth forms of endoplasmic reticulum are known. Ribosomes attach to the former, but not the latter.

The endoplasmic reticulum appears to be an intracellular communications system. The endoplasmic reticulum is also thought to form a connecting network between cells. The cell wall pit pairs are penetrated by plasmodesmata. These appear to link neighboring cells, and it is thought that the endoplasmic reticulum extends through these plasmodesmata. The smooth endoplasmic reticulum is a site of lipid biosynthesis and channels proteins and lipids throughout the cell. The rough form is involved in protein synthesis.

Golgi Apparatus.   Another membranous system is denoted the *Golgi apparatus* (collective body of *dictyosomes*). This structure appears as a complex organization of net-like tubules and smaller vesicles (Figure 3–17). Evidence is strong that this structure is involved in secretion, cell wall biosynthesis, and intracellular transport of materials secreted in the Golgi apparatus and other cellular areas, such as the endoplasmic reticulum.

Nucleus.   The *nucleus* is a spherical or elliptical organelle bounded by the nuclear membrane (Figure 3–17). Numerous pores are present on the nuclear membrane that permit passage of products from nuclear biosynthesis into the surrounding cytoplasm. The nucleus contains the *chromosomes*, which are composed of *deoxyribonucleic acid* (DNA) in association with protein. These contain the genetic information for the cell. This information is transmitted to the cytoplasm by *ribonucleic acid* (RNA) for subsequent protein biosynthesis at the ribosomes. Bodies composed of RNA nucleoproteins are also found in the nucleus. They are called *nucleoli* and are involved in the formation of ribosomes. The role of DNA in determining the physiological functions of the cell through its control of morphology and metabolism in present and future generations will be discussed further in Chapters 5 and 9.

Ribosomes.   *Ribosomes* are the "factories" for protein synthesis. These structures are not membrane bound. Ribosomes, when present in aggregates on the rough endoplasmic reticulum, are called *polysomes*. Numerous ribosomes can also be found free in the cytoplasm; smaller numbers are present inside the nuclei, mitochondria, and chloroplasts. Proteins made by ribosomes are either enzymic or structural proteins. Unattached ribosomes tend to make enzymes used in the metabolic processes that occur in the cytoplasm, whereas the ribosomes on the endoplasmic reticulum make proteins and enzymes associated with membranes.

Plastids.   Plastids are cytoplasmic organelles bounded by a double membrane; they are quite numerous. These disk-shaped structures can be colorless (*leucoplasts*) or pigmented (*chromoplasts*). The *amyloplast* is a leucoplast involved in starch formation and storage and the elaioplasts store oils. The potato tuber contains numerous amyloplasts. Many orange, red, and yellow flowers and fruits owe their color to the presence of pigmented chromoplasts. Examples are the yellow flower of the forsythia and the red tomato.

The most important plastids are the *chloroplasts* (Figure 3–17), which contain protein, lipids, and pigments such as carotenoids and chlorophyll. Chlorophyll gives the chloroplasts and the leaves their green color. Dying chloroplasts in the fall give leaves their orange and yellow colors. Chloroplasts are surrounded by a limiting double membrane, the *envelope membrane,* which is selectively permeable. Chloroplasts contain a series of disklike, chlorophyll-containing membranes, the *thylakoids.* Thylakoids are embedded in a protein matrix called the *stroma.* The importance of the chloroplast lies in its production of food and energy during the photosynthetic process that is discussed in Chapter 4. Starch grains that are formed from the photosynthetic process can accumulate in the chloroplasts.

Two kinds of chloroplasts exist. Some plants, such as spinach, have one type only, the *mesophyll chloroplast* that contains thylakoids packed in stacks called *grana.* In addition to this type, other plants, such as pigweed, sugarcane, and sweet corn, have a *bundle sheath chloroplast* in which the lamellae are not stacked, but rather extend the length of the chloroplasts. Mesophyll chloroplasts are found free in the cells of mesophyll tissue, while bundle sheath chloroplasts are found in cells that ring the vascular bundle (bundle sheath cells). These two plant groups show a difference in their photosynthetic pathways that will be discussed in Chapter 4.

Chloroplasts also contain DNA, RNA, and ribosomes. Some proteins are synthesized within the chloroplast, but most chloroplastic proteins are synthesized in the cytoplasm. Chloroplasts develop from proplastids that are triggered into chloroplast formation by light. Proplastids pass from generation to generation in the egg's cytoplasm. This act is known as cytoplasmic maternal inheritance. Both plastids and mature chloroplasts can reproduce by division. All of these unique features tend to support the theory that plant chloroplasts arose from separate organisms (likely a cyanobacterium) that formed a symbiotic arrangement with early plants.

Mitochondria. Just as chloroplasts are the centers of food production, the *mitochondria* (Figure 3–17) are the centers of energy production. Mitochondria contain lipids and protein and are bounded by an outer membrane. An inner membrane extends into the mitochondrial matrix through invagination to form folds called *cristae*. Enzymic processes related to oxidative metabolism occur in the mitochondria that result in the formation of an energy carrier, adenosine triphosphate (ATP). DNA, RNA, and ribosomes are present in mitochondria. Evidence suggests the synthesis of some mitochondrial protein in the mitochondria, but most of the protein comes from the cytoplasm. Again, like chloroplasts, these organelles likely arose from some early symbiotic combination between a plant-like ancestor and maybe a bacterium.

Vacuoles. *Vacuoles* (Figure 3–17) contain liquid inclusions of the cytoplasm (*cell sap*) bound by a membrane, the *tonoplast*. The membrane is relatively impermeable, so solutes must be "pumped" into it at an energy cost (active transport). Vacuoles have functions such as food storage, waste repository, and degradation of complex macromolecules and even organelles. Vacuoles containing pigments in their cell sap also contribute to the color of various plant tissues, such as the red and pink petals of the rose and tulip. The red and blue colors of certain vegetables and fruits, such as grapes and radishes, also derive from pigment storage in the vacuole. In a mature cell the vacuole occupies a large portion of the cell. The water content of the vacuole also helps maintain the turgidity of the cell.

Microbodies. Other minor membranous structures are present in the cytoplasm. These include aleurone grains, glyoxysomes, oleosomes, peroxisomes, protein bodies, and spherosomes. *Aleurone grains* contain protein stores and enzymes and are found in some seeds, such as wheat. The remaining ones are lumped under the category of *microbodies;* some are associated with the endoplasmic reticulum, chloroplasts, or mitochondria, or are free in the cytoplasm. Microbodies are bound by a single membrane and are found in some cells, but not others. *Spherosomes* contain lipids and enzymes; they likely function in lipid synthesis and storage. *Oleosomes (oil bodies)* appear to be storage sites for oil droplets. *Protein bodies* appear to be storage sites for proteins. *Glyoxysomes* seem to be involved in the enzymic conversion of fats to carbohydrates. *Peroxisomes* are involved in photorespiration (see Chapter 4) and the conversion of fats to sucrose.

Tissues

Masses of cells arise through the process of cell division. These cells enlarge and undergo differentiation (change to a more specialized cell). The end result is an organism consisting of cells that differ in their structure and physiological functions. Despite these cellular differences, groups of similar cells exist in ordered aggregates. These cellular aggregations are known as *tissues*.

Tissues can consist of *undifferentiated cells* that are undergoing active growth and division, or *differentiated cells* that are no longer carrying out division. The first are denoted as *meristematic tissues* and the second as *permanent tissues*. Permanent tissues may be *simple,* composed of only one kind of cell, or *complex,* composed of more than one type of cell.

Meristematic Tissues. The cells of *meristematic tissue* are physiologically young and capable of continual division. Generally, these cells have thin primary cell walls, rather large nuclei and abundant cytoplasm; they initiate the formation of other tissues and also maintain themselves. Under some conditions, permanent tissues achieve meristematic activity, so the distinction between the two tissues is not absolute.

Meristematic tissues (Figure 3–5), or meristems, are classified into two groups: apical meristems and lateral meristems. *Apical meristems* are found at the tips of shoots (growing points) and roots. Tissues initiated from these meristems are *primary tissues. Lateral meristems* lie along the sides of roots and stems. In addition, *intercalary meristems* are zones found in elongating areas below the apex and not restricted to the sides of roots and stems. These have prolonged, but not permanent, meristematic activity and may be found in plants such as grasses. Tissues arising from the activities of lateral meristems are *secondary tissues.*

Permanent Tissues: Simple and Complex.   *Permanent tissues* result from meristematic activity. Those tissues of the one-cell type, or simple tissues, include the collenchyma, parenchyma, and sclerenchyma (Figure 3–19).

Collenchyma is a primary living tissue composed of elongated cells with tapering ends. These cells start to develop thick primary walls of cellulose and pectin at an early age, mainly in the corners of the walls. Their function of mechanical support is most characteristic of flexible, actively growing herbaceous dicots, such as sunflower (*Helianthus annuas*). Collenchyma tissue is the primary component of the ridges visible on the surface of celery petioles.

Parenchyma tissue is the least highly specialized and most abundant tissue. This abundance and the frequent presence of chloroplasts make parenchyma tissue the site of the largest part of the plant's metabolic activity. Parenchyma cells with chloroplasts are referred to as *chlorenchyma cells.* Parenchyma cells often have similar widths and lengths (isodiametric). Their walls are thicker than those of meristematic cells, they have large vacuoles, and are living cells at maturity. The edible tissues of most fruits consist mainly of parenchyma tissue.

Sclerenchyma cells are not alive at maturity when functioning as rigid supporting tissue, such as in the nongrowing stem parts of corn (*Zea mays*). These cells are essentially elongated cells with tapering ends and thick secondary walls. Their size and shape differ, but their tissue form is either *fibers* or *sclerids.* The latter tend to be more isodiametrical than the fibers. Fibers are important in the manufacture of rope and cloth from plants such as hemp, jute, and cotton. Hard tissues of plants, except for wood, are composed partly or entirely of sclerids. Examples are nut shells, hard seed coats, and fruit pits. The hard grains of pears are compact clusters of sclerids known as stone cells.

Complex tissues are another form of permanent tissues. These complex or combined simple and specialized tissues are phloem, xylem, epidermis, and periderm.

Phloem may be either primary or secondary tissue; the former arises from the apical meristem and the latter from a lateral meristem, the *vascular cambium.* The function of phloem is translocation of sugars derived from photosynthetic activity. The functional life of phloem is not long lasting in many plants; it can collapse from pressures of surrounding cells. New phloem is constantly added to replace damaged areas.

The essential cells of the phloem are the *sieve elements,* regarded as the most specialized plant cells. In angiosperms, sieve elements (also called *sieve tube members*) exist in an end-to-end arrangement in the form of a long, multicellular tube called the *sieve tube* (Figure 3–19). Numerous pores exist at the cell ends of sieve tube members in what is known as a *sieve plate.* As sieve elements mature, the nucleus disappears. However, the cells may remain alive and active for up to a year afterward. *Sieve cells* are found in gymnosperms and lower vascular plants; they do not form a definite sieve tube. Instead these cells have pits on the side walls and cells overlap at the pit areas.

Sieve tube members are associated with a form of parenchyma cells, the *companion cells* (Figure 3–19). Their role is unclear, but they likely keep the sieve cells alive after the disappearance of the nucleus. Sieve cells are associated with an *albuminous cell.* Other parenchyma cells and sclerids and fibers may be part of the phloem.

Xylem, like phloem, may be either primary or secondary tissue; the origin of the primary form is the apical meristem, and the secondary form arises from the vascular cambium. Its function is water conduction. Xylem is a long-lasting tissue composed of living and nonliving cells; it is present at higher levels in woody plants than in herbaceous ones. The wood of woody plants is mostly xylem.

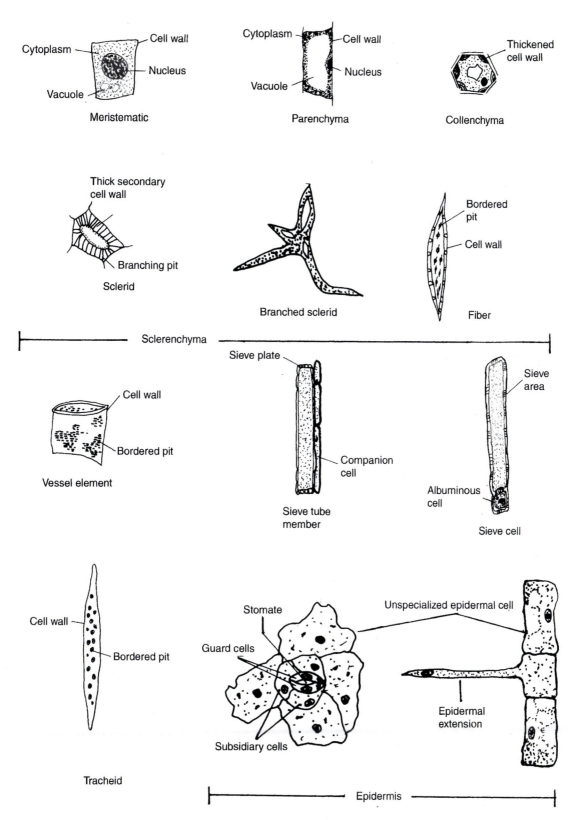

FIGURE 3–19 Various cells involved in the formation of simple and complex tissues. The meristematic cell is found in areas of active cell division (meristems) and through differentiation matures into the other cells that compose the various simple and complex tissues. Epidermal extensions can be either root hairs or leaf/stem hairs (trichomes) depending upon where they occur.

Tracheids and *vessel elements* (Figure 3–19) are the main cells of xylem. The dominant water-conducting cells of angiosperms are vessel elements. Tracheids may be present in some angiosperms, but their role is strengthening and support. Gymnosperms and the lower vascular plants do not have vessel elements in their xylem. Instead water is conducted through tracheids. Fibers and parenchyma cells may also be present in xylem.

Tracheid cells are elongated, with pointed ends and not especially thick walls. *Pits* are found in the cell walls, which aid the flow of water (as cell sap) through the *lumens,* or the space formerly filled by the protoplasts. Vessel elements are shorter and wider than tracheids and have truncated ends. Their end-to-end arrangement and the eventual loss of their protoplasts and end walls lead to tube formations (*vessels*).

Disease organisms and their slimy excretions can clog xylem and phloem; the subsequent disruption of water and food movement leads to wilting and eventual death. Functional phloem usually exists in the outer portions of woody plants, just underneath the bark. Sufficient girdling disrupts the phloem, halts food movements, and results in eventual death. If only minor phloem disruption occurs, such as by ringing or scoring (see Chapter 14), the phloem regenerates. The temporary stoppage of food movement may produce dwarfing, earlier flowering and fruiting, higher yields, or higher sugar contents of fruits.

Epidermis consists of epidermal and accessory cells, trichomes (hairs), guard cells, and sometimes sclerenchyma; periderm contains cork and cork cambial cells. These are both outer, protective tissues. Epidermis is found on the outer part of all soft tissues, whereas periderm is found as the outer tissue of woody parts.

Tissue Systems

Levels of organization in the multicellular plant obviously exist beyond the cell and ordered aggregates of cells discussed in the previous section. If not, the coordination of cells into a structurally organized, functional organism would not result. This section is concerned with plant structures and functions at the supracellular level of organization. Tissue systems at this level include the ground, dermal, secretory, and vascular tissue systems.

Ground Tissue System.   In leaves the ground tissue system consists of only mesophyll tissue that is found between the upper and lower epidermal layers. The upper part of the mesophyll tissue (Figure 3–9) can have elongated cells at right angles to the surface. These cells, called *palisade parenchyma,* are rich in chloroplasts. Below the palisade parenchyma, going toward the lower epidermis, are irregularly shaped cells and large intercellular spaces. These cells constitute the *spongy parenchyma,* and the intercellular spaces there provide for gaseous exchange and transpiration. These cells contain chloroplasts also, but in fewer numbers. Collectively, the palisade and spongy parenchyma cells or undifferentiated parenchyma in other leaves constitute the *mesophyll tissue,* and their chloroplasts are *mesophyll chloroplasts.* Vascular bundles run through the mesophyll cells. A sleeve of parenchyma cells surrounds the bundle sheath. These *bundle sheath cells* may contain chloroplasts that are designated as *bundle sheath chloroplasts.* These chloroplasts are found in plants that possess greater photosynthetic efficiency (see Chapter 4).

The ground tissue system in stems of conifers and woody and herbaceous dicots is divided into an inner part (*pith*) and outer part (*cortex*) by the symmetrical arrangement of the vascular system (Figures 3–3 and 3–4). In the monocots the unsymmetrical arrangement of the vascular system (Figures 3–3 and 3–4) produces no clear-cut division, so the tissue is simply called *ground tissue.* In the root of horticultural plants the ground tissue system consists only of cortex, except for some monocots that have a pith in the center of their vascular system.

The pith is entirely parenchyma cells, and the cortex is mostly parenchyma cells. Some chlorenchyma cells may be present on the outermost part of the cortex of woody twigs and herbaceous stems. This may give a greenish color to the cortex. In addition, collenchyma and/or sclerenchyma cells may occur as strands in the outer regions of stem and petiole cortex.

Dermal Tissue System.   The outermost layers of the cortex give rise to a lateral meristem called the *cork cambium (phellogen)* that to a lesser extent can arise in the epidermis, as discussed shortly. The cork cambium initiates the formation of *cork* (*phellem*) on the outside of stems on woody dicots and gymnosperms. Cork is also produced by woody roots. Most of the cork cells are produced outward, although some may be produced inward (*phelloderm*). Collectively, the cork cambium and cork cells are known as the *periderm* (Figure 3–4), which when present forms part of the dermal system. Essentially, the bark on trees consists of cork.

In some woody plants the formation of cork leads to the death and disappearance of the cortex and epidermis. As the outer cork cells mature, their cell walls are impregnated with *suberin,* a mixture of waxes and other lipids. This waterproof layer prevents excessive water loss from woody stems, but also cuts off the epidermis from its food and water supply, causing it to die and scale away (see Figure 3–4). Cork cells laid inward with time also disrupt the cortex. Outermost layers of cork and included cells die as the area of cork cambial activity shifts. Because they are unable to increase with increases of diameter, these outer corky layers crack from internal pressure to form a rough bark. *Bark* includes all tissues outside the vascular cambium. In other trees with smooth bark, the cortex is not disrupted, and the area of cork cambial activity remains stationary.

Rough, raised areas seen on the bark of some trees, such as cherry, are *lenticels,* or loosely arranged cells produced by the cork cambium that provide for gaseous diffusion. Lenticels are also found on stems, roots, and other plant parts.

The *epidermis* is a tissue that is usually only one cell thick (see Figures 3–3, 3–4, and 3–9). It comprises the surface area (dermal tissue system) of young or herbaceous plant parts, except for woody stems and roots, where periderm formation results in a gradual disappearance of epidermal tissue. In woody parts of plants, the dermal system may eventually consist of only periderm. Several types of cells are found in the epidermis. Cell size and shape vary considerably among species.

Epidermal cells, except for the *guard cells,* have no chloroplasts. A pair of guard cells surrounds a pore; these structures are *stomata* or *stomates.* Stomates are bounded by accessory cells and then epidermal cells. All of these cells are considered to be part of the epidermis. Diffusion of water vapor, carbon dioxide, and oxygen into and out of the plant occurs through the stomates.

Water loss is restricted to the stomates (see Figure 3–9), found mostly on leaves and in lesser numbers on stems, flower parts, and fruit. Loss of water through other parts of the epidermis is prevented by the *cuticle,* a layer of waxes and other lipids secreted by the epidermal cells. This secreted material is called *cutin.*

The cuticle is absent on the epidermal cells of roots. Long projections of the root epidermal layer are denoted as *root hairs;* they are involved in absorption of water and nutrients. Some epidermal cells of stems and leaves may also become hairs called *trichomes.*

Secretory Tissue System.   Some trichomes found on the epidermis are glandular. They secrete materials characteristic of a species. In addition, other glands arise from the epidermis or tissues below it. These structures secrete complex metabolic products: nectar, mucilages, gums, rubber, resins, and essential oils.

Many of these secretions are used commercially. *Resin canals* of conifers produce a resinous substance that yields turpentine and resin. *Lactifers* are found in some vascular plants; they secrete latex, which is a milky to clear liquid. Latex from the rubber tree yields rubber; poppy latex contains well-known alkaloids; and pawpaw latex contains papain, a proteolytic enzyme used in meat tenderizers. *Essential oils,* produced by various glands, are valued for their odors and flavors.

Vascular Tissue System.   Xylem and phloem are the major tissues of the vascular system, whose function is water and food conduction and, to a lesser extent, support. The structural arrangement of the stem vascular system differs among the herbaceous monocots, herbaceous dicots, woody dicots, and conifers.

Vascular systems appear as scattered or circularly arranged bundles in cross sections of stems from herbaceous monocots and dicots, respectively (Figure 3–3). Vascular bundles of herbaceous monocotyledons have no secondary phloem or xylem, since there is no vascular cambium. In herbaceous dicots a cross section would show the vascular bundles arranged in a ring or as a cylinder in longitudinal sections (Figure 3–3). The ring of vascular bundles becomes a continuous cylinder of vascular tissues in some mature herbaceous dicotyledons that possess vascular cambial activity. The woody dicotyledons and conifers, because of vascular cambial activity, have a cylindrical arrangement of vascular tissues (Figure 3–3) that appears as a ring in cross section. Ribbon-like sheets of parenchymous tissue can be observed to extend radially through the vascular cylinder in woody plants. These are *vascular rays* and are important in food storage and lateral conduction of water and food. The vascular cylinder is arranged around a parenchymous tissue called the *pith* (Figure 3–3). The pith is not present in herbaceous monocotyledons or roots.

The vascular systems of roots are more clearly separated from the cortex than are those of stems (Figures 3–3 and 3–4). In roots the *pericycle* and *endodermis* (innermost limit of cortex) surround the vascular cylinders. The pericycle, a parenchymous tissue, can become meristematic and give rise to branch roots, or to lateral meristems that can increase girth and cork production in roots with secondary growth. The endodermis cells have a strip of lignin and suberin in their walls called a *Casparian strip*. This strip plays a regulating role in the movement of water and solutes between the cortex and vascular tissues.

The vascular system is not restricted to roots and stems, but is continuous into the petiole (leaf stalk) through the *leaf trace*, or vascular bundle that connects the vascular tissue of the stem and leaf. The vascular bundles continue into the leaf blade as veins. The vascular system is also present in flowers and fruits; the arrangement in these plant parts is quite complex.

SUMMARY

Plants have two basic parts: roots below ground and shoots above ground. Roots absorb water and minerals from the soil through root hairs found near root tips. Roots also conduct water and minerals to the stem and anchor the plant. Some roots also store food. Roots grow at their tips. Roots growing mainly downward are primary roots, while roots growing laterally near the soil surface are lateral roots. If the plant's primary root dominates, the plant has a taproot system. If lateral roots dominate, a fibrous root system. Plants with taproot systems fare better during droughts, but transplant poorly because of root loss during digging. Plants with fibrous root systems respond quickly to fluctuations in water and nutrient additions and transplant easily. However, all plants suffer some shock during the transplanting process because of damage to root hairs. Roots that occur at abnormal places, such as at the stem, are called adventitious roots. These roots can occur naturally (corn prop roots) or be induced to form such as in the rooting of stem cuttings. Many roots also have a symbiotic arrangement with certain fungi (mycorrhizae). Mycorrhizae improve absorption by roots, especially in areas with no root hairs. Woody plants eventually develop woodiness in their roots.

Shoot functions include support, food manufacture, reproduction, conduction (food, water, and minerals), and sometimes food storage. Shoots consist of stems, leaves, buds, and eventually reproductive structures such as flowers or cones. Stem growth occurs at the tips. Stems can have either soft (herbaceous) or hard (woody) tissues. Herbaceous stems increase in length, while woody ones increase both in length and diameter. Buds are embryonic stems (stem buds), flowers (flower buds), or both (mixed buds). Buds are found at either stem tips (terminal buds) or at points where leaves are attached (lateral buds). Adventitious shoot buds can also arise on roots naturally or be induced such as with root cuttings. This possibility explains why dandelions keep returning when you pull them out, because adventitious shoot buds arise on the taproot part left behind.

A number of stem modifications beyond the normal upright stem exist as a result of evolution. Stem modifications exist both above (climbing stems, succulent stems, offsets, runners, and spurs) and below ground (bulbs, corms, rhizomes, stolons, tubers, and suckers). Climbing stems are adapted to outcompeting other plants by growing over them (vines and lianas), while others survive in dry areas (succulent stems) through water storage and reduced or absent leaves such as cacti. Still others enable the plant to self-propagate asexually such as offsets and runners. Buried stem modifications enable plants to survive unfavorable growing conditions (cold or dryness) with their stored food reserves. All provide useful ornamental and edible plants commonly found in horticulture and they are easily propagated commercially through their stem modifications.

Leaves are essentially the food factories (photosynthesis) for plants and the parts where plants lose water by evaporation (transpiration) through pores called stomates. Leaves vary considerably in form. Some leaves have a flattened part (blade) that is attached to the stem by a leaf stalk (petiole). Others have no petiole (sessile), while others are needle-like or scale-like as in conifers. In ferns, leaves are called fronds. The vascular system is often clearly visible in leaves as veins. Veins in leaves provide an identification feature to tell monocots (parallel venation) apart from dicots (net-like venation). Leaves with one blade are termed simple, while a leaf composed of several blades is called compound.

As with stems, evolution has resulted in modifications. These modifications include fleshy petioles (rhubarb) or fleshy blades (cabbage) or fleshy leaves such as the bulb scales of onions where food reserves are stored. Other modifications are for climbing (tendrils on peas) or defense (spines on cacti). Not all climbing or defensive aids are leaf modifications though. The tendrils on grape and the thorns on hawthorns came from branches while the prickles on roses came from outgrowths of surface tissues (epidermis) on stems. Not all leaves are leaves either. The delicate, leafy looking parts of asparagus plants are actually modified stems. The real leaves are the scaly parts on the spear that you eat.

Flowers and cones are reproductive structures in angiosperms and gymnosperms, respectively. Flowers can contain sepals that protect the flower bud from drying, petals that lure pollinators, and the stamens and pistils that produce pollen (sperm) and ovules, respectively. Flower features that attract pollinators such as bees and butterflies include color, fragrance, and nectar.

Evolutionary processes have also produced flower modifications such as fused or reduced parts. Petals can be fused to form tube-like flowers such as the petunia. Unlike parts can also fuse such as petals with stamens. Flowers with all four parts are termed complete flowers. If sepals, petals, stamens, or pistils are missing or greatly reduced, these flowers are called incomplete. Incomplete flowers lacking petals are usually wind pollinated as in the grasses. If the stamens or pistils are reduced, flower sexuality is affected. Flowers with both stamens and pistils are bisexual flowers, while flowers missing one of these structures are unisexual flowers. If the plant has both male (staminate flowers) and female flowers (pistillate flowers) present, its condition is known as monoecious. Corn is a good example with its staminate flowers, the tassel, and its female flowers, the silk. If two sexed plants exist, one with only staminate flowers and the other with only pistillate flowers, we call this condition dioecious. Holly is a dioecious plant, which explains why two plants are needed in proximity to each other in order to get the red berries. Another modification is associated with the symmetry of the flower's shape. Flowers with a circular shape such as roses and petunias are said to have radial symmetry. Flowers such as orchids with noncircular shapes divisible into two identical halves by only one plane have bilateral symmetry. Fusion, reduction, and bilateral symmetry are considered to be more recent events in evolutionary history.

After pollination and fertilization, flowers produce fruits with seeds. Fruits are found with flowering plants only. Conifers have cones with seeds borne on the scales of the cone. Fruits are the ripened ovary or ovaries of the pistil. If adjacent parts of the ovary end up as part of the fruit, it is termed an accessory fruit. Simple fruits came from the ovary of one pistil and can be either fleshy or dry. The tomato and grape are simple, fleshy (botanical) fruits. Simple fleshy fruits are known as a berry, drupe, or pome. The berry has a fleshy interior with a thin skin such as the grape. The botanical term berry should not be confused with the food term *berry* as in raspberry. If the berry's skin is a hard rind or leathery, it is called a pepo (pumpkin) or hesperidium (orange), respectively. Drupes are stone fruits such as cherries. Pomes have a core such as the apple. Simple dry fruits are dehiscent if they open on maturity and release their seeds while still attached to the plant. If seeds are released after the fruit detaches from the plant, the term indehiscent is used. Fruits derived from several pistils of one flower are called aggregate fruits (blackberry). If the fruit came from many pistils from several flowers clustered together, it is called a multiple fruit (pineapple). Many of the above categories also have accessory forms. Fruits without seeds are parthenocarpic fruits and are highly valued in horticultural production. Most are naturally occurring.

Seeds are essentially embryonic plants with nutritive tissues (endosperm) and a protective structural coat (seed coat). Seeds result from sexual processes and provide the next generation of plants naturally and in commercial propagation. Many seeds are also used as foods because of their oil-rich or protein-rich (or both) endosperm. The embryo consists of cotyledon(s) or seed leaves (one for monocots and two for dicots). Depending on the plant, the cotyledons are either a food source or transfer vector for food during germination. In some plants the cotyledons become green and appear leaf-like. Other parts of the embryo develop into the first shoot and root.

Plants are composed of cells that collectively form the various tissues that make up the plant. Plant cells are complex, having many membranous structures and organelles that are involved in metabolism. For example, one organelle, the chloroplast, is involved in food manufacture. Another structure, the nucleus, contains the genetic material involved in heredity and it also directs the cell's activities. Tissues involved in growth are termed *meristematic tissues* and are usually found at the tips of roots and shoots. Tissues that comprise the body of the plant are either simple (one cell type only) or complex (two or more cell types) permanent tissues.

Tissues can be grouped into four systems, the dermal, ground, vascular, and secretory tissue systems. The dermal tissue system is essentially the "skin" of the plant and helps protect it from dehydration and pathogen entry. The vascular tissue is involved with transport of water, minerals, and sugars. The bulk of the plant tissue in which the vascular system is embedded is the ground tissue system. The secretory tissue system is involved with secretion of various compounds such as nectars, resins, and oils.

EXERCISES

Mastery of the following questions shows that you have successfully understood the material in Chapter 3.

1. How do taproot and fibrous root systems differ? What advantages and disadvantages are associated with each? Does either relate to horticultural practices and, if so, how?
2. What is an adventitious root? Why is it of value to horticulturists?
3. What are the functions of roots? What do root hairs do and do root hairs have any horticultural implications? What are mycorrhizae? What implications do mycorrhizae have in horticulture?
4. What are the functions of the shoot? Name the various parts found on shoots and their functions.
5. What is an adventitious shoot bud? Why is it of value to horticulturists?
6. What are the various stem and leaf modifications? How do they differ? What horticultural value do they have?
7. Name, define, and discuss the functions of the various flower parts. How do flowers attract pollinators?
8. What evolutionary modifications of flowers exist and how do these changes relate to floral sexuality?
9. What are the basic four types of fruits? How do they differ? What is the difference between a cone and fruit?
10. What are the parts found in seeds? What functions do they serve? What is the horticultural value of seeds?
11. What is the relationship between cells and tissues and tissue systems? What do chloroplasts and mitochondria do? What is the connection between plant disease and xylem and phloem?

INTERNET RESOURCES

Botanical Society of America Online image collection. (anatomy, morphology, organography)

> http://images.botany.org/cgi-bin/image-search.pl

Glossary of botanical terms

> http://www.ucmp.berkeley.edu/glossary/gloss8botany.html

Reproductive characteristics of flowering plants

> http://www.csdl.tamu.edu/FLORA/tfplab/reproch.htm

Vegetative characteristics of flowering plants

> http://csdl.tamu.edu/FLORA/tfplab/vegchar.htm

RELEVANT REFERENCES

Bell, Adrian D., and Alan Bryan (illustrator) (1991). *Plant Form: An Illustrated Guide to Flowering Plant Morphology.* Oxford University Press, New York. 341 pp.

Bowes, Bryan G. (2000). *A Color Atlas of Plant Structure.* Iowa State University Press, Ames. 192 pp.

Dickison, William C. (2001). *Integrative Plant Anatomy.* Academic Press, San Diego, CA. 512 pp.

Fahn, A. (1990). *Plant Anatomy* (4th ed.). Pergamon Press, New York. 588 pp.

Gartner, Barbara L. (1995). *Plant Stems.* Academic Press, San Diego, CA. 440 pp.

Mauseth, James D. (1988). *Plant Anatomy.* Benjamin/Cummings, Menlo Park, CA. 560 pp.

Raven, P. H., R. F. Evert, and H. Curtis (1999). *Biology of Plants* (6th ed.). Worth Publishers, New York.

Smit, A. L., A. G. Bengough, C. Engels, M. V. Noordwijk, S. Pellerin, and S. C. V. Geijn (2000). *Root Methods: A Handbook.* Springer Verlag, New York. 520 pp.

Starr, Cecie, and Ralph Taggart (2001). *Plant Structure and Function* (9th ed.). Brooks/Cole, Stamford, CT. 112 pp.

CLASSIC REFERENCES

Bierhorst, D. W. (1971). *Morphology of Vascular Plants.* Macmillan, New York.

Esau, K. (1977). *Anatomy of Seed Plants* (2nd ed.). John Wiley & Sons, New York.

Heyward, H. E. (1938). *The Structure of Economic Plants.* Macmillan, New York.

VIDEOS AND CD-ROMs[*]

Plants (CD-ROM, 2000). Also useful for Chapter 4.
The Leaf (CD-ROM, 1995).

[*]Available from Insight Media, NY (www.insight-media.com/IMhome.htm)

Chapter 4

Plant Processes

OVERVIEW

The life of a plant from sexual or asexual propagation through maturity is a complex process. Growth occurs by cell division and enlargement that increases plant size. Simultaneously with growth, the plant develops its mature form using cellular differentiation to produce organs such as stems, leaves, roots, flowers, fruits, and seeds. Both of these steps are dependent on genes that direct a series of complex and well-integrated biochemical changes. These metabolic and other biochemical processes are reasonably understood through the continuing efforts of plant physiologists and biochemists. A basic understanding of essential plant processes—photosynthesis, respiration, photorespiration, water absorption, nutrient absorption, translocation, transpiration, and other metabolic processes—will heighten the horticulturist's knowledge of crops and provide a better understanding of the basis for specific horticultural practices.

CHAPTER CONTENTS

Photosynthesis
Respiration
Photorespiration
Water Absorption

Nutrient Absorption
Translocation
Transpiration
Other Metabolic Processes

OBJECTIVES

- ❀ Understanding the photosynthetic process and its relationship to horticultural practices, productivity, environment, and evolution
- ❀ Understanding the respiration and photorespiration processes and their connection to water and oxygen needs and crop productivity
- ❀ Learning how plants absorb water and nutrients and which nutrients are essential to plants
- ❀ Tracing the movement of water, food, and nutrients through plants (translocation) and how plants lose water (transpiration) and wilt
- ❀ Understanding basic plant metabolism and biochemistry

PHOTOSYNTHESIS

Few chemical events in our biosphere equal plant photosynthesis in importance and scope. Without photosynthesis the evolution and maintenance of terrestrial life as we know it would not be possible. Each year plants are estimated to produce about 100 billion metric tons of sugar during the photosynthetic process. To accomplish this feat, plants remove huge amounts of carbon dioxide from the air, while releasing oxygen. Plants are the major organisms for maintaining the breathable air so essential to human life. In addition, plants sequester carbon dioxide, thus playing a role in reducing global warming caused by too much carbon dioxide in the atmosphere. If we were to plant a forest the size of Australia, we could slow the process of global warming way down.

Photosynthesis is the process that plants utilize to make their food. This process is unique to plants and also a small group of bacteria. Photosynthesis requires light for energy, gaseous carbon dioxide from the air to provide food building units, a green pigment called chlorophyll that essentially acts as a catalyst, and water, which aids the food construction process. The product of photosynthesis is carbohydrate, and oxygen is also released. The chemist would write photosynthesis as a chemical equation:

$$6CO_2 + 12H_2O \xrightarrow{\text{Light}} C_6H_{12}O_6 + 6O_2 + 6H_2O$$

The equation states that 6 molecules of carbon dioxide and 12 molecules of water in a plant exposed to sunlight produce 1 molecule of glucose and 6 molecules each of water and oxygen. This equation is actually an oversimplification, because photosynthesis is a complex process (Figure 4–1) that requires expertise in radiation physics, biochemistry, physiology, and even ecology to be fully understood. In its simplest sense, photosynthesis is a series of reactions whereby light energy is converted to chemical energy, which in turn is used to construct carbohydrates for the plant to use as food and fuel.

To examine the complex process called *photosynthesis,* we need to treat it like a jigsaw puzzle. First we examine the parts and then put them together into the final picture. Let's start with the most basic question: Where does photosynthesis take place? Photosynthesis takes place in green tissues only, because of its requirement for a green pigment called chlorophyll. With most plants, photosynthesis takes place primarily in the leaves. To a much lesser degree, you will find that some photosynthesis takes place in green stems. The

FIGURE 4–1 Simplified diagram of photosynthesis for most horticultural crops. See text for details. (From Raymond P. Poincelot, *Horticulture: Principles & Practical Applications,* (1980). Reprinted by permission of Prentice-Hall, Upper Saddle River, NJ.)

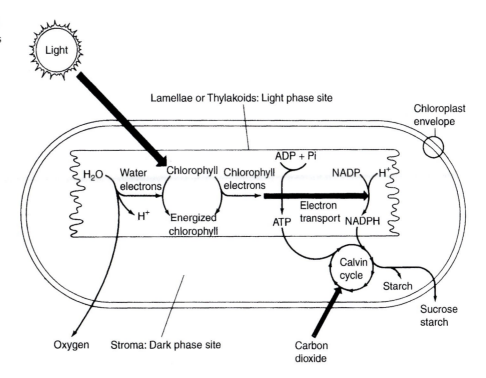

exception to this rule is seen with cacti and other arid-region plants that have lost their leaves through time as they evolved in dry areas. In these plants, photosynthesis is found only in the stems.

The next question is where in the leaf (or stem) does photosynthesis occur? To answer that question, we need to look at a leaf under a light microscope. You will observe that a leaf appears to be composed of brick-like units called cells. If you examine a single cell at 1,000 magnification, you will note tiny, green bodies in the cells. These cellular structures are called *chloroplasts*. They are green because they contain the green pigment chlorophyll. Chloroplasts are essentially the photosynthetic or food manufacturing factories of plants.

Now we need an electron microscope to look at the chloroplast, since the limit to light microscopy is about 1,000 magnifications. To see more minute details, we need a magnification of 60,000 times. At this magnification, you will see that details become visible in the chloroplast. It appears roughly football shaped and has stacks of connected disk-like objects (Figure 4–1). The disks are called thylakoids; these structures contain the green pigment chlorophyll. The material in which the disks are embedded is termed the *stroma*. That's nice, but why do we need to know such details?

As it turns out, photosynthesis is a two-part process. It is analogous to an assembly line where in part one the workers build the engine and drive train, and in the second part, the car's body. At the end, you put everything together and have a new car. The actions of the first part of photosynthesis require light and chlorophyll and are called the light-dependent reactions. The actions that occur in the second part do not require light and are named the light-independent reactions. The chloroplast has two "assembly sites," the thylakoids and the stroma. The light-dependent reactions occur in the thylakoids where you also find the chlorophyll. The light-independent reactions occur in the stroma. These two sets of reactions occur in sequence, first the light-dependent reactions and then the light-independent reactions.

Light-Dependent Reactions

So what happens in each part and how does light energy power this process? Let's look first at what happens. In the light-dependent reactions, light energy is converted to chemical energy, water is broken down to release oxygen, electrons, and protons; and the products are ATP and NADPH. ATP (adenosine triphosphate) is essentially chemical energy; this chemical powers all organisms, including humans. NADPH (nicotinamide adenine dinucleotide phosphate) is essentially an electron carrier; it supplies electrons to ongoing chemical reactions that require them. So what purpose do ATP and NADPH serve in photosynthesis? The light-independent reactions require both ATP and NADPH in order to convert carbon dioxide to carbohydrate. The ATP and NADPH become degraded to ADP (adenosine diphosphate) and $NADP^+$ (oxidized form of NADPH) after their role in carbohydrate construction is completed. ADP and $NADP^+$ then return to the light-dependent reactions for conversion back to ATP and NADPH. So the light-dependent and light-independent reactions are linked by the ATP/ADP and NADPH/$NADP^+$ exchange.

The breaking down of water to oxygen, protons, and electrons during the light-dependent reaction is a result of a process termed *photolysis*. This word roughly means "splitting by light." Somehow light energy is harnessed to break water apart. The oxygen is not a product of photosynthesis; it is essentially a by-product of photosynthesis. Unlike ATP and NADPH, oxygen is not needed for the rest of photosynthesis to proceed. This difference is subtle, but think in terms of need. The material produced is a product, if needed; if not, a by-product.

Photolysis is very important, both for photosynthesis and us. The electrons and protons serve as raw ingredients for some parts of the light-dependent reactions. The oxygen is released as a gas and helps photosynthesis to replenish the earth's supply of breathable air. We exhale carbon dioxide, plants use our waste gas for photosynthesis, and we use their "waste gas," oxygen. This recycling arrangement works well for all involved.

Light as Energy.   So how do the light-dependent reactions operate? To understand this process, we need to back up a bit. Light is a form of energy. This concept is easily illustrated by taking a magnifying glass and concentrating sunlight into a small point. The point of concentrated light can set paper on fire or burn your skin. Lasers also demonstrate the power of light energy when they are used to melt metal or cut through tissues during surgery. The problem when using light energy directly is that its power is intense and somewhat uncontrollable on a cellular level in living plants. Chemical energy, on the other hand, is more easily modulated and usable in organisms. Hence the main purpose of the light-dependent reactions is to convert light energy into a more controllable form. The light-dependent part of photosynthesis is somewhat analogous to using a photovoltaic cell to convert sunlight into electrical energy, which is more useful and controllable for human applications.

Using Light Energy.   To understand how this energy conversion works, we need to look at the level of the atom, the most basic building block of all matter. An atom has a central body called a nucleus with electrons that move around it in orbitals. You could think of the atom as somewhat like our solar system, with the sun being the nucleus and the electrons being the planets. In our solar system, increasing amounts of energy are required to move further from the earth outward or when rocketing from the earth to the moon and then Mars. So what happens when light energy intercepts an electron on the molecular level?

Light energy is thought of in terms of photons. If you are a *Star Trek* fan, you are certainly familiar with the photon torpedoes used as weapons on the star ship *Enterprise*. When a photon slams into an electron, its energy is transferred to the electron. This added energy causes the electron to leap to a higher energy level, or to orbitals that are further out from its original position. This process of energy acceptance and leaping outward is referred to as *excitation*. The electron is said to be "excited." This excitation is only temporary, and the electron eventually returns to its original level, releasing the captured energy as it returns. The released energy comes off as heat. Something very similar happens when you are exposed to the sun. Skin atoms are excited and heat is released, warming or burning you, depending on the exposure time. The problem is that heat energy is not very useful to a plant, because unlike a mammal, it is not in need of maintaining a certain body temperature. Plants over time evolved a way to capture the energy released from excited electrons before it becomes heat. The energy is captured through a transfer process and then tapped to charge or fill up molecules (ATP) that serve as sources of chemical energy.

How Plants Do It.   So how do plants pull off this stunt? To be perfectly honest, scientists don't know how this happens in any detail. One problem is the speed at which the process occurs. Things happen at speeds that exceed the limits of our current technology. Still, scientists have proposed a reasonable model that appears to explain the process. The process is quite complex and best left to more advanced books (see the reference section at the end of the chapter for suggestions). I will attempt to simplify the model so that you get an idea of what we think happens.

The first thing to understand is that sunlight, essentially white light, is actually composed of several colors mixed together. This concept is easily understood when one looks at a rainbow, where droplets of water break up the white light into its spectrum or composite colors: red, orange, yellow, green, blue, indigo, and violet. A prism, a specially ground piece of glass, will also do the same thing. Incidentally, it is easy to remember the colors and their order in a rainbow: just think of the name, Roy G. Biv, where each letter stands for a color. Why is this important? One question is does a plant use all the colors of white light? The answer is no. The photosynthesis process uses only the red and blue part of sunlight. The leaf absorbs the red and blue and the other colors are reflected back. Your eye picks up these reflected colors and sees a green color since the dominant color of the reflected light is the green.

This knowledge is also important for designing lamps for growing plants in offices and malls, where available light is insufficient. The light source has to be rich in red and blue emissions. Special lamps for plant growth are designed to maximize these colors. The ordinary fluorescent lamp can be used to grow plants, because it emits reds and blues and thus is able to make photosynthesis work. The incandescent, regular light bulb can't support photosynthesis, because it emits red, but very little blue.

So what absorbs the red and blue part of light? The important molecules for trapping and transferring light energy are chlorophyll and the carotenoids. Chlorophyll is the green pigment mentioned previously. Carotenoids are yellow and orange pigments also found in the thylakoids of chloroplasts. One carotenoid is the principal source of the vitamin A required by humans. Carotenoids are considered to be accessory or helper pigments in the light-dependent reactions. Both types of molecules on the atomic level are capable of capturing photons of light such that electron excitation occurs. The interesting part is that an excited molecule of chlorophyll can transfer the captured energy to any nearby, "unexcited" molecule of chlorophyll. In turn, this molecule can then transfer its energy onward. Any molecule of chlorophyll not in the excited state or passing energy is available for a "photon hit." Essentially, it is somewhat like a "bucket brigade" line at a fire. Carotenoids can also intercept and pass on light energy, with one catch. Carotenoids can only pass energy to chlorophyll (chlorophyll helper) and chlorophyll can't pass energy to carotenoids.

Where will the energy end up and how will it become chemical energy? Here we need an analogy to answer this question. The best analogy is a satellite dish used for capturing television signals and passing them on to your television set for your viewing pleasure. Think of the molecules of chlorophyll and carotenoids assembled together to form a satellite dish with each molecule acting as an antenna. Each antenna molecule is capable of receiving a photon ("television signal") and passing the captured energy down the dish via other antenna molecules, until the concentrated energy (or signal) reaches the feedout collector at the bottom of the dish. The difference is that instead of a collected signal, we have a concentrated package of energy resulting from millions of photon hits. At the bottom of the "chlorophyll" dish is a molecule of chlorophyll attached or bonded to a molecule of protein. The energy blast that hits this strengthened collector chlorophyll is so intense that the excited electron reaches escape velocity and leaves the chlorophyll molecule, never to return. Essentially, this part is analogous to a rocket ship reaching escape velocity, becoming free of earth's gravitational field and able to move to the planets or even stars.

Now multiply this event by a million or so and have it happen in a millionth of a second. The signal from the "chlorophyll" dish becomes a steady stream of electrons. Electricity running through a copper wire is also a steady stream of electrons that supplies energy to run your computer or microwave oven. This stream of electrons follows a carefully controlled route through different molecules after it leaves the "chlorophyll" dish. Like electricity, the electron flow can be tapped and some of its energy transferred to depleted ATP (ADP), thus regenerating ATP. Perhaps you are wondering how this stream of electrons can be maintained? Surely, the special chlorophyll molecule will run out of electrons. Yes, you are right, but the molecule can be reloaded with more electrons, like an "electronic gun." These electrons are supplied from the photolysis of water. Why water? Water is the most abundant chemical in plants and is essentially an economic source of electrons.

Are we done yet? Not quite, we have an additional complication. There are actually two "chlorophyll" dishes working in tandem. The first dish produces ATP (actually called photosystem II) and releases oxygen through photolysis. The second dish (actually called photosystem I) operates in a similar manner, but its stream of electrons is fed into an "electron trap," $NADP^+$, to restore NADPH. The other difference is that its source of electrons for reloading the special chlorophyll is the spent stream of electrons from dish one that are not totally used up during the formation of ATP. Of course, there are untold numbers of these tandem dish pairs operating throughout the thylakoids.

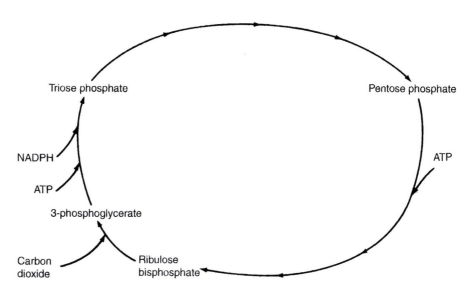

FIGURE 4–2   Calvin cycle is shown here in simplied form; *i.e.*, a number of intermediates are left out. The cycle in C_3 plants is initiated in mesophyll chloroplasts by the addition of CO_2 (carboxylation) to a five-carbon phosphorylated sugar (ribulose bisphosphate). This unstable six-carbon unit splits into two 3-carbon compounds (3-phosphoglycerate) that undergo a number of reactions, eventually regenerating the starting compound, ribulose bisphosphate. Some triose phosphate is drawn off and metabolized into sucrose and starch, but enough remains to keep the pathway cyclic. (From Raymond P. Poincelot, *Horticulture: Principles & Practical Applications*, (1980). Reprinted by permission of Prentice-Hall, Upper Saddle River, NJ.)

Light-Independent Reactions

The light-independent reactions are less complex. In fact, they are so well understood that scientists can duplicate them in a test tube. These reactions are slow enough that we can measure them with our instruments. This part consists of a series of reactions forming a loop or cycle. The cycle is named the Calvin cycle (Figure 4–2) after the scientist who led the group that discovered it. The cycle requires carbon dioxide, ATP, and NADPH in order to run. The carbon dioxide is assembled via several reactions into glucose. The ATP supplies energy and the NADPH provides electrons to operate the cycle. The spent ATP and NADPH are returned to the light-dependent reactions as ADP and $NADP^+$ for regeneration to ATP and NADPH.

The glucose formed from photosynthesis is rapidly converted to sucrose outside the chloroplast. Starch is made both inside and outside the chloroplast The starch is a storage form of carbohydrate and the sucrose is the active form used for supplying energy and for conversion into other chemicals needed by the growing plant. Further metabolic processes convert carbohydrates into lipids, proteins, nucleic acids, and other organic compounds. These organic molecules are assimilated into the tissues, organs, and ultimately all parts of the plant.

Adaptive Forms of Photosynthesis

The first chemical formed when carbon dioxide enters the Calvin cycle has three carbons. Hence this type of photosynthesis or the plant that does it is referred to as C_3 photosynthesis or C_3 plants, respectively. The descriptive terms derives from the three-carbon product formed at the start of the cycle. This term is actually important. Plants over time evolved some variations on photosynthesis as a result of climatic stresses.

The photosynthetic type common to the majority of horticultural crops is C_3. Plants with only the Calvin cycle have less efficient photosynthesis, one type of chloroplast (mesophyll), and high rates of photorespiration (wasteful of photosynthetic intermediates; see the discussion later in this chapter).

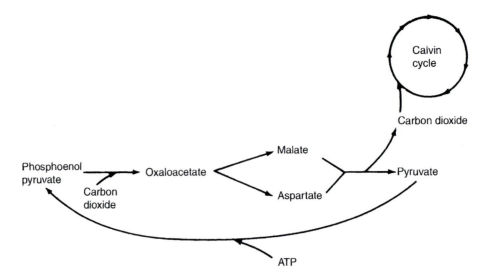

FIGURE 4–3   Hatch and Slack cycle shown in simplified form. This pathway in C_4 plants starts in mesophyll cells by the carboxylation of phosphoenol pyruvate to produce a 4-carbon compound, oxaloacetate, that is converted to malate or aspartate (dependent on plant species). This compound is shuttled to the bundle sheath chloroplasts where it is decarboxylated with the released CO_2 now going through the Calvin cycle (Figure 4–2). The pyruvate is returned and converted to phosphoenol pyruvate to restart the cycle. (From Raymond P. Poincelot, *Horticulture: Principles & Practical Applications*, (1980). Reprinted by permission of Prentice-Hall, Upper Saddle River, NJ.)

A few horticultural crops, of which sweet corn is the most familiar, and many obnoxious weeds have a second pathway, the Hatch and Slack cycle (Figure 4–3). This is in addition to the Calvin cycle. These C_4 plants have both mesophyll and bundle sheath chloroplasts (see Chapter 3), more efficient photosynthesis, and low to zero rates of photorespiration. The first product from carbon dioxide in C_4 plants has four carbons, hence, the term C_4 plants.

Even fewer horticultural plants have the CAM (crassulacean acid metabolism) form of photosynthesis that is essentially a Calvin cycle with carbon dioxide being provided in a different form and time (Figure 4–4). Horticultural crops with this pathway are *xerophytes* like cacti, many succulents of arid regions, and the pineapple. These plants exist under very hot, bright, dry conditions. Their stomates (leaf and stem pores) are closed during the day. Open stomates under hot, dry conditions lose water at a very rapid rate. Carbon dioxide (CO_2) enters the plant at night when the stomates are open, not during the day as in the preceding two groups. The CO_2 is fixed at night into an organic acid (malic acid) that is oxidized internally during the day to produce CO_2. The internally released CO_2 is utilized to produce carbohydrate via the Calvin cycle.

Photosynthesis, Evolution, and Climate.   Plants evolved these three forms of photosynthesis as a consequence of climatic pressures. The result is that some plants (C_3 plants) are best adapted to cooler, wetter climates found mainly in the temperate zones. Others (C_4 plants) do best in hotter, slightly drier climates or the subtropical to tropical areas. Some (CAM plants) exist quite well in very hot and dry areas, the deserts. Having said that, this doesn't restrict their geographical distribution as much as you think. C_4 plants can grow in areas where you find C_3 plants and the reverse is true. CAM plants can also be found in areas other than deserts. These plants are best adapted to their climate of origin, but are not restricted to them. These plants show considerable variation when one looks at their rates of photosynthesis, water usage, and light-independent reactions, as seen in Table 4–1.

Why Crabgrass Wins.   The C_4 plants with their more efficient photosynthesis grow faster and are better adapted than C_3 plants to conditions of high temperature, bright light, and dryness. This explains why many weeds outcompete horticultural crops during July

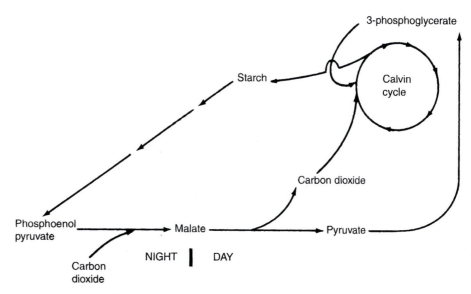

FIGURE 4–4   Simplified CAM pathway. Plants with this pathway take up CO_2 at night to carboxylate phosphoenol pyruvate prior to malate formation. During the day malate is decarboxylated to supply an internal CO_2 source for the Calvin cycle (Figure 4–2). Pyruvate is converted to 3-phosphoglycerate that can be used in the Calvin cycle or elsewhere. Some of the starch produced during photosynthesis can be converted to phosphoenol pyruvate for the night use in the pathway. (From Raymond P. Poincelot, *Horticulture: Principles & Practical Applications*, (1980). Reprinted by permission of Prentice-Hall, Upper Saddle River, NJ.)

TABLE 4–1   *Comparative Facts for Different Forms of Photosynthesis (PS)*

Type	Climate of Origin	PS Rate	Water Needs	Type of Light-Independent Cycle
C_3	Temperate	X	High	Calvin cycle, CO_2 enters during day
C_4	Subtropical to tropical	$2–3X$	Moderate	Hatch–Slack pathway plus Calvin cycle, CO_2 enters during day
CAM	Desert	$0.5X$	Low	Calvin cycle, CO_2 enters during night

and August. The best example of C_4 and C_3 competition takes place in the lawn. The dominant grass is usually some cultivar of bluegrass, a C_3 plant. Crabgrass is a problem weed-grass also found in lawns and is a C_4 plant. In the spring or fall, a time of cool, wet weather in much of North America, lawns look great and grow fast. The main reason is that this type of weather is most favorable for C_3 plants. Come the hot, dry days of July and August, lawns start to brownout and crabgrass thrives. The C_3 bluegrass is at a competitive disadvantage during these times, unlike the C_4 crabgrass that thrives under these conditions.

Switch-Hitters.   Some CAM plants, such as cacti, can convert from C_3 to CAM photosynthesis, when rainfall appears. These plants can go from slow to fast growth, since their photosynthesis rate doubles when this switchover occurs. This ability helps to explain why cacti under cultivation seem to grow faster than their desert relatives. It also helps explain in part why explosions of growth and color take place in arid regions after heavy, sudden spring rains.

RESPIRATION

The maintenance of life requires the continuous use of energy. In plants most of the needed energy has its origin in the photosynthetic process, whereby light energy is converted into a chemical form. This stored energy can be released by the biological oxidation of the organic compounds resulting from photosynthesis, or from compounds synthesized later from the photosynthetic products. The release of energy through the biological oxidation of organic compounds is termed *respiration*.

Respiration in plants provides for a slow, controlled release of energy with a minimal amount of heat as a by-product. In a sense it may be viewed as a slow, controlled form of combustion. Control is only possible through a number of biochemical steps catalyzed by many enzymes. About 40% of the energy released during the respiration of glucose is trapped in the form of the chemical energy carrier, ATP. This energy can be utilized later or elsewhere in cellular processes, when it is released on the conversion of ATP to ADP.

Carbohydrates are the primary substrates of respiration in higher plants. The most important ones are glucose, fructose, sucrose, and starch. Other substances to a much lesser extent may serve as respiratory substrates in certain organs or circumstances. These include the fat reserves stored in the endosperm of some seeds, organic acids, and proteins.

The most common substrate is glucose, which is oxidized in the presence of oxygen (*aerobic respiration*) to produce energy in the form of 38 molecules of ATP per molecule of glucose:

$$C_6H_{12}O_6 + 6O_2 \longrightarrow 6CO_2 + 6H_2O + ENERGY$$

| Glucose | Oxygen | Carbon dioxide | Water | |

This reaction, for all practical purposes, appears to be the reverse of photosynthesis. Of course, this view is a gross oversimplification. The above equation is only indicative of the overall process and gives no idea of the sequence of reactions or intermediates that occurs. Many of the intermediate compounds formed during respiration can be used as precursors for the biosynthesis of numerous cellular constituents. We will come back to this precursor connection later in this chapter.

The sequence of reactions during respiratory metabolism includes phosphorylations, oxidations, hydrations, decarboxylations, group transfers, isomerizations, and cleavages. Respiratory metabolism is summarized in very brief form in Figure 4–5. Pyruvic acid is produced from glucose in a series of enzyme-catalyzed reactions collectively called *glycolysis*. This occurs in the cytoplasm of the cell.

FIGURE 4–5 Simplified aerobic respiration pathway. See text for details. (From Raymond P. Poincelot, *Horticulture: Principles & Practical Applications,* (1980). Reprinted by permission of Prentice-Hall, Upper Saddle River, NJ.)

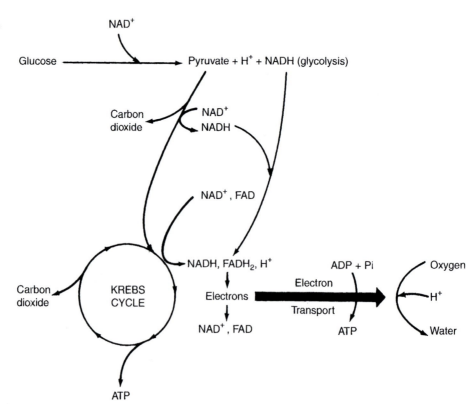

Pyruvic acid is oxidized in the mitochondria through a series of enzymic reactions termed the *tricarboxylic acid cycle* (also known as the *Krebs* or *citric acid cycle*). The net result of this cycle is the production of carbon dioxide and hydrogen atoms; the hydrogen atoms are used in the reduction of the hydrogen acceptor, nicotinamide adenine dinucleotide (NAD^+), to form NADH. Some H^+ is also produced. Electrons are carried by NADH and transferred (some protons transferred at some steps) through a series of cytochromes and other electron acceptors at progressively lower energy levels. The final electron acceptor is oxygen, which in its reduced form combines with protons to form water. The energy lost during the electron transfer is trapped during the formation of ATP from ADP and inorganic phosphate. This process is termed *oxidative phosphorylation* and takes place also in the mitochondria.

The respiratory metabolism described above occurs when oxygen is present. Respiration in the presence of oxygen is known as *aerobic respiration*. When oxygen is lacking, the respiration process becomes *anaerobic*. In that process, pyruvic acid is converted to alcohol or lactic acid. Anaerobic respiration takes place in roots deprived of oxygen. This condition is found in water-logged soil or in plants submerged by floodwaters. Vascular plants generally cannot survive prolonged anaerobic respiration and usually die.

Roots and Respiration

Anaerobic respiration induced by overwatering is the leading cause of house plant death. Overwatering kills the roots, because aerobic respiration ceases. The dead roots cannot conduct water to the shoot, so the shoot wilts. The house plant caretaker thinks more water is needed. Unhappily, all that water accelerates the rotting of the roots and eventually the rest of the plant.

Those plants that thrive even though they have submerged roots (rice, cattails, and willows) probably do not survive on anaerobic respiration, but on their ability to carry out aerobic respiration with the levels of oxygen dissolved in the water. Anaerobic respiration may occur in the center of bulky tissues or in germinating seeds until oxygen becomes available.

Are Plants Oxygen Consumers?

Gaseous exchanges in respiration and photosynthesis occur in opposite directions. In respiration, oxygen is used and carbon dioxide is a by-product, as opposed to the use of carbon dioxide and the production of oxygen in photosynthesis. During darkness, no photosynthesis occurs, but in the light, both processes are taking place. The question arises as to which gaseous exchange dominates.

Evidence indicates that oxygen produced during photosynthesis can be used in respiration, and carbon dioxide from respiration can be used in photosynthesis. Under low light intensities a point can be reached where the two processes are balanced; that is, oxygen and carbon dioxide neither enters nor exits the leaf. The light intensity at which this happens is termed the *light compensation point*. The light compensation point is considerably less than full sunlight, so the rate of photosynthesis is much higher than respiration during the day. Even including the night, when only respiration occurs, the net effect is that plants liberate much more oxygen than they use and release less carbon dioxide to the air than they utilize. If this were not so, the plant would be unable to have food available for immediate use or for storage and ultimately would die.

In earlier, unenlightened times, it was feared that plants placed in a hospital room would deprive the patient of oxygen and sicken them. From the preceding and other studies of oxygen usage and carbon dioxide production, it is clear that plant respiration does not have an adverse effect on the air of rooms used for sleeping or patient care. If anything, flowers are likely to lift the spirits of people recuperating in hospitals.

Several external and internal factors affect the rate of respiration in plants. Temperature, water, and levels of O_2 and CO_2 are the most important environmental factors. Internal factors include injury, the levels of available food, the age or kind of tissues concerned, and levels of intermediates utilized in the respiratory process. Increasing levels of O_2 and CO_2 increase and decrease, respectively, the rates of respiration. An increase in temperature causes an increase in the respiratory rate. The rate of respiration is approximately doubled for each 10°C rise in temperature between 5° and 36°C. These factors and others pose problems in connection with the transportation and storage of fruits, vegetables, and cereal grains that continue to respire after harvest. Refrigeration and sometimes storage in modified atmospheres can slow post-harvest respiration.

PHOTORESPIRATION

The term *photorespiration* was coined to cover a process that appears similar to respiration, but only occurs in daytime with C_3 and CAM plants. This light-induced form of respiration accounts for the higher rates of respiration observed with these plants, but not C_4 plants, in the light as opposed to the dark. The process is extremely wasteful, in that it causes up to a 50% loss of the carbon fixed during photosynthesis. The similarity to respiration is only superficial, because the conversion of photosynthetically fixed carbon to CO_2 does not produce ATP. Photorespiration increases with warmer temperatures and is enhanced as oxygen levels rise and CO_2 levels decrease.

Effects of Photorespiration

C_3 plants (see discussion on photosynthesis) generally have high rates of photorespiration and photosynthetic rates two to three times slower than C_4 species at 25° to 35°C (77° to 95°F) and high light intensities. C_4 plants under similar conditions have very low to zero rates of photorespiration. This fact explains why weeds, many of which are C_4 plants, have a competitive edge. A few species, such as sunflower (*Helianthus annuas*) and cattail (*Typha latifolia*), are exceptions in that they are C_3 species with high rates of photorespiration, yet they also have high rates of net CO_2 fixation.

The primary substrate of photorespiration is glycolic acid ($CH_2OH–COOH$). The reaction responsible for most of the glycolic acid occurs in the first stage of the Calvin cycle. The enzyme that catalyzes the first step of this cycle (ribulose-1,5-bisphosphate carboxylase) is capable of utilizing either CO_2 or O_2 as the substrate. Under normal conditions CO_2 is favored, but when levels drop low, O_2 competes with CO_2 for the active site of this enzyme. During stomate closure, such as in high temperatures and during wilting, CO_2 concentrations drop inside the leaf, thus favoring the use of O_2. Why is this so? It has to do with evolution. This enzyme evolved in a time when CO_2 concentrations were very high and O_2 levels very low, so this dual use posed no problem. As our oxygen levels rose as plants increased in numbers, this flaw in the enzyme became a problem. C_4 plants evolved a process to compensate for this problem, but C_3 and CAM plants did not.

When CO_2 interacts with ribulose-1,5-bisphosphate carboxylase, two molecules of 3-phosphoglyceric acid (3-PGA) are produced and both continue on in the Calvin cycle to produce carbohydrate. However, when O_2 interacts, only one molecule of 3-PGA results. The second molecule is phosphoglycolic acid. Since only one molecule of 3-PGA goes on in the Calvin cycle, its efficiency is cut drastically. The phosphoglycolic acid is converted to glycolic acid and ultimately into CO_2 and serine in a series of reactions that take place in the peroxisomes and mitochondria. The net result is a loss of up to 50% of the carbon fixed in photosynthesis.

Photorespiration and Plant Productivity

Because plant productivity correlates closely with rates of net CO_2 fixation, processes such as photorespiration that diminish the assimilation of CO_2 are considered wasteful. Measurements indicate that plants with high rates of photorespiration are usually less productive than those with lower rates. Many scientists share the belief that photorespiratory processes can be regulated through biochemical means. Recent advances in somatic cell genetics (tissue culture) suggest the usefulness of the technique to selectively screen plant cells so that only those with low rates of photorespiration will survive. Control of photorespiration would appear to be a promising possibility for the future horticulturist.

WATER ABSORPTION

In vascular plants, most water (and mineral) absorption is from the soil through the roots (Figure 4–6). However, epiphytes can absorb water through aerial roots, leaves, or other organs. Most water (and minerals) is absorbed in the root hair zone in young roots with primary tissues a short distance back from the root tip. Here the water transport system, the xylem, is well differentiated, water permeability has not yet been reduced by suberization, and the root hairs present a large surface area for water absorption. Water and mineral absorption also occurs through root-associated fungi, the mycorrhizae.

Older roots with secondary tissues impregnated with suberin can also absorb water through breaks and cracks in the suberized secondary tissues. This process is particularly true for large trees, in which the suberized tissues are a large part of the total root system, and in winter when unsuberized root surfaces are minimal.

Active Absorption of Water

Water may be absorbed by two mechanisms: *active* and *passive absorption*. *Active absorption* of water implies an essential participation of the roots in water absorption. This process occurs when soil moisture is high and the rate of transpiration or loss of water from the plant tissues as vapor is low (see discussion on transpiration rates in this chapter). These conditions occur mostly at night, when the stomates are closed and transpiration rates are very low. Active absorption of water constitutes a smaller part of the total water absorption of the plant than does passive absorption. It also does not account for water absorption during the day. In fact, the process does not occur in all plants, such as many conifers.

Active absorption generates root pressure through osmosis. The inward movement of water across the differentially permeable membranes of the living root cells to xylem by

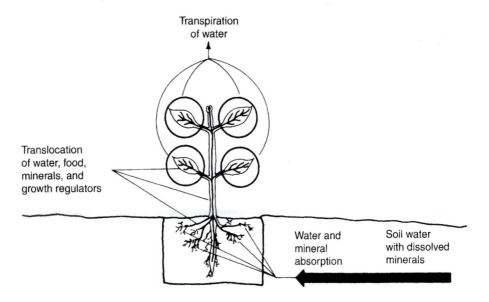

FIGURE 4–6 Translocation in the vascular plant. Water is absorbed at the roots and translocated through the plant via the xylem portion of the vascular system; much of the water is eventually lost by transpiration. The water also serves as a solvent for the translocation of minerals, foods (sugars), phloem translocation, and growth regulators (both internal hormones and externally applied synthetic growth regulators). (From Raymond P. Poincelot, *Horticulture: Principles & Practical Applications*, (1980). Reprinted by permission of Prentice-Hall, Upper Saddle River, NJ.)

osmosis requires a higher solute concentration in the xylem vessels relative to the soil solution. This condition occurs at night, when solutes leak and are secreted from the living root cells into the adjacent xylem tissues. The endodermis around the vascular tissues prevents the back-leakage of ions from the xylem. Because transpiration is essentially zero at night, the ions accumulate, as water does not move away. The resultant diffusion of water into the xylem cells by osmosis is responsible for creating a positive pressure called *root pressure*. Root pressure is responsible for the loss of liquid water drops from the leaf (*guttation*). If the solute level in the soil solution becomes higher than that in the xylem vessels, which can result from de-icing salts or too much soluble fertilizer, the direction of water movement by osmosis will reverse. Unless checked, this will result in wilting, dehydration, and death of the plant.

Passive Absorption of Water

The process primarily responsible for water absorption is passive absorption. *Passive absorption* of water occurs when plants are transpiring rapidly. It does not require participation of the root cells, but comes about from forces arising at the top of the plant. In fact, passive absorption of water will take place with a dead root system. Active absorption of water does not happen during rapid transpiration, since the rapid movement of water from passive absorption dilutes and washes away the solutes in the xylem on which active absorption depends.

The flow of water in the passive mode is dependent on two important properties of water molecules: adhesion and cohesion. Water molecules hold firmly to molecules of other substances (*adhesion*) that contain large numbers of oxygen or nitrogen atoms. Water molecules also hold very firmly to each other (*cohesion*), accounting for the surface tension of water. Because of these water properties, the theory used to explain passive absorption of water is called the *cohesion tension theory*.

Upward movement of water from the root xylem to the leaves, or the transpiration stream, starts with the loss of water vapor through the stomata. This causes local tissue dehydration, and the movement of water into this tissue results, because of the cohesive properties of water. This in turn causes partial dehydration deeper into the plant tissues, bringing about water diffusion from the leaf xylem. Again the cohesive property of water in turn helps to "pull" the water through the xylem of the stem and eventually the root. This generates a pressure potential between the roots and the soil solution. This hydrostatic pressure differential is the driving force for the movement of water into the roots. The adhesive property of water prevents the too rapid removal of water from the xylem and, along with cohesion, maintains continuity of the water column.

NUTRIENT ABSORPTION

Sixteen (or possibly more) elements are necessary for plant growth. Three of these, carbon, hydrogen, and oxygen, are derived from water, oxygen, and carbon dioxide. The remaining 13 (or more) are absorbed by roots as inorganic salts from the soil solution. They are nitrogen, potassium, calcium, magnesium, phosphorus, sulfur, iron, copper, manganese, zinc, molybdenum, boron, and chlorine (Figure 4–6). Some plants, but not all, appear to have additional requirements. For example, some grasses require silica, soybeans need nickel, and C_4/CAM plants use sodium.

All of these mineral elements are present in the soil solution as ions or absorbed onto soil particles. To reach the interior of the root, these ions have to cross several layers of different substances. The mechanisms by which this passage occurs are complex and not as well understood as many other plant processes.

Routes for Ion Movement

The ions first cross the cell walls of the root hairs and other root tissues. Plant cell walls offer little hindrance to the movement of gaseous molecules, dissolved nutrients, or ions, except for having relatively weak attractive forces for *cations* or positively charged ions (K^+,

Mg^{2+}, Ca^{2+}, and others). These cations move across the cell wall in a leapfrog manner. Cations absorbed at a negatively charged site can be displaced by another cation with a higher absorptive affinity for that site. The released cation will move inward to another cation-absorptive site. Negatively charged ions (*anions*), some cations, gases, and dissolved nutrients will diffuse through the cell wall dissolved in water. The area comprised by the cell walls is termed *free space* because of unopposed ion movement. Movement by the cell wall pathway is referred to as the *apoplastic pathway.*

If the ions and water leave the apoplastic route, they must move through a plasma membrane. This pathway is more restrictive and consists of nonfree space. This pathway becomes the only route when the ions reach either an outer exodermis or the endodermis around the root's vascular system. Ion movement across plasma membranes occurs primarily by active transport and is defined as *symplastic*. Movement by the symplastic route involves active transport through the plasma membrane and then movement from protoplast to protoplast by the connecting plasmodesmata. Ions can also move from vacuole to vacuole too, which is termed the *transcellular* route. Evidence suggests that most ions utilize the symplastic route.

Gases, small polar molecules, and fat-soluble molecules can cross the plasma membrane or others by passive transport. Ions must cross by active transport. These two processes are differentiated as follows. Membranes may be visualized as gateways with selective or discriminatory abilities that allow passage of ions, gases, neutral molecules, and the like at rates of passage varying from very low to rapid. *Passive transport* refers to transport resulting from physical driving forces. The movement or diffusion of solutes from an area of higher concentration to an area of lower concentration is a form of passive transport. Some molecules may cross the plasma membrane or other membranes by passive transport. These molecules include gases such as oxygen or carbon dioxide, small polar molecules like ammonia, and fat-soluble molecules. It is speculated that the first two types may move through the membrane if it acts as a *molecular sieve;* the fat-soluble molecules are soluble in the lipid portions of membranes, so it is suggested that diffusion occurs with the membrane acting as a *selective solvent.*

Active Transport of Ions

Active transport of ions is an energy-requiring process. Most of the ions probably cross the plasma membrane or other membranes by active transport. Evidence indicates the involvement of carriers that bind ions, transport them across the membrane, and discharge them on the other side. These carriers have properties that suggest they are large protein molecules analogous to enzymes. Evidence strongly suggests that a carrier that is thought to be adenosine triphosphatase (ATPase), a well-characterized enzyme, transports Ca^{2+} across plasma membranes in roots.

Once into the cytoplasm, diffusion brings about further movement of the ions. Here the ions may undergo several fates that result in their removal. This removal is necessary if ion absorption by the roots is to continue. Ions may be transformed during the course of a metabolic reaction into another substance. They may be absorbed by protein molecules or transported both in the active and passive mode through other cells until they reach the xylem. They may also be accumulated in vacuoles or cellular organelles.

TRANSLOCATION

The movement of water, minerals, and food from one part of a plant to another is called *translocation* (Figure 4–6). Minerals absorbed by the roots are translocated to the stems, leaves, and reproductive organs of the shoots. Once there, minerals may be further translocated up or down the vascular system, such as from older leaves to younger leaves. Fertilizers applied as foliar sprays are translocated to metabolically active tissues. Most minerals are translocated in the form of ions; however, nitrates absorbed in the roots are converted there and translocated as amino acids.

Most food is translocated in the form of sugars. Much of the sugar is translocated as sucrose, except for a few species in which sorbitol, raffinose, stachyose, and verbascose are also present in large quantities. The predominant flow is from leaves to metabolically active tissues, such as meristems or areas of growth, where sugar demand exceeds the synthesized supply. Translocation of sugars from storage organs after hydrolysis of starch also occurs when the leaf supply is insufficient. Leaves and storage sites are termed *sources,* and where sugars are needed, *sinks.* Most sugar translocation is from sinks to sources.

Other substances can be translocated, but to a much lesser extent. Amino acids are translocated; the peak flow is probably from senescent leaves prior to abscission. Auxins, vitamins, and other plant growth substances are translocated through the vascular system. Although their concentration is low, their translocation has a profound effect on the development of the plant.

Translocation in Xylem

Xylem is clearly the part of the vascular system through which water flows. Minerals absorbed at the roots as ions are translocated upward primarily through the xylem. Some lateral movement of minerals occurs from the xylem to the phloem. Minerals can also be translocated from older leaves through the phloem in an upward or downward direction. Some translocation of minerals through the phloem may be important in deciduous trees in the winter, when water flow through the xylem is slow.

Small amounts of sugars may move upward in the xylem during certain seasons, but the bulk of the sugar translocation, both upward and downward, takes place in the phloem. Sugars may also move laterally through the vascular rays. Amino acids, organic acids, soluble proteins, auxins, and vitamins are translocated through the phloem.

Translocation through the xylem is primarily upward and dependent on passive transport. Minerals and small amounts of other solutes are simply carried along with the upward flow of water. In the section on water absorption, it was indicated that water movement was dependent on the loss of water through the stomata and the resulting transpiration stream.

Translocation in Phloem

Phloem translocation has been explained by several hypotheses. The most widely accepted one is the *pressure-flow hypothesis.* Sugar movement from the site of availability, the *source,* to the site of utilization, the *sink,* is dependent on a concentration gradient. First, the sugar leaves the parenchyma cells by active transport adjacent to the phloem and enters the sieve tubes. This process, *phloem loading,* increases the solute concentration in the sieve tubes, causing osmotic movement of water from the xylem into the sieve tubes. Water movement carries the sugar along the sieve tubes to the sink. Adjacent parenchyma cells remove the sugar by active transport, and the reduced solute level causes water movement by osmosis out of the sieve tube.

TRANSPIRATION

Only about 1% or less of the water absorbed by plants is utilized in biochemical processes. Water is used for photosynthesis, hydrolysis, the hydration of cell walls, and maintenance of the swollen conditions of cells (*turgor*) through internal water pressure (*turgor pressure*). Most of the water is lost from the aerial parts of the plant by evaporation, followed by diffusion of water vapor into the air. This process is called *transpiration* (Figure 4–6).

Much water is lost through the stomata and, to a much lesser extent, the cuticle and lenticels. The major part of stomatal transpiration is from the leaves. Under conditions of high soil moisture and humidity, a small fraction of water may be lost by *guttation,* that is, as exuded drops from terminal ends of veins in the leaf margin.

Wilting

When the loss of water by transpiration exceeds that of replacement by absorption, a water deficit occurs within the plant. Water deficits reduce the turgor pressure of the plant cells and, eventually, the leaves and herbaceous stems will droop. A decrease in turgor pressure of the guard cells does cause the stomatal aperture to decrease and slow down the wilting process, but environmental factors may be such that this stomatal response is too slow to prevent wilting. Wilting is not observed in plants with extensive mechanical support, such as in magnolia.

Wilting can be either temporary or permanent. When adequate soil moisture is present, a water deficit can result in *temporary wilting*, which disappears at night when the rate of transpiration slows. This condition is often seen in fields and gardens during the noon hour of a hot, sunny day. Inadequate soil moisture produces *permanent wilting* that can cause death by desiccation if prolonged. Recovery from permanent wilting is possible if water is added to the soil.

Transpiration Rates

Transpiration rates are usually higher in the day than at night. An exception is many succulents with crassulacean acid metabolism (see discussion of photosynthesis), since their stomata are open at night, rather than during the day. Several environmental factors affect the rate of transpiration: internal plant factors, light, humidity, temperature, wind, and soil water content. These factors will be discussed in Chapter 6.

OTHER METABOLIC PROCESSES

Metabolism in the plant is the sum total of all the biochemical reactions that take place in the plant body (Figure 4–7). Most of these reactions are catalyzed by enzymes. Synthetic metabolism is called *anabolism*; degradative metabolism is termed *catabolism*. Photosynthesis

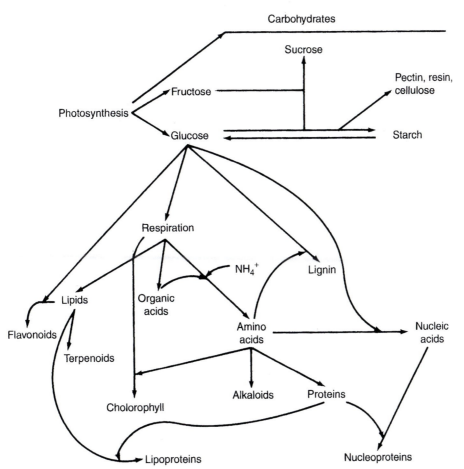

FIGURE 4–7 An overall very simplified view of metabolism in horticultural plants. Intermediates and lesser pathways are left out. (From Raymond P. Poincelot, *Horticulture: Principles & Practical Applications*, (1980). Reprinted by permission of Prentice-Hall, Upper Saddle River, NJ.)

and respiration are examples of anabolic and catabolic processes, respectively. Growth occurs only when anabolism exceeds catabolism. Some of the better known metabolic processes have been discussed already in this chapter. However, other equally important forms of plant metabolism exist, and brief discussions of these follow. No more than a brief, simplistic diagram of these many metabolic pathways is possible, since a complete consideration would require a book in itself. Many of the metabolites involved are important enough to people to be or to have been isolated from plants.

Carbohydrates

The ultimate source of the carbon in carbohydrates is atmospheric carbon dioxide. Simple sugars arise from the fixation of carbon dioxide during photosynthesis and from pyruvic acid by what is essentially a reversal of glycolysis (see respiration). These simple sugars (Figure 4–8), or *monosaccharides* (glucose, fructose, ribose, and others), are produced in the form of monosaccharide phosphates during the two processes just mentioned. These phosphorylated sugars can be converted to free sugars by additional enzymic reactions. The monosaccharide phosphates can undergo several interconversions through further enzymic processes.

Disaccharides (Figure 4–8) are produced when two monosaccharides are linked. Sucrose, the major product of photosynthesis, is a disaccharide composed of glucose and fructose. Sucrose is the major translocation form of carbohydrate in plants. It is a significant storage form in some plants, such as sugarcane and sugar beet, the major sources of sucrose. Common table sugar is also sucrose.

FIGURE 4–8 Some examples of plant carbohydrates. (From Raymond P. Poincelot, *Horticulture: Principles & Practical Applications,* (1980). Reprinted by permission of Prentice-Hall, Upper Saddle River, NJ.)

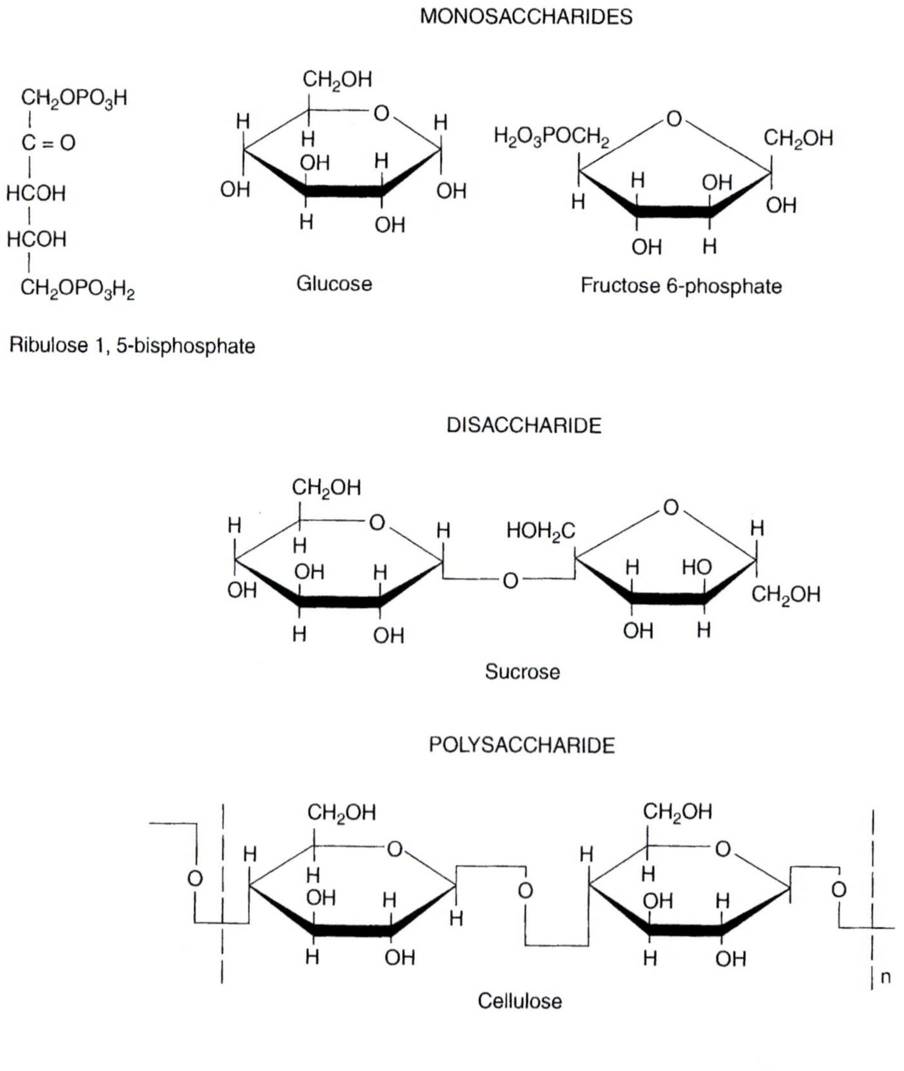

Polysaccharides (Figure 4–8), macromolecules formed from long chains (*polymers*) of carbohydrates, result from the enzymic linkage of many monosaccharides. Starch and cellulose, horticultural products of commercial importance, are the most prevalent polysaccharides. Starch is synthesized in chloroplasts and amyloplasts; it is an important storage carbohydrate. Cellulose is an important cell wall carbohydrate. Starch and cellulose are both formed from glucose, but the bond arrangement differs. Other polysaccharides are hemicellulose and pectin, which are also cell wall constituents.

Storage forms of carbohydrate can be degraded enzymically in the plant by a process called *digestion*. After digestion the resulting soluble foods can be assimilated, respired, or transported elsewhere. Starch is broken down to glucose when needed, such as during seedling germination or development of vegetative tissue. Inulin, a polysaccharide accumulated instead of or in addition to starch in some plants, is degraded to fructose. Pectin is degraded to galacturonic acid.

Amino Acids and Proteins

Nitrogen is taken up by plants mostly as the nitrate ion (NO_3^-). In the legumes and a few other vascular plants, nitrogen is fixed by symbiotic bacteria located in root nodules. These bacteria convert free nitrogen to ammonium ion (NH_4^+), which is readily used in subsequent metabolic pathways. Nitrate, once absorbed by plants, is also enzymically converted to NH_4^+.

Ammonium ion does not accumulate, since it is used rapidly in the enzymic synthesis of amino acids. The initial synthesis of amino acids is the reaction of α-ketoglutaric acid (carbon framework derived from glucose, a photosynthetic product) that arises from the tricarboxylic acid cycle (see respiration), with NH_4^+ to produce glutamic acid:

$$
\begin{array}{cc}
\begin{array}{l}
COOH \\
| \\
C=O \\
| \\
CH_2 \\
| \\
CH_2 \\
| \\
COOH
\end{array} + NH_4^+ \; NADH
&
\Leftrightarrow
\quad
\begin{array}{l}
COOH \\
| \\
H-C-NH_2 \\
| \\
CH_2 \\
| \\
CH_2 \\
| \\
COOH
\end{array} + NAD^+ + H_2O
\end{array}
$$

α-Ketoglutaric acid	Glutamic acid

Other amino acids are synthesized by transamination, or the transfer of an $\sim NH_2$ group:

$$
\begin{array}{l}
COOH \\
| \\
H-C-NH_2 \\
| \\
CH_2 \\
| \\
CH_2 \\
| \\
COOH
\end{array}
+
\begin{array}{l}
COOH \\
| \\
C=O \\
| \\
CH_2 \\
| \\
COOH
\end{array}
\Leftrightarrow
\begin{array}{l}
COOH \\
| \\
C=O \\
| \\
CH_2 \\
| \\
CH_2 \\
| \\
COOH
\end{array}
+
\begin{array}{l}
COOH \\
| \\
H-C-NH_2 \\
| \\
CH_2 \\
| \\
COOH
\end{array}
$$

Glutamic acid	Oxaloacetic acid	α-Ketoglutaric acid	Aspartic acid

Several enzymes, called *transaminases*, catalyze the various transaminations needed for the different amino acids. Amino acid metabolism is dependent on nitrogen and photosynthetic carbohydrate, so both carbohydrate and protein metabolism are linked. These compounds are also the most abundant in plants. Any environmental factor that affects one indirectly affects the other, thereby resulting in an extensive effect on plant development.

About 20 or so amino acids are combined in various ways to synthesize *proteins* (Figure 4–9). A larger number (more than 100) exist only as free amino acids; the metabolic role of the nonprotein amino acids is unclear. The sequence of the amino acids in the proteins is determined by the arrangement of the nucleotides (see other nitrogenous compounds in this chapter) in the chromosomal DNA. Various combinations of three sequential nucleotides code for the different amino acids. A form of RNA, *messenger RNA* or *mRNA*, reads the triplet sequence and carries this information to the cytoplasm. The information is contained in the

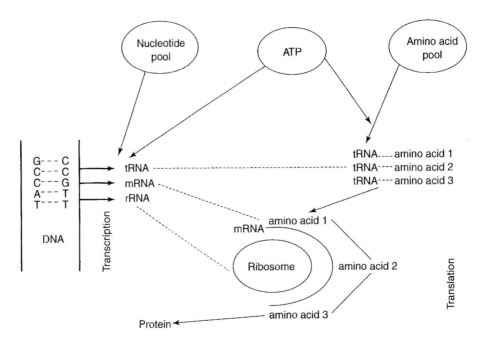

FIGURE 4–9 Events in protein synthesis. The nucleotide sequence code in DNA is transferred to the various forms of RNA (see text). The amino acid carrier (tRNA) brings the amino acid to the right place in the growing protein based on information read (translated).

nucleotide sequence of the mRNA that becomes attached to the ribosomes. Amino acids are carried and oriented to the ribosomal template by another form of RNA, *transfer RNA* or *tRNA*. This form of RNA matches the correct amino acid to the portion of the ribosomal template that codes for it. The amino acids are polymerized through the formation of peptide bonds.

Proteins are utilized as enzymes to catalyze the various metabolic processes that occur in plants. Enzymes are named by adding an *–ase* to the root form of the material or *substrate* that the enzyme acts on or to some specific reaction occuring in the presence of the enzyme. For example, the first enzyme mentioned for the Calvin cycle was ribulose-1,5-bisphosphate carboxylase. This enzyme adds carbon dioxide as a carboxyl group to ribulose-1,5-bisphosphate to change it from a five-carbon to a six-carbon compound. This compound then splits into two molecules of 3-phosphoglyceric acid. Other proteins are used as structural components, such as in membranes.

Proteins can also be enzymically degraded into their constituent amino acids. Both synthesis and degradation probably occur simultaneously. The meat tenderizer papain is an enzyme used by plants for protein degradation. In actively growing tissue, protein synthesis exceeds breakdown. However, protein degradation occurs to a larger extent during dehydration, senescence, and starvation.

Lipids

Lipids (Figure 4–10) are much more diverse than either carbohydrates or proteins, since they are not composed of repetitive, well-defined monomeric units. As a group they are defined more in an operational sense; that is, they are insoluble in water but soluble in "fat" solvents such as chloroform, benzene, and petroleum ether. As such, lipids include the fats, phospholipids, glycolipids, waxes, and sterols.

Fats are found in plant cells, but the highest concentrations are present in the endosperms of seeds. If the fat is a liquid at room temperature, it is called an *oil*. The oils derived from corn, soybean, olive, sunflower, cottonseed, peanut, coconut, and safflower are horticultural products of great value.

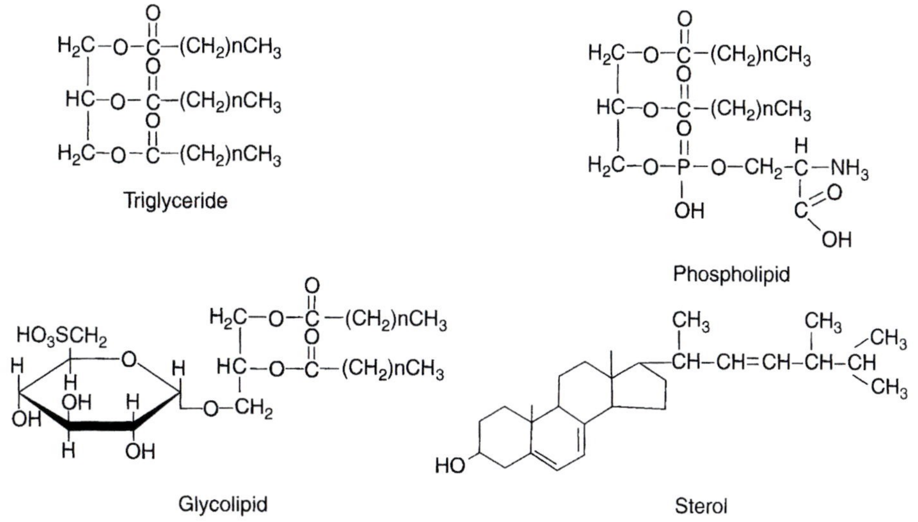

FIGURE 4–10 Some examples of plant lipids. (From Raymond P. Poincelot, *Horticulture: Principles & Practical Applications,* (1980). Reprinted by permission of Prentice-Hall, Upper Saddle River, NJ.)

The synthesis of fats can be summed up in three steps: the production of glycerol from carbohydrates, the production of fatty acids, and the esterification of glycerol with the fatty acids. This fat is correctly called a *triglyceride,* since three fatty acids are esterified to the glycerol.

Phospholipids and *glycolipids* differ from triglycerides in that the third –OH of the glycerol is substituted with a group other than a fatty acid. For phospholipids this group is either phosphoric acid or a substituted phosphoric acid, and for glycolipids it is a sugar or a substituted sugar. Glycerol can be phosphorylated to glycerol phosphate, which in turn can be esterified with two fatty acids to produce phosphatidic acid. Enzymic reactions of glycerol with two fatty acids followed by interactions with nitrogenous compounds or sugars lead to the synthesis of phospholipids and glycolipids.

Plant *waxes* are long-chain esters of fatty acid and alcohols of great complexity. Most of the waxes are found in the cutin and some in suberin. *Sterols* are complex alcohols with a tetracyclic ring structure. The best known plant sterol is erogosterol, which can be converted to vitamin D by irradiation.

Lipids can be enzymically digested to glycerol and fatty acids. Further metabolism of glycerol can produce sugars for respiration or other uses. Fatty acid may be oxidized to produce ATP. For example, lipid digestion occurs during seed germination.

Other Nitrogenous Compounds

Much of the nitrogen in a plant is present in the amino acids and proteins. A few phospholipids, glycolipids, and carbohydrates also contain nitrogen. Nitrogen is also found in a wide variety of other minor compounds. These include purines, pyrimidines, nucleic acids, porphyrins, alkaloids, vitamins, coenzymes, and hormones.

Purines and *pyrimidines* (Figure 4–11) are heterocyclic nitrogenous bases. They are involved in the synthesis of a number of essential compounds, such as plant hormones, nucleic acids, DNA, RNA, ATP, ADP, coenzymes, vitamins, and alkaloids. A considerable number of purines and pyrimidines are synthesized by plants, but the most important are adenine, cytosine, guanine, thymine, and uracil.

Nucleic acids are DNA, the genetic code carrier, and RNA, the genetic code transcriber. Nucleic acids are usually linked to proteins, forming compounds called *nucleoproteins.* The nucleic acid portion is a high-molecular-weight polymer of nucleotides; each *nucleotide* consists of a purine or pyrimidine, a sugar, and phosphoric acid. Both DNA and RNA contain the purines adenine and guanine and also the pyrimidine cytosine. However, DNA and RNA differ, too. DNA has the pyrimidine thymine and the sugar deoxyribose, whereas RNA has the pyrimidine uracil and the sugar ribose.

FIGURE 4–11　Purines and pyrimidines. (From Raymond P. Poincelot, *Horticulture: Principles & Practical Applications,* (1980). Reprinted by permission of Prentice-Hall, Upper Saddle River, NJ.)

Purine

Pyrimidine

Nucleotide

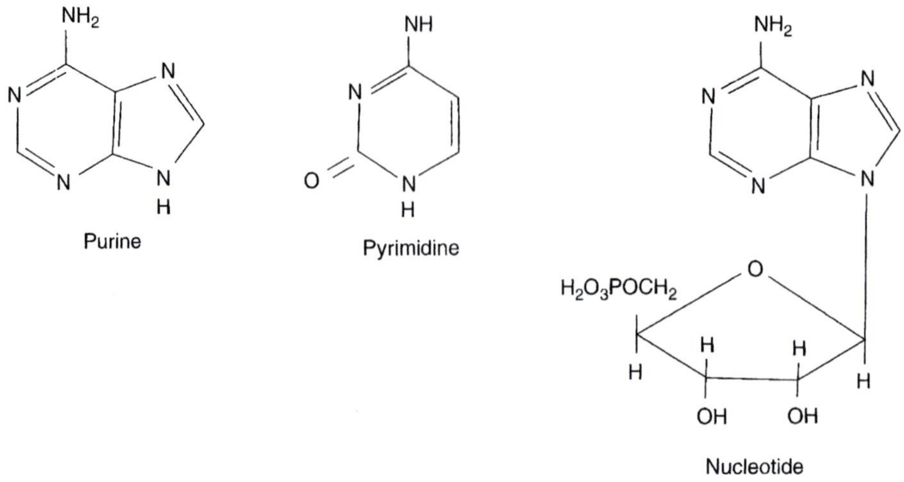

FIGURE 4–12　A plant porphyrin: Chlorophyll b. (From Raymond P. Poincelot, *Horticulture: Principles & Practical Applications,* (1980). Reprinted by permission of Prentice-Hall, Upper Saddle River, NJ.)

Chlorophyll b

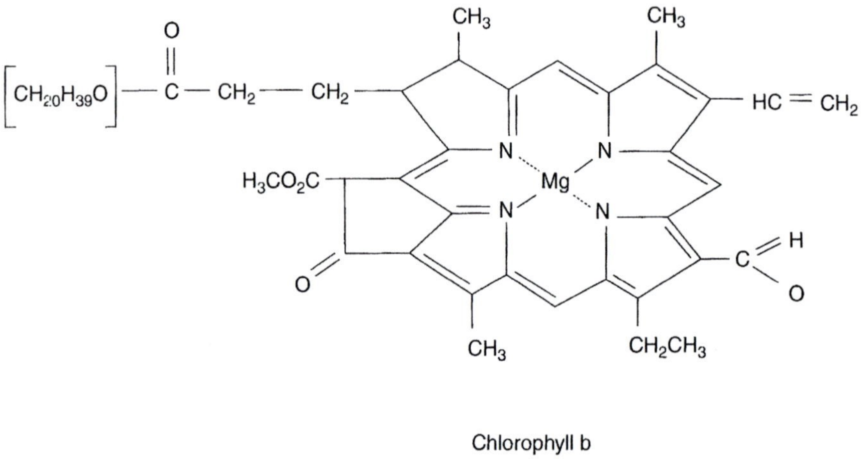

Porphyrins (Figure 4–12) are composed of pyrrole rings. The best known plant porphyrins are chlorophyll and phytochrome. These are involved in photosynthesis and photomorphogenetic responses, respectively. Porphyrins are also found in cytochromes involved in electron transport during photosynthesis, and in enzymes such as catalase, which convert hydrogen peroxide to oxygen and water.

Alkaloids (Figure 4–13) are not synthesized in all plants, but are concentrated in a few families. Many alkaloids are medicinals (atrophine, cocaine, morphine, quinine, and reserpine to name a few), insecticides (nicotine), poisons (strychnine), or stimulants (caffeine and theobromine).

Many *coenzymes* are nitrogenous compounds, such as NADP (see photosynthesis). Vitamins B (thiamine) and B_2 (riboflavin) contain nitrogen (Figure 4–14), as well as plant hormones (see Chapter 5) like indole-3-acetic acid and kinetin.

Other Organic Compounds

Terpenoids (Figure 4–15) are an abundant and diverse group of compounds found in plants. Some have important metabolic functions, and others appear to have no known function. As a group, terpenoids include terpenes, essential oils, sterols, pigments, glycosides, carotenoids, and many other related compounds. The isopentane unit is a basic building block for all terpenoids. Added head to tail, long chains can be produced, and cyclization of these chains can produce the ring structure found in some terpenoids.

Terpenes of note include the resins, essential oils, rubber, hormones, and carotenoids. Terpenes containing up to 20 carbons are volatile oils. Many are valuable horticultural products. Some of these are *essential oils,* like lemon, rose, peppermint, and lavender oil, and others are *resins,* such as those found in pine trees. Resins and essential oils can be found in combined

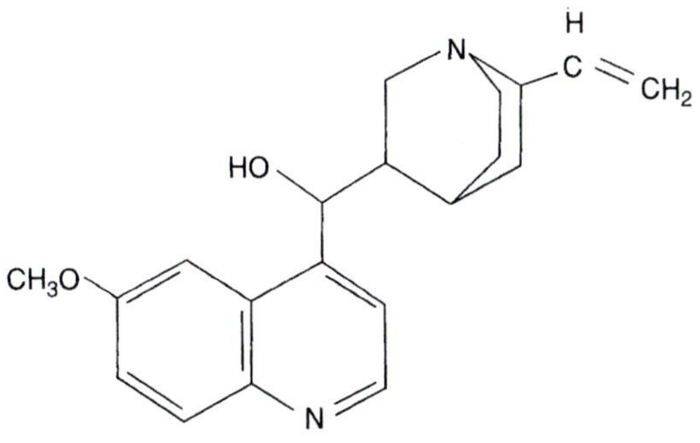

Quinine

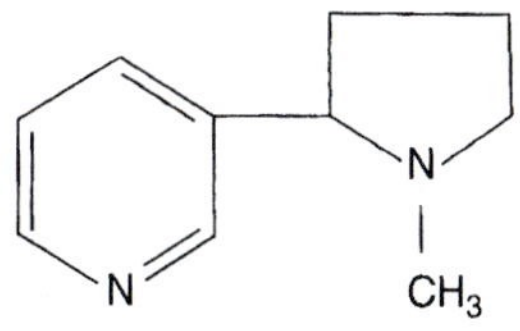

Nicotine

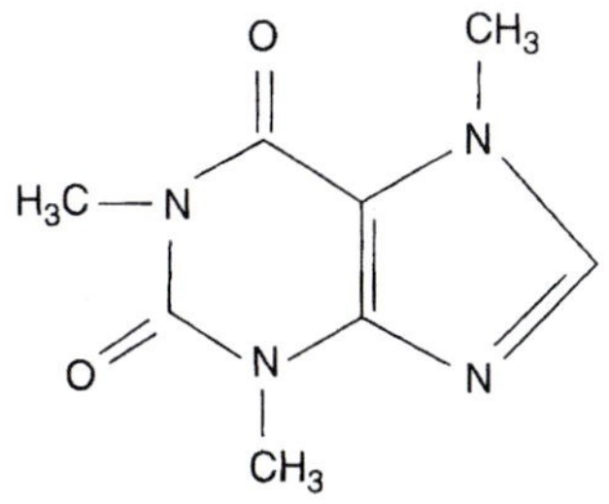

FIGURE 4–13 A few well-known plant alkaloids. (From Raymond P. Poincelot, *Horticulture: Principles & Practical Applications,* (1980). Reprinted by permission of Prentice-Hall, Upper Saddle River, NJ.)

Caffeine

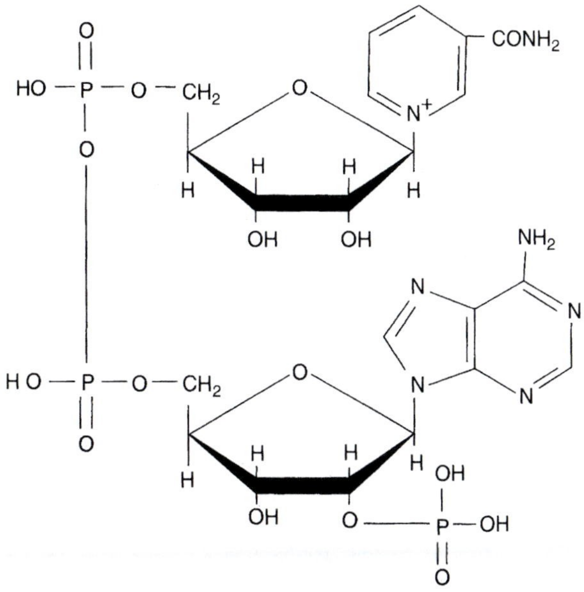

Nicotinamide adenine dinucleotide phosphate

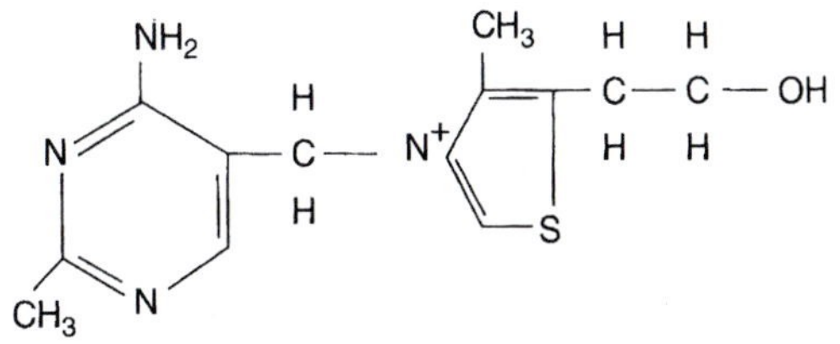

FIGURE 4–14 A few plant coenzymes. (From Raymond P. Poincelot, *Horticulture: Principles & Practical Applications,* (1980). Reprinted by permission of Prentice-Hall, Upper Saddle River, NJ.)

Thiamine

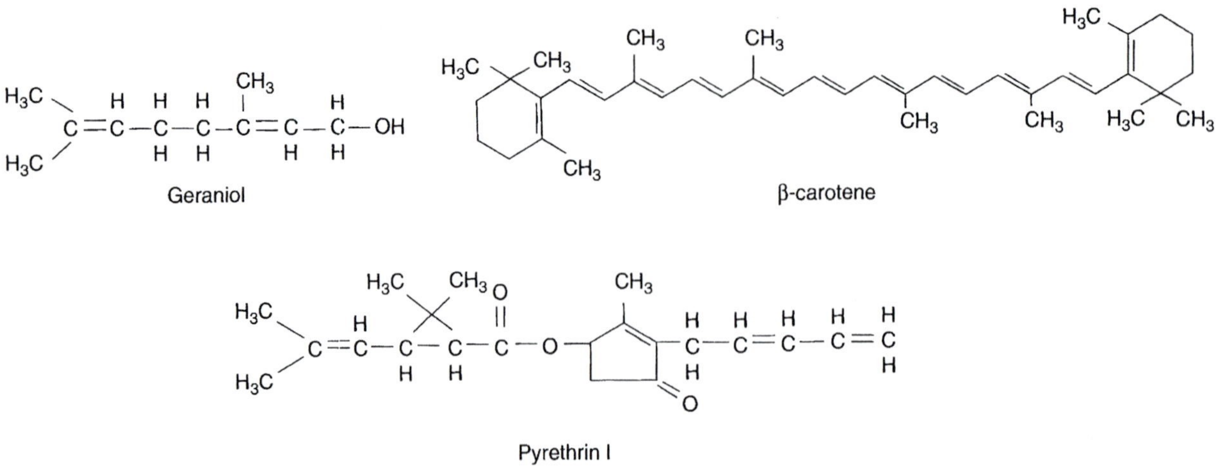

FIGURE 4–15 Terpenoids. Geraniol is an essential oil, β-carotene is a carotenoid, and pyrethrin is a mixed terpenoid. (From Raymond P. Poincelot, *Horticulture: Principles & Practical Applications,* (1980). Reprinted by permission of Prentice-Hall, Upper Saddle River, NJ.)

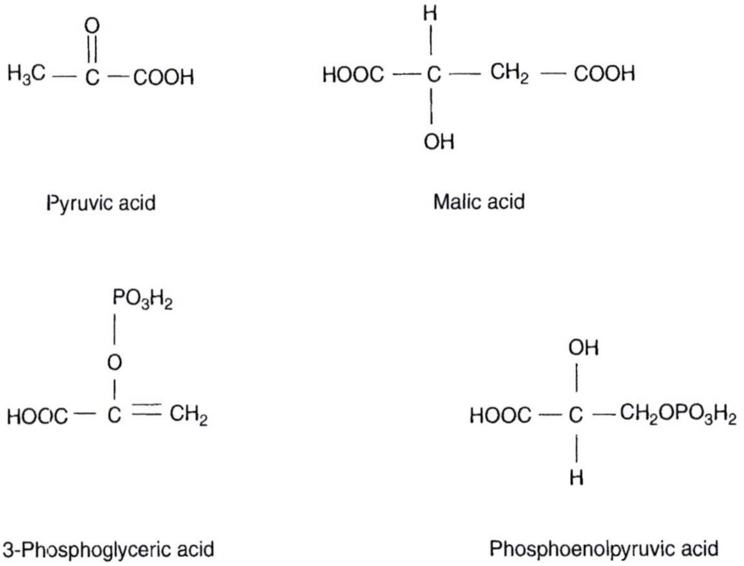

FIGURE 4–16 Some organic acids involved in photosynthetic metabolism (see Figures 3–2, 3–3 and 3–4). (From Raymond P. Poincelot, *Horticulture: Principles & Practical Applications,* (1980). Reprinted by permission of Prentice-Hall, Upper Saddle River, NJ.)

forms like turpentine. Plant *hormones,* such as gibberellins and abscisic acid, are derivatives of terpenes with less than 20 carbons. Terpenes with 40 carbons are termed *carotenoids,* yellow to orange pigments found in chloroplasts and other plant parts. One plant carotenoid, β-carotene, is a source of vitamin A. Terpenes with many thousands of carbons exist. The best known is rubber, which is derived from the latex of *Hevea brasiliensis.*

Other terpenoids are the sterols (see lipids), the glycosides (some are valuable medicinals such as digitalis), some alkaloids like caffeine, and the bitter principles (responsible for the bitter taste in cucumbers and citrus fruit). Mixed terpenoids containing a sugar or fatty acid also exist. Pyrethrin I, an important natural insecticide, and cannabidiol, the active ingredient of marijuana, are well-known mixed terpenoids.

Organic acids (Figure 4–16) arise during some metabolic pathways, such as the tricarboxylic acid cycle (see respiration). Some accumulate in various plant organs, like shikimic and malic acids in apples or citric acid in citrus fruits and grapes. Organic alcohols, ketones, aldehydes, and esters exist; some are part of the characteristic flavor and aroma of fruits.

Another group is denoted *aromatic compounds* (Figure 4–17), based on the presence of a benzene ring in their structure. Those in plants include the phenolics, the flavonoids, lignin, and tannin. Some of the aromatics are derived from acetate and others from shikimic acid, a key intermediate in the synthesis of aromatic amino acids. Some of the simpler *phenols* are re-

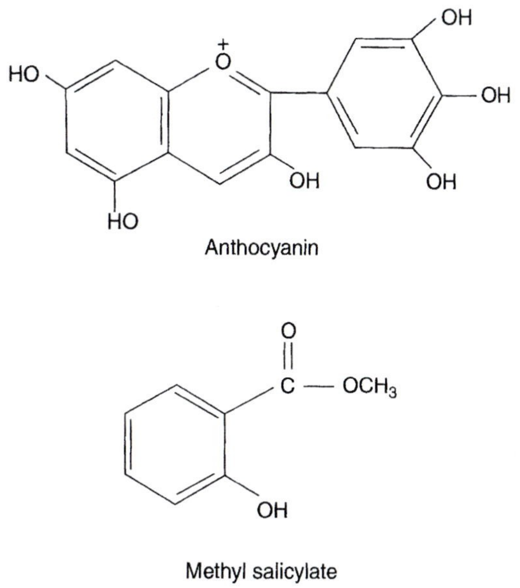

FIGURE 4–17 Some aromatic compounds found in plants. Anthocyanin is involved in many red-based colors in horticultural plants. Methyl salicylate is the aroma of wintergreen. (From Raymond P. Poincelot, *Horticulture: Principles & Practical Applications*, (1980). Reprinted by permission of Prentice-Hall, Upper Saddle River, NJ.)

sponsible for the odors and flavors of plant oils, many of which are valuable horticultural products. The *flavonoids* are a group of pigments consisting of *anthocyanins* and *flavonols*. They contribute color to some plant organs, but their metabolic role is not clear. *Lignin* is a cell wall constituent (see Chapter 3), and the function of *tannin* is unknown. Tannin is used in the tanning of leather. The taste of tannin is bitter, and its higher level in unripe fruits, such as persimmons and plums, and in overbrewed tea gives rise to their astringent tastes.

SUMMARY

A number of processes occur in plants. One process unique to plants (except for cyanobacteria) is photosynthesis. Photosynthesis benefits both plants and humans in that it makes food (sugar and starch) for the plant from carbon dioxide (CO_2) and helps maintain the air we breathe by producing oxygen (O_2) as a by-product.

Photosynthesis in part involves a green pigment (chlorophyll) and the conversion of light energy into chemical energy. The later process during which CO_2 becomes sugar and starch requires this energy. Photosynthesis takes place in green tissues, mostly leaves or sometimes stems as in cacti. The cellular location for photosynthesis is the chloroplast.

The actual process has two steps. The first part requires light (light-dependent reactions) and takes place in the thylakoids, membranous disks in chloroplasts that contain chlorophyll. In step one, light is converted into chemical energy, water is broken down into oxygen, electrons, and protons (photolysis), and the end products are ATP and NADPH. The protons and electrons from water are needed during these light-dependent reactions. ATP and NADPH will be needed later in step two. ATP is a form of chemical energy and NADPH is a source of electrons (electron carrier). The entire process is extremely fast and complicated and is not completely understood.

Sunlight provides the light energy. Being white light, it is composed of several colors (wavelengths) mixed together. Photosynthesis uses only two wavelengths, the red and blue ones. The remaining wavelengths are reflected back. Green, being the dominant reflected wavelength, makes the leaves look green to our eyes. Growth lamps designed for indoor plant growth are rich in red and blue wavelengths. Energy from the red and blue wavelengths causes chlorophyll electrons to be raised in energy levels (excitation). This condition is unstable, so the electron gives up its newfound energy. This energy is normally released as heat, as in sunburns, but in plants it is transferred to another chlorophyll. Carotenoids in the chloroplast can also transfer energy to chlorophyll. Ultimately, all this energy is channeled into a special molecule of chlorophyll bonded to a protein at a terminal site. The blast of energy causes an electron to be forced away from the special chlorophyll molecule. This process keeps on happening very fast in many special chlorophylls, producing a stream of electrons. The electrons lost from chlorophyll are replaced by an abundant, cheap source of electrons produced from water during photolysis. The stream of electrons is energy rich; some of the energy is used to convert ADP (low-energy compound) into ATP (high-energy compound). Eventually a chemical called $NADP^+$ arises that, together with protons (H^+) provided by photolysis traps the electrons, becoming NADPH.

The second step doesn't require light (light-independent reactions) and takes place in the colorless part of the chloroplast, the stroma. This process is slower and better understood. Here during a chemical cycle, the Calvin cycle, CO_2 undergoes a number of reduction steps that require energy (ATP) and a source of electrons (NADPH). The end product of this cycle is glucose, which can be converted into sucrose or starch. The waste products, ADP and $NADP^+$, are sent back to the thylakoids for regeneration into ATP and NADPH.

Evolution has brought about three forms of photosynthesis. These forms differ slightly in the light-independent steps and in being adapted better to specific climates. Plants using the Calvin cycle as described above have C_3 photosynthesis (C_3 plants). This type of plant does best in cooler, wetter climates, essentially the temperate zones. Most of our horticultural crops fall in this category. The term C_3 comes from the first product in the Calvin cycle having three carbons. Another group follows another process, the Hatch–Slack pathway, prior to the Calvin cycle. This group has C_4 photosynthesis (first product has four carbons). C_4 plants do best in hotter, drier climates, the subtropics to tropics. Corn is one of the few food crops with this photosynthetic form. This group has two to three times better photosynthetic rates than C_3 plants. These photosynthetic differences between C_3 and C_4 plants explain why bluegrass (C_3) grows well in the spring and fall and why crabgrass (C_4) dominates the lawn in July and August.

The last group is adapted to arid areas and has crassulacean acid metabolism (CAM). This group (cacti and pineapple) is like the C_3 form, except the CO_2 enters at night and is stored as an organic acid. In the day the CO_2 is released and goes through the Calvin cycle. This mechanism is geared toward water conservation, because much less water is lost from open stomates at night with CO_2 intake as opposed to during the hot and dry day. The price for high water efficiency is a 50% reduction in photosynthesis rate compared with C_3 plants. Some CAM plants can switch to C_3 photosynthesis when enough water is present, thus enabling faster growth when it rains.

All organisms release energy through the biological oxidation of organic compounds, a process termed *respiration*. The process involves many steps and enzymes. The energy is needed to maintain life. Carbohydrates such as glucose, fructose, sucrose, and starch are primary substrates for plant respiration. When glucose is oxidized in the presence of oxygen (aerobic respiration), it is broken down into CO_2 and water and energy is released. Mitochondria and the tricarboxylic acid cycle are involved in this process. The released energy is trapped in ATP molecules that serve as a form of energy currency to be used where and when needed. Respiration can also occur in the absence of oxygen and is then called anaerobic respiration. Such conditions can occur in roots in waterlogged soil where oxygen is deficient. The roots die (root rot) if subjected to prolonged anaerobic respiration. Root rot is a common cause of house plant death from overwatering. While respiration consumes some oxygen, the amount made during photosynthesis far exceeds this use.

Respiration rates during the day are higher than at night in some plants. In others no difference is seen. The extra amount is known as photorespiration and is found in C_3 and CAM, but not C_4 plants. Photorespiration appears to be a wasteful process, causing the loss of carbon fixed during photosynthesis. This waste helps explain why C_4 plants are so much more efficient in photosynthesis than C_3 and CAM plants.

Water and dissolved minerals are absorbed at root tips through root hairs, from mycorrhizae further back (when present), and at breaks or cracks in older roots having secondary tissues and suberin (waterproof wax). Two mechanisms are used to explain water absorption: active and passive absorption. Active absorption occurs usually at night and only when soil moisture is high and water loss (transpiration) from the plant is low (closed stomates). It also does not occur in all plants such as conifers. Solutes can accumulate in the xylem under these conditions and osmosis causes water to enter the roots, creating positive root pressure. This pressure can cause water drops to form at leaf tips (guttation). When deicing salts are used near trees, salt levels outside the root can become higher than inside the roots, leading to reverse osmosis. Water then leaves the roots and can result in damage or death.

Most water enters the roots through passive absorption. The driving force for this process is transpiration and it involves two properties of water molecules, adhesion and cohesion. Water molecules hold onto each other strongly (cohesion) and to molecules of other substances (adhesion). Essentially the loss of water from the top creates a "pull" or "suction," causing lower water molecules to move upward because of cohesion. However, because of adhesion of water to the xylem the water is not ripped out, but removed in a slow continuous stream.

At least 16 elements are needed for plant growth. Some are supplied from air and water, carbon, hydrogen, and oxygen. The remainder enter the plant via the roots as dissolved ions. These nutrients include nitrogen, potassium, calcium, magnesium, phosphorus, sulfur, iron, copper, manganese, zinc, molybdenum, boron, and chlorine. Ion uptake mechanisms are complex. Water can carry the dissolved ions by diffusion through highly permeable cell wall space (apoplastic pathway) from areas of high ion concentrations outside the root to lower ones inside the root. This process requires no energy and is called passive transport. Another pathway requires water and ions to cross cell membranes (symplastic pathway). This route requires energy-driven, active transport of ions across membranes. Once in a cell ions can go from cell to cell via the plasmodesmata (membrane tunnels between cells) without further energy need. Evidence suggests that most ion movement is through the symplastic route.

Water, minerals, and food move throughout the plant by a process called *translocation*. These materials move through the vascular system, water and minerals in xylem tissue and food in phloem tissue. Most food is translocated in the form of sugars such as sucrose. Sugars are often moved from areas of production (source) to areas of active growth (sinks). Other substances such as hormones and vitamins are also translocated.

Water is evaporated from plants through a process called *transpiration*. Most water is lost via the stomates when they open to take in carbon dioxide for photosynthesis. Plant form in soft (not woody) tissues is maintained by internal water pressure (turgor pressure). If transpiration losses exceed water entry at the roots, the loss in turgor results in the collapse of form (wilting). If the wilting is short, it is temporary wilting. If continuous, it becomes permanent wilting and can cause death. As expected, transpiration rates are higher during the day than at night.

Numerous other biochemical reactions take place in the plant (metabolism). Enzymes catalyze most of these reactions. These reactions involve either synthesis (anabolism) or degradation (catabolism) of various substances. These substances include carbohydrates, amino acids, proteins, lipids, nucleic acids, chlorophyll, alkaloids, vitamins, coenzymes, hormones, and numerous other organic compounds.

EXERCISES

Mastery of the following questions shows that you have successfully understood the material in Chapter 4.

1. What purpose does photosynthesis serve in plants? Does it have any value beyond plants? (*Hint:* Think about global warming and our need to breathe.)
2. What part or parts of the whole plant are involved in photosynthesis? Where does photosynthesis occur in the plant cell? Why is this cell organelle green?
3. Which parts of the chloroplast are involved in photosynthesis? What step(s) of photosynthesis occur in each part. What are the names of these steps? How do they differ in terms of light?
4. What events and products occur in the first phase of photosynthesis? What is photolysis? What purpose do the products of this first part of photosynthesis serve? Why is oxygen called a by-product?
5. How do plants convert light energy into chemical energy? What roles do ADP, $NADP^+$, and water play in this process? What parts of visible light from the sun do plants use during photosynthesis? What horticultural application results from this knowledge?

6. What events and products occur in the second phase of photosynthesis? What inputs are needed in this phase? What does the Calvin cycle do? What are the products of the Calvin cycle used for in plants?
7. What connection exists between the first and second parts of photosynthesis? Is this a one-way or two-way connection?
8. Name the three photosynthetic variations. What connection do these three forms have with evolution and climate? How do they differ? What connection is there to crabgrass and cacti?
9. What happens during respiration? What connection does respiration have with root rot and human breathing?
10. What is photorespiration? What connection does photorespiration have to plant productivity?
11. What happens during water and nutrient absorption? How do these events relate to translocation?
12. What happens during transpiration? What connection does transpiration have with wilting?
13. What are the various groups of biochemicals associated with the metabolic pathways of plants?

INTERNET RESOURCES

Photosynthesis Directory; good explanations of photosynthetic process
http://esg-www.mit.edu:8001/esgbio/ps/psdir.html

Photosynthesis links
http://photoscience.la.asu.edu/photosyn/education/learn.html

RELEVANT REFERENCES

Buchanan, Bob B., Wilhelm Gruissem, and Russell L. Jones (eds.) (2000). *Biochemistry & Molecular Biology of Plants*. American Society of Plant Physiology, Waldorf, MD. 1408 pp.

Dey, P. M., and J. B. Harborne (eds.) (1997). *Plant Biochemistry*. Academic Press, San Diego, CA. 554 pp.

Kozlowski, Theodore T., and Stephen G. Pallardy (1996). *Physiology of Woody Plants*. Academic Press, San Diego, CA. 432 pp.

Lambers, Hans, Thijs L. Pons, and F. Stuart Chapin (1998). *Plant Physiological Ecology*. Springer Verlag, New York. 356 pp.

Lawler, D. W. (2000). *Photosynthesis: Molecular, Physiological and Environmental Processes*. Springer Verlag, New York. 340 pp.

Lea, Peter J., and Richard C. Leegood (eds.) (1999). *Plant Biochemistry and Molecular Biology* (2nd ed.). John Wiley & Sons, New York. 384 pp.

Lüttge, U. (1997). *Physiological Ecology of Tropical Plants*. Springer Verlag, New York. 284 pp.

Pallardy, Stephen G., and Theodore T. Kozlowski (1996). *Physiology of Woody Plants*. Academic Press, San Diego, CA. 432 pp.

Salisbury, F. B., and C. W. Ross (1991). *Plant Physiology* (4th ed.). Wadsworth Publishing, Belmont, CA.

Taiz, Lincoln, and Eduardo Zeiger (1998). *Plant Physiology* (2nd ed.). Sinauer Associates, Sunderland, MA.

VIDEOS AND CD-ROMS*

Photosynthesis: Light into Life (Video, 1997).
Plants (CD-ROM, 2000). Useful for both Chapters 3 and 4.

*Available from Insight Media, NY (www.insight-media.com/IMhome.htm)

Chapter 5

Plant Development

OVERVIEW

The collective processes of growth and differentiation over time form the mature plant. Growth is defined as an irreversible increase in size, caused by a combination of cell division and enlargement. Growth alone cannot lead to the formation of an organized plant body. Differentiation denotes the processes involved in the establishment of distinctive differences in the structures and functions of various cells and tissues and organs. The combined, integrated activities of cellular plant growth and differentiation are defined as plant development.

Genes control development by regulating a complex array of biochemical events that are mediated by hormones. Understanding the functions of genes in development is critical to improving plants through genetic engineering. The focus here is to understand development on the macro level and the influence of genes and hormones.

CHAPTER CONTENTS

Development at the Cellular Level
Development of the Plant Body

Factors Affecting Development: Genes
Factors Affecting Development: Plant Hormones

OBJECTIVES

* To learn the basics of development at the cellular level
* To understand events in vegetative and reproductive development and their horticultural implications
* To learn about the dormancy process and its relationship to development
* To understand the relationship between fall foliage changes and dormancy

* To understand the role of genes in development and the applications of this knowledge to horticulture
* To learn about the basic five plant hormones
* To understand the role of plant hormones in development

DEVELOPMENT AT THE CELLULAR LEVEL

In most organisms, cell division consists of two phases: the division of the nucleus (*mitosis* or *karyokinesis*), followed by the division of the cytoplasm (*cytokinesis*). The details of this process are not described, except to point out some events of plant cell division that are unique to plants. Mitotic plant cells generally do not have the asters and centrioles observed during mitosis of animal cells (Figure 5–1). The separation of the cytoplasm in animal cells is caused by the invagination of the plasma membrane from both sides toward the center. However, cytokinesis in plant cells starts with the formation of a bisecting pectic cell plate that becomes the middle lamella after the formation of cell walls and plasma membranes on both sides. Undoubtedly, the biochemical events that trigger cell division in plants and animals differ, since hormones that have a function in plant cell division have no effect on animal cell division.

Enlargement of plant cells can be extensive. For example, the meristematic cells of the shoot apex can increase 20 times in length and 5 times in width during the conversion to palisade parenchyma (in leaf) or cortical parenchyma (in stem) cells. In this case, cell enlargement is essentially cell elongation, but enlargement in other cells comes close to being more uniform.

A plant hormone, auxin, helps loosen (only one of the many events influenced by auxin) the cell wall. Turgor pressure then produces a subsequent enlargement of the cell.

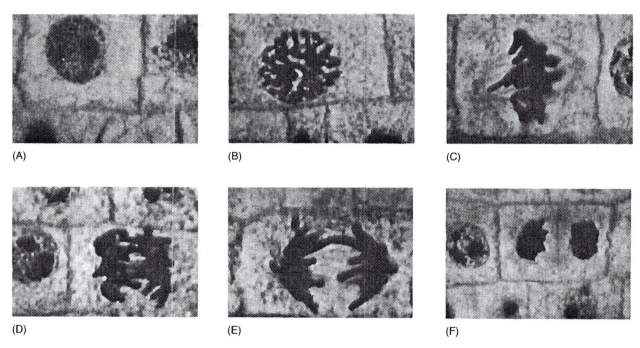

(A)   (B)   (C)

(D)   (E)   (F)

FIGURE 5–1   Mitosis in onion root cells. (A) Interphase. This stage marks a period of extensive cellular activity between mitotic events. (B) Prophase. Chromosomes become visible, shorten and thicken, nuclear envelope breaks down. At high magnifications, each chromosome is seen to actually be two strands (duplicate chromatids). (C) Metaphase. Chromatids are lined up at the equatorial plane of the cell. Spindle fibers are present. (D) Early Anaphase. The chromatids separate and are drawn to opposite ends of the cell by the contracting spindle fibers. (E) Late Anaphase. Each set of chromosomes has completely separated and is at opposite ends of the cell. (F) Telophase and Cytokinesis. Telophase marks the end of mitosis. Chromosomes become indistinct, nuclear envelope forms around each separated group. During cytokinesis a cell plate forms and bisects the cell into two daughter cells and then the two new cells enter interphase. (Author photograph.)

The resulting greater volume produces a decrease in the turgor pressure, which induces the diffusion of more water into the cell. Besides cell enlargement, a concurrent increase in the vacuole size occurs from the entry of water into the vacuole. Mature cells in some instances also have thicker cell walls than the younger meristematic cells from which they were derived. On completion of cell enlargement, cellulose, pectin, and lignin are added to the cell walls to produce this thickening. The volume of the cytoplasm is also greater; hence, some assimilation also occurs during cell enlargement.

During and after cell enlargement, biochemical changes directed by genetic and hormonal events bring about differentiation of the cells at certain times and places. It is much easier to describe these changes than to explain how these complex biochemical changes induce the events of plant development. Some aspects of cell differentiation include development of various cell organelles, alterations in the shape of cells, the production of suberin and cutin in cork and epidermal cells, the deposition of secondary cell walls, the total loss of the nucleus and vacuolar membranes in sieve tube members of phloem, the disappearance of end walls in vessel elements of xylem, and the loss of the protoplast in the fibers and vessel elements. Most of these changes appear after the meristematic cells have enlarged and gone through a parenchyma-like state. Cell differentiation is most extensive during the development of the specialized cells: guard cells, cork cells, collenchyma cells, vessel elements, tracheids, sclerenchyma cells, sieve tube members, and xylem fibers. Development of the whole plant body on the cellular, tissue, and tissue system level is shown in Figure 5–2.

DEVELOPMENT OF THE PLANT BODY

Development is obviously more noticeable to the horticulturist on the whole plant level. Whole-plant development can be divided into two phases: vegetative and reproductive development. Senescence, the aging process that terminates the functional life of the plant, is a natural consequence of vegetative and reproductive development. First, let us turn to the vegetative phase.

Vegetative Development in Seed Vascular Plants

The vegetative phase can be conveniently divided into the following events: embryogenesis, seed dormancy, germination, juvenility, and maturity. Reproduction is only possible in the mature stage. In addition we need to consider dormancy, whether it be whole plant dormancy, bud dormancy, bulb and other modified stem dormancy, or root dormancy for perennial, but not annual, plants.

Embryogenesis. This stage starts right after fertilization and ends with the mature seed. The development events during this stage establish the basic body plan pattern for plants, an axial arrangement. The radical end of the axis is essentially the potential first root and the opposite end the potential first shoot. The plan will be elaborated or undergo further development after seed dormancy is broken and germination commences. In seeds without dormancy, the plan unfolds on germination.

Seed Dormancy. A period of growth inactivity in seeds is termed *seed dormancy*. Dormancy can also occur in whole plants, buds, bulbs, corms, tubers, and other plant organs. Dormant seed is viable, but will not germinate until the physical or physiological cause of dormancy is negated, even when favorable environmental conditions exist. Seed dormancy protects the seed against premature germination. For example, germination shortly after seed maturation prior to winter or a dry season might produce seedlings or young (juvenile) plants that lack the food reserves and tolerance needed to withstand the unfavorable conditions. Many of our cultivated vegetable and flower seeds do not possess seed dormancy, having lost

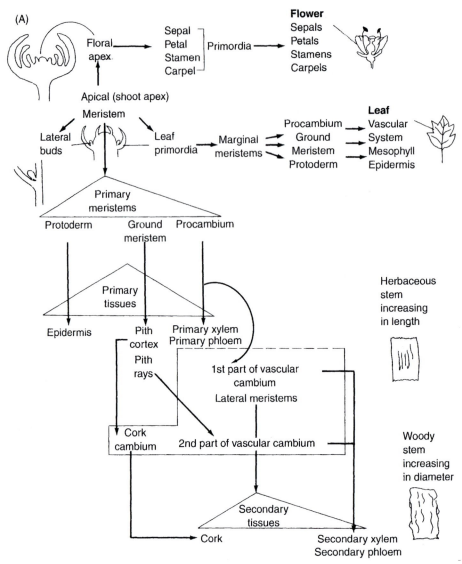

FIGURE 5–2(A)   Development of the vascular plant stem, leaves and flowers. Herbaceous development would be concluded at the point indicated by the herbaceous stem piece mid-way on the right. This generalized sequence would be true for herbaceous dicot angiosperms (flowering plants) and herbaceous gymnosperms (very young conifers). Herbaceous monocot angiosperms would differ slightly in that no vascular cambium would form and the pith, cortex and pith rays would not be clearly distinguished, so they are collectively referred to as ground tissue. Monocots can have additional increases in length from a meristem not found in dicots, the intercalary meristem. Development of woody dicots and woody gymnosperms would go beyond the herbaceous point as indicated. Some thickening of certain monocots, but no true woodiness is possible, because of either a primary thickening meristem or a secondary meristem as in palms. (From Raymond P. Poincelot, *Horticulture: Principles & Practical Applications,* (1980). Reprinted by permission of Prentice-Hall, Upper Saddle River, NJ.)

seed dormancy over many years of plant breeding. In a sense, we have replaced the dormancy of cultivated crop seeds with safe storage until proper planting times.

Seed dormancy can be caused by physical and physiological reasons. Seed coats that are impermeable to oxygen (basswood, *Tilia americana;* Canadian hemlock, *Tsuga canadensis*) or water (clover, *Meliotus alba*) or are mechanically resistant to embryo enlargement (cherry, peach, raspberry) impose physical restraints on germination. Impermeability to oxygen and water results from the impregnation of the seed coat or underlying membranes with waxes or similar substances. Physically induced dormancy is broken naturally

(B)

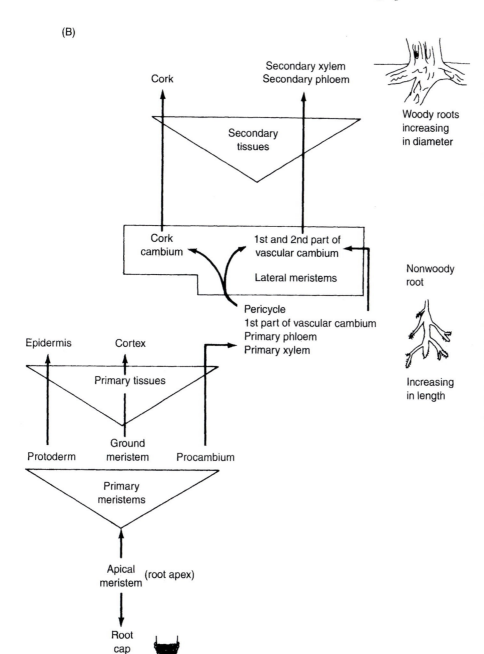

FIGURE 5–2(B)
Development of the root
system of vascular
plants. Again the
sequence would be true
for both herbaceous
angiosperms (monocot
and dicot) and
gymnosperms except
for no vascular cambium
in monocots. Woody
dicots and woody
gymnosperms would
develop beyond that
point as indicated. In
some species the primary
root tissues may be
produced directly from
the apical meristem
rather than indirectly
via primary meristems.
(From Raymond P.
Poincelot, *Horticulture:
Principles & Practical
Applications,* (1980).
Reprinted by permission
of Prentice-Hall, Upper
Saddle River, NJ.)

through actions against the seed coat, such as weathering (alternate drying and wetting or freezing and thawing), fire, external attack by soil microorganisms, and internal attack by enzymes. Artificial means include moist storage at high temperatures, treatment with acids or organic solvents, and mechanical scarification (abrasion of seed coat). Care must be taken to ensure that the embryo is not damaged by these treatments.

Physiological causes of seed dormancy result from partially developed embryos at the time of seed dispersal (*Viburnum;* holly, *Ilex;* pine), morphologically mature but physiologically immature embryos (wild ginger, *Asarum; Trillium grandiflorum*), the inhibitory presence of abscisic acid (corn), and other germination inhibitors (European ash, *Fraxinus excelsa*). Dormancy resulting from partially developed embryos is broken naturally simply by the embryo continuing development after seed dispersal. Physiologically immature embryos develop the enzymes needed to catalyze germination and growth after being subjected naturally to moisture and low temperatures for 6 weeks or more. Physiological changes that take place in dormant seeds after dispersal are termed *after-ripening*. The above two forms of embryo dormancy are broken artificially through refrigeration under moist conditions, a process termed *stratification*.

Germination inhibitors can be produced in the seed coat, the embryo, or the endosperm; they can also arise in the fruit and diffuse into the seed. A number of chemicals have been identified that inhibit germination; many of them inhibit seedling or plant growth. These include organic acids, alkaloids, phenolics, tannins, coumarin and other unsaturated lactones, essential oils, cyanide- or ammonia-releasing compounds, abscisic acid, and aldehydes. The mode of action for these inhibitors varies. Some are hormonal growth inhibitors; others provide an unfavorable pH or interfere osmotically. Germination inhibitors are removed naturally by the leaching action of rain or artificially through soaking and washing. In some cases, light exposure or low temperatures may break the hold of certain inhibitors.

In some instances, seeds may be dormant from both physical and physiological causes. This condition is called *double dormancy*. Seeds of this type often show 2-year dormancy. One year may be needed for weathering or microbial attack of the seed coat and the second year to leach out an inhibitor. The cherry seed has a seed coat that is mechanically resistant, and the embryo is not fully developed on seed dispersal.

Germination.   Seed germination starts with the imbibition of water during favorable environmental conditions. Several morphological and biochemical changes follow. These events include internal hydration, changes in the organization of the embryo and endosperm or cotyledon, light activation of a photoreceptor (phytochrome), and resulting photomorphogenetic phenomena, enzyme activation and synthesis, respiration and digestion of food reserves, synthesis of organic molecules and subsequent translocation, and cellular development.

The first external evidence of germination (Figure 5–3) is the emergence of the rootlike radicle that develops into a root system. Next, depending on the plant, the epicotyl or plumule and sometimes the cotyledon(s) appear, leading to the formation of the young shoot. Germination is considered finished when the plant becomes self-sustaining; that is, when photosynthesis commences. Some prefer to view germination as ending when the radicle appears and to call the remaining period until photosynthesis starts *establishment*.

Juvenility.   The period of extensive vegetative growth when the plant cannot be readily induced into reproductive growth, regardless of environmental effects, is termed *juvenility*. During this stage the plant has high rates of metabolic activity and can increase exponentially in size. The juvenile plant can differ from the mature plant in terms of leaf size and shape, stem growth patterns, and sometimes the timing of leaf abscission (Figure 5–4). One good example is the juvenile and mature forms of several oak species. The juvenile oak retains its dead, brown leaves very late into the fall, whereas the mature form drops its leaves much earlier. Often no visible differences occur with the juvenile and mature stages of many plant species. Other differences occur internally, such as with levels of tissue complexity and hormonal concentrations. Juvenility is a time when the plant competes strongly with other plants in the plant community.

Much descriptive information exists about differences between juvenile and mature plants and their organs, but much less is known about the causes of these differences. The juvenility-to-maturity transition has been attributed to various factors. It has been suggested that progressive aging of the apical meristem results in a loss of the ability to produce juvenile tissue. Others attribute it to the initial presence and subsequent exhaustion of juvenile-inducing hormones, such as gibberellin and auxins. Indeed, gibberellin can produce temporary reversion to or prolong the juvenile form, and the morphological characteristics of the juvenile form suggest a higher auxin content than the mature form. In some biennials, such as cabbage, applications of gibberellin can bring about flowering and seed formation, without the normal environmental signals of long days and cold exposure. The involvement of hormones in the control of juvenility is a most interesting and more likely possibility.

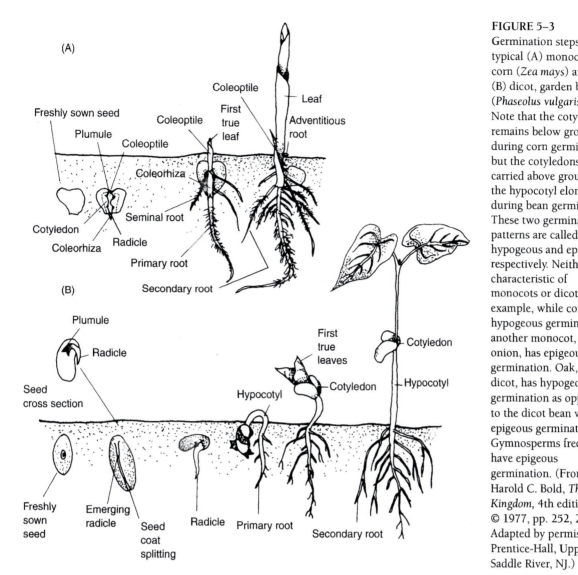

(A)

Freshly sown seed

Plumule

Coleoptile

Cotyledon

Coleorhiza

Coleoptile

Coleoptile

Coleorhiza

Seminal root

Radicle

Coleoptile

First true leaf

Coleoptile

Leaf

Adventitious root

Primary root

Secondary root

(B)

Plumule

Radicle

Seed cross section

Freshly sown seed

Emerging radicle

Seed coat splitting

Radicle

Hypocotyl

First true leaves

Cotyledon

Hypocotyl

Cotyledon

Primary root

Secondary root

FIGURE 5–3
Germination steps for a typical (A) monocot, corn (*Zea mays*) and (B) dicot, garden bean (*Phaseolus vulgaris*). Note that the cotyledon remains below ground during corn germination, but the cotyledons are carried above ground as the hypocotyl elongates during bean germination. These two germination patterns are called hypogeous and epigeous, respectively. Neither is characteristic of monocots or dicots. For example, while corn has hypogeous germination, another monocot, the onion, has epigeous germination. Oak, a dicot, has hypogeous germination as opposed to the dicot bean with epigeous germination. Gymnosperms frequently have epigeous germination. (From Harold C. Bold, *The Plant Kingdom*, 4th edition, © 1977, pp. 252, 255. Adapted by permission of Prentice-Hall, Upper Saddle River, NJ.)

FIGURE 5–4 This deciduous tree exhibits leaf retention differences that can be observed in the juvenile vs. mature stages of development. This juvenile tree, yellow birch (*Betula alleghaniensis*) was photographed in January in Vermont and still retains many of its dead leaves. Mature forms drop all their leaves in the fall. Juvenile oak trees also show similar foliage retention patterns. (Author photograph.)

Maturity.   Juvenility is followed by maturity. This phase is characterized by a decreased rate of vegetative growth, the potential for the development of flowers or other reproductive structures, and, in some cases, morphological differences. The flowering potential will not be realized unless environmental conditions are favorable. The diversion of food, minerals, and metabolites into the reproductive organs, especially the seeds and fruits, leads in part to the reduced growth of the shoot and root in perennials. In annuals, shoot and root growth is terminated when reproduction occurs. Removal of flowers can often enhance vegetative growth, suggesting that substances produced by the reproductive organs may possibly inhibit such growth. As stated previously, some evidence exists that the levels of auxins and gibberellin are reduced during the juvenility-to-maturity transition. This reduction might be one factor involved in reproductive development.

In annuals and biennials the period of maturation may last from a few weeks to a few months. Maturity can be measured in terms of several years with herbaceous perennials or up to hundreds of years with trees. The length of maturity is related to different growth patterns: *determinate* and *indeterminate* growth. In the former, growth occurs over a period of time and then stops once the genetically determined limit is reached. The determinate form, largely under genetic control, is characteristic of leaves, fruits, and seeds. The complete conversion of a bud primordium from the vegetative to reproductive phase on a nonbranching plant, such as the common sunflower (*Helianthus annuas*), results in termination of shoot growth. This event is also an example of whole plant determinate growth. Determinate growth can also occur in branched plants, if all buds become flower buds. This form is seen with certain tomato cultivars that tend to be smaller and produce most of their tomatoes on the outer edges of the plant. Many annuals and biennials exhibit this form of determinate growth, where flowering ends shoot growth.

Indeterminate growth is characterized by continuity and increasing size with age. Vegetative growth is slowed by reproduction, but not ended. Growth ends when the transition from maturity to senescence occurs, but the time to this point is considerably longer than with determinate growth. With herbaceous and woody perennials, the conversion from shoot to floral apices is only partial, so the plant retains both reproductive and growth capabilities. Stem and root meristem growth and cambial growth are also forms of indeterminate growth.

Senescence.   As a plant passes through the various developmental stages, it undergoes various chemical and structural changes that are all part of the aging process. At some point these changes become irreversibly degenerative and lead to the death of the plant. The period from that turning point to the death of the plant is known as *senescence*.

Senescence can be on the whole plant level, such as with the senescence of annuals and biennials that follows reproductive development. It can also be partial, as with the loss of top growth on a perennial, the loss of leaves annually by deciduous trees or over several years by evergreens, and the loss of reproductive organs.

During senescence, destructive metabolism (catabolism) exceeds constructive metabolism (anabolism). Declines occur in photosynthetic and respiratory rates. These factors and subsequent removal of solutes through translocation will cause losses in dry weight. As synthesis slows, losses in protein, lipids, and carbohydrates also occur. Abscission layers will form with leaves and flowers. Most of the changes will be metabolic, rather than morphological. Earlier stages of development, that is, the vegetative and reproductive portions, were extensively characterized by morphological changes.

The causes of senescence are not entirely known. Senescence and flowering are clearly linked in some plants such as annuals where flowering is always followed promptly by senescence. Extensive translocations of foods and minerals to developing seeds, fruits, and vegetative storage organs play some role, since the removal of young fruits can prevent or slow senescence. However, senescence can also be triggered by young flowers prior to solute mobilization. One possible cause of this is the production of inhibitors of a hormonal nature that induce senescence. However, none has yet been identified. Some of the known plant hormones can slow (cytokinin) or accelerate (ethylene and abscisic acid) the symptoms of senescence, but their role is not clear. External environmental signals can initiate senescence, such as happens with leaves in the fall. Certainly, genetic factors are also involved.

Vegetative Development of Bulbs, Corms, Rhizomes, and Tuberous Roots and Stems

Although the formation of underground structures is part of vegetative development, one feature is similar to events in reproductive development. Underground stem (and root) modifications and reproductive structures share similar nutrient movement activities in that the development of bulbs, corms, rhizomes, tuberous roots, and tubers involves the translocation of nutrients and growth stimulants at the expense of all other plant parts. The difference is that the bulbs, corms, rhizomes, tuberous roots, and tubers are food storage organs that enable the plant to survive periods of cold or water stress. Reproductive structures use most of the sugars for growth; only modest amounts are stored in the seed. The development processes of these organs are not well understood. It is apparent that several growth hormones are involved. Environmental signals, such as photoperiod and temperature, play a role in the development of some, but not all, of these structures.

Reproductive Development in Seed Vascular Plants

Much of the plant development that appeals to the horticulturist is associated with the colorful, spectacular events of reproductive development. The transition from vegetative to reproductive growth signals a major change in the life cycle of all plants. The most obvious events of reproduction are flower formation and fruit and seed or cone and seed development. Reproductive development is also a complex process concerned with physiological, biochemical, anatomical, and morphological events.

Flower Development. A major event in the start of the reproductive phase is the formation of floral apices and ultimately flower primordia (Figure 5–2). This takes place after a period of juvenile growth and often only if certain environmental conditions are satisfied. Environmental factors, light and temperature, will be discussed in Chapter 6, where environmental stimuli are related to plant responses. The relationship of climate to flowering and fruiting (phenology) is important for determining the time of harvest and will be covered in Chapter 6. Phytochrome and hormones, such as the long-sought, but elusive, florigen or the well-known gibberellins, auxins, and cytokinins play some part in the initiation of flower primordia through their modification of metabolic reactions.

Development and the maturation of floral bud parts (sepal, petal, pistil, stamen, and others) ensue with the shifts in metabolic processes. Environmental effects can influence the sexual expression of unisexual flowers, that is; the constancy or alteration of the ratio of male to female flowers. This will be considered further in Chapter 6. Changes in levels of phytohormones occur, and the synthesis of protein increases. A redistribution of water, nutrients, growth substances, and many other compounds to the flowering portions occurs at the expense of other plant parts. Internal changes prior to the opening of the mature flower (*anthesis*) prepare the flower for pollination and subsequent fertilization. Flower development covers the period from the initiation of floral primordia through anthesis.

Fruit Development. Fruit development is considered to start after anthesis, with the initial event being pollination followed by fertilization, growth, maturation, and ripening. However, some aspects of fruit development, even though not obvious to the observer, start soon after flower induction. These earlier cellular-level events are significant in establishing the developmental pattern of the fruit. All stages of fruit development are associated with changes on the cellular and metabolic levels. These changes include cell division and enlargement, activities of hormones such as gibberellins and cytokinins, and major translocation of materials to the developing fruit.

Pollination in the flowering plants (angiosperms) involves mostly insect-assisted (or other organisms) and, to a lesser degree, wind-assisted transfer of pollen (encapsulated male gametes) to the pistil's stigma (Figure 5–5). One important physiological process started by pollination is the prevention of fruit or flower drop. The sperm are delivered ultimately to the ovaries via a pollen tube through which sperm descend in a gravity-assisted manner. Fusion

FIGURE 5–5 During pollination pollen is transferred from the anthers to the stigma by a vector, usually a pollinating insect such as a honeybee or by wind. The pollen grain produces a pollen tube (pollen germination) that grows downward through the style and stops upon contacting an ovule. One sperm then fuses with the egg nucleus (normal fertilization) and the second fuses with the two polar nuclei (second fertilization). This double fertilization is unique to flowering plants (angiosperms); seed formation requires both fertilizations. (From Raymond P. Poincelot, *Horticulture: Principles & Practical Applications*, (1980). Reprinted by permission of Prentice-Hall, Upper Saddle River, NJ.)

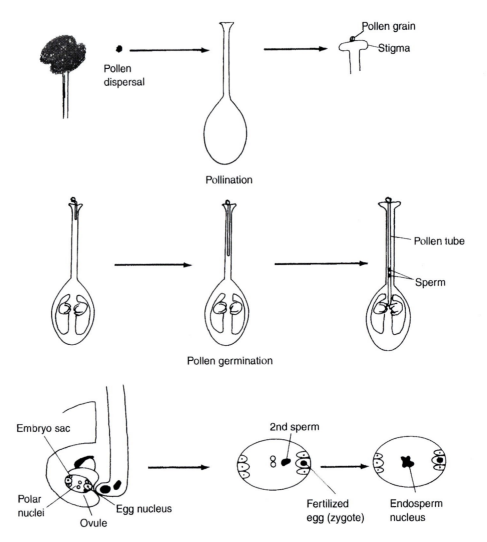

of the male gametes with the egg nucleus and polar nuclei forms the zygote and endosperm nucleus. This process, called *double fertilization* (Figure 5–5), is characteristic of the angiosperms (flowering plants) and also the major impetus to fruit development. Fertilization usually occurs in a matter of days after pollination. The activities of growth substances, such as gibberellins, cytokinins, and auxin, are involved with subsequent development of the zygote and endosperm. The growth of the surrounding ovary and any accessory tissues, if present, into the mature fruit depends on activities of the zygote and endosperm and on the translocation of nutrients and growth substances from the rest of the plant. Early growth involves cell division; cell enlargement is primarily associated with the latter stages of growth.

Abundant evidence exists that seeds can produce regulatory substances that control many aspects of fruit development. Yet seedless or parthenocarpic fruits also grow in a manner similar to that of seeded fruits. The effects of growth substances from seeds range from stimulatory to inhibitory, depending on the stage of fruit growth. In parthenocarpic fruits, other fruit parts may supply the growth signals normally associated with the developing seeds.

Maturation of the fruit is complete when the fruit reaches its full size. Afterward, several subsequent events occur that include softening of the fruit flesh, chemical changes in pigmentation and flavor, and hydrolytic changes of storage materials into sugars. These changes are associated with fleshy fruits that are usually consumed as food. Other fruits enlarge, harden, dry up, and become inedible to humans. These changes are collectively called *ripening*. Such changes are dependent on an increased or *climacteric* rise of respiratory rates.

After the climacteric peak, respiration falls, and the fruit enters the senescent phase. Some fruits, such as citrus fruits, do not appear to undergo a climacteric rise in respiration.

The trigger of ripening for most fruits appears to be the production of the hormone ethylene. Some nonclimacteric fruits, such as the strawberry, are unaffected by ethylene. The mechanisms of ripening are not clearly established. The modern concept is that it is under the control of DNA. A rise in RNA is observed during the climacteric rise, as well as increased protein synthesis that is undoubtedly associated with the production of enzymes utilized during the ripening process.

Seed Development.   The development of the fruit and seed is concurrent. As would be expected from this, the two processes interact and influence one another. Seed development is dependent on nutrients (sugar, salts, and water) translocated from other plant parts. Other substances, such as amino acids, vitamins, and phytohormones, are synthesized within the seed by the endosperm and later by the embryo. As discussed previously, some aspects of fruit development are dependent on phytohormones produced by the enclosed seeds. The embryo is important in seed development, because embryo abortion stops seed development. As the seed develops, stores of starch, hemicellulose, proteins, fats, and, in some species, sucrose are accumulated. Germination inhibitors are also synthesized in many ripening seeds. Seed coats may become desiccated and hardened or impregnated with substances that tend to make them impermeable. Respiration and other processes decrease. These later aspects are involved in seed dormancy and prolonged seed viability.

Seed production in the gymnosperms (Figure 5–6, conifers) differs somewhat from the angiosperms. Their seeds are not enclosed in an ovary, so no fruit develops. Pollen is produced in a male or staminate cone and distributed by wind. Pollination is completed when the pollen reaches an ovulate or female cone. Fertilization takes place shortly (3 days to 3

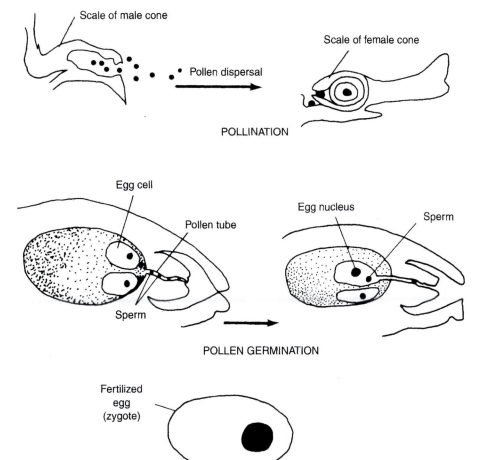

Scale of male cone

Scale of female cone

Pollen dispersal

POLLINATION

Egg cell

Pollen tube

Egg nucleus

Sperm

Sperm

POLLEN GERMINATION

Fertilized egg (zygote)

FERTILIZATION

FIGURE 5–6 In gymnosperms pollen is transferred from the microsporangia found on the scales of male (pollen-bearing) cones to megasporangia on scales of female (ovulate) cones. Usually male and female cones are found on the lower and upper branches, respectively, of cone-bearing trees. Eventually a pollen tube is produced through which sperm move. Only the fusion of one sperm with the egg nucleus is required for seed production in gymnosperms. In the above diagram one scale of a cone is shown with a microsporangium and megasporangium before pollination. Scales portions are left out thereafter. (From Raymond P. Poincelot, *Horticulture: Principles & Practical Applications,* (1980). Reprinted by permission of Prentice-Hall, Upper Saddle River, NJ.)

or 4 weeks) after pollination with most conifers. Pines are an exception in that 12 to 15 months may elapse between pollination and fertilization. After fertilization, the zygote undergoes mitosis. Double fertilization does not occur as in the angiosperms. An endosperm surrounds the developing zygote, but it develops from a reproductive tissue that was not fertilized. The fully developed seed will eventually be dispersed from the ovulate cone.

Vegetative and Reproductive Development in Seedless Vascular Plants

The preceding discussion was concerned with seed and fruit development in seed vascular plants. But a large group of vascular plants are seedless. Those of interest to the horticulturist are the ferns. Ferns reproduce from homospores that produce a bisexual, free-living gametophyte, as opposed to seed vascular plants in which there are two kinds of spores (heterospores), the microspore and megaspore (Figure 5–7). These produce the male and female gametes, respectively. The gametophyte in seed vascular plants is not free living, but dependent on the adult plant. A very few ferns produce heterospores that give rise to separate, free-living male and female gametophytes.

FIGURE 5–7 Seedless vascular plants (ferns) differ from seed plants in their life cycle. Most ferns produce homospores that give rise to a bisexual, independent gametophyte. In seed plants the gametophyte is much reduced, a male and female gametophyte is present, and the gametophytes are enclosed in reproductive structures on the adult plant, on which they are dependent. Sperm and egg unite on the fern gametophyte. The fertilized egg develops into the adult fern with which we are most familiar. The gametophyte breaks down as the fern is maturing. (From Hudson T. Hartmann, Dale E. Kester, Fred T. Davies, Jr. and Robert L. Geneve, *Plant Propagation: Principles and Practices,* 6th edition, © 1997, p. 14. Adapted by permission of Prentice-Hall, Inc., Englewood Cliffs, NJ.)

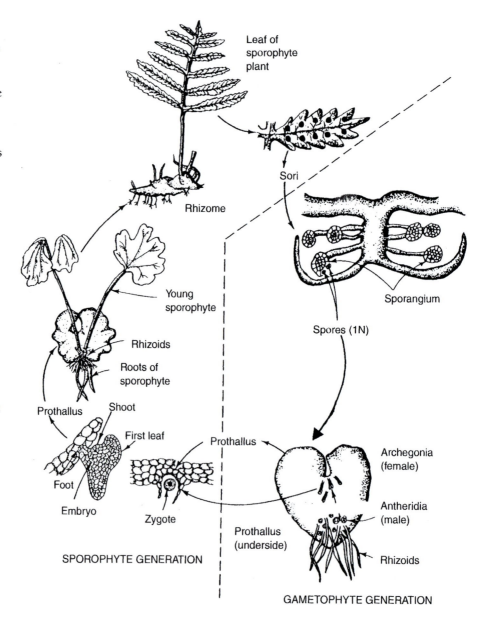

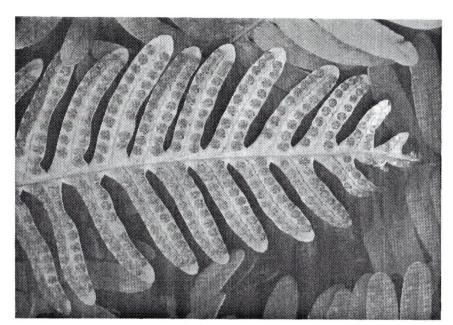

FIGURE 5–8 The brown dots seen on the underside of the fern frond are sori. Each sorus contains numerous clusters of sporangia, which in turn contain countless numbers of spores. (Author photograph.)

The germinating fern homospore produces a tiny green gametophyte that varies from heart shaped to irregular in form (*prothallus*). It is anchored to the soil by root-like structures called *rhizoids*. The sperm and egg develop in specialized structures on the gametophyte, and fertilization occurs when the sperm swims through water to contact the egg. The fertilized zygote undergoes mitosis and develops into the vegetative structure we call the fern, during which time the gametophyte disintegrates. The adult fern produces spores on the frond undersides by meiosis in specialized structures called *sporangia* (Figure 5–8).

Development and Dormancy

During vegetative and reproductive phases, plants usually experience periods of growth inactivity, or *dormancy*. Dormancy can occur on the whole plant level, as with trees and shrubs, and also with specific parts, such as buds, rhizomes, tubers, bulbs, corms, or root systems. Dormancy is seen with biennials and perennials, but not with annuals since the latter only last one growing season.

Dormancy is an adaptive mechanism that allows plants to meet seasonal or environmental limitations (cold, extreme heat, dry periods) on growth. Arrested development can occur throughout the vegetative (and reproductive) phase, but especially during periods of moisture stress and temperature extremes. Some herbaceous perennials may enter a dormant phase after spring flowering, even though conditions are favorable for growth. Woody plants develop vegetative buds at the nodes and shoot tips during active growth and sometimes flower buds that are not developed further as part of the current season's growth. Instead, seasonal environmental cues, such as increasingly colder weather and changes in day length, put these buds into dormant condition, and then subsequent cues, such as warmer weather or longer days, break their dormancy. Similar signals can regulate the dormancy of modified stems and root systems.

The causes for starting or breaking dormancy are not completely clear, but it appears that environmental cues induce changes in the levels of plant hormones, which in turn regulate the nucleic acid system. Because the nucleic acid system directs protein synthesis, this would be an effective on/off switch for growth. The hormones abscisic acid and gibberellin likely play some role in inducing and breaking dormancy, respectively.

Dormancy and Fall Foliage Color

One of the most interesting horticultural phenomena associated with dormancy is the change of foliage colors that occurs with woody plants during the fall. These color changes are at their best when day temperatures are warm and nights become crisp and cool to cold. Hence, New England foliage displays some of the best fall colors found in the United States. This plant phenomenon is the basis of a profitable tourist industry during the fall in states like Maine, New Hampshire, New York (northern part), Vermont, and the provinces of eastern Canada.

So how do the colors come about? Let's start with the signal; that is, how do trees and shrubs know that winter is coming? As the summer winds down and fall approaches, two environmental signals change: The length of the days decreases and the temperature also decreases. Plants somehow respond to these two linked environmental signals and keep track of the data on their biological clocks and calendars. Scientists poorly understand the "time clock" plant functions even today.

We do know that a pigment called phytochrome plays some role in receiving signals related to the change in day length as seasonal change occurs. This pigment changes form in response to red and far-red parts of white light. Because the ratio of red/far-red changes with the length of the day, it causes changes in phytochrome that somehow translate into actions on the cell and biochemical level that induce dormancy in the plant. The thinking is that phytochrome brings about changes by altering the permeability of membranes (thus altering how easily things move through membranes) or by altering synthesis of proteins, such as enzymes. The interaction between the plant and the changing temperature part of the environmental signal is even more poorly understood. The hormone that mediates between the environmental signals and cellular changes is absiscic acid. This dormancy hormone inhibits growth.

Things start to change within the plant as the biological clock/calendar approaches some critical juncture. One of the first processes to change involves the synthesis of chlorophyll. When a plant is actively growing, chlorophyll needs to be made constantly so it can be added to new leaves as they appear and also replace existing chlorophyll as it wears out. As fall approaches and abscisic acid inhibits growth, chlorophyll synthesis slows and then stops. As the chlorophyll breaks down and is not replaced, the leaf color starts to change.

Some leaves go from green to yellow or orange. The yellow and orange colors come from carotenoids, the same pigments covered in our discussion of photosynthesis. These colored pigments are present in green leaves, but are masked by the dominating green of chlorophyll. As the chlorophyll degrades, the yellows and oranges become visible.

Red colors also appear on some plants. The red color comes from a pigment called anthocyanin, the same pigment that colors many flowers red. Anthocyanin is synthesized in response to warm days coupled with cold night temperatures, hence the brilliant reds of New England. What purpose the anthocyanin serves is unknown, but it does help fill out the palette of fall colors.

While the external colors are changing, other events are taking place internally. Nutrients in the leaves are gradually removed and stored in nearby areas, primarily the twigs or small end areas of branches. This process helps to conserve nutrients, because if not removed, the nutrients would be lost when the leaves drop to the ground. This placement helps to explain why deer browse the twig growth of trees and shrubs when their normal food sources are gone. This placement of nutrients is also strategic for spring growth on the breaking of dormancy. The buds for the following year are part of the twiggy areas.

The other change has to do with the water content of the cells. Water is lost, making the cellular fluids more viscous. Some sugars and proteins are also synthesized. The concentration of dissolved solutes goes up as the water amount decreases. Because the freezing point of an aqueous solution is related to solute content, the more solutes, the lower the freezing temperature. In a sense the tree or shrub is making "antifreeze" that will help it to survive the greatly reduced temperatures of winter without freezing. How effective this change is can be easily seen. If you take a small tree growing in a pot during the summer and place it in a freezer, it will freeze and die. Yet this same tree can survive temperatures way below freezing in the winter. The difference is seen in whether or not the dormancy-induced change in cellular fluids has taken place.

After the nutrients are removed, the colored leaves drop. This removal protects the plant from invading microbes that would attack the dead leaves and ultimately infect the rest of the plant. The microbes attack the leaves on the ground. Some of the microorganisms utilize the carotenoids and chlorophyll as nutrient sources. As these pigments break down, the degraded products become somewhat brown in color. A brown pigment called tannin, found in leaf vacuoles, also contributes to the brown color. Tannins and the brown degradation products of the colored pigments are resistant to microbial attack, so the brown color persists. Tannins are used in tanning leather and are also responsible for the bitter taste found in overbrewed tea. Overbrewing releases the tannins from tea leaves.

FACTORS AFFECTING DEVELOPMENT: GENES

The development of plants or any organism is controlled by two factors. The first is an internal factor: the hereditary potential (genes). The second is an external factor: the environment. These two factors influence one another in complex and not always obvious ways. Apparent examples are mutations that induce altered hereditary potentialities; these may result from environmental factors such as ionizing radiation. Natural selection and the influence of long-term environmental conditions on this process is another obvious example. Both heredity and environment together determine the biochemical processes of the plant, and these in turn regulate the pattern of plant development. Simply expressed, heredity determines what the plant can be, and environment determines to what extent these potentials are realized. The effects of heredity on plant development or plant genetics are covered in this chapter, but the effects of environment will be discussed in the following chapter.

Genes and Development

Gene pairs called alleles (see more later on alleles) collectively constitute the *genotype* of the plant. Observable, outward physical appearances and biochemical traits expressed by the genotype are called the *phenotype*. Genes are found on chromosomes. Chromosomes are composed of chromatin that consists of DNA combined with protein. Most of these proteins are called histones. Chromosomes are located in the cell's nucleus and contain the information needed to express the hereditary potential. This information is directly used to instruct the synthesis of structural and enzymic proteins. Each enzyme in turn is associated with the catalysis of a specific biochemical reaction(s). In effect, the control of enzyme synthesis implies biochemical control over development. The *gene* is a unit on the chromosome that directs the synthesis of a given protein. The presence or absence of a particular gene in a plant will determine whether or not the plant possesses a specific enzyme (or structural protein) along with the associated biochemical reaction(s) and products.

Genes consist of coded information in the form of specific sequences of nucleotides in the DNA molecule. The translation of this coded information into a specific type and sequence of amino acids in protein molecules was covered in Chapter 4. Of importance to the developmental process is the fact that any particular cell of the plant at a given time does not synthesize all the potential proteins and enzymes its DNA encodes. Each vegetative cell contains the information required to produce a whole plant, but what is synthesized depends on the stage of development and the location of the cell. This implies that genes are regulated in both a positive and negative manner.

Gene Regulation

A typical gene consists of the gene coding region that specifies the given enzyme or protein and the gene regulatory regions known as the promoter and enhancers. The promoter is the region of the gene that is recognized by RNA polymerase, which binds to the promoter sequence and initiates transcription. Enhancers are the regions of the DNA sequence that are recognized by factors (proteins and hormones) and also influence transcription of a nearby gene by RNA polymerase at the right place and time in development of the plant. As stated previously, all genes are not needed at all times. Enhancers can also act over large distances (several thousand base pairs) from the gene that they regulate.

Regulators of gene expression are usually called *transcription factors*. These factors are proteins or hormones that bind to specific promoter and enhancer sequences. The types of factors that bind to promoters are basal transcription factors. These types help facilitate the proper alignment of the RNA polymerase at the promoter sequence. The factors that bind at the enhancer sequence are called special transcription factors. Interactions among the basal transcription factors, the special transcription factors, and the RNA polymerase are also possible and can regulate the transcriptional activity of that gene. Enhancers have less critical orientation and fewer positional aspects compared to promoters.

Depending on the nature of the transcription factor, the binding will result in either activation or repression of the transcription of a particular gene. In this fashion, genes are regulated so that expression occurs only in the tissues and at the needed times. This tight regulation of gene expression by transcription factors is controlled by internal genetic factors such as stage of development and signaling molecules such as hormones and growth factors. External environmental stimuli such as light, heat, or seasonal changes are also involved.

Genes, Propagation, and Plant Improvement

The influence of heredity on plant development is important to the horticulturist in two applications: propagation and plant improvement. Both of these applications will be covered in Chapters 7, 8, and 9. Propagation and plant improvement may be considered as passively and actively dependent on heredity. The aim of propagation is to increase the numbers of a plant, while maintaining its essential characteristics. Preservation of a plant's unique characteristics requires the transfer and preservation of the original genes or genotype through successive generations. Propagation is then a passive use of the genes. The goal of plant breeding is to produce plants with new characteristics. This application requires a change in the genotype that is being transferred to the next generation. Plant breeding is an active use of genes.

Propagation can be achieved either by *sexual* (seeds) or *asexual* (nonseed parts) means. Each is possible because of reproductive and vegetative development of the plant (Figure 5–9). Propagation based on sexual techniques (seeds) produces offspring that are a product of genetic contributions from both parents. Some genotype alteration is expected, but it is minimized through proper handling by the grower. Propagation by the asexual approach makes use of various vegetative plant parts, all of which preserve the genotype and phenotype of the plant. The applied aspects of propagation will be covered in Chapter 7.

Mitosis and Asexual Propagation.   Each plant cell contains all of the genes necessary to reproduce the entire plant (*totipotency*). The ultimate application of this knowledge is the goal of the genetic engineer. Indeed, whole plants, from petunias to tomatoes and even trees, have been produced from single plant cells. Asexual propagation on any scale is possible because of cell totipotency plus nuclear division, or *mitosis* (Figure 5–1). During mitosis the cell divides, the chromosomes duplicate themselves (DNA replication), and then split into two identical sets, with each set going to one of the daughter cells produced during the cellular division. The end result is two cells from one, each with the same chromosomes and genes as the original parent cell. The propagation of a vegetative part, such as a stem cutting, into a whole plant that preserves the genotype and phenotype of the original plant is made possible by mitosis.

Mitosis is the basic process of vegetative growth, regeneration of plant parts, and wound healing. Heavy mitotic activity is found in meristem regions, such as those in the shoot apex, root apex, the vascular and cork cambiums, and the intercalary zones (see discussion of meristematic tissue in Chapter 3). Mitosis also occurs when callus forms on a wound or when new growth points are started on stem cuttings and other plant parts. These new growth points, such as adventitious roots and shoots, are essential for successful vegetative or asexual propagation (see Chapters 4 and 7).

Meiosis, Asexual Propagation, and Plant Improvement.   Sexual propagation is possible because of another cellular process termed *meiosis* (Figures 5–10 through 5–12). To understand meiosis and its implication to the plant breeder, who may either wish to minimize genetic variation during propagation or maximize it for plant improvement, we must look at meiosis and further aspects of plant genetics. Meiosis takes place in the reproductive structures of both the seedless and seed vascular plants.

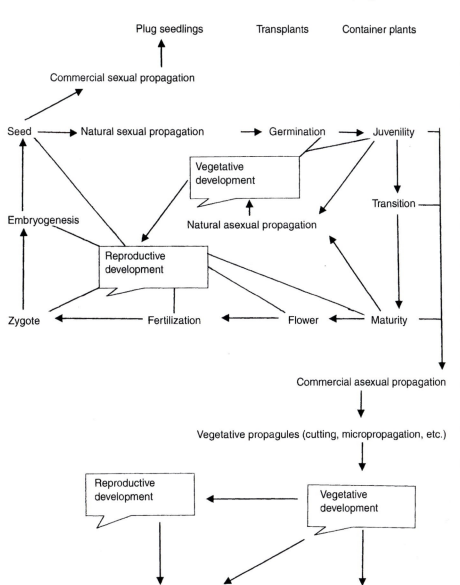

Plug seedlings   Transplants   Container plants

Commercial sexual propagation

Seed → Natural sexual propagation → Germination → Juvenility

Vegetative development

Transition

Embryogenesis

Natural asexual propagation

Reproductive development

Zygote ← Fertilization ← Flower ← Maturity

Commercial asexual propagation

Vegetative propagules (cutting, micropropagation, etc.)

Reproductive development

Vegetative development

Container plants ← Liners

FIGURE 5–9 Natural and commercial propagation and sexual/asexual cycles in plants. Sexual plant propagation is dependent on seeds produced after vegetative development. Some genetic variation occurs among the offspring. After germination the plant passes through the juvenile phase, a transition phase, and the maturation (adult) phase. Cuttings or other vegetative propagules can be taken at these points (best point often depends on plant material in question). These vegetative propagules can be asexual propagated. No genetic variation will occur with asexually propagated plants. Eventually these plants enter the reproductive phase and subsequent seed formation.

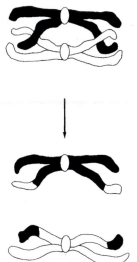

FIGURE 5–10 Crossing-over, an exchange of corresponding chromatid segments between homologous pairs. This exchange is a fairly common occurrence during meiosis. (From Raymond P. Poincelot, *Horticulture: Principles & Practical Applications*, (1980). Reprinted by permission of Prentice-Hall, Upper Saddle River, NJ.)

FIGURE 5–11 Meiosis and fertilization (sexual cycle) in angiosperms. (From Hudson T. Hartmann, Dale E. Kester, Fred T. Davies, Jr. and Robert L. Geneve, *Plant Propagation: Principles and Practices,* 6th edition, © 1997, p. 13. Adapted by permission of Prentice-Hall, Inc., Englewood Cliffs, NJ.)

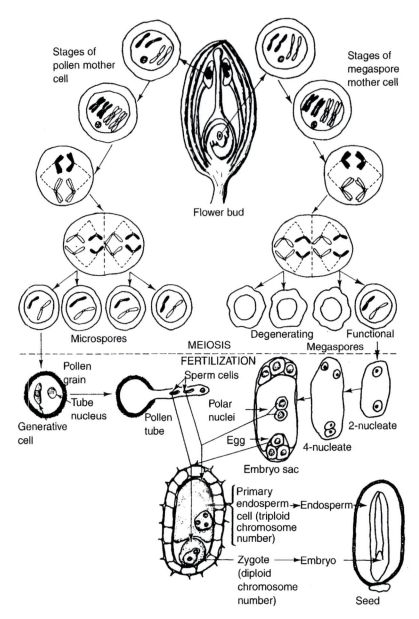

Reproductive structures are collectively called gametangia. In the fern these structures are called the antheridia (male) and archegonia (female). In the flowering plants, their names are the microsporangia (pollen sac), found in the anther (male), and the megasporangia (the nucellus) in the ovary (female). In gymnosperms the microsporangia and megasporangia are structures borne on male and female cones, respectively (Figure 5–13).

The somatic cells (vegetative cells that compose the plant body) of each plant species have a characteristic number of chromosomes referred to as the *2n* or *diploid* number. For example, the Easter lily (*Lilium grandiflorum*) has a diploid number of 24, and corn (*Zea mays*) is 20. The diploid number of chromosomes in corn consists of 10 pairs (total 20) of homologous chromosomes. Each member of the homologous pair was derived from a different parent. Remember that each chromosome actually consists of two duplicates or *chromatids* that are attached.

During meiosis the chromatids of each member of a homologous pair may exchange corresponding segments (termed *crossing-over;* Figure 5–10), and the homologous pairs are randomly divided in the first division of meiosis. During the second division the chromatids divide into single, unduplicated chromosomes that eventually duplicate. The end result of meiosis is four cells, each with 10 chromosomes, the *n* or *haploid* number.

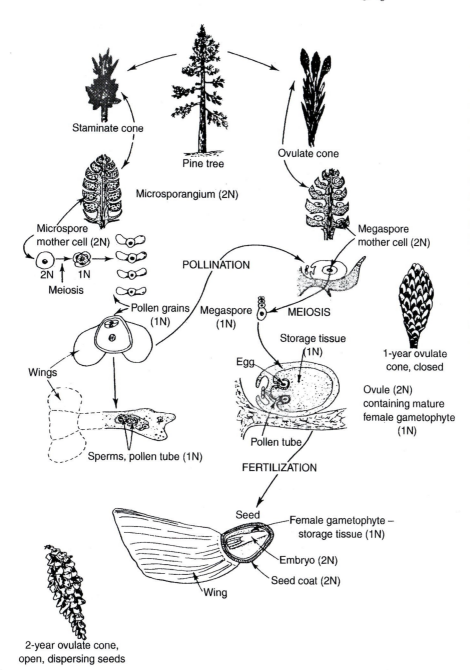

FIGURE 5–12 The sexual cycle in gymnosperms (pine). Note that gymnosperm seeds are borne exposed on cone scales, rather than a protective fruit as found with angiosperms. Cones replace flowers and the food storage tissue is also not a product of fertilization as with angiosperms. (From Hudson T. Hartmann, Dale E. Kester, Fred T. Davies, Jr. and Robert L. Geneve, *Plant Propagation: Principles and Practices*, 6th edition, © 1997, p. 136. Adapted by permission of Prentice-Hall, Inc., Englewood Cliffs, NJ.)

Because of the random division and crossing-over that precede the formation of the four haploid cells, it is highly unlikely that the next generation would be genetically identical to the preceding generation. Each haploid cell develops further by mitosis, such that the adult gamete, sperm or egg, is still haploid. The original diploid number is restored when the chromosomes of the sperm and egg are fused during fertilization. New gene combinations in the offspring (*recombination*) result because of the variance introduced during meiosis. If the fertilization is preceded by cross-pollination, rather than self-pollination, the recombination possibilities are even greater.

Genes and Inheritance

Genes are located on specific places on the chromosome designated as *loci* (singular, *locus*). A given gene can exist in multiple forms (typically two, but possibly more) that affect the same characteristic in different ways (such as flower color, for example, red or white). Alternate forms of the same gene are called *alleles*. Homologous chromosomes can have at

(A)   (B)

FIGURE 5–13   (A) Young ovulate (left) and and mature pollen-bearing staminate cone (right) from pine. (B) Mature pine ovulate cone opened to show scales and seeds (left, 9:00 to 12:00 position). (Author photograph.)

a given locus identical or different alleles of the same gene. The first situation is referred to as *homozygous* (identical alleles) and the second as *heterozygous* (different alleles).

In the heterozygous situation, the expression of the two alleles will differ. When the expression of one allele masks the expression of another, the two alleles are termed dominant and recessive, respectively. A characteristic resulting from the heterozygous condition would be the same as one that resulted from the homozygous condition for the dominant gene. The characteristic controlled by the recessive gene could only be expressed if the homozygous condition for the recessive gene occurred. Alleles may show varying degrees of dominance. When the character is expressed as an intermediate form, with neither allele showing *dominance,* the resulting condition is known as *incomplete dominance.*

If the plant is homozygous for most characteristics, it will breed true if self-pollinated or if pollinated from a genetically similar plant. On the other hand, a plant that is predominately heterozygous will not breed true; the resulting phenotypes may differ from those of the parents as well as each other.

Genes are arranged on the chromosomes in a manner analogous to a string of beads. Therefore, a tendency exists for genes to be inherited as a group; this tendency is called *linkage.* However, linkage is disrupted somewhat by crossing-over during meiosis. Either condition can help or hinder the plant breeder. The genotype and phenotype will be affected minimally or maximally, depending on the interplay between linkage and crossing-over. If the linkage occurs between a desirable and undesirable gene, extensive breeding might be necessary such that only the desirable characteristic is expressed. Linkage maps have been developed for some plant species.

Inheritance becomes more complicated beyond the control of a single character by one gene. For example, genes may act together to produce an effect that neither could do alone. Genes of this type are called *complementary* genes. Several genes may influence each other so that anyone can produce the same or similar character; these genes are termed *duplicate* genes. Some genes may mask or act as dominant modifiers of other genes (control their expression) that are nonallelic; this is known as *epistasis.*

A more complicated situation occurs when numerous genes influence traits of economic importance such as yield. The influence of many genes on a character is called *polygenic inheritance.* The resulting phenotypic variation is expressed in a continuous manner over a range. Traits expressed in this manner are *quantitative* traits. Besides the complication of many genes, this condition is made more complex by the influence of environmental conditions on the phenotypic variation and the occurrence of undesirable genes in some of the linkage groups. The plant breeder must use statistical analysis on large numbers of plants to evaluate quantitative traits.

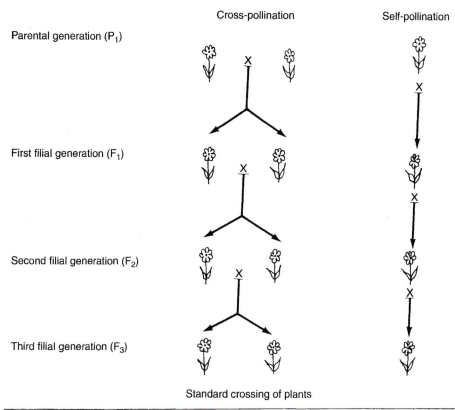

Cross-pollination · Self-pollination

Parental generation (P₁)

First filial generation (F₁)

Second filial generation (F₂)

Third filial generation (F₃)

Standard crossing of plants

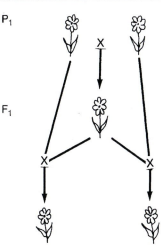

P₁

F₁

Backcrossing of plants

FIGURE 5–14
Terminology used in plant crosses for standard crossing and backcrossing of plants. (From Raymond P. Poincelot, *Horticulture: Principles & Practical Applications,* (1980). Reprinted by permission of Prentice-Hall, Upper Saddle River, NJ.)

Specific terminology is used to denote the crossing of plants to produce new plants (Figure 5–14). The starting point for crossing the two parents to obtain the best qualities of each is the P₁ or *parental generation*. The offspring of this cross is the first filial generation and is expressed as F₁. If the F₁ plants are self-pollinated or intercrossed, the next generation is the F₂ generation. Generations produced in the same manner are called the F₃, F₄, and so on, generations. Sometimes the F₁ generation is crossed with either parent, and this is called a *backcross*. Repeated backcrosses help the breeder to accumulate desirable genes more rapidly when only one parent possesses desirable genes.

An example of inheritance is shown in Figure 5–15. This is a simple case with a single gene pair in a monohybrid cross. A *hybrid* is the progeny of a cross between two parents that differ in one or more genes. A simple example of multigenic inheritance in a dihybrid cross is shown in Figure 5–16.

FIGURE 5–15 Shown is an example of one-trait inheritance involving one gene in a monohybrid cross between a hairy-leaved and non-hairy leaved plant. H stands for the hairy-leaved gene and is dominant, while h stands for the non-hairy form and is recessive.

P_1 (Parental) Generation	HH	hh
Gametes	H	h
F_1 Generation	Hh	
Gametes	H and h	

Pollen from Male Plant

Ovules from Female Plant		H	h
	H	HH	Hh
	h	Hh	hh

F_2 Generation following self-fertilization

HH, Hh, Hh	3 hairy-leaved plants
hh	1 hairless-leaved plant

FIGURE 5–16 Shown is a two-trait inheritance in a dihybrid cross involving garden peas as originally done by Mendel. The yellow and round characteristics of pea seeds are each dominant and each gene is represented by Y and R, respectively. The green and wrinkled characteristics of pea seeds are each recessive and the respective genes are represented by y and r. Mendel was either very insightful or just plain lucky in his choice of these characteristics. These characteristics are independently inherited and not complicated by linkage (see earlier coverage of linkage) with other nearby genes.

P_1 (Parental) Generation	YYRR	yyrr
Gametes	YR	yr
F_1 Generation	YyRr	
Gametes	YR, Yr, yR, yr	

Pollen from Male Plant

Ovules from Female Plant		YR	Yr	yR	yr
	YR	YYRR	YYRr	YyRR	YyRr
	Yr	YYRr	YYrr	YyRr	Yyrr
	yR	YyRR	YyRr	YyRR	YyRr
	yr	YyRr	Yyrr	yyRr	yyrr

F_2 Generation following self-fertilization

1 YYRR 2 YYRr 2 YyRR 4 YyRr	9 with yellow round seeds
1 yyRR 2yyRr	3 with green round seeds
1 YYrr 2Yyrr	3 with yellow wrinkled seeds
1yyrr	1 with green wrinkled seeds

Genes and Hybrid Vigor

Often, increased vigor is noted in the progeny of a cross between different inbred lines or unrelated species, varieties, and forms. This is noted as *hybrid vigor* or *heterosis*. It usually appears in the F_1 generation from parents that are nearly hormozygous. The consequences of continual crossing of closely related parents, or *inbreeding,* can be the opposite of hybrid vigor. These consequences can range from a reduction of size and vigor to a weak, sterile plant. Fortunately, the loss of vigor during inbreeding can often be recovered on crossing inbred lines. Hybrid vigor can be viewed as a benefit from cross-pollination, whereas self-pollination leads to a loss in vigor, but a gain in genetic constancy. Selection of the best

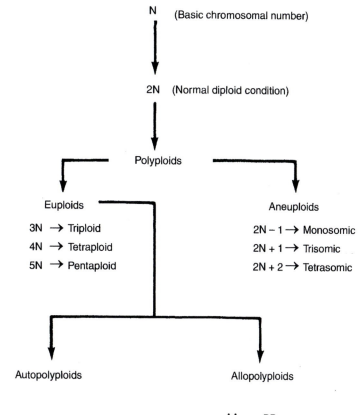

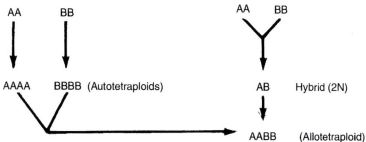

FIGURE 5–17
Terminology and relationships among the various polyploids. Chromosomal doubling within a species (AA→AAAA) is termed autoploidy. Chromosomal doubling of a hybrid produced by crossing two species (AA × BB→AB→AABB) is called alloploidy. Of course, AABB can also result without alloploidy, that is, by hybridization of the two autotetraploids, AAAA and BBBB. (From Raymond P. Poincelot, *Horticulture: Principles & Practical Applications,* (1980). Reprinted by permission of Prentice-Hall, Upper Saddle River, NJ.)

progeny from two inbred lines and their crossing produces plants with hybrid vigor and a good degree of genetic constancy. Continual inbreeding favors the increase of the homozygous condition and hence increases the chances for expression by recessive genes with deleterious characteristics. Crossing two inbred lines increases the heterozygous condition, and the chances for expression by deleterious recessive genes diminish.

Genes and Polyploidy

Natural variations in chromosomal number are possible; this condition is known as *polyploidy* (Figure 5–17). Polyploids arise from mistakes in mitosis, cell division, and meiosis. For example, if the chromosomes fail to separate during meiosis, the resulting gamete (sperm or egg) will be diploid (2n) instead of the normal haploid state (n). These gametes could then fertilize each other to produce a 4n embryo or fertilize a normal gamete to produce a 3n embryo. If a mistake then occurred during meiosis with the 4n plant, the number could become 8n. Polyploids can also be deliberately induced by the use of a chemical called colchicine.

Polyploids that contain multiples beyond the basic two sets of chromosomes are termed *euploids*. Examples are plants with three-chromosome sets known as *triploids*; four sets, *tetraploids*; five sets, *pentaploids*; and so forth. Even-numbered polyploids are usually fertile, whereas uneven ones are not. These qualities can be useful. A triploid marigold flowers continuously until frost. Being sterile, no seeds are produced, so reproduction of

FIGURE 5–18   A bud mutation in an African violet. Bud mutations are reasonably common in African violets. On the left is a normal plant with all one color flowers. On the right is a bud mutated flower (12:00 position) that is all white and considerably larger than the normal flowers below it that are variegated blue and white forms. (Author photograph.)

flowers is not terminated by seed production. A triploid watermelon is seedless because of sterility. This quality is desirable for human eating enjoyment.

Fertile, even-numbered polyploids tend to have bigger flowers and fruits and other desirable horticultural qualities than their diploid counterparts. For example, today's day lilies have larger flowers than day lilies from years ago, because they are tetraploids. Today's strawberries, being octaploids, are much bigger than wild strawberries. Many of our horticultural crops have polyploidy in their lineage. The genetics of polyploids and their use in plant breeding is of considerable interest. As would be expected, the genetics involved in polyploidy are complex; they are not discussed here.

Euploids can be subdivided on the basis of how the added genomes arose. Direct increases (such as doubling) of chromosomal sets within a species produce (fertile) *autopolyploids*. A normal cross between two different species produces a sterile hybrid, because the chromosomes cannot pair during meiosis. If chromosomal doubling occurs in this sterile hybrid, fertility is restored. The resulting fertile hybrid polyploid is termed an *allopolyploid*.

Differences of one or more chromosomes (fractions of the genome) from the basic diploid number are also possible. These are called *aneuploids*; examples are $2n - 1$, *monosomics*; $2n + 1$, *trisomics*; $2n + 2$, *tetrasomics*; and so on. Many aneuploids do not survive. Some do and are capable of reproduction without undergoing fertilization. These types are called apomitic polyploids and have been observed in blackberries, garlic, and bluegrasses. Apomitic cultivars have found their way into commercial production in the grass seed industry.

Genes and Mutations

Sudden heritable changes are known as *mutations*. When mutations occur in plants, the resulting altered form (Figure 5–18) is termed a *sport* by horticulturists. Some sports have been preserved either by asexual or sexual propagation because of useful horticultural qualities. These changes may be spontaneous or deliberate through treatment by the plant breeder with reagents known to cause mutation. These reagents, or *mutagens,* include colchicine (a plant alkaloid) and exposure to radiation.

These changes can occur at a specific location on the chromosome (*point mutations*), as multiple duplications of the genome (*polyploidy*), as large structural alterations of the chromosomes (*inversions, duplications, deletions*), and as gains or losses in some of the chromosomes within the genome (*aneuploidy*). In addition, the chloroplasts and mitochondria contain some DNA and play a role in the determination of plant characteristics (*cytoplasmic inheritance*). Any changes in these organelles may lead to mutations also. An example of cytoplasmic inheritance is seen in certain forms of variation in leaf color. Mutated chloroplasts that have no or low levels of chlorophyll are responsible for this condition.

Proplastids in the cytoplasm of the egg carry the determination of leaf color through control of chlorophyll synthesis and can pass any induced changes to the next generation. Mutations can also occur in part of the meristem and produce mutated tissue in parts of what is otherwise a normal plant. Plants of this type are termed *chimeras*.

FACTORS AFFECTING PLANT DEVELOPMENT: PLANT HORMONES

Changes in the pattern of development are triggered and regulated by plant *hormones,* or *phytohormones.* Phytohormones are biochemical, non-nutrient compounds found in low concentrations. They are usually translocated from the site of synthesis to the site of action, although they can induce responses at the site of synthesis. The broader term, *plant growth regulators,* includes both natural phytohormones and synthetic substances capable of producing phytohormone-like responses. Plant growth regulators are important to the horticulturist in terms of development, plant propagation, plant improvement, and chemical control.

A number of natural or *endogenous* phytohormones and their relationship to development are discussed as are applications of these hormones and synthetic versions for propagation, improvement, and control. Phytohormones can promote or hinder development through their action alone or interaction with other hormones. In addition, environmental factors and phytohormones interact and influence development (see Chapter 6). The action of phytohormones is complex and incompletely understood. Brief coverage of each hormone follows.

Abscisic Acid

Abscisic acid (ABA) is found in most monocots, dicots, conifers, and ferns at levels of 0.01 to 1 part per million (ppm) of fresh tissue. Its name is somewhat misleading, for ABA's actions have little to do with abscission. This phytohormone (Figure 5–19), once termed dormin, is synthesized in mature leaves, especially leaves under water stress. Small amounts can also be made in green stems and green fruit. ABA is synthesized from carotenoids. Transport is primarily from leaves in the phloem. Some movement by the xylem and parenchyma also occurs. Movement in any direction is possible (nonpolar movement). The known effects of ABA include promotion of closure of stomates during water stress, stimulation of seed storage protein production, enforcement of seed dormancy, stimulation of sugar transport from leaves to

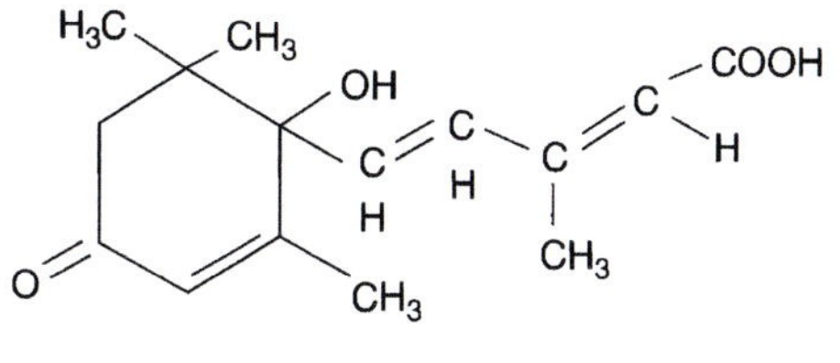

Abscisic acid

FIGURE 5–19   Abscisic acid can act in an inhibitory manner, but is essential for preventing death from dehydration. (From Raymond P. Poincelot, *Horticulture: Principles & Practical Applications,* (1980). Reprinted by permission of Prentice-Hall, Upper Saddle River, NJ.)

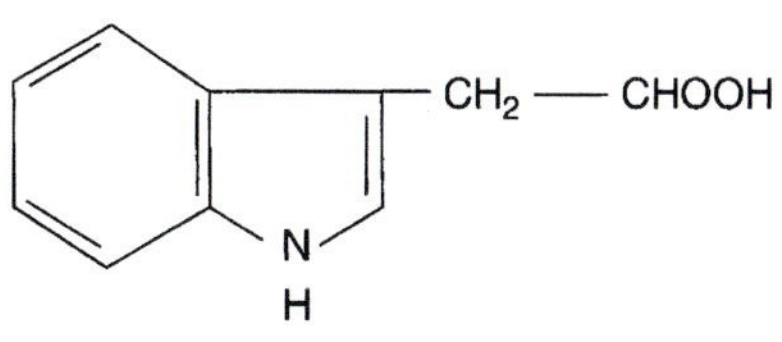

Indole-3-acetic acid

FIGURE 5–20   Widely known as auxin, this phytohormone promotes growth and is the basis for numerous rooting compounds used in horticultural propagation. (From Raymond P. Poincelot, *Horticulture: Principles & Practical Applications,* (1980). Reprinted by permission of Prentice-Hall, Upper Saddle River, NJ.)

developing seeds, and inhibition of growth. ABA often interacts with other hormones in an inhibitory manner that counteracts their stimulatory effects. The effects of ABA suggest some action at the plasma membrane and inhibition of RNA and protein synthesis.

Auxin

Auxin is the best understood and earliest known phytohormone and is assumed to be universal in vascular plants. The auxin found in most vascular plants is indole-3-acetic acid, or IAA (Figure 5–20). IAA is a growth promoter at concentrations near 10^{-9} to 10^{-5} M. High concentrations can lead to growth inhibition. Synthesis of auxin occurs in active meristems, particularly in leaf primordia and young leaves near the shoot tips, buds, fruits, flowers, and developing seeds. Tryptophan is considered to be the precursor of IAA, although some recent evidence questions this fact. Mutants lacking the ability to make tryptophan were still capable of IAA synthesis. Either a second pathway exists or tryptophan is not an intermediate in the synthesis of IAA. Translocation of auxin requires active transport and occurs from cell to cell in one direction only (polar transport). Auxin moves to the base of the plant and in roots to the root tips. Calcium ions are required for the transport of auxin.

At the cellular level, auxin affects cell elongation and differentiation. In turn, certain areas of development are influenced. These include shoot growth caused by cell elongation, prevention and delay of senescence and abscission, differentiation of vascular tissues, adventitious root development, and fruit development. Cytokinins and gibberellins work with auxin to delay and prevent senescence and abscission. Some of these effects cannot be attributed to auxins alone, but are possible through interactions with other phytohormones. For example, gibberellins help auxin to moderarate vascular tissue differentiation. A high auxin/gibberellin level favors xylem development; a low ratio favors phloem. Sugar concentrations also interact with auxin and influence which vascular tissues develops. Auxin also regulates tropisms such as phototropism, geotropism, and thigmotropism (described later in this chapter).

Auxin and Apical Dominance.   Auxin also has a holiday connection. The typical, cone shape of Christmas trees is a consequence of apical dominance in conifers. Auxin is produced at the shoot apex or shoot tip. Buds near the tip have suppressed growth. As you move downward, the auxin effect diminishes and growth of branches increases. The inhibitory effect of the shoot tip on lateral bud growth is known as apical dominance. The current explanation for this phenomenon is based on the fact that auxin stimulates ethylene production. Near the shoot tip, where auxin levels are high, ethylene is produced. The ethylene inhibits the growth of the lateral buds. As you move downward, the concentration of auxin lessens. Less ethylene results and growth of buds increases. An additional factor is that cytokinins moving upward from the root also stimulate bud growth on the lower branches. Again, we see the complicated interactions of hormones.

Horticultural Usage of Auxin.   Synthetic versions of auxin are used frequently for horticultural applications. Propagators use a slightly modified version of auxin to stimulate the production of adventitious roots on cuttings. The cuttings are dipped in a powder or solution containing the growth regulator prior to placement in the propagating medium. Premature fruit drop can also be prevented with applications of growth regulators based on auxin. These products are used on apples, grapefruits, holly, and oranges. Fruit yields at harvest are higher and holly berries stay longer on Christmas greens. Auxin derivatives at higher concentrations can also be used to increase fruit abscission. The higher levels stimulate ethylene production and accelerate fruit abscission over shorter periods. These growth regulators are used to improve the harvesting of fruits such as apples, oranges, and olives. Auxin derivatives are also used as herbicidal products to kill weeds. One of those products, now banned, was the notorious Agent Orange used as a defoliant in the Vietnam War. Auxin in combination with cytokinins is also important for organ development during tissue culture (see cytokinins).

Auxins and Plant Movements (Tropisms).   Movement of plant organs takes place in response to environmental stimuli. Movements in which the direction is determined by the direction from which the stimulus originates are termed *tropisms*. Movements that bear no relationship to the direction from which the stimulus arose are termed *nastic* movements. Examples of tropisms include phototropism, geotropism, thigmotropism, and skototropism. Nastic movements include photonasty, thermonasty, and thigmonasty. The exact causes and mechanisms of these movements are poorly understood. Phytohormones are probably involved in growth movements that stem from uneven rates of growth on different sides of the organ. Turgor movements caused by losses and gains of cellular water are probably also involved.

These growth movements are defined as follows. *Phototropism* is a growth movement in response to one-sided illumination, such as the bending of the plants toward the sun. *Geotropism* is a growth movement influenced by gravity; that is, the root grows downward and the stem upward even if a bulb is planted upside down. *Thigmotropism* is a growth movement toward an object as a result of contact, such as with tendrils of vine. *Skototropism* is the growth movement of a vine along the ground toward an upright dark object, such as a tree. *Photonasty, thermonasty,* and *thigmonasty* are nastic movements in response to light, temperature, and touch, respectively. An example of photonasty is the "sleep" movement of beans: During the day the first pair of leaves above the cotyledons is horizontal, and at night they are folded downward alongside the stems. The sensitive plant, *Mimosa pudica,* exhibits extreme thigmonasty, collapsing completely upon being touched by a hand. It is a good plant for demonstration purposes, especially with children, who often love it to death by constant touching.

The most interesting plant movements appear to be those associated with auxins. Some plants, such as the sunflower, exhibit extreme phototropism, and others, such as trees, exhibit hardly noticeable phototropism. The flower on sunflowers appears to follow or "worship" the sun, facing east in the morning, south at noontime, and finally west in the late afternoon. The common name is quite appropriate, given its round shape, yellow color, and "devotion" to the sun. How does this response work? It is based on the concept of differential growth. When the sunflower stem is exposed to sun in the morning, the auxin migrates to the darkened or opposite side. The sunlit side has no auxin, so it does not increase in length. The dark side, having auxin, grows. This uneven growth causes the stem to bend toward the sun. During the night, the auxin becomes more evenly distributed, which helps straighten the stem until the next day.

Tree leaves also move in response to the light, slowly turning to present as much surface as possible to the light. The process is so slow that we do not see it. If we videotape the leaves over several hours and then run the tape at high speeds, you would see the leaves in constant motion with the changing direction of the light.

A similar response is observed with house plants. You buy a new house plant and place it in your window. You notice a few weeks later that it does not look quite as good. When you pick it up and rotate it, you observe that it has a "good" and "bad" side. The good side faces the window. Phototropic responses have shaped the plant to the point where the leaves face more toward the window than into the room's interior. By rotating the plant every once in a while, you can keep it better looking from your viewpoint.

Bulbs, when planted upside down, have their roots come out the top and then bend downward, while the shoot comes out the bottom and then grows upward. Geotropism or gravitropism helps determine the growth direction of both. Roots always grow toward (positive response) and shoots away (negative response) from the gravity stimulus. Imagine if seeds had to be planted exactly right, such that the root end faced down? Thankfully, it doesn't matter, given that geotropism exists. Vines with tendrils show thigmotropism, response to touch, which enables the tendril to wrap around objects, enabling the vine to climb over objects. Both of these responses involve auxins, but the process is not as well known as that for phototropism.

Cytokinins

These phytohormones are present in most vascular and even some nonvascular plants. Cytokinins are effective at low concentrations of 10^{-7} to 10^{-5} M. Cytokinins exist in more than one form; examples are zeatin (Figure 5–21) and isopentyl adenine. Their synthesis is by the same pathway plants utilize for gibberellins, the mevalonate pathway. Cytokinins are synthesized primarily in root tips and to a lesser degree in young leaves, fruits, and

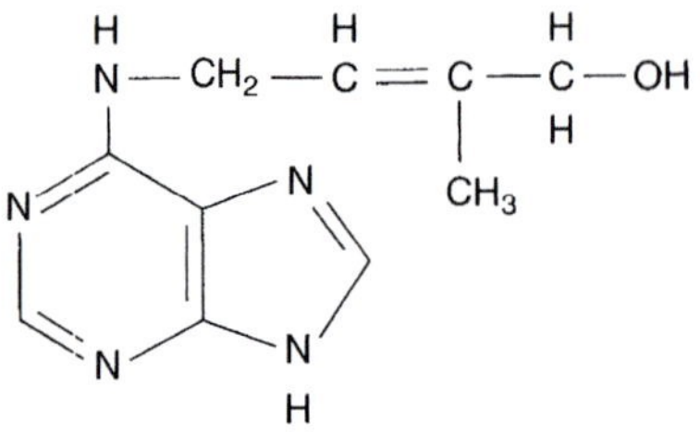

Zeatin

FIGURE 5–21   Zeatin is the major cytokinin. Kinetin is a well-known synthetic analogue. Cytokinins, along with auxins, are essential components of plant tissue culture media. (From Raymond P. Poincelot, *Horticulture: Principles & Practical Applications*, (1980). Reprinted by permission of Prentice-Hall, Upper Saddle River, NJ.)

seeds. Translocation occurs mostly upward from root tips via xylem; some movement in phloem and parenchyma cells occurs in various directions.

On the cellular level, cytokinins primarily enhance cell division in the presence of auxin. Calcium appears to enhance the effect of cytokinin on cell division. The ratio of auxin to cytokinin also helps to determine shoot and root development. This factor is very important in tissue culture and is used to regulate roots and shoot formation from single cells and tissues. A high ratio of cytokinin to auxin favors shoot formation and the reverse favors root formation. Cotyledon expansion is dependent on cytokinins.

Few horticultural applications of cytokinins exist. Cytokinins can delay chlorophyll degradation in cut leaves. This effect is not observed in attached leaves. As such, cytokinins can be used in cut flower preservatives to maintain green coloration. Usage on green edible crops, such as broccoli, is banned. The major role of cytokinins is in tissue culture, as indicated above.

Ethylene

Most plant organs produce ethylene ($H_2C{=}CH_2$). This hormone is derived from the amino acid methionine. Because ethylene is a gas, it moves from cell to cell by diffusion. Ethylene production can be stimulated by high concentrations of auxin. High concentrations of CO_2 and low concentrations of O_2 inhibit ethylene synthesis.

Etyhlene triggers ripening in many fruits; a few fruits, such as grapes and strawberries, show no response to this hormone. Flowering in most plants is inhibited by ethylene; ethylene also promotes flower senescence. A few ornamentals and mangos and pineapples flower in response to ethylene. This hormone also triggers abscission of leaves, flowers, stems, and fruits and causes downward bending of leaf petioles (epinasty). Ethylene in tandem with gibberellin also regulates the sex of flowers on monoecious plants such as cucurbits; female and male flowers form in response to ethylene and gibberellin, respectively. Inhibition of stem elongation also results from ethylene produced in response to mechanical shaking.

Horticultural Uses of Ethylene.   Several commercial applications utilize ethylene. Fruits, such as apples, can be stored in high concentrations of CO_2 and low concentrations of O_2, allowing them to be held unripe for long periods. When ripening is needed, fruits picked prematurely for reasons of storage and shipping can be gassed with ethylene and ripened. This usage makes possible the extended availability of apples, tomatoes, bananas, and other fruits. Most artificially ripened fruits taste as good as fruit ripened by nature. Tomatoes are an exception. Vine-ripened tomatoes develop more flavor components than green tomatoes gassed with ethylene. A few fruits, like strawberries and grapes, do not ripen on ethylene exposure, explaining their more seasonal market appearance. Pineapples are sprayed with a compound that releases ethylene. This treatment initiates flowering, synchronizes fruit production, and makes possible mechanical harvesting. Otherwise, pineapples left on their own produce flowers at different times, making hand harvesting necessary.

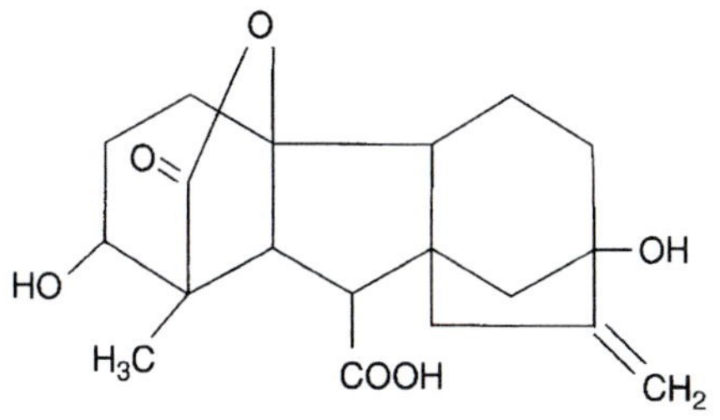

Gibberellic acid

FIGURE 5–22   Gibberellic acid is one of the few hormones other than ethylene where an external application can induce physiological changes. (From Raymond P. Poincelot, *Horticulture: Principles & Practical Applications,* (1980). Reprinted by permission of Prentice-Hall, Upper Saddle River, NJ.)

Ethylene is also used as a fruit-thinning agent on peaches (makes fewer, but larger peaches) and as an aid to mechanical harvesting for blackberries, blueberries, cherries, and grapes.

Rotten Apples.   Finally, most people are familiar with the saying that "a rotten apple spoils the barrel." Apples that rot produce copious amounts of ethylene that speed up the ripening and ultimate spoiling of the other apples in the barrel.

Gibberellins

These phytohormones are present in most vascular and nonvascular plants. Unlike the auxins, for which IAA is the primary auxin, the gibberellins now consist of at least 80 compounds with similar structures and activities (Figure 5–22). They are related to the terpenoids; synthesis is by the mevalonic acid pathway. Gibberellins are effective at concentrations as low as 2×10^{-11} M. Gibberellins are most abundant in developing seeds, apices of roots and shoots, and expanding leaves. Lesser amounts can be found in most plant parts. Translocation is through the xylem and phloem in a nonpolar manner.

On the cellular level, gibberellins enhance cell division and cell elongation. In trees, shrubs, and a few grasses, gibberellins cause shoot elongation in the more mature regions. Auxin is similar in this regard, but shoot lengthening is seen in grass seedlings and herbaceous plants instead. Dwarf plants often lack active forms of gibberellin and can be induced to grow taller by external applications of gibberellins. Other effects include stimulation of seed germination in some plants, maintenance of juvenility, initiation of flowers in certain plants requiring low temperatures followed by long days, and increases in fruit sizes in some plants.

The main horticultural use is as a fruit enhancer on 'Thompson Seedless' grapes. Grape clusters develop larger grapes that are more loosely packed in the cluster. A lesser use is to speed seed production on biennials, such as cabbage.

Other Phytohormone-Like Compounds

More recent plant growth regulators include brassinolides, jasmonates, salicylic acid, and systemin. Their role as possible growth regulators is less well understood than those just discussed, but some evidence suggests that they function as such. Brassinolides are steroids (also called brassosteroids) involved in plant growth through stimulation of cell division and elongation. Jasmonates are volatile fatty acids found in floral fragrances and are involved in the production of defensive proteins, regulation of seed germination, root growth, and protein storage. Salicylic acid, an aspirin-like compound, initiates defenses against pathogens by activating certain genes. Systemin, a peptide, is synthesized in wound tissue and initiates genes activity in defense against plant-eating insects and animals.

SUMMARY

Two processes contribute to plant development. One is growth through cell division and cell enlargement. Cell division involves nuclear division (mitosis) and cytoplasmic division (cytokinesis). Cell enlargement involves the loosening of the cell wall by the hormone auxin and expansion by turgor (water) pressure. In many cells enlargement is mostly a length increase such as in stems and roots. The second is differentiation resulting from the various differences in structure and function imposed on embryonic cells as they mature. Numerous changes result from differentiation such as the development of chloroplasts in some cells (chlorenchyma) or the loss of the nucleus and vacuolar membranes (sieve tube members of phloem) in others. Genes exert developmental control and hormones and other proteins mediate the process as do environmental stimuli.

Whole plant development has two basic stages, vegetative and reproductive development. The former consists of extensive development of the root and shoot systems prior to reproduction, while the latter is concerned with reproductive events. The vegetative stage includes embryogenesis, seed dormancy, germination, juvenility, and maturity. Reproduction becomes possible in maturity. These events prior to reproduction can occur in one growing season for annuals or take several to dozens of years with perennials.

Embryogenesis starts right after fertilization and terminates with the production of the mature seed. It establishes the basic axial body plan of the potential plant. Seeds of most plants, except for many flower and vegetable cultivars, have a growth restraint built into the seed called seed dormancy that is caused by physical or physiological factors. Hard seed coats or germination inhibitors are respective examples for the two. Seed dormancy is a protective mechanism to prevent germination under unfavorable environmental conditions. This dormancy is broken by natural environmental conditions or through artificial means for commercial propagation.

Germination requires favorable environmental conditions and starts with water intake by the seed. Numerous biochemical changes occur involving enzymes, nutrient mobilization, extensive cell division, and other events. The first visible sign is the emergence of the radicle, the first root. Eventually the young shoot appears and germination ends when the seedling is self-sustaining, that is, capable of photosynthesis.

Juvenility is a period of intense growth where the plant establishes itself. Reproduction is not possible at that time. A few plants have a different form as a juvenile plant versus a mature plant. The change from juvenility to maturity involves age and hormonal changes. Reduced vegetative growth and the ability to reproduce characterize maturity. Reproduction slows vegetative growth in perennials, but terminates growth in annuals.

Two different growth patterns are associated with these reproductive patterns: determinate and indeterminate growth. Determinate growth is characterized by genetically limited growth. Leaves, flowers, fruits, and seeds show determinate growth. Annuals also show determinate growth when reproduction starts. The apical meristem stops shoot production and only forms flower buds, thus ending all shoot growth. Indeterminate growth allows for growth and reproduction. Here reproduction slows but does not stop shoot growth. This difference is made possible by the conversion of the shoot apical meristem only partially to flower formation. For this reason perennials show indeterminate growth on the whole plant level.

The final stage, aging and eventual death, is known as senescence. The term can apply to plant parts, such as when leaves fall off, or to the whole plant. In annuals flowering and senescence are linked. Hormonal and genetic factors are likely involved in this aging transition.

The vegetative development of bulbs, corms, rhizomes, and tuberous roots and stems has some similarity to reproductive development. The development of these structures takes priority over all other parts as does reproductive development. However, unlike flowers, these structures are designed for food storage and not reproduction. Their purpose is survival during cold or dry conditions.

The transition from vegetative to reproductive development represents a major change in plant life cycles. Reproduction initiation can involve sufficient aging or environmental stimuli such as temperature and light as well as hormones. In annuals and biennials the shoot apex converts to a floral apex, and in perennials the shoot apex starts producing floral primordia. Energy and nutrients are directed massively to the developmental process that produces the flower bud and its parts (sepals, petals, stamens, pistils). Buds are protected from dehydration by the sepals. When the flower is mature, it opens (anthesis).

Pollen is transferred from the stamens to pistils (pollination) by wind or through insect assistance (pollinators). A pollen tube is produced and two sperm are delivered to the ovules within the ovary. One fertilizes an ovule to form the beginning embryo (zygote) and the other fuses with two polar nuclei to initiate endosperm formation (nutritive tissue). This process, double fertilization, is unique to flowering plants and leads to seed formation.

Eventually the surrounding ovary of the pistil (and sometimes adjacent parts) becomes the fruit enclosing the seed(s). Hormones are involved in both seed and fruit development. Seed development also influences the development of the fruit. As the fruit matures, it softens, colors up, and sweetens (ripening) or some become dry and hard. In most fruits, ripening is initiated by a hormone, ethylene. By the time the fruit is ripe, the seed(s) has developed a protective seed coat around the embryo and the seed is packed with food, usually proteins and oils. Respiration becomes low in the seed and it is usually dormant.

Seed production in conifers differs. Seeds are not enclosed in a fruit, but are borne unprotected on the cone scales. Wind moves pollen from male to female cones. Only one fertilization results. Nutritive tissue is produced, but not through fertilization. Reproduction in ferns does not involve pollen or seeds. Ferns produce spores that germinate into a small green structure (prothallus) that eventually produces sperms and eggs on separate specialized structures. The sperm must swim to the nearby egg, as opposed to the gravity-assisted fall down a pollen tube that occurs in flowering plants and conifers. The resulting embryo develops into the fern and the prothallus withers away.

Dormancy occurs throughout the development of perennials and once for biennials and not at all for annuals. Dormancy is a period of growth inactivity that enables the plant to survive unfavorable cold, hot, or dry conditions. The whole plant and its buds can become dormant such as with trees or shrubs. With some plants the root system becomes dormant (grass) and in other plants underground parts (bulb, corm, rhizome, and tuber) go dormant. Decreasing temperatures and decreasing day length trigger dormancy. Dormancy is usually broken by reversal (increasing) of these signals. Abscisic acid and gibberellin are involved in initiating and breaking of dormancy, respectively.

One of the most colorful events associated with dormancy is fall leaf color change. Leaves become orange or yellow as the underlying carotenoids become visible with the loss of green chlorophyll. Some leaves turn red from synthesis of the pigment anthocyanin. Warm days and cold nights are required for best colors. By the time the leaves fall, most of their nutrients have been removed and stored in the nearby twigs. Water loss increases cell viscosity, thus acting as antifreeze to prevent freezing damage.

Two factors control plant development: genes and environment. Genes determine plant potential and the environment determines to what extent this potential is reached. Hormones mediate the development in response to genetic and environmental commands or signals. The genotype is the sum total of the plant's genes and the phenotype is the plant's outward appearance. Genes are found on chromosomes in the cell's nucleus. Genes direct the synthesis of proteins, usually enzymes or structural proteins. The information for the protein is encoded in the nucleotide sequences of the DNA that compose genes. The genes in any one cell have the coded information to construct the entire plant, but in any given cell at any time only some genes are expressed. Others are turned off.

For development to proceed, genes must be regulated so that they turn on and off as needed. Genes have regulatory regions called promoters and enhancers. Certain transcription factors such as hormones can bind to these sites and activate or repress the transcription of the gene.

Gene effects are important for propagation and plant improvement. With propagation the goal is to increase plant numbers while preserving the existing genotype. This goal is assured with asexual or vegetative propagation, since mitosis and cell division preserve the genotype and phenotype. The opposite is true with plant improvement where the goal is to change the genotype, thus improving the phenotype.

Sexual (seed) propagation is meiosis dependent. During meiosis gene pairs (one from each parent) are randomly divided and genes can also swap segments (crossing-over) thus assuring that the sperm or egg (haploid number) will not be identical to the sperm or egg that produced the current generation (diploid number). When the sperm fertilizes the egg, the diploid number is restored (recombination), but the genotype has changed. This variation can be managed so that it is minimal for propagation or maximized for plant improvement.

Given that genes come from two parents, gene pairs (alleles) that affect the same characteristic could be identical (homozygous) or different (heterozygous). Expression of the heterozygous alleles will differ. One is dominant and the other recessive. Expression in the heterozygous case would be the same as the one resulting from the homozygous condition. The recessive gene will only be expressed if the allele consists of two recessive genes (homozygous). Sometimes the expression can be intermediate between the recessive and dominant gene (incomplete dominance). Plants where most characteristics are homozygous show minimal genetic variation if self-pollinated or cross-pollinated with a genetically similar plant. Plants with mostly heterozygous alleles show genetic variation when bred.

Given the proximity of genes on chromosomes, a tendency exists for genes to be inherited as a group (linkage). Linkage can be disrupted somewhat by crossing-over. The interplay between these two can produce minimal to maximal variation in offspring. When desirable and undesirable genes are linked, much breeding is needed to end up with expression from only the desirable gene.

Parents are the P_1 or parental generation. The offspring of their cross is the F_1 or first filial generation. If F_1's are crossed or selfed, they produce the F_2 generation. Further breeding gives us F_3, F_4, and so on. If an F_1 is crossed with either parent, it is called a backcross. A cross between parents that differ genetically yields progeny denoted as hybrids. Increased vigor (hybrid vigor or heterosis) occurs with the progeny of crosses between different parents. This outcome provides an opportunity to restore vigor to inbred lines.

Crosses are often complicated when characters are controlled by more than one gene. Sometimes genes act together and cause an effect that any one gene alone could not cause (complementary genes). Several genes can influence one another so that any one of them produces the same effect (duplicate genes). Other genes can control expression of other nonallelic genes (epistasis). Even more complex is the situation where many genes influence economic traits such as yield (polygenic inheritance). Phenotypes vary by small amounts over a range (quantitative traits) and can be influenced by environmental effects, making evaluation difficult.

Alterations of chromosomal numbers occur naturally and also artificially with the use of colchicine. The normal diploid number ($2n$) can be changed by mistakes in mitosis, cytokinesis, or meiosis, leading eventually to triploids, tetraploids, and on up to octoploids. These abnormally numbered organisms are known as polyploids or, more specifically, euploids. Even-numbered ones are fertile, while odd-numbered ones are sterile. The latter can be saved only through human intervention. Even-numbered polyploids have bigger flowers and fruits and other improved qualities relative to the normal diploid condition. Triploids are usually seedless or never stop flowering, which also makes them horticulturally desirable. Many of our horticultural crops are polyploids. Polyploidy can occur within a species (autoploidy) or after the cross of two different species (allopolyploidy). The cross is usually sterile, but the occurrence of polyploidy restores fertility. Fractions of the chromosomal number are also possible such as $2n - 1$, $2n + 1$, $2n + 2$, and so on. These types are aneuploids. Most do not survive, but some do and are able to reproduce without undergoing fertilization (apomitic polyploids).

Natural and induced mutations can occur, affecting either the whole plant or organelles such as the chloroplast and mitochondria. These mutations can have horticultural value and be useful in breeding programs.

Changes in development in response to genetic or environmental factors are regulated by phytohormones (plant hormones). Five known hormones each have specific effects, but often interact in complex ways. Abscisic acid often inhibits the actions of other hormones. It is also involved in reducing water stress, enforcing seed dormancy, stimulating translocation of sugars to developing seeds, stimulating the production of seed storage proteins, and inhibiting growth. Auxin stimulates cell and shoot development, delays senescence, and abscission of plant parts (in concert with cytokinins and gibberellins), affects differentiation of vascular tissues in concert with gibberellins), and promotes fruit and adventitious root development. Auxin is also involved with plant responses to environmental stimuli such as gravity (geotropism), light (phototropism), and touch (thigmotropism). The cone shape of gymnosperms also results from auxin effects at the shoot apical meristems (apical dominance). Horticultural uses of synthetic auxins include rooting of cuttings, production of weed killers (herbicides), and prevention of premature fruit drop. At high concentrations auxin derivatives can be used to cause fruit abscission for easier mechanical harvesting. Cytokinins promote cell division in concert with auxin and changes in the ratio of cytokinin/auxin regulate shoot and root development. Cytokinins and auxin mixtures are very important ingredients in tissue culture of plants. Ethylene, the only gaseous hormone, promotes fruit ripening. It also inhibits flowering and promotes flower senescence and abscission of plant parts. The major horticultural use of ethylene is to ripen fruit on demand after shipping. Gibberellins enhance cell division and elongation, promote shoot elongation and fruit development, stimulate seed germination, maintain juvenility, and initiate flowers in some plants. The main commercial use is the production of larger and more loosely packed 'Thompson Seedless' grapes.

More recent plant growth regulators include brassinolides, jasmonates, salicylic acid, and systemin. Brassinolides are steroids (also called brassosteroids) involved in plant growth, jasmonates are volatile fatty acids involved in production of defensive proteins, regulation of seed germination, root growth, and protein storage. Salicylic acid (aspirin-like compound) initiates defenses against pathogens, and systemin initiates defenses against plant-eating insects and animals.

EXERCISES

Mastery of the following questions shows that you have successfully understood the material in Chapter 5.

1. What is the difference between growth and differentiation?
2. By what process can a cell enlarge?
3. What are the differences between vegetative and reproductive development?
4. What are the five basic events that occur in the life of a vascular seed plant? What happens in or defines each one? In which one can reproduction occur?
5. What mechanisms ensure seed dormancy? How is seed dormancy broken in nature and in commercial horticulture? Are all seeds dormant? If not, which ones are not?
6. What events occur in germination from initiation to completion? What environmental factor is the most critical one during germination?
7. During what phase of development is growth most extensive? Least extensive?
8. What are the basic differences between juvenility and maturity? What are the basic differences between determinate and indeterminate growth?
9. What happens during flower and fruit development? What happens during seed development?
10. How do ferns reproduce? How does this process differ from seed plants?
11. Is dormancy restricted to seeds? If not, where in the plant can you find dormancy? What function does dormancy have?
12. What causes the color changes in leaves during the fall?
13. How do genes and the environment influence development?
14. What is a gene? What does a gene do? How are genes regulated?
15. What connection exists between genes and the propagation and breeding of plants?
16. What is an allele and what do homozygous and heterozygous mean in connection with alleles?
17. How does linkage impact plant breeding?
18. What do the terms F_1 and F_2 refer to? What is a backcross?
19. What is a hybrid and what do we mean by hybrid vigor?
20. What is polyploidy? What value does polyploidy have for horticulturists and plant breeders?
21. Name the five basic hormones in plants. What functions does each serve?
22. What is apical dominance, phototropism, geotropism, thigmotropism, and skototropism? What hormone is involved with all of these events?

23. What are nastic movements? Name and describe each type discussed in the text.
24. What horticultural applications have been derived from our knowledge of auxin?
25. What horticultural applications have been derived from our knowledge of ethylene?

INTERNET RESOURCES

Gene regulation

http://www.ultranet.com/~jkimball/BiologyPages/P/Promoter.html

Plant hormones

http://www.rrz.uni-hamburg.de/biologie/b_online/e31/31a.htm

Plant reproduction and development

http://nsccbio.sccd.ctc.edu/jones/bio203/34.html

Polyploidy

http://fletcher.ces.state.nc.us/programs/nursery/metria/metria11/ranney/polyploidy.htm

Review of factors affecting plant growth (some development)

http://www.hydrofarm.com/content/articles/factors_plant.html

RELEVANT REFERENCES

Altman, A., and Yoav Waisel (eds.) (1998). *Biology of Root Formation and Development* (Basic Life Sciences, Vol. 65). Plenum Publishing Corporation, New York.

Davies, Peter J. (ed.) (1995). *Plant Hormones: Physiology, Biochemistry, and Molecular Biology* (2nd ed.). Kluwer Academic Publishing, New York.

Fosket, Daniel E. (1994). *Plant Growth and Development: A Molecular Approach*. Academic Press, San Diego.

Greyson, Richard I. (1994). *The Development of Flowers*. Oxford University Press, New York.

Howell, Stephen H. (1998). *Molecular Genetics of Plant Development*. Cambridge University Press, New York. 300 pp.

Kofranek, Anton, Hudson T. Hartmann, Stephen C. Myers, and Vincent E. Rubatzky (2001). *Plant Science: Growth, Development and Utilization of Cultivated Plants*. Prentice Hall, Upper Saddle River, NJ. 752 pp.

Raghavan, V. (2000). *Developmental Biology of Flowering Plants*. Springer Verlag, San Diego. 354 pp.

Raven, Peter H., Ray F. Evert, and Susan E. Eichhorn (1999). *Biology of Plants* (6th ed.). W. H. Freeman, Worth Publishers, New York. 944 pp.

Singer, Susan (2000). Mechanisms of Plant Development. In Scott F. Gilbert, *Developmental Biology* (6th ed., pp. 621–646). Sinauer Associates, Sunderland, MA.

Steeves, Taylor A., and Ian M. Sussex (1989). *Patterns in Plant Development* (2nd ed.). Prentice Hall, Upper Saddle River, NJ.

Westhoff, Peter (1998). *Molecular Plant Development: From Gene to Plant*. Oxford University Press, New York.

VIDEOS AND CD-ROMS[*]

The Gene (CD-ROM, 2000).
Introductory Genetics (CD-ROM, 1997).
Understanding Basic Genetics (CD-ROM, 2000).

[*]Available from Insight Media, NY (www.insight-media.com/IMhome.htm)

Chapter 6

Plants and Their Environment

OVERVIEW

Plants, like all living organisms, have environmental needs. These requirements must be satisfied if plants are to thrive and be productive. Environmental parameters (Figure 6–1) influencing plants include climate, water, temperature, light, and soil. All play a major role in the development and productivity of plants. A thorough understanding of these relationships is needed before we can appreciate horticultural practices designed for, at best, management of these environmental parameters. In some cases we must be satisfied with practices that help us make the best of what are sometimes less than perfect or uncontrollable environments. Knowledge of environmental parameters also helps us with our attempts to sustain soil and water resources and to minimize pollution of on-farm and off-farm environments.

CHAPTER CONTENTS

Climate
Water
Temperature

Light
Soil

OBJECTIVES

* To understand the determining role of climate on commercial and home horticulture and how skillful usage of microclimates can offset climate to some degree
* To learn how water in the soil and air affects plants and understand the complex relationship between plants and water in the soil
* To understand how temperature impacts both physiological and developmental aspects of

plants and how this knowledge explains certain horticultural phenomena and practices
* To understand how light impacts both physiological and developmental aspects of plants and how this knowledge explains certain horticultural phenomena and practices
* To learn about soil formation, soil physical and chemical properties, soil nutrients, and their impact on soil management and plant growth

FIGURE 6–1
Environmental factors
which influence plant
development.

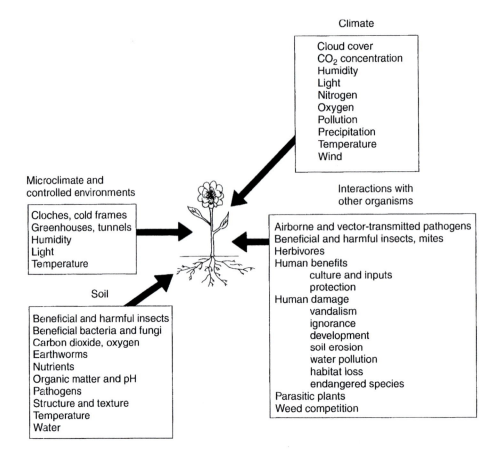

Climate

| Cloud cover |
| CO_2 concentration |
| Humidity |
| Light |
| Nitrogen |
| Oxygen |
| Pollution |
| Precipitation |
| Temperature |
| Wind |

Microclimate and
controlled environments

| Cloches, cold frames |
| Greenhouses, tunnels |
| Humidity |
| Light |
| Temperature |

Interactions with
other organisms

| Airborne and vector-transmitted pathogens |
| Beneficial and harmful insects, mites |
| Herbivores |
| Human benefits |
| culture and inputs |
| protection |
| Human damage |
| vandalism |
| ignorance |
| development |
| soil erosion |
| water pollution |
| habitat loss |
| endangered species |
| Parasitic plants |
| Weed competition |

Soil

| Beneficial and harmful insects |
| Beneficial bacteria and fungi |
| Carbon dioxide, oxygen |
| Earthworms |
| Nutrients |
| Organic matter and pH |
| Pathogens |
| Structure and texture |
| Temperature |
| Water |

CLIMATE

The role of climate is important in plant development both because of its direct effect on plants and its indirect influence on the other elements of the plant environment. Those aspects of climate that affect plants are precipitation, radiation (heat, light, ionizing), air and soil temperature, carbon dioxide concentration, air pressure, wind, moisture, and cloud cover. Variation of one of these factors can extensively alter the effects of the others on the rates of photosynthesis and transpiration, two important parameters of plant development.

Climatic variations can be averaged over long periods of time to yield an average annual profile. Plant formations interact so closely with these varied climates that certain natural plant formations and horticultural crops come to be associated with specific types of climates. This linkage explains why the citrus industry is based mainly in California, Florida, Arizona, and Texas. Profitability is greatest when the crop is grown in the climate that maximizes crop productivity. This association also explains the concentration of apple production in New England and the Pacific Northwest, wine grapes in California, and grains in the Midwest. California has an ideal climate for the production of vegetables and fruits. Consequently, California leads the United States in horticultural production. Some 50% of all U.S. vegetables and fruits are produced in California.

Microclimate

Remember that these concepts of climate are derived from instrumentation at 2 or more meters above ground. However, many of our horticultural crops grow below this level and are subject to influences arising from the surface profile. These influences result from natural and artificial conditions, such as proximity to water bodies, variations in slope, wind exposure, masses of nearby vegetation, soil characteristics, buildings, and paved surfaces. Climatic conditions tend to be modified by these surface influences. The term *microclimate* is used to distinguish the surface-induced climate from the less affected atmospheric macroclimate.

The microclimate is very important to the horticulturist, who must assess its nuances with care. Microclimates can be advantageous, like those areas with a favored exposure, such as land near a large body of water or on the southern slope of a valley. These areas are favorable because in the former winters are milder, and in the latter safe frost-free planting dates arrive earlier than in surrounding areas. Other microclimates can be disadvantageous, such as lower areas that become frost pockets earlier in the fall than adjacent sites. Large stretches of pavement also induce unfavorable areas.

Let's look at some specific microclimates. One of the most advantageous microclimates for horticulture exists where land is adjacent to large water bodies. Examples of this include ocean-induced microclimates in states fronting on the Atlantic, Pacific, and Gulf coastal areas of the United States. Large lakes and rivers can also produce favorable microclimates, such as those found around the Great Lakes.

The advantage of these microclimates is moderated summer heat and winter cold. This advantageous effect can be considerable. For example, in the coastal areas of Connecticut, the growing season ends later and is up to a month longer than land 15 miles inland. Winters are also milder on the coast than inland, so perennial plants have greater winter survival rates in the coastal microclimate as opposed to further inland. In the colonial days, most people were farmers, so this microclimate had major horticultural value. Today, its value is more appreciated by gardeners. Water-induced microclimates in Florida, Louisiana, California, Oregon, Washington, and the states around the Great Lakes still have considerable commercial value to horticultural industries. For example, the microclimate makes for a good fruit-growing region around the Great Lakes and a superb fruit and vegetable area along the coastal regions of California, Oregon, and Washington.

How does this effect come about? The effect results from the fact that land and water absorb, retain, and lose heat at different rates. Land heats up faster and cools quicker than water. Water retains heat considerably better than land. Water is used in some solar energy systems because of its effectiveness as a heat trap. In the summer, the coastal land area heats faster than the nearby ocean. Hot air rises from the land and is displaced by cooler air present over the water. This phenomenon is the so-called "sea breeze" enjoyed at the land/water interface in the summer. Summers are generally cooler near the water than further inland because of this effect. In the winter, the ocean retains heat more effectively than the coast, acting as a "heat sink" that moderates winter temperatures in the coastal area.

Another beneficial microclimate is found on gently sloping land that faces south. The most exposure to sun occurs on southern exposures. The warm air generated during this exposure sweeps up the slope, bathing it constantly. This particular exposure heats up earlier in the spring than other locations, allowing earlier planting. Fruit growers prefer gentle southern slopes. Fruit trees get an earlier start and less danger exists for an unexpected late spring frost damaging the flower buds and blossoms. Such damage reduces fruit yields. Other growers also favor this type of land for similar reasons.

A smaller version of this southern slope microclimate exists in the home garden. Bulbs planted near southern foundation walls of houses flower several days earlier than those on eastern or western walls. The southern wall gets more hours of sun heating the foundation wall. Some of the heat is transferred into the nearby soil. The warmer soil gives the bulbs a head start in the spring. This site is particularly good for getting the early harbingers of spring to show their colors as early as possible. Preferred bulbs for this site include the crocus and snowdrop.

Unfavorable microclimates also exist. Low land at the bottom of a southern slope becomes a frost pocket. As the warm air moves up the slope, the pocket at the bottom fills with heavier, cold air. These low-lying areas with cold air experience frosts before higher areas. Often they have late frosts in the spring and early frosts in the fall. Plants in these areas experience shorter growing seasons. Any depressed areas in the land profile are potential frost pockets. You can often see this phenomenon in the fall when the first frost arrives. Early in the morning, you will likely observe a sporadic cover of frost on the lawn and garden. Some areas have no frost and others have a heavy cover. Often the heavily frosted areas are lower lying lands.

Large paved areas, such as around shopping malls, also produce unfavorable microclimates. These areas are often covered with hot, dry air that does not favor optimal plant growth. The black asphalt surface is a heat trap and warms up much faster and reaches higher temperatures than the adjacent land. In fact, you can feel the difference. When you get out of your car on one of these lots in July or August, you are met with a blast of hot, dry stifling air. If you walk off the lot into a grassy area nearby, you will immediately feel cooler air. In fact, the air temperature is several degrees cooler. The grassy area absorbs less heat and the moisture from the plants and soil helps cool the area. Plantings in these lots fare poorly. When was the last time you saw a beautifully landscaped parking lot? The hot, dry air desiccates the plants, leading to tissue death in leaves and poor growth. Look also at the areas around trees designed to catch rain. Are they large? No, because large areas mean fewer parking spaces. The area of soil should be as large in diameter as the maximal diameter of the foliage ball expected for the mature tree. With small areas of soil around the tree, less water reaches the tree, adding further to the problems.

Big cities often experience some very unfavorable microclimates for similar reasons. Much of the city is paved over. In addition, the large buildings have reflective glass that reflects much of the sun's heat back into the air, in order to reduce the load on the air conditioners. The air conditioners are also heat pumps; that is, they pump heat out of the building into the air. This combined heat load from these sources results in higher air temperatures than would be found in more suburban sites. If you listen to weather reports in the summer, you will note that temperatures are generally higher in big cities when compared to nearby smaller towns.

Finally, small-scale microclimates can exist inside buildings. Often you can use these microclimates to benefit certain house plants. The kitchen and bathroom generally have a moister microclimate than other rooms. The increased moisture results from water usage for cooking and washing. Plants that like high humidity would fare better in these rooms than others. Examples are African violets and ferns. Of course, you also need to place them in the appropriate light exposure in these rooms for best results. Light and temperature microclimates also exist. Northern windows are cooler than southern windows, because they get much less sun. Cacti do well in southern windows because of the high light levels and warmth. Cacti grow poorly in northern windows. Light levels decline (assuming no shading from trees) as you go from southern to eastern, western, and northern windows. Awareness of your home's microclimates as you place your house plants leads to better plant growth.

Wind Effects

Wind influences vegetation through physical impact, acceleration of moisture loss, and convective heat transfer. The uptake of carbon dioxide during photosynthesis is also increased when wind is present, because when there is no wind effect, the lack of air circulation can lead to localized depletion of carbon dioxide in the immediate environment of the leaf. Pollination and seed transfer are also assisted by moderate winds. Severe winds, however, can damage plants, interfere with pollination, cause premature droppage of fruits and nuts, and enhance soil erosion. Plantings adjacent to coastal areas are particularly hard hit by winds because of the combined damage from salt sprays and the drying wind.

Air Quality

Pollutants have adverse effects on plants (Figure 6–2) and their development. Air pollutants include ethylene, lead, sulfur dioxide, ozone, boron, fluorides, hydrocarbons, and photochemical products such as peroxyacetyl nitrate. Most of these pollutants arise from nonagricultural activities, such as automobiles, smelters, electrical utilities, and industrial processes. Plants show different degrees of physiological resistance toward these pollutants. Some appear to suffer no damage, others experience mild leaf damage, and yet others die.

The relationship between air pollution and plants is important to the horticulturist, especially in the urban environment, where the success of a city tree may rest on its pollution

(A)   (B)

FIGURE 6–2   Symptoms of air pollution damage on foliage. (A) Damage from sulfur dioxide, which is a major air pollutant. Chlorophyll-containing cells are destroyed, leading to white areas between leaf veins. (B) Sun acting on nitric oxide, nitrogen dioxide and certain hydrocarbons produced by combustion of fossil fuels yield photochemical oxidants that damage plants. Flecking in leaves is often indicative of such pollution. (U.S. Department of Agriculture photographs.)

tolerance. It is unfortunate that pollution susceptibility must be considered when choosing or breeding plants for use in many urban environments. Crop yields can also be reduced by air pollution. Besides yield reductions, damage from pollutants can also reduce crop quality. For example, spinach with considerable ozone damage is unmarketable. Sites downwind from pollution sources are to be avoided for commercial production.

Phenology

Climate may be related to developmental events in plant life. Climate can also be related in a similar fashion to peak insect outbreaks and serious weed problems. The science dealing with this relationship is *phenology*. Phenological data can be employed to predict or forecast the best times for planting crops, and when fruits, vegetables, and ornamentals will flower, and when fruits and vegetables will ripen and be ready to harvest. Similarly, such data can also predict when specific insect and weed outbreaks will occur for numerous crops. These dates are determined by monitoring climatic factors prior to and up to the time of the event. These data can be represented on maps on which an isophene connects phenological events that take place on the same date. These data can also be tailored to specific localities and regions. For example, maps with frost date lines are based on this concept.

Phenological data are based on degree-days. Heating oil firms also use this same concept to calculate when deliveries should be made to households. One degree-day is defined as a day when the mean daily temperature is 1 degree above the minimum temperature for growth (zero temperature) of a plant or of an insect for that matter. Zero temperatures vary according to the crop considered. For example, pea, a cool season crop, has a zero temperature of 38°F (4°C), sweet corn, a warm season crop, has a zero temperature of 50°F; potatoes, 44.5°F; and most deciduous fruit and nut trees, 41°F. A day when the mean daily temperature was 48°F would provide 10, 0, 3.5, and 7 growing degree-days, respectively, for peas, corn, potatoes, and deciduous fruit and nut trees. From accumulative research through various regions of the United States, researchers have determined how many degree-days (sometimes referred to as heat units) are needed for various crops to reach harvest time, or for an insect or weed outbreak to occur.

Some problems exist with this approach. Degree-days cannot differentiate among seasonal variations, such as cold spring–hot summer versus a warm spring–cool summer, among differences in soil–air temperatures, or among unusual extremes of temperature during the growing season or during the diurnal variation of temperature. Certain developmental stages of crops might be more temperature sensitive than other ones or dependent on a particular photoperiod. One such crop with developmental stages that are sensitive to differing temperatures and dependent on the photoperiod is the pea. The earliest possible plantings of peas can require 45% more growing degree-days to harvest than a planting made 5 weeks later because of increases in the mean daily temperature and photoperiod. Unless the horticulturist was aware of this change, the prediction of harvest dates and timing for processing could be off.

Additional refinements of the heat-unit concept to correct for photoperiod and effects of temperature change on developmental stages are possible. The degree-days can be multiplied by the number of daylight hours to give degree-day-length units. A further refinement is to measure the rate of growth at various temperatures in the growth range and to multiply the number of hours of each day at each temperature by the relative growth rate of the plant at that temperature. This concept is known as positive degree-days or growing degree-days. It is also possible to keep tweaking the data on a yearly basis, such as with insect outbreaks. As the predicted outbreak occurs, trapping and counting can be employed. The actual peak outbreak can be compared with the predicted one, and the data can be fined-tuned. After several years of this fine-tuning, predictions based on phenology can become very accurate.

The concept of negative or dormancy degee-days must also be considered for perennial fruit and nut crops that have minimal chilling requirements to break dormancy. This type of data can be used to predict the time of bud swell and to assess the danger of late spring frosts, or the timing of a bud swell pesticide application. Sprinkling of trees near the time when bud swelling occurs cools by evaporation and reduces accumulation of degree-days. By this technique, bloom can be delayed by up to 2 weeks and late frost damage avoided.

The use of phenological data, especially both growing and dormancy degree-days, is feasible. Best results are obtained for those horticultural crops for which development is not closely dependent on photoperiod or differences in temperature sensitivity, or for those insect pests that have narrow periods of peak infestations. Heat units can be useful even for the more difficult crops if appropriate corrections are made. Heat units are best tempered with a working knowledge of your local area's climate and microclimate. A large number of phenological models are used in California for many horticultural crops, especially to control insects, mites, nematodes, and weeds.

Careful utilization of phenological data makes it possible to establish planting and harvesting schedules that are most suitable for the local climate, labor availability, and fresh market processing demands. It may also be possible to schedule crops to avoid seasonal weather phenomena, maximal disease and insect activity, or maximal need for irrigation.

The importance of phenological data for harvesting and food processing should not be underestimated. If a harvest glut occurs within a short period, overtime labor and the use of more mechanized equipment than necessary often result. Crops might even spoil before they are harvested. On the processing end, the factory's capacity might be exceeded, or it might have to be larger than is really efficient, overtime might be required, and some crops could spoil or be at less than optimal quality when processed because of time delays. The efficient and economical use of equipment and labor at planting and especially harvesting times therefore makes it expedient to plant crops in succession to spread the harvest over a longer time. Phenological data can be used to predict the start and finish of the first and last harvest, respectively. By using this strategy, the harvesting, processing, and marketing labor and equipment requirements are not overwhelming in a short time span; equipment and labor are used more efficiently over a longer period each growing season; and processing operations need not be as large. This is especially useful for crops whose duration of harvest period is very short for quality maintenance.

Successive harvesting of one crop can be realized by making several plantings of a single cultivar at spaced time intervals or by making one planting of several cultivars with different maturation times. Different crops can be planted on dates determined from phenological data to ensure a smooth transition, with adequate time between crop changeovers for harvesting and processing. If adverse climatic factors cause a deviation from the predicted harvest date, all fields within the processing plant region should be affected similarly. Therefore, the sequence of harvest is merely shifted, impacting the processing plant operation minimally.

Remember that phenological data determine the *approximate* time of harvest. Times are approximate because some variables cannot be controlled. Harvest windows can be altered somewhat by factors such as the genetic variations between cultivars, the timing of the planting date, and the environmental conditions during the growing season. A number of other criteria can verify the exact harvest window. Again, this input can be used to fine-tune the phenological model. These criteria vary with the crop being considered, but include the number of days from germination, setting out of plants, and flowering; the skin color, degree of softness, sugar levels, sugar–acid ratios; abscission zone formation; differences in sound obtained from tapping; overall appearance; and the taste test.

Phenological data are of considerable importance to sustainable agriculture in terms of pest control. Ideally, the farmer wants to reduce or even eliminate pesticides. The ability to predict outbreaks and then confirm them with trapping offers security and helps reduce the preventive applications of pesticides that often are not needed. It also offers the opportunity to start employing biological and other natural controls prior to the predicted date, such that their effectiveness is maximized.

WATER

Within wide temperature limits, water is probably the most important critical factor for plant production. Water is an essential plant constituent, both for hydration and a medium for biochemical reactions. Water is absorbed and used as a transport medium for sugars, minerals, and phytohormones. Losses of water occur through transpiration; heat losses during transpiration help to control plant temperatures. The amount of water available to the plant is a function of precipitation (Figure 6–3), surface infiltration, moisture retention of the soil, and losses of water through evaporation from the soil. These soil factors are covered later in this chapter.

A water balance exists among the plants, soil, and atmosphere. Certain features of plants minimize water loss: stomates (leaf pores) and waxy layers (cutin or suberin). In extremely arid situations, plants evolved to survive drought. These water-efficient plants (*xerophytes*) include the cacti and other succulents that have thickened cell walls, heavy layers of cutin, mucilage-like cellular contents, and tiny or no leaves. Plants also adapted to very moist or wet environments (*hygrophytes*). Carnivorous bog plants such as pitcher plants (*Darlingtonia*), sundews (*Drosera*), butterworts (*Pinguicula*), and Venus flytrap (*Dionaea*) fall into this category, as do crop plants such as cranberries and rice. The majority of plants with moderate water needs are known as *mesophytes*.

Water and Its Effects on Plants

Certain physiological processes are affected adversely during times of water stress. The most obvious symptom of water stress is wilting (see Chapter 4), a consequence of insufficient water. Wilting can also result from excessive water (see later discussion on excessive water). Water stress also affects photosynthesis, respiration, biosynthesis, nutrient absorption, and translocation. These processes are altered because water is the solvent for reactions, and it is an essential reaction component in many cases. Water is the medium for nutrient absorption and translocation and metabolic reactions. It is a chemical reactant in hydrolyses of molecules, such as the reaction of starch with water to produce glucose. Water is involved with electron transport during photosynthesis and respiration. A decrease of water causes an increase in reactant (biochemicals) concentration in the cytoplasm and a

FIGURE 6–3 Average precipitation maps for the United States. (A) Annual precipitation. (B) Warm season precipitation. Maps of these types indicate the potential water available through rainfall and snow, but natural climatic variations and soil parameters determine how much water is actually available each year for crops. (Reproduced by permission of the United States Department of Agriculture.)

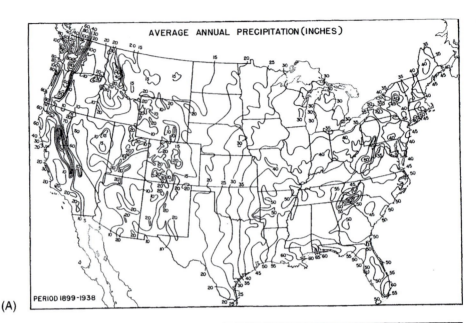

(A)

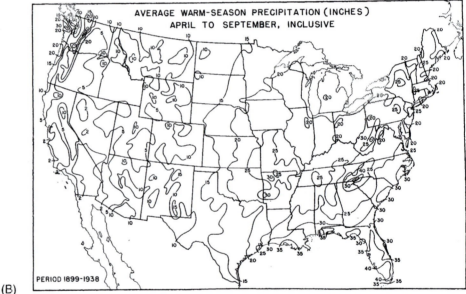

(B)

reduction in turgor pressure (visualized as wilting). There is also evidence that the binding of water molecules to enzymes is altered during water stress, causing configuration changes and hence losses of activity for the enzymes.

Water stress also influences plant development through stimulation of abscisic acid synthesis. This phytohormone induces closure of the stomates to prevent dehydration. Water stress can also cause growth suppression and hasten leaf senescence. Water can relieve some forms of seed dormancy through the leaching of growth inhibitors.

One would not expect water stress to limit photosynthesis, because less than 1% of a plant's absorbed water is used during photosynthesis. However, water stress does quickly reduce the photosynthetic rate and consequently growth. Unless the water balance is restored rapidly, the growth rate may never return to the original rate. Water stress, as indicated, does affect other processes, but the rapid effect on photosynthesis is probably the most important in terms of horticultural productivity. Irrigation is essential during droughts or in dry areas to prevent water stress and the associated decreases in productivity.

Limitations of photosynthesis probably arise from decreased hydration of the protoplast and chloroplast. In turn, this induces molecular disorientation and decreased enzymic activity, which disrupts light absorption and carbon dioxide fixation. However, the major impact on growth is a result of stomate closure in response to water stress. This closure cuts off the supply of carbon dioxide for photosynthesis, thus decreasing the supply of sugars that fuel growth.

Humidity and Its Effects on Plants

Besides the effects of water within the plant, certain effects are associated with atmospheric moisture (*humidity*) around the plant. The influence of humidity depends on the type of plant and on the availability of soil moisture. Xerophytes tolerate reduced humidity, whereas plants of tropical origin often do not. During times of low humidity, the latter exhibit leaf browning and dropping of leaves and flower buds. The primary effect of humidity is a reduction of the transpiration rate as humidity increases, providing the wind force is not overly large. Increased transpiration coupled with low humidity results in wilting and desiccation.

Humidity becomes of prime concern during propagation with cuttings. No water uptake is possible until adventitious roots form, yet the leaves carry on transpiration. Therefore, water loss through the leaves must be minimized until roots form. A high *relative humidity* can greatly reduce this water loss. This measure of humidity refers to the ratio of the concentration of existing water vapor in the air to the total amount the air could hold at the given temperature and pressure. High relative humidity is also necessary to prevent desiccation and to promote callus formation with grafts done in greenhouses. Outdoor grafts are sealed with wax to prevent desiccation.

Symptoms of low humidity stress are observed frequently with house plants of tropical origin during winter. Homes have low levels of humidity, unless a humidifier is used to increase humidity during the heating of homes.

Water in Soil

Precipitation ends up as surface water, soil water, or groundwater (becomes part of the water table). The entry or movement of water into the soil is defined as *infiltration*. This term should not be confused with *percolation,* which merely refers to the movement of water through a column of soil. Excess water over the storage capacity of the soil continues to move downward through capillary forces and gravitation and becomes groundwater. Every soil differs in its *infiltration capacity,* the measure of the amount of water infiltrated per unit of time.

Many soil parameters are involved in the control of water infiltration. One is soil structure. Larger water-stable aggregates shaped like imperfect spheres, the spheroidal type, increase filtration, whereas matted, flattened, plate-like aggregates, the platy type, decrease it. Organic matter content in increasing amounts and coarseness improves infiltration in a fine-textured soil, but decreases it in a coarse-textured soil. The amounts of sand, clay, or silt also play a role. Coarse sands increase infiltration, whereas fine sand, clay, or silt decrease it. Even water content (drier soils increase infiltration), temperature (warmer soils increase infiltration), depth (increasing soil depths to the hardpan increases infiltration), and degree of compaction (decreasing compaction increases infiltration) affect filtration. Surface litter such as from crop residues also improves infiltration.

Available Soil Water.   Water is held by the soil through adhesion and cohesion, and it is lost through gravity and evaporation. Films of water that adhere to soil particles are termed *hygroscopic water.* This water is tightly held and unavailable to plants. Other water molecules are held to the hygroscopic water through cohesion or attraction of water molecules for each other. This form, *water of cohesion* or *capillary water,* is less tightly held and supplies much of a plant's water needs (usually found in micropores; see the section on soil). If precipitation or irrigation is heavy, the water may saturate the soil to the extent that it fills all pore space (micropores and macropores). This water will drain at a rate determined by the soil's percolation capacity. This downward-moving water enters the groundwater and is termed *gravitational* or *free water,* since the soil is unable to retain it against the force of gravity.

Water is retained by the soil with differing degrees of retention energy (indicated above), or *soil-moisture tension.* Water retention or storage capacity depends on the texture (which is a function of amounts of sand, clay, and silt; see soil discussion later) and particle sizes of

FIGURE 6–4   The
influence of soil texture
on water storage capacity.
(From Donald Steila, *The
Geography of Soils:
Formation, Distribution,
and Management,*
© 1976, p. 46. Adapted
by permission of
Prentice-Hall, Inc.,
Upper Saddle River, NJ.)

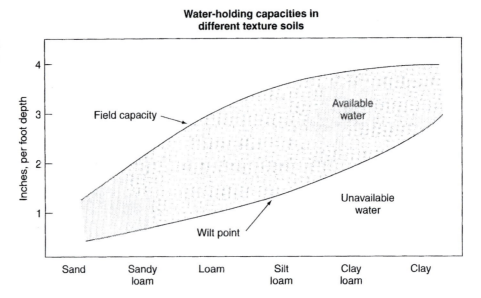

the soil (sand is bigger than silt; silt is bigger than clay) and also on the organic matter content. The smaller the particles are, the greater the surface area per unit volume; this in turn increases the storage capacity. Increasing organic matter content also increases water storage capacity. The effect of texture on water storage capacity is shown in Figure 6–4.

Measures of Water Availability.   Together the soil-moisture tension and soil water retention capacity determine the amount of water available to the plant. After sufficient rain or irrigation, gravitational water drains into the groundwater, leaving only capillary soil water. Some capillary water will evaporate from the surface of the soil, causing it to dry. The remaining capillary water is the maximal amount that can be retained by the soil. This amount of soil moisture is the *field capacity* of the soil and is capillary water with minimal movement. When the energy required for the plant to absorb capillary water becomes too great as the hygroscopic water is approached, the plant starts to wilt. The level of moisture left is called the *wilting point.* The difference between the field capacity and the wilting point is known as *available water* (Figure 6–5). This is the water available to plants, and it consists of capillary water that has little or no capillary movement. The most water-efficient irrigation systems use meters in the soil to measure the wilting point and field capacity. Such systems prevent wastage of water.

Drought and Excess Water

Too much or not enough soil moisture can cause problems for the horticulturist. Soil moisture is lost through evaporation directly from the soil's surface and indirectly through transpiration by horticultural crops and weeds. When these combined losses, collectively called *evapotranspiration,* exceed the amount of available soil moisture, we have a water deficit referred to as *drought.* Drought occurs in areas of arid climate with insufficient rainfall, in areas with distinct periods of greatly reduced rainfall, or in areas with unpredictable variations in precipitation. Localized drought can occur in plantings, such as trees, that are surrounded by large expanses of water-impervious pavement. In addition, factors such as high temperatures, winds, and low humidity can speed drought because they increase evapotranspiration.

Drought can be combatted by the use of irrigation, cultivation to remove weeds, drought-resistant crops, and mulches, and by increasing the soil's field capacity by adding organic matter. Wilting, poor development, decreased yields, and eventual death result from uncorrected drought. Continued drought also reduces soil pH temporarily and increases soluble salts. Fertilizer salts, salts produced by degradation of organic matter, and

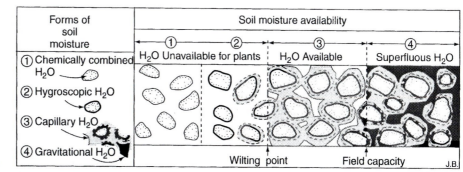

(A)

(B)

FIGURE 6–5   (A) The relationship between forms of soil moisture and water available for use by plants. (From Donald Steila, *The Geography of Soils: Formation, Distribution, and Management*, © 1976, p. 45. Adapted by permission of Prentice-Hall, Inc., Upper Saddle River, NJ.) (B) Plant appearance when wilting point is reached. Unless available water is replenished at this point in sufficient time, the plant will die. (From Raymond P. Poincelot, *Horticulture: Principles & Practical Applications*, (1980). Reprinted by permission of Prentice-Hall, Upper Saddle River, NJ.)

salts brought to the surface by capillary movement and evaporation are no longer leached downward by water infiltration. The soluble-salts level increases in the topsoil during periods of drought. Increased exchange of certain ions (from salts) for hydrogen ions on soil particles frees more H^+ to the soil moisture, which results in a more acidic soil.

Excessive soil moisture can result from excessive rainfall, excessive water retention by the soil, overwatering of container plantings, too much field irrigation, poor drainage of soil in the root zone, or high water tables. Increasing soil moisture means decreasing levels of oxygen in the soil. In turn, respiration decreases in the plant roots, causing poorer plant development and eventual death if the excessive water condition remains uncorrected. As the plant deteriorates, the roots are the first to die, turning from a healthy white to brown in color and rotting from microbial infection. At this stage, water uptake is greatly reduced and the plant wilts. If mistaken for drought-induced wilting, the remedy is the addition of more water, which accelerates the loss of the plant. In addition, excessive soil moisture increases the leaching of nutrients and may cause nutrient deficiencies. Chronic problems can be corrected by improving drainage or by switching to water-tolerant crops, such as cranberries.

Although not often considered, excessive soil moisture that might not be great enough to result in death for the plant can affect the plant by the leaching of growth regulators present in the soil. These leachates from the aboveground portions of the local flora are part of the ecosystem, and their removal undoubtedly has some effect (not yet fully researched) on plant development. If the excessive soil moisture results from excessive rain, an additional effect is to increase the leaching of water-soluble growth regulators from the aboveground portions of the plant. The effect of this is not completely assessed, but one consequence is delayed dormancy due to leaching of abscisic acid (see Chapter 5).

In the previous section we saw that water is an essential reaction component for many of the biochemical processes associated with plant development. Heat in turn is necessary for these processes, because all plants have minimal, maximal, and optimal temperature requirements for all aspects of plant development, as well as for individual biochemical processes. The accumulation of heat over time (degree-days) can be directly related to number of days needed to harvest. This concept, phenology, was covered earlier in this chapter.

Plants, as well as other biological forms, are generally limited to a temperature range from 32° to 122°F (0° to 50°C). Many plants, biennials and perennials, can survive below this range in a dormant stage. Above 122°F plants die because of protein denaturation. Some organisms, such as hot-spring algae, desert lichen, arctic mosses, and arctic lichens, can endure temperatures that exceed this range, but plants cannot.

Effects of Temperature on Physiological Processes

Transpiration shows a marked increase as the temperature rises. This increase is even greater if the temperature increase is associated with a decrease in relative humidity and an increase in air movement. Temperature also affects photosynthesis and respiration. Increases in temperature usually cause higher rates of photosynthesis, respiration, and photorespiration. As expected, decreasing temperatures slow photosynthesis, transpiration, and respiration. As temperatures increase, the photosynthetic rate levels off and then starts to decline. The actual temperature for this changeover varies with species. However, for most plants, the optimum temperature for photosynthesis ranges between 76° and 86°F (25° and 30°C). Plants of tropical and desert origin can have a slightly higher range. Temperatures of more than 122°F cause disruptions of enzyme systems and breakdowns of physiological/biochemical processes.

Effects of Temperature on Plant Development

Normal variations in temperature can affect the various metabolic processes that in turn affect plant development. Temperature influences the quantity and types of food used by the plant, the amount of ATP produced during respiration, the rates of phytohormone synthesis, rates of translocation, internal levels of hydration, and many metabolic processes. In turn, changes in these metabolic processes have considerable impact on plant development.

Growth Temperature Ranges.   Optimal temperatures and suitable temperature ranges for plant development are highly variable. These temperature conditions vary among the species, the time of day, and the particular stage in development. Plants do not often proceed to the next stage of plant development until a specific temperature requirement has been satisfied.

However, some generalizations may be permitted. Horticultural crops of temperate origin do not grow below 41°F (5°C) or above 95° to 104°F (35° to 40°C), and have an optimal growth temperature from 77° to 95°F (25° to 35°C). With tropical plants, these temperatures in respective order are 50°F (10°C), 113°F (45°C), and 86° to 95°F (30° to 35°C). Optimal temperatures for development are generally lower at night than during the day; this phenomenon is termed *diurnal thermoperiodicity*. Of course, this difference varies with species, but the day–night difference is often about 9° to 21°F (5° to 12°C). Change in optimal temperature with the various developmental stages is sometimes called *annual* or *seasonal thermoperiodicity*. Both forms of thermoperiodicity are discussed later.

Seed Germination.   More specific effects of temperature on developmental stages and specific plant parts are of great interest to the horticulturist. Of all the environmental factors that regulate seed germination and seedling growth, temperature is the most important. Seeds have minimal and maximal temperature limits for seed germination, as well as optimal ones. Extensive variation in these temperatures exists among plant species, and they can vary for a particular plant because of climate of origin, seed age or condition, or interaction with light.

Some broad generalizations can be made. The optimal temperature for germination is between 68° and 85°F (20° and 30°C) for many plants. Seedling growth is favored by temperatures somewhat lower than those for germination. A diurnal thermoperiodicity temperature differential of about 18°F (10°C) also appears to be most favorable for germination and seedling growth. Cool season plants, which include many of the flowers and vegetables (carrots, cole crops, and pansy) that originated in the temperate zones, can germinate at temperatures below this range, but some in this group (delphinium, lettuce, and onion) fail to germinate above 77°F (25°C). This high-temperature sensitivity is known as *thermodormancy*. Warm season plants such as the solanaceous crops, cucurbits, bean crops, and sweet corn have a minimal germination temperature of 50°F (10°C) and are highly susceptible to chilling injury. Chilling appears to cause damage and subsequent leakiness of membranes within the chilled cells.

Root and Shoot Development. Root and shoot development are affected by temperature. Both shoot and root growth increase with rising root temperature and constant shoot temperature up to a point, and then decrease. The shoot-to-root ratio increases at the same time. Small changes in root temperature, therefore, can cause extensive changes in top growth. The best root and shoot growth also requires diurnal thermoperiodicity.

Situations of interest include container crops and mulched areas. Many plants have some mature and especially young, roots that are less hardy than the shoot portion. With containers in the fall, the soil is at air temperature rather than the warmer field soil temperature. Therefore, root damage or death can result if plants are unprotected or insufficiently acclimatized by winter's arrival. Applications of cold water to container plantings can also lower the rootball temperature and retard or damage growth. Premature application of mulches to insufficiently warmed soils will slow the warming of soil and can retard top growth and even flowering. One possible cause of reduced top growth may be the reduced uptake of water and nutrients associated with decreasing root temperatures. Root temperatures are also important when rooting cuttings, because higher air than soil temperatures favor shoot over root development.

Reproduction and Cold Temperature Exposure. Flower initiation with some plants, such as many summer annuals and plants of tropical origin, can occur once the plant is physiologically ready (reached the appropriate stage of development). In others further stimulus by other environmental factors, such as temperature or light, may be required. The light requirement (photoperiodism) is discussed later in this chapter.

Flower initiation and earlier flowering, when induced by cold temperature exposure, is a phenomenon known as *vernalization*. Not all cold temperature reproductive responses are promoted by vernalization. Some trees and flowering shrubs, such as apple, dogwood, forsythia, lilac, maple, oak, and pear, do flower after cold temperature exposure. These plants require cold exposure (*chilling period*) to break dormancy so that vegetative and reproductive growth can proceed in the spring. Bulbs, such as hyacinths and tulips, also require cold exposure that allows development of the already formed embryonic flowers in the bulb. In the former, cold temperatures break dormancy, and in the latter, they promote development of the existing bud. In neither case does the cold initiate flower bud formation. Examples of plants that require vernalization include beet, cabbage, carrot, cereal grains, foxglove, lettuce, peach, rye grass, spinach, and tomato.

Note that the cold temperature required is variable and depends on the plant. For example, tomato seedlings exposed to temperatures in the 50°F range for 2 weeks flower earlier than unexposed seedlings. Most plants have a 30°F range requirement. The temperatures and times of exposure vary with species, but a general range is 28° to 50°F (−2° to 10°C) for several weeks. In most cases, the vernalization is only effective with plants where a certain amount of vegetative growth has occurred. Exceptions include the seeds of winter cereal grains, such as winter wheat.

In some plants subject to vernalization, a photoperiod with long days must follow after the low temperatures if flowering is to be completed. Short day lengths (see the photoperiod discussion in the section on light) can substitute for low temperatures in some plants, but long days afterward are still required.

The effects of low-temperature preconditioning can be lost if subsequent exposure to higher temperatures around 86°F (30°C) occurs. This process, called *devernalization,* has some commercial value. Onion sets, small onions produced by seed the first year, are harvested and stored. During the following spring, they are sold and planted to produce large, mature onions. Between seasons, the sets must be stored at cold temperatures to prevent spoilage. This exposure vernalizes the onion, so that on planting, flowering would occur at the expense of bulb enlargement. However, exposure of the sets on removal from storage to 80°F for a few weeks reverses the effects of the vernalization.

Vernalization and Lysenko.   Vernalization has an important historical connection. The vernalization of winter wheat seeds is connected to the Cold War with Russia and the serious setback of its biological sciences. The effects are still visible today. Genetic engineering requires the most advanced knowledge in molecular biology. The United States and Japan lead this technology, with Russia far behind because of this setback. How is this possible?

Wheat cultivars exist in two general groups: winter and spring wheat. Spring wheat is planted in the spring and harvested late in the summer. Winter wheat is planted in the fall and harvested the following summer. If winter wheat is planted in the spring, grain heads develop very late in the season and are likely to be damaged by frosts before harvesting takes place. Planting winter wheat in the fall allows vernalization to occur and promotes earlier flowering.

It was discovered that cold exposure of the seeds of winter wheat could satisfy the vernalization requirement, such that winter wheat could be spring sown. This discovery was made in the late 1800s, but was thought of as a scientific curiosity. In 1929, a Russian scientist, Trofim Lysenko, expanded on these earlier experiments and showed that freezing winter wheat seeds in ice caused it to behave like spring wheat. He named the phenomenon *vernalization,* the origin of the term. He also falsely claimed to be the discoverer of vernalization, conveniently ignoring the work of earlier scientists. Since winter wheat had some merits over spring wheat, Lysenko achieved some prominence in Russia. Lysenko claimed that vernalization caused a change in the inherited qualities. The change, however, was physiological, and not genetic.

He promoted the concept that the environment could change inherited characteristics, which caught the attention of the Russian leader, Stalin. Stalin was impressed with Lysenko and his "discovery." Lysenko used this connection to rise to a leadership post in Russian biology and genetics. Other Russian scientists opposed him because of his false views and later work that could not be reproduced. Lysenko likely falsified experiments to gain further favor with Stalin. His opponents soon learned to be quiet, because they were taken away by the secret police to Siberia, where most died. During his reign over Russian biology, Lysenko imposed his views, silenced the opposition, and effectively destroyed Russian biology. Others were afraid to do science that either showed Lysenko wrong or made them look better than their boss. He eventually was exposed and fell from power during the Khrushchev reign. However, he had done serious harm to Russian biology and genetics, fields to this day that lag behind other developed countries. Given the Cold War, these events worked in the favor of the United States.

Propagation.   Temperature is also of great importance to the plant propagator. It has an effect on bulb development and storage, callus production, rooting cuttings, seed dormancy and germination, seedling growth, and seed storage. As would be expected, maximal, minimal, optimal, and harmful temperatures will show great variation for these phenomena and among plant species. Propagation temperatures will be examined more closely in Chapter 7.

Temperature Extremes.   The influence of temperature extremes on physiological processes has a noticeable effect on plant development. High temperatures usually speed growth up to a point. Damage results primarily from the secondary effects of increased evapotranspiration, followed by dehydration. Increased evapotranspiration and dehydration at high temperatures can be forestalled if the soil and atmospheric moisture are sufficient. Winds may speed the damaging effects of high temperatures by increasing rates of evapotranspiration during the summer.

At the other end of the spectrum, low temperatures and moisture are associated with damaging and killing frosts and freezes. If a tender or improperly acclimatized hardy plant freezes, structural disruption of the tissues and even death can occur. Damage can be severe, especially with newly transplanted materials, if the plant is subjected to alternate freezing and thawing. The latter produces a heaving action that can lift root systems above the surface of the ground. Cold temperatures insufficient to produce freeze damage can still damage or kill plants of subtropical or tropical origin that are sensitive to chilling. For example, a house plant left out too long into the fall can be subjected to night temperatures in the high 30s, possibly resulting in damage.

When temperature extremes are combined with water problems, the effects can have considerable—often negative—impacts on horticultural crops. Examples of these phenomena include physiological chilling and winterkill.

Microclimates can reduce the effects of temperature extremes on development, such as those to be discussed later under frosts. Slopes with low sun exposure (north, east, or west facing) can experience lower temperatures than surrounding level areas, thus permitting the growth of plants sensitive to high temperatures. Bodies of water slow the rise of temperatures on their leeward side and are advantageous for slowing the blossoming of fruit trees until the danger of frost is past. The southern side of a building, because of reflected and radiated heat, can promote more rapid plant development, such as earlier flowering of spring bulbs, than surrounding areas.

Frost. A temperature phenomenon of special interest to the horticulturist is frost. Moisture in the air can drastically affect plant development as the temperature drops and the moisture becomes frost. The last killing frost of the spring and first killing frost of the fall (see frost maps in Chapter 12: Figure 12–1 and Tables 12–1 and 12–2) determine the favorable period of plant development, the *growing season.*

Frosts can be caused in two ways. *Air-mass frost* results from the appearance of a cold air mass with a temperature of less than 32°F (0°C). This frost is more of a freeze and apt to happen during winter. The only threat to horticultural crops from this frost is to plants limited in winter hardiness.

Radiation frost results from rapid radiational cooling of plant material to 32°F or lower. Radiation frosts usually occur in spring or fall. Later-than-normal spring frosts can be a threat to seedlings and budding trees or shrubs. Earlier-than-normal fall frosts can be a threat to plants that have not gone into dormancy and unharvested crops. This frost occurs on clear nights when the sky acts as a black body and absorbs radiant heat from leaves. Cloud covers slow radiant heat loss and thus help prevent frost. Leaves can experience a temperature drop of several degrees below air temperature. Dew forms on the cooling leaf when the temperature is reached at which the relative humidity is 100 percent (*dewpoint*). The dew freezes into ice crystals at 32°F or less; this is termed a *white frost.* If the dewpoint has not been reached and the temperature is 32°F or less, it is called a *black frost,* because no visible signs of frost exist until the plant material is blackened from freeze injuries.

Microclimates and Frost. Microclimate effects on frost can be quite significant. Sloping areas experience different temperature regions or thermal belts. Cold air, being more dense than warm air, settles into the lower portions and can create frost pockets. The downward movement of cold air displaces the warmer air upward, causing the formation of a *thermal belt* on the slopes. Valleys with north-to-south orientation are shaded quicker than east to west ones and have longer times to develop the beneficial thermal belt. Vegetation shades the ground and lowers conduction and convection; consequently, temperatures above bare ground may be up to several degrees higher than above the area with vegetation. An area of clean cultivation could escape a frost, while a nearby vegetated or mulched area would not. Large bodies of water can keep temperatures relatively higher on windward sides. Their heat loss is slower than soil heat loss; thus they can act as a heat sink and slow the advent of killing frosts.

Winterkill.   Growth of most plants generally ceases when the soil temperature drops below 41°F (5°C). At these temperatures, water uptake through the roots is limited. Warm spells during the winter pose no problem for deciduous plants, since they have lost their leaves. However, needled or broad-leaved evergreens have the potential to lose water from their leaves (transpiration) during winter thaws. When this happens, the rate of transpiration might exceed the rate of water uptake. Damage follows from desiccation or from increases in solute concentration caused by reduced levels of water within the plant. Increased winds or warmer-than-normal temperatures (a January thaw, for example) can increase transpiration rates to a point where serious stem and foliage desiccation occurs, because the roots cannot take up water from the still-frozen soil to replace transpiration losses. This type of damage (*winterkill* or *winter-burn*) can be a problem in the northern United States, especially with needled and broad-leaved evergreens. Similar damage, flower petal burn on magnolias in the South, is also another good example.

Damage can be especially bad with rhododendrons, mountain laurels, and andromedas in landscaped areas. Yet these same plants in their native habitats show much less damage. Why? The answer is failure to pay attention to the natural environment of these plants. In their natural habitat, these plants grow in the understory of the forest. As such, they are shaded throughout the year. In the January thaw, the shaded habitat does not lead to photosynthetic activity of any consequence. Hence, water loss from the leaves does not occur or is minimal.

Plants that suffer the most winterkill damage in the landscape are located on the south side in full sun. During the January thaw, such plants can have considerable photosynthetic activity. If the plants were placed on the north side of the house, or in the shade of trees on the south side, winterkill damage would be minimized. Wrongly placed plants can be protected from winterkill by the application of antitranspirants. These materials are waxes or plastics sprayed on the leaves after frost. The plastic seals the stomates, preventing water loss under sunny conditions. By spring, the plastic has degraded sufficiently to allow normal stomate function.

Physiological Chilling.   The reduction of water uptake at lower temperatures to levels below transpirational needs of shoots at higher air temperatures can cause physiological drought in some plants that results in the yellowing and browning of leaves. Many tropical foliage plants are highly susceptible to this type of damage, called *physiological chilling*. Chilling damage results by applying cold water to the soil, such as with 35°F (1.8°C) water directly from the tap in January. Besides the possibility of damage, the chilling of the roots slows development of the plant. In the greenhouse fuel energy is wasted on returning the soil temperature to normal levels. The timing of greenhouse plants to coincide with a holiday could be ruined by cold water. Water for plants in the home or greenhouse during the winter should be allowed to warm to room temperature or be mixed with warm water.

Thermoperiodicity

It is known that the optimal development (Figure 6–6) of most plants occurs when the daytime temperature (*phototemperature*) is 9° to 21°F (5° to 12°C) higher than night temperatures (*nyctotemperature*). A possible explanation of diurnal thermoperiodicity is that carbon dioxide assimilation is better (up to a point) with higher temperatures, and respirational losses of fixed CO_2 decrease with decreasing temperature; thus the plant gains maximal retention of fixed CO_2.

The day and night temperatures selected for controlled conditions are a compromise between maximizing photosynthesis and minimizing respirational losses short of depriving the plant of its essential respirational energy. Most plants in the home and greenhouse respond favorably to diurnal thermoperiodicity. Not all plants respond favorably to this regime. Some plants do best when phototemperature and nyctotemperature are similar; the African violet (*Saintpaulia*) succeeds better when the nyctotemperature is about 10°F (6°C) higher than the phototemperature. Unlike vernalization, the effect of diurnal thermoperiodicity is on current development. Some effect is exerted on flower initiation, but vegetative development and fruit set and development are influenced more.

FIGURE 6–6 The effect of different day/night temperatures on the growth of cucumbers is clearly evident. From left to right plants were subjected to day/night temperatures of 75/65, 80/70, 85/75, 90/80, and 95/85 for 20 days. (U.S. Department of Agriculture photograph.)

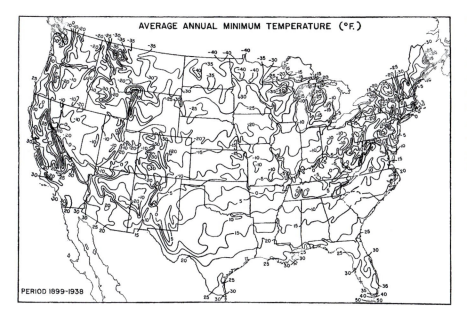

FIGURE 6–7 Map showing average annual minimal temperatures in United States. Plants vary in response to these temperatures. Such temperatures are used in part to determine zones of hardiness for different plant material. (U.S. Department of Agriculture photograph.)

Many plants show responses to longer range fluctuations in temperature, since the various developmental stages have different optimal temperatures. This annual or seasonal thermoperiodicity is most pronounced with temperate-zone plants. In tandem with photoperiodic changes, seasonal thermoperiodicity induces and breaks dormancy. Besides day length, water availability can moderate responses to seasonal temperature changes such as with plants that enter dormancy in areas having a cold, dry season.

Natural Distribution of Plants. Seasonal changes in temperature and the ability of plants to meet them tend to determine the distribution of plants. Average annual minimal temperatures vary (Figure 6–7), and so do the degrees of hardening shown by different species. In addition, the needs of plants vary in terms of chilling requirements. These factors would determine the northern and southern limits of a plant's distribution. On this basis the United States and Canada can be divided into areas based on minimal temperature and the degree of plant hardiness (zones of hardiness, 1–10). A knowledge of a plant's response to low temperatures allows the horticulturist to select appropriate plants for any given geographical areas (see Section Three, Chapter 12: Table 12–3 and Internet Resources in this chapter and in Chapter 12).

Hardening. The most familar response of plants to seasonal thermoperiodicity is *low-temperature preconditioning,* or *hardening.* A plant must undergo this process in order to enter dormancy (see discussion on dormancy in Chapter 5 also) and survive the rigors of winter. Continued exposure to low temperatures affects the development of the plant during the following spring and summer. Breaking of seed and bud dormancy and the blooming of biennials are some of the phenomena influenced by low-temperature exposure. The hardening process occurs in steps on exposure to low temperatures. Development slows, causing simple sugars to accumulate and increase the osmotic potential. A hardiness-promoting phytohormone is thought to be produced on exposure to low temperatures and short days. Exposure to frost causes additional changes, such as reduced hydration, increased membrane permeability, and structural alterations of protein. This gives the protoplasm a property that enables it to resist the stresses of further frosts and freezing. Subzero temperatures induce even greater hardiness.

LIGHT

Light influences several aspects of plant development and physiology. While not the most important limiting factor in the distribution of plants, light is an environmental factor of utmost importance in terms of a given plant in a specific environment. Plants respond to a broader light span, 300 to 800 nanometers (nm), than do our eyes. The term *light* will refer to this broader spectral range. The nature of these responses depends on an interrelationship among the spectral quality, duration, and intensity of light. The actions of light are not always singular, in that effects of temperature and phytohormones are often involved also.

Intensity and Its Effects on Physiology and Development

The degree or amount of light, *light intensity,* varies under natural conditions. Cloud cover, leaf canopies, season, pollution, and latitude cause variations in the light intensity reaching plants. Light intensity is often expressed in terms of foot-candles. For example, a bright, sunny summer day has an intensity of about 10,000 foot-candles as opposed to about 30 for a good reading lamp. However, these units are based on the spectral sensitivity of the eye. A more useful measure of light energy in terms of plants is the absolute energy of the light wavelengths utilized in the photosynthetic process (the photosynthetically active radiation or PAR, 400 to 700 nm) expressed per unit area. Light sensors are available for these measurements. Units of measurement for PAR are microwatts per square centimeter ($\mu W/cm^2$).

Under some conditions, light intensity can be the limiting factor in photosynthesis. At night, in morning when the light is increasing, and in evening when the light is decreasing are obvious examples. Dense cloud cover or heavy shade are also limiting. Light underneath a forest canopy or large shade tree can be limiting, even though the intensity does not appear greatly diminished. Wavelengths effective in photosynthesis have been partially absorbed as the light filters through the leaves. However, the reduced light and lower levels of useful wavelengths provided by large trees or physical obstructions can create a useful microclimate in a sunny area. Plants unable to tolerate high light intensities could succeed under these conditions.

Increasing light intensity raises the photosynthetic rate until a point of light saturation is reached. The light saturation, and hence photosynthetic rate, varies considerably among species. In shade-adapted species or plants not acclimated to high light intensities, levels of light intensity beyond the light saturation point can damage the photosynthetic apparatus. The less photosynthetically efficient plants are unable to make maximal use of full sunlight, unlike the more photosynthetically efficient species (see discussion on photosynthesis in Chapter 4). Photorespiration also increases with increasing light intensity in the less photosynthetically efficient plants, thus further reducing their competitiveness with the more efficient species.

Etiolation.   Plants grown from seed under darkness develop until their stored food reserves are exhausted. These plants are yellow, with spindly stems having long internodes. If you leave a brick on the lawn for several days, the grass under it becomes yellow-white. In both cases, this condition is known as *etiolation.* The yellow color is from carotenoids which are normally masked by green chlorophyll. In the dark-grown plants, chlorophyll fails to appear, as light is required for chlorophyll synthesis. With covered grass, the chlorophyll already present breaks down during darkness and is not replaced, as again light is required for chlorophyll synthesis. The absence of light, which promotes some processes and inhibits others, results in a disruption of normal, balanced development.

Blanching. Etiolation is sometimes deliberately induced in a few vegetables; this process is called *blanching*. This process helps to improve flavor and texture. The blanched area is white to yellow in color. Cauliflower, celery, endive, and Witloof chicory (Belgian endive) are often blanched. Various practices are used for blanching: tying leaves together, covering with soil or paper, placement of boards along rows, and forcing roots in darkness. Some plants, such as cauliflower, have been bred to be self-blanching. These cultivars cover their heads with their leaves.

The term *blanching* is also applied to food processing. This process involves boiling or steaming fruits and vegetables briefly to inactivate enzymes. This process is done prior to freezing and helps to preserve taste and extend storage times.

Sun and Shade Plants. Plants vary in their response to light intensity. Some plants succeed at low light intensity (shade plants) and others do not (sun plants). Ferns, *Hosta, Coleus,* and Impatiens are examples of shade plants. Cacti, corn, tomatoes, sunflowers, and maple trees are sun plants. A variegated cultivar requires more light than the unvariegated plant. The absence of chlorophyll in white, yellow, or pink parts of the leaf reduces the surface area capable of photosynthesis per given amount of light; so photosynthetic rates must be increased for the rest of the leaf.

Sun and Shade Leaves. Certain morphological and physiological changes can occur within a plant to maximize or minimize utilization of light. Leaves of small shade plants or within the interior of a tree canopy differ in appearance from sun-grown small plants and leaves from the exterior of the tree canopy (Figure 6–8). Sun leaves are often thicker, smaller, and have shorter petioles; the thickness usually stems from several extra layers of palisade parenchyma cells. Sun leaves have higher rates of photosynthesis. Such leaves produce more sugars that can be used to build ("fatten") leaf tissue layers.

Shade leaves are often thinner and have a larger surface area. Less light results in less food (thinner) for tissue construction. Leaf surface area is maximal to trap as much of the low levels of light as possible. Shade leaf chloroplasts often have increased granal development that increases the lamellar membrane surface to maximize light interception. Alignment of chloroplasts in shade leaves will often be at the upper cell surfaces, with their long axes at right angles to entering light (maximal area for light interception). In sun leaves the chloroplasts exhibit parallel alignment of their long axes with the vertical cell walls.

This knowledge was useful when horticulturists began to produce tropical trees for use as indoor foliage plants in homes, offices, malls, and other interiorscapes. Growing such plants in field culture under full sun produced spectacular trees. However, these plants fared poorly once placed in interior environments. Most of the leaves fell off, causing consumer complaints. The problem resulted from the cultural conditions. Field-grown plants produced sun leaves. Instead shade leaves were needed, given the relatively low light

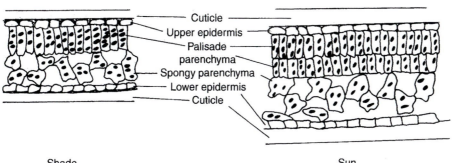

Cuticle
Upper epidermis
Palisade parenchyma
Spongy parenchyma
Lower epidermis
Cuticle

Shade leaf

Sun leaf

FIGURE 6–8 Cross sections through shade and sun leaves. (From Raymond P. Poincelot, *Horticulture: Principles & Practical Applications,* (1980). Reprinted by permission of Prentice-Hall, Upper Saddle River, NJ.)

levels experienced in interior areas. The defoliated plant, if left long enough, would have grown shade leaves to replace the lost sun leaves. The solution was to grow these field plants under shade cloth, such that shade leaves were produced. The plant was then ideally suited for interior use.

Other processes are associated with light intensity. Leaf movements also occur that can expose more or less surface area to light, depending on its intensity. Increasing light intensity also activates the development of pigments, such as the anthocyanins found in leaves and stems. A good example of this process is seen in the well-known house plant, the jade tree (*Crassula argentea*). This plant has green leaves when grown in an eastern or western window, but develops a red blush on leaves when moved to a southern window. Shade-grown leaves may appear less colorful than their sun-grown counterparts.

Effects of Spectral Quality and Duration on Physiology and Development

The effectiveness of light on plant development and physiology varies with the emitted wavelength or *spectral quality*. A number of pigments in plants absorb wavelengths from specific parts of the light spectrum. These pigments include those involved in photosynthesis and photomorphogenesis (the regulation of plant development by light), all of which are discussed in more detail later. The photosynthetic pigment, chlorophyll, absorbs light in the red and blue spectral region. Phytochrome, the carotenoids, and flavins are pigments involved in photomorphogenesis; they absorb blue, red, and far-red wavelengths.

Spectral quality is critical when using artificial light to grow plants. Any light source must have the appropriate spectral quality to initiate and sustain photosynthesis. Chlorophyll absorbs the red (600 to 700 nm) and blue (400 to 500 nm) wavelengths, and lamps designed for plant growth must emit these wavelengths. The application and types of available lamps will be covered in Chapter 13.

Light initiates reactions that control plant development. Light-controlled developmental processes are known as *photomorphogenesis reactions*. These events generally require lower light intensities for initiation than do the photoprocesses of photosynthesis and photorespiration. A very few photomorphogenesis reactions do require high light intensities. Spectral quality is the more important parameter for photomorphogenesis.

Processes in photomorphogenesis have wavelength dependencies on blue (450 nm), red (660 nm), and far-red (730 nm) light. The pigments involved include the carotenoids, flavins, and phytochrome. The photomorphogenesis functions of the carotenoids include the regulation of phototropism in seedlings. Flavins may also be involved in phototropic curvature and in the transfer of light energy from blue to red wavelengths that would mean an indirect participation in the phytochrome functions. Phytochrome, activated by red and far-red wavelengths, is involved in numerous photoresponses. Some of these are shown in Table 6–1. Other factors besides phytochrome are probably involved in photomorphogenesis. These include phytohormones, enzymic alterations and protein syntheses brought about by repression or derepression of genes, and alterations of membrane properties.

Photoperiodism.   The duration of light-hours in relationship to dark-hours in a 24-hour day is called the *photoperiod*. As the seasons change, so does the photoperiod. Many flowering plants respond to these seasonal changes in the photoperiod; this phenomenon is termed *photoperiodism*. These responses include the change from vegetative to flowering development, leaf drop and autumn coloration, tuber formation, grass tillering, and the onset of winter dormancy. Photoperiodism is one of many phytochrome-mediated events of importance to the horticulturist, because of its relation to flowering. The flowering behavior of plants in response to photoperiod varies. Photoperiodic needs for flower production may be sufficient alone, may not be needed at all, or may be needed in addition to vernalization, or can be overidden by exposure to certain temperatures.

TABLE 6–1   *Some Phytochrome-Mediated Photoresponses*

Seedling Development
  Seed germination
  Hypocotyl hook unfolding
  Enlargement of cotyledons
  Hair formation along cotyledons
  Seed respiration
Leaf Development
  Formation of leaf primordia
  Differentiation of primary leaves
  Formation of tracheary elements
  Elongation of leaf and petiole
  Unfolding of monocot leaves
  Differentiation of stomata
  Plastid formation
  Leaf abscission
  Epinasty
Root Development
  Root primordia initiation
Reproductive Development
  Flower induction
  Sex expression
  Photoperiodism

Shoot Development
  Incorporation of sucrose into growing buds
  Elongation of stem
  Bud dormancy
  Internode extension
  Rhizome formation
  Bulb formation
  Succulency
Biochemistry/Physiology
  Synthesis of anthocyanin
  Increase in protein synthesis
  Increase in RNA synthesis
  Formation of phenylalanine deaminase
  Changes in the rate of fat degradation
  Auxin catabolism
  Changes in rate of cell respiration
  Membrane permeability
  Lipoxygenase metabolism
  Degradation of reserve protein
  Geotropic sensitivity
  Phototropic sensitivity
  Electrical potentials

Photoperiodism and Flowering.   The first understanding of the effects of photoperiodism came about in 1920. Scientists in the U.S. Department of Agriculture discovered the process while working with tobacco. Essentially, some plant's flowering behavior was determined by changes in the photoperiod, that is, hours of light versus hours of darkness in each day. At that time it was assumed the critical factor was hours of light.

The classification of flowering behavior in response to photoperiod does not technically depend on day length, but on a factor called the *critical night length* (Figure 6–9). However, because this determination came later in the 1930s, the original photoperiodic group names, based on day length, were retained to avoid confusion. The night duration of the photoperiod required to induce flowering is called the critical night length. The actual value has to be determined experimentally for each species.

Plants that flower during nights longer than the critical night length are *short-day plants* (actually long-night plants), and those that flower with nights shorter than the critical night length are *long-day plants* (actually short-night plants). Those plants unaffected by photoperiod in terms of flowering are called *day-length neutral* or *indeterminate plants*. Plants that require a night longer than the critical night length followed by one of less than the critical night length are *short-long-day plants;* plants requiring the reverse are *long-short-day plants*. Long-day plants usually bloom in spring to summer, short-long-day plants in late spring to early summer, long-short-day plants in late summer to early fall, and short-day plants in the fall.

These groups can be categorized further. Those plants with an absolute dependence on day length for flowering are termed *obligate, absolute,* or *qualitative,* whereas those that appear to flower in various photoperiods, but one photoperiod versus another induces flowering either earlier or later, are called *facultative* or *quantitative*. Some examples are shown in Table 6–2.

Photoperiodism and Temperature Effects.   Photoperiodic response may be altered by temperature (Table 6–3). Some plants may be short- or long-day plants at some specific

FIGURE 6–9
(A) Shown here is the relationship between critical night length, photoperiod and flowering response. Flowering response can be altered by changes in the photoperiod brought about by natural means (seasonal changes) or artificial means. The latter includes darkening plants with black cloth to lengthen the night period or illumination to shorten the night period. (From Raymond P. Poincelot, *Horticulture: Principles & Practical Applications,* (1980). Reprinted by permission of Prentice-Hall, Upper Saddle River, NJ.)
(B) The response of *Celosia* to variations in the photoperiod is shown here. Not only is flowering affected, but also so is stem elongation. (U.S. Department of Agriculture photograph.)

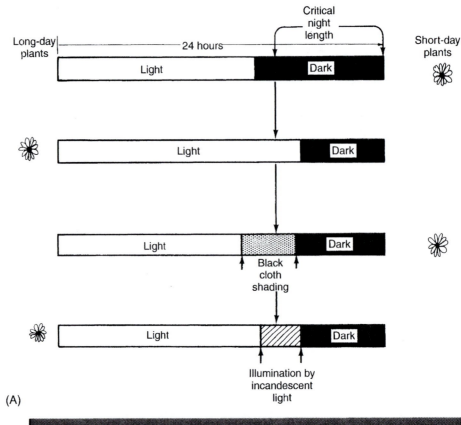

(A)

(B)

temperature range, but day-length-neutral plants below that range. An excellent example is the Christmas cactus, which requires short days to flower and blooms somewhere around Christmas time as a house plant. However, because of lights being turned on and off in houses, the critical night length becomes interrupted. One way to produce flowers is to place them in a darkened area where lights remain off, such as an attic. However, in some cases, the cacti flowered at Christmas time even when lights were turned on. It turns out that exposure of the plant to temperatures ranging from 45° to 55°F for 3 to 4 weeks will cause flower buds to develop. So those cacti left in cold attics, even with the lights on, would flower. The trick then is to leave the Christmas cactus outside for about a month in the fall on your back porch or in your garage and let the cool night temperatures bring about flower buds.

TABLE 6–2 *Photoperiodic Response of Some Horticultural Crops*

Response Unaffected by Temperature

Qualitative Short-Day	Qualitative Long-Day	Quantitative Short-Day
Amaranthus caudatus (lovelies bleeding)	*Anethum graveolens* (dill)	*Capsicum frutescens** (pepper-tabasco)
Cattleya trianaei (orchid)	*Chrysanthemum maximum*	*Chrysanthemum X morifolium**
Chrysanthemum indicum	*Dianthus superbus* (lilac pink)	*Cosmos bipinnatus*
*Chrysanthemum X morifolium**	*Fuchsia**	*Glycine max* (soybean)*
Coffea arabica	*Hibiscus syriacus* (shrub althea)	*Senecio cruentus* (cineraria)
*Cosmos sulphureusa**	*Mentha X piperita* (peppermint)	*Solanum tuberosum* (potato)*
Ipomoea batatus (sweet potato)	*Nicotiana sylvestris* (flowering tobacco)	*Zinnia*
Kalanchoe blossfeldiana	*Phlox paniculata*	**Quantitative Long-Day**
K. pinnata	*Rhaphanus sativus* (radish)	*Brassica rapa;* rapifera group (turnip)
*Phaseolus vulgaris** (kidney bean)	*Rudbeckia hirta* (black-eyed susan)	*Camellia japonica*
Zea mays (corn)*	*Scabiosa ucranica*	*Dianthus barbatus* (sweet william)
Long-Short-Day	*Sedum spectabile* (showy stonecrop)	*Nigella damascena* (love-in-a-mist)
Cestrum aurantiacum	*S. telephium*	*Solanum tuberosum* (potato)*
C. diurnum	*Spinacia oleracea* (spinach)	
Kalanchoe tubiflora		
Short-Long-Day		
Echeveria harmssi		
Symphyandra hoffmannii		

*Does not include all cultivars.

TABLE 6–3 *Photoperiodic Response of Some Horticultural Crops**

Response Affected by Temperature

Day-Neutral	Qualitative Short-Day	Qualitative Long-Day
Browallia speciosa 'Major'	*Chrysanthemum morifolium*[†]	*Beta vulgaris* (beet)
Calendula officinalis	*Cosmos sulphureus*[†]	*Brassica rapa* pekinensis group (Chinese cabbage)
Cucumis sativus (cucumber)	*Euphorbia pulcherrima* (poinsettia)	*Cichorium intybus* (common chicory)
Fragaria vesca (woodland strawberry)	*Fragaria X ananassa* (garden strawberry)	*Delphinium X cultorum*[†]
Cardenia jasminoides	*Ipomoea purpurea*[†] (morning glory)	*Dianthus gratianopolitanus* (cheddar pink)
Helianthus annuus[†]	*Salvia splendens*[†]	**Quantitative Long-Day**
Ilex aquifolium (English holly)	**Quantitative Short-Day**	*Antirrhinum majus* (snapdragon)
Lunaria annua (honesty, annual strain)	*Allium cepa* (onion)[†]	*Begonia semperflorens*
Lycopersicon lycopersicum (tomato)[†]	*Chrysanthemum morifolium*[†]	*Campanula persicifolia* (willow bellflower)
Pisum sativuma (pea)	*Schlumbergera truncata*	*Centaurea cyanus* (cornflower)
Apium graveolens (celery)		*Dianthus caryophyllus*[†]
Daucus carota (carrot)		*Digitalis purpurea* (foxglove)
Geum bulgaricum		*Lactuca sativa* (lettuce)[†]
Vicia faba (broad bean)		*Matthiola incana* (stock)
Short-Long-Day		*Petunia X hybrida*
Poa pratensis (Kentucky bluegrass)		
Campanula medium (Canterbury bells)		

*For classification of photoperiodic response changes induced by temperatures above or below the normal range causing the above listed responses, see Vince-Prue, D. (1975). *Photoperiodism in Plants*. McGraw-Hill, New York.

[†]Does not include all cultivars.

Other Photoperiodic Effects. The photoperiodic effect and temperature relationship are not restricted to flowering. Germination of some seeds can be enhanced or inhibited by long days, but germination of these is also promoted by low temperatures. The formation of dahlia tubers, gladiolus cormels, and potato tubers are promoted by short days, whereas the formation of onion bulbs is a long-day phenomenon. Rooting of cuttings can often be influenced by photoperiod, but other environmental factors complicate the situation, so generalizations are not possible.

Soil

The roots of a plant receive nutrients, water, oxygen, and mechanical support from the soil. The capacity of the soil to supply water is a function of its physical properties, and its capacity to supply nutrients is dependent on chemical properties. These combined properties regulate root growth. The maintenance of soil quality and health in a sustainable manner is critical for crop production and to minimize soil losses from erosion. These aspects are examined next in more detail.

Soil Formation

Soil consists of two basic components: inorganic and organic. The inorganic part consists of various sizes of particles formed from the weathering of rocks and minerals. These particles are the result of long-term processes caused by climate and biological life forces. For example, when water freezes in the pores and cracks of rocks, it expands, often producing enough pressure to split rock. Constant cycles of freeze–thaw action over time reduce the rock to fine particles. Early colonists used this freeze–thaw process to quarry rocks. By drilling holes and adding water, the freezing and thawing cycles that occurred over winter cleaved the rock into manageable blocks. Temperature extremes of heat and cold over time can also cause rocks to fracture, especially when mixed materials are present. One material may expand and contract at a different rate, leading to internal pressures that gradually split the rock. Water and chemicals in it, such as acids and oxygen, along with lichens and plant roots over long times cause decomposition of rocks into fragments.

The other part of soil, organic matter, consists of a dynamic mixture of living, dead, and decomposing life-forms (Table 6–4). Numerous living organisms reside in the soil, including bacteria, fungi, algae, protozoans, plants, worms, insects, mites, and burrowing animals. These living organisms die, decompose, and add to the organic matter. Their excretory wastes and other by-products also add to the mix. The end product of decomposing organic matter is a reasonably stable product called *humus*. The use of pesticides often alters and even destroys the organisms that inhabit the soil ecosystem. From Table 6–4, we can see that the loss of some or all life-forms can dramatically alter, usually in a detrimental manner, the processes that benefit both soil and plant health. Sustainable agriculture seeks to preserve soil health and life through minimal to no pesticide usage.

The time to produce soil from these materials varies. It can take anywhere from 220 to 1,000 years to convert bedrock into 1 inch of topsoil, the richest uppermost soil layer. The average is roughly 500 years. Normally, this slow process is not a problem in natural areas, such as forests and meadows. A rough balance exists between soil formation and soil loss.

Soil Erosion

Soil is lost through a process called *erosion*. Soil can be carried away by wind and water, such as during a windstorm or heavy rains. Most soil erosion in the United States is caused

TABLE 6–4 *Life-Forms Found in Soils*

Groups	Activities Affecting Plants and Soils
Bacteria (actinomycetes, cyanobacteria, free-living bacteria, mycoplasms, symbiotic bacteria)	Conversion of nitrogen, sulfur, and iron into forms absorbed by roots; decomposition of organic matter and maintenance of humus levels; nutrient recycling of macro-nutrients and trace elements; fixation of atmospheric nitrogen into plant useful form (some are symbiotic in roots, others are free living); contributions to global carbon and nitrogen cycles; decomposition of pesticides; building water-stable, soil structure; pathogens cause plant disease (some); source of organic matter in death.
Viruses	Plant diseases.
Fungi (free-living and symbiotic forms)	Decomposition of organic matter and maintenance of humus levels, nutrient recycling of macronutrients and trace elements, improvement of nutrient and water uptake by roots (mycorrhizae), cause plant disease (some), nematode control (some), source of organic matter in death.
Algae (diatoms, green algae, yellow-green algae)	Source of organic matter in death.
Protozoans	Keep populations of bacteria, fungi, and nematodes from getting out of control, source of organic matter in death.
Nematodes	Organic matter recycling, keep populations of bacteria, fungi, and algae (and even other nematodes) from getting out of control, plant disease (some), source of organic matter in death.
Earthworms	Aeration of soil, recycling of organic matter, maintenance of humus and nutrient levels, increased nutrient availability, source of organic matter in death.
Snails and slugs	Recycling of decaying vegetative matter, plant damage by eating, source of organic matter in death.
Arthropods (centipedes, grubs, millipedes, mites, soil insects, sowbugs, springtails)	Recycling of decaying vegetative matter, soil aeration, plant damage by eating, source of organic matter in death.
Burrowing animals (armadillos, badgers, chipmunks, gophers, groundhogs, mice, moles, prairie dogs, rabbits, rats, shrews, snakes, voles)	Soil aeration, increased nutrients (from waste products), plant damage by eating, source of organic matter in death.

by water. Scientists recognize three types of water-induced soil erosion: sheet, rill, and gully. Water moving in a wide flow down a slope or across a field peels off soil in a thin layer or sheet, leading to sheet erosion. This form is not noticeable over the short term, because of the uniform, thin layer of topsoil that disappears. Fast-moving, narrow bands of water cutting small channels into the soil cause rill erosion. When streams are large enough, wider and deeper channels result, causing gully erosion. Rivers and other natural, moving bodies of water can also erode soil. Estimates place the loss of productive cropland in the United States from our founding to date at about 25%. U.S. soil is currently eroding 16 times faster than it forms.

When soils have active, growing plant communities, erosion is minimal, because the plant cover protects the soil from wind and water erosion. Exposed soils are usually found in crop production systems that currently prevail in horticultural practices. Bare soil is found between rows. Rows were designed for the passage of farm machinery and are kept free of competitive weeds, both for reasons of crop productivity and ease of machinery passage. Fields are often left bare in the off-season. Exposed soil is at much greater risk for erosion. Of course, the use of artificial and living mulches along with winter cover crops can greatly reduce the erosion problem.

Another complicating factor is the loss of organic matter. Organic matter is part of the "glue" that holds soils together. Our current system of production tends to shortchange the maintenance of soil organic matter. Prior to the Civil War, the source of fertilizer was manure. Manure not only added nutrients, it also added fresh organic matter. After the Civil War, farmers increasingly relied on chemically manufactured superphosphate and imported Chilean nitrate. In the early 1900s manufactured nitrogen fertilizers started to appear. Farmers rapidly adopted chemical fertilizers for many reasons: ease of handling, no unpleasant smell, and higher nutrient levels. Farmers also abandoned green manure, cover crops, and crop rotation, valuable practices for building organic matter, fertility, and reduced insect and pathogen levels. Exceptions to this change included organic farmers who continued to farm in a more sustainable manner.

This widespread changeover in fertilization practices meant less organic matter was added back to soils. Organic matter tends to be gradually lost in cultivation, because it moves upward through soil turned by plows and cultivators. Although stable in the soil, it does oxidize and break down when exposed to oxygen in the upper layers of topsoil. As organic matter decreases, the aggregates tend to come apart. Wind and rain can then more easily blow or wash away individual soil particles. The smaller and hence lighter particles, clay and silt, are gradually removed. The sand, being heaviest, is left behind; ultimately the topsoil is turned into desert (desertification). A further complication is that erosion not only destroys valuable soil, it also causes water pollution. The chemicals in the soil, such as fertilizers and pesticides, are carried along in the water and on soil particles. Some of these end up in water bodies, leading to pollution by soil and agricultural chemicals.

Soil erosion by water is also accelerated by decreases in infiltration and percolation processes. Surface litter improves infiltration. Soils kept clean of litter because it may harbor insects and pathogens have less water infiltration. Water that does not enter the soil tends to flow over the surface, seeking the lower lying areas. This surface movement of water is referred to as *runoff*. In many areas, runoff is the primary source of lost soil and chemical pollutants found in surface water bodies near agricultural areas. Of course, water percolating through the soil can also contribute to groundwater pollution by farm chemicals.

Runoff can also be increased by soil compaction. Heavy farm machinery causes soil compaction. Compaction produces a densely packed layer of soil with few pores. This layer of soil, called *hardpan*, interferes with water percolation, causing water to puddle and increasing surface runoff. Sometimes, even with high-quality soils, runoff occurs. Long periods of heavy rain lead to water-saturated soils, such that further rain cannot infiltrate/percolate, resulting in runoff. A very heavy rain in a short period can also produce runoff, as rain falls faster than the soil can absorb it. The consequences of soil erosion for the long-term outlook in crop production were covered in Chapter 1.

Physical Properties and Composition

Because soil contains both inorganic and organic materials of different sizes and shapes, a number of measurable physical properties are associated with it. You can measure composition and packing arrangements for the various components, including air volume. Soil texture refers to composition, soil structure to packing, and pores to spaces that hold air and water.

Soil Texture.   The inorganic portion of soil consists of particles called sand, silt, and clay. These categories are based simply on size. Sand (very fine to very coarse) ranges in size from 0.05 to 2.0 millimeters (mm), silt from 0.002 to 0.05 mm, and clay is less than 0.002 mm. Larger particles are sand, medium silt, and fine ones clay. The varying percentages of these particles determine *soil texture*.

Soil textures are named according to the component present in the greatest amount. For example, soils where clay, sand, or silt are the dominant components are termed clays, silts, and sands. A desert would have a soil texture called sands. Soils with an approximately equal amount of clay, sand, and silt have a soil texture called loam. Loams are the best soils for growing plants. Loams are also the best soils for heavy construction, hence the conversion of farmland into housing projects and malls.

Loams with a little too much sand have a soil texture called sandy loam. The soils of New England are primarily sandy loams. One driving force for the westward movement of colonists was the search for better soils than the sandy loams of New England. The majority of citizens were farmers in the early days of our country. The Midwest turned out to be rich in loams. Likewise, too much clay produces clay loams such as those of the Southeast, and too much silt, silt loams. Other soil textures are also known, such as silty clays and sandy clays. Some 20 different soil textures are recognized (Figure 6–10).

Aggregates and Soil Structure. Humus, soil microorganisms, and their secretions act like "glue" to hold soil particles together. Soil particles, both inorganic and organic, can be stuck together in varying degrees into larger, secondary particles termed aggregates.

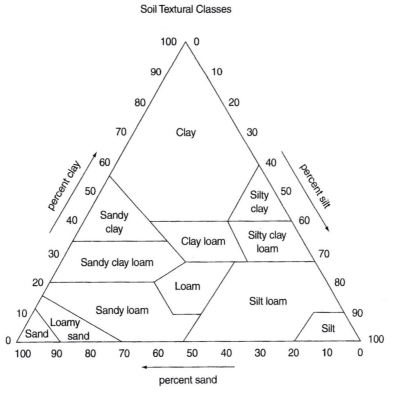

Soil Textural Classes

FIGURE 6–10 Chart showing the percentage of clay (below 0.002 mm d.), silt (0.002 to 0.05 mm d.) and sand (0.05 to 2.0 mm d.) in the basic soil textural classes. (From Donald Steila, *The Geography of Soils: Formation, Distribution, and Management,* © 1976, pp. 15–16. Adapted by permission of Prentice-Hall, Inc., Upper Saddle River, NJ.)

Sands. Soil material that contains 85% or more of sand; the percentage of silt plus 1.5 times the percentage of clay shall not exceed 15%. Sands are rated on sand particle size and percentages. Sand types from largest to smallest particles include coarse sands, sand, fine sand, and very fine sand.

Sandy loams. Soil material that contains either 20% clay or less, and the percentage of silt plus twice the percentage of clay exceeds 30%, and 52% or more sand: or less than 7% clay, less than 50% silt, and between 43 and 52% sand. Variants based on sand particle size and percentages include coarse sandy loam, sandy loam, fine sandy loam, and very fine sandy loam.

Loamy sands. Soil material that contains at the upper limit 85 to 90% sand, and the percentage of silt plus 1.5 times the percentage of clay is not less than 15%. At the lower limit it contains not less than 70 to 85% sand, and the percentage of silt plus twice the percentage of clay does not exceed 30%. Variants based on sand particle size and percentages include loamy coarse sand, loamy sand, loamy fine sand, and loamy very fine sand.

Loam. Soil material that contains 7 to 27% clay, 28 to 50% or more silt and less than 52% sand. Variants based on the percentages of sand, silt, and clay include silt loam, silt, sandy clay loam, clay loam, silty clay loam, sandy clay, silty clay, and clay.

FIGURE 6–11 Various types and classes of soil structure. (From Donald Steila, *The Geography of Soils: Formation, Distribution, and Management,* © 1976, p. 17. Adapted by permission of Prentice-Hall, Inc., Upper Saddle River, NJ.)

Soil structure: Types and classes

Type of Ped	Platelike	Prismlike	Blocklike	
	Platy	Prismatic or columnar	Blocky: Angular or subangular	Spheroidal: Granular or crumb

Ped size classes		Platelike	Prismlike	Blocklike	
	Very fine or Very thin	1 mm	10 mm	5 mm	1 mm
	Fine or thin	1–2 mm	10–20 mm	5–10 mm	1–2 mm
	Medium	2–5 mm	20–50 mm	10–20 mm	2–5 mm
	Coarse or thick	5–10 mm	50–100 mm	20–50 mm	5–10 mm
	Very coarse or very thick	10 mm	100 mm	50 mm	10 mm

The extent of aggregation varies with many factors: organic matter content, chemical nature and amount of clay, nature of soil microorganisms, cultivation, temperature, and rainfall. *Soil structure* refers to the packing arrangement of these aggregates and individual particles.

Peds are natural aggregates that are reasonably water stable; *clods* are artificial aggregates that are not water stable. Peds are categorized (Figure 6–11) according to type (shape and arrangement), class (size), and grade (degree of distinctness). Soil structure influences water infiltration, water-holding capacity, and aeration. Peds are the best soil structure for growing plants. Clods are found in poorly or improperly managed soils.

Soil Pores. Spaces or *pores* are found between soil particles and aggregates. These pores comprise 30% to 60% of the soil volume. Pores result from the packing of irregularly shaped particles and aggregates. Micropores are small pores generally filled with water. Macropores are large pores filled with air alone or with air and water. Larger particles, like those found in sandy soils, produce mostly macropores. Soils high in sand, and hence macropores, have good aeration and rapid water absorption and they warm up fast in the spring. The downside is that such soils have poor water retention. Smaller particles like clay (and organic matter/humus) are micropore rich. Clay soils contain mostly micropores and have poor aeration and drainage, but high water retention. They are slower to warm up in the spring. Clay (and humus) also has the ability to hold and release certain nutrients in the soil. This property is called cation exchange and will be discussed later under chemical properties.

Soil atmosphere is also very important. Roots respire and need oxygen. Root activities and soil animal life deplete the oxygen and raise the concentration of carbon dioxide in the soil. Inadequate macropore space results in poor diffusion of oxygen into and carbon dioxide out of the soil. On the other hand, insufficient numbers of micropores mean very little capillary water is available for plant roots. Both low levels of oxygen and water cause plants to grow poorly or die. For these reasons, the best soil structure contains a good balance of both macropores and micropores, such as is found in loams.

Organic Matter. Soils also contain organic matter in addition to the inorganic materials already discussed. *Organic matter* includes all the living (Table 6–4) and dead matter associated with the soil. The nondegraded twigs, leaves, or stalks on the surface are referred to as *litter.* Litter increases water infiltration by providing more entry points into the soil. It also reduces the impact of rain. The resulting reduction of surface runoff and erosion means more water for soil entry. *Humus* includes the stabilized fraction of organic matter left after enzymic decomposition by microorganisms of the plant and animal residues. Humus is rich in micropores. It acts like a sponge to hold water. Humus, like clay, also regulates the availability of certain nutrients in the soil. Humus improves the cation exchange capacity of the soil. Under most conditions, the capacity of humus for cation exchange exceeds that of clay.

Organic matter content is desirable for horticultural purposes. Current crop production practices deplete the reserves of organic matter, and it becomes necessary to replace it by the techniques discussed in Chapter 11. Organic matter added beyond the naturally present amounts can be used to improve certain soils. Addition of organic matter to a sandy soil (mineral soil with at least 70% sand; particle size varies between 0.05 and 2.0 mm, making it coarse textured) improves water retention, mainly because of micropores in the organic matter. An incidental improvement to that of water retention is an improvement of soil structure. The improvement of soil structure in fine-textured soils (particle size of less than 0.05 mm), consisting mostly of clay or silt-clay (40% or more), is the main reason for adding organic matter to these soils. The workability or *tilth* of these fine-textured soils is improved by the resulting soil structure improvement, as is aeration and water movement.

In all soils the decomposition of organic matter contributes to the nutrients and trace elements needed by plants. Organic matter also acts in a buffering capacity against rapid chemical changes when lime or fertilizers are added. The humic fraction is involved in cation exchange (see chemical properties next). Decomposing organic matter releases chemicals that free some of the normally unavailable phosphorus. Decomposing organic matter also promotes numerous microorganisms, many of which are beneficial for pest and disease reduction. The decomposition products of organic matter, the secretions of microorganisms involved in decay, and the hyphae of fungal decomposers all contribute to the formation of good aggregates and ultimately good soil structure. A more detailed explanation of the horticultural value of organic matter is given in Chapter 11.

Chemical Properties

The chemically active portion of soil consists mostly of humus and clay particles with diameters of less than 0.002 mm. Particles of this size are called *colloids.* The structure of colloidal clay is crystalline, and humus is amorphous. Two chemical properties are active in soil:cation exchange and pH.

Cation Exchange. Ions carry either negative (anions) or positive (cations) charges. Examples are NO_3^- and Ca^{2+}, respectively. Most anions are in solution, that is, dissolved in water found in the soil pores. Under certain conditions small amounts of anions can be held in an exchangeable form. These include nitrate and sulfate ions. Phosphate is an exception, in that it is firmly held chemically with little being present in soil solution. As water percolates downward through the soil, some of the dissolved anions are carried to deeper soil layers. This process is called *leaching.* Cations are mostly held or adsorbed by two soil components. Colloidal clay or humus is negatively charged and attracts and holds positively charged cations. As such, cations are much less susceptible to leaching than anions. The retention of cations by clay and humus helps to reduce the loss of some nutrients through leaching.

Plant roots can only absorb ions that are free in solution, such as anions. Cations must first be released into the soil water from their retention sites before they can be absorbed. The release process involves the exchange of the adsorbed cation by another cation by a process called *cation exchange* (Figure 6–12). Cation exchange can take place between cations adsorbed on colloidal particles and cations in soil solution, cations released by plant roots (usually H^+), and cations on other colloidal particles. Cations are not randomly exchanged, but

FIGURE 6–12 Various types of cation exchange that can occur in soil solution. (A) Cation exchange between soil particles and soil solution caused by cations released by roots or indirectly by cations produced from CO_2 released by roots. If the particle is in contact with the root, direct exchange (contact exchange) in theory could occur with neither cation going into soil solution. (B) Cation exchange between cations in soil solution and those adsorbed on soil particles. (C) Cation exchange between soil particles. (From Raymond P. Poincelot, *Horticulture: Principles & Practical Applications,* (1980). Reprinted by permission of Prentice-Hall, Upper Saddle River, NJ.)

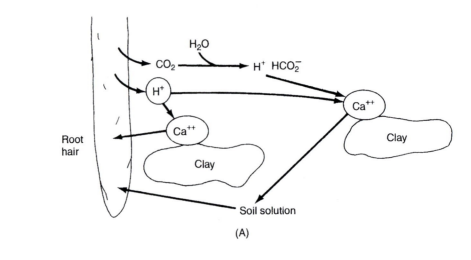

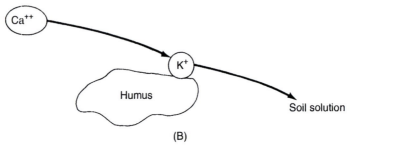

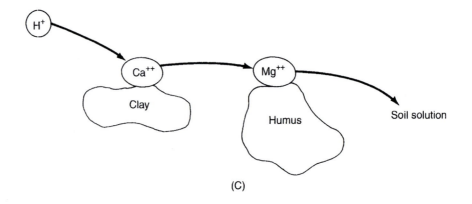

are displaced by another cation with a higher absorptive affinity for that site. The displacement strengths are in decreasing order: H^+, Ca^{2+}, Mg^{2+}, K^+, and Na^+.

Uptake of cations by roots appears to be both by direct exchange of root-released H^+ for another cation bound on an adjacent soil particle (contact exchange) without the cations entering the soil solution and by H^+ from the root exchanged for a cation that enters the soil solution and is then taken up. The H^+ may either be directly released from the root or formed indirectly from respiratory CO_2. The predominant mode depends on several soil variables.

Soils, of course, vary in their cation exchange capacity, depending on the amounts and types of clay and amounts of humus present. The cation exchange capacity of the colloidal particles in increasing order is kaolinite, illite, montmorillonite (all clays), and humus. For those low levels of the few anions that are adsorbed, organic matter, kaolinite, and iron can increase anion exchange. Anions for the most part exist free in the soil solution and are taken up by roots directly from the soil solution.

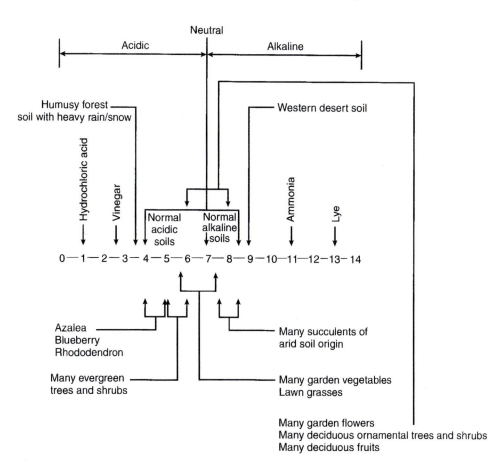

FIGURE 6–13 The pH scale showing acceptable pH ranges for certain horticultural crops, natural pH of certain soils and the pH of some common chemicals for purposes of comparison. (From Raymond P. Poincelot, *Horticulture: Principles & Practical Applications*, (1980). Reprinted by permission of Prentice-Hall, Upper Saddle River, NJ.)

Soil pH. The term *soil reaction* or *soil pH* refers to the degree of acidity or alkalinity (basicity) present in the soil. At neutral pH the number of H^+ and OH^- ions is equal, such as in freshly distilled water. If the amount of H^+ ions becomes increasingly larger, the solution increases in acidity. If the OH^- ions increase instead, the solution increases in alkalinity. The actual expression is $pH = \log 1/(H^+)$, where (H^+) is the active concentration of H^+ ions in grams per liter. The pH scale runs from 0 to 14, and soil pH generally runs from 4 to 8. Extremes include the acid forest humus layers in the Northeast that can be as low as pH 3.5, and the alkaline desert areas in the West may reach pH 9. In general, soils are acidic in areas of high rainfall in the eastern half of the United States and the Pacific Northwest. Dry areas are usually neutral to alkaline, such as the arid western states. The pH scale is shown in Figure 6–13 for both horticultural and other examples.

Plants have an optimal pH range for development. Higher or lower values result in less than optimal development. The explanation is that soil pH regulates the availability of nutrients. Native plants have coevolved with soil pH, so we find acid-loving plants (blueberry, cranberry, and pines) in the New England soils and alkaline-loving plants (cacti and sagebrush) in the alkaline soils of the Southwest deserts.

Crop and many landscape plants pose a pH problem in that most originally came from soils between the acidic and basic extremes associated with many native U.S. plants. Situations occur where cultivated plants are mixed in with native plants in landscaped areas, and soil pH needs to be adjusted specifically for each plant group. For example, many evergreens do well at pH 5.0, but the vegetable garden does better at pH 6.5. Lime is used to raise pH (make it less acidic) and sulfur is used to reduce pH (make it more acidic).

Buffers enable solutions to resist changes in pH. Soils have some degree of buffering capacity that increases with rising clay and humus contents. Clays vary in their buffering capacity. Alteration of pH by addition of lime or sulfur would require larger amounts in a soil with increased buffering capacity.

Plants and Soil Nutrients

Essential elements needed by plants and their uptake by roots were discussed in Chapter 4 under nutrient absorption. Uptake can only occur if the nutrients are present in available forms (ions). In addition, optimal plant development is dependent on sufficient levels. Nutrient forms and levels are influenced by solubility properties, soil texture, organic matter content, cation exchange capacity, and pH. Some of these have been discussed previously. The effect of pH on nutrient availability is shown in Figure 6–14.

General-purpose fertilizers add nitrogen, phosphorus, and potassium. These nutrients are required in large amounts by plants. Normally these nutrients are recycled back to the soil by microbial processes after the plant dies. However, the harvesting of crops removes these nutrients from the soil, thus necessitating the need for fertilizers. Calcium and magnesium, when deficient, are often added as limestone. Regular limestone contains calcium,

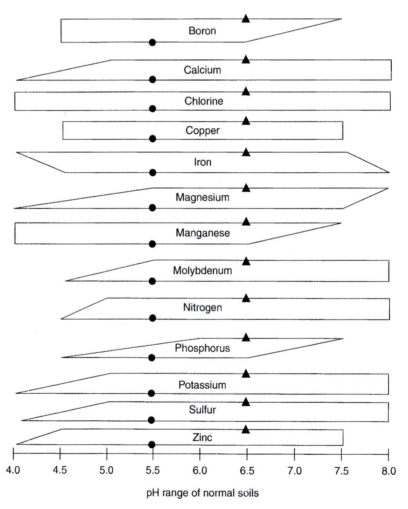

FIGURE 6–14 Nutrient availability in the pH range of most normal soils. The top of the bar line is for mineral soils (less than 20% organic matter) and the bottom bar line is for organic soils (greater than 20% organic matter such as peat and muck soils). The optimal nutrient availability in mineral soils occurs around pH 6.5 (▲) and around 5.5 in organic soils (•). Nutrient availability does not stop at the indicated cutoff points, but is substantially below (at least 40% less) that which occurs at the maximum. The optimal pH does not in itself guarantee sufficient nutrient availability, since the actual concentration of nutrient present as well as the plant's requirement for that nutrient must be considered. Some plants develop well at a pH where others fail, simply because their requirement for a nutrient limited by solubility at that pH is greatly reduced compared to the other plant. (From Raymond P. Poincelot, *Horticulture: Principles & Practical Applications,* (1980). Reprinted by permission of Prentice-Hall, Upper Saddle River, NJ.)

and dolomitic limestone contains both calcium and magnesium. Sulfur, when needed, is added as powdered sulfur. Other nutrients are required in small or trace amounts. As such, applications to the soil are seldom required.

Nitrogen. Nitrogen is a constituent of proteins, amino acids, purines, pyrimidines, chlorophyll, and many coenzymes. As such, the vegetative growth of plants (leaves, stems, and roots) is especially dependent on nitrogen. For example, lush green lawns require the addition of nitrogen. The atmosphere contains 78% nitrogen by volume, yet it is the element that is most often lacking for plant development. This lack happens because plants cannot absorb nitrogen directly as a gas from the atmosphere. Plants in the legume group are an exception, since they possess nitrogen-fixing bacteria on their roots. Nitrogen must exist in the soil in a form suitable for nutrient absorption. These forms are NH_4^+ and NO_3^-. Reactions such as these are part of the nitrogen cycle shown in Figure 6–15.

The conversion of atmospheric nitrogen into NO_3^- and NH_4^+ in the soil is called *nitrogen fixation*. Some nitrogen fixation occurs through the incorporation of atmospheric NO_3^- and NH_4^+ into the soil by rainfall and snow. However, some nitrogen fixation results from bacterial action. Nitrogen-fixing bacteria of the genus *Rhizobium* live in nodules on the roots of plants in the family Leguminosae. Other nitrogen-fixing microorganisms include unidentified species of actinomycetes in a symbiotic relation with several nonleguminous plants of horticultural interest, such as *Alnus* sp. Through bacterial action, atmospheric nitrogen is converted into NH_4^+ that is utilized by the bacteroids and plants. Nitrogen fixation also occurs in the soil through some free-living bacteria and to a much lesser extent through some actinomycetes and fungi. These microorganisms degrade organic material into amino acids by enzymic means. On the death and decay of these microorganisms, other bacteria convert the amino acids into NH_4^+ and NO_3^-.

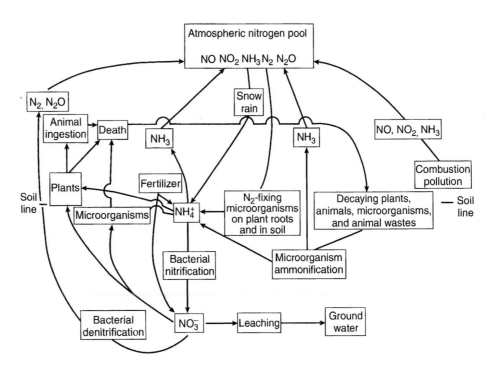

FIGURE 6–15 Nitrogen cycle. Losses are incurred by the leaching of highly soluble NO_3^-, the erosion of soil, the removal of most plant material during harvest, fires, and by certain microbial processes. The loss of nitrogen while in the ammonium form (NH_4^+) is considerably less than in the nitrate form (NO_3^-). Soil biochemistry, however, favors the formation of NO_3^- from NH_4^+. NH_3 can be lost from warm soils with an alkaline pH and from decaying organic matter with a carbon/nitrogen ratio below 30. Loss of NO_3^- through denitrification occurs in wet, anaerobic soils. (From Raymond P. Poincelot, *Horticulture: Principles & Practical Applications*, (1980). Reprinted by permission of Prentice-Hall, Upper Saddle River, NJ.)

If NH_4^+ and NO_3^- are limited in the soil, competition for these compounds takes place between plants and microorganisms. The microorganisms win, and symptoms of nitrogen deficiency appear in the plants. This occurs if the soil carbon : nitrogen ratio by weight is greater than 10 : 1, such as happens when incompletely degraded organic matter is added to the soil, or nitrogen is low and a heavy organic mulch is used. Addition of external sources of nitrogen will correct the ratio.

Nitrogen is easily lost from the soil by leaching, utilization by plants, and the conversion of nitrates into atmospheric nitrogen. The latter, a bacterial process, is favored by a lack of soil oxygen, such as with a waterlogged soil. Excessive nitrogen promotes vegetative growth at the expense of reproductive growth. Addition of nitrogen is often decreased as the reproductive cycle is nearing.

Phosphorus.   Phosphorus is found in a number of compounds in the plant, such as nucleic acids, phospholipids, ATP, and NADP. Phosphorus is important in cell division and growth, especially in areas of rapid development, such as the meristem. Phosphorus is often limiting, even though sufficient amounts are present, because it is present in unavailable forms with iron and aluminum below pH 5 and calcium above pH 7 (Figure 6–16). Losses of phosphorus by leaching are low. The availability of phosphorus is increased by organic matter that produces organic acids during decomposition that complex more readily with the iron and aluminum than phosphorus. Organic matter is also a source of phosphate released by microbial action.

Potassium.   Potassium is very important as an enzyme activator in plants. It is also involved in membrane permeability and translocation. Much soil potassium is present in minerals that dissolve slowly, thereby limiting its availability. The readily available potassium is regulated by cation exchange. Potassium leaching increases as the amounts of clay and humus decrease.

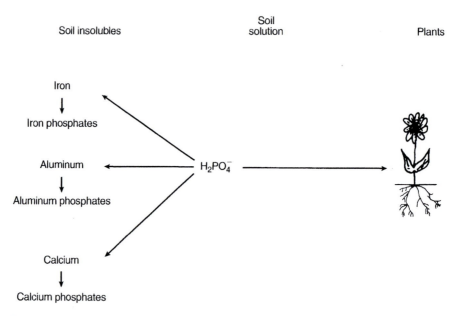

FIGURE 6–16   Soil pH limitations on the availability of phosphorus to plants. The optimal pH for formation of insoluble iron, aluminum, and calcium phosphates is pH 3, 5, and 8, respectively. Not only does acid pH tie up the phosphorus, but it also ties up another essential element: iron. Chelating agents can be used in soils to reduce the loss of iron and other trace elements such as zinc. These form a specific soluble complex with the trace metal, thus preventing loss incurred by the formation of insoluble compounds. Iron and zinc chelates are widely used in areas where deficiencies are encountered. (From Raymond P. Poincelot, *Horticulture: Principles & Practical Applications*, (1980). Reprinted by permission of Prentice-Hall, Upper Saddle River, NJ.)

Calcium. Calcium is needed for the synthesis of calcium pectate in the cell wall. It is also involved in mitosis and is present in membranes as the calcium salt of phosphatidylcholine. Calcium is reasonably soluble and exists as an exchangeable cation on clay and humus. Calcium is derived from organic matter and soil minerals.

Magnesium. Magnesium is a constituent of chlorophyll, ribosomes, and magnesium pectate. The latter is present in cell walls. It is also an enzyme activator. Magnesium is also available as an exchangeable cation, but in amounts less than calcium. It is derived both from organic matter and minerals present in the soil.

Sulfur. Sulfur is present in certain amino acids that are involved in enzymic activities and cross-bonding of proteins. The B vitamins thiamin and biotin contain sulfur. Organic matter and insoluble minerals are the principal sources of sulfur released by microbial action and weathering, respectively.

Other Nutrients. The remaining essential elements supplied in the soil are iron, copper, manganese, zinc, molybdenum, boron, chlorine, and possibly nickel. These are usually classed as micronutrients. Some are present as parts of enzymes, such as iron, copper, zinc, and molybdenum. Others may be required for the synthesis of certain compounds. Most are derived from weathering of minerals and from organic matter to a lesser extent. Many of these can be toxic to plants if present in more than trace amounts.

Mycorrhizae

Nutrients often become more available to plants because of the activities of soil microorganisms called *mycorrhizae*. These fungi are associated with the roots of as many as 97% of all plants, including many deciduous and evergreen trees, shrubs, and herbaceous plants. These fungi colonize the roots, and are especially valuable in areas that no longer contain root hairs in that mycorrhizae function somewhat like root hairs. Mycorrhizal fungi benefit plants and sustainable horticulture. In exchange, the plant supplies carbon as sugars to the fungi.

Ion (and water) uptake, especially for phosphate and micronutrients such as zinc and copper, is enhanced when mycorrhizal fungi infect roots. Absorption is improved for several reasons. The hyphae of the mycorrhizal fungi greatly increase the root's absorbing surface area by up to 80%. Their diameter is narrower than root hairs and diffusion improves as the diameter of the absorbing structure decreases. These hyphae also extend beyond the root hair zone where nutrients with low soil mobility are readily depleted, such as phosphate and micronutrients. Finally, the hyphae release certain chemicals that increase the availability of both inorganic and organic phosphorus. Because plants with mycorrhizal fungi need less phosphate, fertilizer applications can be reduced, leading to cost savings and less pollution of water.

Other benefits are attributed to mycorrhizae such as improvement of plant health. Good plant health means fewer pest problems and reduced applications of various pesticides. Again costs and pollution are lower. Plant health is better because the mycorrhizae increase the plant's uptake of water and nutrients and they also help protect roots from pathogens by their presence. Mycorrhizae also improve the soil by increasing water retention, releasing nutrients from organic matter by enzymic means, trapping nutrients, and in death providing organic matter.

Two basic types exist: ectotrophic mycorrhizae and endotrophic mycorrhizae. The former only penetrates intercellular spaces and the latter penetrates the cortical cells of the root. These fungal partners are very important for plants growing on soils with low fertility. Their absence explains why some tree and shrub species grown from seed in artificial soils fail to grow as well as their counterparts in the forest. Adding the natural strains of mycorrhizae found in the forest soils around these plants improves their growth noticeably.

Given the importance of mycorrhizae, more research should be encouraged in this area. Because we need to reduce inputs of fertilizers for reasons of cost, pollution, and resource strain, mycorrhizae could become a valuable part of sustainable horticulture.

SUMMARY

Climate is a major factor in determining where horticultural commercial activities are to be found and also what we can grow in our gardens. Modifications of climate are possible from localized influences that result from surface features. These surface-induced climatic variations are called microclimates. Some can be beneficial and others harmful to horticultural pursuits.

One excellent microclimate occurs where land meets a large water body, such as a lake or ocean. Summers tend to be cooler and winters milder, usually resulting in a slight increase in growing season length and increased plant choices in terms of winter hardiness. The effect arises from the ability of water and land to heat and cool at different rates. Commercial fruit growers benefit from this type of microclimate around the Great Lakes as do fruit and vegetable growers out on the West Coast. Southern slopes are bathed by rising heat and receive the most hours of sun, thus allowing earlier planting and earlier flowering on fruit trees and less danger of a frost causing bud or flower damage. Gentle, southern slopes are favored by fruit growers. The southern-facing foundations on homes also offer earlier warming and are favorite spots for the earliest possible flowering bulbs. Valleys are not favorable, since cool air sinks, bringing late spring planting dates and early autumn frosts. Such areas are often referred to as frost pockets. Large, paved areas also have unfavorable horticultural conditions, being hot and dry with little water infiltration. Big cities with massive buildings and much paved area also tend to be hotter and drier. Small microclimates can be found in homes, such as in high humidity areas in the kitchen and bathroom. House plants requiring high humidity do well in such areas. Southern windows are also the warmest and brightest of any window exposure, making them ideal for house plants requiring warmth and much sun.

Wind has both positive and negative impacts on plants. Air circulation can prevent localized depletion of carbon dioxide around plant leaves during photosynthesis. Wind also facilitates pollen transfer and seed dispersal for a large number of plants. Severe winds can cause plant damage and increase soil erosion. Plants in coastal areas are damaged by a combination of drying winds and salt sprays.

Many pollutants in air have a negative effect on plants. Plants intended for urban use must be chosen carefully to ensure a degree of pollution tolerance. Airborne pollutants also reduce crop yields and marketability.

Climate is related to developmental events of plants and also to insect outbreaks and major weed infestations. The science of this relationship is phenology. Phenological data can be used to forecast times for planting, flowering, ripening, and harvesting of fruits and vegetables. It can also predict times of insect and weed outbreaks. Such data are accumulated over many seasons and correlated time-wise with the above events. The units of data are degree-days, the same unit that oil delivery services use to determine where to deliver oil so your tank does not run dry. One degree-day is defined as a day when the mean daily temperature is 1 degree above the minimum temperature for growth (zero temperature) of a plant or insect. Each crop or insect has a specific zero temperature. This approach is not perfect, because certain conditions can throw it off track such as seasonal changes in seasonal variations or unusual temperature extremes. Certain refinements are possible to account for the effect of photoperiod and temperature changes on development. Many years of fine-tuning phenological data can also increase the accuracy of predictions. Negative degree-days (dormancy degree-days) can also be used to predict bud swell on fruit trees to assess the likelihood of damage from frost or pesticide sprays. Phenological approaches are especially valuable in integrated pest management programs for determining when to apply natural or chemical controls and also for gauging factory and labor needs for processing harvested fruits and vegetables.

Water is essential for plants for hydration, transport, and as a support medium for biochemical reactions. Plants possess features to reduce water loss such as stomates (leaf pores) and waxy layers. Some plants are especially adapted to conserving water and surviving in arid regions (xerophytes such as cacti and other succulents). Other plants have adapted to survival in extremely wet areas (hygrophytes such as cranberries). Most plants fall between these extremes and are known as mesophytes. Wilting is usually a symptom of water stress, although wilting can also be indicative of dead or nonfunctional roots. Water stress affects numerous physiological and developmental plant processes in a negative way, including photosynthesis and growth. Photosynthetic shutdowns from water stress reduce yields, making irrigation a critical input in horticultural production.

Water in the air (humidity) also affects plants through its impact on transpiration, which tends to decrease as humidity rises. Plants adapted to high humidity such as many of our house plants show leaf browning and dropping of leaves and flower buds on exposure to low humidity. Humidity also plays a major role in vegetative propagation. For example, high humidity levels are maintained around cuttings until root development is complete. Low humidity can also damage new graft unions.

Water enters the soil (infiltration) and then moves downward by gravity (percolation). Infiltration is increased by spherical aggregates, coarse organic matter in fine-textured soils, surface litter, coarse sand, soil dryness, and soil warmth, while compaction decreases it. Some water is lost in soils to gravity (gravitational water). The water held by the soil is capillary water, which plants use. The amount of capillary water or retention capacity is influenced by texture (clay increases it, while sand decreases it) and organic matter (which acts as a sponge with its many pores).

After a heavy rain and loss of water to gravity followed by a slight loss of capillary water at the surface, the soil is maximally charged with water. This maximal storage is the field capacity. Plants then absorb water to the point where absorption fails and the plant wilts (assuming no rain); this point is known as the wilting point. The difference between these two states is called available water (what plants use). Irrigation is applied just before the wilting point and stopped at field capacity.

Water is lost both by plants (transpiration) and by soil (evaporation): the combined loss is called evapotranspiration. When these losses exceed the amount of soil moisture, the water deficit is called drought. Drought is common in arid areas, but can also occur even in places with moderate rainfall. Irrigation is used to eliminate drought to prevent reduced yields and ultimately crop loss. Alternatively, crops with low water requirements better suited to arid regions can be grown. Adding organic matter can also increase water retention and water evaporation can be reduced with mulches. Excessive soil water can also be a problem, leading to root damage. If the condition is chronic, improving the drainage or planting of water-tolerant crops can be helpful.

Temperature has a major effect on plant development and physiology. Transpiration rates rise with increasing temperature, as does photosynthesis, respiration, and photorespiration. The reverse is also true. Optimal photosynthesis for most plants occurs between 76° and 86°F. Tropical and desert plants can exceed this somewhat. Most plants have a range of temperature that is best for development. While variable with plants, some generalizations are possible. Plants that originated in the temperate areas can grow between 41° and about 100°F, while for tropical origins, the range is 50° to 113°F. Optimal development temperatures for each group, respectively, are 77° to 95°F and 86° to 95°F.

Good root and shoot growth is associated with diurnal thermoperiodicity. Growth of both also increases with rising root temperature and a constant shoot temperature, but the shoot outgrows the root. Consequently, small changes in root temperature can cause extensive change in top growth. Horticultural situations of importance include container crops in fields and mulches. In the fall, cold air can drop root temperatures below that of soil, which is warmer at that time. Damage can result unless the plants are protected. Applications of cold water drop the root ball temperature and slow or damage root growth. Mulches placed before the soil warms up can retard shoot growth, given the connection between root and shoot growth. Cuttings also root better when the rooting medium is warmer than the air.

In some plants, exposure to cold temperatures is required to initiate flowering (vernalization). In other plants (certain tree fruits and spring flowering shrubs), cold is required to break dormancy, and in others, to promote development of existing flower buds such as in bulbs. Plants where vernalization initiates reproduction include certain grains and vegetables. The cold temperature range and duration for these processes vary according to the plant. Temperature is of great importance for propagation. It has an effect on bulb development and storage, callus production, rooting cuttings, seed dormancy and germination, seedling growth, and seed storage.

Temperature extremes have horticultural implications. Very high temperatures with drying winds and low soil moisture can lead to lower yields, poor growth, and even death. Very cold temperatures (frost and freezing) kill annuals and even improperly acclimatized perennials. Newly planted perennials, unless mulched, can be heaved out of the soil by freeze–thaw cycles. House plants left outside can be damaged by above freezing temperatures (high 30s).

Frost dates are important. The period from the last killing spring frost to the first killing fall frost is known as the growing season, the maximal annual time for plant growth. Later than usual frosts can harm new plants, damage fruit tree blossoms, and reduce fruit yields. Early than usual frosts can end harvests prematurely and damage plants not yet far enough into dormancy. Microclimates offer both advantages and disadvantages in terms of frost events.

Certain phenomena, known as winterkill and physiological chilling, are also associated with low temperatures. Winterkill is associated with broad-leaved evergreens in northern landscapes when planted in southern exposures. Warm temperatures in the January thaw cause growth in southern exposures. The soil is frozen and the roots cannot replace the lost water, so damage to leaves and bark occurs. Physiological chilling damage occurs with house plants of tropical origin when watered with cold tap water in the winter. The roots are chilled below their temperature range for absorption, but the shoot still transpires in the warm air. Leaves yellow and fall from this physiological drought.

Temperatures show distinctive temporal variations both in terms of day and night and by season. These variable events are known as diurnal and seasonal thermoperiodicity, respectively. Both are essential to plant development. Seasonal thermoperiodicity (along with changes in day length) is involved with entering and breaking dormancy. Minimal winter temperatures also determine the northern and southern limits of plants, the former in terms of hardiness to cold (winter survival) and the latter in terms of chilling requirements needed to break dormancy so growth and reproduction can resume. Hardening is the process through which plants become adjusted so they can survive temperatures below freezing.

Light influences many aspects of development and physiology. The effects of light result from its intensity, duration, and spectral quality. Light intensity can be measured either in foot-candles or photosynthetically active radiation. Intensity is directly related to photosynthetic activity. Plants adapted to shady environments need less light (shade plants) than plants adapted to full sun (sun plants). Seedlings and plants deprived of light develop into spindly, yellow plants, a condition known as etiolation. When deliberately induced in certain plants such as celery to improve flavor and texture, it is known as blanching. Leaves on larger plants receive more light on their exterior foliage ball and less inside. Leaves in these two locations have definite morphological differences and are known as sun and shade leaves, respectively. Field-grown foliage plants in Florida are shaded with shade cloth to produce only shade leaves in order to be better adapted to low-light interiors.

In terms of spectral quality, the red and blue wavelengths are used in photosynthesis. Other light-influenced processes involved in development (photomorphogenesis) are controlled by blue, red, and far-red wavelengths. Lights designed for plant growth indoors need to be enriched with red and blue for photosynthetic requirements. Photomorphogenesis involves different pigments than photosynthesis (chlorophyll): carotenoids, flavins, and phytochrome. For horticulturists, the most important event of photomorphogenesis is photoperiodism, in that it impacts the timing of flowering in many plants. Being able to control flower timing is especially important for the floral trade in holiday plants. The photoperiod is based on the hours of day and night in a 24-hour day and the change of this ratio with season. Photoperiodic changes along with temperature changes initiate and break dormancy. Some plants flower when the nights are long and days short (short-day plants) such as chrysanthemums, poinsettias, and Christmas cacti. Other plants flower when nights are short and days long (long-day plants) such as certain begonias and cornflowers. Photoperiod does not affect flowering in other plants (day-length-neutral plants). These groups can be refined further depending on whether flowering is absolutely dependent on photoperiod (qualitative) or only made slightly earlier or later by photoperiod (quantitative). In some plants photoperiodic effects can be altered by exposure to certain temperatures such as the Christmas cactus where temperatures between 45° and 55°F for a few weeks induce flowering regardless of photoperiodic effects.

Soil contains inorganic materials derived from weathering of rocks (sand, silt, and clay) and organic matter, both living and dead (soil organisms, decomposing organic matter, humus). The average time to form 1 inch of topsoil from bedrock is 500 years. In natural ecosystems the loss of soil is usually kept in balance with soil formation. Soil is lost through erosion brought about by wind and water. Exposed or uncovered soils between crop rows and between cropping seasons are especially prone to erosion. Often the rate of erosion exceeds the natural rate of soil replacement. Losses of organic matter also exacerbate soil erosion, because organic matter holds the soil together. Unlike old pre–Civil War farming or contemporary organic farming systems, today's systems with chemical fertilizers no longer rely on organic-rich manure, mulch, and cover crops. Consequently, organic matter lost through tillage followed by oxidation is often replaced at insufficient rates with crop residues. Less plant litter was also left on fields (because it may harbor insects and diseases), resulting in decreased infiltration of water. Less litter means increased water runoff, which carries soil, nutrients, and pesticides to nearby water bodies. Runoff is the leading cause of water pollution as well as a major avenue of soil loss. Heavy farm machinery also compacts soil, reducing percolation of soil water and leading to more runoff.

Soils have both physical and chemical properties. The varying amounts of rock-derived particles (sand, silt, and clay) determine the property known as soil texture. Sand, silt, and clay are defined by size with sand being the largest and clay the smallest. Because sand is the largest and heaviest, losses of organic matter followed by erosion leave the heaviest particles behind, sand, the basis for desert formation. Soil textures are classified based on the percentages of the three particles. A loam is the best soil texture for crop production. Loams with higher amounts of sand are sandy loams; with too much clay, clay loams. Some 20 different soil textures are recognized. Microbial excretions and humus form a glue-like substance that causes soil particles, both inorganic and organic, to stick together to varying degrees into larger particles known as aggregates. Soil structure refers to the packing arrangement of these aggregates and individual particles. One form, peds, is water stable and provides the best structure for crop growth. Clods denote a soil structure brought about by poor management of horticultural soils.

Spaces or pores can be found in soils as a result of the packing of irregular soil particles and aggregates. These pores comprise some 30% to 60% of the soil's volume and are named based on size, with smaller ones called micropores and larger ones, macropores. Micropores usually hold water, while macropores hold either air alone or air and water. Increases in sand tend to increase macropores, while clay and organic matter/humus tend to increase micropores. Macropores tend to favor soil aeration and drainage, while micropores favor water retention. Clay and humus also have the ability to retain and release certain soil nutrients (cation exchange). Given the importance of aeration, drainage, and water retention for roots and plant growth, a balance of both pore types is critical. Loams are best because this group has the proper balance of both types of pores.

Organic matter includes living organisms in the soil, their secretions/excretions, dead organisms including plant parts, decaying organic matter, and the end product of decay, humus. Nondegraded plant remains on the surface are called litter. Litter increases water infiltration and reduces runoff and erosion. Humus is spongy and contributes micropores and cation exchange properties like clay. Organic matter increases can improve water retention in sandy soils and soil structure in fine-textured soils (high clay or silt content). Organic matter, through decomposition, contributes some nutrients and trace elements, buffers the soil against sudden pH changes, and helps improve phosphorus availability. Organic matter and its microbial decomposers contribute to healthy, beneficial soil ecosystems and the formation of aggregates.

The chemical properties of soil include cation exchange and pH. Clay and humus, being negatively charged, hold water-soluble cations such as K^+ and Ca^{2+}. These cations can be displaced by or exchanged for other cations with a greater affinity for the retention site. Protons (H^+) have the greatest replacement ability and plant roots release protons directly and indirectly that free up bound cations into solution whereupon cation absorption by the roots occurs. Anions such as NO_3^- are not held by soils, with the exception of chemically bound phosphate. Anions are more prone to removal by water (leaching).

Soil pH is important in that it regulates the availability of soluble nutrients. Soil pH varies from acidic to alkaline (pH 4 to 8). The most acidic soils are found in the Northeast and the most alkaline in the Southwest. In general areas with high rainfall east of the Mississippi River and in the Northwest are acidic, while arid regions such as the western states are neutral to alkaline. Native plants are adapted to the existing soil pH. Crop and many landscape plants often require a different pH from the existing one in order to develop optimally. Soil pH is made less acidic with ground limestone and more acidic with sulfur.

Nutrients in soils must be replaced in soils where crops are cultivated. Harvesting removes nutrients in the plants that would normally return to the soil through recycling. Nitrogen, phosphorus, and potassium are required in large amounts and are added with most general-purpose fertilizers. Dolomitic limestone not only adjusts pH, but also adds calcium and magnesium. Sulfur, too, not only adjusts pH, but also is a source of the essential plant nutrient, sulfur. Many nutrients are only needed in trace amounts. Applications are seldom needed, because soil levels are adequate and often maintained by recycling of organic matter and weathering or oxidation of rock minerals. Symbiotic fungi in many plant roots (mycorrhizae) increase nutrient availability and uptake of ions by roots. Mycorrhizae are especially helpful with phosphate absorption.

EXERCISES

Mastery of the following questions shows that you have successfully understood the material in Chapter 6.

1. How does climate influence commercial horticultural production?
2. What is a microclimate? What are the microclimatic advantages and disadvantages for horticulture?
3. What is phenology and what value does it offer for horticulture?
4. What role do wind and air quality play in horticulture?
5. Plants have different water requirements and can be grouped into three categories. What are the three groups? How do they differ and what are a few plant examples in each group?
6. What plant processes and events are associated with the need for water? Do any water-related events impact horticultural production? If so, how?
7. What is the relationship of humidity to the rooting of cuttings?
8. How do infiltration and percolation differ? What factors influence the infiltration of water into soil?
9. Which form of soil water is available to plants? What do field capacity and wilting point have to do with water in the soil?
10. What is the relationship of water to wilting and root rot?
11. Temperature influences a number of physiological events in plants. What are they specifically and how do they impact horticulture?
12. Temperature influences a number of developmental events in plants. What are they specifically and how do they impact horticulture?
13. Temperature extremes can be harmful to plants. How so? Give specific examples involving plants both outdoors and indoors.
14. What is thermoperiodicity? Are there different kinds? If so, what are they?
15. What is the relationship of zones of hardiness to plant usage?
16. What is hardening? Why is it needed in horticulture?
17. What is the difference between light intensity and spectral quality?
18. What is etiolation and blanching? What connection do they have to horticulture?
19. What are sun and shade leaves? How does knowledge of the different types of leaves translate to practical value in the production of foliage plants for interiorscapes?

20. What is photomorphogenesis? What is photoperiodism? What relationship exists between temperature and photoperiodism? How does the knowledge of photoperiodism translate to practical value in horticulture?
21. How is soil formed? What are the basic components of soil? How do these two components differ?
22. What is soil erosion? What are the causes of soil erosion? Why is soil erosion bad?
23. What is soil texture? What is soil structure? How do these physical soil properties relate to horticulture?
24. What types of pores are found in soils? What functions do they serve in the soil and for plants?
25. Why is organic matter so important to soil function and productivity?
26. What is cation exchange and soil pH? What is their relationship to plant growth?
27. What are the essential plant nutrients?
28. What are mycorrhizae? What is their value in horticulture?

INTERNET RESOURCES

Examination and description of soils; many soil topics
http://www.statlab.iastate.edu/soils/ssm/chap3toc.html

Frost dates
http://www.victoryseeds.com/frost/index.html
http://cdmplanning.hypermart.net/frost.html

Mycorrhizae and plant health
http://www.agroforester.com/overstory/overstory8.html

Mycorrhizal symbioses
http://dmsylvia.ifas.ufl.edu/mycorrhiza.htm

Phenology and insects
http://spectre.ag.uiuc.edu/cespubs/hyg/html/199803e.html

Plant hardiness zones
http://www.usna.usda.gov/Hardzone/ushzmap.html

Plant responses to light: phototaxis, photomorphogenesis, and photoperiodism
http://www.rrz.uni-hamburg.de/biologie/b_online/e30/30.htm

Review of factors affecting plant growth
http://www.hydrofarm.com/content/articles/factors_plant.html

Soil basics
http://www.umassvegetable.org/soil_crop_pest_mgt/soil_nutrient_mgt.html

Soil biology primer
http://www.statlab.iastate.edu/survey/SQI/SoilBiology/soil_biology_primer.htm

RELEVANT REFERENCES

Barber, Stanley A. (1995). *Soil Nutrient Bioavailability: A Mechanistic Approach* (2nd ed.). John Wiley & Sons, New York. 414 pp.

Boul, S. W., F. D. Hole, R. J. McCracken, and R. J. Southhard (1997). *Soil Genesis and Classification* (4th ed.). Iowa State University Press, Ames. 544 pp.

Brady, N. C., and R. R. Weil (1999). *The Nature and Properties of Soils* (12th ed.). Prentice Hall, Upper Saddle River, NJ. 881 pp.

Davies, G., and E. A. Ghabbour (1998). *Humic Substances: Structures, Properties and Uses.* Springer Verlag, New York. 272 pp.

Dubbin, W. (2001). *Soils.* Iowa State University Press, Ames. 110 pp.

Elsas, Jan Dick van, Jack T. Trevors, and Elizabeth M. H. Wellington (eds.) (1997). *Modern Soil Biochemistry.* Marcel Dekker, New York. 683 pp.

Franks, F. (2000). *Water: A Matrix of Life* (2nd ed.). Springer Verlag, New York. 192 pp.

Harpstead, Milo I., Thomas J. Sauer, and William F. Bennett (1997). *Soil Science Simplified* (3rd ed.). Iowa State University Press, Ames. 220 pp.

Hock, B. (2000). *Fungal Associations.* Springer Verlag, New York. 250 pp.

Hopkins, William G. (1998). *Introduction to Plant Physiology* (2nd ed.). John Wiley & Sons, New York. 414 pp.

Jones, Hamlyn G. (1992). *Plants and Microclimate: A Quantitative Approach to Environmental Plant Physiology* (2nd ed.). Cambridge University Press, New York. 428 pp.

Kramer, Paul J., and John S. Boyer (1995). *Water Relations of Plants and Soils.* Academic Press, San Diego. 495 pp.

Marschner, Horst (1995). *Mineral Nutrition in Higher Plants* (2nd ed.). Academic Press, San Diego. 889 pp.

Miller, Raymond W., and Roy L. Donahue (1990). *Soils in Our Environment.* Prentice Hall, Upper Saddle River, NJ. 768 pp.

Pessarakli, Mohammad (ed.) (1995). *Handbook of Plant and Crop Physiology.* Marcel Dekker, New York. 1024 pp.

Pfleger, F. L., and R. G. Linderman (eds.) (1994). *Mycorrhizae and Plant Health.* APS Press, St. Paul, MN. 360 pp.

Podila, G. K., and D. D. Douds (eds.) (2000). *Current Advances in Mycorrhizae Research.* APS Press, St. Paul, MN. 214 pp.

Redlin, S. C., and L. M. Carris (eds.) (1996). *Endophytic Fungi in Grasses and Woody Plants.* APS Press, St. Paul, MN. 231 pp.

Rengel, Z. (ed.) (2002). *Handbook of Plant Growth: pH as the Master Variable.* Marcel Dekker, New York. 472 pp.

Smith, A. Ian, and David J. Read (1996). *Mycorrhizal Symbiosis* (2nd ed.). Academic Press, San Diego. 640 pp.

Soil and Water Conservation Society (2000). *Soil Biology Primer* (rev. ed.). Soil and Water Conservation Society, Ankeny, IA.

Sumner, M. E. (1999). *Handbook of Soil Science.* CRC Press, Boca Raton, FL. 2148 pp.

Vince-Prue, D. (1975). *Photoperiodism in Plants.* McGraw-Hill, New York.

Waisel, Y., and A. Eshel (eds.) (2002). *Plant Roots: The Hidden Half* (3rd ed.). Marcel Dekker, New York. 1136 pp.

Wilkinson, R. E. (ed.) (2000). *Plant–Environment Interactions* (2nd ed.). Marcel Dekker, New York. 480 pp.

VIDEOS AND CD-ROMS[*]

Cation Exchange Properties in Soils (Video, 1974).

The Earthworm (CD-ROM, 1997).

From Rock to Sand to Muck: All the Dirt on Soils (Video, 1996).

How Dare They Call it Dirt? (Video, 1993).

Microbiology of Turf Soils (CD-ROM, 1998).

New Methods in Soil Ecology: Combining Biology & Computation (Video, 1997).

Soil (Video, 1981).

Soils (Video, 1996).

Soil and Tissue Testing (Video, 1983).

Soil Organic Matter (Video, 1983).

[*]Available from Insight Media, NY (www.insight-media.com/IMhome.htm)

INTRODUCTION TO HORTICULTURAL PRACTICES

*T*he competent horticulturist knows that his or her applied science (technology) is only as good as the science behind it. He or she is also aware that technological advances only arise as science improves. Basic horticultural science and applied science may be likened to the artist and the tools of his or her craft. Science or the artist holds forth the knowledge of what is possible, but unless the science is applied or the artist uses the tools correctly, the potential will not be expressed. And indeed, like the artist, the applied science or technology of horticulture can lead to works that vary from bad to practical to very beautiful.

Today we need to introduce another aspect of technology, sustainability. Is the technology in harmony with the resources that feed it and the environment in which it will be used? If so, is the technology economically feasible? Do the products satisfy public needs safely and completely? Only when all of these parameters are satisfied can we say that the technology of these horticultural practices is sustainable.

The key to near-term sustainability is found in the soil. The critical factor is soil health. Good soil health is found with soils where organic matter content is maintained or increased through various practices and chemical inputs (fertilizers, pesticides) are minimal, or better yet, organic materials. Healthy soil has a thriving ecosystem of bacteria, fungi, and earthworms. The majority of the microbial populations are beneficial types where different species recycle nutrients, help make nutrients available for plants and even fix atmospheric nitrogen into forms usable by plants. Many are competitive with or openly antagonistic to pathogenic microorganisms that harm plants, thus reducing the need for fungicides. Such soils produce more vigorous, healthier plants that require less water, less fertilizer and have fewer pest problems when compared to soils with poor soil health. Lowered inputs and good organic matter content in turn mean less loss of soil by erosion and fewer off-site pollution problems from nutrients and pesticides.

The long-term key to horticultural sustainability is the development of renewable, alternative forms of energy that end our dependence on finite supplies of fossil fuels. Fuel cells are likely to provide the answer, both in the mid- and far-term time frame. Earlier types will likely use natural gas or propane. Later types will be fueled by hydrogen gas produced by the splitting of water into hydrogen and oxygen. The energy for the splitting process will be solar energy. Not only will horticultural sustainability be achieved, so will the sustainability of our society, once it is freed from oil and gasoline and all the political baggage that goes with dependence on foreign oil sources.

The chapters in the following section examine the practices and technology behind horticulture. Sustainability will be the guiding principle whether looking at the older practices of yesterday such as propagation, conventional plant breeding, cover crops, and irrigation or exploring the cutting-edge practices of today and tomorrow such as biotechnology, integrated pest management, and precision farming.

Chapter **7**

Plant Propagation

OVERVIEW

The propagation of plants and ultimately their improvement are fundamental to the horticulturist. Success in both areas requires a basic knowledge of the science behind plant propagation. This science includes plant structure, development, and metabolism, all of which were covered in Section One. Knowledge of technical propagation skills is even more essential. These techniques, once mastered, must be matched with the appropriate plant materials.

Issues of sustainability with propagation are concerned with environmental and sustainable management of inputs and resources. Basic resources include the materials in propagation media, such as vermiculite and peat moss, and inputs include fertilizers, pesticides, and fungicides. Environmental management of propagation resources and inputs is covered here. Propogation resources and inputs in the greenhouse and field are covered later in chapter 13 and 11, respectively.

CHAPTER CONTENTS

Propagation Basics
Propagation Media
Propagation Needs and Equipment

Sexual Propagation
Asexual Propagation
Propagation Aftercare

OBJECTIVES

- To understand the purpose of propagation, the types of basic propagation, and the reasons for choosing each type
- To understand propagation media in terms of raw materials, formulation, physical/chemical properties, and pasteurization
- To learn about sustainability issues involved with raw materials used in propagation media

and sustainability issues with greenhouse water use
- To understand the basics of sexual propagation from seed production through propagation by seeds
- To learn the basics of asexual propagation from cells to plant pieces
- To understand propagation aftercare

PROPAGATION BASICS

Propagation is essentially the multiplication or increasing of plants in numbers and also involves the maintenance of their essential qualities (genetic characteristics). If absolute continuity is required in the genotype, then propagation utilizes asexual (cloning) techniques. If slight to essentially very little change is tolerable, propagation is accomplished with sexual (seeds) methods.

A number of requirements drive the need for propagation. One need is to replace plants that are harvested for food or ornamental reasons and do not continue into the next growing season, such as annual flowers and vegetable crops or cut flowers. Another reason is to replace diseased, damaged, old, or dead plants that persist for many years, such as herbaceous and woody perennials used for food or landscaping purposes and plants used as house plants in homes or commercial and institutional interiorscapes. Another need is to provide new plants for landscaping around new homes, new businesses, new parks, new arboreta, new gardens, and new institutional buildings. New plants are also required as tokens of appreciation and for gifts, weddings, funerals, and holidays.

Many different types of propagation exist, ranging from sexual (seeds) to asexual (vegetative or cloning) methods. Asexual propagation includes numerous different practices from conventional approaches involving stem, leaf, or root cuttings, whole plant division, or air and soil layering of stems and branches to modern techniques that are highly technical and use small parts, such as cells or pieces of tissue (micropropagation).

The method of choice is determined by cultural, marketing, ease of doing, and economic factors. Seed propagation is widespread in nature and is an efficient and economical method of commercial propagation. Propagation by seed is the largest form of commercial and home propagation and is the choice where large numbers of plants are required. Cost per plant tends to be modest. Seeds are also the best mechanism for storing plants over time and are also good for starting plants free of disease, because most fungal and viral diseases are not transmitted via seeds. The disadvantages of seed propagation are a longer time to flowering or harvest and sometimes minor variations in genotype.

The use of seeds is the propagation method of choice for annual bedding plants, both vegetables and flowers, destined for home gardens and landscapes. Many vegetables are also propagated by seed in commercial horticultural operations, both by direct field sowing and for provision of vegetable transplants. Seed propagation is also the method of choice for flowers and vegetables that are sown directly in fields or garden soil, such as carrots, radish, and sweet corn, where a headstart or difficulty in transplanting rule out the use of bedding plants. Seeds are widely utilized for the establishment of new lawns and commercial sod production. Rootstocks for grafted fruit trees are usually grown from seed, as are some common conifers, shrubs, and trees.

When no variation in genotype is acceptable, such as in fruits or high-quality landscaping cultivars, asexual propagation is the method of choice. Such factors are important for cultural and economic reasons. Farmers expect uniform plant size, uniform ripening, continual resistance to diseases and insects, consistency in time of harvesting, and other aspects that contribute to standardization of cultural practices and minimization of operating costs. Consumers expect no changes in their apples, oranges, grapes, and bananas, so consistency of product is a marketing need. The disadvantage of asexual propagation is cost, which exceeds that for sexual propagation. The bottom line is that economics dictate that asexual reproduction be used to produce higher value plants than is possible with sexual reproduction.

Vegetative propagation is widely used for the production of woody cultivars having high landscape value, such as deciduous and broad-leafed evergreen flowering shrubs, needled conifers, deciduous shade trees, fruit trees, roses, herbaceous perennials, foliage and flowering house plants, florist crops, bulbs, and perennial vegetables. The choice of which asexual method to use is based on ease of rooting, ease of handling, marketing needs in terms of numbers and quality, labor requirements, and unit costs versus market value. For example, grafting and micropropagation are more costly and labor intensive, but qualities desired in the final product result in grafting being used for fruit trees and roses, whereas the huge volumes needed for certain foliage plants, such as ferns, require that micropropagation be used. Cuttings are the most widely used form of asexual propagation.

A number of specialized horticultural industries and institutions are involved in the support of propagation activities, both on the nonprofit and commercial levels. These include state agricultural experiment stations, arboreta, botanical gardens, land-grant universities, seed and plant repositories, seed producers, seed certifiers, nurseries, bedding plant producers, foliage plant growers, rootstock producers, micropropagation laboratories, various trade associations, various professional societies, and the U.S. Department of Agriculture.

PROPAGATION MEDIA

Numerous materials have been used alone or in mixed formulations for propagation. Soil would appear to be a natural material for propagation under most circumstances, but its role is primarily restricted to direct outdoor propagation in farm fields or gardens.

Most house and bedding plants produced by commercial propagators, and even most transplants raised by gardeners, are grown in artificial soils. Why? Natural soils show considerable variation; no standard soil exists. The propagation industry prefers to work with standardized inputs. Artificial soils offer the advantage of consistency, being produced from standard materials and specific formulations.

Shipping costs are another factor. Soil is moderately heavier when dry or wet relative to artificial soils in the dry and wet condition. While soil weight raises shipping costs somewhat, the main problem is soil's lack of compressibility. Artificial soils can be compressed in the dry state into much smaller volumes than soil. Bags and bulk lots of artificial soils are less costly to ship than soil. Artificial soils are light in weight. The average person can easily lift a large bag containing several bushels.

Lastly, propagators must work with relatively sterile media to reduce the incidence of disease during propagation. Soil can harbor pathogens that harm seedlings and chemical or heat treatments of soil (pasteurization) can lead to undesirable changes in its chemical and physical properties and unpleasant odors. Artificial soils are easily prepared in a disease-free form. The bottom line is that artificial soils have all the beneficial effects of soil when it comes to propagating plants, but none of the undesirable ones.

Properties of Propagation Materials

Certain chemical and physical properties are required of materials used for propagation media. In terms of chemical properties, salt content must be low and the material must remain unaffected by pasteurization or sterilization. The last requirement becomes important for media that are not naturally sterile or are to be reused. Sufficient cation exchange capacity and nutrient retention is also needed in media used for extended growth periods beyond propagation.

Physical properties include sufficient density to support seeds or vegetative propagules and sufficient porosity to provide a suitable balance among aeration, water drainage, and water retention. Wetting and drying cycles should not cause any changes in these physical properties. The material should also be free of organisms harmful to plants.

Materials Used for Propagation Media

Basic starting materials can be inorganic (Figure 7–1), such as perlite, pumice, sand, Styrofoam, and vermiculite. Organic forms (Figure 7–1) include compost, peat moss, sphagnum, and wood by-products. Costs for all have been increasing, but especially for perlite and vermiculite. Both require energy inputs as heat to prepare a suitable product

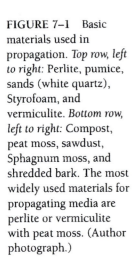

FIGURE 7–1   Basic materials used in propagation. *Top row, left to right:* Perlite, pumice, sands (white quartz), Styrofoam, and vermiculite. *Bottom row, left to right:* Compost, peat moss, sawdust, Sphagnum moss, and shredded bark. The most widely used materials for propagating media are perlite or vermiculite with peat moss. (Author photograph.)

for horticultural use. The chemical and physical properties of these materials are summarized in Table 7–1.

The inorganic materials are usually added to mixtures to improve aeration and drainage, and the organic materials are utilized to increase water retention and cation exchange capacity. At present, peat moss, perlite, and vermiculite are the most widely used. Perlite is used more than vermiculite, because the latter has poorer physical stability when handled wet or mixed in large containers where bottom pressures cause compression. As used here the term *peat moss* refers to moss peat, the moderately degraded remains of hypnum and sphagnum mosses. Sphagnum moss refers to undegraded, fresh, dried remains of the *Sphagnum* species.

Styrofoam is an inexpensive substitute being used in place of perlite, but it cannot withstand steam sterilization or chemical sterilization by certain treatments, such as chloropicrin or methyl bromide. It also has an annoying tendency to float to the surface when pots or containers are watered. Bark and sawdust are becoming increasingly popular. These materials must be treated by adding enough nitrogen to supply the needs of microorganisms during decomposition as well as the needs of the plants, and at least 30 days of aging are needed to minimize phytotoxicity problems with hardwood barks. Pine bark has been used fresh successfully. For example, redwood sawdust is amended with 0.5% ammonium nitrate on a dry weight basis (3 pounds per cubic yard).

Sustainability Issues with Propagation Media

Sustainability issues associated with propagation media include resource limitations and energy inputs during product manufacture and sustainability problems with water, fertilizers, and pesticides. Media issues are covered here and water issues later in this chapter. All of the materials are essentially natural products used either as is or processed to some degree. Supplies range from very limited to very high, as do energy inputs during processing. Of the inorganic materials, pumice (a volcanic mineral) is the most limited resource, both in terms of stocks and replacement by natural processes. Pumice particles break down easily and are not used much in propagation media, the one exception being growing media for cactus and succulents. Sand exists in plentiful supply and does provide good properties for propagation media, except for its weight. As such, use is somewhat limited, except for cuttings rooted in sand.

The current favored inorganic components, perlite and vermiculite, are porous products produced by heat from naturally occurring obsidian and micaceous minerals, respectively. These resources have reasonable availability for the near future, but given the geological time needed for production and the rate of current usage, they are limited in the long term. Styrofoam and other plastic particles, used as substitutes for perlite and vermiculite, are even less acceptable, given that plastics are derived from petroleum.

TABLE 7-1 *Basic Materials for Propagation Media*

	Source	Dry Wt.[a]	Sterile	pH	Nutrients[b]	Buffering Capacity[c]	Cation Exchange[d]	Water Retention[e]	Aeration Drainage	Particle Size[f]	Use[g]	Comments
Inorganics												
Perlite	Heated obsidian silicate	VL	Yes	6.5–7.5	None	None	None	Fair	Very good	1/16–1/8 D_i 1.6–3.2 D_m	R, M	Nasal irritant
Pumice	Volcanic silicate	M	No	7.0	Slight (K)	Low	Slight	Fair	Good	1/16–0.4 D_i 1.6–10 D_m	R, M	Limited resource
Sand	Soil mineral	H	No	7.0	None	None	None	Poor	Good	0.002–.079 D_i 0.05–2 D_m	R, M	Quartz sand best
Styrofoam	Polystyrene	VL	No	7.0	None	None	None	Poor	Very good	0.16–0.47 D_i 4–12 D_m	M	Pasteurizing problems
Vermiculite	Heated micaceous mineral	VL	Yes	6.5–7.5	Slight (K, Mg)	Moderate	High	Fair	Good	0.003–0.12 D_i 0.075–3 D_m	S, R, M	Mix with perlite for long use
Organics												
Bark	Softwood tree waste	L	No	3.5–6.5	Slight (N, P, K, Ca, Mg)	Moderate	Low–medium	Good	Good	1/16–5/10 D_i 1.6–13 D_m	M	Age 30 days with nitrogen
Compost	Plants, manures	L	No	5.5–8.5	Low (N, P, K, Ca, Mg)	Moderate–low	High	Good	Good	1/16–5/10 D_i 1.6–13 D_m	M	Slow release of nitrogen
Peat moss	Peat bogs[h]	L	No	3.8–4.5	1% N	Moderate	High	Good	Good	1/16–0.4 D_i 1.6–10 D_m	M	Use wetting agent
Sawdust	Lumber wastes	L	No	3.5–6.8	Slight (N, P, K, Ca, Mg)	Moderate	Low–medium	Good–very good	Fair	1/16–0.4 D_i 1.6–10 D_m	M	Age 30 days with nitrogen
Sphagnum	*Sphagnum* sp.	VL	Yes	3.5	None	Moderate	High	Very good	Fair–good	1/16–3/8 D_i 1.6–9.5 D_m	S, M	Contains natural fungicides

[a] VL, under 8 lb/ft³ (128 kg/m³); L, 8–14 lb/ft³ (224 kg/m³); M, 15–30 lb/ft³ (480 kg/m³); H, 100 lb/ft³ (1,600 kg/m³).

[b] None have sufficient nutrients for long-term use.

[c] The higher the buffering capacity, the more lime needed for pH adjustment.

[d] The higher the cation exchange capacity, the fewer problems with salinity and nutrient leaching.

[e] Fair: absorbs 3–4 times its weight in water; good: 5–10 times; very good: 10–20 times.

[f] P, particles, D_i, diameter in inches; D_m, diameter in millimeters.

[g] R, rooting medium alone; M, added to mixtures; S, for seeds alone (smaller particles).

[h] Moderately decayed remains of hypnum and sphagnum, (i.e., moss peat).

Trade-offs are inevitable. Perlite and vermiculite are lightweight porous materials that provide good aeration and drainage for propagation media. Their short-term drawback is that energy inputs are needed to process the raw materials (vermiculite at 1090°C and perlite at 760°C). As such, these energy-intensive products prepared from limited mineral resources do not rate high on the sustainability index. Sand, while ideal in terms of resource stocks and limited processing energy, weighs more and increases shipping costs and energy requirements for transportation. Basalt rock and clay can also be heat-processed (rockwool and calcined clay, respectively) and used as substitutes. While potential supplies are high, energy inputs are still a consideration.

Based on these considerations, perlite and vermiculite are the materials of current choice. Rockwool and calcined clays are likely the future materials, especially after further research and improved manufacturing processes. In the sustainability view, alternate materials are clearly needed. Recycling improvements, both in terms of amounts recycled and methods, can extend the time for current materials until new choices are available.

If new materials are not forthcoming, better ways to utilize sand can help. One possibility is to mix either perlite or vermiculite with sand. This strategy buys time. A move to total replacement by sand would mean a dramatic change in marketing, from national to regional. Currently, commercial propagation for some products, such as house plants, is concentrated in California and Florida. Some very specialized products come from other areas, such as bedding plants from Connecticut. Sand raises transportation costs and its use would tend to decentralize the industry to more regional operations. However, regional production would bring about increased energy costs for greenhouse operations in colder climates. Good systems analysis is needed to determine what approach is best in terms of sustainability.

Organic materials of choice include peat moss and sphagnum moss for reasons of standardization and minimal disease problems. Peat moss is a natural resource that is formed after thousands of years of natural processes. Peat moss is mined from peat bogs and current usage for fuel and propagation exceeds natural production, so sustainability is a problem. However, given the reserves of peat moss worldwide, peat moss extraction will hold up for a long time. Sphagnum moss is a more renewable resource in that *Sphagnum* species can be grown and dried to supply the needs for propagation media. Both peat and sphagnum moss are coming under self-regulation, as their costs rise.

Compost and wood by-products (shredded bark and wood shavings) are also potential renewable replacements. The long-term direction should be a shift to increased use of compost and wood by-products. However, problems with variability, processing problems involved in pasteurization, and phytotoxicity dictate a need for further research. In addition, much bark, especially pine, is now used as a fuel at many sawmills, thus reducing supplies. Hardwood barks need composting prior to use. A promising bark product comes from *Melaleuca quinquenervia*, a weed tree from Australia that is widespread in Florida. This bark might replace peat and sphagnum moss in the future, because it is renewable, given the rapid growth rate of this weed tree.

Formulation of Propagation Mixtures

Propagation media are often composed of mixtures of the preceding materials for germination of seeds and for vegetative propagation, such as the rooting of cuttings. The materials for these mixtures should be uniform and of correct size; if not, they are screened for uniformity and the elimination of large particles. Some form of fertilizer may be added as an amendment. The ingredients are moistened before mixing, either by hand or mechanized mixers. Peat moss is difficult to wet, and the use of a wetting agent is advisable when this component is used. Mixtures are allowed to stand for at least 24 hours to equilibrate moisture content. Mixtures are usually pasteurized (see later section), unless such treatment will cause harmful alteration.

The oldest propagation mixtures were those based on soil for seed germination and sand for rooting cuttings and other asexual propagation. An early soil mixture was developed around 1939 at the John Innes Horticultural Institution in England through the research of W. J. C. Lawrence and J. Newell. The use of this soil-based media and others has declined for a number of reasons. Problems with soil mixtures include increasing scarcity of loam, lack of standardization (variations in chemical and physical properties of soil), and heavy weight. Sand when wet is heavy and retains water poorly under hot, dry conditions. An exception occurs when sand is used for rooting in shallow containers that are not well supplied with drainage holes; in this case, sand can have excessive water retention.

Formulations for Seed Germination. Artificial soil mixtures were developed in the late 1950s and early 1960s. These formulations provided a standardized mixture with little variation between batches, which could be prepared on a large-scale basis. This media is easily produced and maintained on a pathogen-free basis. Some popular mixes were those developed at Cornell University by J. W. Boodley and R. Sheldrake and at the University of California at Los Angeles by K. F. Baker and colleagues (see the Relevant References at the end of this chapter). The Cornell "Peat-Lite" mixes require minimal decontamination, other than the pasteurization of the peat moss. The complete University of California mixtures can be steam or chemically pasteurized without any harmful effects.

Many commercial variants of these artificial mixtures are available today, such as Jiffy Mix®, Metro Mix®, and Pro Mix®. These materials come ready to plant, including nutrients. Wetting agents to improve water absorption are usually included, too. Various formulations are available for seed germination, vegetative propagation, and growing of container or pot plants. Most commercial propagators utilize bulk amounts of these commercial formulations. Prefilled propagation containers are also available.

The Cornell Peat-Lite and the University of California mixtures used for seed germination follow:

Peat-Lite Mix C

1 bushel horticultural-grade vermiculite (fine, no. 4)
1 bushel shredded peat moss (at least 70% *Sphagnum* sp.)
4 level tablespoons (3 oz) ammonium nitrate
8 level tablespoons (12 oz) superphosphate (20%); best is powdered
10 level tablespoons (7 ½ oz) ground limestone (best form is dolomitic)

University of California Mix B

75% sand (0.05 to 0.5 mm in diameter)
25% peat moss base

To each cubic yard add the following:

Mixture to Be Stored

6 oz potassium nitrate
4 oz potassium sulfate
2 ½ lb single superphosphate: $CaH_4 (PO_4)_2 H_2O$
4 ½ lb dolomite lime
1 ¼ lb calcium carbonate lime
1 ¼ lb gypsum

This provides a moderate amount of available nitrogen for about 2 weeks after germination.

Mixture to Be Used within One Week

2 ½ lb hoof and horn meal or blood meal
6 oz potassium nitrate
4 oz potassium sulfate
2 ½ lb single superphosphate
4 ½ lb dolomite lime
1 ¼ lb calcium carbonate lime
1 ¼ lb gypsum

This formulation provides available nitrogen plus a moderate reserve. It cannot be stored, because the organic nitrogen will break down during storage. With all of the previous mixtures, care must be observed when wetting the peat moss during preparation. A nonionic wetting agent, 3 ounces in 5 to 10 gallons of water, is helpful for wetting 1 cubic yard of mix.

Seeds can also be germinated in straight shredded or milled sphagnum moss, 0.75 to 1.0 mm vermiculite or sand. Pasteurization is needed for the latter two. Some fertilizer can be added initially to prolong the need for supplemental fertilization. The better results obtained with the previous mixtures, plus their commercial availability and the need for no supplemental fertilization until transplanting, account for the preferred use of mixtures over straight sand, milled sphagnum moss, or vermiculite.

Formulations for Vegetative Propagation.   Vegetative propagation, such as the rooting of cuttings, is possible in straight perlite, pumice, rockwool, sand, or vermiculite. Perlite or pumice is often used to root leafy cuttings under mist because of their good drainage properties. These materials are more likely to be mixed in varying proportions with organic materials (shredded bark, compost, peat, or sphagnum moss). Clean, sharp sand (builder's sand) is used to root cuttings, especially those of evergreens (yews, junipers, arborvitaes). Peat moss is often added to sand to improve water retention, such as 2 parts sand and 1 part peat moss. One part sand and 1 part shredded sphagnum moss is also used. Vermiculite is also used alone, but more often on a 1:1 basis with perlite or sand, since long-term use of straight vermiculite results in particle collapse and reduced aeration and drainage. Soil is sometimes used for rooting deciduous hardwood cuttings and root cuttings. A well-aerated sandy loam appears to be best. Pasteurization of these mixtures is recommended and necessary if they are to be reused.

The following University of California mixture can be stored indefinitely and is useful for rooting cuttings:

University of California Mix C

50% sand (0.05 to 1 mm in diameter)
50% peat moss

To be added to each cubic yard

4 oz potassium nitrate
4 oz potassium sulfate
2 ½ lb single superphosphate
7 ½ lb dolomite lime
2 ½ lb calcium carbonate lime

Pasteurization of Media

Nematodes, fungi, bacteria and weed seeds can inhabit some or all of the following: soils, soil mixtures, bark, compost, peat moss, sand, and any recycled propagation media. These organisms can harm plants. For example, the loss of recently germinated seedlings can occur from a disease termed damping-off. This disease results from attack by fungal species such as *Rhizoctonia, Phytophthora, Fusarium,* and *Pythium.* Pasteurization is used to prevent the damaging effects of these organisms. Fresh materials heat-treated during manufacture (perlite, pumice, and vermiculite) and materials containing natural antibiotics do not require pasteurization.

Steam Pasteurization.   Harmful organisms can be destroyed in propagation media by heat or chemical treatment. From a sustainable horticulture perspective, steam pasteurization is preferred. Heat treatment can be regulated to bring about either pasteurization or sterilization. Pasteurization primarily destroys harmful organisms, and sterilization destroys both harmful and beneficial organisms. Pasteurization is more often used in actual practice.

The moist material is covered in a bin or bench and steamed through perforated pipes 6 to 8 inches below the surface. A temperature of 180°F (82°C) for 30 minutes will destroy nematodes, pathogenic fungi and bacteria, insects, and most weed seeds. Straight steam has a temperature of 212°F (100°C). By mixing air with steam (4.1:1), a temperature of 140°F (60°C) can be attained. This temperature for 30 minutes will destroy many pathogenic organisms, but not weed seeds. However, it will not harm many beneficial organisms that have an antagonistic effect toward pathogens that may appear through recontamination. This lower temperature also reduces the risk of toxicity problems caused by heat-induced breakdown of chemical compounds, such as occurs with some nutrients.

Chemical Techniques.   Chemical treatment may also be utilized to kill organisms in the propagation media. Less physical and chemical disruption occurs, but the toxicity of these fumigation chemicals poses problems. Wetting agents are often added to the chemicals to increase their effectiveness. Safe practices and caution are required with chemicals during storage and application. If chemicals are used, sustainable usage is best accomplished by minimal use of chemicals in the context of an integrated pest management (IPM) program designed for propagation. IPM programs will be covered in Chapter 15.

Better results occur when chemicals are applied to moistened propagation mixtures at temperatures of 65° to 75°F (18° to 24°C). The prescribed waiting time (up to 2 weeks) after fumigation must be observed to allow escape of fumes from the media. The chemicals currently used include methyl bromide, chloropicrin (tear gas), and methyl bromide–chloropicrin mixtures. These treatments vary in their effectiveness, depending on the choice of chemical and observation of proper treatment conditions.

Chemical drenches can be added to the growing media (and to established plants) to inhibit the growth of many pathogenic fungi, such as *Botrytis, Pythium, Fusarium, Phoma, Phytophthora, Rhizoctonia,* and *Sclerotinia.* These chemicals include pentachloronitrobenzene (Terraclor, PCNB, or Quintozene), 5-ethoxy-3-trichloromethyl-1, 2, 4-thiadiazole (Truban, Terrazole, or Etridiazole), Banrot, and compounds containing the systemic fungicide thiophanate methyl (Cleary 3336, Domain, SysTec 1998, and Topsin M).

Fungicides should be rotated to reduce the possibility of induced resistance to pathogens. Given usage choices, the least toxic compound should be used preferentially. As with any chemical treatment, the user must be aware of phytotoxicity problems, proper usage, and safe disposal methods, including disposal of chemical containers.

PROPAGATION NEEDS AND EQUIPMENT

Conventional propagation activities require extensive support facilities, good water supplies, and ancillary equipment. Micropropagation requires much more sophisticated facilities and supplies. The former uses a greenhouse as the main facility; the latter a high-tech laboratory facility stocked with expensive scientific equipment and high-quality biochemicals. Both need good-quality water supplies. Skills required of managers and workers are considerably higher at tissue culture operations, and start-up and maintenance costs are also higher. Field propagation of some materials is also done. Many container plants are grown in field nurseries, especially in the Southeast and Southwest.

Water Supplies for Propagation

Seedlings, cuttings, and plant tissue cultures are more sensitive to salts than older plants. Electrical conductivity is used as a measure of salt levels in water. An instrument called the solubridge or salinity-measuring meters are utilized to measure electrical conductivity. Levels of soluble salts in water (or propagation media) should be ideally below 0.75 millisiemens per centimeter. Under no conditions should soluble salts exceed 1,400 ppm in water used for propagation. Water containing a high proportion of sodium to calcium and magnesium is also unacceptable. The water pH should range from 5.5 to 7.0.

Although hard water is not harmful to plants, the high levels of magnesium and calcium can cause buildup problems. Deposit buildup from hard water can harm mist propagation units and evaporative water-cooling systems. Hardness over 6 grains (100 ppm) generally necessitates water softening. However, water should not be softened with the usual ion-exchange resins like zeolite. Such resins produce high levels of sodium that are toxic to plants and affect the water absorption rates and physical structures of soil. An alternate way to soften hard water (deionization) is to use ion-exchange resins that replace calcium, magnesium, and sodium by hydrogen ions, and chlorides, sulfates, and carbonates by hydroxyl ions. Reverse osmosis can also be used, but is more expensive.

Municipal water is usually treated with chlorine or sodium fluoride. The amounts usually are not high enough to cause phytotoxicity. However, if fluoride exceeds 1 ppm, some leaf damage results with certain foliage plants, such as spider plants (cultivars of *Chlorophytum comosum*). Fluoride damage is mainly a problem in long-term growth, so it is not a factor during propagation, because exposure is short. Boron should not exceed 1 ppm. No good method exists for removing boron. Another water supply will be required if boron is a problem, because excess boron is phytotoxic.

In addition, the water temperature should be monitored, because water that is too cold will retard growth of plants, seed germination, and even cause chilling injury if cold enough. During the winter, unheated tap water may have a temperature of 34° to 40°F (1 to 4.5°C). Propagation mixes watered thusly could take several hours to return to initial temperature. Mixer valves can be used to blend hot water with the incoming cold tap water. Water pH should range from 5.5 to 7.0.

Propagation and Water Sustainability

Two problems face the propagation industry in terms of water sustainability: water pollution and water scarcity. Methods to alleviate these problems exist. Water pollution can be minimized with reduced inputs of fertilizers and pesticides. Fertilizers should be broken into two programs: fertilizers for propagation (preplant fertilizer) and fertilizers for growing in containers and pots (postplant fertilizer). Preplant nutrients usually consist of very low nutrient levels (NPK), micronutrients, and dolomitic limestone.

Nutrient levels are increased for postplant fertilization programs. Postplant fertilizers (and micronutrients) should be applied via the watering system at lower levels with more frequent applications as needed, using a proportioner to inject liquid concentrate. For example, a 10-4-6 NPK soluble formulation at 80 to 100 ppm nitrogen five times weekly at 0.5 inches is useful for container plantings. These rates can be raised at certain developmental and seasonal times as needed. Soluble salt levels in the containers should be monitored and flushed as needed. Drainage water should be handled carefully to prevent leaching into water supplies.

Controlled-release fertilizers can also be used, usually on high-value container plantings, to maximize fertilizer efficiency and reduce water pollution for plants in long-term growth containers. However, during propagation, slow-release fertilizers should be avoided because leaching occurs when plants are small and have low nutrient needs. Rates of application should be adjusted to compensate for different levels of nutrient needs by plants at various stages of development.

Irrigation systems for watering are also factors. Applications of soluble fertilizers in irrigation water by overhead sprinkler systems are the most economical. However about 60% to 70% of the fertilizer does not reach the containers, but ends up in runoff. From a pollutant view, controlled-release fertilizers work better with overhead sprinklers, because runoff nutrient levels are considerably less. An alternative approach is to use drip irrigation with soluble fertilizers, which also greatly reduces nutrients in runoff water. The downside is that both of these alternatives cost more. In areas with water scarcity issues, drip irrigation also offers the plus of being the most water-efficient form of irrigation. If overhead sprinklers are used, better water efficiency, reduced runoff, and less pollution results from short runs several times a day rather than one long run daily. Containers absorb the majority of their water in the first 5 minutes, so three or four 5-minute runs daily are recommended.

Runoff can be channeled through constructed wetlands that help to reduce pollutants and sediments. Constructed wetlands have already been demonstrated as very effective for the cleaning of sewage plant effluent. Ongoing research is also showing value for the treatment of effluent from propagation facilities and nurseries. This partially purified water can be treated further and used as recycled irrigation water.

Further treatments include the use of chlorine or bromine to reduce algae and plant pathogens. Straining and filtration is needed to remove particulates, followed by charcoal filtration to remove herbicides and other phytotoxic chemicals. A final treatment with ultraviolet radiation will remove any surviving pathogens. If salts pose a problem, either deionization or reverse osmosis will be necessary. The recycled water can then be mixed with freshwater, if necessary, to achieve the desired water quality.

The adoption of additional practices termed best management practices (BMP) can also reduce problems. Fertilizers should be applied only when good growth response is expected such as in spring and summer, but not autumn for field containers. Even with greenhouse plants, winter growth is slow and lower nutrient levels are required. If top-dressing containers, aim well (do not broadcast) and do not top-dress containers outdoors that are prone to blowover.

An IPM program should be adopted to minimize the levels of pesticides. Greenhouse and field IPM are covered later in Chapter 15.

Facilities Used for Propagation

Successful propagation depends on control of environmental factors. The most widely used propagating structure is the greenhouse. Other useful structures include the hotbed and propagating chambers. Light levels are adequate within these structures, and the control of relative humidity and temperature is feasible. Plants propagated in these structures cannot be set directly outdoors without prior conditioning. Structures used for conditioning recently propagated plants include the cold frame and lathhouse.

Greenhouses, Hotbeds, Cold Frames, and Lathhouses.   Greenhouses are either free standing with even-span, gable-roof construction or attached (lean-to) structures surfaced in polyethylene, fiberglass, or glass. Glass is the most permanent and most expensive. Fiberglass, the most widely used, is less expensive and less breakable, but its light transmission deteriorates with age (some better grades will last up to 20 years); thus replacement or refinishing is required. Polyethylene is very inexpensive, but requires replacement often at yearly intervals. However, recent advances appear to be increasing its useful span. Polycarbonate and acrylic double-wall plastics have become the glazing materials of choice, because they offer reasonable costs and energy efficiency.

Hotbeds and cold frames are much simpler structures of a non-walk-in type and are usually covered with polyethylene, fiberglass, or glass. Propagating chambers are even smaller units where a high humidity level can be maintained. These structures are enclosed in polyethylene, and the light source can be natural, fluorescent, or high-pressure metal halide lamps. Hotbeds and propagating chambers are particularly useful for starting seedlings or rooting leafy cuttings.

The cold frame and lathhouse are primarily hardening or conditioning areas for plants in transition (see Chapter 13) from the ideal propagating conditions to the more severe conditions outdoors. The cold frame is also used to propagate some of the hardier plants, such as cole crops. Lathhouses consist of a shading material, such as wood strips, aluminum strips, or a woven plastic (saran fabric) supported by wood or pipe members. The lathhouse, another "halfway house," is used to provide protection from the sun and wind for recently propagated plants held for sales or landscaping use and, in some instances, a comfortable outdoor sales area in hotter climates. The sustainable management of the environment within the above propagation structures and more specific details will be covered in Chapter 13.

Sanitation.   The areas where propagation takes place need to be kept clean and sanitary. Spent propagation media and plant debris must be swept up, floors washed, and any working surfaces and tools disinfected. Surfaces can be washed down with bleach solutions, benzylkonium chloride (Greenshield®, Physan®) or even pine disinfectant. Propagating tools, such as pruning shears and knives, can be dipped in Physan®.

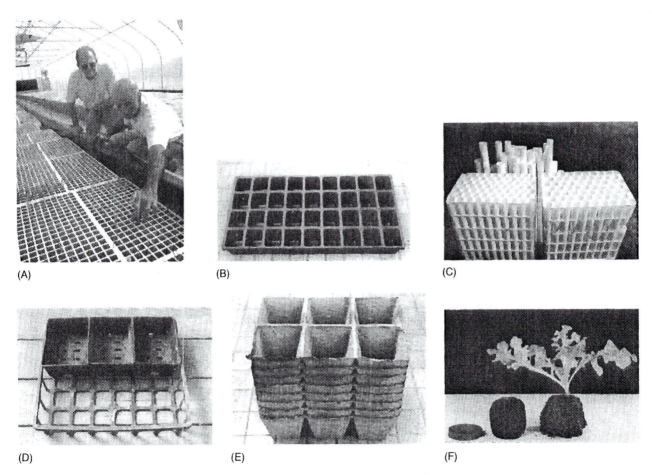

(A)   (B)   (C)

(D)   (E)   (F)

FIGURE 7–2   Various types of propagation containers. Plug trays are very popular for starting seeds and cuttings and come in many sizes. Shown here are the 200 plug (A) and the 36 plug (B) sheets. Wild sea oats are shown in (A). Sea oats are used for beach erosion control projects around the country. The tray holding the plug sheet shown in (B) confers stability and ease of transport. Deeper plug sheets (C) are used for propagating trees and shrubs. Plastic flats (D) in tray were popular prior to plugs, but are used increasingly less. Peat pots (E) and peat pellets (F) are used to a lesser degree. (A and F are photographs from the USDA-NRCS and USDA, respectively. B, C, D, and E are author photographs.)

Containers

During propagation, some form of container is necessary (Figure 7–2), unless the plants are being propagated directly in the ground. Types of containers include pots of clay, plastic or fiber, plastic or wooden flats, plastic plug trays, propagating blocks of fiber or peat, containers of metal or asphalt-coated felt paper, cups of Styrofoam or paraffined paper, and even polyethylene bags.

Container Sanitation.   Recycled containers of any kind should be thoroughly washed and disinfected. Heat treatment with steam is preferred for the least environmental impact. Second best is a ninefold-diluted solution of bleach. Spent solutions should be allowed to sit for several days to permit reduction of chlorine to acceptable levels if water is discharged onto soil. If discharged to a drain, first check the local and state water quality regulations. All tools and working surfaces coming into contact with containers should also be sanitized.

Types of Containers.   Flats or shallow trays are used for rooting cuttings and germinating seeds. Composition can be wood, plastic, or metal. Drainage is provided through bottom holes. An advantage is ease of mobility and storage, especially for the lightweight plastic flats that nest and have nearly completely replaced wood and metal flats.

Pots are usually round or square and come in many sizes and materials. Clay pots are fragile, prone to surface accumulations of salts that can become toxic, expensive, and heavy. However, they are recyclable, easily decontaminated by steam, and their porosity allows good water and oxygen diffusion. Plastic pots offer the advantages of reduced weight, ease of stacking, low cost, reusability, varied colors, and ease of cleaning. Their nonporosity requires different handling, that is, a reduction in watering and fertilization as opposed to clay. Heavy-duty or flexible plastics resist breakage. They cannot be steam sterilized, but can be sterilized by chemical treatment or semipasteurized by hot water for 3 minutes at 158°F (70°C). Chemicals that react with plastic should be avoided.

Lightweight plastic square pot packs in attached units of 8 or 12 are convenient and easy to move. Fiber pots are made of pressed peat, wood fiber, and fertilizer, which is suitable for short-term use. They are biodegradable, easy to store, and the joined unit packs are convenient. When wet they tear easily and dryness of the pot walls discourages good root growth. Propagating blocks composed of peat or wood fiber mixed with fertilizer are a combined "pot and propagation media." Like fiber pots, the whole container can be set directly into the ground. Plastic plugs (similar to egg cartons) come in various cell sizes and numeric counts. Plugs have become extremely popular with seed propagators and are widely seen in the wholesale and retail markets for bedding plants.

A number of containers serve well for temporary use. Asphalt-coated felt paper containers are lightweight, inexpensive, tough, and storable. Styrofoam or paraffined paper cups and even polyethylene bags provided with drainage holes are inexpensive containers for propagating plants.

Bottom Heat and Misting

Often the optimal temperature for germinating seeds or rooting cuttings exceeds the optimal temperature for subsequent development. It may be uneconomical to raise the temperature of the entire propagating area, or areas with mixed plants at various developmental stages may preclude such a temperature increase. In such cases lead- or plastic-coated electric soil-heating cables with thermostatic controls are useful for providing bottom heat in the propagating media (Figure 7–3). Low-voltage systems are available to reduce the hazard of electric shock.

The loss of water through transpiration can lead to severe and even fatal wilting of cuttings, until roots are formed and water uptake commences. One way to reduce water losses by transpiration is to keep the relative humidity of the air around the cuttings at a high level. A simple way to provide these conditions is to place a glass jar over the cuttings or to use a propagating chamber covered with polyethylene. More sophisticated control is provided with intermittent misting equipment or fog generators. Fog droplets are considerably smaller than mist droplets. Dips of antitranspirant chemicals have also been used.

Misting or fog increases the relative humidity, as does a glass jar or plastic frame, but the film of water supplied by the mist has the additional advantage of reducing air and leaf temperature, which in turn reduces the rate of transpiration. The propagating media can be maintained at a warmer temperature than the air with a heating cable. Temperature control with misting allows direct placement of cuttings in full sun, unlike those in an enclosed jar or plastic frame. The resulting increased photosynthesis speeds development of the cutting, especially root formation. In addition, substances important to rooting (carbohydrates, auxins, and flavonoid substances) seem to be increased in cuttings that are under mist compared to those that are not. Intermittent misting provides this kind of environment and is less wasteful of water. Disease problems under mist appear to be minimal. An example of an effective propagating setup is shown in Figure 7–4. Misting can be used indoors and outdoors, both for sexual and asexual propagation.

A solenoid (magnetic) valve regulates water flow through the misting system. These valves should be hooked up so that in the event of a power failure they will be in open or "misting" mode to prevent damage to the cuttings. The solenoid valve in turn is controlled by various mechanisms. These controls include the very reliable dual variable timers that

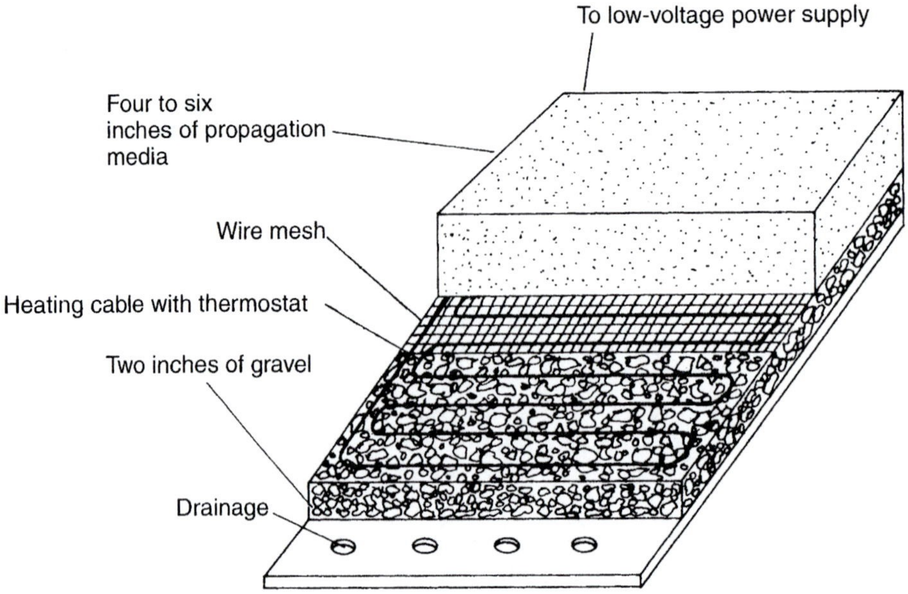

FIGURE 7–3   Construction of propagation bed supplied with bottom heat through a soil- or propagation media-heating cable. This method is useful for a hotbed. Flat plastic mats with built-in heating cables can also be used alone to supply bottom heat directly to propagation containers filled with media that are placed on the mats. The latter approach is used in greenhouses. (From Raymond P. Poincelot, *Horticulture: Principles & Practical Applications,* 1980. Reprinted by permission of Prentice-Hall, Upper Saddle River, New Jersey.)

turn the entire system on and off at preset times, for example, morning and night, and a second timer that determines the length of the misting period in minutes.

Another device consists of a thermostat that turns on the system when a preset temperature is reached. It has proved especially effective when used in conjunction with the timer control described previously, since sudden rises in temperature might dry cuttings prior to the timer-activated mist. Another type of control, the "electronic leaf," sits in the cutting bed, and wetting or drying of it breaks or makes electrical contact. A weight-sensitive device can turn the mist on or off when losses or gains in water weight occur from evaporation or misting. This device is especially useful in areas for indoor or outdoor use that experience extensive, rapid weather changes, thus causing rapid variations in the evaporative power of the air (Figure 7–4). Another control with a photoelectric cell turns mist on in proportion to light intensity. Controls and solenoid valves can be run on low voltage to reduce shock hazards.

Filters in the water line may be necessary to prevent clogging of the mist nozzles. Nutrients can be dispensed through the misting system to counteract nutrient leaching and to improve root formation and subsequent plant growth. Algae growth can be a problem because it is unattractive and slippery. Overnight drying, Bordeaux mixtures, or chemical treatment can reduce algae. Water quality must be watched (see earlier discussion in this chapter) because deposits can cause clogging or salts can hinder rooting.

SEXUAL PROPAGATION

Sexual propagation consists of propagation by seed. Annuals and biennials destined for sales as bedding plants for flower and vegetable gardens are propagated in this manner. Commercial vegetable growers also use these bedding plants for early harvests of certain vegetables, such as tomatoes. Many other vegetables and annuals are direct-seeded in gardens and fields. Lawn grasses are seeded by homeowners and by sod producers. Some herbaceous and woody perennials may be sexually propagated, especially when used for grafting rootstocks and restoring forests and wetlands.

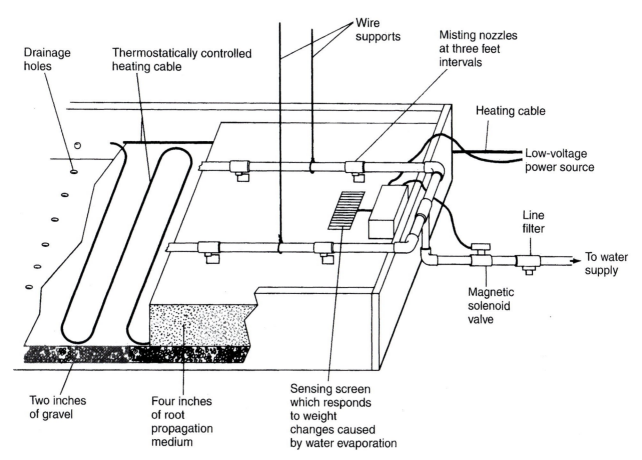

FIGURE 7–4   Construction of a misting propagation unit supplied with bottom heat. The cable is overlaid with heavy plastic mesh to prevent possible cable damage or shock hazard if a tool accidentally comes in contact with it. A highly recommended additional safety feature would be a low-voltage system. The extent of leaf coverage and not height is the determining factor for the misting water lines. Each line covers about 1.5 feet on each side, so a bench 4 to 6 feet wide would need two lines. (From Raymond P. Poincelot, *Horticulture: Principles & Practical Applications,* 1980. Reprinted by permission of Prentice-Hall, Upper Saddle River, New Jersey.)

Generally, sexual propagation is the method of choice, unless seedling variability cannot be controlled, marketing needs require absolute genetic consistency (fruits), or alternate asexual (vegetative) methods are more rapid, practical, and economical. It is also the best means for producing new cultivars for plant improvement, which, along with the production of genetically pure seed, will be covered in Chapter 8.

Seed Production and Handling

The production of seeds for horticultural use is an important industry on which plant propagators are dependent, unless they produce their own seed. Generally, commercial seed is raised in areas where the climate is most favorable for the production and harvesting of the desired seed crop. Much seed production is located in the western coastal states (California, Oregon, and Washington). Other important areas of seed production for the United States include Central America, South America, and Southeast Asia, where seed production is favored both for climatic and economic reasons. In addition, seed is collected from naturally grown plants, such as trees or shrubs, and from fruit-processing industries.

Seed Harvesting.   Harvesting of seeds is best done when the seed is mature or can be removed without reduction of its germination potential. Harvesting too early will yield an immature seed with undesirable qualities, and harvesting too late may mean the loss of seed as a result of dropping to the ground, blowing by the wind, or being removed by animals and birds. Harvesting may be by hand or mechanical equipment. Cleaning is by hand or mechanical threshing, followed by air blasts, screening, or gravity separating if needed. These techniques depend on physical differences between the seeds and associated debris. Some drying may be required after harvesting and prior to cleaning. If a fleshy fruit encloses the seed, it must be removed. Mechanical macerators can be used for fruit disruption and the seeds then separated by fermentation and floating, screening, or other mechanical means. Many seeds will require drying after harvest to moisture contents of usually 8 to 15 percent.

Seed Pretreatments.   Seeds are often pretreated by either the seed producer during or after the harvest or by the farmer prior to sowing. Pretreatments are used to either maximize germination rates or to facilitate mechanical sowing. Some treatments, termed seed protectants, protect the seeds from conditions that reduce germination potential. Seeds can be treated with chemicals (disinfectants and fungicides), beneficial microorganisms (biocontrols such as *Pseudomonas* and *Trichoderma*), heat treatments, microwaves, and ultraviolet radiation to protect the seed from pathogens. Biocontrols, heat, and radiation are preferable from the viewpoint of sustainability.

Seed sizing and priming are techniques used to enhance germination. With sizing, larger seeds are selected and marketed for their increased vigor and longer shelf life. Priming depends on osmotic solutions or moist inert substrates (matrix priming) to bring about partial internal germination of the seeds. Growth enhancers, such as phytohormones or seaweed extracts, can be added to further enhance germination. The seeds are then dried. Primed seeds show higher vigor. From the sustainability perspective, matrix priming is better, because its chemical disposal does not produce any environmental problems.

Techniques are also available to bring about partial external germination. This method is called *pregermination* to distinguish it from the above procedure, termed *germination enhancement*. Pregermination depends on either fluid drilling or pregerminated seeds. With fluid drilling, the seeds are first partially germinated to the point of radicle emergence. This task can be performed with wet blotters or aerated aqueous solutions. As with germination enhancement, growth enhancers can be added to the solutions to further promote germination. The seeds are then suspended in a gel and sown with specialized machinery. Pregerminated seeds are also partially germinated (radicle emergence), but then dried and sown with conventional machinery. In effect, pregermination removes seeds that are not viable or vigorous, such that the sowings give a very high stand that approaches 100%.

Phytohormones have been used to stimulate seed germination (and seedling growth) for various seeds. The phytohormones and treatments are (1) gibberellins at 100 to 10,000 ppm for 24 hours, (2) cytokinins at 100 ppm for 3 minutes, and (3) ethylene for various times. Other chemicals used to stimulate germination in some species are potassium nitrate, thiourea, sodium hypochlorite, and seaweed extracts. Preliminary trials of chemicals and germination tests on small batches of seeds prior to large-scale use are suggested.

Some seeds, either because of size (*Begonia* sp.) or irregular shape (marigolds) are hard to sow with mechanical sowers. Seeds in this group can be coated with inert materials held together by binders. Tumbling methods produce easily handled pelletized seeds, such as lettuce or petunia seeds. Thinner coats are achieved with a process called film coating. The latter offers the advantage that beneficial microbes or fungicides can be added as part of the film.

Seed Testing.   Labeling of seeds is required of producers by various state and federal laws. Procedures for seed testing are used to provide some of the information on the labels. The best indicators of seed quality are germination rates, purity, vigor, seed health, and level of noxious weed seed contamination. The label also must indicate the name of the plant including the cultivar name, place of origin, and amount.

Tests are useful for determining the germination percentage and the number of days required to reach a given germination percentage, the germination rate. In addition, seeds should be examined for purity. Contaminants such as other crop seed, weed seed, and inert material (chaff, broken seed, soil, sand, gravel, and the like) reduce the value of the seed.

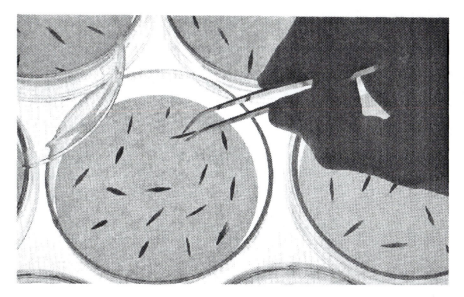

FIGURE 7–5 Grass seeds are placed in dishes for germination testing at the NRCS National Plant Materials Center. (USDA-NRCS photograph.)

Together, the purity, germination percentage, and germination rate are required to establish rates of sowing for the seeds to be planted.

Methods of germination testing are as follows. In a greenhouse, seeds may be sown on flats of sterile sand or peat moss. Tests vary in length from 10 days to 4 weeks, unless the seed is known to germinate slowly. Seeds may be placed between two blotters or on top of one blotter (Figure 7–5). These blotters, held in trays or petri dishes, are placed in germinators with controlled optimal conditions of moisture, temperature, and light. Maximal care in handling is used to avoid conditions suitable for growth of disease microorganisms. Seeds may also be placed on rolled moist paper towels and put in a germinator. Seeds from woody trees or shrubs that need extensive preconditioning prior to germination are tested by excising the embryo, which is then tested directly for signs of germination. The chemical 2,3,5-triphenyltetrazolium chloride, which changes from colorless to red as a result of aerobic respiration, is also useful for germination testing. X-ray analysis can be helpful for checking seed viability.

Seed purity is influenced by physical and genetic contaminants. Physical ones include soil, weed seeds, plant debris, and other inert material. Genetic contaminants are likely to be other similar cultivars or inbreds when one deals with hybrid seeds. Microscopy and flotation can be used to detect physical contaminants. Genetic contamination requires sophisticated chemical tests, protein electrophoresis, and DNA fingerprinting.

Seed vigor provides useful information for predicting germination percentages under conditions that are less than optimal. Generally, tests for seed vigor measure germination percentages under various environmental stresses, such as temperature and humidity extremes. Seed health is lowered by the presence of contaminating organisms (insects, nematodes, and pathogens) in seed lots. Tests to measure seed health rely on microscopy, biochemistry, and incubation.

Seed Storage. Usually, some form of seed storage is required after harvest, except for seeds whose viability is very short unless allowed to germinate quickly. Seeds should be of the highest quality when stored and also be mature and undamaged. Some physiological decline will occur during storage, and this affects viability. The amount of decline depends on the kind of seed, the length of storage, and environmental storage conditions such as temperature and humidity. Storage conditions are selected to minimize respiration and other metabolic processes within the seed. These conditions will vary, depending on the plant species. In general, low moisture storage under refrigerated conditions is best for many seeds. Seeds that are short lived are usually stored under near-freezing (temperate origin) or cool (tropical origin), moist conditions. Cryopreservation offers promise as an economical storage method for long-term preservation. Further research is needed to perfect this technology.

Seed Propagation Techniques

Seeds selected for propagation should be of the best possible quality, often expressed in terms of viability. The number of seedlings that can be produced from a given number of seeds, the germination percentage, is a useful indicator of viability. Additional features of viability are prompt germination, vigorous seedling growth, and normal appearance.

Breaking Seed Dormancy.   Most vegetable and flower garden seeds have been bred free of seed dormancy. Such seeds, said to be quiescent, can be germinated at any time. Some seeds, such as the seeds of woody plants and wildflowers, are in a state of dormancy that must be broken prior to germination. The various forms of seed dormancy and examples were discussed in Chapter 5.

Some forms of dormancy are broken during storage, but others require physical or chemical preconditioning to stimulate germination. Hard or impermeable seed coats may be abraded, but the embryo must not be injured, to improve permeability to gases and water. This abrasion or mechanical scarification can be done with a file, sandpaper, or mechanical tumblers lined with sandpaper or filled with sand or gravel. Hard or impermeable seed coats are also modified by acid treatment, such as a soak in two times their volume of concentrated sulfuric acid. The length of time is determined experimentally. A water soak may be helpful in softening the seed coat or removing inhibitors; it also stimulates germination in some cases. The seeds are placed in about five times their volume of water at 170° to 212°F (77° to 100°C). The water is allowed to cool and the seeds removed after 12 to 24 hours. Seeds should be planted soon after the water soaking process is completed.

Certain seeds must be subjected to cold temperatures prior to germination. This can be done under controlled conditions of moist-chilling (stratification). Dry seeds are first soaked in water for up to 24 hours and then placed in a plastic (polyethylene) bag containing a moisture-retaining medium (peat moss, sphagnum moss, vermiculite, or weathered sawdust). These bags are held at 35° to 45°F (2° to 7°C) for periods of 1 to 3 months (Figure 7–6).

In some cases, two treatments may be necessary to overcome the double dormancy of many woody shrub and tree seeds, such as a hard coat and a chemically inhibited or dormant embryo. The former can be broken by mechanical or acid scarification followed by water soaking and the latter by either form of scarification followed by stratification. An ef-

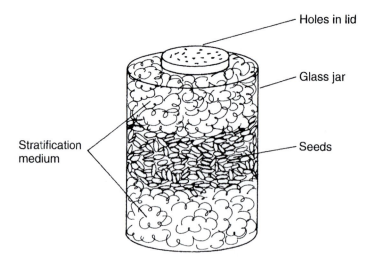

FIGURE 7–6   Seeds prepared for stratification in a glass jar. Seeds and moist stratification media layers are equally thick, that is, 0.5 to 3.0 inches. The lid must be vented because the seeds require oxygen. Polyethylene plastic bags, which are highly permeable toward oxygen, also make good stratification containers. (From Raymond P. Poincelot, *Horticulture: Principles & Practical Applications*, 1980. Reprinted by permission of Prentice-Hall, Upper Saddle River, New Jersey.)

fective treatment for many seeds having double dormancy consists of moist, warm conditions to enhance microbial degradation of hard seed coats and stratification afterward to break embryo dormancy.

Timing. The timing for seed planting is dependent on several variables. These factors include the number of days until maturity of the plant, the length of the growing season as determined by the average date of the last spring killing frost and the first fall killing frost, soil temperatures, other climatic variables, the temperature preference of the seed and developing plant, the desired date of harvest or sale of plants, the cold or heat hardiness limits of the plant, the number of days required for seed germination, and the timing of setting plants outdoors. Some of these factors apply only to seedlings started indoors or only to seedlings started outdoors. Recommended dates are usually available from state agricultural experiment stations or state extension services.

Controlling Disease. Disease control during sexual propagation is extremely important. Extensive losses of seeds, seedlings, and young plants can result from a group of fungi that cause a disease syndrome collectively referred to as damping-off. Symptoms of damping-off include seed decay or seedling rot prior to emergence, stem rot of the seedling at soil level and subsequent falling over, stem girdling and eventual death of the seedling or rot of the root system, and eventual death of young plants. The major fungi responsible for this disease are *Rhizoctonia solani* and *Pythium ultimum*. *Phytophthora* sp. and *Botrytis cinerea* may be involved to a lesser degree.

Pasteurization of the propagation media, discussed earlier in this chapter, will eliminate the microorganisms. Seeds should also be treated to prevent reintroduction of the microorganisms. Treatments include protectants that prevent attack by microorganisms from the propagation media, chemical disinfectants that eliminate microorganisms on the seed surface, and disinfectants that destroy organisms within the seed. Fungicides, such as captan, thiram, and zineb, are used as protectants and are sold under various trade names. Many fungicides come under federal and state regulations that must be observed by the user. Calcium hypochlorite at 1% to 2% strength for 5 to 30 minutes is used as a disinfectant, and hot water at 120° to 135°F (49° to 57°C) for 15 to 30 minutes serves as a disinfectant.

Environmental Factors. A number of environmental factors affect germination: water, temperature, gases, and light. Conditions should be optimal for germination and early growth; often, optimal germination conditions produce unfavorable conditions for damping-off. Insufficient moisture supply adversely affects the germination percentage, germination rate, and the rate of seedling emergence from the seedbed. In addition, as the moisture supply decreases, the actual concentration of soluble salts increases, which may be harmful if the concentration becomes high enough. Excess water can also be harmful because the aeration of the propagating media is reduced, which favors damping-off; the buildup of algae and mosses is also favored.

Water is usually supplied gently by hose, such as with a fogging nozzle or automatic misting systems. Alternatively, bottom watering by placing the propagating container in a tub of water (not submerged) is feasible for small numbers of containers. The reduction of excessive evaporation and water needs outdoors is helped by light mulching and indoors by the use of well-drained media with good water retention and the covering of flats or containers with glass or clear plastic wrap (gas permeable). Containers should be kept out of the sun and the cover removed when germination commences.

Minimal, maximal, and optimal temperatures have been established for seed germination. These temperatures are influenced by diurnal cycles, seasonal cycles, and interactions with other environmental factors. Broad categories are helpful in deciding which temperature regimes are to be provided. Cool season crops, vegetables, and flowers (cabbage, broccoli, cauliflower, lettuce, radish, pea, spinach, and pansy) that originated in the temperate climates generally germinate poorly above 77°F (25°C). Warm season crops (eggplant, pepper, squash, melon, cucumber, dahlia, sweet corn, tomato, and beans) originating from subtropical or tropical areas have a minimal germination temperature of 50° to 60°F (10° to 15°C). Many other cultivated plants have a temperature range for germination from as low as 40°F (4.5°C) to as high as 104°F (40°C). In many instances a night drop in temperature favors germination and seedling growth and results in savings in energy costs. The drop should be around 18°F (10°C). After seeds germinate, somewhat lower temperatures favor more vigorous development of the seedling and young plant. More detailed temperature requirements of many more species are found in Section Four.

Oxygen, carbon dioxide, and, in some instances, ethylene, can affect seed germination in the propagation medium. Oxygen is required for respiration of the germinating seed and carbon dioxide is a product of that respiration. Anything that restricts oxygen influx and carbon dioxide efflux can be harmful to germination. A surface crust or excess water in the medium can hinder gaseous movement and availability. Ethylene is released by some seeds and may stimulate germination.

Light is required for the germination of some seeds (*Begonia,* coleus, and lettuce), inhibits the germination of others (*Allium,* phlox, and portulaca), and is not required for others. Some seeds respond to photoperiod; birch (*Betula*) and California poppy (*Eschscholzia californica*) are long- and short-day seeds, respectively. These light-dependent processes are controlled through phytochrome responses to light. The light requirement can be eliminated in some seeds by dry storage, chilling, alternating temperatures, chemical treatment, or phytohormone treatment. Light-sensitive seeds are usually tiny, and the best germination takes place at or near the soil surface, so that the seedlings sprout and commence photosynthesis quickly. Light-inhibited seeds often originate in dry desert environments. Light requirements of many seeds can be found in Section Four.

Low light intensities produce etiolated seedlings with low rates of photosynthesis. Trees or buildings must not shade fields and gardens. Optimal light intensity produces sturdy, vigorous plants that carry out higher rates of photosynthesis. However, very high light intensities can damage tender new seedlings. Therefore, some shading of certain seedlings may be necessary in greenhouses. If insufficient or no natural light is available, supplementary or completely artificial light may be used. Fluorescent lights are used to provide 1,500 to 2,500 or more foot-candles for about 16 hours, followed by 8 hours of darkness. More details on the use of artificial light can be found in Chapter 13.

Sowing Seeds Indoors.   Indoor seedling production (Figure 7–7) is carried out in greenhouses and other controlled environment facilities to produce plants for subsequent transplanting to the field nursery and vegetable or flower garden or for use in the greenhouse or home. The bedding plant industry is especially dependent on the production of seedlings indoors. Some container-grown stock is produced in this manner. Trees and shrubs for use in landscapes, reforestation, and as rootstocks are sometimes started indoors.

In large operations, filling containers with propagation media, media wetting, leveling, and seed sowing are completely automated with specialized machinery. Mechanization is essential to keep costs down and to produce competitively priced products. Mechanical seeders need to produce precise spacing of seeds, because labor costs for the nonmechanized tasks of thinning and transplanting are high. Gardeners, of course, perform these steps manually.

Plugs have become the most popular container types for mechanized production of bedding plants, having almost completely replaced flats. Growing media and seeds are added to plug flats mechanically. Plugs grown by large growers are usually kept in special germinators where environmental conditions can be carefully controlled. Smaller growers often place plugs under mist or fog with bottom heat in the greenhouse. After germination, regular watering and fertilization practices are employed.

Nonplug containers (flats) of propagating media are leveled and gently tamped (except for vermiculite where compaction destroys structure) to reduce the level to about ½ inch below the container top. The propagation media is thoroughly wetted prior to seeding. Fine seeds are sown on the surface without any covering media, and medium to large seeds are covered with two to three times their minimal diameter of soil.

(A)

(B)

(C)

(D)

FIGURE 7-7 Manual seed sowing for bedding plants. (A) Start off with wetted propagation media. It should be wet enough that it holds a molded ball shape when you release your hand. (B) After the flat is filled and leveled, holes or rows are made for larger seeds, while smaller seeds can be broadcast sown. Seeds are covered with media to the proper depth. Fine, broadcasted seeds can be lightly pressed in with the flat of the hand. (C) Seeds are labeled and lightly watered by gentle sprinkling or by mist from a misting nozzle. (D) Seeds must not dry out. Placing the seed flat in a plastic bag or covering with plastic sheeting can reduce watering frequency. Seed flats are germinated in indirect light. Most large commercial growers now use plugs that are machine-filled with media and mechanically seeded. Plug trays are then germinated in specialized germinators (essentially environmental chambers) where the temperature, humidity, and light are carefully controlled. (A, B, and C are author photographs. D is a USDA photograph.)

Careful attention is paid to maintaining moisture, temperature, and light (if required). Bottom heating is supplied when needed. Misting or a glass or gas-permeable plastic cover (polyethylene) can be used to maintain moisture. If covered, containers are kept out of direct sunlight. Covers are removed during germination. After germination, temperatures are often lowered somewhat and light exposure increased to produce strong, vigorous seedlings. Fertilizer is provided if necessary. In commercial operations, fertilizer is applied with the irrigation water (*fertigation*). This procedure is usually automated. Surface dryness of the media is allowed once the root system is well established. Seedlings are transplanted after they develop at least four true leaves. Mechanical transplantors exist for transplanting plugs. With flats, this step must be performed manually. The young plants are usually transplanted to larger joined cell packs for eventual wholesale and retail sales.

Ferns (see reproduction cycle in Chapter 5) can be propagated from spores in a manner similar to that of seeds. Spores are sown on the surface and kept covered the same as for seeds. Propagation temperatures vary from 65 to 75°F (18° to 24°C), and bottom heat may be helpful. A moss-like growth that consists of numerous prothallia is produced; prothallia are removed by tweezers and transplanted to wider spacing. When these reach about ½ inch in size, tiny sporophyte plants are produced. With further transplanting these sporophytes subsequently develop into mature ferns. Many ferns are no longer propagated in this manner. Most ferns sold as house plants today are micropropagated (see later in this chapter). Ferns sold for outdoor use are often propagated by vegetative means, although some are propagated from spores.

Sowing Seeds Outdoors.   Seeds of many annual, biennial, and some perennial plants (herbaceous and woody) are seeded directly outdoors in farm fields (vegetables, herbs, and cut flowers), field nurseries (for growing various woody plants and vegetables as transplants or container stock for the commercial trade), gardens, and new lawns. This method eliminates losses or slowed growth resulting from transplanting, but seed germination and density are more difficult to control. Often, thinning of seedlings is needed, a considerable expense in commercial operations. Seed tapes that consist of plastic tape with properly spaced seeds are less wasteful of seed and can reduce or eliminate the need for thinning in gardens. Pelleted seed improves the handling of small seeds. Mechanical precision seeders are also available for mechanized spacing of seeds and are widely used in commercial operations.

Seedbed preparation is most important for seeding outdoors. Good contact between the seed and soil is needed for moisture supply, but pore space and good drainage are also required for adequate soil oxygen. Aggregate size is optimal if three-quarters of the particles range from 1 to 12 millimeters in diameter. Smaller particles, if too numerous, can eventually cause problems through the formation of surface crusts. Small gardens are prepared with a spade, fork, or rototiller. Plows or disks are used on larger areas in commercial operations. Seedbeds should be worked 6 to 10 inches in depth. If moisture retention is poor, organic matter can be added or mulches used to conserve moisture. Soil tillage, nutrient levels, and organic matter should be maintained in a sustainable manner through sound practices (see Chapter 11) and by soil tests. Special attention to weeds, pests, pathogens, and water needs are required for good seedling development.

Time of outdoor sowing is determined by the seed and plant's temperature requirements, time to maturity, time of desired harvest, the length of the growing season, the date of the last killing frost in the spring, and the earliest time of soil workability. Your local agricultural experiment station or extension service should be consulted for the time of sowing. Special attention to soil management is essential to maintain soil sustainability. More details on soil management will be provided in Chapter 11.

ASEXUAL PROPAGATION

The principle behind asexual propagation is the potential for regeneration possessed by the vegetative parts of plants (see Chapter 5). For example, adventitious roots arise from stem cuttings and, in turn, root cuttings can produce a new shoot system from adventitious shoot buds. New shoots and roots can also develop adventitiously from leaves.

Asexual propagation is the method of choice for growing cultivars that produce no seeds, for preserving clones (most fruit and nut crops, trees and shrubs used in landscaping, bulbs, and many herbaceous perennials), and for the production of high-value individual plants. It is possible to bypass certain undesirable characteristics or a particular developmental stage through asexual propagation of vegetative material from a later developmental stage. With some plants it may be more rapid and more economical than sexual propagation and even result in earlier flowering. Finally, the production of virus-free or mycoplasma-free plant material is dependent on propagation of heat-treated vegetative parts or certain other methods of asexual propagation. As a general rule, asexual propagation costs more than using seeds. For that reason, seed propagation is used with field crops where large numbers count more than individuals.

Sources used for asexual propagation need careful management. Correctness of cultivar (identity or true-to-name), verification of no genetic shifts (true-to-type), and the absence of pathogens (pathogen-free) are important criteria. True-to-name tests are basically molecular biology approaches (DNA fingerprinting) and involve electrophoresis of DNA fragments alone or subsequent hybridization with labeled probes (Southern blotting). These techniques are explored in detail in Chapter 9. For true-to-type determinations, DNA fingerprinting is helpful, but visual inspection by cultivar experts is more widely used. Pathogen presence is determined in various ways including visual inspection and laboratory tests. These tests involve culturing and identification on agar-based media, the use of test plants that show easily recognized infection signs, and the use of pathogen-specific antibodies and gel electrophoresis. If pathogens are detected, tissue culture of shoot tips and apical meristems can produce pathogen-free (including viruses) plants. Heat treatments are also used.

Rooting media is used in conjunction with misting or fog and bottom heat for asexual propagation. Media is highly variable, depending on the plant, type of plant part, whether fog or mist is used, the time of year, and economic considerations. Some media are similar to those used for seedlings, having an organic (shredded bark, compost, peat, or sphagnum moss) and inorganic component (perlite, pumice, rockwool, sand, vermiculite). Inorganic materials alone are also used, but less often.

Only the briefest generalizations are possible when discussing asexual propagation techniques. Great differences exist in the ease and methods of propagation among the large numbers of plant species. Empirical trials were needed to establish optimal propagation conditions; these have been conducted for most horticultural plants of economic interest. Others have yet to be investigated. A comprehensive book on plant propagation (see Relevant References) will provide the conditions for plants that have been examined. In addition, the asexual propagules for a large number of species can be found in Section Four.

Propagation by Cuttings

A part of a stem, root, or leaf can be cut from a stock plant, and this "cutting" can form roots and shoots under suitable environmental conditions. A low nitrogen to high carbohydrate ratio or balance in the stock plant favors rooting in many cases, so fertilization prior to removal of cuttings should be avoided. Cuttings taken during the vegetative rather than the reproductive phase root better. Cuttings are the most widely used type of asexual propagation. Many plants can be produced from a few mother plants in a small area. Cuttings are also less costly, require minimal technique, and are simpler and faster relative to other asexual methods.

Types of Cuttings. Cuttings (Figure 7–8) can be classified as stem cuttings, modified stem cuttings (sections of bulbs, corms, rhizomes, tubers), leaf cuttings, leaf-bud cuttings, or root cuttings. Stem cuttings, based on timing, location, and plant type, can be further divided into deciduous hardwood (currant, *Euonymus,* grape, *Hibiscus,* and highbush blueberry), narrow-leafed evergreen (*Arborvitae, Cupressus,* Monterey pine, *Taxus*), semihardwood (*Allamanda, Azalea, Buxus, Clematis, Pittosporum, Pyracantha,* and *Rhododendron*), softwood (*Acer, Cryptomeria, Hydrangea, Lantana,* and *Viburnum*), herbaceous (*Begonia sempervirens,* chrysanthemum, geranium, and philodendron), and stem (cane) sections (*Dieffenbachia* and *Dracaena*).

Cutting to Plant. New or adventitious roots and shoots develop from the cutting in three stages. Many cells can return to the meristematic condition (dedifferentiation) and initiate meristematic cells as root or shoot initials. In turn, these cells become root or shoot primordia that grow and differentiate into roots and shoots. The sites of origin for adventitious roots in stem cuttings are usually within or near the vascular bundle. Sites of origin vary for root or leaf cuttings.

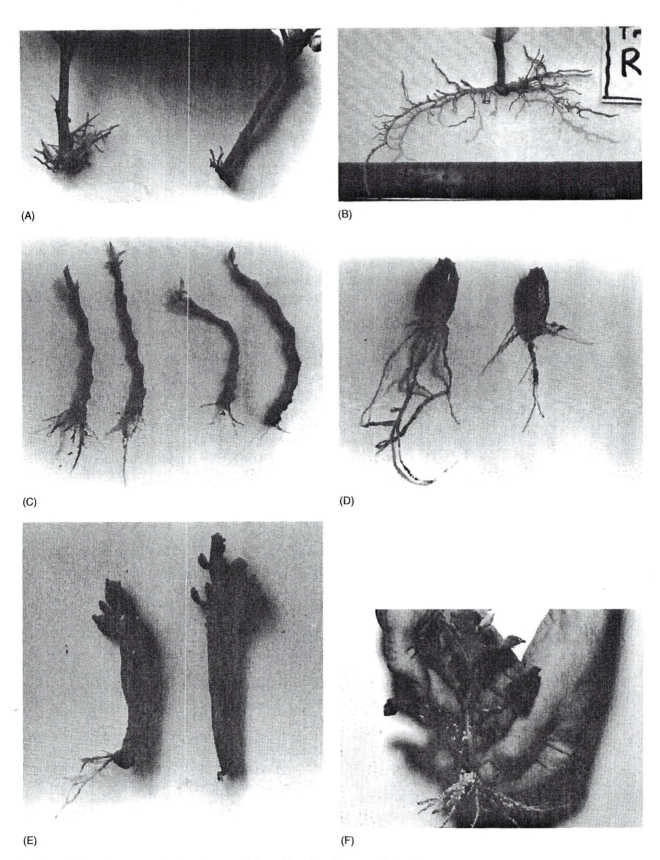

FIGURE 7–8 Examples of cuttings. (A) Broad-leaved hardwood cuttings (citrus) showing early root formation (*right*) and later root formation (*left*). (B) Citrus cutting with well-developed root system. Herbaceous cuttings are shown in C through E. Early root formation (*right*) and later root formation (*left*) are shown for *Pedilanthus tithymaloides variegatus* (C), *Stapelia* sp. (D), *Senecio deflersii* (E), and Chrysanthemum (F). (All author photographs, except for F, which is a USDA photograph now in National Archives.)

Cutting Treatments. After taking cuttings, they should be treated to prevent disease problems. One effective method is to dip cuttings in an oxidizing biocide, such as Agrobrom. This usage is preferable from the sustainable perspective as opposed to the more toxic fungicides. Even more promising in terms of sustainabilty are biocontrol agents. These are beneficial microorganisms that are antagonistic to media pathogens and also enhance rooting. These materials need additional research before they can be perfected.

Certain phytohormones and rooting cofactors, as well as synthetic root-promoting chemicals, favor the formation of adventitious roots by some cuttings. Such cuttings are often dipped into synthetic root-promoting chemicals (0.4% indolebutyric acid) to enhance rooting. Fungicides are often mixed with indolebutyric acid as a precaution against fungal infection. Treating root cuttings with cytokinins stimulates adventitious shoots, and mixtures of auxins and cytokinins enhance adventitious shoots and roots on leaf cuttings.

Environmental Conditions. Cuttings are frequently rooted under mist and with bottom heat (75° to 80°F or 24° to 27°C), as described earlier under propagation aids. Some cuttings may be rooted directly outdoors. General air temperatures for rooting are 70° to 80°F (21° to 27°C) in the day and about 60°F (15°C) at night. Light intensity during rooting can be low, 150 to 200 foot-candles, but should be higher once the cutting is rooted and growing. Optimal light intensity varies according to plant species. Photoperiod appears to have some influence. The photoperiod under which the stock plant is grown, as well as during the rooting of the cutting, can influence the rooting and shoot development of some, but not all, cuttings, depending on the plant species.

Stem Cuttings. Hardwood or narrow-leafed evergreen stem cuttings are taken from the wood of the previous or older season's growth during the dormant season. Semihardwood stem cuttings are taken during the summer from woody, broad-leaved evergreens or partially matured wood of deciduous plants. Softwood stem cuttings are derived from soft, succulent new spring growth (greenwood) of deciduous or evergreen species. Herbaceous stem cuttings come from the succulent tissue of herbaceous plants. Stem sections produce shoots from hidden buds that can be removed and rooted, or the section can be left to produce shoots and adventitious roots.

Sometimes it may be desirable to take leafy cuttings during the pruning of woody plants or pinching of herbaceous plants. If propagation is not convenient at that time, cuttings can often be held in cold storage for several weeks. Cuttings are placed in polyethylene bags and stored at 31° to 40°F (−0.5° to 4.5°C). Low-pressure storage (1/30 atmosphere) at these temperatures extends the life of the cuttings. Examples of cuttings treated this way include azalea, privet, chrysanthemum, carnation, and geranium.

Leaf Cuttings. Plants consisting of mostly leaves and little or no stem parts are propagated with leaf cuttings. With leaf cuttings the intact leaf blade (*Echeveria, Gloxinia, Streptocarpus,* and *Sedum*), leaf sections with primary veins (*Begonia rex* and *Sansevieria*), and leaf blade plus petiole (*Cyperus, Saintpaulia,* and *Tolmiea*) are used. Adventitious roots and an adventitious shoot are formed. Leaf bud cuttings consist of the leaf blade, petiole, and a short piece of the stem with the attached axillary bud. Examples of plants propagated with leaf bud cuttings include *Crassula argentea* and *Camellia*.

Root Cuttings. Root cuttings are sectioned roots usually taken in late winter or early spring when the root's stored food is high (*Bouvardia, Cordyline, Daphne, Ligularia, Plumbago,* and *Rubus*). Root cuttings produce stems from adventitious shoot buds. Orientation of the cutting is important. The end nearest the crown should be placed uppermost when cuttings are rooted in a vertical position. If in doubt about which end was near the crown, place the cutting in a horizontal position in the propagating media. Plants with prickles, such as blackberry and raspberry, are preferentially asexually propagated with root cuttings to avoid the unpleasant handling of prickly stem cuttings.

Propagation by Layering

Layering is an old method of asexual propagation. With layering, adventitious roots are formed on a stem while still attached to the parent plant. After roots are established, the stem is severed and planted. Unlike a cutting, there is no critical maintenance of the attached stem. Layering may be natural, as with *Forsythia* or trailing blackberries, or it may be artificially induced. Layering is simple and usually successful, but it is expensive on a large scale since it is not easily mechanized and requires much hand labor. Gardeners and homeowners find it useful for the propagation of the occasional woody plant that they wish to increase in number or to give away. Nurseries do still use this method for difficult-to-root cultivars and high-value plants.

Bending a stem downward, fastening it in place, and covering it with soil or a rooting medium can induce layering (Figure 7–9). The darkening of the stem causes etiolation,

FIGURE 7–9 Types of layering. Arrows indicate where plant is severed after rooting is completed. The rooted daughter plant is then dug and planted elsewhere. (From Raymond P. Poincelot, *Horticulture: Principles & Practical Applications,* 1980. Reprinted by permission of Prentice-Hall, Upper Saddle River, New Jersey.)

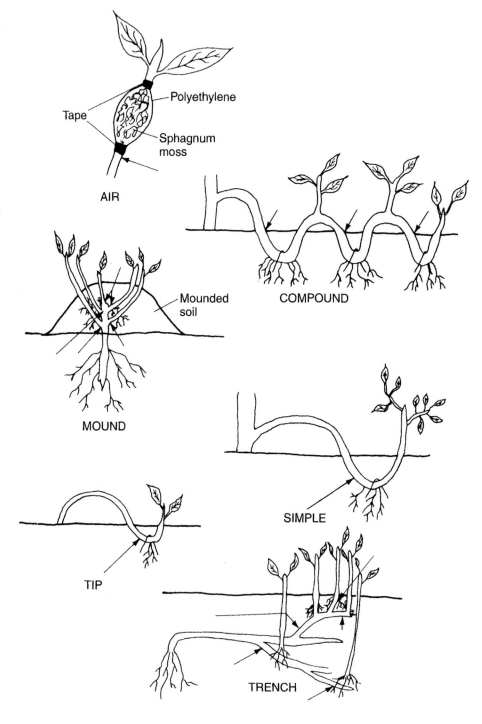

which increases the formation of adventitious roots. If only the region of the stem near the current season's shoot is covered, this is termed tip layering. This form of layering is used for *Forsythia* and *Rubus* sp. When the submerged bend is 6 to 12 inches back from the tip, it is termed simple layering. This procedure is useful for hard-to-root shrubs, such as *Rhododendron, Jasminum,* and *Weigela*. If the stem is alternately covered and exposed along its length, such that each exposed section has at least one bud, it is termed compound or French layering. When long, flexible vines (*Philodendron, Wisteria*) are propagated this way, it is termed serpentine layering. Other forms of ground layering include mound (stool) layering for clonal apple stocks and quince and trench layering for propagating difficult-to-root fruit tree rootstocks. In the former, several shoots are cut back and mounded with soil. The latter technique requires that the plant be laid horizontally in a trench and covered with soil.

Layering done on aerial stems is known as air layering. This procedure is useful for some tropical and subtropical shrubs, such as *Ficus, Codiaeum,* and *Coccoloba,* and for temperate woody trees and shrubs whose stems are too rigid to be bent to the ground. It is especially useful for the house plant that gets lanky and leafless at the base, such as the rubber tree. A stem, usually woody, has a strip of bark removed. A cut is made partially through this area. The cut is prevented from closing with either a piece of toothpick or wad of sphagnum or peat moss. Next a wet ball of sphagnum moss is wrapped around the area and then covered with polyethylene plastic film. After roots become visible, the rooted part is cut off and transplanted.

Propagation by Crown Division

A crown is usually that part of the plant at ground level from which new shoots are produced. The crown is an arbitrary zone of demarcation between the shoots (aerial plant parts) and roots. With many plants the crown increases with age by the formation of new shoots and adventitious roots. Crowns in a sense develop through layering and, as such, provide a means for propagation. Propagation of crowns from herbaceous plants (chrysanthemum, *Cyperus, Spathyphyllum, Maranta,* and *Campanula*) and some woody shrubs (*Sorbaria, Viburnum, Spiraea*) is by crown division.

Many herbaceous plants require crown division every 2 or 3 years to prevent overcrowding, and the divisions provide some increase of plant numbers. This method is easy and generally successful, but of importance mostly for gardeners, because of limited returns. Generally, smaller crowns are cut apart with a knife, whereas larger ones may require a hatchet or shovel. Spring- and summer-blooming plants are usually divided in the fall; those that bloom during late summer or fall are divided in the spring. Indoor potted plants are usually divided whenever necessary to prevent overcrowding.

Propagation of Specialized Stems and Roots

Other plants have growth habits or structures that lead to propagation naturally. Some of these are essentially natural forms of layering, although they can be accelerated, for example, by pinning the vegetative structure to the soil. Examples of the structures adapted to natural layering include runners, stolons, crowns, offsets, and suckers. These specialized stem structures were described in Chapter 3.

Certain specialized stems and roots are involved in food storage. These fleshy organs, already described in Chapter 3, include bulbs, corms, tubers, tuberous roots, rhizomes, and pseudobulbs. These structures contain buds that enable the plant to survive dormancy and, after dormancy is broken, to produce shoots and to grow. As such, the organs are well suited to natural and induced propagation. Generally, these specialized organs are found in herbaceous, perennial plants that undergo dormancy as a result of warm-to-cold or wet-to-dry cycles found in temperate and tropical-to-subtropical areas, respectively. If the specialized organ is naturally detachable, the propagation method is termed separation, as opposed to division when the structure is cut into sections.

Runners and Stolons. Runners are specialized stems that grow horizontally along the ground. Their point of origin is from the axil of a leaf at the crown of the plant. The strawberry,

Chlorophytum sp., *Episcia* sp., and *Saxifraga sarmentosa* (strawberry geranium) are examples of plants propagated from runners (Figure 7–10). The daughter plants that form on the runner at nodes are cut and dug after rooting. Some plants produce a similar form of specialized modified stem, the stolon. Aboveground prostrate or sprawling stems, such as with woody species like *Cornus stolonifera* or herbaceous species like Bermuda grass, are often considered to be stolons. However, some prefer to restrict the term stolon only to underground stolons, such as those found with mint (*Mentha* sp.) or those that develop into tubers, like the potato. After natural layering, stolons can be cut from the parent plant and planted elsewhere. Runners of house plants, such as the spider plant, can be pinned with a clip to a pot of soil and rooted easily.

Offsets.   An offset, sometimes called an offshoot, is a lateral shoot or branch that arises from the base of the main stem. Bulbs also produce offset bulblets at their base. Often, offsets have a short, thick, rosette-like appearance. Offsets may also be lateral shoots arising from rhizomes. Rooted offsets are cut close to the main stem and planted, or if rooting is incomplete, offsets are treated like leafy stem cuttings. Many succulents (Figure 7–11) and the pineapple can be propagated by offsets.

Suckers.   Suckers are shoots that arise below ground from an adventitious bud on a root. Shoots that arise from the vicinity of the crown are termed suckers, although by definition they are not, because they arise from stem and not root tissue. Technically, those shoots arising from an adventitious stem bud are water sprouts. Suckers may be dug and cut after natural rooting or treated as cuttings. The red raspberry produces true suckers.

Bulbs.   Bulblets develop from parent bulbs and at maturity are called offsets. These can be propagated by separation and will eventually reach flowering size. Bulbils (aerial bulblets formed in the leaf axils of some *Lilium* species) can also be propagated in the same manner. When it is desired to increase the production beyond offset numbers, three processes are used. Individual bulb scales are removed from the mother bulb, and under favorable growing conditions adventitious bulblets form at the base of the scales. This method, called scaling, is particularly useful for lilies.

Other bulbs, such as the hyacinth, can be propagated through basal cuttage, a form of wounding that causes multiple development of adventitious bulbs. Other bulbs may be

FIGURE 7–10 *Chlorophytum comosum,* the very popular hanging basket houseplant known as the spider plant, produces runners that are easily rooted. This plant is very effective at removing various gaseous air pollutants found indoors. (Author photograph.)

FIGURE 7–11 Offsets shown on *Hoodia gordonii* (*left*) and *Greenovia aurea* (*right*). (Author photograph.)

propagated with bulb cuttings, such as *Hippeastrum* (amaryllis). Bulbs are cut into portions that contain fractions of three or four scale segments and part of the basal plate (short, fleshy vertical stem axis). The cut areas are dusted with fungicide and callused at 70°F (21°C). They are subsequently treated as leaf cuttings. Bulblets appear and are grown to maturity.

Corms.   Propagation of corms (gladiolus and crocus) is primarily through natural formation of new corms and offsets or cormels (Figure 7–12). Both of these types are separated and cured at 95°F (35°C) for 1 week. Big corms with multiple buds can be sectioned, so that each section contains a bud. These sections will produce a new corm. Fungicide dusting is needed to prevent decay of the exposed tissue.

Tubers.   Whole tubers can be propagated by planting. Larger numbers can be produced by cutting into sections, each of which contains a bud or "eye" (Figure 7–13). Cut sections are allowed to heal or suberize at 68°F (20°C) and 90% relative humidity for 2 or 3 days. Alternately, the cut surface can be treated with a fungicide. Examples of plants propagated by tubers include *Arisaema, Eranthis, Gloriosa, Solanum tuberosum* (Irish potato), *Caladium,* and *Helianthus tuberosus* (Jerusalem artichoke).

Rhizomes.   Division is possible for plants with a rhizome structure (Figure 7–14). Rhizomes can also be sectioned, with each section having a lateral bud, as is done with bamboo, bananas, German iris, certain grasses, and *Agapanthus*. Shoots produced on large rhizomes can also be removed and treated as cuttings.

FIGURE 7–12
Gladiolus corm showing formation of recent small cormels, new well-developed corms (*upper right and left*), and the old withered corm (*bottom*). (Author photograph.)

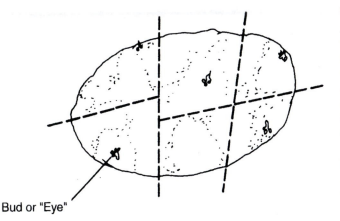

Bud or "Eye"

FIGURE 7–13
A diagram of a potato tuber indicating cuts by dotted lines. Each section should contain one node and weigh about 1 to 2 ounces. (From Raymond P. Poincelot, *Horticulture: Principles & Practical Applications*, 1980. Reprinted by permission of Prentice-Hall, Upper Saddle River, New Jersey.)

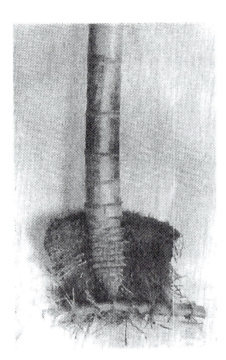

FIGURE 7–14 Rhizome propagation. A rhizome section from bamboo has produced adventitious roots and a well-developed shoot. (USDA photograph.)

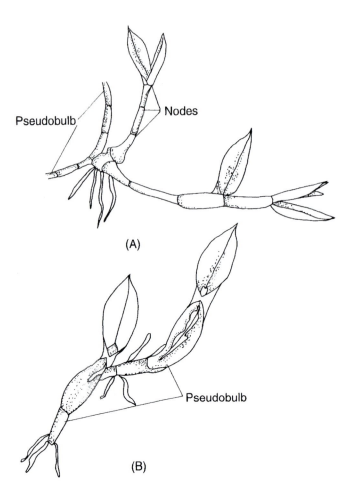

FIGURE 7–15 Orchid pseudobulbs. (A) *Dendrobium* pseudobulb, which has formed a rooted offshoot. (B) A rooted rhizome section from *Catteleya* showing new growth at the pseudobulbs. (From Raymond P. Poincelot, *Horticulture: Principles & Practical Applications,* 1980. Reprinted by permission of Prentice-Hall, Upper Saddle River, New Jersey.)

Pseudobulbs. Pseudobulbs are produced by many orchids. Some pseudobulbs form rooted offshoots that can be cut and potted (Figure 7–15). In others, the rhizome with attached pseudobulbs is planted in rooting media. Growth will begin at the pseudobulb base.

Tuberous or Storage Roots. Tuberous roots can be propagated either from adventitious shoots or division. Sweet potatoes (*Ipomoea batatus*) can be forced to sprout at 81°F (27°C), and these sprouts will eventually form roots. The rooted shoots can be removed and planted. Division (Figure 7–16) is used for tuberous root plants such as the dahlia. Other plants with tuberous roots are *Alstroemeria* and *Ophiopogon*.

Propagation by Grafting and Budding

Grafting is an ancient horticultural art in which parts of plants are joined such that the tissues unite and growth follows as for one plant. The upper part of the graft union is the scion, and the lower is the rootstock or understock, or simply the stock. Budding is a special aspect of grafting in which the scion is a small piece of bark or wood with a single bud. Most fruit tree and hybrid rose cultivars are sold as grafted plants.

Not all plants can be grafted successfully, because grafting incompatibility causes failure of the graft union of many combinations. As the botanical relationship becomes more

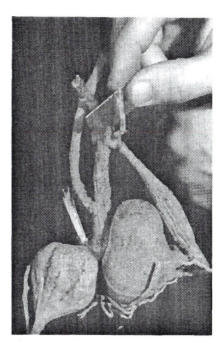

FIGURE 7–16
Divisions of a garden dahlia tuberous root must include one adventitious shoot per section. (USDA photograph.)

distant, the chances of a successful graft diminish. Grafting within a clone is assured, and grafting between clones within a species is almost always possible. The chances of successful grafting between species within a genus are less and are remote to nil for grafting between genera within a family and grafting between families.

Reasons for Grafting. Many reasons exist for the use of grafts. Grafting is the economically feasible method for plants that are difficult to propagate by usual means, such as cuttings. Desirable cultivars that have poor root systems can be grafted onto superior rootstocks. Grafting rootstocks can provide increased tolerance to unfavorable soils or plant pathogens in the soil. Other rootstocks induce increased vigor or dwarfing in the scion. Grafting is useful for replacing undesirable cultivars on good, established trees, using the existing rootstock. This approach is termed top-grafting or top-budding, or even top-working. Certain forms of tree damage can be repaired through grafts. Grafting is also useful for the creation of unusual or novelty plant forms, such as "tree" roses, combination fruit trees, or colorful mutant cacti. Fruiting may also be improved by grafting a scion from a staminate plant onto a pistillate plant. Virus detection is also possible if a plant suspected of carrying a virus is grafted onto a susceptible plant.

Sometimes a third, different plant section may be inserted between the rootstock and the scion. This section is designated as an intermediate stem section, an intermediate stock, an interstem, or an interstock. This approach, called double-working, is used to overcome incompatibility between the scion and rootstock or to provide desired qualities not present in the conventional graft.

Grafting Methods. The most important part of grafting is the formation of a successful graft union. The required close matching of the callus-producing tissues of the cambium layer tends to restrict grafting to the dicotyledons in the angiosperms and to the gymnosperms. Both have a continuous tissue of vascular cambium between the phloem and xylem. Temperatures must be favorable to high cell activity (55° to 90°F or 12.8° to 32°C). High moisture content is essential to prevent desiccation of the newly arising callus tissue. Waxing the graft union or wrapping it in a moist medium helps it maintain moisture. Prompt waxing also helps to prevent infection by plant pathogens. The graft union must also mesh tightly. Wedging, wrapping, nailing, or tying helps create a tight fit. Tightness will prevent the dislocation of cellular growth, which could destroy the graft union. Finally, the scion and rootstock must be at the proper physiological stage. Generally, the scion is dormant and the rootstock is dormant or active, depending on the grafting method.

There are essentially two general forms of grafting: detached scion and approach. The first is the most commonly used; roots are present only on the rootstock and the scion is detached from the donor plant. The second method involves the grafting of two plants, both of which are self-sustaining. After the union is formed, the scion is severed from the donor plant. This method is employed with plants that are difficult to graft. Numerous grafting and budding techniques are used to make either of these graft types. Some are shown in Figure 7–17. These differences result from the differing ways on which pieces can be joined or meshed.

Micropropagation

Micropropagation (Figure 7–18) is essentially another tool for the plant propagator. It differs from the previous vegetative or cloning methods in terms of scale and technological sophistication. Seeds and large parts of plants, such as cuttings or leaves, are not used. Instead, small parts of plants (explants) are used to propagate new plants. Embryos, stem tissue, shoot tips, apical meristems (frequently free of pathogens), cotyledons, leaf tissue, seed parts, root tips, bud tips, bulb scale sections, various flower parts, callus (parenchyma), one to many cells, and pollen grains are examples of explants. Another difference is the propagation media. Instead of artificial soils, agar or liquid cultures are used. Contamination with microorganisms is an even more serious problem with micropropagation than the other types of propagation, so facilities and all materials must be pathogen free (aseptic culture). Because of these differences, costs are higher for micropropagation. Generally, high-value plants are the most likely candidates for micropropagation. Foliage plants are often micropropagated, with about 50% of these plants produced by this technique.

The terms *micropropagation* and *tissue culture* can be confusing. Tissue culture refers to the maintenance and growth of plant tissues in an aseptic culture. Tissue culture can be utilized as the starting point for micropropagation of plants, but it can also be used as a tool of genetic engineering to produce either genetically altered plants (plant breeding) or to produce biomass for the harvest of biochemicals. Micropropagation, tissue culture, genetic engineering, biomass, and harvesting of plant-produced biochemicals are all included under the umbrella term *biotechnology.*

Advantages.   One advantage of micropropagation is the rapid, vegetative propagation of numerous plants. With conventional propagation, you are limited by how many cuttings a plant yields or how many divisions can be made or how many seeds are produced. Under optimal conditions with micropropagation, a million plus plants of a given cultivar can be produced in a year. Although expensive, a number of situations do justify the costs. Candidates include high-value plants that grow slowly (bulbs, ferns, herbaceous perennials, foliage house plants, orchids, and palms), new cultivars having limited stock and high market demand (a new cultivar produced by conventional plant breeding or genetic engineering), very high-value cultivars where increased costs are justified (certain woody landscape plants), cultivars that are hard to root by conventional techniques (mountain laurel), and endangered plants (to save species on the verge of extinction or to satisfy market demand to prevent overcollecting).

Another plus is the ability to produce pathogen-free plants. This technique can produce certified, virus-free plants that command higher prices (many berries) or rid a valuable clone of various pathogens. Pathogen-free plants are also essential for the movement of plants in international trade. Micropropagation can also be used for the clonal propagation of parental plants used for the production of hybrid seed (flower and vegetable hybrids used for gardens and farms). Another advantage is that explants can be stored frozen (cryopreservation) and micropropagated later on thawing, thus being useful for germplasm preservation. Lastly, micropropagation is adaptable to year-round production, whereas many propagating nurseries are more seasonal.

Disadvantages.   The main disadvantage is the high cost. Micropropagation requires sophisticated and highly technical facilities and professional workers. College degrees in the sciences are essential for the propagators. Start-up costs are high and labor costs and

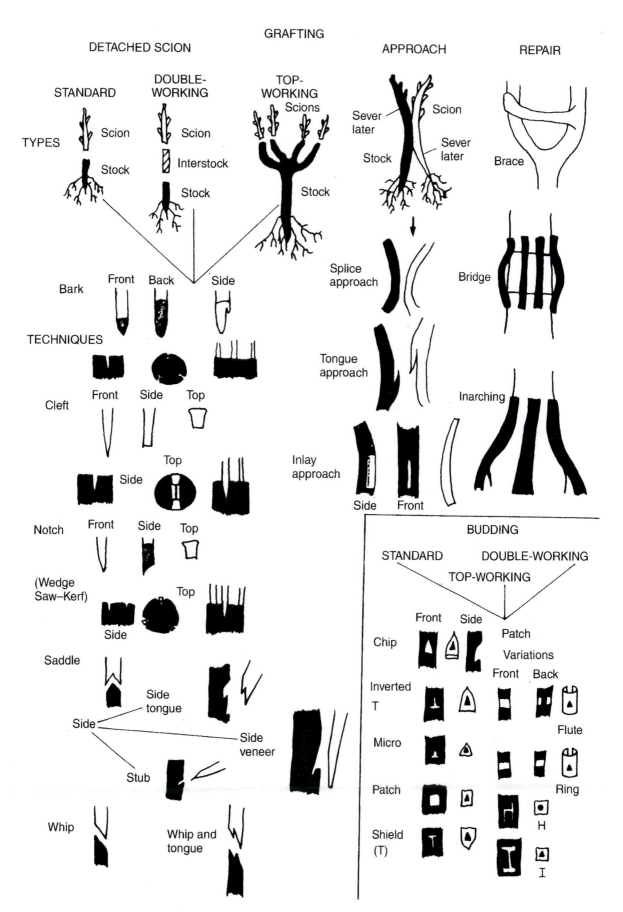

FIGURE 7–17 Grafting techniques used during detached scion (standard, double-working, and top-working), approach and repair grafting are illustrated. Budding techniques are also shown. Detached scion grafting and budding are used for trees that are relatively easily grafted. Approach grafting is for more difficult trees. Repair grafting is used for damaged trees with sufficient value to warrant saving them. (From Raymond P. Poincelot, *Horticulture: Principles & Practical Applications,* 1980. Reprinted by permission of Prentice-Hall, Upper Saddle River, New Jersey.)

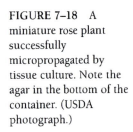

FIGURE 7–18 A miniature rose plant successfully micropropagated by tissue culture. Note the agar in the bottom of the container. (USDA photograph.)

supplies are continuously high. The cost for lab equipment alone in a small operation can run about $50,000. Losses can be severe if disease or insects contaminate the facility or plants. Quality control needs to be rigorous, given the possibility of off-type plants being produced during micropropagation. Management skills must be very good in terms of cost control and marketing.

Facilities and Equipment.   Facilities for micropropagation vary considerably in size from the small hobbyist and nursery setup to the major research laboratory to the very large commercial micropropagation factory. All three, regardless of size, must have preparation, transfer, and growing areas. Because of the need for aseptic conditions throughout the entire facility, the building should be separate from other horticultural operations and have limited entry. All areas must be kept extremely clean, including all equipment. Workers usually wear protective clothing to minimize contaminant introduction.

The preparation area functions include glassware cleaning, media preparation, media sterilization, and storage for chemicals, supplies, and media. Media is usually either agar (solid) or liquid culture. Various biochemicals are also incorporated into the media, such as inorganic salts, sugars, vitamins, hormones, and growth regulators. High-quality dishwashers, various types of glassware, autoclaves, balances, refrigerators, mixers, pH meters, filters, water purification equipment, and automatic pipettes are essential components.

The transfer area is where explants are transferred into the prepared, sterilized media. A special laboratory hood is essential for the transfer area, a laminar flow hood. The hood uses air filtered through high-efficiency particulate air filters and ultraviolet radiation to remove all contaminants and a positive pressure is maintained in the hood. This feature ensures that air blows toward the user and into the room as the explant transfer is performed. Any airborne contaminants are swept away from, not into, the media. Of course, all the media, working surfaces in this room (and other areas), and manipulating tools must be sterile. Hairnets, lab coats, and latex gloves are also needed.

The growing area is where the media and the explants (cultures) are placed for growth under controlled light, temperature, and humidity conditions. Various forms of fluorescent lights are used to provide the needed light. Cultures are usually in test tubes, petri dishes, or flasks. These glass or plastic containers are sealed with cotton plugs or plastic caps or plastic film. Covers need to allow for gas exchange, but also retain moisture and prevent contaminant entry. Cultures usually require additional transfers to different media in order to stimulate first shoot formation followed by root stimulation. Eventually cultured plants have to be moved to greenhouses for hardening or acclimatization and even cold frames, if they are destined for the outdoor market.

Media.   No standardized, single medium is used for micropropagation. The medium varies with the plant being micropropagated, the medium changes with shoot and root inducement, and the basic support component can be either solid (agar) or liquid based. Media can be prepared or purchased commercially. Two of the better known media formulations are the Murashige and Skoog and the Woody Plant Medium. The former is useful for herbaceous plants and the latter, woody plants.

Some common features do exist. All media formulations usually incorporate certain chemicals, whether agar or liquid systems are used. Macroelements and microelements needed by plants are supplied through the inclusion of inorganic salts. These include boron, calcium, cobalt, copper, iodine, iron, magnesium, manganese, nitrogen, phosphorus, potassium, and zinc. Some form of carbohydrate is also required, since photosynthesis is not operational during micropropagation. Sucrose is the usual choice, although a few other sugars, such as glucose or fructose, have been used. Vitamins are also added. Thiamine appears to be critical, and pyridoxine and nicotinic acid are often needed. Other helpful vitamins include biotin and pantothenic acid.

Hormones are also critical for the development of shoots and roots during micropropagation. Auxin and cytokinin are critical. Sometimes gibberellins are added. The ratio of auxin/cytokinin is varied to induce shoot (low/high) and root (high/low) formation (organogenesis). Therefore formulations need to be changed as micropropagtion moves forward. Lastly, antioxidants, such as citric acid, or ascorbic acid, are added. Other occasional components include inositol, myo-inositol, and glycine.

Certain plants that are hard to micropropagate do not respond to known formulations. Additives have been used successfully when this situation occurs. Hydrolyzed casein, yeast extract, coconut milk, orange juice, tomato juice, and malt have been used with good results.

Methods.   Many techniques exist for the micropropagation of plants by the tissue culturing of different plant tissues and organs. The techniques can produce seedlings, plants, callus (undifferentiated tissue), and somatic embryos (vegetative embryo). Regardless of which technique is used, all have four basic steps in common: establishment and stabilization, shoot multiplication, root formation, and acclimatization.

During establishment and stabilization, explants are chosen, removed, disinfected, transferred to the culture media, and readied for shoot multiplication. Stock plants used to supply explants must be free of internal bacteria, fungi, and viruses. If plants are not known to be free of microbial contaminants, then special tests must be conducted on some explants to verify that the stock plants are contaminant free. Even with "clean" plants, disinfestation of the explants is done to remove surface contaminants. Chemical treatments used for this purpose include diluted bleach, various types of alcohols, hydrogen peroxide, silver nitrate, benalkonium chloride, and mercuric chloride. From the sustainable perspective, one should work with the first three listed compounds. The explant is then transferred to the appropriate media, using the laminar flow hood and sterile equipment. Usually the media will be higher in cytokinin than auxin. The period from transfer to shoot production (microshoots must be stable enough to undergo further culture to produce more shoots) is called stabilization. Many herbaceous explants need only 1 or 2 months to achieve stabilization. Woody plants can require up to 12 months.

The next step, shoot multiplication, starts with the division of the multiple microshoots into individual microshoots. Microshoots are transferred to new media that can be the same media as before, although sometimes the levels of some biochemicals are increased. This process, called subculturing, is repeated every 1 or 2 months. Limits are usually set until some number is reached or it is known that beyond a certain amount of subculturing, the risk of abnormal microshoots increases.

The next procedure, root formation, requires the transfer of the microshoots to a new medium, usually one that is auxin rich relative to cytokinin. In a few cases, microshoots form roots without this treatment. In some cases, microshoots can be treated with auxin and rooted in normal artificial soil media. In the last step (acclimatization), plantlets are removed from culture media, washed, and transplanted into normal artificial soil media. These plantlets are grown first under high humidity conditions and gradually "weaned" from high to low humidity and from fluorescent to natural light in the greenhouse. From this point on, the plants are treated like any normal plant.

PROPAGATION AFTERCARE

Generally, the sexual propagation mixes covered earlier in this chapter or outdoor seedbeds have been amended with moderate levels of nutrients. Nutrients are usually sufficient to maintain the seedlings up to the transplanting stage. Rooting media often has little or no nutrient content. Most asexual propagation media contains little or no available nutrients compared with outdoor propagation beds. The vegetative propagules, such as cuttings or bulbs, are usually dependent on stored tissue nutrients only, which are usually adequate to maintain them during the time of root formation. Under some conditions, nutrients may become self-limiting, and supplementary soluble fertilization will become necessary. Conditions include slow-rooting propagules, a delay in transplanting of the rooted propagule, and the fact that rooting of some species is improved by supplemental nutrients.

If plants must be held in the propagation media or outdoor seedbed past transplanting time, it will be necessary to add additional fertilizer at levels recommended for seedlings and rooted propagules. Fertilizers in soluble form can be added by hand watering or more often through automated means, such as a nutrient mist or fertigation. This aspect was covered earlier under postplant fertilization in the propagation and water sustainability section. Supplementary fertilization for outdoors will be covered later in Chapter 11.

Events following propagation, although not part of the process, seriously affect the successful establishment of the recently propagated plant.

Transplanting and Hardening

Transplanting of seedlings (Figure 7–19), ferns, or asexually propagated materials is normally not a problem if root disturbance is kept minimal, unless there are associated environmental changes. Examples of environmental changes are transplanting indoor seedlings outdoors, moving outdoor foliage and flowering plants indoors, or transplanting cuttings from under mist to an unmisted site. For success, the plant material must be conditioned to accept the different conditions of the new environment.

Plants moved from indoors to outdoors must be hardened off (acclimatization or conditioning). Hardening involves lowering the temperature, withholding water, and introducing a gradual change in the environment. This slows growth and builds a carbohydrate reserve that supplies the needed sugars for rapid reestablishment of the rooting system. Cold frames (see Chapter 13) are used to harden plants for about 7 to 10 days prior to transplanting outdoors.

Field-grown plants, such as recently propagated foliage and flowering plants destined for indoor use, must be conditioned or acclimated to the reduced light levels indoors. One way is to transplant the field plants into containers and then place the containers in a lathhouse, where the development of shade leaves will allow it to survive in a lower light situation. Even better is to produce field-grown foliage plants (except for large-trunk-producing types) under shade right from the start.

Plants propagated under mist must be hardened-off to enable them to survive lower levels of relative humidity. Misting periods may be gradually decreased, and then the plants are transplanted; or the plants may be transplanted and placed in a humid, shaded area or another misting system operated at reduced levels.

Conditions that favor reduced transpiration after transplanting increase the chances of success. Favorable environmental factors include lower temperatures, lower light intensity, low or no wind speed, and high relative humidity. Best transplanting occurs during cloudy days, days of light rain, periods of no wind, and late afternoons. Plants should be watered well afterward. Shade can be provided after transplanting, if possible. A soluble fertilizer (starter solution, high in phosphorus) is sometimes added to aid the reestablishment of the plants, but not if soil moisture is low. The increase in soluble salts might damage the plant under conditions of low soil moisture.

(A)   (B)   (C)

(D)   (E)   (F)

(G)

(H)

FIGURE 7–19   Seedling transplanting steps. (A) Seedlings should be transplanted at about the two true leave stage up to when leaves of seedlings begin to contact each other (*left*). Seedlings on the right have been allowed to grow well past the transplanting stage. (B) Seedlings can be transferred with a spatula or by hand. (C) If by hand, seedlings should be held only by the top leaves and not the stem, which is too easily damaged. (D) Seedlings are carefully handled to minimize root damage and placed into a hole made by a finger or dibble. Next media is gently firmed around the seedling. If flats, rather than individual containers, are used, adequate space for development should be provided. (E) Adequate cultural care (water, fertilizer, temperature, and humidity control) is required. Plants on the left were improperly cared for as opposed to good care on the right. (F) These 8-week-old seedlings are ready to be hardened off prior to outdoor planting. (G) Timing of bedding plant production is very important to achieve proper size and developmental stages for marketing. Plants shown here are at maximal acceptable sizes. One reason for widespread use of plugs is the simplicity of transplanting. The plug is easily removed and inserted in a cell pack (see Figure 7–20). No root damage or transplanting setback occurs and the operation is much faster. (H)Excellent root formation and no damage is achieved with cell pack plants. Note the healthy white roots showing no damage. (All photos, except C, G, and H (author photographs), are USDA photographs.)

In some instances recently rooted plants can be held in cold storage prior to transplanting. Cuttings rooted in late summer or early fall for spring planting can be protected over the winter if outdoors or allowed to grow on indoors. An alternative is to dig them and put them bared-rooted in polyethylene bags at 32° to 40°F (0°C to 4.5°C). Reduced atmospheres (1/30 atmosphere) seem to improve storage life.

Containers for Growing-On Transplants

Transplanting may be done directly into outdoor sites or containers maintained indoors or outdoors. Containers described in an earlier section in this chapter may be used. In addition, larger containers are available that are more suitable for container stock. These include plastic and metal containers (Figure 7–20) in sizes from 1 quart to 15 gallons. These are available commercially or reclaimable from restaurants, canneries, and bakeries. They are useful for nursery plants and easily adapted to mechanized techniques.

(A)   (B)   (C)

(D)   (E)

FIGURE 7–20 Growing-on containers. (A) Cell packs in tray designed for easy handling and sales. Most garden bedding plants (flowering annuals and vegetables) are now sold in this form. Production is easier and the gardener sees better transplanting results. (B) and (C) Square and round pot packs are popular for sales of herbaceous perennials, high-end flowering annuals such as geraniums, and late spring, larger flowering annuals. (D) Plastic containers shown here are long and narrow for formation of taproots. These trees will eventually be moved upward to larger plastic or metal trees (*see E*). (E) Various plastic, metal, and clay containers. The larger plastic and metal containers (*middle and right back, side right*) are used for marketing trees and shrubs and larger herbaceous perennials. The large and smaller plastic pots (*left corner and middle left*) are used for marketing of house plants, as is the plastic hanging basket in the left foreground. Clay pots (*right foreground*) are used much less for house plants, but are still popular for cacti and other succulents sold as house plants. (Author photographs.)

Growing Media

Growing medium is needed for transplants going into containers that will be placed in an indoor or outdoor site. Materials utilized to prepare the media and their pasteurization are similar to those described earlier for propagation media. The main difference is increased nutrient levels. Even so, all of these mixtures will need supplemental fertilization with time. Sometimes particulate gels based on starch are incorporated into the media to increase water-holding ability. These water-retaining polymers can hold 150 times their dry weight in water. Their costs restrict usage to high-value container crops.

Winter Protection

Container plantings left outdoors during the winter often need protection. Container and ball-and-burlap nursery stock are more susceptible to low-temperature root kill in a reduced soil volume as opposed to in-ground sites. Damage can also occur through bark splitting and desiccation. Colder areas in the Northeast and Northcentral regions are at greater risk.

Storage and protection should be started from mid-October through early November. The storage structure is usually erected right over the container plants at their growing sites, because the labor involved in plant movement is not economical. The structure usually consists of pipe hoops covered by a sheet of clear or milky polyethylene (Figure 7–21). Supports may be needed in areas that experience heavy snowfalls. White polyethylene is suggested when early flowering or growth is disadvantageous and clear polyethylene when it is not. Containers should be irrigated prior to storage.

EXTERIOR

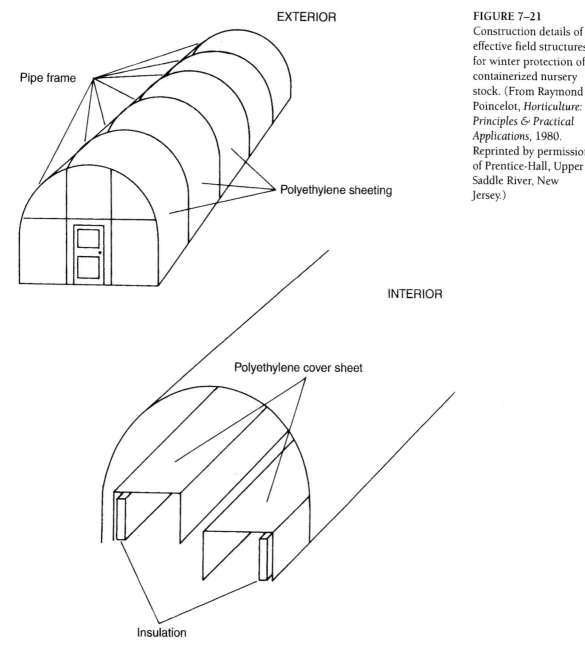

Pipe frame

Polyethylene sheeting

INTERIOR

Polyethylene cover sheet

Insulation

FIGURE 7–21
Construction details of
effective field structures
for winter protection of
containerized nursery
stock. (From Raymond P.
Poincelot, *Horticulture:
Principles & Practical
Applications,* 1980.
Reprinted by permission
of Prentice-Hall, Upper
Saddle River, New
Jersey.)

Plants with tender roots subject to root kill may require additional protection. This is usually a polyethylene blanket that is draped over the plants and tucked at the edges. Styrofoam sheets may be placed between the outermost containers and the polyethylene sheet. Desiccation injury is minimal with the polyethylene sheet.

SUMMARY

The increasing of plant numbers is done through propagation. Annual plants used as flowers and vegetables are continually propagated because of harvest and their demise in one growing season. Plants also die from pests, diseases, and old age and need to be replaced. Plants are also needed to fill holiday needs, special occasions, and installations of new landscapes.

When minor genetic change is acceptable, many plants are needed and cost is a factor, propagation by seed is the choice. When no genetic variation is tolerable, asexual propagation by nonseed parts such as cuttings is used. Choices of methods are based on cultural, marketing, and economic factors. Seeds are widely used for vegetable and flowering bedding plants in home and commercial applications, as well as for direct sowing of vegetables and flowers in gardens and fields that are difficult to transplant. Other seed applications include lawns, rootstocks for grafts, and common conifers, trees, and shrubs. Fruits and high-value landscape tree, shrub, and perennial cultivars are usually asexually propagated. Numerous specialized horticultural industries and institutions are involved in the support of propagation activities.

Most propagation operations use artificial soils as their propagation media. These artificial soils are based on formulations developed at the University of California at Los Angeles and Cornell University and now available in many commercial forms. These materials are standardized products that are light in weight and available in pathogen-free forms. Most products for seed germination and growing on of container/potted plants contain either vermiculite or perlite as the inorganic component and peat moss as the organic part. Nutrient formulations and wetting agents are added. Products for rooting of cuttings and other asexual techniques usually consist only of an inorganic material such as vermiculite or perlite. These media have properties very similar to soil. The main sustainability issues are resource limits, and both vermiculite and perlite are energy-intensive products. Possible future components include rockwool, calcined clay, compost, shredded bark, and wood shavings.

When certain materials are used alone (bark, compost, peat moss, sand), formulated into media on site, or recycled, treatment to remove pathogens and possible pests is necessary. Heat treatments by steam are best from the sustainability perspective. Depending on temperature, treatments can kill all organisms (sterilization) or only harmful ones (leaving beneficial ones alive). The latter process, pasteurization, is preferable. Chemical fumigants and drenches are also available, but their use should be limited and in the context of an overall IPM program.

Propagation is usually carried out in a greenhouse, although some high-technology forms are carried out in special laboratory facilities. Some propagation and growing on of container plants is done in a field nursery. Good nearby water supplies are critical to most propagation operations, because seedlings, cuttings, and plant tissues are more sensitive than older plants to water contaminants. Soluble salt levels and boron must be low and hardness must be reduced with ion-exchange resins. The first two are harmful to plants and the last damages misting and irrigation equipment.

Water has some sustainability issues, scarcity and pollution. Drip irrigation and water recycling are helpful for the former. For the latter, reduced inputs of fertilizers and pesticides are essential. Very low levels of fertilizer should be used during propagation. At later growth stages soluble fertilizers should be fed through the irrigation system at dilute but frequent applications. Slow-release fertilizers can also be used. Overhead sprinklers cause more polluted runoff than drip irrigation when soluble fertilizers are used. Slow-release forms are a better choice with overhead sprinklers. Fertilizers should be used only when active growth is in progress. Pesticide use should be minimal and only in an IPM context.

Runoff can be purified of nutrients and pesticides by treatment through constructed wetlands. Further treatment of this water can result in water that can be recycled in the propagation facility.

The most used propagation structure is the greenhouse, in which careful control of environmental factors for propagation is possible. Greenhouses are designed with minimal support structures and maximal glazing with glass, plastics, or fiberglass to allow maximal light entry. Equipment to provide heat, humidity, irrigation, and ventilation is also included. Plants propagated in greenhouses need conditioning before outdoor use. Cold frames and lathhouses are often used to condition recently propagated plants. Careful sanitation is required in these facilities as well as with propagation containers and propagating tools.

Numerous types of containers are available for propagation. Flats or shallow trays for many years and more recently (and highly popular) plugs made from plastic are used for seed propagation. Flats are also used for asexual propagation. Plastic pots and larger plastic containers are most popular for growing on foliage plants and nursery stock, respectively. Plastic plugs in their egg carton patterns are widely used for bedding plant production.

Bottom heating and misting are essential to propagation. Cuttings and seeds are best propagated at temperatures warmer than those needed for development. Raising the temperature to these higher levels in the greenhouse is energy wasteful and costly. Mixed plantings with different temperature requirements may also prevent temperature changes. The most economical approach is to heat the propagating containers from their bottoms with electric cables, hence the term *bottom heating*. Once germinated or rooted, the cables can be turned off.

The rooting of cuttings and other plant parts is difficult under bright or sunny conditions. Water lost from these plant parts cannot be replaced, because the plant parts have no roots. Severe and even fatal wilting results. Maintaining high humidity around the parts to be rooted solves the problem. Transpiration is essentially stopped and roots are formed under sunny conditions, which favor photosynthesis and, in turn, fast and successful root formation. Fog generators or misting nozzles can be used to maintain a fog or mist around the cuttings. The mist does not need to be constant, so intermittent misting is used.

The production and handling of seeds is an important horticultural industry. Seeds are raised in areas with favorable climates such as California and Central and South America. Most seeds are harvested when mature, preferably by machine harvesters. Cleaning after harvest to remove nonseed dry parts or even fruit parts is usually necessary. Seeds are usually dried after harvest to reduce moisture contents for better storage. Many seeds are pretreated prior to being sold. Some treatments include fungicides, heat, UV radiation, or beneficial microorganisms to protect against pathogens. Other pretreatments include those to improve germination such as sizing and priming. Larger seeds have increased vigor and longer shelf life. Priming uses osmotic solutions and sometimes growth enhancers as a seed soak. Primed seeds have higher vigor. Some seeds are partially pregerminated (radicle appearance) often with growth enhancers, suspended in a gel, and sown by machine. Others are partially germinated and then dried and sown in a conventional manner. Pregerminated seeds have germination rates at nearly 100%. Some very tiny seeds are coated with inert materials such as clay to increase their size for use in mechanical sowers. Phytohormone treatments and seaweed extracts are used to improve germination.

Seed labels are legally required. Tests for germination rates, physical and genetic purity, vigor (germination ability under less than favorable conditions), seed health (presence of insects, nematodes, and pathogens), and contamination by weed seeds are conducted to supply some of the label requirements. The label also indicates the plant name and cultivar, origin, and amount.

Seed storage is needed between harvests and marketing of seeds. High-quality seeds store best. Some decline during storage is expected. The decline is a factor of the type of seed, storage conditions, and the length of storage. Low-moisture cold storage is best for many seeds. Short-lived seeds are stored in cool, moist conditions. Cryopreservation in liquid nitrogen offers future promise.

Some seeds such as many flower and vegetable cultivars do not require breaking of seed dormancy prior to sowing them. Seeds of many woody plants and wildflowers have seed dormancy that must be broken prior to planting. Seeds with hard coats can be treated with abrasives or acid. Water soaks can soften hard coats and remove germination inhibitors. Cold, moist storage works with other seeds. Some seeds need two treatments because of double dormancy conditions.

Numerous factors affect the timing of seed propagation. These include frost dates, soil temperatures, marketing windows, length to harvest, and temperature preferences of the seeds and developing plants. Some factors apply to indoor sowing and others to outdoor sowing.

Disease control, especially for damping-off, is very important during germination and seedling development. Pasteurization of the growing media is necessary. Seeds can also be treated with fungicides or disinfectants such as calcium hypochlorite or hot water. Environmental control is especially important, too. Water is critical during germination; seeds must not dry out. Proper temperatures and good oxygen levels during germination are also important. Some seeds require light during germination. In others light inhibits germination or has no effect. Good light intensity is needed for seedling development, except for certain plants where some shading in the greenhouse may be needed. Lowering temperatures 5° to 10°F in greenhouses after germination improves seedling and young plant development.

Most bedding plants are greenhouse produced from seeds. The preferred container has become plugs. Growing media and seeds are added to plugs mechanically. Plugs are often germinated in special environmentally controlled chambers (germinators). Plugs can also be placed under mist or fog. Flats are used much less and are mechanically filled and sown. Bottom heat and adequate moisture (misting) are needed. Proper fertilization practices are followed, usually dilute soluble fertilizers are added through irrigation systems. When seedlings have reached the four or more leaf stage they are transplanted to larger cell packs for wholesale and retail sales. Mechanical transplantors are now available for transplanting plugs to larger cells. Flats still require hand transplanting.

Seeds are also sown outdoors in farm fields, field nurseries, gardens, and new lawns. Timing of sowing is important for successful germination and seedling development. Placement of seeds is important to eliminate thinning of seedlings. Mechanical precision seeders are used commercially, and seed tapes or hand sowing are used on the garden level. Seedbeds need careful tillage to produce good contact between the seed and soil. Soil fertility and organic matter should be good and maintained with sustainable practices and amendments based on soil tests. Careful attention to weeds, pests, pathogens, and water is critical for good seedling and early development.

Asexual or vegetative propagation is widely used for genetic preservation of seedless cultivars, most fruits, nuts, and ornamentals for indoors and landscape usage (woody and herbaceous perennials). Faster results and virus-free plants are pluses. Cost is not. Stocks must be maintained true in terms of identity or name, genetic composition, and free of pathogens. Microbiological, molecular biology techniques, and visual inspection all play a role in stock maintenance. Micropropagation of shoot tips and heat treatments can produce pathogen-free materials when necessary. Rooting requires a media, usually a mixture of inorganic/organic components similar to those used for seed production. Inorganic components alone are sometimes used. Bottom heat and either misting or fog are also needed.

Cuttings (stem, modified stem, leaf, leaf bud, and root) taken prior to reproduction are a widely used form of asexual reproduction. Many plants can be produced from a few mother plants in a small area. Propagation via cuttings is also less costly, simpler, faster, and requires minimal technique relative to other asexual methods. Stem cuttings can be either herbaceous or woody. Cuttings are often dipped into biocides to kill harmful pathogens and rooting hormones to speed formation of adventitious roots. Leaf cuttings are useful with plants that have little or no stem. Root cuttings are preferred with certain plants with prickles such as raspberries.

Propagation by layering is an old technique that is easy and highly successful. It is expensive because of hand labor, slow, and few plants result. No maintenance or special equipment is needed. Stems are bent to the ground, held in place with soil or a metal clip, and allowed to root naturally at the soil/stem contact area. Once rooted, it is cut and transplanted. The procedure is very useful for hard-to-root woody plants. If the shoot is covered near the tip or further back, it is called tip and simple layering, respectively. A stem covered at alternate points over some distance is called compound layering. If multiple shoots are cut back and mounded with soil, it is called mound layering. Plants bent flat into trenches and covered with soil are termed trench layering, a procedure especially useful for very difficult-to-root plants. Plants can also be air-layered. A strip of bark is removed, a cut is made in the debarked area and held open with a toothpick piece. Wet sphagnum moss is wrapped at the area and then wrapped in turn with polyethylene wrap. After roots form, the rooted stem is cut off and transplanted.

Many woody shrubs and herbaceous perennials produce numerous shoots annually at their root/shoot juncture (crowns). Division by knife or spade can provide new plants. While easy and highly successful, only a few plants are provided, making this method mostly of interest to gardeners. Herbaceous perennials also get overcrowded with time, so crown division is needed eventually and offers new plants in the process.

Many plants have modified stems and roots that are involved in natural asexual propagation. These same structures can also be commercially propagated. Plants with runners and stolons propagate by natural layering by rooting at nodes along these structures. Once rooted, the daughter plant can be cut and transplanted. Offsets of plants such as cacti can be cut and rooted as stem cuttings or, if naturally rooted, cut and transplanted. Shoots arising from the roots through adventitious buds (suckers) can also be dug, cut, and transplanted after natural rooting.

Bulbs and similar structures (corms, rhizomes, tubers, tuberous roots, and pseudobulbs) are also easily propagated. Bulbs produce offsets called bulblets that can be separated and grown on elsewhere to flowering size. Some species of *Lilium* also form bulblets on their stems known as bulbils. These can be removed and treated the same as bulblets. When greater numbers of new bulbs are required, other methods are used. Scaly bulbs can be separated and placed in growing media whereupon they produce adventitious bulblets (scaling). Cutting or wounding some bulbs (basal cuttage) can cause them to produce numerous bulblets. Other bulbs can be cut into pieces with three or four scales and part of the basal plate (stem tissue) and then the bulb cuttings are treated as leaf cuttings. Corms are propagated through formation of new corms and also offsets called cormels. Big corms with multiple buds can be sectioned with each section containing a bud. These sections will produce new corms. Division can propagate both rhizomes and tuberous roots. Rhizomes can also be cut into sections, each of which has a bud. Shoots produced on large rhizomes can be cut and treated as cuttings. Tubers can be planted whole or sectioned into pieces with buds. Tuberous roots can also be propagated from adventitious shoots that can be induced to root while on the tuber. Some orchid pseudobulbs form rooted offshoots that can be cut and potted. In others, the rhizome with attached pseudobulbs is planted in rooting media, with growth beginning at the pseudobulb base. Fungicide dusting is needed to prevent decay of cut tissue in all instances above, except for those suberized.

Grafting involves the joining of plant parts such that the tissues unite and grow as one plant. The upper part of the graft union is the scion, and the lower is the rootstock or understock, or simply the stock. Budding is a special graft in which the scion is a small piece of bark or wood with a single bud. Most fruit tree and hybrid rose cultivars are sold in grafted forms. Grafting is the method of choice for plants that are difficult to propagate by usual means, such as cuttings. Desirable cultivars with poor root systems can be improved with grafted rootstocks. Grafting rootstocks can offer increased tolerance to unfavorable soils or soil-borne pathogens. Other rootstocks can induce increased vigor or dwarfing in the scion. Grafting is useful for replacing undesirable cultivars on good, established trees, using the existing rootstock. This is termed top-grafting or top-budding, or both may be referred to as top-working. Certain forms of tree damage can be repaired through grafts. Grafting is also useful for the creation of unusual plant forms, such as "tree" roses or combination fruit trees. Fruiting may also be improved by grafting a scion from a staminate plant onto a pistillate plant. Virus detection is also possible if a plant suspected of carrying a virus is grafted onto a susceptible plant.

Not all plants can be grafted successfully. Grafting within a clone is assured, and grafting between clones within a species is almost always possible. The chances of successful grafting between species within a genus are less and are remote to nil for grafting between genera within a family and grafting between families.

Successful grafting depends on a good graft union. The callus-producing tissues of the cambium layer must match closely, restricting grafting to dicotyledons and gymnosperms, which have continuous vascular cambiums. Temperatures must be 55° to 90°F (13–32°C) and high moisture prevents desiccation of the forming callus tissue. Waxing the graft union helps maintain moisture and also helps to prevent infection by plant pathogens. The graft union is meshed tightly by wedging, wrapping, nailing, or tying, which prevents the dislocation of cellular growth. The scion and rootstock must be at the proper physiological stage: Scion is dormant and the rootstock is dormant or active, depending on the grafting method.

Two general forms of grafting exist: detached scion and approach. The first is preferred. Roots are present only on the rootstock and the scion is a cutting from the donor plant. The second method involves the grafting of two self-sustaining plants. After the union is formed, the scion is severed from the donor plant. This method is employed with difficult-to-graft plants. Numerous grafting and budding techniques are available to make either of these graft types. These differences result from differing ways in which pieces can be joined or meshed.

Micropropagation differs from the previous cloning methods in terms of scale and technological sophistication. Small parts of plants (explants) are used to propagate new plants: embryos, stem tissue, shoot tips, apical meristems, cotyledons, leaf tissue, seed parts, root tips, bud tips, bulb scale sections, various flower parts, callus (parenchyma), one to many cells, and pollen grains. The propagation media consists of agar or liquid cultures. Contamination with microorganisms is a problem, so facilities and all materials must be pathogen free (aseptic culture).

The terms *micropropagation* and *tissue culture* differ. Tissue culture refers to the maintenance and growth of plant tissues in aseptic culture. Tissue culture can be utilized as the starting point for micropropagation of plants, but it can also be used as a tool of genetic engineering to produce either genetically altered plants (plant breeding) or to produce biomass for the harvest of biochemicals.

Conventional propagation is limited by how many cuttings or divisions a plant yields or how many seeds are produced. Micropropagation can yield a million plus plants of a given cultivar in a year. Cost restricts the process to high-value, slow-growing plants (bulbs, ferns, herbaceous perennials, foliage house plants, orchids, and palms), new cultivars from conventional plant breeding or genetic engineering having limited stock and high market demand, very high-value cultivars where increased costs are justified (certain woody landscape plants), cultivars that are hard to root by conventional techniques, and endangered plants. Other pluses include the ability to produce pathogen-free plants and the clonal propagation of parental plants used for the production of hybrid seed. Another advantage is that explants can be stored frozen (cryopreservation) and micropropagated later on thawing (germplasm preservation). Lastly, micropropagation is adaptable to year-round production, whereas many propagating nurseries are more seasonal. Disadvantages include expensive high technology, sophisticated laboratory facilities, and highly educated workers. Start-up and labor costs and supplies are costly.

Agar or liquid culture media formulations usually incorporate certain chemicals. Macroelements and microelements needed by plants are supplied through the inclusion of inorganic salts. Some form of carbohydrate is also required, such as sucrose. Plant vitamins are also added. Auxin and cytokinin are also critical for root/shoot development. Sometimes gibberellins are added. The ratio of auxin to cytokinin is varied to induce shoot (low/high) and root (high/low) formation (organogenesis). Therefore formulations need to be changed as micropropagtion moves forward. Lastly, antioxidants, such as citric acid or ascorbic acid, are added. Other occasional components include inositol, myoinositol, and glycine. Hydrolyzed casein, yeast extract, coconut milk, orange juice, tomato juice, and malt have been used with good results for plants that are difficult to micropropagate.

Many techniques exist for the micropropagation of plants by the tissue culturing of different plant tissues and organs. The techniques can produce seedlings, plants, callus (undifferentiated tissue), and somatic embryos (vegetative embryo). Regardless of which technique is used, all have four basic common steps: establishment and stabilization, shoot multiplication, root formation, and acclimatization.

During establishment and stabilization, explants are chosen, removed, disinfected, transferred to the culture media, and readied for shoot multiplication. The explant is then transferred to the appropriate media, using laminar flow hoods and sterile equipment. Usually the media will be higher in cytokinin than auxin. The period from transfer to shoot production is called stabilization. Many herbaceous explants need only 1 or 2 months to achieve stabilization. Woody plants can require up to 12 months.

The next step, shoot multiplication, starts with the division of the multiple microshoots. Microshoots are transferred to new media that can be the same media as before, although sometimes levels of some biochemicals are increased. This process, called subculturing, is repeated every 1 or 2 months. Limits are usually set until some number is reached or it is known that beyond a certain amount of subculturing, the risk of abnormal microshoots increases.

The next procedure, root formation, requires the transfer of the microshoots to a new medium, usually one that is auxin rich relative to cytokinin. In the last step (acclimatization), plantlets are removed from culture media, washed, and transplanted into normal artificial soil media. These plantlets are grown first under high humidity conditions and gradually "weaned" from high to low humidity and from fluorescent to natural light in the greenhouse.

If plants must be held in the propagation media or outdoor seedbed past transplanting time, it will be necessary to add fertilizer. In greenhouses soluble fertilizers are added through the irrigation system (fertigation). Conventional fertilization is used outdoors.

Transplanting of seedlings, ferns, or asexually propagated plants requires minimal root disturbance. Transplanting indoor seedlings outdoors, moving outdoor foliage and flowering plants indoors, or transplanting cuttings from under mist to an unmisted site must be hardened off (acclimatization or conditioning). Hardening involves lowering the temperature, withholding water, and a gradual change in the environment. This slows growth and builds a carbohydrate reserve that supplies the needed sugars for rapid reestablishment of the rooting system. Cold frames are used to harden plants for about 7 to 10 days prior to transplanting outdoors. Field-grown plants, such as recently propagated foliage and flowering plants destined for indoor use, must be conditioned or acclimated to the reduced light levels indoors, usually in a lathhouse. Even better is to produce field-grown foliage plants under shade right from the start. Plants propagated under mist must be hardened by gradually decreasing misting periods.

Conditions that favor reduced transpiration after transplanting (lower temperatures, lower light intensity, low or no wind speed, and high relative humidity) increase the chances of success. Best transplanting occurs during cloudy days, days of light rain, periods of no wind, and late afternoons. Plants should be watered well before and afterward. Shade can be provided after transplanting, if possible. A soluble fertilizer (starter solution, high in phosphorus) is sometimes added to aid the reestablishment of the plants.

Transplanting can be done directly into outdoor sites or containers maintained indoors or outdoors. Containers include plastic and metal containers in sizes from 1 quart to 15 gallons. Growing medium is needed for transplants in containers. Materials are similar to propagation media. The main difference is increased nutrient levels. Outdoor container plantings in colder areas benefit from winter protection.

EXERCISES

Mastery of the following questions shows that you have successfully understood the material in Chapter 7.

1. Why are artificial soils preferred over natural soil as propagation media?
2. What chemical and physical properties are essential for an artificial soil?
3. What inorganic and organic materials can be used in artificial soils? What properties do both provide to the artificial media? Which materials are generally used the most in artificial soils?
4. What sustainability issues occur with regard to propagation media?
5. Are there any differences in propagation media used to raise seedlings versus the rooting of cuttings? If so, what are they?
6. What reason(s) is there to pasteurize propagation media? How is pasteurization accomplished? Is one method preferred from the sustainability perspective? Why?
7. What qualities are needed in water used for propagation? What issues of sustainability exist relative to propagation water?
8. What facilities and equipment are needed for plant propagation? Do any of these require sanitation steps? If so, which ones and by what method?
9. What is bottom heating and misting? Why and how are they used in propagation?
10. What is sexual propagation? What types of plants are usually propagated in this manner?
11. What types of pretreatments are used with seeds prior to propagation? What is seed priming? Pregermination? What are the reasons both are done?
12. What is seed testing? What is its purpose? How is testing of seed germination done?
13. How are seeds stored?
14. What is seed dormancy? Do all seeds possess this condition? If not, which seeds are free of seed dormancy? How is seed dormancy broken in nature? In commercial propagation?

15. What environmental factors affect seed germination? What can go wrong?
16. What are the basic steps in seed sowing indoors and outdoors?
17. What is asexual propagation? For what plant types is it the method of choice?
18. What are the many and various plant parts used in asexual propagation?
19. What types of treatments are used in asexual propagation? What are the favorable environmental conditions?
20. What is grafting? Why is grafting used?
21. What is micropropagation? What are its advantages? Disadvantages?
22. What special facilities and media are needed for micropropagation? What are the basic steps involved in micropropagation?
23. What basic care is needed after propagation?
24. Why is transplanting necessary? Why is hardening after propagation needed?

INTERNET RESOURCES

Budding

http://muextension.missouri.edu/xplor/agguides/hort/g06972.htm

Grafting

http://muextension.missouri.edu/xplor/agguides/hort/g06971.htm

Growth media for container-grown ornamental plants

http://edis.ifas.ufl.edu/CN004

Media and Soils Links

http://bluestem.hort.purdue.edu/plant/FMPro?-DB=plant.fp3&-Format=category.html&-Max=100&-SortField=Description&category=Soils%20and%20Media&-Find

Plant propagation by leaf, cane, and root cuttings: instructions for the home gardener

http://www.ces.ncsu.edu/depts/hort/hil/hil-8700.html

Plant propagation by stem cuttings: instructions for the home gardener

http://www.ces.ncsu.edu/depts/hort/hil/hil-8702.html

Plant tissue culture information exchange

http://aggie-horticulture.tamu.edu/tisscult/tcintro.html

Propagation Links

http://bluestem.hort.purdue.edu/plant/FMPro?-DB=plant.fp3&-Format=category.html&-Max=100&-SortField=Description&category=Propagation&-Find

Rose cuttings

http://www.hortus.com/rose.htm

Seed germination and soil temperature

http://www.ext.vt.edu/departments/envirohort/articles2/sdgrmtmp.html

Seeds and germination propagation

http://www.thompson-morgan.com/guide_germination_introduction.html

Vineyard propagation by cuttings

http://www.cahe.nmsu.edu/pubs/_h/h-322.html

Woody cuttings

http://www.hortus.com/hudson.htm

http://edis.ifas.ufl.edu/BODY_EP030

RELEVANT REFERENCES

Anderson, Peter, A. Toogood, and the American Horticultural Society. (1999). *American Horticultural Society Plant Propagation: The Fully Illustrated Plant-by-Plant Manual of Practical Techniques.* DK Publishing, London. 320 pp.

Baker, C. J., K. E. Saxton, and W. R. Ritchie (1996). *No-Tillage Seeding: Science and Practice.* CAB International, New York. 258 pp.

Baker, K. F. (1957). *The U.C. System for Producing Healthy Container-Grown Plants.* California Agricultural Experiment Station Extension Service Manual 23, Berkeley.

Ball, Vic (ed.) (1997). *Ball Redbook* (16th ed.). Ball Publishing Co., St. Charles, IL. 802 pp.

Baskin, Carol C., and Jerry M. Baskin (1998). *Seeds: Ecology, Biogeography, and Evolution of Dormancy and Germination.* Academic Press, San Diego. 672 pp.

Basra, A. S. (ed.) (2000). *Hybrid Seed Production in Vegetables.* Food Products Press, Binghamton, NY. 135 pp.

Boodley, J. W., and R. Sheldrake, Jr. (1964). *Cornell "Peat-Lite" Mixes for Container Growing.* Department of Floriculture and Ornamental Horticulture, Cornell University Mimeo Report, Ithaca, NY.

Collin, Hamish A., and Sue Edwards (1998). *Plant Cell Culture: Introduction to Biotechniques.* Springer Verlag, New York. 176 pp.

Doijode, S. D. (2001). *Seed Storage of Horticultural Crops.* Food Products Press, Binghamton, NY. 340 pp.

Garner, R. J. (1993). *The Grafter's Handbook* (5th ed.). London: Faber & Faber Ltd. 323 pp.

Hartmann, Hudson Thomas, Dale E. Kester, Fred T. Davies, and Robert L. Geneve (1997). *Plant Propagation: Principles and Practices* (6th ed.). Prentice-Hall, Upper Saddle River, NJ. 770 pp.

Havis, J. R., and R. D. Fitzgerald (1976). *Winter Storage of Nursery Plants.* Cooperative Extension Service, University of Massachusetts Publication No. 125, Amherst.

Holcomb, E. Jay (ed.) (1995). *Bedding Plants IV: A Manual on the Culture of Bedding Plants as a Greenhouse Crop.* Ball Publishing Co., St. Charles, IL. 452 pp.

Justice, O. L., and L. N. Bass (1978). *Principles and Practices of Seed Storage* (USDA Handbook No. 506). U.S. Department of Agriculture, Washington, DC.

Kyte, Lydiane, and John G. Kleyn (1996). *Plants from Test Tubes: An Introduction to Micropropagation* (3rd ed.). Timber Press, Portland, OR. 240 pp.

MacDonald, Bruce (1986). *Practical Woody Plant Propagation for Nursery Growers.* Timber Press, Portland, OR. 660 pp.

Nau, Jim (1996). *Ball Perennial Manual: Propagation and Production.* Ball Publishing Co., St. Charles, IL. 512 pp.

Nau, Jim (1993). *Ball Culture Guide: The Encyclopedia of Seed Germination.* Ball Publishing Co., St. Charles, IL. 144 pp.

Schopmeyer, C. S. (1974). *Seeds of Woody Plants in the United States* (USDA Handbook No. 450). U.S. Department of Agriculture, Washington, DC.

Sheldrake, R., Jr., and J. W. Boodley (1965). *Commercial Production of Vegetable and Flower Plants.* Cornell Extension Bulletin 1056, Cornell University, Ithaca, NY.

Styer, Roger C., and David S. Koranski (1997). *Plug & Transplant Production: A Grower's Guide.* Ball Publishing Co., St. Charles, IL. 374 pp.

Thompson, P. (1992). *Creative Propagation: A Grower's Guide.* Timber Press, Portland, OR. 220 pp.

U.S. Department of Agriculture (1961). *Seeds. Yearbook of Agriculture.* U.S. Government Printing Office, Washington, DC.

Young, J. A., and C. G. Young (1986). *Collecting, Processing, and Germinating Seeds of Wildland Plants.* Timber Press, Portland, OR. 236 pp.

Chapter 8

Conventional Horticultural Crop Breeding

OVERVIEW

Horticultural productivity and quality have continuously improved ever since the cultivation of horticultural crops by humans began. Much of this improvement was a result of deliberate, artificial selection over many years, essentially saving seeds from the best field plants for use next season. The genotypes of our current food crops owe their heritage to farmers' seed-saving choices over thousands of years.

With the advent of plant genetics in the early 1900s, a scientific basis for plant improvement was established (see Chapter 5). The resulting technology allowed the plant breeder to manipulate plants. Knowledge about genetics brought about the more rapid production of plants with a directed selection for better characteristics. Crop improvement was no longer dependent on the random appearance of different individuals with desirable features. The only drawback to conventional plant breeding was the limitation of sexual compatibility and the slowness of the process. Some 10 to 20 years of work are needed to bring new cultivars into production and market availability.

The advent of molecular biology techniques, especially tissue culture and recombinant DNA technology, ushered in a new era of plant improvement possibilities. Neither sexual compatibility nor growouts and selection limited choices of traits any longer. Production times were reduced to about 5 years. These more advanced techniques for plant improvement are covered in the next chapter.

CHAPTER CONTENTS

OBJECTIVES

- To learn which desirable characteristics plant breeders strive for
- To learn about sustainability issues involved with plant breeding
- To understand the basics of gene pools in terms of geography and time

- To learn the basics of plant breeding techniques from crosses to selection
- To understand the basics of seed production
- To learn about strategies to protect and promote new plant products

DESIRABLE CHARACTERISTICS

Many aspects of horticultural productivity and quality can be realized through plant improvement directed by the plant breeder. All efforts are generally aimed toward improving productivity either directly or indirectly. Those aspects resulting in increased marketability receive more attention for obvious reasons. Improvements in horticultural crops are not necessarily confined to the production of higher yielding vegetables and fruits. Improvements also include uniform ripening for purposes of mechanical harvesting, increased vitamin and/or protein content, better taste, improved keeping and processing qualities, and resistance to insects and disease. With ornamentals it may be desirable to obtain aesthetic improvements such as larger or double flowers, improved form or scent, or even improved foliage appearance. Dwarfing may be desirable to minimize pruning and reduce the space requirements of ornamentals and fruit trees. The tolerance of lawn grasses to shade and abuse can be improved. Resistance to insects and diseases would be a desirable trait for all horticultural crops, because it minimizes pesticide needs. Other desired characteristics might be increased tolerance toward heat, cold or drought, more efficient absorption of soil minerals, insensitivity to photoperiod, more adaptability to mechanical harvesting, or a shorter period to the production of flowers, fruits, or vegetables.

SUSTAINABILITY ISSUES

One key issue with sustaining plant improvement is genetic erosion or the loss of genes. Gene loss can occur in a number of ways. One major way is the loss of habitat in centers of crop origin. More than 90% of the food crops grown in the United States originated in other countries. These sites are disappearing as development alters the face of the land through deforestation, conversion to cropland, urbanization, and desertification. Other genes are lost as new cultivars replace older ones or the modernization of agriculture brings about the replacement of folk varieties honed by thousands of years of use. Genes for the future are also lost as the tropical rain forests give way to slash and burn agriculture, urban resettlement, mineral and oil exploitation, and logging. Such losses threaten the ability of the plant breeder to alter plants either by conventional or biotechnological means. The genetic engineer can only manipulate and move genes that are based on the discovery and mapping of known genes. If a wild plant or folk cultivar is lost before the genome is characterized, it is truly a loss.

SOURCES OF GENETIC DIVERSITY

Most of our major horticultural food crops belong to few families: Brassicaceae (Cruciferae; cabbage and relatives), Cucurbitaceae (cucumber, pumpkin, and squash), Fabaceae (Leguminosae; beans, peanuts, and peas), Lilliaceae (asparagus, onion), Rosaceae (apples, pear, plum, raspberry, etc.), Rutaceae (citrus), and Solanaceae (eggplant, peppers, potatoes, and tomatoes). Even if we expand the list to include the agronomic crops, cereals and grains, we need add only one more family: Graminae (grasses). Ornamentals and minor food crops expand the list to a bit under 200 families, but these crops represent only a small percentage of the species found in these families.

Not only are the families limited, but the origin of horticultural food crops grown in the United States derives from habitats found in other countries. The regions of origin are also not highly species rich and are usually characterized by seasonal differences. Seasonal changes (hot to cold, wet to dry) bring about dormancy and usually result in plants that store food and are primed and ready to go when favorable growth conditions return. The plant's stored food in turn becomes our food. Most of these sites (centers of origin) are either temperate or semitropical. Few food crops of major commercial value originated in the tropical rain forests.

Given this historical context, several things become apparent regarding plant breeding. One, the potential for new food crops in horticulture is immense. The tropical rain forests are underutilized in terms of current food crops. If we add plants used for ornamental or medicinal applications, the number of potential horticultural crops that might originate from tropical rain forests is easily over 2,000! The second conclusion is that the germplasm to maintain current crops and to produce future crops lies beyond the United States. Finally, the centers of origin for both current crops and future crops are threatened by increased population and agricultural/urban development. Preserving germplasm access for breeding requires political, social, and economic attention from the United States and other developed nations, global cooperation, and fair compensation for usage of germplasm.

Often a wide diversity of genotypes exists for many cultivated crops. Some are well known and usually found currently or recently in cultivation. Others have been neglected or fallen out of favor, and still others remain only in isolated areas (centers of origin) of the world. The latter is especially true for the primitive ancestors of our cultivated crops that were discarded as more favored genotypes were produced by natural and artificial selection.

The more diverse the genetic background, the more scope the plant breeder has to work with. The more primitive ancestors or closely related, but uncultivated, species may possess genes for qualities the plant breeder wishes to incorporate into today's cultivated plants. Accordingly, the plant breeder must assemble a collection of these plants. Some are more readily obtainable than others. For example, the lineage of the strawberry and the garden dahlia are reasonably well known, because of the economic value of the strawberry and the enthusiasm of the dahlia specialist and collector. Knowledge about the lineages of other horticultural crops, especially those of limited economic value, is very limited.

Most horticultural crops of economic value have existing collections of genotypes maintained by governments in both developed and Third World nations. Many of these seed banks and clonal repositories (Figure 8–1) are associated with an international network, the Consultative Group on International Agricultural Research (CGIAR). Collections of genotypes for crops are on the increase, since most food crops run the risk of genetic vulnerability toward plant pest epidemics, because of limited diversity in the gene pool of cultivars that are widely grown.

In the United States the National Plant Germplasm System (NPGS) and Germplasm Resources Information Network (GRIN) provide both germplasm and information needed to conduct plant breeding. The NPGS helps plant breeders acquire crop germplasm by preserving, evaluating, documenting, and distributing crop germplasm. GRIN aids these programs by enhancing communication with scientists regarding the location and characteristics of germplasm in the above programs, permitting flexibility to users in storing and retrieving information, reducing unneeded redundancy of data, and relating pertinent information about germplasm accession.

Germplasm comes to NPGS either through donation, purchase, or collection. Sometimes it becomes necessary to go to the geographic center of origin to locate normally unobtainable genotypes. These centers of origin are often the source of a rich gene pool. For example, past plant expeditions to New Guinea have yielded a rich gene pool for those plant breeders interested in hybridizing New Guinea impatiens and rhododendrons.

(A)

(B)

(C)

(D)

FIGURE 8–1 (A) Staff at the National Seed Storage Laboratory in Fort Collins, Colorado, preserve more than 1 million samples of plant germplasm. Here, a technician retrieves a seed sample from the −18°C storage vault for testing. (B) Apples in the Agricultural Research Service (ARS) germplasm collection at Geneva, New York, vary widely in size, shape, and color. (C) A technician packages Brassica seeds for distribution at the North Central Regional Plant Introduction Station in Ames, Iowa. (D) At the National Plant Germplasm Repository in Corvallis, Oregon, a plant pathologist examines a dwarf blueberry mutant—just one example of clonal germplasm that is available to members of the Farmer Cooperative Genome Project for evaluation or further development. Note the drip irrigation lines going to each container. (USDA photographs.)

Specialized plant societies, particularly for ornamental plants, are another source. Sometimes historical societies are involved in genotype preservation. For example, through the creation of a historical orchard the Sturbridge Village group in Massachusetts has preserved many old apple cultivars no longer grown. Commercial seed companies, botanical gardens, arboreta, and plant research centers are also sources.

Centers of Origin

Most horticultural and agricultural crops originated outside the United States. The few food crops of national origin include mostly minor crops such as blackberry, black walnut, blueberry, chestnut, cranberry, certain grapes, hazelnut, hickory nut, Jerusalem artichoke, pecans, persimmon, plum, raspberry, sunflower, wild rice, and tepary bean. More than 90% of U.S. horticultural crops are dependent on genes in species introduced from outside the country.

Horticultural crops originating in Central and South America include avocado, beans (dry, lima, and string), papaya, peanut, peppers, pineapple, pumpkins, squash, sweet corn, sweet potato, and tomato. The Mediterranean area was likely the original source for asparagus, broccoli, brussels sprouts, cabbage, cauliflower, celery, collards, kale, leek, some mustards, parsley, and radish. Almonds, apples, apricot, banana, broadbean, chick pea, citrus, coconut, date, fig, garlic, kiwifruit, lentil, lettuce, mango, onions, peach, pear, pomegranate, quince, rhubarb, turnip, and spinach came from various parts of Asia. The cucumber, eggplant, and endive derived from India, the carrot and horseradish from Europe, and the macadamia nut from Australia. Artichokes, muskmelon, okra, and watermelon arose in Africa. The origin of peas is uncertain, but was probably the Mediterranean area or Asia. Chives probably came from North America and northern Eurasia. Parsnips and plums are native to Europe and Asia. Cherries and grapes arose in Europe and parts of Asia. Strawberries arose from ancestors in North and South America.

Crops today are still highly dependent on the ancestral plants found in the above-mentioned centers of origin. Given that agriculture and crops developed in these centers and the density of native species, these areas are also centers of diversity. These centers contain germplasm for insect resistance, disease resistance, drought resistance, and restoration of vigor. Given the ability of insects and disease to overcome the genetic barriers produced by plant breeding and their ability to become resistant to pesticides, germplasm from these centers of diversity will be needed far into the future.

All of these centers of origin are under environmental assault, primarily through urban and agricultural development. Most developed nations try to maintain germplasm collections from these areas, either in seed banks or field collections (clonal preservation). However, these collections are not complete replacements for the germplasm of all our food crops. Some protection through parks and botanical reserves is crucial.

The tropical rain forests are likely to be the future centers of diversity, although the rapid rate of their destruction threatens this outcome. These habitats have potentially a few thousands of new fruits and vegetables. Ornamentals and medicinals could easily bring the total to some several thousand potential crops from the rain forests. Global cooperation is necessary to save these centers of diversity for future generations of horticultural crops. Only major efforts will stem the destruction caused by political and socioeconomic pressures found in rain forest countries. These pressures result in transfers of urban poor populations into the rain forest where the new farmers slash and burn the rain forests. These nations also exploit the rain forests in an environmentally unacceptable manner through logging, mining, and oil production. Most do this to maintain trade and to finance their national debt. Only global cooperation to provide financial support and rational conservation efforts can save these genetic resources for our future.

Folk Varieties

Folk varieties, also known as *landraces*, can be very valuable sources of germplasm. Local farmers developed folk varieties through selection over long periods of time. Essentially, folk varieties evolved through artificial selection during the last 10,000 years that humans practiced agriculture. Selection was generally for improved production under local, often adverse, conditions for plants grown essentially under sustainable agricultural conditions. Although many of these landraces are agricultural crops, such as the various grains, some are also horticultural crops.

These plants usually possess considerable genetic diversity and are often adapted to more rigorous habitats than are many of our current horticultural crops. For example, some are adapted to arid environments, such as maize and bean folk varieties used by the Hopi and Zuni peoples in the arid Southwest. Similarly, native peoples in the Andes region of South America have amassed numerous potato folk varieties. Most of these folk varieties have been displaced by modern cultivars, except in a few isolated areas. Government and other institutions have collected and preserved some folk varieties in seed and clonal repositories. Others are lost forever.

Folk varieties are faced with extinction when modern horticulture and its cultivars are accepted. For example, the tepary bean (*Phaseolus acutifolius*) is highly adapted to the arid Southwest. However, it all but vanished with the advent of modern agriculture and irrigation. Under irrigation, its yields were less than modern cultivars bred for areas with more available water. Tepary germplasm could become important to genetic engineers designing crops that use water in a more sustainable manner. Such future crops might have improved drought resistance or be better adapted to dryland farming. Should folk varieties have commercial value, the issue of fair compensation must be considered in the context of the indigenous peoples responsible for their origin.

Folk Varieties and the Zuni Case Study

Native American Hopi and Zuni in Arizona and New Mexico, respectively, developed a number of bean, corn, and squash folk varieties during the last 1,000 years as farmers. These folk varieties include colorful red and blue corns, speckled beans, and wonderfully shaped, colorful squashes. Many of these folk varieties are now threatened as new cultivars from the present day take their toll. Fewer and fewer Zuni and Hopi make their living as farmers. Those who do farm increasingly adopt new cultivars that offer higher income as cash crops. Despite community pride in the cultural and horticultural heritage of these folk varieties, market forces and societal cultural inroads threaten their existence. The threat is both from abandonment and cross-pollination from modern cultivars.

Groups such as The Center for People, Food and Environment in Tucson, Arizona, work with Native Americans to preserve folk varieties. One project involves the Hopi and preservation of blue maize folk varieties best adapted to sustainable horticulture. The growing season in the Hopi area runs from 120 to 160 days in their high desert climate. Drying winds, high summer temperatures, and little water from surface sources or precipitation make the environment for crops difficult. Folk varieties are generally tough, drought and heat resistant, and quick to reach maturity. These genetic traits are ideal for the breeding or engineering of crops for arid environments and dryland farming.

The Hopi use several blue maizes. Their folk varieties include standard blue (sakwaqa'o), hard blue (huruskwapu), and gray-blue (maasiqa'o). The separation between these folk varieties was on the verge of loss. Hard blue came about because of its resistance to insect pests during storage. However, commercial availability of foods today made storage qualities less important. The gray-blue form was soft, making it easier to grind into flour. However, as machine grinding replaced hand grinding, this quality became less important. Genetic erosion also came about through the introduction of modern blue maizes from seed companies and subsequent cross-pollination. These cultivars were introduced in response to consumer demand for blue tortilla chips. Some Hopi farmers tried the new bluer cultivars. Folk varieties tend to be abandoned as the natural or social environment changes and the adaptability or stability of local folk varieties lose their appeal.

Given this rapid demise of blue maize folk varieties, the Zuni Folk Varieties Project was established. The project involves working with the Zuni community to better understand the values of folk varieties, how to maximize their usage, and to help the community preserve their folk variety heritage. Considerable educational outreach, documentation, and social networking with the community is required. Efforts to preserve and distribute the folk varieties are also essential. Finally, steps are needed to help the Zuni protect their intellectual property rights to these folk varieties. Sustainable horticulture will benefit from the genetic traits of folk varieties, and the indigenous peoples who brought them about should benefit financially from their germplasm heritage.

Heirloom Varieties

Heirloom varieties arose more recently in the early kitchen gardens of most countries. The early colonists started the process in the United States with European vegetable and flower varieties. Each further wave of immigrants added to the heirloom varieties. By the late 1800s, numerous vegetable, fruit, and flower heirloom varieties existed in the United States, having passed through several generations of families. The advent of modern cultivars, especially hybrids, brought about their replacement and disappearance. Groups exist today that seek to preserve and distribute these heirloom varieties, such as Brandywine tomatoes and Russett apples. One such group is the Seed Savers Exchange, Inc., in Decorah, Iowa.

Lines and Cultivars

Lines are populations of seedling plants whose genetic composition can be maintained relatively unchanged from generation to generation. Lines with similar phenotypes and differing traits, such as degree of insect resistance, can be blended to create a multiline. Lines can arise through self-pollination, cross-pollination, or hybridization.

Cultivars are lines altered through breeding to produce improved, reproducible characteristics that set the cultivar apart from both the source line and other lines of the particular crop. Cultivars are afforded legal protection and assigned unique names (see Chapter 2). Lines formed from self-pollinating plants usually consist mostly of recognized cultivars, referred to as line cultivars. Cross-pollinating plants usually require enforced self-pollination brought about by isolation to produce lines. These lines are known as inbred lines. Inbred line cultivars tend to be fewer in number. Instead inbred lines are crossed to produce hybrid lines that generally become hybrid cultivars. Crops propagated asexually are usually sold as clonal cultivars.

Lines and cultivars have come to dominate horticultural crops in the 20th century. Most offer improvements over the previous open pollinated, single parent folk and heirloom varieties, especially for highly mechanized, commercial horticulture. Home gardeners are also attracted to their qualities. The advent of legal protection for both seed and clonal cultivars increased their market share further, as seed companies sought to maximize their markets. More recently, legal protection was extended to genetically engineered plants. As such, lines and cultivars have essentially replaced folk and heirloom varieties, leading to the need for preservation of these displaced sources and their genetic diversity. Given that lines and cultivars are also replaced over time with newer versions, older lines and cultivars also need preservation for similar reasons.

NATURAL AND ARTIFICIAL OCCURRENCES OF PLANT VARIABILITY

Mutations in reproductive plant tissues can occur by chance or be artificially induced to produce an altered plant (*sport*) of horticultural value. *Sports* arise from a change on the chromosome or gene level. The nucleotide sequence of the gene may be rearranged, a portion of a chromosome may be relocated within itself or to another nonhomologous chromosome, a chromosomal portion may be lost or duplicated, or entire chromosome sets may be duplicated. Examples of sports are the pink-fleshed grapefruit and the seedless 'Washington Navel' orange. Sports can be preserved through seed and clonal stocks.

Spontaneous mutations can also occur in tissues that do not give rise to sex cells (somatic tissue). These somatic mutations, or chimeras, are sometimes stable enough to be propagated asexually. The chimera can be propagated asexually and become the source of improved plants. Examples of chimeras are the thornless blackberry and many variegated plants. Occasionally, a natural hybrid between two species or even two different genera may arise as a result of chance cross-pollination by insects or wind. Even if the hybrid is self-sterile, it may backcross with one or both parents, thus introducing additional genetic variability.

The horticulturist is ever alert for these natural occurrences of variability and selects altered forms that have horticultural value for purposes of propagation. However, the plant breeder has no control over these accidental events. Increases in hereditary variation can be artificially induced through the use of radiation such as X-rays or chemical mutagens such as colchicine. These procedures are useful for developing new characteristics in horticultural crops.

Often, desired characteristics can be found scattered throughout varieties, cultivars, and species. Crossing these plants to produce offspring with desirable characteristics from both parents might be possible. This process can occur naturally through pollination by insects or can be forced to occur by the plant breeder under controlled conditions. The latter produces the bulk of today's acceptable hybrids.

TECHNIQUES OF PLANT BREEDERS

The conventional plant breeder employs a number of techniques designed to move genes around in sexually compatible plants. These techniques involve the deliberate and careful control of pollination. Sometimes genetic variation is given an assist through the use of mutagens. A thorough understanding of the benefits and shortcomings of the various ways to breed plants is essential. Monitoring the changes and preserving the desired change are crucial for success.

Hybridization

Hybridization is the crossbreeding of sexually compatible plants having differences in genotypes to create offspring with altered characteristics. The parents can be two different species such as two species of clematis, or two different cultivars such as two tomato or two petunia cultivars. The offspring are referred to as *hybrids*. Hybrids exhibit uniform, vigorous improvements compared to either parent. This improvement is known as *heterosis* or *hybrid vigor* (discussed in the following section). Improvements are generally valuable to horticulturists and include characteristics such as higher yields, larger flowers and fruits, and/or earlier maturation. If the cultivar is a perennial, these improvements are permanent. If the hybrid is an annual, the seeds from the hybrid will not produce these improvements in the next generation. The hybrid of these types needs to be preserved by obtaining seed by repeating the original crossing of the two parents. Many popular flowers and vegetables grown commercially and in home gardens are hybrid cultivars. Examples of best-known hybrid flowers include begonia, impatiens, lisianthus, marigold, nicotiana, petunia, seed geranium and snapdragon. Vegetable hybrids of note include broccoli, cabbage, pepper, sweet corn, summer squash and tomato.

Besides care and skill in setting breeding goals and selecting appropriate plants for cross-pollination, one needs to understand and be skillful at hand pollination. Controlled hand pollination is a basic technique in plant breeding (Figure 8–2). The technique used depends on flower structure and pollination characteristics of the different species, but we can state some general principles. While hand cross-pollination is essential for investigating potential hybrids, once a hybrid is successful and destined for market, the cross needs to be achieved on a large scale without hand pollination. Fortunately, systems exist for just such purposes and are covered under the discussion of mass seed production later.

Contamination by unwanted pollen must be minimized. Alcohol can be used to wash the hands and instruments prior to collecting and transferring pollen. Anthers should be taken from flower buds just prior to bloom and before they open or dehisce or, if they are dehisced anthers, only from caged plants or bagged flowers. These precautions will minimize unwanted contamination by foreign pollen.

Anthers can be separated from about-to-open flower buds by pulling gently with forceps, by pressing the flower between the fingers, or by scraping the bud across a wire screen. Anthers are dried in a warm room on paper until they can be seen to dehisce under a magnifying glass. Next pollen is taken from the detached anthers. Removal is done indoors to prevent contamination by foreign pollen. The pollen can be sieved to remove impurities. Alternatively, pollen may be collected when ripe on a pocket knife blade directly from flowers that were caged or bagged to prevent contamination.

If the seed parent is not ready to receive the pollen because of different blooming times or environmental effects, it is either necessary to synchronize the flowering of the seed and pollen parents or to store the pollen. Storage is relatively easy, but success depends on the plant groups involved, because two types of pollen exist. Pollen either has two or three nuclei (binucleate or trinucleate). Different storage conditions are required. Binucleate types store well at 10% to 50% relative humidity and 0°C (32°F) or lower; trinucleate types store better at higher relative humidity and temperatures just above freezing. Binucleate pollen is common for plant families containing ornamental plants. Trinucleate pollen has

FIGURE 8–2 Steps in hand pollination. (A) Removal of stamens from a flower. Their anthers will serve as a pollen source. However, if the flower to receive pollen is self-fertile, stamens are removed (emasculation) to prevent self-pollination. (B) A piece of soda straw is used to protect a stigma that has not become receptive yet. (C) Pollen is transferred to a receptive stigma via a soft camelhair brush. (D) The pollinated flower is bagged to eliminate additional pollination and properly labeled. (From Raymond P. Poincelot, *Horticulture: Principles & Practical Applications,* 1980. Reprinted by permission of Prentice-Hall, Upper Saddle River, New Jersey.) (E) Pollen transfer shown with sunflowers to create new inbred lines that produce oil in the midoleic range. (USDA photograph.)

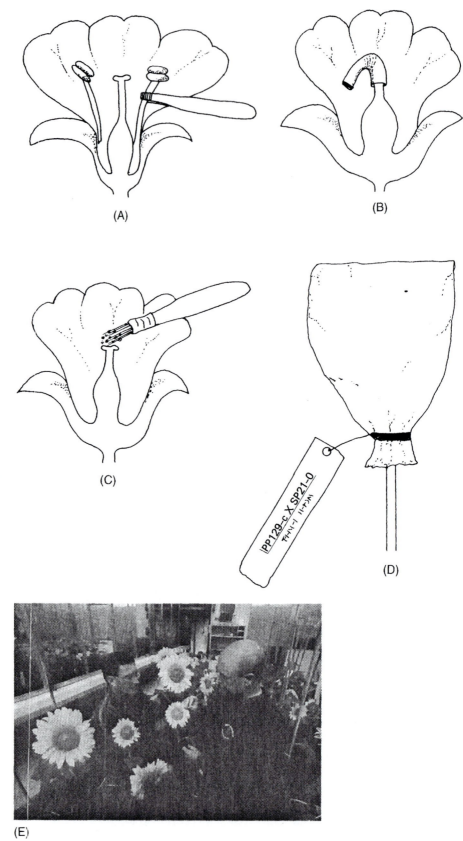

(A)

(B)

(C)

(D)

(E)

been reported in the Cactaceae, Compositae, Geraniaceae, and Caryophyllaceae. Some families, such as Campanulaceae and Labiatae, have both kinds.

If the seed parent is self-fertile, the stamens should be removed with a fine-pointed forceps or scissors just before the flower opens (usually late bud stage) to prevent self-pollination. This process is called emasculation. It may be necessary to remove some flower parts (petals, sepals) in order to remove the stamens; care must be taken that only parts known not to affect the pistil are removed. For example, the pistil may dry on removal of the calyx in some species.

The stigma will need protection at this time if it has not become receptive (usually fully expanded and often sticky). Protection prevents accidental foreign pollination, since pollen remains alive for a while after landing on the stigma. The stigma can be protected in several ways. The petals, if large enough, can be tied together or the stigma can be covered with aluminum foil, a length of soda straw with the tip bent over, or half of a gelatin capsule. If the flower is self-sterile, it need not be emasculated, but only enclosed to prevent insect or wind pollination. Whole plants can be caged with screens, or large numbers of plants can be protected in insect-proof greenhouses.

Once the stigma is receptive, the pollen may be applied with a camel hair brush, the fingertip, the anther itself, a pencil eraser, a pocket knife, or a glass rod. Care must be used to not damage the stigma. If the two parents differ greatly in the length of the style, the long-styled parent should be the pollen source, since pollen from short-styled plants is sometimes not able to grow down a long style. Earliest possible pollination will also give the pollen tube maximal time to grow down the style.

After pollination the seed parent flower should be immediately enclosed in a paper or closely woven cloth bag or caged to eliminate the chance of further pollination by insects or wind. The flower or flower branch should also be labeled with seed parent and pollen parent identification, along with the date. Bags may have to be removed from some species prior to complete seed ripening to prevent damage by excessive moisture. Records should be kept and observations recorded.

Hybrid Vigor

Heterosis or hybrid vigor is an important tool for the plant breeder, who can use it to restore the vigor lost through continual inbreeding (self-pollination) of a line of plants (inbred line). The original line was heterozygous and normally cross-pollinated. Continued inbreeding produces homozygosity (genetic consistency) for desired qualities, but at the expense of vigor. If two inbred lines are cross-pollinated, the F_1 generation often (not always) can show an increase in vigor (hybrid vigor), yet still retain the desirable characteristics existing in the two parental inbred lines. The effects of hybrid vigor may be expressed as increased growth rates, increased size, increased yield, earlier germination, uniformity, and resistance to unfavorable environmental factors.

The most popular use of hybrid vigor by plant breeders is the single cross (the crossing of two inbred lines). The resulting seeds, once shown to produce an F_1 generation (or F_1 hybrid) with hybrid vigor and a desirable phenotype, can be marketed depending on the ease of emasculation and pollination (see later discussion on mass seed production). Pollination control is needed to ensure hybrid seed is not mixed with inbred line seed.

Hybrid vigor has been most successful with sweet corn. It has also been found and utilized to varying degrees with many other horticultural crops, such as Brussels sprouts, cabbage, carrot, eggplant, onion, pepper, tomato, ageratum, marigold, petunia, snapdragon, and zinnia. Hybrid vigor is used frequently in the commercial production of horticultural crop seed.

Backcrossing

Backcrossing is a technique used by the breeder to accumulate desirable genes in a rapid manner. A wild species or another cultivar (nonrecurrent parent) might possess a highly desirable characteristic, such as insect or disease resistance, but have many horticulturally undesirable characteristics. The nonrecurrent parent is crossed with a cultivated species or cultivar (recurrent parent) that possesses many desirable horticultural characteristics, but is highly susceptible to insects or disease. The resulting F_1 hybrid is selected for the desirable characteristic and then is crossed back with the recurrent parent. The most desirable F_1 hybrid of this cross is in turn crossed with the recurrent parent. This process is repeated until the desirable characteristic is bred in and the many desirable characteristics of the recurrent parent are retained.

If the desired characteristic is recessive, it will not show in the hybrid. The breeder must resort to self-pollinations in each cycle to select the plant(s) in the next generation that possess the desired characteristic; these plants can then be used for the backcross. Obviously, the technique of backcrossing, when dealing with a desirable, recessive gene, will easily work with plants that can be self-pollinated. The process can be used with cross-pollinating plants, if they are bagged or isolated in some other way and then artificially self-pollinated.

Regardless of whether the desired characteristic is dominant or recessive, the backcross procedure works well with cross- or self-pollinating plants. The desired plant could be obtained strictly through selection during a selfing program after initial hybridization, but much larger plant numbers are required. Use of the backcross produces the desired plant in fewer generations with fewer plants. For best results the desired character should be controlled by a single or few genes, be simply inherited, and easily identified. Best results have been observed with increasing insect and disease resistance or height changes.

Artificial Mutagens

Polyploids occur naturally (see Chapter 5), and some can be produced artificially with the use of chemical mutagens. Some plants having polyploid forms are raspberry, apple, pear, cherry, plum, strawberry, cabbage, potato, aster, daylily, rose, chrysanthemum, delphinium, iris, narcissus, and marigold.

The chemical mutagen of choice, colchicine, is used to double the chromosome number. Colchicine is an alkaloid derived from the autumn crocus (*Colchicum autumnale*). Because colchicine is a poison, it should be handled with care. With colchicine treatments, some of the plants derived from diploids are tetraploids. Occasionally, a polyploid is produced with more than double the number of chromosomes, such as an octaploid from a colchicine-treated diploid. Technically, these various polyploids (triploids, tetraploids, and so on) are termed euploids (see Chapter 5).

The technique is simple, but frequently large numbers of plants (or sometimes seeds) must be treated to ensure production of a desirable polyploid. Seeds are germinated on water-moistened filter paper in a closed petri dish. When the cotyledons expand, the seedlings are placed upside down (to avoid colchicine treatment of the roots) in another petri dish containing an aqueous colchicine solution. The concentration of colchicine varies from as low as 0.01% to as high as 1.0%. The concentration is based on the nature and maturity of the seedlings, as well as the length of exposure. A good starting point is 24 hours with 0.5% colchicine solution. The colchicine-treated seedlings should be placed under fluorescent lights for around 24 hours and then transferred to a propagation medium.

The colchicine solution is applied directly to the growing tip when treating larger seedlings or plants. A wetting agent may be necessary if the growing tip has young leaves with a very waxy cuticle. The length of treatment varies. With some plants it may be necessary to keep the growing tip in contact with the colchicine solution for several days.

About 5% of the surviving plants will exhibit evidence of tetraploidy. Signs include thicker, broader leaves and flower parts. Flowers, fruits, leaves, and stems of some tetraploids, being larger, are often more attractive or desirable. Other signs are slower growth or larger cell, pollen, and stomate size; however, an actual count of the chromosomes is required to verify tetraploidy. If no tetraploids result, it may be necessary to increase the colchicine concentration and/or exposure time. Plants distorted in appearance are sometimes produced. These usually have an unbalanced chromosomal number (aneuploidy; see Chapter 5) and are discarded.

Colchicine is also useful for other reasons, such as overcoming hybrid sterility. From Chapter 5 we know that euploids can be either autopolyploids (direct increase of entire genomes within a species, ABC→ABCABC) or allopolyploids (multiples of genomes derived from two different species, ABCA'B'C'→ABCABCA'B'C'A'B'C'). Even-numbered autopolyploids (such as autotetraploids) are usually fertile (ABCABC), while hybrids between species are sterile. If a hybrid is sterile (ABCA'B'C') from a failure of chromosomal pairing at meiosis, the treatment of the sterile hybrid with colchicine will produce a fertile allopolyploid (ABCABCA'B'C'A'B'C') that is a double diploid hybrid (amphidiploid). Sterility is overcome because each chromosome will now have an identical pairing partner in meiosis. An alternative route to the interspecies tetraploid hybrid is to cross the tetraploids of both species (ABCABC × A'B'C'A'B'C'→ABCABCA'B'C'A'B'C'). These fertile tetraploids also often breed true from seed.

If the tetraploid and diploid of the same species are crossed, the resulting triploid is sterile. This outcome is only possible with a few plants (banana, citrus, watermelon, marigold), since most tetraploid–diploid crosses abort due to endosperm failure. Triploids are often valued for their increased vigor or flowering/fruiting, such as with "mule" marigolds. Triploid sterility often means an extended blooming period, since there is no seed production to terminate flowering. Other triploids produce seedless fruits, such as watermelon and citrus, that are valued for food purposes.

To induce mutations, plant breeders have used X-rays, gamma rays, and chemical mutagens other than colchicine. Fewer useful genetic traits have been produced with these techniques in comparison to the wide genetic variability available naturally in the higher plants plus those produced with colchicine. For plants the X-ray dosage is usually from 35 to 400 roentgen units. Neutrons and alpha particles have also been used. These mutagens have been used with limited success, and most have been used with agricultural and not horticultural crops.

SELECTION

The testing process leading to the final selection is time consuming and a laborious part of the plant breeder's task. Because the creation of new cultivars and the adaptation of existing plants to new areas or needs are the plant breeder's goals, success is highly dependent on proper selection. This conclusion follows from the fact that the desired characteristics result from changes in the genotype. Such changes, being on the molecular level, can only be recognized and preserved through visual selection of those phenotypes showing the desired new characteristics.

Natural versus Artificial Selection

Artificial selection differs from natural selection in several aspects. Artificial selection is carried out under the direction of the breeder, unlike random natural selection in the wild. For example, in the artificial selection of insect-resistant plants, the breeder may grow plants under natural infestation in a field or in a temperature-controlled greenhouse with the specific insect introduced on the plants. Artificial selection is faster and more favorable for genotypes that would not survive in nature, such as ornamental triploids, and more specific for desired alterations. Natural selection is slower and favors those altered genotypes with enhanced chances of survival. Natural selection might or might not be desirable for human purposes. Artificial selection occurs with controlled crossing of a relatively small number of selected plants, in contrast to the uncontrolled, random crosses among relatively large numbers of chance plants. Artificial selection also provides more uniform conditions for selection in each cycle and thus more rapid genetic gain for desired characteristics.

Methods of Selection

The success of plant breeders in achieving genetic improvements ultimately rests on their ability to indirectly select for desirable genotypes by being able to recognize the expression of desirable trait(s) among the phenotypes that make up the test population. Selection (Figure 8–3) involves perception, judgment, and assessment of relative merits. A thorough knowledge of the morphological and physiological characteristics of the crop is required. Often, selection becomes difficult because the phenotype can be greatly influenced by environmental factors. Some view selection as an art in which the breeder with intuition does best.

Individual plants can be selected on the basis of individual merit; this approach is called individual or single plant selection. If the seeds of several or more individual selections are pooled and subsequently intercrossed, it is termed mass selection. Different procedures and degrees of record keeping are needed in these two approaches.

Selection in Self-Pollinating Plants

Single plant selection with self-pollinating plants is essentially the selection of a large number of superior phenotypes from a genetically variable population of a horticultural crop. These selections are each self-pollinated, and the resulting seeds of each plant are propagated. These progeny, often in different environments, are selected for best lines and their selfed-progeny in turn are selected until no observable differences are found in the remaining lines. The remaining lines are compared with established cultivars for two or more seasons and often in more than one location until a single, pure line is retained. This type of selection is also called pure-line selection.

Characters determined by one or a few genes, such as disease resistance, are more easily selected in the preceding manner than are characters inherited quantitatively (see Chap-

(A)

(B)

FIGURE 8–3 Selection involves field trials and comparisons with other cultivars and the newly bred plant. These trials take place in many locations and by several professionals to account for climatic effects and to reduce bias and subjectivity. (A) Scientists catalog bags of potatoes grown at the Western Regional Research Center in Albany, California. They will send the experimental potatoes to ARS colleagues in Aberdeen, Idaho, for field evaluation.
(B) Evaluations don't always end in the field, as food crops, unlike ornamentals, need to be tested for postharvest factors and food qualities. Entomologists examine sweet potato breeding lines for maintenance of quality during long-term storage. (USDA photographs.)

ter 5). Characters selected as described become fixed through continual selfing in homozygous progeny, that is, a pure line.

Mass selection with self-pollinated crops consists of culling or roguing out undesirable phenotypes. Care must be exercised not to overly select off-types; otherwise, the number of lines is reduced and the genetic base is narrowed. The resulting mass-selected variety consists of several closely related lines (not one pure line) that impart maximal adaptability to the crop in terms of environmental variability. Mass selection can be used to retain the best characteristics of a crop in terms of adaptability, but in less time than with pure-line selection. Pure-line selection can be used to further improve mass-selected lines.

In established self-pollinating crops, individual and mass selection for variability has been utilized nearly to its fullest. Plant breeders use mass selection now mostly for the propagation of established cultivars (both old and new) as a means of preserving identity and variability.

The most widely used selection technique by plant breeders for self-pollinated crops is pedigree selection after hybridization. After hybridization of two parents having characteristics the breeder desires to see in one plant, the F_1 generation is examined, and obvious off-types are removed. In the F_2 generation the best individuals are selected. In the F_3 generation the progeny (family) derived from the individual F_2 selections are examined, and the best individuals are selected. This procedure is continued through the F_4 and F_5 generations.

In the F_6 generation the selection changes from individual to family. With family selection the plant breeder judges the phenotypes of the individual and its siblings collectively. Family selection involves the choice or rejection of entire families with one or both parents in common, since individual selection within the family is no longer possible due to homozygosity. If the families have both parents in common, they are full sibs and have at least 50% of their segregating genes in common. If the sibs have one parent in common, they are half-sibs with at least 25% of their segregating genes in common. From this we may conclude that the more closely related two individuals are, the more reliably we may consider the evaluation of one to be representative of the other. For this reason, family selection is quite useful with inbreeding, in which one expects a high genetic relationship.

In the F_7 generation the best families (now from replicates) are selected. At this point, variability is greatly reduced, and we use the term *line* or *selection* to denote reduced variability. The best lines are chosen and sown in replicates through the F_{10} generation. In the F_{11} and F_{12} generations, strip-tests and seed accumulation of the final chosen line are undertaken. This entire procedure of pedigree selection may take 10 to 16 years.

Another type of selection for handling hybrids is bulk selection. Instead of retention of individuals in the F_2 generation, undesirable types are rogued out and the entire seed crop is retained. This process is continued, and individual selection is not attempted until the F_5 generation at the earliest or as late as the F_8 generation. In addition to roguing out, natural selection will help somewhat to reduce inferior types. Growing the generations in different environments will help to ensure that survivors adapted to those several environments become predominant. After the individual selections are made, they are essentially treated as lines in pedigree selection. Bulk and pedigree selections are also used in combination.

In summary, the pedigree technique permits early removal of off-types, allows evaluation over several years and hence different conditions, and achieves rapid homozygosity. Bulk selection is usually the choice when extensive numbers of plants from complex or several crosses are involved, and natural and/or artificial selection is such that large numbers of off-types can be quickly eliminated.

Selection with Cross-Pollinating Plants

Single plant selection with cross-pollinated plants is possible, but to a more limited extent than with self-pollinated plants. Continued inbreeding leads to elimination of undesirable recessive genes and genetic consistency, but also a loss of vigor, fertility, and productivity. Inbreeding and selection of the best phenotypes is practiced over fewer generations, and then crossing between phenotypes may be necessary to introduce hybrid vigor. Single plant selection opportunities are essentially zero when the plant in question is self-incompatible.

Mass selection with cross-pollinated plants consists of selecting desirable phenotypes and pooling their seed. Such selection is most effective for easily observed or measured characteristics; its greatest contribution is in the development of cultivars for special purposes or adaptation of cultivars to new areas. With established crops, its effectiveness decreases.

Progeny selection can be used to choose between apparent improvements in the phenotype that are environmentally and not genetically induced. Seeds from individual plants selected from mass selection are not pooled, but evaluated under different environmental conditions. Plants are then selected on the basis of performance, and each may become a progeny line. The pooling of several progeny lines is termed *line breeding*.

Recurrent selection is widely used with cross-pollinated plants. In its basic form, recurrent selection consists of selecting superior phenotypes from a cross-pollinated population. These phenotypes are then self-pollinated, and the following generations derived from these selfed selections are intercrossed in all possible combinations. Further selection and intercrossing follow.

A more complex method is as follows. From two separate, genetically distinct populations, selections with desirable characteristics are made. These two selected groups are self-pollinated and each is cross-pollinated with a random sample of pollen from the other population. The selfed seed is kept in reserve. The crossed progeny from the two sources are evaluated in replicated field plots. The parents of superior phenotypes among the crossed progeny are noted. From the selfed seed reserve, the parents that have demonstrated superiority are chosen, based on the field trials of crossed progeny. These are sown in separate blocks and each intercrossed among themselves. The seeds from each group are sown to produce two groups of plants that may be cross-pollinated to produce a final seed source, or the whole cycle may be repeated with these seeds.

Selection and Quantitative Inheritance

Selection as outlined previously and in conjunction with hybrid vigor or backcrossing obviously works best with characteristics that are easily observed or measured, such as insect resistance or flower color change. Other characteristics, such as fruit size and color, resistance to drought or cold, or time to maturity are regulated by several genes (quantitative inheritance; see Chapter 5) and are not easily recognized. The problem is the difficulty of sorting out small contributions from individual genes against the backdrop of larger effects resulting from environmental factors. Selections for these traits can only be analyzed by statistical analysis of considerable data over many generations of the plants in question. Details are beyond the general treatment here, but most plant breeders find themselves dealing with this concept at some time. A good plant breeding textbook provides all the details needed to understand this evaluation process.

Tissue Culture Selection

A newer selection technique for the plant breeder holds great promise because of reduced labor and costs, as well as less need for test land. This is based on the aseptic method of propagation covered in Chapter 7. Basically, single cells of plants are maintained in tissue culture. One such petri dish of cells could be equivalent to a field containing tens of thousands of plants. These cells are grown in the haploid state whenever possible, because it is easier to select mutants from them than from diploid cells. In a haploid cell the one copy of a mutant gene can be clearly expressed, without interference from the other gene copy (allele) present in a diploid cell.

The cells can be treated with mutagens, such as radiation or colchicine, and then placed in the presence of appropriate screening agents. For example, the screening agent might be a fungal toxin produced by a plant pathogen; the cells would be subjected to mutagens and then the fungal toxin added. Those cells that survive may possibly be converted into plants resistant to the plant disease caused by the fungus. These haploid survivors sometimes change spontaneously to diploids, or colchicine may be added to increase diploid production. The diploids have identical gene copies and thus are instant inbreds that will breed true when self-pollinated.

This approach assumes that the diploid cells can be induced to develop roots and a shoot after the addition of phytohormones and growth regulators. Limited production of mutated plants has been achieved to date. Techniques for regeneration of whole herbaceous and woody plants from cells exist for many horticultural crops.

Another possibility with cultured cells is the creation of hybrids by cell fusion. This technique would be especially valuable for producing hybrids that cannot be obtained by sexual means because of incompatibility. Cells of two different species are stripped of their cell walls by enzymic treatment and the resulting protoplasts fused. The fused cell is generated into a plant by the techniques mentioned previously for cell cultures. The resulting plant is a hybrid, except it is not produced by conventional hybridization.

Although exciting, this plant breeding tool will only complement classical plant breeding techniques. It will reduce the work and offer some approaches that cannot be handled with classical techniques. However, there will always be a need for field selection. The next chapter explores the genetic engineering approach to plant improvement.

MASS SEED PRODUCTION

The plant breeder's work is not over once a selection has been made. The selection, if it is to have commercial horticultural value, must be reproduced on a large scale. If the plant can be propagated asexually, the production of an unaltered genotype (clone) is straightforward.

If one desires to maintain the genetic purity of a cultivar propagated by seed, this variability must be controlled. The degree of difficulty in controlling genetic variation varies according to the pollination mode of the cultivar. Regardless of the pollination mode, off-type plants should be removed, especially before pollination. Weeds should be controlled. Testing must be conducted to ensure genetic purity and to detect any genetic drift that may occur if conditions controlling pollination should fail to any degree or if environmental pressures over time favor adaptability changes.

Seeds from Self-Pollinated Plants

In plants that are only self-pollinated, the pollen arises only from the same flower, from a different flower on the same plant, or from different plants of the same clone. Self-pollinated plants are largely homozygous; consequently, their offspring are also homozygous and have the characteristics of their parents. Cultivars of these plants can be preserved quite easily, even in the vicinity of closely related cultivars. Self-pollinated plants that can also accept pollen from a different plant or clone (such as tomatoes) can produce cross-pollinated seed. Maintenance of these cultivars requires the prevention of cross-pollination, usually by isolation (see below).

Isolation is the method of restricting unwanted cross-pollination by wind or insect vectors, as well as preventing accidental mechanical mixing of harvested seeds. The distance required depends on the species, the type of pollination, and the desired degree of seed certification for genetic purity. The distance for production of seed from self-pollinated plants need not be as great as with cross-pollinated plants. Minimal spacing between self-pollinating cultivars is 10 to 15 feet.

Seeds from Cross-Pollinated Plants

Seeds produced from cross-pollinated plants may be variable, depending on how inbred the parents are. Most individuals are heterozygous. We have seen that heterozygosity is important to the plant breeder as a source of new characteristics. Heterozygous plants produce another generation of plants with some characteristics that segregate out (such as tall versus dwarf) and, through continued control and self-pollination, can be developed into nearly true breeding lines. Selection to a standard and prevention of cross-pollination from unwanted sources can produce a cultivar that, although not necessarily homozygous, is uniform for the important traits. Cross-pollination can be restricted by isolation or using staggered planting times to prevent simultaneous pollen production. Distances vary from 165 to 5,280 feet, depending on species and desired degree of seed certification.

Hybrid Seeds

Continual self-pollination of normally cross-pollinated plants produces an inbred line that has phenotypic uniformity, but often at the expense of vigor. Vigor may be restored by crossing with another inbred line (single-cross) to produce a hybrid cultivar with uniform phenotype and restored hybrid vigor. A desirable hybrid may be realized from other crosses, such as between two species. Other hybridization alternatives include two single-cross parents (double-cross), an inbred line and an open pollinated cultivar (top-cross), or a single cross and an inbred line (three-way cross). These latter hybridizations are used for few horticultural crops of value. Hybrid cultivars cannot be saved for seed in the field because of extremely variable progeny, so the hybrids must be continually reproduced from the parent inbred lines.

A problem arises in hybrid production on a large scale since a mode of assuring cross-pollination without any self-pollination is needed to produce pure hybrid seed. Emasculation and hand pollination, used to produce hybrids during the selection process, are of limited value in commercial hybrid production. The prohibitive labor cost makes it feasible only for high-value seed such as begonia, melon, petunia, pansy, tomato, or snapdragon. Commercial production of hybrids requires some system of sterility that favors cross-pollination over self-pollination.

The removal of male flowers of monoecious species followed by natural pollination by the other parent, such as the detasseling of sweet corn, reduces labor and enforces cross-pollination. However, considerable labor is still involved in removing tassels from female parents chosen to be the seed producers.

Fortunately for the plant breeder, a genetic factor for male sterility exists in many crop plants. This factor is responsible for inhibition of pollen production or nonviability of produced pollen. Male sterility may arise from a cytoplasmic factor (cytoplasmic male sterility, CMS) or a genetic factor (genetic male sterility). Male sterility has been observed in beet, carrot, corn, cucumber, lima bean, melon, onion, potato, and tomato. Plant breeders have been able to breed male sterility into a number of hybrid cultivars and have worked out the methods to maintain the male-sterile parental line. The most useful system involves sterility produced from both cytoplasm and nuclear gene activity. This approach has been used with horticultural crops, such as onion and sweet corn, to prevent self-pollination.

Recently, a restorer gene was discovered in petunias, which restores pollen production in plants with CMS. This restorer gene can be added by conventional breeding (or recombinant DNA technology) to cultivars processing CMS. Consequently, two CMS cultivars can also be large-scale hybridized for seed production if one gets the restorer gene.

Cabbage hybrids have been produced through the use of self-incompatibility. This factor results from the inability of the pollen tube to grow properly in the style of a flower on the same plant or clone (Figure 8–4). The pollen will grow normally on another plant. Hybrids can be easily produced with dioecious plants (male and female flowers on separate plants) such as spinach.

Certain monoecious plants, such as the cucumber, tend to produce female flowers predominantly at the first nodes, and male flowers appear later. Breeders have utilized this trait with cucumbers to produce gynoecious plants, individuals that produce mostly female flowers. If the other inbred line is a standard monoecious type, hybrid seed is easily produced.

Growth regulators have been used to shift sex expression in certain monoecious plants such as the cucurbits (squash, cucumber, and melon). Gynoecious cucumbers treated with 1,000 ppm of gibberellin three times a week from the expansion of the first true leaf can be induced to produce male flowers. Treatment of one parent and not the other makes it easy to produce hybrid seed.

Seeds from Trees and Shrubs

Clones that are heavily heterozygous, such as many herbaceous and woody perennials, are not ideally suited for seed production because of extreme seedling variability. However, depending on the intended purpose, seeds may be desired over vegetative propagation. In that case, some control may be achieved by following recommended seed selection techniques.

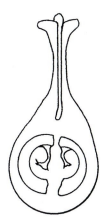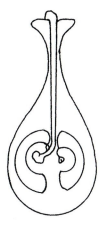

FIGURE 8–4   Self-incompatibility. Pollen from same plant or clone fails to produce pollen tube (*left*) in time to reach receptive ovule, thus preventing self-pollination. The same pollen on another genetically similar but not identical cultivar produces pollen tube in sufficient time to allow fertilization (*right*). (From Raymond P. Poincelot, *Horticulture: Principles & Practical Applications,* 1980. Reprinted by permission of Prentice-Hall, Upper Saddle River, New Jersey.)

Seed should be gathered from areas of comparable environment and from trees, shrubs, and herbaceous perennials showing a desired phenotype. Progeny testing, as discussed under selection, may be required for seed evaluation from these sources.

Seeds by Apomixis

Seed may also arise in some plants through an asexual process, *apomixis,* which provides a means to ensure uniformity. With some plants, seeds can arise through nonsexual processes. These asexually produced seeds preserve the genotype of the parent. Apomixis varies from 100% with some species or individuals (obligate apomicts) to variable percentages mixed with seed produced by normal sexual means (facultative apomicts). Selection for progeny with uniform genotypes relative to the parent line with facultative apomicts can be used to produce apomitic cultivars such as 'Adelphi' Kentucky bluegrass. Seed uniformity is easily maintained in apomitic cultivars. Apomitic seedlings of citrus are usually virus-free, vigorous and uniform; these seedlings are used as citrus rootstocks.

PLANT VARIETY PROTECTION ACT

Existing legislation has improved potential profitability for those who produce new plants. Because it may take 10 to 12 years from initial crosses through selection and final introduction of a cultivar, some form of protection to give exclusive propagation rights to the seed producer is desirable for commercialization. The Plant Variety Protection Act of 1970 did just that. The U.S. Department of Agriculture can issue a Plant Variety Protection Certificate for the seed production of a cultivar that is valid for 17 years. To obtain such protection, the cultivar must meet certain criteria to prove that it is indeed novel. Vegetatively propagated materials may be patented based on 1930 plant patent legislation. A few horticultural crops are exempt from these rules: potato, cucumber, carrot, tomato, pepper, and celery. In the early 1990s, the Supreme Court extended legal protection to cultivars produced by genetic engineering.

The protection and enhanced marketability associated with new improved plants have led to an increasing number of introductions. Hybrid seed is commercially available for more than 40 different flowers grown in the garden and greenhouse and for at least 20 different vegetables. In addition, a great number of new, improved plants are sold in propagating forms other than seed. These clonal cultivars include ornamental shrubs and vines, fruit trees, nut trees, cane and vine fruits, bulbs, corms, and tubers. These offerings would be much reduced in number without plant patents. Some plant breeders, especially in the public sector, have argued that plant patents discourage the exchange of germplasm and cooperation among researchers. This debate continues.

ALL-AMERICA SELECTIONS

The ultimate honor for a flower or vegetable hybrid is to be chosen as a winner in the All-America Selections, or All-America Rose Selections, All-America Daylilies, and Perennial Plant of the Year. Besides the honor, the financial return on such a cultivar is usually higher because of public awareness of the selections. The competition for this annual award is intense among plant breeders from seed companies, botanical gardens, government research stations, and universities.

Over the years hybrids of a number of flowers and vegetables have been chosen. A few recent winners are shown in Figure 8–5.

FIGURE 8–5 2002 All-America Selections and All-America Rose Selections:
(A) Rudbeckia Cherokee Sunset (B) Petunia F_1 Tidal Wave Silver
(C) Petunia F_1 Lavender Wave (D) Cleome F_1 Sparkler Blush
(E) Geranium F_1 Black Magic Rose (F) Vinca Jaio Scarlet Eye
(G) Ornamental Pepper F_1 Chilly Chili
(H) Squash Cornell's Bush Delicata
(I) Cucumber Diva
(J) Basil Magical Michael
(K) Pumpkin F_1 Sorcerer
(L) Starry Night (close-up) (M) Starry Night (whole plant) (N) Love & Peace (close-up)
(O) Love & Peace (whole plant) (A–K, courtesy of All-America Selections; L–O, courtesy of All-America Rose Selections.)

(A)

(B)

(C)

(D)

(E)

(F)

(G)

(H)

(I)

(J)

(K)

(L)

(M)

(N)

(O)

SUMMARY

Plant breeders have improved horticultural productivity and quality. Improvements in edible crops include higher yielding vegetables and fruits, uniform ripening for purposes of mechanical harvesting, increased vitamin and/or protein content, better taste, dwarfing of fruits for mechanical harvesting, and improved keeping and processing qualities. Esthetic improvements of ornamentals include improved form or scent, better foliage appearance, dwarfing to minimize pruning and space requirements, and improved tolerance to shade and abuse for lawngrasses. Other desirable traits for sustainable horticulture are resistance to insects and diseases, increased tolerance to heat, cold, or drought, more efficient absorption of soil minerals, insensitivity to photoperiod, more adaptability to mechanical harvesting, or a shorter period to the production of flowers, fruits, or vegetables.

The key issue with sustaining plant improvement is genetic erosion (gene loss). One problem is loss of plants (and their genes) in centers of crop origins. Genes are lost as new cultivars replace older folk varieties and as development shrinks the tropical rain forests. Gene loss threatens the resources of the plant breeder and the genetic engineer. If a plant is lost before the genome is characterized, it is truly a loss.

Most of our major horticultural food crops belong to a few families that arose outside the United States Regions of origin are not species rich and are characterized by seasonal differences. Seasonal changes produce dormant plants that store food for when favorable growth conditions return. The plants' stored food in turn becomes our food. Most of these centers of origin are either temperate or semitropical. Few current food crops originated in the tropical rain forests. The tropical rain forests are underutilized. The number of potential edible, ornamental and medicinal horticultural crops from tropical rain forests easily exceeds 2,000.

Germplasm to maintain both current and future crops lies beyond the United States in centers of origin threatened by increased population and agricultural/urban development. Preserving germplasm access for breeding requires political, social, and economic attention from the United States and other developed nations, global cooperation, and fair compensation.

Often a wide diversity of genotypes exists for many cultivated crops. Some are well known, and found currently or recently in cultivation. Others have been neglected or discarded, and others remain in isolated areas. The latter is true for the primitive ancestors of our cultivated crops that were discarded as natural and artificial selection produced better genotypes. The more diverse the genetic background, the more scope the plant breeder has to work with. Primitive ancestors or closely related, uncultivated species may possess many desirable genes. The plant breeder must assemble a collection of these plants. Some are more readily obtainable or better known than others.

Most major horticultural crops have genotype collections maintained by governments in both developed and Third World nations. Many seed banks and clonal repositories are part of an international network, the Consultative Group on International Agricultural Research. In the United States, the National Plant Germplasm System (NPGS) and Germplasm Resources Information Network (GRIN) provide both germplasm and information, respectively, to plant breeders. Germplasm comes to NPGS either through donation, purchase or collection at centers of origin. Specialized ornamental plant societies, some historical societies, seed companies, botanical gardens, arboreta, and plant research centers are also sources.

Most crops depend on genes in species introduced from Central and South America, the Mediterranean area, and various parts of Asia, India, Europe, Australia, and Africa. These centers of diversity contain ancestral plant germplasm for insect resistance, disease resistance, drought resistance, and restoration of vigor. All these areas are undergoing urban and agricultural development. Most developed nations try to maintain germplasm collections from these areas. However, these are not complete collections. Some protection through parks and botanical reserves is crucial. Tropical rain forests are the future centers of diversity, although destruction threatens this outcome. These habitats have potentially thousands of new crops. Only global cooperation, financial support and rational conservation efforts can save these future genetic resources.

Folk varieties (landraces) developed through long-term selection by local farmers are valuable sources of germplasm. Selection was generally for improved production under local, often adverse, conditions. These plants possess considerable genetic diversity and are adapted to more rigorous habitats. Some are adapted to arid environments, such as maize and bean folk varieties used by the Hopi and Zuni peoples in the arid Southwest. Similarly, native peoples in the Andes region have amassed numerous potato folk varieties. Most have been displaced by modern cultivars, except in a few isolated areas. Government and other institutions have collected and preserved some folk varieties in seed and clonal repositories. The issue of fair compensation must be considered in the context of the indigenous peoples responsible for their origin.

Heirloom varieties arose more recently in the early kitchen gardens of most countries. The early American colonists came with European vegetable and flower varieties. Later waves of immigrants added to the heirloom varieties. By the late 1800s, numerous vegetable, fruit, and flower heirloom varieties existed in the United States, having passed through several generations of families. The advent of modern cultivars, especially hybrids, brought about their replacement and disappearance. Groups exist today that seek to preserve and distribute these heirloom varieties.

Lines are populations of seedling plants whose genetic composition can be maintained relatively unchanged. Lines with similar phenotypes and differing traits, such as degree of insect resistance, can be blended to create a multiline. Lines can arise through self-pollination, cross-pollination, or hybridization.

Cultivars are lines altered through breeding to produce improved, reproducible characteristics that set the cultivar apart from both the source line and other lines of the particular crop. Cultivars are afforded legal protection and assigned unique names. Lines formed from self-pollinating plants usually consist mostly of recognized cultivars, referred to as line cultivars. Cross-pollinating plants usually require enforced self-pollination brought about by isolation to produce lines. These lines are known as inbred lines. Inbred line cultivars tend to be fewer in number. Instead inbred lines are crossed to produce hybrid lines that generally become hybrid cultivars. Crops propagated asexually are usually sold as clonal cultivars.

Lines and cultivars dominate horticultural crops today. Most offer improvements over the previous folk and heirloom varieties, especially for highly mechanized, commercial horticulture. Home gardeners are also attracted to their qualities. The advent of legal protection for cultivars increased their market share, as seed companies sought to maximize their markets. More recently, legal protection was extended to genetically engineered plants. Lines and cultivars have essentially replaced folk and heirloom varieties, leading to the need for preservation of these displaced sources and their genetic diversity. Given that lines and cultivars are also replaced over time with newer versions, older lines and cultivars also need preservation for similar reasons.

Mutations in reproductive plant tissues on the chromosome or gene level can occur by chance or be artificially induced to produce an altered plant (sport). Sports are preserved through seed and clonal stocks. Spontaneous mutations also occur in tissues that do not give rise to sex cells (somatic tissue). These somatic mutations, or chimeras, are sometimes stable enough to be propagated asexually. Occasionally, a natural hybrid between two species or even two genera may arise. Even if the hybrid is self-sterile, it might backcross with one or both parents, thus introducing additional genetic variability. The horticulturist is ever alert for these natural occurrences of variability and selects altered forms that have horticultural value for purposes of propagation.

Desired characteristics exist in varieties, cultivars, and species. Crosses between plants to produce more desirable offspring are possible, either naturally through insect pollination, or more likely by plant breeders. The conventional plant breeder employs several techniques to move genes around in sexually compatible plants. These techniques involve the deliberate and careful control of pollination. Sometimes genetic variation is given an assist by mutagens. A thorough understanding of the benefits and shortcomings of the various ways to breed plants is essential. Monitoring the changes and preserving the desired change is crucial for success.

Hybridization is the crossbreeding of sexually compatible plants having differences in genotypes to create offspring (hybrids) with altered characteristics. The parents can be two different species or cultivars. Hybrids exhibit uniform, vigorous improvements compared to either parent. This improvement is known as heterosis or hybrid vigor. Improvements include higher yields, larger flowers and fruits, or earlier maturation. If the cultivar is a perennial, these improvements are permanent. If the hybrid is an annual, the seeds produced from the hybrid will not show these improvements. Annual hybrids are preserved by obtaining seed through repetitive crossing of the two parents. Many commercial and garden flower and vegetables are hybrid cultivars.

Besides setting breeding goals and selecting appropriate plants for cross-pollination, skill with a basic technique, hand pollination, is critical. While hand cross-pollination is essential for investigating potential hybrids, large-scale production for mass marketing needs systems other than hand pollination. Contamination by unwanted pollen must be minimized. Anthers are taken from unopened flower buds by various techniques. Anthers are dried and pollen is removed, usually indoors to prevent contamination by foreign pollen. Alternatively, ripe pollen is collected directly from flowers that were caged or bagged to prevent contamination.

If the seed parent is not ready to receive the pollen because of different blooming times or environmental effects, it is sometimes necessary to synchronize the flowering of the seed and pollen parents or to store the pollen. Pollen is stored under conditions of controlled humidity and near freezing temperatures. If the seed parent is self-fertile, stamens are removed (emasculation) just before the flower opens to prevent self-pollination. The stigma will need protection to prevent accidental foreign pollination, if it is not receptive. If the flower is self-sterile, it need not be emasculated, but only enclosed to prevent insect or wind pollination. Plants can be caged with screens, or protected in insect-proof greenhouses.

Once the stigma is receptive, the pollen is applied carefully to not damage the stigma. If the two parents differ greatly in the style length, the long-styled parent should be the pollen source, since pollen from short-styled plants is sometimes not able to grow down a long style. Earliest possible pollination gives the pollen tube maximal time to grow down the style. The seed parent flower should be immediately enclosed in a paper or cloth bag or caged to eliminate further pollination. The flower or flower branch should be dated and labeled with seed parent and pollen parent identification. Bags may have to be removed with some species prior to complete seed ripening to prevent damage by excessive moisture. Records should be kept and observations recorded.

Hybrid vigor is an important tool to restore the vigor lost through continual inbreeding (self-pollination) of a line of plants (inbred line). If two inbred lines are cross-pollinated, the F_1 generation can show an increase in vigor, yet still retain the desirable characteristics of the parental inbred lines. Hybrid vigor effects are expressed as increased growth rates, increased size, increased yield, earlier germination, uniformity, and resistance to unfavorable environmental factors. The most popular use of hybrid vigor by plant breeders is the single cross (the crossing of two inbred lines). The resulting seed, once known to produce an F_1 generation (or F_1 hybrid) with hybrid vigor and a desirable phenotype, can be marketed depending on the ease of emasculation and pollination. Pollination control is needed to ensure hybrid seed unmixed with inbred line seed. Hybrid vigor has been most successful with several ornamentals and vegetables and is used frequently in the commercial production of horticultural crop seed.

Backcrossing accumulates desirable genes rapidly. A wild species or another cultivar (nonrecurrent parent) may possess a desirable characteristic such as insect or disease resistance, but have many horticulturally undesirable characteristics. The nonrecurrent parent is crossed with a cultivated species or cultivar (recurrent parent) possessing desirable traits, but which is highly susceptible to insects or disease. The resulting F_1 hybrid is selected for the desirable characteristic and then is crossed back with the recurrent parent; the most desirable F_1 hybrid of this cross is in turn crossed with the recurrent parent. This process is repeated until the desirable characteristic is bred in, and the many desirable characteristics of the recurrent parent are retained. If the characteristic that is desirable is not dominant (recessive), it will not show in the hybrid. The breeder must resort to self-pollinations in each cycle to select the plant(s) in the next generation that possess the desired characteristic; these plants can then be used for the backcross. Obviously, the technique of backcrossing, when dealing with a desirable, recessive gene, will easily work with plants that can be self-pollinated. The process can be used with cross-pollinating plants, if they are bagged or isolated in some other way and then artificially self-pollinated. The desired plant could have been obtained strictly through selecting during a selfing program after initial hybridization, but much larger plant numbers are required. Use of the backcross produces the desired plant in fewer generations with fewer plants. For best results the desired character should be controlled by a single or few genes, be simply inherited, and easily identified.

Polyploids occur naturally and can be artificially induced with the use of chemical mutagens (colchicine). With colchicine treatments, about 5% of the surviving plants exhibit tetraploidy. Signs include attractive, larger leaves, flowers, and fruits. Other signs are slower growth or larger cell, pollen, and stomate size. Verification of tetraploidy is done by an actual chromosomal count. The colchicine production of tetraploids is also useful for overcoming hybrid sterility or for crossing with diploids to produce triploids. If no tetraploids result, it may be necessary to increase the colchicine concentration and/or exposure time. Plants distorted in appearance (unbalanced chromosomal number) are sometimes produced and are discarded. The resulting polyploids or euploids can be either autopolyploids (direct increase of entire genomes within a species) or allopolyploids (multiples of genomes derived from two different species). If a hybrid is sterile from a lack of chromosomal pairing at meiosis, the production of a tetraploid from the sterile hybrid with colchicine will produce a fertile allopolyploid that is a double diploid hybrid (amphidiploid). Sterility is overcome because each chromosome will now have an identical pairing partner in meiosis. These fertile tetraploids also often breed true from seed. If the tetraploid and diploid of the same species are crossed, the resulting triploid is sterile. This outcome is only possible with a few crops, since most tetraploid–diploid crosses abort due to endosperm failure. Triploids are valued for their increased vigor or continuous flowering/fruiting. Other triploids produce seedless fruits such as watermelon and citrus that are valued for food purposes.

The evaluation process leading to the final plant selection is time consuming and laborious. Since the creation of new cultivars and the adaptation of existing plants to new areas or needs are the plant breeder's goals, success is highly dependent on proper selection.

Artificial selection differs from natural selection. Breeders direct artificial selection, unlike natural selection that occurs randomly in the wild. For example, in the artificial selection of insect-resistant plants, the breeder grows plants under natural infestation in a field or in a temperature-controlled greenhouse with the specific insect introduced on the plants. Artificial selection is faster and more favorable for genotypes that would not survive in nature, and more specific for desired alterations. Artificial selection occurs with controlled crossing of a relatively small number of selected plants, in contrast to the uncontrolled, random crosses among relatively large numbers of chance plants. Artificial selection also provides more uniform conditions for selection in each cycle and thus more rapid genetic gain for desired characteristics.

The success of plant breeders in achieving genetic gains ultimately rests on their ability to recognize through indirect selection desirable genotypes by identifying plant phenotypes with the desirable alterations in characteristic(s). Selection involves perception, judgment, and assessment of relative merits and a thorough knowledge of the morphological and physiological characteristics of the crop. Often, selection becomes difficult because the phenotype can be greatly influenced by environmental factors. Some view selection as an art in which the breeder with intuition does best.

Individual plants can be selected on the basis of individual merit; this approach is called individual or single plant selection. If the seeds of several or more individual selections are pooled and subsequently intercrossed, it is termed mass selection. Different procedures and degrees of record keeping are needed in these two approaches.

Single plant selection with self-pollinating plants involves the selection of a large number of superior phenotypes from a genetically variable population. These selections are each self-pollinated, and the resulting seeds of each plant are propagated. These progeny, often in different environments, are selected for best lines and their selfed-progeny in turn are selected until no observable differences are found in the remaining lines. The remaining lines are compared with established cultivars for two or more seasons and often in more than one location until a single, pure line is retained (pure-line selection). Characters determined by one or a few genes such as disease resistance are more easily selected in the preceding manner than are characters inherited quantitatively.

Mass selection with self-pollinated crops consists of culling (roguing) undesirable phenotypes. The resulting mass-selected variety consists of several closely related lines (not one pure line) that impart maximal adaptability to the crop in terms of environmental variability. Mass selection can be used to retain the best characteristics of a crop in terms of adaptability, but in less time than with pure-line selection. Pure-line selection can be used to further improve mass-selected lines.

The most widely used selection technique by plant breeders for self-pollinated crops is pedigree selection after hybridization. After hybridization of two parents having characteristics the breeder desires to see in one plant, the F_1 generation is examined, and obvious off-types are removed. In the F_2 generation the best individuals are selected. In the F_3 generation the progeny (family) derived from the individual F_2 selections are examined, and the best individuals are selected. This procedure is continued through the F_4 and F_5 generations. In the F_6 generation the selection changes from individual to family. With family selection the plant breeder judges the phenotypes of the individual and its siblings collectively. Family selection involves the choice or rejection of entire families with one or both parents in common, since individual selection within the family is no longer possible due to homozygosity. If the families have both parents in common, they are full sibs and have at least 50% of their segregating genes in common. If the sibs have one parent in common, they are half-sibs with at least 25% of their segregating genes in common. From this we may conclude that the more closely related two individuals are, the more reliably we may consider the evaluation of one to be representative of the other. For this reason, family selection is quite useful with inbreeding, in which one expects a high genetic relationship. In the F_7 generation the best families (now from replicates) are selected. At this point, variability is greatly reduced, and we use the term line or selection to denote reduced variability. The best lines are chosen and sown in replicates through the F_{10} generation. In the F_{11} and F_{12} generations, strip-tests and seed accumulation of the final chosen line are undertaken. This entire procedure of pedigree selection may take 10 to 16 years.

Another type of selection for handling hybrids is bulk selection. Instead of retention of individuals in the F_2 generation, undesirable types are rogued out and the entire seed crop is retained. This process is continued, and individual selection is not attempted until the F_5 generation at the earliest or as late as the F_8 generation. In addition to roguing out, natural selection will help somewhat to reduce inferior types. Growing the generations in different environments will help to ensure that survivors adapted to those several environments become predominant. After the individual selections are made, they are essentially treated as lines in pedigree selection. Bulk and pedigree selections are also used in combination.

Single plant selection with cross-pollinated plants is possible, but to a more limited extent than with self-pollinated plants. Continued inbreeding leads to elimination of undesirable recessive genes and genetic consistency, but also a loss of vigor, fertility, and productivity. Inbreeding and selection of the best phenotypes is practiced over fewer generations, and then crossing between phenotypes may be necessary to introduce hybrid vigor. Single plant selection opportunities are essentially zero, if the plant in question is self-incompatible.

Mass selection with cross-pollinated plants consists of selecting desirable phenotypes and pooling their seed. Such selection is most effective for easily observed or measured characteristics; its greatest contribution is in the development of cultivars for special purposes or adaptation of cultivars to new areas. With established crops, its effectiveness decreases.

Progeny selection can be used to choose between apparent improvements in the phenotype that are environmentally and not genetically induced. Seeds from individual plants selected from mass selection are not pooled, but evaluated under different environmental conditions. Plants are then selected on the basis of performance, and each may become a progeny line. The pooling of several progeny lines is termed line breeding.

Recurrent selection is widely used with cross-pollinated plants. In its basic form recurrent selection consists of selecting superior phenotypes from a cross-pollinated population. These phenotypes are then self-pollinated, and the following generations derived from these selfed selections are intercrossed in all possible combinations. Further selection and intercrossing follow.

A more complex method is as follows. From two separate, genetically distinct populations, selections with desirable characteristics are made. These two selected groups are self-pollinated and each is cross-pollinated with a random sample of pollen from the other population. The selfed seed is kept in reserve. The crossed progeny from the two sources are evaluated in replicated field plots. The parents of superior phenotypes among the crossed progeny are noted. From the selfed seed reserve, the parents that have demonstrated superiority are chosen, based on the field trials of crossed progeny. These are sown in separate blocks and each intercrossed among themselves. The seeds from each group are sown to produce two groups of plants that may be cross-pollinated to produce a final seed source, or the whole cycle may be repeated with these seeds.

Selection as outlined previously and in conjunction with hybrid vigor or backcrossing obviously works best with characteristics that are easily observed or measured, such as insect resistance or flower color change. Other characteristics such as fruit size and color, resistance to drought or cold, or time to maturity are regulated by several genes and are not easily recognized. The problem is the difficulty of sorting out small contributions from individual genes against the backdrop of larger effects resulting from environmental factors. Selections for these traits can only be analyzed by statistical analysis of considerable data over many generations of the plants in question.

Tissue culture selection holds great promise because of reductions in labor, costs, and test land. Basically, single cells of plants are maintained in tissue culture. One such petri dish of cells is equivalent to a field containing tens of thousands of plants. These cells are grown in the haploid state whenever possible, because it is easier to select mutants. In a haploid cell the one copy of a mutant gene can be clearly expressed, without interference from the other gene copy (allele) present in a diploid cell.

The cells can be treated with mutagens followed by appropriate screening agents such as fungal toxins. Surviving cells can possibly be converted into plants resistant to the fungal disease. These haploid survivors sometimes change spontaneously to diploids, or colchicine may be added to increase diploid production. The diploids have identical gene copies and thus are instant inbreds that breed true when self-pollinated. This approach assumes that the diploid cells can be induced to develop roots and a shoot after the addition of phytohormones and growth regulators. Limited production of mutated plants has been achieved to date. Techniques for regeneration of whole herbaceous and woody plants from cells exist for many horticultural crops.

Another possibility with cultured cells is the creation of hybrids by cell fusion. This technique would be especially valuable for producing hybrids that cannot be obtained by sexual means because of incompatibility. Cells of two different species are stripped of their cell walls by enzymic treatment and the resulting protoplasts fused. The fused cell is generated into a

plant by the techniques used for cell cultures. The resulting plant is a hybrid, except it is not produced by conventional hybridization.

Although exciting, tissue culture selection only complement classical plant breeding techniques. It will reduce the work and offer some approaches that cannot be handled with classical techniques. There will always be a need for field selection.

The plant breeder's work is not over once a selection has been made. The selection, if it is to have commercial horticultural value, must be reproduced on a large scale. If the plant can be propagated asexually, the production of an unaltered genotype (clone) is straight-forward.

If one desires to maintain the genetic purity of a cultivar propagated by seed, this variability must be controlled. The degree of difficulty in controlling genetic variation varies according to the pollination mode of the cultivar. Regardless of the pollination mode, off-type plants should be removed, especially before pollination. Weeds should be controlled. Testing must be conducted to assure genetic purity and to detect any genetic drift that may occur if conditions controlling pollination should fail to any degree or if environmental pressures over time favor adaptability changes.

In plants that are only self-pollinated, the pollen arises only from the same flower, from a different flower on the same plant, or from different plants of the same clone. Self-pollinated plants are largely homozygous; consequently, their offspring are also homozygous and have the characteristics of their parents. Cultivars of these plants can be preserved quite easily, even in the vicinity of closely related cultivars. Self-pollinated plants that can also accept pollen from a different plant or clone can produce cross-pollinated seed. Maintenance of these cultivars requires the prevention of cross-pollination, usually by isolation.

Isolation is the method of restricting unwanted cross-pollination by wind or insect vectors, as well as preventing accidental mechanical mixing of harvested seeds. The distance required depends on the species, the type of pollination, and the desired degree of seed certification for genetic purity. The distance for production of seed from self-pollinated plants need not be as great as with cross-pollinated plants.

Seeds produced from cross-pollinated plants may be variable, depending on how in-bred the parents are. Most individuals are heterozygous. We have seen that heterozygosity is important to the plant breeder as a source of new characteristics. Heterozygous plants produce another generation of plants with some characteristics that segregate out (such as tall versus dwarf), and through continued control and self-pollination can be developed into nearly true breeding lines. Selection to a standard and prevention of cross-pollination from unwanted sources can produce a cultivar that, although not necessarily homozygous, is uniform for the important traits. Cross-pollination can be restricted by isolation.

Continual self-pollination of normally cross-pollinated plants produces an inbred line that has phenotypic uniformity, but often at the expense of vigor. Vigor may be restored by crossing with another inbred line (single-cross) to produce a hybrid cultivar with uniform phenotype and restored hybrid vigor. A desirable hybrid may be realized from other crosses, such as between two species. Other hybridization alternatives include two single-cross parents (double-cross), an inbred line and an open pollinated cultivar (top-cross), or a single cross and an inbred line (three-way cross). These latter hybridizations are used for few horticultural crops of value. Hybrid cultivars cannot be saved for seed in the field because of extremely variable progeny, so the hybrids must be continually reproduced.

A problem arises in hybrid production on a large scale since a mode of controlling cross-pollination without any self-pollination is needed to produce pure hybrid seed. Emasculation and hand pollination, used to produce hybrids during the selection process, are of limited value in commercial hybrid production. The prohibitive labor cost makes it feasible only for high-value seed. Commercial production of hybrids requires some system of sterility that favors cross-pollination over self-pollination. The removal of male flowers of monoecious species followed by natural pollination by the other parent such as the detasseling of sweet corn, reduces labor and enforces cross-pollination. However, there still is considerable labor involved in removing tassels from female parents chosen to be the seed producers.

Fortunately, a genetic factor for male sterility exists in many crop plants. This factor is responsible for inhibition of pollen production or nonviability of produced pollen. Male sterility may arise from a cytoplasmic factor (cytoplasmic male sterility) or a genetic factor (genetic male sterility). Plant breeders have been able to breed male sterility into a number of hybrid cultivars and have worked out the methods to maintain the male-sterile parental line. The most useful system involves sterility produced from both cytoplasm and nuclear gene activity. Some hybrids have been produced through the use of self-incompatibility. This factor results from the inability of the pollen tube to grow properly in the style of a flower on the same plant or clone. The pollen will grow normally on another plant. Hybrids can be easily produced with dioecious plants (male and female flowers on separate plants) such as spinach.

Certain monoecious plants tend to produce female flowers predominately at the first nodes, and male flowers appear later. Breeders have utilized this trait with cucumbers to produce gynoecious plants, individuals that produce mostly female flowers. If the other inbred line is a standard monoecious type, hybrid seed is easily produced. Growth regulators have been used to shift sex expression in certain monoecious plants such as the cucurbits. Gynoecious cucumbers treated with gibberellin can be induced to produce male flowers. Treatment of one parent and not the other makes it easy to produce hybrid seed.

Clones that are heavily heterozygous such as herbaceous and woody perennials are not ideal for seed production because of extreme seedling variability. However, depending on the intended purpose, seeds may be desired over vegetative propagation. Some control may be achieved by following recommended seed-selection techniques. Seed should be gathered from areas of comparable environment and from trees, shrubs, and herbaceous perennials showing a desired phenotype. Progeny testing may be required for seed evaluation from these sources.

Seed may also arise in some plants through an asexual process, apomixis, which provides a means to assure uniformity. Plants that can produce seed both by self- and cross-pollination, or by either apomixis and normal pollination, can be maintained as pure lines. Success in maintaining purity requires that the predominate form of seed production be known.

Since it may take 10 to 12 years from initial crosses through selection and final introduction of a cultivar, some form of protection to give exclusive propagation rights is desirable. The Plant Variety Protection Act (1970) allows the USDA to issue a 17-year Plant Variety Protection Certificate for the seed production of a cultivar. To obtain such protection, the cultivar must meet certain criteria to prove that it is indeed novel. Vegetatively propagated materials are patented based on 1930 plant patent legislation. In the early 1990s, the Supreme Court extended legal protection to cultivars produced by genetic engineering. The protection and enhanced marketability associated with new improved plants have led to an increasing number of introductions.

EXERCISES

Mastery of the following questions shows that you have successfully understood the material in Chapter 8.

1. What is the singular driving force behind efforts to improve plants and in what various ways can this goal be realized?
2. What sustainability issue is associated with plant improvement?
3. What various sources provide the genetic material needed for plant improvement?
4. What are centers of origin? Where do you find them? Give examples of crops associated with each one.
5. What are folk varieties and why are they important? What do the Native American Zuni have to do with folk varieties?
6. What are heirloom varieties and why are they important?
7. What are lines and cultivars? How important are they in contemporary horticulture?

8. What causes natural plant variability? How can we induce variability artificially?
9. What is hybridization? What are the basic steps in hybridization?
10. What is polyploidy? How can polyploidy be induced? Why is it important in horticulture?
11. What is hybrid vigor and why is it important to horticulture?
12. What is the single cross? What does F_1 signify? What is a backcross? What is the purpose of a single cross and a backcross?
13. What is the difference between natural and artificial selection?
14. What is the difference between single plant selection and mass selection?
15. How is selection done with self-pollinating plants? With cross-pollinating plants?
16. What is quantitative inheritance? How is selection accomplished with this trait?
17. What is asexual propagation? For what plant types is it the method of choice?
18. How is selection done in tissue culture?
19. How are seeds mass produced for self-pollinating plants, cross-pollinating plants, and hybrids?
20. What steps are necessary in seed production for seeds from trees, shrubs, and apomictic plants?
21. What is the Plant Variety Protection Act? Why is it important in horticulture?
22. What does the term All-America Selections signify? Does it have any horticultural value? If so, what?

INTERNET RESOURCES

All-American Daylilies

http://www.daylilyresearch.org

All-America Selections

http://www.all-americaselections.org

National Plant Germplasm System; aids scientists and the need for genetic diversity by acquiring, preserving, evaluating, documenting, and distributing crop germplasm

http://www.ars-grin.gov/npgs/

Perennial plant of the year

http://www.perennialplant.org/ppy/ppyindex.html

Plant Variety Protection Act

http://www.ams.usda.gov/science/PVPO/pvp.htm

United States and international organizations involved in crop improvement and germplasm preservation

http://www.ars-grin.gov/npgs/ngo.html

RELEVANT REFERENCES

Agrawal, Rattan Lal (1998). *Fundamentals of Plant Breeding and Hybrid Seed Production.* Science Publishers, Enfield, NH.

Allard, R. W. (1999). *Principles of Plant Breeding* (2nd ed.). John Wiley & Sons, New York. 240 pp.

Callaway, D. J., and M. B. Callaway (eds.) (2000). *Breeding Ornamental Plants.* Timber Press, Portland, OR. 359 pp.

Callaway, M. Brett, and Charles A. Francis (eds.) (1993). *Crop Improvement for Sustainable Agriculture.* University of Nebraska Press, Lincoln. 261 pp.

Harlan, Jack R. (1998). *The Living Field: Our Agricultural Heritage.* Cambridge University Press, New York. 287 pp.

Kalloo, G., and B. O. Bergh (eds.) (1993). *Genetic Improvement of Vegetable Crops.* Pergamon Press, New York. 500 pp.

Lesser, W. (1998). *Sustainable Use of Genetic Resources Under the Convention on Biological Diversity: Exploring Access and Benefit Sharing Issues.* CAB International, New York. 218 pp.

McGregor, S. E. (1976). *Insect Pollination of Cultivated Crop Plants.* (Agriculture Handbook No. 496). U.S. Department of Agriculture, Washington, DC.

National Academy of Sciences (1972). *Genetic Vulnerability of Major Crops.* National Academy of Sciences Printing and Publishing Offices, Washington, DC.

Proctor, M., P. Yeo, and A. Leck (1996). *The Natural History of Pollination.* Timber Press, Portland, OR. 487 pp.

Raeburn, Paul (1995). *The Last Harvest: The Genetic Gamble that Threatens to Destroy American Agriculture.* University of Nebraska Press, Lincoln. 269 pp.

Richards, A. J. (1997). *Plant Breeding Systems* (2nd ed.). Chapman and Hall, London. 560 pp.

Simmonds, N. W., and J. Smartt (1999). *Principles of Crop Improvement* (2nd ed.). Iowa State University Press, Ames. 424 pp.

Soleri, D., and D. Cleaveland (1993). "Seeds of Strength for Hopis and Zunis." *Seedling* **10**(4): 13–18.

Stoskopf, Neal C., with Dwight T. Tomes and B. R. Christie (1993). *Plant Breeding: Theory and Practice.* Westview Press, Boulder, CO. 531 pp.

Virchow, D. (1999). *Conservation of Genetic Resources: Costs and Implications for a Sustainable Utilization of Plant Genetic Resources for Food and Agriculture.* Springer Verlag, New York. 275 pp.

Chapter 9

Biotechnology

OVERVIEW

Advancing technology has the potential to play a major role in sustainable agriculture. Its ultimate value depends on how the technology is applied and who benefits from these advances. Biotechnology offers considerable promise, especially in its ability to transfer genes into plants (transgenic plants). Sustainable agriculture can benefit if transgenic plants are created that depend less on pesticides and fertilizers. This scenario is possible through the incorporation of insect and disease resistance and the addition of nitro-gen fixation genes, respectively. If the plant is instead made herbicide resistant such that broad-spectrum herbicides can be applied during the growing season, this scenario may benefit only the chemical manufacturer and not the grower, thus contributing little to sustainable agriculture. The overuse of Bt transgenic plants might also bring about increased pest resistance to Bt applications and reduce Bt's usefulness to organic farmers and home gardeners. Still, many exciting possibilities exist for the future.

CHAPTER CONTENTS

OBJECTIVES

- ❁ To understand the potential of biotechnology, especially genetic engineering
- ❁ To learn what DNA libraries are and how to make them
- ❁ To understand how DNA fragments are manipulated and analyzed

- ❁ To learn how one identifies and selects the right gene for transfer
- ❁ To understand gene mapping and how genetic engineering works
- ❁ To learn about the applications of gene engineering in plant biotechnology

BIOTECHNOLOGY BASICS

Biotechnology is not new. The word itself is new, having surfaced sometime in the early 1970s. Humans have been practicing biotechnology for about 6,000 years, ever since we started making bread, cheese, beer, wine, and yogurt. Breeding plants and animals for improved farm production is another old form of biotechnology. Biotechnology is simply the use of living organisms (especially microorganisms, plants, and animals) to manufacture products to enhance the well-being of the human race. Products are usually food, medicine, and some chemicals.

What is new is the DNA revolution that has given us a nearly complete understanding of genes in terms of function and control and how these genes can be moved between organisms through gene splicing and recombinant DNA technology. This knowledge changed the biotechnology focus from manipulating whole organisms to manipulating DNA. New plants and animals can be bred much more efficiently and microorganisms can be super-enhanced for improved productivity or even turned into biochemical manufacturing plants. No longer does sexual compatibility or the randomness of natural genetic recombination during reproduction limit plant breeding. We can insert genes for selected characteristics directly into a plant, thus bypassing typical physical plant crosses. We can fuse the cells of two sexually incompatible plants and convert the hybrid cell into a new plant. We can even add nonplant genes to a plant.

This book deals with horticulture, thus we will not discuss the medical and animal agricultural applications of biotechnology. Plant biotechnology applications include micropropagation, which was covered in Chapter 7. In this chapter we consider *in vitro* production of pathogen-free plants, *in vitro* germplasm storage, genetic engineering, cellular fusion, tissue culture, transgenic plants for pest control, and biocontrol of pathogens and weeds.

The brave, new world of plant biotechnology holds forth much promise, but like any tool it can be misused. Hopefully, plant biotechnology will be harnessed for the betterment of sustainable horticulture. An enormous debate is currently raging on food safety, escape of engineered genes into the environment, and control of plant germplasm by a few multinational corporations. These challenges need to be resolved before benefits can ensue.

RECOMBINANT DNA TECHNOLOGY, AKA GENETIC ENGINEERING

Today's biotechnology owes its origins to recombinant DNA technology, also popularly known as genetic engineering. With this technology, one can deliberately manipulate genes and bring about desired changes in an organism. A gene or genes can be obtained from any organism and piggybacked onto the DNA of a carrier used to insert the gene(s) into the cells of a target organism. Carriers are usually specially modified bacteria or viruses. In some cases, direct insertion of the gene by physical methods is possible. Gene transfer is more precise and sexual incompatibility is no longer a limiting factor. Processing time is also greatly reduced by eliminating random genetic recombinations and selection for the desired genotype. Gene transfer between virtually any two organisms is theoretically possible.

In genetic engineering, genes in organisms should be identified by their placement on the chromosomes (gene map) and what function(s) they control. Ideally, the functions should be thoroughly understood in a developmental, biochemical, and physiological context. If that were so, we could just select the gene for the desired characteristic we wished to add into another organism and do it. The reality is that gene maps for plants are limited to date. Our knowledge of plant development, biochemistry, and physiology is also not complete, thus limiting our understanding of how a given gene would function in the selected organism. Certainly, such maps and the elucidation of gene function and associated life processes are being worked on, but at least a decade or two will be needed to complete the knowledge gap for the major horticultural crops. In the meantime, we must make do with alternative approaches to finding our way without gene maps, or what I call creative gene splicing. Let's look at the current reality.

DNA Libraries

Gene libraries or, more appropriately, DNA libraries can be created for any organisms, including plants. These libraries contain all of the genome of a particular organism represented by DNA fragments of various lengths. However, these collections are not paginated, alphabetized, or even logically organized. Instead, the library is a pool of DNA clones representing the genome of a given plant species or cultivar. Fortunately, techniques exist for finding and locating genes and even for creating useful maps of gene locations. None of this is easy or inexpensive. Still DNA libraries are useful in that they can be made in multiple copies and stored for indefinite periods and used to retrieve genes at later dates.

Cutting DNA into Fragments

Genomic DNA can be cut into pieces by restriction enzymes. With partial to complete enzymic digests, fragment sizes can be varied from long to short. These enzymes are derived from bacteria. Restriction enzymes cut double-stranded DNA within or near specific areas called *recognition sequences*. These special sequences typically consist of four to six nucleotides (units of nucleic acid that form DNA). Some cut straight, but a few restriction enzymes produce a jagged cut, leaving one strand a few nucleotides longer than the other strand (Figure 9–1). When this uneven cut occurs, the double-stranded DNA is said to have "sticky" ends. These sticky ends have exposed, complementary nucleotides that can pair and be rejoined with other DNA in the presence of an enzyme called DNA ligase.

If the same restriction enzyme is used to cut DNA from two different sources, such as the small circular DNA unit from bacteria called a plasmid and DNA from some plant, the sticky-ended DNA strands from both sources have complementary ends. These DNA fragments from two sources, the plant and the bacterial plasmid, can then be rejoined by DNA ligase (Figure 9–2). The combined DNA now exists as recombinant DNA in the bacterial plasmid. Such altered plasmids are called *vectors*.

Creating the Library

A collection of plasmids, each containing a specific fragment of plant DNA, can now be placed in medium containing growing bacteria. Some bacteria will take up the altered plasmids by a process called *transformation*. Bacteria containing the recombinant DNA, transformed bacteria, can then be cultured to essentially produce millions of copies of each recombinant DNA fragment, a process known as gene or DNA cloning. Of course, for this DNA cloning, one does have to sort out transformed from untransformed bacteria, because not all bacteria take

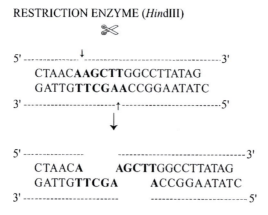

FIGURE 9–1   Restriction enzymes, also called restriction endonucleases, act like molecular scissors. A piece of DNA shown here is treated with the restriction endonuclease called *Hind*III (pronounced hindy3). This enzyme always cuts between the two As in the recognition sequence AAGCTT shown in bold type. The resulting two fragments have "sticky ends."

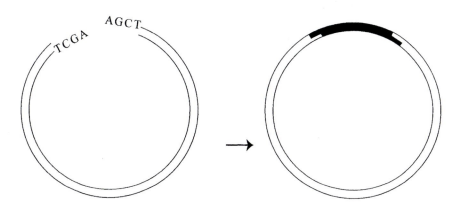

AGCTTGGCCTTATAGCTAACA
ACCGGAATATCGATTGTTCGA

DNA FRAGMENT WITH DESIRED GENE

+ DNA LIGASE

AGCT

TCGA

PLASMID

RECOMBINANT PLASMID VECTOR

FIGURE 9–2 If a bacterial plasmid, which consists of DNA, and plant DNA are treated with the restriction endonuclease *Hind*III, the same sticky ends (AGCT and TCGA) are produced on both. If both of these sticky-ended DNA molecules are placed with another enzyme that can rejoin DNA fragments, DNA ligase, the result is a recombinant plasmid vector. This recombinant product contains plant DNA (*dark area*) inserted into the plasmid's DNA.

up the plasmid vector. This identification process is defined as *screening*, although some prefer to call the process *selection* and reserve the term *screening* for locating specific genes.

Two basic screening techniques are used. One depends on antibiotic resistance and the other on the breakdown of a sugar called lactose. With the latter technique, when a specific restriction enzyme cleaves the recognition sequence in the plasmid, it interrupts the nearby *lacZ* gene. This gene codes for the protein known as β-galactosidase, the enzyme that hydrolyzes lactose. Bacteria transformed with plasmids containing plant DNA inserted at the recognition site are unable to break down lactose. In the absence of the enzyme (caused by interruption of the *lacZ* gene by the inserted plant DNA), colonies with the altered plasmids will be white. Plasmids that do not have plant DNA inserts continue to produce β-galactosidase and in the presence of X-gal (special sugar) will turn blue. In the other method, a genetically altered plasmid containing genes for antibiotic resistance is used. For example, the gene amp^R gives resistance to ampicillin. Bacteria that received this plasmid with its attached DNA fragment are able to grow in cultures containing ampicillin, whereas cultures that did not take the plasmid will not.

Library Variations

The DNA cloning process results in a cloned collection of the various fragments produced from the original total DNA of the organism (genome). This collection is referred to as the genomic library. By using several different restriction enzymes, the same DNA can be cut into different sets of fragments. More than 1,200 restriction enzymes are known, many of which act at different recognition sites. Using different restriction enzymes, one can prepare several versions of a genomic library for any organism's DNA. The probability of finding whole genes of interest among the various fragments is thus ensured.

It is also possible to make another type of library, the complementary DNA (cDNA) library. A cDNA library represents all the individual genes of an organism that are being expressed in the tissue supplying the extracted DNA. This process requires reverse transcriptase, an enzyme produced by retroviruses. When this enzyme is placed with messenger RNA from an organism, it catalyzes the synthesis of a single strand of DNA. This strand can be doubled in the presence of DNA polymerase. This DNA, known as complementary DNA, differs from the genomic DNA of the organism in that cDNA contains no introns. Introns are noncoding regions of DNA that do not translate into proteins.

Why have gene libraries and why have two types? If the genetic engineer had to go to organisms and isolate gene(s) every time they were needed, it would be a difficult and time-consuming process. Instead, these genes need only be isolated once and placed in the gene library. Multiple copies are then made and are readily available as needed.

Two library types are needed, based on where the genes will be inserted for expression. Genes destined for manufacturing a protein in a bacterium must come from a cDNA library. Bacteria and other prokaryotes are unable to properly process DNA from higher organisms (eukaryotes). Eukaryotic DNA, such as from a plant, contains introns and exons. Exons are regions that code for proteins. These regions are not continuous, but are interrupted by regions that do not code for proteins. These noncoding sequences are introns. When eukaryotic DNA is transcribed, the mRNA contains the complement of both the intron and exon regions. The intron part is then removed from the mRNA before it is translated into a protein. Bacteria do not possess introns and have no mechanisms for removing regions transcribed from introns. The cDNA library, containing only the exon parts of DNA, can be readily translated into proteins when inserted into bacteria. The cDNA library is also more efficient for use in eukaryotes, but the genomic library is much easier to make.

The genomic library contains all of the genetic material of an organism, including genes, structural DNA, and other DNA sequences. The genomic library is important for gene mapping and isolating regions of DNA that act to control the expression of a given gene (e.g., promoters and enhancers; see Chapter 5). A knowledge of gene control is necessary for genetic engineering. Without proper control elements, genes will not be expressed.

One last thought here: Both of these libraries are uncataloged in that the locations and functions of the genes within are unknown. Additional screening to locate the desired genes is needed before these libraries can be of any use. We will look at this problem later.

WORKING WITH DNA FRAGMENTS

These same restriction enzymes can be useful for determining the nucleotide sequences of DNA. Knowing the sequence can be one useful tool in terms of screening libraries for specific genes of interest. Three things are critical for sequence determination: workable units of DNA, sufficient amounts of DNA, and primers. The entire genome is not currently a workable unit, given that a single gene could typically have a few thousand base pairs!

Cutting and Copying DNA

Workable units are the fragments of DNA produced with various restriction enzymes. Adjusting the extent of treatment with restriction enzymes carefully controls fragment length. If the reaction time is too short, overly long fragments are produced. If too long, many very short fragments might be produced, thus limiting the efficiency of subsequent uses of DNA fragments. The length of the ideal fragment for genomic library construction, for instance, is about 4,000 bases.

In some cases we might need more copies of the fragments. Several reasons exist for making many copies of the DNA fragments or an individual fragment or even an individual gene. One reason is to increase small amounts of DNA to workable units for purposes of sequencing or identification. Small amounts of DNA might exist at a crime scene or when only a small amount was isolated from an organism, small tissue samples, or even a few plant cells. An individual fragment might be found with a sensitive screening technique, then increased for subsequent protocols. Multiple copies of a gene might be made prior to being added to another organism, as only a small amount actually gets transferred into the working DNA of an organism. The more genes you transfer, the better the chance of successful transfers.

Essentially millions of copies of DNA fragments can be produced via the polymerase chain reaction (PCR). The enzyme involved is *Taq* polymerase. This enzyme can make a copy of single-stranded DNA if the proper primers are applied at the beginning and end of the part you wish to copy. The enzyme is also heat stable, because it is isolated from a hot springs bacterium (*Thermus aquaticus*). For the PCR technique, the double-stranded DNA is heated, which denatures the DNA and separates it into two strands. On cooling, primer DNA binds to the strands of DNA. Primer DNAs are special short segments of DNA with nucleotide sequences that are complementary to the ends of the DNA fragment to be amplified. The *Taq* polymerase works off the primers using supplied nucleotides and copies each strand, resulting in two double-stranded units of DNA that are exact copies of the original DNA fragment. The reaction is initiated again by heat. The two double-stranded molecules then become four intact, double-stranded DNA units. A device called a thermocycler is used to automate the heating and cooling. These steps can occur many times, resulting in up to a million-fold increase of DNA.

Separating DNA Fragments with Electrophoresis

The tool of choice for separating DNA fragments is electrophoresis. Applications include the separation of gene libraries into fragments. The entire library is much too big to use for direct manual or even automated sequencing. Separated fragments can be pooled into smaller batches for sequence determinations. Another use is to separate fragments on a gel, transfer these fragments to a solid matrix, and use DNA or mRNA probes to locate a gene on a specific fragment. If the probe had a fluorescent or radioactive tag, the band can be visualized under UV light or through film exposed to the radioactivity.

For DNA electrophoresis, DNA is loaded onto an agarose or acrylamide gel. An electrical field is applied to the gel using a buffered solution. The fragments move across the gel at different rates based on their differences in length and charge. Smaller fragments move faster than larger ones. The separated fragments can be removed in gel slices. The DNA fragment or fragment with an attached probe can be washed out of the gel slices. If enough is present in the sample, it can then be sequenced. If larger amounts are needed, PCR is used to make additional copies.

Sequencing and Primers

Primers are critical components for sequence determinations. Sequencing requires the synthesis of special DNA sequences (primers) that are complementary to a stretch of the cloning vector just before where the fragment, or target DNA, is attached. These primers act essentially as the "start" or "on" button for the sequencing process. Primers, just as with DNA and mRNA probes (used for screening fragments for specific genes), can be synthesized with any desired nucleotide sequence. These short pieces of DNA, technically oligonucleotides, are synthesized by the phosphoramidite method.

Sequencing Methods

Sequencing is done either manually or with automated DNA sequencing machines. The manual method depends on the presence of dideoxynucleotides (sugar has two fewer oxygens than normal DNA nucleotides). When these nucleotides are incorporated into a replicating DNA strand, replication ends when a dideoxynucleotide is added to the growing chain. Termination occurs because the next incoming, normal nucleotide cannot form the required phosphodiester bond with the dideoxynucleotide. This simple event forms the basis of manual sequencing.

Annealing a primer (special DNA sequence to initiate replication) at a predetermined site on the cloning vector (plasmid DNA) some 10 to 20 nucleotides before the original insertion site of the cloned DNA starts the process. Actually many copies of the cloned DNA are needed for this step. We will come back to this point later. The site DNA sequence is known and is near where the cloned or target DNA had been inserted into the vector when making the library. When this known sequence just after the primer, but before the cloned DNA, is read later, it serves as a reference point for reading the sequence of the cloned DNA.

The primed DNA is divided into four equal volumes and placed in test tubes. Into each test tube is added the four normal deoxyribonucleotides needed for replication (dATP, dCTP, dGTP, and dTTP); one of the nucleotides is labeled with P^{32} or S^{35}. Each of the four tubes has a different nucleotide labeled. Dideoxynucleotide is added to each tube also; one tube gets ddATP, another ddCTP, another ddGTP, and the last ddTTP. The amount of dideoxynucleotide must be carefully controlled so that it is incorporated into every possible site on the growing chain. Many copies of the primed DNA were used at the priming stage so that the dideoxynucleotide has the potential to appear at each site. If only one copy were used, the chain would be stopped at the first incorporation site and all you would get is one short chain. With many copies being made, one will be stopped at the first possible incorporation site, another chain might not be terminated until a dideoxynucleotide is added at the second possible site, the first one having received a normal nucleotide, and so forth.

The reaction is allowed to proceed to completion using DNA polymerase. The reaction is terminated by adding formamide, which interferes with base pairing. You end up with a mixture of many DNA fragments in each tube that vary considerably in length. Some differ in length from each other by only one nucleotide. The mixtures are applied in four lanes on a polyacrylamide gel slab and separated by electrophoresis. An autoradiograph is made (Figure 9–3). The sequence is read from the bottom (shortest fragments) to the top (longest fragments). If lane 1 contains the dideoxynucleotide ddATP and shows the lowest visible band, the sequence starts with base A (adenine). If the next lowest band is found in the ddTTP lane, then the second nucleotide is T (thymine), and so forth. If the mixture was incorrect in terms of components, you end up with multiple bands at the same place. For example, if nucleotide location 15 has four bands, all visible in a horizontal smear across the four lanes, something was wrong with the original mixture.

This method is limited in terms of fragment size. The best results usually give a resolution of about 300 to 400 nucleotides on a typical autoradiograph. This procedure has to be repeated many times before all fragments of the original DNA library are sequenced. Even then the sequence is not complete. All of the parts are there, but what is the order of the fragments? This question is resolved with a comparative, puzzle-solving approach. DNA fragments produced with different restriction enzymes are sequenced and compared. One can now assemble the sequence for the entire genome of the organism in a jigsaw puzzle manner. This approach works because the various sets of fragments from different restriction enzymes have areas of overlap. These overlaps are compared, as shown in Figure 9–4, and help to determine the final sequential order.

FIGURE 9–3 A retooled gene in 'Endless Summer' tomatoes controls ripening to give better flavor and longer shelf life. Note the autoradiographs below the tomatoes from which the sequence of nucleotides can be read. Autoradiographs result from overlaying X-ray film on the polyacrylamide gel. The film becomes exposed by the radioactively labeled nucleotides. The nucleotides appear as dark bands. (USDA photograph.)

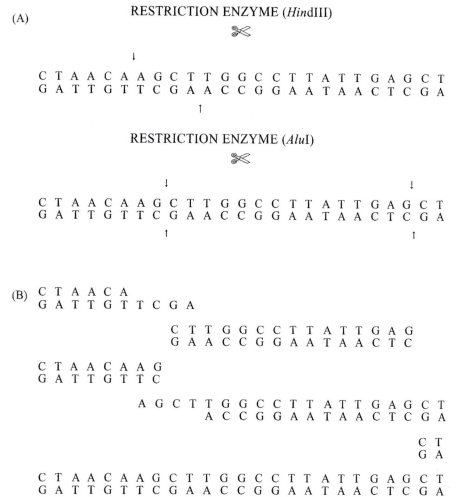

(A)

RESTRICTION ENZYME (*Hin*dIII)

✄

↓

C T A A C A A G C T T G G C C T T A T T G A G C T
G A T T G T T C G A A C C G G A A T A A C T C G A

↑

RESTRICTION ENZYME (*Alu*I)

✄

↓   ↓

C T A A C A A G C T T G G C C T T A T T G A G C T
G A T T G T T C G A A C C G G A A T A A C T C G A

↑   ↑

(B) C T A A C A
     G A T T G T T C G A

                    C T T G G C C T T A T T G A G
                    G A A C C G G A A T A A C T C

     C T A A C A A G
     G A T T G T T C

                    A G C T T G G C C T T A T T G A G C T
                        A C C G G A A T A A C T C G A

                                             C T
                                             G A

     C T A A C A A G C T T G G C C T T A T T G A G C T
     G A T T G T T C G A A C C G G A A T A A C T C G A

FIGURE 9–4   (A) Identical DNA strands are treated by two separate restriction endonucleases, *Hin*dIII and *Alu*I (pronounced a (letter)-loo-1). The former always cuts between the two As in the recognition sequence AAGCTT and leaves sticky ends, while the latter always cuts between the G and C of the recognition sequence AGCT and leaves blunt ends. The first yields two fragments and the second, three fragments. (B) If the sequences are determined and then laid out such that overlaps are lined up, the complete sequence is determined as shown in the last line.

The automated procedure is somewhat similar. Many copies of the DNA fragment are primed and the four normal nucleotides are added. However, none of them are labeled with P^{32} or S^{35}. Four types of dideoxynucleotides are also used, one type per tube. These dideoxynucleotides are each fitted with an attached fluorescent probe. Each of these "fluorescent" nucleotides fluoresces at a different wavelength. The DNA polymerase reaction proceeds and the mixture of various-length units is produced as above in the manual procedure. These samples are scanned with lasers. Each nucleotide substitution has a unique fluorescent wavelength that relates to the original nucleotide. The machine prints out a chromatogram containing four different height peaks. One height corresponds to adenine, another to thymine, and so forth. The base letter (A, C, G, T) is printed across the chromatogram, making the sequence easier to read (Figure 9–5). While the automated method is used to determine longer sequences, the overlap method is still often needed for the final properly ordered sequence.

An alternate approach to the jigsaw overlap is also possible. In fact, it is highly recommended when the piece of DNA exceeds 5,000 base pairs. Sequencing beyond this length is very time consuming because of the need to determine the sequences of numerous fragments. The method of choice here is known as *primer walking*. The vectors used here are different than those used for the overlap procedure. These vectors can carry much larger DNA pieces. The same procedures are used, either the manual or automated approaches described above. This step determines the sequence of the first 250 to 350 bases. Using this just completed sequence, a primer is synthesized for placement at the end of the sequence, usually at the last 20 bases. The sequencing is then done for this section. A primer is made and attached at the last 20 bases of the second known continuing sequence. This procedure

FIGURE 9–5   A sequencer can also be hooked up to a computer, using software to interpret the results. With an automated DNA sequencer, one can quickly obtain a detailed genetic analysis of unidentified DNA. Here, a DNA sequence is inspected from a previous run, while samples are loaded for new determinations. If the picture was in color, you could see the four colors that identify the four different nucleotides. You are looking at the raw data, which can also be printed in an output file that shows the actual peaks and sequence. (USDA photograph.)

is repeated as many times as necessary to determine the total sequence. The procedure gets it name from the fact that a primer is "walked" from place to place on the DNA.

FINDING THE RIGHT GENES

In a perfect world, one would have a completely cataloged gene library for every plant and all other organisms that might be used to supply genes for altering plants. We are a long way from the ideal. Several problems complicate the situation. These include the sheer number of genes and the fact that sometimes one gene can express a single useful trait, like insect resistance, while other characteristics, such as fruit size, are controlled by many genes working together. Sorting through this can be frustrating and tedious. Fortunately, some techniques do exist that can help us locate specific genes in terms of function.

Probes are molecular tools used to screen gene libraries. These tools are able to recognize specific nucleotide sequences in either DNA or RNA. Another type of probe can recognize a protein or other molecule synthesized by a specific gene. Probes are usually used on copies of the gene library, so that the original gene collection remains unaltered by any experimental protocol.

Library Copies and Basic Screening Protocol

Copies of the library are made with a technique called *replica plating*. A piece of sterile velvet cloth is placed on the surface of an agar-cultured library plate. Some cells from each colony present are transferred through contact onto the cloth. The cloth is carefully touched to a new petri dish containing sterile growth media. A new set of colonies is cultured that are copies of the original library. In this manner, the original library remains intact and unaltered by any screening techniques.

While the type of probe varies, the technique used for screening does not vary much. Applied screening protocols are named based on what is being screened: DNA, mRNA, or protein. The first technique was called Southern blotting, named after its discoverer in the mid-1970s, Dr. E. M. Southern. Subsequent protocols were named Northern and Western blotting. The later discoverers' names were not used for these variations of Southern blotting; instead names were chosen to retain or expand on the "directional" context of the first protocol. We will discuss blotting types later.

The protocol starts with the transfer of some material from the library copy. The material can be whole cells, or some part thereof, such as proteins or nucleic acids. The transferred material is placed on special paper called nitrocellulose filter paper. Materials are transferred so as to retain the original spatial relationship found on the starting point, the library copy. The transferred material usually requires some pretreatment prior to analysis. The treatment alters the form so that the analysis can proceed properly. For example, whole cells would be lysed with a dilute basic solution, such as sodium hydroxide. This solution not only lyses the cells, but also converts the double-stranded DNA to the single-stranded form.

A prepared probe is added. Usually a "hot" or radioactive probe is used, although fluorescent probes can also be used. The radioactive or fluorescent probes allow for easy localization later of the areas where the probe reacted. The probe will bind specifically to a complementary DNA or mRNA sequence or, in some cases, to a specific protein expressed by the genes. The binding to a complementary sequence is known as hybridization. Chemical treatments can be varied to control the degree of hybridization. Conditions known as high stringency result in hybridization between two perfectly complementary sequences. Less restrictive methods produce hybridization between partially complementary sequences.

Next a washing step is used to remove excess probe. After the filter paper is dried, it is placed against a sheet of photographic film so that radioactive parts expose the film. After autoradiography, the visible areas correspond to molecules that hybridized. These areas can be traced back to colonies that contained the sought-after (target) genes.

Probe Types

Various types of probes are used, depending on starting material and what is known about the target gene. One type of probe, known as a cDNA probe, is prepared from the cDNA library. First the plasmids containing cDNA inserts are isolated. Next, the double-stranded cDNA is converted into single-stranded form by denaturation. The preparation is then added to nitrocellulose filter paper. A mixture of mRNA, chosen because of its complementarity to many known DNA sequences, is added to the filter paper. Any mRNAs that are bound to a complementary cDNA are left on the filter paper after unbound mRNAs are washed away. These bound mRNAs are removed. The mRNAs are then translated into proteins that are identified. Often the function of an identified protein is known. The cDNA from which the bound mRNA was removed has now been identified as the gene that produces the particular protein. Let's call it gene X. The complement to this known sequence of cDNA can now be used as a probe itself to screen other DNA libraries for gene X. This probe is called a *cDNA probe*.

Known mRNAs that code for specific proteins can also be used as probes. If a mRNA codes for an enzyme or a subunit of an enzyme, such as carbonic anhydrase, this mRNA is used to probe gene libraries for the gene that makes carbonic anhydrase. This probe would be termed a *mRNA probe*.

If the amino acid sequence of a protein is known, another type of probe is used. Based on the amino acid sequence, a sequence of nucleotides coding for production of a segment of the protein is synthesized. The nucleotide sequence (oligonucleotide) is constructed with radioactive components and the resulting product is called an *oligonucleotide probe*. If this probe binds to something in the gene library, you have found the gene (or a gene with a very similar structure) that codes for carbonic anhydrase.

Another type of probe is based on the immune system. Antibodies are specific for and bind to foreign cells or molecules, known as antigens. Gene products for which a specific antibody exists, or can be made, are useful tools for finding the genes producing the antigen. If the antigen is a well-known protein, the antibody can become an antibody probe and verify the presence of the gene indirectly.

Blotting Types

Although the above probes can be used to screen the DNA of gene libraries, these probes can also be used for screening of genomic DNA directly without the step of cloning a library. The technique for screening DNA sequences directly is known as Southern blotting. DNA from some source, such as a plant, is cleaved into fragments by using restriction enzymes as described earlier in this chapter. Fragments are next separated by means of agarose gel electrophoresis. On separation of fragments, the gel is soaked in a dilute solution of sodium hydroxide. This step converts the DNA from double- to single-stranded form. The gel is then neutralized and placed against a solid support, usually filter paper. The filter paper acts as a wick to draw a salt solution into the gel. On top of the gel a sheet of nitrocellulose filter paper is placed and this paper is capped with an absorbent layer, such as paper towels. Between the wicking of the support paper and the absorbent layer, the salt solution moves through the gel via capillary action and deposits single-stranded DNA in the nitrocellulose filter paper. This process creates a replica of the DNA banding pattern originally found on the agarose gel.

The nitrocellulose filter paper is then treated with either a radioactive or fluorescent cDNA, mRNA or oligonucleotide probe. The sheet is then subjected to autoradiography (or viewed under ultraviolet light for fluorescence). The bands, once localized, on the sheet correspond to those on the gel. The gene of interest is excised from the gel and its nucleotide sequence determined. If the fragments are also sequenced, the location of this specific gene can be mapped.

Northern blotting is somewhat similar, but RNA rather than DNA is probed. Agarose gel electrophoresis is used to separate total RNA or purified mRNA preparations. A denaturing agent is employed to prevent the RNAs from hybridizing during electrophoresis. The procedure after this point is similar to that described above for Southern blotting. RNAs complementary to the probes are identified. Northern blotting allows for the determination of whether a given gene is being expressed in a particular tissue.

Western blotting screens for actively produced products of genes, usually proteins, and uses an antibody probe. This technique is more useful for determining whether an inserted gene is functioning in its host organism, although it can be used to screen for the same gene in an unaltered organism or cell preparation. Usually, the host cells are lysed if the product protein is not secreted from the cell. Filter paper treated with the appropriate antibody is placed over the replicate plate containing the lysed or unlysed cells. If the protein specific to the antibody is present, binding will occur. Next the filter paper is treated with a labeled antibody that can bind to the original antibody. Unbound antibodies are washed off and the complexes of antigens/antibodies are visualized either with autoradiography or fluorescence. Once the gene is known to be present and active, the other blotting methods can be used for pinpointing the gene for sequencing and mapping.

Mutations and Transposons

Another way of screening genes involves using mutagen-induced mutations and a natural type of mutation caused by something called a transposable element. The presence of these "elements" was discovered long ago. However, their value, especially as a screening tool, was only recently realized. A Nobel Prize is also involved. Let's take a closer look.

Transposable elements, one form of which is now often referred to as *transposons,* are not new. Such elements, also thought of as "jumping genes," were first identified by Barbara McClintock in maize during the late 1940s. These "jumping genes" sometimes interfered with pigmentation of kernels, resulting in mottling and other abnormalities. Subsequent generations of maize sometimes reverted back to normal kernel pigmentation. McClintock proposed that genetic elements that caused pigment abnormalities could move about on the chromosomes, thus restoring normal functions when they left an affected gene. Her research finding demonstrating gene movements was ignored for many years until recombinant DNA technology was born. The importance of her research was finally realized, resulting in a Nobel Prize in 1983. Transposons are now one of the premier tools for determining gene functions!

In nature, transposable elements cause numerous spontaneous mutations in plants. These mutations result from the duplication and movement of transposable elements among the plant's chromosomes. Two types of transposable elements exist: insertion sequences and transposons. Insertion sequences are smaller, consisting of a few thousand bases. Their function genes code for enzymes that allow for removal and insertion of the element elsewhere in the DNA. Transposons are larger and carry additional genes beyond the removal/insert ones. These genes code for various proteins not involved in the jumping process. Transposons are found in most organisms, including plants.

Once these elements were identified and understood, mutations could be deliberately induced by the insertion of transposons. If these mutations result in visible or biochemical detectable changes in a plant, the function of the affected gene can be determined. For example, if a flower color changes from pink to white, a gene involved in the production of pink pigment was likely damaged. If the transposon can somehow be located in the organism's DNA, we then know the location and function of the gene. Then it can be sequenced, bringing about better genetic maps.

Transposons and insertion sequences from sources such as maize and snapdragons have been isolated and sequenced. Some of these transposable elements have been introduced into plants such as *Arabidopsis*, flax, petunia, rice, tobacco, and tomato. The process is called *transposon tagging*. Transposon tags can be fitted with markers and primers and inserted into whole plants or cultured cells. This form is especially useful for using vectors to introduce transposons into plant cell cultures. The marker can be a gene to confer antibiotic resistance or one that prevents lactose breakdown (see marker discussion earlier under DNA libraries). The marker is used to screen for those cells that have taken up the transposons. These cells can be converted into plants via micropropagation (see Chapter 7).

DNA libraries can then be created for those plants with observable mutations. Those fragments that contain the transposon will have a built-in primer in the vicinity of the mutated gene. This area can be sequenced. Once the sequence is known, a complementary probe is prepared and used to prepare copies of the undamaged gene from DNA libraries of the same plant where transposon tagging was not used. The gene's function, location, and sequence are now known and another piece of the DNA map falls into place.

Transposon tagging in other organisms can also be useful in horticulture. A moth transposon, "piggyBac," has been used to change eye color in fruit flies. Possible horticultural applications under investigation are using these transposons to cause mutations in grain moths and fruit flies in order to reduce grain and fruit damage.

In a similar manner, mutations can be induced either chemically or by radiation. Actual populations of plants or cell cultures can be treated with mutagens. Progeny can be examined for alterations on the cellular, biochemical, or whole plant level. By comparing DNA sequences of the "before" and "after" plants, it is possible to find the damaged gene or genes. For example, dwarf plants produced by mutagens usually are deficient in gibberellins. The damaged gene codes for an essential enzyme involved in gibberellin synthesis.

MAPPING

Two kinds of gene maps exist. One type, called a gene linkage map, gives a rough approximation of certain genes relative to each other. It is a very rough, minimal information map. The other map, the physical map, shows the specific location of all functional genes. The physical map is analogous to a high-quality road map.

Gene Linkage Maps

Certain traits always appear to be inherited together, while others do not or appear together with varying frequency. This phenomenon occurs because of something called *linkage*. If two genes are close together, separation of the two during meiotic crossing over is unlikely. The separation becomes more likely as the distance between them increases. The larger the distance, the more likely one gene may be transferred from one chromosome to another during crossing over, while the other gene remains behind. Once a linked gene is identified and sequenced, it can be used as a marker gene in DNA libraries.

Another marker, the restriction fragment length polymorphism (RFLP, pronounced "rif-lip") can also be used instead of a marker gene. The RFLP is a stretch of functionless DNA that usually remains intact in the presence of restriction enzymes and during crossing over. The area remains intact because at the cleavage site a change in nucleotide base pairs has prevented cleavage by the restriction enzyme. This difference results from single nucleotide polymorphism (SNP) caused by variations in populations at a particular nucleotide site. These multiple forms or polymorphisms are useful as DNA markers.

Genes that are close by are linked with the RFLP during crossing over, so once genes in the vicinity of a RFLP are identified, the RFLP can be used like a marker "gene" in libraries as above with two linked genes. RFLPs are easily identified during analysis by Southern blotting, making them easy-to-use markers.

RFLP studies require some knowledge of partial sequences (so that primers can be cloned) and are time consuming and somewhat difficult, so some other marker types are also used. If no sequence information is available, random amplified polymorphic DNA (RAPD) is used. This approach uses a mix of several short primers that have essentially random sequences. These usually anneal at several sites, and sequences in between can be amplified by PCR. Because of sequence variations in populations such as from SNPs, some areas will be amplified in some DNA samples, but not others. This variable means that the fragment is a polymorphic area that can serve as a marker and can be visualized after electrophoresis and staining and eventually sequenced. This approach is simpler and quicker, but can lead to dark bands that are caused by overlapping fragments of similar sizes and other bands can sometimes be too light to visualize properly.

This problem can be corrected with amplified fragment length polymorphisms (AFLPs). This process yields more uniform band intensities on gels, but is technically difficult. Double-stranded oligonucleotide sequences (primer adapters) that match the primer sequences used later are attached to the jagged overhang of the fragments produced by treatment with restriction enzyme. Primer adapters are complementary to the overhang and are joined by DNA ligase. By varying how well the primer matches, one can vary the number of amplified fragments. A perfect primer match yields so many amplified fragments that separation on the gel becomes difficult. Some alteration of the primer sequence produces a workable number of fragments, some of which will be useful polymorphic markers.

Based on the frequency of paired traits in offspring, rough maps called linkage maps can be created from either marker genes, RFLPs, RAPDs, or AFLPs. Although these maps are not detailed, they can be useful. For example, a rough distance between two or three genes might be determined. It would be analogous to a road map that shows the cities of New York, Chicago, and San Francisco. We get a rough idea of their placement relative to each other, but the exact mileage and what cities are in between are unknown.

These gene linkage maps can be useful for screening once the function and location of one gene of a linked pair is known. This gene becomes known as a marker gene. DNA libraries can be probed for the marker gene or RFLP with various screening techniques. The library fragment that contains the marker gene has an extremely high probability of also having the linked gene. Successful screening and sequencing then puts two specific genes on the map.

Physical Maps

Using the various techniques discussed earlier one can assemble physical gene maps. Probes and transposons are used to screen libraries and the identified genes are placed into previously determined sequences. Although numerous plants have partial maps, only a few plants have a complete map.

One such plant is *Arabidopsis thaliana,* the workhorse of plant genetics, which has a role similar to that of the fruit fly (*Drosophila*). This plant, a weedy member of the Mustard family, has a number of characteristics that endear it to plant geneticists and molecular biologists. One feature is small size, such that thousands can be grown indoors under fluorescent lights on a lab bench. Another is a 30- to 45-day life cycle and the production of several thousands of seeds per plant. Effects of mutations and transposons can be determined over generations in short times. Its genome is small and has no repetitive DNA, consisting of five chromosomes and about 30,000 genes. Screening, sequencing, and mapping are more manageable on this scale.

The map and sequence of *A. thaliana* are the most complete of all plants. In fact, the entire sequence was completed in December 2000. The closest, other nearly finished plant gene map is that for rice. The map and sequence of *A. thaliana* are providing considerable data that are useful for mapping and genetically engineering higher plants. In addition, a number of questions about plant development and evolution are being answered. Genes involved in disease and insect resistance, flowering, and photosynthesis have been discovered. Genes controlling the developmental processes of the embryo, seedling, root, and flower have been found.

While clearly not a crop plant, *A. thaliana* provides a model map that can be extrapolated to provide information on horticultural crops. Certain genes, such as those involved in photosynthesis and development, are likely to be basic to all plants. Once these genes were located on *A. thaliana,* probes could be constructed and used to screen crop plants for similar genes. Even genes that might be somewhat modified when one goes from *A. thaliana* to crop plants can be screened with conditions of lower selectivity such that partial similarities in sequence can be found. Once some genes are mapped on higher plants, they can be used as marker genes, and linked genes can be probed. Transposons and mutations can be used to expand the maps further. Where protein sequences are known, mRNA probes can be constructed to seek out the gene responsible for the protein.

Much of this information, DNA sequences, known genes, and partial maps exist in databases on the Web at the National Center for Biotechnology Information's Genbank. Often the databases can provide clues for other unmapped plant genomes. Probes can be constructed based on existing genes in other plants and then used to screen other plants. Much of this "sharing" is possible because of common ancestry in plants, such that today's plants show high degrees of similarity at the gene and protein level.

GENE TRANSFER TECHNOLOGY

The alteration of plants through genetic engineering has great promise for the future of sustainable horticulture. This technology represents the next step forward. Horticultural production steadily increased as we moved from horse power to mechanical power and then to chemical technology. The latter did cause some problems in terms of environmental consequences. Genetic engineering has the potential to move productivity ahead further and to correct environmental problems. Like any technology, noble purposes can be subverted. Some believe this misdirection is already happening. We will come back to this point later.

Possible applications include the ability to increase resistance to insects and pathogens, thus cutting back on the usage of pesticides. Yields could be increased through the engineering of bigger parts destined for harvest, although this has not yet been done. Quality might be increased through the addition of enhanced vitamin or protein composition and amount. Beneficial microorganisms that benefit certain plants might be altered to work with all plants. For example, nitrogen-fixing bacteria that currently only work with legumes, if adapted to other horticultural crops, could result in reductions in nitrogen applications and less pollution.

Certainly, many of these desirable characteristics have been accomplished with conventional plant breeding. Genetic engineering, to put it simply, offers more with less fuss. Some limitations imposed by conventional plant breeding no longer exist with genetic engineering.

You can move genes between plants that are not sexually compatible and you can get to the desired product directly with gene transfer, rather than randomly and through much selection to find the desired plant. You can also achieve microadjustments not possible with conventional techniques, such as the tweaking of an individual protein's composition.

Two basic tools are needed to accomplish these goals. One is a method to move genes from one place to another, essentially DNA insertion. The second is the ability to take an altered cell with inserted and functional DNA and convert it to a whole plant. The latter technique was covered earlier in Chapter 7 under micropropagation. The procedures involved in micropropagation are routine and possible with most plants. More difficult, but still doable, is DNA insertion.

DNA Insertion

A number of methods exist for the insertion of DNA into other organisms, usually on the cellular level with plants. No one method is the perfect answer, as each has its limitations. Methods include the use of *Agrobacterium tumefaciens* as a vector, ballistic transfer, direct transfer, electroporation, microinjection, silicon whiskers, and viral vectors.

The method of choice is usually to use *A. tumefaciens* as a vector to deliver foreign DNA to a target cell. A second species, *A. rhizogenes,* has been used to a lesser degree. In their natural state, the first causes crown gall disease (tumors) and the second, hairy root disease. After DNA insertion, the target cell can then be micropropagated into a whole plant with altered DNA (Figure 9–6). The limitation is that this bacterium attacks dicots, but not monocots. Another factor is that successful transfer varies considerably by species, the organelle source of the target cells, physiological state of the donor, differences in bacterial strains, and even techniques and skills involved in transfer and plant regeneration.

Strains of *A. tumefaciens* used in transfers are disarmed; that is, the genes responsible for tumor formation have been removed. In their place, one substitutes the gene or genes destined for transfer. The bacterium acts normally, that is, transfer occurs, but the substituted genes are transferred in the place of tumor-causing genes. In addition, the bacterial vector contains other genes, such as marker or reporter genes. These genes are necessary so that successful transfers can be identified. Marker genes usually confer antibiotic resistance to the target cell, such that successfully transformed cells survive exposure to the antibiotic. Commonly used antibiotics include kanamycin and geneticin, and hygromycin to a lesser degree. Marker genes can also be used to confer herbicide resistance. Reporter genes do not involve resistance. Instead, these genes code for some product that can be readily determined by assay. One common example is the luciferase reporter gene system. The assay detects bioluminescence.

FIGURE 9–6   When new genes are inserted into cells that are ultimately micropropagated into a whole plant, one can verify the new gene's presence by doing PCR analysis of the starting and altered plants' DNA or use one of the blotting methods, if a suitable probe exists. Normal and transgenic potato DNA loaded onto an agarose gel electrophoresis chamber will likely reveal the presence of an experimental gene. (USDA photograph.)

Ballistic Transfer

This procedure employs microscopic tungsten or gold projectiles coated with DNA. The projectiles are shot through the cell walls into the cellular cytoplasm. Some of the DNA will be taken up by nuclear and/or organelle genome. Projectiles are propelled with compressed gas now; earlier methods used gunpowder. The system can be used with monocots, such as corn, and conifers, for which *A. tumefaciens* does not work. Its limitation in the successful transformation of cells is 10% or less. Successes with corn, cotton, cranberry, papaya, poplar, rice, sorghum, soybean, spruce, sugarcane, and wheat have occurred.

Direct Transfer

This method of DNA transfer utilizes plant protoplasts. Enzymic digestion of the cell wall is necessary to prepare protoplasts. DNA is delivered via bacterial plasmids that have been modified with added target and marker/reporter genes. DNA can also be encapsulated in liposome vesicles for an alternate insertion method. The plasma membrane around the protoplast becomes permeable in the presence of poly-1-ornithine or polyethylene glycol (PEG), allowing the entry of altered plasmids. Liposome vesicles release their contents into the cytoplasm after fusion with the plasma membrane. The problem with this method is that protocols for regenerating plants from protoplasts are limited.

Electroporation and Microinjection

Protoplasts are also needed for electroporation, which suffers the same limitation as direct transfer. Success has been observed with corn and rice. The genes are attached to gold beads suspended in solution along with the plant protoplasts. An electrical discharge drives the beads of DNA into the protoplasts. This method has also been used with embryos that have undergone partial cell wall digestion.

Microinjection employs a microscopic, needle-like, glass capillary tube with a diameter of 1 or 2 microns. This tube is guided under a microscope with micromanipulators so that it pierces the target protoplast or cell. The plasmid carrying the DNA is injected into the cytoplasm or nucleus. Meristematic plant cells have been used with limited success. This method is tedious, requires considerable skill, and has an extremely low success level.

Silicon Whiskers and Viral Vectors

Microscopic needle-like fibers of silicon carbide are added to suspensions of plant cells or protoplasts. Plasmids carrying desirable genes are also present in the solution. The entire mix is shaken, some of the "whiskers" penetrate the cells or protoplasts, and the plasmids enter. This method is very low technology and has only worked currently with corn.

An ideal alternative to use of the bacterial vector *A. tumefaciens* would be to utilize plant viruses as a vector to deliver DNA. Viruses invade plant tissues and co-opt the plant's DNA. Cells are reprogrammed with viral DNA; these cells then churn out copies of the virus. In a sense, viruses are nature's genetic engineers. The idea is to use a viral vector whose pathogenic genes are disarmed and replaced by target DNA. This process is still under investigation, but a few successes have been reported in the recent literature. For example, cauliflower mosaic virus has been used to add genes for herbicide resistance and insect resistance to soybeans, corn, and a few other crops that are now commercially available. This system has the advantage of working with both monocot and dicot crops.

Gene Surgery with Chimeraplasty

Plants can be altered in other ways than by conventional breeding (see Chapter 8) and gene transfers. The newest alternative technology is called chimeraplasty. This technique is a form of molecular surgery that can be used either to create or repair a mutation in a specific gene. The "molecular knife" is known as a chimera; it consists of a hybrid molecule formed from RNA and DNA.

These segments of RNA and DNA are short and chemically synthesized. The DNA matches a segment of the target gene, except for one base that differs. If the target gene is normal, the base change will result in a mutation. If the gene has a mutation, the correct base can be included in the DNA segment, thus restoring normal function to the gene. The RNA is a perfect mirror (compliment) of the targeted gene's DNA.

The chimera works because cells have DNA repair mechanisms. The chimera pairs up with the target gene, specifically at the chosen segment, and fools the cell's DNA repair machinery into replacing an original base with one supplied by the chimera. The chimera can be delivered into plant cells with either ballistic transfer or electroporation.

This technology is in early stages, but appears to have promise. It offers one advantage over gene transfer. It does not introduce permanent foreign DNA into an organism. The cell degrades chimeras after their purpose is accomplished. Opponents to gene transfer may view this technology with less alarm. Chimeras also have less impact on the genetic machinery. One base change does not alter physical locations or regulatory aspects.

APPLICATIONS IN PLANT BIOTECHNOLOGY

Numerous applications exist in plant biotechnology (Figure 9–7) for horticultural crops. Biotechnology can be used to produce pathogen-free plants, to provide *in vitro* storage of germplasm, to provide new hybrids from plants that are not sexually compatible, and to propagate countless numbers of plants (micropropagation; see Chapter 7). Plants can be genetically altered in numerous ways (Figure 9–8). Plants can be made resistant to various insects, pathogens, and even frost or herbicides. Microorganisms can be genetically altered to provide nitrogen or even insect protection to plants.

Frost Protection

Frost damage limits plant productivity, in that early plantings run the risk of frost damage and even death of plants. Some fruit trees that flower early, such as apricots, are commercially grown only in a few areas where the likelihood of an early frost is minimal. Fall frosts arriving earlier than usual also cause damage, especially to ripening fruit. Citrus fruits in Florida are at high risk, and citrus growers do suffer economic losses. When such years occur, up to 60% of the energy budget for citrus production goes to frost protection. The burning of fuel oil and air circulation with wind machines can be used to protect late ripening fruit. Sprin-

FIGURE 9–7 DNA sequences are unique, so they can be used for identification purposes like fingerprints in humans. To achieve accurate identification of grape cultivars, a geneticist examines a computer image of DNA fingerprints from 36 different cold-hardy grape cultivars grown commercially in the Finger Lakes region of New York. (USDA photograph.)

FIGURE 9–8 A technician checks on futuristic peach and apple "orchards." Each dish holds tiny experimental trees grown from lab-cultured cells to which researchers have given new genes. (USDA photograph.)

klers can be used for minor frost protection during blossoming and fruit ripening. These procedures are energy intensive and some air pollution results from the burning of fuel.

An alternative involves genetically engineered bacteria. This approach is based on the knowledge that certain bacteria act as ice-nucleation centers. Their presence causes dew or condensation to form ice crystals on plants at 32°F. If the bacteria are killed with a bactericide, frost damage does not happen until the temperature approaches 20°F. The most common bacterium involved is *Pseudomonas syringae.* Research demonstrated that the ice-nucleation centers turned out to be under the control of one gene. When this gene was removed, the new strain of *P. syringae* protected plants from frost damage to 23°F. The efficacy of these genetically engineered bacteria was determined with field testing after a difficult and time-consuming legal process. Opponents of recombinant DNA technology brought legal challenges citing safety and regulation issues. After 5 years of arguments, this genetically altered bacterium was field-tested on potatoes and strawberries. To date, this product has not been a commercial success, but does offer a reasonable alternative to energy-intensive forms of frost protection.

Insecticides from *Bacillus thuringiensis*

Bacillus thuringiensis (Bt) has been known since the early 1900s to produce a substance toxic to caterpillars. It has been used since the 1940s as a natural insecticide by organic farmers, gardeners, and by others as part of integrated pest management systems. Suspensions of the bacterium are sprayed on plants needing protection from caterpillars, such as corn. The bacterium produces a potentially fatal protein in crystalline form (protoxin). The protoxin on ingestion becomes activated in the insect's gut through action by a protease that releases a toxic fragment. This natural control, often simply referred to as Bt, has seen increasing use in the last decade, especially by those farmers and gardeners wishing to limit pesticide usage. The one problem is that rain can wash it off, thus limiting the protective period. Bt causes no environmental harm and has not yet been shown harmful to other organisms.

As such, Bt evoked the interest of genetic engineers early on. The ideal solution would be to transfer the gene either to common, plant-associated bacteria or directly to the plant. The genes for the protoxin were identified (*crylA* genes) and soon spliced into the vector *A. tumefaciens.* Results with dicots at first were not exciting. The problem was that the fatal, functional fragment came along as part of a much larger protein and needed gut activation. When gene material for only the fragment was used, results improved. However, expression of the toxic fragment was at unacceptably low levels. An artificial gene was synthesized, based on the natural gene, but minus parts that contributed little or impaired toxin production. The artificial gene was also enriched for two nucleotides, guanine and cytosine. This "tweaking" allowed for better expression in plants such as maize that show strong preferences for enriched areas of these two nucleotides. This final alteration resulted in an expressible gene that produced acceptable and fatal levels of the toxic fragment in host organisms. Field tests determined that cotton, corn, and potatoes could be protected against various caterpillar pests. Crops have appeared in farmers' fields with the added protection conferred by synthetic Bt genes.

Herbicide Resistance

Glyphosate, usually sold as Roundup™, is a popular, biodegradable, broad-spectrum herbicide widely used by farmers and gardeners. Because it essentially kills any annuals and most herbaceous perennials, it cannot be used once a crop is up. However, once genes that conferred resistance to glyphosate were found in *Escherichia coli* and also a plant mutation, it became possible through gene transfer to make crops resistant to glyphosate. Crops with this herbicide resistance include corn, cotton, rape, soybeans, and sugar beets. Once this door was opened, the usage of glyphosate went from only preemergence application to potential for use throughout the growing season. Protection from numerous other herbicides has also been added to other crops, including alfalfa, peanut, potato, rice, sugar beet, tobacco, tomato, and wheat.

Chimeraplasty has also been used to confer herbicide resistance to tobacco. The chimera was used to alter one base in the gene that codes for the enzyme acetolactate synthesis. This change destroyed the enzyme responsible for sensitivity to sulfonyl herbicides. These efforts are in early stages, so no commercialized products have resulted yet.

Another method is to expose cell cultures to mutagens in the presence of a herbicide. Any cells that survived likely have a mutation that confers herbicide resistance. These cells can then be micropropagated into plants, followed by field-testing. The drawback to this method is that mutations are random. Finding the right one is like looking for the needle in the haystack.

Slowing Ripening

Supermarket tomatoes were an early target for the genetic engineers. The problem with tomatoes is that ripe tomatoes ship poorly and have a short shelf life. Consequently, tomatoes are picked green and have been bred for hardness, both for mechanical harvesting and shipping. In fact, these tomatoes can remain damage free after crashing at 35 km/hr! Taste suffers because these tomatoes are not vine ripened and the texture is less pleasing.

An alternative is to pick tomatoes at a later stage, but to alter the genetic controls on ripening so that shelf life and taste are improved. One approach is to interfere with polygalacturonase, an enzyme that causes cell wall softening during ripening. The gene for this enzyme was isolated, a complementary DNA copy was made, and the complement gene was transferred to tomatoes using cauliflower mosaic virus as the vector. When the complement gene was expressed, a complementary mRNA for encoding of polygalacturonase was made. This complementary mRNA combined with the normal mRNA and inactivated it. This approach is known as *anti-sense expression*.

Another method is to interfere with ethylene production, so that hormonal ripening is delayed. The enzyme ACC deaminase can degrade a key intermediate of ethylene synthesis. The gene for this enzyme was transferred into tomatoes (Figure 9–9) using cauliflower mosaic virus as the vector. Ripening for these tomatoes was slowed considerably.

Neither of these tomatoes has captured the market to any significant degree. Problems include some public resistance and slow market penetration relative to existing market tomatoes. For example, major chefs at New York restaurants expressed a no confidence statement regarding these tomatoes. However, given that tomatoes are close to being a $1.5 billion annual crop in the United States, genetic engineered forms are likely to make some inroads.

FIGURE 9–9   A plant physiologist compares Florida-grown 'Endless Summer' tomatoes to his greenhouse-grown fruit. All contain the bio-engineered ACC synthase gene. (USDA photograph.)

Flower Color

Flowers are at least a $1 billion annual industry in the United States. Special efforts are directed in this industry toward improving the appearance of flowers and extending their life as cut flowers. One way genetic engineering could help is to transfer genes for more uniquely colored flowers, such as a true-blue color. Another way would be to alter pathways involved in post-harvest deterioration, such as the introduction of anti-sense genes that inactivate key enzymes involved in post-harvest degradation of cut flowers.

The current approach to manipulating flower color through transgenic means involves flavonoid biosynthesis. Flavonoids are components of flower pigments. Colors are changed by alteration of the side chains. Blue and red colors are enhanced with derivatives of delphinidin and cyanidin, respectively. The anti-sense method discussed for slowing tomato ripening has been used to inactivate one or more genes involved in flavonoid biosynthesis. A number of interestingly colored petunias have resulted with this approach. Before this approach becomes a method of choice over contemporary plant breeding, the biochemistry and genetics of flavonoid pathways needs more research. The lack of knowledge leads to more or less unpredictable results. Another method would be to use genes for certain colors from nonplant organisms.

Better Proteins and Other Nutrients

Traditional plant breeding has been the mainstay for the improvement of nutritional quality of crops. Goals have included the production of a complete plant protein and higher protein and/or vitamin contents. Increased sweetness is another, especially for sweet corn and cherry tomatoes. Transgene technology has opened another approach to these goals.

A more complete protein can be achieved in two fashions. A gene coding for a protein rich in the desired amino acid can be added to the crop. Another method is to alter the pathway to the desired amino acid. For example, a key enzyme might regulate levels by inhibiting the production of the desired amino acid. Interference with the genetic control of this enzyme could result in less inhibition and more amino acid.

Some 20 amino acids are assembled in various ways to create proteins. Humans can synthesize 12 amino acids, leaving 8 essential amino acids that must be consumed in food. Red meat, chicken, and milk can supply these essentials, but no plant protein contains an adequate amount of all eight amino acids. Beans are low in methionine, corn in lysine and tryptophan, soybean in methionine and cysteine, and wheat and rice in lysine. Plants can provide a complete protein if the diet contains a mixture of complementary plants, such as the bean and corn mixture (succotash) of native peoples in the arid southwest United States. Given the lack of meat, poultry, and milk in many diets throughout impoverished areas and dependency often on a single plant such as rice, the value of a plant with a complete protein is considerable.

Some progress has been made. Genes from the Brazil nut, which is rich in methionine, have been moved into other plants, such as *Brassica* and sunflower, resulting in methionine-enriched storage proteins. However, given nut allergies, this approach requires caution. Interference with enzyme systems that inhibit lysine formation has been accomplished with gene transfers to corn, soybean, and canola. Seeds of these crops were enriched in lysine and/or methionine. Other projects are under way with rice. The problem is that considerable testing is needed to ensure that other changes or breakdown products detrimental to humans have not occurred. To date, some of these gene-enhanced proteins are close to commercialization or already available, but only as animal feeds. The enrichment of proteins in horticultural crops for human consumption is not far off.

Major advances have been made in alteration of oil quality, especially in soybeans and peanuts. Gene transfer with soybean, using genes that enhance oleic acid, produced oleic-enriched soybean seeds. This form of fatty acid results in increased stability during frying and cooking applications. Food products containing soy derivatives consequently need less hydrogenation, which benefits health as lower amounts of unhealthy trans-fatty acids are present. Gene transfer efforts are also under way to improve palm oils, especially oleic and stearic acid content. The latter is used as a substitute for cocoa butter and as a raw material in soaps.

Gene transfer for enhanced vitamin, mineral, and other healthy phytochemicals is also in progress. This area is now called *nutritional genomics,* a term used to describe genetic engineering where human nutrition, plant biochemistry, and gene transfers are involved. The motivation is to produce healthier foods, especially for areas dependent on one or two plants as dietary mainstays. In richer countries, transgenic crops that fight heart disease and cancer are being investigated.

The former involves increases in vitamin and mineral contents to bring plant-based diets into the U.S. recommended dietary allowance (RDA). This level is the minimum for alleviation of nutritional disorders. Higher levels are actually required for optimal health. This area has an overlap with cancer and heart disease prevention, in that vitamins E and C play a role in this area. Certain plant chemicals (phytochemicals) are targeted for increased levels in transgenic plants. Examples include carotenoids such as lycopene and lutein (tomatoes and spinach, respectively). Others of note include resveratrol in red grapes and wine, glucoraphanin in broccoli, and genistein in soybeans. These phytochemicals are believed to be important in the prevention of cancer, heart disease, and macular degeneration. The complexity of biochemical pathways for vitamins, minerals, and phytochemicals is complex, but healthier vegetables and fruits are likely to appear in the near future.

Microbial Pesticides

Microorganisms can be used to control insects, weeds, and pathogens. Two routes for biocontrol exist. Genes from microorganisms can be utilized in transgenic plants to offer enhanced protection. An example of this was discussed: the use of the Bt gene complex to make plants toxic to caterpillars. Another approach is to harness microorganisms with enhanced genetic properties to provide protection in the plant's immediate environment. Some beneficial microbes compete with pathogens, thus reducing or displacing their populations. Others produce antibiotics that help control pathogens. Still others can act as parasites on pathogens. These beneficial organisms can operate in soil, although some can be applied aerially to control airborne diseases. Biocontrols appear to offer promise as an alternative to fungicides, bactericides, and herbicides.

Genetic improvements can be made on these organisms. For example, genes involved in antibiotic production or proteases can be tweaked to increase production of these control agents. These genes might be transferred to microbial partners that normally inhabit the root zone of crop plants, but don't carry genes for the expression of antibiotics or proteases. The approach is to make multiple copies of the gene to increase production of antibiotics or proteases, or to alter the regulation of the existing genes. Another possibility is to use molecular techniques to turn virulent microbes into nonvirulent forms. Pathogens that attack weeds can have increased virulence through gene multiplication or gene transfer. Several successes have been demonstrated.

Transposons have been used to alter the genome of *Pseudomonas solanacearum*, a bacterium that causes tomato wilt. The altered form invaded tomato tissue, but provided some protection against the virulent form. Genes for chitinase have been introduced into various bacteria, such as *E. coli*, *Pseudomonas*, and *Rhizobium* spp. and used to control pathogen fungi that contain chitin in their cell walls. Genes for antibiotics have been multiplied in *Pseudomonas fluorescens* and transferred to other *Pseudomonas* sp. that lack the ability to produce antibiotics. Genes for proteases have been multiplied and moved about in various *Trichoderma* sp.; this fungus appears to offer promise as a biocontrol. The latest angle is to use molecular techniques to add two or more useful traits to a biocontrol agent, such as antibiotic, chitinase, and protease production. Much more work remains to be done with genetic alteration of weed-specific pathogens.

Nitrogen Fixation

The transfer of nitrogen fixation properties (ability to convert atmospheric nitrogen to a user-friendly form for plants, NH_4^+) from leguminous crops to other crop plants will likely be the most significant event in the history of genetic engineering, as well as to sustainable agriculture. In a sense, this direction is the "holy grail" of biotechnology. Untold riches will befall the group that patents the N-fixation "gene module" that can be plugged into crops. These riches won't come easy, however. The genetic complexities are extreme. Even with the incorporation of nitrogen-fixing genes, today's cultivars with their high nitrogen needs might not be supplied with enough nitrogen. Levels might be increased by inserting multiple copies of the nitrogen-fixing genes, or we might have to settle for adding small amounts of nitrogen fertilizer (but much less than is required for plants having no nitrogen-fixing genes). However, for agriculture to be sustainable on an industrial level, we need to move away from energy-intensive, manufactured fertilizers to utilizing free nitrogen from the air.

Nitrogen fixation occurs in several microorganisms that live in the soil. These include *Azobacter*, *Azospirillum*, *Bradyrhizobium*, cyanobacteria, *Frankia*, and *Rhizobium*. Most of these microbes bring about nitrogen fixation in the soil and help supply soil nitrogen to plants. Some form a symbiotic relationship with plant roots and bring about nitrogen fixation and nitrogen transfer directly to the plant. Many of the *Rhizobium* sp. are involved in the latter approach, but only work with alfalfa, beans, clover, peas, and other legumes. *Bradyrhizobium japonicum* forms a symbiotic arrangement with soybeans. Still some 50% of the nitrogen used in food production must be supplied with synthetic fertilizers. This amount totals about 50 million tons annually.

The two bacteria likely to form the basis of future attempts to genetically engineer bacteria capable of nitrogen fixation with all crops are *Bradyrhizobium* and *Rhizobium*. Currently these microorganisms only form symbiotic relationships with legumes. These microbes invade through the roots, where they bring about the formation of root nodules through developmental alterations. Inside these nodules, the bacteria utilize an enzyme, nitrogenase complex (consists of nitrogenase and nitrogenase reductase), to bring about nitrogen fixation. The transformation from N_2 to NH_4^+ requires nitrogenase complex, magnesium, ATP (chemical energy), and reducing equivalents (electron donor). In turn, the plant supplies some photosynthetically fixed carbon that facilitates bacterial growth.

One approach to achieving nitrogen fixation in plants is to isolate the genes responsible for nitrogen fixation (*nif* genes). Such genes have been found and isolated from *Klebsiella pneumoniae*. Because of many complex interactions, attempts to add these genes to crop plants have failed to date. Another problem is oxygen sensitivity of nitrogenase. Normally, oxygen levels are low in nodules, but the expression of *nif* genes alone without nodules places the nitrogenase in plant tissues that are too oxygen rich. It is unlikely that simple *nif* gene transfer will work.

Another approach was to find the genes involved in nodulation. The soil contains many species and strains of *Rhizobium* that compete, such that some are more effective at bringing about nodulation in leguminous crops. Unfortunately, the more competitive strains were less efficient at nitrogen fixation. It was reasoned that elucidation of the gene differences relative to competition could be used to enhance the competitiveness of commercial strains currently used with legumes. Essentially, the added commercial strain could be made more competitive than indigenous ones, thus increasing nitrogen fixation.

Nodulation genes (*nod* genes) were found and characterized in *R. meliloti*. Some 15 to 20 genes are involved in the nodulation process. It was observed that some of the genes were involved in host specificity. These genes produced a product that bound specifically to flavonoids, phenolic derivatives that are involved with pigmentation and defense. Once this bound complex is formed, promotion of nearby genes occurs that brings about the formation of lipochitin oligosaccharides (Nod factors) involved with root hair alterations. These compounds are host specific and essentially function like hormones to bring about developmental changes. The plant appears to respond to Nod factors by producing several nodule-specific proteins (nodulins). These root hair changes make the roots susceptible to bacterial invasion. Then other genes produce proteins involved with forming nodules on the roots. Plant nodulins also contribute to nodule formation. To date the complexity of nodulation has hindered efforts to improve the competitiveness of commercial strains of *Rhizobium*. A minor success has occurred with *R. meliloti* that was enhanced with extra copies of a few genes involved in regulation of nitrogen fixation and metabolite transport from the plant to the bacteria. Biomass of alfalfa was increased about 12% with the enhanced bacteria, but this effect was only observed in soils low in nitrogen and with indigenous populations of *Rhizobium*.

A recent break may have occurred indirectly. A pea cultivar was created that formed nodules extremely well with a highly efficient, nitrogen-fixing strain of *R. leguminosarum* nodule in the presence of other indigenous strains. Instead of genetically engineering more competitive stains of bacteria with high nitrogen fixation ability, the pea plant was altered so that it would work with an existing, favorable natural strain.

A third approach is to identify genes involved in conferring host specificity. The idea is that such genes on either the bacteria or plants could be altered, so that *Bradyrhizobium* and *Rhizobium* species could be made to form nodules on nonleguminous crops. While mechanisms have been determined, involving flavonoids and Nod factors, this knowledge has not been translated into changes needed to make nitrogen-fixing bacteria work with nonleguminous crops.

Another avenue toward reducing manufactured nitrogen dependency is to work with free-living, nitrogen-fixing microorganisms. These soil microorganisms do contribute some fixed nitrogen to the soil and eventually to plants. The main concern here would be to genetically enhance the nitrogen fixation ability so that more soil nitrogen would result. Little research appears here and more is clearly warranted.

Pathogen-Free Plants

Certain horticultural crops can be pathogen prone. While seeds can be surface decontaminated, vegetative propagules can harbor internal pathogens. The best possible method of producing pathogen-free, especially virus-free, propagating stock for horticultural crops is to propagate from shoot apices and to raise plants through tissue culture (see discussion of micropropagation in Chapter 7). Plants produced in this manner can be officially certified. Such plants command higher prices and offer increased yields.

The shoot apex is highly likely to be free of bacterial, fungal, and viral pathogens. This fact applies even with infected plants. The risk of contamination approaches zero as the size of the removed shoot apical explant decreases. The problem is that the smaller the explant, the less likely that micropropagation will be successful. The best choice, in terms of balancing these opposing trends, is to use pieces in the 0.25- to 1.00-mm-long range. Techniques exist to check whether pathogens are present in the explant and during micropropagation. Culture indexing is used to detect bacteria and fungi, and virus indexing for viruses. The former involves culture and identification on petri dishes, and the latter uses grafting or budding onto very sensitive indicator plants.

Pathogens and viruses have been eliminated in numerous crops with this approach. Floral crops include amaryllis, carnation, chrysanthemum, dahlia, freesia, geranium, gladiolus, iris, lily, narcissus, nerine, and orchid. Fruits include apple, banana, cherry, citrus, gooseberry, grape, peach, pineapple, plum, raspberry, rhubarb, and strawberry. Vegetables include brussels sprouts, cauliflower, garlic, potato, and sweet potato.

Viral Resistance

Plant viruses can cause considerable damage to crops, and effective chemical treatments are few. The standard approach has been to rely on introducing virus resistance by conventional plant breeding techniques and by destruction of virus-infected plants. These measures are, at best, holding actions. Genetic engineering has altered the battle in the producer's favor (Figure 9–10). The current strategies involve a form of vaccination and the anti-sense approach to inactivate production of critical viral components.

Adding genes to plants that code for viral coat proteins has proven to be the best current technology. The anti-sense approach has resulted in less resistance because it is only effective at low viral concentrations. When the plant expresses these genes and produces viral coat proteins, it appears to block the ability of that particular virus to infect the plant. The method by which this occurs is not yet well understood. Tolerance or resistance to a number of viruses has been achieved. These viruses include alfalfa mosaic virus, cucumber mosaic virus, grape fan-leaf virus, potato virus X, potato virus Y, rice stripe virus, tobacco etch virus, tobacco mosaic virus, tobacco rattle virus, and tobacco streak virus. Horticultural crops made virus resistant by this approach or heading that way include cucumbers, gladiolus, grapes, melons, potatoes, squash, and tomatoes. Field tests have demonstrated higher yields with these altered crops versus normal crops in the presence of viruses. Some concerns are that the added viral genes in the plant might recombine with invading viral genes, resulting in changed viruses that are more troublesome than the original ones. Others have expressed fear that this approach might accelerate the production of more viruses capable of overcoming natural resistance.

Bacterial Resistance

Diseases caused by bacteria result in economic loss and can make the cultivation of some crops in certain regions economically unfeasible. While some antibiotics are cleared for horticultural applications, costs can be prohibitive. The current strategy is to use pesticides

FIGURE 9–10 Plum pox virus is vectored by the green peach aphid (*Myzus persicae*) and is a troublesome disease. These transgenic plums contain a gene that makes them highly resistant to plum pox virus. (USDA photograph.)

to control the insects that spread bacterial diseases. The ideal solution is to produce culti-vars possessing genetic resistance to bacterial diseases. Success has been limited with conventional plant breeding, but genetic engineering offers greater promise.

A good example involves Pierce's disease of grapes, which is caused by *Xyllela fastidiosa*. This unusual bacterium (see Chapter 15) is a fastidious xylem-limited bacterium that lives only in the water conducting tissue (xylem). It eventually clogs the xylem, causing the plant to die. This organism causes millions of dollars in losses for Californian grape growers. Many popular table and wine grapes can't be grown in the Southeast because of this disease.

Recently a joint patent was issued to the USDA and the University of Florida for a gene group expected to confer resistance against Pierce's disease. This gene group is a synthetic analogue of genes found in various organisms such as silkworm larvae. In the silkworm these natural genes produce a protein that kills bacteria and fungi. Research is now in progress to produce genetically modified grapes that are resistant to Pierce's disease. Resistant grape cultivars will reduce or eliminate the need for pesticides to control insects that spread Pierce's disease, save millions of dollars in control costs and damage, and lead to a more sustainable and profitable grape growing industry in the Southeast.

New Hybrids

Conventional plant breeding is limited by sexual compatibility. The ability to maintain tissue cultures ushered in a new era. The cell walls of somatic plant cells can be removed with enzymic treatment, using cellulase. The resulting cell suspensions, devoid of cell walls, are known as protoplasts. Protoplasts, say, from tomatoes and potatoes, when exposed to PEG or electrofusion can fuse. The resulting fused cell can be micropropagated into a plant that shows potato and tomato characteristics. These types of fusion overcome sexual compatibility. The fusion of potato and tomato has been done, but the resulting hybrid has yet to reach commercialization. Little progress has been made commercially in this entire area.

Another advantage of this form of fusion is that cytoplasm merges from both partners, whereas in conventional breeding, only the female contributes cytoplasm. This method can be used to transfer extrachromosomally inherited characters such as cytoplasmic male sterility. This characteristic is especially important for hybrid seed production, such as in corn. Using somatic cell fusion, cytoplasmic male sterility (CMS) has been transferred to *Brassica* sp. and sugar beets. Restorer genes can also be transferred to CMS cultivars to restore pollen production, thus two CMS cultivars can be hybridized (see Chapter 8).

Germplasm Protection

Germplasm preservation is needed to protect genetic diversity and to maintain gene pools for future needs. Seed storage has been the method of choice for many horticultural crops. Maintaining living collections in facilities devoted to their preservation preserves crops not readily propagated by seed, such as potato or apple. The advent of tissue culture and micropropagation now offers an alternative. Tissue samples that can be readily micropropagated can be preserved through storage in liquid nitrogen (cryopreservation). Frozen samples can be thawed and micropropagated, thus making the original plants available. Numerous plants have been successfully preserved in this manner, including apple, carnation, carrot, chicory, coffee, grape, mulberry, pear, potato, and walnut. While not yet routinely employed, cryopreservation is likely to increase in the future. One exception is oil palm, which is already cryopreserved.

Pharmaceutical and Industrial Plants

This area is in its infancy. Currently, the production of pharmaceuticals and other industrial chemicals of various types are conducted with recombinant bacteria. These cultures are maintained in large fermenters, harvested, and separated from the culture medium; then the product is separated and purified. Products include human proteins, such as growth hormone or interleukins, vitamin C, various amino acids, antibiotics, enzymes, and biopolymers, such as xanthan gum. These systems are expensive and complicated, requiring sophisticated equipment and highly skilled personnel. Plants could be genetically programmed to produce these same products and other novel pharmaceuticals. Growing plants is easy and considerable biomass can be produced. The downside is the difficulty of expressing sufficient product so that it can be easily isolated and purified, given the large number of metabolites produced in plants. Of course, costs need to be similar or less than comparable microbial systems.

Products currently expressed in plants include bioactive peptides, such as Leuen-kephalin with opiate activity, human serum albumin, α-amylase, hepatitis B surface antigen (hepatitis B vaccine), a vaccine to prevent infection from *E. coli*, poly-D- (-)-3-hydroxybutyrate (biodegradable thermoplastic), and cancer antibodies. The term *plantibodies* has been coined for plant-produced antibodies. Extensive research by numerous companies is under way in these areas. Some of the pharmaceutical products have entered human clinical trials.

A recent patented technique by the Agricultural Research Service for altering future plants through pollen has brought "pharming" closer to reality. This method is known as pollen electrotransformation. It is essentially a variant of the electroporation method discussed earlier in this chapter. An electrical discharge is used to temporarily open the pollen cell walls, allowing genes to be transferred in. The pollen is used to pollinate unaltered plants and some of the seeds will carry the new gene. This approach eliminates the necessity for micropropagation. Success has been achieved with alfalfa, corn, and tobacco. Commercial plans are under way to use this technique with alfalfa. Alfalfa, with altered genes, will be programmed to produce enzymes, proteins, and other compounds for human and animal medicines, vitamins, skin care products, food additives, and other products. Other companies are looking to use this method to "pharm" other crops, such as ornamentals, soybeans, stone fruits, and sugar beets. In the future, your fruits and vegetables might supply you with vaccines, all vitamins, and even prescribed medications!

SUSTAINABILITY AND PLANT BIOTECHNOLOGY

Some problems have occurred in the application of biotechnology. Crop failures have resulted with cotton genetically engineered with Bt genes and herbicide-resistant genes (see next section). Some research studies have indicated that a possible linkage between Bt crops and immunological and organ damage exists with humans, based on studies with rats. Lab studies have also shown that pollen from Bt corn contains enough Bt toxin to kill the caterpillars of Monarch butterflies. These butterflies, already under environmental pressure, prefer milkweed as their food source. Milkweed is often found on the perimeter of wild areas around cornfields. A more recent field study showed that pollen from Bt corn is naturally deposited on milkweed nearby and could possibly kill the butterfly larvae. Still, another study says the danger is minimal. A definitive answer is needed, because it would be a tragedy if this benign butterfly were wiped out by windblown pollen from Bt corn that is now widely planted throughout the United States.

Resistance toward field trials and consumption of genetically engineered crops has surfaced in Europe, especially England, and Asia, especially India. Some backlash has surfaced in the United States from groups such as the Center for Food Safety. This group has sued the FDA to force the removal of genetically engineered foods from the marketplace. The suit stems from the lack of labels, such that the consumer cannot identify genetically engineered foods, and insufficient safety testing.

In 1999, some 35%, 45%, and 50% of U.S. corn, cotton, and soybean acreage, respectively, were genetically engineered with either herbicide resistance or Bt properties. The recall of several food products containing genetically modified corn (Starlink) approved for animal feeds and not human consumption points out the consumer problems. Clearly, better testing is needed in terms of food safety for humans and environmental pressures on beneficial or benign insects. The National Research Council in 2000 recommended a number of ways to improve the process.

While the majority of the current plantings of genetically engineered crops are not horticultural crops, the horticultural sector can expect major inroads. At present, less than 5% of potatoes are genetically engineered with Bt, and small inroads have been made with tomatoes. Transgenic crops in horticulture production and development now include bananas, melons, papaya, peanuts, peppers, pineapple, potato, raspberry, squash, strawberry, sugar beet, sunflower, sweet corn, and tomato. The question of whether sustainable horticulture has a place for genetically engineered crops remains open, as does whether the consumer has a right to know, and whether more safety/environmental regulations are forthcoming. At least one part of the sustainable horticultural community has no genetically engineered foods in its future. New organic standards from the U.S. Department of Agriculture specify no use of bioengineered crops.

The question has been answered in part for the sustainable horticulture community. Organic practitioners want no part of genetically engineered crops. Certainly, some aspects of genetically engineered crops are likely to be acceptable to the nonorganic part of the sustainable horticulture community. Some examples include crops with gene-added insect and disease resistance, where the genetic change has been proven safe for humans and the environment, or just safe for the environment with ornamental crops or with edible crops when the gene alteration does not appear in the edible part. Another example is the addition of genes that give crops the ability to fix nitrogen directly from the air, thus decreasing the need for fertilizer. Clearly, these examples are sustainable, in that pesticide and fertilizer usage is reduced. Under the right circumstances, these genetic alterations might even be acceptable to the organic community.

Unfortunately, current agricultural applications are not sustainable. The increased usage of herbicide-resistant crops is a case in point. Increased chemical use is not a sustainable goal. Although only agricultural crops are currently involved, the potential for herbicide-resistant horticultural crops is open. Recent figures from the USDA show that glyphosate usage on soybeans rose 72% once herbicide resistance was introduced. In 1998, 71% of genetically engineered crops patented were designed to be resistant to herbicides such as glyphosate. Even of more concern is recent research from Sweden that shows a link between glyphosate and non-Hodgkin's lymphoma, a form of cancer. This cancer has also been reported as increasing 80% since the 1970s. Increased allowed tolerances of glyphosate on soybeans have also been recently permitted. Herbicide resistance clearly is not a trait that is a desirable goal for sustainable horticulture. Given the increased profits for pesticide manufacturers with genetic engineering operations, such products might be offered to horticultural producers in the near future. The sustainable horticulture community should just say no!

SUMMARY

Humans have practiced biotechnology for 6,000 years of making bread, cheese, beer, wine, and yogurt and breeding plants and animals. Biotechnology is the use of living organisms to produce products to enhance our lives. The DNA revolution has rapidly advanced biotechnology. We have a nearly complete understanding of gene function and control and how to move genes between organisms (gene splicing and recombinant DNA technology). This knowledge changed the focus from manipulating whole organisms to manipulating DNA. No longer does sexual compatibility or the randomness of natural genetic recombination during reproduction limit plant breeding. We can directly insert genes for desired characteristics into a plant. Cells of two sexually incompatible plants can be fused and the hybrid cell converted into a new plant. We can even add nonplant genes to a plant. Plant biotechnology applications include micropropagation, *in vitro* production of pathogen-free plants, *in vitro* germplasm storage, genetic engineering, cellular fusion, tissue culture, transgenic plants for pest control, and biocontrol of pathogens and weeds.

The brave, new world of plant biotechnology holds forth much promise, but like any tool it can be misused. Hopefully, plant biotechnology will be harnessed for the betterment of sustainable horticulture. An enormous debate on food safety, escape of engineered genes into the environment, and control of plant germplasm by a few multinational corporations is ongoing. These challenges need to be resolved before benefits can ensue.

With recombinant DNA technology or genetic engineering, one can manipulate genes. A gene or genes from any organism is piggybacked onto the DNA of a carrier used to insert the gene(s) into the cells of a target organism. Carriers are usually modified bacteria or viruses. Direct insertion of the gene by physical methods is also possible. Gene transfer is more precise and overcomes the limitation of sexual incompatibility. Processing time is greatly reduced by eliminating random genetic recombinations and selection afterwards. Genetic transfer between virtually any two organisms is possible.

Ideally genes are identified by their placement on the chromosomes (gene map) and what function(s) they control. Functions should be understood in a developmental, biochemical, and physiological context. Then we could just select the gene for the desired characteristic and add it into another organism. Much research is in progress, but the reality is that plant gene maps are limited. Our knowledge of plant development, biochemistry, and physiology is also not complete. For now, we must make do with alternative approaches to finding our way without gene maps.

Plant gene libraries (DNA libraries) contain DNA fragments of various lengths. These collections are not paginated, alphabetized, or even logically organized. Fortunately, techniques exist for finding and locating genes and even for creating useful maps of gene locations. DNA libraries are useful because multiple copies can be stored for indefinite periods and used later to retrieve genes.

DNA is cut into pieces by bacterial restriction enzymes. Using partial to complete enzymic digests, fragment sizes can be varied. Restriction enzymes cut double-stranded DNA within or near specific areas called recognition sequences. These special sequences typically consist of four to six nucleotides. Some cut straight, but a few restriction enzymes produce a jagged cut, leaving one strand a few nucleotides longer than the other strand. When this uneven cut occurs, the double-stranded DNA is said to have "sticky" ends. These sticky ends have complementary exposed nucleotides that can pair and be rejoined with other DNA using an enzyme called DNA ligase.

If the same restriction enzyme is used to cut DNA from two different sources, such as the small circular DNA unit in bacteria (plasmid) and plant DNA, the sticky-ended DNA strands from both sources have complementary ends. These DNA fragments can then be rejoined by DNA ligase. The combined DNA now exists as recombinant DNA in the bacterial plasmid (vector).

These plasmids are placed in medium containing growing bacteria. Some bacteria take up the altered plasmids (transformation). Transformed bacteria containing the recombinant DNA can then be cultured to produce millions of copies of the recombinant DNA fragment (DNA cloning). Transformed bacteria must be sorted from untransformed bacteria, because not all bacteria take up the plasmid vector. This identification process is defined as screening, although some prefer to call the process selection and reserve the term screening for locating specific genes.

Two basic screening techniques exist: antibiotic resistance and lactose breakdown. When a specific restriction enzyme cleaves the recognition sequence in the plasmid, it damages the nearby *lacZ* gene. This gene codes for β-galactosidase, the enzyme that hydrolyzes lactose. Bacteria transformed with the plasmid are unable to break down lactose. When a special sugar (X-gal) is present, β-galactosidase produces a blue color. In the absence of the enzyme, colonies with the altered plasmids remain white. In the other method, a genetically altered plasmid containing genes for antibiotic resistance is used. For example, the gene *amp*^{R} gives resistance to ampicillin. Bacteria that received this plasmid with its attached DNA fragment are able to grow in cultures containing ampicillin.

The DNA cloning process results in a cloned collection of the various fragments produced from the original organism's total DNA (genome). This collection is called the genomic library. By using several different restriction enzymes, the same DNA can be cut into different sets of fragments. Some 1,200 restriction enzymes are known, many of which act at different recognition sites. Using different restriction enzymes, one can prepare several versions of a genomic library for any organism's DNA. The probability of finding whole genes of interest among the various fragments is thus ensured.

One can make another library, the complementary DNA library. This process requires a retrovirus enzyme (reverse transcriptase). When this enzyme is placed with messenger RNA, it catalyzes the synthesis of a single strand of DNA. This strand can be doubled in the presence of DNA polymerase. This DNA, known as complementary DNA or cDNA, differs from the genomic DNA of the organism, in that cDNA contains no introns. Introns are noncoding regions of DNA that do not translate into proteins.

Why have gene libraries and why two types? If the genetic engineer had to isolate gene(s) every time they were needed, it would be a difficult, time-consuming process. Instead, these genes need only be isolated once and placed in the gene library. Multiple copies are then made and are readily available as needed. Two library types are needed, based on where the genes will be inserted for expression. Genes destined for manufacturing a protein in a bacterium must come from a cDNA library. Bacteria and other prokaryotes are unable to process DNA from higher organisms (eukaryotes) because of the presence of introns. Bacteria possess exons but not introns and have no mechanisms for removing regions transcribed from introns. The cDNA library, containing only the exon parts of DNA, can be readily translated into proteins when inserted into bacteria.

The genomic library works fine with eukaryotes, because they possess the cellular machinery to work with introns. However, even here, a cDNA library is efficient, because the processing of intron regions on foreign DNA (and possible complications) can be eliminated. So why not just make cDNA libraries? The genomic library is much easier to make. Both of these libraries are uncataloged in that the locations and functions of the genes within are unknown. Additional screening to locate the desired genes is needed before these libraries can be of any use.

Restriction enzymes can be used for determining the nucleotide sequences of DNA. Knowing this sequence is a useful tool for screening libraries for specific genes. Three things are critical for sequence determination: workable units of DNA, sufficient amounts of DNA, and primers. The entire genome is not currently a workable unit, given that a single gene could typically have a few hundred thousand base pairs!

Workable units are the fragments of DNA produced with various restriction enzymes. Adjusting the time of treatment with restriction enzymes carefully controls fragment length. Ideal fragment length is about 4,000 bases. Sometimes multiple copies of fragments are needed. One reason is to increase small amounts of DNA to workable units for purposes of sequencing or identification. Multiple copies of a gene might be made prior to being added to another organism, as only a small amount actually gets transferred into the working DNA of an organism.

Millions of copies of DNA fragments can be produced via PCR (polymerase chain reaction) using *Taq* polymerase. This enzyme copies single-stranded DNA if the proper primers are applied at the beginning and end of the part you wish to copy. The enzyme is heat stable. The double-stranded DNA is heated, denaturing and separating it into two strands. On cooling, primer DNA that codes for "start" and "stop" binds to the strands of DNA. The *Taq* polymerase reads the start and stop signal and, using supplied nucleotides, copies each strand, resulting in two double-stranded DNA units. The reaction is initiated again by heat. The two double-stranded molecules then become four intact, double-stranded DNA units. A device called a thermocycler is used to automate the heating and cooling. These steps can be repeated many times, resulting in up to a million-fold increase of DNA.

Electrophoresis is the tool of choice for separating DNA fragments. Applications include the separation of gene libraries. The entire library is too big to apply directly for manual or even automated sequencing. Separated fragments can be pooled into smaller batches for sequence determinations. Another use is to separate fragments after DNA or mRNA probes are used to locate a gene. If the probe had a fluorescent or radioactive tag, the band can be visualized under UV light or through film exposed to the radioactivity.

The library or "probed" collection of fragments is added in a buffered solution as a band on special polyacrylamide gels. An electrical field is applied and fragments move across the gel at different rates based on their differences in length and charge. Smaller fragments move faster than larger ones. The separated fragments can be removed in gel slices. The DNA fragment or fragment with an attached probe can be washed out of the gel slices. If enough is present in the sample, it can then be sequenced. If larger amounts are needed, PCR is used to make copies.

Primers are critical for sequence determinations. Sequencing requires special DNA sequences (primers) that are complementary to a stretch of the cloning vector just before where the fragment, or target DNA, is attached. These primers act essentially as the "start" or "on" button for the sequencing process. Primers can be synthesized with any desired nucleotide sequence. These short pieces of DNA, technically oligionucleotides, are synthesized by the phosphoramidite method.

Sequencing is done either manually or with automated DNA sequencing machines. The manual method depends on the presence of dideoxynucleotides. When these nucleotides are incorporated into a replicating DNA strand, replication ends. Termination occurs because the next incoming, normal nucleotide cannot form the required phosphodiester bond with the dideoxynucleotide. This simple event forms the basis of manual sequencing.

Annealing a primer on the cloning vector (plasmid DNA) some 10 to 20 nucleotides before the original insertion site of the cloned DNA starts the process. Many copies of the cloned DNA are needed for this step. The site DNA sequence is known and is near where the cloned or target DNA had been inserted into the vector when making the library. When this known sequence just after the primer, but before the cloned DNA, is read later, it serves as a reference point for reading the sequence of the cloned DNA.

The primed DNA is divided into four equal volumes in test tubes. Into each test tube is added the four normal deoxyribonucleotides needed for replication (dATP, dCTP, dGTP, and dTTP); one of the nucleotides is labeled with P^{32} or S^{35}. Which one is labeled changes for each tube. Dideoxynucleotide is added to each tube also; one tube gets ddATP, another ddCTP, another ddGTP, and the last ddTTP. The amount of dideoxynucleotide must be carefully controlled so that it is incorporated into every possible site on the growing chain. Many copies of the primed DNA were used at the priming stage so that the dideoxynucleotide has the potential to appear at each site. If only one copy were used, the chain would be stopped at the first incorporation site and all you would get is one short chain. With many copies being made, one will be stopped at the first possible incorporation site, another chain might not be terminated until a dideoxynucleotide is added at the second possible site, the first one having received a normal nucleotide, and so forth.

The reaction is allowed to proceed to completion using DNA polymerase and terminated by adding formamide. The mixture of many DNA fragments in each tube varies considerably in length. Some differ in length from each other by only one nucleotide. The mixtures are applied in four lanes on a polyacrylamide gel slab and separated by electrophoresis. An autoradiograph is made. The sequence is read from the bottom (shortest fragments) to the top (longest fragments). If lane 1 contains the dideoxynucleotide ddATP and shows the lowest visible band, the sequence starts with base A (adenine). If the next lowest band is found in the ddTTP lane, then the second nucleotide is T (thymine), and so forth. If the mixture was incorrect in terms of components, you end up with multiple bands at the same place. For example, if nucleotide location 15 has four bands, all visible in a horizontal smear across the four lanes, something was wrong with the original mixture.

This method is limited in terms of fragment size. The best results usually give a resolution of about 300 to 400 nucleotides on a typical autoradiograph. This procedure has to be repeated many times before all the fragments of the original DNA library are sequenced. Even then the sequence is not complete. All of the parts are there, but in what order? This question is resolved with a comparative, puzzle-solving approach. DNA fragments produced with different restriction enzymes are sequenced and compared. One can now assemble the sequence for the entire genome of the organism in a jigsaw puzzle manner. This approach works because the various sets of fragments from different restriction enzymes have areas of overlap. These overlaps are compared and help to determine the final sequential order.

The automated procedure is somewhat similar. Many copies of the DNA fragment are primed and the four normal nucleotides are added. However, none of them are labeled with P^{32} or S^{35}. Four types of dideoxynucleotides are also used, one type per tube. These dideoxynucleotides are each fitted with an attached fluorescent probe. Each of these "fluorescent" nucleotides fluoresces at a different wavelength. The DNA polymerase reaction proceeds and the mixture of various-length units is produced as above in the manual procedure. These samples are scanned with lasers. Each nucleotide substitution has a unique fluorescent wavelength that relates to the original nucleotide. The machine prints out a chromatogram containing four different height peaks. One height corresponds to adenine, another to thymine, and so forth. The base letter (A, C, G, T) is printed across the chromatogram, making the sequence easier to read. While the automated method is used to determine longer sequences, the overlap method is still often needed for the final properly ordered sequence.

An alternate approach to the jigsaw overlap is possible and highly recommended when the piece of DNA exceeds 5,000 base pairs. Sequences beyond this length take a long time to process because of the need to determine the sequences of numerous fragments. The method of choice here is known as primer walking. The vectors are different than for the overlap procedure. These vectors carry much larger DNA pieces. The same procedures are used, either the manual or automated approaches. This step determines the sequence of the first 250 to 350 bases. Using this just completed sequence, a primer is synthesized for placement at the end of the sequence, usually at the last 20 bases. The sequencing is then done for this section. A primer is made and attached at the last 20 bases of the second known continuing sequence. This procedure is repeated as many times as necessary to determine the total sequence. The procedure gets its name from the fact that a primer is "walked" along the DNA.

Ideally one has a completely cataloged gene library for every plant and all other organisms used to supply genes for altering plants. Several problems keep us from the ideal. These include the sheer number of genes and the fact that sometimes one gene can express a single useful trait, like insect resistance, while other characteristics, such as fruit size, are controlled by many genes working together. Sorting all this can be frustrating and tedious. Fortunately, some techniques do exist that can help us locate specific genes in terms of function.

Probes are molecular tools used to screen gene libraries. These tools are able to recognize specific nucleotide sequences in either DNA or RNA. Another type of probe can recognize a protein or other molecule synthesized by a specific gene. Probes are usually used on copies of the gene library, so that the original gene collection remains unaltered by any experimental protocol.

Library copies are made by replica plating. A piece of sterile velvet cloth is placed on the surface of an agar-cultured library plate. Some cells from each colony present are transferred through contact. The cloth is carefully touched to a new petri dish containing sterile growth media. A new set of colonies is cultured that are copies of the original library. The original library is intact and unaltered by any screening techniques.

While the type of probe varies, the technique used for screening does not vary much. Applied screening protocols are named based on what is being screened: DNA, mRNA, or protein. The first technique was called Southern blotting, named after its discoverer in the mid-1970s, Dr. E. M. Southern. Subsequent protocols were named Northern and Western blotting.

The protocol starts with the transfer of some material from the library copy. The material can be whole cells, or some part thereof, such as proteins or nucleic acids. The transferred material is placed on special nitrocellulose filter paper. Materials are transferred so as to retain the original spatial relationship found on the starting point, the library copy. The transferred material usually requires some pretreatment prior to analysis. The treatment alters the form so that the analysis can proceed properly. For example, whole cells would be lysed with a dilute basic solution, such as sodium hydroxide. This solution lyses the cells, and converts the double-stranded DNA to the single-stranded form.

A prepared "hot" or radioactive probe is added. Fluorescent probes can also be used. The radioactive or fluorescent probes allow for easy localization later of the areas where the probe reacted. The probe will bind specifically to a complementary DNA or mRNA sequence or, in some cases, to a specific protein expressed by the genes. The binding to a complementary sequence is known as hybridization. Chemical treatments can be varied to control the degree of hybridization. Conditions known as high stringency result in hybridization between two perfectly complementary sequences. Less restrictive methods produce hybridization between partially complementary sequences. Washing is used to remove excess probe. After drying, the filter paper is placed against a sheet of photographic film so that radioactive parts expose the film. After autoradiography, the visible areas correspond to molecules that hybridized. These areas can be traced back to colonies that contained the sought-after (target) genes.

Various probes are used. One type, a cDNA probe, is prepared from the cDNA library. Plasmids containing cDNA inserts are isolated. Next, the double-stranded cDNA is denatured into single-stranded form. The preparation is then added to nitrocellulose filter paper. A mixture of mRNA, chosen because of its complementarity to many known DNA sequences, is added to the filter paper. Any mRNAs bound to a complementary cDNA are left on the filter paper after unbound mRNAs are washed away. Bound mRNAs are removed and then translated into proteins that are identified. Often the function of an identified protein is known. The cDNA from which the bound mRNA was removed is now identified as the gene that produces the particular protein. Call it gene X. The complement to this known sequence of cDNA can now be used as a probe to screen other DNA libraries for gene X. This probe is called a cDNA probe.

Known mRNAs that code for specific proteins are also used as probes. If a mRNA codes for an enzyme or a subunit of an enzyme, such as carbonic anhydrase, this mRNA is used to probe gene libraries for the gene that makes carbonic anhydrase. This probe is termed a mRNA probe.

If the sequence of a protein is known, another type of probe is used. Based on the amino acid sequence, a sequence of nucleotides coding for production of a segment of the protein is synthesized. The nucleotide sequence (oligonucleotide) is constructed with radioactive components and the resulting product is called an oligonucleotide probe. If this probe binds to something in the gene library, you have found the gene (or a gene with a very similar structure) that codes for carbonic anhydrase.

Another type of probe is based on the immune system. Antibodies are specific for and bind to foreign cells or molecules, known as antigens. Gene products for which a specific antibody exists, or can be made, are useful tools for finding the genes producing the antigen. If the antigen is a well-known protein, the antibody can become an antibody probe and verify the presence of the gene indirectly.

While the above probes can be used to screen DNA of gene libraries, these probes can also screen genomic DNA directly by a technique called Southern blotting. DNA is cleaved into fragments by using restriction enzymes and followed by separation with agarose gel electrophoresis. Several steps later, a radioactive or fluorescent cDNA, mRNA, or oligonuclotide probe is used. Once identified, the gene of interest is excised from the gel and its nucleotide sequence determined. If the fragments are also sequenced, the location of this specific gene can be mapped.

Northern blotting is somewhat similar, but RNA rather than DNA is probed. Agarose gel electrophoresis is used to separate total RNA or purified mRNA preparations. A denaturing agent is employed to prevent the RNAs from hybridizing during electrophoresis. The procedure after this point is similar to that described above for Southern blotting. RNAs complementary to the probes are identified.

Western blotting screens for actively produced products of genes, usually proteins, and uses an antibody probe. This technique is more useful for determining whether an inserted gene is functioning in its host organism, although it can be used to screen for the same gene in an unaltered organism or cell preparation. Once the gene is known to be present and active, the other blotting methods can be used for pinpointing the gene for sequencing and mapping.

Another way of screening genes involves using mutagen-induced mutations and a natural type of mutation caused by something called a transposable element. Barbara McClintock discovered the presence of these "elements" in maize during the late 1940s. However, their value, especially as a screening tool, was only recently realized. Transposons are now one of the premier tools for determining gene functions!

In nature, transposable elements cause numerous spontaneous mutations in plants. These mutations result from the duplication and movement of transposable elements among the plant's chromosomes. Two types of transposable elements exist: insertion sequences and transposons. Insertion sequences are smaller. Their function genes code for enzymes that allow for removal and insertion of the element elsewhere in the DNA. Transposons are larger and carry additional genes beyond the removal/insert ones. These genes code for various proteins not involved in the jumping process.

Once these elements were identified and understood, mutations could be deliberately induced by the insertion of transposons. If these mutations result in visible or biochemical detectable changes in a plant, the function of the affected gene can be determined. For example, if a flower color changes from pink to white, a gene involved in the production of pink pigment was likely damaged. If the transposon can somehow be located in the organism's DNA, we then know the location and function of the gene. Then it can be sequenced, creating better genetic maps.

Transposons and insertion sequences from sources such as maize and snapdragons have been isolated and sequenced. Some of these transposable elements have been introduced into plants by a process called transposon tagging. Transposon tags are fitted with markers and primers and can be inserted into whole plants or cultured cells. This form is especially useful for using vectors to introduce transposons into plant cell cultures. The marker can be a gene to confer antibiotic resistance or one that prevents lactose breakdown. The marker is used to screen for those cells that have taken up the transposons. These cells can be converted into plants via micropropagation.

DNA libraries can then be created for those plants with observable mutations. Those fragments that contain the transposon will have a built-in primer in the vicinity of the mutated gene. This area can be sequenced. Once the sequence is known, a complementary probe is prepared and used to prepare copies of the undamaged gene from DNA libraries of the same plant where transposon tagging was not used. The gene's function, location, and sequence are now known and another piece of the DNA map falls into place.

In a similar manner, mutations can be induced either chemically or by radiation. Actual populations of plants or cell cultures can be treated with mutagens. Progeny can be examined for alterations on the cellular, biochemical, or whole plant level. By comparing DNA sequences of the "before" and "after" plants, it is possible to find the damaged gene or genes.

Two kinds of gene maps exist. One type, called a gene linkage map, gives a rough approximation of certain genes relative to each other. It is a very rough, minimal information map. The other map, the physical map, shows the specific location of all functional genes. The physical map is analogous to a high-quality road map.

Certain traits appear to be inherited together, while others do not. This phenomenon results from linkage. If two genes are close together, separation of the two during meiotic crossing over is unlikely and becomes more likely with increasing distance. Once a linked gene is identified (location and function) and sequenced, it can be used as a marker gene in DNA libraries.

Another marker, the restriction fragment length polymorphism (RFLP), is similar to a marker gene. The RFLP is a stretch of functionless DNA that remains intact in the presence of restriction enzymes and during crossing over. Genes close by are linked with the RFLP during crossing over, so once genes in the vicinity of a RFLP are identified, the RFLP can be used like a marker "gene" in libraries as above with two linked genes. RFLPs are easily identifiable markers during electrophoresis.

Based on the frequency of paired traits in offspring, rough maps called linkage maps can be created from either marker genes, RFLPs, RAPDs, or AFLPs. These maps are not detailed, but are useful for screening once the function and location of one gene of a linked pair is known. This gene becomes known as a marker gene. DNA libraries can be probed for the marker gene or RFLPs, RAPDs, or AFLPs with various screening techniques. The library fragment that contains the marker gene has an extremely high probability of also having the linked gene. Successful screening and sequencing then puts two specific genes on the map.

Using the various techniques discussed earlier one can assemble physical gene maps. Probes and transposons are used to screen libraries and the identified genes are placed into previously determined sequences. Although numerous plants have partial maps, only a few plants have a complete map. One is *Arabidopsis thaliana*, the workhorse of plant genetics. A weedy member of the Mustard family, it has a number of characteristics that endear it to plant geneticists and molecular biologists. One feature is small size, such that thousands can be grown indoors under fluorescent lights. Another is a 30- to 45-day life cycle that produces several thousand seeds per plant. Effects of mutations and transposons can be determined over generations in short times. Its genome is small and has no repetitive DNA, consisting of five chromosomes and about 30,000 genes. Screening, sequencing, and mapping are more manageable on this scale.

The map and sequence of *A. thaliana* are the most complete of all plants. They are providing considerable data that are useful for mapping and genetically engineering higher plants. A number of questions about plant development and evolution are also being answered. Genes involved in disease and insect resistance, flowering, and photosynthesis have been discovered. Genes in embryo, seedling, root, and flower developmental processes have been found.

Although not a crop plant, *A. thaliana* provides a model map that can be extrapolated to provide information on crops. Certain genes, such as those involved in photosynthesis and development, are basic to all plants. Much of this information, DNA sequences, known genes, and partial maps exist in Web databases. Often the databases provide clues for other unmapped plant genomes. Probes can be constructed based on existing genes in other plants and then used to screen other plants. Much of this "sharing" is possible because of common ancestry in plants, such that today's plants show high degrees of similarity at the gene and protein level.

The alteration of plants through genetic engineering has great promise for the future of sustainable horticulture. Genetic engineering has the potential to move productivity ahead further and to correct environmental problems. Like any technology, noble purposes can be subverted. Some believe this misdirection is already happening. Possible applications include increased resistance to insects and pathogens, increased yields, and enhanced quality (better vitamin or protein composition and amount). Beneficial microorganisms in certain plants might be altered to work with all plants.

Certainly, many of these desirable characteristics have been accomplished with conventional plant breeding. Genetic engineering offers more with less fuss. Some limitations imposed by conventional plant breeding are removed. You can move genes between plants that are not sexually compatible and you can get to the desired product directly with gene transfer, rather than randomly and through much selection to find the desired plant. You can also achieve microadjustments not possible with conventional techniques, such as the tweaking of an individual protein's composition.

Two basic tools are needed to accomplish these goals. One is a method to move genes from one place to another (DNA insertion). The second is the ability to take an altered cell with inserted and functional DNA and convert it through micropropagation into a whole plant. Several methods exist for DNA insertion into plant cells.

The best choice is usually *Agrobacterium tumefaciens*, the bacterium that causes crown gall disease (tumors). Strains of *A. tumefaciens* used in transfers are disarmed by removing the genes responsible for tumor formation. In their place, one substitutes the gene or genes destined for transfer. The bacterium acts normally, that is, transfer occurs, but the substituted genes are transferred in the place of tumor-causing genes. In addition, the bacterial vector contains marker or reporter genes. These genes are necessary so that successful transfers can be identified. Marker genes usually confer antibiotic resistance to the target cell, such that successfully transformed cells survive exposure to the antibiotic. Reporter genes do not involve resistance. Instead, these genes code for some product that can be readily determined by assay. After DNA insertion, the target cell can then be micropropagated into a whole plant with altered DNA. The limitation is that this bacterium attacks dicots, but not monocots. Another factor is that successful transfer varies considerably by species, the organelle source of the target cells, physiological state of the donor, differences in bacterial strains, and even techniques and skills involved in transfer and plant regeneration.

Ballistic transfer employs microscopic tungsten or gold projectiles coated with DNA. The projectiles are shot through the cell walls into the cellular cytoplasm. Some of the DNA will be taken up by nuclear and/or organelle genome. The system can be used with monocots, such as corn, and conifers, for which *A. tumefaciens* does not work. Its limitation in the successful transformation of cells is 10% or less.

Direct transfer of DNA utilizes plant protoplasts prepared by enzymic digestion of the cell wall. DNA is delivered via bacterial plasmids that have been modified with added target and marker/reporter genes. DNA can also be encapsulated in liposome vesicles for an alternate insertion method. The plasma membrane around the protoplast becomes permeable in the presence of poly-1-ornithine or PEG, allowing the entry of altered plasmids. Liposome vesicles release their contents into the cytoplasm after fusion with the plasma membrane. The problem with this method is that regenerating plants from protoplasts is difficult. Protoplasts are also needed for electroporation, which suffers the same limitation as direct transfer. The genes are attached to gold beads suspended in solution along with the plant protoplasts. An electrical discharge drives the beads of DNA into the protoplasts. This method has also been used with embryos that have undergone partial cell wall digestion.

Microinjection employs a microscopic, needle-like, glass capillary tube with a diameter of 1 to 2 microns. This tube is guided under a microscope with micromanipulators so that it pierces the target protoplast or cell. The plasmid carrying the DNA is injected into the cytoplasm or nucleus. Meristematic plant cells have been used with limited success. This method is tedious, requires considerable skill, and has an extremely low success level. Microscopic needle-like fibers of silicon carbide (silicon whiskers) can be added to suspensions of plant cells or protoplasts. Plasmids carrying desirable genes are also present in the solution. The entire mix is shaken, some of the "whiskers" penetrate the cells or protoplasts, and the plasmids enter. This method is very low technology and has very limited success.

Plant viral vectors are an ideal alternative to bacterial vectors. Viruses invade plant tissues and co-opt the plant's DNA by reprogramming with viral DNA; these cells then churn out copies of the virus. Viruses are nature's genetic engineers. The idea is to use a viral vector whose pathogenic genes are disarmed and replaced by target DNA. A few successes have been reported. This system works with both monocot and dicot crops.

The newest alternative technology to alter plants is called chimeraplasty. This technique is a form of molecular surgery that can be used either to create or repair a mutation in a specific gene. The "molecular knife" is known as a chimera; it consists of a hybrid molecule formed from RNA and DNA. These segments of RNA and DNA are short and chemically synthesized. The DNA matches a segment of the target gene, except for one base that differs. If the target gene is normal, the base change will result in a mutation. If the gene has a mutation, the correct base can be included in the DNA segment, thus restoring normal function to the gene. The RNA is a perfect mirror (compliment) of the targeted gene's DNA.

The chimera works because cells have DNA repair mechanisms. The chimera pairs up with the target gene and fools the cell's DNA repair machinery into replacing an original base with one supplied by the chimera. The chimera can be delivered into plant cells with either ballistic transfer or electroporation. This technology is in early stages, but appears to have promise. It offers the advantage of not introducing permanent foreign DNA into an organism. The cell degrades chimeras after their purpose is accomplished. Opponents to gene transfer may view this technology with less alarm. Chimeras also have less impact on the genetic machinery. One base change does not alter physical locations or regulatory aspects.

Numerous applications exist in plant biotechnology for horticultural crops. Some are in use, others are being field-tested, while others are being researched for the future.

Frost damage limits plant productivity. Citrus fruit growers in Florida burn fuel oil and use air circulation with wind machines to protect blossoms and late ripening fruit or use sprinklers. These procedures are energy intensive and some air pollution results from the burning of fuel. An alternative involves using genetically engineered bacteria to displace normal bacteria that serve as ice-nucleation centers for frost. To date, this product has not been a commercial success, but offers a reasonable alternative to energy-intensive forms of frost protection.

Bacillus thuringiensis (Bt) has been known since the early 1900s to produce a substance toxic to caterpillars and used since the 1940s as a natural insecticide. The bacterium produces a potentially fatal crystalline protein (protoxin) that upon ingestion becomes activated in the insect's gut through action by a protease that releases a toxic fragment. As such, Bt evoked the interest of genetic engineers early on. The ideal solution would be to transfer the gene either to common, plant-associated bacteria or directly to the plant. The genes for the protoxin were identified (*crylA* genes) and soon spliced into the vector *A. tumefaciens*. Field tests determined that cotton, corn, and potatoes could be protected against various caterpillar pests. Crops have appeared in farmers' fields with the added protection conferred by synthetic Bt genes.

Glyphosate, usually sold as Roundup™, is a biodegradable, broad-spectrum herbicide widely used by farmers and gardeners. Because it kills any green plant, it cannot be used once a crop is up. However, once genes that conferred resistance to glyphosate were found in *E. coli* and also a plant mutation, it became possible through gene transfer to make crops resistant to glyphosate. The usage of glyphosate went from only preemergence application to use throughout the growing season. Most of these herbicide-resistant plants are not horticultural crops, but other forms of herbicide resistance has been added to a few horticultural crops. Chimeraplasty and exposing cell cultures to mutagens in the presence of a herbicide have also been used to confer herbicide resistance.

Supermarket tomatoes were an early target for the genetic engineers. The problem with tomatoes is that ripe tomatoes ship poorly and have a short shelf life. Consequently, tomatoes are picked green and have been bred for hardness, both for mechanical harvest and shipping. An alternative is to pick tomatoes at a later stage, but to alter the genetic controls on ripening so that shelf life and taste are improved. One approach is to interfere with polygalacturonase, an enzyme that causes cell wall softening during ripening. Another method is to interfere with ethylene production, so that hormonal ripening is delayed. The enzyme ACC deaminase can degrade a key intermediate of ethylene synthesis. Neither of these genetically altered tomatoes has captured the market to any significant degree.

Flowers are at least a $1 billion annual U.S. industry. Special efforts are directed toward improving flower appearance and extending cut flower life. One way genetic engineering could help is to transfer genes for more uniquely colored flowers, such as a true-blue color. Another way would be to alter pathways involved in post-harvest deterioration, such as the introduction of anti-sense genes that inactivate key enzymes involved in post-harvest degradation of cut flowers. A number of interestingly colored petunias have resulted with this approach.

Traditional plant breeding has improved the nutritional quality of crops. Goals are the production of a complete plant protein and higher protein and/or vitamin contents. A more complete protein can be achieved in two fashions through transgene technology. A gene coding for a protein rich in the desired amino acid can be added to the crop. Another method is to alter the pathway to the desired amino acid. No plant protein contains an adequate amount of all eight amino acids needed in the diet. Some progress has been made. The problem is that considerable testing is needed to ensure that other changes or breakdown products detrimental to humans have not been introduced. The enrichment of proteins in horticultural crops for human consumption is not far off.

Major advances have been made in alteration of oil quality, especially in soybeans. Gene transfer efforts are also under way to improve palm oils, especially oleic and stearic acid content. Gene transfer for enhanced vitamin, mineral, and other healthy phytochemicals is also in progress. This area is now called nutritional genomics, a term used to describe genetic engineering where human nutrition, plant biochemistry, and gene transfers are involved. The motivation is to produce healthier foods, especially for geographical areas dependent on one or two plants as dietary mainstays. The complexity of biochemical pathways for vitamins, minerals, and phytochemicals is complex, but healthier vegetables and fruits are likely to appear in the near future.

Microorganisms can be used to control insects, weeds, and pathogens through two routes. Genes from microorganisms can be utilized in transgenic plants to offer enhanced protection. Another approach is to harness microorganisms with enhanced genetic properties to provide protection in the plant's immediate environment. Some beneficial microbes compete with pathogens, thus reducing or displacing their populations. Others produce antibiotics that help control pathogens. Still others can act as parasites on pathogens. These beneficial organisms can operate in soil, although some can be applied aerially to control airborne diseases. Biocontrols appear to offer promise as an alternative to fungicides, bactericides, and herbicides. Several successes have been demonstrated.

The transfer of nitrogen fixation properties from leguminous crops to other crop plants is the most sought-after goal of genetic engineering and of proponents of sustainable farming. The genetic complexities are extreme. Even with the incorporation of nitrogen-fixing genes, today's cultivars with their high nitrogen needs might not be supplied with enough nitrogen. Levels might be increased by inserting multiple copies of the nitrogen-fixing genes, or we might have to settle for adding small amounts of nitrogen fertilizer (but much less than with plants having no nitrogen-fixing genes). The two bacteria likely to form the basis of future attempts to genetically engineer bacteria capable of nitrogen fixation with all crops are *Bradyrhizobium* and *Rhizobium*. Success is many years away.

Certain horticultural crops are pathogen prone. While seeds can be surface decontaminated, vegetative propagules can harbor internal pathogens. The best possible method of producing pathogen-free, especially virus-free, propagating stock for horticultural crops is to propagate from shoot apices and to raise plants through tissue culture. Plants produced in this manner can be officially certified. Such plants command higher prices and offer increased yields. Pathogens and viruses have been eliminated in numerous flower, fruit, and vegetable crops with this approach.

Plant viruses cause considerable damage to crops, and effective chemical treatments are few. The standard approach has been to rely on introducing virus resistance by conventional plant breeding techniques and by destruction of virus-infected plants. The current genetic engineering methods involve a form of vaccination and the anti-sense approach to inactivate production of critical viral components. Adding genes to plants that code for viral coat proteins has proven to be the best current technology. When the plant expresses these genes and produces viral coat proteins, it appears to block the ability of that particular virus to infect the plant. The method by which this occurs is not yet well understood. Tolerance or resistance to a number of viruses in horticultural crops has been achieved.

Conventional plant breeding is limited by sexual compatibility. The cell walls of somatic plant cells can be removed with cellulase to produce protoplasts. Protoplasts, say, from tomatoes and potatoes, when exposed to PEG or electrofusion can fuse. The resulting fused cell can be micropropagated into a plant that shows potato and tomato characteristics. These types of fusion overcome sexual compatibility. The fusion of potato and tomato has been done, but the resulting hybrid has yet to reach commercialization. Little progress has been made commercially in this entire area.

Another advantage of this form of fusion is that cytoplasm merges from both partners, whereas in conventional breeding, only the female contributes cytoplasm. This method can be used to transfer extrachromosomally inherited characters such as cytoplasmic male sterility. This characteristic is especially important for hybrid seed production, such as in corn. Using somatic cell fusion, cytoplasmic male sterility has been transferred to *Brassica* sp. and sugar beets. Much work remains to be done in this area.

Germplasm preservation protects genetic diversity and maintains gene pools for future needs. Seed storage is the method of choice. Maintaining living collections in facilities devoted to their preservation preserves crops not readily propagated by seed, such as potato or apple. The advent of tissue culture and micropropagation now offers an alternative. Tissue samples that can be readily micropropagated can be preserved through storage in liquid nitrogen (cryopreservation). Frozen samples can be thawed and micropropagated, thus making the original plants available. While not routinely employed, cryopreservation is likely to increase in the future.

Currently, the production of pharmaceuticals is conducted with recombinant bacteria. These cultures are maintained in large fermenters, harvested, and separated from the culture medium; then the product is separated and purified. These systems are expensive and complicated, needing sophisticated equipment and highly skilled personnel. Plants could be genetically programmed to produce these same pharmaceutical products. Growing plants is easy and considerable biomass can be produced. The downside is the difficulty of expressing sufficient product so that it can be easily isolated and purified, given the large number of metabolites produced in plants. Of course, costs need to be similar or less than comparable microbial systems. Some pharmaceutical and antibody products are currently expressed in plants. The term *plantibodies* has been coined for plant-produced antibodies. Extensive research by numerous companies is under way in these areas. Plants, with altered genes, will be programmed to produce enzymes, proteins and other compounds for human and animal medicines, vitamins, skin care products, food additives, and other products. In the future, your fruits and vegetables might supply you with vaccines, all vitamins, and even prescribed medications!

Some problems have occurred with biotechnology. Crop failures have resulted with cotton genetically engineered with Bt genes and herbicide-resistant genes. Some research studies have indicated that a possible linkage between Bt crops and immunological and organ damage exists with humans, based on studies with rats. Lab studies have also shown that pollen from Bt corn contains enough Bt toxin to possibly kill the nearby caterpillars of Monarch butterflies. Resistance toward field trials and consumption of genetically engineered crops has surfaced in Europe and Asia. The Center for Food Safety in the United States has sued the FDA to force the removal of genetically engineered foods from the marketplace. The suit stems from the lack of labels, such that the consumer cannot identify genetically engineered foods, and insufficient safety testing.

Although the majority of the current plantings of genetically engineered crops are not horticultural crops, the horticultural sector can expect major inroads. At present, less than 5% of potatoes are genetically engineered with Bt, and small inroads have been made with tomatoes. The question of whether sustainable horticulture has a place for genetically engineered crops remains open, as does whether the consumer has a right to know, and whether more safety/environmental regulations are forthcoming. At least one part of the sustainable horticultural community has no genetically engineered foods in its future. New organic standards from the USDA specify no use of bioengineered crops.

The question has been answered in part for the sustainable horticulture community. Organic practitioners want no part of genetically engineered crops. Certainly, some aspects of genetically engineered crops are likely to be acceptable to the nonorganic part of the sustainable horticulture community. Some examples include crops with gene-added insect and disease resistance, where the genetic change has been proven safe for humans and the environment, or just safe for the environment with ornamental crops or with edible crops when the gene alteration does not appear in the edible part. Another example is the addition of genes that give crops the ability to fix nitrogen directly from the air, thus decreasing the need for fertilizer. Clearly, these examples are sustainable, in that pesticide and fertilizer usage is reduced.

Unfortunately, current applications are not sustainable, such as the increased usage of herbicide-resistant crops. While only agricultural crops are involved, the potential for herbicide-resistant horticultural crops is open. Recent figures from the USDA show that glyphosate usage on soybeans rose 72% once herbicide resistance was introduced. In 1998, 71% of genetically engineered crops patented were designed to be resistant to herbicides such as glyphosate. Even of more concern is recent research from Sweden that shows a link between glyphosate and non-Hodgkin's lymphoma, a form of cancer. This cancer has also been reported as increasing 80% since the 1970s. Increased allowed tolerances of glyphosate on soybeans have also been recently permitted. Herbicide resistance clearly is not a desirable trait for sustainable horticulture. Given the increased profits for pesticide manufacturers with genetic engineering operations, such products might be offered to horticultural producers in the near future. The sustainable horticulture community should just say no!

EXERCISES

Mastery of the following questions shows that you have successfully understood the material in Chapter 9.

1. What is the difference between old forms of biotechnology and the current applications?
2. What are the advantages of recombinant DNA technology?
3. What are DNA libraries? How are these DNA libraries made? How do we know that the library actually contains fragments of DNA?
4. How many types of DNA libraries are there? How do they differ? Why do we need different types?
5. How do we make multiple copies of DNA fragments? Why would we need multiple copies?
6. What are probes? What does hybridization mean in reference to probes? What are probes used for?
7. Why do we make copies of DNA libraries prior to using probes? Why do we use primers?
8. What is *primer walking*?
9. What are probes and what can probes tell us?
10. What are the different types of probes and what do they screen for in DNA libraries?
11. How do the three screening techniques of Northern, Southern, and Western blotting differ? What advantage do they offer as a screening technique?
12. What is hybridization? What are the basic steps in hybridization?
13. What is polyploidy? How can polyploidy be induced? Why is it important in horticulture?
14. What is a transposable element? What do mutations and transposons have to do with screening? What is transposon tagging?
15. What is the difference between a gene linkage map and a physical map? How do RFLPs relate to gene linkage maps? Which plants have the most complete physical maps?
16. Gene transfers into plants have desirable goals. What are they?
17. What is chimeraplasty?
18. Many possible applications or products can be developed with plant biotechnology. What are they and what is the basic premise with each? Which ones have shown some level of success? Which ones appear to be successful further in the future?
19. What are some of the downsides of biotechnology from the standpoint of horticultural sustainability?

INTERNET RESOURCES

Arabidopsis information resource

http://www.arabidopsis.org/home.html

Breeding Distrust: An Assessment and Recommendations for Improving the Regulation of Plant Derived Genetically Modified Foods

http://www.biotech-info.net/Breeding_Distrust.html

How genetic engineering differs from conventional breeding, hybridization, wide crosses, and horizontal gene transfer

http://www.biotech-info.net/wide_crosses.html

International Plant Genetics Resources Institute

http://www.ipgri.cgiar.org/

National Center for Biotechnology Information; allows search of GenBank for plant DNA sequences

http://www.ncbi.nlm.nih.gov/

Plant genome data and information center
http://www.nal.usda.gov/pgdic/

Transgenic crops: An Environmental Assessment
http://www.winrock.org/transgenic.pdf

RELEVANT REFERENCES

Academy of Natural Sciences (2000). *Genetically Modified Food: The Science and the Controversy.* Environmental Associates of the Academy of Natural Sciences of Philadelphia, PA. 5 pp.

Alcamo, I. Edward (2000). *DNA Technology: The Awesome Skill* (2nd ed.). William C. Brown, Dubuque, IA. 348 pp.

Altman, Arie (ed.) (1998). *Agricultural Biotechnology.* Marcel Dekker, New York. 770 pp.

Brush, Stephen B. (2000). *Genes in the Field: On-Farm Conservation of Crop Diversity.* Lewis Publishers, Boca Raton, FL. 288 pp.

Chawla, H. S. (2000). *Introduction to Plant Biotechnology.* Science Publishers, Inc., Enfield, NH.

Chopra, V. L., V. S. Malik, and S. R. Bhat (eds.) (1999). *Applied Plant Biotechnology.* Science Publishers, Inc., Enfield, NH.

Christou, Paul (1996). *Particle Bombardment for Genetic Engineering of Plants.* Academic Press, San Diego. 173 pp.

Geneve, R. L., J. E. Preece, and S. A. Merkle (eds.) (1997). *Biotechnology of Ornamental Plants* (Biotechnology in Agriculture Series, vol. 16). CABI Publishing.

Hammond, J., Peter McGarvey, and Vidadi Yusibov (eds.) (1999). *Plant Biotechnology: New Products and Applications.* Springer Verlag, New York.

Khachatourians, G. G., A. McHughen, R. Scorza, W. Nip, and Y. Hui (eds.) (2001). *Transgenic Plants and Crops.* Marcel Dekker, New York. 888 pp.

Marvier, Michelle (2001). "Ecology of Transgenic Crops." *American Scientist* 89 (March–April): 160–167.

McHughen, Alan (2000). *Pandora's Picnic Basket: The Potential and Hazards of Genetically Modified Foods.* Oxford University Press, New York. 277 pp.

National Research Council (2000). *Genetically Modified Pest-Protected Plants: Science and Regulation.* National Academy Press, Washington, DC.

Oksman-Caldentey, Kirsi-Marja, and Wolfgang Barz (2002). *Plant Biotechnology and Transgenic Plants.* Marcel Dekker, New York. 720 pp.

Paterson, Andrew H. (1996). *Genome Mapping in Plants.* Academic Press, San Diego. 330 pp.

Rissler, Jane, and Margaret Mellon (1996). *The Ecological Risks of Engineered Crops.* The MIT Press, Cambridge, MA.

Taji, A., P. Kumar, and P. Lakshmanan (2001). *In Vitro Plant Breeding.* Food Products Press, Binghamton, NY. 168 pp.

Tokar, B. (ed.) (2000). *Redesigning Life? The Worldwide Challenge to Genetic Engineering.* Zed Books, London.

VIDEOS AND CD-ROMS[*]

Introduction to Biotechnology (Video, 2001).
Patent on Life (Video, 1998).
Understanding Biotechnology (CD-ROM, 2001).

[*]Available from Insight Media, NY (www.insight-media.com/IMhome.htm)

Chapter 10

Energy and the Food System

OVERVIEW

The horticultural industry provides a large and varied array of fresh and processed vegetables, fruits, and nuts for our enjoyment. A similar situation exists for ornamentals. This abundance comes at a price—a considerable usage of energy to reduce labor—to provide chemical inputs for production and to provide processed food in markets and restaurants, and ornamentals at florists and nurseries. Of the total energy utilized in the United States, the food industry consumes about 17%. When we look at a per capita basis, we use three times as much energy for food production in the United States than is used for all energy-consuming activities in developing countries. This usage is a major factor in our consumption of 25% of the world's resources, even though we have only 5% of its population.

A number of reasons exist as to why we should be concerned about energy consumption in sustainable horticulture. One is clearly a moral issue that is beyond the scope of this book. Other reasons have to do with resources and sustainability. Much of the energy consumed in horticultural production and subsequent processing and distribution of products derives from fossil fuel energy, such as oil, natural gas, and coal. These materials are not renewable resources, given the geological conditions and periods of time needed to produce them in nature. Based on current consumption, we will run out of economically accessible oil and natural gas at some future point (see later discussion about oil dependency). Coal will last for another 100 years or more beyond that time.

Granted, any projected dates are educated guesses and new technology or discoveries might extend the lifetimes of fossil fuels. New technology might bring about much more efficient usage of fossil fuels. It might also lead to new ways to discover unknown reserves or economical methods to extract fuels from oil shales and sands or from the deep ocean floors. Unknown variables also cloud the picture. Energy demands from rising Second and Third World nations might exhaust energy supplies faster than predicted. Still, an end to these fuels is inevitable.

Two approaches toward energy sustainability exist. One is to buy time with conservation. Using more energy-efficient farm machinery, switching to conservation tillage, and using fewer inputs can postpone the day of reckoning. Energy conservation in horticulture has been successful. Later in this chapter we look at conservation and examine how it worked with Florida vegetable production. Another, longer range effort is to switch to renewable and alternative sources, such as solar energy, wind power, geothermal power, hydroelectric power, and biomass fuels.

CHAPTER CONTENTS

OBJECTIVES

- To learn why we need to move toward renewable energy sources for horticultural production
- To understand the role of direct and indirect energy inputs into horticulture
- To learn the effects of energy conservation in horticulture
- To understand the patterns of energy usage in horticulture and how we can reduce usage
- To learn about food energy returns as compared to energy inputs
- To understand the patterns of energy usage of crops from the farm to the consumer
- To learn about strategies to reduce our dependence on fossil fuels and their replacement by alternative forms of renewable energy

ENERGY CONSUMPTION IN PRODUCTION HORTICULTURE

Energy use for crop production has been tracked since the 1970s, primarily in response to the OPEC oil embargo of that decade. Tracking has enabled the spotting of trends and excesses and has clearly helped with energy conservation. Today, we rely on computer models, such as the Florida Agricultural Energy Consumption Model (FAECM, developed by the University of Florida), to determine energy consumption for various crops. Energy budgets, from the view of sustainability, must take into consideration primary, indirect, direct, on-farm, and after-the-farm energy consumption.

Specific Types of Energy Consumption

Energy consumption in horticultural operations occurs on several levels. Primary energy is the sum total of all energy used on the farm. It includes both direct and indirect energy inputs. Farming activities require fuels (diesel, gasoline, LP gas, natural gas), lubricants, and electricity. These can be thought of as direct energy inputs. Much of the direct energy input is used for labor performed by machines.

Energy is also needed to produce these direct forms of primary energy through drilling activity, refining, power plant construction, transportation, and other processes before the direct energy input is used on the farm. This production energy is described as indirect energy inputs. Nonenergy forms of farm inputs, such as fertilizers, pesticides, and machinery, also consume energy in the production and delivery phase prior to use on the farm. These farm inputs utilize indirect energy, too. For example, 1 pound of nitrogen fertilizer represents the input of 33,000 Btus of indirect energy. Another way of thinking is that it requires about 1 gallon of number 2 fuel oil to make 4 pounds of nitrogen fertilizer, not including the energy used to construct the factory.

Sustainable horticulture places greater emphasis on the direct energy inputs that go into crop production, an area where farmers have more control over applications and practices, than on the indirect inputs. Indirect energy inputs and postproduction energy usage are controlled more by the business and manufacturing sectors than the farmers. These areas need to become sustainable, too, if we are to become a sustainable society.

Overall Energy Conservation in Production Horticulture

How have we fared with energy consumption in horticulture? Has conservation worked at all? The FAECM computer model of energy usage in Florida provides some answers. Primary energy needs in Florida production agriculture declined by about 15% from the oil embargo in the 1970s to 1990. Much of the agricultural activity in Florida is involved in horticultural production, so this decline is probably a reasonable value for this sector. Direct energy input values on the farm have held relatively constant during this time period.

Therefore, because primary energy inputs consist of indirect and direct energy, the decline is mostly attributable to a decrease in the use of indirect energy. Conservation has worked here because of increased energy efficiency during the manufacture of agricultural inputs, such as fertilizers, pesticides, and farm machinery.

During this same time period (1974–1990), total energy consumption for the entire state of Florida increased by about 66%. Most of this rise resulted from an increasing population. When we look at total primary energy in agricultural production during this period as a percentage of total state energy consumption, we observe a drop from about 7.8% to 3.9%, nearly a 50% decrease. Conservation has helped moderate crop production energy usage.

Again, most of this success is attributable to the manufacturers of agricultural production inputs. Granted, their motivation may be to cut energy costs in the face of rising energy prices, so that profit margins can be maintained and consumer sticker shock for farmers can be controlled. Even so, these industries have done a great job. This fact also lends considerable support to the need for a sustainable societal approach to energy consumption. Energy conservation by horticultural systems in the isolated sense looks good, but when viewed in the context of societal energy usage, we realize that much more remains to be done.

Energy Consumption of Horticultural Crops

Computer models take both indirect and direct energy inputs into consideration for crop production (Tables 10–1, 10–2, and 10–3). Crop production can be expressed on an energy-consumed basis per acre of apples or tomatoes or, in its preferred form, kilocalories per hectare (kcal/ha). One can also convert the crop yield into food energy and get some idea of energy in and energy out. The return on energy ratios shown in Table 10–4 give you some idea of energy efficiency for the production of various crops. One can also compare the ratio of energy costs to crop production value. All of this energy is essentially crop production energy that does not take into consideration energy costs beyond the farm gate. Energy needs for crops after production include inputs for transportation, processing, packing, distribution, marketing, and even food preparation in homes, restaurants, and institutions. This energy consumption can be thought of as postharvest or postproduction energy.

TABLE 10–1   *Energy Inputs in Fruit and Vegetable Production*

	Apple (NY)	Orange (FL)	Brussels Sprouts (US)	Potato (NY)	Spinach (US)	Tomato (CA)
			Inputs (kcal/ha)			
Labor	179,000	98,000	27,900	16,275	26,040	76,725
Machinery	1,029,000	432,000	480,000	480,000	480,000	480,000
Fuel	19,120,000	2,006,000	2,881,065	4,484,576	2,970,000	9,156,296
Seeds	—	—	16,120	1,088,700	135,000	20,000
Nitrogen	662,000	3,308,000	2,646,000	3,013,500	6,909,000	2,469,600
Phosphorus	627,000	339,000	135,000	1,044,000	1,062,000	168,000
Potassium	231,000	360,000	64,000	316,800	217,600	153,600
Limestone	1,438,000	406,000	12,600	—	143,010	15,750
Irrigation	—	—	—	—	69,500	1,010,900
Pesticides	4,850,000	1,363,000	1,433,650	4,882,050	373,640	2,631,985
Electricity	57,000	—	300,000	14,561	300,000	200,000
Transportation	787,000	125,000	63,993	577,993	73,759	177,587
TOTAL	28,980,000	11,862,000	8,060,328	16,038,455	12,759,849	16,560,443

Source: Adapted from David and Marcia Pimentel (eds.), *Food, Energy and Society* (rev. ed.), University Press of Colorado, Niwot, 1996.

TABLE 10–2   *Energy Inputs in Florida Fruit and Vegetable Production*

Input	Grapefruits	Oranges	Peanuts	Bell Peppers	Tomatoes
			Energy Inputs (%) Into Crops		
Diesel for irrigation	13.5	18.8	5.6	1.2	3.2
Diesel for nonirrigation	37.7	32.0	25.3	20.1	23.4
Gasoline	5.8	8.9		1.7	2.0
Electricity for irrigation	1.2	1.7	18.7	1.0	3.9
Lubricants	1.8	3.1	1.3	2.7	2.4
Labor	6.1	11.2	3.1	24.2	13.7
Nitrogen	9.4	10.2	3.0	5.3	6.3
Potash	2.1	3.7	3.2	1.0	0.8
Phosphorus	0.4	0.8	2.1	1.3	1.3
Lime	0.2	0.2	2.2		1.3
Insecticides	5.6	3.9	2.5	1.2	1.0
Herbicides	4.4		1.6	0.1	0.2
Fungicides			3.5	1.9	3.2
Other pesticides				6.5	8.4
Other chemicals	0.2	0.2			
Other costs	10.4	5.1	28.0	31.7	28.9
Return on primary energy[*]	$30	$31	$43	$44	$78
Return on direct energy[†]	$60	$58	$134	$199	$263
% of primary agricultural energy in Florida	18.1	5.5	1.3	4.6	4.6
% of direct agricultural energy in Florida	8.6	30.3	1.4	4.3	4.3
% of cropped land	5.0	23.0	4.2	1.0	2.3

*Dollar return per million Btus of primary energy.
†Dollar return per million Btus of direct energy.
Source: Adapted from data of Richard C. Fluck, University of Florida, Institute of Food and Agricultural Sciences, http://edis.ifas.ufl.edu/MENU_AGCROP.

Fuel Usage in Mechanized Production Horticulture.   If we step into our time machine and return to the early 1600s, we see that our first agricultural activities in colonial America were dominated by human labor. More than 80% of farm work was provided by hand labor, with a small amount from draft animals. Moving ahead to the Revolutionary War and independence, we see that in 1776, draft animals provided nearly 80% of the farm power, with a small amount from hand labor. In effect, animals took the place of humans. Today, more than 90% of the labor on farms is done by machines and only a few percent by animals and humans. We have now traded off animal energy for farm machines. The price for less human and animal labor and increased productivity is high inputs of fossil fuels.

Looking again at Tables 10–1, 10–2, and 10–3, we can see that the production of fruits, vegetables, and ornamentals consumes a considerable amount of fossil fuel energy. Much of it goes to providing energy for production machinery, although in the more detailed Tables 10–2 and 10–3, we see that irrigation also consumes moderate amounts of fuel. Given the mechanized aspect of production, little possibility exists for reducing energy consumption. The importance of fuels to horticultural production becomes even more apparent when we look at some facts gleaned from Florida. The FAECM figures indicate that diesel and gasoline fuels account for roughly four-fifths of the direct energy consumption of Florida vegetable and fruit production.

How critical is the need to reach sustainability of energy sources for farm machinery? Fuel efficiency with farm machinery has likely reached technological limitations and conservation efforts have been maximized, so direct energy consumption remains constant.

TABLE 10–3   Energy Inputs in Florida Ornamental Production

	Energy Inputs (%) Into Crops			
Input	Field Nurseries	Woody Ornamentals (Containers)	Foliage Plants	Cut Greens and Flowers
Diesel for irrigation	3.1	4.4	0.4	1.3
Diesel for nonirrigation	23.3	19.4	21.4	19.8
Gasoline	8.8	7.5	2.1	
LP gas			7.9	
Electricity		8.3		
Electricity for irrigation	2.6	0.4	0.1	1.1
Electricity for nonirrigation	5.6		12.4	
Lubricants	2.2	1.9	0.5	1.0
Labor	19.7	20.1	17.0	34.1
Nitrogen	6.9	3.1	0.9	1.2
Potash	1.2	0.3	0.1	0.1
Phosphorus	1.6	0.4	0.1	0.1
Lime	0.4	0.1		0.2
Insecticides	0.5	0.7	0.4	0.4
Herbicides	0.7	1.0	0.5	0.1
Fungicides	0.1	0.2	0.1	1.3
Other pesticides	0.1	0.1	0.1	2.9
Other costs	23.1	32.2	36.0	36.3
Return on primary energy[*]	$78	$34	$43	$43
Return on direct energy[†]	$229	$112	$144	$220
% of primary agricultural energy in Florida	1.4	2.4	9.5	2.4
% of direct agricultural energy in Florida	1.5	2.4	9.4	1.5
% of cropped land	0.7	0.1	0.2	0.4

[*]Dollar return per million Btus of primary energy.
[†]Dollar return per million Btus of direct energy.
Source: Adapted from data of Richard C. Fluck, University of Florida, Institute of Food and Agricultural Sciences, http://edis.ifas.ufl.edu/MENU_AGCROP.

TABLE 10–4   Energy Outputs in Fruit and Vegetable Production

	Apple (NY)	Orange (FL)	Brussels Sprouts (US)	Potato (NY)	Spinach (US)	Tomato (CA)
	Outputs					
Yield (kg/ha)	54,743	40,370	12,320	12,320	11,200	49,616
kcal/ha (Food Energy)	30,656,080	19,781,000	5,544,000	19,702,032	2,912,000	9,923,200
Protein (kg/ha)	109	404	604	722	358	496
	Inputs[*]					
kcal/ha	28,980,000	11,862,000	8,060,328	16,038,455	12,759,849	16,560,443
	Return on Energy					
kcal output/kcal input	1.1:1	1.7:1	0.69:1	1.23:1	0.23:1	0.60:1

[*]From Table 10–1.
Source: Adapted from David and Marcia Pimentel (eds.), Food, Energy and Society (rev. ed.), University Press of Colorado, Niwot, 1996.

This energy source is not sustainable. The need to reduce fuel consumption and to move toward renewable energy sources is critical, given the high fuel consumption, our need to maintain food supplies, and our dependence on diminishing global supplies of fossil fuels. One solution, a return to an agrarian society where most of the population farms use animal labor, is not an attractive one. The only foreseeable alternative is renewable alternative fuels, which are discussed later in this chapter.

Fuel Usage in Irrigated Production Horticulture.   Considerable fossil fuel energy, not to mention water, is consumed during the irrigation of crops. These activities are certainly critical to the production of horticultural crops in the arid Southwest, and even in wetter areas, such as for citrus production in Florida and California. Decreased irrigation or more efficient irrigation is needed from an energy standpoint. The usual impetus is more likely to be economical.

Irrigation costs are often government subsidized, in that taxpayers foot the bill for dams, waterways, aqueducts, pipelines, and pumping associated with the movement of large volumes of water, such as moving water from the Colorado River to California. If these indirect energy, construction, and maintenance costs were factored in, horticultural producers in arid regions could not compete nearly as well with growers in wetter areas. Even with subsidized water, direct costs will eventually become prohibitive, caused by the need to pump groundwater for the production of crops such as citrus, grapes, and avocados from ever increasing deeper depths. Some low-value irrigated crops are now less likely to be produced in the arid Southwest. Reductions in irrigation are also likelier as groundwater and surface water supplies become dangerously depleted and city dwellers clash with irrigation farmers over limited water supplies.

More efficient irrigation systems can be utilized (see Chapter 11), leading to more efficient water usage and reduced energy costs. The best systems for efficiency are trickle or drip irrigation systems. However, initial costs, in terms of energy and capital expenditures, are high for such systems. Trickle irrigation is generally only used for high-value crops and in areas where only deep pumping of groundwater is possible.

An irrigation study in Florida with tomatoes showed that switching from subsurface irrigation (ditch system) to micro-irrigation could result in considerable energy savings. The latter uses a system of pipes to apply water directly at the root zone. This system could result in energy savings of 70,000 kWh per acre, and could reduce water consumption by up to 80%. This system needs about 50% less nitrogen fertilizer to produce comparable yields to crops with subsurface irrigation. The reduction in fertilizer, if practiced statewide, could save 761,000 gallons of gasoline or some 95 billion Btus annually. Initial costs are high, but operating costs are lower. If energy prices climb, these systems could be economically viable with a short cost payback period. Such systems are more likely to be adopted as water consumption and pollution problems increase. If the true costs of irrigation and water pollution were borne by growers, these systems would be rapidly adopted.

Fertilizer and Energy in Production Horticulture.   The next highest energy input usually involves nitrogen fertilizer. With a few crops, nitrogen use places third behind pesticide energy use. Apples and potatoes are major examples of crops for which pesticide energy use is higher than that used in nitrogen (Tables 10–1, 10–2, and 10–3). Management practices to reduce fertilizer input can lead to energy reductions. However, such practices are usually undertaken in response to reducing off-site water pollution. These practices include better soil tests, more precise applications of fertilizers, more attention to timing of applications, and the use of legumes as cover crops and green manures. These practices result in reduced fertilizer use and lower losses from leaching and volatilization. Practices to reduce fertilizer applications to manage pollution are covered in detail in Chapter 11.

Organic fertilizers such as manure are an alternative to chemical nitrogen fertilizer. Manure is much less energy intensive, is renewable, offers the additional benefits of more organic matter, and results in less pollution when properly managed. While this solution is ideal for a mixed crop/livestock organic farm, it is not suitable as a universal alternative. Supplies are limited in most areas, nutrient levels can be insufficient for certain crops, and transportation costs become unacceptable for sites not close to feedlot manure sources. Manure also loses nutrients during storage, so prompt usage is important for maximizing nutrient value. Manures are discussed further in Chapter 11.

Future advances may come if genetic engineers are able to produce nitrogen-fixing bacteria tailored for nonleguminous crops. Such genetically modified bacteria could greatly reduce dependency on chemical nitrogen sources and result in less energy usage and less water pollution. Nitrogen inputs will be supplied from atmospheric nitrogen, essentially an endless supply. This topic is covered in Chapter 9.

Pesticides and Energy in Production Horticulture.   Pesticides are energy-intensive products, and they consume petroleum feedstocks that could be better used for other applications. Pesticides vary from number three to number two in terms of energy consumption during crop production in horticulture. While the primary driving force for pesticide reduction is environmental and health based, moving in this direction also helps reduce energy consumption. A move from pesticides to integrated pest management (IPM) or, even better, biological control could result in energy savings. The reduction for biological control would be higher than for IPM, because the latter still involves some dependency on pesticides. Biological control will be considerably improved as the genetic engineers move new genes into plants that enhance tolerance and resistance to insects and diseases.

The success experienced in switching to IPM or total biological control varies. Economics and level of difficulty motivate switches. Major horticultural crops offer more motivation than minor ones in terms of economics. The more pests there are, the more difficult it becomes to create a successful IPM or biological control program. A switch for tomatoes is more feasible than for potatoes and even more so than for apples. Given that about 400 insects and diseases attack apples, a working IPM system or complete method of biological control for apples is extremely difficult. Methods of biological control and IPM are covered in detail in Chapter 15.

The above three targets—fuel, nitrogen fertilizer, and pesticides—account for the majority of energy expenditures in production horticulture. Controlling these three parameters, admittedly for environmental and energy considerations, should be the major task for those interested in achieving a sustainable horticulture. For the near future, the best hope is basically a conservation approach. Reaching a sustainable level through the use of renewable fuels, inexhaustible nitrogen supplies, and plant resistance to insects and diseases depends on technological advances in alternative fuels, genetically engineered microbes for agricultural nitrogen fixation, and genetic modifications to induce high levels of plant resistance.

Energy Returns with Horticultural Crops.   The return on energy invested in horticultural crops can be looked at from two perspectives. One is the trade-off between energy inputs versus the amount of food energy offered by a particular crop. Another approach is to consider economics, or how much is earned for each dollar invested in energy for a crop. In the event of energy shortages or rapidly increasing energy prices and consumer resistance to price increases, certain crops might become candidates for reduced production. The return on dollars spent for energy would be helpful in decision making about the worth of certain crops. Farmers and corporations motivated by sustainability might also find this information useful. Of course, such information must be considered in context with other factors, such as food value (protein, vitamins, and minerals), acreage devoted to the crop, and consumer loyalty.

The FAECM computer model from Florida yields useful comparative data. The comparison is between the dollar value of the crop versus the dollar value of the energy used to produce the crop. These values are indicated in Tables 10–2 and 10–3. These values are better viewed in the context of averages. In Florida, the average crop had a value of $136 when compared to the cost of 1 million Btus of production energy of direct energy. When viewed in the context of primary energy, the return is $44. As a group, vegetables gave the best returns, followed by ornamentals and then fruits. However, fruits offer food value, whereas ornamentals offer aesthetic value. From a strictly practical view, land devoted to ornamentals should be turned over to vegetables and fruits. However, the percentage of land devoted to ornamentals (Table 10–3) is quite low, so there is little potential for gain. On the other hand, foliage plants are the number two consumer of primary and direct energy in Florida's horticultural production. Another factor to consider is the social, aesthetic, and psychological value of landscape and indoor foliage plants.

Based on crop acreage and the return on primary and direct energy, the crop to target is the orange. Some 23% of Florida cropland is devoted to oranges and the returns of $58 and $31, direct and primary energy, respectively, are low compared to the state averages of $136 and $44. Oranges also consume more primary and direct energy than any other Florida crop. As energy prices rise, farmers could find it harder to justify oranges and to sell them at acceptable market prices. Oranges are high in vitamin C, which might help to counter costs. Still, potatoes, because of their high consumption in the American diet, provide about twice as much vitamin C, even though their tissue levels are about 50% of that in oranges. The value of taste and fruit fiber for health reasons might counter a switch to only potatoes. Another factor is that Florida cropland in orange production might not be suitable for potato production, leading to difficult choices for replacement crops. The entire choice might be decided by transportation costs as energy prices rise. Oranges could become a local staple in Florida and a luxury fruit in the North. Tough choices! It would be so much better to develop competitive alternative fuels and keep the horticultural production system intact.

Based strictly on energy efficiency, certain vegetables don't fare nearly as well as fruits (Table 10–4). For example, spinach, tomatoes, and brussels sprouts provide dietary energy equivalent to only 23%, 60%, and 69%, respectively, of the energy provided during their production. Apples, potatoes and oranges offer 10%, 23%, and 70%, respectively, more dietary energy than the energy used for their production. The latter gives another reason not to abandon oranges to potatoes based strictly on energy costs as discussed above.

Greenhouse Energy Usage

Fossil fuel usage by greenhouses is considerable. In northeast Ohio, an acre of greenhouse in 1 year would consume 100,000 gallons (15,336 decaliters/ha) or 14 million ft^3/acre (160,567 m^3/ha) of number 2 fuel oil and natural gas, respectively. This amount of fuel would maintain night temperatures between 55° and 65°F (12.8° and 18.3°C). As one moves south, fossil fuel needs decrease. However, even in warmer climates, greenhouse production requires considerable energy. Coverage of technologies and practices to reduce greenhouse energy inputs is found later in Chapter 13.

In central Florida, the production of greenhouse vegetables and other greenhouse crops (bedding and foliage plants) to harvest requires about 1.25 billion Btus of direct energy per acre. Number 2 fuel oil yields 100,000 Btus for each gallon at a boiler efficiency of 80%. The conversion would yield a consumption of 12,500 gallons in central Florida as opposed to 100,000 in northeast Ohio. Moving further to southern Florida, the need for direct energy per acre with foliage plants drops to 0.17 billion Btus. This difference results in large part from the transition from greenhouse production to field production under shade cloth. Another perspective is that orange production in Florida needs about 21 million Btus of direct energy per acre, as opposed to 1.25 billion Btus of direct energy per acre for greenhouse vegetables. Oranges, as we saw earlier, are not the most energy-efficient crops. If we look at an acre of the average crop in Florida, 13.3 million Btus of direct energy are required. Greenhouse production is hard to justify from an energy perspective, even in warmer climates.

ENERGY CONSUMPTION IN POSTPRODUCTION HORTICULTURE

While fresh vegetables and fruits are good choices in terms of taste and optimal levels of vitamins, minerals, and various beneficial phytochemicals, a need does exist for processed products. Home cooking, for example, can improve the texture and digestibility of some vegetables. Whereas I enjoy a tomato and lettuce in my salads, raw rutabaga does not excite me. Once it is cooked and seasoned, I enjoy it much more. Commercially processed foods travel and store well over far longer periods than fresh produce, overcome seasonal limitations of fresh produce, enable us to utilize surplus production, and also offer convenience. For example, while I can make a nice tomato sauce, it does require time. It is much easier to open a jar of good-quality tomato sauce.

Commercial processing requires energy, because heating to certain temperatures is necessary to destroy microorganisms that would cause decomposition of food and human illness, even death. A considerable amount of transportation is required to get horticultural products to processors, then to wholesale distributors, then to retail markets, and finally to homes. Transportation will be treated as a combined activity later in the discussion that follows. Even after we get the processed food, additional energy is consumed during storage in refrigerators and freezers, and finally by cooking.

Vegetable and Fruit Processing

A number of methods exist for processing fruits and vegetables, including canning, drying, and freezing. The commercial canning of fruits and vegetables requires around 261 kcal/lb (575 kcal/kg) of produce. The manufacture of the can also consumes indirect energy, as covered later under packaging. Home canning is even more energy intensive and requires up to 30% more energy.

Fruits and vegetables can be frozen, producing a better tasting product and retaining higher levels of vitamins. Freezing requires considerably more energy, in that energy must not only be supplied as heat, but also for cooling and then freezing. The storage also consumes additional energy, because the frozen food requires freezer storage, unlike the pantry shelves used for canned goods. True, freezing foods does require less heat. Only enough is needed for the deactivation of enzymes that would alter the taste and color. Additional heat to kill microbes, unlike canning, is unnecessary in that the cold temperatures and lack of liquid water prevent microorganisms from spoiling the food. However, the additional energy needed for cooling and freezing results in a need for 825 kcal/lb (1,815 kcal/kg) of frozen food. Storage of the frozen food in supermarket and home freezers consumes 120 kcal/lb (265 kcal/kg) and 482 kcal/lb (1,060 kcal/kg) monthly, respectively. Frozen foods usually move fairly quickly out of supermarkets, but can be stored for many months in home freezers before they are cooked. The manufacture of containers used for frozen foods do require less energy than the manufacture of cans. Still, freezing foods is the most energy-intensive way to process fruits and vegetables.

The drying process is used to dehydrate fruits, herbs, and some vegetables. Examples include dried fruits such as apples, bananas, pears, and pineapples; dried herbs such as dill, oregano, parsley, rosemary, sage, and thyme; dried vegetables such as onions and garlic; grapes into raisins; and plums into prunes. The low moisture content in dried fruits does not support microbial growth. When drying is accomplished by using the sun, the process uses an essentially unlimited energy source and the price is right. However, much drying in the United States is done with large inputs of heat. This dehydration process consumes about 1,610 kcal/lb (3,542 kcal/kg).

Some fruits are used to make beverages and ice cream, such as wine, orange soda, and strawberry ice cream. Fermentation processes are used mainly to process grapes into wine. One liter of wine requires about 830 kilocalories for processing. This figure does not include the glass bottle. A liter of soda consumes about 1,425 kilocalories. About 400 kilocalories are needed to make about 1 pint of ice cream.

Packaging

Processed foods must be packaged. Canned fruits and vegetables are put into steel cans. The typical can used for 1 pound of vegetables requires about 1,000 kilocalories during the manufacturing process. The paper box around 1 pound of frozen vegetables or fruits uses 722 kilocalories. The glass jar around 16 ounces of jelly consumes about the same number of kilocalories as the same size steel can. Packages are also needed even for fresh produce in some cases. Sixty-nine kilocalories are consumed to make the typical wooden berry basket. A size 6 Styrofoam or paper tray found with brussels sprouts or husked sweet corn uses 215 and 384 kilocalories, respectively. The polyethylene pouch around 1 pound of fresh cranberries requires 559 kilocalories to manufacture.

Food Preparation

Energy usage is not over once fresh and processed fruits and vegetables are brought home. While some fruits are eaten fresh without further processing, some might be baked into a pie, such as apples or berries. Fresh vegetables are often cooked, and canned or frozen vegetables require warming or even further cooking. It is estimated that roughly 2,100 kilocalories (4,700 kcal/kg) of energy are used to cook 1 pound of food per person in the average American home. While the amount used specifically for the part involving horticultural products is unclear, given dietary variations and taste preferences, I estimate that about one-fourth (525 kcal//person/day) of that 2,100 kilocalories is consumed in the preparation of vegetables and some fruits.

Transportation

Transportation is crucial to the distribution of horticultural products. This need results from the centralization of agricultural and industrial activities in the United States. In earlier days, when farming was the dominant activity, food production was localized on mixed crop/livestock farms and so was most business/industrialized activity. Over time the economics of mass production and the desire for annual contracts with few major producers brought about consolidation of agriculture and horticulture into intensive, crop-specialized, large-scale farms in areas with the most favorable climates, good soils, abundant or subsidized water, and cheap farm labor. Industries also tended to concentrate around large urban populations because of power sources, large markets, and big labor pools.

Transportation linked these two activities, because the different requirements for each activity tended to favor separate and distant locations for each. Horticultural production is concentrated in California and Florida (ignoring agronomic crops) and heavy industry in the Northeast and North Central states. Of course, exceptions exist in that all states produce some horticultural products for interstate transport, such as bedding plants from Connecticut, peaches from Georgia, and cranberries from Massachusetts. High-tech industries can also be found in proximity to horticultural operations in California. Another transport factor is that certain crops are imported when U.S. fresh fruit and vegetable production is limited or unavailable because of seasonal constraints.

Transportation works both ways between these groups, because farms need machinery and supplies from industrialized centers and industrialized centers need food products. Horticultural products are mostly transported by truck and rail. Good numbers for horticultural crops are hard to find, because existing data deal with agricultural goods as a whole. Horticultural products constitute some percentage of this, but exactly how much is unclear.

The transport of agricultural goods (food and fiber) from farms to processors and packagers and then to wholesale distribution centers and finally to retail locations consumes considerable energy. The average distance that food travels from farms to final destinations, retail stores and homes, is uncertain. Numbers as low as 620 miles and as high as 1,200 miles have been reported (1,000–2,000 kilometers). The annual transport of agricultural goods consumes 348×10^{12} kcalories or 640 kcal/kg. This transportation is equivalent to about 280 million barrels of petroleum annually. My best guess is that about one-fourth of the agricultural products are horticultural goods, so the horticultural cost is about 70 million barrels of petroleum each year. That exceeds the 50 million barrels of petroleum each year used to move machinery and supplies to all U.S. agricultural farms.

Shopping for food adds additional energy costs, because most people in the United States shop by car. Best estimates are that the U.S. average for grocery shopping is about equivalent to the expenditure of energy used to move the food and fiber to the stores. So the transport of groceries to the home consumes another 280 million barrels of petroleum annually. Horticultural products undoubtedly form a considerable part of most grocery bags brought to the home.

It wasn't always so. When I was in the eighth grade, I had an after-school and Saturday job with the corner grocery store. I delivered groceries via a rather strange bike with a small front wheel and huge basket. I believe it was a World War II product resulting from the rationing of gasoline. While it was old, it functioned well and I delivered groceries to a large number of families. By the time I graduated from high school, the corner store was dying as it struggled against the supermarket chains. Unless we want to return to such conditions, we must develop alternative fuels.

These transportation numbers are averages. Extreme examples can be found. For example, fresh strawberries can be found in supermarkets in the winter, having been grown in Central America and jet freighted to the United States. The energy costs are considerable. Another example is oranges grown in Spain, purchased by Great Britain, and made into orange marmalade, some of which is sold in the United States. Clementines from Spain have become a highly marketed fruit in the United States from late November to early February. These extravagant energy costly products are the antithesis of sustainable horticulture. When one considers that the strawberries were grown on Central American farms and that shipping them out results in a reduced food supply for the native population, it becomes worse. One also needs to consider ethical and moral questions here. Who benefits from these international horticultural luxuries, the workers or the multinational corporations? Clearly the latter, as the development of local horticultural products for regional markets would benefit workers and the surrounding communities more. The exploitation of developing countries is another factor. Given the exploitation and lack of sustainability, can't we do without such products? If you'll pardon the pun, there is much food for thought here.

Some of these extreme examples can be self-correcting. The U.S. sales of clementines from Spain have been gradually increasing for some years. This increase has been at the expense of Californian navel oranges. The switch to clementines resulted from a slight deterioration in navel orange quality over time, primarily in terms of less sweetness resulting from premature harvesting procedures. The response in California has been twofold. One is better quality control with navel oranges. The second is the attempt to grow clementines in California. From a sustainability viewpoint, Californian clementines are more sustainable energy-wise than the ones from Spain.

ENERGY SUSTAINABILITY IN THE FUTURE

Our current horticultural (and agricultural) production relies heavily on the use of fossil fuels. These fuels are not a renewable resource, given the eons of time and geological events needed to produce them from biomass. Being finite resources, fossil fuels will run out. Unless alternative fuels that are renewable and economically sound are adopted, our mechanized horticultural production system will slowly grind to a halt. Realistically, our current system is but a blip or short-term experiment on the 10,000-year time line of agricultural activities. In the interim, we need to conserve energy, determine the realistic number of oil and natural gas fuel years that remain, and continue the research and development efforts needed to produce feasible alternative fuels. The other concern is that we might have to switch even before the fossil fuels run out, given the consequences of unchecked global warming.

Fossil Fuel Limits

The year 1973 stands out as a monument to oil dependency. Oil shipments to the United States were cut as a result of an OPEC oil embargo during the 1973 war in the Middle East. Prices for gasoline soared and waiting lines at gas stations became common. That year also stands out in that the energy conservation movement was finally taken seriously. Horticultural production was hit hard by increasing prices, both for fuels and products that needed high energy inputs (fertilizer) or were manufactured from petroleum feedstocks (pesticides).

The prices of fuel, fertilizer, and pesticides soared. Between 1973 and 1981, the prices of gasoline and diesel increased about 400% and 700%, respectively. Fertilizer prices rose about 250% during that same period. Inflation also rose dramatically during that same time period, and interest rates on farm loans went up sharply. Farmer income did not keep pace with cost increases. Many small family farms folded, accelerating the conversion to mega-industrial farms. The greenhouse industry in the Northeast and upper Midwest was hit hard, because large amounts of oil and natural gas were required to maintain production temperatures during the winter. A number of greenhouse operations went bankrupt.

Predictions of oil shortages became common. Most experts back then agreed that by 2000 we would have severe shortages and very high prices. It didn't happen. Economic and technological forces altered energy costs and production in a positive way. The Worldwatch Institute currently estimates that oil will run out by 2100. Natural gas and coal could last longer. Others believe that technological advances could stall the end for anywhere from 200 to a 1,000 years. While no one can predict with certainty when fossil fuel energy will run out, all agree that it will at some finite point in the future.

A number of technological advances are in progress or on the drawing boards. Extraction efficiency with existing wells has been improved considerably with computerized sensors and three-dimensional seismic imaging. Major oil discoveries in Colombia, Russia, and West Africa have occurred. Technology for oil exploration under water has been enhanced with the availability of seismic imaging of seafloors. Underwater oil deposits can be huge, such as the one under the Caspian Sea in Kazakhstan, Russia, estimated to hold 200 billion barrels. Other potential sites include the coast of Brazil and areas off the coast of West Africa and the Gulf of Mexico. Unmanned submarines with robotic arms and deep-sea crewed drilling submarines have managed to create exploratory wells in 8,000-foot-deep waters and have sunk wells 3 miles down into the seafloor. Further technological advances will likely push exploration to water depths of 10,000 feet.

In the past, huge amounts of natural gas produced during oil extraction were wasted by burning or flaring. This waste has stopped, as the technology for converting natural gas to oil has greatly improved. Natural gas at oil wells is no longer a waste, but a valuable product because of more favorable economics. Estimates are that this new conversion technology, which reduces flaring, could add an additional 60 years to oil supplies. Potential gas reserves might extend this to about 200 years. This time line could be extended even further with hydrates, ice crystals containing trapped natural gas found on seafloors. Technology to recover hydrate gas is currently under development. While the cost is more per barrel than oil, future improvements and rising prices in the oil industry are likely to bring the cost into a competitive range. Technologies to recover oil from tar sands are being used now in Canada and Venezuela. Future technology is likely to be employed to convert vast coal deposits to oil. As of 2002, this technology hadn't been perfected. Tar sands and coal conversion could bring about oil supplies that would last for several hundred to a thousand years.

This rosy outlook is offset by political and social problems. Oil producers could band together, such as they did with the OPEC oil cartel. Such organized efforts can lead to artificially high prices and oil shortages. Sudden changes in oil pricing and supplies can lead to severe economic woes for horticultural producers highly dependent on fossil energy. Another unpredictable fact is how fast developing nations will move toward industrialization and U.S. diets and increased demands for fossil fuels. Fossil fuel needs could easily increase five- to sixfold were this to happen. Fossil fuel supplies would disappear more quickly.

Another factor is global warming. The combustion of fossil fuels contributes carbon dioxide (CO_2) to the atmosphere. Already atmospheric CO_2 levels have climbed some 30% since the beginning of the Industrial Age. Summers have been warmer, as have been winters. Ice cover in the polar regions has been declining steadily. While some of this warmth can be attributed to El Niño cyclic weather events, this source does not totally account for the warming trend observed during the last several years. As the levels of CO_2 increase, this heat-trapping gas will accelerate the effects of global warming. Climatic changes and sea-level rises could have devastating effects on horticultural production, not to mention societal consequences.

Pressure to use less fossil fuel so that global warming might be slowed or reversed could mount. Horticultural production of foliage plants in greenhouses could become hard to justify. Even bedding plant and vegetable production in greenhouses could become socially unacceptable, first in the North and later in the South. Most agree that a return to human labor instead of mechanized labor would not be in the best interests of the economy or society. However, unless viable alternative energy is developed, our choices might be limited.

Should the horticultural industry move toward sustainability in terms of energy consumption? The answer is a definite yes for many reasons! Even if one assumes the best case scenario in terms of oil supplies, the resource is finite and not renewable. In the shorter term, cartel decisions and global warming problems might bring about disastrous consequences for horticultural production that is highly dependent on fossil fuels.

Conservation

The path to sustainability has two steps: conservation and alternative fuels. A third possible step is a drastic change in the human diet. Conservation helps reduce energy dependency, slows global warming, and minimizes the effects of price increases and production cuts. Since the oil scare in the 1970s, efforts at conservation have produced some good results. However, the limits to conservation are being approached. Earlier in this chapter we saw that the manufacturers of horticultural goods had brought about considerable decreases in energy consumption, but direct energy usage in crop production showed little change. Farm machinery likely has reached the point where little, if any, increased fuel efficiency can be teased out of combustion engines or further weight reductions can be achieved without power losses or reduced durability.

Some energy conservation remains possible beyond the limits of machinery, through changes in practices such as adoption of conservation tillage. A typical conventional plowing system, using the moldboard plow, followed by disking, herbicide applications, fertilization, planting, cultivating, and some machine maintenance/repairs, consumes 7.66 gallons of diesel fuel per acre. Changing to a chisel plow brings it to 6.95 gallons, and a further change to the disk instead of a plow, 5.71. Disking systems for horticultural crops can cut fuel use and costs by 25%. The extent to which conservation tillage has been adopted in horticultural production currently has not been accurately assessed. Thirty-seven percent of agricultural cropland used conservation tillage by 2002, but the fraction of this acreage devoted to horticultural production is unknown. Adoption has been motivated mostly by cost reduction, followed by decreased soil erosion, and lastly by energy considerations. Best estimates are that conservation tillage is used on about one-third of the acreage used in horticultural production, leaving some room for further energy conservation.

Lower inputs of herbicides and fertilizers can also reduce energy consumption, as can the more efficient irrigation systems discussed earlier. These changes toward a more sustainable horticulture are usually motivated by the desire to reduce soil erosion and pollution, rather than energy savings. The ultimate sustainable farm in terms of reduced inputs is the organic farm. Studies of energy usage on organic farms show that up to 40% less energy per acre is needed compared with the conventional farm. Energy needs can be reduced as farms adopt more sustainable practices and use fewer chemicals.

Energy Sustainability and Human Diets

Another possible pathway for energy conservation is diet modification. Should fossil fuels start to decline because of diminishing resources, dietary modifications could buy time until alternative fuels become commonplace. As fossil fuel supplies decrease and costs rise, dietary modifications will result simply from increasing food costs. If fuels truly become scarce, dietary modifications will become a necessity for survival and are likely to be enforced by government decree and market forces.

The typical U.S. diet is meat dependent to some degree (i.e., a nonvegetarian diet). It is an energy-intensive diet in that meat production gives about a 10% return on the food energy contained in the plants used to provide feeds for cattle. A switch to plant-based diets without meats could reduce energy requirements from fossil fuels considerably. Although the fossil fuel savings would not be tenfold, in that meat is only part of a balanced diet, a saving of some 50% would result from a switch to a vegetarian diet. A switch to a lacto-ovo diet that allows eggs, milk and milk, products from animals would produce savings in fossil fuel of around 30%.

Alternative Energy Sources

Conservation and dietary changes in themselves are not the total answer. Alternative renewable energy sources that are economically feasible must be developed. Only then can horticultural production (and the global society) become sustainable and immune to fossil fuel problems. Numerous alternative energy sources are possible. These alternatives include solar energy, wind energy, hydroelectric, geothermal, biomass, and hydrogen.

No one alternative source is likely to dominate because geography, climate, and economics will influence choices in various ways. For example, considerable numbers of bright, sunny days in the arid Southwest would likely favor solar energy as the best choice, whereas wind energy would be favored in the North Central plains. A major change might also be necessary in that energy would become decentralized; power *to* the people would be replaced by power *of* the people. Instead of having power delivered long distances by electrical conduit and pipelines, horticultural operations would be run by on-site power sources purchased and maintained by the site owners. Economic transitions could be considerable and bring about major changes in corporate control and the alteration of global politics and economics.

Solar Energy. Solar energy (Figure 10–1) utilizes the sun to provide direct and indirect sources of energy. The most promising technology in this area is photovoltaics. This system utilizes special light-trapping cells that convert sunlight to energy. Solar photovoltaics have an annual growth rate close behind wind power. Many of these cells are used by homes too far removed from power grids to justify economical connections. Others are used to provide electrical power to irrigation and water pumps and communication devices in isolated areas, even watches and calculators, and power sources for satellites. Houses and outbuildings on horticultural acreage can also generate their electrical needs from photovoltaic cells made into roofing shingles.

A major U.S. player in the development of this technology is Energy Conversion Devices. Major corporations pumping money into the development of photovoltaics include British Petroleum (BP) and Enron (U.S.). In 1997, BP committed $1 billion to exploring solar and other renewable energy sources. It is unfortunate that the collapse of Enron in 2002 has ended a major U.S. funding source for the development of solar energy technology.

The current drawback with this alternative, renewable energy system is cost. Photovoltaic (PV) cell manufacturing costs need to drop for this source to compete with fossil fuel-fired utilities that produce electricity. Costs need to decrease by 50% to 75%. Costs will drop as cell orders increase (mass production), manufacturing technology improves, cell efficiency increases, and cell longevity rises. Photovoltaics could also play a major role in the rise of fuel cells and hydrogen fuels (discussed later). An area of PV cells about the size of West Virginia and located in Arizona could essentially supply all electrical power needs in the United States. Photovoltaics don't produce CO_2 during the production of electricity, unlike the burning of fossil fuels in utility plants. Some PV cells do produce hazardous materials when disposed, because some contain gallium arsenide and

(A)   (B)

FIGURE 10-1   The sun can be used to provide direct energy with solar driers. (A) Solar grain drier uses a plastic glazed, flat plate air-type collector. The hot air from the collector is pumped over the grain to dry it. Such devices can also be used to dry peanuts, other nuts, and in the production of raisins or prunes. (B) Similar devices, but with pumped water as the collecting medium, can be used to provide hot water for heating and washing purposes. Different types of solar collectors are shown here on the roof of a milking parlor and are used to provide hot water for washing and cleaning. Such systems could also be used to provide hot water for various horticultural operations. Photovoltaic cells look similar, except the covering is a sturdier, more rigid plastic or glass and there is no air space. A crystalline semiconductor sheet such as silicon or copper indium gallium diselenide converts the sun's radiant energy into electricity, thus supplying indirect energy. (USDA photographs.)

cadmium sulfide. However, the more promising PV cells made from silicon pose fewer environmental hazards.

Wind Energy.   Wind energy can be harnessed through the use of wind turbines that generate electricity. A wind turbine is essentially a two- or three-bladed propeller mounted on a high metal column (Figure 10–2). A typical large, commercial wind turbine can generate 2 million kWh of electricity each year. Wind energy is the fastest growing renewable energy. However, despite that fast growth rate, wind-generated electricity is still a minor fraction of the electricity generated in the United States. Currently, 70% of the wind-produced electricity in the United States is generated in California. Small amounts are also produced in Iowa, Minnesota, and Texas. However, the potential for more exists. For example, Denmark produces 7% of its electricity with wind power and is also home to the leading technological corporation (Vesta) working on wind turbine technology. Three states, North Dakota, South Dakota, and Texas, have sufficient wind resources to produce enough electrical energy for the entire United States. Other promising areas include coastal regions.

The advantage of wind turbines is their flexibility. Sizes exist for powering a water pump for irrigation and drinking water. Larger sizes can power the farmhouse and outbuildings. Pooling many together on a "wind farm" can provide electricity for a small town or even a city. The windmill can also be used for lifting water: It runs a mechanical pump without electricity. A PV cell can run an electric pump for irrigation. In general, mechanical windmills and solar pumps are preferred for pumping small quantities of water and low pumping heads. Wind turbines are better for large-scale watering applications, such as field-irrigated vegetable crops.

(A)   (B)

FIGURE 10–2 (A) At the USDA-ARS Conservation and Production Research Laboratory in Bushland, Texas, wind turbines generate power for submersible electric water pumps that are far more efficient than traditional windmills (background). (B) The vertical-axis Darrieus-type wind turbine in Texas harvests at least 30% of the wind's energy. (USDA photographs.)

If the wind turbine is used where electrical grids are available, the grid can serve as a source of backup power in the event that the wind speed drops too low. In many cases, surplus power can be sold to the utility company via the grid connections. If no grid exists, a backup battery system, charged by the turbine, can provide power during downtimes.

Geothermal. In some areas, hot springs and steam vents lie near or at the surface. These sources are potential energy supplies that can be tapped. Iceland is one of the leading users of geothermal power. The major sites in the United States for possible geothermal power usage are primarily in the West. A few sites have been tapped for residential heating and crop drying. Other potential uses include heating greenhouses and farmhouses and food processing (drying of fruit and vegetable products). All of these uses rely on geothermal hot water sources. In fact, some eight western states have greenhouses that are heated with geothermal sources. Geothermal sources, usually steam or high-temperature water, can also be used to generate electricity, such as with some electrical plants in California, Hawaii, Nevada, and Utah. Geothermal power lags behind wind and solar power in annual growth. As of 2002, geothermal power was being used in California, Colorado, Idaho, Nevada, Oregon, Texas, Utah, and Wyoming.

Although geothermal power is not expected to be a major player in the future of alternative renewable energy sources, it will have important localized roles in certain horticultural operations. Most likely are greenhouses, crop drying, and drying of fruit and vegetable products in western states where geothermal power can be readily tapped. The one disadvantage is that geothermal power sources are often not close to markets or major population centers, in which case transportation costs and energy inputs can offset geothermal energy savings.

Hydroelectric. Water power was used long ago to power water wheels that were harnessed to grinding stones to mill grain. A later use was in water turbines to generate electricity. Hydropower supplies roughly 7% of our electricity. Its main disadvantage is the need

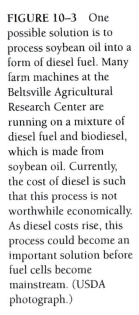

FIGURE 10–3 One possible solution is to process soybean oil into a form of diesel fuel. Many farm machines at the Beltsville Agricultural Research Center are running on a mixture of diesel fuel and biodiesel, which is made from soybean oil. Currently, the cost of diesel is such that this process is not worthwhile economically. As diesel costs rise, this process could become an important solution before fuel cells become mainstream. (USDA photograph.)

to dam rivers and flood large amounts of land in order to produce sufficient hydro head. The construction of these water reservoirs causes environmental problems. Often, good farmland is lost. Another problem is that sediments from soil erosion accumulate in these water reservoirs, reducing the effectiveness of the hydro head over time. Ecosystems are also greatly disturbed. While current sites will remain operative, it is doubtful that much expansion of hydroelectric power will occur.

Biomass Fuels.   Biomass is another old form of renewable energy (Figure 10–3). One of the oldest forms is the burning of wood to provide heat and energy for cooking. Other forms are the use of animal dung and crop residues as fuel sources for fires. Biomass presently supplies slightly more U.S. energy than hydroelectric power. The major form is wood for fuel. An indirect use is the conversion of crop biomass into ethanol that is used as a fuel. Brazil is currently the largest ethanol producer from sugarcane. At first glance, biomass, particularly when used to produce ethanol, looks promising as a future form of renewable energy. Unfortunately, some problems are likely to prevent great expansion in this area, such as the large cost in fossil fuel to produce biomass and convert it to fuels.

The production of firewood as a fuel requires time and land. Present forests are already harvested to a high degree for paper production, house construction, and furniture manufacturing. It is unlikely that more wood would be diverted to use as firewood than is presently done. Wood also grows slowly, requiring more land in rotational succession to keep supplies at constant levels. Little high-quality land remains in the United States for increased tree plantation cropping, in that most of the good cropland is already in use for food production. One possibility, although not for the near future, is that a shift to a more vegetarian U.S. diet could free up cropland normally used for feed grain production. Increased use of acreage for wood production would also increase the detrimental effects on the environment, such as increased soil erosion and water pollution from sediment, pesticides, and fertilizers. Another problem with burning wood is air pollution. Wood smoke contains a number of carcinogens, irritants, and particulates that are released to the atmosphere. Proper pollution controls can reduce these problems.

The same considerations also apply to ethanol production, either from sugarcane or corn. In the United States the diversion of cropland to ethanol production via corn would be limited by the same constraints listed for firewood. Another problem is that with current corn production systems, it takes roughly two times as much energy to produce a liter of ethanol compared to the energy contained in the liter of ethanol. Although the energy used for production can be decreased with improved technology and the dried distiller's

FIGURE 10–4   While ethanol from corn doesn't appear economically feasible, other biomass crops might improve the picture. Switchgrass can yield almost twice as much ethanol as corn, estimates this geneticist, who is conducting breeding and genetics research on switchgrass to improve its biomass yield and its ability to recycle carbon as a renewable energy crop. (USDA photograph.)

grain can be recycled as cattle feed, the energy budget still remains negative. Ethanol burns cleaner than gasoline in terms of carbon dioxide and various sulfur oxides, but it does contribute other serious air pollutants, such as various aldehydes and nitrogen oxides. When one factors in the carbon dioxide produced during the production of ethanol, air quality then becomes a concern. On the whole, ethanol does not look like the fuel of the future (Figure 10–4).

Hydrogen and Fuel Cells.   What if you had a battery that never lost power and never required recharging? Such a device does exist. It is called a fuel cell. Fuel cells supplied with appropriate fuels generate electricity and heat. A fuel cell consists of two electrodes around an electrolyte. With one type of fuel cell, you supply hydrogen at one electrode and oxygen at the other. The result is the generation of electricity and heat; water is the waste product. This type is essentially nonpolluting. This particular fuel cell has been used in satellites and inhabited space stations. Another type of fuel cell uses a fuel reformer or reprocessor. This type can extract hydrogen from essentially any hydrocarbon fuel, such as natural gas, ethanol, methanol, and even gasoline. This type does produce some pollution, but it is considerably less than that from a combustion engine.

A number of fuel cell types exist, based on the nature of the electrolyte. Cells employed in space by NASA use alkaline potassium hydroxide as the electrolyte. These cells approach efficiencies of 70%, quite high in terms of power generation efficiency. Their drawback is high cost, but efforts are under way to bring costs down. Another kind uses phosphoric acid as the electrolyte. This cell has seen the most commercial development. Its power generation efficiency is about 40%, but if the waste steam is harnessed (cogeneration), the efficiency approaches 85%. Steam, instead of water, is produced because the operating temperature is near 400°F. For purposes of comparison, the internal combustion engine operates at about 30% efficiency. These cells have been used in buildings and larger vehicles, such as buses and locomotives. The phosphoric acid-based fuel cell might power the farm tractor of tomorrow. It might also provide electricity for water pumps, farmhouses, outbuildings, and the greenhouse. The steam component, perhaps with the electrical backup, could heat and humidify the greenhouse and even the farmhouse.

Solid oxide-based fuel cells offer promise for applications requiring high power, such as electrical generating plants and industrial activities. They also provide considerable waste heat that could heat buildings. Usually some type of solid ceramic is used as the electrolyte. Temperatures in these cells approach 1800°F and generating efficiencies 60%. These cells are relatively new, but foreseeable uses include providing electricity to horticultural operations and replacing conventional fossil fuel-powered utility plants. These fuel cells could offer localized electricity, thus bringing about decentralization of the electrical industry and the need for much less infrastructure, such as transformers and transmission towers and wires.

The one drawback with fuel cells is that they are not currently competitive economically. If photovoltaics were used to make hydrogen, instead of the current energy-intensive methods, costs could come way down. In this regard, the most promising fuel cell technology in the development stage is the regenerative fuel cell. Photovoltaic cells (solar-powered electrolyzers) are used to provide energy that electrolyzes water into hydrogen and oxygen. These cells then use the two materials to generate electricity and heat. The wastewater formed by the cell is then returned for electrolysis again. This closed-loop cell utilizes solar (unlimited) energy and essentially pollution-free power is produced. This scenario is rapidly becoming reality. In 2002, Honda's Research and Development Center in Torrance, California, launched a hydrogen production, storage, and fueling station. The facility uses solar power to produce hydrogen from water. The hydrogen is pressurized with a compressor and then stored in tanks.

Fuel cells based on proton exchange membranes offer a number of advantages for automobiles and light trucks. Such fuel cells might power the farm car and pickup truck some day. These cells run at low temperatures (175°F), offer high power density, and can respond rapidly to shifts in power demand, such as in quick first starts and rapid acceleration of automobiles. A variation on this cell is the direct methanol fuel cell. It uses a similar polymeric material for the electrolyte, but does not need a fuel processor. The anode serves this purpose, drawing out the hydrogen from the methanol.

As of 2002, some 12 or more U.S. companies were selling fuel cells of various types, and more than 120 companies, universities, research labs, and governmental agencies were researching developing fuel cells worldwide. The only drawback is cost. However, technological improvements and the economies of mass markets are anticipated to lower prices in time. Fuel cells are currently used in some automobiles, some buses, space satellites, space habitats, buildings, and utility plants. Although limited at present, fuel cells are expected to play a major role in the future energy picture, both in horticulture and society.

In the long run, the fuel most likely to replace fossil fuels is hydrogen. It can be utilized to power fuel cells and to generate electricity. As a fuel, hydrogen is relatively pollution free. While fuel cells can use other fuels, such as natural gas or ethanol, these must be converted to hydrogen by a fuel processor. This conversion is an inefficient usage of the fuel. For example, the conversion of natural gas to hydrogen discards 30% of the natural gas fuel value. A fuel processor also increases fuel cell manufacturing costs and the fuel doesn't burn as cleanly as when starting with hydrogen. The one advantage with natural gas conversion is that natural gas and its pipelines and distribution centers could play a major role as an intermediary technology during the conversion to pure hydrogen fuels.

A number of problems remain before hydrogen fuels replace fossil fuels. The current costs for making hydrogen are more expensive than gasoline. Costs can be brought down with better technology, economies of scale, and less costly photovoltaics and wind power. Once this plateau is reached, electricity from wind and solar power can be used to drive the electrolysis of water into hydrogen and oxygen. The value of the oxygen can be used to reduce costs and it can also be used in fuel cells with hydrogen to produce electricity. Another approach being researched is the use of catalytic materials added to water that, when exposed to sunlight, produce hydrogen.

Another problem is storage. Large volumes are needed for gas storage. Volume reductions are possible, such as through compression to a compressed gas and even liquefaction. Then storage vessels need to be stronger for the compressed gas form and strong and cold for liquid hydrogen. The gas can also be treated with certain metals and stored as a metal hydride. Active research efforts are under way to achieve more efficient and less expensive storage options. Hydrogen and fuel cells are the most likely future energy sources for horticultural operations—and society, too.

SUMMARY

Horticulture consumes energy to reduce labor, to provide chemical inputs, and to produce food and ornamentals. The food industry consumes about 17% of the total energy used in the United States. Much of the energy consumed in horticultural production and subsequent processing and distribution of products derives from fossil fuel energy, nonrenewable resources such as oil, natural gas, and coal. Two approaches toward energy sustainability exist. One is to buy time with conservation using more energy-efficient farm machinery, conservation tillage, and fewer chemicals. Another, longer range effort is to switch to renewable and alternative sources, such as solar energy, wind power, geothermal power, hydroelectric power, hydroelectric, and biomass fuels.

Energy use for crop production has been tracked since the 1970s and has helped with energy conservation. Now we rely on computer models to determine energy consumption for various crops. Energy budgets must take into consideration primary, indirect, direct, on-farm, and after-the-farm energy consumption. Primary energy is the sum total of all energy used on the farm, including both direct and indirect energy inputs. Direct energy sources are used on the farm as fuels (diesel, gasoline, LP gas, natural gas), lubricants, and electricity. Much of the direct energy input is used for machine labor.

Energy is needed to produce direct forms of primary energy through drilling activity, refining, power plant construction, transportation, and other processes. This production energy is described as indirect energy inputs. Nonenergy forms of farm inputs such as fertilizers, pesticides, and machinery also consume energy in the production and delivery phase prior to use on the farm. These farm inputs utilize indirect energy, too. Farmers have some control over direct energy inputs through manipulation of applications and practices. Indirect energy inputs and postproduction energy usage are controlled by the business and manufacturing sectors.

The FAECM computer model provides some answers regarding energy conservation. Primary energy needs in Florida production agriculture from the 1970s to 1990 declined by about 15%. Considerable agricultural activity in Florida is involved in horticultural production, so this decline is probably a reasonable value for this sector. Direct energy inputs on the farm have held relatively constant during this time period. The decline is attributable to a decrease of indirect energy. Conservation has worked here because of increased energy efficiency during the manufacturing of agricultural inputs.

The computer models take both indirect and direct energy inputs into consideration. Crop production can be expressed on an energy-consumed basis per acre of crop, or in its preferred form, kilocalories per hectare. One can also convert the crop yield into food energy and get some idea of energy in and energy out. This ratio gives you some idea of energy efficiency for the production of various crops. One can also compare the ratio of energy costs to crop production value. This model does not take into consideration energy costs beyond the farm gate.

Our earliest agricultural activities in colonial America were dominated by human labor. In 1776, draft animals provided nearly 80% of the farm power. Today, more than 90% of the farm labor is done by machines. We have traded off animal energy for farm machines. The price for less human and animal labor and increased productivity is high inputs of fossil fuels.

The production of fruits, vegetables, and ornamentals consumes a considerable amount of fossil fuel energy. Much goes to production machinery, and some to irrigation. Given the mechanized aspect of production, little possibility exists for reduced energy consumption. The importance of fuels to horticultural production is high in Florida. Diesel and gasoline fuels account for roughly 80% of the direct energy consumption.

Fuel efficiency with farm machinery has reached technological limitations and conservation efforts are limited, so direct energy consumption remains constant. The need to reduce fuel consumption and to move toward renewable energy sources is critical, given the high fuel consumption, our need to maintain food supplies, and our dependence on diminishing global supplies of fossil fuels. A return to an agrarian society using animal labor is not possible. The only foreseeable alternative is renewable alternative fuels.

Considerable fossil fuel energy (and water) is consumed during crop irrigation. Irrigation is critical in the arid Southwest, and even in wetter areas (Florida and California). Decreased irrigation or more efficient irrigation is needed from an energy standpoint. The impetus is more likely economical. Irrigation costs are government subsidized. Our taxes pay for dams, waterways, aqueducts, pipelines, and pumping associated with the movement of large volumes of water, such as from the Colorado River to California. If these indirect energy, construction, and maintenance costs were factored in, horticultural producers in arid regions could not compete. Even so, direct costs can become prohibitive, because the pumping of groundwater for the production of crops from increasingly deeper depths is costly. Some low-value irrigated crops are now less likely to be produced in the arid Southwest. Reductions in irrigation are also likelier as groundwater and surface water supplies become dangerously depleted and city dwellers clash with irrigation farmers over limited water supplies.

More efficient drip irrigation systems can be utilized, leading to more efficient water usage and reduced energy costs. However, initial costs, in terms of energy and capital expenditures, are high for such systems. Trickle irrigation is generally only used for high value crops and in areas where only deep pumping of groundwater is possible. An irrigation study with tomatoes showed that switching from subsurface irrigation to micro-irrigation reduced water consumption by up to 80% and nitrogen fertilizer use by 50%, while producing yields comparable to those of subsurface irrigation. While initial costs were high, operating costs are lower. If energy prices climb, these systems could be economically viable with a short cost payback period. If the true costs of irrigation water and water pollution were borne by growers, these systems would be rapidly adopted.

The next highest energy input usually involves nitrogen fertilizer. Management practices to reduce fertilizer input can lead to energy reductions. However, such practices are usually undertaken to reduce off-site water pollution. These practices include better soil tests, more precise applications of fertilizer, attention to timing of applications, and legume cover crops. These practices result in less fertilizer use and lower losses from leaching and volatilization.

Organic fertilizers are an alternative to chemical nitrogen fertilizer. Manure is less energy intensive, renewable, offers the benefits of organic matter, and results in less pollution when properly managed. Whereas this solution is ideal for a mixed crop/livestock organic farm, it is not a universal alternative. Supplies are limited, nutrient levels can be insufficient, and transportation costs become unacceptable for sites not close to feedlots. Manure also loses nutrients during storage, so prompt usage is important for maximizing nutrient value.

Future advances might come from nitrogen-fixing bacteria tailored for nonleguminous crops. Such genetically modified bacteria could greatly reduce dependency on chemical nitrogen sources and result in less energy usage and less water pollution. Nitrogen inputs would be supplied from air, essentially an endless supply.

Pesticides consume petroleum feedstock and vary from number three to number two in horticultural energy consumption. While the driving force for pesticide reduction is environmental and health-based, moving in this direction also helps reduce energy. Moving from pesticides to IPM or even better, biological control results in energy savings. The successful switch to IPM or total biological control varies. Economics and level of difficulty motivate switches. Major horticultural crops offer more motivation than minor ones in terms of economics. The more pests there are, the more difficult it becomes to create a successful IPM or biological control program.

Fuel, nitrogen fertilizer, and pesticides account for the majority of horticultural energy expenditures. Controlling these three parameters is the major task for those interested in achieving a sustainable horticulture. The best hope now is conservation. Reaching a sustainable level through the use of renewable fuels, inexhaustible nitrogen supplies, and plant resistance to insects and diseases depends on technological advances in alternative fuels, genetically engineered microbes for agricultural nitrogen fixation, and genetic modifications to induce high levels of plant resistance.

The return on energy invested in horticultural crops can be looked at two ways. One is the trade-off between energy inputs versus the amount of food energy. Another approach is to consider how much is earned for each dollar invested in energy. Farmers motivated by sustainability might find this information useful. Such information must be considered in context with food value (protein, vitamins, and minerals), acreage devoted to the crop, and consumer loyalty.

The FAECM computer model yields useful data. In Florida, the average crop had a value of $136 when compared to the cost of 1 million Btus of production energy of direct energy. When viewed in terms of primary energy, the return is $44. Vegetables gave the best returns, followed by ornamentals and then fruits. Fruits have food value, ornamentals do not. From a strictly practical view, land devoted to ornamentals should be turned over to vegetables and fruits. The percentage of land in ornamentals is quite low, so little is to be gained. However, foliage plants are the number two consumer of primary and direct energy. Another factor is the social, aesthetic, and psychological value of ornamental plants.

Energy usage by greenhouses is considerable. In northeast Ohio, a greenhouse acre consumes 100,000 gallons of number 2 fuel oil yearly. In central Florida, the production of greenhouse vegetables and ornamentals per acre requires 12,500 gallons yearly. Another perspective is that orange production in Florida needs about 21 million Btus of direct energy per acre, as opposed to 1.25 billion Btus of direct energy per acre for greenhouse vegetables. Greenhouse production is hard to justify from an energy perspective, even in warmer climates.

A need exists for processed fruits and vegetables. Commercial processing requires energy (heat). Energy for transportation is required to get horticultural products to processors, wholesale distributors, retail markets, and finally homes. Additional energy is consumed during storage in refrigerators and freezers, and finally by cooking.

Canning, drying, and freezing are used to process fruits and vegetables. Commercial canning requires around 261 kcal/lb of produce. Home canning requires 30% greater energy. Freezing requires considerably more energy, in that energy must not only be supplied as heat, but also for cooling, freezing, and then storage. Freezing is the most energy-intensive way to process fruits and vegetables.

Drying is used to dehydrate fruits, herbs, and some vegetables. The low moisture content in dried fruits does not support microbial growth. When solar drying is used, the process uses an essentially unlimited energy source and the price is right. However, much drying in the United States requires inputs of heat that consume about 1,610 kcal/lb.

Processed foods must be packaged. The typical 16-ounce can of vegetables requires about 1,000 kilocalories to manufacture. The 16-ounce paper box of frozen vegetables or fruits used 722 kilocalories. The 16-ounce glass jar of jelly consumes about the same number of kilocalories as the same size steel can. Sixty-nine kilocalories are consumed to make the typical wooden berry basket. A size 6 Styrofoam or mould paper tray found with fresh vegetables used 215 to 384 kilocalories, respectively. The polyethylene pouch around 1 pound of fresh cranberries required 559 kilocalories.

Energy usage is not over once fresh and processed fruits and vegetables are brought home. Although some fruits are eaten fresh, some might be baked into a pie, such as apples or berries. Fresh vegetables are often cooked, and canned or frozen vegetables require warming or cooking.

Transportation is crucial to the distribution of horticultural products. This need results from the centralization of agricultural and industrial activities in the United States. Transportation links these two activities, as the different requirements for each activity tended to produce separate locations for each. Another transport factor is that certain crops are imported when U.S. fresh fruit and vegetable production is limited or unavailable because of seasonal constraints.

Truck and rail transport is used for most horticultural products. The transport of agricultural goods from farms to processors and packagers and to wholesale distribution centers and finally to retail locations and homes consumes considerable energy. Estimates for this transport are 620–1,200 miles (1,000–2,000 kilometers). The annual transport of agricultural goods consumes about 280 million barrels of petroleum annually. My best guess is that about one-fourth of the agricultural products are horticultural goods, so the horticultural cost is about 70 million barrels of petroleum yearly. That exceeds the yearly 50 million barrels of petroleum used to move machinery/ supplies to farms. Shopping for food by car adds additional energy costs. Best estimates are that the U.S. average for grocery shopping is about equivalent to the expenditure of energy used to move the food and fiber to the stores.

Extreme transportation examples exist. Fresh strawberries are in our supermarkets in the winter, having been grown in Central America and jet freighted to the United States. Oranges grown in Spain are purchased by Great Britain and become orange marmalade, some of which is sold in the states. Clementines from Spain are now a United States staple. These extravagant energy costly products are the antithesis of sustainable horticulture. When one considers that the strawberries were grown on Central American farms, reducing the local food supply there, it becomes worse. One also needs to consider ethical and moral questions here. Who benefits from these international horticultural luxuries, the workers or the multinational corporations? Clearly the latter, because the development of local horticultural products for regional markets would benefit workers and the surrounding communities more. The exploitation of developing countries is another factor. Given exploitation and unsustainability, can't we do without such products?

Our current horticultural production relies heavily on fossil fuels. These fuels are not renewable resources, given the eons of time and geological events needed to produce them from biomass. Being finite resources, fossil fuels will run out. Unless alternative fuels that are renewable and economically sound are adopted, our mechanized horticultural production system is limited. Realistically, our current system is only a short-term experiment on the 10,000-year time line of agricultural activities. In the interim, we must conserve energy, determine the realistic number of oil and natural gas fuel years that remain, and continue the research and development efforts needed to produce feasible alternative fuels. The other concern is an even earlier switch imposed by global warming.

Most experts 30 years ago agreed that by 2000 we would have fuel shortages and very high prices. It didn't happen. Economic and technological forces controlled energy costs and production. The Worldwatch Institute currently estimates that oil will run out by 2100. Natural gas and coal could last longer. Others believe that technological advances could stall the end for anywhere from 200 to 1,000 years. While no one can predict with certainty when fossil fuel energy will run out, all agree that it will at some finite point in the future.

A number of technological advances are in progress or on the drawing boards. Extraction efficiency with existing wells has been improved considerably with computerized sensors and three-dimensional seismic imaging. Major oil discoveries in Colombia, Russia, and West Africa have occurred. Technology for oil exploration under water has been enhanced with the availability of seismic imaging of seafloors, unmanned submarines with robotic arms, and deepsea crewed drilling submarines.

Technology for converting natural gas to oil is improving. Huge amounts of natural gas at oil wells wasted through flaring can now be converted into oil. Estimates are that this new conversion technology could add an additional 60 years to oil supplies. Potential gas reserves might extend this to about 200 years. This time line could be extended even further with hydrates, ice crystals containing trapped natural gas found on seafloors. Tar sands extraction and coal conversion could bring about oil supplies for several hundred to a thousand years.

This outlook is offset by political and social problems. Oil producers could band together and jeopardize supplies. Another unpredictable fact is how fast developing nations will move toward industrialization and United States diets and increased demands for fossil fuels. Another factor is global warming. Pressure to use less fossil fuel so that global warming might be slowed or reversed could mount. Horticultural production of foliage plants in greenhouses could become hard to justify. Even bedding plant and vegetable production in greenhouses could become socially unacceptable, first in the North and later in the South.

Should the horticultural industry move toward sustainability in terms of energy consumption? The answer is a definite yes for many reasons! Even if one assumes the best case scenario in terms of oil supplies, the resource is finite and not renewable. In the shorter term, cartel decisions and global warming problems might bring about disastrous consequences for horticultural production that is highly dependent on fossil fuels.

The path to sustainability has two steps: conservation and alternative fuels. Conservation reduces energy dependency, slows global warming, and minimizes the effects of price increases and production cuts. The limits to conservation are now being approached. The manufacturers of horticultural goods have brought about considerable decreases in energy consumption, but direct energy usage in crop production shows little change. Farm machinery likely has reached the point where little, if any, increased fuel efficiency can be teased out of combustion engines or further weight reductions can be achieved without power losses or less durability.

Some energy conservation remains possible beyond the limits of machinery, through changes to conservation tillage. The extent to which conservation tillage has become adopted in horticultural production currently has not been accurately assessed. Best estimates are that conservation tillage is used on about one-third of the acreage used in horticultural production, leaving some room for further energy conservation.

Lower inputs of herbicides and fertilizers and more efficient irrigation systems can also reduce energy consumption. All these changes toward a more sustainable horticulture are usually motivated by the desire to reduce soil erosion and pollution, rather than energy savings. The ultimate sustainable farm in terms of reduced inputs is the organic farm. Studies of energy usage on organic farms show that up to 40% less energy per acre is needed relative to the conventional farm.

Another possible pathway for energy conservation is diet modification. Dietary modifications could buy time until alternative fuels become commonplace. If fuels truly become scarce, dietary modifications could become a necessity for survival. The typical U.S. diet is meat dependent to some degree (nonvegetarian diet) and is energy intensive in that meat production gives about a 10% return on the food energy contained in the plants used as animal feeds. A switch to plant-based diets without meats could reduce energy requirements from fossil fuels considerably. Although the fossil fuel savings would not be tenfold, in that meat is only part of a balanced diet, a saving of some 50% would result from a switch to a vegetarian diet. A switch to a lacto-ovo diet that allows eggs, milk, and milk products from animals would produce savings in fossil fuel around 30%.

Conservation and dietary changes in themselves are not enough. Alternative economical renewable energy sources must be developed. These alternatives include solar energy, wind energy, hydroelectric, geothermal, biomass, and hydrogen. No one source is likely to dominate because geography, climate, and economics will influence choices in various ways. For example, considerable numbers of bright, sunny days in the arid Southwest would likely favor solar energy as the best choice, whereas wind energy would be favored in the North Central plains. A major change might also be necessary in that energy would become decentralized. Instead of having power delivered long distances by electrical conduit and pipelines, horticultural operations would be run by on site power sources purchased and maintained by the site owners. Economic transitions could be considerable and bring about major changes in corporate control and the alteration of global politics and economics.

Solar energy utilizes the sun. The most promising technology is photovoltaics. This system utilizes special light-trapping cells that convert sunlight to energy. Many of these cells are used by homes far removed from power grids. Others are used to provide electrical power to irrigation and water pumps. Houses and outbuildings on horticultural acreage can generate their electrical needs using photovoltaic roofing shingles. The current drawback is cost. Photovoltaic cell manufacturing costs need to decrease by 50% to 75%. Costs will drop as cell orders increase (mass production), manufacturing technology improves, cell efficiency increases, and cell longevity rises. Photovoltaics could also play a major role in the rise of fuel cells and hydrogen fuels. Photovoltaics don't produce CO_2 during the production of electricity.

Wind turbines harness wind energy to generate electricity. A typical commercial wind turbine can generate 2 million kWh of electricity each year. Wind energy is the fastest-growing renewable energy. However, wind-generated electricity is but a minor fraction of the electricity generated in the United States. Currently, 70% of the wind-produced electricity in the United States is generated in California. Three states, North Dakota, South Dakota, and Texas, have sufficient wind resources to produce enough electrical energy for the entire United States. Other promising areas include coastal regions.

The advantage of wind turbines is flexibility. Sizes exist for powering a water pump for irrigation and drinking water and even to power the farmhouse and outbuildings. Pooling many turbines together on a "wind farm" can provide electricity for a small city. The windmill can also be used for lifting water: It runs a mechanical pump without electricity. A PV cell can run an electric pump for irrigation. In general, mechanical windmills and solar pumps are preferred for pumping small quantities of water and low pumping heads. Wind turbines are better for large-scale watering applications, such as field-irrigated vegetable crops. If the wind turbine is used where electrical grids are available, the grid can serve as a source of backup power. In many cases, surplus power can be sold to the utility company. If no grid exists, a backup battery system, charged by the turbine, can provide power during downtimes.

In some areas, hotsprings and steam vents lie near or at the surface and can be tapped. The major sites for possible geothermal power usage are in the West United States. A few sites have been tapped for residential heating and crop drying. Other potential uses include heating greenhouses and farmhouses and food processing (drying of fruit and vegetable products). All of these uses rely on geothermal hot water sources. In fact, some eight western states have greenhouses that are heated with geothermal sources. Geothermal sources, usually steam or high-temperature water, can also be used to generate electricity such as with electrical plants in California, Hawaii, Nevada, and Utah. Geothermal power lags behind wind and solar power in annual growth.

Geothermal power is not expected to be a major player in the future of alternative renewable energy sources, it will have important localized roles in certain horticultural operations. Most likely are greenhouses, crop drying, and drying of fruit and vegetable products in western states. The one disadvantage is that geothermal power sources are often not close to markets or major population centers, in which case transportation costs and energy inputs can offset geothermal energy savings.

Water power was used long ago to power water wheels that were harnessed to grinding stones to mill grain. A later use was in water turbines to generate electricity. Hydropower supplies roughly 7% of our electricity. Its main disadvantage is the need to dam rivers and flood large amounts of land in order to produce sufficient hydro head. The construction of these water reservoirs causes environmental problems. Often, good farmland is lost. Another problem is that sediments from soil erosion accumulate in these water reservoirs, reducing the effectiveness of the hydro head over time. Ecosystems are also greatly disturbed. Although current sites will remain operative, it is doubtful that much expansion of hydroelectric power will occur.

Biomass is an old form of renewable energy. Uses include the burning of wood, animal dung, and crop residues to provide heat and energy for cooking. Biomass presently supplies slightly more U.S. energy than hydroelectric power. The major form is wood. An indirect use is the conversion of crop biomass into ethanol for fuel.

The production of firewood as a fuel requires time and land. Present forests are already highly harvested for paper production, house construction and furniture manufacturing. Wood grows slowly, requiring more land in rotational succession to keep supplies at constant levels. Little high-quality land remains in the United States for increased tree plantation cropping; most good cropland is already in use for food production. One possibility is that a shift to a more vegetarian U.S. diet could free up cropland normally used for feed grain production. Increased use of acreage for wood production would also increase detrimental effects on the environment such as increased soil erosion and water pollution from sediment, pesticides, and fertilizers. Another problem with burning wood is air pollution.

The same considerations also apply to ethanol production, either from sugarcane or corn. In the United States the diversion of cropland to ethanol production via corn would be limited by the same constraints listed for firewood. Another problem is that with current corn production systems, it takes roughly two times as much energy to produce a liter of ethanol when compared to the energy contained in the liter of ethanol. Although the energy used for production can be reduced with improved technology and the dried distiller's grain can be recycled as cattle feed, the energy budget still remains negative. Ethanol burns cleaner than gasoline in terms of CO_2 and various sulfur oxides, it does contribute other serious air pollutants. When one considers the CO_2 produced during the production of ethanol, air quality becomes a concern. On the whole, ethanol does not look like the fuel of the future.

What if you had a battery that never lost power and never required recharging? Such a device, the fuel cell, does exist. Fuel cells supplied with appropriate fuels generate electricity and heat. A fuel cell consists of two electrodes around an electrolyte. With one type of fuel cell, you supply hydrogen at one electrode and oxygen at the other. The result is the generation of electricity and heat; water is the waste product. This type is essentially nonpolluting. This particular fuel cell has been used in satellites and inhabited space stations. Another type of fuel cell uses a fuel reformer or reprocessor. This type can extract hydrogen from essentially any hydrocarbon fuel, such as natural gas, ethanol, methanol, and even gasoline. This type does produce some pollution, but it is considerably less than that from a combustion engine.

Both of these types work in a similar manner. The hydrogen, whether directly or indirectly supplied, enters at the anode. A catalyst favors the splitting of the hydrogen atom into an electron and proton. Both migrate toward the cathode, but by different routes. The proton passes through the electrolyte, and the electron passage to the cathode generates a current that can be tapped. At the cathode, oxygen enters and unites with the proton and electron to form water.

A number of fuel cell types exist, based on the electrolyte material. Current ones vary in power efficiency and operating temperature: alkaline potassium hydroxide (70% efficient), phosphoric acid (40% efficient), solid oxide (60% efficient), and proton exchange membrane fuel cells. For comparison, the internal combustion engine is 30% efficient. Some (phosphoric acid and solid oxide) supply waste heat as well as electricity and can be used to power and heat buildings, including greenhouses. The efficiency of the phosphoric acid fuel cell approaches 85% if the waste steam is harnessed (co-generation). Solid oxide-based fuel cells offer promise for applications requiring high power, such as electrical generating plants and industrial activities. Fuel cells based on proton exchange membranes offer applications for farm pickup trucks and tractors.

The one drawback with fuel cells is that they are not currently economically competitive. If photovoltaics were used to make hydrogen, instead of current energy-intensive methods, costs would drop. The most promising fuel cell technology in development is the regenerative fuel cell. Photovoltaic cells (solar-powered electrolyzer) are used to provide energy that electrolyzes water into hydrogen and oxygen. These cells then use the two materials to generate electricity and heat. The wastewater formed by the cell is then returned for electrolysis again. This closed-loop cell utilizes solar (unlimited) energy and essentially pollution-free power is produced. Although limited at present, fuel cells are expected to play a major role in the future energy picture.

The fuel most likely to replace fossil fuels is hydrogen. It can power fuel cells and generate electricity. As a fuel, hydrogen is relatively pollution free. Although fuel cells can use other fuels, such as natural gas or ethanol, these must be converted to hydrogen by a fuel processor. This conversion is inefficient. For example, the conversion of natural gas to hydrogen discards 30% of the natural gas fuel value. A fuel processor also increases fuel cell manufacturing costs and the fuel does not burn as clean as hydrogen. The one advantage with natural gas conversion is that natural gas and its pipelines and distribution centers could play a major role as an intermediary technology during the conversion to pure hydrogen fuels.

A number of problems remain before hydrogen fuels replace fossil fuels. Current costs for hydrogen are higher than gasoline. Costs can be brought down with better technology and cheaper photovoltaics and wind power. Once this plateau is reached, electricity from wind and solar power can be used to drive the electrolysis of water into hydrogen and oxygen. The value of the oxygen can be used to reduce costs and it can also be used in fuel cells with hydrogen to produce electricity. Another approach being researched is the use of catalytic materials added to water that when exposed to sunlight, produce hydrogen. Another problem is storage size. Volume reductions are possible through compression. Then storage vessels need to be stronger for the compressed gas form and strong and cold for liquid hydrogen. The gas can also be treated with certain metals and stored as a metal hydride. Active research efforts are under way to achieve more efficient and less expensive storage options.

EXERCISES

Mastery of the following questions shows that you have successfully understood the material in Chapter 10.

1. How do primary energy, direct energy, and indirect energy differ in terms of horticultural energy budgets? Give some examples of direct and indirect energy inputs in horticultural production.
2. In what type of energy input into horticultural production has conservation been most successful?
3. Which type of direct energy input into horticultural production is the largest? Why is it so difficult to reduce this energy input?
4. Irrigation uses a lot of energy and water. In what ways can we reduce this usage?
5. Manufactured fertilizers are energy intensive. In what ways can we reduce this form of energy consumption?
6. Manufactured pesticides are energy intensive. In what ways can we reduce this form of energy consumption?
7. What are the two ways one can look at the return on energy invested in horticultural crops? What crop is hard to justify in Florida based on energy returns?
8. What are the possible various energy needs for horticultural crops after they leave the farm?
9. Which form of food processing for fruits and vegetables, canning or freezing, consumes more energy? Why?
10. Fossil fuel supplies are finite. In what technological ways might these supplies be stretched as far as possible? What problems might arise with this approach?
11. Conservation and diet modifications can reduce fossil fuel energy usage. Which do you favor and why?
12. What are the various forms of alternative energy? Which one do you think has the most promise? Why?
13. What do you think future farms will look like in terms of alternative energy uses?

INTERNET RESOURCES

Energy and water efficiency in vegetable production
    http://edis.ifas.ufl.edu/EH208

Energy efficiency in crop production (menu)
    http://edis.ifas.ufl.edu/MENU_AGCROP

Energy for Florida oranges

http://edis.ifas.ufl.edu/EH181

Energy in Florida agriculture

http://edis.ifas.ufl.edu/EH179

Fuel cell information

http://www.fuelcells.org

http://www.ott.doe.gov/fuel_cells.shtml

Geothermal education office

http://geothermal.marin.org/

Hydroelectric power facts

http://www.eia.doe.gov/kids/renewable/water.html

http://www.altenergy.org/2/renewables/hydroelectric/hydroelectric.html

Renewable Energy

http://www.eren.doe.gov/

http://ab.mec.edu/curriculum/student/energy.htm

Solar energy information

http://www.eren.doe.gov/pv

http://www.hrel.gov/clean_energy/solar.html

Wind energy: economic development, environmental benefits, and distributed generation (industry)

http://www.windustry.org/

http://www.eren.doe.gov/wind/web.html

RELEVANT REFERENCES

Bassam, N. El, R. K. Behl, and B. Prochnow (eds.) (1998). *Sustainable Agriculture for Food, Energy and Industry: Strategies for Achievement* (two volumes). James & James Science Publishers, London.

Bourne, Geoffrey H. (1990). *Aspects of Food Production, Consumption and Energy Values*. S. Karger AG, Basel, Switzerland. 212 pp.

Hayes, Jack (ed.) (1980). *Cutting Energy Costs: The 1980 Yearbook of Agriculture*. U.S. Department of Agriculture, Washington, DC.

Kleinberg, R. L., and P. G. Brewer (2001). "Probing Gas Hydrate Deposits." *American Scientist* 89: 244–251.

Pimentel, David, and Marcia Pimentel (eds.) (1996). *Food, Energy and Society* (rev. ed.). University Press of Colorado, Niwot. 363 pp.

Poincelot, R. P. (1986). "Crop Energy Conservation." In *Toward a More Sustainable Agriculture*, pp. 33–62. AVI Publishing Co., Westport, CT. 241 pp.

Poincelot, R. P. (1986). "Post Prodution Energy Conservation." In *Toward a More Sustainable Agriculture*, pp. 99–115. AVI Publishing Co., Westport, CT. 241 pp.

Poincelot, R. P. (1986). "Future Technology." In *Toward a More Sustainable Agriculture*, pp. 207–233. AVI Publishing Co., Westport, CT. 241 pp.

Chapter 11

Sustaining Soil, Water, and the Environment

OVERVIEW

Sustaining soil, water, and the environment lies at the heart of sustainable horticulture. No nation will endure for long without proper maintenance of these critical farming and societal resources. The industrialization of horticulture has moved us away from this goal. Returning to resource sustainability need not mean the end of the productivity achieved with our highly mechanized farming systems. These changes need not affect profitability either. Organic farming has already made these changes and is profitable. These changes need not totally convert conventional horticulture to organic horticulture. Systems in between can be reached in gradual increments. For example, a farmer could elect to increase soil organic matter, but decide to rely on IPM methods for pest control. These changes involve practices from the past, but also newer ones, such as precision farming.

CHAPTER CONTENTS

Site Selection
On-Site Protection: Wind, Sun, Cold, and Frost
Soil Fertility
Soil Organic Matter
Sustainable Management of Soil Resources

Reduced Tillage
Sustainable Water Management
Mulches, Soils, and Water
Overview of Precision Farming

OBJECTIVES

- ❀ To learn the desirable site characteristics needed for horticultural production
- ❀ Understanding how to provide on-site protection against wind, sun, cold, and frost
- ❀ Understanding the role soil tests play in sustaining soil nutrients
- ❀ Understanding the key role organic matter plays in sustaining soil

- ❀ Learning the best practices for using fertilizers to achieve maximal productivity without environmental damage
- ❀ Understanding the basics of using crop residues, cover crops, organic amendments, and reduced tillage to sustain soil
- ❀ Learning how to use terraces, contour tillage, dryland farming, and proper management of irrigation to sustain soil and water

❧ Understanding the benefits of mulches in terms of reducing erosion and conserving water

❧ Using precision farming to sustain soil, water, and the environment

SITE SELECTION

The selection of an area for commercial horticultural operations or gardening is of utmost importance in terms of environmental management. Any one of several factors—poor climate, shallow soil, shade from nearby trees and buildings, poor drainage, insufficient water resources, or pollutants—can reduce the horticultural productivity of a site.

Climate

Obviously, the climate and the horticultural crop(s) to be grown must be compatible. Factors such as the length of the growing season (see Tables 12–1 and 12–2 in Chapter 12), seasonal temperature fluctuations, days to maturity, and relative hardiness of the plant material must be considered. The correlation between climate and crop(s) need not be as close for the home gardener as for the commercial horticulturist. With the latter, less than optimal conditions can mean financial disaster. The backyard horticulturist does have the option of altering the environment on a small-scale basis. Microclimates (see Chapter 6) must also be carefully weighed. The localized effects of natural and artificial surface phenomena on frost dates, soil temperatures, air temperatures, wind movement, and pollution levels might alter the desirability of a site for horticultural production or landscape horticulture.

Site Factors

Other environmental factors besides climate must be considered. The soil should be deep, fertile, easy to work, and well drained. Gravel, rock, or a subsoil layer of hardpan is a feature to avoid. Soil of less than ideal conditions can be improved through addition of fertilizers and soil conditioners, but the expense and labor must be weighed against the attributes or desirability of the site. The site should have ample water of good quality, whether from rainfall or irrigation. Direct sunlight should be available all day long. As the available light decreases, the types of crops that can be grown and their favorable development decrease. Shade from trees or buildings is to be avoided. In some cases, tree roots can produce root chemicals that inhibit nearby plants and also compete more successfully than smaller plants for nutrients and water.

Surface grades must be considered. A level site is acceptable if drainage is good; it is desirable if it has sandy or gravel-type soils that increase water infiltration and is preferable to sites with grades of more than 3%. A southerly slope of not more than 1.5% is favorable for early crops. Grading modifications between 0.5% and 1% are sometimes needed to improve drainage and reduce surface water accumulation, especially with fine-textured soils that have poor structure.

The site should not have low places where water can accumulate. Water from surrounding areas should not drain onto the site, and areas prone to flooding from streams are best avoided. Good drainage is essential. Drainage can be improved through the singular or combined use of agricultural tile, ditching or channeling to divert nearby water flow into the area, breaking up of hardpan subsoil plus incorporation of physical amendments such as sand, and modification of surface grades.

The site might also require protection against wind and animals. A windbreak, such as a hedge, a shelterbelt of trees, or a board fence, can provide wind protection. Fences can keep out stray or wild animals. In addition, the proximity and future encroachment of suburbia must be considered by those involved in the commercial aspects of horticulture.

In today's age of energy shortages, the local availability and proximity of alternate energy sources is something to consider. For example, a greenhouse industry in the Midwest about to have its natural gas curtailed in the dead of winter obviously has a need for alternative energy sources.

Pollution also plays a role in site selection. An area subject to atmospheric pollution, such as ozone or ethylene pollution, would not necessarily be a good choice for commercial horticulture. A home gardener would not want to locate a garden in the leaching field of the septic tank, because of the risk of pathogenic contamination of root crops. Areas known to contain lead from old paint residues scraped off of houses are to be avoided. Areas containing hazardous chemicals in the soil are absolutely unacceptable. Questionable areas need to undergo a thorough environmental audit, including soil and water tests for hazardous chemicals.

Drainage

Soil improvements usually deal with the physical condition and fertility of the soil. One physical condition that can be improved is drainage. Poorly drained soils are unsuitable for horticultural applications for many reasons, such as the decrease of soil oxygen caused by excess soil water or the slower warming of wet soils in the spring. Generally, some form of artificial drainage is required when the water table is within 3 feet of the surface, or when excess surface water cannot infiltrate fast enough to keep from seriously depleting soil oxygen.

Sites must be examined carefully to ascertain whether drainage problems exist or have the potential to occur, and whether the problems are associated with the surface, the root area, or the subroot zone area. This analysis will in effect determine the probable modifications required to alleviate the drainage problem. A further determination is necessary to ascertain whether the probable solution is actually possible. This determination is the *drainage capacity* of the soil, which is equal to the percent of total pore space minus the percent of pore space occupied by water at field capacity. The actual technique for making this determination can be found in the texts suggested in the Relevant References section at the end of the chapter.

Soils with surface drainage problems are often level, fine-textured soils with poor structure found in areas of high rainfall. Such soils usually have low drainage capacity, ruling out the use of agricultural tile drainage in the subsoil. Modifications to surface drainage include slight grading (0.5% to 1%) and digging of surface drainage ditches (Figure 11–1).

Drainage problems in the root area can be corrected by breaking up subsoil hardpan (if present) and incorporating physical amendments, such as 70% (by volume) sharp sand and

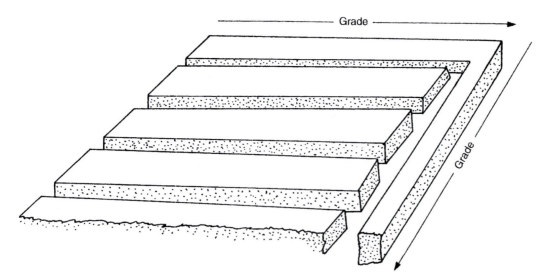

FIGURE 11–1   Trenching to produce feeder ditches that terminate at a main drainage ditch can alleviate surface drainage problems. Assuming proper grading, the main ditch will drain the area by gravity-produced water flow. (From Raymond P. Poincelot, *Horticulture: Principles & Practical Applications*, 1980. Reprinted by permission of Prentice-Hall, Upper Saddle River, New Jersey.)

FIGURE 11–2
Drainage problems in the subroot zone are eliminated by the installation of plastic or tile drainage systems. A machine for installing plastic drainage system is shown here. (USDA/NRCS photograph.)

30% partially decayed organic matter to improve aeration. Drainage problems in the subroot area (i.e., the soil above the root area has adequate drainage capacity) are alleviated by the installation of a tile drainage system (Figure 11–2). The placement of the system and its depth depend on several soil parameters, the types of plants to be grown, and the economical feasibility.

ON-SITE PROTECTION: WIND, SUN, COLD, AND FROST

For various reasons, sites can be less than perfect because of unexpected weather abnormalities such as late spring or early fall frosts. Sometimes a climate might be almost perfect except for prevailing winds. A mild winter climate might have the trade-off of a hot, sunny summer. Other places might have great summers but very cold winters. Sometimes new plantings in their first year might need winter protection, but not in later years when they are more developed. Awareness of these problems is the first step. The next step is the use of certain horticultural practices to solve the problems, such as planting trees as windbreaks or the placement of trees in specific locations to control temperatures.

Protection from Wind and Sun

In some areas the prevailing winds cause problems such as wind damage, snow accumulations, temperature drops, and soil erosion. These areas can occur near the seashore, in level exposed areas, or on hilltops. The unsymmetrical shaping of isolated trees is a good indicator of wind problems that might require modification. In the same or other areas, the sun might make temperature control within structures more difficult. Both problems can be modified through the judicious use of appropriate plant materials. The landscape horticulturist must choose plants not only on the basis of appeal, but also for their modifying influences on the nearby environment. Commercial fruit and vegetable growers and greenhouse operators also need to understand how properly placed plants can modify the environment.

Windbreaks (Figure 11–3) are used to reduce or deflect excessive winds. Most of us are familiar with the moderating influence of windbreaks on open spaces, such as the reduction of the windchill factor. Crops can be protected from wind damage, drying winds, snow accumulations, and sudden temperature drops. Soil erosion by wind can also be reduced by the proper placement of a windbreak. The effect of a windbreak on horticultural and other buildings is important. Greenhouses and homes exposed to extensive prevailing

(A)

(B)

FIGURE 11–3   (A) A cherry orchard in Van Buren County, Michigan, is protected by a Poplar windbreak (*right*). (B) A Resource Conservationist in a flower crop field with a windbreak in background in Allegan County, Michigan. (USDA/NRCS photographs.)

winds require greater energy input for heating than those not exposed to such winds. Unheated facilities can also benefit from wind protection, since their period of usefulness can be extended by the judicious placement of a windbreak. Windbreaks can also deflect wind to another location; for instance, they can be used to guide summer breezes through an outdoor patio. Proper placement of plants can also be used to reduce the speed of wind, rather than deflect it elsewhere.

Windbreaks consist of natural materials, such as hedges or trees, or artificial materials, such as the lath fencing used to control snowdrifts. If the prevailing winds are a winter problem, evergreen plantings would deflect them, whereas deciduous trees would only reduce the wind speed somewhat during the winter. In general, windbreaks are useful for a distance that is 10 to 12 times their height; the maximal protection occurs at a distance up to five to six times the height of the barrier.

Trees can also be used to moderate temperatures in nearby areas. During the summer, the proper placement of deciduous trees (Figure 11–4) can provide shade for homes, greenhouses, and farm outbuildings, thus reducing the amount of energy consumed by an air conditioner or an evaporative cooler or making an uncooled building more comfortable. During the winter, the nature of the deciduous tree would permit the sun's rays to warm the home and provide light to the greenhouse. During a hot summer, the shade of a tree over a patio is most welcome. In areas where sun is a year-round problem, the use of evergreen trees is preferable over deciduous ones.

Windbreaks have even greater future potential for sustainable horticulture. Future possibilities include the incorporation of woody crops into windbreaks so that the "protection" also generates income. Irrigation wastewater could be funneled through windbreak areas and be partially cleaned by the trees and understory plantings. Global warming might be reduced as more windbreaks are planted and carbon is sequestered by the growing trees and accumulated organic matter in the soil. Optimizing plant diversity in the windbreak might attract insects more (trap cropping) and reduce damage to nearby crops.

Winter Protection

Plants in exposed locations, particularly in the northern and northeastern United States, might require some form of winter protection to improve their survival and development rates (Figure 11–5). The protection is mainly against heavy winds that can desiccate foliage;

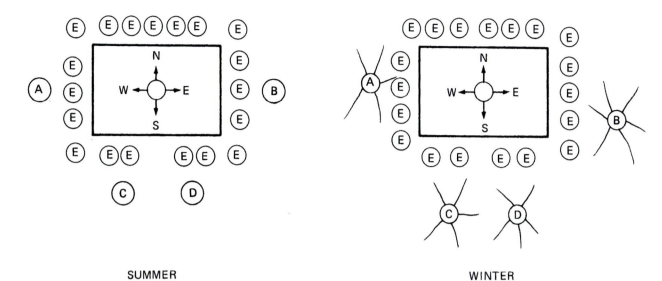

SUMMER   WINTER

FIGURE 11–4   The placement of trees and shrubs can make a home more comfortable and save energy. A high-crowned deciduous tree (A) on the west, a low-crowned deciduous tree on the east (B), and other deciduous trees to the south (C, D) shade the house in the summer. The cooling effect of the shade reduces interior heat buildup, thus minimizing the need for fans or air conditioners. In the winter, the deciduous trees permit warming of the house by the sun, thus reducing heating costs. Additional energy efficiency is provided with low height needled and broad-leaved evergreens (E) around the house, which create a still airspace. This airspace reduces heat loss in the winter and heat entry during the summer through the house foundation and lower walls. (From Raymond P. Poincelot, *Horticulture: Principles & Practical Applications*, 1980. Reprinted by permission of Prentice-Hall, Upper Saddle River, New Jersey.)

extremes of heat and light during winter that increase problems of desiccation; alternate freezing and thawing that could heave a plant from the ground; and protection against heavy snow that can cause physical damage to the plant. The foremost plan against winter threats is the optimal use of horticultural practices to build up and maintain good plant vigor.

Evergreen plants are susceptible to desiccation during the winter, especially when a combination of winds, bright sun, and moderate temperatures occurs (see Winterkill, Chapter 6). The leaves transpire, but the roots cannot take water from the frozen soil; hence leaf desiccation is the result. Protection includes the erection of burlap screens or the use of antitranspirant sprays.

Young, unestablished plants can be carried through the first winter by erecting a burlap square around them and filling it loosely with Styrofoam chips. Young trees often need a wrapping of burlap or tree-wrapping tape around their trunks to prevent winter sun scald the first winter after being transplanted. Soil can be mounded around tender woody plants, such as roses, to protect crowns.

During the period of the year that includes late winter to early spring in the North, newly planted ground covers and newly planted or shallow rooted herbaceous perennials and biennials are prone to heaving when soils (particularly fine-textured soils) undergo thawing and freezing sequences caused by succeeding periods of bright sunny days and cold nights. Applying a light, loose organic mulch to these plants after the ground freezes in midwinter helps to conserve the movement of heat and thereby maintain the soil in a constant frozen condition. This mulch is removed in early spring when air temperature fluctuations are not so extreme. Mulches can be pulled over low, tender plants for additional winter protection. The mulch must be loose and fluffy to ensure good air circulation. Winter mulches usually require some stabilization against winds. Snow damage can be minimized by placing a lath or board structure over the tops of susceptible plants, by tying them together with cord, or by shaking the snow off before it freezes. The labor of winter

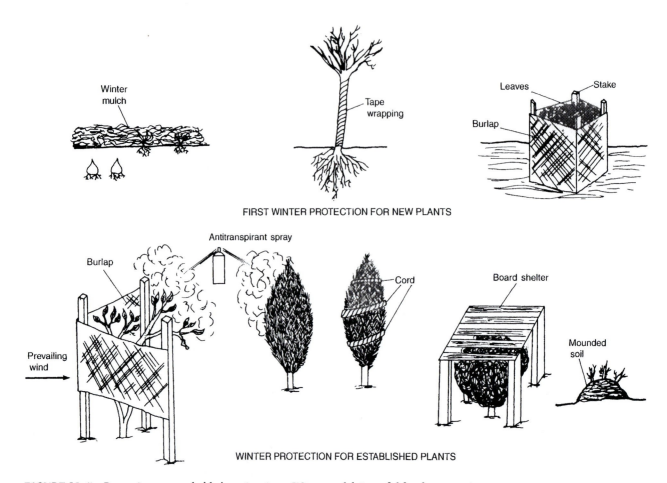

FIRST WINTER PROTECTION FOR NEW PLANTS

WINTER PROTECTION FOR ESTABLISHED PLANTS

FIGURE 11–5   Protecting new and old plants in winter. Winter mulch is useful for the protection of newly planted bulbs and herbaceous perennials and for these same plants once established, if they exhibit borderline hardiness in your area. Tape wrapping protects newly planted trees during their first winter and is not necessary after that time. Antitranspirant sprays can be used to reduce damage from winterkill for broad-leaved evergreens, although it would be better to place them in the proper location to avoid such problems (see Chapter 6). Burlap screens can be used to reduce winter desiccation from winds in windy or seashore areas. Cord tying or board shelters can reduce damage from ice and snow in areas of heavy snowfall, ice storms, or under roof edges from drop off. Mounded soil can protect tender roses and other tender perennials in areas with no winter snow cover or if the plants show borderline hardiness. A borderline hardiness shrub can be well protected with burlap or plastic mesh containment filled with loosely packed leaves or even better, Styrofoam chips. (From Raymond P. Poincelot, *Horticulture: Principles & Practical Applications,* 1980. Reprinted by permission of Prentice-Hall, Upper Saddle River, New Jersey.)

protection can be minimized by avoiding sites with exposure problems (for example, winds or falling snow from roofs) and avoiding plants that have borderline hardiness in your hardiness zone (see Table 12–3 and Internet sources in Chapter 12).

Frost Protection

The propagation and production of many annual flowers and vegetables is keyed to the variable known as the first frost-free date (See Figure 12–1, Tables 12–1 and 12–2 in Chapter 12). It is a variable because of yearly seasonal fluctuations and because garden and field plants themselves vary with respect to their tolerance to frost and freezing temperatures. The bedding plant industry must be able to calculate for each species, variety, and cultivar sufficient lead time so that plants will have time to reach "landscape size" at the time of sales or planting after the last spring frost date, yet still have time to mature before the fall frost. Failure to pay heed to the local environment means poor timing and financial losses in the marketplace.

If at all possible, tender plants should always be planted after the frost-free date. Plants susceptible to frosts at certain stages, such as the blossoming period of fruit trees when a frost would decrease fruit set, should be planted in microclimates such as thermal belts or near large bodies of water (see Chapter 6). A southern exposure can offer better frost protection than other exposures.

The possibility always exists that a late, unexpected frost will damage tender plants recently planted or established plants that are at a critical developmental stage. Here it might become necessary to protect against frost by conserving or adding heat. The gardener can protect against a light frost by laying newspapers, polyethylene sheets, bushel baskets, or cloth over the threatened plants the evening before the frost. This acts to conserve the soil heat and keeps the plants from direct contact with cold air. Often the gardener might wish to risk the chance of frost to get an early start with tender plants such as tomatoes.

A hot cap, a caplike container made of translucent paper, can also be set over the plants. It acts like a greenhouse to increase the soil temperature and conserves heat at night. Plastic tunnels on wire horseshoe-shaped inserts and lightweight spun polyester without hoops are also utilized for this purpose. Hot caps are now seen only in gardens. Polyester and plastic tunnels are used on a commercial scale for high-value horticultural crops to produce earlier harvests, while offering frost protection (see Chapter 13). These structures are also used in gardens for similar reasons.

At one time, smudge pots were used frequently to combat frost in citrus groves and fruit orchards (apple, peach, pear). However, air-quality standards now make them illegal in many areas. Basically, orchard owners burned a cheap oil that gave some heat at tree level, and the black smoke cloud reduced the loss of heat by radiation. More efficient, clean-burning heaters are used now that depend on heat output and convection currents to reduce the danger of frost. Fuels are usually a good grade of oil or propane. Solid petroleum wax candles are also used.

During a radiation frost (see Chapter 6), the plant temperature and the air immediately adjacent could be cooler than that of the air above. Mixing of the air above by fans increases the temperature around the plant and protects it from frost if the radiation frost is not too severe. Fans and heaters together are even more effective for orchards.

Water fogs can also reduce the danger of frost. When water turns to ice, energy, termed heat of fusion, is released. This heat release might be enough to protect against a mild frost. The formation of a white frost on a plant can be protection against frost damage if the air temperature does not decrease too much. Fog also provides a cover to reduce the heat lost through radiation. On occasion, plants touched lightly with frost can be saved by sprinkling them with water before sunrise. A newer technique involves the insulation of the plant with a spray foam. The foam is nontoxic and usually disperses the following day.

SOIL FERTILITY

In terms of soil sustainability, fertility rates a high priority, given its direct relationship (Figure 11–6) to crop production, profitability, and environmental well-being. If fertility is limiting, crop production and profitability decline. If fertility is too high (but not to the point of phytotoxicity), environmental pollution can result. The excess nutrients also decrease profits. It is in the best interest of all farmers and growers, whether organic, sustainable, or conventional, to maintain soil fertility within acceptable limits.

Nutrient Deficiency

An imbalance in nutrients in the soil can cause plant stresses that result in reduced growth and quality. Insufficient nutrients over a period of time produce typical plant-nutrient deficiency symptoms. These symptoms result from an actual deficiency of the nutrients or the presence of the nutrients in chemical forms unavailable to plants. The latter, as discussed in Chapter 6, is a function of the soil pH, cation exchange capacity, nitrogen fixation, solubility

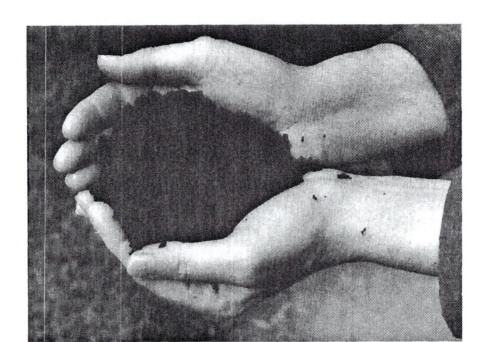

FIGURE 11–6   The key to sustainability is soil. Soil with maintained levels of organic matter, modest nutrient levels, and few or no pesticide residues exhibits good soil health, thus supports a thriving microbial and worm ecosystem. This soil would have minimal soil disease problems, as the antagonistic and competitive effects of the ecosystem's organisms exert a modifying or controlling effect over pathogens and even nematode predation. Some of these same organisms also help moderate nutrient availability and improve nutrient uptake by roots, thus maintaining crop productivity with minimal inputs of water and nutrients supplemented with leguminous cover crops. In turn, plants produced with continuous nutrient availability tend to be healthier, unstressed plants that exhibit fewer pest problems. (Author photograph.)

properties, and mycorrhizae. Excess nutrients can also have detrimental effects, such as phytotoxicity or abnormal growth excesses. For example, excess boron would result in plant death, or excess nitrogen could cause luxuriant leaf growth at the expense of flowers and fruit.

By the time the appearance of nutrient deficiency or excess symptoms is noted, plant growth and probably yield have been reduced. At this stage it is not always likely that the correction of the nutrient deficiency will make up for the lost growth. These symptoms are not always easily diagnosed. For example, purple coloration on leaves can indicate a phosphorus deficiency, but in tomatoes it might also be indicative of a cold soil. Yellowing leaves are not easily diagnosed. Nitrogen deficiency appears as yellow-green leaf color on a normally dark green plant, potassium deficiency appears first as yellow areas and these areas become dead and brown later, and iron deficiency (but also a magnesium deficiency or manganese deficiency or excess) appears as yellowing leaves with green veins. It is obvious that nutrient deficiencies are best avoided by early determination of soil fertility. If symptoms do appear, it might be necessary to resort to analyses of nutrient concentrations in the plant tissue coupled with soil analyses to determine the exact cause.

Nutrients removed by crops must be replaced if the soil fertility is to be maintained. Cover crops, such as legumes, returned to the soil cannot always supply enough nutrients for some of today's high crop yields, so some supplemental fertilization is needed. A soil test will help determine what nutrients, secondary and micronutrients, and lime and soil conditioners are required, as well as the amount required. An application of the appropriate analysis fertilizer will be recommended.

Soil and Water Tests

Both soil and water should be tested before putting new sites into commercial horticultural production. For home gardens, soil tests are important too. Soil tests should be conducted at least annually for existing fields and gardens. If the water source is tap water or well water tested and approved for drinking quality, no further water tests are needed. A certified water testing laboratory should check new water sources for hazardous chemicals, metal content, boron, hardness, salinity, pH, and microbial contaminants.

The actual nutrient requirements of horticultural crops are based on several parameters, including soil testing to determine the total nutrients, the available nutrients and the factors contributing to a nutrient's unavailability; organic matter levels, soil pH, and plant tissue analyses to determine the actual amounts of nutrients absorbed by the plant. Together these are correlated to establish a relationship between plant development and the nutrient concentration of the plant tissue, as influenced by the levels of various nutrients in the soil. Standard curves derived in an empirical manner can be defined for each nutrient. These curves differ depending on the plant species. For example, the optimal range of nitrogen for a leaf crop, such as spinach (*Spinacia oleracea*), would be excessive for a non-leafy crop such as peppers (*Capsicum annuum*).

Additional tests are required for sustainable farming systems. The additional testing aspect has to do with determinations of soil quality (the term preferred by scientists) or soil health (the term preferred by producers). The following questions need to be assessed:

How well does the soil receive, retain, and release water and nutrients to crops?
How effectively does the soil promote root growth?
What is the biotic, especially the microbiological and earthworm, status of the soil?
How does the soil respond to management practices, especially in terms of texture, structure, and erosion change?

These tests are more difficult. Some can be done on farm with a little training. Others need to be conducted by professionals at places where soil testing for sustainable farming is well understood. An excellent reference explaining methods of soil quality testing is given in the Relevant References section at the end of this chapter (Doran and Jones, 1996).

Conventional soil diagnosis or testing services are usually available at moderate cost through your state agricultural experiment station or state extension service. Commercial labs offer similar services. Portable chemical test kits can be used by individuals, but they are not as reliable as the more extensive instrumental methods of analysis used by soil testing laboratories. However, they can be a valuable aid in indicating problems that can be confirmed or eliminated by a more comprehensive soil diagnosis. Analyses of nutrients in plant tissue (frequently leaves) are usually conducted in commercial or university laboratories.

In actual practice, the horticulturist relies mainly on soil analyses before planting each crop. The nutrient needs of most horticultural crops have been established through the plant tissue analyses. Soil analyses provide a base point, and the analyst can recommend the nutrients required to produce good results with the crops to be planted. Computers are increasingly being used to interpret soil test results and to make fertilizer recommendations.

Sampling. A key step in soil analysis (Figure 11–7) is to obtain a soil sample that adequately represents the root zone area. For uniform mixtures of prepared propagation or growing media, a composite sample from two or three samplings is sufficient. For small areas (such as a garden), or larger areas up to 10 or 15 acres with uniform features, a composite sample should be prepared from not less than 20 samplings. Twenty samples are ideal based on statistical studies. Features to consider in terms of uniformity include soil productivity, soil texture, color of topsoil, topography, tillage, drainage, and past management. If large areas show variation in the indicated features, fields should be divided into smaller areas with uniform features. Each smaller area should be represented by 20 samples.

Each of the 20 samples should be a vertical slice or column of unfrozen soil of moderate moisture content taken at plow or spading depth (or deeper if the root zone will exceed this depth) and about 1.27 centimeters (0.5 inch) in cross section. If the lab suggests a depth, follow the suggestion. If not, 12 inches deep is reasonable. The 20 samples should be mixed thoroughly in a clean container. Use a plastic bucket, because metal buckets can lead to false readings with certain nutrients. This composite sample then provides the material for the soil test. About 0.5 to 1 pint of soil is required for the analyses. Subsoil samples

FIGURE 11–7 An ARS physical scientist (*left*) checks a soil sample while an ARS soil scientist (*background*) withdraws another. In the center, a University of Nebraska graduate student operates a GPS device that allows them to relate aerial images to the sampling sites. Thus information, when used with a GPS system at application time, allows for custom applications of fertilizers across fields with differing fertility needs (see precision agriculture later in this chapter). (USDA photograph.)

taken at a depth of 46 to 61 centimeters (18 to 24 inches) can provide additional input for making lime recommendations.

Information should also be supplied with the composite sample for the best fertilization recommendations. Desirable information includes the crop to be planted, the crop planted previously, the yield goal, previous liming and fertilization history, plow depth, and any unusual problems or conditions noted by the grower. One condition to note is drought, which can temporarily increase soluble salt levels and depress pH (see Chapter 6). Because the pH and soluble salts will revert to predrought levels when the drought is ended, more lime than is needed could be falsely indicated. Feel free to provide as much field information and history as possible, because results are improved with additional information.

Test Recommendations.   Reports will furnish the amateur and commercial horticulturist with information on fertility needs, pH needs, and organic matter needs. The response will include specific fertilizer, limestone (or sulfur), and organic amendment recommendations and amounts. Tests for soil quality can involve things like respiration rates (an indicator of microbial activity), organic matter levels (relates to erosion, nutrient and water retention, tilth, microbial activity, and plant root penetration), and aggregate stability (indicator of erosion, water movement, and root growth potentials).

The actual recommendation in soil test results could be based on optimal plant response or on the most economical plant response. This point might not matter much to the gardener, but it certainly does to the commercial horticulturist who wants to minimize costs. This situation results from the nonlinear response of yield versus fertilization cost (Figure 11–8). As fertilizer applications (and hence costs) are increased to reach optimal plant response, the incremental yield increase realized decreases as optimal plant response is approached. In other words, the dollar return measured against fertilization costs can be greater at some point short of the optimal plant response. The situation can be altered by factors such as density of planting, because a planting of high density could make higher applications of fertilizer more profitable than a planting of low density. Factors such as plant variety, insects, disease, weeds, inadequate water, and unfavorable temperatures can make the yield response less than optimal for a given amount of fertilizer.

Another point in favor of the prudent use of fertilizer is to prevent the accumulation of soluble salts with continuous, excessive use. These accumulations appear as a white surface crust. Excessive salinity can cause foliar burn, retarded growth (and profits), and even plant death. Overfertilization should be avoided. Periodic irrigation might be required to

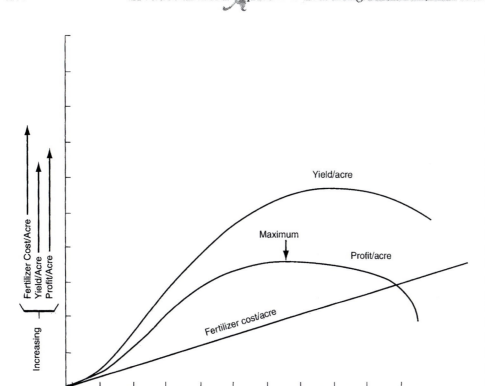

FIGURE 11–8 Fertilizer yield response graph. The maximal profit per acre as a function of fertilizer input occurs somewhat before the yield per acre maximum. This outcome results from the combined effect of the incremental leveling off and decline of the yield/acre versus the linear climb of fertilizer costs. In plain English, the profit returned for each dollar invested in fertilizer between the maximal profit/acre and maximal yield/acre is a losing expenditure. (From Raymond P. Poincelot, *Horticulture: Principles & Practical Applications,* 1980. Reprinted by permission of Prentice-Hall, Upper Saddle River, New Jersey.)

reduce excessive salinity. From the sustainability perspective, a compelling reason is that decreased fertilizer use reduces pollution of watershed areas.

Soil Tests and Sustainable Agriculture

Soil tests have some faults. Variability from lab to lab with the same sample is common. Differences in recommendations result from choice of chemical analytical techniques, technician experience, interpretation of results, the extent of field correlation data used to evaluate soil test data, and whether all factors are considered. Many results are expressed in a conservative fashion; that is, overrecommendations are built in to reduce the risk of inadequate nutrient levels. Some places fail to consider the nitrogen contribution provided by legumes used in rotations or as cover crops. The latter especially concerns organic and sustainable farmers, because it means that excess nitrogen will be used. Not only is this costly, but it increases the likelihood of water pollution. Many tests also fail to address criteria that are important for assessing soil health.

The best solution is to be an informed client. Utilize university, extension, or commercial labs that serve substantial numbers of organic or sustainable farmers. A good clue regarding the existence of a lab is whether the university offers a program in sustainable agriculture. If it does, it probably has a lab. Ask questions, such as do your results take into consideration the amount of nitrogen contributed by legumes?

SOIL ORGANIC MATTER

Organic matter is a critical link in the soil ecosystem. Organic matter plays a role in soil fertility, crop productivity, erosion, water retention, nutrient availability, soil tilth, soil health, and even some pest problems. Maintenance of organic matter runs close behind fertility maintenance as a critical component of soil sustainability. From the organic farming viewpoint of sustainability, the need to maintain soil organic matter is as critical as maintaining fertility.

Functions in Soil

As it decomposes, organic matter provides nitrate for plants. The nitrate is produced by the microbial transformations of complex organic molecules, such as proteins, to ammonium and then to nitrate. Phosphorus, sulfur, and many trace elements are also provided by this conversion process that we call *mineralization*. This contribution to soil fertility is especially important when fertility is maintained by adding complex organic fertilizers and plant residues to the soil. The nutrients in these materials need to be mineralized before becoming available to plants.

Organic matter also contributes indirectly to soil fertility by its regulation of nutrient availability. Organic matter, especially humus, has many negative charges. These charges attract positively charged nutrients (cations); this holding action reduces the losses of nutrients such as calcium (Ca^{2+}) and potassium (K^+) from soils by the leaching process. These held nutrients are released as needed by the cation exchange process described in Chapter 6. Clay also exhibits this property, cation exchange capacity (CEC). While the CEC of organic matter equals or exceeds that of various clay types on an equal weight basis, clay usually contributes 50% to 90% of the CEC because of its higher amounts in most soils utilized for horticultural production. Even so, it is important to keep levels of organic matter properly maintained to ensure optimal CEC along with the other beneficial contributions of organic matter. An exception is oxidized tropical soils with kaolinite clay where organic matter maintains the majority of the CEC.

Another indirect contribution to fertility results from the chelating ability of organic matter. Certain trace elements, such as iron, manganese, and zinc, can be held in chelated form by organic matter. Chelation keeps these trace elements from forming chemical complexes in the soil with other nutrients. Such complexes tend to make trace elements unavailable to plants. Often when nutrient deficiencies appear for trace elements, the real cause turns out to be low organic matter. Enough trace elements are present, but they have become unavailable. Also by forming chelates with iron and aluminum, organic matter reduces the ability of these two chemicals to chemically bond with phosphorus. Phosphate complexes with aluminum or iron make phosphate unavailable to plants. Finally, soluble aluminum is toxic to plants and can be present in acidic soils. The chelation of aluminum prevents toxicity.

Soil health is better when a reasonable amount of organic matter is present. Microbes of all types thrive in the presence of decomposing organic matter. Many of these microbes are beneficial. For example, the ones that release nutrients from organic matter help reduce the need for fertilizers. Others can be helpful in insect, disease, and nematode control; the predatory fungus that attacks nematodes thrives in soils rich in organic matter. Mycorrhizal fungi improve nutrient uptake, plant health, and soils (see Chapter 6). Other microbes help keep populations of pathogens under control. Still others can decompose pesticides held by organic matter, thus reducing the threat of water pollution. Some soil bacteria are capable of fixing nitrogen from the atmosphere. Legume crops can directly utilize this fixed nitrogen, when such nitrogen-fixing bacteria live in their roots. Other nitrogen-fixing bacteria are free living, and their fixed nitrogen becomes released through mineralization. As organic matter declines and soil pesticides are added, these beneficial microbes decline. While this loss can be offset with increased fertilizer, water, and more pesticides, increased costs and pollution become the penalty. It is far better to encourage this biological diversity of soil microorganisms and reap the benefits of lower fertilizer, pesticide, and water needs as well as less pollution.

In Chapter 6, we learned that organic matter decomposition results in numerous microbial excretions and fungal hyphae. These products help bind and stick particles of sand, silt, clay, decaying organic matter, and humus into aggregates. High levels of aggregates result in soil structures having fluffy, loose matrices loaded with channels and pores. The soil is more easily worked, less prone to compaction, aerates well, and holds water well. Surface water infiltrates more readily, reducing both erosion and the need for irrigation. These properties, so beneficial for horticultural crop production, diminish rapidly as organic mat-

FIGURE 11–9 One of the gases causing the greenhouse effect over the Earth is carbon dioxide or CO_2. ARS soil scientists are searching for ways to collect, or sequester, this CO_2 with vegetation. Here two ARS soil scientists discuss regions where winter cover crops or other high-biomass crops could be used to sequester carbon. (USDA photograph.)

ter declines. The organic matter tends to darken the soil's color, thus increasing heat retention. This warming means a slightly early planting date and improved germination. This same organic matter also acts as a buffer and moderates sudden changes in soil pH, such as when lime or chemical fertilizers are added. Sudden shifts in pH can reduce crop productivity as nutrient availability is heavily influenced by pH. Some evidence also exists that humus stimulates root growth and root hair formation.

Organic matter also ties up a lot of carbon in the soil. This carbon came from carbon dioxide (CO_2) removed by plants and passed through the food chain to animals and humans. When organic matter is oxidized in horticultural soils, it can be a major contributor to global warming through the release of CO_2. Practices that increase the addition of soil organic matter, such as cover crops, or reduce the oxidation of organic matter, such as conservation tillage, can reduce the increase of CO_2 in the atmosphere (Figure 11–9). Not only is accumulation and retention of organic matter good for horticultural crop production, it is also good for the environment, both in terms of slowing global warming and reducing the water pollution caused by soil erosion.

Relation to Soil Sustainability

An ideal loam contains about 5% to 6% organic matter by weight, while some sandy soils might have as little as 1%. Sandy soils in hot, dry areas such as California have low levels of organic matter. These soils pose difficulties in terms of organic matter maintenance, and raising the levels of organic matter is quite hard. Even at lower levels, the contributions of organic matter to productivity in these sandy soils are considerable. As such, efforts to maintain low organic matter levels should be a high priority. Continuous cropping reduces this amount of soil organic matter, perhaps by up to as much as 2% of the total organic matter per year. Unless the organic matter is replaced by some form of soil conditioner or cultural practice in addition to fertilizer, the long-term economy of plant production will suffer. Organic matter losses can be offset by increased fertilizer and water inputs, but only for the short term.

The key to sustaining soil for horticultural purposes is organic matter. This statement is based on the considerable importance of organic matter to productive soils and the fact that organic matter is the most fragile, easy-to-lose component in soils. Organic matter plays an important role in maintaining good soil condition and health and it also affects fertility, as discussed in Chapters 1, 6, and this chapter. Maintenance and the increase of organic matter require an annual commitment to good practices.

SUSTAINABLE MANAGEMENT OF SOIL RESOURCES

Soils in horticultural operations need special attention to maintain their soil health and optimal crop productivity. Why? To answer that question, let's look at soils in natural ecosystems. In prairie and rainforest ecosystems, soil health is maintained naturally through effective nutrient recycling and ongoing organic matter formation. As plants, animals, insects, and microorganisms live and die both in and on the soil, they contribute excretions and tissue that are transformed by microbial activity into nutrients and organic matter. Very little is wasted with this closed loop. When we use soils to grow crops, we alter this recycling loop. Most of the plants are harvested and so are the soil nutrients, which are removed with them. In addition, there is less plant mass to contribute organic matter back to the soil. Another factor is that the crops often have different pH requirements than the original, native plants.

We can compensate for the disruption of the nutrient and organic matter cycle by adding fertilizer and organic matter. We can alter pH by adding limestone or sulfur. How this cycle is restored depends on the management style of the farmer and grower. Conventional horticulturists rely heavily on adding chemical fertilizers and primarily adding back organic matter with crop residues. Organic horticulturists substitute organic fertilizers for chemical fertilizers and use multiple approaches to add organic matter. For example, the use of manure contributes both nutrients and organic matter. Animal manure might be supplemented with green manure and winter cover crops to add additional nitrogen and more organic matter. Those sustainable horticulturists who don't use total organic management techniques use practices that lie somewhere between these two examples. Sustainable management here might consist of using smaller, but more frequent applications of chemical fertilizers in conjunction with crop residues or organic fertilizers and a practice or two to increase organic matter. Of course, these are generalized examples, and considerable variability within each group exists.

Fertilizers

Fertilizers are composed of one or more nutrients in varying proportions. If a fertilizer contains the major plant nutrients, nitrogen, phosphorus, and potassium, we call it a *complete fertilizer.* Secondary elements and micronutrients are added in varying amounts to *specialty fertilizers.* Alternatively, such nutrients can be applied separately to counteract a known deficiency. Secondary elements are calcium, magnesium, and sulfur; micronutrients include boron, copper, iron, manganese, molybdenum, zinc, and chlorine, usually as the chloride form of one of the elements. Calcium and magnesium are usually applied through liming.

The nutrient content or the *analysis* or *grade* of most fertilizers is listed as a set of three numbers expressing the percentage by weight of total nitrogen, available phosphorus (expressed as P_2O_5), and soluble potassium (expressed as K_2O). To convert P_2O_5 into elemental phosphorus, multiply by 0.43, and to convert K_2O to potassium, multiply by 0.83. For the reverse, that is, element form to oxide, multiply by 2.33 and 1.20, respectively. Generally, fertilizers are expressed in terms of nitrogen and the two oxides for historical reasons. If the analysis is divided by the lowest common denominator, the result is known as the *fertilizer ratio.* To summarize, a 5-10-5 fertilizer contains by weight 5% total nitrogen, 10% available P_2O_5 (or 4.30% P), and 5% soluble K_2O (or 4.15% K) and has a fertilizer ratio of 1-2-1.

High-analysis or low-analysis fertilizers have available nutrients higher or lower than 30%, respectively. Examples of high-analysis fertilizers are ammonium polyphosphate (15-60-0) and urea ammonium phosphate (34-17-0). In terms of cost, the high-analysis fertilizers are a better buy, but they can cause fertilizer burns, so care must be used to apply only what is required.

Chemical and Organic Fertilizers.   Fertilizers derived from living or dead organisms and natural materials are termed *organic fertilizers;* those from chemically processed materials are *inorganic (chemical) fertilizers.* Examples of materials used to formulate fertilizers (or sometimes used alone) and their nutrient contents are shown in Tables 11–1 and 11–2.

TABLE 11-1 *Chemically Processed Nutrients*

Nitrogenous Materials	Nitrogen Content (% dry wt)
Anhydrous ammonia	82
Urea	46
Urea formaldehyde	38
Ammonium nitrate	33.5
Liquid ammonia	20–24
Ammonium sulfate	20.6
Diammonium phosphate	18
Sodium nitrate	16
Ammonium phosphate sulfate	16
Calcium nitrate	15.5
Potassium nitrate	13
Monoammonium phosphate	11
Magnesium ammonium phosphate	7–8

Phosphorus Sources	Phosphorus Content (P_2O_5)
Monocalcium phosphate	55
Monoammonium phosphate	48
Diammonium phosphate	46
Triple superphosphate	45
Magnesium ammonium phosphate	40
Single superphosphate	20
Ammonium phosphate sulfate	20

Potassium Sources	Potassium Content (K_2O)
Potassium chloride	60
Potassium sulfate	50
Potassium nitrate	44
Potassium frits	36
Potassium magnesium sulfate	22

Some confusion results from these terms. For example, it is easy to see that cow manure from a living cow and blood meal from slaughterhouse animals are organic fertilizers. But what is rock phosphate, a natural ore? If it is ground up to a powder and applied to soil, it is an organic fertilizer, by virtue of being derived from a natural material. A chemist would take affront at this use, though, since the material is an inorganic chemical compound. To the chemist, *organic* implies containing carbon, hydrogen, and oxygen, certainly not phosphate. If the rock phosphate is treated with sulfuric acid to produce superphosphate, the latter is considered a chemical fertilizer because the material was chemically processed.

Chemical fertilizers are easy to apply and their soluble nutrients are rapidly available. However, soluble nutrients are prone to leaching, and fertilizer burning can result with excessive amounts. Chemical fertilizers are either formulated in a dry, granular form for direct soil application, or formulated as a water-soluble powder for application in liquid form. Chemical fertilizers are energy-intensive products and use natural resources that are not sustainable in the long term. As such, the sustainable horticulturist uses chemical fertilizers sparingly (low-input sustainable horticulture) or not at all. Wherever feasible and economically sound, organic fertilizers and green manures are preferred.

Organic fertilizers have some of their nutrients in forms not immediately available to plants; however, the enzymic action of soil microorganisms converts the unavailable forms

TABLE 11–2   Biological and Natural Nutrients

	% Nitrogen	% Phosphorus	% Potassium
		(% dry wt)	
Bat guano	10.0	2.0	1.7
Blood meal	14.0	0.7	0.7
Bone meal (steamed)	4.0	10.1	0
Cattle manure	2.0	1.0	2.0
Compost	2.0	1.0	1.0
Cottonseed meal	6.0	1.1	2.3
Fish meal	10.0	2.6	0
Granite dust	0	0	4.1
Greensand	0	1.5	4.1
Hoof and horn meal	12.0	0.9	0
Horse manure	1.7	0.3	1.5
Olive pomace	1.2	0.4	0.4
Poultry manure	4.3	1.6	1.6
Rock phosphate (concentrated)	0	13.0	0
Seaweed	0.6	0	1.1
Sewage sludge (activated)	5.0	3.0	0.4
Sheep manure	4.0	0.6	2.9
Soybean meal	7.0	0.5	2.3
Winery pomace	1.6	0.7	0.7
Wood ashes (unleached)	0	0.9	5.6
Wool wastes	7.5	0	0

to usable nutrients. These fertilizers are not ideal when nutrient buildup in a hurry is needed. Organic fertilizers offer the advantage of nutrient availability over a period of time and less danger of leaching or burning. Organic fertilizers are much less likely to result in water pollution, unless excessive amounts are applied. Organic fertilizers are available for dry or liquid application. Organic farmers utilize organic fertilizers exclusively. Sustainable farmers vary from those that have total or partial dependence on organic fertilizers to those that only use chemical fertilizers, but in judicious amounts (low input).

Mixtures of organic and inorganic fertilizers that offer the advantages of both are becoming more common. An example of this type is a lawn fertilizer that contains inorganic nitrogen for quick use and organic nitrogen to extend nutrient availability, thus reducing the frequency of fertilizer application. Chemical fertilizers are also being produced in forms that extend nutrient availability and reduce the danger of leaching or burning. These slow-release forms include soluble nutrients coated with a resinous or membranous material or nutrients possessing reduced solubility. An example of a "synthetic organic" is ureaform. These synthetic organic mimics tend to cost more than the straight inorganic forms. They are often incorporated into growing media to reduce the need for supplemental fertilization of container plants.

Fertilizer Application Practices.   Several techniques are used for applying fertilizer (Figure 11–10). *Broadcast scattering* is the uniform scattering of fertilizer by hand or fertilizer spreader over the ground surface (suitable for gardens or small operations). *Plow-sole application* is the dropping by mechanical means of fertilizer behind the plow in the opened furrow. *Band placement* or *soil injection* is the application of fertilizer bands under or to the sides of the planted seeds. *Top* or *side dressing* is the application of fertilizer over the tops or sides, respectively, of growing crops. Soluble fertilizers can also be applied in irrigation water or as water solutions during transplanting. The former is called *fertigation* and the latter is termed a *starter solution*.

FIGURE 11–10 Techniques for applying fertilizer. (A) This tractor's attachment is planting and dropping granular fertilizer all in one operation. Note the no-till planting on the contour in a field in northwest Iowa. While the crop is likely field corn, it could just as easily be sweet corn. (B) This scene shows the application of anhydrous ammonia fertilizer at planting time in Cedar County, Iowa. (C) Water-soluble nitrogen fertilizer being applied to growing corn in a contoured, no-tilled field in Hardin County. Applying smaller amounts of nitrogen several times over the growing season rather than all at once at or before planting helps the plants use the nitrogen rather than have it enter water supplies. (D) Mounted on a high-clearance sprayer, crop canopy sensors monitor plant greenness, which is translated into a signal by an onboard computer that controls the application rate of nitrogen fertilizer to the soil. This high-tech approach uses fertilizer in an efficient manner. (Photographs A–C from USDA/NRCS, photograph D from USDA.)

Fertilizers can also be applied through the foliage instead of the soil. This method is termed *foliar nutrition* and is used to deal with special problems that cannot be solved readily through fertilization of the soil. Not all plants can tolerate foliar feeding; those that can generally have a heavy layer of cuticle wax, such as orchard trees. Situations where this approach is useful include the following. If a quick response is needed, such as with a sudden deficiency of a nutrient or a very rapid use of nutrients during a period of intense growth, foliar feeding responses are more rapid than conventional soil fertilization. The application of micronutrients as a foliar feeding is more effective than the application of chelated forms to soil, where in some cases soil reactions hinder the usefulness of the chelated form or the deepness of roots slows the effect. Foliar feeding can be more economical for some crops, such as forest trees, or serve as a useful alternative when soil fertilization is impractical, such as during a dry spell.

Factors Affecting Placement, Amounts, and Timing.   The use of the same amount of fertilizer and the same application practice and schedule for all crops is to be avoided from the sustainability viewpoint. Decisions on the amounts, timing of initial and supplementary applications, and placement are important if fertilizer usage is to be efficient. These decisions depend on several factors: the nutrient(s) needed, the type of crop and its changing nutrient requirements as a function of its development, the soil type and fertility levels, and the climate. It is especially important to relate applied amounts to specific crops. For example, some vegetables such as sweet corn or broccoli need more nitrogen than others, such as asparagus or carrot. Cool season crops often have higher nitrogen needs than warm season crops, because cool temperatures decrease the conversion of organic forms of nitrogen and even ammonium inorganic forms to nitrates. Levels of soil nitrate are also likely to be lower in the spring, because the previous year's warm season crops uses much of the nitrates mineralized from crop residues applied in the previous year. Leftover nitrates are likely to be leached during the winter, especially in areas with mild winters, sandy soils, and moderate to high rainfall, such as in Florida and California. Recommended fertilization amounts exist for all horticultural crops and are readily available through extension publications (see the Relevant References section).

It is especially important to apply nutrients closer to times in development when they are needed the most. Timing of nutrient needs differs among crops. Fruiting crops such as eggplants, peppers, and tomatoes don't require the bulk of fertilizer nutrients until the stage of reproductive development. Nonfruiting crops and cool season crops such as broccoli, celery, and lettuce take up nutrients slowly during the first half of the season and then suddenly increase nutrient uptake until just prior to harvest. Applying most of the nitrogen early at the sowing or seedling stage rather than later could bring about increased nitrogen leaching and pollution. Fruit trees have much lower nitrogen needs when very young as opposed to during reproductive maturity. Nitrogen needs also increase gradually each year as the tree approaches maturity. In the nursery, container stock nitrogen needs are lower in the earlier, cooler part of the growing season and higher as warmer temperatures prevail and growth becomes more active. Some of the many points to keep in mind with regard to timing and placement are as follows.

So when is the best time for fertilizer applications? Traditional times are just prior to or at the time of planting. Applications of fertilizers in the fall are another option, if the objective is to buy fertilizer at a lower price and to reduce the labor load at planting time. Fertilizers frequently cost more in the spring, when the demand peaks and the supply decreases. While this timing has been widely practiced for many years, other options may be more in harmony with sustainability. Ideally, fertilizers should be applied as close as possible to the time when the developing plant needs fertilizer nutrients, so as to minimize leaching during slow demand times.

The primary danger with fall applications of fertilizer is the loss of nitrogen (mainly as nitrate) through leaching, especially where percolation losses of water are great. This loss can be especially high and contribute to water pollution in areas where winter rains rather than snows are common. If economics dictate fall applications, the loss of nitrates through leaching can be minimized by using an ammonium form, such as gaseous ammonia or an NH_4^+ salt, and applying it to a soil that is 10°C (50°F) or colder. Ammonium nitrogen can be applied in the fall to acid soils, because conversion to the leachable form, nitrate, is slow. Nitrogen fertilizers can also be applied safely in the fall where a cool season crop is growing vigorously and will absorb the nitrogen, or where nitrogen will be utilized by decomposition of crop residues, thus tying up nitrogen and preventing loss by leaching. Fall applications of phosphorus pose no problem, and fall applications of potassium are only a problem on peat and muck soils where leaching losses are great.

One application of fertilizer at planting time, while preferred from the labor reduction and economics aspects, is not always optimal from the sustainability perspective. During early growth, as explained previously, many crops have only low to moderate needs for nitrogen. Losses of unused nutrients during periods of low demand, especially nitrogen, occur by leaching. Therefore, more frequent or supplemental applications (*split applications*) might be better. Split applications of nitrogen are recommended for fruit trees, grapes, container ornamentals, landscape ornamentals, turf, and vegetables. This approach is also recommended when the amount of nutrients required for an unproductive soil is so large as to damage tender seedlings if added in one application. Short season crops can be an exception. One application at planting time with string beans or potatoes is used fast enough that leaching is minimal. Even with multiple (split) applications, in most cases one can still apply all the phosphorus and potassium in the first application, as these two nutrients experience minimal leaching losses. The remainder of the nitrogen can be applied later when needed the most.

Supplemental applications of nitrogen are favored for many crops with changing nitrogen needs during development, in areas with high rainfall, in areas with long growing seasons, for multicropping, with long season crops (garlic), and with crops requiring high nitrogen. The amounts needed can be reduced by following the practices mentioned previously to reduce leaching. Soil testing and tissue analyses (leaf or petiole-sap analysis) are also suggested so that supplemental amounts are kept as low as possible. For best results, testing should be done in the laboratory environment. However, quick soil tests and more recently petiole-sap nitrate tests can be done easily on the farm. The latter test is done with a battery-operated Cardy meter (nitrate-selective electrode).

Nitrogen and potassium are more likely to be needed than phosphorus. The exception for phosphorus would be with soils formed from marine deposits having a pH range from 7.5 to 8.5 (calcareous soils) in cool periods that tend to fix phosphate. One exception for nitrogen is with peat and muck (organic) soils, where the oxidation of organic matter supplies sufficient nitrogen during warm weather. During cool weather, some extra nitrogen would be needed with heavy users of nitrogen, such as lettuce and sweet corn.

Soil textures known as sands, sandy loams, and loamy sands (mineral soils) require supplemental applications, because nitrogen and potassium tend to be highly mobile and easily lost through leaching. The challenge with mineral soils is that fertilizer applications require careful management to provide sufficient nutrients, but not too much to avoid pollution. Mineral soils needing careful management for sustainable horticulture are found in two major horticultural production areas, Florida and California. Applications need to be banded down into the soil close to the outermost reach of the root system. Care must also be used to not damage the crop's root system. A suggested placement is no less than 2 inches from seeds or plant rows and 4 to 7 inches deep. In supplemental applications, liquid fertilizer is easier to apply, especially with mulched crops where an injection wheel can be used. An alternative method is to use mulches and drip irrigation, through which soluble fertilizer can be applied (*fertigation*). Fertigation saves energy, conserves water, and reduces pollution caused by leaching (see later discussion under irrigation and previous discussion in Chapter 10).

If broadcast applications are used with nitrogen fertilizers, some methods to reduce losses and pollution from leaching, especially in sandy soils, exist. Plastic mulches reduce losses from heavy rainfalls, so less fertilizer is needed with this practice. When using plastic mulch, broadcasting or banding of fertilizers is done prior to placing the mulch. With nonmulched soils, required levels of nitrogen should be applied in two or more applications (split application) as opposed to one. This method reduces losses from leaching. One portion is applied at or near planting time and the second or third is added subsequently during the growing season.

Another choice, instead of additional applications, is to use a mixture of organic and chemical forms for extended and quick availability, respectively. Slow-release forms of fertilizer can also be used in place of the organic ones. Low to moderate, but steady supplies of nutrients are often sufficient in the earlier developmental stages of crops. Because nitrogen demand peaks during later development, chemical fertilizers can be used to supply higher levels of nutrients quickly during the window of high need. Slow-release forms are also particularly useful for container nursery stock in the field, where their addition reduces the frequency and labor of application. Another approach is to load organic nitrogen at high levels through legumes, so that by the time of peak demand mineralization should be releasing sufficient amounts of nitrates. Another possibility with drip-irrigated crops is to add soluble fertilizer through the irrigation system. Soluble fertilizers can be applied frequently in dilute forms, and the levels increased or decreased as developmental need dictates.

Leaching is essentially zero to low, because nutrients are present in amounts sufficient for crop needs, leaving little excess for leaching. This approach works well for high-value field crops with plastic mulches, nursery container plants, and greenhouse crops.

Nitrogen, Phosphorus, and Potassium Considerations.   Nitrogen can be taken up as ammonium or nitrate by plants, although nitrate is the preferred form. Ammonium can be held as an exchangeable, readily available form on clay and humus, but it is easily converted into nitrate by bacteria in warm soils. Nitrates are easily lost by leaching, especially in sandy soils. Therefore, high levels of nitrogen cannot be maintained in cropping systems without supplementary additions. Nitrogen can become tied up and unavailable to plants in soils where active decomposition of organic matter is occurring until the carbon to nitrogen ratio is reduced to 12:1. Nitrogen availability can become reduced in cold and/or wet soils, such as in the spring, because of unfavorable bacterial actions that decrease levels of nitrogen.

Nitrates are the preferred form for application in cold or strongly acidic soils, given the slowing of conversion of ammonium forms to nitrate. In Florida, it is recommended that 25% to 50% of the fertilizer for cold or strongly acidic soils be in the nitrate forms. In warm soils, either form can be used. However, several factors favor the use of ammonium forms or urea, which hydrolyzes rapidly to ammonium nitrogen in soil. Ammonium forms are less costly than nitrates. Nitrate applied directly is more likely to be lost through leaching and denitrification. Less nitrate pollution of ground and surface water results with ammonium forms. Ammonium forms or urea are the recommended forms in California vegetable production, given that nitrates in drinking water are an environmentally sensitive issue.

The solubility and toxicity of nitrogen compounds limit the amounts that can be applied at any one time. Excess nitrogen can burn seedlings; cause vegetative development at the expense of flower and fruit in ornamental, fruit, and vegetable crops; and cause poor-quality fruit on trees. Excesses at the season's end can produce lush vegetative growth that is susceptible to winter injury. Surface applications of nitrate forms are preferred over ammonium or amine forms (such as urea) on alkaline soils, since the soil pH favors production of ammonia gas (NH_3) from the latter two and subsequent losses through volatilization. Ammonia should be applied at a 6-inch depth to maximize soil absorption and to minimize losses. Nitrates or ammonium salts can be applied at the surface, and then rain or irrigation water will leach them to root depths.

Phosphorus mobility and availability in field soils is low in most alkaline and acidic soils. It is important to build up soil levels of phosphorus prior to planting because of limited availability, but the low mobility of phosphorus reduces the need for supplemental applications. If one initial application of phosphorus is used, organic matter should be maintained, because the presence of decomposable organic matter and mycorrhizal fungi in the soil increases phosphorus availability.

Because of low mobility, the recommended placement of phosphorus is close to the potential or existing root zone of seeds or transplants, respectively. This choice is better from a sustainability viewpoint, in that less localized applications require higher amounts to achieve the same level of phosphorus near the root zones. Higher amounts are more costly and lead to increased pollution. As such, banding applications provide more efficient use of phosphorus than broadcasting methods. If broadcasting is used, it is highly recommended that the fertilizer be incorporated into the soil prior to planting. Starter solutions containing 1,500 parts per million of phosphorus are especially helpful with transplants and subsequent establishment of good root systems. Their effects are especially noted with early growth in cool soils. Supplemental fertilization is often accomplished with a fertilizer lower in nitrogen and higher in phosphorus than the first applied fertilizer in order not to favor vegetative over reproductive development.

Most sources of phosphorus are equally effective. One possible exception is diammonium phosphate. Reductions in yield of certain crops have been noted with this material on sandy soils that have low levels of micronutrients when banded in mixtures containing micronutrients. Diammonium phosphate applications in unlimed soils having acid pH and large amounts of aluminum and iron are not recommended, because phosphate availability is reduced.

Potassium mobility is greater than that of phosphorus, but less than that of nitrogen. The solubility and toxicity of potassium suggest a placement that is not too close to seeds or plants. Injury is most likely with sandy soils. Potassium needs are less than nitrogen needs and slightly more than phosphorus needs, so the need for supplemental applications is low. One application is reasonable in most soils, except for conditions that favor leaching. Split applications as with nitrogen are best practices for the management of leaching losses where heavy rains and poor water retention favor potassium leaching.

Fertilizers and Sustainability.   Soil sustainability in the long haul will not be achieved with total dependence on chemical fertilizers. For one, chemical fertilizers are energy-intensive products and the raw materials are also limited natural resources, especially high-grade rock phosphate. These fertilizers also do not replace organic matter lost through cultivation, thus reducing overall soil health. The sustainable horticulturist will decrease dependence on chemical fertilizers by using them in smaller amounts and following the recommended timing/application practices (see previous sections). The next step is a gradual shift or even complete switch to renewable organic fertilizers and alternative practices that supply additional nitrogen and organic matter to the soil. Such practices include green manures, cover crops, rotations, and organic amendments.

The use of ground rock minerals such as rock phosphate and green sand to supply phosphorus and potassium in a slow gradual fashion through microbial action is more sustainable and offers the attraction of lower nutrient losses and less pollution. These materials, along with the above approaches to supply nitrogen and organic matter, need special management in order to be highly effective when used in low levels. To ensure yields at acceptable levels, levels of nitrogen, phosphorus, potassium, and organic matter must be at slightly higher levels than needed when nutrients are supplied in a soluble chemical fertilizer. For this reason, the transition from chemical to organic horticulture requires a transition period while nutrient and soil microbial levels are built to acceptable levels so that less soluble, low applications of organic and rock mineral fertilizers can be effective.

Adjusting Soil pH

A soil test is needed to determine the pH (see Chapter 6) of a specific soil. Soils can be naturally acidic as a result of rain leaching the natural, basic elements of soil and lime downward, or because soils were formed from acidic parent materials (granite). Generally, acidic soils in the United States are found in areas of high rainfall, that is, the eastern half of the United States and the coastal Pacific Northwest. Acid rain can add additional acidity to the soil. Neutral-to-alkaline soils occur in the arid western and intermountain areas, where soil cations can accumulate because of limited leaching by rain. Another factor, often overlooked, is that irrigation water from limestone aquifers (such as those found in Florida) can contain lime and this contribution needs assessment through water testing. Horticultural practices also contribute to soil acidity, in that plant roots secrete hydrogen ions, and most nitrogen fertilizers acidify the soil. Fertilizers ranked from the most to the least acid contributing are ammonium sulfate, anhydrous ammonia, ammonium nitrate, urea, and sodium nitrate. An indication of the acidity of many fertilizers is provided by instructions on how much lime is needed to neutralize them.

The pH preference of plants varies. Most horticultural crops in temperate areas grow reasonably well in mineral soils with a soil pH between 6.0 and 7.0. Organic matter content can influence this range, decreasing it as organic matter content increases. However, organic matter must exceed 10% before much effect is noted. For example, in Histosols (peat and muck), the ideal pH range for most crops drops to 5.0 to 5.5. Plant exceptions to the range of 6.0 to 7.0 include a number of succulents from arid regions that grow best in slightly alkaline soils (pH 7.5 to 8.0). Evergreens and a number of plants in the Heath family *(Ericaceae)*, including such plants as the gardenia, rhododendron, azalea, cranberry, camellia, blueberry, *Andromeda,* and mountain laurel, require a pH from 4.5 to 5.5.

Soils with pH values below 6.0 generally have decreasing availability of nutrients such as nitrogen, phosphorus, potassium, calcium, magnesium, and molybdenum. Other nutrients,

FIGURE 11–11 Lime reduces soil pH by replacing acid-causing H^+ on clay and humus particles. The released H^+ reacts with OH^- ions to form water, thus removing the contribution of H^+ to soil acidity.

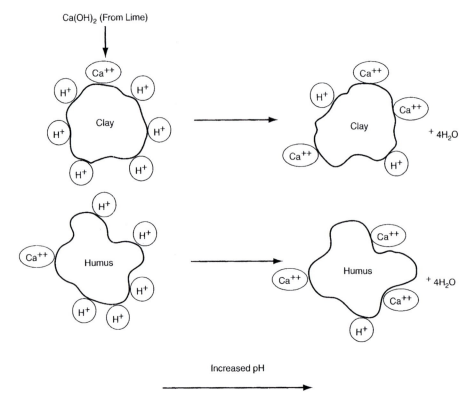

such as manganese, aluminum, and boron, become increasingly available and cause phytotoxicity. The usual recommended soil pH is 6.5. Acidic soils need to have their pH raised for most horticultural crops. Soil pH is raised by the addition of lime. Common liming materials include ground limestone or calcic limestone ($CaCO_3$) and dolomitic limestone [$CaMg(CO_3)_2$]. Less common forms are quicklime (CaO) and hydrated lime [$Ca(OH)_2$], which must be handled with more caution. Lime can also be applied in fluid form. The mode of action of lime in acid soils is shown in Figure 11–11.

Lime has other values besides the regulation of soil pH. These include the addition of the essential element calcium, magnesium if dolomitic forms are used, improvement of phosphorus availability, a beneficial regulation of potassium uptake, an increased availability of nitrogen as a result of accelerated decomposition of organic matter, and general improvement of physical conditions over a period of years.

The amount of lime to apply for the correction of acid pH is dependent on several factors. These factors are the soil pH as determined by a soil analysis, the pH preference of the crop(s) to be grown, and the purity of the lime. The particle size of the lime is a consideration, since the finer the particle the faster the reaction. For example, limestone of 100 mesh or finer will react within a few months of application, 60 mesh in 1 to 2 years, and 20 to 60 mesh in 2 to 4 years. The solubility of the different forms of lime also affects the reaction. The higher the organic matter and clay in the soil, the greater the amount of lime needed to adjust the pH, since their buffering capacity resists changes in pH. The amount of lime is not the same per change in pH unit, but changes with different ranges of pH. For example, less lime is needed to go from pH 5.5 to 6.5 than from 4.5 to 5.5. Remember that pH is a log function, so that the relative strength increases 10-fold for each unit change in pH.

The most efficient way to add lime is in small amounts yearly. However, the increased labor makes this a more costly approach. Besides the initial adjustment of pH, lime must be added afterward to replace losses due to leaching, removal by harvested crops, erosion, and the acidifying properties of fertilizer. Spreaders are the most effective means of liming, and the lime should be thoroughly mixed into the soil. Lime is usually applied in the spring or fall as the soil is cultivated. While lime can be left on the surface, best results are obtained if it is incorporated into the soil. Even with no-till practices, lime should be plowed,

chiseled, or disked in periodically. Lime and all fertilizers can be safely applied together and mixed into the soil. If they are not mixed into the soil, such as on a lawn, fertilizers containing ammonia derivatives of nitrogen should not be applied with the lime, because ammonia formation and subsequent volatilization will waste much of the fertilizer. However, fertilizers not containing ammonia nitrogen can be applied concurrently, or ammonia derivatives could follow the lime safely after at least a 2-week interval.

If it is desirable to acidify a soil, as in the Southwest, the most effective method is to apply sulfur. It is the least expensive acidifier and is long lasting. This might be desirable for growing acid-loving plants. For more rapid changes, aluminum or iron sulfate can be used. These are quick acting, but less effective over the long term than sulfur.

In summary, the maintenance of a desired soil pH is a yearly, ongoing process that requires annual fall soil testing for monitoring. This is necessary, in part, because of the influences of the environment (rainfall, plant materials, plant residues, soils), as well as cultural influences, including irrigation, mulching, and fertilization. These influences cause soil pH to change, resulting in a need for soil pH monitoring and modification.

Other Sustainable Practices for Soil Management

The minimally acceptable practices for sustainable maintenance of soil fertility include the judicious use of chemical fertilizers and organic fertilizers as already discussed. Crop residues represent the minimal approach to maintaining soil organic matter and partially eliminating soil erosion. Additional practices include adding green manure, animal manure, and various organic amendments. All of these are sources of organic matter, but they also add varying amounts of nutrients. Other practices conserve organic matter and reduce soil erosion by providing a covering on the soil, such as with cover crops. The cover prevents the loss of soil by water and wind, but also minimizes the oxidation of organic matter that can lead to increased soil losses. Cover cropping can also be managed to increase soil organic matter.

These practices are not new, having roots in our earlier agricultural history and even reaching back a few thousand years before our nation arose. Such practices, often used by organic farmers, are again increasing in popularity as farmers search for a more sustainable future. The choice must ultimately be based on site specificity, management goals, and the awareness that resource-sparing technologies require much more sophisticated management than resource-consuming technologies. In some instances, trade-offs can be required. For example, green manure can require energy and water inputs, but energy inputs are partially to completely offset by the need for less fertilizer. Growing green manure in the fall/winter fallow period and relying on fall rains can minimize irrigation needs and winter snows for water or, alternatively, a green manure that has low water requirements can be grown.

Crop Residues

Residues are often left after crops are harvested. Crop residues are often roots, but can also be stems and leaves. Crop residues (Figure 11–12) left on the soil after harvest and subsequently only partially plowed under the following season control soil erosion and maintain/increase organic matter in the soil. Other soil properties are also improved, such as water infiltration rates, water retention, and soil aggregation. Soil crusting is also reduced. Crop residues also provide nutrients: nitrogen, phosphorus, potassium, calcium, and trace elements. Their one disadvantage is that crop residues can introduce pathogens into the soil. Residues from crops known to have diseases should not be incorporated into the soil. Crop residues are most effective when soil is not exposed during any part of the year. Cover crops might be needed at some point.

The most widely used method of maintaining organic matter, especially in soils farmed by commercial horticulturists, is the crop residue approach. In general, soil fertility is maintained at high levels through the frequent applications of fertilizers, based in part on frequent soil tests. Plantings are spaced to achieve high density, and the crop stubble left after harvest is plowed partially into the soil at planting time. All field operations are often done

FIGURE 11–12 Crop residues are being plowed under with a chisel plow fitted with a blade chopper. While most of the residue is buried, some is left on the soil's surface. The buried residues will help maintain the soil organic matter, while the surface residues will increase water infiltration and reduce erosion. (USDA photograph.)

at one time: plowing soil and incorporating residues, seeding, fertilizing, applying herbicides, and applying pesticides. Minimizing tillage not only reduces soil compaction and saves labor and energy, but also reduces the oxidation of organic matter.

However, crop residues are not always the answer. With some horticultural crops, very little residue remains, such as with root crops, salad greens, and herbs. Sometimes alternative uses of residues compete, such as using residues for animal feeds or as mulches for another crop elsewhere. Another problem is that residues are often inadequate for restoring or maintaining organic matter on badly managed soils or on soils where rapid oxidation of organic matter occurs. Residues also don't provide enough cover to prevent unacceptable rates of erosion on highly susceptible soils. In these situations, other approaches are needed.

Cover Crop Background

Cover crops are not new. Their history goes back some 2,000 years to southern Europe and China. In our own country, Thomas Jefferson used cover crops at Monticello. Cover crops enjoyed a rebirth because of the organic farming movement and are even more attractive now with the growing interest in sustainable agriculture.

Depending on the management goal, the term *cover crop* sometimes is used interchangeably with other terms. For example, when cover crops are used to raise nitrogen availability and to maintain organic matter, the alternative term is *green manure*. When the goal is to use the crop to remove nutrients left over after a cash crop and thus reduce the potential to lose these nutrients by winter leaching, the term *catch crop* can be used. If these terms are allowed, the term *cover crop* might be more appropriately used in the following manner: *Cover crops* are grown mainly for preventing soil erosion through the formation of a protective soil cover and a soil-holding root mat. Such crops also conserve soil organic matter by preventing the loss of topsoil, which contains more organic matter than the subsoil. To add further confusion, even though the management system might be aimed at a specific goal, certain plants will also accomplish the other two goals. For example, legumes are planted between crops to provide nitrogen for the following crop (green manure), but their presence as a soil cover will also conserve soil organic matter and reduce soil erosion. Conversely, a legume grown primarily over the winter to reduce soil erosion will also add nutrients and organic matter. Cover crops can also reduce insect or weed problems.

No profit is generally expected from the cover crop, since it will not be harvested for any purpose. Some exceptions exist with certain cover crops harvested for forages or seed. The farmer usually manages the cover crop with the expectation of some or several soil ben-

FIGURE 11–13   An ARS plant physiologist in Beltsville, Maryland, checks thickness of killed hairy vetch—a mulch crop that dramatically boosts tomato production. Hairy vetch—a legume—provides nitrogen and reduces weed growth. A dense layer of hairy vetch lowers summertime soil temperature and reduces water loss. (USDA photograph.)

efits. Although the main benefit is the prevention of erosion, the other benefits can be increased with changes in management practices.

Green manure and cover crops are usually sown between cash crops when fields are idle. If the idle period is summer fallow, growth is usually not a limiting factor for either green manure or cover crops. More commonly, the idle period is late fall through winter. In the South, growing seasons are longer and winters milder, thus many cover crops can be successfully raised to mature growth during the off-season. However, as one moves northward, decreases in the growing season and winter hardiness restrictions begin to limit plant choice and the attainment of sufficient growth. In short season, cold areas, green manure and cover crops are often best managed by sowing them in place before the cash crop is harvested. This strategy allows for more choice in cover crops and increased growth. The cash crop must be mature enough to tolerate competition from the emerging green manure or cover crop. Seeds can be sown over the crop when young or later between rows when the cash crop is cultivated for the final time. The latter is often referred to as *intercrop* or *living mulch*. Overseeding would work with sweet corn, while the living mulch has been used successfully with beans, broccoli, cabbage, cauliflower, peppers, potatoes, sweet corn, and tomatoes planted in rows alternating with annual ryegrass strips.

Killing of any mature cover crop, whatever its purpose, is necessary prior to planting any crop into the field. The exception is for permanent cover crops, such as orchards cover-cropped with sod. Cover crops are either killed naturally by winter injury or by mowing, plowing under, or with herbicides. The latter is not recommended from the perspective of sustainability. A 1- to 2-week wait is needed before the following crop is planted after death of the cover crop. This wait eliminates any problems that might interfere with seedbed preparation, seed germination, or seedling establishment. The green manure crop also needs this time to break down, since much nitrogen will go to the decomposition process and not to the new crop.

Killing of the cover crop or green manure is also influenced by moisture considerations. In arid or drought-stricken areas, late killing can result in moisture limitations for the following cash crop. The cover crop or green manure should be killed earlier before too much soil moisture is consumed, if soil incorporation is intended. If the planting is going to be left as mulch (Figure 11–13) for no-till systems, earlier killing is not advised. Later killing will result in more residues, hence better mulch and increased water conservation during the following crop growth.

Crops Managed as Green Manure

Cover crops, when managed as green manures, can supply considerable nitrogen for cash crops. Green manures are plants grown primarily as summer or winter annuals and then plowed under to release a steady supply of nitrogen. Legumes, such as vetch (*Vicia*), clover (*Trifolium, Melilotus*), and Austrian winter pea (*Pisum arvense*) are often used. Nonleguminous plants, such as rye (*Secale cereale*) and ryegrass (*Lolium multiflorum* and *L. perenne*), are also commonly mixed with legumes as cover crops. Green manure is less effective at increasing the organic matter content of the soil. Research results with green manure and soil organic matter are variable, from no effect to some effect. The normal expectation with green manure should be enhancement of soil nitrogen, with little expectations for gains in soil organic matter.

Soil bacteria in the genus *Rhizobium* form symbiotic relationships with legumes. These bacteria inhabit the root hairs of legumes in swollen areas called nodules. Both the bacteria and legumes benefit from this living arrangement. The bacteria capture or fix nitrogen from the atmosphere and transform it into forms used by the legume. In return, the bacteria receive metabolites from the legumes, which enable the bacteria to grow and reproduce better than their alternative lifestyle form, free-living soil bacteria. As a result, legumes are a useful alternative to chemical fertilizers as nitrogen sources for soil enrichment.

Legumes can supply reasonable amounts of nitrogen to the soil. For example, hairy vetch can supply more than 100 pounds of nitrogen per acre to the crop that follows it. Crimson clover is almost as efficient. Nitrogen fixation rates are shown for a number of legumes in Table 11–3. Legumes can be grown as annual summer crops, such as beans, peas, and soybeans, or as winter annuals. The latter are planted in the fall and wintered over. These include berseem clover, crimson clover, and hairy vetch. Biennial and perennial legumes are also grown as longer term cover crops in rotations. These types include alfalfa, crown vetch, and red, sweet, and white clover (Figure 11–14).

TABLE 11–3   *Nitrogen Fixation by Legumes and Legume-Grass Mixtures*

Crop	lb/acre/year
Alfalfa (*Medicago sativa*)	47–196
Alsike clover (*Trifolium hybridum*)	60–70
Annual sweetclover (*Melilotus alba annua*)	44–53
Austrian winter pea (*Pisum arvense*)	53–100
Berseem clover (*Trifolium alexandrinum*)	88
Biennial sweetclover (*Melilotus alba*)	120
Bigflower vetch (*Vicia grandiflora kitaibeliana*)	13–90
Black medic (*Medicago lupulina*)	39–61
Chickpea (*Cicer arietinum*)	24–78
Crimson clover (*Trifolium incarnatum*)	19–114
Dry bean (*Phaseolus vulgaris*)	83
Faba bean (*Vicia faba*)	25–105
Field lupine (*Lupinus albus*)	51
Field peas (*Pisum sativum arvense*)	33–103
Hairy vetch (*Vicia villosa*)	33–145
Hairy vetch + oats	98–145
Hairy vetch + rye	50–100
Hairy vetch + wheat	50
Ladino clover (*Trifolium repens* f. *lodigense*)	90–100
Lentil (*Lens culinaris*)	37–95
Mammoth red clover (*Trifolium pratense perenne*)	60–70
Medium red clover (*Trifolium pratense*)	60–80
Soybean (*Glycine max*)	58–98
Subterranean clover (*Trifolium subterranean*)	11–92
Sweetclover (*Melilotus alba* and *officinalis*)	70–116
White clover (*Trifolium repens*)	100

FIGURE 11–14   Red clover is a short-lived perennial legume that can be used in rotations and as a cover crop to supply nitrogen. (USDA photograph.)

Grasses are often mixed with legumes. Grasses such as rye grass (*Lolium multiflorum, L. perenne*) or cereals (rye, *Secale cereale*) can capture leftover soil nitrogen (catch crop), thus reducing winter leaching losses. Some of the captured nitrogen will be returned when the grass or cereal is disked into the soil in the spring. These mixtures can improve nitrogen fixation by the legume partner under some conditions, as available soil nitrogen can reduce legume nitrogen fixation (see later). This mix also increases the amount of residues and offers the best opportunity to increase the supply of soil organic matter. Grasses alone, when managed for maximal residues, offer minimal amounts of soil nitrogen. The reason is that mature grasses, while giving maximal residues, also have high carbon-to-nitrogen (C/N) ratios and thus have little available nitrogen. In fact, C/N ratios above 32 mean that existing soil nitrogen is consumed during the decomposition process. Harvesting grass earlier eliminates the nitrogen problem, but then minimal residues are the trade-off. By mixing grass with legumes, you can get both high levels of nitrogen and residues. Some evidence also exists that the legumes supply some of their fixed nitrogen to adjacent grasses, thus improving grass production and cutting fertilizer costs.

Green Manure Choices

The choice of legume or legume/grass must be carefully made. The choice must take into consideration the amount of nitrogen required and the goal of the grower. The advantage with summer legumes is that nitrogen fertilizer needs are minimal and they offer some value as a cash crop. However, much of the fixed nitrogen is removed with the harvest, leaving little for the crop that follows. For example, soybeans fix about 30% of their nitrogen needs and about 70% of this is removed with the harvest. Moreover, a horticultural operation might not be interested in handling and marketing what are considered agricultural crops, that is, soybeans and cowpeas.

Winter annual legumes have the advantage of being planted in the fall after a crop is harvested and with sufficient growth can contribute much nitrogen to the following crop. The winter legumes possibly offer a better choice in terms of more retained nitrogen and reduced soil erosion during the winter months. The main problem is with winter hardiness. Severe winters limit growth and hence the amount of fixed nitrogen. Most winter legumes work quite well in the South. Perennial legumes offer reasonable amounts of fixed nitrogen for subsequent crops, but tie up cropland for some part of the rotation. However, some compensation is possible because of their value as forages.

Climate also plays a major role in species choice. Nitrogen capture with legumes in the fall and early winter becomes progressively less as one moves northward or into higher elevations. Some of the ranges of nitrogen fixation for several legumes shown in Table 11–3 illustrate the effects of differing climatic. A legume in the South will fix considerably more nitrogen there than in the extreme northerly range where winter hardiness becomes a factor. Cold areas in the North and upper Midwest have fewer choices than the South; hairy vetch and bigflower vetch appear to be the only realistic choices in cold areas.

The amount of soil water and rainfall are also factors. Insufficient moisture can restrict germination and early growth, and reduce the amount of nitrogen captured by the legume. In arid to semiarid areas, legumes with lower water requirements should be grown. One should consult with local or state extension professionals, state universities, other sustainable farmers with legume experience, and Web pages that provide information with regional or local value.

Maximizing Nitrogen and Organic Matter with Green Manure

The most difficult aspect of green manure management is the timing aspect. It does matter at what developmental point you plow down the legume or use it as a mulch. Up to a point, legume nitrogen content increases, so you want to achieve sufficient growth. If you terminate too early, not only is this inefficient in terms of capturing nitrogen, but the plant material is very succulent and prone to rapid decomposition. As a consequence of rapid nitrogen mineralization, some nitrogen might be lost through leaching and volatilization. If you wait too long, certainly the plant material will have maximal nitrogen, but due to its toughness (lignification), it will decompose slowly and not release enough nitrogen for the crop that follows. In dry areas, waiting too long might also deplete the soil's moisture reserves for the crop that follows. Another problem is that waiting for the legume to achieve full nitrogen content can interfere with the timing of the crop that follows. The ideal compromise is to kill the legume at maximal top growth that occurs just before full flowering. Earlier killing is advisable if crop timing or limited moisture is a problem.

The actual amount of nitrogen returned to the soil and the rate of availability is not only dependent on the legume choice and the developmental stage at which it is killed, but also on the preexisting nitrogen content of the soil and the effectiveness of the *Rhizobium* species in root nodules. Research has indicated that available soil nitrogen reduces the legume's rate of nitrogen fixation. On a per acre basis, each pound of soil nitrogen results in about a 2-pound reduction in nitrogen fixation. A crop with high nitrogen demands should precede the legume crop, so that as much nitrogen as possible will be removed. One can ensure an effective bacterial relationship by adding *Rhizobium* inoculate to seeds prior to planting. This procedure is especially important when legumes are first planted in a new field, or if more than 2 years have elapsed since the last legume sowing, or if you are changing from one legume to another.

In the past, the moldboard plow was used to incorporate legumes deep into the soil. However, this treatment left no surface residue and brought deeper organic matter closer to the surface. These conditions increase soil erosion and loss of organic matter. Conservation tillage is preferable from the sustainability viewpoint. The implement of choice has become the heavy-duty disk. Disking disturbs less organic matter and leaves some surface residue, thus reducing erosion and organic matter losses. Disking also produces even mixing in the uppermost soil layers, thus favoring even mineralization of nitrogen near the surface. This zone is most important for subsequent uptake of nitrate by plants.

Soil nitrogen levels must be watched. Microorganisms will use soil nitrogen during decomposition of the green manure, if the C/N ratio of the plowed under green manure exceeds 30:1 to 35:1. This problem can occur if the green manure was grown past the blossom stage and allowed to produce extensive brown matter.

The amount of the incorporated nitrogen that becomes available over time for crops varies. Three factors affect the availability: the amount of soil organic matter, the rate of mineralization, and the weather. One would think that the more organic matter, the more mineralization, but the rate of mineralization tends to be higher with lower organic matter and lower as organic matter increases. The mineralization rate depends also on soil health, that is, the activity level of microorganisms involved in the mineralization process. Weather is also a factor. Rates are higher under warmer and less moist conditions as opposed to cold, very wet conditions. Soils with high clay contents experience less mineralization of organic matter as opposed to sandy soils. The former tends to retain water and warm slowly, whereas the latter drains well and warms faster. As such, we do not have a reliable way to predict the amount of nitrogen produced by the mineralization process. Consequently, it is not surprising that the amount of nitrogen from legumes available for the following crop varies from 20% to 50%, with additional increments over the next year or two.

Considerable knowledge and management skills are necessary to get the most effective returns with green manure. It is highly advisable that management techniques be based on information and experience accumulated on a regional and local basis. Ideas from local, experienced farmers and land grant institutions are critical for success.

Cover Crops

When the main goal with crops is to control soil erosion and to maintain or improve organic matter, we usually refer to the practice as cover cropping or sometimes winter cover cropping. Given that cover crops can be used in the summer on fallow fields, the latter term is less inclusive. A good definition is that cover crops are legumes, grasses, or small grains grown between the production periods of horticultural crops to protect and improve soils. Because the cover crop can consist of overwintered legumes that are tilled into the soil in the spring prior to conventional planting or killed and used as mulch for no-till planting, one can also get some green manure benefits. As with green manures, plowing should be done with disks and not moldboard plows. The most common cover crops include legumes sown in late summer or early fall, such as Austrian winter peas, bigflower vetch, crimson clover, hairy vetch, and others used also as green manure (Table 11–3). Fall-seeded cereals, such as rye or wheat, and ryegrass and oats are also used.

Cover crops are either temporary (Figure 11–15) or permanent. Temporary cover crops, such as grasses or legumes, prevent winter or summer (fallow) erosion of soil and are plowed under in the spring. Grasses, perennial legumes, or mixtures of both can also be used as permanent cover crops. Permanent cover crops are frequently used in orchards. Oxidation of soil organic matter is minimal because of the constant soil cover and lack of cultivation. Organic matter levels are maintained by the constant degradation of grass roots. This practice has become important in orchards. Prior to fruit harvest, these cover crops are harvested and removed, often for forages or beaten down into a mat. This practice makes it easier to harvest the fruit, especially with mechanical harvesters.

FIGURE 11–15 An NRCS soil specialist examines a cover crop of annual Lana vetch (*Vicia villosa* ssp. *Dasycarpa*) in an old date orchard in California's Coachella Valley. (USDA photograph.)

Rotations

Growing crops in a rotational sequence has many benefits. Insect, nematode, and disease problems are reduced, especially compared to monoculture. If a legume is used as part of the rotation, nitrogen can be supplied to sequential crops. Even when legumes are not part of the rotational cycle, yields can be higher relative to monoculture under similar fertility levels. This yield benefit from rotations is known as the rotation effect, and it is not completely understood.

Rotations are often viewed from agricultural perspectives, that is, corn followed by soybeans followed by forage crops. Horticultural production can also benefit from using rotations. The most common limitation is land for crop production. Most vegetable and herb growers don't have enough field capacity to tie cropland up in grass or legume crops for any period of time. Expensive irrigation and other infrastructure need to be paid for. Still, rotations are possible. One can rotate different vegetable crops, use green manure and cover crops during the off-production times (late fall through winter), and realize important benefits from rotations.

With horticultural crops, one should keep certain facts in mind when rotations are practiced. Rotational crops that follow off-season green manure or legume cover crops should be high nitrogen consumers. Sweet corn and eggplants would be good examples. Later sequential crops should have lower nitrogen needs, such as peppers or carrots. Try not to repeat the same or related crop in any one area, because disease and insect problems are more likely to appear. For example, following tomatoes with peppers, eggplants, or potatoes is not recommended. Some crops also appear to do better when following certain other crops. Examples include potatoes after sweet corn or cruciferous crops following onions. Also rotate root crops with foliage crops whenever possible, because nutrient usage is somewhat complementary and root crops help to open the soil for infiltration of water. Long-term rotations appear to be more beneficial than short-term ones.

A good example is a 10-field, 10-year rotation. In year 1, all 10 fields are planted to the same crops. The main consideration here is crop volume for commercial production. In year 2, the crop is changed, and it is followed by a winter cover crop grown primarily for organic matter contribution. Crops are changed again in years 3 and 4. A green manure follows the fourth year crop in the off time. In the fifth year, a high nitrogen feeder is grown. Crops are changed for the successive years and the winter cover crop and green manure are repeated in the off-seasons following the seventh and ninth years. The tenth year is again for a high nitrogen feeder. If erosion is a worry, cover crops might be used every winter. To increase sustainability, alternate year cover crops could be green manure. These same 10 crops could also be adapted to one field subdivided into 10 areas with all 10 grown each year, but rotated from one area to the next. Gardeners could adopt the latter strategy, since variety and not volume is the consideration.

Organic Amendments

Organic matter in the soil can also be increased by the addition of organic amendments such as manures, compost, sewage sludges, and peat. Many of these amendments depend on proximity, availability, and farm type. Horticultural crops grown on farms also raising animals have manure available. Farms near municipal suppliers of leaf compost have compost availability. Composted or digested sewage sludge free of heavy metals can be an option for other horticultural producers. Home gardeners depend heavily on organic amendments to maintain and improve levels of organic matter in their garden soils.

Manure.   A number of manures are available, such as cow, poultry, horse, and rabbit manure. Besides their value as an organic amendment, they also contain about 1.3% to 5.0% nitrogen (dry weight basis), 0.3% to 1.9% phosphorus, and 0.1% to 1.9% potassium. Wet manure averages about 25% organic matter. Manure is somewhat difficult to handle and can easily lose nutrients through leaching and microbial activity. However, those growers who use manure, such as organic farmers, recognize its value as a soil conditioner and fertilizer. Nitrogen and phosphorus (phosphorus content is low) from manure tends to have extended availability.

FIGURE 11–16 Soil organic matter can be increased by the addition of either liquid manure or sewage sludge. An effluent spreader shown here is used for application. (USDA/NRCS photograph.)

One caution is advised with manure applications (Figure 11–16). Overuse of manure, like any nutrient, can lead to leaching and water pollution. It just takes more and longer, but it can happen. This problem has been observed in organic vegetable production and by other sustainable approaches using manure. With new land, the first application is often heavy to supply nutrients and organic matter. In subsequent years, applications should be based on nitrogen requirements of crops. Even with this approach, the amounts of phosphorus and potassium at this level are more than the crop needs, because crops use considerably less phosphorus and potassium relative to nitrogen. If excess manure is applied, leaching and volatilization of considerable nitrogen can be lost, just as with chemical fertilizers. Phosphorus and potassium are held in unavailable forms usually, so they present minimal problems.

So how do we raise organic matter levels in the soil, but not apply excess nutrients? The answer is that a soil with modest levels of organic matter (2.5% to 3.0%) can be a healthy soil having good biological activity, quality, and fertility. If judicial applications of manure don't produce the desired level of organic matter, as reflected in soil health, then approaches that raise organic matter without the additional application of nutrients are advised. Rotations, especially those involving sod, can build organic matter without raising nutrients too high. If phosphorus and potassium get too high, you can switch from manure to legume cover crops that will add nitrogen and organic matter, but not much phosphorus and potassium.

Compost. Compost, sometimes called synthetic manure, is a soil conditioner that requires some preparation (Figure 11–17) prior to use. It is probably more useful on a small-scale basis, such as with gardens, because the other organic matter sources are less labor intensive. Several misconceptions about compost still prevail, and the composting process is often made much more complex than is needed.

Accordingly, the composting procedure is covered here. *Composting* of collected organic wastes starts when the indigenous microorganisms are subjected to favorable environmental conditions. Bacteria and fungi and other soil biota assimilate some carbon, nitrogen, phosphorus, and other nutrients made available by their enzymic activities during the process of aerobic decomposition. Heat is produced from these biological oxidations and much of it is retained, since organic matter acts as an insulator. The temperature rises to 104°F (40°C), whereupon the population of mesophilic microorganisms is succeeded by thermophilic and thermotolerant microbes. The temperature increases to 158°F (70°C), during which time the rate of degradation accelerates. Decomposition eventually slows and the temperature returns to that of the ambient atmosphere. Changes induced by microbial degradation have converted

FIGURE 11-17
Compost windrow
undergoing
decomposition on a farm
in Hamden Connecticut.
(Author photograph.)

FIGURE 11-18
Compost pile
construction shown here
with alternating 8-inch
and 4-inch layers of
carbonaceous/nitrogenous
layers, respectively,
assures good composting.
A proper carbon/nitrogen
ratio will be achieved,
compaction and hence
anaerobic conditions
will be minimized, but
heat retention will be
sufficient. The latter two
are functions of surface
area and volume.

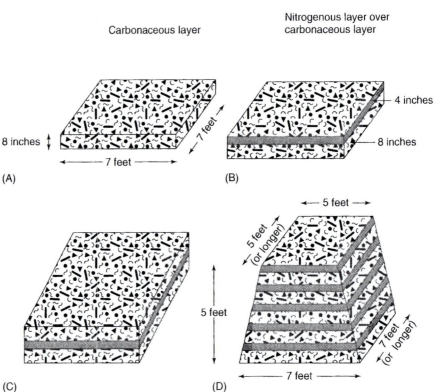

the original wastes into a humus-like material. The volume and weight of the heap will have decreased by as much as 50%, and losses of carbon can be as high as 40%.

A simplified method for the construction of a compost pile is shown in Figure 11-18. The size of a pile is determined by the relationships among heat loss and surface area, heat production and volume, and oxygen content and compaction. While additives and limestone have been shown to be unnecessary for compost production, both carbonaceous and nitrogenous materials are needed during decomposition by microorganisms that utilize about 30 parts of carbon per 1 part of nitrogen. In practice, the optimal C/N ratio is approximated by combining two parts of carbonaceous wastes with one part nitrogenous materials.

The carbonaceous materials include dry leaves, hay, straw, sawdust, wood chips, chopped cornstalks, or other dry plant residues, and small amounts of shredded paper. Nitrogenous materials can be food wastes, manures, or fresh green plant residues such as grass, weeds, and garden remains. Nitrogen content can also be increased by adding fertilizers. Dry leaves from deciduous trees, except for oak leaves that are resistant to degradation, can be composted alone to form leaf mold, since their C/N ratio is sufficient to support microbial growth. The pile is kept moist to the touch, such as the feel of a wrung-out sponge.

Although not necessary for decomposition, shredding and aeration are often practiced. Aeration is achieved by turning or mixing the outer edges with the center of the pile. If the oxygen content is too low, anaerobic decomposition results and is associated with foul odors and slower degradation. Shredding produces greater surface area that increases susceptibility to microbial invasion and availability of oxygen. Extensive shredding and frequent aeration are the basis of rapid composting techniques.

Generally, compost production in small containers such as cans or plastic bags is inefficient, because the container puts some constraints on volume and aeration. Production of compost under these conditions is slowed. Loose-fitting containers, 40 by 40 inches or larger and 5 feet in height, are preferred.

Most plant and animal pathogens, weeds, and insects can be destroyed during composting if the temperature is maintained at 133°F (56°C) for a few days. More resistant pathogens require 149° to 158°F (65° to 70°C) for 3 weeks. Unless rigid control over temperature and turning (to thoroughly expose all material to these temperatures) is practiced, it is not advisable to use materials suspected of containing pathogens.

Dry garden compost contains 1.0% to 3.5% nitrogen, 0.5% to 1.0% phosphorus, and 1.0% to 2.0% potassium. When added to soil, it improves aeration, soil tilth, and moisture retention. Compost piles are usually employed for small-scale use, such as in a garden. Large-scale needs for organic matter in soil can be met with sheet composting or green manuring (see previously under Cover Crops).

Sewage Sludge.   This amendment has value both as a fertilizer and soil conditioner. Digested sewage sludge that is prepared by anaerobic digestion contains 2.0% nitrogen, 0.6% phosphorus, and traces of potassium on a dry weight basis. Similarly, activated sewage sludge that is prepared by aerobic decomposition, has about 5.0% nitrogen, 4% phosphorus, and traces of potassium. These materials have been used successfully, but have not been generally accepted because of possible risks involving human pathogens and the phytotoxicity problems associated with heavy metals that contaminate many sewage sludges. Another factor in the United States is the psychological one. Other countries such as China are not deterred by that factor. Some municipalities are preparing an acceptable compost from sewage sludge and bark.

Peat.   Available peats were discussed in Chapter 7. The best forms for use as a soil conditioner are moss peat, followed by reed-sedge peat. Peat is especially used as a soil amendment in landscape horticulture, where it is necessary to improve the water-holding capacity of coarse textured soils and to improve the structure of fine-textured soils. It must be thoroughly wetted before incorporation into the soil. It is also necessary to add a complete fertilizer to compensate for the nutrients needed for decay, as well as to replace the soil nutrients diluted by its incorporation into the soil. Some lime (see next section) might be needed to counter its acidic soil reaction.

REDUCED TILLAGE

The working of soil by plowing and cultivating is termed *tillage*. The purposes of tillage are weed control, good seedbed preparation, and incorporation of crop residues or organic amendments. In small-scale situations, such as a garden, the soil is dug in the spring when the soil temperature and water content are correct (soil is ready when it does not pack into a hard ball, but crumbles readily). Sandy soils can be tilled anytime, but clay soils are better tilled in the fall to allow alternate freezing and thawing to improve soil structure. Gardens are tilled by digging with a spade, spading fork, or a mechanized tiller (see Chapter 12). Subsequent cultivating is done with a hoe or cultivating fork. Commercial operations use mechanized plows and cultivators.

Conventional Tillage

The moldboard plow was the earliest tillage implement of choice in colonial America, especially with the tough sods found in our Midwestern prairies. Before this time, the moldboard plow was used in Europe and as far back as 3,000 years ago in China! Tillage with the moldboard plow followed by disking and cultivation is referred to as *conventional tillage*. This form of tillage is not favored for sustainable horticulture.

The moldboard plow disturbs soil to a depth of 8 to 10 inches and buries surface crop residues. The loss of surface residues reduces water infiltration and increases runoff, thus making the soil more susceptible to soil erosion. The deep plowing exposes lower reserves of organic matter to oxygen, thus speeding up oxidation of soil organic matter. Continued reduction of organic matter over time reduces soil health, increases susceptibility to erosion, and more water and fertilizer are required to maintain crop productivity. Deterioration of the soil structure also occurs, because the continual use of heavy machinery causes soil compaction. The plow also causes pan formation where its bottom slides over and compresses the soil, resulting in reduced root penetration and less water percolation. Deep soil disturbance also requires heavy energy inputs into plowing operations.

The use of a chisel plow can temporarily eliminate the compaction and pan problems, but the incorporation of soil amendments for long-range relief is needed. The extensive use of machinery is also high in energy costs. These soil structure problems are minimal in home garden operations, since much of the work is done by hand or light machinery. On a larger scale, these problems can be reduced by changing plow types, cutting back on the number of cultivating steps, or eliminating them entirely by mulching or chemical weed control.

Conservation Tillage

The trend has been toward reduced tillage, or what is termed *conservation tillage*. Switching from the moldboard plow (conventional tillage) to conservation tillage can reduce soil erosion by 50% to 90%. Organic farmers and other types of sustainable farmers favor conservation tillage for horticultural crop production. In 2000, 36% of the nation's cropland was under conservation tillage. Reduced tillage can be taken to the extreme to what is called a *no-till* practice. Minimally acceptable conservation tillage leaves 30% of the soil's surface covered with crop residues, while no-till leaves the highest possible amount of residues on the surface.

Several tillage implements for conservation tillage are available, but the most widely used is the chisel plow (Figure 11–19). Other implements include subsoilers, disks, culti-

FIGURE 11–19 A chisel plow is shown turning the soil and partially burying the crop residues. (USDA/NRCS photograph.)

vators, mulch incorporaters, strip rotary tillers, and no-till planters. These implements have been adapted to various methods of conservation tillage. One of these is chisel planting, in which the bed is prepared with a chisel plow (Figure 11–19), which leaves crop residues in the top 2 inches of soil and on the surface. This form tends to rank at the lower end of conservation tillage in that considerable soil disturbance results and some residues are buried. Disk planting (Figure 11–20) is very similar, except the seedbed is prepared by disking the soil. Less soil disturbance and fewer residues are buried. With ridge-till (Figure 11–21),

FIGURE 11–20 Soils that work very easily can be prepared with disking only. Disks can be followed by ring rollers to both work and roughly level the soil. Disks can also be used after chisel plows for further working and smoothing of the soil. If a really fine seedbed is needed, final preparation consists of harrowing (somewhat like raking) with a spring-toothed harrow or drag harrow. This step gives a leveling and pulverizing effect. The downside is that it removes more of the surface crop residues. If the soil contains a lot of crop residues, the drag harrow is preferable, as tangling is less. (USDA photograph.)

FIGURE 11–21
Cultivating corn on a ridge-till field in north central Iowa. Note the crop residues still in the soil surface between the raised rows (ridges). (USDA/NRCS photograph.)

(A)

(B)

(C)

FIGURE 11–22   (A) Planting corn following conservation tillage on terraced land on the Gil Winter farm in Plymouth County. (B) Southern Iowa farmer uses a grass drill to interseed native grasses into a cool season field of grasses. These seeders are useful for sowing cover crops. (C) Mechanical tree planter being used to plant trees in the field. Similar devices can be used to place transplants such as tomatoes into fields. (USDA/NRCS photographs.)

plowing and planting is done in one operation with crop residues mixed into the soil surface between raised rows (ridges). The strip till process also involves one step for plowing and planting of strips, with undisturbed crop residues left in place between the strips. Planting can be carried out at the same time as plowing or later (Figure 11–22).

The least disturbance and maximal surface residues result with the no-till process, in which only the immediate row is disturbed for planting by slotting or slicing through the undisturbed crop residues. In the no-tillage system, herbicides are used to kill weeds prior to planting and during the growing season. The earth is left covered by a mulch that can be the residue of the previous crop, a cover crop grown as a mulch, or a crop grazed by animals. Planting and fertilization are done with a machine that slices through the mulch and soil, deposits fertilizer and seed, and then presses the soil over the seed (see Figure 11–10A, Figure 11–23). Advantages include reduced losses of organic matter, a reduction of soil erosion, less destruction of soil structure, decreased labor, and less energy consumption. Disadvantages include the use of herbicides that some consider objectionable and unsustainable, and increased losses from insect and rodent damage because of the attractiveness of the mulch.

FIGURE 11–23 No-till pumpkin patch at Stahlbush Island Farms, Oregon. While no-till is associated with field or sweet corn, it is ideal for vine crops where clean cultivation becomes impossible quite early as the vines spread rapidly beyond the planting row. (USDA/NRCS photograph.)

SUSTAINABLE WATER MANAGEMENT

Sustainable horticulture requires that water and soil be conserved, because the supply of each is limited. The two are interrelated, because steps taken to conserve soil often aid in the conservation of water. Improper management of soil and water can lead to floods, droughts, erosion (Figure 11–24), water pollution, and other conditions unsuitable for horticultural practice. Water conservation also greatly reduces off-farm pollution, because runoff water can carry considerable amounts of sediment and farm chemicals.

Water conservation is based on the control of soil runoff and improvement of water absorption by the soil. In most horticultural production areas, the aim is to minimize runoff and maximize water absorption. However, where the objective is to collect water for irrigation, the aim is to maximize runoff while minimizing erosion. Practices in the previous sections that reduced soil erosion and increased surface and soil organic matter are also ideal for minimizing runoff, while maximizing water infiltration and water retention. Soils enriched in organic matter also encourage high populations of earthworms, up to 1 to 2 million per acre. The high number of channels created by earthworms can greatly increase the infiltration and percolation of water.

On slopes, contour tillage is essential for controlling runoff that in turn reduces soil erosion (Figure 11–25). Terraces can reduce water runoff even more. Terraces and contour tillage run at right angles to the slope. These features act to trap the rainfall until it can be soaked up by the soil. Conservation tillage is also needed to maximize the amount of surface residues for water infiltration. On level land, practices such as conservation tillage and cover crops or organic amendments are effective ways to conserve water.

Winter cover crops and leftover crop residues are very effective at trapping and retaining water from fall rains and later trapping snow, thus increasing storage water. Frozen soils can take up some water when sufficient organic matter is present. The bulk of the water would be absorbed in the spring when the snow melted.

Nonplastic mulches greatly increase the infiltration of rainwater and reduce soil erosion. Mulches are popular with home gardeners for water conservation. In commercial operations, plastic mulches are preferred for reasons of mechanized application and costs. Slotted or perforated plastic mulches are best for water entry and slowing the evaporation of soil water. Living mulches also control runoff and help to conserve water.

Permanent cover crops in orchards, such as sod, improve water infiltration and prevent runoff. Sodded orchards need to be mowed, especially in dry periods. Mowing

(A)

(B)

(C)

(D)

FIGURE 11–24 Types of soil erosion. (A) Wind erosion on an Iowa farm field. Note the wind-blown dust (soil) cloud on the horizon. Wind (and sometimes water) causes the uniform removal of soil in thin layers on sloping land (sheet erosion). Sheet erosion by rain, if unabated, usually leads to first rill and later gulley erosion (B) Early stage of rill erosion sometimes called shoestring erosion. (C) Advanced rill erosion, also known as gully erosion. (D) Critical gulley erosion on summer-fallowed land in the Palouse Basin (border of Washington and Idaho). (USDA/NRCS photographs.)

FIGURE 11–25
Contour tillage on this slope helps hold rainfall and results in a more even distribution of the water. (USDA/NRCS photograph.)

increases the amount of soil water available to the fruit trees, because less grass surface area means less transpiration.

In areas where water is a limiting factor, the issue of sustainability becomes a greater problem. In many of these areas, we still grow horticultural crops. Production is made possible by the use of irrigation that draws water from surface and groundwater supplies. For true sustainability, withdrawals should be based on rates that are near or below the natural recharge rates of the water resources. In areas where this approach is not possible, horticultural acreage should be reduced to the sustainable level by conversion to dryland farming or abandonment. Short of that, the most water-efficient irrigation should be adopted and crops with lower water consumption should be selected. In addition, practices that increase surface residues, build up organic matter, and reduce erosion are essential.

Terraces

These constructed landforms found on sloping farmland are constructed along the contour to reduce water runoff and subsequently soil erosion and off-site water pollution. Terraces today are constructed to include a channel bordered by a ridge on the lower side. This feature is designed to trap runoff. Terraces can either be graded or level. Graded terraces have a graded channel that slows runoff and guides it to a special discharge area, such as a vegetated or wooded area or some waterway. This type's main function is to reduce water erosion. When this type is built on steep slopes so that the final appearance is step-like, it is termed a *bench terrace* (Figure 11–26). When used on moderately to slightly sloping fields, fewer are needed and the look is not step-like. These types are simply called *graded terraces*. The other type is the *level terrace,* which has no channel grade, thus trapping water behind the ridge and allowing for infiltration into the terrace soil. This type's main function is to trap water for dryland farming and indirectly reduce water erosion. Two types of level terraces are used, the level ridge type and conservation-bench type.

Terraces are not new, but date from the B.C. period in the Middle East. Farmers in the United States in the past constructed terraces, but fewer terraces are built today because construction costs are high and they can interfere with the efficient operation of field machinery. If the indirect costs of soil erosion and lost water are factored in, terraces are a long-term plus for sustainable farming where eroding, sloping land is intensely cropped.

The most effective terraces for steep slopes are bench terraces. For moderate to slight slopes, the best choice to reduce water erosion is the graded terrace. For water conservation on slopes of less than 5%, level ridge-type terraces are recommended. A special form

FIGURE 11–26 A well-constructed bench terrace is shown here. (USDA/NRCS photograph.)

of terrace, the conservation-bench terrace, alternates constructed flat areas at intervals on the slope. It is designed to trap runoff from the sloped parts; the runoff water is used for crop production in the flat areas. In dry areas, the conservation-bench type has the potential to allow annual cropping with minimal irrigation.

Contour Tillage and Contour Strip Cropping

Tillage at right angles to the slope along contours is called *contour tillage*. Contour cultivation (Figure 11–27) includes contour tillage, planting, and cultivating along the contours. The ridges formed in this manner help slow and increase water infiltration. This practice can also control erosion on moderate slopes of up to 8% during moderate rainfalls. Increased effectiveness occurs when this practice is combined with contour strip cropping and/or terraces. Contour strip cropping (Figure 11–28) involves contour tillage, but in-

FIGURE 11–27
Hillside contour cultivation of strawberries in Monterey County, CA. (USDA/NRCS photograph.)

FIGURE 11–28
Contour strip cropping is shown here as found on a farm in Minnesota. (USDA/NRCS photograph.)

stead of planting one crop, strips of row crops are alternated with sod-forming crops, such as hay or grass. The sod strips are effective at trapping water and sediment from the row crops. Unless a market exists for hay, or the horticultural operation includes cattle, this practice has less appeal because crop production is half of that achieved with contour cultivation or from terraces with row crops alone.

Dryland Farming

In areas where water consumption exceeds the recharge capacity of aquifers and surface water, serious consideration should be given to dryland farming. The concept is not new. Native Americans used this practice in the arid Southwest, as did other farmers who settled the Great Plains. The practice declined with the advent of irrigation in some areas, but extensive dryland farming still occurs in the western rim of the Great Plains.

One method of dryland farming requires alternate years of crop production and summer fallow. During summer fallow, water is stored in the field from natural rainfall. This stored soil water is used for crop production the following year. The disadvantage is that crop production is only possible every 2 years. The fallow period is not all that efficient, in that only about 25% of the fallow year's rain and snow are stored in the soil. Another approach is termed the eco-fallow system, as seen in Nebraska. This system is based on no-till or minimal tillage and two crops in 3 years or three crops in 4 years are possible.

Few, if any, horticultural crops are grown in dryland farming systems. Modern cultivars of horticultural crops are often not suitable for dryland farming. Old cultivars from the native Americans, such as the tepary bean (*Phaseolus acutifolius*), are possible candidates either in themselves or as genetic sources for the genetic engineering of future dryland crops. The fruit of the *Opuntia*, prickly pear, a favorite of Luther Burbank, is another possibility.

Irrigation

Much of the irrigation technology used for horticultural applications is derived from agricultural practices. The need for irrigation by horticulturists in the arid West is obvious. However, horticulturists in the East and South can also benefit from irrigation. Although rainfall can be adequate, irregularities in the distribution pattern and timing sometimes cause a number of drought days. Many failures with horticultural crops result from lack of rainfall or poor timing and quantity of irrigation.

Irrigation Parameters.   The amount and frequency of irrigation can be determined from calculations or actual measurements. This step is necessary so that water is not wasted and excess water does not harm plants. Calculations are used to determine the *consumptive use* of water by the plant, that is, the water lost through transpiration and evaporation. Consumptive use is affected by a number of variables including temperature, wind, humidity, sunshine levels, percentage of plant cover, root depth, the developmental stage of the plant, and even the available amount of moisture. Fortunately, a reasonable approximation of the consumptive use can be arrived at from average meterological data, empirical constants available for horticultural crops, estimates of the total plant cover, and measurements of evaporation determined with a Bellani black-plate atmometer. The difference between the consumptive use and the available moisture, the *water deficit,* is the amount of water that must be supplied.

The other approach is the direct measurement of soil moisture. Either a gravimetric procedure that involves weighing a soil sample before and after oven drying or instrumentation is used. The instruments required are a tensiometer or a Bouyoucos bridge. The tensiometer measures soil-moisture tension. The drying of soil increases the soil-moisture tension. This tension causes a suctioning of water from the soil tensiometer that can be observed on the tensiometer gauge. Tensiometers are usually installed in several places and at varying depths in the zone of rooting. The Bouyoucos bridge measures the electrical resistance of a sorption block, such as gypsum, situated in the soil. The electrical resistance changes with the amount of absorbed water in the block, a direct function of the amount of moisture in the surrounding soil. Generally, the amount of soil moisture should be supplemented by irrigation when the level of available water in the root zone reaches 50% of field capacity.

When the water is applied, remember that not all the irrigation water will be available for use by the plant. Some will be lost by runoff, evaporation of free capillary water, adherence to soil particles (hygroscopic water), or sinking below the root depth (gravitational water). The latter can be minimized by applying water only to the depth of rooting, a depth that is well known for most horticultural crops (see Chapter 6). Other points to consider are that soil fertility levels must be good for efficient water utilization and calculations and measurements are only a useful starting point. This must be tempered with experience, personal observations, and variations that prevail in localized sites.

Certain soils in arid and semiarid regions, classified as saline, alkali, or saline-alkali, pose special considerations with regard to irrigation. Saline soils have surface salt crusts. These soils are productive if their drainage is either naturally good or improved by artificial means such that low-salt irrigation water will leach the salts below the root zone. Problems result when only high-salt irrigation water is available or the water table is high. Alkali soils have a pH between 8.5 and 10.0. These soils require applications of sulfur to reduce pH, and gypsum (calcium sulfate) to reduce the levels of exchangeable sodium. Saline-alkali soils have the problems of both saline and alkali soils and need all of the preceding treatments. Horticultural crops also vary in their salt tolerance, and this variation should be considered when selecting crops for these soils. Water for irrigation purposes must be tested to ensure that it meets certain standards. Excesses of soluble salts, sodium, bicarbonate, boron, lithium, and other minor nutrients could cause phytotoxicity.

Irrigation Systems.   Several types of irrigation systems are used: surface irrigation, sprinkler irrigation, subsurface irrigation, trickle or drip irrigation, and ooze irrigation. *Surface irrigation* is used in arid and semiarid areas with level surfaces. The four types of surface irrigation are *furrow, flood, corrugation,* and *border* irrigation. Under the best conditions, efficiency of water usage approaches 60% with surface irrigation. Advantages of surface irrigation include a low initial investment (if the land requires no grading), low energy consumption, and low water evaporation. Disadvantages include uneven distribution of water by gravity flow (costly grading might be required), and the weight of the water can reduce soil aeration and drainage.

Furrow irrigation (Figure 11–29), the oldest known type, depends on water flowing by gravity from a main ditch down into furrows between ridges where crops are planted.

FIGURE 11–29   Laser-leveled land is furrow-irrigated near Phoenix, AZ. Laser-leveling is done by earth movers equipped with laser receiving equipment and is done to optimally manage the flow of irrigation water across a field. (USDA/NRCS photograph.)

Slopes must be less than 2% to avoid soil erosion. Some piping can be used to minimize water losses by seepage and evaporation. Water gates and siphons are used to improve water distribution. In flood irrigation the water is distributed as a continuous sheet. Crops tolerant of excess water, such as rice or cranberries, are generally irrigated by the flood method. Corrugation irrigation (Figure 11–30) can be used on slopes up to 5%. Many small furrows are used to guide rather than carry the water. Border or contour levee irrigation (Figure 11–31) is used on gentle slopes. Narrow, leveled strips are enclosed by levees, and irrigation consists of flooding these areas.

Sprinkler irrigation consists of pumping water through pipelines and distributing it from rotary heads (Figure 11–32). Water use efficiency approaches 75% with sprinkler irrigation. The result is a rainfall-like application. This system is popular for supplementing

FIGURE 11–30
Corrugation irrigation (think how corrugated cardboard looks) is shown here with a potato crop in Idaho. (USDA/NRCS photograph.)

FIGURE 11–31   Border irrigation is being used in this orchard. Note the levees to the right and left of the leveled area with the fruit trees. (USDA/NRCS photograph.)

(A)

(B)

(C)

(D)

FIGURE 11–32   (A) Center pivot irrigation system on wheels advances across the field using LEPA (Low Pressure, Precision Application) system. This type of application uses less water and reduces evaporation. (B) A close-up of a sprinkler head from the LEPA system. (C) Potato field with fixed sprinkler line irrigation in Grand Traverse County, Michigan. (D) Low sprinkler irrigation is used to provide water for citrus trees in this California orchard. (USDA/NRCS photographs.)

natural rainfall during periods of inadequacy, especially in the humid regions of the East and South. Advantages of this system include reductions in land leveling, drainage problems, erosion, and the need for special skills. These pluses result from the even, controlled rate of sprinkler irrigation. The initial investment, evaporation losses, and energy consumption, however, are higher than those for surface irrigation, but the operational costs are less. In addition, winds can interfere with uniform distribution. Newer drip or splash nozzles are available for pivots and are effective at reducing evaporation.

Subsurface irrigation depends on an impervious layer below the soil surface. This layer will stop percolation losses and allow the maintenance of an artificial water table by pumping water through ditches and laterals. This irrigation system depends on unusual soil conditions and is not widely utilized in the United States.

The previous irrigation systems are used mainly on large-scale horticultural operations in the field. The home gardener can use a hose and sprinkler during periods of inadequate rainfall, and this would be a simple form of sprinkler irrigation. The use of misting systems for propagation and automatic-watering sprinkler systems in the greenhouse is another form of sprinkler irrigation.

(A)

(B)

(C)

(D)

FIGURE 11–33 Drip irrigation lines with only holes are used for field and orchard crops. Pipes with narrow bore feeder/emitter lines are used with nursery container crops as the flexible "spaghetti" line can be easily placed into the container's surface. (A) Soil cut away to expose a buried drip irrigation line in a tomato field. (B) Drip irrigation delivers small amounts of water to young citrus trees in California's Imperial Valley. This type of irrigation is highly efficient because it delivers small amounts of water over a long period of time and greatly reduces evaporation of irrigation water. (C) Drip irrigation on grapes in one of California's productive valleys. (D) Ooze irrigation hose showing water oozing out. (A and D, USDA photographs; B and C, USDA/NRCS photographs.)

Other types of irrigation have become popular for greenhouse use, small nursery container operations, home gardens, high-value fruit crops, and small field operations. These types are known as trickle or drip irrigation (Figure 11–33) and ooze irrigation. Drip irrigation is extremely efficient, approaching 90% efficiency. However, these systems cost more initially relative to other irrigation systems. With drip irrigation, water is carried by polyethylene pipes into which are tapped either small holes or smaller feeder lines that terminate with small emitters (valves). Water trickles from these holes or valves. Valves can be shut off individually, or the holes in the hose can be plugged as needed. Drip irrigation is particularly adapted to potted plants, nursery container stock, small vegetable and flower gardens, and orchard operations. California is the leading state in the usage of drip irrigation for commercial horticulture. Ooze irrigation employs polyethylene pipe for carrying water and the feed lines are porous polyethylene that allows water to ooze through micron-sized holes.

The reasonable cost and efficient water use of these systems suggests their increased use in all sustainable horticultural operations. Drip irrigation offers several other advantages for sustainable horticulture. Less energy and fertilizer are needed relative to other irrigation systems. Less cultivation and fewer herbicides are needed, because weed growth is reduced. Greater crop uniformity results, because seedling mortality is reduced. In some cases, crops mature earlier and yields are increased. Less erosion results, so irrigation on more sloping land becomes feasible. Computerized controls are also possible with drip irrigation. These advantages help offset initial costs.

Of particular interest is a demonstration project in Florida with winter tomato production. Irrigation water was supplied from a well and brought to the field via PVC pipes. Drip tapes were placed under plastic mulch about 10 inches from the rows. The system was termed *micro-irrigation*. Fertilizer was applied by the irrigation system. Compared with normal, seepage irrigation by ditches, micro-irrigation with fertigation reduced water consumption, energy consumption for pumping, and fertilizer needs. An earlier project showed that micro-irrigation resulted in up to an 80% water savings relative to seepage irrigation. This later study showed that micro-irrigation versus seepage irrigation resulted in 7,000 kWh of energy savings yearly per acre when the water table was 100 feet down. A 50% reduction in nitrogen fertilizer use was also observed. Initial costs and management skills are higher, but operating costs are less. From the sustainability view, micro-irrigation can provide much value to the sustainable horticultural operation.

MULCHES, SOILS, AND WATER

The cultural practice of mulching has several advantages. Water is conserved, because mulches reduce the evaporation of soil moisture by acting as a barrier and lowering the soil temperature. Water absorption by a mulched (non-plastic) soil is greater than by an unmulched soil, because of physical properties of the mulch and prevention of the formation of impervious soil crusts. Consequently, soil losses from heavy washing and blowing are decreased. In effect, mulches are excellent conservation agents. Mulches also save on the labor of cultivation, since emerging and small weeds perish under their dark barrier. Therefore, they reduce tillage and the use of weed control chemicals. The disadvantages are cost and installation labor.

The insulating property of mulch averts drastic fluctuations of soil temperature, keeping the soil cooler in summer and warmer in winter. During the summer this improves both root growth and nutrient availability. Winter mulches reduce the risk of root damage. Soils are improved with organic mulches as lower layers decompose and become incorporated into the soil. At the end of the growing seasons, organic mulches can be tilled into the soil to further increase the organic matter content. Finally, mulches impart a neat, trim look to gardens, reduce the incidence of mud-splashed flowers and vegetables after heavy rains, and decrease the frequency of vegetable rot caused by soil contact.

Mulches do have some disadvantages. The application of a mulch requires labor of varying degrees, depending on the material selected. However, the labor saved from the decreased need for cultivation usually outweighs the initial input. Mulch and moisture around newly emerging seedlings or perennials can provide an ideal environment for diseases, such as damping-off or crown rot. The potential of disease is increased by long periods of rain. Diseases can overwinter in the mulch. Insects and rodents, who find it an attractive habitat, can cause plant damage. These problems can be minimized by avoiding direct contact between the plants and the mulch. Premature applications of organic and black plastic mulches can retard soil warming, and hence the growth of plants preferring warmer soils, such as tomatoes, peppers, and eggplants. An exception is clear plastic mulch, which will accelerate the warming of soil and produce earlier yields with warm season crops such as corn or muskmelons, but it supplies less weed control than black plastic mulches, unless a herbicide is put down prior to the clear plastic. A mulch that is applied too thickly or one prone to caking can impede the uptake of water and air by the soil, leading to possible plant damage. Mulches with a C/N ratio greater than 30:1 can steal nitrogen from the soil during decomposition, causing nitrogen deficiency in the mulched plant. Prior applications of nitrogen fertilizers can prevent this problem. Dry mulches that are combustible can be a fire hazard near buildings or in public places. Woody mulches near buildings can be a vector for termites.

A practical mulch should be easily obtained, inexpensive, and simple to apply. Availability and cost vary from region to region. Mulching materials can be found in yards, garden centers, lumberyards, sawmills, dairy farms, tree-service firms, breweries, and food-processing plants. A suggested depth is 2 to 4 inches, bearing in mind that too little will give limited weed control and too much will prevent air from reaching roots. Mulches should be applied prior to active weed growth and summer droughts or before the ground freezes if it is a winter mulch. For warm season crops, such as tomatoes, the mulching should be delayed until blossoms appear. A discussion of mulching materials follows with specific emphasis on advantages and disadvantages.

Crop and Food Wastes

A number of wastes from crop and food processing have been used as mulches. These include buckwheat hulls, cocoa shells, coffee grounds, ground corncobs, ground tobacco stems, licorice roots, peanut shells, spent hops, and sugarcane. Buckwheat hulls are light, fluffy, and black. Because buckwheat hulls are prone to caking, they should be applied no deeper than 2 inches. Forceful watering will cause scattering of buckwheat hulls. Cocoa shells are brown, light, easy to handle, and relatively noncombustible. Cocoa shells have some value as a fertilizer and resist blowing in the wind. Their high potash content harms some plants, so they should not be applied to a depth greater than 2 inches. Coffee grounds cake badly; a depth of 1 inch is recommended. Coffee grounds contain some nitrogen. Ground corncobs make good mulch. Some find their light color objectionable. Other competing uses for ground corncobs, such as in feeds and mash, tend to limit the supply for mulching. Ground tobacco stems make a coarse, good mulch with some nitrogen value. Availability is limited, and they should not be used on plants susceptible to tobacco mosaic virus, since they can be a source of infection. Licorice roots are especially good on slopes, because they resist floating and blowing. Licorice roots also have an attractive appearance. Peanut shells are attractive and easy to apply. Peanut shells also contain nitrogen and are long lasting. Spent hops are resistant to blowing and have some nutrient value. However, when fresh they have some odor and can decompose and retain heat, thus causing some plant damage. Sugarcane is a good mulch when available in the crushed form. It has a moderately acidic pH, making it useful around acid-loving plants.

Garden, Household, Municipal, and Yard Wastes

Compost makes good mulch, because it has value as a fertilizer and a soil-like appearance. It is also a good organic amendment for tilling into the soil after the growing season ends. Grass clippings (Figure 11–34) contain nitrogen. Grass clippings cannot be applied thickly when green, because they heat rapidly and form a dense, matted layer that restricts the flow of air and water. They should be applied in thin layers and allowed to brown between each application. Leaves are free, readily available in many areas, release some nutrients on decomposition, and spread easily. However, they have a tendency to form a soggy, impenetrable mat. This problem can be overcome by mixing leaves with fluffy materials, such as hay or straw, or by shredding the leaves. Oak leaves are especially good for acid-loving plants, such as azaleas and rhododendrons. Newspaper is certainly readily available and

FIGURE 11-34   Tomatoes in this garden are mulched with grass clippings. (USDA/NRCS photograph.)

FIGURE 11-35   These tomatoes are mulched with black plastic. (USDA photograph.)

economical, but somewhat difficult to apply. The high C/N ratio necessitates the prior application of nitrogen fertilizer. A good use for newspaper is as undermulch; that is, place four or five sheets under a thin layer of an attractive, more expensive mulch. Pine needles have an esthetic appeal and are not prone to forming a soggy mat as are leaves. They are especially good for acid-loving plants. Tire fiber mulch is extremely durable and effective. It also reduces the disposal problems for tires, offering a useful recycling choice. Given its synthetic nature, it can be used around foundation plantings of homes, because it is not a vector for termites.

Manufactured Materials

Polyethylene film (Figure 11-35) is one of the few mulches that is readily available, adaptable to machine installation, and economical enough to be used in large-scale commercial applications. Polyethylene allows passage of gases such as nitrogen, oxygen, and carbon dioxide. Holes or slits facilitate the planting of seeds or plants and water entry. Some types come with perforations. It can last several years if undamaged by machinery. It is usually black. Clear film is sometimes used, but it offers limited weed control (unless herbicide is applied before mulching), since light passes through it. Earlier crops can be produced with the clear, and to a lesser degree, black plastic mulches because of the warming of the soil.

Natural Products

Peat moss is attractive, easy to handle, but somewhat expensive mulch. Dry peat moss requires considerable time and water to become moist, so it should be applied only to a 3-inch or less depth and avoided in areas subject to drought. Given its acidic pH, peat moss is especially desirable for acid-loving plants. Straw, hay, and salt-marsh hay are lightweight and easy to apply, but their appearance restricts their application mostly to vegetable gar-

dens. These materials are available in spoiled form from animal feed dealers at cheaper prices. They are used more frequently as a winter mulch for protection. They are not long lasting and frequently contain weed seeds. Stone in crushed or gravel form is durable as a mulch with an attractive appearance. It is used often around trees or shrubs. Do not use crushed limestone around acid-loving plants.

Wood By-Products

These by-products include bark, sawdust, and wood chips. Bark is a widely sold, popular horticultural mulch. Small pieces of bark are preferred over large chunks. Bark mulches vary, but all are attractive, durable, and suitable for foundation shrub plantings. Contact with wood framing is to be avoided, because bark can be a termite vector. The high C/N ratio of bark requires prior application of nitrogen fertilizer. Aged or partially rotted sawdust makes satisfactory attractive mulch that lasts a long time. Because it is prone to caking and has a high C/N ratio, apply it only 2 inches deep after adding nitrogen fertilizer to the soil. Wood chips are moderately priced or free, attractive, readily available, and easy to apply. They make excellent mulch. Their high C/N ratio requires an application of nitrogen fertilizer prior to placement of the mulch. Wood chips can last about 2 years. As with a bark mulch, one must consider termites.

OVERVIEW OF PRECISION FARMING

Advancing technology can play a major role in sustainable agriculture. Of course, the value depends on how the technology is applied. One new technology, precision farming, depends on global positioning satellites and global information systems. Precision farming offers fewer conflicts than other new technologies, such as genetic engineering, and generally benefits sustainable agriculture because of its potential lower inputs of chemical or organic fertilizers and of pesticides. Two main limitations of precision farming are economy of scale and complexity of application. Very large producers will find this technology more useful than would smaller farmers. One criticism is that precision agriculture will set back the conversion to sustainable horticulture because it encourages chemical dependence. The technology is also difficult to apply, and actual changes in practices to date have been limited, even on large farms.

Precision farming, also called precision agriculture, traces its roots to the same decade that gave birth to sustainable agriculture, the 1980s. This approach to agriculture makes use of computers and electronic information technology. During the 1990s, precision farming made considerable advances.

Like many technologies, this one owes its existence to military technology. Two components of satellite reconnaissance are the cornerstones of the precision farming technology. They are the global positioning system (GPS) and geographic information systems (GISs). GPS allows the military to locate and "target" specific installations around the globe with pinpoint accuracy. GIS allows placement of geographic features and overlays of terrain-associated data (climate, soil type, water quality, buildings, transportation arteries, population density, land prices, etc.) into maps with extremely high accuracy. In fact, GIS can be an excellent tool in its own right for the selection of optimal land for commercial horticultural production using the indicated terrain overlay data. After the end of the Cold War, these two technologies expanded beyond military uses into new areas such as agriculture, boat navigation, and environmental research.

So exactly what is precision farming? This technology is based on the fact that on any given large farm, a number of factors are variable across the fields. Practices, however, are geared to the overall or average picture of the crop area and do not account for variability in field conditions. Applications of fertilizer, irrigation water, and pesticides are applied at constant rates across fields. Conditions that vary include soil, fertility, temperature, aeration, water retention, and drainage. These variations influence environmental and biological interactions among plants, livestock, insects, pathogens, weeds,

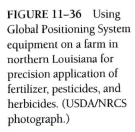

FIGURE 11-36 Using Global Positioning System equipment on a farm in northern Louisiana for precision application of fertilizer, pesticides, and herbicides. (USDA/NRCS photograph.)

and soil. These variations ultimately affect crop yields in different ways in different fields, or even areas within a given field.

Precision farming allows for the precise adjustments of farming practices and inputs (Figure 11-36) that reflect the needs of fields both in the sense of spatial and seasonal differences. These precise adjustments are based on data collected from various sources through the latest information technologies. The potential result is lower input of chemicals and water, yields that reflect the potential of each area, and less environmental degradation. Resources can be better conserved and profits per acre are improved. While most attention has been paid to crop production, precision farming could be applied to animal production or mixed crop/livestock farms. Here we consider horticultural production.

Precision farming sounds like an exact match for the goals of sustainable horticulture. Criticism has arisen, however. One criticism is that the technology, while it does reduce pesticide applications, it does not eliminate them. The reduced amounts of pesticides also do not result in reduced costs for the farmer, because the technology of precision farming is currently costly. Another criticism is that nonrenewable resources are still used and the switch to renewable ones is postponed. Other problems have been noted. We examine this area in more detail later.

Basic Factors Underlying Data Collection

Precision farming depends on understanding and managing field variability. Three aspects of variability must be considered: spatial, temporal, and predictive. Changes across the field or entire farm, such as soil organic matter or nutrient content, are examples of spatial variability. Temporal variability involves changes over time, such as daily, seasonal, or annual changes. For example, yields can change yearly due to climatic variation. Discrepancies between the predicted and actual values are known as *predictive variability.* If precision farming is to work, all forms of variability must be carefully measured (data collection) and then interpreted. Only then can changes in management become realistic.

Precision farming in its infancy is an ongoing technology under rapid development. Existing techniques are undergoing modification and new ones are being added to eventually create optimal tools for the farm manager. The technology is highly sophisticated, involving computerized equipment and software applications. Far more than localized applications of input are involved. The precise monitoring of the localized operation by data collection is essential. The data must be thoroughly assessed and then completely understood. Only then can management practices be changed so that a particular goal, such as increased yield or reduced input cost, is reached. This goal might not be for maximal yield, but targeted for maximal financial advantage within environmental constraints. For sustainable agriculturists, precision farming must be viewed from the context of sustainable resource and environmental management.

Data Collection

Precision agriculture is only as good as the data collected. First data can be collected through a number of means: field sampling (see Figure 11–7), laboratory analyses, remote sensors, in-place sensors, and soil tests. Sensors in soil can be used to detect numerous soil parameters: nitrate, organic matter, salinity, total nitrogen, and water content. Canopy sensors can monitor temperatures, leaf water potential, and leaf chlorophyll content, and combine sensors can monitor yields. Data from leaf water potential and soil water content can be used to adjust irrigation rates. Data on soil nitrate and total nitrogen plus leaf chlorophyll content can be used to modify fertilizer applications.

Sensor technology is in flux, varying from reliable to unreliable. Soil tests and lab analyses are important supplements. Remote sensors, such as those used for satellite imagery or aerial photography, have great promise, but the technology needs considerable further development and cost reduction. Satellite sensors detect microwaves, infrared, ultraviolet, thermal, and multispectral radiation that can be interpreted in many ways, such as to reveal the density and health of vegetation cover. Plans are under way in the business community to dedicate considerable satellite coverage for agriculture and to provide data in GIS format to farmers.

Satellite imaging data for agricultural and horticultural purposes became a commercial reality in 2000, making precision farming more likely to be adopted by farmers. Images and other data are available in black and white, color, and infrared. The latter makes possible quantitative analyses of crop vigor. Details down to certain trees and trucks can be made visible. Precise growing areas can be pinpointed in the field. Possible applications include measurements of field sizes, boundaries of planted fields, and sizes of problem areas; tracking of plant density and health over the growing season; identification of areas under insect and disease attacks; enabling of fertilizer applications to only those field areas where it is needed; monitoring of field moisture and irrigation efficiency; determination of storm damage; evaluation of damage areas caused by floods or droughts; estimation of yields; and better planning of future crop production.

Two types of data collection are possible. The data can be collected, placed on a map using GIS, and then interpreted. Some action is then directed. For example, if soil tests and soil nitrate levels indicate more nitrogen is needed in certain fields or areas within fields, the corrective information can be written as a "prescription" into the GIS. The GIS attached to applicators then can vary the rate of nitrogen in areas needing correction, based on positional information provided by the onboard GPS system. This system does have a time lag inherent in the interpretation of data and programming of corrective action.

Another approach would be to utilize sensors that instantly translate incoming data into action. These sensors would send signals to systems or equipment that would then provide the changes needed. For example, suppose a tractor-driven unit is applying an herbicide. An onboard, digital videocamera looks ahead and compares plants to files in its memory. If a "weed" image is found, a signal is sent to the trailing herbicide unit, which then sprays the offending plant. In this system, GIS and GPS are not essential, but can be useful for record keeping. This scenario is becoming possible, but given the rapid advances in technology, it is not yet a reality. Most precision farming is based on the former approach involving lag time between sensing and action.

Global Positioning Satellites

The GPS system consists of 24 orbital satellites conceived and controlled by the U.S. Department of Defense. These satellites broadcast one-way signals to ground-based GPS receivers that in turn translate the signal into positional information, that is, elevation, latitude, and longitude. These satellites are used as reference points. Essentially the system is used to triangulate (three points determine a location) locations on the surface of earth. At any given location on earth, enough satellites are available to allow for triangulation. Twenty-four satellites are needed to make any earth point "readable" and to allow for some margin of overkill.

Locations can be determined with accuracy to meters or even centimeters, with appropriate corrections and advanced instrumentation. An initial error is deliberately introduced by the military so that the system is not too useful to hostile foreign powers or terrorists intent on delivering weapons to pinpoint targets. However, technology is available to get around this aspect, at considerable cost and expertise. The system is so good that essentially every square meter on earth can be given its own coordinates, or unique address. We won't get into the explanation of how the process, including corrective approaches to greater accuracy, works here. Those who are mathematically inclined can check out the GPS Web sites listed at the end of the chapter.

Ground-based receivers can be permanent or mobile installations. In fact, receivers have become miniaturized to the point of being handheld. For precision farming, GPS receivers can be mounted on tractors and other farm machinery, and even laptop computers. Costs have also become reasonable.

Geographic Information Systems

The GIS tool is essentially a device for the capture, manipulation, and application of a database referenced to specific locations, such as a farm field. Data can be handled in numerous ways: captured, stored in memory, edited, updated, retrieved, analyzed, modeled, and displayed. The display often takes the form of a map with specific features overlaid, such as nutrient levels or salinity or yields or profit (yield and profit map, respectively). The actual tool is a piece of computerized hardware containing GIS software. Users must be highly skilled in computer applications and database manipulation. The technology is relatively new and undergoing rapid changes.

Numerous data layers can be overlaid on field or orchard maps. These overlays can include numerous soil parameters: textures, organic matter content, nitrate levels, total nitrogen, total and available phosphorus, potassium levels, trace element levels, pH, water content, yields, and field topographical features. Features can be compared to see if any correlation exists, such as yields versus nitrate levels, or more complex combinations of factors. These field maps also form the basis for deciding where specific inputs need to be applied.

Databases can also track herbicide applications, irrigation schedules, yields, changes in crop cultivars, fertilizer and lime applications, and pesticide usage. These databases can help predict future needs and show patterns, such as where weed and insect problems are high and likely to need more attention, or where disease outbreaks occur more frequently. These databases can also track where resources are used most heavily and whether dollars are being wisely spent.

Variable-Rate Technology

In order for precision farming to work, machinery working in concert with GPS and GIS must be able to apply inputs across the fields at variable rates. For example, one area might need limestone applied at rates of 2 tons per acre, whereas another area might need 4 tons per acre. The input machinery must be able to change input rates as it rolls across the field. Changes are based on data either obtained earlier or "on the go" through the combined GPS/GIS systems. Variable-rate applicators are specialized pieces of equipment that can be costly to own; farmers might decide to instead rent or lease such equipment.

Sustainability and Precision Farming

Precision farming is currently in early development. The major effort has been in agricultural applications, especially for grains. Most of the effort has been directed toward yield mapping, profit mapping, and input adjustment to maximize productivity. Some horticultural applications are in early stages, such as vegetable farming in Florida. Large orchards and vineyards are other potential horticultural targets. The cost of the technology is currently limiting applications to large farms of single crops, such as tomatoes or sweet corn. The smaller farmer is essentially priced out of the technology. Whether costs will decline over time sufficiently to allow smaller farms to use precision farming remains uncertain. Some data acquisition and processing might be done in the public sector, thus leading to cost reduction.

Is precision farming beneficial for sustainable horticulture? This question has been answered differently, depending on whom you ask. On the plus side, any technology that reduces unneeded inputs of fertilizer and pesticides does increase resource sustainability and does reduce pollution. On the negative side, any continuing use of chemical inputs is not sustainable in terms of resources and environmental damage. The question arises as to whether this quick fix will become a brake on the conversion to sustainability. Also on the negative side is that the cost will prevent smaller farms from improving productivity and sustainability. Small farms would become less competitive, which would further speed their demise. This decline in the number of family farms will accelerate the weakening of many agricultural rural communities.

SUMMARY

When selecting an area for horticultural use, one must consider climate, soil, shade, drainage, water resources, and pollutants. Localized microclimatic effects can affect the usefulness of a site. Soil should be deep, fertile, easy to work, and well drained without hardpan. The site should have ample, good-quality water available and direct sunlight all day. A level site with good drainage is acceptable, it is desirable with sandy or gravel-type soils to improve water infiltration, and is preferable to sites with grades over 3%. Slight southerly slopes favor early crops. Sites might require wind (windbreak) and animal protection. Commercial horticulturists must consider suburban proximity. Local availability and proximity of alternate energy sources may also be critical.

Poorly drained soils require artificial drainage when the water table is high or when surface water infiltrates poorly. Modification of surface drainage includes slight grading or digging of surface drainage ditches. Drainage problems in the root area are corrected by breaking up hardpan and incorporating physical amendments to improve aeration. Drainage problems in the subroot area are corrected with a tile drainage system.

Windbreaks reduce or deflect excessive winds, reduce soil erosion by wind, and protect crops from wind damage, drying winds, snow accumulations, and sudden temperature drops. Windbreaks reduce heating needs for greenhouses and homes and can deflect wind to other locations, such as summer breezes through the patio. Windbreaks consist of trees, hedges or artificial materials and are useful for a distance 10 to 12 times their height; maximal protection distance is five to six times the height.

Trees also control temperature. During summer, proper placement of deciduous trees provides shade, thus reducing energy consumption by cooling devices or making an uncooled building or patio more comfortable. During winter, deciduous trees permit the sun's rays to warm the home and provide light to the greenhouse. In areas where sun is a year-round problem, evergreen trees are preferable over deciduous ones.

Future possibilities include the incorporation of woody crops into windbreaks to generate income. Irrigation wastewater could be funneled through windbreak areas and be partially cleaned by the plantings. Global warming might be reduced as more windbreaks are planted and carbon becomes sequestered by the growing trees and accumulated organic matter in the soil. Optimizing plant diversity in the windbreak might attract insects (trap cropping) and reduce damage to nearby crops.

Plants in exposed locations might require winter protection against winds, heat and light extremes, freezing and thawing cycles, and protection against heavy snow. The erection of burlap screens or antitranspirant sprays can protect evergreens from winter desiccation. Young trees often need tree-wrapping trunk tape to prevent winter sun scald. Soil can be mounded around crowns of tender woody plants. Newly planted herbaceous perennials can be protected from soil heaving by applying light, loose organic mulch after the ground freezes. Snow damage can be minimized by placing lath or board structures over the tops of susceptible plants, or by tying them together with cord.

The propagation and production of many annual flowers and vegetables is keyed to the variable known as the first frost-free date. Sometimes the danger of a late, unexpected frost threatens tender plants recently planted or established plants that are at a critical developmental stage. The gardener can protect against a light frost by laying newspapers, polyethylene, bushel baskets, or cloth over the threatened plants. Polyester and plastic tunnels are used commercially for high-value horticultural crops. Clean-burning heaters use heat convection currents to reduce frost damage in citrus and fruit orchards. Mixing of the air by fans can increase the temperature around orchards and protect from light frosts. Fans and heaters together are even more effective. Water fogs can also reduce the frost danger.

Fertility relates directly to crop production, profitability and environmental well being. Insufficient soil nutrients result from deficiency or their presence in a chemically unavailable form. Excess nutrients cause detrimental effects, such as phytotoxicity or abnormal growth excesses. By the time nutrient deficiency or excess symptoms are noted, plant growth and yield are probably already reduced. These symptoms are not easily diagnosed. Nutrient deficiencies are best avoided by early determination of soil fertility.

Nutrients removed by crops must be replaced to maintain soil fertility. An annual soil test can determine what nutrients, micronutrients, lime, and soil conditioners are required and how much. A key need in soil analysis is a soil sample that adequately represents the root zone area or prepared growing media. Desirable information includes the crop to be planted, crop history, yield goal, previous liming and fertilization history, drought history, plow depth, and any unusual problems or conditions. Computers are mainly used to interpret the test results and to make fertilizer recommendations. Reports supply information on fertility, pH, and organic matter needs, including specific fertilizer, limestone (or sulfur), and organic amendment recommendations and amounts. Plant tissue analyses to determine the actual amounts of nutrients absorbed by the plant supplement soil tests. Additional tests are required for sustainable farming systems in terms of soil quality. Tests for soil quality include respiration rates (an indicator of microbial activity), organic matter levels (relates to erosion, nutrient and water retention, tilth, microbial activity, and plant root penetration), and aggregate stability (indicator of erosion, water movement, and root growth potentials). A certified water-testing laboratory should check new water sources for hazardous chemicals, metal content, boron, hardness, salinity, pH, and microbial contaminants.

Soil tests have some faults. Variability from lab to lab is common. Many results are expressed in a conservative fashion; overrecommendations are built in to reduce the risk of inadequate nutrient levels. Some fail to consider the nitrogen contribution from legumes. The latter concerns organic and sustainable farmers, because excess nitrogen costs more and pollutes. Many tests fail to also address criteria important to soil health.

Organic matter plays a critical role in soil fertility, crop productivity, erosion, water retention, nutrient availability, soil tilth, soil health, and even some pest problems. Decomposing organic matter provides nitrate by the microbial transformations of complex organic molecules. This conversion or mineralization process also provides phosphorus, sulfur, and many trace elements. Mineralization is important when fertility is maintained with organic fertilizers and plant residues. The nutrients in these materials need to be mineralized before becoming available to plants. Organic matter also contributes indirectly to soil fertility by regulating nutrient availability. Organic matter (especially humus) attracts positively charged nutrients (cations), thus reducing leaching of nutrients such as calcium (Ca^{2+}) and potassium (K^+). The cation exchange process releases cations. Clay also exhibits cation exchange capacity.

Another indirect contribution to fertility results from the chelating ability of organic matter. Certain trace elements can be held in chelated form by organic matter. Chelation keeps these trace elements from forming insoluble chemical complexes that make trace elements unavailable to plants. Often trace element deficiencies are really a result of low organic matter. Also by forming chelates with iron and aluminum, organic matter reduces the ability of these two chemicals to chemically bond with phosphorus, thus rendering it unavailable to plants. The chelation of aluminum prevents aluminum toxicity.

Soil health is better when a reasonable amount of organic matter is present. Beneficial microbes thrive where organic matter decomposes. Some microorganisms release nutrients from organic matter, others help control insects, diseases and nematodes; yet others (mycorrhizae) improve nutrient uptake, plant health, and soils. Some soil bacteria fix nitrogen from air. Legume crops with nitrogen-fixing bacteria can directly utilize this fixed nitrogen. Other nitrogen-fixing bacteria are free living, releasing fixed nitrogen through mineralization. As organic matter drops and soil pesticides are added, these beneficial microbes decline. While this loss can be offset with more fertilizer, water and pesticides, increased costs and pollution result.

Organic matter decomposition results in numerous microbial excretions and fungal hyphae. These products help bind and stick particles of sand, silt, clay, decaying organic matter, and humus into aggregates. High levels of aggregates result in soil structures having fluffy, loose matrices loaded with channels and pores. The soil is more easily worked, less prone to compaction, aerates well, and holds water well. Surface water infiltrates more readily, reducing both erosion and the need for irrigation. These beneficial properties diminish rapidly as organic matter declines. Organic matter also darkens the soil's color, thus increasing heat retention. This warming results in a slightly earlier planting date and improved germination. Organic matter also buffers and moderates sudden harmful changes in soil pH when lime or chemical fertilizers are added. Humus likely stimulates root growth and root hair formation.

Organic matter ties up carbon in the soil. When organic matter is oxidized and releases CO_2, it is a contributor to global warming. Practices that increase soil organic matter, such as cover crops, or reduce the oxidation of organic matter, such as conservation tillage, can reduce atmospheric CO_2. Accumulation and retention of organic matter is good for horticultural crop production and good for the environment, both in terms of slowing global warming and reducing water pollution caused by soil erosion.

Organic matter maintenance in sandy soils is difficult and raising the levels of organic matter is even harder. Still, the contributions of organic matter to productivity in sandy soils are considerable. Efforts to maintain low organic matter levels merit a high priority. Continuous cropping reduces organic matter by up to 2% of the total organic matter yearly. Unless the organic matter is replaced by soil conditioners or cultural practices in addition to fertilizer, the long-term economy of plant production suffers. In the short term organic matter losses can be offset by increased fertilizer and water inputs.

Horticultural soils require maintenance of soil health and crop productivity. We can compensate for nutrient and organic matter losses by adding fertilizer and organic matter and altering pH by adding limestone or sulfur. Conventional horticulturists rely heavily on chemical fertilizers and crop residues. Organic horticulturists use organic fertilizers and multiple practices to add organic matter. Other sustainable horticulturists use practices that fall somewhere between these two approaches.

Fertilizers containing nitrogen, phosphorus, and potassium are called complete fertilizers. Secondary elements and micronutrients are added in varying amounts to specialty fertilizers or applied separately. Secondary elements are calcium, magnesium and sulfur, and micronutrients include boron, copper, iron, manganese, molybdenum, zinc, and chlorine. Calcium and magnesium are usually applied through liming.

Organic fertilizers are derived from living or dead organisms and natural materials, and chemical fertilizers from chemically processed materials. Chemical fertilizers are easy to apply with rapidly available soluble nutrients, but are prone to leaching and fertilizer burning. Chemical fertilizers come in dry, granular forms for soil application, or as a water-soluble powder for liquid application. Chemical fertilizers are energy-intensive products and use unsustainable natural resources. Organic fertilizers in dry or liquid forms have some nutrients in unavailable forms that soil microorganisms convert to usable nutrients and are not ideal for rapid nutrient buildup. Organic fertilizers offer the advantage of extended nutrient availability, less leaching or burning, and are less likely to cause water pollution.

Mixtures of organic and inorganic fertilizers offer the advantages of both. An example is lawn fertilizer that contains inorganic nitrogen for quick use and organic nitrogen to extend nutrient availability. Chemical fertilizers are also produced in forms that extend nutrient availability and offer less leaching or burning. Slow-release forms are based on resinous coatings or reduced solubility. These synthetic forms tend to cost more and are often used on container plants to reduce the need for supplemental fertilization.

Broadcast scattering is the uniform scattering of fertilizer by hand or fertilizer spreader. Plow-sole application uses mechanical means to place fertilizer behind the plow in the opened furrow. Band placement or soil injection is the application of fertilizer bands under or to the sides of the planted seeds. Top or side dressing is the application of fertilizer over the tops or sides, respectively. Soluble fertilizers can also be applied in irrigation water (fertigation) or as water solutions during transplanting (starter solution). Fertilizers can also be applied through the foliage (foliar nutrition).

Decisions on the amounts, timing of initial and supplementary applications, and placement are important for efficient fertilizer usage. One must relate applied amounts to specific crops and even to developmental stages of the crop. Some vegetables need more nitrogen than others. Fertilizers are traditionally applied just prior to or at planting time. Fall applications are cheaper and the labor load at planting time is reduced. From the sustainability viewpoint, fertilizers should be applied as close as possible to the time when the developing plant needs nutrients to minimize leaching. Fruiting vegetables use more nutrients during reproduction. Nonfruiting and cool season crops use nutrients slowly the first half of the season and then increase nutrient uptake until just prior to harvest. Young fruit trees have lower nitrogen needs that rise gradually and peak at the stage of reproductive maturity. Nursery container stock nitrogen needs are lower in the earlier cooler part of the growing season and higher as warmer temperatures prevail.

The primary danger with fall applications of fertilizer is nitrate loss through leaching, which contributes to water pollution where winter rains are common. If economics dictate fall applications, nitrate leaching is minimized by using ammonium forms and applying to cold soil (50°F or less). Ammonium nitrogen can be applied in the fall to acid soils, since nitrate conversion is slow. Nitrogen fertilizers can be applied safely where a cool season, fall crop is growing vigorously and will absorb the nitrogen or where nitrogen will be utilized by decomposing crop residues. Fall applications of phosphorus and potassium are acceptable, except on peat and muck soils where potassium leaching occurs.

One fertilizer application, while preferred labor- and cost-wise, is not sustainable. During early growth, many crops have low to moderate nitrogen needs; unused nitrates are leached. Supplemental (split) applications are better and recommended for vegetables, ornamentals, fruit trees, grapes, and turf. Short season crops are an exception. Even with multiple applications, one can still apply all the phosphorus and potassium (less to no leaching losses) in the first application, while splitting nitrogen applications. Soil testing and tissue analyses helps keep supplemental amounts low.

Mineral soils require split applications, because nitrogen and potassium tend to be highly mobile and easily leached. Later applications need to be banded close to the outermost reach of the root system without damaging the crop's root system. In supplemental applications, liquid fertilizer is easier to apply, especially with mulched crops where an injection wheel is used. An alternative method is to use mulches and drip irrigation, through which soluble fertilizer can be applied. If broadcast applications are used with nitrogen fertilizers, methods to reduce losses and pollution from leaching should be used. Plastic mulches reduce losses from heavy rainfalls when placed after broadcasting fertilizer.

Another choice is a mixture of organic or slow-release fertilizer with chemical forms for extended and quick availability, respectively. Steady, low levels of nutrients are often sufficient in earlier developmental stages of crops. As nitrogen demand peaks later, chemical fertilizers can be used to supply the higher levels of nutrients quickly. Slow-release forms are useful for container nursery field stock to reduce the frequency and labor of application. Another approach is to load organic nitrogen at high levels through legumes, which at peak demand release sufficient nitrates through mineralization. Another possibility is to add soluble fertilizer through drip irrigation systems. Leaching is essentially nil, as nutrients are present in amounts just sufficient for needs. This approach works well for high-value field crops with plastic mulches, nursery container stock, and greenhouse crops.

Nitrogen is taken up as ammonium or nitrate (preferred form) by plants. Ammonium is held as an exchangeable form on clay and humus, but is easily converted into nitrate by bacteria in warm soils. Nitrates are lost easily by leaching, making it difficult to maintain nitrogen without supplementary additions. Nitrogen can become unavailable to plants in soils

where active organic matter decomposition is occurring until the C/N ratio reaches 12:1. Nitrogen availability is reduced in cold and/or wet soils in the spring, because unfavorable bacterial actions decrease levels of nitrogen. Nitrates are preferred in cold or strongly acidic soils, given slow conversion of ammonium to nitrate. In warm soils, either form can be used. Several factors favor the use of ammonium forms or urea, which hydrolyzes rapidly to ammonium in soil. Ammonium is less costly than nitrate. Nitrate applied directly is more likely to be lost through leaching and denitrification. Less nitrate pollution of ground and surface water results with ammonium forms. Surface applications of nitrate forms are preferred over ammonium or amine forms (urea) on alkaline soils, since the pH favors production of ammonia gas from the latter two and subsequent losses. Ammonia should be applied at a 6-inch depth to maximize soil absorption and to minimize losses. Nitrates or ammonium salts applied at the surface will be leached to root depths by rain or irrigation water.

Phosphorus mobility and availability in field soils is low in most alkaline and acidic soils. Phosphorus needs are maximized prior to planting because of limited availability, and low mobility reduces supplemental needs. If one initial application of phosphorus is used, organic matter and mycorrhizal fungi in the soil increase availability. Because of low mobility, the recommended phosphorus placement is close to the potential or existing root zone of seeds or transplants, respectively. This choice is better from a sustainability viewpoint; less localized applications require higher amounts to achieve the same level of phosphorus near roots. Higher amounts are more costly and lead to increased pollution. Banding applications provide more efficient use of phosphorus than broadcasting methods. If broadcasting is used, the fertilizer should be incorporated into the soil prior to planting. Starter solutions containing 1,500 ppm of phosphorus are especially helpful with transplants and subsequent establishment of good root systems, especially with early growth in cool soils. Supplemental fertilization is often lower in nitrogen and higher in phosphorus than the first applied fertilizer in order not to favor vegetative over reproductive development.

Potassium mobility is greater than that of phosphorus, but less than nitrogen. Solubility and toxicity suggest placement not too close to seeds or plants. Injury is most likely with sandy soils. Potassium needs are less than nitrogen and slightly more than phosphorus, so the need for supplemental applications is low. One application is reasonable in most soils, except for conditions that favor leaching. Split applications as with nitrogen are best practices for the management of leaching losses where heavy rains and poor water retention favor potassium leaching.

Soil sustainability will not be achieved with total dependence on chemical fertilizers. Chemical fertilizers are energy-intensive products, their raw materials are limited natural resources, and they do not replace lost organic matter, thus reducing soil health. The sustainable horticulturist will use less chemical fertilizer and follow the recommended timing/application practices. The next step is a gradual to complete switch to renewable organic fertilizers and alternative practices that supply additional nitrogen and organic matter to the soil. Such practices include green manure, cover crops, rotations and organic amendments.

The use of ground rock minerals such as rock phosphate and green sand to supply phosphorus and potassium slowly by microbial action is more sustainable and pollutes less than other fertilizers. These materials, along with the above approaches to supply nitrogen and organic matter, need special management in order to be highly effective at low levels. To assure yields at acceptable levels, amounts of N, P, K, and organic matter must be at slightly higher levels than needed when nutrients are supplied in a soluble chemical fertilizer. For this reason, the transition from chemical to organic horticulture requires a transition period while nutrient and soil microbial levels are built to acceptable levels so that less soluble, low applications of organic and rock mineral fertilizers can be effective.

A soil test can determine soil pH. Acid soils exist in the eastern half of the United States and the coastal Pacific Northwest. Neutral-to-alkaline soils are found in the arid western and intermountain areas. Horticultural practices lower soil pH, in that most fertilizers acidify the soil. Most horticultural crops in temperate mineral soils grow well at a soil pH of 6.0–7.0. Organic matter content can influence this range, decreasing it as organic matter content increases. Succulents from arid regions prefer slightly alkaline soils, while evergreens and Heath family plants require a acidic soils.

Soils with pH values below 6.0 generally have decreasing availability of some nutrients, while others become increasingly available and phytotoxic. Acidic soils have their pH increased by the addition of lime for most horticultural crops to pH 6.5. Common liming materials include ground limestone ($CaCO_3$) and dolomitic limestone [$CaMg(CO_3)_2$]. The amount of lime to apply depends on several factors: existing soil pH, pH preference of the crop(s), and lime purity. Particle size is also a consideration. The solubility of the different forms of lime affects the reaction. The more organic matter and clay in the soil, the greater the amount of lime needed. Less lime is needed to go from pH 5.5 to 6.5 than from 4.5 to 5.5. The efficient way to add lime is in small amounts yearly. Increased labor makes this approach more costly. Besides initial pH adjustment, additional lime is needed to replace losses due to leaching, removal by harvested crops, erosion, and fertilizer acidifying properties. Lime is usually applied in the spring or fall with spreaders. Best results are obtained with incorporation into the soil. Even with no-till, lime should be plowed, chiseled, or disked in periodically. Lime and all fertilizers can be safely applied together in most cases and mixed into the soil.

The most effective method to acidify a soil, as in the Southwest, is to apply sulfur. Sulfur is also desirable for growing acid-loving plants. For more rapid changes, aluminum or iron sulfate can be used, but are less effective over the long term.

Crop residues provide the minimal approach to maintaining soil organic matter and partially eliminating soil erosion. Green manure, animal manure and organic amendments supply organic matter (and some nutrients). Cover crops conserve or increase organic matter and reduce soil erosion by covering the soil. Such practices, often used by organic farmers, are increasing in popularity. The choice must ultimately be based on site specificity, management goals and the awareness that resource-sparing technologies require much more sophisticated management than resource-consuming technologies. In some instances trade-offs might be required.

Crop residues left after crops are harvested and subsequently plowed under control soil erosion and maintain/increase soil organic matter. This widely used practice reduces soil crusting, provides some nutrients and improves water infiltration rates, water retention and soil aggregation. One disadvantage is that crop residues can introduce pathogens into the soil. Crop residues are most effective when soil is not exposed during any part of the year. Crop residues are not always the answer. With some crops, very little residue remains. Alternative uses of residues compete. Residues are often inadequate for restoring or maintaining organic matter on badly managed soils or on soils where rapid oxidation of organic matter occurs. Residues also don't provide enough cover to prevent unacceptable rates of erosion on highly susceptible soils.

Cover crops are grown to prevent soil erosion. Such crops also conserve soil organic matter by preventing the loss of the organic matter-enriched topsoil. Legumes are planted between crops to provide nitrogen for the following crop (green manure), but their presence as a soil cover will also conserve soil organic matter and reduce soil erosion. Conversely, a legume grown primarily over the winter to reduce soil erosion will also add nutrients and organic matter. Cover crops can also reduce insect or weed problems. No profit is generally expected, since the cover crop is not harvested. Some exceptions exist with certain cover crops harvested for forages or seed. The farmer usually manages the cover crop with the expectation of soil benefits.

Green manure and cover crops are sown on idle fields. In summer fallow, growth is not a limiting factor. More commonly, the idle period is late fall through winter. In the South, cover crops can mature during the off-season. In the North, a shorter growing season and colder winter limit plant choices and ability to reach maturity. In short-season, cold areas green manure and cover crops are best managed by sowing them in place before the cash crop is harvested. The cash crop must be mature enough to tolerate competition. Seeds can be sown over the crop when young or later between rows (intercrop or living mulch) when the cash crop is last cultivated.

Killing of mature cover crops is necessary prior to planting crops. The exception is for sod cover crops used in orchards. Cover crops either die from winter injury or by mowing, plowing under, or treating with herbicides. Green manure crops need this time to break down to avoid nitrogen depletion in the new crop. Killing is also influenced by moisture considerations. In arid or drought-stricken areas, earlier killing is recommended with soil incorporation, because late killing can result in moisture limitations for the following crop. If the planting is instead left as mulch for no-tillage systems, later killing results in more mulch residues and increased water conservation for the following crop.

Cover crops used as green manures supply considerable nitrogen for cash crops. Green manures consist of summer or winter annuals plowed under to release a steady supply of nitrogen. Legumes are often used. Rye and ryegrass are also commonly mixed with legumes as cover crops. The normal expectation with green manure should be enhancement of soil nitrogen, with little gain in soil organic matter. Soil bacteria (*Rhizobium*) form symbiotic relationships with legumes. The bacteria capture or fix nitrogen from the atmosphere and transform it into forms used by the legume. Legumes can supply reasonable amounts of nitrogen to the soil. Legumes can be grown as annual summer crops or as winter annuals planted in the fall and wintered over. Biennial and perennial legumes are also grown as longer term cover crops in rotations.

The choice of legume or legume/grass is made based on the amount of nitrogen required and the goal of the grower. Summer legumes have minimal nitrogen needs and have value as a cash crop. However, much of the fixed nitrogen is removed at harvest, leaving little for the crop that follows. Moreover, a horticultural operation might not be interested in handling and marketing what are considered agricultural crops.

Winter annual legumes can be fall-planted after a crop is harvested and with sufficient growth can contribute much nitrogen to the following crop. Winter legumes offer a better choice in terms of more retained nitrogen and reduced soil erosion during the winter months. Perennial legumes offer reasonable amounts of fixed nitrogen for subsequent crops, but tie up cropland for some part of the rotation. However, some compensation results from their forage value.

Climate plays a major role in species choice. Nitrogen capture with legumes in the fall and early winter is reduced progressively as one moves northward or into higher elevations. Cold areas in the North and upper Midwest have fewer choices than the South. The amount of soil water and rainfall are factors. Insufficient moisture restricts germination and early growth and reduces the capture of nitrogen by the legume. In dry areas, legumes with lower water requirements should be grown.

Timing is difficult. If you kill legumes too soon, not enough nitrogen is present and the plant material is very succulent and decomposes rapidly. Rapid nitrogen mineralization can lead to the loss of nitrogen by leaching and volatilization. If you wait too long, certainly the plant will have maximal nitrogen, but will be tough and will decompose slowly and not release enough nitrogen for the crop that follows. The ideal compromise is to kill the legume at maximal top growth, which occurs just before full flowering. Earlier killing is advisable if crop timing or limited moisture is a problem.

The actual amount of nitrogen returned to the soil and the rate of availability depends on the legume choice, the developmental stage at which it is killed, the preexisting nitrogen content of the soil, and the effectiveness of the *Rhizobium* species. On a per acre basis, each pound of soil nitrogen results in about a 2-pound reduction in nitrogen fixation. A crop with high nitrogen demands should precede the legume crop, so that as much nitrogen as possible will be removed. One can add *Rhizobium* to inoculate seeds prior to planting.

The moldboard plow, when used to incorporate legumes into the soil, leaves no surface residue and brings deeper organic matter upward, which increases soil erosion and organic matter loss. Disking disturbs less organic matter and leaves some surface residue. Disking produces even mixing in the upper-most soil layers, thus favoring even mineralization of nitrogen near the surface. This zone is most important for subsequent uptake of nitrate by plants. Soil nitrogen levels must be watched. Microorganisms use soil nitrogen during decomposition of the green manure, if the C/N ratio of the plowed under green manure exceeds 30:1 to 35:1. This problem can occur if the green manure was allowed to produce extensive brown matter.

Three factors affect the availability of incorporated nitrogen: the amount of soil organic matter, the rate of mineralization, and the weather. The rate of mineralization decreases as organic matter increases. The mineralization rate depends on the activity of microorganisms involved in the mineralization process. Rates are higher under warmer and less moist conditions. Soils with high clay contents have less mineralization of organic matter. No reliable way exists to predict the amount of nitrogen produced by the mineralization process. The amount of available legume nitrogen for the following crop varies from 20% to 50%, with additional increments the next year or two.

If the main goal is to control soil erosion and to maintain/improve organic matter, the practice is termed cover cropping. A good definition is that cover crops are legumes, grasses or small grains grown between successive horticultural crops to protect and improve soils. Since the cover crop can be over wintered legumes that are spring-tilled into the soil prior to planting or killed and used as mulch for no-till planting, one can derive some green manure benefits. The most common cover crops include legumes sown in late summer or early fall. Fall-seeded cereals are also used.

Temporary cover crops prevent winter or summer fallow erosion of soil and are spring-plowed. Grasses, perennial legumes, or mixtures of both can be used as permanent cover crops in orchards. Oxidation of soil organic matter is minimal, because of constant soil cover and lack of cultivation. Organic matter levels are maintained by grass root degradation. Prior to fruit harvest, these cover crops are harvested for forages or beaten down into a mat. This practice makes fruit harvest easier.

Growing crops in a rotational sequence reduces insect, nematode, and disease problems. If a legume is rotated, nitrogen can be supplied to sequential crops. Even when legumes are not used, yields can be higher relative to monoculture under similar fertility levels (rotation effect). Horticultural production can benefit from using rotations. Most vegetable and herb growers don't have enough land to tie cropland up in grass or legume crops. Expensive irrigation and other infrastructure need to be paid for. Still, rotations are possible. One can rotate different vegetable crops, use green manure and cover crops during the off-production times and realize important benefits from rotations.

Rotational crops that follow off-season green manure or legume cover crops should be high nitrogen consumers. Later crops should have lower nitrogen needs. Do not repeat the same or related crop in any one area to avoid disease and insect problems. Some crops appear to do better when following certain other crops. Also rotate root crops with foliage crops, because nutrient usage is somewhat complementary and root crops open the soil for water infiltration. Long-term rotations are better than short-term ones.

Adding organic amendments increases soil organic matter. Manure contains nutrients and organic matter, but is difficult to handle and nutrients can leach or volitalize. Nitrogen and phosphorus from manure tends to have extended availability; overuse can lead to leaching and water pollution. Soils with modest organic matter content usually have good biological activity, quality, and fertility. If judicial manure applications don't produce the desired level of organic matter, than approaches to raise organic matter without additional application of nutrients are advised.

Compost is a soil conditioner and low-grade fertilizer. Composting of moist, piled organic wastes starts with indigenous microorganisms that assimilate some nutrients made available by their enzymic activities during aerobic decomposition. Heat is produced from biological oxidations and retained by the insulating organic matter. Decomposition eventually slows and the temperature returns to ambient. Changes induced by microbial degradation convert the original wastes into a humus-like material. While additives and limestone are unnecessary, a C/N ratio of 30:1 is needed during decomposition by microorganisms. Nitrogen content can also be increased by adding fertilizers. Most plant and animal pathogens, weeds, and insects are destroyed with proper aeration and turning during composting.

Sewage sludge is both a modest fertilizer and soil conditioner. These materials have been used successfully, but have not been generally accepted because of possible risks involving human pathogens and phytotoxicity problems associated with heavy metals that contaminate many sewage sludges.

The best forms of peat for use as a soil conditioner are moss peat and reed-sedge peat. Peat is used as a soil amendment in landscape horticulture to improve the water-holding capacity of coarse-textured soils and the structure of fine-textured soils. It must be thoroughly wetted before incorporation into the soil. It will be necessary to add a complete fertilizer to compensate for the nutrients needed for decay, as well as to replace the soil nutrients diluted by its incorporation into the soil. Some lime might be needed to counter its acidic soil reaction.

Tillage is the working of soil by plowing and cultivating for weed control, good seedbed preparation and incorporation of crop residues or organic amendments. Organic and other sustainable farmers favor conservation tillage. Reduced tillage to the extreme is called no-till. Minimal conservation tillage leaves 30% of the soil's surface covered with crop residues, while no-till leaves the most surface residues. The most widely used implement for conservation tillage is the chisel plow. Others include subsoilers, disks (highly recommended), cultivators, mulch incorporaters, strip rotary tillers, and no-till planters. In no-tillage, herbicides are used to kill weeds prior to planting and during the growing season.

Sustainable horticulture requires soil and water conservation. The two are interrelated, since steps taken to conserve soil often aid in the conservation of water. Improper management of soil and water leads to floods, droughts, erosion, and water pollution. Water conservation reduces off-farm pollution, since runoff water can carry sediment and farm chemicals. The aim is to minimize runoff and maximize water absorption, except when the objective is to collect irrigation water. Then the aim becomes to maximize runoff and minimize erosion. Practices that reduce soil erosion and increase surface and soil organic matter are ideal for minimizing runoff, while maximizing water infiltration and retention. Soils enriched in organic matter encourage high earthworm population. Numerous earthworm channels increase water infiltration and percolation.

On slopes, contour tillage and terraces reduce soil erosion. On level land, practices such as conservation tillage and cover crops or organic amendments are effective ways to conserve water. Winter cover crops and leftover crop residues are very effective at trapping and retaining water from fall rains and later snow, thus increasing storage water. Frozen soils can take up some water when sufficient organic matter is present. The bulk of the water would be absorbed in the spring when the snow melts.

Nonplastic mulches increase rainwater infiltration and reduce soil erosion. In commercial operations, plastic mulches are preferred for reasons of mechanized application and costs. Slotted or perforated plastic mulches are best for water entry and slowing the evaporation of soil water. Living mulches also control runoff and help to conserve water. Permanent cover crops in orchards improve water infiltration and prevent runoff. Sodded orchards are mowed, especially in dry periods. Mowing increases the amount of soil water available to the fruit trees, because grass transpiration is reduced.

In areas where water is a limiting factor, the issue of sustainability becomes a greater problem. Production is possible by irrigation that draws water from surface and groundwater supplies. For true sustainability, withdrawals should be based on rates that are near or below the natural recharge rates. In areas where this approach is not possible, horticultural acreage should be reduced to the sustainable level by conversion to dryland farming or abandonment. Alternatively, the most efficient irrigation should be adopted and crops with lower water consumption should be selected. In addition, practices that increase surface residues, build-up organic matter, and reduce erosion are essential.

Terraces found on sloping farmland are constructed along the contour and include a channel bordered by a ridge on the lower side. This feature helps trap runoff. Some terraces have no channel grade. This type's main function is to trap water for dryland farming and indirectly reduce water erosion. If the indirect costs of soil erosion and lost water are factored in, terraces are a long-term plus for sustainable farming where eroding, sloping land is intensely cropped.

Tillage at right angles to the slope along contours is called contour tillage. Contour cultivation includes contour tillage, planting, and cultivating along the contours. The ridges formed in this manner help slow runoff, increase water infiltration, and control erosion. Increased effectiveness occurs when this practice is combined with contour strip cropping or terraces. Contour strip cropping involves contour tillage and the alternation of strips of row crops with sod-forming crops. The sod strips are effective at trapping water and sediment from the row crops.

In areas where water consumption exceeds the recharge capacity of aquifers and surface water, serious consideration should be given to dryland farming. The practice declined with the advent of irrigation, but dryland farming still occurs in the Great Plains. One method of dryland farming requires alternate years of crop production and summer fallow. During summer fallow, water is stored in the field from natural rainfall. The stored soil water is used for crop production the following year. Another approach is termed the eco-fallow system, as seen in Nebraska. This system is based on no-till or minimal tillage and two crops in 3 years or three crops in 4 years are possible. Few horticultural crops are currently grown in dryland farming systems.

The need for irrigation in the arid West is obvious. However, horticulturists in the East and South also benefit from irrigation. Even with adequate rainfall, irregularities in the distribution pattern and timing can cause drought days. Many failures with horticultural crops result from lack of rainfall or poor timing and quantity of irrigation. The amount and frequency of irrigation is determined from calculations or actual measurements. The latter involves direct measurement of soil moisture. The amount of soil moisture is supplemented by irrigation when available water in the root zone reaches 50% of field capacity. Calculations and measurements are only a starting point and must be tempered with experience, personal observations, and variations that prevail in localized sites. Not all the irrigation water applied will be available for use by the plant.

Certain soils in arid and semiarid regions pose irrigation problems. Saline soils have surface salt crusts. These soils are productive, if their drainage is naturally good or improved by artificial means such that low-salt irrigation water will leach the salts below the root zone. Problems result with high-salt irrigation water or a high water table. Alkali soils require applications of sulfur to reduce pH and gypsum to reduce the levels of exchangeable sodium. Saline-alkali soils have the problems of both saline and alkali soils and need all the preceding treatments. Horticultural crops vary in their salt tolerance; this variation should be considered when selecting crops for these soils. Water for irrigation purposes must be tested. Excesses of soluble salts, sodium, bicarbonate, boron, lithium, and other minor nutrients could cause phytotoxicity.

The four types of level surface irrigation used in arid and semiarid areas are furrow, flood, corrugation, and border irrigation. Efficiency of water usage approaches 60%. Advantages include a low initial investment (if land requires no grading), low energy consumption, and low water evaporation. Disadvantages include uneven distribution of water by gravity flow (costly grading might be required), and the weight of the water can reduce soil aeration and drainage.

Sprinkler irrigation consists of pumping water through pipelines and distributing it from rotary heads. Efficiency approaches 75%. This system is popular for supplementing natural rainfall, especially in the East and South. Advantages of this system include reductions in land leveling, drainage problems, erosion, and the need for special skills. The initial investment, evaporation losses, and energy consumption are higher than those for surface irrigation, but the operational costs are less. Trickle or drip irrigation and ooze irrigation have become popular for greenhouse use, small nursery container operations, home gardens, high-value fruit crops, and small field operations. Drip irrigation efficiency approaches 90%. These systems cost more initially. With drip irrigation, water is carried by polyethylene pipes into which are tapped either holes or smaller feeder lines that terminate with small emitters (valves). Water trickles from these valves. Ooze irrigation employs polyethylene pipe for carrying water and the feed lines are porous polyethylene that allows water to ooze through micron-sized holes. Drip irrigation offers several other advantages for sustainable horticulture. Less energy and fertilizer are needed, relative to other irrigation systems. Less cultivation or herbicides are needed, as weeds are reduced. In some cases, crops mature earlier and yields are increased. Less erosion results, so irrigation on more sloping land becomes feasible. Computerized controls are also possible. These advantages help offset initial costs.

Mulching has several advantages. Water conservation is increased and water absorption is improved. Consequently, soil losses from heavy washing and blowing are decreased. Mulches are effective weed barriers. Mulch averts drastic soil temperature fluctuations, keeping soil cooler in summer and warmer in winter. Winter mulches reduce the risk of root damage. Soils are improved with organic mulches through decomposition and also when tilled into the soil at the end of the growing season.

Mulches have disadvantages. Mulching incurs labor and materials costs. Reduced labor for weeding usually outweighs the initial input. Mulch and moisture around newly emerging seedlings or perennials can provide an ideal environment for diseases. Diseases can overwinter in the mulch. Insects and rodents, who find mulch attractive, can cause plant damage. These problems are minimized by avoiding direct contact between plants and mulch. Premature applications of organic mulches can retard soil warming. A thick mulch or one prone to caking can impede the uptake of water and air by the soil. Decomposing mulches with a C/N ratio over 30:1 steal nitrogen from the soil. Prior applications of nitrogen fertilizers can prevent this problem. Combustible dry mulches and woody mulches near buildings are a fire hazard and termite vector, respectively.

Precision farming depends on global positioning satellites (GPS) and global information systems (GIS). GPS allows farmers to locate and target specific problem land areas with pinpoint accuracy. GIS allows placement of geographic features and overlays of terrain-associated data (climate, soil type, water quality, buildings, transportation arteries, population density, land prices) into maps with extremely high accuracy. GIS can be an excellent tool in its own right for the selection of optimal land for commercial horticultural production using the indicated terrain overlay data.

Precision farming is based on the existence of farm field variability. Practices are geared to the average crop area and do not account for variability in field conditions. Applications of fertilizer, irrigation water, and pesticides are applied at constant rates across fields. Conditions that vary include soil, fertility, temperature, aeration, water retention, and drainage. These variations influence environmental and biological interactions among plants, livestock, insects, pathogens, weeds, and soil. These variations affect crop yields in different ways in different fields, or even areas within a given field.

Precision farming allows for the precise adjustments of farming practices and inputs that reflect the needs of fields both in the sense of spatial and seasonal differences. These precise adjustments are based on data collected from various sources through the latest information technologies. The potential result is lower input of chemicals and water, yields that reflect the potential of each area, and less environmental degradation. Resources can be better conserved and profits per acre are improved.

One criticism is that the technology reduces pesticide applications, but it doesn't eliminate them. The reduced amounts of pesticides also don't result in reduced costs for the farmer, as the technology is currently costly. Another criticism is that nonrenewable resources are still used and the switch to renewable ones is postponed. The cost also precludes its use on small farms and the practice requires sophisticated management.

Satellite imaging data for precision farming exists. Images and other data are available in black and white, color, and infrared. The latter makes possible quantitative analyses of crop vigor. Precise growing areas can be pinpointed in the field. Two types of data collection are possible. The data can be collected, placed on a map using GIS, and then interpreted. Some action is then directed. Another approach would be to utilize sensors that instantly translate incoming data into action. These sensors would send signals to systems or equipment that would then provide the changes needed.

EXERCISES

Mastery of the following questions shows that you have successfully understood the material in Chapter 11.

1. What factors are important in the selection of commercial horticultural production sites?
2. How does one correct drainage problems?
3. What does a conventional soil test determine? What does soil quality testing examine?
4. What type of sample is needed for a soil test? How are soil samples gathered? What type of information is usually presented with the soil sample?
5. What parameters are usually presented in a conventional soil test? What are some possible parameters for a soil quality test?

6. What are some of the functions of organic matter in the soil? Why is it important to soil sustainability?
7. Why do soils under cultivation need more attention in terms of sustainability than soils in natural plant communities?
8. What do the terms *complete, specialty, grade* or *analysis, fertilizer ratio, chemical,* and *organic* mean in connection with fertilizers?
9. When should one apply fertilizer? What steps should be taken to minimize nutrient loss when fertilizers are applied at nonoptimal times?
10. What are the various ways in which one can apply fertilizer to the soil? What advantage does foliar nutrition offer?
11. What factors affect fertilizer timing and application? What does this mean in practical terms when dealing with actual applications?
12. Can soil sustainability be maintained solely with chemical fertilizers? Why or why not?
13. How do we adjust soil pH? What affects how much input is needed?
14. What practices can we use beyond chemical fertilizers to maintain soil sustainability? Which ones add organic matter? Nutrients? Organic matter and nutrients?
15. Why are crop residues not always sufficient for the maintenance of soil organic matter?
16. What is the relationship among the terms *cover crop, green manure,* and *catch crop?*
17. Does climate or geography limit cover crops in any way?
18. How are cover crops managed differently when the goal is controlling soil erosion versus increasing soil nitrogen?
19. What kinds of plants are used as green manure? What factors affect the choice of green manure plant?
20. Why is timing of concern when determining when to plow down green manure?
21. What plants are used for cover crops? What are temporary and permanent cover crops used for?
22. What benefits are offered by rotations? What are some good examples of rotation sequences?
23. What materials can be used as organic amendments in soils? Is manure applied to soils pollution free under all conditions? If not, why?
24. What is compost? How do you prepare compost?
25. What are the downsides of using sewage sludge?
26. What is conservation tillage? What advantages does it offer?
27. What methods can be employed to reduce runoff and conserve water?
28. What is dryland farming? What benefit does it offer?
29. What problems arise with irrigation? How can we increase irrigation efficiency?
30. What advantages does mulching offer? What disadvantages exist? What types of materials can be used as mulches?
31. What is precision farming? What does GIS and GPS mean? What are the advantages and disadvantages of precision farming from the viewpoint of sustainability?
32. What kinds of information can satellites and other data sources provide for precision farming?
33. What benefits can a windbreak offer?
34. Why is winter protection needed? How can we provide it?
35. Under what circumstances is frost protection needed? What practices are used to provide frost protection? Which one is best from the sustainability view?

INTERNET RESOURCES

Basic Resources for Sustainable and Organic Farmers/Growers:

http://www.attra.org

http://www.sare.org

Composting:

Cornell Composting (advanced, includes science and engineering)
http://www.cfe.cornell.edu/compost/Composting_Homepage.html

Composting resources
http://www.attra.org/attra-pub/farmcompost.html
http://www.epa.gov/epaoswer/non-hw/compost/index.htm
http://www.nal.usda.gov/afsic/AFSIC_pubs/qb9712.htm

Cover Crops and Green Manure:

Cover crops and green manures: overview
http://www.attra.org/attra-pub/covercrop.html

Cover crops: California
http://www.sarep.ucdavis.edu/ccrop/

Cover crops: interseeding
http://www.uvm.edu/vtvegandberry/factsheets/interseeding.html

Cover crops: Michigan
http://www.kbs.msu.edu/Extension/Covercrops/home.htm

Cover crops: Oregon
http://berrygrape.orst/fruitgrowing/cover.htm

Green manure basics
http://www.eap.mcgill.ca/Publications/EAP51.htm
http://www.agric.gov.ab.ca/agdex/100/2300202.html

Fertilization:

Catch crops: nitrogen removal
http://www.wsu.edu/pmc_nrcs/technotes/agronomy/tntag38.htm

Commercial vegetable crop nutrient requirements in Florida
http://edis.ifas.ufl.edu/BODY_CV001

Commercial vegetable fertilization principles
http://edis.ifas.ufl.edu/BODY_CV009
http://vric.ucdavis.edu/selectnewtopic.fertiliz.htm

Environmental guidelines for nursery and turf industry: fertilizer management
http://www.agf.gov.bc.ca/resmgmt/fppa/environ/nursery/nursry.htm

Environmental guidelines for tree fruit and grape producers: fertilizer management
http://www.agf.gov.bc.ca/resmgmt/fppa/environ/fruit/fruit.htm

Fertilizers: manures
http://www.attra.org/attra-pub/manures.html

Fertilizers: nutrient content of fertilizer and organic materials
http://www.soil.ncsu.edu/publications/Soilfacts/AG-439-18/

Fertilizers: organic fertilization
http://www.agr.state.nc.us/agronomi/sfn12.htm
http://www.uvm.edu/~pass/pss161/problem/handout.html

Fertilizers: nitrogen for vegetables
http://res2.agr.ca/stjean/recherche/azote_e.pdf

Fertigation for pecan orchards

http://www.cahe.nmsu.edu/pubs/_h/h-642.html

Fertilizer for ornamentals

http://www.msue.msu.edu/msue/imp/modop/00001794.html

Fertilizer for young orchards

http://www.ncw.wsu.edu/fert.htm

Fertilizer guide for California vegetable crops

http://vric.ucdavis.edu/veginfo/topics/fertilizer/fertguide.html

Fertilizer recommendations for orchards

http://www.penpages.psu.edu/penpages_reference/29401/29401118.html

Fertilizing

http://bluestem.hort.purdue.edu/plant/FMPro?-DB=plant.fp3&-Format=category.
html&Max=100&SortField=Description&category=Fertilizing&-Find

Fertilizing fruit crops

http://www.msue.msu.edu/vanburen/e-852.htm

IFAS standardized fertilization recommendations for environmental horticulture crops

http://edis.ifas.ufl.edu/BODY_CN011

Sources of organic fertilizers and amendments

http://www.attra.ncat.org/attra-pub/PDF/organicfert.pdf

Mycorrhizae:

http://www.unce.unr.edu/publications/factsheets/FS%2001/FS-01-29.htm

Precision Farming:
Remote sensing and its link to precision farming:

http://www.pioneer.com/usa/technology/agremot.htm

Soil Basics:

http://bluestem.hort.purdue.edu/plant/FMPro?-DB=plant.fp3&-Format=category.
html&Max=100&-SortField=Description&category=Soils%20and%20Media&-Find

http://www.umassvegetable.org/soil_crop_pest_mgt/soil_nutrient_mgt.html

Soil Management:

http://www.extension.umn.edu/distribution/cropsystems/DC7398.html

http://www.attra.org/attra-pub/soilmgt.html

http://anrcatalog.ucdavis.edu/pdf/7248.pdf

Soil Quality:
Earthworms

http://www.sarep.ucdavis.edu/worms/

Information and Links

http://soils.usda.gov/sqi/

http://soils.usda.gov/sqi/nusites.html

http://www.sarep.ucdavis.edu/soil/websites.htm

Soil Testing:
Alternative soil testing labs for sustainable agriculture

http://www.attra.org/attra-pub/soil-lab.html

Chemical soil tests for soil fertility evaluation

http://vric.ucdavis.edu/veginfo/topics/fertilizers/soiltests.pdf

Measuring soil quality

> http://www.ext.vt.edu/pubs/compost/452-400/452-400.html
> http://soils.usda.gov/sqi/SQassessments.htm

Plant tissue analysis of vegetable crops

> http://vric.ucdavis.edu/veginfo/topics/fertilizer/tissueanalysis.pdf

Water:

Conserving water and preserving the environment

> www.TurfGrassSod.org/waterright.html

Drought and vegetable production

> http://www.ces.ncsu.edu/drought/dro-16.html

Dryland Agriculture Institute of West Texas A&M

> http://www.wtamu.edu/research/dryland/

Energy and water efficiency in vegetable production

> http://edis.ifas.ufl.edu/BODY_EH208

Irrigation and water quality: landscape sites

> http://bluestem.hort.purdue.edu/plant/irrigation_waterquality.html

Winter Hardiness and Protection:

> http://bluestem.hort.purdue.edu/plant/FMPro?-DB=plant.fp3&-Format=category.
> html&Max=100&-SortField=Description&category=Winter%Hardiness&-Find

RELEVANT REFERENCES

Bezidek, D. F., and J. F. Powers (1984). *Organic Farming: Current Technology and Its Role in a Sustainable Agriculture.* American Society of Agronomy, Crop Science Society of America, and Soil Science Society of America, Madison, WI.

Brady, Nyle C., and Ray R. Weil (1999). *The Nature and Properties of Soils* (12th ed.). Prentice Hall, Upper Saddle River, NJ. 881 pp.

Chaney, D. E., and Laurie E. Drinkwater (1992). *Organic Soil Amendments and Fertilizers.* DNAR Publication No. 21505, University of California, Davis. 36 pp.

Coleman, Eliot (1995). *The New Organic Grower: A Master's Manual of Tools and Techniques for the Home and Market Gardener,* (2nd ed.). Chelsea Green Publishing, White River Junction, VT. 340 pp.

Coughenour, C. Milton, and Shankariah Chamala (2000). *Conservation Tillage and Cropping Innovation: Constructing the New Culture of Agriculture.* Iowa State University Press, Ames. 420 pp.

Doran, John W., and Alice J. Jones (1996). *Methods for Assessing Soil Quality.* Soil Science Society of America Special Publication No. 49, Madison, WI. 410 pp.

Dougherty, Mark (ed.) (1999). *Field Guide to On-Farm Composting.* Northeast Regional Agricultural Engineering Service, Ithaca, NY 128 pp.

El Titi, A. (ed.) (2002). *Soil Tillage in Agroecosystems.* CRC Press, Boca Raton, FL. 376 pp.

Gajri, P. R., V. K. Arora, and S. S. Prihar (2002). *Tillage for Sustainable Cropping.* Food Products Press, Binghamton, NY. 246 pp.

Gershuny, G., and J. Smillie (1986). *The Soul of Soil: A Guide to Ecological Soil Management* (2nd ed.). AgAccess, Davis, CA. 174 pp. Third edition published in 1995.

Hood, Teresa M., and J. Benton Jones, Jr. (1997). *Soil and Plant Analysis in Sustainable Agriculture and Environment.* Marcel Dekker, Monticello, NY. 864 pp.

International Commission of Irrigation and Drainage (1996). *Irrigation Scheduling: From Theory to Practice.* Food and Agriculture Organization of the United Nations/Bernan Associates, Lanham, MD. 384 pp.

Jackson, William R. (1993). *Humic, Fulvic, and Microbial Balance: Organic Soil Conditioning.* Jackson Research Center, Evergreen, CO. 958 pp.

Jeavons, J. (1979). *How to Grow More Vegetables Than You Ever Thought Possible on Less Land Than You Can Imagine.* Ten Speed Press, Palo Alto, CA. 116 pp.

Kuehn, F., T. V. V. King, B. Hoerig, and D. C. Peters (eds.) (1998). *Remote Sensing for Site Characterization.* Springer Verlag, New York. 256 pp.

Lee, Jasper S. (2000). *Natural Resources and Environmental Technology.* Interstate Publishers, Danville, IL. 456 pp.

Little, Charles E. (1987). *Green Fields Forever.* Island Press, Washington, DC. 192 pp.

Ludwick, A. E. (1990). *Western Fertilizer Handbook: Horticulture Edition.* Interstate Publishers, Danville, IL. 279 pp.

Magdoff, Fred, and Harold van Es (2000). *Building Soils for Better Crops: Organic Matter Management,* (2nd ed.). Sustainable Agriculture Publications, University of Vermont, Burlington. 240 pp.

Miller, P. R. (1989). *Cover Crops for California.* DNAR Leaflet 21471, University of California, Davis. 24 pp.

Miller, Raymond W., and Roy Donahue (1990). *Soils: An Introduction to Soils and Plant Growth* (6th ed.). Prentice Hall, Upper Saddle River, New Jersey. 768 pp.

Opie, John (2000). *Ogallala: Water for a Dry Land,* (2nd ed.). University of Nebraska Press, Lincoln. 475 pp.

Oregon State University (1998). *Using Cover Crops in Oregon.* Oregon State University Pub. No. EM 8704, Corvallis. 50 pp.

Parnes, R. (1990). *Fertile Soil: A Grower's Guide to Organic and Inorganic Fertilizers.* AgAccess, Davis, CA. 190 pp.

Poincelot, R. P. (1975). *The Biochemistry and Methodology of Composting.* Connecticut Agricultural Experiment Station Bulletin 754, New Haven. 18 pp.

Rynk, Robert (1992). *On-Farm Composting Handbook.* Northeast Regional Agricultural Engineering Service, Ithaca, NY. 186 pp.

Sarrantonio, Marianne (1991). *Methodologies for Screening Soil-Improving Legumes.* Rodale Institute, Kutztown, PA. 310 pp.

Sarrantonio, Marianne (1994). *Northeast Cover Crop Handbook.* Rodale Institute, Kutztown, PA. 118 pp.

Schmidt, J. (ed.) (2000). *Soil Erosion: Application of Physically Based Models.* Springer Verlag, New York. 366 pp.

Schwenke, Karl, (1991). *Successful Small-Scale Farming: An Organic Approach,* (2nd ed.). Storey Communications, Inc., Pownal, VT. 134 pp.

Smith, Miranda, Northeast Organic Farming Association and Cooperative Extension (eds.) (1994). *The Real Dirt: Farmers Tell about Organic and Low-Input Practices in the Northeast.* Northeast Region Sustainable Agriculture Research and Education (SARE) Program and Northeast Organic Farming Association. Burlington, VT. 264 pp. Second edition appeared in 1998.

Soil Quality Institute (1996). *The Soil Quality Concept.* U.S. Department of Agriculture, Washington, DC.

Soule, Judith D., and Jon K. Piper (1992). *Farming in Nature's Image: An Ecological Approach to Agriculture.* Island Press, Washington, DC. 286 pp.

Stivers, L. (1999). *Cover Crops for Vegetable Production in the Northeast.* Cornell University Extension Service Bulletin No. 1421B244, Ithaca, NY. 12 pp.

Sustainable Agriculture Network (1998). *Managing Cover Crops Profitably,* (2nd ed.). Sustainable Agriculture Publications, University of Vermont, Burlington. 212 pp.

Sustainable Agriculture Network (1996). *Sustainable Agriculture: Directory of Expertise,* (3rd ed.). Sustainable Agriculture Publications, University of Vermont, Burlington.

U.S. Department of Agriculure (1998). *A Guide to USDA and Other Federal Resources for Sustainable Agriculture and Forestry Enterprises.* USDA, Washington, DC. 159 pp.

VIDEOS

Demonstrating Field Capacity of Soils[*] (1997).
How Chemicals Move through Soil[*] (1997).
Irrigation[*] (Video, 1994).
Landscape Irrigation Systems: Maintenance and Troubleshooting[*] (1999).
Life's Hidden Treasure: Protecting our Groundwater[*] (1995).
Soil and Water Management for Agricultural Production: Surface Water[*] (1997).
Soils and Their Role in Protecting Water Quality[*] (1995).
Creative Cover Cropping in Annual Farming Systems (1993).
University of California, Division of Agriculture and Natural Resources, Davis.
No-till Vegetables (1997). Steve Groff, Cedar Meadow Farm. Holtwood, PA.
Using Cover Crops in Conservation Production Systems (1997).
Seth Dabney, USDA-ARS National Sedimentation Lab, Oxford MS.

[*]Available from Insight Media, NY (www.insight-media.com/IMhome.htm)

Chapter 12

Home Horticulture Basics

OVERVIEW

The homeowner has more latitude than the commercial horticulturist does in that business success is not the homeowner's focus. Lawns and home landscapes are created and maintained for esthetic reasons and to provide pleasurable areas for outdoor activities. The only monetary profit derived from these horticultural activities is an enhanced sales price when the home is sold. Gardening is a hobby done for joy, recreation, and health. The rewards are fresh vegetables, fruits, cut flowers, or herbs, not necessarily profits. The home gardener can grow crops in marginal locations and focus on improving soil in small areas. He or she can try new cultivars, shrug off the vagaries of weather or the unexpected insects and diseases, take horticultural chances, experiment with techniques, and spend as much or little time as desired in gardening activities. Even within this context, the gardener is wisely advised to refer back to Chapter 11 and the discussion about climate, microclimate, site selection, and soil tests. Much of the information of importance to commercial producers is useful to homeowners involved with landscaping, lawns, and gardens, and complements the discussion of similar topics in this chapter.

Basic information for your local area can be gleaned from established gardeners in your neighborhood. Local extension offices in your state, your state university, and your state agricultural experiment station are also good sources of information for the homeowner and gardener. Their locations and telephone numbers are found in the government listings in your telephone book. Some states have a Master Gardener program available through extension offices. Master Gardeners are volunteer, trained gardeners who can answer most gardening, landscaping, or lawn care questions and concerns and are a source of considerable information. Much information is also available from books, Web chat rooms for gardeners, and certain Web sites (see listing at end of chapter). Garden clubs are also a good information source and offer the chance to enjoy the company of other gardeners. Don't forget your local nursery or garden center; these vendors often provide useful information to their customers. Finally, a number of good horticultural magazines offer information and creative ideas. A few basics are all you need to enjoy a lifetime of gardening activity.

CHAPTER CONTENTS

Climatic Considerations
Soil Drainage
Soil Tests
Landscape, Lawn, and Gardening Choices
Landscaping Basics
Lawn Care

Why Garden?
Site Selection
Plant and Garden Choices
Planning the Garden
Getting the Most Vegetables and Herbs
Flower Basics

Starting Seeds Ahead: To Buy or Not to Buy
Tilling the Garden
Sowing Seeds

Using Transplants and How to Establish
Culture

OBJECTIVES

❧ To learn about sources of garden information and site selection

❧ To learn how to sample soil and the value of soil tests

❧ To work with soil drainage, site selection, and climate

❧ To explore the basic approaches to landscape maintenance, lawn care, and gardens

❧ To plan and get the most out of your vegetable garden

❧ To learn how to set up flower gardens

❧ To learn how to work with transplants and seeds and start the garden

❧ To understand garden culture and harvesting

CLIMATIC CONSIDERATIONS

Unlike the commercial horticulturist, who locates production fields where climates are favorable, climate is not the primary consideration for the homeowner and gardener. Hobbyists garden in whichever city or town their home is located. The best way to deal with the climate you are given is to understand it thoroughly in terms of potential and limitations. If you live in the far northern reaches of the United States or high up on a mountain, you might have to accept the fact that you can't grow tomatoes. On the other hand, if you live in southern Florida, your tomatoes will be great, but you won't have tulips unless you force them in refrigerated pots.

Instead of concentrating on what you can't do, you should instead work with those plants that are most adapted to your climatic area. First choice goes to native plants that have evolved hand in hand with the prevailing climate. Second choice goes to plants that are not indigenous, but have been naturalized in your area after past introduction from elsewhere. Third choice goes to those non-native plants that based on experience will succeed in your area. Last choice would be marginal plants (in terms of climate) and difficult plants.

Of course, if you are determined, you can try to grow plants that marginally or rarely succeed in your area. Tricks such as taking advantage of a microclimate (see microclimates in Chapter 6) or extending the growing season with tunnels or cloches can be helpful with problematic plants. For example, spring flowering bulbs planted near the foundation on the south side, assuming a sunny location, will flower a few weeks earlier than bulbs elsewhere. This trick takes advantage of longer sun exposure and trapped heat in the masonry or brick foundation, otherwise known as the southern exposure microclimate.

Other critical information related to climate includes the dates of the last killing spring frost and the first killing fall frost (Figure 12–1), the length of the growing season, and the zone of hardiness for your location. Frost and growing season data are shown for selected cities in Canada and the United States in Tables 12–1 and 12–2, respectively. Further and more localized information can be found through the Internet sources at chapter's end or by consulting your local state extension office or extension Master Gardener program. Going northward or higher in elevation shortens the growing season, killing spring and fall frosts arrive later and earlier, respectively, and zones of hardiness get colder. Microclimates can also modify these parameters. Both of these factors can bring about sudden and dramatic changes in frosts, growing seasons, and zones of hardiness in geographically close sites, hence the need for reliable, localized information.

Frost dates and the length of the gardening season determine the best possible times for planting early cool season crops such as peas, lettuce, and pansies, and warm season crops such as tomatoes, squash, and marigolds. The length of the growing season also

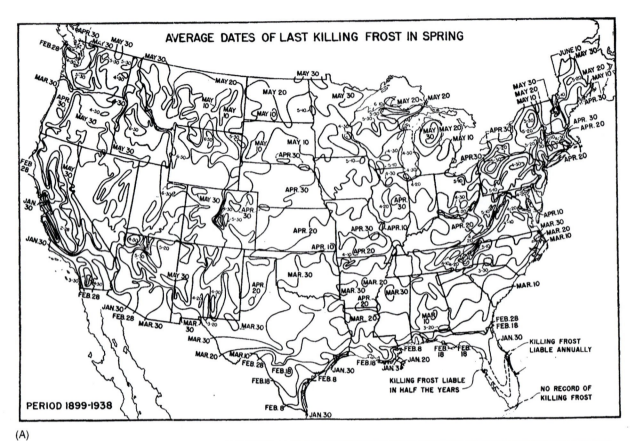

(A)

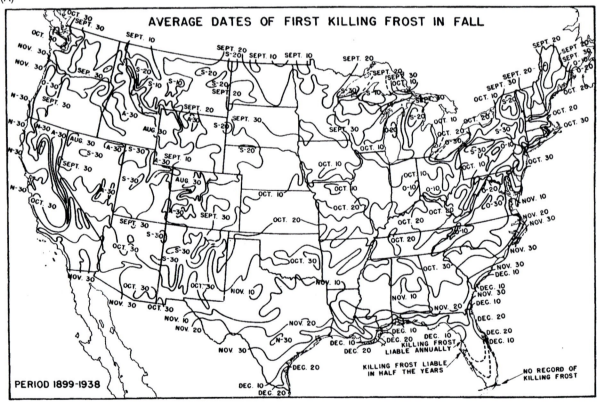

(B)

FIGURE 12–1   (A) Mean date of last killing spring frost. (B) Mean date of first killing fall frost. These two dates determine the growing season length in any given area. (USDA photographs.)

TABLE 12-1   *Frost Dates and Length of Growing Season for Selected Cities in Canada**

Province	City	Spring Frost	Fall Frost	Growing Season (days)
Alberta	Calgary	5/23	9/15	116
British Columbia	Dawson Creek	6/5	8/29	86
Manitoba	Winnipeg	5/25	9/22	121
Newfoundland	St. John's	6/2	10/12	133
New Brunswick	Edmundston	5/28	9/18	114
Northwest Territories	Yellowknife	5/27	9/15	112
Nova Scotia	Yarmouth	5/1	10/18	171
Ontario	Ottawa	5/6	10/5	153
	Toronto	5/9	10/6	151
Prince Edward Island	Charlottetown	5/17	10/14	152
Quebec	Montreal	5/3	10/7	158
	Quebec	5/13	9/29	141
Saskatchewan	Saskatoon	5/21	9/15	118
Yukon Territory	Dawson	6/13	8/17	64

*Data are for a light freeze (32°F). There is a 50% probability that a frost could occur after the spring date and before the fall date.

TABLE 12-2   *Frost Dates and Length of Growing Season for Selected Cities in the United States**

State	City	Spring Frost	Fall Frost	Growing Season (days)
Alaska	Juneau	5/16	9/26	133
Arizona	Phoenix	2/5	12/15	308
California	Eureka	1/30	12/15	324
	San Francisco			365[†]
Colorado	Denver	5/3	10/8	157
Connecticut	Middletown	4/23	10/12	172
Florida	Miami			365[†]
Hawaii	Mana			365
Illinois	Chicago	4/22	10/26	187
Iowa	Cedar Rapids	4/29	10/7	161
Louisiana	New Orleans	2/20	12/5	288
Maine	Portland	5/10	9/30	143
Minnesota	Duluth	5/21	9/21	122
Montana	Helena	5/18	9/18	122
Nevada	Las Vegas	3/7	11/21	259
New Mexico	Los Alamos	5/8	10/13	157
New York	Syracuse	4/28	10/16	170
North Carolina	Fayetteville	4/2	10/31	212
Oregon	Portland	4/3	11/7	217
Texas	San Antonio	3/3	11/24	265
Vermont	Burlington	5/11	10/1	142
Virginia	Richmond	4/10	10/26	198
Wyoming	Casper	5/22	9/22	123

*Data are for a light freeze (32°F). There is a 50% probability that a frost could occur after the spring date and before the fall date.

†Frosts can occur in some years.

TABLE 12–3   *USDA Hardiness Zones and Average Annual Minimal Temperature Range*

Zone	Temperature Range (°F)	Example States and Countries	Respective Example Cities
1	Below −50	Alaska, Northwest Territories (Canada)	Fairbanks, Resolute
2a	−50 to −45	Alaska, Manitoba (Canada)	Prudhoe Bay, Flin Flon
2b	−45 to −40	Alaska, Minnesota	Unalakleet, Pinecreek
3a	−40 to −35	Minnesota, Alaska	International Falls, St. Michael
3b	−35 to −30	Wisconsin, Montana	Tomahawk, Sidney
4a	−30 to −25	Minnesota, Montana, New York	Minneapolis/St. Paul, Lewistown, Malone
4b	−25 to −20	Iowa, Vermont	Northwood, South Burlington
5a	−20 to −15	Iowa, Maine	Des Moines, Bangor
5b	−15 to −10	Missouri, Ohio, Pennsylvania	Columbia, Lima, Mansfield
6a	−10 to −5	Connecticut, Missouri, Pennsylvania	Hartford, St. Louis, Lebanon
6b	−5 to 0	Connecticut, Missouri, Tennessee	Fairfield, Branson, McMinnville,
7a	0 to 5	New York, Oklahoma, Virginia	Queens, Oklahoma City, South Boston
7b	5 to 10	Arkansas, North Carolina, Georgia	Little Rock, Rocky Mount, Griffin
8a	10 to15	Georgia, Texas	Tifton, Dallas
8b	15 to 20	Florida, Texas, Washington	Gainesville, Austin, Bellingham
9a	20 to 25	Florida, Texas	St. Augustine, Houston
9b	25 to 30	Florida, Texas	Fort Pierce, Brownsville,
10a	30 to 35	California, Florida	Eureka, Victorville, Naples
10b	35 to 40	Florida	Coral Gables, Miami
11	Above 40	Hawaii, Mexico	Honolulu, Mazatlan

determines which crops one can grow. For example, if a giant pumpkin requires 120 days to harvest, and the growing season is only 100 days, you are not likely to succeed with pumpkins. Even if the growing season were 120 days, it still might not work, because a warm season pumpkin can't be planted right at the start of the growing season. Growing season length is also useful information when you are planning to use the same area for a succession of crops. Each must be able to mature in the time remaining in the growing season.

The zones of hardiness (Table 12–3 and Internet sources), while not important for annual flowers and vegetables, become very critical for perennial herbaceous and woody crops such as asparagus and apples, perennial ornamentals such as chrysanthemums, and woody ornamentals such as forsythia and azaleas. It also determines when a plant is an outdoor plant or a house plant. For example, the Christmas cactus is a house plant in the United States, but is also a plant native to the tropical rain forests in Central and South America.

SOIL DRAINAGE

Drainage is a critical factor and should be checked thoroughly before buying land or a house. If water pools and sits for hours after a heavy rainstorm on your current site, drainage is not good. This soil will have poor aeration, leading to stunting and rotting of roots. Lawns, landscape, and garden plants will fare poorly. If you have other choices for your garden site, be sure to use other areas with good drainage after a rain. Some things can be done to improve drainage on your grounds and in your garden, although it will be hard work and you might need professional help. All but simple solutions are likely to require the aid of a landscape contractor.

With clay soils, the incorporation of sharp sand can improve drainage. This approach requires a considerable amount of sand to achieve worthwhile results. A typical clay soil might require at least 600 to 1,000 pounds of sharp sand to be tilled in per 100 square feet.

This amount is the minimal amount required to establish a sand content of 10% to 15% to a depth of only 6 to 8 inches.

Copious amounts of organic matter should also be mixed with the sand or applied prior to the addition of the sand. Sometimes organic matter alone can improve drainage sufficiently. Organic matter will improve soil structure and water infiltration. Organic matter, unlike the sand, will require replenishment annually. Manure should be applied only once about every 4 years, because it raises soluble salt levels in clay-based soils to harmful levels, as opposed to sandy soils. The latter drain well and soluble salts are readily leached out. Mature compost is a good source of initial organic matter. A layer of finished compost should be tilled in to at least rooting depth. Ideally, enough compost should be used to form a mixture of one part compost to no more than three or four parts clay soil. Organic mulch can be used annually afterward. This mulch should be tilled in after harvest to maintain organic matter levels and water infiltration.

In some cases, trenches can be dug to divert water that collects from higher areas nearby. Another approach is to install a French drain. A French drain is composed of a gravel-filled trench with embedded perforated drainage pipe. The trench is usually lined with filter fabric to prevent soil-fines from clogging the system. All drainage systems require a point lower than the garden site to which the water drains.

SOIL TESTS

The soil test is critical for all home horticultural activities from landscaping to lawns to gardens. Sustainability is hard to achieve without knowing your soil fertility needs. This need exists whether you are a novice or seasoned veteran when it comes to maintenance of plants in the landscape, lawn, or garden. You can only guess at the potential or nutrient levels of your soil without the results of a soil test. Test results will guide gardeners in their efforts to improve the soil's productivity in order to grow healthy and successful plants. Soil testing is a wise investment. It only takes a few minutes and the payoff is a successful landscape, lawn, or garden. One can buy soil test kits, but a better choice is to take advantage of your tax dollars. Soil tests are a service provided to taxpayers in all states. Either all or one of the following institutions provides these services: extension offices in your state, your state university, and your state agricultural experiment station. These places are usually listed in the government listings section of the telephone book. Some organizations provide this service for free, while others require a modest fee of a few dollars. A quick call can determine the costs involved and any special requirements. Some will provide a mailing carton and directions for the soil test.

One warning is in order: Don't procrastinate until the lab is swamped with soil samples from all those other gardeners who waited until the very last minute. Soil can be tested any time, so do it early. The proper procedure for sampling lawn, landscape, or garden soil, the information required with the soil sample, and how to interpret the soil test results are found in Chapter 11. Make sure you indicate to the testing agency that you want fertilizer recommendations at minimal levels because you are trying to minimize environmental impacts and to practice sustainable horticulture.

LANDSCAPE, LAWN, AND GARDENING CHOICES

Grounds care and gardening, like any other human activity, involves choices. The following questions need to be answered. What kind of yard and garden do you want? Do you want nicely landscaped grounds with a neatly trimmed lawn? Do you want a vegetable or flower garden? What about an herb garden? Or perhaps some fruit or nut trees? Do you want to improve the landscape around your home? Will you do all the work? Will you seek professional help such as from a landscape designer or lawn care or landscape maintenance services? How do you feel about the environment and your family's health? Will you feel strongly enough that no pesticides will be allowed, but perhaps some chemical fertilizers? Or will you learn to be a successful organic gardener and landscaper?

Conventional

Mainstream gardening and landscape care is all about chemicals. The message is to kill the insects with insecticides, diseases with fungicides, weeds with herbicides, and mites with miticides. One pours chemical fertilizers on the lawn, trees, shrubs, and garden. Homeowners, gardeners, and farmers keep the chemical manufacturers of horticultural and agricultural products in business and help them to earn considerable profit. These businesses continue to thrive, because the image of better living through chemistry is hard to shake, given slick advertisements on television and in magazines.

The chemical industries hope you never leave them. Many home, gardening, and farming experts in industry, the service sector, and academia continue to extol the virtues of chemicals for gardens, grounds, and farms. The resulting environmental damage and possible health hazards are downplayed. Chemical-dependent gardening and grounds care is known as conventional or traditional gardening and grounds care. A better, more honest name would be chemical gardening and grounds care, but it doesn't have the respectable ring of conventional or traditional. Given that this book looks for sustainable approaches, two alternatives are offered. One is to use no synthetic chemicals at all, organic gardening. The other choice is to use as few manufactured chemicals as possible, which here is called integrated grounds and gardening management.

Organic

The commercialization and advocacy of organic gardening and farming has its origins in 1945 with J. I. Rodale. Certainly most crop production and gardening prior to the advent of chemical fertilizers and pesticides was essentially organic in practice. Thomas Jefferson undoubtedly practiced organic farming at Monticello. Today he would be labeled an organic farmer. Back then Jefferson was a conventional farmer like everyone else.

The basic premise of organic production is that only naturally occurring materials should be used in the garden and farm. Fertilizers are formulated from various manure and natural products (bone meal, greensand, wood ashes, granite dust), liberal amounts of compost are added to the soil, legumes are grown to add nitrogen to soil, and troublesome weeds are controlled with mulches and hand or mechanical cultivation. Insects and diseases are prevented with the use of resistant cultivars, crop rotations, and natural enemies and products. Natural enemies include beneficial insects and microbes such as predatory or parasitic insects, milky spore disease, and competitive or antagonistic microorganisms against plant pathogens. Natural products include botanical pesticides such as pyrethrum, rotenone, and neem and products such as dormant oils and soap sprays. This approach to insects and disease will be discussed in detail in Chapter 15.

Proponents believe that soil health is optimal when natural materials are used. It is also believed that organically managed soils produce plants that are less stressed and hence less susceptible to insects and diseases. Some research does support the idea that soils not treated with insecticides and herbicides have a healthier ecosystem, especially on the microbial and earthworm level. Plants grown in such soils seem to be more productive and less prone to diseases and insects, which probably results in part from the normal soil microbes being antagonistic to plant pathogens and certain soil inhabitants such as nematodes.

Integrated Grounds and Gardening Management

For various reasons, not everyone is prepared to adopt organic gardening approaches completely. Others, in the heat of a major onslaught from a nasty pest, weaken and resort to nonorganic methods. While the organic concept is one highly visible form of sustainable horticulture, it is still possible to garden in a sustainable manner with minimal loss of resources and to preserve the environment and your family's health. That's what integrated grounds and gardening management or IG²M is all about. IG²M's basic premise is that the guiding principle is to use organic techniques wherever and whenever possible. When these techniques fail or are not acceptable to you, then and only then do you resort to manufactured products. However, these products need to be used in moderation and the least toxic or harmful ones should be chosen.

For example, if you feel the need to use a chemical fertilizer on your lawn, select a slow-release form. This product is less likely to leach nitrogen into the groundwater and adjacent water bodies. Also use it sparingly. Just because the manufacturer says to use so many pounds per square feet and repeat at varying intervals does not mean you have to do it. The manufacturer has a vested interest in your use of the product. The more you use, the greater their profit. These directions are often based on too generous amounts. You would be surprised to see how well your lawn does with less. The manufacturer often errs on the side of too much, because the modest excess won't harm the lawn. However, it can pollute the surrounding environment, especially nearby bodies of water. When you have a soil test (refer back to Chapter 11) to determine fertility for your garden and lawn, make sure you indicate that you want the minimal recommendation for nutrients, as you wish to have a minimal impact on the surrounding environment.

The other consideration is with pesticides. Only after you have truly tried the natural controls used by organic gardeners, such as lady beetles and soap sprays or natural botanical pesticides like neem or pyrethrum, and they have failed should you resort to a chemical pesticide. You should select the pesticide on the basis of least toxicity and persistence in the environment. Examples of the latter include malathion and Sevin®. Your family and future generations will thank you. Details on alternative natural controls and mixed natural/chemical controls (integrated pest management or IPM) are found in Chapter 15.

LANDSCAPING BASICS

Homes, commercial buildings, institutional buildings, and workplaces are all enhanced when properly landscaped (Figure 12–2). A well-landscaped structure appeals to our esthetics and the human psyche. This point should not come as a surprise, given that plants and people have evolved together for countless millennia. More practical reasons for good

FIGURE 12–2 A well landscaped home is not only beautiful, but is also enhanced in value and provides a more pleasant living environment. (USDA photograph now in the National Archives.)

landscaping include increased home values and happier, more productive workers, students, and shoppers. Real estate professionals consider landscaping improvements a good investment. Home value is increased and, at the least, the increased sales value can result in full recovery of landscaping costs. A final reason is that careful landscaping can cut energy costs for nearby buildings (see Chapter 11, Figure 11–4).

Landscaping is a combination of skills, horticultural and artistic. If the former is lacking, the landscaping will likely suffer, resulting in the death of plants. If the latter is lacking, the landscape might thrive, but offer little or no visual pleasure. Landscaping requires careful planning, given cost factors and its long-term permanence. Landscapes also need maintenance and cultural inputs to remain attractive and healthy.

The design of the landscape is especially critical. This cost should not be minimized. It is better to cut expenses by adding plants over time, as funds become available, than to skimp on the design. Landscape designers and landscape architects usually design quality landscapes. Software design packages help both the designer and client to more fully visualize the final product than was possible in the past. Such software can also enable the average person to design landscapes, too. However, this approach is only for those individuals with lots of horticultural expertise and a good artistic sense.

After the designer and you agree on plans, some construction might be needed such as a stone wall or path or even a patio. Usually, a landscape contractor is involved with such construction and the eventual procurement and placement of plants. Often, a garden center or nursery can provide all of these services, from design to final installation. Landscaping, whether for a newly constructed house or office complex, or a remake of an existing landscape, must be taken as a serious exercise, given its value and long-term existence.

Plant choices are critical in landscaping. High-quality plants are recommended, and these plants should be selected with the following criteria in mind. Plants should be rated for your zone of hardiness. Borderline plants that generally survive, except for infrequent abnormally cold winters, are to be avoided. Exceptions to this rule are the availability of microclimatic areas that increase the chance of long-term winter survival or "must have" plants that you understand are at risk. Plants that are known to be prone to insect and disease problems in your area are not recommended, unless you are willing to provide the long-term, required maintenance. From a sustainability perspective, their need for pesticides makes them an unacceptable choice.

Diversity is also recommended. Landscapes are best when they offer something for all seasons. A mix of deciduous and evergreen trees and shrubs is best. Even in the coldest climates, evergreens (Figure 12–3) offer a touch of green and soften the stark, bare bones look

FIGURE 12–3 A dwarf, scaly-leafed evergreen softens the rock mulch all year round. (Author photograph.)

of deciduous specimens. Needled and broad-leafed evergreens mixed with deciduous shrubs and trees also offer an interesting variety of textures (Figure 12–4) in the landscape. Some of the trees and shrubs should be flowering types (Figure 12–5). A few late flowering choices should be on the list so that the beauty and color are not all confined to a few glorious weeks in the spring. Colorful fruits that persist into the fall and even winter are a plus for enhanced late season color and for attracting birds. Herbaceous perennials are desirable for enhancement of spring, summer, and fall colors in your landscape and for cut flowers. Lastly, annuals can be used where fast color or temporary color is needed in the border or on the deck, porch, or patio (Figure 12–6).

Scale and form also need careful consideration. One common error in landscaping has to do with scale. A natural tendency exists for homeowners to not fully appreciate what size

FIGURE 12–5   These flowering shrubs, mountain laurel, offer a bright spot of color on the north side of this house. (Author photograph.)

FIGURE 12–4   The mixture of a scaly-leafed evergreen with these deciduous shrubs and trees offers an array of textures. (Author photograph.)

FIGURE 12–6   Petunias and geraniums offer color and cheerfulness in these window boxes. (Author photograph.)

plants will be at maturity. This failure to project into the future leads to two problems. One, plants are placed at distances that seem reasonable relative to other plants, but become overcrowded when the plants mature. This situation then requires extensive, unnecessary pruning or the removal of valuable plants. These initial plants when properly placed for future growth capabilities make the planting look sparse and unpleasing. The sparseness can be filled in temporarily with herbaceous perennials and annuals until the trees and shrubs fill out with age. The other problem occurs when a nice, compact plant is placed near a smaller home on the front lawn or close to the house. Later when it is 150 feet tall, the plant towers over the house and looks out of scale relative to the house and lot. Such plants only achieve proper scale when the land around the home allows for placing the plant at a great distance from the house where its size can be a focal point, rather than an overwhelming spectacle. Another scale problem occurs with small shrubs planted near or under a window that ultimately reach sizes that obscure the window and block views, thus requiring unnecessary pruning. A quality landscaping contractor or a knowledgeable homeowner will avoid problems with scale through proper plant choices and correct placement.

Form is also important. If all of the chosen plants have the same, natural form such as globular or columnar, the landscaping will look strange. Choosing a diverse mix of plants usually eliminates this problem. Placement of various forms also needs consideration. For example, a columnar evergreen would soften the corner of a house or hide an unattractive drainpipe. On the other hand, it would be a poor choice in front of a window where it would look out of place and partially obscure the view. A globular shrub with a mature height less than the window's bottom sill would be a better choice.

An established landscape will require ongoing maintenance to keep it attractive and healthy. Pruning will be needed annually to maintain flowering, to remove diseased and dead wood, and sometimes for shape or size maintenance. Pruning techniques are covered later in Chapter 14. The use of water porous landscape fabric over which natural (bark, compost, wood chips, peanut shells, etc.) mulch is laid is highly recommended to minimize weeding and watering. Mulch is not only attractive, it also makes for more sustainable water use (mulch slows water evaporation), less herbicide use, and less hand labor. If some weeds appear, you can hand pull, hoe, or spot treat them with safe commercial herbicides containing fatty acids or homemade products such as full-strength vinegar. Even safe herbicides must not contact garden plants to avoid injury or death. Handheld propane weed torches can be used around driveways and walkways (not in mulch). Conventional herbicides should be used as a last resort and in judicious amounts. Chapter 15 provides more information on weed control.

Trees and shrubs in the landscape are deeper rooted than garden plants and grass. As such, less watering is needed. In fact, native trees and shrubs succeed with natural rainfall. The use of mulch also helps reduce water needs. During drought, some water may be needed. If you notice that tree and shrub leaves don't recover from temporary wilting in times of dry, hot weather by late afternoon and early evening or the soil is dry more than a few inches below the surface, watering is needed. Herbaceous perennials will need an inch of water weekly as will grass and gardens. When water is needed, water can be applied sustainably as is done with gardens (see the discussion of garden watering later in this chapter).

Nutrient requirements, both for establishment of the initial landscape and later for maintenance, must be determined on the basis of the soil test as discussed earlier. Nutrients can be applied in a sustainable manner in a number of ways. The annual use of compost mulch will add nutrients slowly as it decomposes. Supplement applications with slow-release fertilizers such as organic products, chemical slow-release granular forms, and chemical slow-release tree and shrub spikes are all good choices. If you use normal chemical fertilizers or water-soluble fertilizers, use them as sparingly as possible. The pH level should be maintained as needed with limestone or sulfur as recommended. Proper pH ensures efficient usage of soil nutrients by plants. More detail on fertility and pH is given in Chapter 11.

Finally, one needs to realize that a lawn is the binding element or the backdrop (Figure 12–7) for perennials, shrubs, and trees that constitute the home's (or other building's) landscape. Good lawn care is essential to keep up the appearance, but not at the risk of undoing sustainability. Fortunately, lawns can be managed more sustainably than they cur-

FIGURE 12–7 The broad sweep of lawn ties all the elements together in this majestic landscape view. (Author photograph.)

rently are. The addition of a garden, whether flower beds/borders or a vegetable garden, also needs to be fitted carefully into the context of the existing landscape. Lawns are covered next, gardens later, and more on landscaping can be found in Chapter 16 on ornamentals.

LAWN CARE

Lawns are not very sustainable ecosystems in that their maintenance requires large amounts of fertilizer, pesticides, and gasoline throughout the growing season. Lawnmowers as currently constructed pollute the air because their engines lack pollution control and fuel vapor recovery systems. The latter is beyond our consideration here. The hope is that congressional legislation will bring about cleaner operating lawnmowers as was done with automobiles. Improved technology might also lead to the development of highly efficient electric lawnmowers that run on clean burning fuel cells (see Chapter 10). It is possible, however, to alter lawn cultural practices such that a higher level of sustainability is achieved.

Step 1 is the earlier mentioned and highly recommended annual soil test. Request minimal fertilizer amounts or ask for organic recommendations if you are an organic practitioner. Lawns can do quite well at fertility levels below the ones suggested by lawn fertilizer manufacturers or those amounts normally given in soil tests, which are based on older, less environmentally sensitive guidelines. Results were generally designed to err on the high side to ensure good plant growth.

Fertilizer applications can be minimized by the following approach. First, maintain the pH at 6.5 using either limestone or sulfur as needed (see Chapter 11). A good rule is to correct pH problems either in the fall or a few weeks prior to applying fertilizer or compost. This pH will ensure efficient usage of nutrients by the grass such that minimal applications of fertilizer will be effective. Highly recommended from a sustainability perspective is an annual early spring or early autumn application of screened compost or dehydrated cow manure. Using a lawn fertilizer spreader, apply compost or cow manure between 0.5 to 1 inch thick. Apply either before a rain or water afterward to settle the materials into the lower grass areas. In conjunction with these materials, mow your lawn with a mulching mower. This type recycles finely chopped grass blades and their nutrient contents back to the lawn. These small pieces decompose fast and do not contribute to thatch buildup.

These combined practices can reduce the subsequent amounts of fertilizer (Table 12–4) that you will need to apply. In some cases, compost and recycled grass clippings alone can provide sufficient nutrients to maintain a healthy lawn or, at the least, reduce the

TABLE 12–4   *Sustainable Applications of Nitrogen for Lawns*

Dominant Grass	Month to Apply	Maximal Amount of Nitrogen/1,000 Square Feet
Bermuda grass	June and July	One pound each month
Fine fescue	October	One pound in October
Tall fescue	September and October	Two pounds each year
Bluegrass		One pound in September
Ryegrass		and one in October
Zoysia grass	June	One pound in June

*Use these amounts only if you don't use compost and recycle grass clippings with a mulching mower. If you do both practices, cut the amounts in half.

supplemental applications of nitrogen annually. If you prefer a greener lawn or a soil test indicates some nutrient deficiencies, then further fertilization will be required. Additional fertilizer should be applied in early fall when grass is actively growing. Do not apply nutrients during the summer, because they are more likely to leach away and harm the environment, given minimal uptake by grass. Preferred fertilizers in terms of the environment include organic and slow-release chemical forms. If more soluble fertilizers are used, use them sparingly. Lastly, products containing seaweed extracts and humic acids have been used to promote better root systems in grass and are likely to improve nutrient uptake and result in greater stress tolerance. Although still under investigation, these products might be worth a try because they are environmentally benign.

Watering grass poses a dilemma in terms of sustainability. Lawns can be left to Mother Nature's sustainable approach. In areas with moderate to high rainfall, grass will survive easily on natural rainfall and enter a dormant (brown up) state in the hot summer months. Re-greening will occur in the autumn as rains increase and temperatures cool. This method works best with older grass cultivars and not as well with improved, modern cultivars. If spring or autumn droughts occur, a green lawn is wanted during the summer, your lawn is in arid regions, or your lawn consists of recent hybrid grass cultivars, some watering is likely to be needed (assuming water restrictions do not prevent it). The same rule of thumb used with gardens applies to lawns: an inch of water a week, including rain, is needed. Sustainable approaches to watering are covered later under gardens.

In terms of weeds, and especially crabgrass, the sustainable approach is to mow high. Grass mowing at a height of 2.5 to 3.5 inches reduces the amount of light reaching the soil through shading. Germination of weed seeds is considerably reduced under these conditions. Higher grass also develops better root systems, leading to more efficient water and nutrient uptake and increased stress tolerance toward insects, disease, and drought. Herbicides should be a last resort and should be organic herbicides (see Chapter 15) such as products made from corn gluten (preemergent) or fatty acids (spotting). If chemical herbicides are used, usage should be kept to a bare minimum and only in lawn areas where small amounts of weeds can't be tolerated. Weeds can be mowed, too, and blend in well with grass. Not everyone needs a golf-course perfect lawn.

Lawns generally can tolerate pests better than gardens and also seem to be less disease prone. If problems develop that need control, try using natural controls or safer "green" insecticides such as citrus-based products (limonene), fatty acids, milky spore disease, neem products, and pyrethrum, or sulfur-based fungicides (see Chapter 15 for greater coverage).

WHY GARDEN?

Gardening has been a popular American pastime over the years. Various polls indicate that more than 40 million American households garden each year, mainly with flowers and vegetables. Several reasons are behind this high hobby activity. One reason is that people like the taste of fresh vegetables and fruits. No store-bought tomato comes close to the taste of a tomato plucked from the garden at its height of ripeness. Another reason is gardening provides an outdoor activity with a moderate level of exercise that is pleasurable (for the most part) and provides a sense of personal satisfaction. The latter feeling comes from raising plants to provide food or enjoyment for your self, family, and others. Some find a sense of "back to nature" with their gardening activity. Gardening is a great form of therapy and is a wholesome family activity. You can also express your personal artistic and esthetic sense with flowers and landscaping plants. Another impetus is economic, since a vegetable garden provides fresh food at reasonable prices. A family can save $400 to $600 and possibly more on their annual fresh vegetable grocery bill if some homegrown vegetables and fruits are processed by freezing or canning. Pesticide-free food is another advantage, because you decide what you put on your food to control pests. Another factor is diversity. You can grow items that are not normally found in the produce department of the large grocery store such as lemon cucumbers, purple peppers, or salsify.

SITE SELECTION

If you are buying your first home or looking for another home and have gardening in mind, you need to pay attention to certain gardening requirements. While some aspects of your property such as climate are determined by geographical location, others are variable and determined by local physical and topographical features. The amount of sun and quality of soil drainage are very important site considerations. Heavily shaded homes and yards may provide poor gardens, and also mean a dreary home interior and higher heating bills. A badly drained soil with standing water also means poor gardening and the possibility of flooding in the basement. Drainage can be corrected with sharp sand and organic matter or drainage tiles (see earlier discussion).

An alternative solution is available and can be easily accomplished with minimal effort and modest expense. Raised beds (Figure 12–8) are essentially the equivalent of a child's bottomless sandbox, with the sand replaced by good garden soil. The sides can be made of wood or cinder block. Avoid the use of pressure-treated lumber with crops destined for consumption, especially root crops, because chemicals that leach from the wood are hazardous. A good alternative is the synthetic wood made from recycled plastics or to use rot-resistant wood such as redwood. You could also use untreated wood that you treat yourself with linseed oil or commercial preservatives that are safe for plant growth such as those containing copper naphthenate. The height for these beds is 12 to 18 inches. Raised beds are also solutions to extremely rocky or stony soils, and also for areas containing only subsoil left behind by builders who sold off the topsoil.

(A)   (B)

FIGURE 12–8   (A) These wooden raised beds are used for a vegetable garden in Texas. (USDA/NRCS photograph.) (B) This stone raised bed is used for an annual flower garden in Connecticut. (Author photograph.)

One very critical question to ask regarding your potential garden site relates to sun exposure. For best growth and highest yields, your site must have at least 6 hours of sun per day, and more is preferable. Also keep in mind that sun exposure from 10:00 A.M. to 2:00 P.M. is more effective than 4 hours starting at 2:00 P.M. If you have less than 6 hours of sun a day, a garden is still possible, but less sun means lower yields and less growth. You will need to experiment with plants to determine which ones can provide acceptable results with your conditions.

North sides of homes and large structures are to be avoided, as well as areas heavily shaded by big trees. If all you have is minimal sun or lots of shade, don't despair. Some ornamentals can be grown in shade, but you should plan to buy your vegetables, fruits, and cut flowers. Of course, you might be able to open an area to sun by removing some trees if a garden is high on your wish list. If you are buying a home and want to garden, always be watchful for sun exposure. The preferred exposure in order of most to least for a potential garden site (assuming no trees or house shadows) is south, southeast, southwest, east, and west.

Plant and Garden Choices

Gardening involves a multitude of plant choices and even style choices. One of the most basic types of gardens is the vegetable garden (Figure 12–9). It is the second most practiced form of gardening in the United States. Even here, choices exist. Many people plant what is known as a salad garden. A salad garden can be simple with lettuce and cherry and slicing tomatoes. A simple salad garden can be placed anywhere that is sunny, from the side of the garage or house to the back yard or even in containers. Salad gardens can be more complex with arugula, cucumbers, lettuce, peppers, radicchio, radishes, and tomatoes. Others prefer a fuller culinary choice in their vegetable garden and add various greens such as mustard, Swiss chard, spinach, herbs, and main-plate vegetables such as broccoli, carrots, eggplants, green beans, and summer squash. Others will add herbs such as chives, basil, and oregano or maintain a separate herb garden. Gardeners with big areas might add asparagus, melons, rhubarb, sweet corn, and winter squash. Personal tastes, available space, and the level of labor one is willing to commit are determining factors for these choices. Chapters 17 and 18 cover vegetables, herbs, fruits, and nuts.

Flower gardens (Figure 12–10) are the most popular type of garden. Some people have a simple cut flower garden, an area where flowers grown for cutting such as dahlias, gladiolas, and zinnias are planted in rows like vegetables. More people prefer to grow some flowers for

FIGURE 12–9 This Connecticut backyard has a typical vegetable garden. The tall poles around the perimeter support a seven-foot tall, black nylon net deer fence that is essentially invisible to the eye from a distance. (Author photograph.)

(A)

(B)

FIGURE 12–10 (A) This flower garden starts off with spring bulbs. (B) The same flower garden later in the season filled with perennials and annuals. The view is shifted somewhat. In (A), the yellow forsythia in the right back corner is now the fully green shrub at 12:00 in the background. (Author photographs.)

FIGURE 12–11 Mixture of prairie plants in eastern Iowa flower garden. (USDA/NRCS photographs.)

color effect such as begonias, coleus, impatiens, marigolds, and others for cutting in special areas known as borders and islands (see later section on flower gardens). Flowers can be used for temporary effects such as various annual bedding plants placed in borders to complement flowering bulbs and other flowering perennials or in landscaped areas to add instant color or to fill in until woody plants mature. Annuals are also widely used in container plantings, hanging baskets, patio pots, and window boxes. Perennials are also used in more permanent arrangements such as borders and islands, where such plantings are expected to stand alone.

Other people prefer to use native flowering plants (Figure 12–11) to create a natural garden. Others use wildflowers (Figure 12–12) in a meadow-like garden. Still others prefer rock gardens with various alpine-type flowers or shade gardens with shade-loving flowers such as begonias, bleeding heart, coleus, hosta, and impatiens. Some find attractive a water or bog garden with water lilies. Gardens can also be designed as butterfly gardens with plants attractive to butterflies (Figure 12–13) such as bee balm, *Buddleia*, butterfly weed, cosmos, milkweed,

FIGURE 12–12   A happy gardener in her wildflower garden in Michigan. (USDA/NRCS photograph.)

FIGURE 12–13   Monarch butterfly sipping nectar at a zinnia. (Courtesy of National Garden Bureau.)

and zinnia. Similar gardens can be designed to attract humming birds with plants like petunias and red salvia. Chapter 16 lists appropriate plants for these various types of gardens.

Flower gardens also present style choices. Flower gardens range from formal to informal. Formal gardens require considerable planning and maintenance. Such gardens were often the mainstay of royalty and the rich, who could afford the staff of gardeners required for these gardens planted in fancy patterns with their mazes and various garden structures known as whimsies and follies. As you might guess such gardens are not very popular these days, given the expense and considerable amount of labor required to maintain them. With busy lives, most of us are content with more informal garden styles with their gentle curves, sweeps of plants, and concentrated, colorful patches here and there. Each one differs, being a reflection of the ideas and artistic talents of the designing individual.

The cottage garden style is typical of informal gardens. Its origins are found in England. The emphasis is not on perfect lawns and picture-perfect flower borders and islands. The idea is essentially an eclectic blend of flowers, vegetables, fruits, trees, shrubs, and other plants around the property. Flowers and vegetables can be mixed in small plots. These plantings tend to look more naturalistic and attract butterflies and birds. While not essential, some purists insist on heirloom and older cultivars for these gardens. The idea is to enjoy beauty, scent, and plants throughout the season. A patch of hollyhock near the back door, some morning glories on a porch trellis, lilacs near the end of the driveway, and a small patch of vegetables and flowers is a good beginning.

Some gardeners also like to grow some fruits and even nuts. Some of the more popular or easier fruits include apples, blackberries, blueberries, grapefruits, grapes, oranges, peaches, pears, raspberries, and strawberries. Almonds and walnuts are popular nuts. This type of gardening requires available sunny space and a commitment to a certain amount of labor beyond the ordinary garden. I grow blueberries and strawberries and consider the effort well worth it when we eat them fresh or frozen or as homemade ice cream. More information on fruits and nuts can be found in Chapter 18.

The concept of sustainability can also influence your choices. You might feel that you are a sustainable gardener if you elect to be an IG^{2}M gardener who uses mostly natural controls, while your neighbor with an organic garden might consider his or her garden to be more sustainable. Is the latter gardener the ultimate sustainable gardener? Not necessarily; it depends to some degree on what plants are used. If you live in an arid area and garden organically with vegetables and flowers that require much water such as corn, cucumbers, impatiens, and marigolds, how sustainable is that? If instead, your flower garden uses indigenous plants that have adapted to arid conditions, your water usage would be considerably less. If you landscaped with native drought-tolerant trees and shrubs, they too would need less water. Many of these same plants might also be more resistant against local pests than those exotics introduced from elsewhere. In fact, this type of gardening and landscaping is recommended from the viewpoint of sustainability in arid regions and is known as *xeriscaping*.

In general, natural or environmental gardening places high emphasis on the use of indigenous plants wherever possible as the best choices in terms of vigor and resource usage. In the arid areas, cacti and succulents are good choices for natural gardening, while in New England a woodland garden filled with native rhododendrons, dogwoods, mountain laurels, and underplantings of trillium, dogtooth violet, and pink ladyslippers would be ideal. An open area in the Northeast or West Coast might become a wildflower meadow garden with Queen Anne's lace, chicory, and yarrow, while in the Midwest a prairie garden might contain purple coneflowers, goldenrod, and wild indigo. All of these are sustainable alternatives to the conventional flower garden.

So what should you do to be a responsible caretaker of your backyard environment and the local watershed? With vegetable gardens the choice is straightforward. Given that hardly any of our vegetables are indigenous plants anywhere in the United States, you would be hard pressed to have a natural vegetable garden. The clear choice is to grow whatever vegetables you wish and to use cultural practices based either on integrated garden management or, even better, organic gardening.

Flower gardens do have the potential to be natural gardens almost anywhere in the United States, given the abundance of native flowering plants. A natural flower garden with cultural practices based on either integrated garden management or organic gardening would be a clear choice for sustainability. However, if your preference is for flowers that came from elsewhere, the conventional garden flowers, you can still be sustainable. Just don't use the conventional cultural practices, instead use either integrated garden management or organic gardening.

PLANNING THE GARDEN

Gardens, like most things in life, turn out best when planned. Planning basics include plant choices and locations. All plant choices are, of course, subject to local climatic conditions. Beyond that, home garden choices of flowering plants are made on the basis of personal pleasure in terms of color, scent, and texture. Vegetables are chosen on the basis of taste preferences and culinary considerations. Site drainage and levels of sun and shade also influence choices. Another factor concerns maintenance levels in terms of cultivation and pest problems. Commercial operations are based on marketing, labor, and economic considerations.

A physical plan can be very useful, especially for vegetable and flower gardens. Plans can be done the old-fashioned way, with paper, pencil, and much thought. However, another option is to use garden planning software. These programs are usually based on your input to specific questions such as garden size, shape, vegetable and flower choices, size of family (for vegetable gardens), length of gardening season, and so forth. Height considerations are also important. Tall plants need to be placed on the north side of vegetable gardens and in the back of flower borders, so that sun is not blocked from reaching smaller plants.

GETTING THE MOST VEGETABLES AND HERBS

Even with a small yard or the limitation of being able to devote only a little space to the gardening area, a vegetable garden can still be productive or one can still have masses of flower color. The trick is to take advantage of certain practices that give you more for less. One thing that you need to do is to forget anything you ever saw on a farm or in big gardens elsewhere.

Those farms and gardens were planned for machines so that labor could be reduced. You are not likely to drive a tractor with various implements up and down your crop rows.

Growing Season and Its Extension

To use some of these space-saving tricks, you need to know the length of your growing season, which can vary from about 3 months in the northernmost United States to 365 days in a few blessed spots in the extreme southernmost parts. Such information is easily available from various sources as discussed earlier in the chapter. Another fact you often need is days to maturity (harvest) for vegetable crops from time of sowing or from time of setting transplants (bedding plants). Maturity times for vegetables are covered in Chapter 17. Another useful piece of information is harvest duration, or how long you have to enjoy those fresh garden beans or leaf lettuce.

Keep in mind that you can extend the season on either end with some techniques. This approach increases your choices for vegetable successions and also leads to earlier harvests in the summer and longer lasting harvests in the fall. These techniques are also useful in northern areas with short growing seasons. Season extenders include hot caps, row covers, tunnels, and cloches (see Chapter 13). Waxed paper hot caps (Figure 12–14) can be purchased, but the more sustainable version is homemade. One-gallon plastic milk bottles or two-liter plastic soda bottles with their bottoms cut off can be recycled as excellent hot caps. Row covers (Figure 12–15) made from polyester material are so lightweight that they are placed over plants without internal supports such as wire hoops. Polyethylene plastic tunnels (Figure 12–16) with wire hoop supports can also be used. Another variation much less used today is the glass or plastic cloche, a box-like or cylinder-like structure.

All of these aids can protect plants from light frosts in the spring or fall. In addition, they increase temperatures by 5° to 20°F over the ambient air temperature, giving many plants a head start of a few weeks. Such structures are especially popular for getting that early tomato! These structures do need to be removed once the weather is settled, because the temperature inside can become high enough to damage plants. Bringing these protective structures back out in the fall can also add a few weeks of extra harvest.

Intensive Gardening

This technique uses space more efficiently by eliminating some rows and moving plants closer together in either bands or beds. It is made possible in the garden because most rows

FIGURE 12–14 A waxed paper hot cap shown here has the vent open to release some heat. Once popular, these aren't used much now. Clear plastic two-liter soda bottles with the bottom cut off provide an inexpensive and effective alternative. (USDA photograph.)

FIGURE 12–15 A close-up of the polyester spun-bonded material used as a floating row cover. It not only provides frost protection and increased heat for faster growth, it also offers insect protection and allows rain through. (Author photograph.)

FIGURE 12–16 This row cover has a built-in square wire mesh and is easily bent into a tunnel shape. Lettuce is shown under it. Other row cover types are shown in Chapter 13. (Courtesy of National Garden Bureau.)

are unnecessary, given that we won't be driving a tractor up and down rows. A few rows must remain, because some space is needed for walking between bands and beds. Even in a commercial truck garden, it is possible to make some use of bands and beds with proper row planning. Bands and beds differ in size, bands being narrower than beds. Beds (Figure 12–17) should be no wider than double your arm length, given the need to reach in for weeding and harvesting. Switching from rows to bands or beds can nearly double the harvest from a given area. Vegetable beds shouldn't be a big surprise, for flowers have been grown in beds for years to achieve color masses.

Bands and beds can also be raised above the soil's surface, as with a raised bed (see Figure 12–8). This form is not necessary, except under certain conditions. If your land consists of subsoil, the builder having sold off the topsoil, you need raised beds. Another reason is if your soil drains poorly and has standing water on it after a heavy rain. If you wish to grow beautiful, symmetrical carrots, like those from California or Massachusetts, you need a sandy soil free of rocks and pebbles. Rocks cause carrots to develop forks and other abnormalities. In many places, such as in the rocky soils throughout New England, the only way to grow good carrots is with a raised bed. In some cases, raised beds have an aesthetic

FIGURE 12–17   A bed containing squash plants is shown here. Note the grass clipping mulch to the right. (Author photograph.)

FIGURE 12–18   Lettuce is interplanted with these squash plants. Most of the leaf lettuce is ready for harvest and will leave only the romaine lettuce (tall ones in middle), whose compact heads will be ready by the time the squash closes in. (Author photograph.)

appeal such as stair-stepped raised beds on a gentle slope or against a high foundation wall such as with a greenhouse.

Interplanting

This method depends on the use of the same space by two different vegetables. These interplanted partners (Figure 12–18) are chosen on the basis of complementary space requirements. Some plants can share space simultaneously because of complementary forms. A good example is the interplanting of broccoli and lettuce transplants. Broccoli grows taller and tends to be umbrella shaped, using little surface row space, but needing more upperstory space to spread its leaves and broccoli heads. Lettuce tends to be low and compact such that it fits nicely between broccoli plants. An added advantage is that the broccoli partially shades the lettuce, producing a better environment for the less heat-tolerant lettuces.

Other vegetables make good space partners because of chronological differences in space use. A good example is the interplanting of radish and lettuce seeds. Radishes grow much faster than lettuce. In about 30 days the radishes can be harvested. The slower growing lettuce then fills in the space freed up by radish removal. Another interplanting is winter squash and corn seed. This mix was practiced by native Americans in New England and, in turn, by the early colonists that adopted this practice. The corn was harvested first and the winter squash later.

Vertical Gardening

This space-saving concept is based on the skyscraper principle. In cities with expensive or limited real estate, the most efficient space use is to build upward. The same is true in the garden. Vine crops are space hogs. Consider the cucumber. Its vines can spread out for several feet in all directions. However, plant the cucumber at a fence, trellis, or net support and it grows upward, using very little garden surface space. There you have the basics of vertical gardening (Figure 12–19). Cucumbers, climbing string and lima beans, peas, small melons,

FIGURE 12–19 Nylon netting supports these pole string beans. (Author photograph.)

FIGURE 12–20 Sweet corn planted in succession at timed intervals is shown here. (Courtesy of National Garden Bureau.)

and small winter squashes are good choices for vertical gardening. Two factors are critical. Make sure your support structure is strong enough to hold the weight of the mature crop and make sure the vertical garden is placed on the north side of your garden. This approach is best for small home gardens where space is a limiting factor. Commercial producers are unlikely to use this technique, because it is not suitable for mechanized harvesting.

Succession Gardening

This method can use the same space, different spaces, the same plants, different cultivars of the same plant, or even successions of the same and different plants (Figure 12–20). One type of succession planting utilizes a timed sequence of plantings in a given area. For example, one could start off with peas, a cool season crop whose seeds can be sown earlier than other crops. Once the peas are harvested, the space could be planted to another fast-maturing cool season crop, rather than being left idle. The second crop could be radishes or perhaps lettuce or maybe broccoli. Later after the second crop is harvested, the same area could be planted to a fast-maturing warm season crop such as bush beans. Once the beans are harvested, the area could be planted with some cool season crop to wind up the season such as turnips or spinach. An alternate scenario is to plant peas, followed by radishes, and then tomatoes to finish out the season. Another scenario might be peas, radishes, a second planting of radishes, and then a corn finish. This approach allows one to harvest a succession of different crops from the same space, thus making efficient and continuous use of the area.

The succession of plants can consist of the same vegetable planted at intervals in different areas. This succession increases the harvest length for a given vegetable. For example, the same sweet corn cultivar can be planted in one area on May 15, another area on May 30, and a third area on June 15. This allows for a continuous corn harvest over several weeks, rather than the much shorter harvest period with only one corn planting. Use of this method allows farm stands to supply corn for several weeks during the summer. Of course, it is also possible with some crops to get the extended harvest in another manner. One can plant the three areas simultaneously to an early, mid, and late season cultivar. This strategy is possible with crops that have differing lengths of time to harvest and several cultivars such as sweet corn and tomatoes.

Succession planting does require one to plan carefully and to know the length of the growing season and days to maturity for crops and also whether a crop can tolerate some frost or is even improved by it. It is also useful to know whether a vegetable is a cool season or warm season crop such as spinach and tomatoes, respectively. For example, if you have used up 120 days of a growing season that lasts 180 days, your next crop(s) must be ready for harvest in 2 months or less. This timing is especially critical for warm season crops that are killed by frost. On the other hand, you could exceed the 2 months with cool season crops that tolerate frost such as broccoli, or with a crop where taste is improved by a touch of frost such as brussels sprouts, parsnip, rutabaga, or winter squash.

FLOWER BASICS

Vegetable gardens provide food, while flower gardens feed the spirit. Flower gardens, unlike vegetable gardens, can satisfy the creative urge. One can express artistic bents, blending colors and textures in pleasing mosaics. Or the gardener can create a single-color garden such as a blue or pink garden. Fragrance can also be emphasized. Flowers can be chosen on the basis of butterfly or humming bird attraction. Emphasis can also be placed on cutting flowers to beautify your indoor living space.

Flowers are pleasing to the sight and bring cheer to the landscape and the viewer. Flowers can be incorporated into the landscape in a number of ways. Some of the permanent woody landscape plants should be chosen on the basis of flowers. Depending on where you live, a few possible choices are azalea, bougainvillea, broom, clematis, dogwood, flowering almond, forsythia, lilac, cotoneaster, magnolia, mahonia, mountain laurel, oleander, rhododendron, rose, spiraea, and viburnum.

Additional flower choices include annuals and perennials. Annuals are economical and offer fast, bold colors and ease of planting, because many are available as bedding plants. As such, annuals are frequently used as filler plants to hide ripening bulb foliage, to fill in areas until more permanent plants take hold, to add color spots to perennial beds during slow periods, and to provide cut flowers. Annuals (Figure 12–21) make excellent choices for beds, borders (along fences, driveways, and paths), containers, islands in the lawn, and window box plantings. The downside is that their life span is but one growing season, so their usage is an annual labor event. The other choice is perennials that are more costly and slower to establish and flower. Fortunately, the latter problem can be minimal, given the large number of perennials available at nurseries as reasonably mature plants. The advantage is that most perennials, once established, come back every year. Another plus is that perennials usually require less maintenance than annuals. Annuals, perennials, and woody plants and their culture are discussed in later chapters.

Stair-Stepping

Stair-stepping of plant heights (Figure 12–22) in a flower bed or border is important for getting sun to all flowers, and it also allows the viewer to see the beauty of all the flowers. Taller flowers are placed in the back and shorter ones in the front facing the sun. For example, perennial phlox and most lilies should be placed in the rear of a perennial bed or border. Similarly, in an annual grouping the old-fashioned cultivars of cosmos that reach 4 feet in height go to the rear. The same would be true of sunflowers. Newer, shorter cultivars of cosmos would go near the middle and portulaca in the front. Most seed packages and plant labels give mature heights for plants. Books and Web sites can help out with unknowns such as incomplete labels or those unexpected gifts of plants from fellow gardeners and your family.

Light and Temperature Needs

Plants in the flower garden need to be selected with temperature and light needs in mind. Best results are obtained when you know your plant cultural requirements and match their needs with appropriate environmental conditions. If you have a shaded area, some flowers are still possible. On the other hand, those flowers would not be good choices for an area

FIGURE 12–21
Annuals are shown in the driveway flower border and in the raised bed island on the lawn. (Author photograph.)

FIGURE 12–22   Low growing petunias are planted in front of tall snapdragons. (USDA photograph.)

FIGURE 12–23   Coleus plants offer colorful foliage in many hues for those shady spots in the yard and are also good container plants. (Author photograph.)

FIGURE 12–24   These gloriosa daisies are ideal, easy care perennials for those sunny spots. (Author photograph.)

receiving sun all day. Some flowers thrive in cool temperatures, so are ideal choices for spring flowers. Others succeed in the heat of the summer. For flowering success, you need to understand the differing needs of flowering plants. Unlike vegetables, it is possible to even have flowers in shaded areas.

Ideal annuals for shady areas include begonias, coleus (Figure 12–23), and impatiens. Coleus is noted for its many-hued foliage, but not for its flowers. Begonias and impatiens have many highly colorful flowers. Some cultivars are available with bronzy foliage. The newer New Guinea cultivars of impatiens are especially attractive. These shade annuals prefer warm summer temperatures and can tolerate some sun. Most annuals such as the ever-popular marigolds and zinnias require lots of sun and warmer temperatures. Water is especially critical for good results with annuals.

Some perennials for partial shade include astilbe, bleeding heart, columbine, daylily, hosta, lily of the valley, and violets. Ferns, while not flowering plants, also thrive in the perennial shade garden. Some do best in the cool of spring such as bleeding heart, lily of the valley, and violets. Others such as astilbe can take the heat. Given the large number of perennials available, it is possible to create a perennial flowering display for your sunny garden (Figure 12–24) from early spring through the first fall frosts. While choices are somewhat limited, it is also possible to create perennial flowering displays in shady to partially shaded areas.

STARTING SEEDS AHEAD: TO BUY OR NOT TO BUY

Every gardener faces this question: Should I use seeds or buy bedding plants? In some cases the decision is made for you. If you want to grow sweet corn, you plant seeds. The same is true for green beans, peas, radishes, and beets. Some vegetables and flowers are simply not available as bedding plants. The reason for this might be economics or the inability to transplant well, such as with sweet corn. In other cases, such as with tomatoes, you have a choice. Many plants are sold in bedding plant form (Figure 12–25). The advantages include faster flowering or vegetable harvesting and less care, because seedlings generally require closer care than bedding plants set out in the garden. For example, with tomato plants, you can harvest tomatoes some 4 to 6 weeks earlier than planting seeds in the garden. This advantage usually outweighs the cost, as bedding plants cost more than seeds.

However, it is possible to buy seeds and raise your own bedding plants and perennials instead of buying them. One reason for doing this approach is greater choice. Bedding plant producers generally sell a limited number of cultivars, usually the more popular ones in terms of marketing. Should you wish another cultivar, you might grow your own bedding plants. The trade-off is work and some skill. If you are doing a very large vegetable garden or perennial bed, cost might be a factor. Bedding plants and nursery perennials, while rea-

(A)

(B)

(C)

FIGURE 12–25 (A) A flat of 'Orchard Daddy' petunias (former AAS winner) offers instant color for the yard, container, or annual flower garden. (B) These 'Sandy White' verbenas (former AAS winner) in their cell pack make a nice choice for the front of the annual flower border. Courtesy of All America Selections. (C) Vegetable bedding plants are also available in similar ways, or as individual transplants as this 'Big Early' tomato. (Author photograph.)

sonably priced given the production costs, can add up. Suppose you wanted to plant 100 perennials and the average price was $6 each (total = $600!). For a few dollars spent on seeds and modest amounts for potting media and containers and a year of sweat equity culture, you could save money. Of course, this assumes you have the space and time. On a smaller scale, you can raise your own vegetable and annual bedding plants for much less over a few months. It is your choice. Most gardeners buy finished plants rather than raising their own. It is hard to beat the convenience factor, given our overly busy lives.

Raising bedding plants at home (Figure 12–26) can be done. Chapter 7 provides basic information about seed propagation. Keys for success are quality seeds, sterile or pasteurized propagation media (commercially available), containers such as plastic, compartmentalized flats, careful watering, fertilizing, and lots of sun. A greenhouse is ideal, but you can make do at a southern window or under high-intensity growth lamps. At some point your seedlings will require transplanting to provide adequate growing room, either in open plastic flats or into individual containers. The latter is best in terms of reducing transplant shock when you move the plants to the garden. Careful attention to watering, fertilization, and possible disease control will be needed throughout the process. You will also need to harden or condition (Figure 12–27) your plants (see Chapters 7 and 13) either in a cold frame or outdoors on a porch or protected area prior to transplanting them into the garden.

FIGURE 12–26   Starting seeds in flats or in other containers is easy and economical. More detail is available in Chapter 7. (Author photograph.)

FIGURE 12–27   Homegrown trays of various annuals and vegetables started in flats and moved to pots are shown here hardening off in the shade of the garage. (Author photograph.)

TILLING THE GARDEN

The most critical step in a garden occurs before you dig the soil: a proper soil test. Sources and methods for soil tests were covered in Chapter 11 and earlier in this chapter. Make sure you indicate in your testing request that you are interested in gardening in a sustainable manner and would prefer that recommendations for fertilizer be based on this consideration. Recommendations from soil tests should be heeded. Proper fertility and pH are essential for best results with all aspects of gardening and landscaping.

To dig or not to dig is a reasonable question when one gardens. The answer depends on a number of factors. If you are converting a sodded area into a garden, the grass does need removal. One way is to till it under, generally with a rotary tiller (Figure 12–28). Doing it by hand with a spading fork or shovel (Figure 12–29) would be hard labor and is advisable only for small areas. You could also use herbicide to kill the grass and plant after the herbicide activity period is over. This solution is the least desirable in terms of sustainability. Another approach is to use organic mulch such as hay or black plastic mulch to kill the grass. Once the grass is dead, you could remove the mulch and then plant bedding plants with a trowel. Seeds could be sown into rows carved out with a hoe corner or a furrow hoe. The mulch should be removed if it was placed early in the spring, since it would slow the warming of the soil to proper germination temperatures. If the mulch was placed after the soil had warmed up, you could leave it.

The mulch, if left in place, can be moved aside or have holes cut into it. You can then place bedding plants, such as tomatoes, into small holes dug with a trowel. Sowing seeds into a mulched area is admittedly harder, but not impossible. Some seeds such as those planted in hills, like squash, can be placed in small holes dug with the trowel. Seeds requiring rows need more effort. One way is to slit the plastic with a knife and then fold over the flap and pin it down with a bent U-shaped wire formed by cutting coat hangers in half or using cut and bent pieces from wire rods used to hold insulation battens in place. Organic mulches can simply be raked aside. A row can then be carved with a hoe or furrow hoe as indicated above.

Once gardens are established, tillage is optional. In some cases, such as in established flower or vegetable perennial beds, tillage ceases. Plants are thinned and new plants added through selective spot digging with trowels or bulb planters. Seeds can be planted in spots with light scratching of soil. Vegetable gardens and cut flower gardens or annual flower beds, borders, and islands can be tilled or not on a yearly basis in the spring. Some gardeners like the smell of freshly turned soil and the smooth raked appearance of newly tilled areas. It does bury surface weed seeds and remove stubble and small weeds, but the downside is that new weed seeds are brought to the surface and some organic matter is lost through oxidation. It is hard work, too. If you must dig, only dig those areas where you will actually plant. You could ignore areas used for paths or in between rows.

FIGURE 12–29 Hand digging of a garden with a spade or fork is pleasant exercise, but can become hard work with large areas. (Courtesy of National Garden Bureau.)

FIGURE 12–28 Rotary tillers are useful for turning sod or weeds under when making a new garden. Large garden areas are also candidates for rotary tillers. Tillers can also be used for cultivation to kill weeds later between rows. (USDA photograph.)

FIGURE 12–30 After tilling or digging, a rake can be used to level soil, remove rocks, and to make a more easily worked seedbed. (USDA photograph now in the National Archives.)

If you opt to dig the garden, you can do it by hand or machine. Rotary tillers are commonly used. These machines are expensive and require maintenance. Your best bet is to rent one, as the annual rental fee won't approach the purchase price for many years. Some lawn and garden operations and private individuals also supply tilling services. Hand labor involves turning over the soil, using either a spade shovel or spading fork. Generally, one removes any large rocks. Fertilizer and limestone can be added (see later section on fertilization and pH adjustment, this chapter) at this point so that tilling or digging will incorporate nutrients and limestone into the rooting zone. Next the soil is raked (Figure 12–30). Lastly, bedding plants and seeds are put into place. Soil nutrients and amendments are discussed in a later section titled Culture.

SOWING SEEDS

Seeds must be fresh, either recently purchased or not saved for too many years. If you save seeds from previous years, their life span is influenced by two parameters: longevity and storage conditions. Some seeds are short lived, even under the best conditions. Examples include lettuce and begonia. Some seeds such as tomatoes can last for several years. Good storage requires cool, dry conditions. If in doubt, purchase fresh seed, because the cost is minimal.

Seed propagation has been covered in detail in Chapter 7. The key to successful seed germination is moisture. If seeds dry out during germination, death usually results. The growing media, whether in an outdoor garden or indoors in flats, needs to be loose and friable, have good drainage and aeration, but be able to retain some water. Indoor germination media are commercially formulated to provide the right conditions. Garden soil fertility and organic matter need to be maintained, based on soil tests and appropriate fertilizers and soil amendments. Soils should be dug and raked or not, as discussed earlier in this chapter. Even if not dug, it is a good idea to loosen the soil in the row or hill where you will plant the seeds. A hoe or some other long-handled cultivator can be used for this purpose.

Seed sowing requires some knowledge and skill. Seeds differ in terms of soil temperature needs during germination. Some seeds can be sown earlier (cool season crops) than others (warm season crops). This topic was covered in Chapter 6. For example, carrot, lettuce, garden pea, and sweet pea are sown earlier than corn, marigold, squash, tomato, and zinnia. With the latter, cold wet soils cause the seeds to rot prior to germination. Sowing times obviously vary based on geographical location. Altitude can also alter planting times even within a given area. Cool season crops can be sown safely a week or two before the last frost-free date, while warm season crops should be delayed at least a few weeks beyond the frost-free date. More specific information can be learned from established gardeners, local extension offices, state universities, and state agricultural experiment stations. Additional information can be found in the later chapters that cover specific floral and vegetable crops.

A few seeds with hard seed coats such as morning glory and parsley often germinate very slowly. Such seeds usually germinate faster if soaked in warm water overnight. Seed packages usually indicate special treatments for problem seeds. Seeds sown in gardens are often sown in rows (Figure 12–31), either with conventional spacing or closer. Closer spacing is usually used in vegetable gardens for beds and bands, as discussed earlier in this chapter. Some seeds such as squash or melons are sown in hills. Other seeds can be sown in a broadcast pattern such as with grass seed for a new lawn or replacement spot. Carrots and salad greens can also be broadcast sown in a bed as an alternative to closely spaced rows. Regardless of what prac-

FIGURE 12–31 Rows can be made with this plow-like tool called a furrower. A rectangular-bladed hoe corner can also suffice. (Author photograph.)

FIGURE 12–32 These spinach seedlings will require thinning to produce good quality plants. The removed seedlings can be either transplanted to other areas or discarded. (Courtesy of National Garden Bureau.)

tice you use, seeds should not be placed too close or be too dense in the beginning. Some allowance is possible here, given that 100% of the seeds are not likely to germinate. Some will also possibly be lost in the seedling stage to insects and disease. Keep in mind that you will need to do considerable thinning of seedlings if your initial sowing density is too high.

Seeds should be sown in moist soil, either after a rain or watering by hose or sprinkler. Seeds should be covered with soil to the depth recommended on the seed package. A good rule of thumb is that seeds are covered to a depth twice to three times their diameter. Very tiny seeds are just surface sown and not covered, but lightly pressed in such as with begonia, coleus, and petunia. Some tiny seeds require light for germination, so covering them can actually be detrimental. A gentle surface watering after the seeds are covered is usually recommended. Soil should be maintained moist and will require watering if you see surface dryness. Careful watch is usually required for a few weeks or more until the seedlings are well established.

Once seedlings are established, some thinning (Figure 12–32) might be required to bring about proper spacing. Plants that are too close develop poorly and yields are often reduced. Crowding can also increase disease problems. You can also transplant small seedlings to fill in spaces where germination was less than optimal. These transplants can also go to other areas or even friends. Transplants require some special handling that will be covered shortly.

While the above deals with seeds sown outdoors, much of the information pertains to sowing seeds indoors to produce transplants (bedding plants) for the outdoor garden. Bedding plants give the gardener a head start and faster color. The head start is especially useful for long maturity vegetables such as tomatoes. Tomato bedding plants produce tomatoes at least 30 days earlier than direct outdoor seed sowing. The same is also true for annuals purchased as bedding plants. Annuals are usually desired for quick color to fill in spots in borders and beds, window boxes, and other containers. In these situations, earlier is better. Bedding plants also extend the already too-short summer of beauty and clearly present much less work for today's harried gardener. Bedding plants are available for many popular garden vegetables such as eggplant, lettuce, pepper, tomato, and various herbs and flowers such as ageratum, begonia, coleus, dianthus, geranium, impatiens, marigold, petunia, salvia, and zinnia. If you are willing to invest the time and energy in doing your own bedding plants, more details can be found in Chapter 7.

USING TRANSPLANTS AND HOW TO ESTABLISH

Certain attention to details is needed for the successful establishment of transplants, whether purchased as bedding plants or home grown. If you purchase bedding plants, make sure you purchase quality plants from a reputable dealer. Look for short, stocky plants with good green color. Pale or yellow-green plants are usually stressed, often due to insufficient nutrients. Try to buy plants that have few or no flowers or buds only. When in doubt, the label or the dealer can give you an idea of the color. Plants in full blossom usually have more difficulty with establishment than those that do not. This trend is especially noticeable with tomato plants. A natural tendency exists for consumers to purchase tomato plants already in flower or even showing little green tomatoes. Such plants actually can take longer to produce red tomatoes and often have lower yields than stocky, shorter plants without open flowers or even buds.

Most bedding plants are now being sold in the plastic plug-type flats, with each plant in its individual "egg carton." Undivided flats are becoming increasingly rarer. Given that such flats produce transplants with tangled roots that are damaged on separation, this trend is better for the consumer. The plug type bedding plants are much easier to transplant and establishment problems are few, if any. Make sure that you keep these purchased bedding plants watered when you bring them home. Place them in a sunny location such as a porch or patio and don't forget their water! Exceptions to the sunny location include shade-loving plants such as begonia, coleus, and impatiens.

Before plug transplants, transplanting was recommended for cloudy or even mildly rainy days. Transplanting on a sunny day was discouraged, because root damage and sun-induced transpiration took its toll on a fraction of transplants. Unless you are using bedding plant flats that need dividing and plant separation, this caveat need not apply. The only reason you might want to consider it is for personal comfort. Transplanting bedding plants on a cloudy or drizzly day might appeal to you over a bright, hot, need-your-sunblock day.

The chore is relatively simple. You need to plan the placement of the bedding plants based on proper distances between plants, making sure that the location is either sunny enough, or shady enough, depending on the plants in question. If you are working with flowering annuals, you also need to consider color compatibility and final heights. These latter points become especially critical if you are transplanting herbaceous or woody perennials to their final placement. Once you have decided where the transplants are to be placed, you dig a hole with a trowel. If you are working with woody or herbaceous perennials, you will probably need a spade and a much bigger hole. The hole should be bigger than the root ball of your transplant, but not deeper, because the root/shoot original line should remain the same. Generally, the size hole produced by one or two applications of the trowel is sufficient for bedding plants, while larger plants might require the removal of several shovels of soil.

Place the transplant in the hole (Figure 12–33) and carefully backfill the hole. Gently tamp down with your palm and add a little more soil to restore the soil level. Next water

(A)

(B)

FIGURE 12–33   (A) This hole was dug with a bulb planter. A trowel is also satisfactory. The hole is dug deeper than needed to accommodate some granular 10-10-10 fertilizer in the bottom, which is covered with a thin layer of soil. This layer will supply nutrients, as the roots grow downward. After the hole is backfilled as needed around the transplant, it will be watered with a phosphorous-rich starter fertilizer to promote rapid root growth and successful establishment. (B) This hole was dug through black plastic mulch with a bulb planter. Granular fertilizer is shown being added as in (A). The process is the same as described in (A). (Author photographs.)

the area with a soluble fertilizer based on recommended package dilutions. A fertilizer richer in phosphorus (starter fertilizer such as 10-15-10 or 5-10-5) is best, because phosphorus promotes root development and soluble forms act rapidly to speed root regeneration and new growth. Some settling might occur later when you water or when it rains. This change is minor and can be ignored. Over the next week or two, you need to pay special attention to watering. Prolonged drying of recent transplants is to be avoided. Never add fertilizer solutions to the hole prior to filling it. The muddy slurry created by adding soil to liquid will harm soil structure and affect root growth adversely.

CULTURE

The successful establishment and maintenance of garden plants is dependent on the careful and timely application of skills and inputs during the garden season. These collective practices are known as *culture.* Cultural practices include the maintenance of several soil parameters: nutrients, pH, organic matter, and moisture. Additional practices include pest control to minimize damage from insects and diseases and competition from weeds. The importance of cultural practices cannot be overemphasized. Your garden is only as good as your cultural skills.

Fertilization and pH Adjustment

Your garden results depend highly on proper nutrient levels. Even with sufficient nutrient levels, results can be poor if the soil pH is too high or low. Soil pH regulates nutrient availability. Both of these areas were discussed in Chapter 6 in terms of science and in Chapter 11 in terms of applications. If you wish to maintain some sense of sustainability in your garden, fertilizers must be applied judiciously and not to excess. The importance of the soil test can't be overemphasized. While it is no guarantee that you aren't applying too many nutrients, it does create a baseline from which to work. Certainly, you should not apply more fertilizer than recommended in the report. Less might be all right, but do this reduction in small increments. From a sustainable viewpoint, organic fertilizers are highly recommended. However, one can use chemical fertilizers also in a sustainable manner and not harm the environment. The key is that less is better with soluble chemical fertilizers. Slow-release forms of chemical fertilizers are preferable in that fewer nutrients leach into ground and surface waters.

Fertilizers are usually added at the beginning of the garden season prior to planting. They can be broadcast by hand or spread mechanically with a lawn fertilizer spreader (Figure 12–34). On freshly dug soil, you can lightly rake in the granules and then water. This process of working the fertilizer in and dissolving some nutrients increases the fertilizer efficiency over just leaving it on the dry surface. Even more efficient fertilizer usage in the garden can be obtained by broadcasting the fertilizer only in areas where the plants will grow such as in rows, beds, or bands. Additional applications of fertilizer are likely to be needed later with established plants, especially if they are heavy feeders. These later applications are referred to as side dressings. Instead of granular applications, foliar feeding is possible with highly soluble fertilizers. Cost is considerably higher. One possibility is to use foliar sprays in place of side dressing, if cost is not a factor.

Limestone in various forms, preferably ground limestone, can be added at any time such as in the fall. It can be added at the same time as fertilizer, but sometimes certain soil conditions can bring about some loss of the fertilizer nitrogen as ammonia gas. This loss is especially critical with urea-based fertilizers. Given this possibility, the best approach is to add limestone either in the fall or about 1 to 2 weeks prior to spring fertilization. The limestone needs to be raked and watered in to enhance soil pH adjustments.

Organic Amendments

Organic matter is a critical component in terms of soil sustainability and plant productivity. It affects the chemical, physical, and microbiological aspects of soil. The functions of

(A)

(B)

FIGURE 12–34   (A) A broadcast fertilizer spreader is useful for spreading limestone or granular fertilizer on lawns and level unplanted garden areas. (USDA photograph.) (B) Fertilizer or limestone can also be spread by hand broadcasting with a scoop. This approach is useful for planted or small areas. (Author photograph.)

organic matter were discussed in Chapters 6 and 11. To be a sustainable gardener, one must maintain and improve wherever possible the organic matter levels in your garden soil.

Several amendments can be used to provide organic matter for garden soils that need more. Levels of application are best determined based on soil test results. Certainly, you would not want to apply more than 1 to 2 inches onto the soil prior to turning it under. One approach is the incorporation of peat moss at levels recommended in your soil test results. Some limestone will also be needed to offset the acidity of the peat moss. Sewage sludge can also be used, assuming the finished product is cleared for home use by health department standards and it contains no heavy metals such as cadmium, lead, or mercury.

Animal manure, both fresh and composted, is a better choice than sewage sludge, especially if you know of a nearby stable or farm that needs to dispose of it. Fresh manure should be added well in advance of planting to ensure decomposition. Fresh manure can sometimes cause plant damage (poultry manure) and might be a source of harmful human pathogens. Fresh manure and sometimes fresh sewage sludge can also contain weed seeds. Composting when properly done destroys harmful pathogens and eliminates viable weed seeds. Composting also eliminates the plant damage seen sometimes with fresh manure.

Compost is another, viable alternative. Many gardeners make their own compost and many communities make it available at minimal cost to gardeners. Methods of composting were discussed in Chapter 11. One useful way to employ compost is to apply it as mulch, then turn it under into the soil at the end of the season.

Watering

The standard garden advice on watering is 1 inch of rain weekly is the minimum. In very hot, dry climates or periods, this number approaches 2 inches. Although good advice, it is not perfect. For example, if you have several light rains in the week totaling 1 inch versus one solid rain yielding 1 inch, the available soil moisture will not be the same. Deeper penetration results with the latter. Surface wetness is rapidly lost through evaporation, especially on hot, summer days. Clearly, if weekly rainfall is less than 1 inch and your garden is not mulched, you should water the garden. Mulches have many valuable functions (see later discussion), one of which is water conservation.

Gardeners can watch for clues beyond rainfall. If plants wilt at midday on a sunny, warm (over 80°F) day, but rapidly become turgid again as the afternoon wears on, soil moisture is likely fine. What you are seeing is temporary wilting, brought on by high transpiration rates for which water intake cannot compensate. If wilting starts to persist longer, some serious watering is in order. Failure to do so reduces the ability of the plant to grow and produce flowers and vegetables. Persistent or permanent wilting results in plant death.

Watering is often done with a hose equipped with a sprinkler head or variable nozzle. This method is fine for small vegetable gardens and flower borders, beds, islands, and window or patio boxes. Patience is required, because a fast pass with the hose is insufficient to wet the soil down to at least an inch or more (1 inch of water is about 60 gallons per 100 square feet). If you can fill a 1-gallon container with your hose in 1 minute, it would take about 1 hour to supply an inch of water to 100 square feet. Until you get the feel for how much time is right, you might put down a flat, small container and water until the container has a 1-inch depth of water. Bigger gardens and sizable lawns are best watered with movable, oscillating, or sweeping sprinklers that you set up and move to various areas. Again, some measuring of water depth is recommended initially so that some idea of watering time needed is gained. Sprinklers can also be piped underground with pop-ups or above ground with fixed sprinkler heads such as for lawns (Figure 12–35) and gardens, respectively. Surface watering is also possible with drip or ooze hoses (Figure 12–36). Drip or ooze watering systems are the most efficient in terms of water usage and, hence, sustainability. Watering systems can be automated with time clocks and other sensing devices.

Timing for watering is also related to sustainability issues. Watering is best done earlier in the day. Less loss of applied water through evaporation occurs then as opposed to midday and early afternoon on sunny days. Water is also then available for plant uptake when high midday demands occur, thus reducing wilting. Less wilting in turn means increased plant productivity since wilting does not disrupt photosynthesis. Foliage also has a chance to dry as opposed to watering late in the day when water persists on foliage. Persistent foliage moisture overnight can increase certain disease problems, which in turn raises fungicide requirements. Water with high but acceptable salt levels can also cause leaf burn when applied in the high heat of the day as the salts become concentrated in rapidly

FIGURE 12–35 A pop-up sprinkler system is used to water this front lawn area of a house in Nevada. (USDA/NRCS photograph.)

FIGURE 12–36 An ooze hose is used in this vegetable garden as seen near this tomato plant. (USDA/NRCS photograph.)

shrinking drops and can enter leaves through stomates that are generally fully open at that time. Watering by hose or sprinkler on days when the wind speed is high should also be avoided, as water loss from wind drift and evaporation can be considerable. Drip or ooze can be used on windy days without worries about water loss through blowing.

Controlling Pests

Solving the pest problems in gardens used to depend on the application of pesticides for insects, mites, diseases, and weeds. Organic gardeners did employ alternative strategies such as natural or biological controls. For example, the lady beetle was employed to mop up aphids, and a bacterium, *Bacillus thuringiensis* or Bt, was used against many caterpillars on vegetables. Today, given the problems associated with pesticides (see Chapter 15), the idea for a sustainable approach is to minimize pesticide usage. The strategy varies from zero usage of pesticides and 100% natural controls (organic gardening) to various mixtures of pesticides and natural controls (sustainable gardening approach based on integrated pest management).

Pesticides, when used, are often used after natural controls fail to alleviate the problem, or when no natural remedies are available. Pesticides are usually selected based on low toxicity and limited environmental persistence. Pesticides made from plants (botanical pesticides) are often preferred over those manufactured from petroleum feedstock. The former include pyrethrum, rotenone, and neem products derived, respectively, from a chrysanthemum-type plant in Kenya, the roots of *Derris elliptica* or *Lonchocarpus* sp., and the neem tree in India. Rotenone should be fresh, because it deteriorates rapidly. Some synthetic chemical pesticides include the popular malathion and Sevin®.

Even with botanical pesticides, caution, safety, and minimal usage are recommended. Although botanicals are usually less toxic to people and animals than chemical ones, caution and safe use is highly suggested, given that these products can kill beneficial insects such as honeybees. Some such as rotenone can kill fish if allowed into water bodies. Bees can be protected by not spraying when flowers are about to or are open. The possibility of adverse reactions with some individuals such as allergies or irritation is always possible. No pesticides, including botanicals, should be sprayed on windy days so as to avoid accidental skin contact. Hot days are to be avoided also, because plant damage can occur on pesticide contact. Future findings might also show that even "safe pesticides" could be involved in cancer initiation such as acting as an endocrine disruptor.

Natural controls (Figure 12–37) are an alternative to pesticides. Some natural controls include the introduction of the previously mentioned lady beetle and Bt, fatty acids for

FIGURE 12–37 This Japanese beetle trap attracts the beetles with either geranium oil or pheromone bait. Once the beetle flies in, it cannot escape. Small infestations can be controlled with a few traps. However, large populations usually overwhelm this system, especially if you are the only one in the neighborhood with these traps. (Author photograph.)

weeds, soap sprays for various insects and mites, and the encouragement of spiders and preying mantises in the landscape. Birds are often encouraged, too, as good forms of pest control. Much more pest control information can be found in Chapter 15.

Weeding and Mulching

Weeding is high up on the list of most disliked garden chores. Imagine a hot, humid, muggy day in July. It has recently rained and you are pulling weeds. Weeds pull out easier from wet rather than dry soil. Mosquitoes are biting and you are slapping, covering your exposed areas with mud while sweat drips down your face into your eyes. You can use two approaches and be sustainable. The usage of herbicides to kill weeds should be avoided, because it ranks dead last in the sustainable race.

The use of a hoe (Figure 12–38) works well in dry soils. With some hoes, you use a chopping motion to "decapitate" weeds at or just below the soil's surface. Other hoes have sharp steel bands shaped to a rectangle or circle and you draw them through the soil just below the surface, slicing weeds as you go.

Even better is mulching, which smothers weeds before they can become a garden presence. Mulches are either synthetic or natural materials (Figure 12–39) applied to the soil as a surface layer. Black plastic is an excellent example of the former and salt marsh hay, the latter. Mulches also reduce the need for water, keep mud from splashing on vegetables, and ameliorate drastic changes in soil temperature that can impair plant growth. Mulches of many types were covered in Chapter 11 in more detail. Bear in mind that some mulches can contain weed seeds such as mixed hay. Pure alfalfa hay is better in that regard.

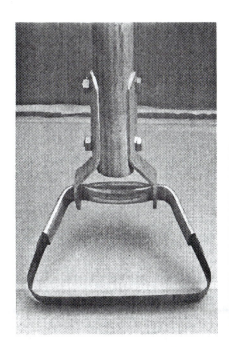

FIGURE 12–38 Many types of hoes exist. This one is an oscillating hoe. The blade is sharp on both sides and can move back and forth in a small arc. It is drawn just under the soil's surface in a back and forth motion. Slicing of weeds at the roots is very effective. Other hoes have rectangular blades sharpened at the base and a chopping motion is used to sever the weeds. (Author photograph.)

FIGURE 12–39 This photo illustrates both black plastic and organic mulch. The eggplants are growing well in the weed-free environment provided by the plastic mulch, which also reduces water loss by evaporation. Organic hay/grass mix mulch is seen on the right. It is being used to provide a weed-free pathway, but could also be used with plants. (Author photograph.)

Staking, Pruning, and Dead-Heading

With many plants some additional chores are required. These chores are connected with visual appearance and flowering. For example, some perennials and certain annuals have mature heights between 2.5 to 5 feet. Combine these heights with many flower heads and a heavy, prolonged rain and even some wind, and the plants will fall over. Plants in this category include peonies, many lilies, delphinium, gladiola, lupines, and certain marigold and zinnias. These taller plants often need staking with thin bamboo green stakes or specialized holders such as peony rings. In some cases, if you are willing to forego the majesty of tall plants, you can opt for shorter cultivars. Staking is also used even in vegetable gardens such as for tomatoes.

Although it can be a time-consuming chore, certain ornamental plants have their spent blooms removed (dead-heading). The benefits are improved appearance such as for geraniums, lilacs, marigolds, mountain laurel, petunias, and snapdragons. An additional benefit is often an additional flush of flowers. This chore can be made easier by pinching off flower heads as you pass by beds or borders on your way to your car as you head off for work. A pruning shear can also be used to cut off stems with spent blooms. Some plants can be cut back extensively when flowering appears to be over. This drastic action often results in a new flush of flowers such as with petunias and snapdragons. Somehow a few minutes here and there, rather than a scheduled hour or two, make this chore easier to bear. Removal of parts beyond flower heads, pruning, is generally only needed for woody plants. Annuals and herbaceous perennials rarely require pruning. Exceptions are occasional tip pinching to promote a bushy appearance in house plants and to increase flower stalks on certain plants such as chrysanthemums. This aspect of pruning is discussed in Chapter 14.

Harvesting

The harvest is the high point of the vegetable gardener's activity. The taste of fresh tomatoes or the crunch of salad greens makes it all worthwhile. Harvesting does apply to flowers in some sense. The use of cut flowers for table arrangements is also a harvest.

Proper harvesting of vegetables ensures peak taste and continuity. For example, sweet corn can be tricky in terms of when to harvest. If harvested too soon, kernels will be watery and lack sweetness; too late and kernels will be cheesy and starchy. When picked at the right time, you can't beat the melt-in-your-mouth sweetness of sweet corn. On the other hand, some vegetables can be picked prior to ripeness or when ripe such as peppers. When picked unripe, or green and not sweet, peppers are edible. When picked ripe, peppers have a sweeter taste and more color such as red, orange, or yellow. Vitamin content is also higher. Depending on the culinary use, peppers can be used at either stage.

Some vegetables can be gently broken off such as tomatoes. Others require cutting such as peppers and eggplants. If you try to break either off, you usually end up removing some of the shoot and reducing future harvest potential. Some vegetables, if partially harvested, will grow back for additional harvesting such as leaf lettuce. Broccoli will also produce side sprouts after the main part is removed. Harvest hints for various garden vegetables are discussed under their respective entries in Chapter 17.

Cutting flowers usually requires a sharp knife. The use of small pruning shears can crush tissues and reduce the stem's ability to take up water. Other ways to increase vase longevity include minimizing the time between cutting and plunging the stems into water and the use of cut flower preservatives. The best way is to recut the flower stem under water and then plunge them into lukewarm water in the vase. One of the oldest tricks is to place aspirin in the water. We now understand that this works because it inhibits jasmonic acid production. This chemical is released on injury to ward off predators. It also promotes aging. Cutting the flower stem releases jasmonic acid, but the aspirin counters it. A number of commercial flower preservatives also exist. The latter often contain ingredients to nourish the cut flower, to prevent bacterial growth, and to prolong life. Such ingredients include sugars for nourishment (sucrose or glucose), bacteriocides such as 8-hydroxyquinoline sulfate or 8-hydroxyquinoline citrate, and chemicals to inhibit ethylene, which speeds aging.

SUMMARY

Information on landscaping, lawn care, and gardening is available from books, garden clubs, extension offices/Master Gardeners, local nurseries and garden centers, magazines, neighbors, state agricultural experiment stations, state universities, and Web sites.

Climate determines which crops you can grow and when to plant. Critical information includes dates of the last killing spring and first killing fall frosts, growing season length, and hardiness zones. Frost dates determine the best times for planting crops. Zones of hardiness determine winter survival of all perennial herbaceous and woody plants. Microclimates can help extend plant choices.

Poor drainage leads to stunted plants and root rot. Incorporation of large amounts of sharp sand and organic matter to clay soil can improve drainage. French and tile drains or the diversion of water that collects from nearby higher areas might be necessary.

Annual soil tests provide a basis for improving your soil and are essential for achieving sustainable home horticulture. You should state that you want sustainable options. Soil tests are a taxpayer service provided in state institutions.

Conventional landscape maintenance, lawn care, and gardening use chemicals (pesticides and chemical fertilizers) can lead to environmental damage and possible health hazards. Organic gardening uses natural materials. Fertilizers are manure and natural products, compost is added to the soil, and troublesome weeds are controlled with mulches and hand/mechanical cultivation. Insects and diseases are controlled by resistant cultivars and by natural enemies and products.

One can also maintain grounds and garden in a sustainable manner and protect the environment and your family's health with integrated grounds and gardening management (IG^{2}M). IG^{2}M's basic premise is to use organic techniques when possible. When organic techniques fail or are not acceptable, then resort to manufactured products. The sustainable strategy varies from zero usage of pesticides and 100% natural controls (organic gardening) to various mixtures of pesticides and natural controls. Pesticides are selected for low toxicity and limited environmental persistence. Botanical pesticides (derived from plants) are often preferred over manufactured chemicals.

Homes and other buildings are all enhanced esthetically and economically when properly landscaped. High-quality shrubs, trees and perennial plants should be selected on the basis of site compatibility, beauty, ease of care, contribution to overall plant diversity, seasonal appeal, and color/texture/form aspects. Careful landscaping reduces energy costs. Landscaping requires horticultural and artistic skills plus careful planning. Established landscapes need maintenance and cultural inputs. Landscape design is critical. Landscape designers and landscape architects design quality landscapes. Software design packages help the client to visualize the final product and can also enable the homeowner to design landscapes. Some construction by a landscape contractor might be needed. Often, a garden center or nursery can provide all of these services, from design to final installation.

Scale and form need careful consideration. A tendency exists to not realize what size plants will be on maturity. Plants are placed at distances that seem reasonable, but become overcrowded later. This situation then requires extensive, unnecessary pruning or the removal of valuable plants. The sparseness of initial placement can be filled in temporarily with herbaceous perennials and annuals. Large plants only achieve proper scale when the land around the home allows for placing the plant at a great distance from the house where its size can be a focal point, rather than an overwhelming spectacle. Small shrubs planted near or under a window that ultimately reach sizes that obscure the window and block views will require unnecessary pruning. Placement and use of various forms also needs consideration. A columnar evergreen can soften the corner of a house or hide an unattractive drainpipe, but be a poor choice in front of a window where it would look out of place and partially obscure the view. A globular shrub with a mature height below the windowsill would be a better choice.

Pruning is needed annually to maintain flowering, to remove diseased and dead wood, and sometimes for shape or size maintenance. The use of water porous landscape fabric overlaid with natural mulch will minimize weeding and watering. During long dry periods, water will be needed. If weeds appear, hand pull, hoe, or spot treat with safe commercial herbicides containing fatty acids or homemade products (full-strength vinegar). Even safer herbicides must not contact garden plants to avoid injury or death. Handheld propane weed torches can be used around driveways and walkways (not in mulch). Conventional herbicides should be used judiciously and as a last resort.

Lawns can be sustainably managed at fertility levels below levels suggested by lawn fertilizer manufacturers or amounts normally given in soil tests. By using certain practices, you can reduce the amounts of applied fertilizer. In some cases, compost and recycled grass clippings alone can provide sufficient nutrients to maintain a healthy lawn or at the least, reduce supplemental applications of nitrogen. Lawns can be left to Mother Nature's sustainable water approach. This method works best with older grass cultivars and not as well with improved, modern cultivars. If spring or autumn droughts occur, a green lawn is wanted during the summer, your lawn is in arid regions, or your lawn consists of recent hybrid grass cultivars, some watering is likely to be needed.

In terms of weeds, and especially crabgrass, the sustainable approach is to mow high. Higher grass develops better root systems, leading to more efficient water and nutrient uptake and increased stress tolerance toward insects, disease and drought. Herbicides should be a last resort and should be organic herbicides or use the bare-minimum of chemical herbicides only in lawn areas where small amounts of weeds can't be tolerated. Lawns generally can tolerate pests better than gardens and also seem to be less disease prone. If problems develop that need control, try using natural controls or safer "green" insecticides such as citrus-based products (limonene), fatty acids, milky spore disease, neem products, and pyrethrum or sulfur-based fungicides.

Gardening provides a pleasurable outdoor activity, moderate exercise, a sense of personal satisfaction, and a back to nature feeling. Gardening offers therapy and a wholesome family activity where you can express your personal artistic and esthetic sense. A vegetable garden provides fresh and potentially pesticide-free food at reasonable prices.

Vegetable and flower gardens need at least 6 hours of sun and good soil drainage. A vegetable garden can be a simple salad garden with lettuce and cherry and slicing tomatoes. Some prefer a fuller culinary choice and add various greens and main-plate vegetables. Others will add herbs in the vegetable garden or maintain a separate herb garden. Some gardeners also like to grow some fruits and even nuts. Personal tastes, available space, and the level of labor you are willing to do determine these choices.

Some people have a simple cut flower garden planted in rows. More grow flowers for color and cutting in borders and islands. Flowers used for temporary effects include various annual bedding plants placed in borders to complement flowering bulbs and other flowering perennials or in landscaped areas to add instant color or to fill in until woody plants mature. Annuals are also widely used in container plantings, hanging baskets, patio pots, and window boxes. Perennials are placed in more permanent arrangements such as borders and islands. Others use native flowering plants to create a natural garden or wildflowers in a meadow-like garden. Still others prefer rock gardens with various alpine-type flowers, shade gardens, a water or bog garden, or butterfly garden.

The concept of sustainability influences choices. Living in an arid area and gardening organically with vegetables and flowers that require much water isn't sustainable. Gardening with indigenous flowering plants adapted to arid conditions and landscaping with native arid-tolerant trees and shrubs uses less water. These same plants are likely more resistant against local pests that exotics from elsewhere. This type of gardening and landscaping, xeriscaping, is recommended in arid regions.

Natural or environmental gardening places high emphasis on the use of indigenous plants wherever possible as the best choices in terms of vigor and resource usage. In arid areas desert plants are good choices for natural gardening, while in New England a woodland garden filled with native woodland plants would be ideal. An open area in the Northeast or West Coast might become a wildflower meadow, while in the Midwest it might become a prairie garden.

Given that hardly any of our vegetables are indigenous plants in the United States, one is hard pressed to have a natural vegetable garden. The clear choice is to grow vegetables with cultural practices based either on IG^2M or organic gardening. Flower gardens do have the potential to be natural gardens pretty much anywhere in the United States, given numerous native flowering plants. A natural flower garden with cultural practices based on either IG^{2}M or organic gardening would be a clear choice for sustainability. However, if your tastes run to the flowers that came from elsewhere, you can still be sustainable with either IG^{2}M or organic gardening.

Garden planning basics include climate, plant choices, and locations. Choices of flowering plants involve color, scent, and texture. Vegetables are chosen on the basis of taste preferences and culinary considerations. Another factor concerns maintenance levels in terms of cultivation and pest problems. Plans can be done with paper, pencil and much thought or with garden planning software.

You can extend the growing season on both ends to increase your choices for vegetable successions and to have earlier and longer lasting harvests. This practice is useful in northern areas with short growing seasons. Season temporary extenders include hot caps, row covers, tunnels, and homemade plastic cloches. All can protect plants from light spring or fall frosts and they increase temperatures, giving plants a few weeks head start.

Intensive gardening uses space efficiently by eliminating some rows and moving plants closer together in either bands or beds (bands being narrower than beds). A few rows remain for walking. Switching from rows to bands/beds can nearly double the harvest in a given area. Bands and beds can also be raised above the soil's surface. If your land consists of subsoil or if your soil drains poorly, you need raised beds.

Interplanting depends on complementary space sharing or chronological differences in space use. A good example is interplanting broccoli and lettuce transplants. Broccoli grows taller and is umbrella shaped, while low lettuce fits nicely between broccoli plants. An added advantage is that the broccoli partially shades the lettuce, producing a better environment for the less heat tolerant lettuces.

Vertical gardening is based on the skyscraper principle. Climbing vegetables on a fence, trellis, or net grow upward, using very little garden surface. Support structures must be able to hold the weight of the mature crop. This approach is best for small home gardens where space is limited.

Succession gardening can use the same space, different spaces, the same plants, different cultivars of the same plant, or even successions of same and different plants. One popular type utilizes a timed sequence of plantings in a given area. Succession planting does require one to plan carefully and to know the length of the growing season and days to maturity and also whether a crop can tolerate some frost.

Flowers are beautiful, bring cheer to the landscape and the viewer, and can be incorporated into the landscape in many ways. Some permanent woody landscape plants should be chosen on the basis of flowers. Additional flower choices include annuals and perennials. Annuals are economical, offer fast, bold colors and ease of planting as bedding plants. The downside is annuals require yearly planting. Perennials are more costly than annuals and slower to establish and flower. Fortunately, perennials are available as reasonably mature plants at nurseries. The advantage is that, once planted, perennials last many years and usually require less maintenance.

Stair-stepping of plant heights in a flower bed or border gets sun to all flowers, and allows an unobstructed view. Taller flowers are placed in the back and shorter ones in the front.

Plants in the flower garden are selected based on temperature and light needs. A few annuals and perennials can grow in the shade. Some prefer cool temperatures, so are ideal choices for spring flowers. Most annuals and perennials require lots of sun and warmer temperatures. Water is especially critical with annuals. Given the large number of annuals and perennials, it is possible to create a flowering display in a sunny garden from early spring through the first fall frosts.

Some vegetables and flowers are simply not available as bedding plants. The reason for this might be economics or the inability to transplant well. In other cases, you have a choice. Many plants are sold in bedding plant form. The advantages include faster flowers or vegetables and less care, as seedlings generally require closer care than bedding plants set out in the garden. This advantage usually outweighs cost, as bedding plants cost more than seeds. It is possible to raise your own bedding plants and perennials, given the time, materials, and proper skills. Most gardeners buy finished plants rather than raising their own.

To dig or not to dig is a reasonable question. When converting a sodded area into a garden, the grass does need removal. One way is to till it under with a rotary tiller or by hand. You could also use herbicide and plant after the herbicide activity period is over (unsustainable). Another approach is to use mulch to kill the grass, remove the mulch and then plant bedding plants with a trowel. Seeds could be sown into rows carved out with a hoe corner or a furrow hoe. The mulch, if left in place, can be moved aside or have holes cut into it. You can then plant bedding plants into small holes dug with a trowel.

Once gardens and flower beds are established, tillage is optional. Plants are thinned and new plants added through selective spot digging. Seeds can be planted in spots with light scratching of soil, or making rows with a hoe. The downside of tillage is that new weed seeds are brought to the surface and some organic matter is lost through oxidation. If you dig, only dig those areas where you will actually plant. Rotary tillers are commonly used. Hand labor involves turning over the soil, using either a spade shovel or spading fork. Fertilizer and limestone are best incorporated during tillage. Next the soil is raked.

Seeds must be fresh. If seeds dry out during germination, death occurs. Seeds differ in terms of soil temperature needs during germination. Some seeds can be sown earlier than others. Cold wet soils cause the seeds of warm season crops to rot prior to germination. Sowing times vary based on geographical location. Altitude can also alter planting times even within a given area.

Seeds with hard seed coats often germinate very slowly, unless soaked in warm water overnight. Seeds in gardens are often sown in rows, either with conventional spacing or closer. Closer spacing is usually used in vegetable gardens for beds and bands. Some seeds are sown in hills. Some seeds can be sown in a broadcast pattern. Regardless of what practice you use, seeds should not be placed too close or too dense in the beginning. Seeds should be sown in moist soil, either after a rain or watering by hose or sprinkler. A good rule of thumb is that seeds are covered to a depth twice to three times their diameter. Very tiny seeds are just surface sown and not covered, but lightly pressed in. Some tiny seeds require light for germination. A gentle surface watering after the seeds are covered is usually recommended. Soil should be maintained moist and will require watering if you see surface dryness. Once seedlings are established, some thinning might be required to bring about proper spacing.

Bedding plants are available for many popular garden vegetables and flowers. Certain attention is needed for the successful establishment of transplants. With purchasing bedding plants, look for quality plants. Short, stocky plants with good green color are best. Pale or yellow-green plants are usually stressed. Try to buy plants that have few or no flowers or buds only. Plants in full blossom usually have more difficulty with establishment and have more delayed maturity. Plug style bedding plants can be transplanted on a sunny day, unlike earlier flats or trays. Once you have decided where the bedding plants are to be placed, you dig a hole with a trowel. With woody or herbaceous perennials, you will need a spade and a much bigger hole. The hole should be bigger than the root ball of your transplant, but not deeper, as you wish the root/shoot original line to remain the same. Place the transplant in the hole and backfill with soil and then water with soluble fertilizer. Prolonged drying of recent transplants is bad.

The successful establishment and maintenance of garden plants depends on the careful and timely application of cultural skills and inputs over the garden season. Cultural practices include the maintenance of several soil parameters: nutrients, pH, organic matter, and moisture. Additional practices include pest control to minimize damage from insects and diseases and competition from weeds.

Your garden results depend on proper nutrient levels. Even with sufficient nutrients, results can be poor at too high or low soil pH. Sustainability in your garden requires that fertilizers be applied judiciously. The importance of the soil test can't be overemphasized. From a sustainable viewpoint, organic fertilizers are preferred, but one can use chemical fertilizers. The key is less is better.

Fertilizers are added during tillage, either by hand broadcast or mechanical spreader. Working the fertilizer in and dissolving some nutrients increases the fertilizer efficiency. Efficient fertilizer usage in the garden can be obtained by broadcasting the fertilizer only in areas where plants grow such as in rows, beds, or bands. Additional side-dressings of fertilizer will be needed later. The annual use of compost mulch will add nutrients slowly as it decomposes. Supplemental applications with slow-release fertilizers such as organic products, chemical slow-release granular forms, and chemical slow-release tree and shrub spikes are good choices. If you use fast-acting chemical or water-soluble fertilizers, use them sparingly. The pH level should be maintained as needed with limestone or sulfur. Proper pH assures efficient usage of soil nutrients by plants. Ground limestone can be applied in the fall or about 1–2 weeks prior to spring fertilization.

Sustainability requires that one maintain and improve organic matter levels in soil. Levels are best determined based on soil test results. Limestone will be needed to offset peat moss acidity. Aged or composted manure is a good choice. Compost is another option.

The standard watering advice is minimally 1 inch of rain weekly. If weekly rainfall is less, nonmulched areas should be watered. If wilting persists or the topsoil is dry a few inches down, watering is needed. Trees and shrubs can last longer without water than lawns and gardens. Watering in small gardens and containers is often with a hose having a sprinkler head or variable nozzle. Bigger gardens and lawns are best watered with movable, oscillating or sweeping sprinklers. Some measuring of water depth is recommended initially to gauge time needed. Sprinklers can also be piped underground with pop-ups or aboveground fixed heads such as for lawns and gardens, respectively. Surface watering is also possible with very efficient drip or ooze hoses. Watering systems can be automated with time clocks and other sensing devices.

No pesticides, even botanicals, should be sprayed on windy or hot days. Bees can be protected by not spraying when flowers are open. Natural controls are an alternative to pesticides. Some natural controls include the introduction of the lady beetle (predator) and *Bacillus thuringiensis* (bacteria causing caterpillar demise), fatty acids for weeds, soap sprays for various insects and mites, and the encouragement of spiders and preying mantises in the landscape. Birds are often encouraged, too, as good forms of pest control.

The usage of herbicides to kill weeds is not sustainable. A hoe or mechanical cultivation works well in dry soils. Mulch smothers weeds. Mulches are either synthetic or natural materials applied to the soil as a surface layer. Mulches also reduce the need for water, and ameliorate drastic changes in soil temperature.

Taller flowers and tomatoes often need staking with thin bamboo green stakes or specialized holders. Removing spent blooms with pruning shears improves appearance and promotes flowering. Pruning is generally needed only for woody plants. Annuals, herbaceous perennials, and house plants might need occasional tip pinching to promote a bushy appearance and to increase flower stalks on certain plants.

Proper harvesting of vegetables assures peak taste and continuity. Some vegetables can be gently broken off such as tomatoes. Others require cutting such as peppers and eggplants. Some vegetables, if partially harvested, will grow back such as leaf lettuce. Cutting flowers with a sharp knife reduces crushed tissues and maintains the stem's ability to take up water. Minimizing the time between cutting and plunging the stems into water/cut flower preservatives raises vase longevity.

EXERCISES

Mastery of the following questions shows that you have successfully understood the material in Chapter 12.

1. What are the differences among conventional, organic, and integrated grounds and garden management?
2. Why should you have a soil test done for your garden soil?
3. What are the benefits of landscaping?
4. Why does landscaping require both horticultural and artistic skills?
5. What are the basic requirements and steps in landscaping?
6. In what ways can you improve lawn sustainability?
7. What are the various reasons that motivate people to become gardeners?
8. What factors should be considered when one selects a site for a garden? How do these factors relate to having a good garden?
9. How does climate influence the selection of plants in a garden? How does the growing season length influence the selection of garden plants?

10. What kinds of plant and garden choices are there with vegetable gardens? With flower gardens?
11. Which types of vegetable and flower gardens have some element of sustainability?
12. Why should you plan your garden? What factors are considered in planning your garden?
13. Explain what the following practices are and how they help you get the most out of your vegetable and herb garden: extension of the growing season, intensive gardening, interplanting, vertical gardening, and succession planting.
14. What are the various ways that one can incorporate flowers into the landscape?
15. Annuals and perennials have pros and cons as flowers. What are the pluses and minuses?
16. Why should flowers be arranged in a stair-step fashion in the flower bed or border?
17. What are the light and temperature needs of flowers. How do these needs influence the selection of plants for the flower garden?
18. Why are some flowers and vegetables not available as bedding plants? What advantages do bedding plants offer? What are the pros and cons of growing your own versus buying bedding plants?
19. How can one have a garden without digging up the soil? How do you prepare the soil bed if you decide to till the garden?
20. What criteria should be used in selecting commercially produced bedding plants?
21. What are the basic steps for planting and establishing bedding plants?
22. What various practices are collectively involved in horticultural culture?
23. How should one approach fertilization, pH adjustment, and organic amendments in terms of sustainability?
24. What is the minimal weekly amount of rain needed in a garden? What clues can alert you to the need to water your garden?
25. What are the various ways one can provide water to a garden? Which one is the most sustainable in terms of water conservation?
26. How should one manage the control of pests in the garden from the sustainable perspective?
27. What two ways are used to control weeds in the garden from a sustainable perspective?
28. Why are staking, pruning, and dead-heading needed in the garden?
29. What are the basics of harvesting vegetables and flowers?

INTERNET RESOURCES

American Nursery & Landscape Association
  http://www.anla.org

Chat room and many gardening links
  http://www.gardenweb.com

Compost
  http://www.ianr.unl.edu/pubs/horticulture/g810.htm

Container gardening
  http://www.windowbox.com

Cottage Gardening Society
  http://www.alfresco.demon.co.uk/cgs/

Gardening links
  http://www.virtualgardener.com

Gardening catalog links
  http://www.qnet.com/~johnsonj/

Garden planning and plant finder
  http://www.bhg.com/home/Free-Garden-Plans.html
  http://www.backyardgardener.com/plant.html

Growing season, planting dates, and frost dates for many cities in the United States
http://www.victoryseeds.com/frost/index.html
http://www.almanac.com/garden/garden.frostchart.html
http://www.floridata.com/tracks/misc/frostdates.htm
http://www.victoryseeds.com/information/plantingdates.html

Growing vegetables in the home garden
http://www.hoptechno.com/book26.htm

Low-input landscaping
http://info.ag.uidaho.edu/Resources/PDFs/CIS1054.pdf

Low-input lawn care
http://www.extension.umn.edu/distribution/horticulture/DG7552.html

National Gardening Association
http://www.garden.org/

Numerous gardening links
http://www.gardenweb.com/vl/
http://GardeningLaunchPad.com/

Organic Gardening
http://www.organicgardening.com/

Plants for butterflies
http://www.butterflies.com

Regional gardening forums
http://www.southerngardening.com
http://forums.gardenweb.com/forums/regional/

Sustainability through xeriscaping
http://forums.gardenweb.com/forums/swest/

Xeriscaping and garden flowers
http://www.ext.colostate.edu/pubs/garden/07231.html

Zones of hardiness by zip codes
http://www.arborday.org/trees/whatzone.html

Zones of hardiness map for U.S. and individual states
http://www.usna.usda.gov/Hardzone/ushzmap.html?

RELEVANT REFERENCES

American Horticultural Society (2000). *American Horticultural Society Gardening Manual.* Dorling Kindersley Publishing, London. 424 pp.

American Horticultural Society (2001). *American Horticultural Society Practical Guides: Bulbs.* Dorling Kindersley Publishing, London. 80 pp.

American Horticultural Society (2002). *AHS Practical Guides: Growing from Seed.* DK Publishing, New York. 80 pp.

American Horticultural Society (2002). *Plants for Places.* DK Publishing, New York. 576 pp.

Anderton, Stephen (2001). *Urban Sanctuaries.* Timber Press, Portland, OR. 144 pp.

Armitage, Allan M. (2000). *Armitage's Garden Perennials.* Timber Press, Portland, OR. 324 pp.

Armitage, Allan M. (2001). *Armitage's Manual of Annuals, Biennials and Half-Hardy Perennials.* Timber Press, Portland, OR. 604 pp.

Asakawa, Bruce, and Sharon Asakawa (2000). *Bruce and Sharon Asakawa's California Gardener's Guide.* Cool Springs Press, North Franklin, TN. 401 pp.

Austin, Richard L. (2002). *Elements of Planting Design.* John Wiley & Sons, New York. 192 pp.

Benjamin, Joan, and Barbara W. Ellis (eds.) (1997). *Rodale's No-Fail Flower Garden: How to Plan, Plant and Grow a Beautiful, Easy-Care Garden.* Rodale Press, Emmaus, PA. 384 pp.

Binetti, Marianne, and Alison Beck (2000). *Perennials for Washington and Oregon.* Lone Pine Publishing, Renton, WA. 352 pp.

Bond, Rich, and Richard H. Bond (1995). *Ortho's Guide to Successful Flower Gardening.* Ortho Books, San Ramon, CA. 352 pp.

Bradley, Fern Marshall (ed.) (1996). *Gardening with Perennials: Creating Beautiful Flower Gardens for Every Part of Your Yard.* Rodale Press, Emmaus, PA. 320 pp.

Bridwell, Ferrell M. (2003). *Landscape Plants: Their Identification, Culture, and Use,* (2nd ed.). Delmar, Albany, NY. 624 pp.

Brookbank, George (1988). *Desert Gardening: Fruits and Vegetables.* Fisher Books, Tucson, AZ. 288 pp.

Brooks, John (2002). *Garden Masterclass.* DK Publishing, New York. 362 pp.

Brooks, John (2002). *Natural Landscapes.* DK Publishing, New York. 192 pp.

Brown, Claud L., and L. Katherine Kirkman (1990). *Trees of Georgia and Adjacent States.* Timber Press, Portland, OR. 384 pp.

Brown, Kathy (2000). *Bulbs for All Seasons.* Anness Publishing, London. 159 pp.

Buchanan, Rita (1997). *Vegetables (Step-By-Step Series).* Better Homes & Gardens Books, Des Moines, IA. 132 pp.

Bush-Brown, Louise, and James Bush-Brown (1996). *America's Garden Book* (rev. ed.). Macmillan, New York. 1,042 pp.

Coleman, Eliot (1999). *Four-Season Harvest: Organic Vegetables from Your Home Garden All Year Long.* Chelsea Green Publishing Company, White River Junction, VT. 212 pp.

Courtright, Gordon (1979). *Trees and Shrubs for Temperate Climates.* Timber Press, Portland, OR. 250 pp. Reprinted in 1984 and 1988.

Cullina, William, and the New England Wildflower Society (2000). *The New England Wild Flower Society Guide to Growing and Propagating Wildflowers of the United States and Canada.* Houghton Mifflin Company, Boston. 322 pp.

Cutler, Karan Davis (1997). *Burpee: The Complete Vegetable & Herb Gardener: A Guide to Growing Your Garden Organically.* Hungry Minds, New York. 416 pp.

Dirr, Michael A. (1997). *Dirr's Hardy Trees and Shrubs: An Illustrated Encyclopedia.* Timber Press, Portland, OR. 494 pp.

DiSabato-Aust, Tracy (1998). *The Well-Tended Perennial Garden: Planting and Pruning Techniques.* Timber Press, Portland, OR. 338 pp.

Druse, Ken (1992). *The Natural Shade Garden.* Random House, New York. 280 pp.

Ellis, Barbara W. (2000). *Annuals, Biennials, and Tender Perennials.* Houghton Mifflin, New York. 400 pp.

Ellis, Barbara W. (2001). *Taylor's Guide to Perennials.* Houghton Mifflin, New York. 400 pp.

Ellis, Barbara W., and Fern Marshall Bradley (eds.) (1993). *Rodale's All-New Encyclopedia of Organic Gardening: The Indispensable Resource for Every Gardener* (reprint ed.). Rodale Press, Emmaus, PA.

Foote, Leonard E., and Samuel B. Jones, Jr. (1989). *Native Shrubs and Woody Vines of the Southeast.* Timber Press, Portland, OR. 255 pp.

Forster, Roy, and Alex M. Downie (2000). *The Woodland Garden: Planting in Harmony with Nature.* Firefly Books, Ontario, Canada. 180 pp.

Foster, F. Gordon (1984). *Ferns to Know and Grow.* Timber Press, Portland, OR. 244 pp.

Foster, Raymond (1982). *Rock Garden & Alpine Plants.* David & Charles, North Pomfret, VT. 256 pp.

Garrett, J. Howard, and Howard Garrett, (2002). *Howard Garrett's Texas Organic Gardening Book* (reprint edition). Gulf Publishing, Houston, TX. 248 pp.

Gelderen, D. M. Van, and J. R. P. Hoey Smith (1996). *Conifers: The Illustrated Encyclopedia,* two volumes. Timber Press, Portland, OR. 706 pp.

Gershuny, Grace (1997). *Start with the Soil: The Organic Gardener's Guide to Improving Soil for Higher Yields, More Beautiful Flowers, and a Healthy, Easy-Care Garden.* Rodale Press, Emmaus, PA.

Grant, John A., and Carol L. Grant (1990). *Trees and Shrubs for Pacific Northwest Gardens.* Timber Press, Portland, OR. 456 pp.

Greenwood, Pippa (1998). *New Flower Gardener.* Dorling Kindersley Publishing, London. 168 pp.

Greenwood, Pippa (2002). *Garden Problem Solver.* DK Publishing, New York. 192 pp.

Grey-Wilson, Christopher, and Victoria Matthews (1997). *Gardening with Climbers.* Timber Press, Portland, OR. 160 pp.

Grissell, Eric (2001). *Insects and Gardens: In Pursuit of a Garden Ecology.* Timber Press, Portland, OR. 345 pp.

Halpin, Anne (ed.) (2001). *Sunset Northeastern Garden Book.* Sunset Books, Menlo, CA. 560 pp.

Halpin, Anne, and Tom Christopher (eds.) (1997). *Rock Gardens: The New York Botanical Garden.* Crown Publishing Group, New York. 192 pp.

Hayes, Jack (ed.) (1977). *Yearbook of Agriculture: Gardening for Food and Fun.* U.S. Department of Agriculture, Washington, DC. 392 pp.

Heger, Mike, and John Whitman (1998). *Growing Perennials in Cold Climates.* NTC Publishing Group, Lincolnwood, IL. 431 pp.

Hillier, Malcolm (1991). *The Book of Container Gardening.* Simon & Schuster, New York. 192 pp.

Hodgson, Larry (2000). *Perennials for Every Purpose: Choose the Right Plants for Your Conditions, Your Garden, and Your Taste.* Rodale Press, Emmaus, PA. 502 pp.

Holmes, Roger (ed.) (1993). *Taylor's Guide to Natural Gardening.* Houghton Mifflin, New York. 507 pp.

Holms, John P. (ed.) (2002). *The Home Depot: Flowering Gardening 1-2-3.* Meredith Books, Des Moines, IA. 256 pp.

Hudak, Joseph (2000). *Design for Gardens.* Timber Press, Portland, OR. 217 pp.

Jeavons, John (1979). *How to Grow More Vegetables: A Primer on the Life-Giving Biodynamic/French Intensive Method of Organic Horticulture* (2nd ed., classic and out of print). Ten Speed Press, Berkeley, CA. 115 pp.

Jeavons, John (1995). *How to Grow More Vegetables: Fruits, Nuts, Berries, Grains, and Other Crops* (5th ed.). Ten Speed Press, Berkeley, CA. 192 pp.

Jeavons, John, and Carol Cox (1999). *The Sustainable Vegetable Garden: A Backyard Guide to Healthy Soil and Higher Yields.* Ten Speed Press, Berkeley, CA. 128 pp.

Jones, David L. (1987). *Encyclopedia of Ferns.* Timber Press, Portland, OR. 450 pp.

Jones, Jr., Samuel B., and Leonard E. Foote (1991). *Gardening with Native Wildflowers.* Timber Press, Portland, OR. 255 pp.

Joyce, David (1996). *The Complete Container Garden.* Reader's Digest Books, Pleasantville, NY. 216 pp.

Knopf, Jim (1991). *The Xeriscape Flower Gardener: A Waterwise Guide for the Rocky Mountain Region.* Johnson Books, Boulder, CO. 192 pp.

Lancaster, Roy (2002). *Perfect Plant, Perfect Place.* DK Publishing, New York. 448 pp.

Lovejoy, Ann, and Leona Holdsworth Openshaw (1999). *Annuals.* Meredith Books, Des Moines, IA. 96 pp.

McClure, Susan (1997). *Easy-Care Perennial Gardens: Techniques and Plans for Beds and Borders You Can Grow and Enjoy (Plus 10 Beautiful Garden Designs).* Rodale Press, Emmaus, PA. 160 pp.

Mineo, Baldassare (1999). *Rock Garden Plants: A Color Encyclopedia.* Timber Press, Portland, OR. 284 pp.

Moran, Neil (1995). *North Country Gardening.* Avery Color Studios, Marquette, MI. 212 pp.

Morse, Harriet K. (1939). *Gardening in the Shade.* Timber Press, Portland, OR. 242 pp. Reprinted in 1962.

Motloch, John L. (2000). *Introduction to Landscape Design* (2nd ed.). John Wiley & Sons, New York. 384 pp.

Nardozzi, Charlie, and the National Gardening Association (eds.) (1999). *Vegetable Gardening for Dummies* (2nd ed.). Hungry Minds, New York. 1,392 pp.

Nottle, Trevor (1996). *Gardens of the Sun.* Timber Press, Portland, OR. 208 pp.

Oudolf, Piet (1999). *Designing with Plants.* Timber Press, Portland, OR. 352 pp.

Perrin, Sandra (2002). *Organic Gardening in Cold Climates* (revised edition). Mountain Press Publishing, Missoula, MT. 160 pp.

Pittenger, Dennis R. (2002). *California Master Gardener Handbook.* ANR Publication 3382, University of California, Davis. 704 pp.

Poincelot, Raymond P. (1986). *No Dig, No Weed Gardening.* Rodale Press, Emmaus, PA. 264 pp.

Poor, Janet Meakin (ed.) (1984). *Plants That Merit Attention: Trees,* volume I. Timber Press, Portland, OR. 349 pp.

Poor, Janet Meakin (ed.) (1996). *Plants That Merit Attention: Shrubs,* volume II. Timber Press, Portland, OR. 349 pp.

Reynolds, Phyllis C., and Elizabeth F. Dimon (1993). *Trees of Greater Portland.* Timber Press, Portland, OR. 216 pp.

Rice, Graham (1988). *Plants for Problem Places.* Timber Press, Portland, OR. 184 pp.

Rice, Graham (1999). *Discovering Annuals.* Timber Press, Portland, OR. 192 pp.

Riffle, Robert Lee (1998). *The Tropical Look.* Timber Press, Portland, OR. 524 pp.

Robinson, Peter (1997). *The American Horticultural Society Complete Guide to Water Gardening.* Dorling Kindersley Publishing, London. 216 pp.

Ross, Robert A., and Henrietta L. Chambers (1988). *Wildflowers of the Western Cascades.* Timber Press, Portland, OR. 204 pp.

Roth, Susan (1995). *Better Homes and Gardens Complete Guide to Flower Gardening.* Better Homes & Gardens Books, Des Moines, IA. 408 pp.

Ruggiero, Michael A., and Thomas Christopher (2000). *Annuals with Style: Design Ideas from Classic to Cutting Edge.* Taunton Press, Newtown, CT. 224 pp.

Ryan, Julie (1998). *Perennial Gardens for Texas.* University of Texas Press, Austin. 400 pp.

Schenk, George (1997). *Moss Gardening.* Timber Press, Portland, OR. 262 pp.

Severa, Joan, and Stan Stoga (eds.) (1999). *Creating a Perennial Garden in the Midwest.* Trails Media Group, Black Earth, WI. 184 pp.

Slocum, Perry D., and Peter Robinson (1996). *Water Gardening, Water Lilies and Lotuses.* Timber Press, Portland, OR. 434 pp.

Small, Ernest, and Grace Deutsch (2002). *Culinary Herbs for Short-Season Gardeners.* Mountain Press Publishing, Missoula, MT. 192 pp.

Smith, Edward C. (2000). *The Vegetable Gardener's Bible: Discover Ed's High-Yield W-O-R-D System for All North American Gardening Regions.* Storey Books, North Adams, MA. 309 pp.

Strom, Steven, and Kurt Nathan (1998). *Site Engineering for Landscape Architects* (3rd ed.). John Wiley & Sons, New York. 328 pp.

Strong, Roy (2001). *Creating Small Formal Gardens.* Conran Octopus, London.

Sullivan, C. (1997). *Drawing the Landscape* (2nd ed.). John Wiley & Sons, New York. 336 pp.

Taylor, Patrick (1996). *Gardening with Bulbs.* Timber Press, Portland, OR. 256 pp.

Thomas, Graham Stuart (1990). *Perennial Garden Plants.* Timber Press, Portland, OR. 535 pp.

Walker, Jacqueline (1996). *The Subtropical Garden.* Timber Press, Portland, OR. 176 pp.

Wasowski, Andy (2000). *Building Inside Nature's Envelope.* Oxford University Press, New York. 176 pp.

Weinstein, Gayle (1999). *Xeriscape Handbook: A How-To Guide to Natural, Resource-Wise Gardening.* Fulcrum Publishing, Golden, CO. 144 pp.

Welch, B. C. (1989). *Perennial Garden Color: For Texas and the South.* Taylor Publishing Company, Dallas, TX. 280 pp.

Westcott-Gratton, Stephen (2000). *Creating a Cottage Garden in North America.* Fulcrum Publishing, Golden, CO. 160 pp.

Wilson, Jim (1993). *Landscaping with Wildflowers: An Environmental Approach to Gardening.* Houghton Mifflin, New York. 244 pp.

Wyman, Donald (1987). *Wyman's Gardening Encyclopedia* (2nd expanded ed.). Simon & Schuster, New York. 1,221 pp.

Ziegler, Catherine (1996). *The Harmonious Garden: Color, Form, and Texture.* Timber Press, Portland, OR. 304 pp.

Chapter 13

Managing Enclosed Plant Environments

OVERVIEW

Horticulturists use several enclosed plant environments, mostly for conventional propagation and, to a lesser extent, for frost protection, growth enhancement, hardening of plants, and plant display. Some depend totally to partially on natural conditions, while others are entirely dependent on controlled environments. All involve enclosed structures requiring energy inputs. Sometimes solar energy suffices, but fossil fuel inputs are usually needed. Alternative environments are an important part of horticulture and will be considered here. Examples include cloches, cold frames, greenhouses, hotbeds, lathhouses, row covers, sun-heated pits, and tunnels. Facilities for micropropagation, high-technology laboratories, were covered in Chapter 10.

CHAPTER CONTENTS

Greenhouses
Greenhouse Environmental Control
Greenhouse Cultural Practices
Energy Conservation
Hotbeds
Cold Frames

Cloches and Hot Caps
Row Covers and Tunnels
Clear Mulch
Sun-Heated Pits
Lathhouses
The Home, Office, and Interiorscapes

OBJECTIVES

- ❁ To learn about the uses and types of greenhouses, structural differences, glazing choices, and interior systems
- ❁ To understand the components of the greenhouse environment and how they are controlled
- ❁ To look at the various aspects of cultural control in greenhouses

- ❁ To learn how to conserve energy in greenhouse operations
- ❁ To learn about other methods of environmental control such as hotbeds, cold frames, cloches, hot caps, row covers, tunnels, clear mulch, sun-heated pits, and Lathhouses
- ❁ To understand the control of home, office, and interior environments

GREENHOUSES

Greenhouses are important plant-growth structures encountered on both the commercial and hobbyist levels of horticulture. Most commercial greenhouses in the United States are used in the wholesale trade for the propagation and production of primarily ornamental crops: bedding plants (annuals and vegetables), cut flowers, and potted plants (foliage and flowering house plants, holiday potted plants). Fewer greenhouses are used for the propagation and production of container nursery stock for landscaping (herbaceous perennials, shrubs, and trees) and vegetable crops such as cucumber, lettuce, and tomato. Home greenhouses and some greenhouses in botanical gardens are usually used for plant displays. The former are also used for living areas and solar heating.

Structural Aspects

Greenhouse structural frameworks vary. The framework can consist of pressure-treated (chromatid copper arsenate or copper napthenate) or rot-resistant wood (cedar, cypress, locust, and redwood), galvanized steel, or extruded aluminum. Heat loss through a metal framework is higher than through wood, but redwood is quite costly. Aluminum alloy comes closest to being maintenance free from rust and corrosion and can be painted or anodized to improve its appearance. Galvanized steel is stronger than aluminum, so frame sizing can be smaller and lighter in weight, leading to lower cost and greater light availability. Galvanized steel is used more than aluminum with commercial greenhouse construction. Home greenhouses are usually manufactured from aluminum or redwood.

Basically, greenhouse structures are either freestanding or lean-to (attached at ridge point to another structure) types. Freestanding greenhouses are usually the choice of commercial horticulturists. Freestanding greenhouses afford maximal light entry and provide the greatest amount of space, but cost more to heat than lean-to greenhouses. The need to maximize space and light availability, however, outweighs fuel cost considerations. In larger commercial operations, several freestanding greenhouses can be joined side to side at their eaves by a structural connector called a gutter (gutter connect), eliminating coverings on sidewalls. The advantages of gutter-connected greenhouses (also known as ridge and furrow greenhouses) include the need for less land, less costly shared equipment for heating and cooling, and easier access to the combined growing area. Since greenhouses have combined water and electrical inputs, all installations should be by a qualified electrician. Ground-fault interrupters and low-voltage systems (where possible) are recommended to prevent possible electrocution.

Freestanding structures are of three basic types: Quonset, bow, and truss greenhouses. The Quonset style offers the advantages of simple, inexpensive construction. The framing consists of piping shaped like a wide, half-circle. These types are usually covered with a single or double layer of polyethylene film (see discussion on glazing later). Quonset styles are popular, especially in areas with mild to moderate winters, for start-up operations, and for rapid expansion operations.

Bow houses cost more than Quonsets, but less than truss types. Both types are even-span greenhouses in that the rafters (roof framing supports) are of equal length on the right and left sides of the ridge point. Both bow and truss types can be constructed as gutter-connected greenhouses (Figure 13–1). A wide choice of glazing exists for bow and truss greenhouses. The structural strength of bow houses is higher than that of Quonsets, but lower than that of truss houses. Bow frames are made from standard framing and not piping. A bowed (arched) or triangular (gable) roof with straight sidewalls characterizes their shape. Truss types are similar in shape to bow houses, but have additional framing support (trusses) for their roofs. Truss greenhouses are the most costly, but offer the greatest strength and the potential to hang equipment or hanging basket plants from their load-bearing trusses. Black-out curtains for photoperiod control, growth lights, humidifiers, misting units, overhead irrigation systems, and thermal blankets for heat retention can be suspended from the trusses.

Lean-to greenhouses (Figure 13–2) are more apt to be the choice of homeowners, who enjoy being able to walk directly into the greenhouse from their homes. Another plus is the

FIGURE 13–1 This greenhouse complex is an example of gutter-connected, arch-style bow greenhouses. The glazing consists of single clear polyethylene at the ends. The roof and halfway down the sides have been covered with a layer of milky polyethylene. (Author photograph.)

FIGURE 13–2 This lean-to greenhouse is characteristic of types attached to homes. The framing is aluminum and the glazing is single-layer glass. This greenhouse is not energy efficient. It was built in the mid-1970s before energy efficiency was a national concern. (Author photograph.)

ability to view the plantings from the living or dining area through sliding glass doors. If used as passive solar energy heating sources, the greenhouse must be of the attached type. Lean-to greenhouses are often seen as part of the serving area in upscale restaurants, being quite pleasant with hanging baskets and some potted tropical foliage plants. Lean-to greenhouses are popular also as sunroom additions.

Floors in greenhouses are usually made of standard or porous concrete and, to a lesser extent, dirt or gravel. Standard concrete is the choice where heavy loads are expected such as areas of heavy traffic or heavy equipment use. It is not recommended for aisles between benches, because water will puddle unless drains have been installed. Porous concrete can be used for aisles between benches because it does drain. However, porous concrete can only be used in areas exposed to light vehicle traffic and people. Porous concrete is also not suggested for propagation areas, because the fine pores can become clogged from water-containing growing media that drains onto the floor. Gravel or dirt floors are inexpensive and sometimes used in home greenhouses (usually gravel), but can't be properly cleaned and disinfected. As such, gravel or dirt floors are not found often in commercial operations. If cost is a factor, areas under benches can be gravel-covered dirt and aisles concrete.

Glazing

Greenhouse covering materials (glazing) consist of glass, fiberglass, and various plastics. Glass has the highest light transmission capability (Table 13–1) and the greatest initial cost. However, its long-lasting qualities and long-term depreciation benefit are advantages, especially in areas with low-light winter conditions. Esthetically, glass is a good choice where greenhouses are primarily for plant display, such as in botanical gardens and private homes. Glass greenhouses are not airtight. As such, they have fewer problems with excessive humidity and roof drips on plants, thus posing fewer disease control requirements. Single-layer glass houses are energy inefficient, but can be improved easily (see section on Energy Conservation). As of now, glass is no longer the dominant glazing material in greenhouse construction.

Fiberglass is often used to cover greenhouses. Panels are either flat or corrugated and fabricated in the clear form specifically for greenhouses. Fiberglass does not last as long as glass and, even when new, transmits less light than glass. Light transmission through fiberglass decreases with age, because of yellowing. Even with special coatings to reduce yellowing, it does not last as long as glass. Another disadvantage is that its translucency reduces views from within and without. Fiberglass is also highly flammable. It has been displaced in popularity by the newer plastics.

Polyethylene film is inexpensive. Because of the lightweight nature of polyethylene, structural support requirements are minimal, adding further to cost reductions. Polyethylene transmits less light than glass, about 85% compared with 90% or slightly higher for quality glass. One disadvantage is that it breaks down on exposure to ultraviolet light, which usually necessitates replacement once a season. The addition of ultraviolet inhibitors results in a plastic film that can last two to four seasons. In the South, with its high ultraviolet levels, less durability is the rule. Roughly half the greenhouses in the United States use polyethylene covering (Figure 13–3), mostly with a double layer (see Energy Conservation section).

Double-layered acrylic and polycarbonate plastics are lightweight, almost as transparent to light as glass, energy efficient, nonyellowing, and strong, making them a very popular choice. Acrylic types are sold as Lucite, Plexiglas, and Exolite. Polycarbonate types are marketed as Cyroflex, Dynaglas, Lexan, and Polygal. Polycarbonate plastics are much less flammable and more easily worked than acrylic types, but can yellow after 10 years.

Older single-layer and newer double-glazed glass-covered houses exist in the trade, but the trend today in commercial ventures is toward the double-layer acrylic and polycarbonate types and polyethylene film. These types are energy efficient and require lower capital investment and construction costs relative to double-glazed glass greenhouses. Homeowners are increasingly attracted to the double-layer acrylic and polycarbonate types for similar reasons, although the esthetics of a quality glass greenhouse still appeals to some.

TABLE 13–1 *Sunlight Transmission through Greenhouse Glazing Materials**

Covering Material	%
Acrylic, double wall	85–92
Fiberglass	80–92
Glass, single layer	92
Glass, double layer	80–85
Polycarbonate, double wall	80–88
Polyethylene, single layer	85
Polyethylene, double layer	70–75

*Numbers vary based on thickness and quality of material.

Bench and Floor Systems

Wooden benches in greenhouses are traditional, but this tradition has experienced serious inroads by other materials in recent years. High-quality, rot-resistant wood and even pressure-treated or preservative-treated wood (chemical treatments compatible with plants must be used) have become expensive. Benches constructed of either metal or strong, rigid plastics have become common.

In the past, benches were spaced primarily with aisles designed for walking purposes. Benches were often permanently fixed in place. Today placement depends on what type of machinery is used to lift and move plants. Today's placement has more to do with making room for electric carts and lift trucks than people walking. Rolling benches are also used and offer considerable flexibility. They can be placed with various aisle spacings to accommodate changes of equipment. They can be rolled together to open up wide aisles for major equipment and returned to position later. Rolling benches also increase usable plant space by 25% to 30% in greenhouses because they can be tightly packed together and only opened up as needed. Then the open area can be moved elsewhere when you are done with a particular area.

Some of today's greenhouses have no benches; they use floor systems instead. These systems use a floor ebb and flood system (flood floor) that is automated. Concrete floors must be carefully laid, leveled, and gently sloped to ensure even flooding, no puddles, and collection of runoff at one end. Flood floors supply both water and fertilizer. Heating pipes and irrigation lines are buried in the ground. These floor systems have found use with plug production, rooting of cuttings, and the growing of tissue-cultured plants. Floor systems increase available growing space relative to fixed benches by about 35%. Floor systems are also environmentally preferred in that the collected runoff can be recycled. Collecting the runoff also ensures that pesticides and fertilizer nutrients don't enter the groundwater. Their disadvantage is seen in the increased and uncomfortable labor required to bend continuously to care and move plants. Another disadvantage is that diseases can spread rapidly in floor systems, making disease prevention critical.

GREENHOUSE ENVIRONMENTAL CONTROL

Environmental control within a greenhouse involves the regulation of temperature (heating and cooling), ventilation, light, and humidity. With today's sophisticated computerized controls, these parameters and even more environmental aspects are easily controlled. Other aspects include automatic watering and fertilization by fertigation, carbon dioxide enrichment, thermal sheet operation, and photoperiodic control with lights and blackout sheets.

Temperature

Temperature control is dependent on regulation of heating, ventilation, and cooling. Some of the heat is supplied by solar radiation, which passes through the transparent glass and warms everything it contacts. Some of the absorbed heat is reradiated from the warmed objects as longer wavelengths, that is, infrared. These longer wavelengths pass back through the glass much less easily than the incoming shorter wavelengths, but are absorbed by carbon dioxide and water vapor in the greenhouse. The enclosed area also has minimal air mixing, which contributes further to warming. Fiberglass is also effective, but polyethylene film is much less so. However, the film of water frequently found on the plastic film does absorb the infrared, so a warming effect can occur even with polyethylene. Double-layer polyethylene greenhouses actually trap more heat than glass types because they are more airtight. The double-wall plastics are more effective than glass at retaining heat. The heat trapped by the greenhouse depends on convection currents or fans for distribution.

However, on cloudy days and winter days, the heat produced by solar radiation is not sufficient to maintain the desired temperature. Some form of generated heat then becomes necessary. This heat source must have a constant high output with uniform distribution, because heat losses through the glass or plastic and framework by conduction and through leaks are rapid. Cold spots may develop because of factors such as cold winds blowing against a wall that increase heat loss in that area compared to other areas. Heat loss can be calculated on the basis of surface area and the rate of thermal transmission, which varies according to the material involved. Heat loss calculations for the structure, the desired running temperature of the greenhouse, heat losses if any by the heating system, expected external winds and temperatures, and the desired safety margin are factors that must be taken into account when determining the number of Btus that must be produced by the heating system.

Numerous types of heating systems (Figure 13–4) and fuels are available for greenhouses. The cost of installation and cost and availability of fuel must be considered. Hot water or steam is used in pipes fitted with fins to radiate heat and start convection currents; these systems are usually dependent on oil-, gas-, or coal-fired boilers. Circulating fans might be required for uniform heat distribution. Forced hot air can be distributed through ductwork. Oil-, gas-, or coal-fired furnaces can be used to produce hot air, or gas or electric fan-forced heaters can be utilized. Hot air can be blown into large polyethylene tubes with holes that produce turbulent outflows of hot air, resulting in good mixing effects. These same tubes can be used also for distribution of fresh air in winter without harming plants through direct cold air contact (see section on Ventilation). During the summer, these tubes can be part of the cooling system and provide fan-forced cooler, outdoor air.

More recent heating technologies include infrared heating and root zone heating. Infrared heating involves heating of plants, but not the air mass or soil below the plants. The main advantage of infrared heating is lower energy use relative to more conventional heating systems. Commercial greenhouse systems use gas-fired radiant tubes housed in reflective shielding. Several of these radiant tubes run the length of the greenhouse in the ridge area usually 6 to 12 feet above the plants. Since the soil is not heated, some changes in cultural practices might be needed such as warmer water for irrigation. Small natural gas- or propane-fired infrared heaters are available for hobby and home greenhouses. Electrical quartz infrared heaters are also used in home greenhouses.

Some problems can occur with infrared heating. Greenhouse crops with deep foliage canopies experience uneven heating. Foliage at the top intercepts most of the infrared heat, while the lower leaves and soil remain cooler. Certain crops such as cucumbers, large foliage house plants, roses, and tomatoes experience some uneven heating problems. Greenhouses also must have adequate height to use infrared heating, because these units operate at high temperatures.

Root zone heating is placed below the plants and heats the root mass, soil, and tops, but not the air mass. Energy costs are considerably lower for this type of heating. Pipes can be

(A)   (B)

FIGURE 13–4   (A) This greenhouse is heated by a ceiling-suspended, forced hot air gas heater. Note the steam jets to the upper left of the heater. These units are used to maintain the humidity level. The dark area above the steam jets is a duct that leads from a wet-pad evaporative cooler on the roof, which helps cool the greenhouse during the summer. The actual greenhouse is to the right and is an attached, rooftop greenhouse used for research purposes. (B) This greenhouse is also an attached greenhouse. You are looking at the foundation part. An oil-fired furnace provides hot water to the double layer of four-inch finned tubing that provides radiated heat. Large, double-finned units are needed to provide rapid and sufficient heating during the winter. (Author photograph.)

placed on the soil surface or below the soil surface with greenhouses where plants are grown directly in soil beds at floor level. With benches, pipes can be placed on the benches. Hot water is pumped through the pipes with a small circulator pump. This heating, essentially bottom heating, is excellent for germination of seeds, rooting of cuttings, and growth of plugs and liner plants. A significant part of the propagation industry now uses root zone heating.

All systems can be controlled with thermostats or, even better, microprocessors (see Automation section), which are often hooked into a temperature alarm system to warn of failure in the heating system. A backup system, even just a simple kerosene heater, can be useful in emergencies or in times of fuel shortages. Backup generators offer more insurance. Fans are often use to ensure even mixing and distribution of heat. These same fans also help ensure even cooling.

Ventilation

Heating makes maximal use of solar radiation and secondary use of conventional heating. Conventional heating turns off when the environment reaches the set temperature, but solar radiation cannot be turned off as readily. Temperatures in an enclosed greenhouse can rise to more than 40°C (100°F) in a short time on a sunny day. The simplest way to control this temperature rise is by ventilation. Sections of the greenhouse, usually along the ridge and sometimes on the sides, can be opened by a motorized drive (Figure 13–5). The motor is thermostatically controlled and set to open the windows at a temperature several degrees above the set heat temperature. A more recent unit designed for use in smaller, home greenhouses uses a heat-activated expanding chemical to operate the ventilation windows. This mechanism saves on electrical costs and is secure from power failures.

Ventilation systems with opening windows depend on the rising of hot air and the entry of cool air to establish convection currents. Such systems can be improved with the use

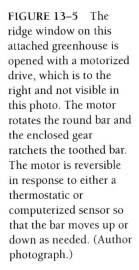

FIGURE 13–5 The ridge window on this attached greenhouse is opened with a motorized drive, which is to the right and not visible in this photo. The motor rotates the round bar and the enclosed gear ratchets the toothed bar. The motor is reversible in response to either a thermostatic or computerized sensor so that the bar moves up or down as needed. (Author photograph.)

of air intakes and exhaust fans to exchange the air at a more rapid rate. Another type of ventilation makes use of an air intake fan and motorized exhaust shutter, which is regulated by a thermostat. A perforated polyethylene convection tube (Figure 13–6) can be connected to the intake to improve air distribution, provide cooling, and reduce plant damage from cold drafts in winter. In addition to heat control, ventilation also serves the useful purpose of replacing air that is depleted of carbon dioxide by plants.

Cooling

During a hot summer, ventilation can prove inadequate to maintain the desired operating temperature. Shading of the greenhouse will provide additional cooling, as well as a reduction (usually around 60% to 70%) of summer light levels, which are often too intense for greenhouse plants. Shading (Figure 13–7) can be provided by spraying shading compounds on the glass, or by the use of roller blinds, green vinyl translucent plastic, or vinyl mesh (Saran). Air can also be exhausted from the greenhouse with air intakes at one end and a large exhaust fan (see Figure 13–6B) at the other end. Even under these conditions, ventilation cannot reduce the interior temperature below that found outside.

Alternate means of cooling might be required, depending on summer temperature conditions. Conventional air conditioning is too costly, but air cooling with fans and wet pads provides a reasonable alternative. Wet-pad evaporative coolers can produce an interior temperature equivalent to an outdoor temperature measured in the shade. High-pressure misting or fog systems are also used to reduce temperatures by evaporative cooling. These systems are used less, because costs are higher than those of wet-pad evaporative coolers and performance is poor in areas having high relative humidity such as the Deep South.

A newer alternative for cooling uses a retractable roof on the greenhouse and retractable benches. During hot weather, the roof can be opened. Under very hot conditions, the benches can be moved completely outdoors. These types of greenhouses save cooling energy costs and are popular for seed and cutting propagation, especially plug production. When night comes or bad weather or frost threatens, the roof can be retracted and benches, if outdoors, can be returned to the greenhouse. Plants produced in these greenhouses need less hardening, because they are exposed more directly to sun, wind, and outdoor temperature variations. In fact, hardening is the main advantage of retractable roofs, because they save on hardening labor and space because crops don't have to be moved to another location. The

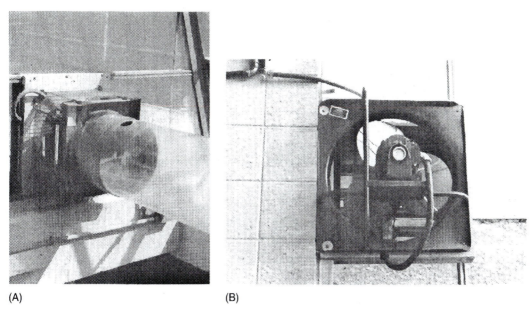

(A)   (B)

FIGURE 13–6 (A) The fan-jet shown here with the perforated polyethylene tube is very useful for providing ventilation for greenhouse plants during the winter. The outer part has a motorized shutter that is activated either thermostatically or by a computer-controlled sensor that also activates the intake fan. The vortex mixing action as cold air rushes from the numerous holes along the tube causes the cold air to mix with the warmer greenhouse air. Plants below then receive warmed air, rather than harmful, direct blasts of winter air. If desired, this perforated tube can also be used to provide hot air from the heating system. (B) When a fan jet is used in conjunction with opening ridge windows (see previous figure) and the exhaust fan shown here, this three-phase system functions both as effective ventilation and cooling system. (Author photographs.)

FIGURE 13–7 Green vinyl translucent plastic is shown here on the end windows and the first, third, fourth, and sixth roof windows. The plastic adheres to the inside glass by a thin water film (capillary action). It is available in different light-reducing grades. This option is more attractive for a home greenhouse than the sprayed-on shading compounds used on commercial greenhouses. Note the end louvered window for supplementing ventilation during the warmer weather. Saran vinyl green mesh is also attractive and easy to apply. It has become popular for both commercial and home greenhouses. (Author photograph.)

energy savings might possibly be offset by slight increases in heating, because retractable roof greenhouses don't seal quite as tightly as conventional greenhouses.

Humidity

Other environmental aspects such as humidity can also be controlled in a greenhouse. Steam humidifiers (see Figure 13–4A) or cool-water humidifiers that operate on the aerosol principle can supply humidity. The latter are much more popular than steam humidifiers, holdovers from the time when greenhouses were heated by steam. Most are heated today with hot water or hot air. Automation can be provided with a humidistat. Ventilation systems can pose control problems, because humidity losses occur with open vents.

Light

Several factors determine the amount of visible radiant energy (light) in a greenhouse. These include variations of the sun's position with time of day and season, the location and orientation of the greenhouse, the slope of the roof, type of glazing, the shape of the greenhouse, and the amount of cloud cover. Because of these variables, the design of a greenhouse is at best a compromise in terms of maximizing the amount of visible radiant energy entering the greenhouse (radiant flux, commonly expressed as watts).

Radiant Flux.   On a daily basis, the available radiant flux is highest at noon, with the start of the morning and the end of the afternoon being lowest. With seasonal changes, the higher the sun is above the horizon, the greater the radiant flux. The minimum occurs during winter when the winter sun's low angle combined with short day-length results in low radiant flux. These factors are also functions of latitude; southern latitudes have a greater radiant flux than northern latitudes in the United States during winter. Of the total radiant flux energy, only the number of quanta or photons (quantum flux) entering the greenhouse in the 400- to 700-nm wavelength range (PAR or photosynthetically active radiation) is important for initiating photosynthesis (see Chapter 6).

Roof Slope.   The photosynthetically useful light within the greenhouse is determined by the angle at which the light strikes the glass. The ideal angle of the glass to maximize transmission of light is a function of latitude and season. Greenhouse roof slope is based on a compromise. The goal is to provide a high (but not maximal) level of transmission, yet still allow snow to slide off and condensation to run off rather than drip on plants. The need for compromise results from the impracticality of a variably sloped roof and the fact that the ideal roof slope is often such as to require an excessively high ridge and large area. These roof slopes are usually 32° for greenhouses up to 25 feet wide and 26° for those over 25 feet.

Shape.   The ideal shape for maximal light transmission as a function of hour of the day and season is the hemispherical dome. This shape is impractical for commercial operations. The Quonset structure, seen frequently with houses made of pipe half-hoops and polyethylene covering, is almost as good a shape as the dome. The other shapes seen in greenhouse construction are not quite as efficient in terms of light transmission.

Orientation.   The directional alignment of the long axis of the greenhouse (orientation) should be such as to maximize radiant flux during the winter. At latitudes above 40°N, an east-to-west orientation is best (long axis faces south). Below this latitude, as the distance to the equator decreases, a north-to-south orientation might be preferred. In terms of structure shape versus orientation, the shape has more potential for maximizing interior light.

Light Reduction.   During the summer, the radiant flux can be too high for most greenhouse plants. Light reduction can be provided with shading, as discussed previously, and with ventilation. Roller shades can be automated through a motorized drive and a photoelectric sensor. During the winter, increases in light levels can be provided with indoor light sources, which can also be automated with photoelectric sensors.

Photoperiod Control. For certain crops, flowering is subject to photoperiodism (see Chapter 6); thus it may become desirable to control the photoperiod. Control of photoperiod could involve lengthening the day with indoor light, interruption of the night period with indoor light, or shortening the day with black shade cloth. Certain horticultural crops of high commercial value such as poinsettias (*Euphorbia pulcherrima*) and chrysanthemums are routinely brought to flowering for holidays or seasonal display through control of the photoperiod.

Chrysanthemums, a short-day plant, experience optimal development of flower buds when the day length is about 12.5 hours. If the natural day is longer, the plants are covered with shade cloth, such as black polyethylene, black polypropylene, or black sateen cloth (64 by 104 mesh or smaller). No more than 2 to 3 foot-candles should penetrate the shade cloth. Temperatures must be watched, since heat buildup is possible, especially during the summer. Some newer materials, such as black vinyl or polyester aluminized on one side, are available for dual purposes: shading and heat retention as a thermal blanket during the winter (see later discussion on fuel conservation).

The day can be lengthened artificially to promote flowering of long-day plants or to maintain the vegetative development of short-day plants. Since low light intensities are effective, it is economical. Incandescent or pink fluorescent bulbs, rich in red radiation, are quite effective for this purpose. Alternately, the night period can be interspersed with cyclic light for 20% of the time every 30 minutes for the duration of the additional hours of daylight needed. Brief light flashes of 4 seconds per minute are also effective. Since the overall light period is reduced, so are the operating costs.

Automation

In earlier greenhouses and still in home greenhouses, thermostats were used to control the heating system, ventilation, and cooling. A humidistat was used to regulate humidity. These systems worked reasonably well, but had their drawbacks. For example, sometimes conflicts arose among heating, cooling, and humidification that these simple controls could not resolve. Thermostats often experience temperature overrides, in that they can't anticipate at what point prior to the set temperature to shut down. Eventually analog systems evolved with electronic sensors that gathered information, enabling the analog system to resolve simple conflicts through various electronic logic systems.

More recent computer automation, especially as first practiced in European commercial greenhouses, came about. These systems are essentially miniature computers in that they use a microprocessor and numerous sensors and can be handled from one preprogrammed console. Various sensors measure temperature, light levels, humidity, and wind speed. Complex situations are easily handled by these systems, as they respond to environmental needs. These systems can also be programmed to handle automatic watering and fertilization by fertigation, carbon dioxide enrichment, thermal sheet operation, photoperiodic control with lights and blackout sheets, and even temperature control in propagating beds. While expensive, these systems save energy and labor.

Some advanced automation is now available for home greenhouses. Although not nearly as complex as commercial operations, it is quite helpful. For example, an automatic control system is available to tie in a wet-pad evaporative cooler and the heating system. It can maintain a set temperature with no drift or overlap of heating and cooling operations.

Air Pollutants

Pollutants in the air (see Chapter 6) are as serious a concern for those managing a greenhouse as they are for those managing horticultural production outdoors. Crops grown in greenhouses in urban or industrial sites can become affected enough to force relocation of the greenhouse or a change to crops less sensitive to air pollution. Some air pollution can even arise from greenhouse operations. Sulfur dioxide and ethylene can be produced from oil- or coal-fired boilers used to heat the greenhouse. Other pollutants may be produced during careless or excessive use of wood preservatives, herbicides, insecticides (especially fumigants), or chemicals used for pasteurization. Mercury vapors may arise from certain paints, which should be avoided in a greenhouse.

GREENHOUSE CULTURAL PRACTICES

Greenhouse success depends on using good growing media and providing water and fertilizer according to plant needs, but in a way that minimizes environmental harm. Media for the propagation and growing-on of plants in greenhouses was covered in Chapter 7. The use of water, fertilizer, and pesticides in greenhouses must be carefully managed so that runoff from greenhouses doesn't pollute nearby water bodies or groundwater. Best management practices for water and fertilizer usage were covered in Chapter 7. Pest control in greenhouses will be covered in Chapter 15, but these aspects are treated briefly here.

Growing Media

Media suitable for growing plants and techniques for pasteurization were described in Chapter 7. Greenhouse growing media consist of soilless mixtures of materials such as peat moss and vermiculite or perlite. The various soil tests and tissue analyses described in Chapter 11 for outdoor cultivation should also be used to monitor growing media and greenhouse plants, respectively. These media must have their nutrient status checked routinely with soil tests. Tissue analyses are useful to confirm suspected nutrient deficiencies or to determine the levels of nutrients needed by greenhouse crops.

The pH level should be monitored and maintained for best growth; ground limestone and sulfur are generally used to raise and lower pH in growing media, respectively. For most flower crops in the greenhouse, a pH of 5.8 to 7.0 is acceptable; for vegetables, 5.8 to 7.4; and for acid-loving crops (azalea, hydrangea, gardenia), 5.0 to 5.5.

Soluble-salt levels should be checked routinely, especially given the small volumes of growing media and the continuous use of fertilizer. Salts accumulate as a white surface crust on the soil surface. Since excessive salinity can cause foliage burn, retarded growth, and even death, it is wise to avoid excessive fertilizer applications, to monitor soluble-salt levels, and to use leaching if necessary to reduce salt levels.

Watering

Handheld hoses equipped with sprinkler nozzles to reduce the force of the water facilitate watering for established plants. Fogging nozzles can be used for recently planted seeds and seedlings. Watering can also be automated for the various irrigation systems used in greenhouses: spray irrigation, trickle or drip irrigation, ooze irrigation, ebb and flood floor systems, or capillary mat irrigation (Figure 13–8). Irrigation systems can be controlled with solenoid valves and time clocks or turned on manually when needed. Drip and capillary mat irrigation are primarily used for potted foliage and flowering plants; ooze and spray irrigation are used for flower beds in greenhouses. Other uses for these systems were discussed in Chapter 11. Water temperature should be watched, especially in the winter, since plant development can be set back (Chapter 6). Wetting of foliage should be minimized or, at the least, foliage should be able to dry off before night to minimize disease incidence. Fewer fungicides and more sustainability result.

Fertilization

Application of nutrients is mostly in solution form. Numerous commercial forms of soluble fertilizers are available for liquid feeding. These formulations vary extensively in dry

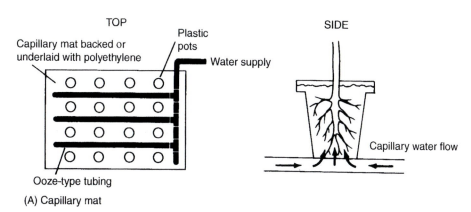

TOP

Capillary mat backed or underlaid with polyethylene

Plastic pots

Water supply

Ooze-type tubing

(A) Capillary mat

SIDE

Capillary water flow

TOP

Header

Water supply

Spaghetti-type tube   Pots

(B) Trickle or Drip

Bulk water flow   Close-up

Raised drainage

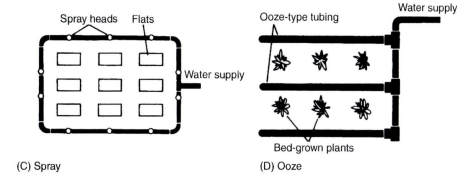

Spray heads   Flats

Water supply

(C) Spray

Ooze-type tubing   Water supply

Bed-grown plants

(D) Ooze

FIGURE 13–8 Various types of automated watering systems used in greenhouses. The spaghetti-type drip irrigation is also popular for watering container plants outdoors in the nursery. (From Raymond P. Poincelot, *Horticulture: Principles & Practical Applications*, 1980. Reprinted by permission of Prentice-Hall, Upper Saddle River, New Jersey.)

weight percentages of N, P_2O_5, and K_2O: 20-20-20, 30-10-10, 25-0-25, 10-6-4, and 9-45-15 are examples. Choices are made on the basis of horticultural crop and developmental stage. Many fertilizers also contain trace elements. The dilution and application schedules vary, but a dilute feeding during every watering or a stronger solution at weekly or biweekly intervals is common.

Soluble fertilizers can also be prepared to any desired nutrient analysis by the grower. Chemicals used include ammonium nitrate or urea (must be biuret free), monoammonium phosphate, potassium nitrate, calcium nitrate, ammonium sulfate, and soluble trace elements if desired. Formulations are often richest in nitrogen, which is the most readily exhausted nutrient.

Liquid feeding is done during watering or sometimes through foliar sprays (see Chapters 7 and 11). For large-scale fertilization, a liquid fertilizer is usually injected into a watering system (fertigation) by the use of a proportioning valve. Controlled-release fertilizers greatly reduce the frequency of application and the risk of injury from excessive applications (Figure 13–9). However, they are more costly than soluble fertilizers. Either a membranous or resinous coating around granules of soluble fertilizer regulates the slow release of nutrients over a 3- to 9-month period. Another approach is to use nutrients in chemical forms that possess reduced solubility or organic forms that require microbial breakdown to release nutrients in available form.

FIGURE 13–9
Controlled-release fertilizer beads can provide long-term nutrient release. Advantages include less labor for applications, a more continuous feeding, and less pollution problems. The disadvantage is higher cost for fertilizer. (Author photograph.)

Pest Control

Pest control in a greenhouse is more difficult than in the field. The greenhouse is an enclosed environment. Pesticides and enclosed environments constitute a dangerous pair. Chemicals must be used with considerable caution, a high regard for safety must exist, and protective measures, including respirators, might be required. From the sustainable perspective, the use of pesticides in greenhouses must be minimal. The best approach is either an organic system with reliance on natural controls or an integrated pest management (IPM) strategy with major reliance on natural controls and minor usage of pesticides. Natural controls and IPM for the greenhouse will be covered in detail in Chapter 15.

Carbon Dioxide Enrichment

Another form of fertilization in an indirect sense is the enrichment of the air with carbon dioxide. Carbon dioxide, a substrate for photosynthesis that accounts for increases in dry weight, is normally present in the atmosphere at 0.033% or 330 ppm. In an enclosed environment with poor air circulation, such as a greenhouse, levels of carbon dioxide can be locally depleted in the immediate vicinity of the plant. Therefore, a fresh air supply and good circulation are important for maintaining normal levels of carbon dioxide. Increasing the level of carbon dioxide from 1,000 up to 2,400 ppm can, under some conditions, produce up to a 200% increase in photosynthesis. This in turn increases the dry weight production of the plants.

This increase can be realized for some, but not all, plants. Temperatures must be high (29.5°C or 85°F), and light levels must be saturating for photosynthesis. As the carbon dioxide levels increase, so does the level of light needed for saturation of photosynthesis. Plant leaves must not be shaded, so plants should not be crowded. The greenhouse must be well sealed.

Carbon dioxide may be derived from dry ice or compressed gas cylinders or by burning natural gas, propane, or fuel oil with fresh air. The latter fuels must be pure to prevent the release of toxic products, and burners must be properly adjusted or harmful amounts of ethylene and carbon monoxide can be produced. The gain in dry weight production must be weighed against the costs of the carbon dioxide supply and its control and distribution. The use of fuel systems tends to offset the costs somewhat as the heat may be used to warm the greenhouse. When the weather warms to temperatures where ventilation is needed for temperature control, the use of carbon dioxide becomes unrealistic.

ENERGY CONSERVATION

Greenhouse operators, whether they are hobbyists or commercial types, face an increasingly serious problem: the cost of energy. It appears that the availability and cost of fuel for heating will continue to get worse, but some steps can be taken to reduce the severity of the problem. In the 1970s, oil price increases put a number of greenhouse operations out of business. In the winter of 2000–2001, natural gas prices rose considerably, causing operating losses for many greenhouses.

Fuels must be examined closely. First, determine your fuel costs on the basis of cost per 100,000 Btus. Another important fuel consideration is the heating cost per square foot over a year; to determine this, divide the total square feet of the greenhouse into the fuel cost per year. These two steps will provide an idea of the heating value per dollar and a cost basis for evaluating changes that reduce energy costs. Economy must be weighed against present and future availability. Heating systems that can burn an alternative fuel appear to be a good choice. One point to consider is where horticulture will stand on the priority scale if fuel is rationed. The Federal Energy Regulation Commission listing places greenhouse operations after hospitals, homes, and small businesses but ahead of big industry. Finally, some of the new heating systems are more efficient than older, existing systems. The installation cost versus projected fuel costs might make a new installation feasible. This reality also suggests placing higher priority on designs that maximize use of solar energy and proper insulation.

Insulation

Insulation offers considerable potential for the reduction of fuel consumption. The main thrust with most existing greenhouses is to utilize an insulation system that forms a dead air space to reduce heat loss. Unfortunately, the production of new glass or fiberglass greenhouses with double glazing that is hermetically sealed is of limited feasibility because of the cost. Many greenhouses today are instead being produced with double glazing of either acrylic or polycarbonate plastics, or fiberglass with inflatable double polyethylene. These materials can be expected to reduce heating costs relative to single glass by 35% to 40%.

A double layer of polyethylene film (4 to 6 mil) can be put over the exterior of older glass or fiberglass greenhouses in a manner that allows normal operation of vents. It is kept inflated by a squirrel cage blower to produce a double bubble over the structure. Experiments indicate a possible reduction in heat loss from 33% to 50% depending on how leaky the original structure was. Up to a 20% light reduction is experienced, which might cause problems with crops such as cucumbers, roses, and tomatoes that require high light levels. This light reduction is especially critical in areas where winter light intensities are close to limiting production. Light loss can be offset with fluorescent or high-intensity discharge lamps, but their cost and energy inputs need to be weighed against the heat energy savings. Given that polyethylene deteriorates on exposure and light losses increase, the double polyethylene should be replaced annually.

A single polyethylene sheet causes less light loss, but savings are less (20% to 30%) if a single layer is attached by stapling it on the inside or outside of the house onto the wooden glazing bars. If the glazing bars are aluminum or steel, a bubble pack plastic is available that can be applied to wet glass and held in place by capillary action. Bubble pack plastic reduces heating needs by 25% and reduces light by about 12%.

Another form of insulation is the retractable heat sheet or movable thermal curtain, sometimes also called a thermal blanket (Figure 13–10). These thermal curtains can cut fuel costs from 20% to 50%, depending on the curtain material, tightness of the greenhouse, and the outside temperature. The heat sheet is basically an insulating blanket of nonporous polyethylene or synthetic material. Some have black bottoms that trap heat and return it by emission. These sheets can also double as blackout sheets for photoperiodic control of crops requiring increased darkness to flower such as chrysanthemums. The disadvantage with black thermal sheets is reduced light below, because black absorbs light and reduces

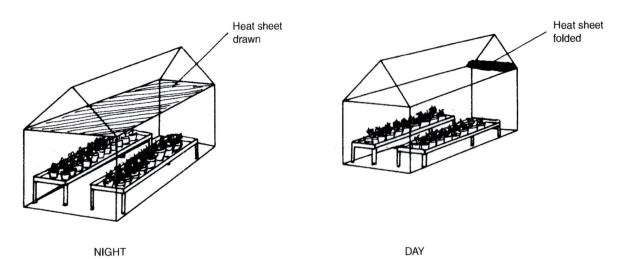

NIGHT   DAY

FIGURE 13–10   At night the thermal curtain is drawn across the greenhouse, usually with a motorized drive. During the day, the thermal curtain is folded, preferably at the north end of the greenhouse to minimize shading of plants. Posts, hanging baskets, and overhead environmental maintenance equipment can preclude the use of thermal curtains. (From Raymond P. Poincelot, *Horticulture: Principles & Practical Applications,* 1980. Reprinted by permission of Prentice-Hall, Upper Saddle River, New Jersey.)

light reflection. Aluminum bottoms increase available light, but aren't as effective at trapping heat. Some sheets have strips of aluminum sewn in among the black as a compromise. Heat sheets can be drawn by hand or by an automated, timed motor across the greenhouse at gutter height. These sheets only work in greenhouses that don't have purlin posts or heaters and pipes below gutter level.

The basic idea is to trap heat under the sheet, and thus to prevent the waste of heat on overhead space that does not contain plants. Care must be taken at temperatures below $-18°C$ (0°F), since heavy frost accumulations can occur and melt so slowly during the day that available light levels becomes a problem. This problem is preventable with a tight-fitting heat sheet that prevents moisture leakage from under the sheet. In addition, heat sheets are not used during heavy snowfalls, since the melting of snow may be slowed enough such that structural collapse from snow weight becomes possible. When not in use, the heat sheet must be tightly folded to prevent light losses caused by large shadows. At present, their use is increasing in some commercial operations.

Other Steps

In older glass houses, injection of a sealant between overlapping glass panes (lapsealing) can result in fuel savings of about 25% to 30%. Other minor (but effective collectively) insulation improvements include recaulking, north wall insulation with fiberglass aluminum-faced insulation (coldest wall and least light), installation of reflector-faced insulation behind fin tubing on perimeter foundation walls, and insulation of heating pipes leading to the greenhouse. Windbreaks can reduce heat losses, which double in exposed areas having winds of 0 to 15 miles per hour. During the zeal to conserve energy, remember that the greenhouse must not be sealed such that it becomes completely airtight, since oxygen is needed for people, plants, and open flames in heating and CO_2 is also needed for plants.

Split Night Temperatures

Another promising approach to fuel conservation is split nighttime greenhouse temperatures. As seedlings, many plants require a uniform high temperature, but as they develop, their requirement becomes thermoperiodic, that is, optimal night temperatures are lower than day temperatures. For how long is the optimal temperature required at night? Can the nighttime physiological processes be completed in a few hours at warm night temperatures and be followed by colder temperatures without affecting development? The answer to the last question appears to be yes, based on research. Development of chrysanthemums, marigolds, petunias, and Easter lilies was normal when the normal optimal night temperature of 15°C (60°F) was maintained only to 11:00 P.M. and then allowed to drop to 7°C (45°F) until 6:00 A.M. To bring the soil temperature back quickly to the normal daytime temperatures, the plants were watered at 6:00 A.M. with warm water 21° to 24°C (70° to 75°F). This practice has not caught on to any great degree, perhaps because of the increased management needed. Another reason is that a savings of only 15% to 20% is likely, and simpler approaches such as adding a polyethylene layer can produce better savings.

The Future

The next energy-efficient generation of greenhouses will involve fuel cells (Chapter 10). The technology exists, but current fuel cell energy prices are not competitive with current energy sources. This economic reality is likely to change in the near future as fossil fuel prices rise and mass production of fuel cells reduces their costs. If one factors in pollution from fossil fuels and its indirect costs, that day is even closer. A number of possible scenarios exist. A dairy farm or sewage plant could use anaerobic digesters to produce methane from cow manure and sewage sludge, respectively. In fact, some already do. A fuel cell exists that can convert methane gas into electricity and heat. These inputs could be tied to greenhouses, making their operation more sustainable. The dairy farm could produce lettuce, tomatoes, cucumbers, or bedding plants for local markets. Sewer plants could produce the same vegetable crops for adjacent big city markets. The environmentally safe waste products from the fuel cell, water vapor and CO_2, could provide humidity for the greenhouse and also a CO_2-enriched atmosphere (see earlier discussion in this chapter) to increase vegetable yields. Greenhouses in proximity to natural gas supplies could use that fuel to accomplish the same outcome. Areas near coal and oil-shale (assuming syn-fuel facilities) could also do the same. Areas with plenty of sun could harness solar energy to photolyze water and use the resulting hydrogen to run fuel cells, too. Some research and prototypic organizations are investigating these possibilities.

Low-technology possibilities include large water tanks in greenhouses in which fish (*Tilapia*) are grown for food and the sun-warmed water helps reduce energy needs for heating the greenhouse. Heat produced by compost piles in greenhouses could also be harnessed and the finished compost used in growing media and soil enrichment. These systems were demonstrated at the New Alchemy Institute in Falmouth, Massachusetts, many years ago. Although now defunct, the Green Center is making the institute's publications available.

HOTBEDS

Hotbeds (also called hot frames) can enhance certain greenhouse functions, such as starting plants and rooting cuttings, at a modest cost (Figure 13–11). A typical hotbed with 6-foot-long window sashes consists of a pressure-treated wood (chromatid copper arsenate or copper naphthenate), rot-resistant wood (cedar, cypress, locust, and redwood), concrete, brick, or cinderblock frame that is 18 inches high in the rear and 12 inches high in the front. The slope height in the back versus front is determined by the rule of thumb that for each foot of sash, you drop 1 inch. The number of standard window sashes (3 feet by 6 feet) used to cover the frame generally determines the length and width. Polyethylene can also be used as a cover, as can greenhouse fiberglass or double plastics. The hotbed is located in a sheltered, well-drained area, and it should be situated in a sunny, southern exposure. The lower front wall should face south. The bed's bottom generally consists of 6 inches of sand or gravel covered with sand.

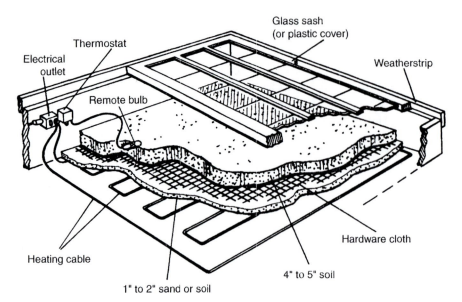

Glass sash
(or plastic cover)

Thermostat

Electrical
outlet

Weatherstrip

Remote bulb

Heating cable

Hardware cloth

4" to 5" soil

1" to 2" sand or soil

FIGURE 13–11   Construction details of an electrically heated hotbed are shown here. Manure was once used to provide heat through biological oxidation as it decomposed. A 24-inch layer of strawy manure after composting for 7 to 8 days was covered with 6 inches of soil. The resulting heat from biological oxidation was capable of maintaining a soil temperature 15° F higher than ambient soil for up to 90 days. This natural source was once the only way of heating a hotbed. A few still use manure today. (USDA photograph.)

A thermostatically controlled heating cable buried in the sand can provide heating quite simply. Alternate heating can be provided by steam, hot water, or hot air forced through pipes. At one time, fermenting manure was used to supply heat for hotbeds. The sand temperature (heating cable) is usually maintained at 21° to 27°C (70° to 80°F), if it is to be used for rooting of cuttings. If the hotbed is to be used to start plants for the garden, air temperatures using the heating source and solar energy can range from 13° to 21°C (55° to 70°F), depending on the requirements of the plants. Plastic flats or plugs can be placed over the sand, or a layer of growing media over the sand can be directly planted. Cooling is provided by manual ventilation or by a mechanized window raiser that is operated by a heat-activated expanding chemical. Shading, as with greenhouses, is used to provide some cooling and a reduction in light levels. Watering is usually done by hand.

COLD FRAMES

Cold frames are the same as hotbeds, but are not provided with heat; they are warmed by solar energy through the greenhouse effect (Figure 13–12). Cooling is by manual ventilation, a mechanized window opener, or shading. Cold frames are used to harden off transplants raised in the greenhouse or hotbed, prior to setting them outdoors. The considerable change in temperature and light levels in going from the greenhouse to outdoors directly is such that the plant is stressed. Death or, at the least, delayed development is the result without the intermediate environment provided by the cold frame, a type of "halfway house." Cold frames can also be used to raise transplants of the more cold-tolerant crops in the spring such as cole crops or lettuce. They also provide a useful place to summer over house plants or to winter over plants that have borderline winter hardiness. For the latter purpose, the cold frame is covered with leaves or straw. A layer of polyethylene tacked over the sash, as in the greenhouse, will provide additional insulation.

(A)

(B)

(C)

FIGURE 13–12 Various types of cold frames.
(A) Polyethylene-covered cold frames are common in commercial use because of ease of construction and low cost. (USDA photograph.) (B) Wooden cold frames with glass sash, once popular commercially, have decreased in numbers. This old glass greenhouse and its attached wooden glass sash cold frames have fallen in disfavor, being replaced by modern plastic-glazed structures. (Author photograph.) (C) Wooden cold frames with a fiberglass cover are economical and favored by homeowners. (Courtesy of National Garden Bureau.)

CLOCHES AND HOT CAPS

The basic principle behind cloches is to provide a greenhouse-like environment at the permanent site of the crops. Cloches are used to produce earlier crops of vegetables and flowers, because air and soil temperatures within a cloche are higher than in the external environment. Originally, cloches consisted of several sheets of glass held together by a wire framework. These units were fitted together end to end to provide a glass tunnel directly over the crop. The practice of using glass cloches in commercial production has decreased because of material and labor costs. More recently, the use of polyethylene film on wire hoops, which can be applied by machine, has replaced cloches (see discussion of sun tunnels). In the home garden, cloches can be as simple as a polyethylene bag over a frame or a gallon plastic bottle with the bottom removed. Tubular polyethylene enclosures that can be filled with water are also used in the garden. Most cloches give a 2-week advantage, but only offer protection from a light frost. Further covering might be needed if a heavier frost is expected. The only temperature control possible with cloches is complete removal when the season has progressed to where the internal temperatures are too high, or gradual perforation of the plastic types to allow ventilation.

Hot caps are made from wax paper laid over a wire frame. Like cloches, protection from light frost is possible and earlier growth of plants is possible. Temperature control is limited. Usually an upside down V is cut into the side away from the wind. This flap can be partially opened if temperatures get too high. Hot caps do not transmit as much light as polyethylene, so growth can be reduced. Better results are obtained with polyethylene or glass cloches or sun tunnels.

ROW COVERS AND TUNNELS

Row covers are essentially field-adapted mini-greenhouses. They are used both commercially and in home gardens to extend the growing season at both ends. Row covers (Figure 13–13) are usually used with vegetable crops, but can be used with flowers. Two basic materials are used: polyethylene and spun-bonded polyester. Polyethylene is available in clear and white forms and requires support by wire half-hoops. The clear form transmits the most light (85% of sunlight) and warms the plants and soil better. White forms pass less light and also reduce heat buildup. White polyethylene can be used with heat-sensitive crops, but its biggest use is for winter storage of container-grown plants. Temperature fluctuations are reduced with white polyethylene, making it ideal for winter use (see Chapter 11). Spun-bonded polyester is so light that the plants become the support. The polyester cover is applied loosely over the plants, which permits it to rise as the plants grow. Polyester also offers a high degree of protection against insects and insect-borne diseases (see Figure 15–19 in Chapter 15). Given its porous nature, it does not require venting and rain can pass through it. Polyethylene offers less pest protection, because some venting is required at times. Rain cannot pass through it either.

Row covers trap heat. Temperatures inside the cover can be considerably warmer than the ambient temperature by 5° to 15°F during the earlier part of the growing season. Warm season crops such as cucumbers, sweet corn, and tomatoes are given a considerable head start by row covers. These covers also offer protection against light and unexpected spring and fall frosts. As such, these covers are also useful for extending plant life and harvests at the end of the growing season. The main driving force commercially is an earlier harvest, which brings a higher price. Row covers are often used to bring in early harvests of sweet corn. Row covers must be removed at some point, because heat buildup can kill plants as the weather warms. The type of crop and the ambient temperature determines the amount of safe time. For example, row covers of polyester can be left on sweet corn, a warm season crop, until the ambient temperature reaches 90°F. Cool season crops such as lettuce or broccoli are harmed more easily than corn, so the cover should only remain on for a few weeks in the spring.

Tunnels come in two forms: sun tunnels and high tunnels. Sun tunnels (Figure 13–14) are low polyethylene tunnels on pipe hoops made from metal or PVC. The covering can be either clear or white polyethylene as with row covers. While larger than polyethylene row covers, you can't walk in them. Some sun tunnels are heated and hence are very similar to a hotbed. Hot water tubing or electrical cables provide heat. Heated sun tunnels are essen-

FIGURE 13–13 Wire hoops covered with this ventilated clear polyethylene cover offer an economical means of producing earlier row vegetable crops and also offering some degree of frost protection. These row covers are used both in gardens and for commercial vegetable crops. (Author photograph.)

FIGURE 13–14 These sun tunnels are being used for a research project to study the effects of various levels of atmospheric carbon dioxide upon plant growth. While these tunnels can also be used to produce earlier crops and offer some degree of frost protection, their main use is for propagation purposes either as hotbeds or cold frames. Sun tunnels are bigger and use sturdier, larger pipe hoops as contrasted with wire hoops in row covers (see Figure 13–13). (USDA photograph.)

tially a less expensive form of hotbed and are used mainly for propagation purposes. Like hotbeds, heated sun tunnels are an inexpensive alternative form of propagation structure compared to the more expensive greenhouse. Unheated sun tunnels are the equivalent of cold frames. Unheated sun tunnels can be used to harden bedding plants and rooted cuttings and for the propagation of cool season crops where heat is not required.

Close attention must be paid to ventilation, shading, and watering. Opening both ends of the tunnel provides ventilation. Placing Saran shade cloth over the tunnel can provide shading. Often both forms of these structures are fitted with trickle irrigation lines for watering the plants.

High tunnels are larger structures suitable for walking in and operating small field machinery and are used for early spring and late fall production of fruits, flowers, and vegetables. These tunnels, also known as French tunnels, have been around for many years in Asia, Europe, and the Middle East. In the United States usage has been mainly as overwintering structures for ornamentals and producing early vegetable crops such as tomatoes. Considerable interest exists regarding their use for early production of salad greens, organic vegetables, figs, and specialty crops. High tunnels depend entirely on solar energy for heat. Warm season crops should not be planted any earlier in tunnels than 2 to 3 weeks prior to normal planting dates, as the frost protection provided is only a few degrees. A propane heater provides good backup insurance before the average last frost date. With tunnels, tomatoes can ripen 3 to 4 weeks earlier and command premium prices. Tunnel costs often can be recovered in 1 year. While harvests can be extended in the fall, more interest rests with earlier spring production and higher market price for early crops.

Covers consist of 6-mil polyethylene (greenhouse grade) that can last for 3 to 4 years. Framing materials can be wood, PVC pipe, or galvanized metal. The preferred frame is metal bows attached to metal posts protruding from the ground. Wooden baseboards are attached to the posts. Hipboards are attached to the bows about 3 to 4 feet above the baseboard. The plastic is attached to the hipboard with a nail or screw fastened batten strip. The lower plastic end is attached to a pipe or board, so that the sidewalls can be rolled up to allow for ventilation. Careful attention must be paid to ventilation to prevent excessive heat and humidity. Tunnels should be oriented perpendicular to prevailing winds to enhance ventilation via the sidewalls. The end walls are usually removable to allow for human and machine access. The tunnel floor consists of black plastic mulch (6 mil) to speed soil warming, control weeds, conserve water, and reduce soil-borne diseases. A trickle irrigation system is usually installed too. Tunnel width varies from 14 to 20 feet, while a good length is 96 feet.

CLEAR MULCH

If one were to flatten a clear polyethylene unheated sun tunnel, you would end up heating just the soil instead of the soil, air, and plants. The use of clear mulch or infrared mulch on soil does the same thing. This type of mulch works especially well during a cool spring, giving essentially a 1- to 2-week head start on the growing season. If the season turns out to be unseasonably warm, your gain is cancelled. Under the right conditions, sweet corn is not only earlier, but increased yields and higher quality can also occur. Increased yield results from improved germination in the warmer soil and higher quality results from less stress and better growth as a consequence of more uniform soil moisture. Given the increased market price for certain earlier crops such as sweet corn, the gamble is often worth doing. Special planters that lay mulch, punch seeds holes, and drop seeds all in one pass are used with clear mulch systems.

Clear polyethylene mulch transmits 85% to 95% of solar radiation, depending on the thickness and clearness. These mulches prevent infrared heat loss from the soil, thus keeping soil warmer by 5° to 9°F at a 4-inch depth and 9° to 14°F at a 2-inch depth. The disadvantage with clear mulch is that weeds can grow under it, but when used with no-till systems this problem is minimal. A compromise is the use of infrared mulch (green in color), which passes infrared but not the blues and reds needed for weed growth.

SUN-HEATED PITS

Sun-heated pits are enlarged cold frames sunk deeply into the earth to provide the dimensions of the greenhouse at a reduced cost. Such a pit has a sloping glass roof that faces south. The sun through the greenhouse effect heats the sun pit and the surrounding earth is an effective insulator. Auxiliary heaters might be used as an emergency measure on severe winter nights. On very cold nights an additional covering, such as blankets or straw mats, might be needed. Plants that tolerate cold temperatures, that is, night temperatures of 2° to 7°C (35° to 45°F), do especially well in sun-heated pits. As energy costs continue to rise, the use of sun-heated pits will most likely increase.

LATHHOUSES

Lathhouses consist of a simple frame of pipes or wood embedded in concrete (Figure 13–15). The frame is covered across the top with thin wood strips about 2 inches wide and spaced to cut out from one-third to two-thirds of the natural light. Alternately, the top may be covered with a plastic woven fabric (Saran), which comes in different densities. Aluminum Lathhouses are also available. The least expensive lathhouse consists of snow fencing attached to a frame.

FIGURE 13–15   A very simple lathhouse constructed with wooden slats is shown here in backyard use. While larger, more elaborate wooden lath ones are seen commercially, the trend is to use plastic woven fabric coverings (Saran) for the roof. (Courtesy of National Garden Bureau.)

Lathhouses provide shade and offer protection for nursery stock in containers. Protection is especially important in regions subject to both high temperatures and excessive light intensities during the summer. Other uses include protection of container plants after transplanting, acclimation of field-grown house plants to reduced light levels prior to selling them, the holding of shade plants, the further hardening of tender plants after the cold frame and prior to field planting, propagation beds, and holding areas for market display. Like the cold frame, the lathhouse may be considered a halfway house.

Light intensities can be controlled by adding or removing wooden laths or by the use of different densities of Saran. Temperatures are generally similar to those found in the shade. Watering frequency is reduced because the shade reduces the water losses incurred by evaporation of soil water or transpiration.

THE HOME, OFFICE, AND INTERIORSCAPES

The home and office do not appear to be plant growth structures at first glance. However, one realizes after some thought that with today's increasing green consciousness and trends toward interior decorating with plants (Figure 13–16) such environments do qualify as controllable plant environments. The design, installation, and maintenance of ornamental plants for interior decoration in homes and commercial sites can be challenging and rewarding to those in the landscape horticulture and floriculture professions. Entire service industries exist to provide plant rentals, plant exchange (rotation from office to greenhouse back to office), and maintenance contracts (watering, pruning, fertilizing, pest control, disease control). More on this topic is covered in Chapter 16.

Environmental Control

Control is limited to a great degree by the dictates of creature comfort. Fortunately, the ideal conditions for health, comfort, and efficiency, that is, day to night temperatures of 20° to 21°C (68° to 70°F) to 16° to 18°C (60° to 65°F) with 50% relative humidity, are suitable for a wide range of foliage and flowering plants. Unfortunately, even with these compelling reasons and the additional problem of energy conservation, many homes and commercial buildings are still overheated and underhumidified. These conditions should be corrected prior to plant installations. The drop in night temperature not only conserves fuel, but is beneficial to plants. At night, plants respire and produce energy for growth. Part of this energy is thought to be wasted, and as the respiratory rates increase with increasing temperature,

(A)

(B)

FIGURE 13–16   (A) Interiorscapes can be as simple as this collection of cacti and succulents on a table under growth lamps. (B) A very elaborate interiorscape is shown here in this estate conservatory. (Author photographs.)

night temperatures that are too high result in a lower net photosynthesis for the day, and in turn a decrease in rates of plant development.

Temperatures within the home and office can be modified to a limited extent through the use of microclimates. Window areas often experience a greater drop in night temperature than other parts of the room, since heat losses through the glass and curtain are usually higher than through the insulated wall. This temperature differential is often localized near the window, since air movement and disturbance become minimal in the house or office at night. A maximum-minimum thermometer placed near the window and monitored over a period of time will give an idea of the temperature range. In the day a southern window area will likely be warmer than a northern one, because it receives more sun. Sun-porches are also cooler at night than other rooms, because large glass areas favor heat loss. If one checks with a thermometer, temperature differentials can be found to occur throughout a house or office. These temperature changes depend on several factors, such as the distance of a room from the heating system, shading from surrounding buildings and trees, wind exposure, and efficiency of insulation.

Air conditioning does not pose a problem, since the temperature is maintained in a range suitable for many plants. As with heating, it is important to avoid a direct draft on the plant. Drafts cause leaf desiccation and eventual plant deterioration. Problems found to occur in air-conditioned offices can usually be traced to the turning off of the air conditioning on a weekend or holiday, which creates a high temperature range that is unsuitable for plants.

Light

Natural light in a home or office is often a limiting factor in the choice of plants for an indoor environment. Indoor light can be utilized to broaden the choice of plants; this topic will be discussed later. Generally, light levels decrease in order for the following window exposures: south, east, west, and north. This order assumes that the exposure is not shaded or obstructed by a large overhang, building, or tree. Light levels also vary with the season. For example, a plant receiving favorable light in the winter at a southern window could receive too much light in the same window during the summer. A lightweight sheer curtain could be used to decrease the light level, or the plants could be located farther back from the window where there is less light. A window or sunroom with a southern exposure is the most versatile, since the range of light from directly at the window to a point farther back is suitable for a number of plants that vary in their light requirements.

Humidity

Ideally, the relative humidity level should be maintained at 40% to 50%. With most indoor environments, this level cannot be maintained constantly without the help of a humidifier. If a humidifier is not available, one can take advantage of humid microclimates. Often the window area over a kitchen sink or in the bathroom has a higher level of humidity. Plants can also be grouped on wetted pebbles. The water level should be below the pot bottom to prevent capillary uptake, possible water-logging of the soil, and salt accumulation on the surface of the growing media. The evaporation of water from the pebbles and transpiration from the grouped plants tend to create a humid microclimate.

Ventilation

Ventilation is important for plants in the home or office, especially in the winter when fresh air movement is considerably less. The level of carbon dioxide in air is about 330 ppm, which is sufficient for plant development under normal environmental conditions. In an enclosed environment with inadequate ventilation, localized depletion of carbon dioxide can occur in the immediate vicinity of the leaf. Ventilation is usually adequate in the spring, summer, and fall. To avoid ventilation problems in winter, plants should be located in traveled areas, and fresh air should be added to the environment, but never as a direct draft on the plant. Poor ventilation is often involved in disease problems, too, since localized increases in water vapor might occur in still air around the leaf and provide an ideal environment for plant pathogens.

Water

Watering house plants is one of the hardest environmental parameters to control and explain. Any explanation is only a generalization; no substitute exists for cultural knowledge and experience. When watered, potted foliage and flowering plants are watered thoroughly until water flows from the drainage hole. Excess water is discarded from the saucer to prevent oversaturation of the soil. The time between watering applications depends on several parameters: temperature, relative humidity, light levels, rate of transpiration, soil properties, type of pot, type of plant and developmental stage of the plant, and current season. All of these conditions interact differently, some in concert and others in opposition. Soil dryness, however, can be seen, felt, and even monitored with a moisture meter.

Water needs vary considerably. Succulents, coming from arid regions, need much less water than tropical plants. Some plants of tropical origin require constant, even soil moisture, whereas others do better when the top inch or two of soil dries out between water applications. Plants with hair-like, delicate fibrous roots and plants with wiry, thick roots are characteristic of the first and second group of tropical plants, respectively.

Indoor Lighting

Indoor light sources can be used in place of or to supplement natural light when the latter is insufficient. The use of indoor light in areas between the natural light and principal viewing point can also prevent the turning of plants toward the natural light. Landscape architects should be consulted for effective placement of plant light sources.

Indoor light sources do not duplicate all of the physical parameters of natural light, but can affect developmental processes similarly by providing the light wavelengths (spectral quality) and intensities most effective for their regulation. The indoor light sources by themselves can maintain photosynthesis and can be used to manipulate the photoperiod. In conjunction with natural light, they can be used to increase photosynthesis and to alter the photoperiod. These aspects were covered in Chapter 6.

Indoor lighting for plant growth is not used to any great extent on a commercial basis, except for plants used in interior decoration of commercial buildings and for greenhouse crops of high value, because of high electric costs. However, this use enjoys great popularity with amateur horticulturists, especially for potted foliage and flowering plants indoors. Control of photoperiod by indoor light is used extensively on a commercial scale.

Indoor light can affect photosynthesis by supplying radiation in the red and blue regions of the spectrum and photoperiodic responses by supplying red, far red, and blue. Different lamps vary in their spectral output. For example, incandescent light is richer in the red and far-red region than fluorescent light. Differences in lamps used for horticultural lighting are as follows.

Fluorescent lamps produce spectral emissions derived from fluorescent phosphors and low-pressure mercury vapor. The more common fluorescent light, called cool white, has the following energy output in the spectral regions given in decreasing order: yellow-green, blue, red, and far-red. This lamp is well suited for human visual purposes and is also useful for growing plants because of the blue and red emissions. By changing phosphors, the energy output of a particular spectral region can be increased. Fluorescent lamps designed expressly for growing plants have phosphors that tend to increase the red and far-red emission at the expense of the yellow-green region. However, these plant growth lamps are not quite as energy efficient as the cool white. Expressed in terms of total energy, they produce about 75% as much as the cool white.

Plants can be grown quite satisfactorily under a one-to-one combination of ordinary cool white and warm white operated for 16 to 18 hours per day. Plant-growth lamps do offer some advantages, even though they cost more, since plants appear to flower a bit earlier and to become bushier and darker green when compared on an equal total-energy basis.

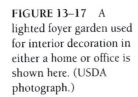

FIGURE 13–17   A lighted foyer garden used for interior decoration in either a home or office is shown here. (USDA photograph.)

The business or homeowner must decide whether these esthetic improvements are worth the extra cost. A setup of a simple fluorescent light arrangement for growing plants is shown in Figure 13–17. Lamps are commonly placed 6 to 9 inches above the plants.

Incandescent lamps are richer in the red and far-red spectral regions than fluorescent lamps. In addition, they are about one-third as energy efficient as fluorescent lamps, their lifetime is about 15 times less, and their light distribution is not as linear. Their spectral output makes them suitable only for photoperiod control, where only low-energy output is required. Fluorescent lamps are also available that produce more red than conventional fluorescent lamps. These types are also useful for control of the photoperiod.

As would be expected from their complementary spectral emissions, a combination of fluorescent and incandescent bulbs is good for photosynthesis and control of photoperiod. A general combination is one or two 15-watt incandescent bulbs for every 80 watts of cool-white fluorescent lamps. With any incandescent bulb in the presence of plants, the heat output (infrared radiation) might be a problem in close, confined quarters with inadequate ventilation.

A third type of lamp is used in horticultural practice, the high-intensity discharge (HID) lamp. These lamps overcome one disadvantage of fluorescent lamps, which at a distance of a few inches can supply only 8% to 10% of the light levels provided by sunlight on a bright, sunny summer day. Light intensities with HID lamps can be as much as five times higher than those of fluorescent lamps. These lamps are more costly than fluorescent or incandescent types. However, the following factors must be weighed. A number of fluorescent lamps must be used to give the same light levels as one HID lamp. HID lamps have a long life, with some giving up to 24,000 hours when lit for five or more hours per start; and some HID lamps have energy efficiencies that exceed those of fluorescent lamps. Collectively, these factors can make a HID lamp cost competitive over the long range.

High-intensity discharge lamps are available in three basic types: mercury vapor, metal halide, and sodium vapor lamps. Mercury vapor lamps, the oldest form of HID lamps, are available in clear and phosphor-coated forms (mercury-fluorescent lamp). The former has spectral emissions in the green and blue and the latter has, in addition, red. Mercury-fluorescent lamps have been used extensively for horticultural lighting. These lamps are generally smaller and have the lowest installation costs, the longest life expectancy, and the best light distribution pattern of all HID lamps. Mercury-fluorescent lamps are also a good choice for home use (Figure 13–18); they are used primarily with large trees in tubs as part of a practical decorative accent.

FIGURE 13–18 A circular mercury-fluorescent lamp is used in this attractive plant growing arrangement. (USDA photograph.)

Metal halide lamps emit light in both the blue and red areas of the spectrum. A phosphor-coated form is even more effective for plant growth. Sodium vapor lamps emit most of their light in the red area. A combination of metal halide and sodium vapor lamps is very effective for plants.

Sodium vapor lamps come in low-pressure (LPS) and high-pressure (HPS) forms. The LPS type is a line source of light (like a fluorescent tube) and is the most efficient of all light sources described here. The HPS form is a point source (such as an incandescent light) and is almost as efficient as the LPS one. Their light levels do not drop rapidly with age, as do the fluorescent or mercury halide lamps, and their life is over 24,000 hours. Even though sodium vapor lamps cost more to install than fluorescent lamps, their efficiency makes them more economical in the long run. Their effectiveness is also better, so they are increasingly the choice in commercial operations for raising seedlings and supplementary lighting. Better growth is produced with LPS or HPS lamps over cool-white fluorescent, and best growth is achieved when supplemented with incandescent light. Suggested installation heights are 5 feet or higher. Since these lamps produce some heat, a fan may be necessary to avoid heat damage to plants.

The best way to determine the number and proper placement of lamps, that is, spacing between lamps and height above plants, is to know the light requirements of the plants and to measure the light levels, following adjustments, until you reach the desired level (see Chapter 6).

SUMMARY

Most United States commercial greenhouses are used for wholesale propagation and production of bedding plants, cut flowers, and potted plants. Fewer are used for the propagation and production of container nursery stock for landscaping and vegetable crops such as cucumber, lettuce, and tomato. Home greenhouses and some in botanical gardens are used for plant displays. Home greenhouses also provide living areas and solar heating.

Greenhouse framework consists of pressure-treated or rot-resistant wood, galvanized steel (preferred commercially), or extruded aluminum. Home greenhouses use aluminum or redwood. Structures are either freestanding (preferred commercially) or lean-to (preferred for home use) types. Freestanding types afford maximal light entry and greatest space and can be joined side to side at their eaves by a structural connector (gutter), eliminating sidewall covering. Advantages include smaller space requirements, less costly shared heating and cooling equipment, and easier access to the combined growing area.

Freestanding structures are Quonset, bow, or truss greenhouses. Quonsets offer simple, inexpensive construction with their half-circular pipe frames and single or double layer of polyethylene film. The structural strength of bow houses is higher and the cost greater than Quonsets, but strength and cost are lower than in truss houses. Both bow and truss types can be constructed as gutter-connected greenhouses. Both are even-span greenhouses in that the rafters are of equal length on the right and left sides of the ridge point. A bowed (arched) or triangular (gable) roof with straight sidewalls characterizes their shape. Homeowners, who enjoy being able to walk directly into the greenhouse from their homes, prefer lean-to greenhouses.

Floors in greenhouses are standard concrete (heavy loads) or porous concrete (aisles, light loads) and less often, dirt or gravel. Porous concrete is not used for propagation areas, because the fine pores become clogged easily. Gravel or dirt floors are inexpensive and sometimes used in home greenhouses, but can't be properly sanitized.

Greenhouse glazing consist of glass, fiberglass, and various plastics. Glass has the highest light transmission and cost, but its long-lasting qualities are best in areas with low-light winter conditions. Glass is no longer the dominant glazing material. Clear flat or corrugated fiberglass is shorter-lived and transmits less light than glass. Light transmission also decreases with age from yellowing. Polyethylene film is inexpensive and lightweight. Structural support requirements are minimal, reducing costs further. Polyethylene transmits slightly less light than glass, but breaks down upon exposure to ultraviolet light. Polycarbonate plastics are much less flammable and more easily worked than acrylic types, but yellow after 10 years. The commercial trend today is toward the double-layer acrylic and polycarbonate types and polyethylene film. These types are energy efficient and require lower capital investment and construction costs relative to double-glazed glass greenhouses. Homeowners are attracted to the double-layer acrylic and polycarbonate types for similar reasons, although the esthetics of glass still appeals.

Wooden benches in greenhouses are traditional. Metal or strong, rigid plastics are now common. Benches once were spaced primarily with walkway aisles and permanently fixed in place. Today placement depends on what type of machinery is used to lift and move plants. Rolling benches offer considerable flexibility and also increase usable plant space.

Some greenhouses have no benches, but use floor systems instead. These systems use an automated floor ebb and flood system. Concrete floors must be carefully laid, leveled, and gently sloped to ensure even flooding, no puddles, and collection of runoff at one end. Heating pipes and irrigation lines are buried in the ground. These floor systems have found use with plug production, rooting of cuttings and the growing on of tissue cultured plants.

Greenhouse environmental control involves regulating temperature, light, and humidity. Temperature control depends on solar energy and regulation of supplemental heating, ventilation, and cooling. Structural heat loss, the desired running greenhouse temperature, heat losses by the heating system, potential external winds and temperatures, and the desired safety margin are factors that determine required Btus.

Numerous types of heating systems and fuels are available. Hot water or steam is used with finned pipes to radiate heat and start convection currents. Circulating fans offer uniform heat distribution. Alternatively, forced hot air is distributed through ductwork. Oil-, gas-, or coal-fired boilers and furnaces produce hot water/steam and hot air. Gas or electric fan-forced heaters can also be utilized.

More recent technologies include infrared and root zone heating. Infrared heating heats plants, but not the air mass or soil. Infrared heating uses less energy than conventional heating systems. Root zone heating (essentially bottom heating) is placed below the plants and heats the root mass, soil, and tops, but not the air mass. Hot water is pumped with a small circulator pump through pipes. Root zone heating is used in a significant part of the propagation industry.

Temperature rise is controlled with ventilation. Air exchange can be increased with air intakes and exhaust fans. A perforated polyethylene convection tube can be connected to the intake to improve air distribution, provide cooling, and reduce plant damage from cold winter drafts. Ventilation also replaces depleted CO_2. Humidity is either supplied by steam humidifiers, or more commonly, cool-water humidifiers that operate on an aerosol principle. Automation can be provided with a humidistat. Ventilation systems can interfere as humidity losses occur when vents open.

Greenhouse shading provides additional cooling and reduces summer light levels. Shading can be provided by sprayed shading compounds, automated roller blinds, and green vinyl translucent plastic or mesh. Wet-pad evaporative coolers produce interior temperatures equal to outdoor shade temperatures. High-pressure misting systems are used in areas of low humidity. A newer alternative uses a retractable roof and retractable benches. During hot weather, the roof can be opened and the benches can even be moved completely outdoors.

Several factors determine the amount of visible radiant energy: variations of the sun's position with time of day and season, greenhouse location and orientation, roof slope, greenhouse shape, and the amount of cloud cover. Given these variables, greenhouse design and shape is a compromise in terms of maximizing the amount of visible radiant energy received per unit of time (radiant flux, commonly expressed as watts).

For certain crops, flowering is subject to photoperiodism. Photoperiodic control involves lengthening the day with indoor light, interruption of the night period with indoor light, or shortening the day with black shade cloth. Certain horticultural crops of high commercial value such as poinsettias and chrysanthemums are routinely brought to flowering for holidays or seasonal display through control of the photoperiod.

In earlier greenhouses and still in home greenhouses, thermostats are used to control the environment and a humidistat to regulate humidity. These systems worked reasonably well, except for some conflicts between humidity needs and ventilation. Thermostats often experience temperature overrides, too. Eventually analog systems evolved with electronic sensors that gathered information, enabling the analog system to resolve simple conflicts through various electronic logic systems. More recent computer automated systems use a microprocessor and numerous sensors and are preprogrammed from a console.

Pollutants are a serious concern. Greenhouse crops in urban or industrial sites can become affected enough to force relocation or a change to crops less sensitive to air pollution. Pollutants can arise in greenhouses and their mitigation must be considered.

Pasteurized media for growing greenhouse plants consist of peat moss and vermiculite or perlite. These media must have their nutrient and pH status checked routinely with soil tests. Tissue analyses are useful to confirm suspected nutrient deficiencies or to determine the nutrient levels needed by greenhouse crops.

Handheld hoses equipped with sprinkler nozzles and fogging nozzles facilitate watering for established plants and recently planted seeds/seedlings, respectively. Watering can be automated with solenoid valves and time clocks. Drip and capillary mat irrigation are primarily used for potted foliage and flowering plants; ooze and spray irrigation are used for flower beds in greenhouses.

Numerous commercial forms of soluble fertilizers are available for liquid feeding, depending on the horticultural crop and developmental stage. The dilution and application schedules vary, but a dilute feeding during every watering or a stronger solution at weekly or biweekly intervals is common. Soluble fertilizers can also be prepared to any desired nutrient analysis. Liquid feeding is done during watering (fertigation) or sometimes through foliar sprays and can be automated. Controlled-release fertilizers greatly reduce the frequency of application and the risk of injury from excessive applications. A form of indirect fertilization is CO_2 enrichment.

Pesticides and enclosed greenhouses constitute a dangerous combination. Chemicals must be used with considerable caution, a high regard for safety must exist, and protective respirators might be required. From the sustainable perspective, the use of pesticides in greenhouses must be minimal. The best approach is either natural controls or an IPM strategy.

Greenhouse budgets face increasing energy costs. Economy must be weighed against present and future availability of fuel. Heating systems that can burn an alternate fuel are good choices. Polyethylene over glass and double-layered plastics can reduce heat loss. The retractable heat sheet or movable thermal curtain (thermal blanket) can cut fuel costs. Heat sheets can be drawn at gutter height by hand or by an automated, timed motor. In older glass houses, injection of a sealant between overlapping glass panes (lapsealing) can save fuel. Other minor insulation improvements include recaulking, north wall insulation with fiberglass aluminum-faced insulation, installation of reflector-faced insulation behind fin tubing on perimeter foundation walls, and insulation of heating pipes leading to the greenhouse. Windbreaks can reduce heat losses. Another promising approach is split nighttime greenhouse temperatures. Seedlings require a uniform high temperature, but as they develop, their requirement becomes thermoperiodic. Night temperatures are dropped several degrees at night and rapidly raised in the morning.

Fuel cells are the next step in energy sources for greenhouses in that they can be fueled with methane, natural gas, syn-fuel, and even hydrogen produced by using solar energy to split water into hydrogen and oxygen. Fuel cells can produce electricity, heat, water vapor, and CO_2, all of which can be utilized in the greenhouse. Pollution is essentially non-existent with fuel cells.

Hotbeds provide certain greenhouse functions such as starting plants and rooting cuttings at a modest cost. The hotbed is heated by electric cables and located in a sheltered, well-drained area with a sunny, southern exposure. Cold frames are similar to hotbeds, but are only solar heated. Cooling is by manual ventilation, mechanized window opener, or shading. Cold frames are used to harden off plants from the greenhouse or hotbed. Another use is to raise transplants of the more cold-tolerant crops in the spring.

Cloches, hotcaps, row covers, and tunnels provide a greenhouse-like environment in the field. Offering only light frost protection, they are used to produce crops of vegetables and flowers about 2 weeks earlier than normal. Row covers are used both commercially and in home gardens for vegetable crops. Two basic materials are used: clear or white polyethylene and spun-bonded polyester. The main driving force commercially is an earlier, premium price harvest.

Tunnels come in two forms: sun tunnels and high tunnels. Sun tunnels are low clear or white polyethylene tunnels on pipe hoops made from metal or PVC. Unheated sun tunnels are used to harden bedding plants and rooted cuttings and for the propagation of cool season crops. Heated sun tunnels are used for propagation purposes. High tunnels are larger structures suitable for walking in and small field machinery. Considerable interest exists regarding their use for early production of salad greens, organic vegetables, figs, and specialty crops. High tunnels depend entirely on solar energy. The use of clear or infrared plastic mulch on soil also offers one to two weeks head start. Special planters that lay mulch, punch seeds holes and drop seeds all in one move are used with clear mulch systems.

Sun-heated pits are enlarged cold frames sunk deeply into the earth to provide the dimensions of the greenhouse at a reduced cost. The sun heats the sunpit and the surrounding earth is an effective insulator. Plants that tolerate cold temperatures do especially well in sun-heated pits. As energy costs continue to rise, the use of sun-heated pits will most likely increase.

Lathhouses consist of a simple pipe or wood frame covered across the top with thin wood or aluminum strips or with a plastic woven fabric. Lathhouses provide shade and offer protection for nursery stock in containers. Other uses include protection of container plants after transplanting, acclimation of field-grown house plants to reduced light levels prior to selling them, the holding of shade plants, the further hardening of tender plants after the cold frame and prior to field planting, propagation beds, and holding areas for market display.

Fortunately, ideal conditions for human health, comfort, and efficiency are suitable for a wide range of foliage and flowering plants. The amount of natural light in a home or office is often a limiting factor for plant growth. Relative humidity level should be maintained at 40% to 50%.

When watered, potted plants are watered thoroughly until water flows from the drainage hole. Excess water is discarded from the saucer to prevent oversaturation of soil. Times between watering depend upon several parameters: temperature, relative humidity, light levels, rate of transpiration, soil properties, pot type, type of plant and developmental stage of the plant, and what season it currently is. Water needs of house plants vary considerably.

Indoor light sources replace or supplement natural light and can also prevent the turning of plants toward natural light. Landscape architects should be consulted for effective placement of plant light sources. Indoor lighting for plants is not used much commercially, except for plants used in interior decoration of buildings and for greenhouse crops of high value, because of high electric costs. However, this use enjoys great popularity with amateur horticulturists, especially for potted plants indoors. Control of photoperiod by indoor light is used extensively on a commercial scale. Fluorescent lamps are useful for growing plants, as are high-intensity discharge (HID) lamps.

EXERCISES

Mastery of the following questions shows that you have successfully understood the material in Chapter 13.

1. What are the various uses of greenhouses in horticulture?
2. What materials are used in greenhouse framing? What is the difference between a freestanding and a lean-to greenhouse? What are the differences among Quonset, bow, and truss greenhouses?
3. What kinds of floors are used in greenhouses? What are the options for coverings or glazing on the greenhouse? What are the pros and cons of each?
4. What choices are available in terms of bench and floor systems in greenhouses?
5. How is a greenhouse heated, ventilated, and cooled? What are some of the newer approaches to greenhouse heating? How is greenhouse humidity maintained?
6. What factors are involved in the amount of available light in a greenhouse? Discuss their relationship to available light.
7. How can computers help control the complex environment within the greenhouse?
8. What parameters should be monitored and maintained with greenhouse plants grown in soilless media?
9. What systems can be used to provide irrigation in a greenhouse? How can these systems, as well as the application of soluble fertilizer, be automated?
10. Why does pest control require more careful attention in a greenhouse compared with field plants?
11. Why would greenhouse growers want to increase carbon dioxide levels?
12. Why should fuels used to provide greenhouse heating be evaluated on a cost basis per 100,000 Btus?
13. What are some approaches to increase greenhouse insulation of the glazing layer? What are some of the advantages and drawbacks of these approaches?
14. What is a hotbed in terms of its structure and usage?
15. What is a cold frame in terms of its structure and usage?
16. What are the differences among cloches, hotcaps, row covers, tunnels, and clear mulch? Why are these structures used in the field?
17. What is a sun-heated pit?
18. What is a lathhouse in terms of its structure? What are its uses?
19. What environmental conditions are best for growing plants in the home, office, and other interiorscapes? How do these conditions relate to human comfort?
20. How can one alter the amount of light in a home or office?
21. Why is the timing of watering plants so difficult to put into a simple rule or explanation?
22. What types of lamps are useful for growing plants?

INTERNET RESOURCES

Benches and greenhouse space
http://www.igcusa.com/greenhouse_benches_by_internati.htm

Caps, mulches, and tunnels
http://www.ag.usask.ca/cofa/departments/hort/hortinfo/veg/hotkaps.html

French and Spanish tunnels
http://www.clovis.co.uk/horticultural/french_spanish_tunnels.htm

Fuel cells and greenhouses

http://www.cnn.com/2000/NATURE/09/18/farm.fuel.enn/

http://www.ansaldoricerche.it/clc/ari.htm

Green Center

http://www.fuzzylu.com/greencenter/home.htm

Greenhouse floors and benches

http://www.agweb.okstate.edu/pearl/hort/greenhouse/f-6703.pdf

Greenhouse glazing links

http://www.h2othouse.com/html/greenhouse_glazing.html

High tunnels

http://www.aginfo.psu.edu/News/october00/tunnel.html

High tunnels for early spring/late fall production

http://www.penpages.psu.edu/penpages_reference/29401/2940170.html

Hobby greenhouse construction

http://www.aces.edu/department/extcomm/publications/anr/anr-1105/anr-1105.htm

Hotbeds and cold frames

http://www.muextension.missouri.edu/xplor/agguides/hort/g06965.htm

National Greenhouse Manufacturers Association

http://www.ngma.com/index.htm

Retractable roof systems and natural ventilation

http://www.agra-tech.com/NatVent.html

Row covers and plasticulture

http://edis.ifas.ufl.edu/BODY_CV106

http://nfrec-sv.ifas.ufl.edu/plasticulture_&_row_covers.htm

Season extenders

http://www.ext.vt.edu/pubs/envirohort/426-381/426-381.html

Sweet corn, plastic mulches, and row covers

http://www.gov.on.ca/OMAFRA/english/crops/facts/swtmulch.htm

RELEVANT REFERENCES

Boodley, James William (1996). *The Commerical Greenhouse* (2nd ed.). International Thomson Publishing, Cambridge, MA. 612 pp.

Freeman, Mark (1997). *Building Your Own Greenhouse*. Stackpole Books, Mechanicsburg, PA. 304 pp.

Hanan, Joe J. (1998). *Greenhouses: Advanced Technology for Protected Horticulture*. CRC Press, Boca Raton, FL. 720 pp.

Hashimoto, Yasushi, Hiroshi Nonami, W. Day, H. J. Tantau, and Gerald P. A. Bot (eds.) (1993). *The Computerized Greenhouse: Automatic Control Application in Plant Production*. Academic Press, San Diego. 329 pp.

Nelson, Paul V. (1997). *Greenhouse Operation and Management* (5th ed.). Prentice Hall, Upper Saddle River, NJ. 637 pp.

Reed, David William (ed.) (1996). *A Grower's Guide to Water, Media, and Nutrition for Greenhouse Crops*. Ball Publishing Co., St. Charles, IL. 322 pp.

Stone, Greg (1997). *Building a Solar-Heated Pit Greenhouse*. Storey Communications, Pownal, VT.

Walls, Ian G., R. A. Martin, A. G. Channon, and J. W. Newbold (1996). *The Complete Book of the Greenhouse* (5th ed.). Ward Lock, West Sussex, United Kingdom. 304 pp.

Chapter 14

Controlling Growth by Physical, Biological, and Chemical Means

OVERVIEW

The horticulturist uses indirect practices to control plant development through the manipulation of the plant's outdoor environment. However, more direct control of plant development through physical, biological, and chemical means is both feasible and highly satisfactory. Physical and biological techniques are among the oldest forms of control, and are still indispensable today, especially from a sustainability perspective. Chemical controls are relatively newer and will become increasingly sophisticated as our understanding of the molecular biology and biochemistry behind plant development expands. The control of plant development is an important tool of the horticulturist, which, if properly mastered, will pay worthwhile dividends.

The importance of the relationship of basic horticultural practices to the success of physical, chemical, and biological control cannot be stressed enough. Plants must have good vigor for best results. Vigor depends on proper identification of all environmental conditions in a site (site analysis). Once these conditions are known, the horticulturist can better select plants suited to that environment, whether the site is to be left natural or modified where possible to maximize plant development. After the plantings are established, cultural practices are directed toward maintaining sound plant development. With good cultural conditions, the physical, biological, and chemical practices of controlling plant development can be expected to yield their best results.

CHAPTER CONTENTS

OBJECTIVES

- To learn the physiological consequences of pruning and reasons why woody plants are pruned
- To understand the tools, types of cuts, cut locations, and timing of pruning
- To learn the basics of pruning used for training of trees and shrubs at the nursery
- To learn the basics of maintenance pruning used with established evergreen or deciduous trees and shrubs

- To understand the basic biological control techniques in the field and greenhouse
- To learn the basics of chemical control using growth regulators
- To understand the various groups of growth regulators, their effects, and their plant-specific applications

PHYSICAL CONTROL BY PRUNING

Physical control of plant development usually involves the selective removal of specific plant parts such as a branch or part of a branch. This practice is known as *pruning*. Pruning is primarily practiced on woody plants: shrubs, trees, and lianas. Some herbaceous plants such as many garden perennials also benefit from minimal pruning to promote branching and reduced height.

Control by pruning has a number of benefits. Pruning can be used to alter the size, shape, and direction of growth of young woody plants for esthetic or practical reasons; this form of pruning is termed *training*. Many ornamental shrubs, ornamental trees, bush fruits, fruit trees, and nut trees undergo training at the nursery prior to being sold.

Pruning can also play a role in plant health such as the removal of a diseased or dead branch to prevent further plant damage. The quality of fruit, flowers, and vegetative parts and the quantity of fruit and flowers on woody plants can be maintained and improved through pruning. Pruning applications include a thorough thinning out of unproductive wood to increase flowering and fruiting, the reduction of fruit-bearing branches to produce larger fruit, and the selective removal of fruit or flowers to increase the size of those left. The balance between growth and flowering can be regulated by the removal of older, mature wood to encourage the production of young wood for growth and/or flowering with some woody plants. Finally, it might be desirable to restrict growth, such as with foundation plants to maintain a proper scale between the house and plantings. All of these goals are accomplished with maintenance pruning.

Pruning and Effects on Development

Pruning is essentially the removal of plant parts. It is the primary means of physical control. Lesser techniques of physical control, such as wiring plant parts, are covered later. The pruning of plants is not a haphazard process. Pruning must be carefully done, based on a thorough knowledge of plant development. To do otherwise results in plant damage, decreased productivity, and poor plant form.

Assuming that the chapters on plant structures and plant development have been mastered, we will proceed to the effects of pruning. First, pruning of the top growth alters the balance between root and shoot, unless compensating root pruning is done. Root pruning is generally done with nursery container plants, field-grown nursery stock, and bonsai and is considered later in this chapter. The root area is usually left untouched and continues to transport water, nutrients, and stored food upward. Since the shoot area is reduced, the upward flow is diverted in part to the buds below the cuts.

These buds, which were dormant because of apical dominance or other genetic factors, are capable of growth when the stem tips are removed. Auxin is produced in stem tips, so stem tip removal temporarily stops its production and, hence, the inhibitory effect of auxin on lateral buds (apical dominance). These buds break dormancy, and the diverted water and nutrients bring about growth and subsequent branching. If, however, the pruning cuts remove entire branches and leave no buds, branching is eliminated. Instead we see strong growth (elongation) of the remaining branches.

With pruning it is apparent that some photosynthetic production is lost. The new growth will compensate partially, but only after a lag time during which new leaves are formed. This lapse causes a temporary setback, but the remaining photosynthetic area and the stored carbohydrates in the untouched roots and older shoot parts carry the plant. Compensation for the lost part, therefore, is never 100%. Dwarfing is the result of cumulative pruning.

Pruning of top growth appears to enhance juvenility in many plants, resulting in vegetative growth. This result is a consequence of the lower, more juvenile buds being forced into growth after pruning. The reasons for the enhanced degree of juvenility are not clear, but might lie in an alteration of phytohormone production or carbohydrate-to-nitrogen ratios. Root pruning, on the other hand, slows down vegetative growth and enhances flowering in many plants. The reason again is unclear, but might also be tied in with phytohormones or carbohydrates.

Pruning and Timing

The effects following pruning can vary, depending on the time of season when the pruning is done. Pruning at the wrong time can produce poor results. The correct time to prune varies with plant material, but some generalizations are possible. Deciduous ornamental trees are usually pruned in late winter and early spring, which makes it easier to see which branches should be removed and gives the new growth time to mature prior to winter. That time of year, when deciduous form is readily apparent, is advantageous both for normal, ongoing maintenance pruning and restoration and rejuvenation pruning of severely neglected deciduous trees and shrubs. Exceptions include *Acer* (maples) and *Betula* (birches), which bleed sap profusely at this time. These trees are usually not pruned until early summer when the sap runs more slowly. Mature fruit trees are pruned while dormant. Although often cold and unpleasant work, it occurs during the time when other normal orchard maintenance is very low. Evergreen trees grown primarily for foliage effects are pruned in early spring just prior to when growth commences or, if they are grown for flowers, right after flowering.

Deciduous shrubs that bloom in the spring (*Forsythia, Syringa*) are pruned soon after flowering in order not to disturb the setting of flowering buds for the following year. Later blooming deciduous shrubs, that is, those that bloom on the current year's wood (*Hydrangea*), are pruned in late winter and early spring. Evergreen shrubs grown mainly for foliage are pruned in early spring just prior to when growth commences. Evergreen flowering shrubs are pruned right after flowering. Woody vines or lianas are treated in the same manner as shrubs; their time of blooming determines when to prune them. These suggestions are meant merely as generalizations. A good pruning guide should be consulted to determine the correct time for pruning particular plants.

Pruning Tools and Techniques

Some basic tools (Figure 14–1) and techniques apply to all forms of pruning. All tools should be disinfected between plants to avoid transmittal of bacterial, fungal, and viral pathogens from one plant to another. This simple act minimizes later need for bactericides and fungicides, which is highly desirable from the viewpoint of sustainability. Various products are available such as those based on *n*-alkyl derivatives of dimethyl benzyl ammonium chloride (Consan Triple Action 20). Tools should also be sharpened periodically and maintained to prevent rusting.

Reference to Figures 14–2 and 14–3 will aid the reader in understanding the following pruning terms. Pruning back growth to a bud is termed *heading back;* complete removal of

FIGURE 14–1 Pruning tools. *Top:* pruning saw with leather scabbard. *Bottom:* left, hedge shears; center, anvil pruning shears; right, lopping shears. (USDA photograph.) Not shown is the extension pole, pruning saw that has a telescoping handle for reaching higher branches. A chain saw for major repairs and removal is also not shown.

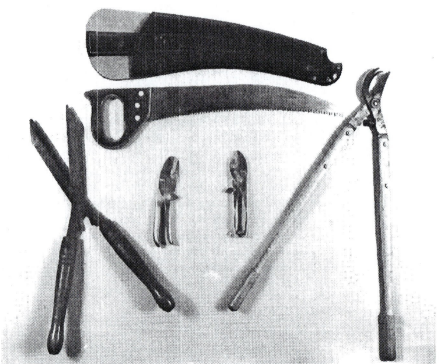

FIGURE 14–2 Pruning back to a bud as shown here with scissors-action pruning shears is known as heading back. The majority of thinning out activity is done with handheld anvil or scissors-action pruning shears. A lopping shear is needed for the occasional thick branch (0.75–1.50 inch diameter) that still has buds. (Author photograph.)

selected growth back to a lateral, the trunk, or the crown is called *thinning out.* Heading back tends to produce a bushy appearance; thinning out results in a more open look. Of the two, heading back alters plant form more and can even spoil form (if excessive); thinning out tends to maintain and even improve plant form. The need for both pruning techniques varies with species, stage of development, overall plant form, and the desired goal expected from the pruning. *Tip pinching* is the removal of herbaceous tissue at shoot ends by pinching off with the thumbnail and forefinger. It is usually used on herbaceous garden perennials such as chrysanthemums to promote branching, more flowers, and reduced height.

The slant and distance of a cut from the bud are important when heading back; an example of a correct cut with pruning shears is shown in Figure 14–4. Cuts on larger branches with lopping shears would be similar. The position of the top bud is important. For example, if the plant has an upright nature, the top bud should be an outside bud such that the branch has room to develop outward and keeps the center open. If you wanted to fill the

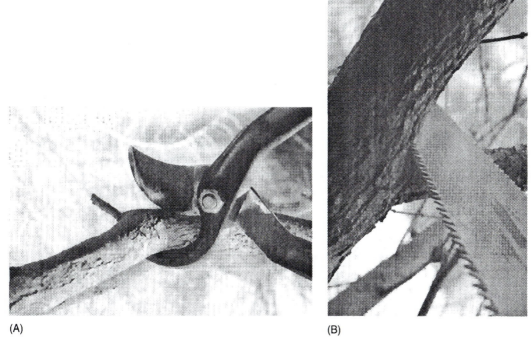

(A)   (B)

FIGURE 14–3   (A) Cutting back to the point of origin of this lateral branch at a fork (other part of fork just slightly visible at bottom left) is called thinning out. The branch is small enough that a lopping shear can do the job. (B) The removal of a lateral branch at the trunk is also an example of thinning out. This larger branch requires a pruning saw. Very large branches would require a chain saw. (Author photographs.)

FIGURE 14–4   Correct heading back cut in relation to buds on woody twigs is shown here. The cut starts even with the bud's base on the opposite side and slants upward, ending even with the top of the bud. In northern areas of extreme cold, the top ends a short distance above the bud, which helps prevent bud desiccation. Cuts ending too high above the bud expose too much wood and can result in twig dieback and bud damage. Cuts too low interfere with the bud's growth. (Author photograph.)

center such as in a storm-damaged tree, you would select for an inside bud. In older trees, top buds may dry out or be crowded out by lower ones, so the choice of top bud or its position can be for naught.

Watersprouts are rapidly growing, vertical unbranched shoots produced from the main stem or head of a tree or shrub. *Suckers* are rapidly growing shoots produced from the plant roots (often incorrectly applied to watersprouts originating from stem tissue near the crown). Both are usually superfluous and are thinned out during maintenance or rejuvenation

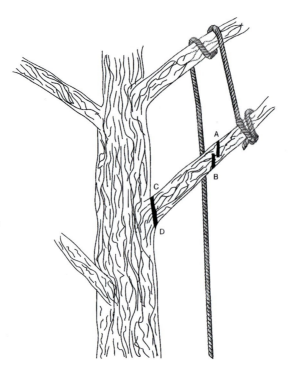

FIGURE 14–5   Larger branches of easily handled weight are double-cut with a pruning saw. Heavy branches are roped and cut first at A, then B. B should be placed to intercept A, but are shown slightly apart here to emphasize the two-cut nature. The rope prevents the heavy branch from crashing down and either hitting the pruning worker or knocking the ladder over. The stub is then removed with a final cut at C to D. This method prevents bark being stripped downward by the weight of the falling branch. Cuts can be made with a hand pruning saw or power chain saw (handle with care and only if trained in its use).

pruning. Watersprouts and suckers may be selectively retained on occasion to fill out a sparse tree or shrub or to replace the loss of a nearby branch caused by disease or accident.

Larger branches that can be handled on a weight basis (easily supported with a free hand) are treated with lopping shears or a pruning saw as shown in Figure 14–3. Branches that are too large to be handled safely are thinned out using first an undercut followed by an overcut as demonstrated in Figure 14–5. These double cuts are necessary to prevent splitting and bark stripping, which could lead to serious tree damage.

The final cut on the tree trunk after the removal of the branch is important in order to promote correct healing and reduce the chance of disease. A correct cut leaves no stub, and is flush with the trunk. However, one should not cut back too far so as to remove bark, as this exposes more heartwood than is necessary, which leads to a longer healing time. This leaves a round to oval area of heartwood. The results of proper and improper cuts are seen in Figure 14–6.

Cuts larger than 2 inches in diameter are often painted with a wound compound. However, research results from the U.S. Forest Service indicate that the use of tree wound paint is more cosmetic than disease preventive. Healing of the wound and disease problems appeared to be statistically similar for tree wounds that were painted and unpainted. The practice still persists, even though it might only be psychologically helpful. It might be useful initially to waterproof the wood, thus minimizing environmental suitability for and entry of pathogens. Growth pressures and weathering lead to cracking of the paint, thus compromising the compound's waterproofing. Yearly, but impractical, applications would be needed to maintain waterproofing.

Caution should be exercised in pruning work, as well as with any horticultural activity where hand or power tools are utilized. Chain saws cause a number of injuries each year when used by the homeowner for pruning chores. When using an extension pole pruner for high branches, beware of possible electrocution from power lines and injury from

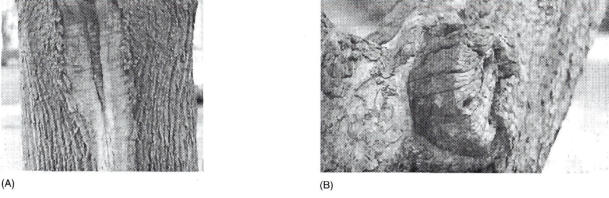

(A)

(B)

(C)

(D)

FIGURE 14–6 (A) The pruning cut was incorrectly done (branch not undercut), leading to excessive tear wounding. (B) This cut was incorrectly done by leaving a stump that prevents proper healing. These poor cuts increase the chances for disease and lengthen the healing period. (C) The pruning cut was correctly done (round to oval) and callus tissue (light gray) is healing the wound. (D) The completed healing process on a correct cut leads to total coverage by callus tissue (lighter gray). (Author photographs.)

falling branches. Unless roped properly, large branches can injure people or nearby structures. A large branch might miss you, but it can spring back on impact and knock the ladder out from under you. Removal of large branches involving climbing and perhaps most pruning of large trees with chain saws are best left to professional arborists.

TRAINING

Young trees and shrubs will eventually flower and fruit without pruning, but training pruning allows the horticulturist to produce a well-balanced plant with a practical and pleasing size and shape. Training also produces a plant with strong, well-placed branches and a plant that is easier to maintain. Training is particularly important to the commercial fruit grower, whose fruit trees must be properly trained to allow for increased productivity, less disease, and mechanical harvesting.

FIGURE 14–7 This ornamental tree was trained to the central leader system. (Author photograph.)

Pruning to train a tree or shrub is started at the nursery and sometimes continued by the purchaser. The nursery generally sells young deciduous trees in the form of a single straight stem (*whip*) or a whip with a few short side branches a few inches in length (*feathers*). If the tree is grafted, it may be called a *maiden*. These forms are sold bare rooted, are relatively inexpensive, and have maximal potential for training by the homeowner. These forms are also sold in containers and, to a lesser degree, balled and burlapped. Container forms are preferred from the perspective of the homeowner who wants to minimize transplant shock and increase survival.

Whips and maidens can be trained further either at the nursery or by the homeowner into *standards* or *central leaders*. The former, used with medium-sized trees, has a clear stem for 6 feet above which the framework is allowed to develop. When the tree reaches about 7.5 feet, the tip is removed to promote branching. Smaller growing trees are usually sold as half-standards (branching starts 3.5 to 4 feet above crown). The central leader system is usually used with the larger trees. These tree forms have a clear stem for 6 to 8 feet and the leader is not pruned. Heading back is done on the other branches to produce a balanced head. The form resulting from the central leader system is shown in Figure 14–7. Pruning may or may not be done to keep the center of the tree open. The above systems with medium to large deciduous trees allow for foot and vehicular traffic (street trees) or mechanized operations such as mowers (lawn trees) underneath. Other trees may be trained by pruning to maximize symmetry and ornamentation instead. These forms are usually destined to be specimen trees.

Evergreen trees, conifers in particular, do not respond favorably to heavy pruning. Therefore, woody evergreens are trained differently than deciduous trees. Training is minimal in the nursery, where the main thrust is to allow development of the natural form. Exceptions include evergreens that will be utilized as hedges (*Taxus, Tsuga*). Broad-leaved evergreen trees such as *Arbutus* are allowed to develop with a central leader, and side shoots are not removed. Surplus leaders are removed. Side branches are thinned out where crowded or crossed. Main branches are kept at 6 to 8 feet or higher. Long-needled conifers are usually only given light heading back of their terminal candle growth when in the candle developmental state of growth. If heading back is done beyond the new growth such as candle growth on *Pinus* or *Picea*, twig dieback can occur. If dieback occurs on a branch, it is thinned out, not headed back as with deciduous trees. Unfortunately, such removal can spoil the symmetry. Surplus leaders are removed on conifers, except for certain shrub-like ones (*Cephalotaxus*), especially if they grow away strongly. Conifers with fine or scale-like foliage such as *Chamaecyparis, Thuja,* or *Juniperus* are sheared or tip-pinched partially on their new growth in order to train them into a desirable shape.

(A)

(B)

FIGURE 14–8 (A) This young apple tree was trained to the modified leader system. (B) These old apple trees in a Vermont orchard were recently pruned to maintain their modified leader system form. (Author photographs.)

FIGURE 14–9 This peach tree was trained to the open center (vase) form. (USDA photograph.)

Young fruit trees are trained by three different systems: central leader, modified leader, and open center systems. The central leader system is the same as that shown in Figure 14–7 for ornamental trees. Once popular in orchards, it has fallen into disfavor because the tall, pyramidal shape makes spraying and picking fruit somewhat difficult. This form still appeals to the homeowner who wants a fruit tree that is both ornamental in appearance and fruit bearing.

The modified leader system is the most popular at present (Figure 14–8), since it produces a well-shaped tree of good structural strength suited for orchard conditions, that is, minimal shading of underlimbs and maximal fruit production. With open center training, the branches all arise from one small area on a shorter, main trunk, and the center is primarily free of branches (Figure 14–9). This type of form has a disadvantage in that the open center might be ideal for a water pocket and subsequent rot. The open center form is also structurally weaker, with all the branches arising from a small area with narrow angles that withstand stress less than wide ones.

Although not hard and fast categories, the central leader system is used for many nut trees and some tropical fruits. The modified leader system is used with apple, apricot, cherry, olive, pear, persimmon, and plum trees. The open center technique is used for almond, fig, nectarine, peach, Japanese plum, and quince trees.

Shrubs as a rule need less training than ornamental and fruit trees. Training produces a shrub with a well-balanced, evenly spaced framework. This training is often done after purchase from the nursery. Deciduous shrubs are initially pruned to three to five strong upright shoots, and side shoots are halved and left with two to three buds. The following year the stems are headed back about one-third and crossed or crowded branches are thinned. Less pruning is done with evergreen shrubs. Three strong shoots are selected and others are only lightly tipped. The following year crowded branches are thinned.

Berries such as blackberries and raspberries are trained according to the growth patterns shown in the area in which they are grown. Where growth is limited such as in the north and northeast United States, berries are trained as unsupported bushes. On the Pacific Coast, growth of berries is so rampant that they are trained on trellises or stakes. Grapes are generally trained on wire and wood supports, trellises, and arbors. Homeowners mostly use arbors.

Some very specialized forms of training are rather labor intensive. They are utilized primarily for esthetic purposes and only to a limited degree on a commercial basis. These forms include topiary, espalier, cordon, and bonsai. *Topiary* is the training of a plant to grow as a certain form, such as an animal or geometric pattern. Topiary was once the high point of formal gardens, but is seldom seen today. Topiary does have some limited appeal on the house plant scale, where one can encounter plants trained on shaped-wire pieces stuffed with sphagnum moss.

Espaliers are fruit trees, shrubs, and vines that have been trained flat against a wall, trellis, or post and wire. If the espalier is restricted to one or two shoots trained in parallel or opposite directions, it is termed a *cordon*. An example of the espaliering technique is shown in Figure 14–10. This technique is used commercially for fruit trees in Europe, to a limited extent in the United States, and worldwide for grapes. For grapes a specialized form of espalier based on the Kniffin system is used.

Bonsai (Figure 14–11) is the art of artificially dwarfing trees into miniatures through top pruning, root pruning, and wiring over a number of years. This horticultural form is ancient, having arisen in Japan. Bonsai was influenced by Penjing, a somewhat similar form that arose earlier in China. Bonsai requires considerable skill, much time, great patience, and very careful cultural care. The final product is often a piece of horticultural art and, like great art, it is fragile.

Root pruning is an essential part of training practiced by nursery people. Without it, the digging of field-grown nursery stock for transplanting and sale could not be done as successfully as it is. Root pruning is a routine, annual cultural activity carried out by hand and

FIGURE 14–10   These vineyard grapes in California are trained in an espalier trellis system known as the Kniffin system. Grapes are one of the few fruits in the U.S. trained as espaliers. The espalier is more common with ornamental trees in the U.S. (Author photograph.)

FIGURE 14–11 A superb example of bonsai is shown here in a bonsai collection. The tree is Japanese Zelkova (*Zelkova serrata*), which at maturity reaches a height of 90 feet. This mature bonsai started in 1949 is only a few feet tall. (Author photograph.)

FIGURE 14–12 Root ball of plant left unpruned and held too long in present size container. Some roots have died, as they do not exhibit the white coloration of healthy roots. This plant will grow poorly when transplanted to its final outdoor site and will likely die eventually. (Author photograph.)

also machines designed for that purpose. The techniques vary, but the simplest example is the forcing of a sharp spade into the ground in a circle around a woody plant about 1 year prior to moving the plant. Most nurseries use mechanized equipment to root prune alternate sides of a row in alternate years. This root cutting forces the development of fibrous roots near the base and facilitates the transplanting operation. Recovery is faster and losses minimal.

Root pruning should also be practiced on container nursery stock as follows. Container stock should be moved to bigger containers in time to avoid producing a constricted root system (Figure 14–12). If constricted roots develop by oversight and remain uncorrected, plant quality will be inferior. The outer, constricted, circular-growing roots must be pruned to eliminate this condition before the plant is moved to another container. When the homeowner removes a container plant, he too must root prune if the plant is to be properly established in its new permanent environment. A sharp knife, hatchet, or spade is drawn through the outermost root area such that it is partially quartered. This step will produce outward root growth in the soil and, hence, better establishment. If the root ball is not broken, root growth will usually remain constricted, and the plant will fare poorly and often will die. Death occurs because the encircling roots girdle each other and the crown, resulting in disruption of food and water movement in the roots.

MAINTENANCE PRUNING FOR ESTABLISHED PLANTS

The preceding section has been concerned with the use of pruning for training of plant material. Much of the training occurs in the nursery. Once plants are trained and established, maintenance pruning, usually by the homeowner, becomes necessary. Maintenance pruning will be needed to restrict growth because of space limitations or deviations from form, to obtain a balance between vegetation and flowers and fruits, to maintain health, and to produce quality improvements of fruits and flowers. Some of these areas apply equally to the various plants we prune; others are handled somewhat differently depending on the plant.

Pruning is but one aspect of the maintenance of plant health. Others will be covered in the following chapter. Weak branches, pests, and diseases impact plant form and can hasten the death of a woody plant. In addition, the falling of large weakened branches can cause human injury and property damage. Diseased or dead wood should always be pruned promptly until only healthy wood remains. Other branches that are potential sources of future health problems or that detract from plant form should be removed before they become problems. Problem branches include watersprouts, suckers, crossed branches, horizontal forks of equal thickness, narrow-angled scaffold branches, and underbranches on a main branch. Suckers and watersprouts (Figure 14–13) tend to grow rapidly and obliterate desirable plant form, are often soft growth, and can rub other branches if left too long. Suckers arising below a graft union must be removed, or the scion will be overtaken by the sucker growth. Under some conditions a watersprout arising above the graft union or on an ungrafted plant can be headed back and left to fill in an opening or to replace diseased material about to be or previously removed.

If branches cross, one should be removed to prevent rubbing and bark damage. Horizontal forks tend to split under wind stress, especially if loaded with fruit, or if one branch is heavier than the other. One branch of the fork should be removed. Underbranches, as do suckers, arise from adventitious buds. Underbranches grow downward and then curve upward or straighten out horizontally. Such branches should be removed because of their structural weakness.

Narrow-angled scaffold branches are weak because of squeezed bark in the crotch and a lack of continuous cambium. An inclination to catch debris and to crowd the crown exists. Branches of this type should be thinned out at an early stage to avoid major tree surgery after a severe storm. Figure 14–14 shows incorrect and correct development of scaffold branches.

Under some conditions a tree might have progressed beyond the point of corrective pruning. This event occurs if the tree is mature and the removal of narrow-angled scaffold branches requires major surgery or the removal of a weakened branch would leave a gaping hole. Under these conditions the scaffolds or weakened branches can be strengthened

FIGURE 14–13 The two straight upright slender branches in the center of this shrub are water sprouts. These branches should have been removed by pruning earlier before they worsened the shrub form. It is not too late to remove these branches. (Author photograph.)

(A)   (B)

FIGURE 14–14 Incorrect (A) and correct development (B) of scaffold branches. The tree in A is more prone to storm damage, more susceptible to rot in the crotch area, and has a less attractive form. (Author photographs.)

FIGURE 14–15 A tree with a weak crotch due to poor scaffold branch development. A cable with two brace bolts is used to strengthen the crotch. (USDA photograph.)

with bracing cables attached to bolts (Figure 14–15). Stresses must be considered carefully if these braces are to work. This treatment may also be used to save a storm-damaged, irreplaceable tree of great value.

After initial training some woody plants will need little or no pruning to maintain their form. Others will need ongoing annual pruning. Pruning will also be required to restrict mature size sometimes. Ongoing pruning will be needed for foundation shrubs, ornamental and shade trees, fruit trees, vines, and even house or greenhouse plants. Heading back and thinning out are the mainstays of size restriction and form retention. These techniques will also be used to maintain a balance between vegetative growth and flowering or fruiting. Maintenance pruning approaches are now covered in a generalized way according to various plant categories.

On deciduous ornamental trees one should remove superfluous branches that develop down low or up near the tip. Branches that spoil the form are thinned. Branches that must be removed because of size restrictions should be headed back to a vigorous side branch. Oldest branches should be cut out and younger wood trained into its place by heading back adjacent branches. Heading back adjacent branches can also fill in gaps caused by storm damage or disease.

Pruning of established deciduous shrubs differs widely, but is based on the time of flower bud development (and flowering time) and the age of the wood that bears the flower buds. Spring-flowering deciduous shrubs bloom on last season's wood; that is, flower buds are formed prior to dormancy. If not pruned, spring-flowering shrubs tend to flower less and less. Pruning is done shortly after flowering and consists of thinning out superfluous growth at the crown such as old wood, young suckers, and watersprouts. Growth that is too excessive in length or detracts from plant form is headed back. Other shoots are thinned out to open the center of the shrub. This combined pruning promotes flowering and tends to keep the shrub in good form and at a desirable size. Deciduous shrubs flowering in the summer and autumn bloom on the current season's growth. These types are dormant pruned. Pruning is the same as for the spring-flowering shrubs just covered, except for the change in time.

The removal of spent flowers (*deadheading*) is a time-consuming pruning practice. Where used, it is primarily with plants that have spent flower clusters that detract from plant esthetics or slow desirable growth because of diversion of food and nutrients to seed and fruit formation. Shrubs of this type include *Syringa* and *Rhododendron*.

In mild climates, deciduous shrubs will often be subjected to insufficient chilling; therefore, buds will not start growth normally in the spring. Irregular leafing and flowering will result. Tip pinching of the first 2 or 3 inches during the summer will often improve leafing and blossoming the following year.

Broad-leaved evergreen shrubs generally require less pruning than deciduous materials. This approach is especially true of shrubs having a slow, even, compact growth, most of which originates from the terminal buds. This produces a dome-like appearance with little interior growth. An example of this type is the *Rhododendron*. Branches that break from form are lightly headed back. Weak, diseased, dead, and crossed wood is thinned. This light pruning can be done in early spring prior to growth; heavier pruning should be delayed until after flowering. Winter-killed branches are headed back to healthy wood in the late spring after it is completely evident that they are dead. If desired, spent flowers are also removed then.

Fast-growing broad-leaved evergreens, such as *Berberii,* that follow the growth habits of the deciduous shrubs are pruned more heavily in the Pacific coastline and Deep South, where they are headed back and thinned out strongly. In other areas, pruning is less severe and not necessarily done on an annual basis. Pruning of evergreen shrubs grown for flowers or berries is delayed until after their appearance; otherwise, pruning is done in early spring prior to active growth.

Coniferous evergreens usually reach the homeowner after minimal shaping and training at the nursery. Fine-foliaged conifers (*Chamaecyparis, Thuja,* and *Juniperus*) are lightly sheared or tip-pinched to maintain shape, usually in the late spring or early summer when growth is completed. Evergreens used as hedges are more heavily sheared (*Taxus* and *Tsuga*). Pruning at this time will give maximal size restriction and minimal upkeep for the remainder of the season. Long-needled conifers can also be pruned in the early to late spring during candle formation by lightly heading back their terminal candle growth either partially or completely. Complete removal results in no growth for the current growing season. Cutting beyond the soft candle growth is to be avoided, because the branch will die back and require complete removal. Older, dying branches in the lower portions should be thinned (see training section on evergreens). Pruning in general should always maintain the inherent natural shape of the evergreen and be kept to a minimum.

Once trained, fruit trees (Figure 14–16) are pruned to maintain maximal fruit yield. Watersprouts, suckers, dead wood, and weak branches are removed on all fruit trees. The trained form is also preserved. For example, with the modified leader form, the center is kept open and the leader is kept dominant. Strong young wood is promoted through pruning to eventually replace the fruit-bearing wood, and spent fruit-bearing wood is removed. With many fruit trees the fruit is produced on spurs. Old spurs are replaced by pruning to encourage constant development of new fruit-bearing spurs. Procedures vary according to the fruit. General pruning information for fruit trees is given in Section Three.

Grapes and berries are pruned to maintain the trained form and to encourage maximal fruiting wood. Pruning varies from the complete removal of all canes that bore fruit right after

FIGURE 14–16   This apple tree was dormant pruned. Extensive thinning out to remove older nonproductive wood was done. Heading back was also done to encourage development of new bearing wood. (Author photograph.)

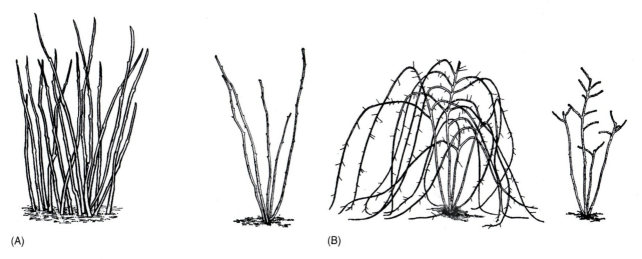

(A)   (B)

FIGURE 14–17   (A) Red raspberries in early spring (*left*) are cut back to five feet and weak or spent fruiting canes are thinned out (*right*). (USDA photograph.) Pruning raspberries. (B) After harvesting black raspberries or in early spring (*left*), laterals should be pruned back to about eight inches (*right*). These laterals have buds that will produce fruit-bearing branches. Weak or spent fruiting canes are thinned out. Later, summer topping will be needed.

harvest to only partial removal (Figure 14–17). Complete removal is practiced with brambles (raspberry, blackberry), which, though perennials, have canes with a biennial fruiting habit.

Partial removal is used with grapes, where the fruit is produced on the current growth from buds set the previous season. Wood older than 2 years is not fruitful. Grapes usually set far more buds than needed, resulting in excessive clusters of fruit with small grapes. Often the vine is unable to ripen all the fruit and produce enough vegetative growth for next year's crop. Grapes are usually pruned (Figure 14–18) in late winter or early spring by a technique called *balanced pruning*. The training system is usually a four- or six-cane Kniffin system, where either two or three canes are left on each trunk side and trained onto trellis wire (see espalier discussion earlier).

At pruning time, four or six 1-year-old canes about the diameter of a pencil are selected to be wire-trained for the current season. Canes on the upper trunk are best. Prune out all other 1-year canes and any older, unfruitful branches. Also form renewal spurs by cutting

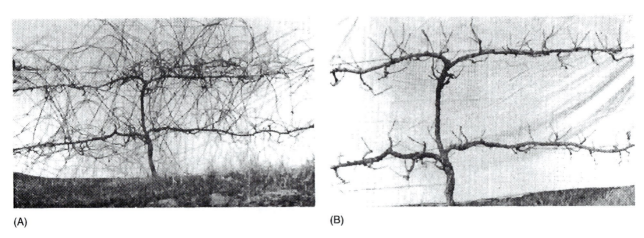

(A)   (B)

FIGURE 14–18   The training trellis system shown here for grapes is a four-cane Kniffin system. (A) Early spring prior to pruning. (B) Pruning both improves productivity and reduces disease problems by removing old wood, encouraging new wood, and opening the vine up to air and light. (USDA photographs.)

back surplus 1-year-old canes to a short branch near the trunk such that two buds are left. These spurs will produce the fruiting canes for next year. Weigh all the removed 1-year canes. For the first pound, retain 30 buds, and anything more than a pound, 10 buds per pound. If the weight is less than 1 pound total, no further pruning is needed. If the canes weighed 3 pounds, you would leave 30 + 10 + 10 or 50 buds total. If you had six canes and four renewal spurs (spurs each had two buds, total is 8), each retained cane would be partially cut back to 7 buds (6 × 7 = 42, plus 8 buds on spurs equals 50 buds). With four canes, each would be cut back to 10 to 11 buds.

Flowering climbers and vines are generally pruned at a time determined by the age of the wood on which flowers are produced, as for flowering shrubs covered previously. In general, climbers and vines that bloom on the current season's wood can be pruned during the winter. Those blossoming on the past season's wood are pruned after flowering is completed.

Pruning is also used to improve the quality of flowers and fruit. This activity involves the removal of some buds, flowers, or fruits to concentrate resources with those remaining. This selective removal increases the final size over that which would have resulted if all were retained. When used to increase the size of fruits, removal is termed thinning, and *disbudding* when flower size increase is desired.

BIOLOGICAL TECHNIQUES

Before plant growth is regulated by chemical means, serious consideration of biological alternatives is needed from the viewpoint of sustainable practices. In organic operations, acceptable alternatives are required. Incentives to adopt biological over chemical practices include the negative image of chemical usage in horticulture held by the public and tighter EPA regulations of plant growth regulators (registration and reentry restrictions).

In some situations, the alternatives require more labor and possibly might cause slight plant damage such as with harvesting of citrus crops without the use of chemical harvest aids. In some cases the alternatives might produce inconsistent results. Alternatives to chemicals can be more costly. Often biological alternatives have been ignored, thus knowledge of their efficacy is limited and minimal research exists to gauge their effectiveness.

Still, organic horticulturists have achieved successful production without plant growth regulators. Some sustainable horticulturists are mixing both biological and chemical techniques, which gives satisfactory crop production and results in the use of fewer chemical plant growth regulators than previously. Total usage of chemical growth regulators is best avoided, except for those situations where no alternatives exist or the chemical approach results in great reductions in labor and energy usage.

Vegetables and Turf

Seaweed extracts combined with humic acids and various enzymes (referred to as biostimulants) are used in foliar applications on some vegetable production in Florida to replace chemical plant growth regulators. These products have various uses such as enhancing and speeding germination, producing earlier flowering and harvest, and improving water and nutrient uptake. To date, the track record has been one of inconsistent results from year to year and from one cultivar to another. Expense is also a factor with these products. Another problem is that, because of past unsupportable claims and product variability, these products tend to have a "snake oil" image. Similar products are also used in turf production and management. Current products appear to work best when plants are stressed and fertility levels are not high. Growers of vegetable crops often mix biological and chemical techniques for best results.

Fruiting and Ornamental Woody Plants

Grafting (covered in Chapter 7) is used mostly for propagation, but can also be used for the biological control of plant growth. For example, modification of growth habits can be achieved through grafting. The choice of a specific rootstock can be used to dwarf the cultivar selected to be the scion, such as with fruit trees (e.g., apple, apricot, and peach). Others can be utilized to give the grafted plant exceptional vigor. Grafting is frequently done with certain highly pigmented forms of cacti, whose vigor is greatly enhanced by the choice of a proper rootstock. Other alterations of growth habits through grafting include the production of tree roses and weeping birches, cherries and needled evergreens. Rootstocks can also be chosen to enable the selected scion cultivar to tolerate unfavorable soil conditions, to resist diseases, to improve the size and quality of fruit, to encourage early bearing of fruit, and to replace rootstocks that are not cold hardy in a particular climatic zone.

Ringing and scoring are forms of girdling. Scoring involves making a thin cut (about one-sixteenth inch) into the bark layer with a knife or saw. Ringing involves a similar cut, but with removal of about a one-eighth- to one-quarter-inch strip of bark. Cuts should go no deeper than the start of the cambial layer. Cuts should not encircle the entire tree trunk, because death might result. If severe scoring or ringing is needed, two overlapping cuts halfway around the trunk from opposite sides can be used. These cuts should be spaced about 4 to 6 inches apart. These techniques are used to reduce excessive vigor and encourage flowering and fruiting with fruit trees and vine fruits. A case in point would be young apple trees that should commence fruiting, but haven't.

These treatments provide only temporary effects and are not used to any great extent. If they are done early in the season with grapes, current fruit production is heavier and the sugar content is higher. With fruit trees such as apple, the setting of fruit buds can be initiated or increased for the following year. The effects result from phloem disruption, but effects are temporary since the phloem regenerates. Infection from fire blight bacteria or other pathogens is a risk if these practices are done when disease is present.

Shaking is another form of biological control, but a natural one that requires minimal output. The stresses of wind movement appear to strengthen plant stems and in some cases cause dwarfing. Nonstaked, widely spaced container nursery stock requires less pruning when transplanted into the landscape than crowded, staked stock with minimal movement. The need to stake new trees, unless in a site exposed to excessive winds and hence shifting of the root ball, would appear to be minimal. Herbaceous material with easily broken stems and of tall height should still be staked. The area of natural stresses and their effect on plant development still needs much research and should produce results of interest in the future.

Greenhouse Ornamental Crops

With greenhouse ornamental crops, alternative choices are somewhat better. The main problem with greenhouse plants is height control. Some simple steps such as not starting plants too early (better scheduling) or using dwarf cultivars (assuming consumer acceptance) are possible. Another simple step is to maximixe light levels in the greenhouse to reduce "stretch." Cleaning glass, replacing aged plastic, and increasing the space between plants can help improve light conditions.

A number of cultural steps can be taken to control plant heights. One is to use a minimal size container such that some root restriction results. Development will be slowed and some nutrient/water stress will result, all of which help to reduce stem elongation. Instead of using minimally sized containers, another option is to keep fertilization levels low, especially nitrogen, and allow slight wilting, which will accomplish the same goal as small containers. Moderate phosphorous deficiencies have also been effective for reducing plant heights. These approaches must be cautiously used to prevent decreased plant quality.

Temperature control can also reduce the need for chemicals to control height. The most useful technique is DIF, which is defined as the difference between day temperature (DT) and night temperature (NT). Stem elongation is less when nights are warmer than days, a situation called *negative DIF*. Warmer days (*positive DIF*) promote stem elongation. A reduction of the difference between DT and NT can bring about less stem elongation and a reduced need to use plant growth regulators for height control. DIF has been shown to be effective with numerous greenhouse crops (Table 14–1), but not all. A warmer night does require more fuel and is less sustainable from the energy side. Fortunately, a reduction in early morning temperatures, which also reduces fuel input, can achieve the desired result. Greenhouse temperatures for 2 to 3 hours prior to dawn need to be reduced by 5° to 10°F relative to the night temperature. Many bedding plants and poinsettias have shorter stems when night temperatures are set at 68°F (20°C), followed by a decrease to 60°F (15.5°C) 30 minutes prior to dawn for 2 hours, and with day venting temperatures set at 65°F (18.3°C). For plants needing warmer temperatures such as begonia, the respective temperatures are 72°, 64°, and 65°F (22.2, 17.7 and 18.3°C, respectively).

DIF does have some drawbacks. It does not work for all crops. Sometimes seasonal temperatures and geographic location make it difficult to drop the predawn temperature. Under these situations, the reduction of heat inputs will not bring about sufficient predawn temperature drop. DIF does alter average daily temperature, which in turn alters development. Raising the average daily temperature speeds development, while reducing the average daily temperature slows development. These changes need to be considered in terms of scheduling impacts and market timing. DIF should not be used late in the developmental

TABLE 14–1   *Response of Greenhouse Ornamentals to DIF*

Significant Response	Minimal or No Response
Asiatic lilies	Aster
Celosia	French marigold
Chrysanthemum	Hyacinth
Dianthus	Narcissus
Easter lily	Platycodon
Fuchsia	Tulip
Geranium	
Gerbera	
Hypoestes	
Impatiens	
Oriental lilies	
Petunia	
Poinsettia	
Portulaca	
Rose	
Salvia	
Snapdragon	
Tomato	

stage either, because flower/bract size can be reduced. Night temperatures should not be dropped more than 10°F, because leaves will droop. DIF use in combination with chemicals is a compromise that reduces chemical usage and can produce better results than DIF alone in certain circumstances.

Light modification has also been used with some success. Stem elongation is increased by far-red wavelengths, which are more prevalent at dawn and dusk. Height reductions of 20% for Easter lilies result when black cloth is drawn over plants from 30 minutes prior to sunset through 30 minutes after sunrise. Other crops might show similar trends, but as of yet have not been fully researched.

Last, mechanical conditioning of plants in the greenhouse can restrict height. Greenhouse tomato growers can draw cardboard tubes or bars suspended from irrigation booms across plant tops starting at the second true leaf stage for up to 18 days. With 10–20 strokes daily, height reductions of up to 20% result. This procedure should be done when plants are dry to prevent disease spread. This method has worked for Easter lilies and pansies, too.

CHEMICAL TECHNIQUES

The growth and differentiation of plants is controlled extensively by endogenous chemical messengers, phytohormones, as discussed in Chapter 5. It would not be unreasonable to expect isolated phytohormones, synthesized versions, and analogs, which are collectively termed plant growth regulators (PGRs), to alter plant development on being applied to the plant. This task has not proved to be as easy to accomplish as first thought.

Advances have been made, and this area holds both promise and peril to sustainable horticulturists who wish to modify plant development for purposes of increased productivity, better marketability, and less labor. PGR-induced results improve economic sustainability through labor reduction and in many cases can reduce fuel consumption by machinery, thus saving energy. Less impact on environmental sustainability is seen with this class of chemicals, as opposed to pesticides. Most PGRs have minimal or no environmental impact in terms of pollution. As such, growth regulators, except some few that also are insecticides, have a role, albeit limited, in sustainable horticulture.

Regulators and the responses of plants to them must be registered for use through the Environmental Protection Agency. In addition, the various states have laws controlling their use to varying degrees. The horticulturist must understand the laws governing PGRs (also insecticides, fungicides, herbicides, and other horticultural chemicals), obtain a license to use them when required by law, watch for changes in the regulations and lists of permissible chemicals, and observe all safety precautions. Care must also be exercised to avoid spray drift to nearby nontarget, but susceptible plants. Sustainable usage of PGRs should be practiced. These products should only be used when no acceptable workable alternatives are available or when the labor/energy savings are significant. Selection should be for the safest product and usage should be as minimal and as infrequent as reasonably possible.

Plant growth regulators can be used with some horticultural crops to increase yield by stimulating flowering and fruiting. More important, however, is their use to modify quality and timing more effectively than other cultural techniques. If this use has the added advantage of reducing labor cost, so much the better. Some regulators are used specifically to reduce labor costs associated with high-value horticultural crops. Labor reduction and the reduced use of any machinery associated with that labor are important considerations in terms of cost and energy usage.

The major problem with achieving the desired results with plant growth regulators undoubtedly lies with the complex interactions that take place between the various plant hormone systems. For example, the application of a PGR might cause changes in the levels of an endogenous phytohormone. In turn, this might have no effect on producing the desired result, it might produce a result other than the one desired, or it might produce the desired result in conjunction with a good or bad side effect. Nevertheless, some results are possible, and they are discussed later in this section.

For the plant growth regulator to be effective, it must enter the plant. A number of possibilities exist. Uptake is possible through the roots, leaves, and stems. The degree of entry varies with many factors such as the plant species, the part of the plant, cuticle thickness and how open stomates are, the chemical nature of the PGR, the temperature, the solubility of the PGR, the relative humidity, and the nature of the carrier (solubilizer). In general, entry through the roots is easier than through the foliage.

After entry, the fate of the plant growth regulator is even more complex. Chemicals can be translocated, they can be nontranslocatable, only certain portions of the translocation system might selectively translocate them, they might be moved out of the plant, and their chemical structure might be unaltered or modified. Any environmental factors that affect translocation will also affect the movement of the PGRs.

In view of the complexities involved, it is sufficient for our purposes to say that entry of plant growth regulators is possible. To obtain a chemical formulation that will enter the plant and produce a beneficial result is often a long, arduous task based on existing experimentation and much trial and error and screening of formulations on large numbers of plants. Concentration is also critical. If it is too low, there is no effect; if too high, plant damage can result.

Application of a plant growth regulator can bring about regulation by three means. The first involves the use of chemicals that alter endogenous phytohormone systems, the second uses chemicals that function as disruptive agents, and the third uses chemicals that bring about regulation through interactions between phytohormone systems.

Alteration of phytohormone systems can be achieved by applying an isolated phytohormone or an analog of it, which brings about regulation directly, or a chemical can be used to stimulate or inhibit the synthesis of an endogenous phytohormone. Disruptive agents disrupt the plant tissue and induce an altered physiological state, which includes changes in the phytohormone systems. Interaction reactions can be induced by applying two combined regulators to produce an enhanced effect or by applying one regulator that interacts with an endogenous phytohormone to produce the desired result. Examples include the use of ethylene to stimulate ripening of fruit, the use of cycloheximide to stimulate abscission of citrus fruits, and the application of auxin to pineapples to induce ethylene production, which in turn causes flowering.

The structures of several plant growth regulators are shown in Figure 14–19. PRGs can be grouped as natural phytohormones or according to function. Each will be treated separately. Trade names are capitalized and common names are not.

Auxin

Auxin (indole-3-acetic acid, see Figure 5–20) derivatives (indole-3-butyric acid or IBA, Hormodin®) are useful for improving the rooting of cuttings (Figure 14–20). A formulation of it with a fungicide has been used to reduce the losses attributable to rotting of cuttings during rooting. Other combinations include a 1:1 mixture of indolebutyric acid and naphthaleneacetic acid, or the preceding two plus naphthalene acetamide (Rootone®). Indolebutyric acid is solubilized in a minimal volume of ethanol and diluted to a final concentration that ranges from 500 to 5,000 ppm. Alternately, the solubilized IBA can be blended with talc to produce a final strength of 0.1% to 1.0%. Generally, much of the cutting is immersed in the liquid; with powder formulations, only the base is dusted. Indolebutyric acid resists breakdown by sunlight and bacteria.

Empirical trials are necessary to determine the optimal concentration for rooting any particular species under a given set of conditions. Too much can inhibit bud development of shoots, and too little will not stimulate rooting. Softwood cuttings of many bedding plants respond favorably to the lower end of the range given.

Indolebutyric acid is not effective in stimulating rooting of all plant species at every developmental stage. Some species that are difficult to root are believed to be lacking or limiting in naturally occurring rooting co-factors. In others it is believed that rooting inhibitors are present. Sometimes success results when the cuttings are leached to remove inhibitors. The identity of the rooting co-factors remains unknown.

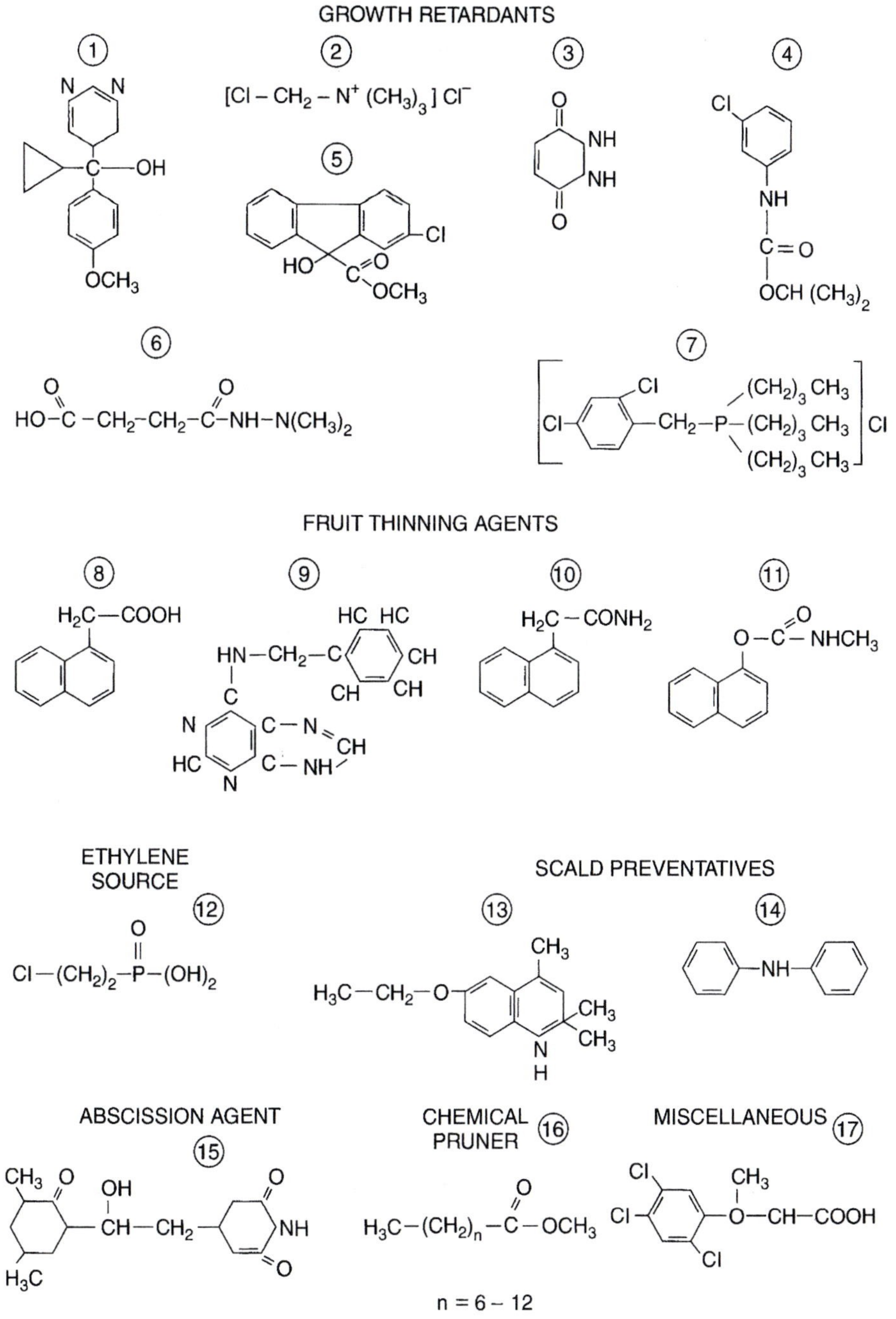

GROWTH RETARDANTS

① ② [Cl – CH₂ – N⁺ (CH₃)₃] Cl⁻ ③ ④ ⑤

⑥ ⑦

FRUIT THINNING AGENTS

⑧ ⑨ ⑩ ⑪

ETHYLENE SOURCE ⑫

SCALD PREVENTATIVES ⑬ ⑭

ABSCISSION AGENT ⑮   **CHEMICAL PRUNER** ⑯   **MISCELLANEOUS** ⑰

n = 6 – 12

FIGURE 14–19
Structures of several growth regulators. (From Raymond P. Poincelot, *Horticulture: Principles & Practical Applications,* 1980. Reprinted by permission of Prentice-Hall, Upper Saddle River, New Jersey.)

Gibberellins

Gibberellins (see Figure 5–22), chemically 2,4a,7-trihydroxy-1-methyl-8-methylenegibb-3-ene-1,10-carboxylic acid → 1-4 lactone (gibberellic acid, Gibberellin, Gib Tabs®, Gibrelo®, Gib-Sole®, Pro-Gibb®, ProVide®, Berelex®, Activolo®, or Grocel®), have been utilized for achieving various plant responses (Figure 14–21). A number of uses have been helpful to the horticulturist. Application on 'Thompson Seedless' and other grapes causes cluster loosening and elongation and increases grape size. Applications at weekly intervals for up to 3 weeks will produce early flowering for many summer-flowering annuals. It is also used to break dormancy in seed potatoes in lieu of low temperatures. Concentrations vary and depend on several factors. The range is critical, and overdosing will produce plants that are horticulturally unacceptable. Aqueous solutions of gibberellins are unstable, so they should be freshly prepared prior to usage.

FIGURE 14–21
(A) Loblolly pine on the
right was sprayed with
400 ppm of gibberellic
acid at one year. Control
is on the left. Plants are
now two years old and
treated plant is taller and
has a double thickness
stem. (B) Chrysanthe-
mum on right was
treated with gibberellic
acid during the fourth
through seventh week of
short days. It shows a
higher stem and flower
stalk and earlier
blooming than the
control on the left.
(USDA photographs.)

(A)   (B)

Numerous other functions exist. On sweet cherries Pro-Gibb can delay harvest and increase the color, firmness, and size of the cherries. With tart cherries, it can help maintain
high fruiting activity. ProVide can reduce russeting of apples and preharvest cracking. Gibberellic acid is especially important to the citrus industry. It is used to control fruit maturity,
fruit set, yield, rind color, and rind aging with citrus (navel and Valencia oranges, lemons,
limes, and minneola tangelos). Other applications include the induction of bolting and the
increase of seed production with lettuce; the increase of yields and as a harvest aid with
forced rhubarb and Italian prunes; the promotion of rapid emergence with beans, peas, and
soybeans; and the extension of the harvest season for sweet cherries and artichokes.

Growth Retardants

Growth retardants are an important group of plant growth regulators. These regulators slow
growth, and many of them produce secondary effects of great value to the horticulturist (Figure 14–22). Most growth retardants are dissolved in water and a surfactant is added, unless one
is present in the growth retardant formulation. Surfactants usually consist of a detergent or a

FIGURE 14–22 The growth retardant Phospfon® was used on the chrysanthemum at right when eight weeks old. Control is on the left. (USDA photograph.)

sulfonated alcohol and are added at a low concentration. Growth retardants are applied as a foliar spray or a soil drench. The various growth retardants will be considered individually.

Butanedioic acid mono-(2,2-dimethylhydrazide), known as daminozide (Figure 14–19:6) (B-Nine®), is used to reduce internode elongation, to induce heat, drought, and frost resistance, to produce darker green foliage and stronger stems, to inhibit undesirable stretching of transplants, and to produce earlier and multiple flowers. It is used on a number of ornamentals (azalea, chrysanthemum, foliage plants, gardenia, hydrangea, marigold, petunia, poinsettia) in enclosed structures such as greenhouses, shade houses, and interiorscapes. Most dicots respond, but monocots do not. Bonzi® (paclobutrazol) also reduces internode elongation, thus producing more compact, uniform, and more marketable plants. Plant usage in enclosed environments is similar to B-Nine, except that bulb crops are also included. Sumagic® (uniconazole-P) has similar effects and usage.

The chemical α-cyclopropyl-α-(4-methoxyphenyl)-5-pyrimidinemethanol (Figure 14–19:1) (ancymidol, A-Rest®, or Reducymol®) is used as a growth retardant on bedding plants, foliage plants, and ornamentals such as chrysanthemums, tulips, poinsettias, and lilies. It is applied as a spray or soil drench. Also used as a growth retardant in the form of a soil drench or spray are (2-chloroethyl) trimethylammonium chloride (Figure 14–19:2) (chlormequat, Cycocel®, or Cycogan®) and tributyl-2,4-dichlorsbenzylphosphonium chloride (Figure 14–19:7) (chlorphonium, or Phosfon®). The first is used on azalea, bedding plants, geranium, herbaceous crops (flowering potted plants, foliage plants, and tropical and temperate perennials), hibiscus, poinsettia, and a few woody crops (fuchsia, hydrangea, lantana, and rhododendron) and the second on chrysanthemums and Easter lilies. Chlormequat is often used with B-Nine.

Maleic hydrazide (Figure 14–19:3) (chemically, 1,2-dihydro-3,6-pyridazinedione), known in the trade by about 20 names (Retard®, Slo-Gro®, Maintain-3®, to name a few), is a growth retardant with several uses. It is used as a growth retardant on turf, as a chemical pruner on trees and ornamental shrubs, as a dormancy inducer for citrus, as a sprout inhibitor for stored onions and potatoes, and as a herbicide in some instances. Methyl-2-chloro-9-hydroxyfluorene-9-carboxylate (Figure 14–19:5) (chlorflurenol) is a growth retardant used in combination with maleic hydrazide on weed growth and turf. Isopropyl N-(3-chlorophenyl) carbamate (Figure 14–19:4) (CIPC®, Chloro-IPC®, Sprout Nip®, or Spud-Nic®) is a growth retardant used to prevent sprouting of potatoes. Dikegulac (Atrimmec®) is also used as a growth retardant (and a chemical pruner) on various ornamentals in Florida nurseries and greenhouses.

Fruit-Thinning Agents

A regulator and also an insecticide, 1-naphthyl-N-methylcarbamate (Figure 14–19:11) (carbaryl, Sevin®) is used to thin apples. Naphthalene acetamide (Figure 14–19:10) (Amid-Thin®) is applied to thin apples (early-maturing cultivars) and pears; it also prevents preharvest fruit drop. Apples, pears, pineapples, and olives can be thinned and preharvest drop prevented by α-naphthaleneacetic acid (Figure 14–19:8) (NAA®, Fruit Tone®, Fruit Fix®). Since one of the above thinners is also a frequently used insecticide (Sevin), care should be exercised to avoid following it with a thinner when used as an insecticide to prevent excessive thinning of fruit. Sevin also harms beneficial insects (honeybees and mite predators) when used as a fruit thinner. As such it is not recommended for frequent use. Accel® is an apple fruit thinner based on cytokinin; the active ingredient is 6-benzyladenine. Best results occur when Accel or NAA is used in combination with either Sevin or Vydate® (another product that serves as both a pesticide and thinner). Accel should not be used in combination with NAA, because this combination can lead to pygmied fruits.

Ethylene Sources

The compound (2-chloroethyl) phosphonic acid (Figure 14–19:12) (ethephon, Ethrel®, Florel®, or Cepha®) is used to release ethylene into plant tissues. Apple responses include increased flowering, thinning of fruit, earlier bearing, and increased red color. With cherries, it loosens fruit and increases color. With other crops it can lead to earlier harvests for filberts, increased yields and earlier uniform ripening for tomatoes, increased fruit set and yield for pickling cucumbers, flower induction on pineapples and ornamental bromeliads, improved harvest efficiency for blackberries and blueberries, faster maturity and color intensity for cranberries, accelerated ripening of lemons and peppers, height reduction for potted daffodils, reduced stem topple for potted hyacinths, loosened fruit and improved color for tangerines, and production of cucumber, pumpkin, and squash hybrid seed. Ethylene is also applied directly as a gas in a confined area. It can be used to de-green and ripen bananas, citrus, honeydew melons, pears, persimmons, pineapples, tomatoes, and walnuts.

Scald Preventatives

Storage scald of apples can be prevented with the plant growth regulators 1,2-dihydro-6-ethoxy-2,2,4-trimethylquinoline (Figure 14–19:13) (ethoxyquin, Stop Scald®, Santoquin®, or Nix-Scald®) and diphenylamine (Figure 14–19:14) (Big Dipper® or Scaldip®). Fruit can also be dipped after harvest in a wax emulsion.

Abscission Agents

Abscission agents are generally used to cause abscission for improving harvest efficiency. Fruit can be loosened for easier picking, or the plant may be defoliated for cleaner harvests. Citrus fruit can be loosened, but not dropped, with 3-[2-(3,5-dimethyl-2-oxocyclohexyl)-2-hydroxyethyl]-glutarimide (Figure 14–19:15), which is also called cycloheximide or Actidione®.

Chemical Pruners

Methyl esters of C_6 to C_{12} fatty acids (Figure 14–19:16) (Off-Shoot-0®) are used as a chemical pruner on some ornamental plants, tomatoes, and woody plants such as azalea, *Cotoneaster,* juniper, *Ligustrum, Rhamnus,* and *Taxus.* Tre-Hold® (an ethyl ester formulation of naphthalene acetic acid) is used to reduce the regrowth of suckers and watersprouts on fruit trees around pruned areas. Prohexadione calcium (Apogee®) can be used to reduce pruning from an annual practice to every other year. Dikegulac (Atrimmec) is also used as both a chemical pruner and growth retardent on ornamentals.

Miscellaneous Regulators

A mixture of cytokinin (6-benzyladenine) and gibberellin (Promalin®, Typy®) is used with tree fruits. With apples, it can be used to increase the length relative to the diameter. Fruits have a more elongated appearance (typiness) such as seen with 'Red Delicious.' This mixture can also be used to increase branch angles and the number of lateral branches with spurs. It is applied to young nonbearing trees in nurseries and sometimes orchards to improve fruit bearing, specifically to apples, cherries, and pears.

ReTain® (amino-ethoxyvinylglycine) aids harvest scheduling. It inhibits the biosynthesis of ethylene. Some uses include delayed ripening of apples and reduced preharvest fruit drop. It also maintains firmness and other qualities that are desirable in ripe apples. The red color also develops more fully. With citrus, fruit drop can be reduced and delayed by 2,4-dichlorophenoxyacetic acid. Blossom set on tomatoes is improved with p-chlorophenoxyacetic acid (Figure 14–19:17) (Fruitone® or Tomatotone®). Deoxy derivatives of gibberellin (Pro-Gibb 47) are sprayed on cucumbers to develop male flowers on gynoecious cucumbers for purposes of seed production.

SUMMARY

Pruning involves the selective removal of specific woody plant parts such as a branch or branch part. Pruning is used to alter size, shape, and growth direction of young woody plants for esthetic or practical reasons (training). Many ornamental shrubs and trees, fruit and nut trees, and bush fruits trees undergo training at nurseries. Flowering, fruiting, health, size, shape, and growth are subsequently maintained through maintenance pruning. Plant health is aided by the removal of diseased and dead branches.

Deciduous trees are usually pruned in late winter and early spring, both for ongoing maintenance pruning and restoration and rejuvenation. Exceptions include maples and birches. Mature fruit trees are dormant pruned. Deciduous shrubs that bloom in the spring are pruned soon after flowering in order not to disturb the setting of next year's flower buds. Later blooming deciduous shrubs bloom on the current year's wood and are pruned in late winter and early spring. Foliage evergreens are pruned in early spring just prior to growth and flowering ones right after flowering. Climbers are treated like shrubs; time of blooming determines when to prune.

Pruning back growth to a bud is *heading back*; complete removal of selected growth back to a lateral, the trunk, or the crown is *thinning out*. Heading back adds new wood and produces a bushy appearance; thinning out results in a more open look. The need for both pruning techniques varies with species, stage of development, overall plant form, and the desired goal. A slanted cut close to the bud is desirable when heading back with handheld pruning shears or on larger branches with lopping shears. The bud should be an outside bud such that the branch grows outward and keeps the center open.

Watersprouts are rapidly growing, vertical shoots produced from the main stem or head of a tree or shrub. *Suckers* are rapidly growing shoots produced near the soil line. Both are superfluous and are thinned out during maintenance or rejuvenation pruning.

Larger branches are removed by either a lopping shear or pruning saw. Branches that are too large to be handled safely are thinned out using support ropes and an undercut followed by a normal cut. Double cuts are necessary to prevent splitting and bark stripping. The final cut on the tree trunk after the removal of the branch is important for correct healing and minimal disease. A correct cut leaves no stub, and is flush with the trunk.

Training pruning allows the horticulturist to produce a well-balanced, easily maintained plant with a practical and pleasing size and shape and strong, well-placed branches. Training is important for fruit trees, in that it allows for increased productivity, less disease and ease of mechanical harvesting. Pruning to train a tree or shrub is started by the nurseryman and sometimes continued by the purchaser.

Young deciduous trees can be trained at the nursery or by the homeowner into *standards* or *central leaders*. The former has a clear stem for 6 feet, and branching above. When the height is 7.5 feet, the tip is removed to promote branching. The central leader system is usually used with larger trees. These tree forms have a clear stem for 6 to 8 feet and the leader is not pruned. Heading back is done on the other branches to produce a balanced head. Pruning may or may not be done to keep the center of the tree open.

Evergreen trees, conifers in particular, respond unfavorably to heavy pruning. Nursery training is minimal; the main thrust is to allow natural form development. Broad-leaved evergreen trees are allowed to develop with a central leader, and side shoots are not removed. Long-needled conifers are usually only given light heading back of their terminal growth. Conifers with fine or scale-like foliage are sheared or tip-pinched partially on their new growth into a desirable shape.

Young fruit trees are trained in three ways: central leader, modified leader, and open center systems. The central leader system has a tall, pyramidal shape that makes spraying and picking fruit difficult. The modified leader system is popular; it produces a well-shaped tree of good structural strength suitable for orchards. With open center training, the branches all arise from one small area on a shorter, main trunk, and the center is primarily free of branches. This type of form has a disadvantage. The open center can become a water pocket that rots and it is structurally weaker. Although not hard and fast categories, the central leader system is used for nut trees and some tropical fruits. The modified leader system is used with apples, apricot, cherry, olive, pear, persimmon and plum trees. The open center technique is used for almond, fig, nectarine, peach, Japanese plum and quince trees.

Shrubs need less training than ornamental and fruit trees. Training produces a shrub with a well-balanced, evenly spaced framework. Berries such as blackberries and raspberries are trained according to the growth patterns shown in the growing area. Grapes are generally trained on wire and wood supports, trellises, and arbors, and pruned by the balanced pruning technique. Some very specialized forms of training are rather labor intensive. They are utilized primarily for esthetic purposes. These forms include topiary, espalier, cordon, and bonsai.

Root pruning is an essential part of training in nurseries. Without it the digging of field-grown nursery stock for transplanting and sale could not be done successfully. Root cutting forces the development of fibrous roots near the base and facilitates the transplanting operation. Recovery is faster and losses minimal. Root pruning is also practiced on container nursery stock. Container stock should be moved to bigger containers in time to avoid producing constricted roots. If constricted roots develop and remain uncorrected, plant quality will be inferior.

Established woody plants need maintenance pruning by the homeowner. Maintenance pruning is needed to restrict growth (space limitations) or correct deviations from form, to obtain a balance between vegetative growth and flowers/fruits, to maintain health, and to produce quality improvements of fruits and flowers. Diseased or dead wood should always be pruned promptly back to healthy wood. Other branches that are potential sources of health problems and/or detract from form should be removed before they become problems. Problem branches include watersprouts, suckers, crossed branches, horizontal forks of equal thickness, narrow-angled scaffold branches, and underbranches on a main branch.

Under some conditions a tree might have progressed beyond the point of corrective pruning. This event occurs if the tree is mature and the removal of narrow-angled scaffold branches requires major surgery or the removal of a weakened branch would leave a gaping hole. Under these conditions the scaffolds or weakened branches can be strengthened with bracing cables attached to bolts. This treatment may also be used to save a storm-damaged, irreplaceable tree of great value.

After initial training some woody plants need minimal pruning to maintain form. Others need ongoing annual pruning. Pruning is also required sometimes to restrict plant size. Ongoing pruning will be needed for foundation shrubs, ornamental and shade trees, fruit trees, vines, and even house or greenhouse plants. Heading back and thinning out are the mainstays of size restriction and form retention. These techniques are also used to maintain a balance between vegetative growth and flowering or fruiting.

On ornamental trees one should remove superfluous branches that develop down low or near the tip. Branches that spoil the form are thinned out. Branches that must be removed because of size restrictions should be headed back to a vigorous side branch. The oldest branches should be cut out and younger wood trained into its place by heading back adjacent branches.

Pruning of established shrubs differs widely, but is based on the time of flower bud development and the age of the reproductive wood. Spring-flowering deciduous shrubs bloom on last season's wood; that is, flower buds are formed prior to dormancy. Pruning is done shortly after flowering and consists of thinning out superfluous growth at the crown. Growth that is too excessive in length or detracts from plant form is headed back. Other shoots are thinned out to open the center of the shrub. Deciduous shrubs flowering in the summer and autumn bloom on the current season's growth. These types are dormant pruned. Pruning is the same as for the spring-flowering shrubs just covered.

Broad-leaved evergreen shrubs generally require less pruning than deciduous materials. This approach is especially true of shrubs with slow, even, compact growth, originating from terminal buds. This produces a dome-like appearance with little interior growth. Light pruning is done in early spring prior to growth; heavier pruning is delayed until after flowering. Fast-growing broad-leaved evergreens that follow the growth habits of the deciduous shrubs are pruned more heavily.

Coniferous evergreens usually reach the homeowner after minimal shaping and training at the nursery. Fine-foliaged conifers are lightly sheared or tip-pinched to maintain shape, usually in the late spring or early summer when growth is completed. Evergreens used as hedges are more heavily sheared. Long-needled conifers can also be pruned in the late spring or early summer by lightly heading back candle growth.

Once trained, fruit trees are pruned to maintain maximal fruit yield. Watersprouts, suckers, dead wood, and weak branches are removed on all fruit trees. The trained form is also preserved. Strong young wood is promoted through pruning to eventually replace the fruit-bearing wood, and spent fruit-bearing wood is removed.

Grapes and berries are pruned to maintain the trained form and to encourage maximal fruiting wood. Pruning varies from the complete removal of all canes that bore fruit right after harvest to only partial removal. Complete removal is practiced with brambles (raspberry, blackberry), which, though perennials, have canes with a biennial fruiting habit. Partial removal is used with grapes, where the fruit is produced on the current growth from buds set the previous season.

Flowering climbers and vines are generally pruned at a time determined by the age of the wood on which flowers are produced. In general, climbers and vines that bloom on the current season's wood can be pruned during the winter. Those blossoming on the past season's wood are pruned after flowering is completed.

Pruning is also used to improve the quality of flowers and fruit. This activity involves the removal of some buds, flowers, or fruits to concentrate resources for those remaining. This selective removal increases the final size over that which would have resulted if all were retained. When used to increase the size of fruits, removal is termed thinning, and disbudding when flower size increase is desired.

Serious consideration of biological techniques for growth regulation is needed from the sustainable viewpoint. Incentives include the negative image of chemical usage held by the public and tighter EPA regulations of plant growth regulators (PGRs). The alternatives are more costly, might require more labor, and possibly cause slight plant damage. In some cases the alternatives might produce inconsistent results. Often biological alternatives are new, thus knowledge of their efficacy is limited and minimal research exists to gauge their effectiveness. Still, organic horticulturists have achieved successful production without plant growth regulators. Some sustainable horticulturists are mixing both biological and chemical techniques, which gives satisfactory crop production and reduces the use of chemical regulators.

Seaweed extracts combined with humic acids and various enzymes (biostimulants) are used somewhat in Florida vegetable production and also in turf production and management. Grafting is used mostly for propagation, but can be used for biological control of plant growth. Rootstocks can be used to dwarf the cultivar selected to be the scion with apple, apricot, and peach. Other alterations include the production of tree roses and weeping birches, cherries, and needled evergreens. Rootstocks can also be chosen to enable the selected scion cultivar to tolerate unfavorable soil conditions, to resist diseases, to improve the size and quality of fruit, to encourage early bearing of fruit, and to replace rootstocks that are not cold hardy.

Ringing and scoring are forms of girdling. Scoring denotes a thin cut into the bark layer with a knife or saw. These techniques are used to reduce excessive vigor and encourage flowering and fruiting with fruit trees and vine fruits. These treatments give only temporary effects and are not used to any great extent. Shaking is another form of biological control, but a natural one that requires minimal output. The stresses of wind movement appear to strengthen plant stems and in some cases cause dwarfing.

With greenhouse ornamental crops, the main problem is height control. Some simple steps such as not starting plants too early (better scheduling) or using dwarf cultivars are possible. Another simple step is to maximize light levels in the greenhouse to reduce "stretch." Cleaning glass, replacing aged plastic, and increased plant spacing can help increase light levels. Cultural steps can control plant heights. One is to use a minimal size container such that some root restriction results. Development will be slowed and some nutrient/water stress will result, all of which help to reduce stem elongation. Keeping fertilization levels low, especially nitrogen, and allowing slight wilting will accomplish the same goal. Moderate phosphorous deficiencies are also effective. These approaches must be cautiously used to prevent decreased plant quality.

Temperature control can also reduce the need for chemicals to control height. The most useful technique is DIF, which is defined as the difference between day temperature and night temperature. Stem elongation is less when nights are warmer than days (negative DIF). Light modification has also been used for height control. Stem elongation is increased by far-red wavelengths, which are more prevalent at dawn and dusk. Height reductions result when black cloth is drawn over plants from 30 minutes prior to sunset through 30 minutes after sunrise. Last, mechanical conditioning of plants in the greenhouse can restrict height. Greenhouse tomato growers draw cardboard tubes or bars suspended from irrigation booms across plant tops to achive height reductions.

Chemical control can be achieved with isolated phytohormones, synthesized versions, and analogs. PGRs are used to modify plant development for purposes of increased productivity, better marketability, and less labor usage. Less impact on environmental sustainability is seen with PGRs, as opposed to pesticides. Judicial and minimal usage of chemical growth regulators is best. PGR uptake occurs through roots, leaves, and stems.

PGRs and the responses of plants to them must be registered for use through the EPA. Various states have laws controlling their use. The horticulturist must understand the laws governing PGRs, obtain a license to use them when required by law, watch for changes in the regulations and lists of permissible chemicals, and observe all safety precautions. Care must also be exercised to avoid spray drift to nearby non-target plants.

Several types of PGRs exist. Auxin derivatives alone, or combined with fungicide or other products are useful for improved rooting of cuttings. Gibberellins are utilized on grapes to cause cluster loosening and elongation and increased grape size. Other uses include earlier flowering for many summer-flowering annuals; dormancy breaking in seed potatoes; delayed harvest and increased color, firmness, and size (sweet cherries); maintaining high fruiting activity (tart cherries); reduced russeting and preharvest cracking of apples, and controlling fruit maturity, fruit set, yield, rind color, and rind aging with citrus.

Growth retardants are used on ornamentals and bedding plants in the greenhouse, shade houses, and in interiorscapes. Applications reduce stem height, induce heat, drought, and frost resistance, produce darker green foliage and stronger stems, inhibit undesirable stretching of transplants, and produce earlier and multiple flowers. Other uses are growth inhibition on turf, chemical pruning on trees and ornamental shrubs, a dormancy inducer for citrus, and a sprout inhibitor for stored onions and potatoes.

Several PGRs are used as fruit thinners and to prevent preharvest fruit drop. Some PGRs are used to release ethylene into plant tissues. Apple responses to ethylene include increased flowering, thinning of fruit, earlier bearing, and increased red color. With cherries, looser fruit and increased color result. Effects on other crops include earlier harvests for filberts, increased yields and earlier uniform ripening for tomatoes, increased fruit set and yield for pickling cucumbers, flower induction on pineapples and ornamental bromeliads, improved harvest efficiency for blackberries and blueberries, faster maturity and color intensity for cranberries, accelerated ripening of lemons and peppers, height reduction for potted daffodils, reduced stem topple for potted hyacinths, loosened fruit and improved color for tangerines, and production of cucumber, pumpkin, and squash hybrid seed. Ethylene is also applied as a gas in a confined area to de-green and ripen bananas, citrus, honeydew melons, pears, persimmons, pineapples, tomatoes, and walnuts.

Storage scald of apples is prevented with PGRs. Abscission agents are generally used to improve harvest efficiency. Fruit can be loosened for easier picking, or the plant may be defoliated for cleaner harvests. Other PGRs reduce the regrowth of suckers/watersprouts on fruit trees at pruned areas. Miscellaneous regulators include a mixed cytokinin/gibberellin. With apples, this mixture increases the length relative to the diameter (typiness). This mixture also increases branch angles and the number of lateral branches with spurs when applied to young nonbearing trees in nurseries and sometimes orchards to improve fruit bearing, specifically to apples, cherries, and pears. Some PGRs inhibit the biosynthesis of ethylene and cause delayed ripening of apples, reduced preharvest fruit drop, maintained firmness, and other qualities that are desirable in ripe apples. With citrus, fruit drop is reduced. Blossom-set PGRs are used to increase tomato fruit set and to develop male flowers on gynoecious cucumbers for seed production.

EXERCISES

Mastery of the following questions shows that you have successfully understood the material in Chapter 14.

1. What is pruning? What types of plants are usually pruned?
2. What does training pruning accomplish? Where does it take place?
3. What goals can be accomplished with maintenance pruning?
4. What happens to buds below pruning cuts? What happens when pruning removes an entire branch and leaves no buds?
5. When should the following groups of plants be pruned: most deciduous ornamental trees, birches and maples, fruit trees, deciduous spring flowering shrubs, deciduous summer flowering shrubs, foliage evergreens (trees and shrubs), flowering evergreens (trees and shrubs), and lianas?
6. What types of tools are used in pruning? Is disinfection needed?
7. What are the differences between the pruning cuts termed *heading back* and *thinning out?* What are the effects of each of these cuts?
8. Does it matter what distance is left to the nearest bud after heading back? Why? Does it matter whether the remaining bud is an inside or outside bud? Why?
9. What are watersprouts and suckers? Should they be removed by pruning? Why?
10. Why do larger branches need to be roped and double-cut with an undercut and overcut?
11. Should large cuts be painted with tree wound paint? Why or why not?
12. What are the likely causes of accidents and injuries during pruning?
13. What type of training is done with various evergreen trees?
14. What are the three training systems used for fruit trees? How do they differ? Which one is preferred? Why?
15. What training is done with deciduous and evergreen shrubs? Berries?
16. What are topiary, espalier, cordon, and bonsai?
17. Is root pruning ever done? If so, on what and why?
18. What types of wood or branches should be removed promptly by pruning?
19. What are some of the basic pruning steps for deciduous ornamental trees and shrubs?
20. What are some of the basic pruning steps for evergreen ornamental trees and shrubs?
21. What is deadheading?
22. What are some of the basic pruning steps for fruit trees, grapes, and berries?
23. What types of biological approaches are available for control of plant growth?
24. How can grafting, scoring, and ringing modify plant growth and reproduction?
25. What are growth regulators? Why should their use be avoided or minimized?
26. What are the three ways in which a growth regulator can bring about results?
27. Name the various groups or classes of growth regulators. What are the basic effects of each one?

INTERNET RESOURCES

Basics and techniques of pruning

http://aggie-horticulture.tamu.edu/extension/pruning/pruning.html

http://www.ext.vt.edu/pubs/nursery/430-455/430-455.html

B-Nine®

http://www3.infolinksa.com/demouniroyal/tablestubs/b-nine_links.asp

Bonsai primer

http://www.bonsaiprimer.com/bonsai/bonsai.html

Bonzi®

http://www3.infolinksa.com/demouniroyal/tablestubs/bonzi_links.asp

Chemical management: pesticides and growth regulators

http://tfpg.cas.psu.edu/part3/part33.htm

Chemical thinning

http://www.umass.edu/fruitadvisor/healthy_fruit/hf0799.html

Citrus and plant growth regulators

http://www.ipm.ucdavis.edu/PMG/r107900111.html

DIF and height control

http://www.for.gov.bc.ca/nursery/fnabc/Proceedings/DIFAndHeightGrowthControl.htm

http://www.open.k12.or.us/mars/etag/mmpht225.html

http://www.umass.edu/umext/programs/agro/floriculture/floral_facts/altpgr.htm

Gibberellic acid and 'Crimson Seedless' table grapes

http://cetulare.ucdavis.edu/pubgrape/tb1298.htm

Gibberellic acid and delayed senescence with citrus

http://www.ipm.ucdavis.edu/PMG/r107900511.html

Growth Regulators in Floriculture

http://www.ext.vt.edu/pubs/greenhouse/430-102/table1.html

Mechanical growth conditioning

http://www.imok.ufl.edu/veghort/pubs/workshop/pdf/latimer2.pdf

Pink coreopsis and growth regulators

http://www.ag.auburn.edu/resinfo/publications/ornamentals99/greenhouse/pgrcoreopsis.html

Poinsettia height control

http://aggie-horticulture.tamu.edu/greenhouse/guides/poinsettia/height.html

Pruning and training fruit trees

http://www.ces.ncsu.edu/depts/hort/hil/ag29.html

http://www.ianr.unl.edu/pubs/horticulture/ec1233.pdf

Pruning berries

http://www.utextension.utk.edu/spfiles/sp284e.pdf

http://www.utextension.utk.edu/spfiles/sp284g.pdf

Pruning evergreens

http://ag.udel.edu/extension/information/hyg/hyg-73.htm

http://www.ces.ncsu.edu/teletip/scripts/5116.htm

http://www.uri.edu/ce/factsheets/sheets/evergreenprune.html

Pruning grapes

    http://www.ent.iastate.edu/ipm/hortnews/1995/3-3-1995/prune.html

Pruning roses

    http://www.ag.ohio-state.edu/~ohioline/hyg-fact/1000/1205.html

Pruning shade trees

    http://www.tree-pruning.com/index.html

    http://www.ces.uga.edu/pubcd/C628-w.htm

    http://muextension.missouri.edu/xplor/agguides/hort/g06866.htm

Pruning shrubs

    http://www.ext.colostate.edu/pubs/garden/07206.pdf

    http://muextension.missouri.edu/xplor/agguides/hort/g06870.htm

Sumagic and hybrid lily cultivars

    http://virtual.clemson.edu/groups/hort/sctop/bsec/bsec-02.htm

Sumagic and mountain laurel

    http://henderson.ces.state.nc.us/newsletters/nursery/sep96/h.html

Survey of chemical use (including growth regulators) in Florida ornamental production

    http://edis.ifas.ufl.edu/BODY_AA242

Topiary at Disney World

    http://www.wdwmagic.com/topiary.htm

RELEVANT REFERENCES

Basra, Amarjit S. (ed.) (2000). *Plant Growth Regulators in Agriculture and Horticulture: Their Role in and Commercial Uses.* Food Products Press, Binghamton, NY. 264 pp.

Brickell, Christopher (ed.) (1996). *American Horticultural Society Pruning & Training.* Dorling Kindersley Publishing, London. 336 pp.

Brown, George E., and John E. Bryan (1995). *The Pruning of Trees, Shrubs and Conifers.* Timber Press, Portland, OR. 374 pp.

Disabato-Aust, Tracy (1998). *Well-Tended Perennial Garden: Planting & Pruning Techniques.* Timber Press, Portland, OR. 269 pp.

Edward, F. Gilman (2002). *An Illustrated Guide to Pruning,* (2nd ed.). Delmar Publishers, Albany, NY. 338 pp.

Ellis, Barbara, and Frances Tenenbaum (ed.) (1997). *Easy, Practical Pruning: Techniques for Training Trees, Shrubs, Vines, and Roses* (Taylor's Weekend Gardening Guides). Houghton Mifflin, New York. 122 pp.

Forshey, C. G., D. C. Elfving, and R. L. Stebbins (1992). *Training and Pruning Apple and Pear Trees.* American Society for Horticultural Science Press, Alexandria, VA.

Gallup, Barbara, and Deborah Reich (1998). *The Complete Book of Topiary.* Workman Publishing, New York. 318 pp.

Hill, Lewis (1998). *Pruning Made Easy: A Gardener's Visual Guide to When and How to Prune Everything, from Flowers to Trees.* Storey Books, North Adams, MA. 224 pp.

Johnson, Eric A., and Scott Millard (ed.) (1997). *Pruning, Planting & Care: Johnson's Guide to Gardening Plants for the Arid West.* Ironwood Press, Tucson, AZ. 160 pp.

Joyce, David (2000). *Topiary and the Art of Training Plants.* Firefly Books, Westport, CT. 160 pp.

Koreshoff, Deborah R. (1997). *Bonsai: Its Art, Science, History, and Philosophy.* Timber Press, Portland, OR. 287 pp.

Liang, Amy (1995). *The Living Art of Bonsai: Principles and Techniques of Cultivation and Propagation.* Sterling Publishing, New York. 288 pp.

Lombardi, Margherita, John E. Elsley (ed.), and Cristiana Serra Zannetti (1998). *Topiary Basics: The Art of Shaping Plants in Gardens and Containers.* Sterling Publishing, New York. 144 pp.

Pessey, Christian, and Remy Samson (1993). *Bonsai Basics: A Step-by-Step Guide to Growing, Training and General Care.* Sterling Publishing Company, New York. 120 pp.

Reich, Lee (1999). *The Pruning Book.* Taunton Press, Newtown, CT. 240 pp.

Somerville, Warren (1996). *Pruning and Training Fruit Trees* (Practical Horticulture). Butterworth-Heinemann, Oxford.

Squire, David (2001). *Pruning Basics.* Sterling Publications, New York. 128 pp.

Yang, Linda (1999). *Topiaries & Espaliers: Plus Other Designs for Shaping Plants* (Taylor's Weekend Gardening Guides). Houghton Mifflin, New York. 128 pp.

Chapter 15

Plant Protection

OVERVIEW

Horticulturists are faced with a number of plant pests that coexist in the horticultural ecosystem. Failure to control these pests can result in consequences that vary in severity. These consequences might only be minor plant damage and some distress for the horticulturist or, worse, loss of plants. A horticulturist responsible for a vegetable crop or a landscape planting that suffers from pest damage because of failure to control pests could quickly lose his or her professional standing.

Control of plant pests, or *plant protection*, involves a number of steps. First, the horticulturist must recognize the possibility or existence of a problem. In this regard it is important to be knowledgeable with respect to plant susceptibility to attacks by plant pests and under what environmental conditions and during which time window a plant is most susceptible. The horticulturist must also be capable of recognizing the presence of a plant pest at its initial appearance, or after it causes a plant response or symptom, and be able to identify the pest once it is discovered. Some pests can only be identified when they are considered in conjunction with the symptoms. Identification is followed by an assessment to determine if control is needed, the control required to eliminate the pest, the method of application, timing of application(s), how much is needed, phytotoxicity and off-site environmental impact (with chemical controls), and effects of control on population balance between beneficial organisms and pests.

CHAPTER CONTENTS

Pests of Animal Origin
Nonanimal Pests
Sustainable Approaches to Plant Protection

IPM Basics
Preventing Pests and Diseases: Proactive Practices
Reactive Practices to Control Insects and Diseases

OBJECTIVES

- To learn to identify and understand the various insects and other pests of animal origin
- To learn to identify and understand the various pathogens that attack plants

- To understand the competitive role of weeds in crop production
- To learn about the usage of proactive controls (barriers, deterrents, cultural practices) to prevent or restrict pests and diseases

❧ To understand how IPM works
❧ To learn about reactive controls (traps, biological controls, cultural controls, and pesticides) and their usage to control pests and diseases

❧ To understand how to use proactive and reactive controls with IPM approaches to achieve sustainable plant protection

PESTS OF ANIMAL ORIGIN

A variety of pests from the animal kingdom attack plants. Pests from the animal kingdom include the well-known arthropods. Arthropods are characterized by segmented bodies and jointed legs. Pests in this group include the insects and noninsect arthropods such as acarids (mites) and centipedes/millipedes. Others pests are the worm-like nematodes and the larger animal pests: slugs, snails, birds, rabbits, deer, mice, moles, other rodents, dogs, cats, and even humans at times. All of the above pests (except for humans) usually injure or kill plants during their feeding activity. Some of the arthropods and nematodes also transmit diseases when feeding.

Insect Arthropods

Insects are members of the same phylum, Arthropoda, as are mites. The class Insecta (Hexapoda) contains the true insects, which differ from mites and other arachnids as discussed in the next section. The species in Insecta constitute the bulk of the plant pests of animal origin and are the number one source of economic damage in horticulture. Insect pests are well adapted to plant destruction because of their feeding habits, small size, protective coloration in some cases, the ability to fly, and their prolific nature.

Insect life cycles are such that we can consider two groups of insects, those that undergo complete metamorphosis and those that undergo gradual metamorphosis. The former group has distinctly different states that give the insect different forms ideally suited for feeding on plants and for reproduction; these are of obvious disadvantage to the horticulturist. In gradual metamorphosis, the change between stages is gradual; hence, the distinction between stages is not obvious. An example of insect metamorphosis is shown in Figure 15–1.

Insecta contains a number of orders with both plant pests and beneficial insects (Table 15–1). There are at least 700,000 species in Insecta and very few plants are completely safe

(A)

(B)

FIGURE 15–1   Stages of insect metamorphosis with the Japanese beetle (*Popilla japonica*). (A) Eggs hatch in the soil to produce a white grub (*left*). Eventually the grub enters an intermediate stage, the pupa, shown right. (B) The adult emerged from the pupa. Both the grub and adult are horticultural crop pests. Grubs do extensive lawn damage and the adults eat many ornamental and vegetable crops. (USDA photographs.)

TABLE 15–1 *Insect Orders*

Order	Insect Types	Comments
Blattaria	Cockroaches	House and food pests.
Coleoptera	Beetles, weevils	Many economically significant plant pests and disease vectors; many important predators.
Collembola	Springtails	Found in debris; annoying pests of potted plants.
Dermaptera	Earwigs	Minor plant damage; found in organic debris.
Diplura	Diplurans	
Diptera	Flies, gnats, mosquitoes	Many predators and parasites; human and plant disease vectors; some plant pests.
Embiidina	Webspinners	Rare.
Ephemeroptera	Mayflies	Found near water.
Grylloblattaria	Rock crawlers	Rare.
Hemiptera	True bugs	Many economically significant plant pests; some carry plant diseases; some are important predators.
Homoptera	Aphids, cicadas, leafhoppers, mealybugs, planthoppers, psyllids, scale, spittlebugs, and whiteflies	Many economically significant plant pests; some carry plant diseases.
Hymenoptera	Ants, bees, ichneumons, sawflies, and wasps	Excellent pollinators; some predators and parasites; few plant pests; some sting humans; some damage wood.
Isoptera	Termites	Wood pests.
Lepidoptera	Butterflies, moths, skippers	Many economically significant plant, food, fabric pests (larvae); many pollinators (adults).
Mantodea	Mantids	Predators of minimal value.
Mecoptera	Scorpionflies	
Microcoryphia	Bristletails	
Neuroptera	Alderflies, antlions, dobsonflies, and lacewings	Some predators.
Odonata	Damselflies, dragonflies	Predators of flies, gnats, mosquitoes, other insects; found near water.
Orthoptera	Crickets, grasshoppers, katydids	Plant pests of economic significance.
Phasmida	Walkingsticks	
Phthiraptera	Lice	Mammal pests.
Plecoptera	Stoneflies	Found near water.
Protura	Proturans	
Psocoptera	Book lice	Found in debris.
Siphonaptera	Fleas	Mammal pests.
Strepsiptera	Twisted-wing parasites	
Thysanoptera	Thrips	Plant pests and disease carriers.
Thysanura	Firebrats, silverfish	Found in debris.
Trichoptera	Caddisflies	Found near water.
Zoraptera	Zorapterans	

from these insects. Damage runs the gamut from microscopic to visible chewed holes in leaves to complete destruction of the plant. A generalized relationship between these symptoms and various source insects is shown in Figure 15–2. In addition, some insects are vectors for certain plant diseases, as covered in a later section.

Noninsect Arthropods

The animal class Arachnida includes spiders, mites, scorpions, and ticks. All differ from the true insects in the class Insecta (Hexapoda). The arachnids have four pairs of legs, as opposed to three pairs for insects and also lack insect antennae, wings, true jaws, and compound eyes.

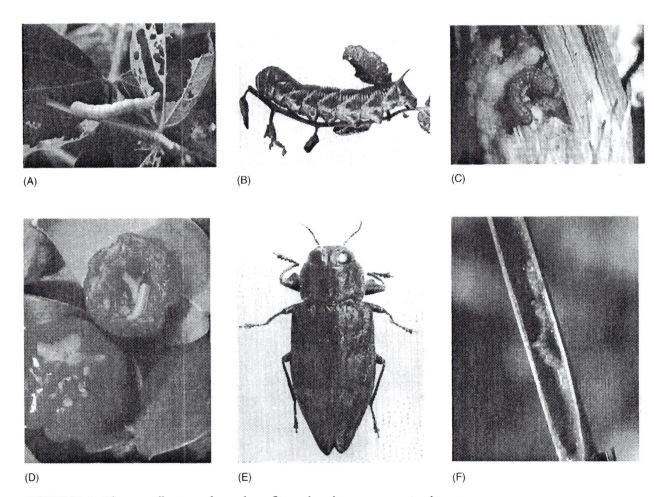

(A)   (B)   (C)

(D)   (E)   (F)

FIGURE 15–2 The caterpillar stage of many butterflies and moths causes extensive damage. Some examples follow. (A) Cabbage looper (*Trichoplusia ni*) eating soybean leaves. This caterpillar is a serious pest on Cole crops. (B) Tomato hornworms (*Manduca quinquemaculata*) can quickly strip tomatoes of foliage and even small green tomatoes. (C) Corn earworms (*Heliothis zea*) are serious pests of sweet corn. Other insects in their larval and/or adult phase eat many plant parts. (D) Fruit flies attack many fruits and vegetables. Fruit fly larvae are feasting on Surinam cherries. (E) The adult form of the flat-headed apple tree borer (*Graphocephala coccinea*) shown here lays eggs on apple trees. The resulting grubs tunnel out wood as they eat and eventually emerge as adults. Borers as a group attack a wide range of deciduous fruit and shade trees. (F) An unidentified stem-boring beetle larva is shown here in this cut-away stem of a herbaceous plant.

In addition, they have only two body regions with no separate head; insects have three body regions, one of which is a separate head. The mites, members of the order Acari, are the only serious arachnid plant pests (Figure 15–3). Mites, along with ticks, are also referred to as acarids. The most common mites are the spider mite, broad mite, and the cyclamen mite. Some mites are also predatory and can be used in biological control. Spiders are beneficial predators around crops and flowers. However, ticks are vectors for human diseases such as Lyme disease or Rocky Mountain spotted fever. Ticks can become a problem for horticultural workers.

Mites are small, usually less than 1/100 inch long. Damage to plants results from their feeding action, the sucking of plant sap. Mites can cause extensive damage since they are prolific, and chemicals used to control insects often eliminate the competitors and enemies of mites, leading to explosive increases of mites. Many insecticides are combined with miticides to prevent such occurrences.

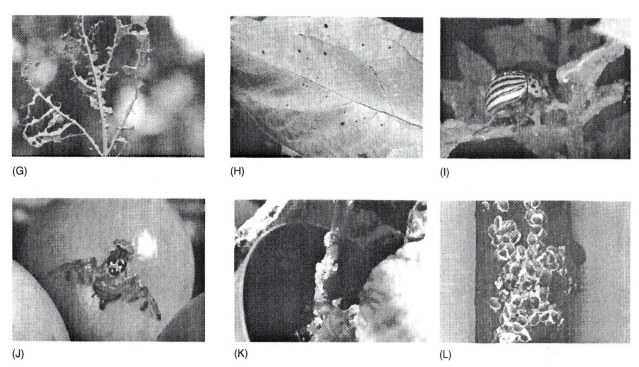

(G)   (H)   (I)

(J)   (K)   (L)

FIGURE 15–2 (*continued*)   (G) Japanese beetles (adults) can cause extensive defoliation as shown here with this green snap bean leaf. (H) These tiny holes in this eggplant leaf show visits from flea beetles. (I) A Colorado potato beetle (*Leptinotarsa decemlineata*) is shown here eating a tomato plant, a close relatives of potatoes. Planting tomato seedlings into a mulch residue of hairy vetch can reduce damage from Colorado potato beetles. (J) The female Mediterranean fruit fly, shown here on a coffee fruit, can deposit eggs 2 to 3 millimeters deep in papayas. Other insects suck plant sap as their food source instead of eating plant tissues. (K) Mealy bugs are sucking insects that damage many plants in the greenhouse, home, and outdoors. The lower stem of this greenhouse plant is covered with mealy bugs. (L) European elm scale (*Gossyparia spuria*) is shown here. Scale are troublesome on ornamentals indoors and outdoors and also on fruit trees. (A-J, L, USDA photographs; K, Author photograph.)

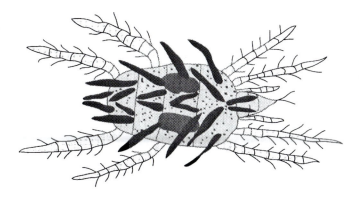

FIGURE 15–3 The two-spotted spider mite (*Tetranychus urticae*) is a common horticultural pest on horticultural crops of all types indoors and outdoors.

Symptoms of mite damage are as follows. Some species cause fine yellow, gray, or white flecking or stippling of the foliage; the whole leaf becomes chlorotic with severe damage and eventually drops off. Webbing can be seen to cover the plant with large populations of certain mites. Other mites cause russetting of leaves, gall production, bud injury, and blistering. Host plants include a very wide range of horticultural material throughout the United States, with some mites being very host specific and others being more far ranging.

Millipedes (class Diplopoda) and centipedes (class Chilopoda) are arthropods characterized by numerous legs (centipedes have considerably fewer legs). Millipedes prefer moist sites, soils with high levels of organic matter and organic debris. Wet conditions and thick organic mulches or compost piles are quite suitable as their habitat. Their numbers are usually minimal, except when warm, wet weather lasts for several or more days. These arthropods cause minor plant damage in the greenhouse and garden. Millipedes feed on decaying organic matter, but can eat plant roots, foliage, and ripening fruits on beans, beets, carrots, corn, cucumbers, melons, parsnips, peas, potatoes, squash, strawberries, and turnips. A few species are beneficial in that they are insect predators.

Centipedes like moist conditions, but are found under stones, boards, or in woodpiles. Centipedes are actually beneficial in that they prey on grubs, other insects, and snails. Unfortunately, they will eat earthworms, too. Another type of centipede called the garden centipede differs in that it is much smaller, has fewer legs, and is white. This species is actually not a centipede, but a symphylan (class Symphala). It feeds on roots of asparagus, cucumbers, lettuce, radishes, tomatoes, and several flowers. It can also be a greenhouse pest.

Nematodes

Nematodes, phylum Nemeta, are roundworms that live in soil and water (Figure 15–4). They range in size from 1/64 to 1/8 inch in length. The typical nematode is 1/20 inch in length. Most nematodes are free living; only a few parasitize plants and animals. Some parasitic types feed mostly on plant roots and a few are leaf feeders. Others parasitic nematodes can be used as predators (entomopathogenic nematodes) in biological control systems. These types will be discussed later in the Biological Controls section.

While feeding, nematodes can introduce diseases into roots that cause wilting or rot. Rootknot nematodes cause galls or swellings on the roots of many plant species. These growths block the flow of water and nutrients, producing a stunted, yellow, wilted plant. Meadow nematodes cause symptoms that could easily be mistaken for root rot. Cyst and sting nematodes cause stunted plants with wilted, curled foliage.

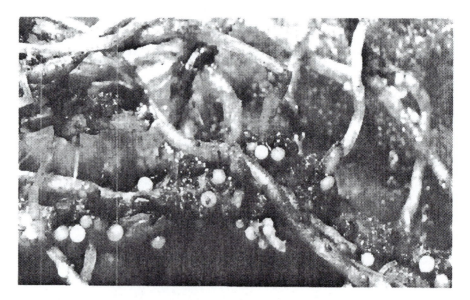

FIGURE 15–4 Nematodes are in the same phylum as hookworms and pinworms that are animal parasites. Nematodes that attack plants are tiny, threadlike worms that mostly feed on roots with a hypodermic-like stylet, although some attack foliage such as the foliar nematodes (*Aphelenchoides* sp.). In some species the female has a more pear-like shape. Some cause knots to form on roots (root knot nematodes, *Meloidogyne* sp.). In others the dead swollen female becomes an egg-containing cyst (cyst nematodes) as shown here in this photograph of roots infected with cysts of the Golden nematode (*Heterodera rostochiensis*). Still others cause root lesions such as the meadow nematode (*Pratylenchus* sp.). (USDA photograph.)

Larger Animal Pests

Slugs and snails damage plants in the garden and greenhouse. Damage varies from holes caused by eating of plant tissues (Figure 15–5) up to partial or total loss of herbaceous plants. For example, slugs eat tomatoes just prior to ripening, leaving holes that make the tomato less appealing. These wounds are also potential areas for pathogen infection. During foraging, slugs and snails leave a slimy trail, which can be used to identify their presence. These pests survive nicely in horticultural habitats because they need moisture, shaded areas, organic debris, and food, all of which are found in garden and horticultural sites.

Some larger animals (birds, rabbits, various rodents, deer, cats, dogs, and humans) can occasionally cause extensive damage to horticultural crops. Damage by larger animal pests can reduce the esthetic qualities of plants, increase their susceptibility to attacks by other plant pests, or kill plants. Many of these creatures are pests simply because horticultural operations and suburban homes are in proximity to rural and forested areas where these animals normally live.

Birds become pests when they eat newly planted seeds and ripe fruits such as sweet corn seed or berries, cherries, and grapes. For example, ripe cherries can be cleaned off of a tree in no time once birds discover them. Ripe strawberries are left partially eaten and riddled with peck holes by early morning. Ripe blueberries also rapidly vanish.

Rabbits and other rodents damage young crops by eating them and often girdle fruit trees when they eat the bark during the winter. Deer can also damage fruit and ornamental trees in the winter by eating the bark. During the growing season, total destruction of some plants such as daylilies, hosta, and tomatoes can occur from deer browsing. Deer also decimate landscape shrubs such as rhodendrons. Deer populations have exploded in many parts of the United States, causing considerable economic damage for farmers and homeowners. Deer are also hosts for Lyme disease-infected ticks. Dogs and cats damage plants by digging activities, sharpening (cats only) of claws on bark, and the use of planted areas as their outdoor toilet.

Humans damage plants through ignorance, carelessness, theft, and vandalism (Figure 15–6). Typical injuries from carelessness include weed whacker damage to lower tree bark and severing of herbaceous perennials, lawn mower and tractor tire injuries from running over plants, ladder and foot damage during painting and construction activities, and careless stepping during walking, playing, and weeding.

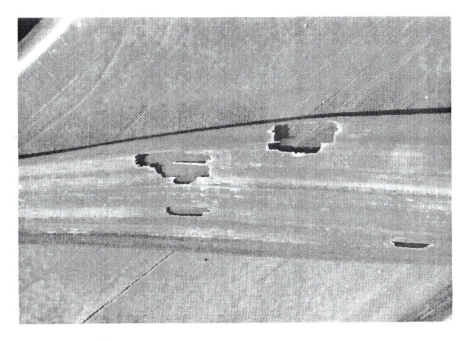

FIGURE 15–5   Slug damage on a Crocosmia leaf. (Author photograph.)

FIGURE 15–6 A street maple planted at a school bus stop. It was vandalized by having its terminal leader broken off. Consequently, sprouting occurred in the lower trunk area, producing an unattractive form. If not corrected by pruning in a short time, the tree will become worthless as a street tree. (Author photograph.)

NONANIMAL PESTS

A number of nonanimal organisms from three nonanimal kingdoms attack plants or interfere with successful cultivation of crops. Pests from the fungal (fungi) and moneran (bacteria) kingdoms injure and kill plants by causing plant diseases. Pests from the plant kingdom, weeds, compete with crop plants for available resources. Others such as viruses and viroids don't conveniently fit into these kingdoms. Nonanimal pests include viruses, viroids, phytoplasmas, spiroplasmas, xylem/phloem-limited bacteria, normal bacteria, fungi, dodder, mistletoe, algae, moss, weeds, strangling vines, and strangling roots.

The microscopic and tinier organisms (viruses, viroids, phytoplasmas, spiroplasmas, xylem/phloem-limited bacteria, normal bacteria, and fungi) cause diseases. The microbial agents are known as *pathogens,* and the result of the microbial infection is termed a *pathogenic disease.* This term is used to distinguish it from nonpathogenic diseases resulting from nutritional or environmental causes. Many prefer to view nonpathogenic diseases as disorders that are physiologically based, rather than diseases. The term *disorder* in this case is more appropriate in that it is more logical and more descriptive of the problem. An example of a disorder is blossom end rot of tomatoes, which, although disease-like in appearance, results from insufficient water (environmental) and calcium deficiency (nutritional).

Viruses and Viroids

The virus is the second smallest pathogen, with dimensions in the millimicron range. As such, a virus can only be visualized with the aid of an electron microscope. Viruses usually consist of a protein coat surrounding an RNA or DNA nucleic acid core. They are not capable of reproduction except within the cells of a living host. Inside the cell they are dependent on the metabolism of the cell, where they direct part of the cell's protein synthesis toward their own reproduction. Viroids are essentially viruses without protein coats and currently are the smallest known plant pathogen. Those viroids associated with plant diseases are of the RNA type only.

Plants of horticultural value affected by viruses include the tomato, cucumber, squash, potato, corn, bean, sugar beet, strawberry, raspberry, apple, peach, and many ornamentals. For a given virus, the host range can be broad or as narrow as a few cultivars within a species. The plant response to viruses, that is, the symptoms (Figure 15–7), provides a basis for division into two viral groups, the *yellows* and *mosaics.* The basic symptom of the former is yellowing

(A)

(B)

(C)

FIGURE 15–7 These plants all exhibit symptoms of viral infections. (A) A healthy dahlia leaf on the right is shown for comparison with a leaf (left) infected with dahlia mosaic virus. (B) These squash leaves are deformed and mottled because of infection with cucumber mosaic virus. (C) This tomato plant exhibits the curly form resulting from infection with curly top virus. (USDA photographs.)

of leaves; mosaic results in a patchwork pattern of normal green and light green to yellow areas in the leaves. Other virus symptoms include spotting, dwarfing, leaf curl, leaf roll, leaf wrinkle, excessive branching, and gall formation. These symptoms result in part from destruction of chlorophyll and, possibly, disruption of hormonal levels and interference with translocation. Viroid diseases include avocado sunblotch, chrysanthemum stunt, and citrus exocortis.

Viruses (Figure 15–8) can be transmitted in several ways among plants. The mosaics group is transmitted by leafhoppers. Other insects such as thrips and whiteflies also spread viruses. Seeds, nematodes, and fungi can also be transmission agents, or *vectors*. Mechanical transmission is also possible, such as rubbing leaves or rubbing viral-infected plant sap into a healthy leaf by hand or with pruning shears or other tools and implements. These same mechanical routes usually spread viroids. Transmission can occur when a viral-infected plant is grafted onto a healthy plant; in fact, this is used as a test to detect suspected viruses in plants where symptoms are few or lacking. The healthy portion of the graft is deliberately chosen to be a highly susceptible stock that will show symptoms rapidly.

Phytoplasmas, Spiroplasmas, and Xylem- and Phloem-Limited Bacteria

A number of unusual prokaryotes related to bacteria cause plant diseases. At first these diseases were thought to be viral based because these organisms produce symptoms that resemble viral diseases (yellowing, sickly green foliage, leaf curl, stunting). Eventually

FIGURE 15–8 Aphids seen on this rose bud are known to spread mosaic viruses. (USDA photograph now in the National Archives.)

organisms that somewhat resembled mycoplasmas and rickettsias were discovered to be the causal agents, hence they were initially termed mycoplasma-like and rickettsia-like organisms. Further research has established that these organisms are neither mycoplasmas nor rickettsias; they merely possess superficial similarities.

Currently the mycoplasma-like organisms are designated as phytoplasmas and spiroplasmas. Phytoplasmas are very small and have no cell wall. They simply consist of a membrane enclosing living protoplasm and, as such, the organism has no constant shape. Spiroplasmas have no cell wall also, but have helical shapes. Both organisms are usually found in the phloem tissue, where they utilize food (sterols for phytoplasmas) from the host, disrupt translocation, and possibly produce toxins or cause hormonal imbalance. Diseases caused by phytoplasmas and spiroplasmas are shown in Table 15–2. Vectors (Figure 15–9) are usually leafhoppers (mostly), plant hoppers, psyllas, spittlebugs, and grafting.

Rickettsia-like organisms are now known not to be rickettsias. They are now designated as either fastidious phloem-limited bacteria (FPLB) or fastidious xylem-limited bacteria (FXLB), depending on their preferred tissue location. Leafhoppers appear to be vectors for both FPLBs and FXLBs and possibly psyllas. Dodder and grafting are also vectors for FPLBs. Damage from FPLBs and FXLBs appears to result from disruptions of phloem and xylem translocation as bacterial growth and excretions block the sieve tubes and xylem vessels, respectively.

Bacteria

Bacteria are single-celled organisms that are larger than viruses; unlike viruses, they can be seen with a light microscope. Some bacteria are plant pathogenic and can enter plants through natural openings or through wounded tissue. Certain insects can also transmit bacteria such as the striped cucumber beetle, which transmits the cucumber wilt bacteria. This beetle's feces contain the bacterium. Entry is through direct deposition on damaged leaf tissue or indirectly by water (rain, irrigation) to the damaged area.

Diseases of bacterial origin produce symptoms that often resemble symptoms caused by fungal attacks. Symptoms include rots induced by enzymatic degradation, wilts caused by bacterial blockage of the vascular system, gall formation, and tissue destruction at various plant parts. Examples of bacterial disease include potato blackleg, Stewart's disease of early sweet corn, crown gall on fruit trees, berry bushes, and roses; fireblight on apple, pear, and quince; cucumber wilt (Figure 15–10), and shot-hole disease of stone-fruit trees. Water-soaked spots on leaves of houseplants are often of bacterial origin.

TABLE 15–2 *Phytoplasma, Spiroplasma, and FPLB/FXLB* Diseases*

Disease	Pathogen	Vector
Ash decline	*Phytoplasma* sp.	?
Aster yellows of carrots, celery, cucurbits, echinacea, lettuce, potato, sage, strawberry, sunflower, tomato	*Phytoplasma* sp.	*Macrosteles fascifrons*
Blueberry stunt disease	*Phytoplasma* sp.	*Scaphytopius magdalensis*
Brittle root of horseradish, mustard	*Spiroplasma citri*	
Elm (phloem necrosis) yellows	*Phytoplasma* sp.	*Scaphytopius luteolus*
Pear decline	*Phytoplasma* sp.	*Psylla pyricola*
X-disease of peaches, nectarine, cherry	*Phytoplasma* sp.	*Scaphytopius acutus*
Citrus stubborn	*Spiroplasma citri*	*Circulifer tenellus*
Corn stunt of field and sweet corn	*Spiroplasma maydis*	*Dalbulus maidis*
Citrus greening	FPLB	*Psylla* sp.
Bermuda grass stunt	FXLB (*Clavibacter xyli*)	?
Elm leaf scorch	FXLB	?
Phony disease of peach	FXLB	?
Plum leaf scald	FXLB	?
Pierce's disease of grape, almond leaf scorch	FXLB (*Xyllela fastidiosa*)	*Draeculacephla minerva* *Graphocelphala atropunctata* *Oncometopia nigricans* *Homalodisca coagulata*
Red maple scorch	FXLB	?
Sycamore leaf scorch	FXLB	?

*Fastidious phloem-limited bacteria/fastidious xylem-limited bacteria.

FIGURE 15–9
Leafhoppers such as this one (*Graphocephala coccinea*) spread phytoplasmas and spiroplasmas. The resulting disease symptoms include yellowing and dwarfing and abnormal shoot production. Flowers are often replaced by green leafy structures. (USDA photograph.)

FIGURE 15-10 The cucumber on the right is healthy, but the left one is infected with cucumber wilt (*Erwinia tracheiphila*). This disease is transmitted by the cucumber beetle (*Diabrotica undecimpunctata* and *Acalymma vittatum*). (Author photograph.)

Fungi

Fungi are multicellular organisms with a thread-like growth pattern, although their fruiting bodies can be elaborate and colorful such as the mushroom, puffball, or stinkhorn. Fungi cause more diseases (Figure 15-11) than bacteria and viruses; in fact, fungal diseases are the number two economic pest problem after insects. All plants and all plant parts are susceptible to fungal attack. Weather can sometimes increase fungal disease problems such as periods of high humidity. Overwatering can also increase root disease problems. Fungi invade plants through natural openings, wounds, and even directly through the cuticle by enzymatic attack. Fungal spores are readily dispersed by air and are carried great distances by the wind. Spores can also be spread by water splashing and by contact. Certain fungal diseases such as Dutch elm disease (Figure 15-12) are also vectored by insects. This disease is spread by the Dutch elm bark beetle.

Symptoms of fungal diseases are often quite pronounced in visual appearance. In fact, many of these specific appearances are the basis for naming several categories of fungal diseases such as wilt or rust (Table 15-3). Other symptoms can include rolling of leaves, fruit deformation, fruit and blossom drop, dwarfing, chlorophyll destruction, tumors, and necrotic tissue.

Dodder, Mistletoe, Algae, and Moss

Dodder is a plant that lacks chlorophyll and parasitizes other plants such as chrysanthemums. Mistletoe is another parasitic plant that is found high in trees. Both are debilitating to the host plant's vigor and can result in death. Dodder can also be a vector for some viral diseases. Algae and moss are simple, nonvascular plants that can be undesirable in certain horticultural situations. Green growths of algae on clay pots or in water bodies are unsightly. Greenhouse floors and garden walks (Figure 15-13) or patios coated with algae can be slippery and dangerous. Moss can be unattractive in lawns and flowerbeds. Moss usually indicates either poor fertility or acid soils.

Weeds

Weeds (Figure 15-14) are highly successful as competitors; in fact, they usually win out over horticultural plants unless steps are taken to control these plant pests. The numbers of seeds produced by weeds are overwhelming. For example, the weed purslane (*Portulaca*

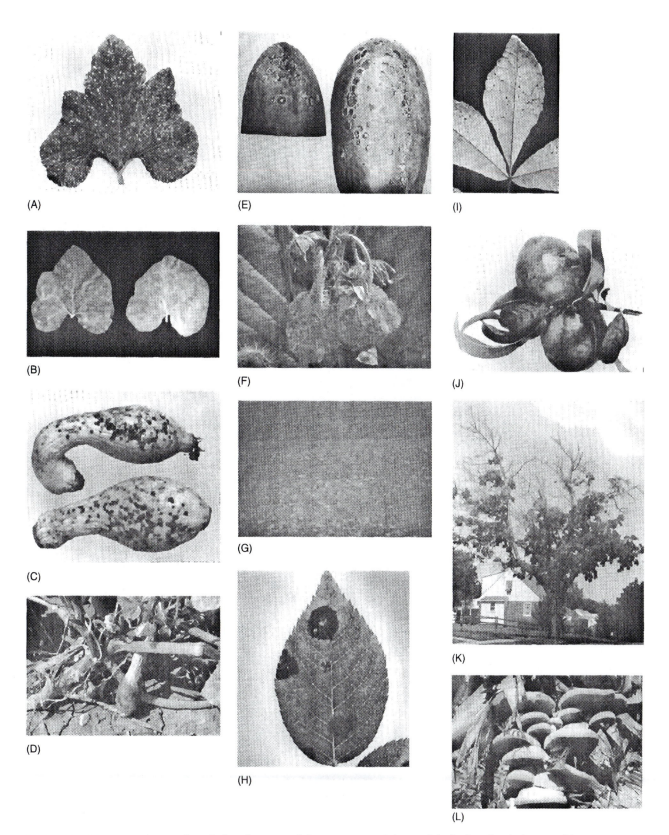

FIGURE 15–11 Some common fungal plant diseases and their symptoms. (A) Squash leaf infected with downy mildew (*Pseudoperonospora cubensis*). (B) Squash leaf infected with powdery mildew (*Erysiphe cichoracearum*). (C) Squash fruit infected with scab (*Cladosporium cucumerinum*). (D) Squash fruit infected with rot (*Choanephora cucurbitarum*). (E) Watermelon infected with anthracnose (*Colletotrichum lagenarium*). (F) A strawberry rachis is completely covered with gray mold (*Botrytis cinerea*). (G) A troublesome disease of golf and other fine turf, dollar spot (*sclerotinia homoeocarpa*), is shown here. (H) Rose leaf infected with black spot (*Diplocarpon rosae*). (I) Pecan scab (*Fusicladium effusum*) lesions can lead to tree defoliation and reduced yields. (J) Nectarines infected with brown rot (*Monilinia fructicola*). (K) Maple dieback or decline observed on a street maple tree. The syndrome results from a two-pronged attack. First the tree is weakened by environmental stress resulting from air pollution, or water stress/aeration (pavement coverage leaving little exposed soil) or insect defoliation. The tree is then susceptible to invasion by fungal pathogens such as *Armillaria* sp. (root rot) that normally doesn't trouble healthy trees. (L) Bracket or shelf fungi growing on a tree trunk is a sure sign of a tree in serious trouble. (Photographs A-J, USDA photographs. K and L are author photographs.) **541**

FIGURE 15–12 This American elm (*Ulmus americana*) is showing severe symptoms of Dutch elm disease (*Ceratocystis ulmi*). The disease is spread by the elm bark beetle (*Scolytus multistriatus*). (USDA photograph.)

TABLE 15–3 *Fungal Plant Diseases*

Group	General Symptoms
Anthracnose	Sunken spots on leaves and fruit; black, shriveled leaves
Blight	Sudden spotting or drying of leaves, flowers; no noticeable wilting
Canker	Initial well-defined lesion, later enlargement to wounds or girdle areas; some oozing
Damping-off	Seeds or seedlings rot in media or soil; stem rots and falls over; stem girdled and seedling stunted
Leaf spot	Spots with clearly defined margins, usually of a different color than interior
Mildew	Powdery coating, usually white
Mold	Colored fungal patches (black, blue, green, pinkish, white)
Rot	Soft pulpy tissue, usually brown or black
Rust	Red or rusty colored spots and patches
Scab	Dark, raised spots
Smut	Large velvety black areas
Wilt	Yellowing foliage and collapsing of stem and leaves

FIGURE 15–13 This patio brick shows a green algal growth. While it looks harmless, it can be the cause of falls and injuries, as it becomes very slippery when wet. (Author photograph.)

(A)

(B)

(C)

(D)

FIGURE 15–14   Many weeds cause trouble. Annual types are controllable by early cultivation, thus preventing seed formation for next year's weed crop. The worst are perennial types that invade viciously by spreading rhizomes such as quack grass. Perennial vines can also be extreme weeds in that spread is rapid and overtaking and strangulation of valued plants can occur. Vines with thorns such as Green brier (*Smilax rotundifolia*) can cause scratches and deep cuts when one tries to remove them. Weeds with deep taproots are also difficult, as the entire taproot must be removed or killed to prevent resprouting. (A) Wild grapes and honeysuckle vines have all but obliterated an ornamental shrub. (B) This corner in a nursery was neglected for a short while and already annual broad-leaved weeds and some grass have moved in. (C) Burdock (*Arctium minus*) has a long taproot, making eradication difficult. (D) Leafy spurge (*Euphorbia esula*) shown here is a noxious perennial weed. It has naturalized in the United States and produces copious amounts of seeds. (Author photographs, A-C. USDA photograph, D.)

oleracea) can produce about 190,000 seeds in one growing season. These seeds are also extremely viable for many years in a natural or induced dormant state. When this dormancy is broken such as by tillage of soil, which exposes seeds to favorable conditions, they can still germinate many years later. Many weed seeds will retain viability for 10, 20, and some even 40 years when buried in soil.

Growth habits of weeds often contribute to their aggressiveness. Seed germination and seedling growth are frequently rapid, and subsequent growth is often extensive. Furthermore, weeds are more tolerant of extremes of heat, cold, and drought as well as plant pests, than are many horticultural plants; the latter seem to have lost some of these characteristics over years of intensive breeding. Some weeds have long taproots that must be completely removed; otherwise, regrowth will occur. A well-known example of this type is the dandelion (*Taraxacum officinale*). Others have vigorous rootstocks that can spread underground and send up new plants elsewhere. Quack grass (*Agropyron repens*) and western sage (*Artemisia ludoviciana*) are examples of this type. In addition, many weeds possess a fibrous surface root system well suited to moisture and nutrient removal.

Weed control is practiced for a number of reasons. The most serious reason is their detrimental effect on crops. In the competition for light, water, and soil nutrients, weeds can seriously weaken or kill horticultural crops. This obviously reduces the crop's economic and esthetic values. In addition, weeds can be alternate hosts for insects and plant diseases that attack horticultural material. They are esthetically unpleasing in lawns, ornamental plantings, vegetable gardens, and roadsides. Some weeds pose health hazards because they are dermal irritants such as Pacific poison oak and poison ivy (*Rhus diversiloba* and *R. radicans*), or allergens like ragweed (*Ambrosia artemisiifolia*) and goldenrod (*Solidago* sp.).

SUSTAINABLE APPROACHES TO PLANT PROTECTION

Plant protection, as is soil health, is a key area in terms of sustainable practices. The most important, primary-use practices should be proactive or preventive practices that minimize pest problems. These practices are nonchemical in nature and are based on physical, genetic, cultural, physiological, and biological methods to create field, garden, or greenhouse environments that are either unfavorable to pest invasion and subsequent colonization or favorable to natural enemies of pests. Many of these practices are in a sense natural practices in that they use naturally occurring organisms (biological controls) or existing genes (genetic control) or work in harmony with nature such as altering crop environments to be more favorable to predators and less favorable to pests.

These practices are widely embraced by the organic horticulturists who have only a few certified techniques or products available to control insects and diseases that become a problem. Other sustainable horticulturists have more control options, but should make proactive practices their first choice such that the need for chemical control is limited. Proactive methods reduce expenditures for pesticides and result in less environmental pollution and health risk.

Should pest levels start to indicate a potential to cause economically unacceptable damage, controls should be moved from proactive to reactive or responsive approaches. Reactive controls include biological controls (predators, parasites, insect diseases, weed diseases, weed feeders, antagonistic and competitive microorganisms), botanical pesticides, natural product pesticides, insect growth regulators, and chemical pesticides. Biological controls straddle both approaches in that some aspects are proactive and others reactive. Organic horticulturists are limited to biological controls and natural botanical (not analogues) pesticides. Other sustainable horticulturists are able to take advantage of all of these strategies.

Pesticides have a number of detrimental consequences associated with their use. As such chemical pesticides should not be used in either a preemptive manner or in large volumes as the sole approach to plant protection. This strategy is clearly not acceptable in the sustainability sense. Pesticides have left a legacy of environmental damage, negative impacts on human health, an ever increasing spiral of induced pest resistance, ever rising costs, and a less than perfect success record with pest control. The public views pesticides with more suspicion and concern than ever before. Other problems include a change of secondary pests to primary pests as pesticides destroy natural controls on secondary pests and explosions in populations of primary pests after pesticides are used as a result of total loss of natural controls. Another aspect is extensive and unnecessary residues that result from only a small amount of the application contacting the pest, leaving the rest to bring about undesirable environmental and human health problems. Details of this track record were covered in Chapter 1 and won't be repeated here. Pesticides also squander energy and petrochemical feedstock that could be put to better usage. For all of these reasons, pesticides are not compatible with sustainable horticulture at their present use levels.

Can crops be produced without pesticides? The answer is a resounding yes! Before the 1940s, crops were produced without pesticides. Farmers relied on natural controls. Crop losses from pests were about one-third. The irony is that present crop losses with pesticides are about one-third. Can we return to pre-pesticide controls? The answer for some farmers is "Been there, done that." Organic horticulturists produce certified organic food in a pesticide-free environment. Organic farmers rely heavily on proactive controls. Clearly, one form of sustainable horticulture in terms of pesticides is organic farming.

The other alternative is to reduce inputs of pesticides to minimal levels, an approach called low-input sustainable horticulture. For example, if herbicides are used, the earlier the better, as young weeds are more susceptible to herbicides. Starting earlier could require less herbicide over the season because of greater initial control. Even better is to make natural controls the primary line of defense against pests and to supplement them with pesticides only when essential. The selected pesticide should also have the lowest possible toxicity to humans and beneficial organisms and pose the least environmental harm and health risk. This approach is called integrated pest management or IPM. The IPM approach is the mainstay of many sustainable horticulturists who can't or aren't ready to adopt organic practices. IPM strategies manage pests with an understanding of their place in the ecosystem relative to other organisms and in a manner that protects the environment. With IPM, pesticides are used as the last resort when pest populations have reached economically damaging levels and all forms of natural controls have failed to bring the pest under control. IPM is a strategy that will increasingly replace single-strategy pesticide applications as off-site environmental damage rises, concerns about human health and food safety increase, and the government applies ever more restrictive regulations on pesticides in response to societal concerns.

A number of criteria form the basis of IPM and also help define the practice. First, the preemptive approach is gone. Pests were often targeted with mass sprayings of chemicals before they constituted a problem or even at first sightings. With IPM, the mere presence of a pest is not automatically labeled as a problem. Criteria and information are examined and a decision is reached as to whether control of the particular pest is warranted or not. It is understood that certain levels of pests are acceptable as part of the normal balance in the functioning ecosystem. The goal with IPM is no longer to totally wipe out the pest species; it is instead to maintain the natural balance between pests and beneficial organisms. The strategy is to let nature or natural controls take their course and to only intervene when necessary and in such a manner as to not disrupt or cause irreparable harm to the ecosystem. Monitoring pest and beneficial organism levels is a critical component of this approach and provides essential information for evaluation of population levels of both in order to make informed decisions regarding control strategies.

All management options are considered in terms of control effectiveness, environmental consequences, economic costs, food safety, effects on the balance of pests and beneficial organisms, and human health considerations. These options include natural control approaches such as biological, cultural, genetic, and physical controls and the final option of chemical control. Use of these controls is based on extensive collection or monitoring of field-level data. These data include pest populations, beneficial organism populations, extent of damage, and the extent of control by natural existing field systems. The subsequent interpretation of the data and projection of added control options in terms of the target pest and the impact on nontarget and beneficial species is carried out prior to control action.

IPM BASICS

Critical to the success of an IPM program (or any control program for that matter) is pest identification. You need to know the enemy with which you are dealing. Numerous resources are available in pest identification. When personal knowledge and experience fail, useful resources include fellow farmers, cooperative extension agents, other university profession-als, private consultants, books, and Internet sites. A wrongly identified pest can cause wasted time, needless effort, and unnecessary expenditures. In some cases, misidentification could even lead to crop losses and unnecessary environmental impact and loss of beneficial insects.

Once the pest is identified, logical questions follow in terms of effective control mech-anisms. What crop is the preferred host for the pest? What field-edge or nearby plants serve as alternative hosts for the pest in terms of food, egg laying, and overwintering? What win-dow of crop vulnerability exists in terms of time? What is the pest's life cycle in terms of destructive, nondestructive, and overwintering forms? Can weather conditions increase or decrease the pest problem? What crop changes can reduce the pest's destructive damage and improve the ability of the natural controls to minimize the pest population?

Monitoring, Scouting, and Record Keeping

The aim of monitoring is to create a dynamic database through record keeping. One might also think of this process as pest mapping. From this database a course of action (or inac-tion) can be determined in terms of pest control, whether it is for insects, mites, diseases, or weeds. Monitoring involves the regular, systematic collection over a number of years of data through a process called *scouting*. Scouting consists of gathering information through visual means and the use of traps for sampling purposes. Scouting information can also be supplemented with real-time data that can be downloaded from the Web (see the entry, IPM resources on the Web, in the Internet Resources list at the end of this chapter).

The key to successful IPM programs is monitoring for pest activity and population levels. Checking for these parameters or scouting need not be complicated. Some labor and time is required, but the trade-off is reduced pesticide costs. Growers can perform scouting themselves or contract out the scouting services to IPM consultants. A number of techniques can be used to assess activity and population levels. Scouting should be done at regular intervals and with increased frequency at times when pest levels have the potential for major destruction.

The basic scouting procedure relies on the gathering of visual information. During scout-ing, one needs to achieve sampling that is representative of the field, garden, greenhouse (Fig-ure 15–15), lawn, or orchard. For example, one could create a zigzag or snaking pattern through the area and follow this same path for each scouting event. Random plants should be chosen along the pathway each time. Plants with various first-time problems should be identified with some marker such as colored yarns so that these plants can be further evalu-ated for changes on subsequent inspections. These plants can serve as indicator plants. For example, one might want to see if introduced beneficial insects or an insect growth regulator or fungicide results in improvement. As many plants as reasonably and economically possi-ble should be inspected. Time and labor are limiting factors. The more inspecting you do, the less pesticide needed to compensate for the switch from conventional to IPM programs. While the numbers and time needed for effectiveness vary considerably depending on crop type, associated pests, developmental stages of both the crop and pest, and weather, your lo-cal extension IPM person or commercial consultant can help in that determination.

Visual inspection requires familiarity with pests, their various life-cycle forms, and their eggs. But one should not place total trust in inspections by eye. Certain insects and mites and their eggs are tiny enough that a $10\times$ magnifying lens or 15 to $45\times$ inexpensive dissecting microscope should be used to enhance your inspection ability. Other tools are also critical to inspection. Not all pests will be visible on plant surfaces. Some might hide under loose or splitting bark or on the undersurface of foliage. In this situation a small pocketknife used as a prying tool can bring bark pests to light as can turning over some leaves. A pruning shear might be useful for removing a suspicious branch for further in-spection. A small spade can be useful for digging around plant bases and root areas in vi-sual searches for soil and root pests. Small zippered plastic bags are useful for saving pests for closer examination under the dissecting scope or for later identification (if unknown) by others or elsewhere by entomologists. Grids the size of one square foot or bigger can be laid out in row crops to provide a basis for comparative weed counts.

FIGURE 15–15 To keep ahead of mite populations in the greenhouse, an entomologist and graduate research assistant recommend that growers examine 38 rose plants for every 10,000 square feet of greenhouse. (USDA photograph.)

During scouting, a number of parameters should be examined. The first step is to conduct an active search for the presence of insects, mites, diseases, and weeds. If pests are found, additional information should be gathered such as the pest identity, its developmental stage, and how many insects are present in a given area or how extensive is the disease or weed presence. Symptoms should be carefully assessed. How severe are the symptoms, how does an infected or damaged plant compare with a normal plant, and is the damage localized or widespread both on an individual and collective crop level? If weeds are present, are nearby plants smaller or less green than areas that are weed free? If symptoms exist without pests being detected, attention needs to be directed toward determining the causal agent.

Other ancillary information also should be gathered during scouting. Besides pests, one should look carefully for the presence of beneficial insects and mites. Assessments of the beneficial insect or mite in terms of its ability to control the pest both in terms of known efficacy and its numeric presence are important data. Weather conditions at that time should also be assessed. Questions to ask include is the weather the same at that time as in past years, did any unusual weather precede the new pest, and is the predicted weather in the next week or two going to have any effect on the pest? For example, if high winds preceded the new pest, it likely blew in from somewhere and might not be a chronic problem. If hot dry weather is predicted, it might favor some pests and not others. Finally, when were pesticides, fertilizer, and irrigation last applied to the affected plants? The appearance of a specific pest repetitively after you have used an insecticide for another pest might indicate a secondary pest becoming a primary pest through removal of its competition or existing predator or parasitoid. Lush plants covered with pests might be connected to overfertilization.

Traps (Figure 15–16) can be a valuable complement to visual activities. A number of trap types (Table 15–4) exist for assessing the presence and population pressures of insects and mites. While a few traps (Figure 15–17) might also serve as control measures, most traps don't. Pheromone traps only capture males, leaving the females free to do damage and possibly reproduce with the aid of as yet untrapped males. Care must be taken to prevent cross-contamination when replacing pheromones in various different traps. Gloves and tweezers should be used to prevent body or clothing contact with pheromones otherwise you will become a "magnet" for that particular pest.

A number of other possibilities exist. One is the use of key plants. Certain plants are much more likely to be attacked by insects and infected with diseases on a continuing basis. For example, summer squash are routinely attacked by squash borers, cucumbers by cucumber beetles that introduce cucumber wilt, geraniums by adult Japanese beetles, turf by Japanese beetle

FIGURE 15–16 (A) The improved version of the McPhail trap uses a combination of three chemicals to attract male and female fruit flies. The older version of the trap used a protein bait that captured large numbers of nontarget insects. (B) and (C) This red sticky sheet and red sticky ball are used in an apple orchard as part of a scouting activity. (D) The high level of attractiveness of the pear ester could be useful in developing "attract and kill" traps that reduce pesticide use while removing moths from orchards before they reproduce. (A and D are USDA photographs; B and C are author photographs.)

(A)

(B)

(C)

(D)

grubs, roses by aphids, eggplants by flea beetles, peaches by peach tree borers, and so forth. These plants are known as *key plants*. These plants should be routinely visually inspected or scouted at the appropriate window of time when problems are known to commence. Similarly, *key pests* lists need to be determined and their window of destruction determined. A key pest is one that causes sufficient damage and economic impact to require control, as opposed to minor pests that cause minimal or inconsequential damage. For example, cucumber beetles introduce cucumber wilt, causing death and economic losses for cucumber producers, so this key pest needs special scouting and visual inspections. On the other hand, aphids on oaks and roses cause little harm and require less inspection or scouting activity.

Records from scouting inspections over time including traps and other activities are used to develop a database, a process known as *pest mapping* in IPM circles. Detailed, careful records must be kept either on the computer or in written form or both. As much time and location-linked data as possible should be recorded, from pests to pesticide use to biological or cultural controls and the weather. Real-time data on major pests and certain crops are also available for some states on the Web and can be an additional, helpful source of record-keeping data. Over time these records help you to hone and fine-tune your IPM

TABLE 15–4 IPM Monitoring Devices

Type	Description	Comments
Bait traps	Bait traps usually use color plus a protein source or volatile chemical that attracts insects. Sometimes actual baits alone are used such as potato pieces for fungus gnat larvae.	Bait traps are useful for insects (usually nonlepidopteran) for which no pheromones are available.
Beating sheets	A cloth sheet stretched on a frame. It is placed under plant foliage. The foliage is struck to dislodge insects and mites that fall downward on the sheet.	Pests are counted to determine levels. Works best with insects that do not fly away readily.
Light traps	Black light traps that attract flying insects.	Experience needed, as these traps can trap high numbers of varied insects in a short time, so interpretation requires skill.
Pheromone traps	These traps are baited with pheromones or attractant chemicals to attract specific pests. Wind traps are for larger insects and delta traps for many insects. The former has a replaceable sticky area while the latter does not.	Good for determining activity periods of specific insects, but limited to those pests where pheromones or other attractants are known and available. Only males are attracted, so these types serve no control purposes.
Pitfall traps	Sunken soil traps consisting of a can or plastic cup. Placed in turf or between rows.	Useful for crawling insects. Careful placement is needed to avoid ankle twisting.
Sticky traps	These consist of flat sheets or round balls that lure insects with specific colors and trap them with sticky coatings.	Red sticky balls are useful for scouting in apple orchards, as are red sticky sheets. Yellow colors also attract insects such as aphids. Other insects are attracted to white.
Sweep sheets	A funnel-shaped net having a frame and handle. It is swept over or around the foliage of plants.	Pests are counted to determine levels. Works best with insects that fly away readily. Weather or time of day can affect number of insects caught, as well as variations in sweep style.
Trap logs	Freshly cut branches, logs or trunks of host tree or shrub placed near intact plants.	Inspected to determine when adults commence laying eggs to determine treatment start. Sometimes the trap logs receive the majority of eggs and can be burned as a control measure.

FIGURE 15–17 This old trap from the 1950s was once used to trap Japanese beetles. It was baited with geranium oil that the beetle found very attractive. Once it flew in, the vanes prevented its exit and it eventually died in a layer of kerosene or oil at the bottom. They were effective at attracting large numbers of beetles whose bodies led to unpleasant smells in the hot summer. They also attracted beetles from the neighbors' yards. Their usefulness as traps was short-lived. (Author photograph.)

program as you compare current data patterns to the "average" picture gleaned from your records. Such information can also be plugged into existing modeling systems for certain key pests and crops, leading to recommendations for appropriate IPM responses.

Phenological Predictions

The concept of phenology was covered in Chapter 6. Phenological models can be used to predict when outbreaks of insects, mites, weeds, and even diseases will occur. These models indicate the timing of outbreaks, not necessarily whether they will occur. However, these models serve a useful purpose in predicting when visual inspections and traps should be applied or examined with greater attention and frequency. These models are based on two approaches: host associations and degree-days. With host associations, some plant activity that seems to correlate with a problem pest is used to predict the timing for the pest's appearance. For example, pest mapping could show that a critical insect or weed appears generally in synchrony with some plant event, either in the crop itself or another plant. An obnoxious weed that occurs with sweet corn might be controllable at a height of 2 inches with minimal herbicide. This height might occur 1 week after the corn has germinated. By careful visual inspection during the early germination period, one can determine whether the weed is present and whether present in sufficient densities to require herbicide treatment. Instead this weed might be found to be at a controllable height when oak leaves were the size of mouse ears. A certain insect might occur when another crop flowers and so forth. This type of model affirms the importance of keeping detailed records of pest problems over many years and why many ancillary observations can be of great value.

The use of degree-days as predictors of pest problems can be a valuable asset in terms of efficiency and economics of IPM programs. While such phenological models can be developed by growers over time (see Internet Resources list, the "Pest control decisions with ornamentals" and "Phenology and pests" entries), these data are available through the Web and provide an excellent starting point. Scouting activities over time can then be used to fine-tune these data until they provide a detailed picture of your own localized operation.

Economic Threshold Level

The point at which the population pressure of a pest, disease, or weed causes plant damage that exceeds the cost of control is known as the economic injury level (EIL). Given human nature and commercial operations, action is usually warranted prior to this point. Ideally, one wants to institute control measures prior to the EIL at the point where control action will prevent the pest population from increasing to damaging levels. This point is known as the economic threshold level (ETL) or sometimes simply as the action threshold. The problem with both of these concepts is that actual EILs or ETLs aren't available for all crops. In addition, those existing static values do not reflect actual changes that occur over the growing season as the nature of the crop ecosystem changes. For example, a single cutworm can do considerable damage to a recently transplanted tomato as opposed to a mature tomato plant. ETLs are also part specific. For example, the ETL will differ for a pest that attacks leaves on broccoli as opposed to the flower head. ETLs are extremely low when the insect in question carries a disease that could destroy the crop. Cosmetic damage is also a factor, especially where consumers views this damage as reducing overall quality, even though it has no impact on the taste or actual food value of the produce. In this case, ETLs will also be set lower. A good analogy is decision making in the health care area, where one has to consider numerous parameters when applying medical care to an injured or diseased patient. When and how to apply treatment needs careful assessment and involves saving the patient and is not independent of economic considerations. It is beyond the scope of this chapter to explain the connections between monitoring data, ETLs, and taking or not taking action. Information from IPM Web sites, publications, your county extension service, and professional growers can help in this area, especially in terms of decision making.

Control Measures

IPM control measures differ considerably compared with the conventional pesticide approach. With IPM, the intent is to reduce or modify pest impact and to reduce pest damage to a tolerable level. Pesticides alone tend to be used with the idea that the pest is eliminated before it can do any damage, thus preemptive use can occur. Some preemptive actions also take place with IPM such as the use of pest-resistant cultivars where major problems are known to exist and the use of cultural practices such as sanitation and those that increase biodiversity of the soil and field ecosystem. These approaches are thought of as proactive IPM. Actual use of biological controls to solve a problem or a green pesticide would be a reactive strategy. The main approach with reactive strategies is to first use natural controls (see earlier sections) such as biological or cultural controls. If these fail to reduce the pests and damage to tolerable levels, then stronger options are used, starting first with "green" pesticides and insect growth regulators. The last resort is the more toxic insecticide, pesticide, fungicide, or herbicide. Another guiding principle is to utilize chemicals that result in the least harm to beneficial organisms and the ecosystem. If the only feasible choice is one that harms beneficial organisms, the method or timing of the application is conducted in a manner that minimizes the impact on the helpful organisms. Throughout all of these steps, detailed record keeping is practiced and becomes part of the database.

PREVENTING PESTS AND DISEASES: PROACTIVE PRACTICES

Proactive practices consist of approaches to and processes for pest and disease prevention that do not involve unnatural or synthetic chemicals or pesticides. Such controls are not new, but the interest in and use of proactive controls is on the increase, especially in view of the increasing concern about damage to our environment and human health from pesticides. In fact, prior to the 1940s, many of these practices were the prevailing method of pest control in crop production.

The advantages of proactive practices are that they are much less of a threat to ecosystems and our environment, and some are self-perpetuating once established. Some disadvantages are that proactive methods are not always capable of completely preventing all pests and diseases. Organic commercial horticulturists and gardeners, but not all nonorganic commercial horticulturists readily accept this disadvantage. The latter group's reluctance stems from the fact that some practices cut into the profit margin and not all are convinced about the efficacy of these practices to prevent problems. As proactive practices become more readily available and more refined, they will be increasingly adopted. Proactive practices are based on mechanical, physical, cultural, genetic, and biological methods (partially).

Barriers and Deterrents

A number of preventive measures are based on the use of physical or mechanical techniques to create a barrier or deterrent, respectively. Some are more expensive or labor intensive than others and these are more likely to be used in a garden rather than commercial operation. Cardboard or newspaper collars (Figure 15–18) can be partially sunk into soil around new transplants such as tomatoes, peppers, and eggplants to form a barrier against cutworms. Polyester floating row covers and plastic tunnels (Chapter 13) serve not only to promote earlier harvests, but to provide considerable barrier protection against most flying insects. Screens keep insects out of greenhouses. Polyester row covers (Figure 15–19) and greenhouse screens will also control some plant pathogens to the degree that the material stops insects that are vectors for plant diseases. In some instances, small waxed paper bags can be placed around apples or pears to prevent insect damage (Figure 15–20). Physical barriers for rabbits, mice and other rodents, and deer (Figure 15–21) include fences (7 to 8 feet high for deer). Screening around tree trunks of fruit and shade trees prevents rabbits, mice and other rodents, and deer from gnawing tree bark.

FIGURE 15–18
Handmade newspaper collars partially buried and placed around tomatoes, peppers, eggplants, and various ornamental bedding plants foil cutworms, as cutworms can't climb over the barrier. (From Raymond P. Poincelot, *Horticulture: Principles & Practical Applications,* 1980. Reprinted by permission of Prentice-Hall, Upper Saddle River, New Jersey.)

Soil

FIGURE 15–19 This floating polyester row cover has been placed over some recently sown vegetable seeds. The cover will trap some heat and speed germination, while its porosity lets rain through. After germination, it offers some degree of frost protection and is translucent enough to allow sufficient sunlight through, while keeping insects at bay. (Author photograph.)

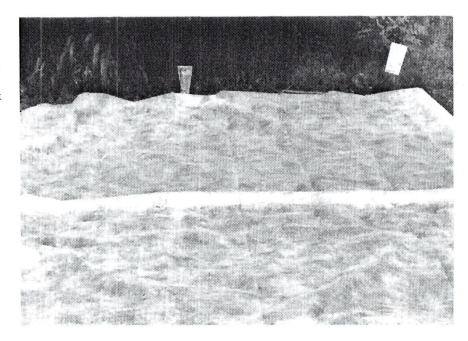

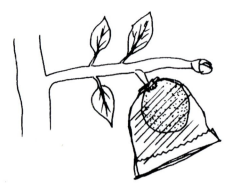

FIGURE 15–20   Waxed bags can be placed around fruits prone to high levels of insect damage such as apples, peaches, and pear. These bags need to be placed quickly after fruit set if they are to be effective. The method shows moderate to good success, but is too labor intensive for commercial orchards. (From Raymond P. Poincelot, *Horticulture: Principles & Practical Applications,* 1980. Reprinted by permission of Prentice-Hall, Upper Saddle River, New Jersey.)

FIGURE 15–21 This inexpensive, black nylon net comes in rolls that are 7 feet tall. When stapled to sturdy poles, its height is an effective deer deterrent. An added plus is that to the human eye, it is inconspicuous. (Author photograph.)

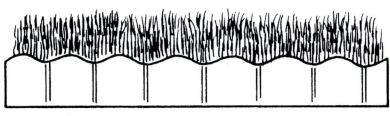

FIGURE 15–22 Plastic edging buried partly into the soil is very effective at preventing grass from encroaching into flowerbeds and foundation plantings. (From Raymond P. Poincelot, *Horticulture: Principles & Practical Applications*, 1980. Reprinted by permission of Prentice-Hall, Upper Saddle River, New Jersey.)

Metallic or plastic strips and edging (Figure 15–22) or brick can be helpful in preventing the entry of weeds or turf into areas such as shrubbery borders or gardens. Netting or twigs placed over new seedbeds discourage dogs and cats. Hot pepper sprays and coyote urine products around threatened plants can keep small animals and deer at bay. Unfortunately, physical barriers for humans who damage plants through vandalism and ignorance are limited. Education appears to be one answer.

Barrier screening (Figure 15–23) can be used to keep birds away from fruits (berries, cherries) and seeds. Barriers include nylon nets or cheesecloth and polyester fabric for small fruit trees and berries and cheesecloth for new seedbeds. Bird deterrents around freshly planted seeds such as sweet corn and ripening small fruits include scarecrows, plastic owls, and noisemakers (carbide cannon, intermittent siren, bird distress calls). Unfortunately, these deterrents lose their effectiveness over time as birds become accustomed to them. Rotation of the various deterrents can extend their effectiveness. Often bird damage is tolerated and compensated for by planting extra material.

Mulches (covered in Chapter 11) control weeds through their action as a physical barrier. Black or other colored polyethylene (Figure 15–24) appears to be the only feasible

FIGURE 15–23 Sturdy cotton netting is a good barrier to prevent birds from eating ripening fruits such as blueberries, cherries, and strawberries. (USDA photograph.)

FIGURE 15–24 A vegetable grower and an NRCS Soil Conservationist inspect a zucchini squash blossom. The grower specializes in unique vegetables. Local restaurants enjoy a white zucchini produced on his farm. Vegetables are growing in black plastic mulch that was machine-laid. While some cultivation is done between rows, no close-in action is needed. (NRCS/USDA photograph.)

mulch that can be mechanically applied for large-scale weed control in commercial operations. Yellow plastic mulches show promise for insect control. The color yellow attracts insects, especially the cucumber beetle. A strip of yellow mulch in various field areas can serve as a monitoring area for insect populations, and lead to effective timing for controls. Another possibility is to alternate one row of yellow mulch with five to six rows of black mulch. Yellow rows would attract insects and those rows could be sprayed as needed, thus reducing the overall use of pesticide on the crop. Another possibility is the use of silver mulch, which repels aphids and could be used on crops highly prone to aphid damage or diseases carried by aphids. Red mulch suppresses nematode damage with tomatoes (also increases yield).

Hay or straw mulches provide a greatly improved environment for some insect predators and spiders relative to bare ground. Considerably less insect damage occurs with a limited number of crops that are provided with straw or hay mulch, mainly from predatory activity. This benefit occurs in addition to weed control. Diseases normally transmitted by the splashing of soil-borne pathogens are also reduced, because the mulch effectively limits soil splashing during rain or sprinkler irrigation. One drawback with organic mulch is the possible introduction of weed seeds from hay or straw (minimal with alfalfa hay) and the possibility of increased pathogen activity by those fungi that thrive in moist environments containing organic matter. Plastic mulches eliminate these problems.

Plant Disease Forecasting Systems

IPM thresholds for insects exist to a much greater extent than for diseases. An alternative is the use of plant disease forecasting systems that rely on somewhat different thresholds, which are not as precise as the economic threshold level of insect IPM programs. These thresholds are more variable and depend on continual monitoring of such components as spore concentrations and relative humidity/temperature profiles. Given the difficulty, time, and cost, these predictive disease models have been developed and implemented for only a few high-value horticultural crops. Examples are potato late blight, apple scab, and tomato early blight. With potato late blight, monitors for temperature and relative humidity (or foliage wetness or rainfall amounts) are placed in strategic fields, usually maintained and monitored by a land grant institution such as an agricultural experiment station. Blight development is favored when the relative humidity goes over 90% and the extent of probability is related to the temperature range during the period of duration over 90%. Data are transmitted on a scheduled, frequent basis. The data are converted into severity indices on a daily basis that are accessible to growers at a Web site. When the severity index reaches a critical value, fungicide applications are indicated. These forecasting systems greatly reduce the regularly scheduled fungicide applications (essentially preemptive) and use fewer, but more effective applications when really needed.

Cultural Practices

Cultural controls involve the selective manipulation of the cropping system, crop production practices, and the crop ecosystem to minimize plant pests and crop damage by plant pests. A number of cultural techniques are practiced in the battle to prevent plant pests. These practices are divided into four basic approaches. One way is to enhance the populations of natural enemies by manipulation of the field and surrounding environments to create optimal conditions for beneficial populations. The field and surrounding environments can also be manipulated to produce the least favorable breeding and survival conditions for chronic plant pests. For example, known alternate hosts for fungi that have life cycles that require two hosts such as apple-cedar rust should not be grown in proximity. Plant breeding or genetic engineering can also produce plants that resist injury or damage from plant pests. Finally, steps can be taken to prevent pests or diseases from colonizing crops. The latter is very important with certain diseases that are difficult or impossible to cure. The best way to control diseases caused by viruses, viroids, bacteria, phytoplasmas, spiroplasmas, and fastidious vascular limited bacteria is to eliminate the insects that infect crops with these diseases.

Sanitation, Surgery, Aseptic Methods, and Inspection. Plants that are suspected of harboring plant pathogens or pests should not be planted. Any plant debris (except that needed to provide organic matter for the soil) should be promptly removed; otherwise, it could provide a breeding area or overwintering material for insects and disease organisms. Plants with uncontrollable plant pest infestations from bacteria, nematodes, phtyoplasmas, spiroplasmas, viruses, and viroids should be destroyed, whether in the field or greenhouse. These practices are termed *sanitation*. For example, culled potatoes should be turned into the soil to decompose, otherwise they can become a source for the late blight fungus. *Surgery* involves the removal of infected plant parts and *roguing* is the removal of diseased plants or seeds. These practices can prevent the spread of insects and disease.

Some viruses can be transmitted through seeds or asexual propagation of plants, so cultural methods have been developed to maintain virus-free stock. The shoot apex is often virus free, and can be propagated by aseptic methods of micropropagation (see Chapter 7) or by grafting onto disease-free rootstock seedlings. Heat treatments from 43.5° to 57°C

(110° to 135°F) for 30 minutes to 4 hours or at about 38°C (100°F) for 2 to 4 weeks, depending on the plant or seed, will often free plants of viruses. A combined heat treatment and shoot apex approach is used for plants where neither method alone is satisfactory. Plants known to be virus prone, such as the strawberry, should be purchased as certified virus free.

Plant diseases, nematodes, and insects can be unwittingly transmitted through shipments of plants. Federal and state regulations exist to control the import of plants and sometimes the shipments of certain plants across state lines in order to prevent such occurrences. These regulations forbid the entry of certain plants and require inspections of others and/or phytosanitary certificates. New plants (preferably certified planting material) being introduced into the greenhouse, nursery, home, or orchard should be inspected carefully for insects and diseases, regardless of their source or paperwork.

Soil Health Maintenance.   Soil health can be a critical component in reducing pest problems, especially in terms of plant-damaging nematodes and soil-borne diseases. Good soil health can lead to the need for fewer pesticides and better results with biological controls. Increasing organic matter through the addition of compost or use of green manure helps to reduce nematode populations. Certain fungi are known to be nematode predators. The success of organic matter in the reduction of nematodes is likely to lie with its ability to create a more hospitable environment for fungal populations that prey on nematodes. A healthy soil is also rich in various beneficial microbes that are antagonistic toward disease-causing microorganisms. Control is achieved through competition and suppression of plant pathogens found in the soil. Beneficial microbes can be added to soils to enhance the soil ecosystem (see the Biological Controls section).

Environmental Alterations.   Field environments often do not provide the conditions needed to maintain populations of natural predators and parasites. These natural enemies of crop pests need food sources nearby such as pollen or nectar or even a second prey not specific to the crop being grown, because crops generally do not supply all their needs. Many natural enemies also need specific plants or host organisms in order to winter over. Natural vegetation surrounding the field might require maintenance or manipulation to maximize levels of natural enemies.

Manipulation of surrounding vegetation can be either destructive or constructive, depending on the prevailing crop and pest dynamics. Destructive situations would take place with pests that overwintered not on the crop, but in specific wild vegetation nearby. An example would be the green peach aphid (*Myzus persicae*), which transmits maize dwarf mosaic virus to sweet corn and cucumber mosaic virus to peppers. This aphid overwinters on peach, wild cherry, and other species of *Prunus*. The aphid tends to stay on these hosts until early July, at which time migration occurs and damage occurs in later plantings of sweet corn and existing pepper crops. Two solutions are available for virus control on the sweet corn. One is the removal of *Prunus* species in wild plantings near the sweet corn fields. A second approach would be to plant only early plantings of sweet corn near cultivated peach trees or near adjacent areas with wild *Prunus* species. Early plantings of sweet corn are unaffected by the virus. Later plantings can be planted at further removed fields, thus escaping the damage as the aphids first stopped at the earlier planting sites before moving further afield. Similar strategies can be adapted for pepper crops. The green peach aphid is also a lettuce pest in the Southwest, where it can be an unacceptable contaminant and also a source of three lettuce viruses: alfalfa mosaic virus, lettuce mosaic virus, and beet western yellows virus. Certain weeds are known to attract the green peach aphid, such as weedy mustards and goosefoot family members. Such weeds should be destroyed not only in lettuce fields, but also in adjacent wild vegetation.

A constructive situation exists with *Anagyrus epos*, a parasitic wasp that infects the eggs of grape leafhopper (*Erythoneura elegantula*). The wasp overwinters in the eggs of an alternative host, as the grape leafhopper overwinters in the adult stage. *Dikrella californica*, the blackberry leafhopper, has eggs available year-round that serve as the winter host for the wasp. The blackberry leafhopper overwinters on wild blackberries (*Rubus* sp.) and is not a troublesome species. Vineyards near wild blackberries achieved early control of grape leafhoppers by the wasps that migrated from their winter hosts. Vineyards further away usually got this natural control later. Construction of blackberry plantings nearby vineyards would seem to be an easy solution, except for one problem, habitat. Wild blackberries grew best in shady, moist habitats provided by tree areas along streams and waterways. The plantings of just California blackberry (*Rubus ursinus*) near vineyards failed for lack of sufficient water, because grape irrigation didn't provide enough. Subsequently, it was found that another blackberry, Himalaya blackberry (*R. procerus*), a naturalized blackberry from Europe, could grow in shade and sun with less water and provided a suitable host for blackberry leafhopper and hence a decent winter environment for the wasp. Later it was found that French prune and its overwintering leafhopper, *Edwardsiana prunicola*, could also provide overwintering for the wasp. Plantings near French prune became another possibility.

Crop Rotations, Cover Crops, and Intercropping. Rotation of crops, preferably of crops not related to each other, can disrupt the life cycles of diseases, insects, and nematodes by eliminating their food supply for a season. This practice works best with insects having long generation cycles and limited migration ability and diseases that are short lived in the absence of hosts. The Colorado potato beetle (*Leptinotarsa decemlineata*) overwinters as an adult in fields of origin or nearby. When it emerges in the spring, it will remain for a week if volunteer crops from the previous year are available for feeding and egg laying. Rotating potatoes with sweet corn and allowing volunteer potatoes to remain buys time (1 to 2 weeks means fewer sprays) for other newer fields to avoid colonization and infestation. Infestations of the Colorado potato beetle are also lower in nearby rotated fields. *Coleomegilla maculata* (lady bug , lady beetle) populations rise in sweet corn as they feed on pollen and aphids and will move into the adjacent potato fields and feed on the eggs of the Colorado potato beetle. Rotate susceptible crops with resistant crops to reduce nematode populations.

Crop rotations can also be effective in weed management. If the same crop is grown continuously in the same field, weeds that have a life cycle similar to the crop will maximize their population. Rotations with different crops can disrupt such weed buildups. Intercropping of two plants together can create an unfavorable weed environment through increased plant density, greater competition, and shading.

Brassica crops (rapeseed, mustard) used in a rotation or as cover crop are effective for reducing nematodes and pathogenic fungi. These plants appear to release allelochemicals such as glucosinolates and isothiocyanates, which on degradation release compounds detrimental to nematodes and fungi. Rapeseed can also be used as a green manure with similar effectiveness.

Timing and Spacing. Timing a crop with regard to sowing, transplanting, or harvesting can be used to avoid peak periods of insect infestations or the time of laying of eggs. Early tillage followed by delayed planting can also eliminate weeds that germinate early. Mechanical weeding just prior to setting vegetable transplants allows crops to grow for a period without weed competition to a point where they can be large enough to shade out weeds that germinate later. Later plantings of potatoes can experience lower infestation by Colorado potato beetles. Less spraying would be needed. Early summer planting of carrots escape damage from the maggots of carrot rust fly more so than early spring plantings. Closer spacing (greater crop density) of plants both between and in rows can reduce weed problems because of faster shading of bare soil, thus reducing weed germination and establishment.

Preventive Plants. Certain plants can be grown to discourage plant pests or serve as trap crops for specific pests. The African and French varieties of marigolds have a root exudate that repels nematodes, thus serving as a repellent plant. The cultivar 'Single Gold' (also sold as 'Nema-gone') is highly effective. Mixed crops can be less susceptible to a particular insect than would a monocrop of the preferred host plant. Insects can prefer other plants. These can be planted as trap plants to draw insects away from what would normally be a host plant in lieu of the preferred species. For example, alfalfa can be used as a trap crop near strawberries to prevent damage from *Lygus* bugs. The *Lygus* bugs can be left to inflict damage on the alfalfa or be eliminated by an insecticide or controlled with certain wasps such as *Anaphes iole.*

Resistance.   Some plants have the ability to resist insects, nematodes, and diseases to varying degrees. This natural or *genetic resistance* can be complete for a particular pest, that is, the plant has immunity, or the resistance might only be *partial resistance*. A complete lack of resistance is termed *susceptibility. Resistance* can take another form called *tolerance;* the plant becomes infected and even injured, but produces an acceptable harvest. The basis of *genetic control* is the introduction through plant breeding of genes for resistance into plants (Figure 15–25) that do not possess natural resistance. Horticulturists should strive to keep abreast of developments in this area and to specify the use of resistant plants wherever possible.

Resistance to insects and diseases is undoubtedly the best possible form of cultural control (or one can think of this approach as genetic control). In crops where genes are available to confer resistance to key pathogens, resistance traits form the mainstay of IPM programs. This form of plant protection involves essentially very little labor and energy inputs relative to most practices used to control insects and diseases. It is also the most environmentally compatible form of protection. Resistance can be based on several forms. Plants can be bred to ripen earlier to avoid windows of vulnerability to insects and disease or susceptible parts can be bred to be tougher or hairier. The former makes it more difficult for insects to penetrate or bite, while the latter because of sticky or sharp hairs creates an unpleasant environment for insects. Specific phytochemicals can be genetically added to thwart insects and even diseases. These phytochemicals can be repellents, toxins, or metabolic inhibitors/disrupters for insects and toxins or disrupters for normal associations between pathogens and plant tissues. Other possibilities are mechanisms to enhance repair or replacement of damaged parts on the plant.

For example, in areas with high nematode populations, use cultivars that are known to have some degree of nematode resistance. Examples are sweet potato 'Nemagold,' white potato 'Russet Burbank,' pepper cultivars 'Charleston Belle' and 'Carolina Wonder,' or tomato cultivars having "VFN" resistance (the N is for nematodes). Use rootstocks, such as Nemagard rootstock for peach, that are nematode resistant for grafting trees that will go into nematode-infested areas. Another example is that butternut squash is more resistant to squash vine borer than summer squashes. 'Mammoth Red Rock' cabbage is less attractive to cabbage loopers than green cultivars. Cultivars of tomatoes resistant to verticillium and fusarium wilts ('VFN' types) should be grown in areas where these diseases are a problem.

FIGURE 15–25
Powdery mildew is a common problem with lilacs. This new lilac cultivar, 'Betsy Ross,' released by the U.S. National Arboretum has fragrant white flowers and shows tolerance to powdery mildew. (USDA photograph.)

Miscellaneous. Other cultural practices include the alteration of soil pH to provide an unfavorable environment for certain disease-causing fungi. Humidity levels and temperatures can be varied in greenhouses to provide less favorable environments for disease microorganisms. Air circulation in greenhouses can be used to speed drying of foliage in order not to provide a good environment for disease initiation. Warm, dry conditions favor mites, so an increase in relative humidity and temperature reduction can reduce their numbers.

REACTIVE PRACTICES TO CONTROL INSECTS AND DISEASES

Reactive or responsive practices come into use when proactive practices have failed to the point that pests or diseases will cause unacceptable economic damage unless controlled. Two strategies exist: nonchemical and chemical (pesticides). Nonchemical techniques are the first choice for all sustainable horticulturists. These practices include mostly cultural controls and biological controls. Pesticides, except for a few botanicals (if allowed by certification), are not available as an option for the organic horticulturist, but are for the remaining sustainable horticulturists. However, the use of pesticides must be in the context of IPM practices (see earlier section on IPM). Once a plant pest is identified, a number of nonchemical control measures are possible.

Traps for Insects

Traps with sex attractants (pheromones) with and without sticky baits or electric blacklight traps are helpful in trapping flying insects. Traps generally work only when small numbers of pests are present. Large numbers of pests require substantial numbers of traps over a wide area. Traps become too expensive for commercial operations and in gardens it is unlikely that everyone in your neighborhood will have traps. Your traps are likely to lure insects from beyond your garden and will become overwhelmed in a short time. Traps work best for monitoring insect population levels (see IPM earlier). Sticky bands can be put on trees to trap climbing caterpillars, but do not deter those species that travel by wind drift. Sticky hanging ribbons can be used to trap flying insects in the greenhouse and home. These traps fare better, given limited space and fewer insects. Boards painted yellow and coated with SAE 90 motor oil trap whiteflies and psylla. Containers of stale beer (traps) are used to attract and drown slugs or snails. Harmless traps can also be used with rabbits and mice and other rodents.

Cultural Practices for Insect Control

Cultural practices include hand picking and destruction of larger insect pests such as the removal of tomato hornworms. Forceful spraying or misting with water can be used on plants that can withstand the force to remove aphids, mealybugs, spider mites, and whiteflies from house and greenhouse plants. Flooding of cranberry fields can control some insects and diseases that attack cranberries. Flooding can also reduce harmful soil nematodes. Overwintering populations of certain pests such as corn earworms (*Heliothis zea*), serious pests of sweet corn, can be controlled by tillage in the fall or spring.

Hot-water dips at 43° to 50°C (100° to 122°F) for 5 to 25 minutes have been employed to destroy plant insects and mites. Heat treatments from 38° to 57°C (100° to 135°F) for varying times have been used to destroy bacteria, fungi, and viruses. For example, a hot-water dip consisting of immersion in water at 44°C (112°F) for 15 to 20 minutes has been used to control cyclamen mite infestations of African violets. Hot-water treatments of 65 minutes at 43°C (100°F) have been used to free sweet potatoes of nematodes prior to planting them. Obviously, this approach is only successful with plants, seeds, and bulbs that can withstand these temperatures.

Cultural Practices for Weed Control

Weeds in the field or garden compete with crop plants for needed water and nutrients. Weeds nearby can be sources of diseases and insects that utilize these plants as alternative

(A)

(B)

FIGURE 15–26   (A) As part of the ridge-tillage system practiced on this farm, an assistant farm manager cultivates for weed control in soybeans. (USDA photograph.) (B) Excellent weed control is possible with cultivation, as seen in this clean-cultivated potato field. (NRCS/USDA photograph.)

hosts. As such, field weed control is required for good crop yields. Nearby weeds, however, should not be blindly destroyed because they can also supply food, refuge, and breeding sites for beneficial insects. Overburning of land and spot flaming with handheld and mechanized equipment have been utilized to destroy weeds. Weeds can be removed through hand picking, hoeing, and tillage. The oldest form of weed control is hand removal, which, although laborious, works well for many annual and biennial weeds. Perennial weeds and others with multiple-shoot rootstocks are apt to reoccur since some roots can be left after hand pulling. The use of hand hoes gives the same results, but with somewhat less work.

Mechanical tillage (Figure 15–26) starting early in the growing season between rows and around beds is the major alternative to herbicides for keeping weeds under control and is a more sustainable option from the environmental impact viewpoint. However, the depth and frequency of tillage need careful oversight to prevent soil erosion and disruption of soil structure. Mechanical weeding between rows can be accomplished with shovels, sweeps, rolling cultivators, and rotary tillers. These implements can dig deep enough between rows (2 to 4 inches) to destroy even big weeds. Small weeds are left in rows. With these implements, shields are needed on the cultivator to prevent burying of young crops by vigorous soil movement. With larger crops, care is needed to prevent damage to roots of larger crops in the inner rows. When crops are larger, sweeps can be used to cause soil-hilling at the rows, effectively burying smaller weeds. Other implements for shallower weeding (2 inches or less) are basket weeders, brush weeders, disk hillers, spyders, and vegetable knives. These devices cause less root damage on crops, can come closer to rows, and throw less soil onto rows, but don't control large or perennial weeds as well as the deeper weeding equipment. Disk hillers and spyders can also be used later in the season to hill-up soil on weeds in rows. Full-field cultivators include harrows and rotary hoes. These implements can be used for full-field weeding prior to planting. Harrows are more likely to be used for seedbed preparation, but are very effective when used on germinating or recently germinated weeds. Rubber-finger weeders, spinners, and torsion-weeders can be used for in-row weeding. Some crop loss occurs, but a slightly higher planting density can compensate. Such devices must be set very carefully and used with skill and precision. Mechanical spring hoes can be used for weeding in orchards and vineyards.

Optimal horticultural practices to produce maximal growth in minimal time can sometimes favor crop competition over weeds. Weeds can be easily prevented with mulches (see Figure 15–24), and weed seeds can be killed by solarization.

FIGURE 15–27   A plant pathologist (*left*) and organic grower inspect the progress of a soil solarization treatment. (USDA photograph.)

Cultural Practices to Control Nematodes and Soil Pathogens

Small batches of propagation media containing soil can be steam pasteurized to destroy nematodes and microbial pathogens. Solarization can also be used to destroy nematodes and microbial pathogens (and weed seeds) in the field (Figure 15–27) and in propagation media. In the field a double layer of polyethylene kept separated by either two by fours or polyvinylchloride pipes is laid over the area. Thirty days of midsummer sun exposure or more is very effective for pasteurization of the field soil. Propagation media is placed in a black plastic trash bag and then double bagged with a transparent plastic bag. The bag is placed on either concrete or asphalt outside and the media is distributed to form a 3-inch layer. From April through August a full day of sun in warmer parts of the United States can raise the temperature beyond 45°C (113°F) for at least 2 hours, the minimal conditions needed to kill nematodes. To be on the safe side, given variations in sun levels and climatic variations, 5 to 7 days should be allowed for certainty.

Cultural Practices to Control Diseases

Alterations of the plant's physiology are possible through such actions as fertilization and pruning. Sometimes these physiological alterations can enable a plant to resist some plant diseases; this cultural control is also known as *physiological control*. Fertilizer can be applied to cause rapid growth, and plant growth can exceed disease progression. Alternatively, fertilization can be reduced to cause slower growth, which would have an adverse effect on those diseases that thrive in vigorously growing plant tissue or insects that prefer lush growth. Avoid overfertilization with plants susceptible to fire blight (*Erwinia amylovora*), because this practice can make the plant more susceptible. Overfertilized plants develop lush growth that is often more attractive to insects. Underfertilization can give plants a more yellowish color that is attractive to aphids.

Biological Controls for Pests and Diseases

Biological controls are the mainstay of organic pest control and a primary method in the remaining types of sustainable horticulture. Most biological controls are essentially reactive practices, but some aspects can also be considered preventive. For example, biological controls such as a predator can be used against a pest, a reactive strategy. Biological controls can also be introduced into an environment adapted for their long-term survival before pests arrive, such that the pest will be controlled when it appears. This approach is preventive. Given the complexity of biological control, it is difficult to separate it into proactive and reactive methods, while still maintaining a cogent explanation of the overall process. Also horticulturists are much more likely to employ reactive strategies, because proactive biological controls require tremendous technological resources and are very expensive. Proactive examples include the release of sterile insects to disrupt pest populations and the search for and evaluation of imported enemies of non-native, but very damaging pests. Proactive biological controls are generally financed and implemented by the government, especially the USDA. As such biological controls will be treated here as mainly a reactive strategy with their preventive aspects pointed out.

Biological control is based on the natural antagonism, competition, and predation existing among organisms. For example, many insects that attack plants have natural enemies. These enemies are *predators* that prey on other insects for food and *parasitoids* that lay eggs in or on other insects or their eggs. The host then serves as a source of food for the young. Predators, parasitoids, and other insects such as pollinators that serve a useful purpose in the horticultural community are collectively termed *beneficial insects*. These helpful insects unfortunately are often casualties when chemical control pesticides are used.

Other forms of biological control are the use of various pathogens to infect insects, mites, and weeds; the use of microbial antagonists to suppress plant diseases and reduce food spoilage; the use of insects to eat weeds; and the introduction of sterile male insects to reduce harmful insect populations. Sterile insect releases have met with some success for the Mediterranean fruit fly (*Ceratitis capitata*) and the Mexican fruit fly (*Anastrepha ludens*), but with little success for Gypsy moth (*Lymantria dispar*).

Parasitoids, Predators, and Other Beneficial Organisms.   Parasitoids (Table 15–5) are usually wasps (Figure 15–28) or flies. The wasps are tiny wasps, not the large stinging type. These insects have an early life stage during development that requires a host organism that eventually succumbs to the parasitic infection. The development proceeds either on or within the host, depending on the specific parasitoid. The adult free-living form in some cases is also a predator. Most parasitoids are species specific, but some can attack several related species. The female finds the host and deposits the eggs either on or within the host.

Predators (Figure 15–29) are usually insects, but can be mites, spiders, other arachnids, and even fungi (Table 15–6). Some are specialized and others are generalists. For example, lady beetles are very effective against aphids, while the praying mantis is such a generalist (they even eat each other!) that their effectiveness is low. Predators vary in their effectiveness. Some alone can control specific crop insects, others offer partial control, and some only late season control. Collectively, predators form an important part of natural control and are important for IPM programs.

Numerous pathogens (Figure 15–30) can be used as biocontrols against pests. Some bacterial, fungal, and viral plant pathogens are shown in Table 15–7. Pathogens cause diseases in the infected pest that lead to decreased feeding and often death. Outbreaks of disease are found in nature when pest populations reach high levels or during periods of highly humid, hot weather. Some pathogens are available from commercial sources and can be sprayed like pesticides. The oldest such pathogen spray is *Bacillus thuringiensis,* used for caterpillar control since the 1940s. Another biological pesticide is *Bacillus popilliae,* which is used on Japanese beetles. Pathogens are generally specific for small groups of closely related pests and usually at certain developmental stages. Beneficial insects are mostly unaffected by applied pathogens.

Microorganisms that compete with other microbes and suppress the latter's populations without killing them are referred to as antagonistic microorganisms or antagonists. Antagonists (Table 15–8) can be used to reduce or limit plant disease and crop or food spoilage. Antagonists can work in the soil at the root level, on surface plant tissues and even as coatings on seeds. *Bacillus subtilis* is a biological antagonist used to suppress root rot and other seedling diseases.

TABLE 15–5 *Examples of Parasitoids in Horticultural Applications*

	Host Pest	Crop Applications
Parasitoid Wasps		
Anagyrus epos	Leafhopper	Vineyards
Anagyrus pseudococci[*]	Mealybugs	Greenhouse crops
Apanteles ornigis	Spotted tentform leafminer	Apple
Aphidius colemani[*] *A. matricariae*[*]	Aphids	Greenhouse crops
Aphytis melinus[*] *Coccophagus lycimnia*[*]	Scales	Greenhouse crops, citrus
Cotesia glomerata	Cabbageworm (*Pieris rapae, P. brassicae*)	Cole crops
Diadegma insulare	Diamondback moth (*Plutella xylostella*)	Cole crops
Encarsia formosa[*] *E. luteola*[*]	Whiteflies (*Trialeurodes vaporariorum*)	Greenhouse crops
Eretmocerus eremicus[*]	*Bemisia tabaci, B. argentifolii*	
Encarsia inaron	Ash whitefly (*Siphoninus phillyreae*)	Trees and shrubs
Eriborus terebrans	European corn borer (*Ostrinia nubilalis*)	Sweet corn
Leptomastidea abormis[*] *Leptomastix daxtylopii*[*]	Mealybugs	Greenhouse crops
Lysiphlebus testaceipes[*]	Aphids	Greenhouse crops
Metaphycus alberti	Cottony cushion scale (*Coccus hesperidium*)	Foliage and flowering tropical plants
Pholetesor ornigis	Leafminers (*Phyllonorycter blancardella, P. crataegella*)	Apples
Trichogramma ostriniae[*]	European corn borer (*Ostrinia nubilalis*)	Sweet corn
Parasitoid Flies		
Trichopoda pennipes	Squash bug (*Anasa tristis*), Southern green stinkbug (*Nezara viridula*)	Squash

[*]Commercially available.

FIGURE 15–28
Parasitoid wasps (*Biosteres arisanus*) inject their eggs into oriental fruit fly eggs. This parasitoid was introduced into Hawaii and has become an established biocontrol. (USDA photograph).

(A)

(B)

(C)

FIGURE 15–29 (A) Lady beetle larva (*right*) and adult (*left*, *Hippodamia convergens*) are feeding on aphids. Lady beetles are available commercially. (B) A Mexican bean beetle larva, a serious pest on snap beans ad soybeans, becomes a meal for the spined soldier bug (*Podisus maculiventris*). It is possible to attract this predator, as the attracting pheromone is now known. (C) This Golden Garden spider (*Argiope aurantia*) is a common predatory spider. (A and B are USDA photographs, C is author photograph.)

Other organisms such as predaceous or parasitic nematodes (Figure 15–31) and protozoa (Table 15–9) can be used as biocontrols for soil and airborne insect pests, respectively. The most effective are the entomopathogenic nematodes. This group invades the insect's body and releases a symbiotic bacterium that kills the insect, whereupon the nematode feeds on the dead insect and bacteria. These nematodes are not species specific, but are generalists that prey on broad ranges of soil insects. Predatory nematodes are more likely to be found in coarser textured soils and soils that are less disturbed. Reduced tillage is better for the maintenance of predatory nematode populations. One protozoan group, microsporidia, infects caterpillars. This infection leads to reduced feeding and reproduction and high mortality.

In some cases certain herbivores such as geese or pigs can be used to control specific weeds. Geese can be utilized to control weed grasses among crops that are not palatable to them. Geese have been used to control grasses early in the production of onion, potato, strawberry, and various perennial nursery crops. Weeds can also be controlled with insect weed feeders (Figure 15–32) and fungal pathogens that cause weed diseases (Table 15–10). The latter can be commercialized as bioherbicides. Biological weed control successes include klamath weed, nodding thistle, and ragwort.

Weeds can also be controlled with natural products. Corn gluten meal has been used as a biological herbicide. Corn gluten meal is a by-product of the wet milling of corn. This natural, patented herbicide has been used on turf, home gardens, field crops, and organic crops. The product inhibits root growth and also contains 10% nitrogen, making it both a "feed and weed" natural product. It is applied as a preemergent herbicide. Hydrolyzed proteins from corn and other grains have higher herbicidal activities and can be applied as sprays. The spray product is less stable than the granular meal.

TABLE 15–6 *Examples of Insect, Arachnid, Mite, and Fungal Predators in Horticultural Applications*

Predator	Prey	Crop Applications

Predatory insects
  Aleochara bilineata
  Rove beetle
	Root maggots (*Hylemya brassicae, H. antiqua*)	Cole crops, corn, radish, onions, turnip

Aphidoletes aphidimyza[*] Aphid midge	Aphids	Berries, cole crops, fruits, gardens, greenhouse crops, ornamentals, potato, vegetables
Chilocorus spp.[*] Beetle	Scales	Greenhouse crops

Chrysopa carnea[*]
  Chrysoperla carnea[*]
  C. rufliabris[*]
  Green lacewing
	Aphids, beetles, leafhoppers, leafminers, moths, small caterpillars, spider mites, thrips, whiteflies	Apples, asparagus, cole crops, eggplant, greenhouse crops, leafy greens, peppers, potato, strawberry, sweet corn, tomato

Cryptolaemus montrouzieri[*]
  Delphastus pusillus[*]
  Beetles
	Aphids, mealybugs, scale, whiteflies	Greenhouse crops

Deraeocois nebulosus *D. brevis*[*]	Aphids, lace bugs, mites, psyllids, scale, whiteflies	Ornamental trees and shrubs, apple, peach, pecan
Geocoris bullatus, G. pallens, *G. punctipes, G. uliginosus*	Aphids, blister beetles, mites, leafhoppers, tarnished plant bugs (*Lygus lineolaris*)	Fruits, garden flowers, vegetables

Lady beetles, 45 species
  Coleomegilla maculata[*]
  Harmonia axyridis[*]
  Hippodamia convergens[*]
  Rhyzobius (Lindorus) lophanthae[*]
  Stethorus punctillum[*]
	Aphids, caterpillars, mites, scale, thrips, whiteflies	Fruits, ornamentals, vegetables, greenhouse crops

Lebia grandis
  Lebia beetle
	Colorado potato beetle
	  (*Leptinotarsa decemlineata*)
		Potatoes

Orius insidiosus[*], *O. tristicolor*[*]
  Minute pirate bug,
  insidious flower bug
	Aphids, caterpillars, mites, thrips	Greenhouse crops, orchards, strawberry, sweet corn

Podisus maculiventris[*] Spined soldier bug	Beet and fall armyworms, cabbage looper, Colorado potato beetle, corn earworm, diamondback moth, European corn borer, flea beetles, imported cabbageworm, Mexican bean beetle	Apple, asparagus, bean, Cole crops, cucurbits, eggplant, onion, potato, sweet corn, tomato

Predatory arachnid
  Phalangium opilio
  Daddy longlegs or harvestman
	Aphids, beetles, caterpillars, leafhoppers, mites, slugs	Apple, cabbage, potato, strawberry, sweet corn

Predatory mites[*]
  Euseius spp.
  Hypoaspis miles
  Iphiseius degenerans[*]
  Neoseiulus fallacis[*]
  Phytoseiulus persimilis[*]
  Typhlodromus pyri
	Aphids, fungus gnats, scales, spider mites, thrips	Fruits, greenhouse crops

Predatory fungus (*Myrothecium verrucaria*)[*]	Nematodes	Greenhouse crops

[*]Commercially available.

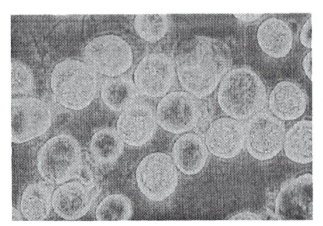

(A)

(B)

FIGURE 15–30   (A) Baculoviruses are available commercially as biocontrol agents. Infected cells from a cabbage looper are shown here. Note that the infected cells have occlusion bodies of the baculovirus AcMNPV within them. (B) Nuclear polyhedrosis viruses are available for use against caterpillars. In this photo, a biological control agent, nuclear polyhedrosis virus, has killed the beet armyworm at top. It is also effective against gypsy moth caterpillars. (C) *Bacillus thuringiensis* (Bt) has been available for several decades as a biocontrol agent. More recently the gene for the Bt protoxin that becomes a toxin in the insect's gut has been genetically engineered into plants such as corn. After only a few bites of peanut leaves with built-in Bt protection, this lesser cornstalk borer larva crawled off the leaf and died. (USDA photographs.)

(C)

Strategies.   Three basic strategies exist in the use of natural enemies to control harmful insects and mites: augmentation, conservation, and importation. Augmentation is a reactive approach widely used by organic and other sustainable horticulturists. Augmentation involves strategies to increase the effectiveness of natural enemies. Natural enemies can be raised in laboratory cultures and released annually in areas where natural enemies fail to establish themselves (inoculative release). Releases can also take place in areas having natural enemies. This release will increase numbers and raise the effectiveness of the natural enemy (inundative release). The drawback with augmentation is that it does not provide permanent control of pests. A good example is the release of *Encarsia formosa*, a parasitic wasp that is used to control greenhouse whiteflies (*Trialeurodes vaporariorum*). Inoculative releases of the parasitic wasp when the first whiteflies are encountered on greenhouse crops can prevent the whitefly from reaching damaging levels.

Conservation methods are essentially a proactive strategy, one of the few in biological control that commercial horticulturists are capable of implementing. With conservation, one looks for factors that reduce the effectiveness of a natural enemy and then attempt to limit these inhibitory factors. Actions are also taken to increase resources needed by the helpful insect. Factors that can be limiting include a lack of food (nectar), reduction in numbers of natural enemies or their host insects by pesticides, and cultural practices (tillage, debris burning) that kill natural enemies or damage their environment. When certain parasitoids are supplied with appropriate nectar-containing wildflowers at the field's periphery, their life span and level of host hits increase considerably. Some crops have

TABLE 15–7   *Examples of Pathogenic Biocontrols in Horticultural Applications*

	Host Pest	Crop Applications
Bacteria		

Bacillus lentimorbus[*]
  B. popilliae[*]
Milky spore disease
  (Doom™, Japademic™)
	Japanese beetle (*Popillia japonica*)	Turf

Bacillus thuringiensis[*]
  (several varieties of Bt for different
  pests: Agree®, Biobit®, Condor®,
  Crymax®, Dipel®, Gnatrol®,
  M-Trak™, Thuricide®)
	Numerous caterpillars, few beetle and fly
	  larvae, fungus gnats
		Cole crops, forest crops, ornamental
		  shrubs and trees, potato, sweet
		  corn, vegetables

Pasteuria penetrans	Nematodes	
Viruses		
Baculoviruses[*] (see Nuclear polyhedrosis virus)		
Granulosis virus[*] (Capex™, Cyd-X™)	Codling moth, leafroller	Apple

Nuclear polyhedrosis virus (NPV)[*]
  Gemstar™, Mamestrin™
	Numerous caterpillars, especially gypsy moth	Apple, cabbage, forests, greenhouse
		  flowers, ornamental trees and
		  shrubs, peach, plum, sweet corn,
		  tomato, walnut

Fungi		

Beauveria bassiana[*]	Aphids, bark beetles, caterpillars, chinch bug,
	  Colorado potato beetle, European corn
	  borer, flea beetle, grasshoppers, Japanese
	  beetle, lygus bugs, Mexican bean beetle,
	  mites, psyllids, thrips, whiteflies
		Fruits, ornamentals, vegetables

(BotaniGard™, Mycotrol™, Naturalis-O™)		
Beauveria bassiana strain GHA[*]	Grasshoppers, locusts, Mormon crickets	Potato, sunflower
Entomophaga maimaiga	Gypsy moth	Ornamentals
Entomophthora muscae	Cabbage, onion, and seedcorn maggot	Cabbage, onion, sweet corn
Hirsutella thompsonii *Neozygites floridana*	Mites	Fruits, ornamentals, vegetables
Myrothecium verrucaria[*] DiTera®	Nematodes	
Nomuraea rileyi	Some caterpillars	Cabbage, sweet corn

Paecilomyces fumosoroseus
  PFR-97[*]
	Aphids, spider mites, thrips, whiteflies	Ornamentals (greenhouses,
		  interiorscapes, outdoors)

Pandora neoaphidis	Aphids	Fruits, ornamentals, vegetables
Verticillium lecanii[*] (Vertalec™)	Aphids, thrips, whiteflies	Greenhouse crops
Zoophthora radicans	Aphids, leafhoppers, several caterpillars, weevils	Fruits, ornamentals, vegetables

[*]Commercially available.

TABLE 15–8   *Examples of Antagonistic Biocontrols in Horticultural Applications*

	Suppression	Crop Applications
Antagonistic Bacteria		
Agrobacterium radiobacter[*] (Diegall™, Galltrol™)	Crown gall	Fruit and ornamental trees
Bacillus subtilis[*] (Epic™, Kodiak™, Quantum 4000 HB™)	Brown rot Soil-borne diseases, some foliar diseases	Field vegetables Peach storage Greenhouse crops
Burkholderia (Pseudomonas) cepacia[*] (Blue Circle™, Deny™)	Soil-borne diseases, nematodes	Greenhouse potting media Vegetable crops
Pseudomonas fluorescens (Blight Ban A506™)	*Erwinia amylovora*	Apple, almond, cherry, peach, pear, potato, strawberry, tomato
Pseudomonas syringae[*] Strain ESC-11	Blue mold (*P. expansum*) Gray mold (*Botrytis cinera*) Mucor rot (*Mucor* spp.)	Red Bartlett pear storage Apple storage Pear storage
Pseudomonas sp. + *Azospirillum*	Brown patch, dollar spot	Turf
Streptomyces griseoviridis[*] (MycoStop™) *S. lydicus*[*] (Acinovate™)	Soil-borne diseases	Greenhouse potting media
Antagonistic Fungi		
Ampelomyces quisqualis[*] (AQ10™)	Powdery mildews	Grape, greenhouse crops, vegetables
Candida oleophila[*] White yeast (Aspire™)	Post-harvest pathogens	Citrus storage Apple storage
Coniothyrium minitans[*] (Contans,®)	Pink rot, water soft rot, white mold	Peanut, sunflower, vegetables
Fusarium oxysporum[*] (nonpathogenic strains) (Biofox C)	*Fusarium* diseases	Basil, carnation, cyclamen, tomato
Gliocladium spp. *Gliocladium virens*[*] (Gliomix™, SoilGard 12G™)	Root diseases	Greenhouse crops
Pythium oligandrum[*] (polyversum)	Soil-borne diseases	Fruits, vegetables
Sporobolomyces roseus Fruit yeast	Blue and gray molds	Pome fruit storage
Trichoderma spp.[*] *T. harzianum, T. viride, T. koningii, T. hamatum* (Binab™, RootShield™, Trichodex™, TopShield™)	Numerous pathogenic fungi causing root diseases, wound diseases, also botrytis and powdery mildew	Numerous crops in field and greenhouse

[*]Commercially available.

harmful chemical defenses that inhibit the natural enemy, but not the harmful insect that has developed immunity.

A good example of conservation occurs with citrus. Extensive tillage deposits dust on the citrus foliage that can kill parasitoids and predators, leading to increases in the harmful pests that are unaffected by the dust. A simple, periodic washing of the foliage resulted in greater control of California red scale (*Aonidiella aurantii*) by parasitoids. Less tillage

FIGURE 15–31 Bacterial-feeding nematodes, *Operculorhabditis* sp. LKC10, frozen in liquid nitrogen. Magnified about 30×. (USDA photograph.) While these types are not biocontrol agents for insects, they are important for maintaining soil health and plant health through nitrogen availability. These nematodes are important in mineralizing, or releasing, nutrients in plant-available forms. When these nematodes eat bacteria, ammonium (NH_4^+) is released because bacteria contain more nitrogen than needed by the nematodes. Biocontrol nematodes and those that attack plant roots look very similar to the nematodes in this picture.

TABLE 15–9 *Examples of Miscellaneous Biocontrol Organisms in Horticultural Applications*

Nematodes	Prey	Crop Applications
Heterorhabditis bacteriophora[*] (Cruiser™, Otinem™, Lawn Patrol™)	Billbugs, root weevils, turf grubs	Berries, container crops in greenhouses and nurseries, cranberries, turf
Heterorhabditis spp.[*] (Larvanem™)	Caterpillars	Evergreens, nursery and greenhouse crops
Steinernema carpocapsae[*] (Bio-Safe™, Ecomask™, Exhibit®, Scanmask™)	Armyworms, black vine weevil, cutworms, fungus gnats, girdlers, shore flies, strawberry root weevil, webworms, wood borers	Artichokes, cranberries, ornamentals, greenhouse crops
Steinernema feltiae[*] (Guardian™, Nemasys™, Otienem™)	Fungus gnats, tipulids, mushroom flies, shore flies	Greenhouse crops, mushrooms
Steinernema riobravis[*]	Citrus root weevil, corn earworm, mole crickets	Citrus, sweet corn, turf
Steinernema scapterisci[*]	Mole crickets	Turf
Protozoan Microsporidia		
Nosema pyrausta	European corn borer	Sweet corn
Vairimorpha necatrix	Armyworms, cabbage looper, corn earworm, fall webworm	Cabbage, corn

[*]Commercially available.

FIGURE 15–32
Tropical Soda Apple (TSA) is a federal noxious weed that has been in Florida for several years and has recently moved into the Southeast. The tortoise beetle, *Gratiana boliviana,* is being considered for release as a biocontrol agent for tropical soda apple. (USDA photograph.)

TABLE 15–10   *Examples of Organisms Used in Biological Weed Control*

Insect Weed Feeders	Target Weed
Rhinocyllus conicus	*Carduus nutans*
	Nodding thistle
Gymnetron tetrum	*Verbascum thapsus*
	Common mullein
Schizocerella pilicornis	*Portulaca oleracea*
	Common purslane
Fungal Pathogens	
Entyloma (Cercosporella) ageratinae	Hamakua Pamakani weed
Colletotrichum coccodes	*Abutilon theophrasti*
	Velvetleaf
Phomopsis amaranthicola	*Amaranthus* spp.
	Pigweed
Phytophthora palmivora[*]	*Morrenia odorato*
(DeVine™)	Strangler vine
Puccinia cardui-pycnocephali	*Carduus nutans*
	Nodding and slender thistle
Xanthomonas campestris[*]	Turf weeds
(X-PO™)	

[*]Commercially available.

would also accomplish the same improvement with this biological control. Improvement of natural resources can also work. Some possible areas include supplying alternate hosts for the natural enemy, overwintering sites, food resources, and microclimatic alterations. A good example is the provision of wildflowers in the cabbage family and Queen Anne's lace as a nectar source for the parasitoid that parasitizes the Diamondback moth (Table 15–5).

Importation of predators and parasites, a proactive strategy, is used against pests that are not indigenous to the United States. This area is largely left to U.S. governmental agencies such as the USDA. Pests such as the Japanese beetle (*Popillia japonica*) arrived mostly by accident through commerce and travel, while the Gypsy moth (*Lymantria dispar*) imported for research escaped accidentally. Having no natural enemies, non-native pests have no checks on their populations and can become serious crop pests. The solution is to import natural enemies from the country in which the pest originated. This solution will not work if the natural enemies are not able to adapt to conditions in the United States.

Importation often requires extensive fieldwork to first identify natural enemies. Permits are required from the USDA for importation of these natural enemies, which are quarantined. The natural enemy is assessed during quarantine to make sure it is free of disease and undesirable parasitoids. Shipment of the natural enemy across state lines and field releases require additional permits. An example would be *Tiphia vernalis* and *T. popilliavora*, parasitic wasps that were imported to control Japanese beetles. *Encarsia inaron*, another wasp, was imported from Italy and Israel to control ash whitefly, an exotic from Europe, the Mediterranean, and northern Africa. This importation was highly successful and the ash whitefly remains controlled to this day in California (Table 15–5).

Chemical Controls

Chemicals used to control plant pests through toxic action are termed *pesticides*. These chemicals usually kill the pest, but in some cases they are only sickened. This toxicity is not necessarily for all life stages of the plant pest, but is often most effective at the most vulnerable stage in the life cycle of the pest. Chemicals that discourage plant pests from attacking plants because of some property that makes the chemical disagreeable to the pest are termed *repellents*. The latest pesticide trend is toward products that are significantly less damaging to the environment. Such products usually have lower toxicity, more selectivity, and less harmful effects on beneficial insects. Some term these new products "green" pesticides. Examples are pesticides made from natural products and fungal metabolites or pesticides that act as insect growth regulators.

Pesticides are usually divided into categories on the basis of which pest they control. These are insecticides, miticides (acaricides), nematicides, bactericides, fungicides, rodenticides, and herbicides. Specific examples of each are discussed in a later section. Pesticides can also be divided into two broad categories: systemics and nonsystemics. *Systemics* are water-soluble pesticides that are absorbed by plant leaves or roots, and then translocated throughout the plant. The plant itself is then toxic to the attacking organism. The advantage is that the plant is protected for several weeks, including new growth to which the systemic is translocated. The disadvantage is that such products can't be applied too close to harvest. Plants not given enough time to detoxify the systemic could sicken and possibly kill people who eat them. Federally established legal waiting periods for detoxification are designed to prevent such occurrences. *Nonsystemics* coat the plant and affect the pest either on contact or after ingestion. These products can be applied closer to harvest, but usually aren't as effective or as long lasting as systemics, unless applied more frequently. Systemics are also not as easily degraded by environmental factors such as rain and sunlight as are nonsystemic pesticides. Systemic pesticides are also less of a threat to beneficial insects than nonsystemic pesticides. Pesticides are also classified for general use or restricted use. The former is a pesticide that usually doesn't have an unreasonable adverse environmental effect. Restricted-use pesticides can cause unreasonable adverse environmental effects and human injury. Regulatory restrictions are imposed by the EPA on restricted-use pesticides; these products can only be applied by licensed, certified applicators.

Pesticides, for the most part, show some degree of selectivity. For example, a miticide would be most specific for mites and would not show any fungicidal activity. Insecticides might only be useful for a small group of insects. Unfortunately, the majority of pesticides are usually not selective enough to spare beneficial organisms. Certain pesticides are designed to eliminate a wide range of plant pests. These are termed *wide-spectrum* or *general-purpose pesticides*. Such products can be a single chemical that shows toxicity to a wide range of organisms, or they can be combinations of chemicals, such as a mixture with an insecticide, fungicide, and miticide. When combining two or more pesticides to achieve multiple control, one must have information on pesticide compatibility. If they are not compatible, plant damage or reduced efficiency of one or more components can result. Pesticides should not be mixed unless such information is available.

The form in which pesticides are sold is termed the *formulation*. The formulation can be ready for direct application, or it could require dilution prior to application. The form that the formulation takes is determined by several factors: storage, handling, method of application, effectiveness, and safety. Formulations can be applied as sprays, dusts, granulars, aerosols, fumigants, or encapsulated materials.

Sprays.   Spraying is the most prevalent mode of application. About 75% of all pesticides are applied as sprays. Formulations sold for use as a spray include water-miscible liquids, emulsible concentrates, wettable powders, water-soluble powders, flowable suspensions of ground pesticide in water, oil emulsions, and ultra-low-volume concentrates.

The bulk of the pesticides sold as sprays are emulsible concentrates. These are concentrated solvent solutions of pesticides (18% to 75%) with added emulsifier, which forms a milky-white emulsion when mixed with water. Only small amounts are mixed in water to produce an effective spray. These emulsions are usually stable for several days and require no agitation when applied as a spray. Certain weather conditions such as high temperatures are avoided with these products because of plant injury from the carrying solvents.

Wettable powders consist of dry pesticide (15% to 80%), a wetting agent and sometimes an inert powder. The wetting agent facilitates mixing with water. The carrier in the pesticide dust is usually clay or talc, which will sink, so the solution will require agitation during spraying. On drying, these pesticides often leave a visible and undesirable residue. Blowing of pesticide during the mixing process can also be a problem. Less plant injury occurs with wettable powders than with emulsible concentrates. Water-miscible liquids and water-soluble powders form solutions readily with water and require no agitation during application.

Pesticides not readily water or oil soluble or flowables are sold as wet dusts containing talc or clay that can be suspended in water and sprayed with agitation. Oil emulsions consist of pesticide (4% to 5%) diluted with oil and sprayed. These products are usually used for nonpest insects because the oil causes plant damage. Ultra-low-volume concentrates are high-strength pesticides used directly for spraying without dilution to eliminate large spray volumes.

Sprays sometimes fail to give good coverage of plant leaves because of the cutin layer, which is water repellent. *Wetting agents* or *spreaders* are therefore added to give more uniform coverage, since their properties are such that they break the surface tension of the droplets sprayed on the leaves. These are usually detergents or sulfonated alcohols. *Stickers* are chemicals added to increase the amount of pesticide that sticks, as well as the length of time that it sticks. Sprays are most effective as a thin, complete coat. Dripping or runoff indicates overspraying.

Spray equipment of many types is available (Figure 15–33). Orchardists use a sprayer with a bank of nozzles, which uses air blasts from a rotating fan as the carrier for the spray. Air-driven sprayers avoid the large water volumes needed as a carrier; these sprayers are ideal for the ultra-low-volume pesticide concentrates. Arborists use a high-pressure, high-volume sprayer with a fire-hose type of nozzle. Boom-type sprayers are often used on row crops. Smaller sprayers are available as tanks on wheels, backpack tanks, and handheld tanks. Many of them rely on hand-operated force pumps for the compressed air needed to deliver the spray. The larger ones use a gasoline-driven engine to produce the compressed air. Siphon sprayers that connect to water hoses and use water pressure to dilute and deliver the spray are also available. A backflush valve should be used with any sprayer connected to a water source to prevent accidental entry of pesticide into the water supply.

Dusts.   Dusts are simple to formulate and easy to apply. Generally, dusts are dry pesticides (0.5% to 10%) mixed with an inert powder such as clay, flour, or talc. However, dusts are less effective and economical than sprays. Considerable amounts are lost from wind blowing and coverage is often incomplete owing to a poor rate of deposit on plant materials. Retention might not be as good as with a spray. Heavy moisture also produces caking of applied dusts, leading to lower effectiveness. Slight moisture on the plant can improve

(A)

(B)

(C)

(D)

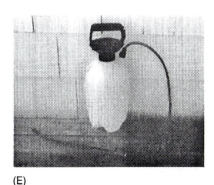

(E)

FIGURE 15–33 (A) This new nozzleless sprayer uses one-half to two-thirds less spray to achieve superior coverage and penetration of the tree canopy. (B) KeyPlex products (a Morse Enterprises, Ltd., product with chelated micronutrient mixture with alpha keto amino acids) stimulate plants to increase production of their own defensive compounds such as the enzyme chitinase, which can break down chitin in insects and fungi. Already used on several crops, they don't require registration by the U.S. Environmental Protection Agency. Some insects and diseases appear to be controllable with this approach. (C) While extracts from exotic plants may thwart whiteflies in the future, other remedies may be right in your kitchen. ARS researchers in Arizona found that a mix of common liquid dishwashing detergent and cooking oil kills sweet potato whiteflies, as well as several common home garden pests. At test plots in Maricopa, Arizona, a technician and biological aide spray plants with a fine mist of vegetable oil and dishwashing detergent. (D) Attached to a tractor, the Mite Meter can be calibrated to dispense from 500 to 20,000 biocontrol mites per acre. (E) This sprayer type comes in 1 and 2 gallon versions and is widely used in gardens and home grounds. Pumping the handle pressurizes it. (Photos A-D, USDA photographs. E is author photograph.)

distribution and adherence. Dust-treated plants are also unattractive. Dusts are of very limited value for trees. Power dusters use gasoline-driven fans to distribute the dust. Smaller power units are available as backpack units. Hand-operated units are also available.

Granular and Encapsulated Materials. Granular pesticides consist of pesticide (1% to 20%) formulated with an absorbent as small pellets of clay, corn cob, fertilizer, limestone, or nut shells ranging in size from 20 to 80 mesh. Drift problems are reduced compared to the drifting experienced with the considerably lighter dusts. Systemic pesticides are available in granular form. Granulars are usually applied with a hand-operated rotary type of distributor or applied directly in the drill at planting time. These products are generally only applied to soils. Encapsulated pesticides use tiny beads of polyvinyl or other plastics to surround the pesticide. These products are slow-release forms.

Aerosols, Fogs, and Fumigants. In an aerosol bomb, which is designed for convenience, the pesticide is sprayed through the release of a compressed gaseous carrier. These bombs are useful for small applications, such as with houseplants, yard plants, or in greenhouses. A fog effect is created by heat volatilization of kerosene or other oil carrier containing a pesticide. Fogs are not usually used for horticultural purposes; they are used mostly for mosquito and fly abatement. Fumigants are pesticides that consist of small volatile molecules that exist as gases above 4°C (40°F). Often they have a density greater than air. They are highly penetrating, which accounts for their use in soils and greenhouses. They are released as a fumigant by heating or, in some cases, by wetting.

Safety.   Pesticides are often toxic to humans (especially infants and children) and, unless handled properly, can cause illness or death to humans who apply pesticides, come into contact with pesticides before or after application, or eat horticultural crops contaminated with pesticides. Some pesticides are also implicated in causing certain cancers (carcinogenic) or can cause damage to reproductive cells and fetuses, resulting in birth defects and miscarriages. Pesticides can also cause soil and water pollution, since some have a long residual life and move through the soil and water cycles. Accordingly, some federal regulation of pesticides came about in the early 1970s when control over pesticide regulation was moved from the USDA to the EPA. Horticulturists using pesticides must be aware of the laws concerning their use.

The provisions of the Federal Environmental Pesticide Control Act are in brief as follows. Pesticide use other than stated on the label is prohibited. Pesticides are classified as general or restricted use. Only professionals that are trained (USDA oversight) and then state certified can apply restricted-use pesticides. Certification is possible in several different categories. Certification is usually recognized by the issue of a license after the applicant has demonstrated proficiency in the following areas: label and labeling comprehension, safety, pesticides, pests, environment, equipment, methods of application, and laws and regulations. All violations are punishable by heavy fines or imprisonment. Other aspects of the law cover the manufacturer and registration procedures for pesticides.

In 1988 the regulations were amended to require considerable chemical, environmental, toxicological, and reentry protection data in order to register pesticides. Experimental use permits and special exemptions were also allowed to allow field testing of pesticides prior to registration and also for use on nonlabeled pests or crops in emergencies when no other registered pesticide is available. Efforts to review pesticides in terms of possible cancellation, suspension, or change of classification also occurred. Suspension essentially puts on immediate stop to the use of a pesticide and its marketing when health and environmental data indicate the pesticide poses unacceptable risks. Cancellation is a slower process and can take years. In all instances, pesticide manufacturers are given notice and can request hearings. Pesticides previously registered before 1984 also had to undergo reregistration, but the process goes slowly.

The most recent change results from the Food Quality Protection Act passed by Congress in 1997. This act made many changes and will have far-reaching implications. Each pesticide will have to undergo reregistration/review every 15 years, even if the previous review gave it an acceptable rating. This process will enable the consideration of new health, safety, and efficacy data that might arise after the initial opproval was granted and could result in loss of registration. Registration for reduced-risk pesticides such as Bt products and antimicrobial pesticides will have a fast track registration process. The ability to use pesticides for minor crops not registered on the label will become easier. Incentives such as extended exclusive rights and fast-track review have been added to encourage pesticide manufacturers to register these minor uses. Minor use will even be allowed on major crops if the use is the only effective alternative or it is important for IPM programs.

Tolerances for pesticides have gone from the zero tolerance of the earlier (1960) Delaney Clause to levels that ensure with reasonable certainty that no harm will result from aggregate exposure for both raw and processed foods. The original clause only covered processed foods, and zero levels eventually became impossible, given our increasingly sophisticated tests that now find pesticides in parts per billion and even parts per trillion. Infants and children, given their reduced tolerance to pesticides relative to the body mass of an adult, are now better protected. The safety factor of 100 used in laboratory animal tests used with adults is increased by another 10-fold factor for infants and children. A right-to-know brochure for consumers concerning pesticides will also be developed in the near future.

Aggregate exposure is also new. Rather than setting tolerance for an individual pesticide, the EPA must weigh aggregate exposure for several or more pesticides having the same mechanism of toxicity. Exposure is also extended beyond food to consider even nonfood exposures such as in lawns, water, homes, and businesses. This tolerance review must be accomplished during the 2000s for all pesticides, starting with the worst first. Pesticides for the first time must also now be tested to see if they have endocrine-disrupting effects. This new test is in addition to all of the other tests such as those for cancer risk, toxicity levels, environmental effects, birth defects, and reproductive system alterations. Already methyl parathion has been banned on all fruits and many vegetables and the amount of allowable residue of Guthion on crops is now lower. Recently, a popular weed killer, atrazine, was shown to disrupt the sexual development of frogs of concentrations 30 times lower than allowed by EPA levels. Prior to this discovery, herbicides were assumed to be safer than pesticides. Now even herbicides are suspect. As expected, challenges to this process for setting aggregate exposure are mounting from all sides to speed (environmental groups) or limit (congress members speaking for farm states) the resetting of tolerances.

A number of safety precautions must be observed when working with pesticides. When purchasing the pesticide, be certain the label is intact and the information therein is up to date. Try not to buy more than is needed for the current season. While transporting the pesticide, use containers that are intact and be alert for leaks of liquid or fumes. If pesticides must be stored, be sure to store them in a secure, locked area away from foodstuffs or water supplies. The storage area should be dry, well ventilated, out of direct sunlight, and kept above freezing. The pesticide storage area should be clearly identified as such. Outdated material or damaged containers should be removed.

Directions should be carefully followed when preparing pesticides for use. A manufacturer's safety data sheet (MSDS) should come with all pesticides. Read it carefully and retain it for future needs such as spills, exposure, contamination, and disposal. Do not smoke or eat while working with pesticides. Mix them in a well-ventilated area away from winds or water supplies. If some is spilled on you, wash immediately with plenty of soap and water. Remove contaminated clothing promptly.

Wear appropriate protective clothing and equipment, such as respirators, when the toxicity of the pesticide warrants it (Figure 15–34). This is especially important during application of the pesticide. Use clean, reliable equipment for application and do not use

FIGURE 15–34 Many of the older pesticides used as sprays or dusts are toxic enough to require a personal cartridge-type respirator. Respirators must be fitted properly and users must receive training in their use. (Author photograph.)

herbicide sprayers to apply insecticides and the like. Follow rates of application, and do not apply pesticides on windy days. Avoid walking in clouds of spray or dust, and keep pesticides away from the body. Try not to knowingly spray when beneficial or pollinating insects are most active; notify nearby beekeepers in advance of any spraying.

After pesticides are applied, observe the required waiting period before reentry or harvest. Be sure the area is clearly posted both in English and any other language needed to cover farm or greenhouse workers. All equipment should be thoroughly cleaned after use. Clothing should be changed afterward, and the operator should bathe completely with soap and water. If an accident occurs, notify a physician at once and have the MSDS available for the emergency responders.

Empty containers should be rinsed and damaged to prevent reuse for any purpose. Containers and unused pesticides should be disposed of in a proper manner. Contact your state extension service or environmental protection agency to determine the recommended means of disposal.

These safety precautions should be carefully observed at all times. Avoid complacency and shortcuts. The risk of injury, death, or lawsuits is high otherwise.

Application Problems.   The timing of pesticide applications is critical; otherwise, the plant pest might not be controlled in time to prevent major damage. Timing is also critical to avoid contamination by residues when harvesting the horticultural crop. The pesticide should be applied before symptoms become very noticeable, that is, not when the plant pest has overrun the plant material. Complete coverage is sometimes difficult because of the plant density or cutin layers on leaves. Pruning or chemical spreaders and stickers can be helpful. Sprays tend to be more effective than dusts.

For reasons of sustainability, IPM principles (see the IPM discussion earlier in this chapter) should pertain, too. Timing of applications should be planned to avoid harming pollinators during the flowering stage when honeybees are maximally active. Periods of maximal activity by predators and parasites should be avoided, too. If spraying is necessary during these times, pesticides that don't harm or are the least harmful to beneficial insects should be used. Pesticides are less harmful to bees when applied in late afternoon (after 3:00 P.M.). Bees generally end plant foraging and return to the hive by late afternoon. Pesticides will decompose somewhat overnight prior to the next day's bee activity. Dusts drift more than sprays and are more dangerous to nearby bees. Granulars are safest for bees in terms of drift. The choice of spray formulation is also a factor. Water-soluble forms are safer than emulsions, fine sprays are less harmful than coarse ones, and diluted sprays are safer than ultra-low-volume concentrates. Microencapsulated pesticides are to be avoided, because their size and appearance makes them look like pollen. Bees can carry these products back to the hive, resulting in deadly consequences.

Pesticides vary considerably in terms of bee toxicity (also for other beneficials). Most insecticides are either highly or moderately toxic. Highly toxic products should not be used when bees are present and never near beehives. They are toxic for at least 24 hours after application. Moderately toxic insecticides should not be applied when bees are active or near hives. These products can be used when bees aren't active assuming dosage, timing, and method of application are correct. As a general rule, older botanicals are relatively nontoxic to bees (but not newer generation analogues). Acaricides, insect growth regulators, natural products (except for oils), fungicides, herbicides, plant growth regulators, and biocontrols are generally nontoxic, too, assuming dosage, timing, and method of application are correct. The various pesticide tables in this chapter indicate bee and beneficial toxicity in some cases. More information is available through the Internet resources (see the list at the end of the chapter). Always read the literature that comes with pesticides to determine whether the product is safe or harmful to bees and other beneficials.

Other timing considerations are to target the pest whenever possible at the developmental stage when it is most susceptible or to time it for periods before the pest is likely to cause economically unacceptable damage (see the earlier discussion of economic threshold level under IPM). Last, applications during windy or rainy weather should be avoided. Drift from wind and washing away by rain result in wasted pesticide and ineffective results. Drift can also harm beneficials on adjacent crops or nearby indigenous plants. Beyond this economic consideration is also the liability that could result, depending on the location to which the pesticide drifts or where runoff ends up.

Phytotoxicity Problems.   Sometimes a plant will show leaf damage after the utilization of a pesticide. The damage appears as marginal burn, spotting, or chlorosis. Sometimes distortion or abnormal growth is observed. These are the symptoms of *phytotoxicity*. The younger leaves and blossoms, or new growth, are most likely to be damaged. In cases like this, the cure can be worse than the original problem.

In many cases it is known that certain pesticides will damage specific plants; the pesticide is phytotoxic to that plant. The horticulturist must be aware of these cases. It is possible to minimize the phytotoxic effect of any pesticide. Apply pesticides in the cooler parts of the day, preferably early morning, which allows the foliage and pesticide droplets to dry before the temperature goes over 28°C (85°F). High temperatures increase phytotoxicity problems. The temperature of the water can also be a factor; water should be tepid, not cold or hot. If a pesticide is available as both a wettable powder and emulsifiable concentrate, the wettable powders are usually less phytotoxic.

Sometimes the carrier, such as the organic solvent in which the pesticide is dissolved, can cause phytotoxicity. Sometimes pesticides show no evidence of phytotoxicity until they are used as a combination spray. Several other factors that influence phytotoxicity are the concentration of the pesticide, the developmental stage of the plant, and the health of the plant.

In many cases the full limits of phytotoxicity of pesticides are unknown or poorly characterized. Whenever possible, the horticulturist should test the pesticide on a few plants under the particular conditions that will prevail during subsequent spray applications. If phytotoxic symptoms appear, another pesticide should be chosen.

Contamination Problems.   The toxicity of pesticides is clearly known. There are several ways of expressing toxicity, but the most common is the LD_{50}, which is the median lethal dose of toxicant (given as milligrams per kilogram of body weight) fatal to 50% of the test animals (usually white albino rats).

On the basis of LD_{50} values, pesticides can be listed in terms of decreasing toxicity to the spray or dust applicator. When a choice is available, select the pesticide with the highest LD_{50}. The LD_{50} values are expressed as acute oral and dermal figures. Both are important, but the dermal value is probably more useful to the horticulturist, who is more apt to contact pesticides dermally rather than orally. Another expression used is LC_{50}, which is similar, except that it is based on fatality through inhalation and is given in terms of milligrams per liter.

Highly toxic compounds have oral LD_{50} values of 0 to 50 milligrams per kilogram, dermal LD_{50} values of 0 to 200 milligrams per kilogram, and inhalation LC_{50} values of 0 to 2,000 milligrams per liter. Moderately toxic compounds have respective values of over 50 to 500, over 200 to 2,000, and over 2,000 to 20,000. The respective LD_{50} oral and dermal values for slightly toxic compounds are over 500 to 5,000 and over 2,000 to 20,000, and for relatively nontoxic compounds, over 5,000 and over 20,000, respectively. The probable oral lethal doses for these categories in decreasing order of toxicity are a few drops up to one teaspoon, one teaspoon to one ounce, one ounce to one pint or one pound, and over one pint or one pound. Labels contain a signal word to relate toxicity in a simple manner. The word "Caution" denotes a pesticide that is either relatively nontoxic or slightly toxic, whereas "Warning" denotes moderately toxic, and "Danger" or "Poison" indicates high toxicity.

All appropriate safety precautions must be taken. Remember also that today's slightly toxic or relatively nontoxic product might turn out to be tomorrow's carcinogen. Some pesticides are suspect even now and could be withdrawn from the market.

Based on their toxicity, some level of pesticide, usually expressed in parts per million, will be allowed in horticultural food crops. Pesticides obviously leave a residual material on the plant. The residual lifetime will vary, depending on several factors, such as chemical stability, water solubility, temperature, and microbial susceptibility. Some pesticides will break down quickly, whereas others will remain relatively unchanged for long periods of time. Given a choice, the horticulturist should choose the pesticide with the least residual lifetime.

The longer the residual lifetime of a pesticide, the more likely it is to appear in the environment as a pollutant. The residual pesticide on the plant can be passed through the food chain to humans, the residual pesticide can wash into the soil from rain, and sprays or dusts will land on the soil. Once in the soil, the pesticide can enter the water table and end up in the food chain. Federal regulations restrict the amount of residues on food crops; other regulations are designed to minimize environmental consequences. The one unpredictable with residues is that our current medical knowledge cannot rule out the possibility that exposure to multiple residues over many years increases our risk for cancer. Tests for safety are feasible for individual pesticides and their residues, but we are limited when it comes to testing and predicting the synergistic effects of multiple exposures over time.

Control Problems. Constant dependence on the same pesticide can lead to less effectual control, since the plant pest can develop genetic resistance. For this reason, it is better to alternate different pesticides capable of controlling a plant pest. Even better, pesticides for a specific pest that vary in their modes of action should be alternated. Mode of action information is not always easy to come by, but manufacturers in compliance with EPA voluntary suggestions might incorporate mode of action information in their labels in the near future.

Another problem is the possible disturbance of the ecosystem. Pests of minor importance can emerge after the elimination of the major pest by the use of pesticides. The pesticide can be toxic to the beneficial organisms that controlled the minor pest, and the killing of the major pest as well has removed all the competition. The horticulturist must be observant and move quickly to control new outbreaks, either by using another form of plant protection to control the new pest or by switching to a form of plant protection that controls both pests.

Types of Pesticides for Specific Pests

This section contains several tables of pesticides. A large number of pesticides are cited to offer maximal flexibility in terms of personal views on sustainability when choosing pesticides for various problems. Availability will vary according to region, supply, and restrictions. Restrictions are imposed by the federal Environmental Protection Agency (EPA). When the tables were prepared, the pesticides listed were not banned outright by the EPA. However, some were under consideration for possible banning or for use under restricted conditions. Others were on waiting lists to be considered similarly. Some pesticides can be purchased by nonlicensed people and used according to their labels, while others are restricted and available only to licensed persons tested by the individual states. Further restrictions on availability and licensed usage beyond those of the EPA vary by state. Some of the pesticides discussed here are designed for use in IPM programs, often because of lower toxicity and minimal or no impact on beneficial insects. It is hoped that, out of the many listed, a few will meet the requirements dictated in the various situations nationwide.

The author and publisher assume no responsibility or obligation for the correctness, legality, or current EPA and state status of the listed pesticides. It is the horticulturist's responsibility to determine such status through local agencies such as the state agricultural experiment station, the state extension service, and state agencies regulating pesticide usage (such as environmental protection agencies). All precautions should be observed, and the previous section on pesticide safety should be learned well. The lists are not meant to be complete, merely indicative of the various types of available pesticides. The omission or inclusion of any pesticide does not imply any rating with regard to its merits.

"Green" Pesticides. A number of older and newer products (Figure 15–35) derived from natural products provide a greener alternative to chemical pesticides, as do synthetic and synthetic modifications of the original natural products. A word of caution is needed; it should not be assumed that all so-called "green" pesticides are nontoxic and safe for beneficial insects, mammals, and humans. These products vary considerably in their toxicity. These products, like all pesticides, should be selected with an eye toward toxicity and effects on beneficial insects, mammals, and humans.

FIGURE 15–35 Untreated Seckel pears (*left*) show insect and disease symptoms while those exposed to the same growing conditions, but treated with a particle film, exhibit undamaged, healthy fruit. The particle film is a white, reflective coating of a specially processed kaolin (a type of clay). Results include control of insects and diseases, reduced sunburn and less heat stress with fruit trees, vines, or vegetable crops. (USDA photograph.)

The oldest are the botanical pesticides, their modern analogues, soaps, and natural oils (Table 15–11). Examples of pests controlled by botanicals include caterpillars, beetles, aphids, mites, leafhoppers, thrips, weevils, Japanese beetles, flea beetles, loopers, and cabbageworms. Natural petroleum oils (relatively nontoxic) are used on dormant woody plants to control insects that winter over, such as scale. Lighter weight oils or summer oils have been used on nondormant plants to control scale, mites, mealybugs, whiteflies, aphids, pear psylla, and various insect eggs. Soap sprays have been used to control soft-bodied insects such as aphids or whiteflies and noninsects such as mites. Fatty acids have been used to control weeds and algae. Homemade weed controls include full-strength vinegar, rubbing alcohol (1 to 5 tablespoons per quart of water), and dishwashing soap (5 tablespoons per quart of water).

Newer natural products used as pesticides are shown in Table 15–12. Most of the lower toxicity pesticides listed in Tables 15–11 and 15–12 are sometimes called biopesticides or biorational pesticides. Another alternative is to use insect growth regulators (IGRs). IGRs (Table 15–13) are another form of least toxic pesticide. These products are very specific in that they disrupt development of the target insects. Hormonal mimics or interference with either chitin synthesis or the molting process brings about disruption. Insects must ingest these products, so good spray coverage is desirable. Mature beneficial insects are unaffected by these products. Both biorational pesticides and IGRs are often used in conjunction with biological controls as part of an IPM approach.

Nematicides. Nematodes have a rather impermeable cuticle, so chemicals with high penetration properties are used. The most widely used are the halogenated hydrocarbons, which, because of their volatile properties, can be used as soil fumigants. Examples of halogenated hydrocarbons used as nematicides in the form of fumigants are shown in Table 15–14. Nematicides other than the halogenated hydrocarbons include the fumigants shown in Table 15–14. Many nematicides also serve other purposes, such as herbicides, insecticides, rodenticides, or fungicides. Sometimes the nematicides shown in Table 15–14 are not used alone, but in combinations of two, such as methyl bromide plus chloropicrin or Telone plus chloropicrin. Some of the nematicides are phytotoxic, so they are applied as preplant materials. A number of compounds of a nonfumigant nature are also used to control nematodes. These compounds are used as combined insecticide/nematicides, and as such will be covered in the later section on insecticides.

TABLE 15–11   *Botanical, Botanical Analogues, and Oils Used as Pesticides*

Compound	Source	Comments
Capsaicin and garlic (Hot Pepper Wax™)	Hot peppers, paraffin, and garlic	Repellent for aphids, loopers, mealybugs, spider mites, thrips, whiteflies. Not compatible with beneficials.
Dormant and summer oils (All Seasons®, Damiol®, Dormant Oil 435®, Omni Supreme Spray, Par F70 Soluble Oil®, Sunspray 6E®, Sunspray Ultrafine®)	Oil	Controls eggs, nymphs, mites, scale. Don't mix with fungicides containing sulfur. Not compatible with beneficials.
Garlic extracts (Garlic Gard®, Garlic Barrier®)	Garlic	Repellent for thrips.
Limonene (*d*-limonene)	Citrus peels, especially oranges	Controls fleas, lice, mites, and ticks. Safe for humans and pets. Blocks sensory nerves in peripheral system.
Neem (azadirachtin)	Neem tree seeds	Controls numerous insect pests; also has insect growth regulator (Azatin®) activity (see Azadirachtin, Table 15–13).
Nicotine	Tobacco	Alkaloid. Broad spectrum, but especially for aphids and caterpillars. Acetylcholine mimic.
Nicotine analogues	Imidacloprid (Admire®, Merit®), acetamiprid, nitenpyram (Bestguard®)	Used in large amounts globally; systemic; long residual control for aphids, Japanese beetles, lace bugs, leafminers, mealybugs, scales, thrips, whiteflies. No effect on predatory mites and nematodes; blocks postsynaptic nicotinergic acetylcholine receptors.
Pyrethrum (mixture of pyrethrins I and II and cinerins I and II)	Species of chrysanthemum grown in Ecuador and Kenya	Fast-acting nerve paralytic agent, with synergist. Kills most insects. Oldest, safest botanical. Photo-unstable and costly for large-scale applications. Minimal bee toxicity.
Pyrethrum analogues	Allethrin, resmethrin, phonothrin, permethrin (Ambush®, Pounce®), bifenthrin (Capture®), cyfluthrin (Countdown®), esfenvalerate (Asana XL®), *zeta*-cypermethrin (Mustang®)	Now in fourth generation, more photostable, longer lasting, and extremely effective at very low level of active ingredient, axonic poisons, broad-spectrum pesticide. Highly toxic to bees. Little residual activity.
Rotenone	Cubé (*Derris* and *Lonchocarpus*)	Kills many insects, but very short residual. Poisonous to fish, leaf-eating caterpillars. Inhibits respiratory enzyme. Minimal bee toxicity.
Ryania	Ground roots of shrub grown in Trinidad	Controls aphids, caterpillars, thrips, and whiteflies. Minimal bee toxicity.
Sabadilla	Extract of seeds from lily family member	Sneezing and skin irritant. Controls leafhoppers, thrips, and true bugs.
Soybean oil (Golden Natur'l Spray Oil®)	Soybeans	Controls aphids, fungus gnats, lace bugs, leafminers, mealybugs, scales, spider mites, and whiteflies.

Miticides (Acaricides) and Insecticides.   Pesticides applied at the peak of mite or insect infestation usually eliminate beneficial parasitoids and predators. For the least damage to plant and beneficial organisms, chemical controls should be utilized before the peak of infestation or in an IPM program based on coordinated natural and chemical controls. Chemical controls against mites are listed in Table 15–15; those against insects include the

TABLE 15–12 *Natural Product and Alternative Pesticides*

Compound	Controls	Comments
Abamectin (Avid®)	Effective against leafminers, spider mites, and other greenhouse pests	Actinomycete (*Streptomyces avermitilis*) product. Local systemic properties (spray on leaf surface, kills underneath leaf). Photodegradable.
Clandosan®	Stimulates soil microbes that attack nematodes	Chitin protein (crab and shrimp shells) mixed with urea.
Corn gluten meal (Concern™, Organic Weedzstop™, W.O.W.™)	Weeds	Contains 10% nitrogen, used as preemergent natural herbicide.
Emamectin benzoate (Proclaim®, Denim®)	Caterpillars	Actinomycete (Abamectin analogue) derivative. Safe for most beneficials. Photodegradable.
Fatty acids (DeMoss®, Scythe®)	Weeds	Avoid contact with crops.
Insecticidal soaps (M-Pede®, Safer®)	Controls soft-bodied insects and mites on contact	Potassium salts of long-chain, biodegradable fatty acids; possibly harmful to beneficial *Stethorus*; can be phytotoxic.
Kaligreen,™ Armicarb™	Flyspeck, sooty blotch	Similar to baking soda (potassium bicarbonate)
Methionine plus riboflavin with detergent and traces of copper sulfate	Flyspeck, sooty blotch	Experimental product showing good control with apples
Remedy™	Controls anthracnose, blackspot, botrytis, leafspot, phoma, phtophthora, powdery mildew, scab	Baking soda product; works best mixed with horticultural oil.
Spinosad (Conserve SC™, Spin Tor®, Success®, Tracer Naturalyte®)	Effective at very low dosage against some beetles and weevils, many caterpillars, leaf miners, thrips, termites	Actinomycete (*Saccharopolyspora spinosa*) metabolite; long residual activity, low toxicity, safe for mammals, birds, fish; beneficials.
Strobilurions: azoxystrobulin (Abound®, Heritage®); kresoximmethyl, (Sovran®); trifloxystrobin (Flint®)	Controls alternaria, anthracnose, bitter rot, black rot, blight, botrytis, Brooks spot, brown rot, downy mildew, flyspeck, leaf spot, lawn diseases, powdery mildew, rusts, scab, soil-borne diseases, sooty blotch, white rot	Derivative of strobilurin found in *Strobiluris tenacellus*. Relatively new; resistance develops with overuse. Classed as a reduced risk compound.
Vinegar (full-strength)	Weeds	Can harm crop or garden plants, too.

nonsystemic insecticides, the systemic insecticides, and combined nonsystemic pesticides (Table 15–16). The nonsystemics kill by acting as a stomach poison, which is effective against chewing insects, or as a contact poison, which is effective against both chewing and sucking insects. The systemics are primarily effective against sucking insects, although some of the systemics also act as contact poisons.

Other Animal Pesticides. Chemical repellents against birds include 9,10-anthraquinone (anthraquinon, slightly toxic) and 4-aminopyridine (Avitrola, highly toxic). Other chemicals are used to kill birds or as birth control agents. Chemical repellents against rabbits, other rodents, and deer include tetramethylthiuram disulfide (Thiram), also used as a fungicide and bone tar oil, (Magic Circle Repellent). These are slightly toxic. The bone tar oil repellent is also used to repel dogs and cats. Many of the chemical insecticides will control millipedes.

Molluscicides for snails and slugs include 4-(methylthio)-3,5-xylyl methylcarbamate (Methiocarb) and metaldehyde (Antimilace). These are moderately and slightly toxic, respectively.

TABLE 15–13 *Insect Growth Regulator Pesticides*

Compound	Controls	Comments
Azadirachtin, Azatin®, Neemazad®, Neemix®	Aphids, armyworms, beetles, fungus gnats, leafhoppers, leafminers, thrips, whiteflies	Interferes with molting and metamorphosis; safe for beneficial insects, earthworms, fish. Breaks down fast. Low toxicity.
Cyromazine, Citation®, Larvadex®	Fungus gnats, leafminers, shore flies	Interferes with molting and metamorphosis; safe for beneficials.
Diflubenzuron, Adept®, Dimilin®	Caterpillars, fungus gnats	Inhibits chitin synthesis; safe for birds, fish. Low toxicity. Breaks down fast.
Diofenolan, Arbor®	Caterpillars and scale	Interferes with molting and metamorphosis; low toxicity.
Fenoxycarb, Insegar®, Precision®, Preclude®	Aphids, fungus gnats, scales, shore flies, thrips, whiteflies	Low toxicity; safe for birds but not fish.
Isomate®	Moths	Pheromone that disrupts mating.
Kinoprene, Enstar II®	Aphids, fungus gnats, mealybugs, soft and armored scales, whiteflies	Rapid photodegradation, Breaks down fast; minimal problems with beneficials.
Lufenuron, Match®	Caterpillars	Interferes with molting and metamorphosis.
Pymetrozine, Endeavor®, Fulfill®	Whiteflies, aphids, leafhoppers, scale, mealybugs, planthoppers, spittlebugs, and psyllids	Technically an anti-feedent specific for homopterous pests; safe for fish and wildlife.
Pyriproxyfen, Admiral®, Esteem®, Knack®, Pyrigo®	Aphids, leafminers, moths, psylla, scale	Interferes with embryogenesis in eggs; adult emergence; toxic to fish.
Tebufenozide, Confirm®	Caterpillars	Interferes with molting and metamorphosis; safe for beneficial insects; *Stethorus,* and predatory mites.

TABLE 15–14 *Nematicides*

Compound	Controls	Comments
Basamid® (dazomet)	Nematodes, soil-borne fungal diseases, weeds	Mechanically incorporated granular; can harm trees planted too soon; low-level fumigant. Ornamentals, turf.
Chlor-O-Pic® (chloropicrin)	Nematodes and soil fungi	Halogenated hydrocarbon. Preplant fumigation. Fruit, ornamental, and vegetable crops.
Mocap® (ethoprop)	Nematodes and soil insects	Vegetable crops.
Nemacur® (fenamiphos)	Nematodes	Organophosphate; needs watering into soil. Turf, citrus, vegetable crops.
Telone II® (1,3-dichloropropene), Telone C-17® (Telone + chloropicrin)	Nematodes, also soil-borne pathogens	Halogenated hydrocarbon. Preplant fumigant. Vegetable crops.
Vapam (meta-sodium)	Nematodes, soil-borne fungal diseases, weeds	Preplant fumigant. Ornamental and vegetable crops.
Vydate (oxamyl)	Nematodes	Carbamate also used as an insecticide, miticide, and growth regulator. Vegetable crops.

TABLE 15–15 *Miticides*

Class	Examples	Comments
Carbamates	Carzol® (formetanate hydrochloride)	Also controls leafminers, leafhoppers, plant bugs, stink bugs, thrips; toxic to bees and predatory mites.
Dinitrophenols	Dinocap (Karathane®)	Used also for powdery mildews; all have been banned.
Organosulfurs	Ovex (Ovotran®), propargite (Omite®), tetradifon (Tedion®)	Relatively nontoxic to honeybees.
Organotins	Cyhexatin (Plictran®), fenbutatin oxide (Vendex®)	Also have fungicidal qualities.
Pyrazoles	Fenpyroximate (Acaban®, Dynamite®)	Fast acting; controls all mite stages; inhibits molting; long residual activity; disrupts ATP formation; has some effectiveness against aphids, psylla, thrips, and whiteflies.
Pyridazinones	Pyridaben (Sanmite®)	Rapid action and long residual activity; has some effectiveness against aphids, leafhoppers, thrips, and whiteflies; safe for mammals and birds.
Quinazolines	Fenazaquin (Magister®, Matador®)	Rapid action; controls all mite stages; ovicidal activity.
Tetrazines	Clofentezine (Apollo®, Acaristop®)	Inhibits mite growth; also has ovicidal activity; low mammalian toxicity; minimal impact on beneficial insects.

TABLE 15–16 *Insecticides*

Class	Examples	Comments
Carbamates	Aldicarb (Temik®), carbaryl (Sevin®), oxamyl (Vydate®), propoxur (Baygon®)	Third-generation pesticides; inhibit cholinesterase. Carbaryl is widely used and has low mammalian toxicity and broad-spectrum action, but harms bees and mite predators; aldicarb (systemic) is very toxic.
Fiproles	Fipronil (Regent®, Icon®)	Systemic; useful on pests resistant to pyrethrum analogues, organophosphates, and carbamates.
Formadines	Amitraz (Mitac®)	Inhibition of monoamine oxidase; used on organophosphates and carbamate-resistant pests.
Organochlorines	DDT, lindane, chlordane, dieldrin, endrin, heptachlor, methoxychlor, mirex, toxaphene	Oldest synthetic organics; many have environmental persistence; problems of bioaccumulation in food chain; few remain in use.
Organophosphates	Acephate (Orthene®), chlorpyrifos (Dursban®), diazinon, dichlorvos (Vapona®), malathion, oxydemeton-methyl (MetaSystox®), methyl parathion	Second-generation organic synthetics, derived from WWII nerve toxicants. Inhibit cholinesterase; toxic to vertebrates; less persistence in environment. In various stages of partial or complete ban by EPA.

Bactericides and Fungicides. The control of viruses by pesticides is not possible at present. Indirect control is achieved by the use of pesticides against insects known to be viral carriers. Antibiotics such as tetracycline inhibit phytoplasmas. Some measure of control has resulted from the injection and translocation of antibiotics within infected woody plants.

Bacterial diseases have been controlled by the use of antibiotic sprays such as streptomycin (Agri-Strep, Agrimycin 17, and Phytomycin). This is relatively nontoxic. Chemicals can be used to prevent fungal infections and to fight those already present. Examples of bactericides and fungicides used on plants and seeds are shown in Table 15–17. Many fungicides are toxic to earthworms and reduce their populations and subsequently soil health.

TABLE 15–17   *Fungicides and Bactericides*

Compound	Controls	Comments
ALIPHATIC NITROGENS: dodine, Cyprex®, Syllit®	Brown rot, cherry leaf spot, scab	Do not mix with lime or sulfur; overuse induces resistance.
ANILINOPYRIMIDINES: cyprodinal, Vanguard®	Brown rot blossom blight, scab	Newer product; do not apply near water; overuse induces resistance.
ANTIBIOTICS: streptomycin (Agri-Mycin®), terramycin (Myco-Shield®)	Fireblight/soft rots and bacterial spot/mycoplasmal diseases (X-disease, decline), respectively	Low mammalian toxicity.
CARBAMATES: ferbam, Carbamate®; mancozeb, Dithane®; maneb, Manex®; metiram, Polyram®; thiram, Vitavax®; zineb, Zinosan®; ziram, Milbam®	Wide variety of diseases in many plants	Widely used foliar protectants; few on seeds and in soil.
BENZIMIDAZOLES: benomyl, Benlate®, thiabendazole, Mertect®; thiophanate-methyl, Topsin-M®	Wide variety of diseases in many plants	Overuse induces resistance in pathogens; systemic.
DICARBOXIMIDES: captan, Captec® and Agrox®; iprodione, Rovral®, vinclozolin, Ronilan®	Captan has broad-spectrum control, others more specific: botrytis, brown rot, lawn diseases, white mold	Resistance more likely to develop with the newer, more specific iprodione and vinclozolin.
INORGANICS: copper (basic copper sulfate, Basicop®; copper oxide and hydroxide, Champ®, Kocide® and Nu-cop®), sulfur (elemental, flowable, lime and wettable (Microthiol Special®, Thiolux)	Sulfur for powdery mildew, copper for several fungal and bacterial diseases	Oldest fungicides. Use carefully to avoid plant injury; sulfur and copper are also essential plant nutrients.
ORGANOPHOSPHATES: fosetyl-Al, Aliette®	Soil-borne diseases, white rust	Systemic.
ORGANOTINS: triphenyltin, Super-Tin®	Alternaria blight, early and late blight, leaf spot, scab	Highly toxic to fish.
PHENYLAMIDES: metalaxyl, Ridomil®	Downy mildew, late blight, soil-borne diseases	Systemic.
PYRIMIDINES: fenarimol, Rubigan®; triforine, Funginex®	Anthracnose, brown rot blossom blight, black rot, powdery mildew, rusts and scab	Systemic.
PYRROLES: fludioxonil, Medallion® and Maxim®	Seed-borne and soil-borne diseases	Low toxicity; safe for bees, birds, and earthworms.
SUBSTITUTED AROMATICS: chlorothalonil, Bravo®, Evade®; chloroneb, Demosan®; dichloran, Botran®	Anthracnose, brown rot blossom blight, leaf curl, cherry leaf spot, coryneum blight, lawn diseases, scab, seed protection, soil diseases	Exhibits cross-resistance with the dicarboximides.
TRIAZOLES*: etridiazole, Truban®; fenbuconazole, Indar®; myclobutanil, Nova®; propiconazole, Orbit®, Topas®; triadimefon, Bayleton®; tebuconazole, Elite®; triflumizole, Procure®	Anthracnose, blossom blight, black rot, brown rot, leaf spot, lawn diseases, powdery mildews, rust, scab, soil-borne diseases, turf diseases	Mostly systemics.

*This group is also known as sterol inhibitors.

(A)   (B)

FIGURE 15–36   (A) This equipment is preparing soil for planting and simultaneously applying a preplanting treatment of herbicide. (USDA/NRCS photograph.) (B) Herbicide is being applied as a postemergence treatment here to a corn crop. (USDA photograph.)

Algicides   Algae can be eliminated through chemical control, such as with copper sulfate penthahydrate and triphenyltin acetate. Another product is bromo-chloro-dimethyl-imidazolidinedione (Agribrom®). Bleach solutions and fatty acid products are also effective.

Herbicides.   Chemicals used to kill weeds are termed herbicides. Weed killers can be divided into several groups based on their functions. Some herbicides are *nonselective;* that is, they destroy any and all plants they touch. These can be useful for treating propagation media or for clearing areas of unwanted vegetation, such as roadsides, railroad right-of-ways, sidewalks, driveways, or areas about to be planted to horticultural crops. Others can be *selective* and kill some plants (weeds), but not others (horticultural crops). Selective herbicides would be useful for weed control in lawns, orchards, or monocrops such as tomatoes. However, selective herbicides are not as effective for mixed horticultural crops, because the various crops will show different susceptibility to being damaged. Selectivity can be brought about by physical means such as avoiding the spraying of crops, or by physiological means, that is, by differences in contact or uptake of the herbicide. The latter might be possible if crops, when compared to weeds, have heavier cuticles, less exposed growing points, different metabolism, less surface area, or deeper roots that limit herbicide effectiveness.

Some herbicides destroy weeds only on areas of *contact;* others destroy even untouched areas since they are absorbed and *translocated* throughout the plant. Herbicides (Figure 15–36) can be applied prior to planting crops (*preplanting treatment*), after crops are planted but before they emerge (*preemergence treatment*), or directly to growing crops (*postemergence treatment*). Whatever treatment is used, care must be exercised to avoid drift of herbicide to susceptible plants. Separate sprayers must be used for herbicides to avoid accidental destruction of plant material. Some herbicides in horticultural use are shown in Table 15–18.

SUMMARY

Control of damaging plant pests involves several steps. First, the horticulturist must recognize the problem's existence. One must be knowledgeable about plant pests and under what environmental conditions and during which season susceptibility occurs. The horticulturist must recognize the plant pest early, or after it causes a plant response or symptom, and be able to identify the pest. Identification is followed by an assessment to determine if control is needed, the control required to eliminate the pest, the method of application, timing of application(s), how much is needed, phytotoxicity and off-site environmental impact, and effects of control on the population balance between beneficial organisms and pests.

TABLE 15–18   *Herbicides*

Class	Examples	Comments
Amides	Napropamide(Devrinol®), pronamide (Kerb®)	Selective preemergents; controls many grasses and some annual broad-leaved weeds.
Aryloxyphenoxy propionates	Fluazifop-*p*-butyl (Fusilade®)	Selective postemergent; controls many grasses.
Benzoic acids	Dicamba (Banvel®)	Effective against difficult broad-leaved weeds such as clover.
Bipyridyliums	Paraquat (Gramoxone®)	Kills all annual broad-leaved weeds and some perennial ones; highly toxic to mammals.
Carbamates	Chlorpropham (Furloe®)	Selective preemergents.
Carboxylic acids	Picloram (Tordon®)	Control perennial broad-leaved weeds and brush.
Cyclohexanediones	Clethodim (Prism®), Sethoxydim (Poast®)	Selective postemergent; controls annual and perennial grasses.
Dinitroanilines	Oryzalin (Surflan®), Pendimethalin (Prowl®)	Selective herbicides; controls grasses and many broad-leaved weeds; widely used. Low toxicity.
Diphenyl ethers	Oxyfluorfen (Goal®)	Selective pre- and postemergents; controls certain broad-leaved and grassy weeds.
Nitriles	Dichlobenil (Casoron®, Norosac®)	Apply only in cool or cold weather; controls broad-leaved weeds.
Phenoxys	Dimethylamine and diethanolamine salts of 2,4-D (Orchard Master®), 2,4-D (Weedar 64®)	Selective for broad-leaved weeds.
Phosphono acids	Glyphosate (Roundup Ultra®, Ranger®, Rattler®, Touchdown®), glufosinate (Ignite®, Rely®)	Broad-spectrum control of annual and perennial grasses plus broad-leaved weeds; moderately toxic; eye irritant.
Triazines	Simazine (Princep 4L®)	Selective for broad-leaved weeds and annual grasses.
Uracils	Terbacil (Sinbar®)	Controls many broad-leaved and grassy weeds avoid use in sandy soils.
Ureas	Diuron (Karmex®)	Preemergent, broad-leaved weeds.

Animal kingdom pests that eat plants include insect arthropods, noninsect arthropods (mites, millipedes, and centipedes), nematodes, and larger animal pests (slugs, snails, birds, rabbits, deer, mice, moles, other rodents, dogs, cats, and even humans). Insects and nematodes also transmit diseases. Insects are the most damaging pests. Insects are well adapted to plant destruction because of their feeding habits, small size, protective coloration in some cases, the ability to fly, and their prolific nature. Mites, being prolific, cause extensive damage. Insecticides often eliminate competitors and mite enemies, leading to mite increases and the need for combined insecticides/miticides.

Millipedes and centipedes have numerous legs. Millipedes prefer moist sites, high levels of soil organic matter and organic debris. These arthropods cause minor plant damage in the greenhouse and garden. Millipedes feed on decaying organic matter, but can eat plant roots, foliage, and ripening fruits and vegetables. A few species are insect predators. Centipedes like moist conditions, but are found under stones, boards, or in woodpiles. Centipedes are beneficial in that they prey on grubs, other insects, and snails.

Nematodes are roundworms that live in soil and water. Most nematodes are free living; a few parasitize plants and animals. Some parasitic types feed mostly on plant roots and a few are leaf feeders. Other parasitic nematodes can be used as predators (entomopathogenic nematodes). While feeding, nematodes can introduce disease organisms that cause wilts or rots into root systems.

Slugs and snails damage garden/greenhouse plants. These pests thrive in horticultural areas with their moisture, shaded areas, organic debris, and food. Some larger animals can cause extensive harm. Birds become pests when they eat newly planted seeds and ripe fruits. Deer, rabbits, and other rodents damage young crops by eating them and can girdle fruit and ornamental trees when eating in winter. During the growing season, total destruction of many horticultural plants can occur from deer browsing. Deer are also hosts for Lyme disease-infected ticks. Dogs and cats damage plants by digging, sharpening (cats only) of claws on bark, and the use of planted areas as a toilet. Humans damage plants through ignorance, carelessness, theft, and vandalism.

A number of nonanimal organisms attack plants or interfere with their successful cultivation. These include disease-causing (pathogenic) microorganisms (viruses, viroids, phytoplasmas, spiroplasmas, xylem/phloem-limited bacteria, bacteria, and fungi) and certain plants (dodder, mistletoe, algae, moss, weeds, strangling vines, and strangling roots). Non-pathogenic diseases are caused by nutritional or environmental causes and are physiologically based disorders.

The virus is the second smallest pathogen. Viruses consist of a protein coat surrounding an RNA or DNA nucleic acid core. They cannot reproduce except within the cells of a living host. Viroids are essentially viruses without protein coats and are the smallest known plant pathogen. Viroids associated with plant diseases are of the RNA type only. Many fruits, vegetables, and ornamentals can be infected by viruses/viroids.

Bacteria cause diseases by entering plants through natural openings or wounded tissue and by insect transmission. A number of organisms related to bacteria cause plant diseases. Formerly, these diseases were thought to be viral-based because their symptoms resembled viral diseases. Eventually organisms that somewhat resembled mycoplasmas and rickettsias were discovered. Currently the mycoplasma-like organisms are designated as phytoplasmas and spiroplasmas. Both organisms are usually found in phloem, where they utilize food from the host, disrupt translocation, and possibly produce toxins or cause hormonal imbalance. Vectors are usually leafhoppers (mostly), plant hoppers, psyllas, spittlebugs and grafting. Rickettsia-like organisms are now designated as either fastidious phloem-limited bacteria or fastidious xylem-limited bacteria, depending on their preferred tissue location. Leafhoppers, psyllas, dodder, and grafting are vectors.

Fungi are multicellular organisms with a thread-like growth, sometimes with elaborate and colorful fruiting bodies. Fungal diseases are the number two economic pest problem. High humidity can increase fungal disease problems. Overwatering also increases root disease problems. Fungi invade plants through natural openings, wounds, insect transmission, and even directly through the cuticle by enzymatic attack.

Dodder is a plant that lacks chlorophyll and parasitizes other plants. Mistletoe is a parasitic plant found high in trees. Both are debilitating to the host plant's vigor and can result in death. Dodder also vectors some viral diseases. Algae and moss are simple, nonvascular plants that can be undesirable in certain horticultural situations.

Weed seeds are numerous and viable for many years. Seed germination and seedling growth are frequently rapid and subsequent growth is often extensive. Weed control is essential. In the competition for light, water, and soil nutrients, weeds can seriously weaken or kill horticultural crops, reducing the crop's economic and esthetic values. Weeds can be alternative hosts for plant pests and diseases. They are esthetically unpleasing and can be dermal irritants and allergens.

Plant protection is a key area needing sustainable practices. The most important, primary-use practices are proactive or preventative practices to minimize pest problems. These practices are biological or natural and based on approaches to create field, garden, or greenhouse environments that are either unfavorable to pest invasion and subsequent colonization or favorable to natural enemies of pests. Should pest levels approach a potential to cause economically unacceptable damage, controls become reactive (responsive). Reactive controls include biological controls, botanical pesticides, natural product pesticides, insect growth regulators and chemical pesticides.

Pesticides have several detrimental consequences. Pesticides should not be used in either a preemptive manner or in large volumes as the sole approach to plant protection. This strategy is clearly not sustainable. Pesticides have left a legacy of environmental damage, negative impacts on human health, an increasing spiral of induced pest resistance, ever rising costs, and a less than perfect success record of pest control. The public views pesticides with suspicion and concern. Other problems include a change of secondary pests to primary pests as pesticides destroy natural controls on secondary pests. Another aspect is extensive and unnecessary residues that result from only a small amount of the application contacting the pest, leaving the rest to bring about undesirable environmental and human health problems. Pesticides also squander energy and petrochemical feedstock.

Organic horticulturists produce organic food in a pesticide-free environment, relying heavily on proactive controls. The other alternative is to reduce inputs of pesticides to minimal levels. Even better is to make natural controls the primary line of defense and to only use pesticides when essential. The selected pesticide must have the lowest possible toxicity to humans and beneficial organisms and pose the least environmental harm and health risk. This approach is integrated pest management (IPM). IPM is the mainstay of many sustainable horticulturists who can't or aren't ready to adopt organic practices. IPM strategies manage pests with an understanding of their place in the ecosystem relative to other organisms and in a manner that protects the environment. With IPM, pesticides are the last resort when pests reach economically damaging levels and natural controls (biological, cultural, genetic and physical controls) have failed.

Several criteria form the basis of IPM and help define it. With IPM, the presence of a pest is not automatically labeled a problem. Criteria and information are examined and a decision is reached regarding whether pest control is warranted. It is understood that certain levels of pests are acceptable as part of the normal balance in the functioning ecosystem. The goal is no longer total wipeout of the pest species, but maintenance of the natural balance between pests and beneficial organisms. The strategy is to let nature or proactive practices function and to only intervene when necessary and in a manner that causes minimal harm to the ecosystem.

The key to good IPM programs is pest identification and monitoring for subsequent activity and population levels. Monitoring creates a dynamic database through record keeping. From this database a course of pest control action (or inaction) is determined. Monitoring involves a regular, systematic collection of yearly data by scouting. Scouting is information gathering by observation and sampling traps plus real time Web data for major pests. Scouting is not complicated. Some labor and time is required, but the tradeoff is reduced pesticide costs. Scouting is done at regular intervals and with increased frequency when pest levels are becoming serious. During scouting, one needs to achieve representative sampling of growing areas.

Assessments of the ability of beneficial insect or mite to control the pest (efficacy, numbers) are important data. Weather conditions should be assessed. Is the weather the same as in past years, did any unusual weather precede the new pest, and is the predicted weather in the next week or two going to have any effect on the pest? Finally, when were pesticides, fertilizer and irrigation last applied to the affected plants? The appearance of a specific pest repetitively after an insecticide for another pest might indicate the insecticide is causing a secondary pest to become a primary pest. Lush plants covered with pests might be connected to overfertilization.

Various traps exist for assessing the presence and population pressures of insects and mites. Other possibilities exist. Certain key plants are much more likely to be attacked by insects and infected with diseases. These plants should be routinely scouted at the appropriate window of time that known problems commence. Similarly, key pests lists need to be determined and their window of destruction ascertained. Phenological models predict pest outbreaks. Near the predicted time visual inspections and traps should be applied or examined with greater attention and frequency. While growers can develop such phenological models over time, some data is available through the Web.

Records from scouting inspections including traps and other activities are used to develop a database or what is known as pest mapping. Over time these records help you to fine-tune your IPM program as you compare current data patterns to the "average" picture gleaned from records. Such information can also be plugged into existing modeling systems for certain key pests and crops, leading to recommendations for appropriate IPM responses.

The IPM intent is to reduce pest damage to a tolerable level. The point at which pest numbers cause damage that exceeds the cost of control is known as the economic injury level (EIL). Control measures are applied prior to the EIL when controls will prevent the pest population from increasing to damaging levels. This point is the economic threshold level (ETL). Some preemptive actions also occur with IPM such as the use of pest-resistant cultivars where major problems exist and the use of cultural practices such as sanitation and those that increase biodiversity of the soil and field ecosystem. These approaches are proactive IPM. Actual use of biological controls to solve a problem or a pesticide would be a reactive strategy. The main approach with reactive strategies is to first use biological or cultural controls. If these fail to reduce the pests and damage to tolerable levels, then stronger options are used, starting first with "green" pesticides and insect growth regulators and ending with toxic pesticides. A principle is to utilize chemicals that result in the least harm to beneficial organisms and ecosystems. If the only choice harms beneficial organisms, the method or timing of the application is controlled to minimize the impact on the helpful organisms.

Proactive practices are approaches to and processes for pest and disease prevention that don't involve pesticides. Preventative measures can be physical or mechanical techniques to create a barrier and/or deterrent. Some are more expensive or labor intensive than others. Proactive practices are less of a threat to ecosystems and our environment, and some are selfperpetuating, once established. Proactive methods are not always capable of preventing all pests and diseases. Organic commercial horticulturists, but not all nonorganic commercial horticulturists, accept this disadvantage. Cost and efficacy are concerns. As proactive practices become more available and refined, they will be increasingly adopted.

IPM thresholds for diseases are limited. Plant disease forecasting systems rely on different thresholds. These thresholds depend on continual monitoring of components like spore concentrations and relative humidity/temperature profiles. These thresholds are expressed as severity indices. Given the difficulty and time/cost, these predictive disease models have been developed and implemented for only a few high-value horticultural crops. When the severity index reaches a critical value, fungicide applications are indicated. Forecasting systems greatly reduce regularly scheduled fungicide applications.

Cultural controls involve the manipulation of the cropping system, production practices, and the ecosystem to minimize pests and damage. One way is to enhance the populations of natural enemies by manipulation of the field and surrounding environments to create optimal conditions for beneficial populations. The field and surrounding environments can also be manipulated to produce the least favorable breeding and survival conditions for chronic plant pests. Plant breeding or genetic engineering can produce plants that resist injury or damage from pests. Steps can be taken to prevent uncontrollable diseases from colonizing crops by eliminating the insects that infect crops with these diseases.

Plants harboring pathogens or pests should not be planted. Plant debris should be promptly removed; otherwise, it could provide a breeding area or overwintering material for pests. Plants with uncontrollable plant pest infestations should be destroyed. These practices are termed *sanitation*. *Surgery* involves the removal of infected plant parts and *roguing* is the removal of diseased plants or seeds. Virus-free stock should be used when available. These practices prevent the spread of insects and disease.

Good soil health leads to the need for less pesticides and better results with biological controls. Increasing soil organic matter reduces nematode populations. A healthy soil is also rich in various beneficial microbes that are antagonistic toward disease-causing microorganisms. Control is achieved through competition and suppression of plant pathogens found in the soil.

Field environments often do not provide conditions needed to maintain natural predators and parasites. Manipulation of surrounding vegetation can be either destructive or constructive. Destructive situations would take place with pests that overwintered not on the crop, but in specific wild vegetation nearby which is destroyed. A constructive situation involves the encouragement and maintenance of nearby vegetation that supplies food or overwintering sites for predators and parasites.

Crop rotation can disrupt the life cycles of diseases, insects, and nematodes by eliminating their food supply. This practice works best with insects having long generation cycles and limited migration ability and diseases that are short-lived in the absence of hosts. Crop rotation is also effective in weed and nematode management. Intercropping of two plants together can create an unfavorable weed environment through increased plant density, greater competition and shading.

Timing a crop in regard to sowing, transplanting, or harvesting can be used to avoid peak periods of insect infestations or the time of egg laying. Early tillage followed by delayed planting can also eliminate weeds that germinate early. Closer spacing of plants both between and in rows can reduce weed problems because of faster shading of bare soil, thus reducing weed germination and establishment.

Certain plants can be grown to discourage plant pests or serve as trap crops for specific pests. Mixed crops can be less susceptible to a particular insect than would a monocrop of the preferred host plant. Insects can prefer other plants. These can be planted as trap plants to draw insects away from what would normally be a host plant.

Some plants resist pests to varying degrees. Horticulturists should strive to keep abreast of developments in this area and to specify the use of resistant plants wherever possible. In crops where genes are available to confer resistance, resistance forms the mainstay of IPM. Resistance involves less labor and energy inputs relative to most practices used to control pests and is environmentally compatible.

Other cultural practices include alteration of soil pH to provide an unfavorable environment for disease-causing fungi. Humidity levels and temperatures can be varied in greenhouses to provide less favorable environments for disease pathogens. Air circulation in greenhouses is used to speed drying of foliage to reduce disease initiation. Increased relative humidity and temperature reduction can reduce mite numbers.

Reactive or responsive practices are used when proactive practices fail such that pests or diseases will cause unacceptable economic damage. Two strategies exist: nonchemical (first choice for sustainable horticulturists) and pesticides. These practices include mostly cultural controls and biological controls. Pesticides, except for a few botanicals are not available as an option for the organic horticulturist, but are for the remaining sustainable horticulturists. However, the use of pesticides must be in the context of IPM practices.

Traps with sex attractants with and without sticky baits or electric blacklight traps are helpful in trapping flying insects. Hand picking and destruction of larger insect pests is possible. Forceful blasts of water from a misting nozzle can remove insects from plants in the greenhouse and home. Flooding of fields can control some insects, nematodes, and diseases. Overwintering populations of certain pests can be controlled by fall or spring tillage. Hot-water dips can destroy plant insects and mites. Heat treatments have been used to destroy bacteria, fungi, and viruses.

Weeds compete with crop plants for water and nutrients and can be sources of diseases and insects. Nearby weeds should not be blindly destroyed as adjacent weeds can also supply food, refuge, and breeding sites for beneficial insects. Overburning of land and spot flaming have been utilized to destroy weeds. The oldest form of weed control is hand removal or by hoe. Mechanical tillage between rows and around beds is the major alternative to herbicides for controlling weeds and is more sustainable. Practices to produce maximal crop growth in minimal time can favor crop competition over weeds. Weeds are prevented with mulches.

Small batches of propagation media containing soil can be steam pasteurized to destroy nematodes and microbial pathogens. Solarization is used to destroy nematodes and microbial pathogens (and weed seeds) in the field and in propagation media.

Alterations of the plant's physiology are possible through fertilization and pruning. Sometimes these physiological alterations enable a plant to resist some plant diseases; this cultural control is also known as *physiological control*.

Biological controls are the mainstay of pest control in sustainable horticulture. Most biological controls are reactive practices, but some aspects are preventative. Biological control is based on the natural antagonism, competition and predation existing among organisms. Many insects that attack plants have natural enemies: Enemies are predators (certain insects, mites, and spiders) that prey on other insects and parasitoids (tiny wasps or flies) that lay eggs in or on other insects or their eggs. Collectively, predators and parasitoids form an important part of natural control and are important for IPM programs. Predators, parasitoids, and other insects such as pollinators that serve a useful purpose are collectively termed beneficial insects. These helpful insects are often casualties when pesticides are used. Other forms of biological control (biocontrols) are the use of various pathogens to infect insects, mites, and weeds; microbial antagonists to suppress plant diseases and reduce food spoilage; insects and geese to eat weeds, the introduction of sterile male insects to reduce harmful insect populations; and predaceous or parasitic nematodes/protozoa for soil and airborne insect pests, respectively.

Three basic strategies exist in the use of natural enemies versus harmful pests. Augmentation involves strategies to increase the effectiveness of natural enemies. Natural enemies are either released annually in areas with no natural enemies (inoculative release) or in areas having natural enemies (inundative release). Conservation methods involve looking for factors that reduce the effectiveness of a natural enemy and then attempting to limit these inhibitory factors. Actions are also taken to increase resources needed by the helpful insect. Importation of predators and parasites is used against pests not indigenous to the United States Importation is left to governmental agencies.

Chemicals used to control plant pests are pesticides. These chemicals usually kill the pest, but in some cases they are only sickened. Chemicals that discourage plant pests from attacking plants are *repellents*. The latest pesticide trend is toward products that are significantly less damaging to the environment. Such products usually have lower toxicity, more selectivity, and less harmful effects on beneficial insects.

Pesticides are divided into categories on the basis of which pest they control: insecticides, miticides, nematicides, bactericides, fungicides, rodenticides, and herbicides. Pesticides can be systemics and nonsystemics. Systemics are water-soluble pesticides absorbed by plant parts and translocated. The plar   ecomes toxic to the pest. Nonsystemics coat the plant and affect the pest either on contact or after ingestion. Systemics are not as easily degraded by rain and sunlight, but are less of a threat to beneficial insects than nonsystemic pesticides.

Pesticides show some degree of selectivity. The majority of pesticides are usually not selective enough to spare beneficial organisms. Certain pesticides are designed to eliminate a wide range of plant pests. These are termed wide-spectrum or general-purpose pesticides. These can be a single chemical that shows toxicity to a wide range of pests, or combinations of pesticides. When combining pesticides to achieve multiple control, one must have information on compatibility. If incompatible, plant damage or reduced efficiency can result.

The form in which pesticides are sold is the formulation. Formulations can be applied as sprays (mostly), dusts, granulars, aerosols, fumigants, or encapsulated materials. Formulations sold for use as a spray include water-miscible liquids, emulsible concentrates (mostly), wettable powders, water-soluble powders, flowable suspensions of ground pesticide in water, oil emulsions, and ultra-low-volume concentrates. Wetting agents or spreaders are added to give more uniform coverage.

Pesticides can be toxic to humans and when improperly handled cause illness or death. Some pesticides are implicated in causing cancers (carcinogenic) or damage to reproductive cells and fetuses, resulting in birth defects and miscarriages. Pesticides can also cause soil and water pollution. Some federal regulation of pesticides came about in the early 1970s when control was moved from the USDA to the EPA. The provisions of the Federal Environmental Pesticide Control Act are as follows. Pesticide use other than stated on the label is prohibited. Pesticides are classified as general or restricted use. Only professionals that are trained and state-certified can apply restricted pesticides. All violations are punishable by heavy fines and/or imprisonment. Other aspects of the law cover the manufacturer and registration procedures for pesticides. In 1988 the regulations were amended to require considerable chemical, environmental, toxicological, and reentry protection data in order to register pesticides. Experimental use permits and special exemptions were also allowed to allow field testing of pesticides prior to registration and also for use on nonlabeled pests or crops in emergencies. Efforts to review pesticides in terms of possible cancellation, suspension, or change of classification also occurred.

More recent changes result from the 1997 Food Quality Protection Act. Each pesticide will have to undergo reregistration/review every 15 years, even if acceptable by previous review. This process will enable the consideration of new health, safety, and efficacy data that arose after approval and could result in loss of registration. Registration for reduced-risk pesticides will have a fast track process. The ability to use pesticides for minor crops not registered on the label will become easier. Incentives such as extended exclusive rights and fast track review are added to encourage pesticide manufacturers to register minor uses. Minor use will even be allowed on major crops, if it is the only effective alternative or is important for IPM.

Tolerances for pesticides have gone from the zero tolerance of the earlier 1960s Delaney Clause to levels that ensure with reasonable certainty that no harm will result from aggregate exposure for both raw and processed foods. The original clause only covered processed foods and zero levels were unattainable, given more sophisticated tests. Infants and children, given their reduced tolerance of pesticides relative to the body mass of an adult, are now better protected. A right to know brochure for consumers concerning pesticides is being developed.

Rather than setting tolerance for an individual pesticide, the EPA must now weigh aggregate exposure for pesticides having the same toxicity mechanism. Exposure is extended beyond food to consider even nonfood exposures such as lawns, water, homes, businesses, and so on. This tolerance review must be done over the next 10 years for all pesticides, starting with the worst first. Pesticides for the first time must also now be tested to see if they have endocrine-disrupting effects. This new test is in addition to all the other tests (cancer risk, toxicity levels, environmental effects, birth defects, and reproductive system alterations).

There are a number of safety precautions to be observed when working with pesticides. These safety precautions should be carefully observed at all times. Avoid complacency and shortcuts. The risk of injury, death, or lawsuits is high otherwise.

The timing of pesticides is critical to control the pest, but not leave excess residue upon harvest. The pesticide should be applied before symptoms become very noticeable. Timing of applications should be planned to avoid harming pollinators during the flowering stage, as well as periods of maximal activity by predators and parasites. If spraying is necessary during these times, pesticides that are least harmful to beneficial insects should be used. Other timing considerations are to target the pest at the most susceptible developmental stage, or to time it for periods before the pest is likely to cause economically unacceptable damage. Avoid applications during windy or rainy weather.

Sometimes a plant shows leaf damage, distortion, or abnormal growth after pesticide use (symptoms of phytotoxicity). To minimize phytotoxic effects of pesticides, apply pesticides in the cooler part of the day, preferably early morning, which allows the foliage and pesticide droplets to dry before the temperatures go over 85°F (28°C). High temperatures increase phytotoxicity problems. The temperature of the water is also a factor; water should be tepid. Wettable powders are usually less phytotoxic than emulsifiable concentrates. Sometimes organic solvent carriers cause phytotoxicity. Sometimes pesticides show no evidence of phytotoxicity until used as a combination spray. Several other factors that influence phytotoxicity are pesticide concentration, the developmental stage of the plant, and the health of the plant.

Pesticide toxicity is commonly expressed as the LD_{50}, which is the median lethal dose of toxicant (mg/kg of body weight) fatal to 50% of test animals. Select the pesticide with the highest LD_{50}. The LD_{50} values are expressed as acute oral and dermal figures (more useful to the horticulturist). Another term is LC_{50}, which is similar, except that it is based on fatality through inhalation and is given in terms of milligrams per liter.

Some level (ppm) of pesticide residue is allowed on food crops. The horticulturist should use pesticides with the least residual lifetime. The longer the residual lifetime, the more likely it is to pollute the environment and be passed through the food chain to humans. Federal regulations restrict the amount of residues on food crops; other regulations are designed to minimize environmental consequences. The one unpredictable with residues is that our current medical knowledge cannot rule out the possibility that exposure to multiple residues over many years increases our risk for cancer.

Constant dependence upon the same pesticide can lead to less effectual control, since the plant pest can develop genetic resistance. For this reason, it is better to alternate different pesticides capable of controlling a plant pest. Even better, pesticides for a specific pest that vary in their modes of action should be alternated.

EXERCISES

Mastery of the following questions shows that you have successfully understood the material in Chapter 15.

1. What pests from the animal kingdom attack plants? What type of damage results from their activity? Which pests can transmit diseases to plants when they feed?
2. What are the differences between insect arthropods and noninsect arthropods? Which group causes the most damage to horticultural crops?
3. Name the various plants pests in the noninsect arthropod group. What is a nematode and how can nematodes harm plants?
4. What larger animals can cause plant damage? How do these animals damage plants?
5. Are there any beneficial organisms in the groups that contain pests of animal origin? If so, which groups contain beneficial organisms?
6. Name the various plant pests of nonanimal origin. Which ones are known as pathogens? Which pathogen causes the most economic damage and how does that compare to insects?
7. What is the difference between a disease and a disorder?
8. How does a virus differ from a viroid? What are the symptoms of the yellows and mosaics? How are viruses transmitted?
9. What are phytoplasmas, spiroplasmas, xylem-limited bacteria, and phloem-limited bacteria? How are these pathogens transmitted?
10. What kinds of diseases do bacteria cause? What kinds of diseases do fungi cause?
11. Why are weeds considered to be a problem when growing horticultural plants? Why are weeds so good at what they do?
12. What do we mean by a proactive practice in terms of pest problems? How do these proactive methods relate to sustainability?
13. What do we mean by a reactive practice in terms of pest problems? Which reactive practice is the least sustainable? Why?
14. What is the best way to use pesticides from the viewpoint of sustainability?
15. What is IPM? What management options for pest control are part of IPM?
16. What is meant by scouting and monitoring? Of what value are these practices? How are these practices accomplished?
17. What role do traps play in IPM?
18. What is the relationship between phenological models and IPM?
19. What is the difference between the economic injury level and economic threshold level? How do these levels relate to IPM?
20. What control measures can be used in an IPM program?
21. What are the types of nonpesticide proactive practices?
22. What kinds of barriers and deterrents can be used to foil pests?
23. What various cultural practices can be used to prevent pest damage?
24. In what way can soil health maintenance reduce pest problems?
25. What are the two basic environmental alterations that reduce pest problems?
26. In what way can crop rotations, timing, and preventive plants reduce pest problems?
27. What role do sanitation, surgery, aseptic methods, and inspection play in reducing pest problems?
28. How can resistance be used to reduce pest problems?
29. What are the various nonchemical and chemical means of pest control?
30. Give examples of various cultural practices for controlling insects, weeds, nematodes, soil pathogens, and diseases.
31. What types of biological controls are available? Briefly describe how each one works.
32. What do we mean by augmentation, conservation, and importation in the context of biological controls?

33. What are the basic categories of pesticides based on the specific pests they control?
34. What do we mean by the term "green" pesticides?
35. What is the difference between a systemic and a nonsystemic pesticide?
36. Do pesticides differ in terms of degree of selectivity? If so, how?
37. What are the various formulations of available pesticides? Which one is the most used?
38. What are the various federal regulations in place to ensure pesticide safety in terms of human health and environmental well-being?
39. What problems are associated with pesticide applications?
40. How can we use pesticides and still protect beneficial insects?
41. Can pesticides exhibit phytotoxicity problems? If so, what are the symptoms?
42. What are the various protocols that we use to assess pesticide toxicity?
43. Does constant dependence on the same pesticide lead to problems? If so, what kinds of problems and how do we minimize these problems?
44. What kinds of pesticides are designated "green" pesticides?
45. How do nematicides differ from other forms of pesticides?
46. What are the various classes of chemicals used as miticides and insecticides?
47. What do we mean by selective, nonselective, preemergent, and postemergent herbicides?

INTERNET RESOURCES

The Internet sites given here are only meant to be a helpful start, to provide basic concepts, and provide links to many other sites. The Web contains an enormous number of sites dealing with pests, diseases, weeds, IPM, and pesticides. A good search engine such as Google can provide you with many choices.

American Phytopathological Society Press
> http://www.shopapspress.org

Bio-Integral Resource Center
> http://www.birc.org/

Biointensive IPM
> http://www.attra.ncat.org/attra-pub/PDF/ipm.pdf

Biological control and IPM
> http://ipmworld.umn.edu/chapters/landis.htm

Biological control guide
> http://www.nysaes.cornell.edu/ent/biocontrol/

California and IPM
> http://ipmworld.umn.edu/chapters/zalom.htm
> http://www.ipm.ucdavis.edu/GENERAL/resources.html
> http://www.ipm.ucdavis.edu/IPMPROJECT/pubs.html

Chemical management: pesticides and growth regulators
> http://tfpg.cas.psu.edu/part3/part33.htm

Christmas tree production and IPM
> http://ipmworld.umn.edu/chapters/mccull.htm

Control of fungus gnats
> http://www.umass.edu/umext/programs/agro/floriculture/floral_facts/fungnat.htm

Control of nematodes by alternative means
> http://www.attra.org/attra-pub/nematode.html

Controlling plant diseases with microbe-induced resistance
http://ipmworld.umn.edu/chapters/zehnder.htm

Cultural control and IPM
http://ipmworld.umn.edu/chapters/ferro.htm

Disease prediction models
http://www.cropinfo.net/AnnualReports/2000/Bliter00.htm
http://www.cdpr.ca.gov/docs/empm/grants/98-99/finlrpts/98-0272.pdf

Entomological Society of America
http://www.entsoc.org/catalog/

Environmental risk and pest management
http://ipmworld.umn.edu/chapters/higley.htm

Farmscaping to enhance biological control
http://www.attra.org/attra-pub/farmscape.html

Future of IPM
http://ipmworld.umn.edu/chapters/cuperus.htm

Insect management for organic crops
http://anrcatalog.ucdavis.edu/pdf/7251.pdf

Insecticides for greenhouses
http://www.agweb.okstate.edu/pearl/hort/greenhouses/f6712.htm

Introduction to fungicides
http://extension.usu.edu/ipm/disfact/upd2.htm
http://www.ent.uga.edu/pest2002/Pesticide_Safety/Fungicides_Bactericides.htm
http://www.msue.msu.edu/ipm/CAT99_fruit/fungicide.htm
http://www.apascc.org/acv/pests/99app06.htm
http://www.caf.wvu.edu/kearneysville/spray/fungbact.html

Introduction to herbicides
http://ipmworld.umn.edu/chapters/wareherb.htm

Introduction to insecticides
http://ipmworld.umn.edu/chapters/ware.htm

IPM for greenhouse crops
http://www.attra.org/attra-pub/gh-ipm.html

IPM overview and links
http://www.attra.org/attra-pub/ipm.html
http://www.reeusda.gov/ipm/

IPM resources on the Web
http://ipmworld.umn.edu/chapters/macrae.htm

Lettuce IPM in the Southwest
http://ipmworld.umn.edu/chapters/kerns.htm

Monitoring and IPM
http://www.ipmalmanac.com/basics/monitor.asp

Monitoring and scouting in the greenhouse
http://www.agweb.okstate.edu/pearl/hort/greenhouses/f6711.htm

Novel organic and natural product insecticides
http://ipmworld.umn.edu/chapters/larson.htm

Pest control decisions with ornamentals (includes methods of calculating degree days)
http://www.ag.ohio-state.edu/~ohioline/b504/

Pesticide information*
http://www.ent.uga.edu/pest2002/Pesticide_Safety/Fungicides_Bactericides.htm
http://www.ent.uga.edu/pest2002/Pesticide_Safety/Herbicides_PGR.htm
http://www.ent.uga.edu/pest2002/Pesticide_Safety/insecticides.htm
http://pmep.cce.cornell.edu/profiles

Pesticide regulations
http://ipmworld.umn.edu/chapters/willson.htm
http://ipmworld.umn.edu/chapters/fqpa96.htm

Phenology and pests
http://www.ipm.ucdavis.edu/PHENOLOGY/models.html

Phenology and pests
http://www.attra.org/attra-pub/phenology.html

Phytoplasmas, spiroplasmas, and fastidious vascular-limited bacteria
http://www.ppws.vt.edu/~bacteria/ENDOPHYT99.pdf

Plant disease management for organic crops
http://anrcatalog.ucdavis.edu/pdf/7252.pdf

Plant resistance and IPM
http://ipmworld.umn.edu/chapters/teetes.htm

Plastic mulches, insects, and diseases
http://www.cstone.net/~agmulch/ref5.html
http://www.hort.uconn.edu/ipm/veg/htms/colrmlch.htm

Prey specialization in insect predators
http://ipmworld.umn.edu/chapters/obrycki.htm

Relative toxicity of pesticides to honey bees
http://ipmwww.ncsu.edu/agchem/chptr5/501.PDF

Semichemicals and IPM
http://ipmworld.umn.edu/chapters/flint.htm

Sterile insect release method
http://ipmworld.umn.edu/chapters/bartlett.htm

Sticky traps for monitoring in floriculture and nurseries
http://www.ipm.ucdavis.edu/PMG/r280390411.html

Strawberries and IPM
http://ipmworld.umn.edu/chapters/rao.htm

Sweep nets
http://www.ipmalmanac.com/tipsheets/tip5.asp

Traps for insect monitoring in fruit
http://www.msue.msu.edu/vanburen/trapsweb.htm

*The date changes yearly in these uga.edu sites

Weed management for organic crops
http://anrcatalog.ucdavis.edu/pdf/7250.pdf

RELEVANT REFERENCES

Agrios, George (1997). *Plant Pathology* (4th ed.). Academic Press, San Diego, CA. 635 pp.

Aldrich, R. J., and R. J. Kremer (1997). *Principles in Weed Management* (2nd ed.). Iowa State University Press, Ames. 472 pp.

Altieri, Miguel (1994). *Biodiversity and Pest Management in Agroecosystems.* Food Products Press, Binghamton, NY. 185 pp.

Barbosa, Pedro (ed.) (1998). *Conservation Biological Control.* Academic Press, San Diego. 424 pp.

Boland, Greg J., and L. David Kuykendall (eds.) (1998). *Plant–Microbe Interactions and Biological Control.* Marcel Dekker, New York. 464 pp.

Bowman, Greg (ed.) (1997). *Steel in the Field: A Farmer's Guide to Weed Management Tools.* Natural Resource, Agriculture and Engineering Service, USDA Sustainable Agriculture Network, University of Vermont, Burlington. 128 pp.

Brooks, G. T., and T. Roberts (eds.) (1999). *Pesticide Chemistry and Biosciences.* Springer Verlag, New York. 400 pp.

Chase, A. R., Margery L. Daughtrey, and Gary W. Simone (1995). *Diseases of Annuals and Perennials: A Ball Guide.* Ball Publishing, St. Charles, IL. 208 pp.

Daughtrey, Margery L., and A. R. Chase (1992). *Ball Guide to Diseases of Greenhouse Ornamentals.* Ball Publishing, St. Charles, IL. 224 pp.

Davidson, R., and W. Lyon (1987). *Insect Pests of Farm, Garden, and Orchard* (8th ed.). John Wiley & Sons, New York, NY. 640 pp.

Dixon, A. F. G. (2000). *Insect Predator–Prey Dynamics: Lady Bird Beetles and Biological Control.* Cambridge University Press, New York. 257 pp.

Fisher, T. W., Thomas S. Bellows, L. E. Caltagirone, D. L. Dahlsten, Carl Huffaker, and G. Gordh (eds.) (1999). *Handbook of Biological Control.* Academic Press, San Diego. 1200 pp.

Foster, Rick, and Brian Flood (eds.) (1995). *Vegetable Insect Management: With Emphasis on the Midwest.* Meister Publishing Co., Willoughby, OH. 206 pp.

Gnanamanickam, Samuel S. (ed.) (2002). *Biological Control of Crop Diseases.* Marcel Dekker, New York. 480 pp.

Higley, Lyon G., and Larry P. Pedigo (eds.) (1996). *Economic Thresholds for Integrated Pest Management.* University of Nebraska Press, Lincoln. 327 pp.

Hoffman, Michael P., and Anne A. Frodsham (1993). *Natural Enemies of Vegetable Insect Pests.* Cornell Cooperative Extension Service, Ithaca, NY. 63 pp.

Hokkanen, Heikki M. T., and James M. Lynch (1995). *Biological Control: Benefits and Risks.* Cambridge University Press, New York. 304 pp.

Horst, R. Kenneth (2002). *Westcotts Plant Disease Handbook* (6th ed.). Kluwer Academic Publishers, Norwell, MA. 1032 pp.

Howard, Ronald, J. Allan Garland, and W. Lloyd Seaman (eds.) (1994). *Diseases and Pests of Vegetable Crops in Canada.* Canadian Phytopathological Society and the Entomological Society of Canada, Ottawa, Ontario. 554 pp.

Ishaaya, I., and D. Degheele (eds.) (1997). *Insecticides with Novel Modes of Action.* Springer Verlag, New York. 320 pp.

Julien, M. H., and M. W. Griffiths (eds.) (1998). *Biological Control of Weeds: A World Catalogue of Agents and their Target Weeds* (4th ed.). Oxford University Press, New York. 223 pp.

Kennedy, George G., and Turner B. Sutton (eds.)(2000). *Emerging Technologies for Integrated Pest Management.* APS Press, St. Paul, MN. 544 pp.

Khan, Jawaid A., and Jeanne Dijkstra (eds.)(2001). *Plant Viruses as Molecular Pathogens.* Haworth Press, Binghamton, NY. 530 pp.

Khetan, Sushil K. (2000). *Microbial Pest Control.* Marcel Dekker, New York. 320 pp.

Kohli, Ravinder K., Harminder Pal Singh, and Daizy R. Batish (eds.) (2001). *Allelopathy in Agroecosystems.* Haworth Press, Binghamton, NY. 447 pp.

Leight, W., and M. Fitzner (eds.) (updated frequently). *Directory of State Extension Pest Mangement Coordinators.* Cooperative State Research, Education & Extension Service, USDA, Washington, DC.

Leslie, Anne R. (ed.) (1994). *Handbook of Integrated Pest Management for Turf and Ornamentals.* Marcel Dekker, New York. 672 pp.

Liebman, Matt, Charles L. Mohler, and Charles P. Staver (2001). *Ecological Management of Agricultural Weeds.* Cambridge University Press, New York. 532 pp.

Mahr, Daniel L., and Nino M. Ridgeway (1993). *Biological Control of Insects and Mites: An Introduction to Beneficial Natural Enemies and their Use in Pest Management.* North Central Region Publication No. 481. Cooperative Extension Service, University of Wisconsin, Madison. 91 pp.

Page, B. G. and W. T. Thomson (2002). *Insecticide, Herbicide, Fungicide Quick Guide.* Thomson Publications, Fresno, CA. 216 pp.

Page, Steve, and Joe Smillie (1995). *The Orchard Almanac: A Seasonal Guide to Healthy Fruit Trees.* AgAccess, Davis, CA. 154 pp.

Perry, A. S., I. Yamamoto, I. Ishaaya, and R. Perry (1997). *Insecticides in Agriculture: Retrospects and Prospects.* Springer Verlag, New York. 300 pp.

Pimentel, David (ed.) (2002). *Encyclopedia of Pest Management.* Marcel Dekker, New York. 903 pp.

Powell, Charles C., and Richard K. Lindquist (1997). *Ball Pest and Disease Manual: Disease, Insect and Mite Control on Flower and Foliage Crops.* Ball Publishing, St. Charles, IL. 448 pp.

Prakash, Anand, and Jagadiswari Rao (1997). *Botanical Pesticides in Agriculture.* Lewis Publishers, Boca Raton, FL. 480 pp.

Rao, V. S. (2000). *Principles of Weed Science* (2nd ed.). Science Publishers, Enfield, NH. 555 pp.

Rechcigl, Jack E., and Nancy A. Rechcigl (eds.) (2000). *Insect Pest Management: Techniques for Environmental Protection.* Lewis Publishers, Boca Raton, FL. 392 pp.

Roberts, T. (ed.) (1998). *Metabolic Pathways of Agrochemicals. Part 1. Insecticides and Fungicides.* Springer Verlag, New York. 1500 pp.

Roberts, T. (ed.) (1999). *Metabolic Pathways of Agrochemicals. Part 2. Herbicides and Plant Growth Regulators.* Springer Verlag, New York. 869 pp.

Ruberson, John R. (ed.) (1999). *Handbook of Pest Management.* Marcel Dekker, New York. 864 pp.

Smith, Albert E. (ed.) (1995). *Handbook of Weed Management Systems.* Marcel Dekker, New York. 736 pp.

University of California (1998). *Natural Enemies Handbook: The Illustrated Guide to Biological Pest Control.* Publication No. 3386B4. University of California Statewide Integrated Pest Management Project, Davis, CA. 164 pp.

Vorley, William, and Dennis Keeney (eds.) (1998). *Bugs in the System: Redesigning the Pesticide Industry.* Earthscan Publications, London. 222 pp.

Ware, George W. (1996). *Complete Guide to Pest Control: With and Without Chemicals* (3rd ed.). Thompson Publications, Fresno, CA. 388 pp.

Ware, George W. (2000). *The Pesticide Book* (5th ed.). Thompson Publications, Fresno, CA. 418 pp.

Wheeler, Willis B. (ed.) (2002). *Pesticides in Agriculture and the Environment.* Marcel Dekker, New York. 360 pp.

Zimdahl, Robert (1998). *Fundamentals of Weed Science* (2nd ed.). Academic Press, San Diego. 556 pp.

CD-ROMs

Learning Plant Pathology, APS Press, St. Paul, MN.

INTRODUCTION TO HORTICULTURAL PRODUCTION

*T*he science and technology of sustainable horticulture must be mastered, but unless they are put into practice, the practitioner is an incomplete horticulturist. Many commercial and noncommercial avenues are open and these are covered in following chapters.

According to the 1997 Census of Agriculture, total crop sales in the United States totaled slightly over $98 billion dollars. Fruits, nuts, and berries accounted for 12.9% of these sales, while vegetables, sweet corn, and melons totaled 8.5%. Greenhouse, nursery, and other ornamental crops comprised 11.2% of the total sales. Collectively, horticultural production accounted for one-third of all crop sales in agriculture.

The satisfaction of having a horticultural career or a recreational and hobby interest can be experienced in any or all the horticultural divisions: ornamental horticulture, olericulture (vegetable production), and pomology (fruit production). The rewards are there, whether they be beauty experienced in the flower garden, the good taste of fresh fruits and vegetables, the satisfied feeling experienced on viewing one's landscaping efforts, the satisfaction that comes from knowing that horticulture and the environment can be complementary, or the financial reward of a horticultural career. Germane to all this is the production of horticultural crops, and special interest is directed toward that topic in the chapters that follow.

Some of my greatest joys have been derived from being a horticulturist: working with tropical plants in my greenhouse on a snowy day, the first flowering bulbs of spring, the beauty of trees framed by the sky, the first bloom of a newly obtained species, the good conversation with other horticulturists. I could go on and on, but the fact that you are using this book leads me to suspect that you have experienced or hope to experience these and many other exciting aspects of horticulture.

Chapter 16

Ornamental Horticulture

OVERVIEW

The ornamental horticultural industry (crop production, not services) had gross sales of $9.3 billion in 1998. Commercial production activities involve the propagation, production, and marketing of floricultural, garden, nursery, turf, and miscellaneous crops. The service sector is involved with landscape design and the selection, installation, and maintenance of ornamental plants in the landscape. Although outdoor landscaping services dominate the market, considerable maintenance of interiorscapes in offices, malls, and businesses occurs. The ornamental horticultural industry can be broken down into several specialized sectors. Some are involved with crop production: floriculture, nursery, unfinished plants and propagation materi-

als, seeds, bulbs, Christmas trees, and turf crops. Others are involved with service aspects: landscape design industry, landscape maintenance industry, and interior plantscaping industry. Each area is considered here on its own merit as a viable, working concept, but the blurred distinction sometimes introduced by commercial consolidation should be kept in mind. For example, bedding plant producers often produce both flowering annuals and vegetables, such as tomatoes and peppers. Production nurseries can sell fruit and nut trees in addition to woody landscape plants. The underlying concept is still the plants propagated, grown, and sold. Data for ornamental crop sales and rank are shown in Table 16–1.

CHAPTER CONTENTS

OBJECTIVES

- ❀ To understand floriculture in terms of crops, economics, and geography
- ❀ To understand nursery production in terms of crops, economics, and geography

- ❀ To understand other specialty crops in terms of economics and geography
- ❀ To learn about the landscape service industry in terms of products and economics

❖ To understand the basics of landscape design
❖ To learn about careers in ornamental horticulture

❖ To learn the basics of annuals, biennials, herbaceous perennials, bulbs, ferns, grass, ground covers, woody perennials, house plants, and greenhouse plants

FLORICULTURE

Floriculture is concerned with the propagation, production, and wholesale/retail distribution of cut flowers, greenery, potted foliage and flowering plants, bedding plants (both flats and pots), and herbaceous perennials. Gross sales in 1998 for floriculture were close to $4.5 billion, making it the largest commercial production sector in ornamental horticulture. Wholesale levels in 1998 were at $3.93 billion and in 2000, $4.57 billion. A significant part of floricultural production involves meeting seasonal markets. These time-dependent events include cut flowers and potted flowering and foliage plants for special holidays; bedding plants, perennials, and potted flowering plants for the spring garden and outdoor summer living; bulbs and perennials for fall planting; and foliage and flowering plants for interiorscapes, especially in the fall and winter. The top five states (in order of importance) for floricultural production are California, Florida, Michigan, Texas, and Ohio. In 2000 these five accounted for 53% of total sales. California and Florida alone accounted for 36%.

California tends to dominate much of the floricultural industry (not to mention fruits and vegetables), followed by Florida. In fact, California generated 21% of the commercial sales in ornamental horticulture. Factors such as climate are part of this dominance. The climates in these two states are favorable ones, both in terms of long growing seasons and moderate winters. Climate is a major factor in potted flowering and foliage plant production. Energy costs are minimal in Florida and California with regard to foliage plant production, as opposed to the ever-increasing energy costs of operating greenhouses in the north for fo-

TABLE 16–1   Ornamental Production: Crops, Sales, and Rank

Crop Category and Subcategories	Gross Sales (in thousands of dollars)	Rank (%)	
Floriculture:	4,462,860	48.22	
Bedding/garden plants:		52.8	
Annual bedding/garden plants			73.4
Herbaceous perennial plants			26.6
Potted flowering plants		19.4	
Foliage plants		13.3	
Cut flowers		11.5	
Cut cultivated greens		2.9	
Nursery plants	3,096,723	33.46	
Turfgrass: sod, sprigs, or plugs[†]	835,212	9.02	
Unfinished plants and propagation materials	493,049	5.33	
Cut Christmas trees	256,161	2.77	
Dried bulbs, corms, rhizomes, or tubers	55,389	0.60	
Aquatic plants	23,635	0.25	
Flower seeds	19,480	0.21	
Short-term woody crops	12,642	0.14	
Total	9,255,151		

[†]Grass seed not included.
Source: Adapted from the 1998 census of horticultural specialties (see Internet Resources list at end of chapter).

liage plant production. Many foliage plants can also be grown directly outdoors, either in full sun or under shade. Other factors include low labor costs, much of which results from the use of illegal and legal immigrants, migrant workers, and retirees.

Approximately 11,000 growers produced floricultural crops in 2000. The covered area devoted to U.S. production of floricultural crops in 2000 was 911 million square feet. Greenhouses accounted for 57% of this space, of which 68% consisted of plastic film-covered structures. The remaining space (43%) was devoted to shade and temporary cover. Some 36,868 acres of open field area were also used for production.

The distribution of growers in 2000 relative to total sales is $10,000–19,999 (10.79%), $20,000–39,999 (14.02%), $40,000–49,999 (6.95%), $50,000–99,999 (24.43%), $100,000–499,999 (28.34%), and $500,000 or more (15.47%). The majority of growers tended to be either individual (mostly) or partnership operations. Most sectors of floriculture were not corporate operated (one-third or less), except for potted flowering and potted foliage plants, for which corporate operation approached the 50% level.

Bedding and Garden Plants

Roughly 50% of floricultural production is involved with the production of bedding and garden plants. Gross sales in 1998 were $2.36 billion. By 2000 the wholesale value alone had reached $2.12 billion. Growers involved in production totaled 3,225. Plants are sold in flats (mostly plug-produced) and pots (conventional pots, large containers, and hanging baskets) for use in the garden and patio, deck, and other aboveground areas.

These plants are sold most heavily in the spring during a 4- to 6-week period and to a lesser extent during late spring through early summer. Plants sold during the latter period tend to be mature, full-blooming plants. Many are sold in large containers for instant color in commercial operations such as corporations and malls. Late-starting gardeners are also buyers. More recently sales have experienced a modest fall burst. Plants in this category tend to be cold hardy annuals such as ornamental kale and pansies, and herbaceous perennials such as chrysanthemums, irises, and daylilies.

Annuals in 1998 dominated in terms of sales (73%), but herbaceous perennials capture a reasonable market share at 27%. The dominant annuals and perennials grown in flats and conventional pots are shown in Table 16–2. The five major hanging basket plants in

TABLE 16–2 *Top-Selling (Flats and Pots Combined) Annual Bedding/Garden Plants and Herbaceous Perennials in the United States*

Annuals		Perennials	
Geraniums (from vegetative cuttings)	Snapdragon	Chrysanthemum (hardy/garden)	Salvia (perennial)
Impatiens	Combination planter/ color bowl	Hosta	Peony
Petunia	Coleus	Ornamental grasses	Columbine
Pansy/Viola	Portulaca	Daylily	Delphinium
Marigold	Dusty miller	Ferns (hardy/garden)	Coral bells
Begonia	Ageratum	Coreopsis	Iris
Geraniums (from seeds and plugs)	Verbena	Phlox	Bleeding heart
Impatiens (New Guinea)	Gerbera daisy	Rudbeckia	Veronica
Vinca (*Catharanthus roseus*)	Celosia	Astilbe	Clematis (nonclimbing)
Salvia (annual)	Zinnia	Sedum	
Alyssum, sweet	Fuchsia	Purple coneflower	
	Gazania		
	Nicotiana		

Source: Adapted from the 1998 census of horticultural specialties (see Internet Resources list at end of chapter). Hanging baskets excluded; ranked in descending order.

TABLE 16–3   *Top 10 States in Sales of Floricultural Crops (% of Sales)*

Annual Bedding Plants	Herbaceous Perennials	Potted Flowering Plants	Foliage Plants
California (11.02)	California (12.23)	California (20.31)	Florida (59.85)
Michigan (8.92)	Michigan (8.47)	Florida (10.16)	California (18.37)
Texas (7.38)	Ohio (8.34)	Texas (7.32)	Texas (4.56)
Ohio (6.18)	Illinois (4.70)	North Carolina (5.08)	Hawaii (2.37)
New York (5.67)	New York (4.11)	Pennsylvania (4.93)	North Carolina (1.52)
Florida (4.00)	Connecticut (4.03)	New York (4.65)	New Jersey (1.39)
Pennsylvania (3.76)	North Carolina (3.95)	Michigan (3.87)	Pennsylvania (1.00)
Washington (3.49)	New Jersey (3.60)	Ohio (3.56)	Ohio (0.71)
New Jersey (3.38)	Pennsylvania (3.60)	New Jersey (3.31)	Massachusetts (0.68)
North Carolina (3.15)	Oregon (3.20)	Virginia (2.72)	Michigan (0.67)

	Combined Percentage of Total Sales		
56.95	56.23	65.91	91.12

Source: Adapted from the 1998 census of horticultural specialties (see Internet Resources list at end of chapter).

descending order are geraniums (from cuttings), New Guinea impatiens, standard impatiens, petunia, and fuchsia. Hanging baskets account for 10% of sales. The top 10 states for annual bedding/garden plant and herbaceous perennial sales are indicated in Table 16–3. California ranked number one in both areas of production.

Bedding plant production continues to grow and can often be a good first introduction to horticultural business or to add to an existing operation. Homeowners are the largest market, but other buyers include landscape operations, large supermarket chains, garden centers, home maintenance/hardware centers, and discount chains. Choice of plants is influenced nationally by top-selling ones (Table 16–2), but can be influenced to some degree by local conditions. For example, if your local market has hot, dry summers, top-selling annuals that are not heat tolerant are not good choices. Similarly, if you ship to distant markets, your local product needs to be compatible with their specific local climate.

Most bedding/garden plants are now started as plugs. Specialist plug growers exist and supply many bedding/garden plant producers, especially new growers and the majority of small to medium-size growers. Large growers have the capital to invest in the equipment and skilled workers needed to produce their own plugs. The cost is considerable and large growers sometimes also sell plugs to other smaller growers to justify equipment and labor costs.

One area needing careful control is growth and timing when meeting the spring market. Spring light conditions can be low, resulting in stretching, causing poor quality plants. Sometimes inclement weather also delays the start of the marketing season. Height control and flowering during these times can be regulated with biological (preferably) or chemical methods (growth retardants) as covered in Chapter 14. Postproduction in the sales area is especially important to maintain quality. Frequent watering and provision of shade are essential for maintaining quality during this time. Elevated plants (i.e., those not sitting in water) have fewer disease problems and are more convenient for customer selection. Artistic or neat arrangements also tend to increase consumer interest and sales.

Potted Flowering and Foliage Plants

Potted flowering and foliage plants, widely utilized as house plants and greenhouse plants and for business interiorscapes, accounted for $868 and $595 million, respectively, of floricultural gross sales in 1998. The top 10 states for potted flowering plant sales are indicated in Table 16–3. Californian producers alone accounted for 20% of national sales. The top state for foliage production is Florida. It accounted for 60% of sales in 1998 and by 2000, 69%. Foliage production is a very centralized sector of floriculture, much more so than bedding/garden plants and potted flowering plants. There were 1,554 foliage producers in 2000. Corporate operations account for half. Growers operate heavily on the wholesale and less so on retail levels.

Production area in 1998 for potted flowering plants was 2,624 open acres, 38 acres of natural shade, 20,242,000 square feet of shade structure, and 141,762,000 square feet of greenhouses. Production area in 1998 for potted foliage plants was 3,240 open acres, 81 acres of natural shade, 94,540,000 square feet of shade structure, and 77,786,000 square feet of greenhouses.

Top-ranked potted flowering plants (excluding baskets) and potted foliage plants are shown in Table 16–4. The five major hanging basket potted flowering plants in descending order are poinsettia, orchids, begonia, spring flowering bulbs, and combination baskets. The five major hanging basket potted foliage plants in descending order are ferns, pothos, ivy, palms, and philodendron.

Production is heavily greenhouse oriented, with minimal field production. Initial capital investment is high due to greenhouse requirements. Expenses for plant materials are also high, as are production inputs both on the material and energy levels. Operations are heavily scheduled and frequent, and monitoring of plant conditions and culture is critical for successful production.

Breaking into the business is best on a part-time, small-scale approach using an inexpensive greenhouse on the order of 3,000 square feet. Greenhouses need to be as energy efficient as feasible, especially in the North. Fuel expenditures can be a significant expense. Marketing is critical and chances for success can be enhanced by providing interiorscaping, using your own grown plants.

Cut Flowers and Cut Greens

Sales of cut flowers in 1998 totaled $513 million. Cut flower production is concentrated in California, which accounted for 62.9% and 67% of national sales in 1998 and 2000, respectively. Sales of cut greens in 1998 reached $130 million. Florida accounted for 80% and 81% of the national sales of cut greens for floral arrangements in 1998 and 2000, respectively. Top-ranked cut flowers are shown in Table 16–4. The top five cut greens in order are leather-leaf ferns, eucalyptus, asparagus tree fern, holly, and pittosporum. In 2000 there were 503 and 245 producers of cut flowers and cut greenery, respectively.

TABLE 16–4 *Top-Ranked Potted Flowering Plants (Excluding Baskets), Potted Foliage Plants, and Cut Flowers*

Flowering	Foliage	Cut
Poinsettia	Palms	Roses
Orchids (combined)	Spathiphyllum	Lilies
Chrysanthemum (tender)	Combination planters	Gladiolus
Spring flowering bulbs	Ivy	Chrysanthemums
Azalea	Dracaena	Snapdragon
Easter lily	Bromeliad	Tulip
Rose	Ficus	Carnations
African violet	Cacti and succulents	Gerbera daisy
Cyclamen	Dieffenbachia	Delphinium
Kalanchoe	Pothos	Lisianthus
Hibiscus	Schefflera	Alstroemeria
Hydrangea	Ferns	Anthurium

Source: Adapted from the 1998 census of horticultural specialties (see Internet Resources list at end of chapter).

Cut flowers are a high-value, specialized crop. In the last 10 years, cut flowers have appealed to horticulturists as a way to diversify operations. Demand has risen as more consumer disposable income has been spent on cut flowers. Cut flowers are no longer the domain of florists, but are increasingly being sold in supermarket chains, green grocers, farmers' markets, and even roadside farm stands. Cut flower growing is highly labor intensive and involves considerable horticultural skill and business acumen. Marketing of cut flowers can take more time and effort than the growing of them.

Some cut flower growers operate as pick-your-own operations, either as a stand-alone or supplementary business to pick-your-own strawberries, blueberries, or other crop. These operations often specialize in field-grown cut flowers such as gladiolas, Gloriosa daisies, snapdragons, sunflowers, and zinnias. Others sell to retail florists or wholesale cut flower distributors. Many cut flowers are now field grown, as opposed to former greenhouse-intensive operations. Greenhouses are now used more for starting flowers for earlier starts prior to transplantation into fields. Another increasing use is for early cut flowers, followed by field cut flowers, followed by winter greenhouse cut flowers. This approach expands the marketing period and produces some flowers when supplies are low and market price is high. Sales of cut flowers for Valentine's Day and Mother's Day can be especially profitable.

The choice of cut flowers is determined by several factors such as market demand, vase life, color retention, production cost, length of harvest season, long and multiple stems, flower presentation (location of stem and angle of view), ease of cultural requirements, and insect/disease problems.

Postharvest appearance and longevity are especially critical with cut flowers. Repeat business is highly dependent on cut flower quality. Conditions prior to harvest, conditions at harvest, and conditions after harvest influence postharvest quality. Prior to harvest, optimal cultural care is essential to produce high-quality cut flowers as would be true of any crop. At harvest time, flowers should not be cut during the heat of the day. Best times are early morning or evening. Flowers, once cut, should be kept shaded and should be hydrated (preferably with a floral preservative solution) as soon as possible. Storage conditions prior to transport should be short, cool, and ventilated.

NURSERY PRODUCTION

The production nursery (Figure 16–1) is concerned mainly with the production and wholesale (and retail to a lesser degree) sales of ornamental woody plants such as trees, shrubs

FIGURE 16–1 The production nursery shown here specializes in container production of shrubs and trees utilized in foundation plantings and yard landscaping. (Author photograph.)

and vines, herbaceous plants used as ground covers, and turf grass sod. The woody material can be deciduous or evergreen. Production can go beyond ornamentals in that fruit and nut trees may be included, as well as smaller fruits such as strawberries and grapevines and perennial vegetables like asparagus and rhubarb. The breakdown of production and rank are shown in Table 16–5. Fruit and nut trees are covered in Chapter 18. The top-selling plants for each nursery category are found in Table 16–6. Sales tend to be seasonal, being most heavy in the spring with light summer sales and modest fall sales.

Of the nearly $9.3 billion retail value of ornamental crops in 1998, about $3.1 billion was accounted for by production nursery crops (woody and herbaceous plants). California alone accounted for nearly 23% of U.S. nursery production in 1998 and 28% in 2000. California, Oregon, and Florida together nearly produced 50% in 1998 and by 2000, 57%. The top 10 states and their sales are shown in Table 16–7. Production area in 1998 was 270,668 open acres, 835 acres of natural shade, 101,149,000 square feet of shade structure, and 69,326,000 square feet of greenhouses. The total acreage in 1998 was 311,000 and 370,000 in 2000, and the number of nurseries was 6,535.

In terms of specific crops, California leads in the production of combined deciduous shrubs, ground covers, vines, and other associated minor ornamentals with 33% of sales in 2000. Florida was a close second at 28%. California in 2000 also led in sales of broadleaf evergreens, deciduous flowering trees, and fruit/nut trees at 20%, 20%, and 73%, respectively.

TABLE 16–5 Nursery Crops: Sales and Rank

Crop	Sales ($)	Rank (%)
Deciduous shade trees	491,708,000	15.88
Deciduous flowering trees	335,613,000	10.84
Broadleaf evergreens	651,737,000	21.05
Coniferous evergreens	607,935,000	19.63
Deciduous shrubs	441,314,000	14.25
Ground covers	83,576,000	2.70
Vines	39,994,000	1.29
Palms	78,166,000	2.52
Fruit and nut plants	254,115,000	8.21
Miscellaneous crops	112,566,000	3.64
Total	3,096,723,000	

Source: Adapted from the 1998 census of horticultural specialties (see Internet Resources list at end of chapter).

TABLE 16–6 Top-Selling Nursery Plants in the United States

Deciduous			Evergreen	
Shrubs	**Shade Trees**	**Flowering Trees**	**Broad-leaved**	**Needled (Coniferous)**
Roses	Oak	Crabapple	Holly	Juniper
Spireas	Red maple	Crapemyrtle	Azaleas	Pine
Hibiscus	Ash	Callery pear	Rhododendron	Spruce
Hydrangeas	Japanese maple	Dogwood	Euonymus	Arborvitae
Viburnum	Birch	Flowering cherry	Boxwood	Yew
Lilacs	Honey locust	Magnolia	Viburnum	Cypress (*Cupressus*)
Buddleias	Norway maple	Flowering plum	Privet	Hemlock
Weigelas	Other maples	Redbud	Pittosporum	Fir
	Sugar maples	Hawthorn	Magnolia	Cedar (*Cedrus*)
	Linden	Amelanchier	Cotoneaster	

Source: Adapted from the 1998 census of horticultural specialties (see Internet Resources list at end of chapter). Ranked in descending order.

TABLE 16–7   *Top 10 States in Sales of Nursery Crops*

	(% of Sales)
California	22.32
Oregon	12.20
Florida	11.93
Ohio	4.53
Texas	3.71
Michigan	3.37
North Carolina	3.19
Illinois	3.10
Tennessee	2.91
Connecticut	2.58
Combined Percentage of Total Sales	69.84

Source: Adapted from the 1998 census of horticultural specialties (see Internet Resources list at end of chapter).

Oregon led in sales for deciduous shade trees and coniferous evergreens at 25% and 21%, respectively. Oregon was also second for broadleaf evergreens and deciduous flowering trees at 15% and 18%, respectively.

Nursery Types

Nurseries are divided into wholesale, retail, landscape, and mail-order types. Wholesale nurseries usually specialize in large-scale growing of a few crops, with certain firms noted for their specialties (for example, roses, broad-leaved evergreens, or shade trees). Often the specialties are determined by the climate where the nursery is situated. These nurseries often start production from lining-out stock, which is grown on to marketable size, a process that can require up to several years. The liners and other propagation materials are raised by essentially specialized propagation nurseries (see Other Specialty Production section). Wholesale nurseries sell to other nurseries, landscapers, lawn and garden centers, large chain discount stores, large chain home hardware/building supplies stores, and retailers. Sales are sometimes contract based. A limited number of wholesale nurseries also sell finished products directly to retail consumers.

Many retail nurseries have stopped growing nursery stock and act only as a distributor between the wholesalers and the consumer. Some of the larger retail nurseries still grow some or all of their plants on their own acreage. Landscape nurseries usually sell retail plants and offer a broad range of landscape services such as design, installation, and maintenance. Many produce their own plants. Some small farms also produce nursery crops in addition to vegetables and fruits and sell their nursery products at farmers' markets. Mail-order nurseries sell on the national level through catalog and Internet sites.

Trends

Trends are toward container-grown nursery stock (80%) because less labor is required, growers have more control over soil quality, higher plant densities per area are possible, they are accepted better by consumers, they recover quicker after transplanting, and they are marketable year-round (Figure 16–2). Balled and burlapped stock is still used, but the labor costs and more difficult transplanting task for consumers have reduced the use of this form (Figure 16–3). Bare-root stock is used for small deciduous plants, but losses can run higher with this form because of failure to control temperature and humidity at all storage sites. Container plants are not problem free. Watering and fertilization need careful attention, winter protection can be required, containers and potting labor are costly, some hand weeding is required, and root stress and root-bound conditions can be a problem. The latter can be minimized with copper-treated containers (paint).

FIGURE 16–2 Most nursery stock is sold in container form because of easier production and higher consumer satisfaction in terms of planting and survival. Homeowners are the main buyers. (Author photograph.)

FIGURE 16–3 Ball and burlap production has been heavily replaced by containerized production. Ball and burlap usage is primarily for larger, premium tree and shrub stock. Much of this market is directed toward commercial landscapers and more expensive landscape projects, where mature plantings take precedence over cost. Students at a middle school are planting a large, already flowering balled and burlapped tree as part of an improvement project on the school grounds. (USDA/NRCS photograph.)

The biggest driving force for the switch to container nursery crops is that container-grown crops yield 10 times as many sales per acre as field crops. Consumers tend to want smaller container plants in the 1- to 3-gallon sizes, while landscapers want larger 3- to 5-gallon container plants. Landscapers also buy most of the balled and burlapped stock that is currently sold.

Most nurseries utilize some form of irrigation, either overhead, drip, or sand capillary beds. Drip is best from the water sustainability view, but is more costly to install. Runoff from all of these systems needs to be managed properly to reduce off-site pollution from nutrients and pesticides.

Nursery production offers opportunities for start-up operations. About half of the U.S. nurseries are owned by individuals and most tend to be regional operations. Operations tend to be state specific in their customer draw. Small nurseries sell about 80% in state, and large ones, 60%. Good areas to begin a business include regions undergoing major construction activities, especially when associated with areas experiencing rapid population growth, much start-up housing, and low unemployment. Areas that are already built up offer less opportunity for start-ups, but do offer the possibility of nursery production jobs or buying an ongoing nursery. New nurseries often require 5 to 7 years before they show a profit. New nursery owners often have another full- or part-time job until the nursery is successful.

OTHER SPECIALTY CROPS

Several other crops are produced as horticultural specialty plants by specialized producers: turf grass (sod, sprigs, or plugs), unfinished plants and propagation materials, cut Christmas trees (Figure 16–4); dried bulbs, corms, rhizomes, and tubers; aquatic plants, flower seeds, and short-term woody crops. Collectively, this group accounts for roughly 18% of ornamental horticultural production (Table 16–1).

California and Florida were the top two states for turf production in 1998 at 15.46% and 14.90%, respectively. These two states plus the next four ranking states, Alabama, Georgia, Texas, and Colorado, accounted for 52.13% of turf production. During 1998 326,108 acres were involved in turf production. Oregon is the leader in grass seed production. In 1997, Oregon growers harvested 430,000 acres of grass seed valued at $320 million. Cool season grass seed is produced primarily in the Pacific Northwest. Warm season grass seed is produced on a much smaller scale in the Southern Plains.

Florida (24.01%), California (21.74%), and Michigan (10.00%) produced slightly more than half of the unfinished plants and propagation materials in 1998. Production categories include cuttings, bedding/flowering plant liners, foliage plant liners, nursery plant liners, plug seedlings, tissue cultured plantlets, and prefinished plants. Sales totaled $493 million. Rankings and leading products are shown in Tables 16–8 and 16–9, respectively.

In 1998 just two states, Oregon (33.91%) and North Carolina (18.05%), accounted for slightly more than half of the cut Christmas tree production. Around 191,000 acres were used for cut Christmas tree production. The leading tree was Douglas fir (24.28%) followed by Fraser fir (19.79%), Scotch pine (18.00%), and Noble fir (10.81%). Individuals (75%) dominate Christmas tree production. Many operations are part-time to supplement incomes from a normal nonhorticultural job or to supplement other horticultural operations.

Over half of the dried bulbs, corms, rhizomes, and tubers were produced by California (29.19%) and Oregon (20.02%) in 1998. Some 7,259 acres were involved in production.

FIGURE 16–4 A Christmas tree farm in Iowa. Christmas tree farms are often a part-time operation. Many of these trees are old enough for the next harvest. (USDA/NRCS photograph.)

TABLE 16–8 Unfinished Plants and Propagation Materials: Sales and Rank

Category	Sales ($)		Rank (%)	
Cuttings	95,746,000		19.42	
Liners	237,222,000		48.11	
Bedding/flowering plants		90,258,000		38.05
Foliage plants		25,812,000		10.88
Nursery plants		121,153,000		51.07
Plug seedlings	107,073,000		21.72	
Tissue cultured plantlets	34,558,000		7.00	
Prefinished plants	18,450,000		3.74	
Total	493,049,000			

Source: Adapted from the 1998 census of horticultural specialties (see Internet Resources list at end of chapter).

TABLE 16–9 Top-Selling Unfinished Plants and Propagation Materials in the United States

Cuttings	Bedding/Flowering Plant Liners	Foliage Plant Liners	Nursery Plant Liners
Chrysanthemums	Herbaceous perennials	Spathiphyllum	Coniferous evergreens
Geraniums	Bedding plants	Ferns	Deciduous shrubs
Nursery plants	Flowering plants	Dieffenbachia	Deciduous shade trees
Poinsettias	Poinsettias	Dracaena	Deciduous fruit/nut trees
New Guinea impatiens	Geraniums	Philodendron	Broadleaf evergreens
Foliage plants	New Guinea impatiens		Deciduous flowering trees
Herbaceous perennials	African violets		Landscape roses
African violets			

Plug Seedlings	Tissue Cultured Plantlets	Prefinished Plants
Annual bedding/garden plants	Foliage plants	Orchids
Herbaceous perennials	Herbaceous perennials	Poinsettias
Nursery plants	Nursery plants	Potted roses
Cut flowers	Potted flowering plants	Foliage plants
Foliage plants		Easter lily

Source: Adapted from the 1998 census of horticultural specialties (see Internet Resources list at end of chapter). Ranked in descending order.

Florida alone accounted for 50.35% of the aquatic plant sales. Open areas in production numbered 624 acres, while areas under protection were 1,863,000 square feet.

California handled 56.51% of 1998 flower seed production in the United States. Sales of U.S. flower seeds in 1998 approached $20 million. This figure, however, is low for total flower seeds in the United States, because many flower seeds sold in the United States are now produced in areas outside the United States, such as Costa Rica, Mexico, and Thailand. These sales were not tracked in the 1998 horticultural census.

Plant growing and seed harvesting are important aspects of ornamental seed production. Seed processing includes milling, cleaning, packaging, and storing. Much seed production is under contract with seed houses. As with nurseries, there are wholesale, retail, and mail-order parts to the seed industry.

Short-term woody crops were under production in 20,927 acres. The leading state in 1998 sales was Oregon at 26.29%. These crops are fast growing trees used to supply firewood, pulp for paper, and fibers for various products.

LANDSCAPING SERVICES

Landscape horticulture can be broken into landscape design, landscape installation/ construction, lawn/landscape maintenance, and tree care. Landscape horticulture is a multibillion dollar industry. Gross receipts for landscape design, landscape installation/construction, lawn/landscape maintenance, and tree care totaled $17.4 billion in 1999 according to the American Nursery and Landscape Association. These horticultural services ranked second only to animal husbandry services in all the agricultural services. The breakdown in terms of percentages was landscape design (11.5%), landscape installation/construction (35.1%), lawn/landscape maintenance (38.5%), and tree care (14.9%). Firms providing landscape horticultural services tend to be concentrated in and around major urban areas, particularly the Eastern and Western seaboards, but also around big cities in the interior of the United States.

The above services may be provided from several sources. Landscape architects or designers (see Careers section) are involved in landscape design and landscape contractors in landscape construction. Lawn care services, arborists, and full-service landscape maintenance firms provide lawn, tree, and other landscape maintenance. In some instances a retail nursery (see Nursery Production section) may attempt to provide design, construction, and even maintenance services in addition to retailing ornamental plants. Combined operations of this sort are referred to as landscape nurseries.

Landscaping is rewarding in both the sense of esthetics and of value. According to the American Nursery and Landscape Association, the following values can be attributed to plants and landscaping. The overall value of a home is increased by 7% to 15% with landscaping. The quality of the landscaping is also relevant. If two homes are equivalent, but one has excellent as opposed to good landscaping, the one with excellent landscaping can sell for 6 to 7% more. Just increasing the landscape from average to good can raise the selling price by 4% to 5%. The recovery value of landscaping varies from 100% to 200% as opposed to kitchen remodeling at 75% to 125%. Nearly 100% of real estate agents believe that landscaping enhances the appeal of a house when sold.

Trees also are an important component of the landscape. A mature tree can have a value of $1,000 to $10,000. Real estate agents also believe that mature trees can influence the salability of homes (83% for homes under $150,000 and 98% for homes over $250,000). Trees also offer direct value to the homeowner. Shade cast from trees, especially deciduous trees on the south side, can reduce air conditioning costs by 50%. Evergreen trees would work, except in the winter the heating effect from sunlight would be reduced. Trees can also reduce urban noises by acting as a barrier and masking noises with pleasant wind noises. Outdoor areas under trees can be cooler by 9°F. A single city tree can provide $273 annually in terms of value when one considers air conditioning, erosion control, pollution reduction, storm water control, and wildlife shelter.

Design

Landscape design and architecture are core elements of ornamental horticulture. Design is intimately associated with humans, land, plants, structures, and the environment. Its aim is to balance the requirements of each, yet present an esthetically pleasing image. The final product must be economical, functional, beautiful, safe, durable, and environmentally responsible. Good design should maintain or enhance the existing environment. Bad design can even be harmful if physical alterations are not done with foresight. The landscape designer or architect should work with and not separately from the building architect.

The concepts involved are quite extensive. Entire texts are devoted to landscape concepts. The area of concern to us is the relationships possible among ornamental plants, horticultural structures (walls, fountains, pools, and the like), and the site and how to integrate them with nonhorticultural structures (home, church, school, industrial building). Only minimal detail is possible here.

The basic elements of design are color, form, line, and texture. Plant color varies considerably. All parts must be considered: foliage, stems, bark, flowers, and fruit. Seasonal variations in color and leaf retention patterns are also a factor. A few examples include the dark green, seasonal color of a deciduous tree changing to red followed by no leaves versus the year-round blue-green of certain evergreens. On the other hand, certain deciduous woody plants exhibit reddish, attractive bark in the winter. Flowering shrubs add color in the spring. Holly has shiny green foliage all year and red berries from the fall through winter. Chrysanthemums add color in the fall. Spring bulbs start color early in the spring. Colors involved with nearby structures (walls, pools, fountains, paving, patios, decks, and the house) must be considered as part of the overall picture.

Plant form is essentially the visual impact of the total mass of a plant or, collectively, of a group of plants. A low, spreading plant and an upright, narrow plant have horizontal and vertical forms, respectively. A long hedge composed of numerous plants has a horizontal form. A large evergreen might be said to have a conical form with some species or a columnar form with others. A large maple tree could have a globular form. A weeping cherry has a weeping form.

Collective plant masses or horticultural structures can be used to guide or direct eye movement; this constitutes the concept of line. A hedge, walk, or wall can be used to direct the eye toward a focal point such as a specimen tree, an attractive fountain, or the entrance to a home. The visual impact of the surfaces of plants and other horticultural surfaces offer various textures. Grass (both lawn types and ornamental forms) and pea gravel have fine textures, whereas a large-leafed tree would have a coarse texture. Coniferous evergreens have a needle-like texture. A Japanese maple might be said to have a delicate, fine cut texture.

These design basics must be artistically blended such that plantings, horticultural structures, and buildings form an overall, appealing picture. Plantings and structures should be mixed to produce variety, instead of a monotonous sameness. Some repetition, however, is necessary to bring order into the overall image. Skillful use of plantings and structures can be used for emphasis of a focal point. The overall arrangement must be considered for balance, whether it is exact on either side of an axis, as in a formal garden, or a bit asymmetrical to give an informal appeal.

In design, the environmental requirements of the planting must also be known. Failure of plantings would seriously harm the reputation of the landscape designer or landscape architect. The mature form and heights of plants must be known to maintain the proper scale of plantings with time. A large tree that in time obscures the facade of a building and is not amenable to pruning obviously distorts scale. For example, a Colorado blue spruce planted close to a Cape Cod or ranch-style house will look out of scale in no time.

The effects of microenvironment, especially in urban situations, must not be neglected. On the whole, the environment can appear suitable for a planting, but the microclimate might be sufficiently adverse to cause failure of the planting. One example in an urban situation will suffice (see Chapter 6 for others). Landscape installations in high-density urban sites are often under the detrimental influences of microclimates. This can be manifested by daily late summer–early fall temperature fluctuations. During this season, in particular, air temperatures can be halved in a matter of minutes. Late afternoon sun, perhaps intensified by building proximity (particularly buildings with reflecting surfaces), can produce very high desiccating air temperatures (120°F plus). Moments later the sun is shielded by buildings and air temperatures often drop suddenly (60°F plus or minus). This rapid change can be detrimental to plant development.

Microclimates in these sites often influence air temperatures in spring or fall such that growing seasons are extended. This influence can result in late autumn growth that is unable to harden off and therefore is very susceptible to cold temperature injury in the form of twig dieback. This environmental influence can also produce early season growth that can be very susceptible to cold injury produced by early spring nocturnal freezes.

Design as stated before is a unifying concept in ornamental horticulture that involves close cooperation with activities in other areas, especially site preparation and subsequent maintenance. Necessary modifications of the site for purposes of design and to provide a satisfactory environment (grading, drainage, berms, etc.) can only be achieved through cooperation between the designer and the landscape contractor. A mistake could cause loss of plants (improper drainage) or an incorrect design (misplacement of wall or a plant). Close attention to planting is critical, because improper planting can mean an unrealized final design. Maintenance personnel must know what the designer had in mind. Formal pruning of a woody plant could destroy the informal effect the designer had intended for the landscape.

Construction and Maintenance

Landscape installation/construction is involved with the total preparation of the site for the planting, which might be for a school, home, apartment complex, condominium, shopping mall, housing development, business, industrial complex, church, temple, mosque, and so on (Figure 16–5). These activities include the procurement of the necessary plants (trees, shrubs, perennials, sod, vines, etc.); the modification of the site, which can involve grading, soil modification, walls, terraces, pools, paved surfaces, drainage systems, and excavation; plant installation; and site maintenance until construction is complete. Landscape and lawn maintenance involves maintaining established ornamental plantings, which includes fertilization, minor pruning and shearing, mowing, and plant protection against insects, disease, and weeds. Tree care involves major pruning, repair of storm damage, tree removal, and insect/disease protection. Established ornamental plantings include lawns, shrubs, trees, flower borders and islands, and gardens.

CAREERS IN ORNAMENTAL HORTICULTURE

A large number of career positions are associated in whole or part with ornamental horticulture. Some of the general areas for careers include these:

- Arboriculture
- Environmental management
- Horticultural therapy
- Interiorscape design and management
- Landscape construction and installation
- Landscape design
- Landscape maintenance
- Nursery management
- Pest management
- Plant propagation and production
- Turf management in parks or golf courses.

Brief coverage of these careers follows that is not all inclusive and is not meant to imply that 100% of a person's activity will be concerned with some aspect of ornamental horticulture.

Floriculture

The production area in larger floricultural operations involves four basic positions: production superintendent, grower, marketing manager, and an inventory controller. The production supervisor coordinates production and oversees the growers, who are actually responsible for the production of floral crops. The inventory controller handles scheduling of crop production for correct timing and maximal profits. The marketing manager controls the handling of the finished products and the sales (wholesale) and deliveries that follow. Computer skills are essential in terms of greenhouse operational controls, inventory management, and marketing. In smaller operations, these jobs might be combined and spread out over fewer staff.

The wholesale commission florist is an intermediate firm between the production and retail aspects. There is usually an overall manager, buyer, sales manager, and salespeople whose positions are self-explanatory. This group usually handles cut flowers and hard goods. Retail

(A)

(B)

(C)

(D)

(E)

FIGURE 16–5 Examples of residential and commercial landscape projects. (A) A well-landscaped house has enhanced the visual appeal of this modest home. (B) The landscaping of this penthouse overlooking New York City reduced the big city atmosphere for its occupants. (C) Landscaping of this mall in California provides a more pleasant atmosphere for shoppers and likely increased sales. (D) Landscaped grounds at this industrial headquarters provide a pleasant working atmosphere and warm weather lunch area and likely contribute to increased productivity. (E) A landscaper-installed pond is the focal point of a conservation scene in this Iowa backyard and provides many hours of viewing enjoyment for its owners. (Photographs A–D are USDA photographs: A is now in the National Archives. Photograph E is from the USDA/NRCS.)

florists are the main outlet for retail sales of floricultural crops and products, but garden centers also offer a fair share. Supermarkets and discount stores also sell some products. The florist shop generally has a manager, often the owner, sales personnel, and floral designers. There may be a managerial position associated with the floricultural portion of the garden center. Generally, little growing, other than holding plants until sold, is involved in the retail end.

Nursery Production

Positions in the production nursery include the manager, propagator, inventory controller, field foreman, field superintendent, sales manager, shipping foreman and traffic manager, salesperson, and broker.

The main manager is responsible for the overall operation of the nursery. The propagator is responsible for propagation to produce marketable stock and is usually in charge of a propagation crew. The field foreman oversees the crews that bring the recently propagated stock to marketable size. The field superintendent is actually responsible for all production stages, from time of propagation through reaching marketable size and must make many decisions involved with fertilization, pest control, and the like.

The inventory controller coordinates order requests with stock. The sales manager, salesperson, and shipping foreman and traffic manager are self-explanatory positions. The broker (usually self-employed) is a middleman between retail outlets, such as garden centers, and the production nursery.

Landscape Horticulture

Landscape architects have a degree in landscape architecture and are licensed in most states. Landscape designers are usually unlicensed and do not have a degree in landscape architecture, but do have sufficient training and education. Both specialize in landscape design either as self-employed individuals or as employees of firms involved in landscape horticulture.

Landscape contractors are usually self-employed and direct the people and equipment needed to provide the wide range of landscape construction activities described previously. These services include both site alterations and building of structures.

A manager oversees the landscape nursery operation. A landscape designer working on commission for a nursery usually handles landscape designs for homes and smaller buildings. Larger projects (industrial complexes, housing developments, schools) are mainly handled by a landscape architect, who is self-employed in many cases. The construction superintendent coordinates construction jobs, crews, plant material, and construction equipment. The construction foreman oversees construction activity. Services are sold by a salesperson, who sometimes doubles as a landscape designer.

Some specialists may be employed by the landscape nurseries, or they may depend on subcontractors. The latter include landscape contractors involved with site preparation, site modification, and construction of horticultural structures. Other specialists might be required for irrigation installations and lawn construction. These services must be provided by trained people, otherwise the reputation of the nursery may suffer.

Landscape maintenance firms are run by a manager, who might also oversee assignments in small firms. Larger firms employ a superintendent of operations for the latter task. The crew foreman oversees the work crews, and services are sold by a salesperson.

Firms may specialize in one maintenance activity or several. Some of the specialties are as follows. One specialty is grounds maintenance for private individuals, public and private institutions, businesses, cemeteries, arboreta, parks, botanical gardens, and others. This service may include planting and transplanting, and even plant improvement, depending on whether it is involved with landscape contractors or with a specialty firm. When plant improvement is included, the specialty is more correctly named plantation maintenance; this form is more apt to be found at arboreta, botanical gardens, and private estates. Another specialty is arboriculture, which deals with maintenance and management of woody plants. Turf care is another common specialty. Chemical control with pesticides, growth regulators, and fertilizers is still another.

Garden Centers and Retail Nurseries

Garden centers and retail nurseries are basically retail outlets for production nursery crops, floricultural crops, nonornamental crops, horticultural hard goods such as fertilizers and tools, and sometimes horticultural services such as designing. Development and maintenance of the home landscape by the homeowner are made possible by such outlets, which may be individually owned, a partnership, an incorporated chain, or part of a horticultural or other conglomerate. They generally have a manager and buyer, and sometimes a designer and plant doctor. Although these operations are not completely involved in ornamental horticulture, it usually constitutes a large share of the business.

Arboreta and Botanical/Horticultural Gardens

Horticultural institutions are involved with plant collections (not completely ornamental plants) and require people knowledgeable in ornamental horticulture. Some of these positions are director of the institution, superintendent of horticultural operations, propagator, curator (involved with planning, obtaining, labeling, and so on, of plant material), greenhouse manager, librarian, writer, educational director, and various researchers.

Education and Research

Teachers of ornamental horticulture and other areas are needed in high schools, vocational training or skill centers, community colleges, colleges, and universities. Cooperative extension service agents, dealing in whole or part with ornamental horticulture, are found in all states. They are educational consultants operating on a governmental level and dealing with homeowners, commercial horticulturists, and others in need of their services. Private self-employed consultants may also provide similar services. Writers and lecturers (usually self-employed) provide ornamental horticultural information in magazines, books, newspapers, radio, television, and lectures.

Researchers dealing with various specialties within or associated with ornamental horticulture can be found in universities, state agricultural experiment stations, federal research laboratories, industry, arboreta, and botanical gardens.

Seed Production

Commercial seed firms dealing with ornamentals need plant breeders, propagators, growers, sales managers, catalog and Web site designers, market specialists, and researchers. Some seed companies specialize in a few seed product lines, while others are generalists. Some are strictly wholesalers, providing seeds to other seed retailers.

Horticultural Therapists

Horticultural therapists work with people who have physical or mental disabilities. Much of the activity is involved with ornamental horticulture, especially house plants and greenhouse activities. These therapists offer services at hospitals, rehabilitation facilities, senior centers, and nursing homes.

Common Link

All of the positions just discussed have a common link, some aspect of the production and culture of the various groups of ornamental plants that make ornamental horticulture the enjoyable field it is. Many opportunities exist for individuals who wish to go into ornamental plant production (Table 16–10). These groups of plants or crops are annuals, biennials, herbaceous perennials, bulbs, ferns, grass, ground covers, woody perennials, and house/greenhouse plants. The next section covers these plant groups.

ANNUALS

Annuals are plants that pass through the vegetative and reproductive cycles and senescence in one growing season (Figure 16–6). Plants that pass through the vegetative and reproductive cycles—but not senescence—and do die as a result of killing frosts, are also treated as annuals. Many vegetables are also annuals, but they are covered in Chapter 17.

TABLE 16–10 *Type of Business Organization for Ornamental Crop Producers*

Crop	Individual	Partnership	Corporation	Other
Annual bedding/garden plants	61.55	9.10	28.27	1.08
Herbaceous perennial plants	59.11	8.12	31.16	1.61
Potted flowering plants	43.33	7.94	47.56	1.16
Foliage plants	45.03	6.80	48.06	0.11
Cut flowers	56.26	10.81	31.90	1.03
Cut cultivated greens	66.49	11.17	22.34	0
Nursery plants	49.59	9.08	40.07	1.26
Turfgrass sod, sprigs, or plugs	37.84	9.96	51.37	0.83
Unfinished plants, propagation materials	50.58	7.60	38.68	3.14
Cut Christmas trees	74.30	14.40	10.22	1.07
Dried bulbs, corms, rhizomes, or tubers	54.62	9.24	35.29	0.84
Aquatic plants	56.91	7.53	35.55	0
Flower seeds	51.79	16.07	32.14	0
All producers (average)	54.41	9.83	34.82	0.93

Source: Adapted from the 1998 census of horticultural specialties (see Internet Resources list at end of chapter). Based on each crop counting for at least 50% or more of the operation's total sales.

(A)   (C)   (E)

(B)   (D)   (F)

FIGURE 16–6   A few examples of annuals. (A) Sunflower F_1 'Ring of Fire,' a 2001 All America Selections winner. (B) Nicotiana F_1 'Avalon Bright Pink,' a 2001 All America Selections winner. (C) This porch pot contains shade loving coleus and impatiens. It adds a nice touch of summer color to the porch. (D) The deck rail starkness is softened with this deck rail window box filled with cascading 'Purple Wave' petunias. (E) Breeders have improved cosmos, as shown by this double-petal example. (F) This geranium is one of the newer seed geraniums in the Orbit series. (Photographs A and B are courtesy of All America Selections. Photographs C–F are author photographs.)

Annuals are the mainstays of the flower garden, since they give a greater effect for less money and labor than other flowers, and their exciting, long-term flowering makes them ideal choices for summer color in the landscape. Annuals are easily mixed with herbaceous perennials and bulbs as complementary colors and to fill in bare spots or hide spent foliage or finished blooms. Annuals are also readily available in flats as almost flowering plants during the spring.

Annuals such as impatiens, marigolds, nicotiana, petunia, and salvia are used for masses of color in flower borders and islands. Some such as sunflowers and zinnias make excellent cut flowers. Nicotiana (flowering tobacco) and dianthus (pinks) have pleasing fragrances. Shorter ones such as ageratum, alyssum, dwarf marigolds, and portulaca (rock moss) serve as flower edging in borders. Morning glories and sweet peas (vines) can be used for colorful screening. Statice and strawflowers can be dried for fall and winter arrangements. Many annuals are also excellent choices for window boxes, as well as hanging baskets and container plantings for patios, porches, and decks. Uses and placement are determined by such factors as seasonal time of bloom, height, frost tenderness, response to hot or cool weather, appearance, color, disease resistance, insect susceptibility, and lasting qualities as cut flowers.

Given their versatility, beauty, availability, and reasonable price, annuals are valued both by the home gardener and commercial horticulturists. Annuals are an established part of the bedding plant industry because of their appeal. Annuals typically sold in flats include ageratum, alyssum, celosia, dianthus, marigold, nicotiana, petunia, portulaca, salvia, seed geranium, snapdragon, verbena, and zinnia. These are the flowering bedding plants that succeed in a sunny to partly sunny location. Those adaptable to shadier areas include begonia, coleus, and impatiens. There are many more annuals than the top-selling ones (Table 16–11; see also Table 16–2). Good seed houses sell at least 150 annuals, and there are many varieties of each of these from which to choose. Surprisingly, many of these are seldom seen in gardens. There is no excuse for being bored with annuals.

The growing treatment of annuals provides a basis for two groups: (1) those that are sown in the place where they are expected to bloom and (2) those that should be started early in a hotbed or greenhouse in order to provide sufficient time for blooming during the growing season. Some of those sown directly in place, if sown earlier indoors, can be expected to have an extended blooming period because of the earlier start.

Scheduling of propagation dates is of extreme importance to the floriculturist involved in the production of annual bedding plants. Such plants must attain a size sufficient to ensure flowering at or close to the correct marketing and planting time range. Size must be held if weather delays the selling season.

Most should not be started outdoors until all danger of frost is past. Some annuals are hardy and will tolerate a light frost. Those started indoors are started 6 to 10 weeks prior to setting out in the garden, depending on their speed of development. Some need more time such as seed geraniums (10 to 12 weeks). Starting seeds indoors and transplanting were discussed in Chapters 7 and 12. Depth to sow seeds varies and is indicated on seed packets. A good rule of thumb is to cover a seed two to three times its thickness, and very fine seeds such as petunia or begonia should not be covered, but lightly pressed into the propagating media or soil. Seeds that require light for germination such as begonia and coleus should be pressed into the media and not covered with growing media, paper, or anything opaque during germination.

A word or two about the annuals is in order. Many are not true annuals, but perennials or biennials treated as annuals. If they are used in areas where they are winter hardy, their perennial habit would be observed. For example, some years in my Connecticut garden, snapdragons and dusty millers survive the winter and resprout the following year. In addition, some of these annuals can self-sow in your area and return year after year such as cornflower (bachelor buttons), *Cleome,* and *Nicotiana.*

All annuals, even those of a perennial nature, can be expected to bloom the first year if they are given an early enough start. This is accomplished by starting them indoors if the growing season in your area is too short or if earlier blooming is desired. Alternately, if their seeds are of sufficient hardiness, they can be started early outdoors directly after the danger of heavy spring frost is past or planted in late fall just before the ground freezes. Gardeners can start

TABLE 16–11   *Some Popular Annuals*

Common Name	Scientific Name*	Comments
Ageratum	*Ageratum houstonianum*	Shorter ones good for edging; tall blue form for cut flowers
Balsam	*Impatiens balsamina*	Can self-sow
Batchelor buttons	*Centaurea cyanus*	Can self-sow, nice blue
Blanket flower	*Gaillardia pulchella*	Good cut flower
California poppy	*Eschscholzia californica*	Can winter over
Cockscomb	*Celosia cristata*	Crested to feathered; can be dried
Coleus	*Coleus × hybridus*	Colorful foliage; good for shade
Coreopsis	*Coreopsis tinctoria*	Dwarf and tall types
Cosmos	*Cosmos bipinnatus*	Tall forms dramatic
Dusty miller	*Centaurea cineraria*	Silvery-white, hairy foliage plant; can winter over
Flowering tobacco	*Nicotiana alata*	Fragrant
Geranium	*Pelargonium × hortorum*	Popular; colorful; good container plant
Impatiens	*Impatiens wallerana*	Colorful; good in shade
Lobelia	*Lobelia erinus*	Nice blue; good edging and container plant
Marigold	*Tagetes erectus, T. patula, T. tenuifolia*	Carefree; short edging and taller cut flower types
Morning glory	*Ipomoea purpurea*	Vine; nice blues
Nasturtium	*Tropaeolum majus, T. minus*	Colorful; good in containers and window boxes
Petunia	*Petunia × hybrida*	Popular; colorful; good container plant
Pinks	*Dianthus chinensis*	Fragrant; can winter over; good in rock gardens
Portulaca	*Portulaca grandiflora*	Needs warmth and sun
Pot marigold	*Calendula officinalis*	Good cut flowers
Salvia	*Salvia splendens*	Humming birds like red cultivars
Scarlet runner bean	*Phaseolus coccineus*	Vine; fast grower
Snapdragon	*Antirrhinum majus*	Good cut flower; can winter over
Spider flower	*Cleome hasslerana*	Tall; nice background plant
Statice	*Limonium carolinianum*	Nice blue; good for dry arrangements
Strawflower	*Helichrysum bracteatum*	Good for dry arrangements
Sunflower	*Helianthus annus*	Tall; attracts birds
Sweet alyssum	*Lobularia martima*	Shorter ones good for edging
Sweet pea	*Lathyrus odoratus*	Vine; fragrant
Verbena	*Verbena × hybrida*	Colorful edging; can winter over
Wax begonia	*Begonia × semperflorenscultorum*	Waxy leaves; many flowers; good in shade
Zinnia	*Zinnia elegans*	Rich colors; good cut flower

*Many are actually grown as cultivars of the indicated genus or species.

annuals indoors or outdoors, but most prefer to buy established annuals, leaving the planting to commercial growers. Because of the extensive variation of growing conditions in the United States, it would be wise to check with your local agricultural extension service for exact treatment in your area. Not all annuals will be suitable for your area either, and suitability should be checked with a local source. For example, some annuals succeed in hot, dry sunny areas such as marigolds, portulaca, torenia, and zinnias. Impatiens on the other hand can survive in heat with some shade, but only when kept well watered. Sweet peas prefer cooler weather.

BIENNIALS

Biennials are plants that require two growing seasons to complete their life cycles. Generally, the first year is vegetative in habit, and the second reproductive. Many biennial ornamental plants are treated as annuals by giving them an early start indoors or by seeding in late fall or early spring outdoors (if hardy enough). Some perennials are also so short lived

FIGURE 16–7 The pansy is a popular springtime biennial that flowers best in the cooler weather. Shown here is Pansy F$_1$ 'Ultima Morpho,' a 2001 All America Selections winner. (Courtesy of All America Selections.)

as to be regarded as biennials such as *Dianthus barbatus* (sweet william). Biennials are few in number and are less popular than annuals, but a few are sold as bedding plants sufficiently old enough to bloom the first year.

Examples of biennials (Figure 16–7) include the following ornamentals: *Alcea rosea* (hollyhock), *Campanula medium* (Canterbury bells), *Daucus carota* (Queen Anne's lace), *Digitalis purpurea* (foxglove), *Lunaria annua* (honesty), *Meconopsis betonicifolia* (blue poppy, only suitable for Pacific Northwest), *Verbascum blattaria* (moth mullein), and *Viola* × *wittrockiana* (pansy). Pansies are one of the more popular biennials and are frequently sold as bedding plants for early spring color.

HERBACEOUS PERENNIALS

Perennials are plants that persist for more than 2 years (Figure 16–8). Many of our favored ornamentals such as shrubs, trees, ferns, and bulbs are perennials. However, horticultural usage of the word *perennial* has come to imply herbaceous, nonbulbous flowering perennials used in beds or borders. This approach will be used here, and the ferns, woody, and bulbous perennials will be treated in separate sections. However, ornamental grasses and bamboos are usually included with the herbaceous flowering perennials. Some popular perennials are shown in Table 16–12.

Many perennials can be started from seed, and it is economical if extensive numbers are required. Drawbacks include the care of plants until they reach blooming size, usually about 2 years, and that many of the more modern varieties of perennials are genetically variable when grown from seed. The latter are propagated asexually. The old-fashioned perennials can be profitably raised from seeds, but plants of the newer ones should be obtained as container stock from a nursery. Once established, it is always possible to increase their numbers through such asexual means as division and cuttings from stems and roots.

The perennial border must be carefully planned, given its longevity. A poorly planned one wastes money and will give years of dissatisfaction. The advantage over annuals is that they are planted once, not every year. True, some initial installation is required and maintenance is needed to divide perennials that become crowded. However, over the long term, perennials are less labor intensive then annuals, giving years of colorful viewing pleasure and cut flowers.

Since there is a reasonable amount of work involved in establishing a perennial planting, the arrangement should certainly be right the first time. Things to consider in the plan are height, spacing, color, hardiness, time of bloom, and required cultural conditions. The ideal perennial border or bed provides a succession of bloom throughout the growing season, but such an effect requires good planning and moderate maintenance.

FIGURE 16–8   Some examples of herbaceous perennials. (A) Two easily grown, carefree perennials are growing well at the back corner of a house. In the foreground are Gloriosa daisies (*Rudbeckia hirta*), and purple coneflowers (*Echinacea purpurea*) are in the background. (B) Daylilies (*Hemerocallis fulva*) are easy to grow, offering color in the late spring through early summer. (C) This variegated *Hosta* cultivar offers interesting foliage for those shady spots, but will need protection if deer are around. (D) The Oriental poppy (*Papaver orientale*) is an old favorite for color and dependability. (E) Pinks (*Dianthus* sp.) are appealing with their low mounds of colorful, fragrant flowers. (F) A clump of variegated grass produces an interesting accent in the perennial bed. (G) Delphiniums are good perennials for those interested in blue or purple flowers. (H) Bamboo offers an interesting accent all by itself. (I) The chrysanthemum is a reliable perennial that offers fall flowers when most perennials are done. (J) Two container perennials purchased from a nursery are shown here. On the left is a Montauk daisy (*Nipponanthemum nipponicum*) and on the right, a sedum (*Sedum spectabile* 'Autumn Joy'). Both extend the flowering time into the fall. (All photographs except for H and I are author photographs. H and I are USDA photographs.)

Care should be observed in considering the hardiness zones for perennials. A perennial in California may very well be an annual or pot plant in the North. In fact, a large number of plants that are perennials in the warmer areas of the South are grown as annuals in the North. In addition, zones are not absolute. Factors such as microclimates must be considered, as well as additional environmental conditions that may eliminate a plant for your

TABLE 16–12 *Some Popular Perennials*

Common Name	Scientific Name*	Comments
Aster	*Aster hybridus, A. novae-angliae*	Autumn bloomer
Astilbe	*Astilbe × arendsii, A. × crispa*	Flowers in partial shade
Balloon flower	*Platycodon grandiflorus*	Long-lived, nonspreading
Bamboos	Species of genera *Arundinari, Bambusa, Chimonobambusa, Dendrocalamus, Phyllostachys, Pseudosasa, Sasa, Semiarundinaria, Shibataea*	Most species and cultivars of these genera do best in warmer areas of the U.S; only a few can succeed around New York
Bee balm	*Monarda didyma*	Attracts hummingbirds
Blanket flower	*Gaillardia × grandiflora*	Good cut flower
Blazing star	*Liatris spicata*	Attracts butterflies; blooms top down
Bleeding heart	*Dicentra formosa, D. spectabilis*	Flowers in partial shade
Blue sage	*Salvia azurea*	Blue
Bugbane	*Cimicifuga simplex*	Flowers in partial shade
Butterfly weed	*Asclepia tuberosa*	Attracts butterflies; OK in dry soil
Campanula	*Campanula carpaica, C. latifolia, C. persicifolia, C. rotundifolia*	Showy; cultivars for beds, rock gardens
Catmint	*Nepeta cataria, N. × faassenii*	Attractive to cats
Chinese lantern	*Physalis alkekengi*	Colorful fruit
Christmas rose	*Helleborus niger*	Early bloomer
Chrysanthemum	*Chrysanthemum frutescens, C. × morifolium, C. × superbum, C. zawadskii*	Autumn bloomers
Cinquefoil	*Potentilla aurea, P. nepalensis, P. nitida, P. recta*	Rock garden and flower bed
Columbine	*Aquilegia caerulea, A. chrysantha, A. vulgaris*	Rock garden and borders
Coral bells	*Heuchera sangiunea*	Easy, trouble-free
Coreopsis	*Coreopsis auriculata, C. lanceolata*	Easy, trouble-free
Cranesbill	*Geranium cinereum, G. pratense, G. sanguineum*	Rock garden and borders
Daylily	*Hemerocallis fulva*	Easy care; mostly tetraploid cultivars
Delphinium	*Delphinium × belladonna, D. elatum, D. grandiflorum*	Especially good blues
Eryngium	*Eryngium alpinum, E. planum*	Rock garden and borders
Globeflower	*Trollius chinensis, T. europaeus*	Good in wet soil
Gloriosa daisy	*Rudbeckia hirta*	Easy care; tetraploid cultivar; good cut flower
Goat's beard	*Aruncus dioicus*	Flowers in partial shade
Hosta	*Hosta fortunei, H. lancifolia, H. sieboldiana, H. sieboldii, H. undulata*	Good in shade; attractive foliage
Japanese anemone	*Anemone × hybrida*	Autumn bloomer
Lady's mantle	*Alchemilla alpina, A. vulgaris*	Low grower
Lily-of-the-valley	*Convallaria majalis*	Low, fragrant speader; OK in shade
Lungwort	*Pulmonaria angustifolia*	Easy grower; OK in shade; blue
Lupine	*Lupinus polyphyllus*	OK in poor soils; Russell hybrids best (uncertain parentage)
Masterwort	*Astrantia major*	Can be planted in wet places
Oriental and Iceland poppy	*Papaver orientale, P. nudicaule*	Oriental cultivars long-lived when left undisturbed
Ornamental grasses, pampas grass, uva grass, silver grass	*Cortaderia selloana, Gynerium sagittatum, Miscanthus sinensis*	Ornamental grasses are usually tall; planted in clumps and noted for plumes; some are variegated.
Peony	*Paeonia lactiflora*	Showy, fragrant flowers
Phlox	*Phlox paniculata, P. stolonifera, P. subulata*	Varies from creepers; mats to tall
Pinks	*Dianthus deltoides, D. plumaris*	Rock gardens and beds; fragrant
Purple coneflower	*Echinacea purpurea*	Easy, long-lasting; good cut flower

(continued)

TABLE 16–12 *Some Popular Perennials* *(continued)*

Common Name	Scientific Name*	Comments
Sedum	*Sedum acre, S. album, S. rosea, S. rupestre, S. spurium, S. spectabile*	Rock gardens and beds; some bloom in spring; others autumn
Sneezeweed	*Helenium autumnale*	Good cut flower; late summer–autumn bloomer
Solomon's seal	*Polygonatum biflorum, P. commutatum*	OK in shade
Speedwell	*Veronica longifolia, V. spicata*	Easy to grow
Spiderwort	*Tradescantia* × *andersoniana*	Easy culture
Starry eyes	*Omphalodes cappadocica*	Blue flowers with white center
Turtlehead	*Chelone glabra*	OK in wet, shaded areas
Violet	*Viola cornuta, V. nuttallii, V. odorata, V. tricolor*	Rock garden; partial shade; can be weedy
Yarrow	*Achillea millefolium*	Easy care; good cut flower
Yellow loosestrife	*Lysimachia punctata*	OK in wet areas
Yucca	*Yucca filamentosa*	Good for hot, dry areas

*Most are actually grown as cultivars or hybrids of the indicated species.

area, even though it would be of sufficient hardiness. An example of such a condition would be the arid-alkaline region found in the Southwest.

Planting time (and hence finished production time in a nursery) for perennials varies according to region. Many perennials can be planted in the North as soon as the ground can be worked in early spring. This appears to be better than early fall planting in the North; however, early fall planting is equally possible in warmer sections. This gives the perennial root systems time enough to become established. Fall planted perennials should be mulched to prevent freezing and thawing cycles that can heave newly planted perennials from the soil. Late fall planting is to be avoided in the North, because the root systems either become poorly established or not at all.

Perennial plantings require some additional efforts beyond those of watering, weeding, fertilizing, and prevention of pests and diseases. Tip pinching is required with some perennials to produce shorter, stockier plants with multiple blooms such as chrysanthemums. Some of the taller perennials will require staking. Perennials become crowded through natural propagation and will require division of the clump at intervals. The frequency of division varies according to the species and existing cultivation. If this is not done, flowering will become sparser and even nonexistent. Division can usually be done most favorably in the early fall. Certain perennials can also be divided early in the spring. With spring division, care must be taken not to damage the young shoots. Some perennials need division every few years to keep them at peak blooming. Still others considered low-maintenance perennials would only need little attention over several years.

BULBS

The term *bulb,* when used in a horticultural sense, does not have the narrow connotation attributed to it by botanists. To the horticulturist the word *bulb* suggests such plant parts as corms, tubers, and thickened rhizomes, as well as actual bulbs. Most of the plants arising from these structures are simple stemmed perennials. Many are monocots, but not all. For example, dahlias are dicots as are tuberous begonias. Generally, bulbs are divided into two groups: spring and summer bloomers. Examples of spring and summer flowering blooms are shown in Tables 16–13 and 16–14, respectively.

Most spring flowering bulbs (Figure 16–9) require a cold period and are limited to areas of minimal to extensive frost. Such bulbs do not flower in the more extreme southerly parts of the United States. On the other end, extreme cold can also be a limiting factor. For example, hyacinths are less cold hardy than are daffodils. Winter mulch can help extend

TABLE 16–13   *Some Popular Spring Flowering Bulbs*

Common Name	Scientific Name*	Comments
Avalanche lily	*Erythronium grandiflorum*	Rock garden; partial shade; winter mulch helpful
Butterfly tulip	*Calochortus* sp.	Mulch to prevent harmful freeze/thaw cycles and to protect winter foliage; rock garden and border
Crocus	*Crocus* sp.	Good for naturalizing, borders, early flowering; can be forced, some nice blues; small botanical kinds (Snow crocus) and larger (Dutch) types
Crown imperial	*Fritillaria imperialis*	Tall, colorful, and imposing; repels mice, moles, other rodents
Daffodils	*Narcissus*	Trumpet narcissi cultivars; good for naturalizing; safe from deer; good cut flower, smaller ones for rock gardens and forcing
Dutch (English, Spanish) iris	*Iris* sp.	Cultivars of mixed bulb-type iris species
English bluebells	*Endymion nonscriptus*	Popular in England; good for naturalizing; nice blue
Glory of the snow	*Chionodoxa luciliae*	Early bloomers; need moisture; replace every 3 years
Grape hyacinth	*Muscari armeniacum, M. azureum, M. botryoides*	Easy culture; good blues; long lasting; good for naturalizing
Grecian windflowers	*Anemone blanda*	Nice blues; good for naturalizing; long blooming period
Guinea-hen tulip	*Fritillaria meleagris*	Good for moist soil; rock garden and naturalizing; borders
Hyacinth	*Hyacinthus orientalis*	Colorful, including blues; fragrant; easily forced
Iris	*Iris*	Cultivars of mixed rhizome iris species grouped as beardless (Japanese and Siberian), bearded, crested, regalia (mild climates)
Narcissus	*Narcissus*	Small-, double-, and large-cupped narcissi cultivars; good for naturalizing; safe from deer; good cut flower; smaller ones for rock gardens; some forced
Ornamental onion	*Allium caeruleum, A. christophii, A. cyaneum, A. flavum, A. giganteum, A. neapolitanum*	Colorful, globe-shaped flowers; some can be used as cut or dried flowers; easy culture
Pink buttercups	*Oxalis adenophylla*	Mild climate; can be difficult
Siberian squill	*Scilla siberica*	Easy culture; good for naturalizing; some good blues
Snowdrops	*Galanthus nivalis*	Easy culture; very early flowering
Snowflake	*Leucojum aestivum*	Best left undisturbed; good for naturalizing
Spring starflower	*Ipheion uniflorum*	Blues; good for naturalizing
Striped squill	*Puschkinia libanotica*	Rock gardens; borders; naturalizing
Tulips	*Tulipa*: botanical (*fosterana, greigii, kaufmanniana*), single early, double early, triumph, darwin hybrid, single late, lily-flowered, fringed, viridiflora, parrot, double late	Colorful, can be forced; good cut flowers; troubled by rodents and deer; not long lasting; many types in terms of blooming time and size; botanical ones good for rock gardens and last longer; Darwin hybrids live longer than others
Wild hyacinth	*Camassia leichtlinii, C. scilloides*	Some blues; good with daffodils; best not disturbed
Winter aconite	*Eranthis cilicica*	Good for naturalizing; early flowering
Wood hyacinth	*Endymion hispanica*	Easy culture; good in woodland garden

*Many are actually grown as cultivars or hybrids of the indicated species.

the range of the less cold hardy bulbs. Microclimates, especially areas adjacent to the house foundation on the southern side, can also extend bulb hardiness ranges.

Summer blooming bulbs (Figure 16–10) are usually even more limited in cold hardiness and are grown as annuals or pot plants in the colder areas of the United States. In more southerly areas, summer flowering bulbs can be left in the ground to survive for many

TABLE 16–14   *Some Popular Summer Flowering Bulbs*

Common Name	Scientific Name*	Comments
Baboon flower	*Babiana stricta*	Good for rock gardens, borders, and containers; winters over in more southerly areas
Caladium	*Caladium* × *hortulanum*	Noted for colorful foliage; good in partial shade; use in beds and containers
Calla lily	*Zantedeschia aethiopica, Z. albomaculat, Z. elliottiana, Z. rehmannii*	Start indoors in North; use in beds and containers
Canna	*Canna* × *generalis, C. orchiodes*	Start indoors in North; good in beds and borders; tall, but dwarf hybrids available
Chinese lantern lily	*Sandersonia aurantiaca*	Use in beds, containers, and greenhouse; winters over in more southerly areas
Corn bells	*Ixia maculata, I. monadelpha*	Warmer climates; pot plant in North
Crocosmia	*Crocosmia* × *crocosmiiflora*	Good cut flowers; good near seaside; winters over further north than most; flowers over long period; long lived
Dahlia	*Dahlia:* single, anemone, colarette, peony, formal decorative, informal decorative, ball, pompon, incurved cactus, straight cactus, semicactus, miscellaneous	Numerous hybrid cultivars; colorful, vary from short to tall cultivars with huge flowers; good in beds, borders, and containers (smaller ones); good cut flowers; need lots of water; very frost sensitive
Freesia	*Freesia* × *hybrida*	Good cut flower; start indoors in North; good bedding and pot plant
Gladioli	*Gladiolus* × *hortulanus*	Good cut flower; colorful; frost sensitive; winters over in more southerly areas
Glory of the sun	*Leucocoryne ixioides*	Good container plant; can be forced; outdoors in warmer climates
Hardy cyclamen	*Cyclamen hederifolium*	Start indoors in North; outdoors after all danger of frost past; good in rock garden and partial shade
Harlequin flower	*Sparaxis tricolor*	Front of border; winters over in more southerly areas
Homeria	*Homeria breyniana*	Flowers for several weeks; easy culture; winters over in more southerly areas
Lily	*Lilium*	Easy culture; long-lasting; can be forced; many colors; plant in fall or early spring; numerous species and many cultivars; very popular
Mexican shell flowers	*Tigridia pavonia*	Easy culture; nice flowers; winters over in more southerly areas
Nerine	*Nerine bowdenii*	Good cut flower; early autumn flowering; frost sensitive (needs long growing season); can be forced; containers and beds; winters over in more southerly areas
Peacock orchid	*Acidanthera bicolor*	Fragrant; easy culture; winters over in more southerly areas
Persian buttercup	*Ranunculus asiaticus*	Good cut flower; can be forced; winters over in more southerly areas
Pineapple lily	*Eucomis autumnalis*	Long-lasting flower; easy culture; beds and containers; winters over in more southerly areas
Queen fabiola	*Triteleia laxa*	Nice blue; good cut flower; long-lasting flower; needs good drainage; rock garden; winters over in more southerly areas
Scarborough lily	*Vallota speciosa*	Good container plant; can be forced; winters over in more southerly areas
Spider lily	*Hymenocallis* × *festalis*	Easy cultivation; long-lived and winters over in more southerly areas
Tritonia	*Tritonia crocata*	Flowers late summer; colorful; winters over in more southerly areas

TABLE 16–14 *Some Popular Summer Flowering Bulbs* (continued)

Common Name	Scientific Name*	Comments
Tuberous begonias	*Begonia × tuberhybrida*	Good container; hanging basket, and bedding plant; continuous flowering; numerous groups; start indoors in North; semishade; lots of water; lift and save (in North before frost, in South when leaves yellow and drop)
Windflowers	*Anemone coronaria*	Can be forced; good cut flower; winters over in more southerly areas
Wonder flower	*Ornithogalum*	Good cut flower; can be forced; beds and containers; winters over in more southerly areas

*Many are actually grown as cultivars or hybrids of the indicated species.

(A)

(B)

(C)

(D)

FIGURE 16–9 Examples of spring flowering bulbs. (A) Hyacinths (bulbs) are both colorful (pink, rose, red, blue, purple, white, yellow, and peach) and very fragrant spring flowers. (B) Daffodils (bulbs) in yellows and whites offer cheerful trumpet-shaped, deer-proof flowers that are ideal vase flowers. (C) Crocuses (corms) are colorful purple, white, and yellow harbingers of spring. (D) Irises come in many hues, make great cut flowers, and easily grown from rhizomes. (Author photographs.)

(A)   (C)   (E)

(B)   (D)

FIGURE 16–10   Examples of summer flowering bulbs. (A) Lilies (bulbs) are large and colorful in the early summer garden. (B) Dahlias (tuberous roots) produce many large, various colored flowers of great attraction both outdoors and in the vase. (C) Gladiolas (corms) are beautiful in the yard and the vase. (D) Crocosmia (corms) offers long-lasting periods of bright red or orange flowers. (E) The caladium (tuber) is a shade loving plant noted for its brightly colored variegated foliage. (Author photographs.)

years. In parts of the South, summer flowering bulbs can be planted in the fall or spring. Fall planting results in earlier flowering such as with Baboon flower.

Spring flowering bulbs are quite popular, since their blooming period occurs prior to that of the annuals and bedding perennials covered previously. Summer blooming bulbs provide excellent cut flowers and offer color and diversity in formal plantings beyond annuals and bedding perennials. The versatility of bulbs leads to their use in naturalized or formal plantings. Various bulbs are suited to rock gardens, woodland gardens, beds, borders, and even container gardens.

Culture of either bulb group is relatively straightforward. Bulbs are planted at various depths, depending on species and varieties. The depth is meant as the distance from the top of the bulb to the soil level. Planting depths also vary according to soil texture. Soils containing much sand (sandy loam) require a planting depth about one-third greater than a sandy clay loam, and a clay loam would require a depth about one-third less than that in the sandy clay loam. A general rule of thumb is that bulbs 2 inches or larger in diameter are planted at a depth of two or three times their diameter, and smaller bulbs are planted at depths three or four times their diameter. Approximate spacing is about twice the planting depth. Squirrels can be troublesome in that they dig up and eat certain bulbs. Deer and rabbits also eat young bulb growth such as tulips, but leave daffodils and narcissus alone.

A well-drained soil with fertility maintained at a moderate level is essential, since most bulbs will be left undisturbed for many years. Some bulbs such as most tulips, except for the botanical types, are short lived and need replacement more frequently. Still others will become crowded with decreasing blooms, and lifting, dividing, and replanting will be required. Those treated as annuals and discarded have lower soil fertility requirements, as nutrients will not be needed to produce stored food for next year's flowers. If these bulbs are dug and stored, their fertility needs are higher.

Spring blooming or hardy bulbs are planted in midfall up to just before the ground freezes too hard to be dug. Small numbers can be planted with a trowel or bulb planter. Large numbers require a shovel. Mulch is often used on beds where the ground is subject to alternate freezing and thawing to prevent heaving of bulbs and to increase the survival of half-hardy bulbs. If squirrels are troublesome, hardware cloth can be placed just below the soil's surface as you backfill with soil.

After flowering is over, the foliage should be left until it yellows and dies or is frost damaged. The foliage is involved in manufacturing food, which will be translocated to the bulb and stored for the following year's growth. Good soil fertility is essential at this time, and supplemental fertilization might be required. Quite often annuals or other perennials may be used to hide the unsightly or uninteresting foliage of spring bulbs at this stage.

Summer flowering or tender bulbs are usually planted when the danger of spring killing frosts is past. Frost-tender bulbs after a light frost ($-1°$ to $0°C$; $30°$ to $32°F$) should be trimmed of dead and damaged foliage and lifted from the soil in the North. They are next cured of wounds or bruises under conditions that minimize shriveling and disease problems. These curing conditions are 2 to 4 weeks at $16°$ to $21°C$ ($60°$ to $70°F$) and 40% to 50% relative humidity. After curing, a dusting with sulfur is often used to eliminate disease problems and waxing is sometimes used to prevent shriveling. Storage conditions for the various species vary; the relative humidity must be high enough to prevent shriveling, but not high enough to encourage disease, and cool temperatures below $10°C$ ($50°F$) are best. Tender bulbs are often packed in sand, sawdust, or peat moss and inspected periodically for disease or shriveling. Moisture is added as required. Commercial horticulturists use temperature and relative humidity-regulated storage facilities.

On the arrival or purchase of spring flowering bulbs, they can be stored in cool, well-ventilated areas out of direct sunlight. Temperature will vary according to the type of bulb, but the homeowner may use a range between $4.5°$ and $10°C$ ($40°$ and $50°F$). Cool, dry storage is needed for fall bulbs on receipt by the homeowner if they are not to be planted soon. Long-term storage by commercial horticulturists after harvesting bulbs, of course, requires temperature and relative humidity regulation as determined by bulb type.

Spring flowering bulbs may be forced indoors for early winter color or for holidays such as Easter. Bulbs that are usually forced and sold potted include *Hyacinthus orientalis* (hyacinth), *Narcissus* (includes daffodil, narcissus, jonquil), *Tulipa* (tulip), *Crocus*, *Muscari* (grape hyacinth), *Iris danfordiae,* and *I. reticulata.* Daffodils and tulips are also sold as cut flowers after being forced from bulbs. Forcing is the only way some of the warmer areas can use these bulbs, because of their cold requirement. The procedure is relatively easy, but strict attention to a time and temperature program is essential if the grower desires blooming bulbs for a specific day. This type of treatment is beyond the scope of this text, but information can be obtained from the references at the end of the chapter.

The art of forcing, for those who are not concerned with a specific day, is as follows (Figure 16–11). First, the cold requirement of the bulbs must be satisfied. Cold temperatures are needed to promote flower stem elongation. This can be done prior to potting or after potting them, and the cold treatment can be naturally or artificially supplied. A temperature between ($5°$ and $9°C$) ($41°$ and $48°F$) supplied through refrigeration can be used with potted or bare bulbs. Hyacinths may also be forced in special hyacinth bulb glasses filled with water at similar temperatures. The treatment lasts for about 8 to 12 weeks, and

(A)

(C)

(E)

(B)

(D)

FIGURE 16–11   Steps in bulb forcing. (A) Place crocking over drainage hole in clean pot to prevent soil leakage. (B) Place growing media in pot to height such that top of bulb will be about 1/2 inch below top of pot. (C) Fill with growing media around bulbs leaving bulb tip slightly exposed or barely covered. Water and place in cold storage (see text). (D) When shoots appear and root growth is abundant, put pots in cool area with indirect light. (E) If growing hyacinths in hyacinth glasses, move to a cool, indirectly lit area when appearance is similar to glass on right. (F) After a few days the shoots become green and pots are moved to a bright sunny area with cool temperatures. (USDA photographs.)

(F)

no natural light should be provided at this time. Rooting also occurs with potted bulbs during the cold storage period, but precooled bulbs can be rooted at warmer temperatures. Potted bulbs are generally ready for flower forcing when roots are seen to come from the drainage hole of the pot. This type of treatment can be done whenever bulbs are available. If it is done in November or December, bulbs can be flowered during the winter.

Potted bulbs can also be placed outdoors in the late fall in a cold frame and covered with several inches of hay, dry leaves, or peat moss. An unheated garage may also be used, with the pots placed in boxes and covered in the same manner. Whichever form of cold treatment is used, one thing must be remembered: The soil must not dry out. This should be checked periodically. A growth of grayish mold may be observed on the soil. It can be ignored, as it will disappear when the pots are brought into a sunny, warm environment.

Once past the cold treatment, bulbs can be moved into a warmer but not sunny area to allow acclimation of the new sprouts. Temperatures may range from 10° to 20°C (50° to 68°F). The higher the temperature, the more rapid the flowering and the longer the flower stalk. Pots should be moved into the direct sun after a few days in dim light or after the whitish sprouts become green.

FERNS

Ferns, although not a substantial part of ornamental horticulture, are still of value for three reasons (Figure 16–11). First, they are an important group of nonflowering, foliage plants that can be useful in perennial borders, rock gardens, or naturalized plantings of a woodsy nature. Second, florists value their foliage for use as greenery in floral arrangements. Finally, they are attractive pot plants found in greenhouses and homes that can be grown under low-light conditions.

Horticulturists generally recognize two groups of ferns. The first consists of hardy ferns (Figure 16–12) of native and exotic origin that can be cultivated outdoors throughout much of the United States and Canada. The second group includes ferns originating in tropical areas. The latter group, except for the southern extremities, is restricted in most of the United States to use in the greenhouse and home. Of the approximately 11,000 species of ferns, about one-third are found in the temperate areas. Many tropical ferns are epiphytes such as the well known, majestic Staghorn fern (*Platycerium bifurcatum*, Figure 16–13) grown in greenhouses on bark or boards. A few such as *Azolla* and *Salvina* are aquatic and can be grown in aquaria. Some ferns have solid fronds (*Asplenium nidus*, bird's nest fern),

(A)

(B)

FIGURE 16–12
Example of a temperate fern. (A) Close-up of New York fern (*Thelypteris noveboracensis*). New York fern used as a base planting around a white birch in Vermont. (Author photographs.)

(A)

(B)

FIGURE 16–13
Example of subtropical and tropical ferns. (A) A magnificent Staghorn fern (*Platycerium bifurcatum*) is growing well as a hanging basket fern in a greenhouse. (B) This Tasmanian tree fern (*Dicksonia antarctica*) was photographed in San Francisco, California. (Author photographs.)

TABLE 16–15   *Some Popular Ferns*

Common Name	Scientific Name*	Comments
Bulblet bladder fern	*Cystopteris bulbifera*	Small bulb-like bodies on fronds can produce new ferns; does well in limestone soils
Christmas fern	*Polystichum acrostichoides*	Evergreen; leathery
Cinnamon fern	*Osmunda cinnamomea*	Some fronds look like cinnamon sticks
Common polypody	*Polypodium vulgare*	Evergreen; fancy cultivars
Crested shield fern	*Dryopteris cristata*	Woodland garden fern
Delta maidenhair fern	*Adiantum raddianum*	Greenhouse or tropical fern
Ebony spleenwort	*Asplenium platyneuron*	Evergreen; rock garden
Fragile bladder fern	*Cystopteris fragilis*	Rock garden
Glade fern	*Diplazium pycnocarpon*	Tall, woodland garden fern
Hay-scented fern	*Dennstaedtia punctiloba*	Fragrant when crushed; good for naturalizing
Interrupted fern	*Osmunda claytoniana*	Fertile segments "interrupt" the barren segments; easy culture
Lady fern	*Athyrium felix-femina*	More tolerant of sun and dryness
Maidenhair fern	*Adiantum pedatum*	Dainty, attractive native fern
Maidenhair spleenwort	*Asplenium trichomanes*	Evergreen; rock garden
Marginal shield fern	*Dryopteris marginalis*	Slow to spread; nice specimen plant
Marsh fern	*Thelypteris palustris*	Good in wet areas
Narrow-leaved chain fern	*Woodwardia areolata*	Woodland garden; wet areas
New York fern	*Thelypteris noveboracensis*	OK in sun
Ostrich fern	*Matteucia pennsylvanica*	Large, wet areas
Rabbit's foot fern	*Polypodium aureum*	Hairy rhizomes look like rabbit's feet; greenhouse or indoor fern
Rattlesnake fern	*Botrychium virginianum*	Very hardy
Rusty woodsia	*Woodsia ilvensis*	Rock or alpine garden; rusty appearance
Sensitive fern	*Onoclea sensibilis*	Wet areas; weedy
Southern maidenhair fern	*Adiantum capillus-veneris*	Subtropical
Virginia chain fern	*Woodwardia virginica*	Woodland garden; wet areas
Walking fern	*Camptosorus rhizophyllus*	Evergreen; roots at tips on ground contact

*Many are actually grown as cultivars or hybrids of the indicated species.

but most have finely divided fronds (*Dryopteris goldiana,* Goldie's wood fern). Some such as the Christmas fern (*Polystichum acrostichoides*) have evergreen fronds. Many ferns of interest to the horticulturist are listed in Table 16–15.

Transplanting of ferns requires about the same effort as for flowering perennials. Soil and light conditions should be similar to the natural environment of the ferns' natural habitat. Many ferns thrive in full or partial shade when grown outdoors and require filtered sunlight or lower intensities of light indoors. Ferns often grow in soils having a high humus level and a covering of organic litter such as leaves. Since the rhizomes are often near the soil–litter interface, it is best not to cultivate in areas planted to ferns. Most ferns require added soil moisture during dry periods and usually high humidity levels in the greenhouse or home. Fertilizer requirements are minimal for ferns, particularly since the organic matter and litter provide small amounts of nutrients.

GRASS

Lawns are generally regularly mowed plantings of grass (Figure 16–14). Their purpose is to enhance the visual appeal of open spaces around buildings and parks, to provide useful outdoor living space for recreation and enjoyment and to improve environmental quality. The

(A)   (B)

FIGURE 16–14   Nothing beats the sweep and integrating effect of a well-manicured lawn on the landscape, whether it is a suburban landscape (A) or an estate (B). (Author photographs.)

latter consists of soil improvement (soil health and erosion prevention) and runoff reduction (better water infiltration), as well as a modifying influence on drastic changes in soil–air temperatures. Areas carpeted with lawns usually have a cooler microclimate than paved areas.

Grasses are generally used for lawns, since their basal meristems are adapted for mowing in that plant damage or loss does not occur. One exception is *Dichondra micrantha,* used as a grass substitute in the southwest United States and especially California. Ground covers other than these are utilized, but not in areas requiring mowing. Their uses will be covered in the next section.

The use of grass species is determined by climate. The main two climatic areas for grasses are the North and South. The first climatic region consists of the northern two-thirds of the United States. The dividing line between the northerly and southerly regions extends from Washington, D.C., through northern Tennessee and Arizona to San Francisco, California. The perennial cultivars of choice for the North are derived from Kentucky bluegrass, perennial ryegrass, fine fescue, and bent grasses. Grasses of choice for the south include Bermuda grass, zoysia, St. Augustine grass, Bahia grass, and centipede grass. Characteristics of each grass are shown in Table 16–16. In addition, some native grasses utilized for difficult or specialized situations are indicated. Native grasses are often good choices for low maintenance and more sustainable lawns (see the Relevant Resources list at the end of the chapter).

Clovers, especially species of *Trifolium,* are common companion plants in lawns and are often tolerated in home and park areas, but not golf courses. Given the nitrogen-fixing abilities of clover, it should be left in lawns because it improves sustainable lawn care in that less fertilizer is needed. Other companion plants are considered weeds. Weeds include, but are not limited to, crabgrass (*Digitaria sanguinalis*), dandelion (*Taraxacum officinale*), ground ivy (*Glechoma hederacea*), knotweed (*Polygonum aviculare*), milky spurge (*Euphorbia maculata*), plantain (*Plantago lanceolata*), and quackgrass (*Agropyron repens*).

Lawns can be planted either from seed or by vegetative means, such as sod, sprigs, or plugs. Seeding is more economical and, unlike with sod, there is minimal danger of introducing diseases, weeds, or insects. Sod offers the advantage of quick, but costly cover. Sprigs, individual stems, and plugs (biscuits of sod) are economical means of establishing grasses that exhibit genetic variability when grown from seed.

In the North, seeding is done either in early spring or, better still, from mid-August in the northernmost states to mid-September for the lower reaches of the North. Vegetative establishment is the method of choice in the South. Seedbed preparation is the same for either seed or vegetative means. Basically, the bed must be cultivated to at least 3 inches such that soil lumps larger than a golf ball and vegetative and inorganic debris are not evident. Excessive cultivation is avoided to minimize breakdown of soil structure. Fertilizer and lime are added according to soil tests (see Chapter 11). Organic matter can be added,

TABLE 16–16   *Northern and Southern Lawn Grasses*

Common Name	Scientific Name*	Comments
Northern		
Bent grass	*Agrostis* sp.	Very fine texture. Wide use in golf courses. Needs lots of moisture and quality care. Heavy maintenance grass. Weak in hot muggy weather.
Fine fescue	*Festuca rubra*	Quick to germinate. Fine companion to blue grass. Adaptable to shade, poor soils, dry periods, low nutrient levels. Some weakness in warm, humid weather.
Kentucky bluegrass	*Poa pratensis*	Excellent, attractive, nicely textured grass. Widely adaptable. Easy care. Forms dense, strong sod. Best in cool weather, but survives hot, dry conditions periodically. Tolerates traffic and light shade. Quick recuperation. Some weakness with persistent high-temperature weather.
Perennial ryegrass	*Lolium perenne*	Very quick to germinate. Reasonably attractive. Weak during weather extremes. Does not mow cleanly. Doesn't form dense sod.
Southern		
Bahia grass	*Paspalum notatum*	Highly adaptable and modest care needed. Can take heat and shade. Somewhat coarse and cold intolerant.
Bermuda grass	*Cynodon dactylon*	Attractive texture and color. Fast growing, can be aggressive. Recuperates well. Performs well under warm, humid conditions. Some thatch problem. Needs frequent mowing, fertilization, and watering. Growth slows at 50°F and dormancy starts near freezing. Fares poorly in shade.
Centipede grass	*Eremochloa ophiuroides*	Requires minimal care. Tolerates heat and shade. Sensitive to high fertility and alkalinity. Slow to establish. Cold intolerant. Sensitive to chlorosis.
St. Augustine grass	*Stenotaphrum secundatum*	Attractive. Tolerates shade. Needs modest attention. Good in Deep South. Coarse. Thatch problem. Susceptible to cinchbug.
Zoysia grass	*Zoysia* sp.	Dense, attractive. Requires minimal attention. Tough and durable. Minimal mowing needed. Slow to establish. Needs heavy-duty mower. Thatch problem. Recuperates slowly. Long dormancy period. Billbug susceptible.
Native Grasses for Problem Areas		
American beach grass	*Ammophila breviligulata*	Good for stabilizing beach dunes and coastal areas in the North.
Buffalo grass	*Buchloe dactyloides*	Good in dry plains area where lawns are not irrigated.
Carpet grass	*Axonopus affinis*	Good for poorly drained, southern soils.
Fairway wheat grass	*Agropyron cristatum*	Good in dry, cool plains and mountains. Holds soil well.
Rough bluegrass	*Poa trivialis*	Good for moist, shaded areas in North where traffic is light to none.
Tall fescue	*Festuca elatior*	Good in border states where neither northern nor southern grasses are optimal.
Western wheat grass	*Agropyron smithii*	Like fairway wheat grass, but can adapt to more alkaline soils.

*Most grasses are grown as cultivars of the indicated species and many products are blends of two or more cultivars such that varying conditions (sun and shade, heavy and light traffic) can be accommodated.

especially if the soil structure is poor. Keep in mind that once grass is established, organic matter will build up. The soil is settled by watering, not rolling, which compacts the soil. If light rolling is deemed necessary, it should be done when the soil is not wet.

Seed is either hand or mechanically sown. Unless the surface is not crumbly, no light dragging or raking is required after sowing. The seed can be covered with a light straw or excelsior mulch or woven net. This prevents washing away of seed by rain, conserves moisture, and reduces seed losses to birds. Watering is especially important during germination and subsequent establishment of seedlings or sod. Sections of established lawns can be reseeded as follows. Vegetation is removed by either mechanical or chemical means. The surface is roughed such as by hand raking or scarifying machines. Seeding and subsequent care are the same as outlined previously.

Established lawns require regular mowing, periodic fertilization and adjustment of soil pH, irrigation when needed, and pest control. Fertilization, pH control, watering, and weed control in a sustainable manner are covered in Chapter 12 under lawn care. Fertilization and liming are best done according to soil tests. Mowing height varies according to grass type as follows: northern bent and southern Bermuda grasses, 1 inch; Kentucky bluegrass, perennial ryegrass, fescue, Bahia, and St. Augustine grasses, 2 or 3 inches; improved varieties of Kentucky bluegrass, most zoysias, centipede, and common Bermuda grasses, 1.5 to 2 inches. Mowing is required when the grass exceeds 50% of its cut height.

GROUND COVERS

Plants selected for ground covers are herbaceous perennials or low woody plants requiring minimal care (Figure 16–15). They are usually used in place of lawns on areas where lawns fare poorly or not at all, or on sites where lawn care is difficult. Such locations include very sandy soils (dunes and arid areas) and heavily shaded areas under trees and slopes. Ground covers on dunes, slopes, and banks also help prevent soil erosion. As a rule, ground covers are not suitable for walking or recreation. Some are grown mostly for foliage (evergreen are better than deciduous) and others offer attractive flowers as a bonus.

(A)

(B)

(C)

FIGURE 16–15   Examples of ground covers.
(A) Wintercreeper (*Euonymus fortunei*). (B) Gold moss stonecrop (*Sedum acre*). (C) Periwinkle (*Vinca minor*). (USDA photographs.)

Beds for ground covers should be prepared the same as for lawns, except that they should be prepared to 10 inches deep if possible. Spacing is determined by several factors: cost, size of area to be covered, rate of growth, and the importance attached to the time for achieving full cover. Mulch is important for minimizing weed competition and maximizing water retention during the establishment of the ground cover. Once a ground cover is established, weeding is minimal. Fertilization and watering are done as needed. A number of plants suggested for ground covers are listed in Table 16–17.

WOODY PERENNIALS: TREES, SHRUBS, AND VINES

Woody perennials include deciduous and evergreen trees, shrubs, and vines. Woody plants have great value in the overall landscape because of their long-lasting, substantial year-round effect. As such, they are mainstays in landscaping planning, whether utilized for background, screening, foundation plantings, visual impact, shade, ecosystem maintenance/improvement, energy conservation (see Chapter 11), or windbreaks.

Obviously, the woody plant group is very diverse, but some factors are common to all that should be considered when choosing plants. Foremost is the habit of growth, which determines the ultimate shape, height, and width of the mature plant. The placement of a plant, both in terms of mature scale and shape, in a landscape will ultimately be determined by this fact. Hardiness is also important for determining the ultimate success of a plant in a given area. As before, it must be stressed that hardiness zones are at best a guide to relative hardiness. Too often, local environmental conditions will increase or decrease the hardiness of a particular plant in a given area, or the hardiness limitations of a plant may not be truly known, but only guessed at. There is no substitute for localized knowledge of the plants that are hardy in your area.

Two other factors are important: permanence and foliage effect. The plant should be long lived in your locality, considering the investment of time and money put into woody plants. If you know a plant is generally not long lived, avoid it. For example, a plant might be hardy in your area, but require cool, moist summers. If your summers are hot and dry, that plant is not a wise choice. Other plants might have borderline hardiness in your area and thrive, until that one cold winter that occurs every so many years arrives. Some plants might be highly susceptible to certain pests or diseases that prevail in your area.

The latter is a critical point beyond longevity in terms of both sustainability and maintenance perspectives. Plants prone to infection from insects and diseases of your area require high maintenance to keep them healthy. Are you prepared to give the time or pay someone to control pests? The usage of pesticides is also more prevalent than with plants less prone to pests. Plants needing little or no pesticides are a better choice for environmental sustainability. For example, lethal yellowing is a serious threat to palm trees. Some 27 species of palms are impacted. Certain species such as the Christmas palm (*Veitchia merrilli*), Fiji fan palm (*Pritchardia pacifica*), and Canary Island date palm (*Phoenix canariensis*) are highly susceptible and should not be planted unless resistant selections are found. Instead, one should choose highly resistant palms such as the Cabada palm (*Chrysalidocarpus cabadae*) and Chinese fan palm (*Livistona chinensis*) in the Florida landscape. Similarly, Eastern hemlock (*Tsuga canadensis*) and Carolina hemlock (*Tsuga caroliniana*) are to be avoided in the landscape in the Northeast and Southeast, given the high rate of attack by wooly agelids. Dogwoods also are faring poorly in New England because of borers and dogwood blight. Your local extension office and regional horticultural books can help you to avoid woody plants plagued with pests in your locality.

Foliage effect (Figure 16–16) can be quite important, since you view the plant throughout the changes of seasons year after year. Several factors are involved in foliage effect: size, texture, deciduous versus evergreen, summer and fall color, arrangement on the stem, and appearance and condition throughout the seasons. The bark color and texture, flowers and fruit, and the branching pattern are also factors that can further contribute to the overall effect of appearance and condition throughout the year. Other selection factors involve use factors such as shade, a specimen for the lawn or backyard, foundation plantings, property line plantings, and vista views.

TABLE 16–17   *Some Popular Ground Covers*

Common Name	Scientific Name*	Comments†
Aaron's beard	*Hypericum calycinum*	Flowers; roots on contact; evergreen; purple fall foliage; good in shade and poor soil
Alleghany spurge	*Pachysandra procumbens*	Evergreen in South; use in shade
Bearberry	*Arctostaphylos uva-ursi*	Evergreen; very cold hardy; bronze fall foliage; red berries; good in poor soil and shore areas; sun; holds banks
Bearberry cotoneaster	*Cotoneaster dammeri*	Evergreen; roots easily in moist soils on contact; red fruits
Carpet bugleweed	*Ajuga repens*	Spreads fast; flowers; mostly evergreen; good in sun and shade
Creeping lilyturf	*Liriope spicata*	Evergreen; flowers; grass-like; not suitable near grass as grass invades and outcompetes
Creeping mahonia	*Mahonia repens*	Evergreen; stoloniferous shrub
Crown vetch	*Coronilla varia*	Good for covering banks; can be very weedy
Dichondra	*Dichondra micrantha*	Good grass substitute; can take light traffic; only in warmer, more southerly areas
Dwarf bamboos	*Sasa veitchii, Shibataea kumasaca*	Spreads fast; best in sun, but grows (taller) in shade; only in warmer, more southerly areas
Ferns	See Table 16–15	Good woodland and rock garden ground cover; moist and shady areas; ferns available for most climates; some evergreen
Ground ivy	*Glechoma hederacea*	Mat-like; grows in sun and shade; does well in moist soils; can become weedy
English ivy	*Hedera helix*	Evergreen; grows in sun and shade, rich and poor soils, flat areas and banks
Ice plant	*Mesembryanthemum crystallinum*	Good for sandy, hot areas with sun; only in warmer, more southerly areas
Japanese spurge	*Pachysandra terminalis*	Widely (overused) grown in North; evergreen; use in shade; spreads easily; good near shore
Moss pink	*Phlox subulata*	Semi-evergreen; nice flowers; good slope cover; sun; very cold hardy
Partridge-berry	*Mitchella repens*	Moist, acid soils with partial shade; evergreen; red berries; dainty; woodland/rock gardens
Prostrate junipers	*Juniperus chinensis procumbens, J. communis depressa, J. conferta, J. horizontalis, J. virginiana 'Horizontalis'*	Evergreen; woody; long lived; good in hot and dry, shore and urban areas; attractive foliage; nice greens and blues (scale-like needles)
Salal	*Gaultheria shallon*	Good in acid soils; flowers; fruits; vigorous grower; evergreen
Scotch heather	*Calluna vulgaris*	Evergreen; nice flowers; forms thick mats; good cold hardiness; acid, moist, well-drained soils
Spring heath	*Erica carnea*	Good in acid soils; evergreen; flowers
Stonecrop	*Sedum sarmentosum, S. spurium*	Good in hot, dry, sandy conditions; flowers; evergreen to semi-evergreen
Sweet fern	*Comptonia peregrina*	Good for covering banks; aromatic; difficult to transplant; very good cold hardiness; moist to dry, sandy soils; woodland edges, shrubby
Trailing periwinkle	*Vinca minor, V. major*	Good near shore; V. minor better in North, evergreen; flowers; grows in sun and shade
Trailing seafig	*Carpobrotus chilensis*	Good for sandy areas; only in warmer, more southerly areas, flowers
Wandflower	*Galax urceolata*	Good in acid soils; shade; high cold hardiness
Wedelia	*Wedelia trilobata*	Flowers; only in warmer, more southerly areas
Weeping lantana	*Lantana montevidensis*	Woody; trailing, flowers; sun; evergreen; only in warmer, more southerly areas
White lilyturf	*Ophiopogon jaburan*	Grass-like; evergreen; flowers; good near shore; easy culture
Wineleaf cinquefoil	*Potentilla tridentata*	Evergreen; sun; good in dry to moist acid soils; tiny flowers; rock gardens
Yellowroot	*Xanthorhiza simplicissima*	Good for moist shady areas; deciduous; fall color

*Some are actually grown as cultivars or hybrids of the indicated species.
†Unless indicated otherwise, moderate cold hardiness.

FIGURE 16–16 (A) The foliage of the extremely hardy deciduous shrub, the cork bush (*Euonymus alata*, also known as the flame bush and winged euonymus) is a light green in the spring, dark green in the summer, and a flaming red in the fall. A compact cultivar is available (*E. alata* 'Compacta'). (B and C) These shrubs are both cultivars of *Euonymus japonica* ('Albo-marginata' and 'Aureo-marginata,' respectively). Both offer variegated leaves (green/white and green/yellow, respectively) and the extra bonus of evergreen foliage. These variegated leaves offer a break from the solid green of many needled, scaly, or broad-leaved evergreens in months when deciduous trees and shrubs have no leaves. (D) The needles of this evergreen, Colorado blue spruce (*Picea pungens*), have a soft blue-green color that contrasts nicely with the greens of other trees and shrubs. (E) The shiny green leaves of this holly add their sheen in the dark of winter and offer a nice foil to needled and scaly conifers. (Author photographs.)

FIGURE 16–17 A large, mature deciduous shade tree such as this English elm (*Ulmus procera*) offers a cooling patch of shade during the summer. (USDA photograph.)

Some generalized principles apply to the selection of woody plants. For purposes of shade, large deciduous trees (Figure 16–17) are useful. These trees provide ample shade in the summer, both in the yard and along the street. In the winter, these trees allow sun to warm houses and adjacent property. Smaller deciduous trees are more useful as specimen plantings on the lawn (Figure 16–18) or in the backyard. These plants are chosen mostly

(A)   (B)

FIGURE 16-18   (A) This small deciduous tree (Eastern Redbud, *Cercis canadensis*) is ideal as a lawn tree, especially in the spring when covered with rose-colored flowers that appear before the foliage. (Author photograph.) (B) A crab apple offers a double bonus in the yard: colorful spring flowers varying from white to pink followed by the small ornamental crab apples in the summer. (USDA photograph.)

FIGURE 16-19
(A) This dwarf conifer cultivar of the Colorado blue spruce has excellent blue shades and is suitable for viewing from a window in the winter (*Picea pungens* 'Moerheimi'). (B) This scaly dwarf evergreen (cypress, *Chamaecyparis obtusa* 'Nana Gracilis') has dark green color and its foliage exhibits a fan-shaped pattern, making it a very interesting specimen. (Author photographs.)

(A)   (B)

on the basis of some outstanding characteristic such as flowers, fruits, or attractive summer and autumn foliage.

Evergreens, both broad-leaved and needled, are useful for maintaining some color throughout the year. Form and texture are also other good points for evergreens. The conifers (Figure 16–19) make ideal unique specimens as well as background plantings, both for colorful foreground plants and for privacy. Conifers also hide unattractive views and act as a windbreak year-round. Given their wide size range, it is always possible to find suitable ones for any landscape or park. Broad-leaved evergreens (Figure 16–20) also offer the added benefits of flowers and fruits. This group is widely useful as a complement to deciduous flowering and fruiting shrubs (Figure 16–21). A good landscape combines both deciduous and evergreen plantings in the landscape (Figure 16–22) to provide visual appeal year-round. Interest is created from flowers appearing in the spring and summer followed by colorful fruits. This mix also provides attractive foliage color, texture, and form (needled, scaly, and broad leaves) during the growing season, colorful changes in deciduous leaves at the fall transition, and evergreen foliage and persistent fruits during the dormant season.

(A)

(B)

(C)

(D)

FIGURE 16–20 (A) Broad-leaved evergreen shrubs such as this Rhododendron (a P. J. M. hybrid, *Rhododendron dauricum sempervirens* × *R. carolinianum*) are literally completely covered with pink to rose-colored flowers. This feature makes it an excellent choice for a foundation or border specimen. (B) Mountain laurel (an unnamed "red-bud" cultivar of *Kalmia latifolia*) is a good yard, border, or foundation broad-leaved, flowering evergreen shrub. Its habit is tidy (other than deadheading), flowers are prolific, pruning is minimal, and it thrives in partial shade. (C) This Southern magnolia (*Magnolia grandiflora*) broad-leaved, flowering evergreen tree is a fine specimen for the Southeast. (D) Palms are excellent broad-leaved evergreens for warmer climates such as this *Washingtonia filifera* seen widely in California and Arizona. (A and B are author photographs; C and D are USDA photographs.)

Vines (Figure 16–23) can offer a unique form, thus becoming a focal point, and the benefits of flowers, fruits, and either deciduous or evergreen foliage. Vines are useful for covering stone walls, trellises, and arbors. Care should be exercised when one places the vine. Certain vines can be destructive of house exteriors and are also a problem when painting or other maintenance needs to be done behind them.

The major means of producing woody perennials in the production nursery is in containers. Balled and burlapped stock is still produced, especially for larger evergreen or deciduous trees and shrubs. Deciduous trees, to a lesser degree, are also sold in bare-root form. Very small-needled evergreens such as would be used to establish a wood lot or forest are sometimes sold bare root. Most evergreens cannot be handled successfully in bare-root form.

(A)

(B)

(C)

FIGURE 16–21 (A) The double-file viburnum (*Viburnum plicatum* f. *tomentosum*) is a deciduous shrub noted for its horizontal branches filled with white flowers in a double-file followed by reddish fruits that eventually turn black. (B) The border forsythia (*Forsythia* × *intermedia*) with its bright yellow covering of spring flowers is a dependable flowering deciduous shrub. Flowers can be forced in the late winter on cut stems in water. (C) Lilacs (*Syringa vulgaris*) are relatively carefree, flowering deciduous shrubs other than a minor problem with powdery mildew. Their fragrance in the spring is noteworthy. (Author photographs.)

(A)

(B)

FIGURE 16–22 Mixtures of deciduous and evergreen woody plants provide textural and all-season appeal in the landscape, whether the planting be simple (A) or sophisticated (B). (Author photographs.)

FIGURE 16–23 Vines are highly versatile as shown in this backyard trellised hybrid clematis (A, *Clematis* 'Will Goodwin'), this variegated ivy used as a house plant (C, *Hedera helix* 'Argenteo-variegata'), this greenhouse-grown passion flower (B, *Passiflora* × *alatocaerulea*), or this split-leaf philodendron used as an office plant (D, *Monstera deliciosa*). (Author photographs.)

(A)

(B)

(C)

(D)

Container stock (Figure 16–24) can be transplanted successfully anytime from early spring through fall. Balled and burlapped stock can also be planted during this time period. The best planting times for both in the North are early spring while in the dormant stage prior to growth, and in the fall and winter in the South after cessation of growth just as dormancy is about to start. Bare-root stock should only be planted at these times while dormant. These preferred planting times allow the plant to recover from transplanting shock without the added stress produced by high rates of transpiration during warmer weather. Container-grown stock is most likely to recover sooner than other types, since their roots experience less transplanting shock.

If the nursery stock cannot be planted within 2 days, some precautions are necessary. Bare-root stock should be heeled in (Figure 16–25); that is, a trench is dug and the plants laid in on a 45° angle, watered, and then the roots covered with soil. Balled and burlapped or container stock can be placed temporarily in a shady location and the root ball kept moist.

When ready to plant the shrub, tree, or vine, dig a hole adequate to accommodate the root ball or roots without crowding. If necessary, organic matter can be incorporated into the backfill soil to improve it or sand or gravel may be incorporated to improve drainage. If the root ball is in natural burlap, slit it with a knife to aid root penetration. Plastic burlap should be removed after the ball is in place. Container-grown plants should have the root mass quarter-scored with a knife or spading shovel. Otherwise, the roots will continue to grow in a constricted pattern with detrimental effects on the long-range adjustment of the

FIGURE 16–24
Container stock from nurseries offers the highest success rate for do-it-yourselfers around the home. (Author photograph.)

FIGURE 16–25
Heeled-in bare-root stock is shown here. (USDA photograph.)

plant. The plant should be placed at the same soil depth it grew in previously. New plants should be watered and mulched to carry them through the critical establishment period. Once they are established, fertilization, pruning, watering, and other cultural inputs should be applied as indicated for normal culture.

Because of dryness, failure is more apt to occur with fall-planted rather than spring-planted material. Often the growing media around the root ball drains faster and dries quicker than the site soil; this differential drying bears close watch during the establishment period. Adequate moisture should be supplied if needed right up to a hard freezing. An antidesiccant for broad-leaved evergreens may help to slow transpiration on certain bright warm days in winter when water losses occur. However, frozen soil prevents compensating water uptake, so the antidesiccant may not be sufficient. A tree wrap can prevent drying, splitting, and scalding of bark on trees until they are established. Staking, based on recent evidence, would appear to be unnecessary, as stronger trees are produced without it. Staking would be in order if a specimen were susceptible to wind damage or shifting of the root ball.

Some popular deciduous trees and coniferous evergreens are listed in Tables 16–18 and 16–19, respectively. Popular deciduous shrubs and broad-leaved evergreens are found in Tables 16–20 and 16–21. Woody vines are listed in Table 16–22. The tables are only meant

TABLE 16–18 *Some Popular Deciduous Trees**

Common Name	Scientific Name[†]	Comments
Ash, European	F. excelsior	Rapid growing shade trees. Good street or lawn tree.
Ash, flowering	F. ornus	Transplants easily. White and green ash are very cold hardy and
Ash, green	F. pennsylvanica	widely used in North, while Oregon, Shamel, and velvet ashes
Ash, Oregon	F. latifolia	are popular in the Pacific Coast, California, and the Southwest,
Ash, Shamel	F. uhdei	respectively. Flowering ash has the best flowers, while European
Ash, velvet	F. velutina	ash has dwarf, variegated, and weeping cultivars.
Ash, white	Fraxinus americana	
Bald cypress	Taxodium distichum	Deciduous conifer. Best near a pond or stream in low, wet areas. Lower Northeast through Southeast and South Central U.S.
Beech, American	Fagus grandifolia	Majestic, very large shade trees. European Beech has many
Beech, European	F. sylvatica	cultivars; some have coppery, purple, or variegated foliage, some weeping branches. Good cold hardiness. American Beech: small, edible nuts.
Birch, canoe or white	Betula papyrifera	Graceful, short-lived trees for North, pendulous branches.
Birch, European	B. pendula	Canoe and European birch has notable white bark. Sweet
Birch, sweet	B. lenta	birch yields oil of wintergreen.
Bradford pear	Pyrus calleryana	Noted for heavy covering of white flowers. Good lawn and street tree. Moderate cold hardiness.
Buckeye, Ohio	Aesculus glabra	Small to medium trees. Yellow buckeye has nice flowers, Ohio
Buckeye, yellow	A. octandra	one orange fall leaves. Very good cold hardiness.
Cape chestnut	Calodendrum capense	Profusion of flowers, short deciduous period, grown in California and Florida. Tolerates some dryness.
Cherry, bird	Prunus padus	Notable for spring flowers, interesting bark and fall color.
Cherry, Japanese	P. serrulata	Japanese cherry has moderate cold hardiness, others more
Cherry, Sargent	P. sargentii	hardy.
Coral tree	Erythrina coralloides	Brilliant, showy flowers. Small tree. Best in California and Arizona.
Cork tree	Phellodendron amurense	Medium shade tree, great form and bark. Shiny summer leaves, yellow in fall. Easy culture.
Crab apple, showy	M. floribunda	Notable for spring flowers. Siberian crab apple very cold hardy,
Crab apple, Siberian	Malus baccata	showy one moderate. Small trees.
Crape myrtle	Lagerstroemia indica	Showy flowers. Small tree. Best in southern half of U.S. Interesting bark; many cultivars.
Dawn redwood	Metasequoia glyptostroboides	Ancient lineage, deciduous pyramidal conifer. Large specimen tree. Good cold hardiness.
Dogwood, flowering	Cornus florida	Showy spring flowers. Good cold hardiness. Small trees.
Dogwood, Japanese	C. kousa	Attractive red fruits and red fall foliage.
Elm, American	Ulmus americana	Problems with Dutch elm disease. Few cultivars with some
Elm, Chinese	U. parvifolia	resistance. Large shade tree. American elm has vase shape and
Elm, Holland	U. × hollandica	very good cold hardiness.
European mountain ash	Sorbus aucuparia	Very cold hardy. Notable for red fruit clusters. Red foliage in fall. Easy culture, tolerates dry soils.
Fringe tree, Chinese	Chionanthus retusus	Notable for profuse spring flowers, small shrubby tree. Best
Fringe tree, common	C. virginicus	with hot summers, moderate winters.
Golden trumpet tree	Tabebuia chrysotricha	Colorful flowers. Best in Florida and California. Likes heat. Medium-size tree.
Hackberry, sugar	Celtis laevigata	Good, large shade tree. Yellow fall leaves. Good street and park tree in South. Birds like fruit.
Hawthorn, cockspur	Crataegus crus-galli	Small trees with flowers and red fruits. Bad thorns. Attractive
Hawthorn, English	C. laevigata	red fall foliage. Easy culture. Few thornless cultivars. Moderate
Hawthorn, single seed	C. monogyna	cold hardiness.

TABLE 16–18 *Some Popular Deciduous Trees* * (continued)

Common Name	Scientific Name[†]	Comments
Hornbeam, American Hornbeam, European	*Carpinus caroliniana* *C. betulus*	Difficult to transplant. Medium-size tree. Good cold hardiness. Thrive in moist soils. Red fall leaves for American hornbeam.
Jacaranda	*Jacaranda mimosifolia*	Showy flowers. Deciduous in spring. Best for southern California and Florida. Medium-size tree.
Japanese pagoda tree	*Sophora japonica*	Medium, nice late flowering tree. Foliage dark green in summer. Easy culture. Interesting fruit. Good cold hardiness.
Japanese Stewartia	*Stewartia pseudocamellia*	Medium, nice flowering bushy tree. Foliage dark green in summer and orange/scarlet in fall. Nice bark. Good cold hardiness.
Katsura tree	*Cercidiphyllum japonicum*	Large, nice specimen tree. Pest resistant. Best in rich, moist soil.
Kentucky coffee tree	*Gymnocladus dioica*	Large legume tree, does well in Midwest.
Korean evodia	*Evodia danielli*	Small, flowering specimen tree. Easy culture. Good cold hardiness.
Larch, European Larch, Japanese	*Larix decidua* *L. kaempferi*	Deciduous conifer. Good yellow fall color. Very good cold hardiness. Large tree.
Linden, American Linden, Crimean Linden, littleleaf Linden, silver	*Tilia americana* *T.* × *euchlora* *T. cordata* *T. tomentosa*	Easy culture, fares poorly in drought. Very cold hardy. Large shade tree. Good bee plant. Littleleaf linden is widely planted as street and lawn tree. Crimean favored in Midwest. Silver one has silvery-white leaf undersides.
Locust, black	*Robinia pseudoacacia*	Easy culture, large legume tree. Fragrant flowers. Bee plant. Very cold hardy. Good in poor soils.
Magnolia, cucumber tree Magnolia Magnolia, saucer Magnolia, star	*Magnolia acuminata* *M. campbellii* *M.* × *soulangiana* *M. stellata*	Beautiful flowers, colorful fruits. Small to large trees. Fertile, well-drained soil, good soil moisture needed. Transplant poorly. Many cultivars. Some magnolias are also evergreen (see Table 16–19). Moderate cold hardiness.
Maidenhair tree	*Ginkgo biloba*	Large street and lawn shade tree. Good cold hardiness. Avoid female tree, fruit stinks. Interesting leaf. Good yellow fall foliage.
Maple, amur Maple, box elder Maple, Japanese Maple, Norway Maple, Oregon Maple, paperbark Maple, red Maple, Rocky Mountain Maple, southern sugar Maple, sugar Maple, sycamore	*Acer ginnala* *A. negundo* *A. palmatum* *A. platanoides* *A. macrophyllum* *A. griseum* *A. rubrum* *A. glabrum* *A. barbatum* *A. saccharum* *A. pseudoplatanus*	Widely used street, lawn, park, and backyard shade or specimen trees. Small, medium, to large trees, good fall colors (yellows, oranges, and reds). Easy culture. Some have very good cold hardiness, others moderate. Variable shapes. Smaller ones can look shrubby. Numerous cultivars. Red maple good red fall color, sugar maple yields maple syrup, Japanese maple has finely divided, attractive all-season red to purple red leaves. Shallow-rooted, some prone to storm damage. Wide range of cold hardiness.
Oak, black Oak, bur Oak, California black Oak, English Oak, pin Oak, red Oak, sawtooth Oak, scarlet Oak, white	*Quercus velutina* *Q. macrocarpa* *Q. kelloggii* *Q. robur* *Q. palusris* *Q. rubra* *Q. acutissima* *Q. coccinea* *Q. alba*	Medium to large, majestic shade trees. Very good to good cold hardiness. Some are evergreen in South (see Table 16–19). Good street, lawn, yard, and park trees. Often have good fall red color. Numerous species and cultivars. Oaks with taproots transplant poorly. Acorns add interest. Long-lived trees. Good timber. Most thrive in rich, moist soil.
Orchid tree	*Bauhinia variegata*	Very showy flowers. Small to medium tree. Best in southern California, Florida, and Hawaii.
Pepperidge	*Nyssa sylvatica*	Large tree, glossy summer and orange/scarlet fall leaves, birds like fruit. Difficult to transplant. Very good cold hardiness.

(continued)

TABLE 16–18 *Some Popular Deciduous Trees** *(continued)*

Common Name	Scientific Name[†]	Comments
Persian silk tree	*Albizia julibrissin*	Notable powdery puff flowers. Medium tree. Withstands dryness and alkalinity. Best in California and Southwest. Nice umbrella shape.
Plane tree, American Plane tree, California Plane tree, London	*Platanus occidentalis* *P. racemosa* *P. × acerifolia*	Large street and big yard tree. Best in rich, moist soil. Interesting bark. London plane tree good city tree. Good cold hardiness, except for California one.
Poplar, Eastern Poplar, Freemont Poplar, Lombardy Poplar, white	*Populus deltoides* *P. fremontii* *P. nigra* *P. alba*	Easy cultivation. Fast growth. Short lived. Roots can damage sidewalks, drains. Large trees. Very good to good cold tolerance. Freemont good in dry, alkaline soils of Southwest. Lombardy has columnar shape, but prone to canker. White poplar best of large tree ones. Eastern poplar good in Midwest.
Redbud, Eastern Redbud, Chinese	*Cercis canadensis* *C. chinensis*	Notable spring flowers before leaves. Medium tree. Chinese one more showy, but less cold hardiness.
Sassafras	*Sassafras albidum*	Easy culture. Good cold hardiness. Medium to large tree. Tolerates poor soils. Good orange/red fall color. Bluish berries. Difficult to transplant.
Serviceberry, downy Serviceberry, shadblow	*Amelanchier arborea* *A. canadensis*	Showy, early spring flowers, often before leaves. Edible fruit. Good cold hardiness. Small, shrubby to medium tree.
Shagbark hickory	*Carya ovata*	Large tree, interesting bark. Edible nuts. Transplants poorly. Good cold hardiness.
Snowbell, fragrant Snowbell, Japanese	*Styrax obassia* *S. japonicus*	Small, nice flowering specimen tree. Good cold hardiness.
Sourwood	*Oxydendrum arboreum*	Medium specimen tree. Bees like it. Glossy summer leaves. Scarlet fall foliage. Good cold hardiness.
Sweet gum	*Liquidambar styraciflua*	Large, shade tree. Needs room. Good cold hardiness. Star-shaped leaves. Scarlet fall foliage. Difficult to transplant. Interesting fruit. Good street, lawn, yard, and park tree. Widespread use.
Tulip tree	*Liriodendron tulipifera*	Quite large, popular, long-lived handsome tree. Tulip-like flowers, nice yellow fall leaves.
Varnish tree	*Koelreuteria paniculata*	Nice fragrant, bee flowers. Small tree. Good cold hardiness. Easy culture. Interesting pod fruits.
Wild olive	*Halesia carolina*	Small tree. Good cold hardiness. Nice flowers, bark.
Willow, florist's Willow, pussy Willow, Wisconsin weeping Willow, weeping Willow, white	*Salix caprea* *S. discolor* *S. × blanda* *S. babylonica* *S. alba*	Numerous species and cultivars. Good near water. Rapid growers. Graceful, weeping cultivars. Some can be messy and damage water/drain/septic pipes. Weak wood. Small to medium trees. Hold wet banks in place. Good to moderate cold hardiness.
Witch hazel	*Hamamelis virginiana*	Small, shrubby tree. Moist soils. Good cold hardiness. Fall flowers. Yellow autumn color. Tolerates shade.
Yellowwood, American	*Cladrastis lutea*	Medium legume tree. Good cold hardiness. Fragrant flowers. Good yellow fall foliage. Easy culture.

*While most are deciduous, a few trees can be semi- to totally evergreen given very southerly locations or unusually mild winters, for example, *Fraxinus uhdei* and *Calodendrum capense* in Southern California.

[†]Many are actually grown as cultivars or hybrids of the indicated species.

TABLE 16–19 *Some Popular Broad-Leaved, Needled, and Scaly Leaved Evergreen Trees*

Common Name	Scientific Name*	Comments
Arborvitae, American	*Thuja occidentalis*	Compact growth. Easy culture. American one very cold hardy, while Oriental one best in South and in areas with hot, dry summers. Many dwarf cultivars. Medium to large. Does not age well in terms of form.
Arborvitae, giant	*Thuja plicata*	
Arborvitae, Oriental	*Platycladus* (*Thuja*) *orientalis*	
California incense cedar	*Calocedrus decurrens*	Columnar growth. Scale-like leaves. Large tree. Few pests. Good cold hardiness. Dwarf cultivar available. Avoid areas with hot, dry summers.
California nutmeg	*Torreya californica*	Moderate cold hardiness. Pyramidal to rounded medium tree. Yew-like leaves and yew berry-like covering.
Cedar, atlas	*Cedrus atlantica*	Excellent specimen, large needled trees. Cultivars with blue, dark green, yellow, and silvery foliage, pendulous and dwarf ones, too. Moderate cold hardiness.
Cedar, deodar	*C. deodara*	
Cedar, Lebanon	*C. libani*	
China fir	*Cunninghamia lanceolata*	Medium, needled tree, pyramidal, slightly pendulous. Moderate cold hardiness. Good in warmer, moister parts of U.S. Lawn tree.
Cypress, Hinocki	*Chamaecyparis obtusa*	Scale-like leaves. Good cold hardiness. Numerous cultivars, dwarf and foliage colors. Lawson best in mild areas with much moisture.
Cypress, Lawson	*C. lawsoniana*	
Cypress, Nootka	*C. nootkatensis*	
Cypress, Sawara	*C. pisifera*	
Cypress, Arizona	*Cupressus arizonica*	Scale-like leaves. Arizona and Modoc, good cold hardiness; rest best in Southwest. Medium to large trees. Many foliage colors, varying shape and size cultivars.
Cypress, Italian	*C. sempervirens*	
Cypress, Modoc	*C. bakeri*	
Cypress, Monterey	*C. macrocarpa*	
Cypress, Portuguese	*C. lusitanica*	
Douglas fir	*Pseudotsuga menziesii*	Large needled tree. Moderate cold hardiness, except for var. Glauca. Some nice foliage; weeping and dwarf cultivars.
Eucalyptus	*Eucalyptus*	Fast growing, broad-leaved medium to large tree. Shallow-rooted, few pests. Bee plants. Numerous species and cultivars. Best grown in California.
Fir, balsam	*Abies balsamea*	Medium to large pyramidal, needled trees. Not for areas with hot, dry summers. Good to very good cold hardiness. Nice foliage; weeping and dwarf cultivars. Few pest problems.
Fir, Himalayan	*A. spectabilis*	
Fir, Nikko	*A. homolepis*	
Fir, noble	*A. procera*	
Fir, Spanish	*A. pinsapo*	
Fir, Veitch	*A. concolor*	
Fir, white	*A. veitchii*	
Giant sequoia	*Sequoiadendron giganteum*	Gigantic, scale-like, needled, long-lived tree. Dwarf cultivar exists. Moderate cold hardiness. Needs lots of rain and rich soil.
Hemlock, Canada	*Tsuga canadensis*	Large, pyramidal needled trees, graceful look. Good cold hardiness. All can make nice hedge. Numerous interesting cultivars. Not good in areas with hot, dry summers.
Hemlock, Carolina	*T. caroliniana*	
Hemlock, Japanese	*T. diversifolia*	
Hemlock, Siebold	*T. sieboldii*	
Holly, Altaclara	*Ilex* × *altaclarensis*	Glossy broad-leaved medium tree with great red berries (need male and female). Bird food, Christmas decoration. Numerous cultivars, spineless ones better. Moderate cold hardiness, American furthest north range.
Holly, American	*I. opaca*	
Holly, English	*I. aquifolium*	
Japanese cedar	*Cryptomeria japonica*	Large, pyramidal needled specimen tree. Moderate cold hardiness, some foliage browning. Several nice cultivars.
Juniper, Chinese	*Juniperus chinensis*	Medium, scaly-needled tree, berry type (need male and female). Good to very good cold hardiness. Numerous, interesting cultivars, some shrub-like (see also Table 16–21). Widely planted.
Juniper, common	*J. communis*	
Juniper (Eastern red cedar)	*J. virginiana*	
Juniper, Rocky Mountain	*J. scopulorum*	

(continued)

TABLE 16–19 *Some Popular Broad-Leaved, Needled, and Scaly Leaved Evergreen Trees* (continued)

Common Name	Scientific Name*	Comments
Oak, California live	*Quercus agrifolia*	Medium to large, long-lived, broad-leaved trees of moderate to
Oak, canyon live	*Q. chrysolepis*	low cold hardiness. Favored in South and southern California.
Oak, cork	*Q. suber*	Cork oak is cork source.
Oak, holly	*Q. ilex*	
Oak, laurel	*Q. laurifolia*	
Oak, live	*Q. virginiana*	
Magnolia, Southern	*Magnolia grandiflora*	Medium to large, long-lived, broad-leaved trees of moderate to
Magnolia, sweet bay	*M. virginiana*	low cold hardiness. Notable flowers. Sweet bay grows furthest north, but becomes deciduous.
Palms	*Acoelorrhaphe* sp.	Most are medium size and tree-like. Few are very large or
	Brahea sp.	shrubby. Palms are restricted to Arizona, California, Florida,
	Coccothrinax sp.	Hawaii, and Texas. Majority grown in California, Florida, and
	Phoenix sp.	Hawaii. The genera listed here are only a few of those from
	Pseudophoenix sp.	which species and cultivars are grown horticulturally. Some
	Rhapidophyllum sp.	produce dates or coconuts.
	Rhaphis sp.	
	Roystonea sp.	
	Sabal sp.	
	Serenoa sp.	
	Thrinax sp.	
	Washingtonia sp.	
Pine, Austrian	*Pinus nigra*	Small, medium, and large needled trees widely planted and
Pine, bristlecone	*P. aristata*	highly popular. Cold hardiness varies from very good (Eastern
Pine, Eastern white	*P. strobus*	white, red, Scotch, Swiss mountain) through limited (Italian
Pine, Himalayan	*P. wallichiana*	stone, Monterey, shore). Numerous cultivars in terms of size,
Pine, Italian stone	*P. pinea*	form, foliage color. Japanese black excellent near seashore. Easy
Pine, Japanese black	*P. thunbergiana*	to moderate culture. Bristlecone pines can live for 3,000 to
Pine, Japanese red	*P. densiflora*	4,000 years! Many uses in landscape.
Pine, Japanese white	*P. parviflora*	
Pine, Korean	*P. koraiensis*	
Pine, lace-bark	*P. bungeana*	
Pine, Monterey	*P. radiata*	
Pine, Swiss mountain	*P. mugo*	
Pine, Norway	*P. resinosa*	
Pine, Scotch	*P. sylvestris*	
Pine, shore	*P. contorta*	
Pine, Swiss stone	*P. cembra*	
Pine, Western white	*P. monticola*	
Pine, Western yellow	*P. ponderosa*	
Pittosporum, narrow-leaved	*Pittosporum phillyraeoides*	Small to medium tree. Mostly in South and Pacific Coast. Nice
Pittosporum, Queensland	*P. rhombifolium*	foliage, flowers, and fruit.
Pittosporum, Victorian box	*P. undulatum*	
Redwood	*Sequoia sempervirens*	Gigantic, long-lived needled tree. Dwarf cultivar exists. Limited cold hardiness. Needs lots of rain and rich soil. Pacific Coast best.
Southern yew	*Podocarpus macrophyllus*	Small to medium, shrubby to tree-like needled specimen with yew-like berries. Moderate cold hardiness, mostly in the South.
Spruce, Colorado blue	*Picea pungens*	Large, sharp-needled trees with very good cold hardiness.
Spruce, Engelmann	*P. engelmannii*	Numerous cultivars in terms of foliage color, form, and size.
Spruce, Norway	*P. abies*	Best as specimen plant with lots of room.
Spruce, Oriental	*P. orientalis*	
Spruce, Serbian	*P. omorika*	
Spruce, white	*P. glauca*	

TABLE 16–19 *Some Popular Broad-Leaved, Needled, and Scaly Leaved Evergreen Trees* (continued)

Common Name	Scientific Name*	Comments
Umbrella pine	*Sciadopitys verticillata*	Large, needled, pyramidal specimen tree. Few pest problems. Interesting bark. Moderate cold hardiness.
Yew, English Yew, intermediate Yew, Japanese	*Taxus baccata* *T. × media* *T. cuspidata*	Medium, needled with red, berry-like arils (neeed male and female). Widely planted. Respond well to pruning. Poisonous plant. Good cold hardiness, English has least. Numerous interesting cultivars, including dwarves.

*Many are actually grown as cultivars or hybrids of the indicated species.

TABLE 16–20 *Some Popular Deciduous Shrubs*

Common Name	Scientific Name†	Comments
American elder	*Sambucus canadensis*	Flowering shrub, edible fruit. Good cold hardiness. Several cultivars including yellow leaves. Best near wet areas and naturalizing.
Beautybush	*Kolkwitzia amabilis*	Medium flowering shrub, good cold hardiness. Easy culture.
Black jetbead	*Rhodotypos scandens*	Small tough, flowering shrub with persistent fruits. Good cold hardiness.
Broom, Scotch Broom, Warminster	*Cytisus scoparius* *C. × praecox*	Notable medium legume shrub with nice flowers, small scaly leaves. Easy culture. Tolerates dryness and sandy soil. Green twigs in winter. Good cold hardiness. Few cultivars.
Buckeye, bottlebrush Buckeye, red	*Aesculus parviflora* *A. pavia*	Nice flowers, bottlebrush tolerates shade. Good cold hardiness.
Bush cinquefoil	*Potentilla fruticosa*	Small, flowering shrub with very good cold hardiness. Several cultivars. Few pest, disease problems. Flowers in summer when other shrubs finished.
Butterfly bush	*Buddleia davidii*	Medium, late summer flowering shrub, good cold hardiness, but often dies back in North. Butterflies love it.
Carolina allspice	*Calycanthus floridus*	Medium shrub with flowers mostly noted for fragrance. OK in shade, easy culture. Very good cold hardiness.
Cherry prinsepia	*Prinsepia sinensis*	Medium flowering shrub with red fruit that birds love. Spiny. Good hedge. Good cold hardiness. Few pest problems.
Clethra, summersweet	*Clethra alnifolia*	Summer fragrant-flowering medium shrub, easy culture. OK near seashore. Very good cold hardiness.
Cotoneaster	*Cotoneaster multiflorus*	Medium graceful shrub noted for both flowers and fruits. Good cold hardiness.
Cutleaf stephanandra	*Stephanandra incisa*	Small to medium shrub noted mostly for interesting leaves and red fall color. Easy culture.
Deutzia	*Deutzia gracilis*	Small graceful, flowering shrub, very good cold hardiness. Good in masses or as hedge.
Dogwood, bloodtwig Dogwood, gray Dogwood, Redosier Dogwood, silky Dogwood, Tartarian	*Cornus sanguinea* *C. racemosa* *C. sericea* *C. amomum* *C. alba*	Small to medium flowering shrubs with fruits. Fall foliage usually attractive. Redosier and bloodtwig have red twigs in winter. Redosier also good in wet soils. Very good to good cold hardiness. Many cultivars.
Double-flowered pomegranate	*Punica granatum*	Cultivars of pomegranate often used in South as a notable flowering shrub. Limited cold hardiness.

(continued)

TABLE 16–20   *Some Popular Deciduous Shrubs* *   (continued)

Common Name	Scientific Name[†]	Comments
Euonymus, winged	*Euonymus alata*	Medium shrub with interesting winter twigs, branches. Very cold hardy. Nice red fall foliage. Prunes well, hedge use or shrub border. Compact cultivar available.
Fiveleaf aralia	*Acanthopanax sieboldianus*	Thorny, medium shrub, OK in shade. Very good cold hardiness. Best used as a hedge.
Flowering plum	*Prunus triloba*	Large shrub noted mostly for masses of flowers. Good cold hardiness.
Flowering quince	*Chaenomeles speciosa*	Medium shrub, spiny branches, nice flowers. Very good cold hardiness. Good in hedges. Fruits OK for preserves. Numerous cultivars.
Forsythia, border Forsythia, weeping	*Forsythia × intermedia* *F. suspensa*	Easy culture large, fast-growing shrub. Flowers nice, appear before leaves. Popular. Flowers can be forced. Very good cold hardiness. Several cultivars.
Fothergilla, dwarf Fothergilla, large	*Fothergilla gardenii* *F. major*	Small to medium flowering shrub, nice fall color. Good cold hardiness.
Hardy orange	*Poncirus trifoliata*	Large tree-like, spiny shrub noted for fragrant flowers and bitter fruits. Limited cold hardiness. Good hedge.
Harry Lauder's walking stick	*Corylus avellana* 'Contorta'	Hazelnut large shrub noted for its twisted, contorted branches in winter. Good specimen. Very good cold hardiness.
Hibiscus, Chinese Hibiscus (Rose-of-Sharon)	*Hibiscus rosa-sinensis* *H. syriacus*	Late summer flowering large tree-like shrub. Chinese one limited to subtropical U.S., other good cold hardiness. Many cultivars.
Holly, finetooth Holly (possum haw) Holly (smooth winterberry) Holly (winterberry)	*Ilex serrata* *I. deciduas* *I. laevigata* *I. verticillata*	Large, flowering shrubs with glossy leaves, red berries. Birds love them. Good to very good cold hardiness. Many cultivars. Need both sexes for fruits.
Honeysuckle, Tartarian	*Lonicera tatarica*	Large shrub, fragrant flowers, red fruit. Very good cold hardiness. Birds love fruit. Several cultivars.
Hydrangea, French Hydrangea, hills of snow Hydrangea, oakleaf	*Hydrangea macrophylla* *H. arborescens* *H. quercifolia*	Notable flowers. Small to medium shrub with good cold hardiness. Can die back in North, but regrows and flowers fast. Can be evergreen in deeper South. Likes moisture, can tolerate some shade.
Japanese barberry	*Berberis thunbergii*	Thorny medium shrub notable for flowers, red fruits, and red fall leaves. Easy culture. Grows in poorest soils. Good cold hardiness. Makes good hedge. Several cultivars.
Japanese beautyberry	*Callicarpa japonica*	Small shrub mainly noted for its persistent purple fall fruit. Moderate cold hardiness, but in North can die back, resprout, and produce fruit.
Japanese kerria	*Kerria*	Medium flowering shrub, good cold hardiness. Nice winter stem color. Few cultivars. OK with some shade.
Lilac, Chinese Lilac, common Lilac, Meyer's Lilac, summer	*Syringa × chinensis* *L. vulgaris* *L. mayeri* *L. villosa*	Very popular, fragrant flowering small to large shrub. Good to very good cold hardiness. Numerous cultivars. Summer one flowers later.
Mockorange, lemoine Mockorange, sweet	*Philadelphus x lemoinei* *P. coronarius*	Large shrub noted for fragrant flowers. Easy culture. Few pests. OK in dry soil. Good cold hardiness. Several cultivars.
Oriental photina	*Photina villosa*	Large flowering shrub with red fruit. Good fall color. Can get fire blight disease. Good cold hardiness.

TABLE 16–20   *Some Popular Deciduous Shrubs**   *(continued)*

Common Name	Scientific Name[†]	Comments
Rugosa rose	*Rosa rugosa*	Very rugged, medium shrubby, prickly plant. OK in sandy soil, near coast. Nice flowers and reddish orange fruit for jellies and tea. Very good cold hardiness. Easy culture. Pest free.
Russian olive	*Elaeagnus angustifolia*	Large, tree-like shrub noted mostly for attractive gray foliage. Can be spiny. Very good cold hardiness. OK near coast.
Seabuckthorn	*Hippophae rhamnoides*	Thorny, large shrub with grayish leaves and orange fruits that persist. OK near coast. Need both sexes for fruits. Good hedge. Very good cold hardiness.
Shrubby St. John's Wort	*Hypericum prolificum*	Small shrub, nice flowers, good winter twig color. Good cold hardiness.
Siberian peashrub	*Caragana arborescens*	Large tough, leguminous shrub, nice flowers, very good cold hardiness. Good for hedge, screening.
Silver buffaloberry	*Shephardia argentea*	Large, thorny, flowering, and fruiting shrub. Fruit used for jelly. Very good cold hardiness. Easy culture. OK in poor soils.
Smokebush	*Cotinus coggygria*	Early summer flowering large shrub with great plumose blooms (need female). Good fall color. Good cold hardiness.
Snowberry	*Symphoricarpos albus*	Medium shrub with white fruits. Easy culture. Very good cold hardiness.
Spirea, bridalwreath Spirea, Bumald Spirea, crispleaf Spirea, Japanese Spirea, Thunberg Spirea, Vanhoutte	*Spiraea prunifolia* S. × *bumalda* S. *bullata* S. *japonica* S. *thunbergii* S. × *vanhouttei*	Very small to medium shrub with masses of flowers. Some have arching habit. Good cold hardiness. Easy culture. Few pests. Several cultivars for variegation and textured foliage.
Star magnolia	*Magnolia stellata*	Large shrub with fragrant flowers before leaves. Good cold hardiness. Popular specimen plant. Flowers sometimes injured by frost in North.
Tamarisk, five stamen Tamarisk, small-flowered	*Tamarix ramosissima* T. *parvifolia*	Large flowering shrub with scaly leaves, overall feathery look. Very good to good cold hardiness. Easy culture. OK near coast.
Weiglia	*Weiglia florida*	Medium flowering shrub. Easy culture. Several cultivars. Good cold hardiness.
Witch hazel	*Hamamelis vernalis*	Medium shrub noted for early spring flowers before leaves. OK in moist soil. Tolerates some shade.
Viburnum, American cranberrybush Viburnum, Burkwood Viburnum, Chinese snowball Viburnum, European cranberrybush Viburnum, fragrant Viburnum, fragrant snowball Viburnum, Japanese snowball Viburnum, Judd Viburnum, Linden Viburnum, Withe-Rod	*Viburnum trilobum* V. × *burkwoodii* V. *macrocephalum* V. *opulus* V. *carlesii* V. × *carlcephalum* V. *plicatum* V. × *juddii* V. *dilatatum* V. *cassinoides*	Popular medium to large shrub noted for flowers (some very fragrant) and fruits (black, blue, red, yellow) and fall color. Easy culture. Birds love fruit, some edible for jellies. Few pests. Cold hardiness very good to good. See evergreen ones in Table 16–21.

*Unless indicated otherwise, spring flowering.

[†]Many are actually grown as cultivars or hybrids of the indicated species.

TABLE 16–21   *Some Popular Broad-Leaved and Needled Evergreen Shrubs*

Common Name	Scientific Name*	Comments
Abelia	Abelia × grandiflora	Glossy, leafy foliage turning purplish-bronze in fall. Good cold hardiness, becomes deciduous or dies back in colder areas. Nice flowers. Good in masses.
Andromeda, Japanese Andromeda, mountain	Pieris japonica P. floribunda	Medium broad-leaved flowering shrub, nice foliage. Good cold hardiness. Good foundation shrub.
Azalea (see Rhododendron)		
Barberry, Chenault Barberry, warty Barberry, wintergreen	Berberis × chenaultii B. verruculosa B. julianae	Thorny, small-leaved medium shrub notable for flowers, colorful fruits. Easy culture. Grows in poorest soils. Good cold hardiness. Makes good hedge. Several cultivars.
Bayberry, California Bayberry, northern	Myrica californica M. pennsylvanica	Large to very large, broad-leaved shrub, purple fruit (need male and female). Very good (Northern, semievergreen further north) to moderate cold hardiness. Bayberry (aromatic) candles made from Northern one. Easy culture. Good in sandy soils, near coast.
Bog rosemary	Andromeda polifolia	Low, broad-leaved flowering shrub for rock garden with very good cold hardiness.
Boxwood, common Boxwood, littleleaf	Buxus sempervirens B. microphylla	Glossy, leafy foliage. Small to large shrub. Prunes well. Good for hedge. Good cold hardiness (common boxwood best).
Camellia	Camellia japonica	Broad-leaved, nice late-summer onward flowering large shrub. Limited cold hardiness. Best in Southeast and Pacific Coast. OK in partial shade. Many cultivars.
Cotoneaster, brightbead Cotoneaster, rock spray Cotoneaster, small-leaved	Cotoneaster glaucophyllus C. horizontalis C. microphylla	Small to large broad-leaved shrubs, flowers, nice fruit and form. Limited cold hardiness. Best in Southeast and Pacific Coast.
Euonymus, evergreen	Euonymus japonica	Large, broad-leaved tree-like shrub, lustrous leaves. Limited cold hardiness, best in South. Some variegated cultivars.
Firethorn	Pyracantha coccinea	Medium broad-leaved shrub noted for red berries in fall. Good cold hardiness, becomes deciduous in lower North. Easily espaliered. Good hedge. Several cultivars. Fire blight problem.
Gardenia	Gardenia jasminoides	Medium broad-leaved shrub with very fragrant flowers. Limited cold hardiness. In North, greenhouse plant.
Holly, inkberry Holly, Chinese Holly, Japanese Holly, Meserve Holly, Yaupon	Ilex glabra I. cornuta I. crenata I. × meserveae I. vomitoria	Glossy broad-leaved medium to large shrub with colorful berries. Bird food. Numerous cultivars, spineless ones better. Good to moderate cold hardiness (inkberry furthest north). Need both sexes for fruits.
Honeysuckle, box	Lonicera nitida	Medium broad-leaved shrub with fragrant flowers and purple fruit. Moderate cold hardiness. Several cultivars.
Indian hawthorn	Rhaphiolepsis indica	Small, broad-leaved flowering, fruiting (purple/black) foundation shrub. Limited cold hardiness (southern California).
Japanese Aucuba	Aucuba japonica	Large, glossy broad-leaved flowering and red fruited (need male and female) shrub. Several cultivars, including variegated. Moderate cold hardiness.
Juniper, creeping Juniper, Japanese garden Juniper, savin Juniper, singleseed	Juniperus horizontalis J. chinensis procumbens J. sabina J. squamata	Small to medium scaly-needled shrub, berry type (need male and female). Good to very good cold hardiness. Numerous, interesting cultivars. OK near coast.
Lantana	Lantana camara	Small, broad-leaved flowering, fruiting shrub. Moderate cold hardiness. Good in South.
Laurel cherry	Prunus laurocerasus	Large, tree-like broad-leaved shrub noted for flowers, dark purple fruit. Moderate cold hardiness. Several cultivars. Good in South.

TABLE 16–21   *Some Popular Broad-Leaved and Needled Evergreen Shrubs*   (continued)

Common Name	Scientific Name*	Comments
Leucothoe, drooping	*Leucothoe fontanesiana*	Small, broad-leaved fragrant flowering shrub, graceful form. Bronze leaves in fall, semievergreen in upper North. Very good cold hardiness.
Mahonia, leatherleaf Mahonia, Oregon holly-grape	*Mahonia bealei* *M. aquifolium*	Medium to large broad-leaved, flowering and fruiting (blue/black) shrub. Good cold hardiness. Few cultivars. Bronze/purplish winter foliage. Foundation plant, masses.
Malpighia holly	*Malpighia*	Small, broad-leaved, flowering and fruiting holly-like plant. Limited cold hardiness. Useful foundation plant.
Mountain laurel	*Kalmia latifolia*	Large broad-leaved flowering shrub, requires acid soil. Very good cold hardiness. Several cultivars.
Oleander	*Nerium oleander*	Large, broad-leaved shrub, nice flowers. Easy culture, stands dryness. Some sweet-scented cultivars. Tub plant in North.
Photina, Japanese	*Photina glabra*	Large broad-leaved, flowering shrub with red fruit. Moderate cold hardiness, best in Southeast. Useful as hedge.
Pink powder puff bush	*Calliandra*	Large, broad-leaved, tree-like shrub with unusual puffy flowers. Limited cold hardiness (southern California and Florida).
Pine, mugo	*Pinus mugo* var. *mugo*	Small, needled shrubby pine widely used. Good foundation plant. Very good cold hardiness.
Pittosporum, Japanese	*Pittosporum tobira*	Large tree-like shrub. Limited cold hardiness. Good in seaside plantings. Mostly in South and Pacific Coast.
Privet, California Privet, Japanese	*Ligustrum ovalifolium* *L. japonicum*	Large, broad-leaved shrub often used as hedge. Good to moderate (Japanese) cold hardiness. Easy culture. Variegated cultivar.

Rhododendron (and Azalea)
Rhododendron hybrid groups:
  Catawba, Caucasicum, Fortunei, Dexter, Griffithianum, Javanicum, Thomsonii.
Azalea hybrid groups:
  Gable, Ghent, Glen Dale, Indian, Kaempferi, Knap Hill, Kurume, Molle.
	Rhododendron cv.	Small, medium, and large broad-leaved flowering shrubs of great popularity throughout U.S. Most are hybrid cultivars, although a few species are used. Many make good foundation plants, others nice specimen. Includes both rhododendrons and azaleas. Some azaleas are deciduous, especially further North. Cold hardiness varies from very good to limited, depending on the group.

Rockrose	*Cistus incanus*	Small, broad-leaved flowering shrub, nice flowers, fire resistant. Easy care. Limited cold hardiness. In California and South.
Tea tree	*Leptospermum scoparium*	Large, tree-like broad-leaved shrub, nice mass of flowers. Limited cold hardiness (California, Oregon coast, and Florida). Few cultivars.
Viburnum, David Viburnum, Henry Viburnum, Japanese Viburnum, Laurustinus Viburnum, sweet	*Viburnum davidii* *V. henryi* *V. japonicum* *V. tinus* *V. odoratissimum*	Popular small, medium to large broad-leaved shrub noted for flowers (some fragrant) and fruits (black, blue, red yellow). Easy culture. Birds love fruit, some edible for jellies. Few pests. Cold hardiness moderate to limited.
Yellowbells	*Tecoma stans*	Medium to large broad-leaved fall flowering shrub. Limited cold hardiness (California, Florida).
Yew, Canadian	*Taxus canadensis*	Medium, needled shrub with red, berry-like arils (need male and female). Widely planted. Respond well to pruning. Poisonous plant. Very good cold hardiness. Takes some shade. Few cultivars.

*Many are actually grown as cultivars or hybrids of the indicated species.

TABLE 16–22   *Some Popular Flowering Vines and Lianas*

Common Name	Scientific Name*
Very Good to Good Cold Hardiness[†]	
American bittersweet	*Celastrus scandens*
Boston ivy	*Parthenocissus tricuspidata*
Bower actindia	*Actinidia arguta*
China fleece vine	*Polygonum aubertii*
Clematis, Jackman	*Clematis × jackmanii*
Climbing hydrangea	*Hydrangea anomola*
Dutchman's pipe	*Aristolochia durior*
English ivy	*Hedera helix*
Fiveleaf akebia	*Akebia quinata*
Honeysuckle, Japanese	*Lonicera japonica*
Honeysuckle, trumpet	*L. sempervirens*
Japanese hydrangea-vine	*Schizophragma hydrangeoides*
Porcelain Ampelopsis	*Ampelopsis brevipedunculata*
Trumpet creeper	*Campsis radicans*
Virginia creeper	*Parthenocissus quinquefolia*
Wintercreeper	*Euonymus fortunei*
Wisteria, Chinese	*Wisteria sinensis*
Wisteria, Japanese	*W. floribunda*
Moderate Cold Hardiness[**]	
Algerian ivy	*Hedera canariensis*
Bengal clockvine	*Thunbergia grandiflora*
Carolina jessamine	*Gelsemium sempervirens*
Cat's claw vine	*Macfadyena unguis-cati*
Chinese trumpetcreeper	*Campsis grandiflora*
Clematis, Armand	*C. armandii*
Japanese staunton vine	*Stauntonia hexaphylla*
Jasmine, common white	*Jasminum officinale*
Limited Cold Hardiness[††]	
Allamanda	*Allamanda cathartica*
Bleeding heart glorybower	*Clerodendrum thomsoniae*
Bougainvillea	*Bougainvillea × buttiana*
Cape honeysuckle	*Tecomaria capensis*
Combretum	*Combretum grandiflorum*
Cup of gold	*Solandra guttata*
Easter lily vine	*Beaumontia grandiflora*
Guinea gold vine	*Hibbertia scandens*
Jasmine, goldcoast	*Jasminum dichotum*
Lilac vine	*Hardenbergia comptoniana*
Orchard trumpet vine	*Clytostoma callistegioides*
Passionflower, blue	*Passiflora caerulea*
Passionflower, red	*P. coccinea*
Phanera	*Bauhinia corymbosa*
Thunbergia, laurel-leaved	*Thunbergia laurifolia*
Trumpet vine, red	*Distictis buccinatoria*
Trumpet vine, royal	*D. × riversii*
Trumpet vine, vanilla	*D. lactiflora*

*Many are actually grown as cultivars or hybrids of the indicated species.
[†]Can grow pretty much anywhere in the U.S.
[**]Not reliable north of the South, southern part of the Southwest, and the Pacific Coast.
[††]Limited to Florida, Gulf Coast, southern Arizona, and California.

to show examples for each category and are not meant to be a comprehensive listing. The Relevant Resources and Internet Resources lists at the end of the chapter can lead you to more comprehensive listings. When species are given, as for maples or oaks, it should not be assumed that the selection is inclusive. Again, only examples are indicated. Choices were made on the basis of popularity, usefulness, and reliability. An effort was also made to select examples for all regions in the United States.

HOUSE AND GREENHOUSE PLANTS

Success with house or greenhouse plants (Figure 16–26) depends on the ability of the horticulturist to provide conditions similar to those found in the native habitat of the plant. Obviously, these environmental conditions are more easily managed in a greenhouse than in the home or business environment. The latter locations allow for only minimal adjustments of light levels, temperature, and humidity in view of the restrictions imposed by the

(A)
(B)
(C)
(D)
(E)
(F)

FIGURE 16–26   Examples of house and greenhouse plants. (A) Red-veined prayer plant (*Maranta leuconeura erythroneura*) thrives in low-light conditions and modest humidity levels on this end table in a room receiving indirect light. (B) Cultivars of the African violet (*Saintpaulia ionantha*) offer colorful flowers in indirect light, but require reasonable levels of humidity given their rain forest origins. (C) Orchids, except for a few exceptions, are best grown in a greenhouse. *Cattleya* species shown here is a reliable choice for the beginning orchid grower. (D) Cacti are easy to grow in sunny southern windows or in greenhouses, providing they are not overwatered or grown in high humidity conditions. (E) Ponderosa lemon (*Citrus limon* 'Ponderosa') can be grown in sunny conditions with reasonable humidity in the home or greenhouse. (F) Christmas cactus is an epiphytic cactus that produces colorful blossoms under moderate light conditions in the home near Christmas. If humidity is too low, the buds fall off before flowering. Flowering is photoperiod-dependent, but can be overridden by temperature. Exposure to 40–50°F for 7 to 10 days on the back porch can cause buds to be set. (Author photographs.)

comfort level of human occupants. Plants that succeed outside the greenhouse usually have wider tolerance for deviations from the environmental norm than do those that are only successfully grown in the greenhouse. Therefore, the key to growing house, conservatory, or greenhouse plants (Figure 16–27) lies first with the ability of the horticulturist to associate normal environmental conditions with a given plant, and then to provide these conditions as closely as possible within the limits of accepted cultural practices.

The regulation of conditions within a greenhouse or home was discussed in Chapter 13. The use of artificial lighting in these environments was also covered there. Potting soils and plant protection were discussed in Chapters 7 and 15, respectively. Great caution must be observed with the selection and application of pesticides in enclosed environments such as the home or greenhouse. Examples of popular plants used in the home and greenhouse are given in Table 16–23. These are the most common plants found in houses and greenhouses. It should be realized that many other plants found in the other tables such as bulbs, ferns, and shrubs can be cultured under greenhouse, home, or container conditions if the prevailing climate rules them out for outdoor consideration. In fact, many house plants are outdoor plants in subtropical, tropical, and arid environments.

The three most likely areas for failure with these plants are overwatering, failure to observe dormancy, and lack of insect control. Frequent overwatering results in an increasing loss of soil aeration, thus reducing root respiration to a harmful level. It also enhances environmental conditions favorable to the growth of pathogens involved in rotting of roots. Failure to recognize dormancy (often late fall through winter) leads to overwatering and excessive fertilization during a portion of the plant's cycle when these requirements are minimal. Over the long run, this excess is detrimental to the plant. Constant debilitation of plants by insect attack often increases susceptibility to secondary infections by pathogens. If left unchecked, the plant eventually dies.

(A)

(B)

(C)

FIGURE 16–27 (A) This simple house plant collection is growing in a heated sunroom on a house. Bright light with slightly cooler night temperatures because of window heat loss provides ideal growing conditions. (B) This conservatory-style restaurant in a San Francisco hotel offers elegant dining with the simple addition of some palms. Bright light is provided through the overhead glassed-in area and tougher palms are used. (C) This warm temperature greenhouse with a southern exposure and controlled humidity provides ideal conditions for numerous different plants of tropical and desert origin. (Author photographs.)

TABLE 16–23 *Some Popular House and Greenhouse Plants*

Common Name	Scientific Name*	Common Name	Scientific Name*
		Require Bright Light[†]		
Aloe	*Aloe barbadensis*	English ivy	*Hedera helix* cv.
Amaryllis	*Hippeastrum* cv.	Flowering maple	*Abutilon* × *hybridum*
Asparagus fern	*Asparagus sprengeri*	Gardenia	*Gardenia jasminoides*
Bird of paradise	*Strelitzia reginae*	Ghost plant	*Graptopetalum paraguayense*
Cacti	*Astrophytum* sp.	Inca lily alstroemeria	*Alstroemeria pelegrina*
	Cephalocereus sp.	Italian jasmine	*Jasminum humile*
	Cereus sp.	Jade tree	*Crassula argentea*
	Coryphantha sp.	Jerusalem cherry	*Solanum pseudocapsicum*
	Echinocactus sp.	Kalanchoe	*Kalanchoe blossfeldiana*
	Echinopsis sp.	Lantana	*Lanatana camara*
	Espostoa sp.	Living stones	*Lithops* sp.
	Ferocactus sp.	Oxalis	*Oxalis rubra*
	Gymnocalycium sp.	Ponderosa lemon	*Citrus limonia* 'Ponderosa'
	Mammillaria sp.	Pussy ears	*Cyanotis somaliensis*
	Notocactus sp.	Sea onion	*Urginea maritima*
Calamondin orange	× *Citrofortunella mitis*	Strawberry saxifrage	*Saxifraga stolonifera*
Chenille copperleaf	*Acalypha hispida*	Velvet plant	*Gynura aurantaica*
Cigar flower	*Cuphea ignea*	Umbrella tree	*Brassica actinophylla*
Croton	*Codiaeum variegatum*	Wax vine	*Hoya carnosa*
		Require Moderate Light[**]		
Aluminum plant	*Pilea cadierei*	Japanese aucuba	*Aucuba japonica*
Begonias	*Begonia* sp. and cv.	Kangaroo vine	*Cissus antarctica*
Bromeliads		Monkey puzzle plant	*Araucaria araucana*
Foster's favorite	*Aechmea fosteriana*	Nerve plant	*Fittonia verschaffeltii*
Earth stars	*Cyptanthus zonatus*	Norfolk Island pine	*Araucaria heterophylla*
Quill tillandsia	*Tillandsia fasciculata*	Orchids	
Burro's tail	*Sedum morganianum*		*Cattleya* sp. and cv.
Ceriman	*Monstera deliciosa*		*Cymbidium* sp. and cv.
Christmas cactus	*Schlumbergera*		*Cypripedium* sp. and cv.
Climbing onion	*Bowiea volubilis*		*Dendrobium* sp. and cv.
Drunkard's dream	*Hatiora salicornioides*		*Oncidium* sp. and cv.
Dumb cane	*Dieffenbachia amoena*		*Vanda* sp. and cv.
Euphorbia		Palms	
African milk tree	*Euphorbia trigona*	Lady palm	*Rhapis excelsa*
Baseball plant	*Euphorbia obesa*	Parlor palm	*Chamaedorea elegans*
Crown-of-thorns	*Euphorbia milii*	Sentry palm	*Howeia forsterana*
Milk bush	*Euphorbia tirucalli*	Staghorn fern	*Platycerium bifurcatum*
Poinsettia	*Euphorbia pulcherrima*	Peperomia	*Peperomia* sp. and cv.
False aralia	*Dizygotheca elegantissima*	Piggy-back plant	*Tolmiea menziesii*
Gesneriads		Pothos	*Epipremnum aureum*
African violet	*Saintpaulia ionantha*	Prayer plant	*Maranta leuconeura*
Flame violet	*Episcia cupreata*	Rubber tree	*Ficus elastica*
Gloxinia	*Sinningia speciosa*	Spider plant	*Chlorophytum comosum*
Lipstick plant	*Aeschynanthus radicans*	Swedish ivy	*Plectranthus oertendahlii*
Magic flower	*Achimenes longiflora*	Wandering Jew	*Tradescantia fluminensis*
Grape Ivy	*Cissus rhombifolia*		*Zebrina pendula*

(continued)

TABLE 16–23   *Some Popular House and Greenhouse Plants*   (continued)

Common Name	Scientific Name*	Common Name	Scientific Name*
		Require Low Light††	
Cast iron plant	*Aspidistra elatior*	Maidenhair fern	*Adiantum hispidulum*
Chinese evergreen	*Aglaonema modestum*	Mother fern	*Splenium bulbiferum*
Ferns		Rabbit's foot fern	*Davallia canariensis*
Bird's nest fern	*Asplenium nidus*	Philodendron	*Philodendron scandens*
Boston fern	*Nephrolepsis exaltata*	Snake plant	*Sanseveria trifasciata*
Holly fern	*Cyrtomium falcatum*	Spathe flower	*Spathiphyllum floribundum*

*Many are actually grown as cultivars or hybrids of the indicated species.

†Light found at an unobstructed southern exposure window or in a greenhouse.

**Light found at either an unobstructed eastern window, a sheer-curtained southern exposure window, further back in the room near an unobstructed southern window, a partially shaded southern window, or under larger plants in a greenhouse.

††Light found at an unobstructed western window or a sheer-curtained eastern exposure window or under shelves in a greenhouse.

SUMMARY

The ornamental horticultural industry consists of crop production and service industries. Crop production involves floriculture, nursery, unfinished plants and propagation materials, seeds, bulbs, Christmas trees, and turf crops. Service aspects include landscape design, landscape maintenance, and interior plantscaping.

Floriculture involves the propagation, production, and wholesale/retail distribution of cut flowers, greenery, potted foliage and flowering plants, bedding plants, and herbaceous perennials. Floricultural production involves meeting seasonal markets. Time-dependent events include cut flowers and potted flowering and foliage plants for special holidays; bedding plants, perennials, and potted flowering plants for the spring garden and outdoor summer living; bulbs and perennials for fall planting; and foliage and flowering plants for interiorscapes, especially in the fall and winter.

Roughly 50% of floriculture involves the production of bedding and garden plants in flats and pots for use in the garden and patio, deck and other above ground areas. Most bedding plants are started as plugs. Plants are sold heavily in the spring and less during late spring to early summer. Bedding plant production continues to grow and is a good introduction to horticultural business or an add-on to an existing operation. Homeowners are the largest market, but other buyers include landscape operations, large supermarket chains, garden centers, home maintenance/hardware centers, and discount chains.

Foliage production is very centralized, more so than bedding/garden plants and potted flowering plants. Production is heavily greenhouse oriented, with minimal field production. Initial capital investment is high due to greenhouse requirements. Expenses for plant materials are also high, as are production inputs both on the material and energy levels. Operations are heavily scheduled and frequent and monitoring of plant conditions and culture is critical for successful production.

Cut flowers are a high-value, specialized crop. Recently, cut flowers have appealed to horticulturists as a way to diversify operations. Demand has risen as consumer disposable income has been spent on cut flowers. Cut flower growing is highly labor-intensive and involves considerable horticultural skill and business acumen. Postharvest appearance and longevity are critical with cut flowers.

The production nursery is concerned mainly with the production and wholesale (and sometimes retail) sales of ornamental woody plants such as trees, shrubs and vines, herbaceous plants used as ground covers, and grass sod. The woody material can be deciduous or evergreen. Sales tend to be seasonal, being most heavy in the spring with light summer sales and modest fall sales. Wholesale nurseries usually specialize in large-scale growing of a few crops, with certain firms noted for specialties. Many retail nurseries no longer grow nursery stock, but are a distributor between wholesalers and consumers. Landscape nurseries usually sell retail plants and offer landscape services such as design, installation, and maintenance. Mail-order nurseries sell on the national level through catalogs and the Internet. Nursery production is a good start-up business.

Trends are toward container-grown nursery stock because of less labor, more control over soil quality, higher plant densities per area, better acceptance by consumers, quicker recovery after transplanting, and year-round marketability. Balled and burlapped stock is still used, but the labor cost and more difficult transplanting for consumers has reduced its use.

Several other crops are produced as horticultural specialty plants by specialized producers: turf grass (sod, sprigs, or plugs), unfinished plants and propagation materials, cut Christmas trees; dried bulbs, corms, rhizomes, and tubers; aquatic plants, flower seeds, and short-term woody crops. Plant growing and seed harvesting are important aspects of ornamental seed production. Seed processing includes milling, cleaning, packaging, and storing.

Landscape horticulture consists of landscape design, landscape installation/construction, lawn/landscape maintenance, and tree care. Landscape architects or designers are involved in landscape design and landscape contractors in landscape construction. Lawn, tree, and other landscape maintenance are provided by lawn care services, arborists, and full-service landscape maintenance firms.

Landscape design and architecture are core elements of ornamental horticulture. The basic elements of design are color, form, line, and texture. Plant color varies considerably. Plant form is essentially the visual impact of the total mass of a plant or, collectively, of a group of plants. Collective plant masses or horticultural structures can be used to guide or direct eye movement; this constitutes the concept of line. The visual impact of the surfaces of plants and horticultural surfaces offer various textures.

These design basics must be artistically blended such that plantings, horticultural structures, and buildings form an overall, attractive picture. Plantings and structures should be mixed to produce variety. In design, the environmental requirements of the planting must be known. The mature form and heights of plants must be considered in order to maintain the proper scale of plantings with time. Design is a unifying concept in ornamental horticulture that involves close cooperation with activities in other areas, especially site preparation and subsequent maintenance.

Landscape installation/construction is involved with the total preparation of the site for the planting. These activities include the procurement of all plants, site modification, plant installation, and site maintenance until construction is complete. Landscape and lawn maintenance involves maintaining established ornamental plantings. Tree care involves major pruning, repair of storm damage, tree removal and insect/disease protection.

A large number of career positions are associated in whole or part with ornamental horticulture. Some of the general areas for careers include: arboriculture, environmental management, horticultural therapy, interiorscape design and management, landscape construction and installation, landscape design, landscape maintenance, nursery management, pest management, plant propagation and production, and turf management in parks or golf courses.

Many plants are used in ornamental horticulture: annuals, biennials, herbaceous perennials, bulbs, ferns, grass, ground covers, woody perennials, house plants and greenhouse plants.

Annuals are plants that grow, flower, and die naturally in one growing season. Annuals are the mainstays of the flower garden and their long-term flowering makes them ideal choices for summer color in the landscape. Annuals are used for masses of color in flower borders, islands, and containers. Some make excellent cut flowers. Others have pleasing fragrances. Shorter ones serve as flower edging in borders. Given their versatility, beauty, availability and reasonable price, annuals are valued both by the home gardener and commercial horticulturists.

The growing treatment of annuals provides a basis for two groups: (1) those sown in the place where they are expected to bloom, and (2) those started early in a hotbed or greenhouse in order to provide sufficient time for blooming during the growing season.

Biennials are plants that require two growing seasons to flower. Generally, the first year is vegetative and the second reproductive. Many biennial ornamental plants are treated as annuals by giving them an early start indoors or by seeding in late fall or early spring outdoors (if hardy enough). Some perennials are also so short-lived as to be regarded as biennials. Biennials are few in number and are less popular than annuals, but a few are sold as bedding plants sufficiently old enough to bloom the first year.

Perennials are plants that persist for more than 2 years. Many of our favored ornamentals such as shrubs, trees, ferns, and bulbs are perennials. However, horticultural usage of the word *perennial* has come to imply herbaceous, nonbulbous flowering perennials used in beds or borders.

The long-lived perennial border must be carefully planned. A poorly planned one wastes money and will give years of dissatisfaction. The advantage over annuals is plant once, not every year. True, some initial installation is required and maintenance is needed to divide perennials that become crowded, as crowding reduces flowering. However, over the long term, perennials are less labor intensive then annuals, giving years of viewing pleasure and cut flowers. Things to consider in the plan are height, spacing, color, hardiness, time of bloom, and required cultural conditions. The ideal perennial border or bed provides a succession of bloom throughout the growing season. Perennials are usually planted in the spring in the North, and either spring or fall in the South.

The versatility of bulbs leads to their use in naturalized or formal plantings. Various bulbs are suited to rock gardens, woodland gardens, beds, borders, and even container gardens. Bulbs are divided into two groups: spring and summer bloomers. Most spring flowering bulbs require a cold period and are limited to areas of minimal to extensive frost/freezes. Such bulbs do not flower in the more extreme southerly parts of the United States. Extreme cold can also be a limiting factor. Spring flowering bulbs are quite popular, since their blooming period occurs prior to that of the annuals and bedding perennials covered previously. Summer blooming bulbs are more limited in cold hardiness than spring ones and are grown as annuals or pot plants in the colder areas of the United States. In more southerly areas, summer flowering bulbs can be left in the ground to survive for many years. Summer blooming bulbs provide excellent cut flowers and offer color and diversity in formal plantings beyond annuals and bedding perennials.

Spring blooming or hardy bulbs are planted in midfall up to just before the ground freezes too hard to be dug. Summer flowering or tender bulbs are usually planted when the danger of spring killing frosts is past. In parts of the South, summer flowering bulbs can be planted in the fall or spring. Spring flowering bulbs may be forced indoors for early winter color or for holidays such as Easter. Some bulbs are forced to provide cut flowers.

Ferns, although not a substantial part of ornamental horticulture, are still of value for three reasons. First, they are an important group of nonflowering, foliage plants that can be useful in perennial borders, rock gardens, or naturalized plantings of a woodsy nature. Second, florists value their foliage for use as greenery in floral arrangements. Finally, they are attractive pot plants that can be grown under low-light conditions.

Horticulturists generally recognize two groups of ferns. The first consists of hardy ferns of native and exotic origin that can be cultivated outdoors throughout much of the United States and Canada. The second group includes ferns originating in tropical areas. The latter group is restricted in most of the United States to use in the greenhouse and home. Many ferns thrive in partial to full or partial shade when grown outdoors and require filtered sunlight or lower intensities of light indoors. Ferns often grow in soils having a high humus level and a covering of organic litter such as leaves.

Lawns are generally regularly mowed plantings of grass. Their purpose is to enhance the visual appeal of open spaces around buildings and parks, to provide useful outdoor living space for recreation and enjoyment, and to improve environmental quality. The latter consists of soil improvement (soil health and erosion prevention), runoff reduction (better water infiltration), as well as a modifying influence on drastic changes in soil–air temperatures. Areas carpeted with lawns usually have a cooler microclimate than paved areas. Grasses are generally used for lawns, since their basal meristems are adapted for mowing in that plant damage or loss does not occur. The use of grass species is determined by climate. The main two climatic areas for grasses are the North and South. The perennial cultivars of choice for the North are derived from Kentucky bluegrass, perennial ryegrass, fine fescue, and bent grasses. Grasses of choice for the south include Bermuda grass, zoysia, St. Augustine grass, Bahia grass, and centipede grass. Native grasses are often good choices for low maintenance and more sustainable lawns.

Lawns can be planted either from seed or by vegetative means, such as sod, sprigs, or plugs. Seeding is more economical, and, unlike with sod, there is minimal danger of introducing diseases, weeds, or insects. Sod offers the advantage of quick, but costly cover. Sprigs, individual stems, and plugs (biscuits of sod) are economical means of establishing grasses that do not come true from seed. Established lawns require regular mowing, periodic fertilization and adjustment of soil pH, irrigation when needed, and pest control. Fertilization and liming are best done according to soil test.

Plants selected for ground covers are herbaceous perennials or low woody plants requiring minimal care. They are usually used in place of lawns on areas where lawns fare poorly or not at all, or on sites where lawn care is difficult. Such locations include very sandy soils (dunes and arid areas) and heavily shaded areas under trees and slopes. Ground covers on dunes, slopes, and banks also help prevent soil erosion. Ground covers are not suitable for walking or recreation. Some are grown mostly for foliage (evergreen over deciduous) and others offer attractive flowers as a bonus.

Woody perennials include deciduous and evergreen trees, shrubs, and vines. Woody plants have great value in the overall landscape because of their long-lasting, substantial year-round effect. As such, they are mainstays in landscaping planning, whether utilized for background, screening, foundation plantings, visual impact, shade, ecosystem maintenance/improvement, energy conservation, or windbreaks.

The woody plant group is very diverse, but some factors are common to all and should be considered when choosing plants. Foremost is the habit of growth, which determines the ultimate shape, height, and width of the mature plant. The landscape placement of a plant, both in terms of mature scale and shape, will ultimately be determined by this fact. Hardiness is also important for determining the ultimate success of a plant in a given area. Hardiness zones are only a guide to relative hardiness. There is no substitute for localized knowledge of the plants that are hardy in your area.

Two other factors are important: permanence and foliage effect. The plant should be long lived in your locality, considering your investment of time and money. Some plants might be highly susceptible to certain pests and/or diseases that prevail in your area. The latter is a critical point beyond longevity in terms of both sustainability and maintenance perspectives. Plants prone to insects and diseases of your area require high maintenance to keep them healthy. Plants needing little or no pesticides are a better choice for environmental sustainability.

Foliage effects are important, since you view the plant throughout the changes of seasons. Several factors are involved in foliage effect: size, texture, deciduous versus evergreen, summer and fall color, arrangement on the stem, and appearance and condition throughout the seasons. The bark color and texture, flowers and fruit, and the branching pattern are also factors that contribute to the overall appearance and condition throughout the year. Other selection factors involve use factors such as shade, a specimen for the lawn or back yard, foundation plantings, property line plantings, and vista views.

For purposes of shade, large deciduous trees are useful. These trees provide ample shade in the summer, both in the yard and along the street. In the winter, these trees allow sun to warm houses and adjacent property. Smaller deciduous trees are more useful as specimen plantings on the lawn or in the backyard. These plants are chosen mostly on the basis of some outstanding characteristic such as flowers, fruits, or attractive summer and autumn foliage.

Evergreens, both broad-leaved and needled, are useful for maintaining some color throughout the year. Form and texture are also other good points for evergreens. The conifers make ideal unique specimens as well as background plantings, both for colorful foreground plants and for privacy. Conifers also hide unattractive views and act as a windbreak year-round. Given their wide size range, it is always possible to find suitable ones for any landscape or park. Broad-leaved evergreens also offer the added benefits of flowers and fruits. This group is widely useful as a complement to deciduous flowering and fruiting shrubs.

Vines can offer a unique form, thus becoming a focal point, and the benefits of flowers, fruits, and either deciduous or evergreen foliage. Vines are useful for covering stone walls, trellises, and arbors. Care should be exercised when one places the vine. Certain vines can be destructive of house exteriors and are also a problem when painting or other maintenance needs to be done behind them.

Container and balled and burlapped stock can be transplanted successfully anytime from early spring through fall. The best planting times for both in the North are early spring while in the dormant stage, and in the fall and winter in the South after cessation of growth just as dormancy is about to start. Bare-root stock should only be planted at these times while dormant. Container-grown stock is most likely to recover sooner than other types, since their roots experience less transplanting shock.

Success with house or greenhouse plants depends on the provision of conditions similar to those found in the plant's native habitat. Environmental conditions are more easily managed in a greenhouse than in the home or business environment. Plants that succeed outside the greenhouse usually have wider tolerance for deviations from the environmental norm. Great caution must be observed with the selection and application of pesticides in enclosed environments such as the home or greenhouse. The three most likely areas for failure with these plants are overwatering, failure to observe dormancy, and lack of insect control.

EXERCISES

Mastery of the following questions shows that you have successfully understood the material in Chapter 16.

1. What commercial production areas constitute the ornamental horticultural industry? What is their rank in terms of sales?
2. What specific crops are produced in the floriculture area? What are the top five states for floricultural production?
3. What seasonal markets are associated with floriculture?
4. What are bedding and garden plants? How are they used in horticulture? When are they sold? In what form are they started and finally sold? Give some specific examples of the two types of bedding plants.
5. What area of bedding and garden plant production needs careful attention? Why?
6. What are potted flowering and foliage plants used for? Why do you think Florida is the number one producer of potted foliage plants?
7. What factors go into determining what cut flowers are grown commercially? What stage is especially critical for cut flower production?
8. What crops are produced by production nurseries? What types of nursery operations exist. Do any provide services? What type of services?
9. Why does container-grown nursery stock dominate the market?
10. What other specialty crops are produced in ornamental horticulture?
11. What areas constitute landscape horticulture? What services or products does each provide? Does landscaping have any value for the homeowner? If so, what?
12. What are the basic elements of landscape design? Explain each.
13. What types of construction and installation might be needed for landscaping a site?
14. Which horticultural career might appeal to you? Why?
15. What is an annual? What uses do they have? Why are they so important to consumers? Give examples of some popular annuals.
16. Why are some annuals started ahead in greenhouses?
17. What is a biennial? Name a few.
18. What is a herbaceous perennial in terms of horticultural usage? What advantage over annuals do herbaceous perennials have? What disadvantages do herbaceous perennials have? Give some examples of popular herbaceous perennials.
19. What factors should be considered when planning a perennial border?
20. What maintenance beyond normal culture do herbaceous perennials require?

21. What is a bulb in the sense of horticultural use?
22. What are the differences between spring and summer flowering bulbs? Give examples of each bulb type.
23. What are the basic steps for forcing bulbs?
24. What are the horticultural uses of ferns? What are the two groups of ferns? Give some examples of each.
25. What types of grass are used in the North? In the South?
26. Does clover have any value as a companion plant to grass? If so, what?
27. When is a ground cover used in place of grass? Why?
28. What types of plants constitute the group called woody perennials?
29. What factors come into play when one chooses woody perennials?
30. What generalized principles apply when using deciduous and evergreen trees and shrubs?
31. Where should one use a vine or liana?
32. What are the conditions needed to successfully grow house or greenhouse plants? What mistakes are commonly made?

INTERNET RESOURCES

American Nursery & Landscape Association
   http://www.anla.org

Annuals
   http://muextension.missouri.edu/xplor/agguides/hort/g06629.htm

Bulb information source
   http://www.bulb.com

Care (culture) of potted flowering plants
   http://muextension.missouri.edu/xplor/agguides/hort/g06511.htm

Care of potted flowering plants from greenhouse to market
   http://www.penpages.psu.edu/penpages_reference/29401/2940140.HTML

Census of horticultural specialties, 1998
   http://www.nass.usda.gov/census/census97/horticulture/horticulture.htm

Christmas tree production
   http://www.gov.on.ca/OMAFRA/english/crops/facts/info_xmastreeprod.htm
   http://www.exnet.iastate.edu/Publications/CRP19.pdf

Cut flower handling for florists
   http://www.oznet.ksu.edu/library/hort2/mf2323.pdf

Cut flower harvesting
   http://www.oznet.ksu.edu/library/hort2/MF2155.PDF

Deciduous shrubs
   http://muextension.missouri.edu/xplor/agguides/hort/g06830.htm

Evergreens with broad leaves
   http://muextension.missouri.edu/xplor/agguides/hort/g06820.htm

Evergreens with needles
   http://muextension.missouri.edu/xplor/agguides/hort/g06815.htm

Ferns
   http://www.ces.uga.edu/pubcd/B737-w.htm

Ground covers

    http://muextension.missouri.edu/xplor/agguides/hort/g06835.htm

Growing and marketing bedding plants

    http://www.aces.edu/department/extcomm/publications/anr/ANR-0559/anr-559.html

Guidelines for utilization of fall-planted spring and very early summer flowering bulbs in North American landscapes and gardens

    http://www.ces.ncsu.edu/depts/hort/consumer/factsheets/bulbs-summer/bulletin32/key_steps.html

Guidelines for utilization of flowering bulbs as perennial (naturalized) plants in North American landscapes and gardens

    http://www.ces.ncsu.edu/depts/hort/consumer/factsheets/bulbs-perennial/bulletin37/perennial_bulbs.html

Herbaceous ornamentals

    http://www.hcs.ohio-state.edu/mg/manual/herbac.htm

House plant information

    http://aggie-horticulture.tamu.edu/plantanswers/publications/houseplant/houseplant. html

Lawn care

    http://www.TurfGrassSod.org

Links to annuals

    http://www.ces.ncsu.edu/depts/hort/consumer/hortinternet/flowers-annual.html

Links to production of potted flowering and foliage plants

    http://flowers.hort.purdue.edu/web/pottedpltguides.htm

Nursery production references

    http://www.umassgreeninfo.org/fact_sheets/plant_culture/nursery_references.htm

Perennials

    http://muextension.missouri.edu/xplor/agguides/hort/g06650.htm

Potted flowering plants for home and office

    http://www.ca.uky.edu/agc/pubs/ho/ho51/ho51.htm

Small-scale foliage plant production

    http://www.sfc.ucdavis.edu/pubs/brochures/foliage.html

Small-scale greenhouse business

    http://www.ces.uga.edu/pubcd/b1134-w.html

Sustainable cut flower production

    http://www.attra.org/attra-pub/cutflower.html

Sustainable nursery production

    http://www.attra.org/attra-pub/nursery.html

Transport of potted flowering and foliage plants

    http://www.ams.usda.gov/tmd/Tropical/pottedplants.htm

Tree Planting

    http://hort.ifas.ufl.edu/woody/planting

Trees for flowers

    http://muextension.missouri.edu/xplor/agguides/hort/g06805.htm

Trees for shade

    http://muextension.missouri.edu/xplor/agguides/hort/g06800.htm

Updated economics, floriculture

    http://usda.mannlib.cornell.edu/reports/nassr/other/zfc-bb/

Updated economics, nursery crops

    http://usda.mannlib.cornell.edu/reports/nassr/other/nursery/index.html

Vines

    http://muextension.missouri.edu/xplor/agguides/hort/g06840.htm

RELEVANT REFERENCES

Readers should also refer to the relevant resources found in Chapters 2 and 12.

Aclyn, Jan (1995). *Handbook of Landscape Palms.* Great Outdoors Publishing Company, St. Petersburg, FL. 70 pp.

Armitage, Alan (1993). *Specialty Cut Flowers: The Production of Annuals, Perennials, Bulbs, and Woody Plants for Fresh and Dried Cut Flowers.* Timber Press, Portland, OR. 372 pp.

Armitage, Alan (1997). *Herbaceous Perennial Plants. A Treatise on Their Identification, Culture, and Garden Attributes* (2nd ed.). Stipes Publishing Co., Champaign, IL. 1141 pp.

Arnold, Michael A. (2002). *Landscape Plants for Texas and Environs* (2nd ed.). Stipes Publishing, Champaign, IL. 1088 pp.

Ball, Vic (1998). *Ball Redbook* (16th ed.). Ball Publishing, Batavia, IL. 816 pp.

Boodley, James W. (1996). *The Commercial Greenhouse* (2nd ed.). Delmar Publishers, Albany, NY. 624 pp.

Booth, Norman K. (1990). *Basic Elements of Landscape Architectural Design.* Waveland Press, Prospect Heights, IL. 315 pp.

Booth, Norman K., and James E. Hiss (1998). *Residential Landscape Architecture: Design Process for the Private Residence* (2nd ed.). Prentice Hall, Upper Saddle River, NJ. 448 pp.

Botanica staff (eds.) (1999). *Botanica's Trees and Shrubs.* Advantage Publishing Group (Laurel Glen), San Diego, CA. 1008 pp.

Bryan, John E. (2002). *Bulbs* (rev. ed.). Timber Press, Portland, OR. 524 pp.

Buchanan, Rita (2000). *Taylor's Master Guide to Landscaping.* Houghton Mifflin Co., Boston. 400 pp.

Byczynski, Lynn (1997). *The Flower Farmer: An Organic Grower's Guide to Raising and Selling Cut Flowers.* Chelsea Green Publishing Company, White River Junction, VT. 208 pp.

Camenson, Blythe (1998). *Opportunities in Landscape Architecture, Botanical Gardens & Arboreta.* McGraw Hill-NTC, New York. 150 pp.

Chin, Wee Yeow (1998). *Ferns of the Tropics* (reprint). Timber Press, Portland, OR. 190 pp.

Clarke, Graham (1997). *Indoor Plants: The Essential Guide to Choosing and Caring for Houseplants.* Reader's Digest Association, Pleasantville, NY. 240 pp.

Daniels, Stevie (1997). *The Wild Lawn Handbook: Alternatives to the Traditional Front Lawn.* Hungry Minds, New York. 224 pp.

Daniels, Stevie (ed.) (1999). *Easy Lawns: Low-Maintenance Native Grasses for Gardeners Everywhere.* Brooklyn Botanic Garden, New York. 111 pp.

Darke, Rick (1999). *The Color Encyclopedia of Ornamental Grasses, Sedges, Rushes Restios, Cattails, and Selected Bamboos.* Timber Press, Portland, OR. 325 pp.

DeHertogh, August (1996). *Holland Bulb Forcer's Guide.* Ball Publishing, Batavia, IL. 597 pp.

DeWolf, Gordon P. (ed.) (1987). *Taylor's Guide to Ground Covers, Vines and Grasses.* Houghton Mifflin Company, Boston. 495 pp.

Dirr, Michael A. (1997). *Dirr's Hardy Trees and Shrubs: An Illustrated Encyclopedia.* Timber Press, Portland, OR. 493 pp.

Dirr, Michael (1998). *Manual of Woody Landscape Plants. Their Identification, Ornamental Characteristics, Culture, Propagation and Uses* (5th ed.). Stipes Publishing Co., Champaign, IL. 1250 pp.

Dole, John, and Harold Wilkins (1998). *Floriculture: Principles and Species.* Simon & Schuster, New York. 613 pp.

Dwelley, Marilyn J. (2000). *Trees and Shrubs of New England* (2nd ed.). Down East Books, Camden, ME. 276 pp.

Edinger, Philip, and staff of Sunset Books (eds.) (1999). *Vines and Ground Covers.* Sunset Books, Menlo Park, CA. 112 pp.

Evans, John (1994). *The Complete Book of Houseplants: A Practical Guide to Selecting and Caring for Houseplants.* Penguin Studio Books, New York. 256 pp.

Everitt, James H., J. H. Everitt, and D. Lynn Drawe (1992). *Trees, Shrubs and Cacti of South Texas.* Texas Tech University Press, Lubbock. 216 pp.

Foote, Leonard E., and Samuel B. Jones (1998). *Native Shrubs and Woody Vines of the Southeast: Landscaping Uses and Identification.* Timber Press, Portland, OR. 199 pp.

Gilman, Ed, and R. J. Black (1999). *Your Florida Guide to Shrubs: Selection, Establishment and Maintenance.* University Press of Florida, Gainesville. 116 pp.

Godfrey, Robert K. (1989). *Trees, Shrubs, and Woody Vines of Northern Florida and Adjacent Georgia and Alabama.* University of Georgia Press, Athens. 734 pp.

Grant, John A., and Brian O. Mulligan (eds.) (1994). *Trees and Shrubs for Pacific Northwest Gardens* (2nd ed.). Timber Press, Portland, OR. 456 pp.

Griffith, Jr., Lynn (1998). *Tropical Foliage Plants: A Grower's Guide.* Ball Publishing, Batavia, IL. 318 pp.

Griner, Charles P. (2000). *Floriculture: Designing & Merchandising* (2nd ed.). Delmar Publishers, Albany, NY. 496 pp.

Hamrick, Debbie (1996). *Grower Talks on Plugs II.* Ball Publishing, Batavia, IL. 214 pp.

Harlan, Michael, and Linda Harlan (2000). *Growing Profits: How to Start & Operate a Backyard Nursery* (2nd ed.). Chelsea Green Publishing Company, White River Junction, VT. 216 pp.

Harris, Richard Wilson, James R. Clark, and Nelda P. Matheny (1998). *Arboriculture: Integrated Management of Landscape Trees, Shrubs, and Vines* (3rd ed.). Simon & Schuster, New York. 687 pp.

Heffernan, Cecelia (2001). *Flowers A to Z: Buying, Growing, Cutting, Arranging.* Harry Abrams, New York. 160 pp.

Hessayon, David G. (1992). *The House Plant Expert.* Sterling Publishing Company, London. 255 pp.

Hessayon, David G. (1998). *The Evergreen Expert.* Sterling Publishing Company, London. 128 pp.

Hightshoe, Gary L. (1997). *Native Trees, Shrubs, and Vines for Urban and Rural America: A Planting Design Manual for Environmental Designers.* John Wiley & Sons, New York. 819 pp.

Hill, Lewis, and Nancy Hill (1995). *Lawns, Grasses, and Groundcovers* (Vol. 1). Rodale Press, Emmaus, PA. 160 pp.

Hillier, Malcolm (2000). *Flowers: The Book of Floral Design.* Dorling Kindersley Publishing, London. 468 pp.

Holcomb, E. (1994). *Bedding Plants IV. A Manual on the Culture of Bedding Plants as a Greenhouse Crop.* Ball Publishing, Batavia, IL. 452 pp.

Hunter, Norah (2000). *The Art of Floral Design* (2nd ed.). Delmar Publishers, Albany, NY. 450 pp.

Jerome, Kate (1999). *Ortho's All about Houseplants.* Meredith Books, Des Moines, IA. 96 pp.

Jones, David L. (1992). *Encyclopaedia of Ferns* (Vol. 1). Timber Press, Portland, OR. 450 pp.

Jozwik, Francis X. (1992). *The Greenhouse and Nursery Handbook: A Complete Guide to Growing and Selling Ornamental Container Plants.* Andmar Press, Mills, WY. 511 pp.

Jozwik, Francis X. (2000). *Plants for Profit: Income Opportunities in Horticulture.* Andmar Press, Mills, WY. 304 pp.

Knopf, Jim (1999). *Waterwise Landscaping with Trees, Shrubs and Vines: A Xeriscape Guide for the Rocky Mountain Region, California and the Desert Southwest.* Chamisa Books, Boulder, CO. 389 pp.

Larson, Roy A. (ed.) (1992). *Introduction to Floriculture* (2nd ed.). Academic Press, San Diego, CA. 636 pp.

Laurie, Alex, D. C. Kiplinger, and Kennard S. Nelson (1979). *Commercial Flower Forcing: The Fundamentals and Their Practical Application to the Culture of Greenhouse Crops* (8th ed.). McGraw-Hill Higher Education, New York. 438 pp.

Lerner, Joel (2001). *Anyone Can Landscape!* Ball Publishing, Batavia, IL. 472 pp.

MacKenzie, David S. (1997). *Perennial Ground Covers.* Timber Press, Portland, OR. 452 pp.

Martin, Tovah (1996). *Window Boxes: How to Plant and Maintain Beautiful, Compact Flowerbeds.* Houghton Mifflin, Boston. 122 pp.

Mathias, Mildred E. (ed.) (1976). *Color for the Landscape: Flowering Plants for the Subtropical Climates.* California Arboretum Foundation, Arcadia, CA. 210 pp.

McHoy, Peter (2000). *The Complete Houseplant Bible.* Anness Publishing, London. 256 pp.

Mickel, John T. (1996). *Ferns for American Gardens: The Definitive Guide to Selecting and Growing More than 500 Kinds of Hardy Ferns.* Macmillan Publishing Company, New York. 334 pp.

Nash, Helen, and Steve Stroupe (1998). *Plants for Water Gardens: The Complete Guide to Aquatic Plants.* Sterling Publications, London. 224 pp.

Nau, Jim (1996). *Ball Perennial Manual: Propagation and Production.* Ball Publishing, Batavia, IL. 512 pp.

Nell, Terril (1993). *Flowering Potted Plants.* Ball Publishing, Batavia, IL. 96 pp.

Nelson, Kennard S. (1991). *Flower and Plant Production in the Greenhouse.* Interstate Publishers, Danville, IL. 220 pp.

Nelson, Paul V. (1997). *Greenhouse Operation and Management* (5th ed.). Prentice Hall, Upper Saddle River, NJ. 637 pp.

Ouden, P. den, and B. K. Boom (1978). *Manual of Cultivated Conifers* (English ed.). Martinus Nijhoff, Boston. 526 pp.

Perry, Leonard P., and Cathleen Walker (1998). *Herbaceous Perennials Production: A Guide from Propagation to Marketing* (NRAES-93). Northeast Regional Agricultural Engineering Service, Cooperative Extension, Cornell University, 152 Riley-Robb Hall, Ithaca, NY 14853-5701. 220 pp.

Pfahl, Peter Blair, and P. Blair Pfahl, Jr. (1994). *The Retail Florist Business* (5th ed.). Interstate Publishers, Danville, IL. 336 pp.

Pleasant, Barbara (1997). *Lawns and Ground Covers.* Oxmoor House, Birmingham, AL. 128 pp.

Poor, Janet Meakin (ed.) (1984). *Plants That Merit Attention: Trees* (Vol. 1). Timber Press, Portland, OR. 349 pp.

Poor, Janet Meakin, and Nancy Peterson Brewster (eds.) (1984). *Plants That Merit Attention: Shrubs* (Vol. 2). Timber Press, Portland, OR. 363 pp.

Reid, Grant W. (1993). *From Concept to Form: In Landscape Design.* John Wiley & Sons, New York. 162 pp.

Royer, Kenneth (1998). *Retailing Flowers Profitably.* Ball Publishing, Batavia, IL. 495 pp.

Smith, Katie LaMar (2001). *Ortho's All About Ground Covers.* Meredith Books, Des Moines, IA. 96 pp.

Snyder, Leon C. John (2000). *Trees and Shrubs for Northern Gardens.* University of Minnesota Press, Minneapolis. 320 pp.

Stephens, H. A. (1998). *Trees, Shrubs, and Woody Vines in Kansas.* University Press of Kansas, Lawrence. 250 pp.

Stevens, Alan B. (1997). *Field Grown Cut Flowers: A Practical Guide and Sourcebook: Commercial Field Grown Fresh and Dried Cut Flower Production* (2nd ed.). Avatar's World, Edgerton, WI. 395 pp.

Stresau, Frederic B., and James N. Baker (ed.) (1991). *Florida, My Eden: Exotic and Native Plants for Use in Tropic and Sub-Tropic Landscapes.* Florida Classics Library, Hobe Sound, FL. 300 pp.

Stuart, John David, and John O. Sawyer (2001). *Trees and Shrubs of California*. University of California Press, Berkeley. 467 pp.

Sturdivant, Lee (1994). *Flowers for Sale: Growing and Marketing Cut Flowers: Backyard to Small Acreage*. San Juan Naturals, Friday Harbor, WA. 200 pp.

Styer, Roger C., and David S. Koranski (1997). *Plug & Transplant Production: A Grower's Guide*. Ball Publishing, Batavia, IL.

Taylor, Patricia A. (1996). *Easy Care Native Plants: A Guide to Selecting and Using Beautiful American Flowers, Shrubs, and Trees in Gardens and Landscapes*. Henry Holt & Company, New York. 336 pp.

Taylor, Ted M. (2000). *Secrets to a Successful Greenhouse and Business*. Greenearth Publishing Company, Melbourne, FL. 280 pp.

Thomas, Graham Stuart (1997). *Trees in the Landscape* (2nd ed.). Sagapress/Timber Press, Portland, OR. 216 pp.

Viertel, Arthur T. (1970). *Trees, Shrubs and Vines: A Pictorial Guide to the Ornamental Woody Plants of the Northern United States Exclusive of Conifers*. Syracuse University Press, Syracuse, NY. 600 pp.

Whitcomb, Carl E. (1984). *Plant Production in Containers*. Lacebark Publications & Research, Stillwater, OK. 638 pp.

CD-ROMs

Dirr, Michael, A. (2002). *Interactive Manual of Woody Landscape Plants*. Plant America: www.plantamerica.com.

Chapter 17

Vegetables and Herbs

OVERVIEW

Olericulture is the branch of horticulture that deals with the production, storage, processing, and marketing of vegetables. What amount of vegetables and herbs are grown in a sustainable manner is unknown, because current statistics from the USDA's National Agricultural Statistics Service do not differentiate among differing systems of vegetable and herb production. Part of the problem is that the exact limits of what constitutes sustainable horticulture have not been clearly defined. Only one part of sustainable horticulture, organic farming, has been clearly defined such that a certification process exists. Individual states and recent federal government standards allow for the tracking of organic crops and products. According to the latest data released in 2000 by the Economic Research Service of the USDA, certified organic cropland doubled in the 1992–1997 period. The number of certified organic farmers in the United States jumped to 7,800 for an increase of 18% over 1999. The organic market (all aspects and crops) is now at $7.7 billion. The average U.S. certified organic vegetable farm is roughly 12 acres as opposed to 70 acres for the conventional vegetable farm. *Trends in Agriculture 2000* indicates that roughly one-fourth of U.S. farmers believe that they are operating as sustainable farmers. About 10% of vegetable farmers thought their practices were sustainable.

CHAPTER CONTENTS

OBJECTIVES

❀ To learn about the economics of and the contribution of sustainable and organic vegetable operations to commercial vegetable and herb production and vegetable gardening

❀ To learn about the acreage, greenhouse space, production, value, states, and vegetables involved in fresh/processing vegetable markets

❀ To learn about herb, vegetable seed/transplant, and mushroom production

❀ To explore possible careers

❀ To understand basics of vegetable and herb production, harvest, post-harvest, and marketing

❀ To learn the cultural and culinary basics of most vegetable, mushroom, and herb crops

VEGETABLE GARDENS

Vegetable production on the family or noncommercial level is substantial. In 1999, 33.74 million American households had a vegetable garden (see Chapter 12) for a total of 35% of the 96.4 million households in America, as indicated by Gallup polls. Retail sales for vegetable-oriented products and services reached $1.21 billion. The number one garden plant was the tomato, which was raised in 85% of all vegetable gardens.

COMMERCIAL FIELD VEGETABLE AND MELON CROPS

The federal government currently tracks 34 vegetables and melons under the vegetable category. White and sweet potatoes are tracked separately under crops. The current vegetables and melons tracked are artichokes, asparagus, beans (lima and snap), beets, broccoli, brussels sprouts, cabbage, cantaloupes, carrots, cauliflower, celery, cucumbers, eggplant, escarole (endive), garlic, greens (collards, kale, mustard, and turnip), honeydews, lettuce (head, leaf, and romaine), okra, onions, peas, peppers (bell and chile), pumpkins, radishes, spinach, squash, sweet corn, tomatoes, and watermelons. Tracking includes acreage, production, and value on the individual crop and state levels as well as the federal level.

Acreage

On the commercial level, vegetables are big business. During 2000 about 2.10 million acres were harvested for fresh market vegetable and melon production (34 selected crops) and 1.45 million acres for 10 selected processed vegetables. California led in fresh vegetable production, accounting for 42.6% of the U.S. acreage. Florida was next with 9.7%. The leading states for acreage of harvested processed vegetables were California (21.6%), Wisconsin (16.1%), and Minnesota (15.7%).

Certified organic vegetable acreage in 1997 was 1.3% of the combined fresh/processed acreage. When viewed on the organic level, California in 1997 led with nearly 48% of the total certified organic U.S. acreage for all vegetables combined, but this amount was only 2.1% of California's total vegetable production. Vermont led the nation with 23.6% of its vegetables in 1997 being certified organic, but its total U.S. contribution was 1.4% relative to California's 48%.

For all harvested fresh vegetable acreage combined, the leading vegetables in decreasing order were sweet corn, head lettuce, romaine lettuce, and watermelons. These four vegetables accounted for 37% of the fresh harvest acres. Sweet corn, tomatoes, green peas, and snap beans were the leaders for harvested processed vegetable acreage and accounted for 86% of the tracked processing acreage. For certified organic vegetables, the leading vegetables in terms of acreage were lettuce, tomatoes, and carrots, accounting for about 25% of the organic acreage. Mixed vegetable organic farms accounted for 35% of the organic acreage.

Production

Fresh vegetable production was 24.1 million tons in 2000 and during that same year, processed vegetables totaled 17.1 million tons. For all harvested fresh vegetable production combined, the leading vegetables in decreasing order were head lettuce, onions, watermelons, tomatoes, and carrots. These five accounted for 51.8% of the combined production. In terms of fresh market vegetable/melon production, the five leading states were California

(48.4%), Florida (9.9%), Arizona (7.7%), Georgia (5.3%), and Texas (4.3%). Similarly, for vegetables utilized in processing, the leading vegetables in decreasing order were tomatoes (63.2%), sweet corn (18.4%), snap beans (4.8%), cucumbers (3.6%), and green peas (3.1%). The leading states were California (62.3%), Wisconsin (7.3%), Washington (6.7%), Minnesota (5.6%), and Oregon (3.2%).

Value

Fresh vegetable production value in 2000 totaled $9.33 billion. During that same year, processed vegetables were valued at $1.43 billion. In terms of fresh market vegetable/melon value, the five leading states in 2000 with percentages of total production value were California (53.0%), Florida (13.1%), Arizona (6.4%), Georgia (4.3%), and New York (3.6%). Similarly, for vegetables utilized in processing, the states were California (46.8%), Wisconsin (8.2%), Washington (7.5%), Minnesota (7.2%), and Oregon (4.7%).

In the United States the three most important fresh vegetables based on the total crop value in 2000 are as follows in decreasing order: head lettuce, tomato, and onion. Together these three accounted for 34% of the total value for the 34 tracked vegetables and melons (excluding white potatoes). The next seven accounted for another 34% of the value: bell peppers, broccoli, sweet corn, carrot, cantaloupe, cabbage, and celery. For processed vegetables, the leading three were tomatoes, sweet corn, and snap beans, which accounted for 74% of the value of the 10 major processed crops.

COMMERCIAL FIELD WHITE AND SWEET POTATO CROPS

Some 1.35 million acres of white potatoes were harvested in 2000. Only 0.3% of potatoes were certified organic. The leading states in terms of potato acreage were Idaho (30.6%), Washington (13.3%), and North Dakota (8.1%). California led in organic potatoes with 25% of the organic U.S. potato acreage. The total potato yield was 25.8 million tons and was valued at $2.6 billion. The top two potato producers in terms of yield were Idaho (29.5%) and Washington (20.9%). A little over half of the fresh potato crop was processed into frozen french fries and other frozen potato products (60%), potato chips (18.2%), and the remainder went into canned potato products, dehydrated potatoes, potato starch, and potato flour. About 94,000 acres of sweet potatoes were harvested in 2000 with a production of 681,000 tons. The leading sweet potato producers production-wise were North Carolina (40.7%) and Louisiana (22.9%).

COMMERCIAL GREENHOUSE VEGETABLE AND MUSHROOM CROPS

Greenhouse food crops had a value of $223 million in 1998. Some 32 million square feet of greenhouse space was devoted to these crops with tomatoes accounting for slightly over half. The other major crops in descending order were cucumber, lettuce, and peppers (sweet and hot). The two leading states were California (32%) and Colorado (16%). Florida, Pennsylvania, and Massachusetts were the next three. Cultivated mushrooms in 1998 were worth $864 million. The three leading states were Pennsylvania (39.5%), California (20.6%), and Texas (6.0%).

COMMERCIAL FIELD AND GREENHOUSE HERB CROPS

Greenhouse space devoted to herb production in 1998 totaled 4.73 million square feet and brought in sales of $31 million. The leading producer was California with about 64% of the herb-producing greenhouse space. About 8% of the greenhouse-grown herbs were certified organic. Vermont led in the production of certified organic herbs with 73.6%! Herbs (culinary and medicinal) were also cultivated on roughly 17,000 acres, of which nearly 40% were certified as organic. Medicinal (mostly) herbs were also gathered from natural stands (wild-crafting) on some 84,000 acres, essentially all of which are certified as organic given no inputs other than natural ones.

COMMERCIAL VEGETABLE SEED AND TRANSPLANT CROPS

Most vegetable seeds sold by firms are produced under contract to independent growers. Many of these are situated on the West Coast, where climatic conditions are favorable for seed production. In 1998 about 54,000 acres of seeds were harvested with a combined production value of $93 million. Seeds grown outside the United States for U.S. markets are not included. The three leading states in terms of production value were California (46%), Washington (29%), and Oregon (10%).

Production of vegetable transplants had a wholesale value of about $130 million in 1998. Protected areas in transplant production were 23 million square feet and 4,054 open acres. The leading state was California (68%) followed by Florida (18%).

CAREERS IN OLERICULTURE

Seed firms need people experienced in olericulture, as well as in ornamental horticulture. Positions include a breeder who develops improved cultivars of vegetables, a propagator in charge of producing seed and vegetative propagules, an independent grower who contracts with seed firms for seed production, and sales personnel such as managers, dealers, and salesmen.

Vegetable production takes place on farms, some of which are family owned and operated. Others are owned on the corporate level. All require a person experienced in olericulture to manage all stages of vegetable production. Vegetable growers and managers are also part of the greenhouse vegetable production industry. Processing companies usually have a person in the field who acts as a liaison and consultant between independent growers and the processing company. Sales personnel knowledgeable in vegetables are found with companies that produce horticultural machinery and inputs. Financial specialists that deal in vegetable-related activities are also needed.

There are numerous positions within the vegetable industries, all of which require varying degrees of knowledge of olericulture. A number of them occur with sales agencies that handle sales of vegetables for growers. These agencies require an overall manager, a storage supervisor, a supervisor for grading and packing, a field person who is a liaison and consultant between the agency and growers, and a buyer. Brokers resell vegetable wholesale lots obtained from growers or sales agencies, and buyers for retail chains obtain such lots from growers, agencies, or brokers. There are also promotional agencies on the state or national level for specific vegetables that would need a director.

Extension personnel, professors, and researchers dealing with vegetables are often part of state land-grant universities and the USDA. Integrated pest management (IPM) consultants with vegetable backgrounds can consult and provide services to farmers or work in research or extension positions.

VEGETABLE PRODUCTION BASICS

Success with vegetables on the commercial and garden levels is dependent on a number of factors covered in previous chapters such as propagation, soil and water management, gardening, greenhouse management, and plant protection. Having mastered those chapters, only generalizations will be given here, because details are available in these referral chapters (Chapters 7, 11, 12, 13, and 15).

Crop and Cultivar Selection

Vegetable crops are selected by the home gardener on the basis of climate, ease of culture, personal taste, and desirability. Commercial vegetable growers choose crops on the basis of expected productivity in their climatic region, marketability, profitability, length of growing season, soil factors, water needs, and extent of local pest pressures. For example, carrots are misshapen when grown in soils with rocks and stones. The best carrots are produced in sandy soils free of rocks and stones such as those in certain areas of California, where the majority of carrots are grown. California, with its excellent and variable climate, long growing seasons, good soils and water availability is the main vegetable producer. Cultivars are often chosen on the basis of such factors as good yields and pest resistance, heat or cold tolerance, and requirements imposed by commercial processors or fresh market demands.

Soil Preparation

The soil must be prepared properly for sowing of seeds and setting out transplants. Soil fertility, organic matter content, drainage, aeration, and pH must be properly maintained. From a sustainability viewpoint, organic matter content and pH are critical. Organic matter plays a vital role in reducing soil erosion and maintaining soil health, especially with regards to microbiological processes that contribute to the availability of nutrients derived from decaying organic matter. These biologically processed nutrients not only reduce the need for easily leached chemical fertilizers, but also provide nutrients on a continuous basis long after chemical fertilizer nutrients are exhausted. In addition, organic matter contributes to the maintenance of beneficial microorganisms that protect plants from soil-borne diseases through competitive and antagonistic processes. Soil pH is critical in that less than optimal ranges can lead to decreased availability of nutrients, thus leading to decreased yield and the need for more chemical fertilizers.

Soil pH should be monitored through at least annual soil testing, and limestone or sulfur added as recommended. Organic matter can be maintained through a number of approaches such as crop residues, cover crops, green manure, and organic amendments such as manure or compost. Manure is an especially attractive option for vegetable growers (if available locally) in that it not only maintains organic matter, but also provides some nitrogen, phosphorus, potassium, and several other nutrients that can decrease the levels of applied chemical fertilizers.

Fertilization

The amount of fertilizer to apply to vegetable crops is highly variable. For each vegetable, the amount of removed nutrients (NPK) is known. One approach would be to simply apply that amount plus a margin of safety amount to account for losses from leaching. Essentially, that is almost always what is done with nitrogen, as conventional soil tests for nitrogen vary in reliability and often don't account for some forms of "banked" nitrogen that will be released over time. Nitrogen tests done by labs that specialize in soil tests for sustainable agriculture provide more accurate nitrogen needs (see Chapter 11). Ideally, one should apply enough nutrients to bring soil levels up to the vegetable's needs. Essentially, that is done with phosphorus and potassium, because soil tests given good accounting for these two nutrients.

Ranges of nutrient requirements for various vegetables are shown in Table 17–1. These numbers are expressed in pounds per acre. On a gardening level, these ranges are somewhat like the directions to apply 3 to 5 pounds of 5-10-5 per 100 square feet. Two questions arise when one views these numbers: (1) Why the range? (2) What amount do I use if I wish to practice sustainable vegetable production? Several factors contribute to the range. One factor is soil type. Mineral soils need more nitrogen than muck soils or other soils with high organic matter. A soil with considerable organic matter has good soil health leading to the availability of nitrogen from decaying organic matter, so less is needed. With phosphorus and potassium (expressed as P_2O_5 and K_2O), the range results from correlation with soil test data for both nutrients. The high number is an amount to apply if either nutrient is very low. The lower amount results from readings of high phosphorus and potassium (and very high with certain crops). Another factor is planting systems. High-density plantings need more nutrients than low-density plantings. For perennial crops, new plots need more nutrients than established plots. Another consideration is that fields with histories of legume cover crops have lower nitrogen needs than those without such crops.

TABLE 17–1   *Seasonal Nutrient Needs of Vegetable Crops*

Vegetable Crop	Nitrogen (lb/acre)	Phosphorus (lb P_2O_5/acre)	Potassium (lb K_2O/acre)
Artichokes	50–200	100–200	50–200
Asparagus	50–150	50–300	50–250
Beans	40–100	50–150	50–150
Beets, chard	75–200	50–150	60–150
Broccoli, brussels sprouts, cauliflower	100–200	50–200	50–200
Cabbage	80–125	50–200	50–200
Carrots	50–150	50–150	50–200
Celery/celeriac	150–300	100–200	50–200
Chinese cabbage, Chinese mustard	100–175	50–200	50–200
Collards, kale	100–125	50–150	50–150
Corn (Sweet)	75–300	50–200	50–200
Cucumber	75–150	50–200	50–200
Eggplant	75–125	100–250	50–250
Horseradish	100–200	50–150	50–150
Kohlrabi	100–150	50–150	50–150
Lettuce, chicory, dandelion, endive, escarole, radicchio	80–150	50–200	50–200
Muskmelons, watermelons	75–150	50–200	50–200
Mustard, arugula, New Zealand spinach	60–120	50–150	50–200
Okra	25–50	50–100	25–100
Onion, garlic, leek, shallot	80–250	50–250	50–200
Parsley/parsley root, parsnips	75–100	50–150	50–150
Peas	20–40	50–150	50–150
Peppers	75–250	50–150	50–200
Potato	120–250	50–250	50–350
Pumpkin, winter squash	75–150	50–200	50–200
Radish	30–150	50–150	50–200
Rhubarb	70–80	70–80	140–160
Rutabaga, turnip	50–80	50–150	50–200
Salsify	75–100	50–150	50–150
Spinach	60–120	50–150	50–150
Summer squash	75–130	50–150	50–150
Sweet potato	50–100	30–150	60–180
Tomato	30–175	50–200	50–200

*Splitting, timing, and placement of fertilizer varies with each crop. First plantings on new land can have larger nitrogen needs than those of established crops shown here. For example, plantings of asparagus on new land can need as much as 280 lb/N per acre.

Source: Adapted from Swiader, Ware and McCollum, 1992 and Commercial Vegetable Production Guides (http://www.orst.edu/Dept/NWREC/vegindex.html).

The use of these amounts will certainly give you high yields, but are not necessarily what you should use with a sustainable system. A first step in achieving this goal is to have soil tests done by labs that have experience with tests that account for nutrient contributions based on soil health and the use of practices such as green manure, legume cover crops and animal manure. A list of these types of labs is available through the Internet (see the Internet Resources at the end of the chapter). Generally, these soil tests would recommend lower amounts than conventional soil testing. Other things to do are to use practices that promote more efficient use of nutrients such as banding and split applications (see Chapter 11). Also faster maturing cultivars generally need less nutrients than later maturing ones.

Planting

Crops, whether direct-seeded or by transplants, must be put in at favorable times based on knowledge of several factors. These factors include dates of local last killing spring frost and first killing fall frost, days needed for maturity, crop responses expected from the variations in weather known to occur during the growing season, and marketing and processing demand. Propagation of many vegetables is by seed, which should be obtained from a reliable seed firm and be clean, properly labeled, fresh, and insect and disease free. Seed should be viable, whether fresh or stored. Some vegetables are propagated asexually such as asparagus, garlic, horseradish, onion, potato, rhubarb, shallot, sweet potato, and taro.

Some vegetable crops are grown from transplants (Figure 17–1) for reasons of early harvest (especially in the North) and for economical use of space. The transplants are usually produced in a greenhouse or outdoor seedbed at seeding rates three to six times those practiced in directly seeded fields. This leaves some field space free for early crops, which can be followed by transplants. Plants commonly grown from transplants include celery, cole crops, herbs, onion, and solanaceous crops. Removal of excess seedlings (thinning) or proper spacing of transplants reduces plant crowding. Precision seeding with mechanical seeders can be used to eliminate the labor and costs associated with thinning. Mechanical transplanters are also available.

Optimal efficiency can be realized in vegetable production by the use of certain horticultural practices. Restricted areas, such as small city gardens, as well as commercial olericultural operations, can benefit from some of the following practices. Keep in mind that increased productivity per unit area increases the loss of nutrients also. Therefore, soil fertility must be maintained if increased productivity is to be realized. These methods are intensive gardening, vertical gardening, interplanting, and succession planting. These methods were discussed in Chapter 12 with regard to gardening. Commercial vegetable producers would not make use of all of these gardening methods, but succession planting is one that is widely used commercially to assure continuous harvests. Another is high-density planting to increase yields, the equivalent of intensive gardening.

Irrigation

Water must be supplied if rainfall is insufficient (see Chapter 11). While uniform soil moisture throughout the growing season is generally required for optimal yields of most

FIGURE 17–1 An employee shows the NRCS District Conservationist seedlings of a few of the crops such as cantaloupe, collards, tomatoes, and watermelon grown on the farm. Half of the farm's crops are raised from transplants propagated in the greenhouse. (NRCS/USDA photograph.)

vegetables, certain stages are more critical than others. Vegetables subjected to periods of water stress during early development have lower yields and delayed maturity. Water stress during fruiting or maturation can produce poorer quality fruits such as with cucumbers (misshapen) or tomatoes (growth cracks). Too much water, however, with muskmelons and hot peppers at fruit development can result in reduced soluble solids and capsaicin, respectively. It should also be kept in mind that irrigation during germination can improve germination, resulting in improved stands. The upper layer of soil at seed planting depth dries out faster than below seeding depth.

Requirements also vary somewhat by crop. Crops grown for enlarged parts such as fruits (beans, peas, peppers), seeds (sweet corn), bulbs (onions), tubers (potato), tuberous roots (sweet potato), and fleshy leaf heads (cabbage, head lettuce) need the larger part of their water budget during enlargement of the harvested structures. Fruiting vegetables also have critical water needs prior to enlargement, from flowering time through fruit set. Water stress during this period can result in incomplete pollination and less fruit. Crops grown for their leaves such as salad crops and greens need uniform moisture throughout their development. Excessive water at any time can also cause problems such as increased diseases or result in poorer quality fruits or tubers grown for their dry matter, if applied excessively during fruit enlargement.

Vegetables tend to be shallow-rooted, so can exhibit drought stress in sandy loam after 1 inch of rain in as little as 3 days as observed in tomatoes. Some vegetables are more shallow-rooted than others such as cole crops, celery, corn, cucumber, leek, New Zealand spinach, onion, potato, and radish. These crops tend to need irrigation sooner than medium- to deeper-rooted crops. Crops with medium rooting depth are snap and soy beans, beet, carrot, eggplant, greens (kale, mustard, and turnip), peas, peppers, rutabagas, summer squash, and turnip. Deep-rooted vegetables are asparagus, lima beans, lettuce, okra, parsnip, pumpkin, rhubarb, winter squash, sweet potato, tomato, and watermelon.

Rooting depth also needs to be weighed against soil type. Sandy soils need more water than loams and clays. These soils also absorb water at different rates. For example, the same shallow-rooted crop in a sandy soil, a loam, and a clay soil might have the following irrigation rates/frequency, respectively: 1 to 2 inches/4 to 6 days, 2 to 3 inches/7 to 10 days, and 3 to 4 inches/10 to 12 days. Medium- and deep-rooted crops would have different numbers for these same soil variations. The trend would be more inches of water, but at greater intervals. Drought tolerance and periods of critical water needs are other factors that affect irrigation rates and frequencies. For example, asparagus has deep roots, high drought tolerance, and critical water needs during crown set and transplanting. On the other hand, tomatoes are deep rooted, but have only moderate drought tolerance and benefit from continuous water in terms of fruit size and minimizing blossom end rot.

Given all these variables, it is difficult to generalize about irrigation needs. Irrigation of 0.50 to 1.0 inch just before or right after seeding is helpful, unless weather is rainy. Similar amounts are helpful with vegetable transplants. A rule of thumb is that vegetable crops with a spread of 12 inches or more need 1.5 inches of water weekly during hot weather, but only 0.75 inches weekly in cooler weather. Application rates are 0.40, 0.30, and 0.20 inches per hour for sandy, loamy, or clay soils. Higher rates exceed absorbency rates and result in runoff and subsequent erosion and pollution. Deeper and less frequent irrigation is preferred over light, frequent ones, which tend to cause shallower root systems in vegetables. There is no substitute for local knowledge and frequent measurements of soil moisture. Tensiometers are used to measure soil moisture as soil tension in centibars (0.001 bars) and are preferred for sandy soils. Soil blocks are best with loams and clays: these measure available soil moisture (ASM) as a percent. ASM is the percentage of soil water between field capacity (−0.1 Bar) and permanent wilting point (−15 Bars). The latter section with individual vegetables will report preferred minimum soil moistures in both units.

Weed and Pest Control

Weeds must be sustainably controlled through either cultivation or mulching. Herbicides are the last resort. Cultivation depth needs to be matched to the crop. Deeper, more effective cultivation can be used with deep-rooted crops. Shallow cultivation is used with

FIGURE 17–2 Black plastic mulch is used commercially with tomatoes such as in Florida, both for reasons of earlier harvest and weed control. (Author photograph.)

shallow-rooted crops such as lettuce and later in the season with most crops. Closeness of cultivation to plants also changes from closer earlier to further away later. Mulches (Figure 17–2) are very effective and are popular with gardeners and with some commercial crops such as peppers. Mulch use is more likely if, besides weed control, higher yields and earlier harvests are a consequence of mulch usage. Preemergent herbicides are frequently used to give young plants a better start and to reduce later weeding problems. Herbicides are also used for weeds that are difficult to eradicate with shallow cultivation.

Insects and diseases must be controlled through rotations, resistant cultivars, and with biological controls supplemented with judicious use of pesticides when necessary (or without chemical pesticides in organic production). These two areas have been covered in Chapters 11 and 15. From a sustainable viewpoint, the best approach to pests is proactive, not reactive. Proactive approaches include sanitation, rotations, and the use of resistant cultivars. Plants suspected of disease or insect problems should not be planted, plants in the field or greenhouse with uncontrollable diseases must be destroyed and unused crop residues should be removed. Many hybrid vegetable cultivars are available with resistance to diseases. While more costly, this benefit plus higher yields can compensate for cost.

Rotations are critical to the prevention of soil disease and weed and pest (insects, mites, and nematodes) buildups in your fields. Rotations disrupt the feeding and breeding chains of these organisms. Rotations also improve nutrient status and soil health in that different vegetables contribute differing amounts of crop residues (root crop, no roots: salad crops, roots left) and different vegetables remove varying amounts of nutrients (sweet corn removes high nitrogen, radish low nitrogen). Each crop that follows later here includes the family name. Crops in the same family should not replace one another. For example, broccoli should not follow cabbage, as both are members of the Brassicaceae family. Potatoes should not follow tomatoes, as both are in the Solanaceae family. Peas could follow broccoli or cabbage, because peas are in the Leguminosae family. Weeds are also a factor in family considerations. Wild mustard and wild radish are weeds in the same family (Brassicaceae) as broccoli, cabbage, and other cole crops, so the presence of these weeds count when considering rotation times.

Rotation times vary. For example, club root of cabbages can persist in soil for 4 to 5 years, while *Fusarium* root rot lasts 2 to 3 years. Most rotation plans use a 4- or 5-year cycle. Besides time, another factor is choice. While it might seem always acceptable to follow one crop with another from an unrelated family, consideration of growth form is important. A deep-rooted crop better follows a shallow-rooted crop than another shallow-rooted crop. Deeper penetration means more earthworm channels and more organic matter residue. A wider-form crop (greater shading and killing effect on weed seeds left from previous crop) best follows a narrow-form crop. Nutrient uptake is also a consideration. A light nutrient-requiring plant better follows a heavy nutrient-requiring plant than another heavy nutrient-requiring plant. Some vegetables interfere with the subsequent growth of others. For example, lettuce that follows sweet corn is inhibited early growth-wise from decomposition products resulting from corn stubble decomposition. These same factors also play a role in choosing plant sequences for succession plantings and intercropping (see Internet Resources).

The best preventive practices for pest control are either the use of IPM systems or IPM systems without synthetic pesticides (organic systems). Many commercial biological controls are available against insects, mites, nematodes, and diseases. Examples are parasitoid wasps, predatory insects, predatory mites, predatory fungi, predatory nematodes, and antagonistic fungi. Specific examples for use for protection of field and greenhouse vegetables and mushrooms are found in the tables of Chapter 15. In addition, many botanical, natural product, and environmentally safe (green) pesticides can also be found there.

HARVESTING VEGETABLE CROPS

Determining when to harvest is a critical part of vegetable production. For the home gardener, the choice is clear. Harvesting usually occurs at peak quality, because the picked crop is either going to be used quickly, refrigerated only for a short period, or processed soon after harvest. For commercial operations, harvest is determined by a number of factors. For vegetables destined for processing, harvest can be close to peak maturity, because processing happens shortly after and better quality products result from using peak maturity vegetables as opposed to under-ripe vegetables. For fresh market, harvest can be close to peak maturity or somewhat less, because ripening will continue in many cases. Dead ripe vegetables are generally avoided, as bruising can occur during shipping and decreased quality can result during the time from harvest through sales. In some cases, gassing with ethylene is used prior to harvest to initiate uniform ripening; this is done with vegetables such as peppers and tomatoes. The determination of the correct picking time can be tricky. Days to maturity from seeding or transplanting are known and can be helpful predictors of maturity. Degree-days or heat units are even more reliable indicators. Other indicators are color (tomatoes and peppers), stem slippage (muskmelons), hollow "thunk" tap noises (watermelons), dry silk and milky kernel juice (sweet corn), rinds easily indented with fingernails (summer squash), or rinds not easily dented with fingernails (winter squash).

Harvesting equipment usually saves time and cuts the harvesting cost considerably. Many agricultural engineering departments in universities, the U.S. Department of Agriculture, and industrial research laboratories are involved in developing new harvesting machinery. At one time commercial operations did not have suitable mechanized harvesting equipment for many crops or the equipment was of limited feasibility because of variations in the time of ripening. However, harvesting machinery is under continual development and improvement even for easily bruised crops. Plant breeders have also aided mechanized harvesting by developing crop varieties with uniform ripening habits, compact growth, less easily bruised fruits, more easily detached vegetables (or fruits), or an angle of hang better suited for mechanized removal. In addition, more and more growth regulators are used as mechanized harvest aides to promote uniform ripening, color, and abscission. The one downside is that texture and taste are sometimes reduced in favor of a more harvest-friendly cultivar as has happened with tomatoes.

The oldest harvesting technique is hand harvesting, which is still practiced. The home horticulturist makes extensive use of this approach for his or her vegetable, fruit, and nut crops. Commercial operations also use this hand harvesting (Figure 17–3) for easily damaged vegetables and fruits that are intended for fresh market sales. Hand harvest is also employed for many other fresh market vegetables because of nonuniform ripening of various vegetables. Less easily damaged crops such as root crops and thick-skinned fruits can be mechanically harvested for fresh markets to reduce labor costs. Crops intended for processing such as tomatoes to juice (or apples to apple juice) are mechanically harvested, be-

FIGURE 17–3 Much of the commercial lettuce crop continues to be harvested by hand in the United States. (USDA photograph.)

cause damages or blemishes are not of concern. Even hand harvests are facilitated to some degree by aids such as conveyor belts or mechanized undercutting of carrots (to be sold with tops on) to facilitate hand removal. Other reasons for not using mechanized equipment are that suitable harvesting equipment is not available or is bypassed in favor of pick-your-own operations, and the cost payback or feasibility of the equipment is questionable.

The mechanization of harvesting varies as to degree and the number of steps involved. Some crops are highly mechanized in terms of harvest such as fresh or processed potatoes. First vines are killed as a necessary first step prior to harvests. Machines that have rubber or steel flails (mechanical beaters) go over rows. An alternative is to spray chemical desiccants to kill the vines. Mechanical diggers next dig up the potatoes and carry them directly to trailers or trucks. For processing tomatoes, machines cut and lift the vines, then shake them over conveyor belts while separating vines from fruits. Next the tomatoes are delivered to trailers off to the side. More sophisticated harvesters can sort fruits by color with electronic sensors. Humans are still needed to cull unacceptable fruits. Peas utilized for canning are whole-plant mechanically harvested by mowing machines. Vines are taken elsewhere where machines remove the vines and then shell the peas. Other vegetable crops are not as mechanized as these two. Some operations might be handled in a partially mechanized manner, for example, the hand picking of vegetables followed by mechanized grading and packing. Fresh market crops such as broccoli, cabbages, cauliflower, peas, and peppers would be handled in this manner. More on harvesting can be found later under discussions of the individual vegetables.

POSTHARVEST CHANGES OF VEGETABLES

When a horticultural crop is harvested, ongoing metabolic processes continue to produce changes in the picked crop. These changes constitute the science of *postharvest physiology*. The biochemical and physiological changes can enhance or detract from the appeal of the crop. The appeal of the crop is determined by three factors: quality, appearance, and condition. Quality is used in reference to flavor and texture, as determined by taste, smell, and mouth-tongue touch. The appearance refers to the visual impression of the crop, whereas the condition refers to the degree of departure from a physiological norm or the degree of bruising or disease.

Postharvest alterations are numerous and include water loss, the conversions of starch to sugar and sugar to starch, flavor changes, color changes, toughening, vitamin gain and loss, moisture loss, sprouting, rooting, softening, and decay. These processes may improve or detract from the quality, flavor, and texture of the crop. Most people are familiar with the loss of quality associated with the rapid enzymatic conversion of sugar to starch that occurs with harvested sweet corn, and the reverse reaction, starch to sugar, that takes place in potatoes. On the whole, postharvest changes produce more detrimental results than beneficial ones. If the crops are harvested for subsequent processing, the time between harvesting and marketing or processing should be as short as possible to minimize postharvest alterations.

No one set of conditions will achieve this end, because postharvest alterations are a function of the crop type, and each crop varies as to the optimal air temperature, relative humidity, and concentrations of oxygen and carbon dioxide required for stabilization. Some form of air circulation is usually beneficial in reducing disease problems and removing any released gases that might accelerate harmful physiological and biochemical changes. The level of disease pathogens should also be kept to a minimum.

The application of postharvest preventive measures will also vary as to when they should be applied. For example, if a postharvest change is both rapid and seriously detrimental, some treatment might be required from the moment the fruit is picked. This would be a problem with broccoli, which is highly perishable, but not with turnips. The treatment might be either ice packing or hydrocooling (see later discussion). Then it might have to be shipped to a grading and packing area in a refrigerated truck. The grading and packing area might even need to be temperature controlled; yet the comfort of the workers must be considered.

The most important aspect of postharvest treatments is in storage facilities designed for long-term crop storage, so that the crop can be sold at a reasonable price over an extended period of time. The best storage conditions slow the life processes but avoid dehydration, tissue death, and storage diseases, which produce gross deterioration and drastic differences in the quality, appearance, and condition of the stored crop. Although conditions vary, it is possible to make some generalizations.

Vegetables (also fruits and nuts) intended for storage should be free from mechanical, insect, or disease injuries and at the proper stage of maturity. Storage for the less critical crops, in terms of better postharvest longevity, usually consists of common (unrefrigerated) storage, which lacks precise temperature and humidity control. Examples of common storage include insulated storage houses, outdoor cellars, and mounds. Root crops such as rutabaga and carrots can be stored in such facilities, as well as other crops with good keeping qualities such as winter squash. More reliable storage consists of cold (refrigerated) storage with precise control of temperature and humidity. An improvement on this form of storage is the addition of a controlled atmosphere (CA storage) by which the concentration of oxygen and carbon dioxide are kept at an optimal level. A low temperature decreases the rate of respiration, the controlled humidity level slows water losses through transpiration and evaporation, and the controlled atmosphere brings about an overall reduction in the rates of chemical reactions occurring in the stored materials. The overall change in reactions is reflected in the inhibition of respiration and ethylene production. Apples and pears are examples of fruit stored under controlled atmosphere, temperature, and humidity conditions. Vegetables generally are stored under controlled temperature and humidity conditions.

Storage temperatures and humidity, of course, differ for each vegetable and fruit, but some generalizations are possible. A number of vegetables can be held near 0°C (32°F) and 90% to 95% relative humidity. Others need higher temperatures and lower relative humidity. Storage conditions for each vegetable are covered later in the chapter.

PREMARKETING OPERATIONS

Premarketing operations include washing, trimming, waxing, curing, precooling, grading, prepackaging, and shipping. Some steps are part of preventing postharvest damage; others are designed to improve or enhance the marketing value of the vegetable. Still others are used to remove unwanted materials. These steps can be done in the field, at specialized facilities, or at a combination of both places. Each place has advantages and disadvantages. Field operations reduce transport of and damage to the crop and cost less. The downside is that quality control is less, weather can cause problems, large machines can disrupt field activities and cause soil compaction, and cooling of small lots is less energy efficient than bulk lot cooling.

FIGURE 17–4   Harvested celery that has been trimmed to improve its appearance prior to marketing. (USDA photograph.)

FIGURE 17–5   Field of recently mechanically harvested and piled (windrowed) onions. After curing by air-drying, the onions will be removed. (USDA photograph.)

Washing, Trimming, Waxing, and Curing

At the receiving end, vegetables are transported either by dry conveyors or running water troughs. The more fragile crops and those needing washing are placed in the latter. Early on, some form of elimination is done to remove rotten or undersized produce and foreign objects such as rocks. Water is often chlorinated to reduce disease problems. Washing, often in conjunction with brushing, is needed especially for root crops, since they have adhering soil particles that reduce consumer appeal and might increase disease problems. A similar problem exists with leafy greens, especially after mud-splashing rains. Drying by air or giant sponge rollers follows.

Trimming (Figure 17–4) is used to cut back roots or discolored leaves or to cut back excess green tops for vegetables such as beet, carrot, celery, lettuce, radish, rutabaga, spinach, and turnip. Waxing is used to improve appearance and to slow moisture loss, which causes shriveling. Waxing has been employed to varying extents on cucumber, pepper, tomato, rutabaga, and turnip to reduce water loss and to improve appearance. Some sort of machine or hand sorting is next to remove off-quality materials. Vegetables are also sized mechanically or electronically and manually when necessary. Sizing is based on weight, volume, or some other physical dimension. Curing (Figure 17–5) is used to heal cuts or bruises incurred during harvesting of crops such as onions, potatoes, and sweet potatoes. Potatoes are cured by applying heat with humidity control.

Precooling

Precooling is used for rapid removal of heat from recently harvested fruits and vegetables. In some cases this might be done at the field or later at the packinghouse. For some crops cooling can be introduced at the washing step. With other crops cooling might occur after packaging. Precooling permits the grower to harvest at maximal maturity and yet be assured that the crop will have maximal quality when it reaches the consumer. Precooling slows the rate of respiration in the fruit or vegetable, slows wilting and shriveling by decreasing water loss, and inhibits the growth of decay caused by microorganisms. Precooling is an energy-requiring process; harvesting at night and in the early morning, when the temperature of the crop is lowest, can minimize its cost.

Major precooling methods include hydrocooling, contact icing, vacuum cooling and air cooling. In immersion hydrocooling, cold or iced (chlorinated) water flows around the bulk or bin containers before packaging has occurred. With spray hydrocooling, cold water is sprayed over the crop as it passes below on conveyor belts. The cold water absorbs heat from the vegetables and fruits. Crops that are suitable for hydrocooling include artichokes, asparagus, broccoli, cauliflower, celery, leeks, lettuce, muskmelon, peas, radish, scallions, spinach, and sweet corn. Contact or package icing is simply the placing of crushed ice in or around the packed produce. This technique has declined in use. One disadvantage is that packing containers must be able to withstand wetting. Crops can also be top-iced during rail car shipping. Crops suitable for icing include stem and root vegetables, brussels sprouts, carrots, melons, and green onions.

Vacuum cooling cools the crop through the rapid evaporation of water under reduced pressure, that is, evaporative cooling. Leafy vegetables are suitable for vacuum cooling. Some water is lost from the crop, but can be minimized by using fine water sprays during the vacuum cooling process. Room cooling and forced-air cooling is simply cold air blown over the crops. This technique can be used on fruit-based (bean, cucumber, eggplant, pepper, and tomato), tuber-based, and bulb-based vegetables.

After precooling the crop, it must be kept cool in transit by shipment in refrigerated trucks, railroad cars, and cargo holds in planes. If stored, cold storage is used, and refrigerated or water-mist display cases are preferred. Storage conditions are discussed briefly earlier in this section and later under individual vegetables.

Grading

Uniformity in quality, size, shape, color, condition, and ripeness is usually preferred (Figure 17–6). This is for reasons of consumer appeal, freedom from disease, and ease of handling by automated machinery. To achieve this end, vegetables and fruits are often graded according to standard grades, which form a basis of trade. The requirements of grading vary depending on factors such as whether the crop is destined for the fresh market or for processing. Grades have been established by federal law and are strongly suggested; however, they are not enforced legally unless the federal grades are used to describe the produce. Government grades include such designations as Fancy or Extra Fancy, No. 1, No. 2, and so on. Crops not meeting grading standards are usually processed or employed in products using lesser grades and off sizes. Grading used to be done manually, but automated video systems with computer-assisted packages are increasingly being used.

Packaging

At least half of all vegetables and fruits are packaged in retail units at some point prior to reaching the retail store. This type of packaging is called *prepackaging* (Figure 17–7). Products are placed in bags of transparent plastic film or in trays and cartons covered over by transparent film. Transparent film is usually made of polyethylene, which can be made to have a differential permeability toward oxygen and carbon dioxide. Often a controlled at-

FIGURE 17–6 Permissible shapes allowed for U.S. No. 1 Grade parsnips. (USDA photograph.)

FIGURE 17–7 Prepackaged cauliflower heads are wrapped in plastic wrap and shipped in this cardboard carton. (USDA photograph.)

mosphere can be reached in these containers covered in polyethylene, which consists of less oxygen and more carbon dioxide than is present in normal air. The retention of water vapor is also improved with plastic film, which is usually beneficial for the quality of the fruits and vegetables. Paper and mesh bags are also used for prepackaging products such as onions. Trays and cartons used for the prepackaging unit are usually made of Styrofoam or wood pulp. The master containers for the prepackaged units are usually made of paperboard. Wooden crates and bushel baskets are seen less and less. Master containers are designed for easy stacking, easy shipping, and minimal bruising. Packaging is automated in many cases. These packaged units are bulk-packed into larger shipping containers.

Shipping

Fruits and vegetables can be shipped a number of ways. Large trucks with refrigerated containers, which sometimes have a controlled atmosphere, are used extensively (for roughly 70% to 75% of vegetables). Refrigerated railroad cars are also used. Crops are also air freighted by large cargo planes. Cargo boats are another source of transportation. Loading and unloading are often facilitated by the use of wooden pallets and forklifts.

Preservation

Vegetables, fruits, and nuts destined for fresh consumption can be made available over a longer time by storage, which was discussed earlier. However, crops destined for processing can be preserved by a number of techniques. These include canning and freezing of many vegetables, dehydration (onion and parsley flakes), pickling (cucumbers to pickles, pickled beets), fermentation (cabbage to sauerkraut), and juicing (tomatoes). Chemical preservatives are often added to maintain quality in terms of taste, texture, and color. Coloring agents are frequently added to improve visual appeal. Home growers are more likely to only use freezing and sometimes canning. Some vegetables such as salad greens are not processed. Minor vegetables are also not commercially processed such as kohlrabi. Some crops are also processed into starch and flour such as the sweet potato.

Crops grown for processing usually cost less per unit area of land or per ton. One factor is that, unlike crops grown for the fresh market, appearance is not of major concern. However, size, quality, and uniformity are important. An extended harvest through successive plantings or the use of varieties with different maturation dates is needed to maintain a constant supply. This enables the processing factory to operate at an even flow over a reasonable period of time.

VEGETABLE CROPS A TO Z

Generally, vegetables can be grouped in the following categories, which are not always absolute. Salad crops include leafy vegetables that are mainly eaten raw, as in a salad. Arugula, celery, chicory, dandelion, endive, escarole, frisee, lettuce, parsley, raddichio, and watercress are often included in this group. Certainly, other vegetables are eaten raw in salads, but their main use is for other purposes. The legumes (family Leguminosae) or pulse crops can form a symbiotic relationship with *Rhizobium* sp., nitrogen-fixing bacteria. The edible legumes include chickpeas, cowpeas, fava beans, lima beans, mung beans, peas, snap beans, and soybeans. The solanaceous crops (family Solanaceae) include the eggplant, peppers, potato, and tomato. Cucurbits are vine crops (family Cucurbitaceae), which include the cucumber, melon, pumpkin, squash, and watermelon. The cole crops (family Brassicaceae or Cruciferae) encompass many *Brassica* species, such as broccoli, brussels sprouts, cabbage, cauliflower, Chinese cabbage, Chinese mustard, kohlrabi, kale, collards, and mustard. The last three are often included in the group of crops called greens, which are the immature leaves and stems of plants prepared by boiling. Chard, New Zealand spinach, spinach, and turnip greens are also in the group called greens. The onion group includes the common onion and its close relatives, garlic, leek, and shallot. The root crops are grown for their underground edible portions. They include the beet, celeriac, carrot, horseradish, parsnip, radish, rutabaga, salsify, sweet potato, taro, and turnip. Artichoke, asparagus, and rhubarb are included in the perennial crop group, whereas some vegetables such as mushrooms (not really a vegetable but treated as such), sweet corn, and okra are classed as miscellaneous vegetables.

Vegetables are a valuable component of the human diet. They contribute vitamins A and C, various B vitamins, and several minerals (calcium, iron, magnesium, phosphorus, and potassium). Vegetables are also valuable sources of fiber and phytochemicals that help prevent and fight various cancers.

An individual listing of the vegetables more commonly grown both commercially and in vegetable gardens follows in alphabetical order. A similar treatment for herbs follows the vegetable listings.

Artichokes (Globe and Jerusalem)

Globe artichokes are half-hardy, cool season perennial herbaceous, thistle-like plants. First-time eaters are often puzzled about how to eat them. Jerusalem artichokes are perennials cultivated as annuals. Both types of artichokes are in the Compositae or Sunflower family (Compositae or Asteraceae). Globe artichokes and Jerusalem artichoke cultivars are derived from *Cynara scolymus* and *Helianthus tuberosus,* respectively. Cultivars of globe artichoke are 'Green Globe Improved,' 'Imperial Stars,' and 'Purple Sicilian.' Cultivars of Jerusalem artichoke are 'Brazilian White,' 'Brazilian Red,' 'Mammoth French White,' 'Stampede,' 'Sun Choke,' and 'Sunray.'

Globe artichokes are essentially harvested for their flower buds, which have edible scales and receptacle. Globe artichokes thrive in areas with cool, mild climates. California is the leading commercial producer. Manure can be used at the rate of 20–30 tons/acre. Yields are 5 to 6 tons per acre. Their main food value is carbohydrate and some vitamin C and K. Aphids and plume moths can be troublesome, as can *Botrytis* and *Fusarium* root rot.

Jerusalem artichokes (a.k.a. girasole or Sunchoke®) are harvested for their potato-like tubers. Tubers store sugars as inulin, which can be isolated. Tubers are also good carbohydrate sources for diabetics. Jerusalem artichokes can be grown in most of the United States. Commercial and garden cultivation are minimal. This plant is one of the very few U.S. food crops indigenous to North America. Flavor is sweet and water chestnut-like. It can be grated and used in salads and main dishes. Modest amounts of iron, phosphorus, and B vitamins are present. Although trouble free for the most part, Jerusalem artichoke can become weedy if not controlled by harvesting.

Globe artichokes are asexually propagated, as seeds show variability. Either rooted offshoots from the base or crown division are used. Fall planting 6 inches deep is best, with rows spaced 8 feet apart and in row plants, 4 to 8 feet. Harvesting is possible over 5 to 8 years. A good plant can yield 40 buds. Hand harvesting is usually employed. Harvest starts in the fall and peaks in late spring. Smaller buds are sold as "baby" artichokes. Short-term storage at 0°C (32°F) and 90% relative humidity is possible.

Jerusalem artichokes are asexually propagated from the tubers in either the spring or fall. Tubers are planted 3 inches deep in rows about 3 feet apart with in-row spacing of 3 to 4 feet. Maturity takes about 5 months. Long-term storage of Jerusalem artichokes is possible at 0°C (32°F) and 90% relative humidity.

Arugula (*see* Mustard)

Asparagus

Asparagus (Figure 17–8) is a hardy, cool season perennial crop grown for its edible early spring shoots (also known as spears). Male and female plants appear, given its dioecious nature. Male plants are preferred, because they are longer lived and don't produce volunteer seedlings that reduce yield through competition. Mature height of the unharvested,

FIGURE 17–8 Asparagus is the leading supplier among vegetables of folic acid. It is also one of the richest sources of rutin, a bioflavonoid that strengthens capillary walls. Rutin also aids the absorption and use of vitamin C and decreases the oxidation of vitamin C within the body. A small amount of premium white asparagus is produced commercially in the United States under black plastic row tunnels (blanched). New plantings of asparagus can require up to 280 pounds nitrogen per acre. The use of green manure legumes can cut the need for chemical nitrogen in half. Established asparagus has modest nitrogen needs (Table 17–1). Asparagus water needs during early growth through harvest are roughly 5 to 6 inches monthly and during fern growth, 7 to 8 inches. (USDA photograph.)

fern-like parts is 3 to 6 feet. Cultivars of asparagus are derived from *Asparagus officinalis*. This species is found in the lily family (Liliaceae). As such, asparagus is one of the few monocot vegetables. Asparagus is also the most important of the perennial vegetable crops grown in the United States. The leading commercial producer in terms of crop value by far is California, followed by Washington and Michigan. Average yields vary, but for the West Coast yields average 1.4 tons/acre. Washington yields approach 1.8 tons/acre.

The early shoots themselves are somewhat tender and easily frost damaged. Spears are sold in fresh, canned, and frozen forms. Culinary use of asparagus is in soups, salads, and as a hot vegetable, especially with sauces such as hollandaise. It is noted for its reasonable amounts of vitamins A, C, and K, folate, riboflavin, calcium, and phosphorus.

There is a southern limit for asparagus, since it does best where the ground freezes to at least a few inches. This requirement would preclude the Deep South. Best spear quality occurs at a soil pH of 6.5 to 7.5 and when the temperature for 4 to 5 nights preceding harvest is 15.6° to 18.3°C (60° to 65°F), and daytime temperatures are moderate. Temperatures below 12.8°C (55°F) at this time produce a purple, tougher spear. First time plantings of asparagus on new land might need 10 tons or more barnyard manure per acre.

Direct seeding is usually limited to the commercial production of 1-year-old crowns, which are the propagule of choice for commercial and home plantings of asparagus. Seeds have a viability of 3 years. Seeds germinate (10 to 15 days) best at soil temperatures between 15.5° and 29.4°C (60° and 85°F) and are planted about 1.0 to 1.5 inches deep. Certified, disease-free crowns are initially planted 6 to 8 inches deep in either furrows or holes, with soil being gradually added as growth advances upward. Spacing for mechanical cultivation varies from 4 to 6 feet between rows and 10 to 24 inches between plants within each row. In gardens rows can be 18 to 24 inches apart. Weeds are best controlled by shallow cultivation early in the season before spears appear and afterward by mounding of soil. Asparagus has high drought tolerance, being able to grow well with 1 inch of rain in 20 days. The preferred minimal soil moisture is −0.70 Bars or 40% available soil moisture (ASM).

Harvesting should not be started with new crowns until the second (light) or third spring in order to establish strong crowns. Once established, an asparagus bed can be harvested for up to 25 years. The harvest period lasts from 5 to 12 weeks, depending on the prevailing climatic conditions, age, and vigor of the bed. Harvest periods are longer where the growing season is long and cool. Harvesting should be stopped when spear diameter diminishes noticeably. A row of 50 to 100 feet is sufficient for a family of six persons. Spears are harvested by knife or snapping when 5 to 8 inches above the ground surface. Generally larger diameter spears are best, as they have less fiber than smaller spears. Fresh market asparagus is usually hand harvested, while asparagus intended for processing is mechanically harvested. Asparagus should be used promptly or kept cool (0°–2.2°C, 32°–36°F) during storage and/or transit. For commercial situations, hydrocooling is preferred and relative humidity should be high, (i.e., 95%).

The main diseases affecting asparagus are rust (*Puccinia asparagi*), root rot (*Fusarium oxysporum* f. *asparagi*), and stem/crown rot (*F. moniliforme*). Insect pests include the common and spotted asparagus beetles (*Crioceris asparagi* and *C. duodecimpunctata*, respectively), asparagus aphid (*Brachycolus asparagi*), and the garden symphalen (*Scutigerella immaculata*).

The use of disease-resistant cultivars is recommended. Early cultivars are 'Mary Washington' and 'Waltham Washington.' More recent ones include 'Jersey Knight Hybrid,' 'Jersey Giant Hybrid,' and 'Waltham Hybrid,' all male, higher yielding cultivars possessing disease resistance. Several other all-male cultivars in the 'Jersey' line are also available. 'Apollo,' 'Atlas,' 'Grande,' and 'UC 157' are high-yield cultivars for milder areas such as the West Coast. Two novelty cultivars are also known, 'Purple Passion F_1' and 'Larac Hybrid.' These have purple, sweet tasting, and mild tasting white (self-blanching) spears, respectively.

Beans

Basically, three types of major beans are grown: the green or yellow (wax) snap bean (Figure 17–9) with edible pods (also called string beans), the full-sized but immature lima (Figure 17–10) and green shell beans (French horticultural) not eaten with pods, and the mature, dry shell beans (black bean, red kidney beans, navy beans) not eaten with pods. Lesser grown beans include asparagus beans (a.k.a. Chang dou, Dau gok, and yard-long beans), cowpeas (a.k.a. black-eyed peas, crowder beans, southern peas), fava beans (a.k.a. broad beans), garbanzo beans (also called chickpeas), garden soybeans (Figure 17–11), mung beans, and scarlet runner beans.

The immature pods of snap beans, asparagus beans, soybeans, and scarlet runner beans are usually steamed, microwaved, or sautéed as a main vegetable dish. Lima and other green shell beans (no pod) such as immature fava, French horticultural, garbanzo beans, or soybean are simmered and buttered. Dry shell beans (no pod) require soaking followed by boiling/simmering and are often made into baked beans, soups, or combined with rice. Dry garbanzo beans and cowpeas after boiling/simmering are used in salads, soups, stews, rice dishes, and in Middle Eastern cooking such as hummus (garbanzo only). Mung beans are usually used as bean sprouts. Snap beans are good sources of vitamins A, C, and K, folate, and potassium. Asparagus beans are rich in potassium and vitamins A and C. Fava beans supply protein, iron, fiber, vitamins A and C, and potassium. Lima beans are high in protein, fiber, potassium, and vitamins A, B_1, and B_6 and C. Garden soybeans, cowpeas, and garbanzo beans are high in protein.

All three types occur normally as twining vines, but bush (dwarf) cultivars are available. Bush cultivars are grown mostly in commercial operations, being much more adaptable to once-over, destructive mechanical harvest than vine types. Bush types tend to bear over a short period versus a long period for vines. Vine types can be intercropped with sweet corn. Snap beans are very popular both as a fresh/processed commercial crop and as a garden crop. Lima beans, shell beans, dry beans, and other minor beans are less popular in gardens and commercially.

Cultivars of snap beans, green shell beans, and dry shell beans are derived from *Phaseolus vulgaris*, large lima beans from *P. limensis*, and smaller lima beans (also called butter beans)

FIGURE 17–9   String beans should be rotated to minimize damage from soil-borne diseases. Avoid fields previously planted to cabbage, carrots, cucurbits, lettuce, parsnips, potatoes, and tomatoes, as they can harbor sclerotinia white mold. Research at Oregon State University with many green snap bean varieties suggests that 36 square inches/plant (174,000 plants/acre) is optimal for yield. Two critical times for water needs during bean development are bloom and pod set. Peak water use for green beans use is about 0.20 and 0.16 inches per day for April and June plantings, respectively. On most soils, weekly irrigations during peak periods are best, but sandy and sandy loam soils may need irrigation as often as every 3 to 5 days. (USDA photograph.)

FIGURE 17–10   Baby lima beans are planted at 3-inch row intervals and rows are 22 inches apart. For large-seeded lima beans, the respective numbers are 4 and 36 inches. The rule of thumb for harvesting lima beans is that a field where 10% of the pods are "dry" will produce good quality yields. Rows are cut using either rotary cutters or knives, windrowed, and then picked up by mobile viners for threshing. New "pod-picking" harvesters used for peas can also be used for harvesting processing limas. These harvesters end the need for swathing and windrowing. Fresh market lima beans are usually hand harvested, but new mechanical harvesters exist that harvest intact pods. (USDA photograph.)

FIGURE 17-11
Soybeans are essentially an agricultural crop, so are not covered much here, except that they are mentioned in the context of a garden vegetable. For a complete history on soybeans, refer to http://www.agron.iastate.edu/soybean/history.html. (USDA photograph.)

from *P. lunatus.* Bush cultivars of snap, lima, and butter beans are derived from *P. vulgaris humilis, P. limensis limenanus,* and *P. lunatus lunonanus,* respectively. Lesser grown bean cultivars are derived from the following species: asparagus beans (*Vigna unguiculata* subsp. *sesquipedalis*), cowpeas (*Vigna unguiculata*), fava beans (*Vicia faba*), garbanzo beans (*Cicer arietinum*), garden soybeans (*Glycine max*), mung beans (*Vigna radiata*), and scarlet runner beans (*Phaseolus coccineus*). All are members of the pea or pulse family (Leguminosae or Fabaceae).

Commercial production for fresh snap beans is extensive in Florida, Georgia, New York, and North Carolina. Wisconsin, Oregon, Michigan, and New York are noted for processed snap beans. Average yields per acre are 3.0 and 3.7 tons for fresh and processed snap beans, respectively, although rates as high as 5 and 6 tons are reported in California and Oregon for fresh and processed types, respectively. Fresh lima beans are limited in production, most are processed. Georgia and North Carolina grow most of the fresh supply and California plus Delaware supply processed lima beans. Lima bean yields are around 1.3 tons/acre for both fresh and processed crops. Dry bean yields are 0.8 tons/acre with the leading states being North Dakota, Nebraska, Michigan, and Minnesota. This group includes large and baby lima, navy, great northern, small white, pinto, kidney, pink, small red, cranberry, black, blackeye, and garbanzo beans.

Cowpeas are best grown in the South and Southwest. Garden soybeans have cultivars adapted to either the North or South. Garbanzo beans are best grown in dry areas of California and mung beans in the South. Scarlet runner beans grow best where summers are warm and fava beans are best grown on the Pacific Coast.

Beans, being legumes, can form a symbiotic relationship with bacterial species of the genus *Rhizobium.* Since these are nitrogen-fixing bacteria, it is good horticultural practice to encourage their growth. Inoculants of *Rhizobium* sp. are available, and their use is desirable if the soil has no *Rhizobium* present. Once introduced into the soil, they should persist, especially if legumes are continued in the soil. Soil pH should be in the range of 5.5 to 7.0.

Beans are tender, warm season annual legume crops that are not frost tolerant. The lima bean is less cold tolerant (very tender) than the snap bean. Lima beans need a longer, warmer growing season than snap beans, which makes their culture somewhat difficult in the far northern parts of the United States. Snap beans typically mature in 50 to 70 days, while limas need 75 to 100 days. Dry beans, cowpeas, garbanzo beans, fava beans, and soybeans also need a longer growing season to properly mature.

Bean seeds, since they decay quickly, should not be sown until the soil has warmed to at least 10°C (50°F). Seeds have a viability of 3 years. Seeds germinate in 4 to 8 days best at soil temperatures between 60° and 85°F and are sown about 1.0 to 1.5 inches deep. Bush bean rows can be as close as 12 inches apart to up to 3 feet, depending on whether hand or machine cultivation is used. Plants in the row can be placed 2 to 4 inches apart. High-density plantings with close rows are popular with multiple-row mechanical harvesters, but need more fertilizer and careful weed and disease control. Pole beans require more space (rows 3 to 4 feet apart and in row spacing of 6 to 12 inches). If hills are used with several plants, hills should be 3 to 4 feet apart each way. Preemergent and postemergent herbicides are often used to control early and later weeds, respectively. Cultivation needs to be shallow, because beans are surface rooted.

Beans have low to moderate drought tolerance. Snap beans, dry beans, and lima beans need 1 inch of water weekly; their preferred minimal soil moisture is −0.45 Bars or 50% ASM. Pole beans and cowpeas/edible soybeans need 1 inch of water in 5 (−0.34 Bars, 60% ASM) and 14 (−0.70 Bars, 40% ASM) days, respectively. The critical water need is during flowering and pod fill. No irrigation is needed with dry beans after the pods begin to dry.

Pole bean bearing period is longer than the bush forms. However, succession plantings of the bush forms at 2-week intervals, until only the minimal time required for harvest remains before the first killing frost, can extend the harvest period. Pole beans are ideal for nets on the garden's edge. Bush forms are especially useful for following early maturing crops, such as spinach, radish, peas, lettuce, beets, or early potatoes. Snap beans are harvested when the pods are almost full sized with the beans one-quarter to one-third developed. Lima beans and other green shell beans are picked with full-sized green pods and nearly full-sized beans. Dry beans are picked when pods are mature and bean seed moisture is 15% to 20%. Snap beans and lima beans are perishable. As such after harvest they are washed, culled, and quick-cooled, preferably hydrocooling, but vacuum cooling and forced-air cooling are also used. During storage and shipping, snap beans require temperatures of 4.4° to 7.2°C (40° to 45°F) and lima beans, 2.8° to 5.0°C (37° to 41°F); relative humidity should be maintained at 90% to 95%. Dry beans require storage conditions of 4.4° to 10.0°C (40° to 50°F) and 40% to 50% relative humidity.

A number of insects and diseases attack beans. Insects include the Mexican bean beetle (*Epilachna varivestis*), leafhopper (*Empoasca fabae*), bean leaf beetle (*Cerotoma trifurcata*), aphid (*Aphis fabae*), bean weevil (dry beans only, *Acanthoscelides obtectus*), and Japanese beetle (*Popillia japonica*). Nematodes can be a serious problem with cowpeas, although some cultivars have resistance to nematodes. The cowpea aphid (*Aphis crassivora*) is another pest. Diseases include anthracnose (*Colletotrichum lindemuthianum*), bacterial blight (*Xanthomonas phaseoli, Pseudomonas phaseolicola*), bean mosaic virus, downy mildew (*Phytophthora phaseoli*), rust (*Uromyces phaseoli*), and sclerotinia white mold (*Sclerotinia sclerotiorum*). Disease problems can be reduced by the use of resistant varieties. Beans in rotation should not follow cabbage, carrot, cucurbits, lettuce, parsnips, potatoes, strawberries, and tomatoes, given possible disease similarities.

Suggested varieties of bush green snap beans include 'Astro,' 'Blue Lake 274,' 'Blue Wonder,' 'Bush Kentucky Wonder 125,' 'Early Contender,' 'Early Bush Italian,' 'Jade,' 'Oregon Trail,' 'Roma II,' 'Tendercrop,' and 'Tenderpick'. Pole forms of green snap beans include 'Blue Lake,' 'Cascade Giant,' 'Kentucky Wonder,' and 'Romano Italian Pole.' Yellow bush snap beans of choice are 'Brittle Wax,' 'Cherokee,' 'Gold Mine,' 'Pure Gold,' 'Rocdor,' and 'Slenderwax'. 'Goldmarie' and 'Kentucky Wonder Wax' are pole forms of yellow snap beans. Green shell beans include 'Cannellini,' 'Dwarf Horticultural,' 'Flambeau,' and 'French Horticultural.' Baby bush lima bean (butter bean) cultivars include 'Eastland Lima' and 'Baby Fordhook.' Regular bush limas are 'Dixie Butterpea,' 'Fordhook No. 242,' 'Henderson,' 'Kingston,' and 'Thorogreen,' whereas pole varieties include 'Burpee's Best,' 'King of the Garden,' and 'Large Speckled Christmas.' Fava bean cultivars include 'Acme,' 'Aprovecho Select,' 'Broad Improved Long Pod,' 'Express,' 'Jumbo,' and 'Windsor.' 'Black Coco,' 'Black Turtle,' 'Maine Yellow Eye,' 'Pinto,' 'Red Kidney,' 'Trout,' and 'White Marrowfat' are used for dry shell beans. Cowpea cultivars include 'Brown Crowder,' 'California Blackeye No. 5,' 'Dixilee,' 'Magnolia Blackeye,' 'Monarch Blackeye,' 'Pinkeye Purplehull,' 'Queen Anne,' and 'White Acre.' Suggested garden soybean cultivars include 'Butterbaby,' 'Early Hakucho,' 'Envy,' and 'Shironomai.' 'Garbanzo' and 'Scarlet Runner' are chickpea and scarlet runner bean cultivars, respectively. 'Asparagus Yardlong' is an asparagus bean cultivar.

Beets

Beet cultivars are derived from *Beta vulgaris*. Beets (Figure 17–12) are in the Goosefoot family (Chenopodiaceae). The beet is both a root and greens crop that is a half-hardy, cool season crop that can be grown in all parts of the United States. Root types are found in the Crassa group; roots are used as a cooked (boiled) or pickled vegetable, for forage, and as a commercial source of red food pigment and sugar. Beet tops are also used as leafy salad greens or cooked like spinach. Other names for beet tops are beet leaves, chard, and Swiss chard; these cultivars are from the Cicla group. Beets are usually red, but can be orange, yellow, or nearly white. Beet roots are good sources of vitamin C and carbohydrates. Beet tops are rich in iron and vitamins A, C, and K. Commercial processed production is highest in Wisconsin and New York, and Texas leads in fresh production. Yields per acre average 8 to 10 tons and 15.6 tons for fresh and processed beets, respectively. Rates as high as 22 tons/acre have been reported for the latter.

Beets grow poorly in strongly acid soils and are highly sensitive to boron deficiency. A well-tilled soil at pH 6.0 to 6.8 and devoid of rocks is best; otherwise, beet roots will be misshapen. Since the beet matures quickly, successive sowings at 2-week intervals are possible until insufficient time remains for full maturity prior to a killing frost. Beet seeds (actually seedballs containing several seeds) are sown between 0.5 to 1.0 inches deep in rows 1 to 3 feet apart depending on hand or machine cultivation. Seed viability is 4 years; germination takes 6 to 10 days. Optimal soil temperature range for germination is 10° to 29.4°C (50° to 85°F). Fresh beets usually require thinning to 6 to 8 beets per foot of row. Processing beets can be left at 12 to 20 plants per row-foot.

Beets are deep rooted and require minimal irrigation, except in dry areas. Beets require 1 inch of water in 14 days. Their critical period is during root expansion, when insufficient water causes growth cracks. Their preferred minimal soil moisture is −2.00 Bars or 20% ASM. Shallow cultivation is critical at seedling stage, because beet seedlings fare poorly

FIGURE 17–12   Hybrid beets have better seedling vigor, improved top growth, and resistance to certain diseases. The downside is that their vigor can result in beets moving past grade more rapidly when poor weather or other causes delay the harvest. Cylindrical beets possess greater uniformity and are best for sliced beet production, a major packing requirement by processors. About 6–8 and 15–25 lb of beet seed per acre are required for fresh and processing beet sowings. Beet greens can also be found in the Orientalis group, where the leafy parts are dominant as opposed to the large, pronounced petiole with Swiss chard. Peak water consumption (0.21 inches per day for beets) occurs in July. Weekly irrigation during peak use is adequate, but with sandy and sandy loam soils, irrigation frequency can increase to every 3 to 4 days. (USDA photograph.)

against weeds. If herbicides are used, care is urged, given beet sensitivity to some herbicides, especially on light, sandy soils.

The best quality beets are produced during the cooler part of the growing season and are about 1.25 to 1.50 inches in diameter. Fresh beets are harvested at the 1.25- to 2.5-inch-diameter stage (55 to 80 days after planting) and processing beets when the roots size average is around 2.0 inches in diameter (90 to 110 days). Machines are available for harvesting and topping beets. Beets sold with tops on (bunch beets) are hand harvested. Both types can be stored at 0°C (32°F) at 98% relative humidity. Bunch beets can be stored for only 1 to 2 weeks, while topped beets keep for 4 to 6 months.

Insects that attack beets include the beet leafhopper (*Circulifer tenellus,* vector for curly top virus), beet leafminer (*Pegomya betae*), beet webworm (*Loxostege sticticalis*), and cutworms. Fungal pathogens include both leaf spot (*Alternaria* and *Cercospora* sp.) and scab (*Streptomyces scabies*). Black spot is a physiological root disorder caused by boron deficiency, especially in dry, alkaline soils with high levels of calcium. Popular early beets are 'Chicago Red Hybrid' and 'Early Wonder.' Other popular cultivars are 'Charlotte,' 'Cylindra,' 'Detroit Dark Red,' 'Golden Beet,' 'Perfected Detroit,' 'Ruby Ball,' 'Ruby Queen,' and 'Red Ace Hybrid.' 'Big Top' and 'Crosby's Greentop' are good for bunched beets.

Broccoli

The nomenclature of broccoli is somewhat confusing. The typical broccoli (Figure 17–13) in the supermarket, especially in the spring, is sprouting (a.k.a. asparagus, Calabrese, or Italian) broccoli. Its cultivars are derived from *Brassica oleracea,* Italica group. A second type, called heading or Romanesco broccoli, is a lighter green, more cauliflower-looking broccoli. In effect, it is a form of cauliflower. Broccoli rabe (a.k.a. brocoletto, brocoli raab, and turnip broccoli) is a nonheading, loosely sprouting, more pungent, broccoli derived from *Brassica rapa,* Ruvo group. Adding further to the confusion is broccoflower, a cross showing the physical features of cauliflower, but the green color of broccoli. Further confusion results from broccolini, a hybrid of broccoli and Chinese Kale.

Broccoli, a cole crop in the Mustard family (Brassicaceae or Cruciferae), is a hardy, cool season vegetable grown for its large, edible terminal and later lateral green buds and stems.

FIGURE 17–13 Broccoli provides the best vegetable nutrition available. A medium size stalk of broccoli provides 220%, 100% and 20% of the daily requirement of vitamins C, K, and A, respectively. Broccoli is rich in the anticarcinogens indole carbinol and sulforaphane and has as much calcium, ounce for ounce, as milk. When planting broccoli, the field should have been free of crucifer crops (and crucifer-related weeds) for at least 2 years. Most broccoli for fresh and processing markets is direct-seeded with precision seeders. Some early market fresh broccoli is from transplants. (USDA photograph.)

It is somewhat more heat resistant than cauliflower. Since it is a cool season crop, it is best grown in the spring and late summer. Commercial production is heavily concentrated in California, both for fresh and processed crops. Broccoli is served steamed, microwaved, or sautéed as a main vegetable or used as a raw dipping component in dips. Broccoli is considered to be one of the more healthful vegetables, being rich in cancer-fighting phytochemicals, vitamins, fiber, iron, calcium, and potassium.

Seeds (4-year viability) are usually started indoors (4 to 8 days for germination, optimal soil temperature range for germination, 7.2° to 35°C (45° to 95°F) for the spring crop in areas with cool, but short springs such as the North and Northeast. Areas with longer, cool springs such as California use direct seeding. Seeds are early-spring sown about 0.5 inch deep in rows 24 to 36 inches apart, depending on whether hand or mechanical cultivation is used. Plants in rows are 10 to 24 inches apart. Transplants in early spring about 4 to 5 weeks old are set out in the soil, which should be at pH 6.0 to 7.0. A lower pH can lead to molybdenum deficiency in broccoli (and cauliflower). Fall crops are also possible. Placement of seeds or transplants in fields is highly mechanized. Yields average 7.0 tons per acre, although yields as high as 10 have been reported in Arizona.

Broccoli has low drought tolerance, especially during head development. Insufficient water causes strong flavor. The preferred minimal soil moisture is −0.25 Bars or 70% ASM. Water needs are one inch in five days.

Broccoli should be harvested prior to the opening of the green buds. Many broccoli cultivars are ready in 40 to 60 days. Broccoli raab matures in 50 to 60 days. Broccoli is often harvested by hand because cultivars do not usually mature uniformly. Later sprouts are often gathered in gardens, but are usually not cost effective for commercial operations. Broccoli is usually ice-packed or hydrocooled, and stored/shipped at 0°C (32°F) and 95% relative humidity. Broccoli needs to be kept away from ethylene-producing apples and pears. Broccoli needs to be moved fast, because it is highly perishable. Fresh broccoli has a dark green color (a touch of purple is really good). Yellowing or light green broccoli is past good quality.

Insects and diseases associated with broccoli are those that attack cabbage. Popular cultivars include 'Arcadia,' 'Bonanza Hybrid,' 'Emperor,' 'Gallant Hybrid,' 'Green Comet Hybrid,' 'Italian Green Sprouting,' 'Mariner Hybrid,' 'Packman Hybrid,' 'Premium Crop,' and 'Southern Comet Hybrid.' Raab cultivars include 'Broccoli Raab,' 'Di Rapa,' 'De Brocoletto,' and 'Sessantina Grossa.'

Brussels Sprouts

Brussels sprouts (Figure 17–14), a cole crop in the Mustard family (Brassicaceae or Cruciferae), is a hardy, cool season vegetable grown for its edible lateral buds produced in series along the upright stem in the form of miniature heads. Cultivars are derived from *Brassica oleracea*, Gemmifera group. Brussels sprouts contain cancer-fighting phyochemicals, vitamins A, C, and K, thiamin, folate, iron, potassium, and phosphorus. They can be served steamed, microwaved, or sautéed as a main vegetable. Fresh ones have a sweet, pungent taste, but old ones tend to be bitter. Sometimes the whole stalk with still-attached sprouts is marketed as assurance of freshness. Most of the fresh and processed crop is grown in California.

In the North they can be grown as a spring or late summer crop, and in addition to these times also a winter crop in the South. Brussels sprouts mature in 80 to 160 days, with cultivars for the North having shorter maturity times. However, the best quality brussels sprouts are obtained when days are sunny and light frosts occur during the night. Because of this, brussels sprouts are favored as a fall crop. Both direct seeding and transplants are used. Seeding depth, germination times, viability, spacing, and culture are very similar to broccoli and cabbage. Transplants are usually produced in an outdoor seedbed or cold frame. Placement of seeds or transplants in fields is highly mechanized. Soil pH should be between pH 6.0 and 7.0. Seeds are sown in early spring through early summer and transplants are put in the permanent field in late spring through summer. Placement of seeds or transplants in fields is highly mechanized. Timing is determined by whether the intent is for a spring, late summer, or fall crop. Yields average 8 to 9 tons per acre.

FIGURE 17–14 When planting brussels sprouts, the field should have been free of crucifer crops (and crucifer-related weeds) for at least 3 years. Most commercial production is done with transplants. Up to 15 to 20 inches of water are needed seasonally with this crop. (USDA photograph.)

Brussels sprouts have moderate drought tolerance; insufficient water during sprout formation causes poor sprout production. Water needs are 1 inch in 5 days. The preferred minimal soil moisture is −0.25 Bars or 70% ASM.

The harvest period is about 6 to 8 weeks, since the harvesting of the mature lower sprouts (1 to 2 inches in diameter) and the removal of the associated leaves does not end the usefulness of the plant. Further sprouts higher up the main stem develop, since the terminal crown maintains the vigor of the plant. Fresh sprouts are often hand harvested. Treatment with growth regulators produces a larger number of simultaneous sprouts on a more compressed stem for mechanical harvest, especially with processed sprouts. An alternative favored from the sustainability point is to cut off the top growing point when lower sprouts are about 0.5 inch in diameter. Tip removal promotes a stem full of uniformly mature sprouts in 4 weeks. Some cultivars are better adapted to this method.

Disease and insect problems are those of cabbage. Care is required for rotations to ensure pest control such that cole crops or related weeds should not be in one place more than once in 3 to 4 years. Sprouts can be ice-packed and stored/shipped at 0°C (32°F) and 95% relative humidity for about 30 days. Sprouts need to be kept away from ethylene-producing apples and pears.

Suggested cultivars include 'Jade Cross Hybrid,' 'Lunet,' 'Oliver,' 'Prince Marvel Hybrid,' 'Trafalgar Hybrid,' and 'Tasty Nuggets Hybrid.'

Cabbage

Cabbage (Figure 17–15), a cole crop in the Mustard family (Brassicaceae or Cruciferae), noted for its heads of tightly folded, fleshy leaves, is a hardy, cool season crop. It is a biennial grown as an annual crop. Heads vary in shape from conical (Wakefield) to round (Copenhagen Market), round/oval (Danish Ballhead), and flat globe (Flat Dutch). Red cabbages also exist. Cultivars are derived from *Brassica oleracea*, Capitata group. The Savoy cabbage, a cultivar with puckery, blistered leaves is *B. oleracea bullata*. The main states for commercial fresh production are New York, California, Texas, Georgia, and Florida. Leading states for cabbage processed into sauerkraut are Wisconsin and New York. Yields average 15.7 and 27.0 tons per acre, respectively, for fresh and processed (kraut) cabbage. Fresh yields on the order of 23.5 tons/acre have been reported in Colorado.

Fresh cabbage is usually boiled or steamed and served perhaps most notably as corn beef and cabbage. Raw cabbage is also made into cole slaw and processed by fermentation into sauerkraut. Cabbage contains reasonable amounts of vitamins C and K, and considerable fiber.

FIGURE 17–15   Cabbage was one of the earliest cultivated crops. Northern Europe was the origin for wild cabbage, which was loose leafed like collards. It was one of the first crops brought by colonists to the New World, where it grew well. When planting cabbage, the field should have been free of crucifer crops (and crucifer-related weeds) for at least 3 years. In the Pacific Northwest, yields of processing cabbage can be 40 tons/acre. Water needs can be as high as 23 inches in the growing season. (USDA photograph.)

Cabbage as a garden crop is easily overdone, as heads often ripen uniformly and you end up with too many heads and few recipes.

Cabbage is generally treated as a spring, early summer, and fall crop in most of the United States. In the South it can be carried over as a winter crop. Seed viability is 4 years; seeds are sown 0.5 inch deep in rows 24 to 36 inches apart, depending on whether hand or mechanical cultivation is used. The optimal soil temperature range for germination is 7.2° to 35.0°C (45° to 95°F) and germination takes 4 to 8 days. Plants in rows are 12 to 24 inches apart. The spring crops of cabbage are frequently grown from transplants started 4 to 6 weeks earlier to allow for maturity before warm weather arrives, especially in the North. Transplants can be set out when hard frosts cease. Later cabbage is either seeded directly in place (thinning required) or grown from transplants produced in an outdoor seedbed. Cabbage destined for processing is usually direct seeded. Field placement of seeds or transplants is highly mechanized. Late cabbage can be started between rows of potatoes prior to harvest, or it can follow early potatoes, spinach, beets, peas, radishes, and other early crops. Soil pH is best from 6.0 to 6.8. If club root exists, a slightly alkaline pH can help control this disease, although rotation should be tried first.

Cabbage has moderate to high drought tolerance. The critical period is during head development, where too little water causes growth cracks. The water needs are one inch in 10 days; preferred minimal soil moisture is −0.34 Bars or 60% ASM.

Preplant herbicides are often used for weed control, followed by cultivation. Cultivation can start out deep, but needs to become shallower because cabbages are shallow rooted. Winter cabbage crops generally have fewer weed problems. Cultivation should stop when heads begin to form. Cabbage is reasonably drought tolerant. However, too much dryness not alleviated by irrigation followed by heavy rain leads to head splitting.

Heads are ready for picking anywhere from 55 to 125 days, depending on cultivar and when planted. Some cabbages planted as winter crops can take up to 180 days. Cabbage for fresh market is harvested when the head is firm and white and weighs 2 to 5 pounds. Cabbages for processing are harvested at higher weights up to 12 pounds. Mechanized harvest is usually done for processing cabbage, while hand harvest is employed for fresh market. Cabbages can be stored at 0°C (32°F) and 98% relative humidity for about 60 days (early cabbage) or 180 days (late cabbage). Cabbages should be kept away from ethylene-producing apples and pears.

Several insects attack cabbage and other cole crops. These include the cabbage worm (*Pieris rapae*), cabbage looper (*Trichoplusia ni*), diamondback moth caterpillars (*Plutella xylostella*), aphid (*Brevicoryne brassicae*), root maggot (*Hylemya brassicae*), thrips, and black or red harlequin bugs (*Murgantia histrionica*). Diseases include the fungal-caused clubfoot (*Plasmodiophora brassicae*), black rot (*Xanthomonas campestris*), black leg (*Phoma lingam*), leaf spot (*Alternaria* sp.), and cabbage yellows (*Fusarium oxysporum*). Some cultivars possess varying degrees of disease resistance.

Cabbage (and all related cole crops) should never follow another cole crop or cole related weed for at least 3 years.

One physiological disorder called internal tipburn is seen with fast-growing, high-density cultivars when transpiration is restricted, leading to lower calcium mobility in the plant. Internal tissue damage results. Uniform moisture availability can limit this problem.

Cabbage cultivars for early crops include 'Copenhagen Market,' 'Early Jersey Wakefield,' 'Everlast Hybrid,' and 'Stonehead Hybrid.' Midseason and later cabbages include 'Bravo Hybrid,' 'Cardinal,' 'Castello Hybrid,' 'Grand Prize Hybrid,' 'King Slaw,' 'Red Jewel Hybrid,' 'Super Red Hybrid,' 'Savoy Express Hybrid,' 'Savoy King Hybrid,' 'Surprise Hybrid,' and 'Sweet Surprise Hybrid.' The late fall or winter cabbages are 'Danish Roundhead,' 'Late Flat Dutch,' 'Marabel Hybrid,' 'Savoy Siberia Hybrid,' 'Tundra Hybrid,' and 'Winter Star.'

Carrots

Carrots (Figure 17–16) are root crops grown for their edible, fleshy orange (sometimes yellow) taproots that vary from long and pointed through blunt and cylindrical. Carrot cultivars are derived from *Daucus carota sativus* and are found in the parsley or carrot family (Umbelliferae or Apiaceae). They are a half-hardy, cool season crop and make a good fall, winter, and spring crop in the South, and a good summer through fall crop in the North. Best growth occurs when the temperatures range between 15.5° and 21.2°C (60° and 70°F). California is the main commercial producer of fresh carrots and Washington, California, and Wisconsin are the primary suppliers of processed carrots. Fresh yields average 15.3 tons/acre, although tonnage above 30 has been reported. Carrots for processing have average yields near 25 tons per acre. Carrots are excellent sources of vitamins A and K, and fiber. Carrots can be eaten raw, in salads, or as a dipping vegetable. As a main vegetable, carrots are usually steamed, microwaved, or boiled.

Seed viability is 3 years, the optimal soil temperature range for germination is 7.2° to 29.4°C (45° to 85°F) and seeds are covered with 0.25 to 0.50 inch of soil. Germination is slow, taking 10 to 18 days. Rows, or even better, beds for hand cultivation can be placed 12 inches apart. For machine cultivation, 24 to 36 inches is required. In-row spacing for processing carrots is 1.5 to 2.0 inches. Fresh carrots can be placed at 0.5 to 1.0 inch. Thinning is a prohibitive expense for carrots, so precision mechanical seeding is required. Soil should be in the pH range of 6.0 to 8.0 and of a deep, friable nature free of debris and rocks. If not, an 8-inch raised bed is recommended. Otherwise, the roots will be of poor shape. Carrots can be started in early spring and sown at 3-week intervals for successive harvests.

Carrots have critical water needs during germination and root expansion. Poor stands, growth cracks, and misshapen roots result from water stress. Drought tolerance is moderate to high; one inch of water in 21 days is needed. The preferred minimal soil moisture is −0.45 Bars or 50% ASM.

Carrots can be harvested from finger size to maturity. While some "baby carrots" are young-harvest, high density plantings, most are produced by cutting and peeling of large carrots. Fresh carrots are harvested short of maturity and those for processing at maturity. Days to maturity vary from 60 to 100, with fresh carrots being at the lower end. Carrots are all

FIGURE 17–16   Carrots originated in South Asia. Carrots can provide up to 220% of the required daily allowance of Vitamin A through the transformation of their beta-carotene (a powerful antioxidant) into vitamin A in the body. Carrots also contain calcium pectate, a substance that reduces cholesterol by binding to bile acids and preventing their reabsorption into the body. While most carrots today are orange, other colors exist. Most "baby carrots" are actually pieces cut from larger carrots, but real baby carrot cultivars are available such as Amca, Amstel, Babette, Colora, Caropak, Indu, Minicore and Verona. Popular commercial cultivars for fresh and processed carrot markets include Apache, Caro-Best, Caropak, Gold Pride, Legend, Plato, and Orlando Gold. Carrots for the premium fresh market benefit from the use of floating row covers. Water needs seasonally for carrots run 20 to 30 inches. (USDA photograph.)

machine harvested with either tops on (bunch carrots) or off (bulk carrots). Bunched carrots require some hand labor, but command a higher market price. Mature, topped carrots can be stored at 0°C (32°F) and 98% relative humidity for about 7 to 9 months. Bunched carrots, if hydrocooled, top-iced, and packed in polyethylene bags can be held for 30 to 45 days. Ethylene-producing produce such as apples and pears can cause bitterness to develop in carrots.

Weed control during early carrot growth is critical. Pre- and postemergent herbicides are often used until carrot foliage is large enough to shade out weeds. Shallow cultivation between rows is effective and often near the end some soil is hilled over the tops to prevent top-greening.

The carrot rust fly (*Psila rosae*) is a serious insect pest. Proper timing as to sowing can minimize damage by the larvae of this fly; otherwise, control by insecticides is necessary. Diseases include leaf blight (*Cercospora carotae, Alternaria dauci*) and carrot (aster) yellows. The latter is a phytoplasma transmitted by a leafhopper (*Macrosteles fasifrons*).

Four basic cultivar groups of carrots are used, based on root size and shape. These are 'Chantenay,' 'Danvers,' 'Imperator,' and 'Nantes.' A few carrot cultivars of choice are 'A-Plus' (high vitamin A), 'Apollo,' 'Bolero,' 'Caroline Hybrid,' 'Caropak' (often used for baby carrots), 'Danvers Half Long,' 'Early Nantes,' 'Imperator 58,' 'Little Finger,' 'Nantes Half Long,' 'Red Cored Chantenay,' 'Royal Chantenay,' 'Scarlet Nantes,' 'Short 'n Sweet,' 'Sweet Sunshine,' 'Sweet Treat Hybrid,' and 'Fly Away Hybrid.'

Cauliflower

Cauliflower (Figure 17–17), a cole crop in the Mustard family (Brassicaceae or Cruciferae), is not quite as hardy as cabbage, being a half-hardy, cool season crop. It is also not as heat tolerant as cabbage and will not form heads if the temperatures are too high. Cauliflower cultivars are derived from *Brassica oleracea,* Botrytis group. Cauliflower is noted for its large edible,

FIGURE 17–17 Cauliflower does best with uninterrupted growth. Growth delays cause the plants to prematurely form small heads of no value. Soil must have good fertility and should be high in organic matter to retain moisture. Cauliflower crops that follow a good clover sod perform well. Well-manured soils also give good results with cauliflower. While most cauliflowers are white, Romanesco and broccoflower ('Alverde' and 'Macerata') cultivars are green. When planting cauliflower, the field should have been free of crucifer crops (and crucifer-related weeds) for at least 3 years. Moisture stress, especially at the 6 to 7 leaf stage, can cause cauliflower to form unmarketable button heads prematurely. (USDA photograph.)

whitish "snowball" head (curd), which consists of a condensed, thickened malformed flower cluster. A second later or winter type, called heading or Romanesco broccoli, is a lighter green cauliflower. This type requires cold exposure and a long growing season and sees limited production in the United States. Both should be harvested before they discolor and become loose.

Cauliflower is grown both for the fresh and processing market. The major producing state is California. The average yield is 8 tons/acre. Cauliflower is served raw as a dipping component or in salads and steamed, stir-fried, or microwaved as a main vegetable. It is also pickled. It is a good source of vitamins C and K, and cancer-fighting phytochemicals.

Temperature requirements make it a spring, fall, and winter crop in the South, and a spring or fall crop in the North. Cauliflower is often started from 4- to 6-week-old transplants set out in the spring or by direct seeding in the late spring through early summer. Soil pH can range from 6.0 to 7.0. Placement of seeds or transplants in fields is highly mechanized. Seed viability, planting depth, germination temperature, germination time, and other cultural information is similar to broccoli, so will not be repeated.

Cauliflower has low drought tolerance, especially during head development. Water-stressed heads show buttoning (small heads) and ricey curds. Preferred minimal soil moisture is −0.34 Bars or 60% ASM.

Heads of early cultivars are protected from sun injury by tying the leaves over the heads or buttons as they begin to form. Later cultivars often have in-curving inner leaves, which makes this practice unnecessary. Harvest is usually by hand, but has been reduced in frequency by using cultivars with more uniformity for maturation. Fresh market cauliflower is either hydrocooled or vacuum cooled and ice-packed. Storage for 21 to 28 days at 0°C (32°F) with a relative humidity of 95% is possible.

Insects and disease are the same as described for cabbage.

Recommended cultivars include 'Andes,' 'Apex,' 'Candid Charm,' 'Early White Hybrid,' 'First White Hybrid,' 'Ravella Hybrid,' 'Snowball,' 'Snow Crown Hybrid,' and 'Hybrid Snow King.' 'Shannon' and 'Minaret' are late or winter Romanesco cultivars. A broccoflower (cross between broccoli and cauliflower) cultivar of note is 'Green Harmony.'

Celery and Celeriac

Cultivars of celery (biennial harvested as an annual) and celeriac are derived from *Apium graveolens dulce* and *A. graveolens rapaceum,* respectively. Both are found in the parsley or carrot family (Umbelliferae or Apiaceae). Celery (Figure 17–18), a salad and lesser soup crop, is noted for its edible fleshy petioles. Celeriac is a crop noted for its thickened, edible root/crown. Both are half-hardy, cool season crops grown as winter and spring crops in the South and as spring or fall crops in the North. California is the primary commercial producer of celery. Celeriac is mostly a home garden product and grown to a very small extent by specialty growers. It is also used in salads, dips, stir-fries, stuffing, and soups. Both have minimal nutritive value other than fiber and vitamin K. Average yields of celery are 34 tons/acre. Most celery goes to fresh market use. Celery seeds are also used in cooking.

Soil should be rich, moist, and well drained. Soil pH can be from 5.5 to 6.6, with the lower end better for organic soils and the upper end, mineral soils. Celery or celeriac is usually started outdoors with transplants that are 8 to 10 weeks old. Optimal germination soil temperature is 15.5 to 21.1°C (60° to 70°F). Germination of seeds can be speeded by overnight soaking in water. Direct field seeding is also used, but labor costs for thinning make direct seeding costs almost equivalent to transplants. Seeds have a 5-year viability, germinate in 12 to 20 days, and are covered with one-eighth inch of soil. Spacing between rows for hand and mechanical cultivation is 18 to 24 and 30 to 36 inches, respectively. In-row spacing is 5–8 inches. Celery and celeriac from transplants take 100 to 140 days to reach maturity.

Celery has low drought tolerance and a continuous need for water. Moisture deficits can result in small petioles and even irreversible growth stoppage. Water needs are one inch in five days. Preferred minimal soil moisture is −0.25 Bars or 70% ASM.

Early celery is sometimes blanched in the field by mounding soil. Late celery may be partially field blanched and the final blanching completed in storage. Self-blanching cultivars and high-density plantings can reduce blanching labor. Blanching reduces the vitamin A content somewhat and green celery is now preferred, so blanched celery is less common today. Because of slow growth, weeds can be troublesome. While some herbicides are used, shallow cultivation is effective. Celery has high water needs, and irrigation is often needed.

Celery can be harvested before it reaches full size if desirable. Mechanical harvesting of celery is used; some celery is hand harvested. Celery requires washing and trimming. Hydrocooling is preferred. Celery should be shipped and stored at 0°C (32°F) and 98% relative humidity. Storage for 60 to 90 days is possible.

FIGURE 17–18 Celery does best with a uniform and regular supply of water. Frequent irrigations are preferred, as irregular or infrequent applications can increase black heart problems. Celery is often harvested well into the fall. Given that frost can occur, irrigation is sometimes used for frost protection. The following cultivars might have *Fusarium* resistance: Deacon (tall), Matador, Picador, Promise, UC8-1, UC10-1, UC26-1, Vicar, and XP-85. (USDA photograph.)

Insects that attack celery include the carrot rust fly (*Psila rosae*) and the tarnished plant bug (*Lygus lineolaris*). Diseases include early and late blight (*Cercospora apii* and *Septoria petroselini*, respectively), bacterial leaf spot (*Pseudomona apii*), root rot (*Phoma apiicola*), yellows (*Fusarium oxysporum*), and pink rot (*Sclerotinia sclerotiorum*). Other diseases include aster yellows (a phytoplasma, see carrot entry) and western celery mosaic virus. The former is spread by leafhoppers and the latter by aphids. Physiological disorders include black heart, chlorosis, and cracked stem. The first results from calcium deficiency and nutrient imbalance, the second from magnesium deficiency, and the third from boron deficiency.

Good celery cultivars include 'Florida 683,' 'Giant Pascal,' 'Golden Plume,' 'Golden Self Blanching,' 'Matador,' 'Picador,' 'Slow Bolting Green,' 'Summer Pascal,' 'Tall Utah 52-70 Improved,' 'Tango Hybrid,' 'Tendercrisp,' and 'Victoria F$_1$ Hybrid.' 'Alabaster,' 'Brilliant,' 'Diamant,' and 'Monarch' are celeriac cultivars.

Chard

Chard or Swiss chard (Figure 17–19) is one of the vegetables termed greens. It is a type of beet grown for its edible leaves. If the outer leaves only are removed, the harvest period can be extended throughout the summer, since chard is well adapted to hot weather. Usually, only one sowing is necessary. Seeds are usually used, although transplants are possible. Chard has moderate drought tolerance. Maturity is in 40 to 65 days. Leaves and stalks are cooked by steaming, boiling, and stir-frying. Chard is a minor commercial crop. It is of easy garden culture. Chard is a good source for vitamin A and iron.

The nomenclature, planting, culture, and pests for chard are the same as for beet. The only difference is the in-row spacing is slightly larger (4 to 6 inches). Cultivars include 'Bright Lights,' 'Burpee's Fordhook Giant,' 'Burpee's Rhubarb Chard,' 'Lucullus,' and 'Vintage Green.'

Chickpeas (*see* Beans)

Chicory (*see* Endive)

Chinese Cabbage and Chinese Mustard

Chinese cabbage, a cole crop in the Mustard family (Brassicaceae or Cruciferae), noted for either cos lettuce-like heads of loosely folded, fleshy leaves, or a nonheading type similar in growth form to celery or chard. Both forms are half-hardy, cool season crops, primarily

FIGURE 17–19 Chard tolerates hot weather and provides greens after spring spinach and asparagus are finished. With partial harvest, chard will continue to produce into the fall. It can withstand frosts and some freezing, thus it is available even into early winter. (USDA photograph.)

FIGURE 17–20   Napa cabbage (shown in figure) yields are about 22 tons/acre and pak-choi about 11 tons/acre. (Courtesy of National Garden Bureau.)

FIGURE 17–21   Pak-choi (shown in figure) yields are about 11 tons/acre. (Courtesy of National Garden Bureau.)

grown as a fall crop in the North and as a winter crop in the South. The heading form (Figure 17–20) is known as pe-tsai or Napa cabbage (*Brassica oleracea,* Pekinensis group) and the nonheading form (Figure 17–21) as pak-choi (*Brassica oleracea,* Chinensis group). Other names for pak-choi are bok choi, Chinese mustard, or celery mustard. Both types can be used in salads or cooked, either by steaming or stir-frying.

Planting is either by seeding directly outdoors or from 5- to 6-week-old transplants. Drought tolerance is low and water needs are continuous at 1 inch in 5 days. Water stress causes tough leaves. Preferred minimal soil moisture is −0.25 Bars or 70% ASM. If planted too early, hot weather will force flowers before the head is fully developed. In addition, hot weather may adversely affect texture and flavor. Soil pH should be between 6.0 and 7.0. Seed information, planting, culture, and other information is similar to that for cabbage. Heading types reach maturity in 65 to 100 days and nonheading types in 35 to 60 days. Neither type keeps long and should be used fresh.

Insects and diseases are similar to those of cabbage. Suggested head-forming cultivars are 'Jade Pagoda Hybrid,' 'Joi Choi Hybrid,' 'Michihli,' 'Nagaoka,' and 'Wong Bok.' 'Crispy Choy' and 'Pac Choi' are nonheading cultivars.

Collards

Collards (Figure 17–22), a cole crop in the Mustard family (Brassicaceae or Cruciferae), are noted for its cabbage-like cluster of leaves that do not form heads. Cultivars are derived from *Brassica oleracea,* Acephala group. Collards are a hardy, cool season crop, primarily grown as a fall crop in the North and as a winter crop in the South. Collards are somewhat more tolerant of heat than cabbage. Georgia, North Carolina, and South Carolina are the leading producers. Yields average 7 tons/acre. Collards are good sources of vitamins A, C, and K, folic acid, and calcium. Collards are treated as greens, generally boiled or sautéed.

Collards have continuous water needs; water stress causes tough leaves. Water needs are one inch in 14 days, as drought tolerance is moderate. Preferred minimal soil moisture is −0.45 Bars or 50% ASM.

Soil pH should be 6.0 to 7.0. Seeds are usually sown directly in place, although 5- to 6-week-old transplants can be used. Seed information, planting, culture, and other information is similar to that for cabbage. Leaves may be partially removed to extend the har-

FIGURE 17–22
Collards are planted at rates of 12,000 to 14,000 plants per acre. When planting collards, the field should have been free of crucifer crops (and crucifer-related weeds) for at least 3 years. The cultivar, 'Green Glaze,' has been reported to exhibit some resistance to imported cabbage worm, diamondback moth, and cabbage looper. Harvest is done both by hand and machine. (USDA photograph.)

vest. Leaves or whole rosettes should be harvested before becoming tough and woody (60 to 90 days from seed). Insects and diseases are similar to those of cabbage. Suggested cultivars are 'Blue Max,' 'Champion,' 'Georgia,' 'Green Glaze,' 'Morris Heading,' and 'Vates.'

Corn (Sweet)

Sweet corn (Figure 17–23) or sugar maize is a tender, warm season, frost-sensitive plant grown in most parts of the United States. It is noted for its edible, sweet, immature seed ears. Kernels can be yellow, white, or mixed yellow and white. Sweet corn is a monocot found in the grass family (Graminae). Cultivars are derived from *Zea mays rugosa* (= *saccharata*). Yields per acre of fresh sweet corn are 5–6 tons. Florida, California, Georgia, and New York are the leading fresh producers. Yields per acre of processing sweet corn are 7–9 tons. Washington, Minnesota, and Wisconsin are the leading commercial states for processed sweet corn. Sweet corn is served as a vegetable alone on the cob or as removed kernels alone or mixed with other vegetables by steaming, roasting, boiling, or microwaving. Sweet corn supplies mostly carbohydrates and modest amounts of protein and vitamins A and C.

Corn is usually directly seeded 2 inches deep for a summer and fall crop. Seed should be fresh each year and is sown 1 inch deep. Different cultivars should be isolated to prevent cross-pollination, which can have a detrimental effect on kernel quality. Germination takes 6 to 8 days, but can be longer with cool soils. Soil temperature must be at least 10.0°C (50°F). The optimal soil temperature range for germination is 15.5° to 35.0°C (60° to 95°F). Rows are spaced 24 to 48 inches apart with in-row spacing of 6 to 18 inches. Exact spacing is based on cultivar and types of mechanization and whether for fresh or processed markets. Garden spacing can be at the lower end. Soil pH should be between 6.0 and 7.0.

Corn has moderate to high drought tolerance. Irrigation prior to the critical water need period, silking, has little benefit. Water stress during silking causes poor ear fill. Water needs are one inch in 14 days. Preferred minimal soil moisture is −0.45 Bars or 50% ASM.

The harvest season is brief because of texture changes and enzymatic conversion of sugar to starch. Some sowings can be made shortly before or after harvest of early crops such as peas, beets, radishes, lettuce, and others. An extended harvest can be obtained through the simultaneous use of early, midseason, and late cultivars or succession plantings of a cultivar at 2-week intervals or times determined by degree-days. Early sweet corn commands premium prices and is often produced with soil covers such as polyethylene or spun-bonded polyester (see Chapter 13). Corn is shallow rooted, so adequate water must be supplied and cultivation must be shallow and often discontinued when corn is 24 inches

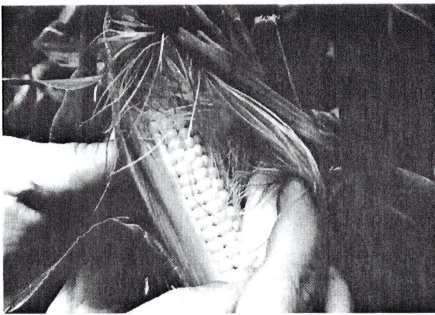

FIGURE 17–23 Sweet corn cultivars need to be isolated by one of three methods: distance, timing of pollination, or accepting some outer rows contamination and discarding that portion (blocking). Distance is the preferred method. White corn is usually separated by 500 feet from yellow or bicolor types. Similar color corns that differ in sweetness (normal, sugary-extended, and supersweet) should be isolated by 250 feet. Supersweet corns should also be located upwind of other types. Similar distances (250 feet) are used with popcorn, ornamental corn, and regular corn. Some popcorn cultivars are 'White Cloud' and 'Crookham 1084.' A few popcorn ornamentals are Chinook, Indian Ornamental, and Fiesta (standard size) and Little Indian, Strawberry, and Wampum (miniature types). Corn information can be found at http://maize.agron.iastate.edu/general.html#article. (USDA photograph.)

tall. Herbicides are often used on sweet corn. Harvesting is highly mechanized. Harvest is best when the silk first browns/dries and the kernel juice is milky.

Sweet corn must be picked fresh and promptly used, because sugar converts to starch fast. Newer, supersweet cultivars can hold sugar contents for several days, making them popular with supermarkets. Other gene-altered corns are available with increased sugar holding power. The various types each have pluses and minuses. All sweet corn should be cooled promptly, preferably hydrocooled, kept cool during shipping, and moved rapidly in the marketing chain.

A number of insects attack corn: corn earworm (*Heliothis zea*), European corn borer (*Pyrausta nubilalis*), southern corn borer (*Diatraea crambidoides*), corn (northern, southern, and western) rootworm (*Diabrotica longicornis, D. undecimpunctata howardii,* and *D. virgifera*), wireworm (*Agriotes mancus*), corn flea beetle (*Chaetochema ectypa*), army worm (*Pseudaletia unipuncta*), and chinch bug (*Blissus leucopterus leucopterus*). Birds, such as crows and starlings, frequently dig and eat freshly sown seed. Diseases include bacterial wilt (*Bacterium stewartii*), corn smut (*Ustilago maydis*), corn blight (*Helminthosporum turcicum*), leaf rust (*Puccinia sorghi*), and maize dwarf mosaic virus. Resistant cultivars are available. Racoons can be especially troublesome.

There are many hybrids of sweet corn. Maturities vary from 60 to 100 days. A few of the regular, older yellow hybrids (rapid sugar loss) are 'Early Sunglow,' 'Golden Queen,' 'Golden Bantam,' 'Seneca Chief,' 'Seneca Horizon,' 'Sundance,' and 'Merit.' Older white hybrids are 'Silver Queen' and 'Stowell's Evergreen Hybrid.' An older mixed yellow-white hybrids is 'Honey and Cream.' New, gene-enhanced sweet corn cultivars are:

1. *Yellow:* 'Bodacious,' 'Florida Staysweet,' 'Illni Xtra-Sweet,' 'Kandy Corn,' 'Northern Xtra-Sweet,' 'Sugar Ace,' 'Summer Sweet,' and 'Sweet Riser';
2. *White:* 'How Sweet It Is,' 'Silver King,' 'Silver Princess,' and 'Sweet Ice';

3. *Bicolor:* 'Brocade,' 'Butterfruit Bicolor,' 'Fantasy,' 'Honey and Pearl,' 'Seneca Dancer,' and 'Sun and Stars.'

Numerous corn cultivars are also available for the production of popcorn ('Crookham,' 'Robust,' and 'White Cloud') and ornamental corn ('Chinook,' 'Fiesta,' 'Indian Ornamental,' 'Little Indian,' 'Strawberry Popcorn,' and 'Wampum') used decoratively during autumn and Halloween. While not discussed here, these items can offer extra income to sweet corn producers. If grown, these two corns need to be isolated from each other and also from sweet corn by 250 feet.

Cowpeas (*see* Beans)

Cucumber

The cucumber (Figure 17–24) is a very tender, warm season annual vine crop or cucurbit that does not withstand frost. Cucumbers are grown for their edible fruits. Cucumber cultivars are derived from *Cucumis sativus,* a member of the gourd family (Cucurbitaceae). Mostly served raw in salads, cucumbers can also be cooked. Cucumbers are also processed into pickles. Cucumbers have only modest amounts of vitamin A and C, plus small amounts of iron and calcium.

Cucumbers can be grown throughout most of the United States. Fresh cucumber average yields are 10 tons/acre, with most being produced in Florida, California, and Georgia.

FIGURE 17–24   The slicing cucumber shown here is a fresh market type. Slicing cucumbers can be monoecious (separate male and female flowers on the same plant) and gynoecious (female flowers only) and predominantly female (PF) types. Others can be parthenocarpic types (also called burpless or seedless). Oriental (slicing) cucumbers are longer and narrower and of the burpless type. Cucumbers should not follow cucumbers or the other curcurbits such as muskmelons, pumpkins, squash, or watermelons. English or European cucumbers are seedless (parthenocarpic) or all female (gynoecious) varieties that are grown as a greenhouse crop. Greenhouse production requires careful management and good growing skills to be profitable. Some cultivars are 'Corona,' 'Factum,' 'Fidelio' (powdery mildew tolerant), 'Femfrance,' 'Femspot,' 'Fertila,' 'LaReine,' 'Pandorex,' 'Pepinex '69,' 'Pepinova,' 'Sandra' and 'Santo.' Pickling cucumbers are usually direct-seeded to yield 75,000–85,000 plants/acre for machine harvest and 18,000 to 20,000 for hand harvest. Most pickling cucumbers are machine harvested with 'Calypso' being most popular. Cucumbers yields can be enhanced with plastic mulch over trickle irrigation and floating row covers. (Author photograph.)

Processing cucumber yields average 5.9 tons/acre. Yields of 16 tons per acre have been reported in California. Michigan, North Carolina, Florida, and Wisconsin are leading commercial processing producers. There is also a parthenocarpic (burpless or seedless) cucumber with edible skin known as the European type. Cucumbers are the second most important vegetable in terms of commercial greenhouse production.

Cucumbers can be sown directly after all danger of frost is past, or 3-week-old transplants are used, but peat pots are suggested such that minimal root disturbance occurs. Seed viability is 5 years. Germination (6 to 10 days) is poor below 15.5°C (60°F). The optimal soil temperature range for germination is 15.5° to 35.0°C (60° to 95°F). Seeds are sown 0.5 to 1.0 inch deep. Rows are spaced 30 to 60 inches apart with in-row spacing of 8 to 12 inches for hand cultivation. Rows for mechanical cultivation are closer at 12 to 30 inches with in-row spacing of 2 to 6 inches. Choices of cultivars are critical with these high-density plantings. If space is limited, cucumbers can be grown on fences, trellises, nylon net, and the like. Soil pH can range from 5.5 to 8.0; 5.8 to 7.0 is better. A second sowing 4 to 5 weeks later can be used to extend the harvest season, or successive sowings based on degree-days can be used. Manure can be applied up to 10 tons/acre.

Cucumbers can be monoecius, with male flowers appearing first, then equal percentages of male and female flowers, followed mostly by female flowers. Fruit set and ripening tend to not be uniform, making these cucumbers more suitable to gardens than for mechanized horticulture. Newer cucumbers have either complete or mostly gynoecious flowering, having either all female or mostly female flowers, thus assuring more uniform fruit set and uniform maturity for mechanized harvest. A percentage of male pollinators are often included in the seed mix as a pollen source. Seedless (parthenocarpic) cucumbers are used for greenhouse culture, but require spraying with a fruit-setting growth regulator. Cucumbers for mechanized harvest mature earlier (40 to 50 days) than older cultivars (55 to 70 days).

Cucumbers have low drought tolerance and need a constant moisture supply to prevent malformed development. Water stress during flowering and fruiting causes pointed and cracked fruits. Earlier water stress results in reduced yield and quality. One inch of water in 7 days is needed. Preferred minimal soil moisture is −0.45 Bars and 50% ASM.

Weed control can be difficult, in that cucumbers show more senstivity to herbicides than most crops. Cultivation can also be difficult with high-density crops, although eventually foliage shades out new weeds. Cultivation needs to be earlier and close to plants up to the time vines start to run. After that point, cultivation becomes shallow and further out from plants. Black plastic mulch is effective and promotes earlier crops.

Cucumbers should be picked when dark green (any yellow color indicates the seeds will be hard) and not overly large; otherwise, seeds become more noticeable. Cucumbers for pickling are picked at a less mature stage. Fresh cucumbers are handpicked; mechanical harvest is used for processing cucumbers. Cucumbers are hydrocooled and held at 10.0° to 12.8°C (50° to 55°F) and 95% relative humidity for 10 to 14 days. Cucumbers are often waxed to reduce moisture loss.

A number of insects attack cucumbers. They are the striped cucumber beetle (*Acalymma vittata*), which carries bacterial wilt and cucumber mosaic; spotted cucumber beetle (*Diabrotica undecimpunctata*), which carries bacterial wilt; aphid (*Aphis gossypii*); and pickle worm (*Diaphania nitidalis*). Diseases include powdery mildew (*Erysiphe cichoracearum*), anthracnose (*Colletotrichum lagenarium*), scab (*Cladosporium cucumerinum*), angular leaf spot (*Pseudomonas lachrymans*), bacterial wilt (*Erwinia tracheiphila*), mosaic (a virus spread by aphids mostly), downy mildew (*Pseudoperonospora cubensis*), and leaf blight (*Alternaria cucumerina*). Some cultivars have multiple disease resistance.

Cultivars for pickling include 'Bounty' (gynoecious hybrid), 'Burpee Pickler,' 'Catalina Hybrid, 'County Fair Hybrid' (mostly female flowers), 'Miss Pickler' (gynoecious hybrid), 'National Pickling,' 'Regal' (mostly female flowers), and 'Wisconsin SMR 18.' Cultivars for slicing are 'Burpee Hybrid II,' 'Celebrity' (gynoecious hybrid), 'Dasher II,' 'Diva' (burpless), 'FanFare,' 'Marketmore 76,' 'Oriental Express' (oriental, gynoecious hybrid), 'Palace King Hybrid' (oriental), 'Raider' (mostly female flowers), 'Salad Bush Hybrid' (a bush cultivar), 'Straight Eight,' 'Streamliner Hybrid,' 'Sweet Slice' (burpless), and 'Sweet Success' (gynoecious hybrid). Greenhouse cultivars include 'Aidas F_1 Hybrid,' 'Carmen F_1 Hybrid,' 'Corona,' 'Fidelio,' 'Femspot,' 'Kyoto,' and 'Toska 70.' Novelty yellow cucumbers are 'Lemon' and 'Sunsweet F_1 Hybrid.'

Dandelion (*see* Endive)

Eggplant

The eggplant (Figure 17–25) is a very tender, warm season perennial crop cultivated as an annual that does not withstand frost. Eggplant or aubergine is grown for its edible, usually purplish-black fruits. Eggplant cultivars are derived from *Solanum melongena esculentum*, a member of the potato or nightshade family (Solanaceae). Cooked eggplant is usually served in combination with other foods in Italian and Middle Eastern dishes or grilled alone as marinated slices. Eggplants have only minimal nutritive value other than fiber. Fresh eggplant production is a minor commercial (and garden) crop mainly grown in Georgia, Florida, California, and New Jersey. Yields average 12.6 tons/acre.

The longer growing season and warmer temperatures of the South are ideally suited to the culture of eggplant. There the seeds may be sown directly in an outdoor seed bed and transplanted in about 6 to 8 weeks to start both fall and spring crops. In the North, eggplants must be started indoors about 8 weeks before the transplants are set in place. All danger of frost must be past, and the daily mean temperature should be 15.5°C (60°F) or higher. Row tunnels and plastic mulch can be used for early crops for the premium market. Soil pH should be 5.0 to 6.5.

Seed viability is 5 years. Germination (10 to 15 days) is poor below 10°C (60°F). The optimal soil temperature range for germination is 23.9° to 35.0°C (75° to 95°F). Seeds are sown 0.5 inch deep. Rows are spaced 36 to 60 inches apart with in-row spacing of 18 to 24 inches for mechanical cultivation with the lower end favored for hand cultivation. Water is especially needed to keep yields up during flowering and fruiting. Eggplants are deep rooted, so periodic cultivation can be used for weed control. Plastic mulch is another option. Herbicides are sometimes used in addition to cultivation.

Eggplant has moderate drought tolerance; water needs are critical during flowering and fruiting. Water stress causes blossom end rot and misshapen fruit. Preferred minimal soil moisture is −0.45 Bars or 50% ASM.

Maturity is 50 to 75 days. Fruits should be harvested when shiny, as opposed to dull, to avoid hard seeds. Hand harvesting with stem pieces left on is used. Eggplants store poorly; short-term storage is at 7.8° to 12.2°C (46° to 54°F) and 90% relative humidity.

FIGURE 17–25 The eggplant likely became a culinary item in China around 500 A.D. At that point eggplants were strange looking, thorny fruits with a bitter taste. The Chinese thought eggplants were poisonous unless properly prepared by a trained chef. Eggplants should not follow other solanaceous crops such as peppers, potatoes, and tomatoes. (USDA photograph.)

Insects that attack eggplant are the Colorado potato beetle (*Leptinotarsa decemlineata*), flea beetle (*Epitrix fuscula*), and aphid (*Aphis gossypii, Myzas persicae,* and *Macrosiphum euphorbias*). Troublesome diseases include wilt (*Verticillium, Fusarium*) and blight (*Phomopsis vexans*). Some cultivars have disease resistance.

Some eggplant cultivars include 'Black Beauty,' 'Black Enorma F₁ Hybrid,' 'Black Magic,' 'Burpee Hybrid,' 'Dusky,' 'Early Bird Hybrid,' 'Florida High Bush,' 'Nadia Hybrid,' and 'Park's Whopper Hybrid.' Longer, thinner oriental cultivars are 'Ichiban Hybrid,' 'Little Fingers,' and 'Pingtung Long.' Cultivars with white, lavender, or green skin include 'Cloud Nine Hybrid,' 'Ghostbuster,' 'Green Goddess Hybrid,' 'Lavender Touch,' 'Neon Hybrid,' 'Purple Blush,' 'Purple Rain,' and 'Red Egg.'

Endive plus Chicory, Dandelion, Escarole, and Radicchio

This group of salad greens, sometimes used as cooked greens, leads to much confusion in common names. These crops are all characterized by a bitter taste and are mixed with other salad greens such as lettuce. All are in the Compositae or Sunflower family (Compositae or Asteraceae). Endive is a half-hardy, cool season salad crop noted for its rosette of edible lettuce-like leaves. It is somewhat less sensitive to heat than lettuce. Both fringed, fine-cut, slightly curled and curly, broad-leaved cultivars exist. The latter is often referred to as escarole. Frisee, found in fancy salad mixes, is another form of endive that has extremely fringed, deeply cut leaves and less bitterness than standard endive. These endive cultivars are derived from *Cichorium endiva.* California, Florida, and New Jersey are the leading producers. Yields average 9.4 tons/acre. Endive is rich in vitamin K.

Chicory (a.k.a. Belgium endive, French endive, green chicory, witloof chicory, and radichetta) is taller with a more open rosette of leaves; its roots are used as a coffee substitute. Some forms also form heads. One semiheading form is also grown as a forced crop whose large initial leaf bud (chicon) is blanched to reduce bitterness and picked before it opens. This whitish bud form is frequently called witloof chicory or French/Belgium endive. Certain head forms with intense red or burgundy color are called radicchio. These cultivars of chicory are derived from *Cichorum intybus.* While only a minor, specialty crop, their use in upscale restaurants is increasing their popularity with consumers and gardeners. States that produce endive also produce the chicory crops.

Many cultivars of endive, escarole, and chicory take 60 to 70 days to mature, although some of the endive/escarole/radicchio cultivars can take 90 to 100 days. Witloof chicory requires about 1 month of forcing. All require good continuous moisture to avoid tough leaves, reduced yields, and extreme bitterness.

In the South, endive is mainly a winter crop, and in the North, a spring, summer, and fall crop. Chicory is more a spring crop. Endive and chicory are usually sown directly in place, and successive sowings at 2-week intervals are used to extend the harvest period. Transplants are also used. Culture and seed information is similar to that for lettuce. Endive can be blanched (if bitterness is objectionable) by loosely tying the leaves together. Insect and disease problems are minimal with endive and chicory.

Dandelions are a very minor commercial perennial crop also grown for greens for salads and sometimes cooked greens. Cultivated dandelions (*Taraxacum officinale*) are less bitter than those gathered in the wild. Leaves are picked young, from 60 to 95 days from seeding. After hand harvesting the rosette of saw-toothed, barbed leaves, the root is left to regenerate the plant for next spring's harvest. Rows are spaced 12 to 18 inches apart and plants are thinned to 8 to 12 inches apart in the rows. Culture is very similar to that for spinach. Dandelions are rich in vitamin A and iron.

The main escarole cultivars with fringed or curled leaves are 'Bianca Riccia,' 'Green Curled,' and 'Salad King.' Broad-leaved cultivars are 'Broad Leaved Batavian,' 'Coral,' 'Perfect,' and 'Full Heart NR 65.' 'Frizz E' and 'Neos' are frisee cultivars. Radicchio cultivars include 'Ambra,' 'Augusto,' 'Chioggia,' 'Indigo Hybrid,' 'Red Surprise,' 'Treviso,' and 'Versuvio.' Witloof chicory cultivars include 'Flash' and 'Roelof.' A dandelion cultivar is 'Amelioré.'

Florence Fennel

Florence fennel or finocchio cultivars are derived from *Foeniculum vulgare* var. *azoricum*. It is a cool season perennial grown as an annual crop. It is noted for its feathery dill-like foliage and lower, aboveground bulb-like structure. The species produces mostly leaves and leave stalks and is the herb called fennel (see herbs). The bulb is served as a vegetable after steaming or grilling or eaten raw in salads. It has an anise-like taste. Seed is sown about 1/8 inch deep in rows spaced 18 inches apart. Plants are thinned 6 inches apart. In the North seeds are planted in the spring for a summer crop and in the South seeds are sown near the end of summer for a winter crop. At about egg-size, soil is mounded around the base to blanch the lower parts. Maturity takes about 80 days. Cultivars include 'Florence' and 'Zefa Fino.'

Garlic

Garlic cultivars are derived from *Allium sativum* and garlic is in the amaryllis family (Amaryllidaceae). Garlic is a hardy, cool season perennial (monocot) grown as an annual crop. Garlic is noted for its bulb, which can be broken apart into cloves. Garlic is used cooked as a seasoning addition to many foods and can be eaten raw. Some is also processed into garlic powder, dehydrated flakes, and chopped garlic in olive oil.

It can be grown successfully in both the North and South. Propagation is by cloves in the early spring. Cloves are covered with 1 to 2 inches of soil and spaced 2 to 6 inches apart in rows (based on cultivar). Rows are set 12 to 16 inches apart for hand cultivation and twin rows of similar spacing with 30 to 36 inches between each set are use for mechanical cultivation. Light, sandy soils are preferred. Maturity is in 150 to 180 days. California is the primary producer of this minor specialty crop. Yields average 8.3 tons per acre. Culture and pests are similar to that for onion. Harvest is after the leaves have wilted. Mechanical and hand harvesting are used. After harvest, bulbs are sun cured or shed cured (humid or wet areas) for 7 to 14 days with light cover (usually garlic tops) to prevent sunscald. Garlic can be stored for half a year at 0°C (32°F) and 65% relative humidity. Storage for 100 to 120 days without special conditions in well-ventilated storage is possible. Antisprout growth regulators are often used.

Early cultivars are of the white or Mexican type, and late ones the pink or Italian type. Later cultivars are preferred for reasons of quality and storage. Examples of cultivars are 'California Early,' 'California Giant White,' 'Creole,' 'Early Italian,' 'Elephant,' 'German Red,' 'Giant French Mild,' 'Pioneer Softneck,' 'Spanish Roja,' and 'Spanish Rose.'

Globe Artichoke (*see* Artichoke)

Horseradish

The name "horseradish" is likely derived from an English adaptation of its German name, "Meerrettich." The translation is "sea radish," so named because it grew wild in coastal areas. The German word for sea, meer sounds like the English word: mare. Possibly over time, the English "Mareradish" became "Horseradish." Horseradish (*Armoracia rusticana*) is a very hardy perennial usually treated as an annual, in that the enlarged taproots are harvested in the fall. It is in the Mustard family (Brassicaceae or Cruciferae). Mature height is 1.5 to 3 feet.

Propagation is by root cuttings made from small, lateral roots gleaned from the harvested crop. These root cuttings are stored under cool to cold conditions until spring. Root cuttings in the early spring are placed either by hand or machine at a slant (works better, reason unknown) with the flat end up and covered with 3 inches of soil. In-row spacing is 18 inches. Rows are spaced 3 feet apart for mechanical cultivation and less for home gardens. Yields average 2 to 4 tons/acre and Illinois is the leading producer of this minor crop.

Insect pests are few. Flea beetles and beet leafhoppers are minor pests. Diseases include brittle root (*Spiroplasma citri*), turnip mosaic virus, and white rust (*Albugo candida*). Brittle root is transmitted by the leafhopper. Controlling leafhoppers is recommended. Some degree of resistance is available with cultivars such as 'Big Top Western.'

Harvest is after the tops have been frost killed. Roots are dug by machine. Roots can be stored in plastic bags in root cellars, pits, or barns. Alternatively, roots can be left in the ground and dug in the early spring. The majority of the mature roots are grated to prepare the familiar processed, pungent relish. Horseradish is also used as a substitute for *Wasabi japonica*. The later is used in the product called wasabi. It can become a weed if roots are allowed to remain more than 1 year. Horseradish contains some vitamin C. Cultivars include 'Big Top Western,' 'Bohemian,' 'Improved Bohemian,' and 'Maliner Kren.'

Jerusalem Artichoke (*see* Artichoke)

Kale

Kale, a cole crop in the Mustard family (Brassicaceae or Cruciferae), is noted for its cabbage-like cluster of leaves that do not form heads. Kale is closely related to collards. However, leaves are not flattened, but are either gray-green and very curly and crumpled (Scotch kale) or blue-green with less curliness (Siberian kale). Scotch kale (Figure 17–26) cultivars are derived from *Brassica oleracea,* Acephala group. Siberian kale (Figure 17–27) cultivars are derived from *Brassica napus,* Pabularia group. Kale is used as greens, generally boiled or sautéed. Kale is a good source of vitamins A, K, and C.

Kale is a hardy, cool season crop, primarily grown as an early spring or fall crop in the North and as a late fall/winter crop in the South. Kale is somewhat more tolerant of heat than cabbage. Maturity is in 40 to 65 days from seed. California and Georgia are the leading producers of both fresh and processing kale. Yields average 11 tons/acre.

Kale can follow early vegetables such as peas, radishes, early potatoes, and green beans. Soil pH should be 6.0 to 7.0. Seeds are usually sown in place and thinned, although 5- to 6-week-old transplants can be used. Kale has continuous water needs at one inch weekly, otherwise tough leaves result. Drought tolerance is low. Preferred minimal soil moisture is −0.25 Bars or 70% ASM. As with collard, kale may be partially or completely harvested. Generally

FIGURE 17–26   Scotch kale. After harvesting, satisfactory precooling is accomplished by vacuum cooling or hydrocooling. Kale should be held as close to 32°F as possible with relative humidity of at least 95%. These leafy greens are commonly shipped with package and top ice to maintain freshness. (Courtesy of National Garden Bureau.)

FIGURE 17–27   Siberian kale. When planting kale, the field should have been free of crucifer crops (and crucifer-related weeds) for at least 3 years. (USDA photograph.)

FIGURE 17–28
Kohlrabi potassium content is 245 grams for one-half cup. While some think this vegetable is a cross between cabbage and turnip, it is not. This misnomer likely arises from the fact that the German names for cabbage and turnip, respectively, are "kohl" and "rabi." Yields are 6–8 tons/acre. (Courtesy of All America Selections.)

harvest is best before the plants become large and tough. Seed data, culture, and other care are similar to that of broccoli and cabbage. Insect and disease pests are similar to those of cabbage.

Recommended cultivars include 'Blue Curled Scotch,' 'Curled Vates,' 'Dwarf Blue,' 'Dwarf Blue Scotch,' 'Dwarf Green Curled,' 'Dwarf Siberian,' 'Green Glaze,' 'Hi-Crop,' 'Redbor Hybrid,' 'Red Russian,' and 'Winterbor Hybrid.'

Kohlrabi

Kohlrabi (Figure 17–28) is a cole crop grown as a hardy, cool season vegetable. It is a biennial grown as an annual crop is in the Mustard family (Brassicaceae or Cruciferae). Kohlrabi or cabbage turnip is noted for the edible swollen portion of the stem near ground level, which should be harvested before it becomes tough and stringy (at most 3 inches in diameter). Leaves can be used in salads or cooked as greens. The stem bulb can be consumed raw or cooked like turnip. Kohlrabi is high in vitamin C and iron. Green and purple cultivars are derived from *Brassica oleracea*, Gongylodes group.

Commercial production is very limited. In the North, kohlrabi is grown as a spring or fall crop, and as a fall and winter crop in the South. Seeds are usually sown in place, although 5- to 6-week-old transplants can be used. Soil pH should be in the 6.0 to 7.0 range. Maturity is 42 to 85 days from seeds. Smaller 3-inch cultivars are at the 42- to 65-day end, while large cultivars harvested at the 8- to 10-inch size are 85 days. Seed and cultural information is very similar to that for broccoli and cabbage. Harvest is by hand. Insect and disease problems are the same as for cabbage.

Cultivars include 'Blusta,' 'Early White Vienna,' 'Express Forcer Hybrid,' 'Grand Duke,' 'Kolibri Hybrid,' 'Kossack Hybrid' (large cultivar), 'Purple Danube F$_1$ Hybrid,' 'Purple Vienna,' 'Sweet Vienna,' and 'White Vienna.'

Leek

Leeks (Figure 17–29) are hardy, cool season biennials (monocots) grown as an annual crop found in the amaryllis family (Amaryllidaceae). Cultivars are derived from *Allium ampeloprasum*, Porrum group. The leek is hardier than its relative, the onion. Leeks are noted for their slight, soft bulbs and sheaf of leaves. Leeks can be used in soups, salads, combined with other vegetables, alone or with a cream sauce. Leeks have moderate amounts of vitamins A and C plus potassium.

FIGURE 17–29   Leeks need an inch of water every 5 days during active growth. Most leeks are machine-dug and then harvested, cleaned, and packed by hand. Yields are 18–19 tons/acre. (USDA photograph.)

In areas with long growing seasons, leeks are usually sown directly in place. Shorter season areas use transplants. Leeks are early spring planted in the North and West, being long season crops (100 to 200 days). In the South and Southwest, leeks are winter crops that are spring harvested. Soil pH should be 6.0 to 7.0. Leeks are usually blanched by mounding soil at their base. Leeks have continuous water needs; water stress causes thin scales. Drought resistance is low to moderate; an inch of water every five days is needed. Preferred minimal soil moisture is −0.25 Bars or 70% ASM.

They can be harvested when they become 1 to 2 inches in diameter. Mechanical and hand harvests are used. Leeks are hydro-, ice, or vacuum cooled. They can be stored at 0°C (32°F) and 95% relative humidity for 60 to 90 days. Seed and cultural information is similar to that for onions, as are insects and diseases.

Suggested cultivars are 'American Flag,' 'Blue Leaf,' 'Dawn Giant,' 'Electra,' 'Jolant,' 'Kilima,' 'King Richard,' 'Leekool,' 'Lyon,' 'Monstrous Carentan,' 'Toledo,' 'Upton F_1 Hybrid,' and 'Winora.'

Lettuce

Lettuce is a half-hardy, cool season salad crop with poor heat resistance. Heat can produce bitterness and sometimes cause reproductive growth (bolting), making the lettuce not marketable. Lettuce is noted for its edible green and sometimes red leaves, primarily used as the main salad green. Lettuce is an annual found in the Compositae or Sunflower family (Compositae or Asteraceae). Cultivars are derived from *Lactuca sativa*. There are three basic types (Figure 17–30): head, which varies from firm and hard (crisphead or iceberg) to loose and soft (butterhead); looseleaf or bunching, clusters or bunches of leaves; and cos (romaine), long upright cylindrical heads. These are listed in order of increasing tolerance toward heat. Cos and looseleaf lettuce are an excellent source of vitamins A,C, K, and folate (much more so than the other types). California is by far the leading producer of all types of lettuce. Arizona is next. Yields are 17.5, 11.4, and 19.8 tons/acre for head, looseleaf, and romaine lettuce, respectively.

Mesclun is the term applied to a blend of leaf lettuces plus several other greens combined to create a variety of textures, colors, and flavors. Leaves are partially harvested and regrowth is encouraged. Other greens besides lettuce are generally several of following in various combinations: arugula, beet leaf, chard, chicory, (edible) chrysanthemum, corn

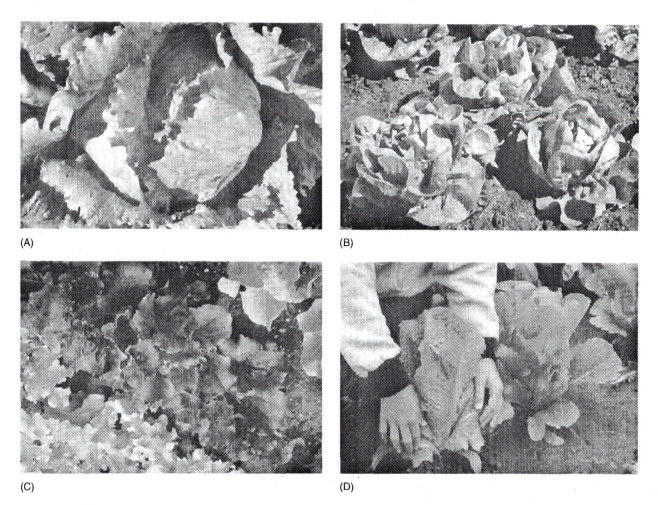

(A) (B) (C) (D)

FIGURE 17–30 Lettuce types shown are: (A) Crisphead. (B) Butterhead. (C) Leaf. (D) Cos. Lettuce is 95% water by weight. These high cellular water levels cause the crunchiness of the lettuce when eaten. Leaf lettuce dates back to the Greek and Roman times, where it was used as an early course (the salad) to their huge feasts. Heading lettuce did not appear until the late 1500s. The heading variety is widely used in the United States; its consumption is over four billion heads a year. Americans consume about 30 lbs of lettuce *per capita* yearly. "Mesclun" is a mix of fresh, tender greens notable for their varying of textures, flavors, and colors. Leaves are partially harvested and plants are allowed to regrow. Ingredients vary, but always contain leaf lettuces (sometimes cos) plus other greens such as arugula, beet leaf, chard, chicory, corn salad, dandelion, endive, escarole, frisée, mizuna, mache, mustard tips, radicchio, sorrel, spinach, and watercress. Crisphead cultivars 'Alpha' and 'Target,' butterhead cultivars 'Divina' and 'Little Gem,' and cos cultivar 'Valmaine' show resistance to downy mildew. Harvestors are now available for leaf lettuce and mesclun. (Courtesy of National Garden Bureau.)

salad, dandelion, endive, escarole, frisée, kale, mache, mizuna, mustard tips, nasturtium leaves, orach, parsley, plantain, purslane, radicchio, sorrel, spinach, tango, tat-soi, travissio, and watercress. In addition, certain herbs and flowers are added to the mix. Herbs include basils, borage, chervil, chives, fennel, and salad burnet, while blossoms can be borage, calendula, nasturtium, violas, and violets.

In the North, lettuce is a spring and fall crop, and a fall, winter, and spring crop in the South. It can be grown as a summer crop at high altitudes and in the far North during the summer. Lettuce can be sown directly in place or 3- to 5-week-old transplants are sometimes used. Depth of planting is shallow, from 0.25 to 0.50 inch. Seed should be fresh, because viability of lettuce seed is short. Seed germination is optimal for soil temperatures from 4.4° to 26.6°C (40° to 80°F). Some cultivars germinate poorly or not at all at high temperatures (thermodormancy). Others require light for germination. Growth regulators are often used by commercial seed producers to remove light and temperature effects. Soil pH can vary from 6.0 to 7.0. Spacing varies with lettuce type and cropping system. Both row and bed systems are used. Spacing between rows or beds is 12 to 18 inches with crisphead types at 18 inches. In-row or plant-to-plant spacing varies from 4 to 10 inches with crisphead at 10 inches. Thinning by hand is often required, because thinning machinery is not satisfactory.

Successive plantings of the leaf type at 2-week intervals are often used to extend its harvest period. Succession crops such as cabbage, beet, celery, or green beans can follow early lettuce. Sometimes lettuce is grown as a companion crop with low-growing, longer season crops such as the cole crops. Lettuce is the third largest commercial vegetable crop grown in greenhouses.

Weeds are often controlled by preplant herbicides followed by cultivation. Shallow cultivation above 2 inches is needed. Lettuce has moderate to high drought tolerance. Water needs are one inch per week. The critical periods for heading and leaf lettuce are the head expansion and rosette formation, respectively. Water stress causes tough leaves and bitterness, especially in warmer weather. Preferred minimal soil moisture is −0.34 Bars or 60% ASM.

Crisphead types are harvested when heads become firm and full (50 to 120 days for spring-planted lettuce). Looseleaf types are ready in 40 to 50 days, while butterhead and cos cultivars require 55 to 70 days. All should be picked before the onset of hot weather, which causes bolting and bitterness. Harvest is by hand, followed by field packing and vacuum cooling. Refrigerated shipping is used. Storage time must be brief and at 0°C (32°F) and high humidity. Ethylene can cause brown spotting on lettuce during shipping and storage (russetting). Excessive CO_2 and very low O_2 develop brown stains (watery lesions) and pink discoloration (pink rib).

Insects that damage lettuce include aphids, caterpillars (armyworms: *Spodoptera exigua* and *S. ornithogalli*; cabbage looper, *Trichoplusia ni*), leafhoppers, plant bugs, whiteflies, and wireworms. Aphids carry some diseases. Troublesome diseases include big vein (possible virus), bottom rot (*Rhizoctonia solani*), downy mildew (*Bremia lactucae*), gray mold rot (*Botrytis cinerae*), lettuce drop (*Sclerotinia sclerotiorum*), mosaic (virus carried by aphids), and yellows (a phytoplasma carried by leafhoppers). Tip burn is a physiological disorder resulting from high temperatures and low soil moisture.

Popular cultivars are as follows. Head-lettuce cultivars include:

1. *Butterhead:* 'Balisto,' 'Bibb,' 'Buttercrunch,' 'Divina,' 'Louisa,' 'Tania,' and 'White Boston';
2. *Crisphead:* 'Alpha,' 'Beatrice,' 'Bullseye,' 'Calamar,' 'Great Lakes 659,' 'Iceberg,' 'Ithaca,' 'Imperial,' 'Nevada,' 'Pacific,' 'Salinas,' 'Simpson Elite,' and 'Summertime';
3. *Looseleaf lettuce:* 'Black-Seeded Simpson,' 'Grand Rapids,' 'Green Ice,' 'Oak Leaf,' 'Red Salad Bowl,' 'Red Sails,' 'Royal Oak,' 'Ruby Red,' 'Salad Bowl,' and 'Slobolt';
4. *Cos:* 'Dark Green,' 'Green Towers,' 'Little Ceasar,' 'Little Gem,' 'Olga,' 'Parris White,' 'Sierra,' and 'Valmaine.'

Lima Beans (*see* Beans)

Melons

Melons (Figure 17–31) are very tender, warm season vine crops (cucurbits) noted for their fruits. Melons are in the gourd family (Cucurbitaceae). Melons are grown for their edible fruits and are treated here because their culture is similar to vegetables and is also tracked under vegetables by the USDA. The major commercial melons are muskmelons and honeydews. Minor melons include casabas, crenshaws, canary melons, and Persian melons. Muskmelon and Persian melon (netted melons) cultivars are derived from *Cucumis melo,* Reticulatus group. Honeydew, casaba, crenshaw, and canary melon (winter melons) cultivars are derived from *Cucumis melo,* Inodorus group. The watermelon is treated separately. The term *cantaloupe* is often incorrectly used to indicate muskmelons. True cantaloupes are seen more in European horticulture. Yields per acre are 10.5 and 9.7 tons for muskmelon and honeydew, respectively. California vastly leads in melon production, followed by Arizona. Most melons are good sources of vitamin C and orange-fleshed ones are good sources of vitamin A.

FIGURE 17–31 Muskmelons (photo) are a common market melon. Higher melon yields are obtained with drip irrigation under plastic mulch as opposed to plastic with sprinkler irrigation. Some less common specialty melons include Charentais ('Acor F_1,' 'Alienor F_1,' 'Charentais Improved,' 'Ido,' 'Panchito'), Mediterranean ('Casablanca'), Ogen ('Galia,' 'Gallicum') Rochet ('Verdol F_1'), Chinese "Hami" ('Red-Pink Hami,' 'Snow Charm,' 'Tiger-Skin Hami'), Japanese melons ('Tokyo King,' 'Emerald Jewel,' 'Emerald Pearl,' 'Ginryu' and 'Zuikoh'). (Author photograph.)

Melons are seeded directly in place in the warmer areas. In the North, where the growing season is short, 3- to 4-week-old transplants can be used, if transplanting shock is minimal (preferably with peat pots or large plugs). Machines are available for direct seeding and placing of transplants including through plastic mulch. Soil temperature should be at least 15.5°C (60°F) and optimal soil temperature range for germination is 23.9° to 35.0°C (75° to 95°F). Seeds are covered with 0.5 to 1.5 inches of soil. Rows are spaced 5 to 8 feet apart and in-row spacing is 12 to 24 inches. Soil pH can be between 6.0 and 7.0. Sandy soils appear to be best, as they warm faster. Clear or black (slower) plastic mulch can be used in the North to produce a soil temperature better suited for melon culture earlier than would occur naturally. Even better is the use of row tunnels or row covers coupled with plastic mulch. Weeds are highly competitive, and plastic and cultivation methods are best for control.

A constant water supply is needed, so melons are often irrigated. The critical period is during flowering and melon development. At least an inch of water every 10 days is needed. Water stress causes poor melon development. Preferred minimal soil moisture is −0.34 Bars or 60% ASM. Drought tolerance is moderate.

The best muskmelons are harvested when the stem separates easily and cleanly from the melon. Such melons are picked in gardens or for local markets. Melons for shipping are picked when the stem separates, but shows a little resistance. Honeydew melons do not show stem separation. Best indicators are a change of rind color from lime green to creamy white and a slight softening at the blossom end. Ethylene gassing at either the shipping or receiving end can be used to improve honeydew ripening. Fruit color changes and slight softening are also useful criteria for the other specialty melons. Melons are harvested by hand. Some mechanization of melon collection and movement is used in large operations. For shipping, melons should be precooled by hydrocooling, ice, packs or forced-air cooling. Refrigerated shipping is used. Melon storage is usually not practiced.

Insect pests include those that attack cucumbers such as aphids, pickleworm, and striped cucumber beetle. Diseases include those that attack cucumbers (anthracnose, bacertial wilt, downy mildew, cucumber mosaic virus, and powdery mildew) plus *Alternaria* leaf spot (*Alternaria cucumerina*), black rot (*Phoma cucurbitacearum*), *Fusarium* wilt

(*Fusarium oxysporium*), gummy stem rot (*Didymella bryoniae*), squash mosaic virus, and watermelon mosaic virus.

Muskmelon cultivars include 'Ambrosia Hybrid,' 'Burpee Hybrid,' 'Early Dawn,' 'French Orange,' 'Hales Best Jumbo,' 'Harper Hybrid,' 'Hy-Mark,' 'Saticoy,' 'Super Star,' and 'SuperSun Hybrid.' 'Early Crisp Hybrid,' 'Earli-Dew Hybrid,' 'Honeybrew,' 'Honey Moon Hybrid,' 'Milky Way,' and 'Sweet Delight' are honeydew cultivars. 'Golden Beauty' and 'Sungold' are casaba cultivars, and 'Burpee Early Hybrid,' 'Carnival,' and 'Goleten Crenshaw' are crenshaw cultivars. Canary melon cultivars include 'Gold King,' 'Sweet Yellow Canary,' and 'Tenerife.'

Mushrooms

Mushroom production (Figure 17–32) is a highly specialized, very competitive operation, whether intended for fresh or processed markets. While not a plant and certainly not a vegetable or fruit, production is more closely allied to vegetable production than any other horticultural industry. Mushroom kits are also available for home operations for interested gardeners or for learning about mushrooms while investigating the launch of a commercial operation. Mushrooms are also gathered in the wild on both a commercial and personal basis, which can be a risky to deadly venue, unless you know exactly what mushrooms you pick. Mushroom identity for wild gathering will not be discussed here.

Mushrooms sold commercially include California brown and white button (*Agaricus bisporus* and *A. bisporus hortensis,* respectively, immature stage), chanterelle (*Cantharellus cibarius*), ink cap (*Coprinus* spp.), Crimini (almost fully mature California brown mushroom), ear (*Auricularis*), enoki or winter (*Flammulina velutipes*), lion's mane (*Hericium erinaceum*), morel (*Morchella esculenta, M. angusticeps, M. crassipes, M. hybrida*), Nameko (*Pholiota nameke*), oyster (*Pleurotus ostreatus*), pioppini (*Stropharia rugosoannulata*), porcini (*Boletus edulis*), portobello (a fully mature California brown mushroom), shiitake (*Lentinus edodes*), straw (*Volvariella volvacea*), truffle (*Tuber aestivum, T. melanosporum*), and white jelly (*Tremella* spp.) mushrooms. Organic certified mushrooms are an option as with all vegetables. Pennsylvania and California lead in the production of commercial mushrooms.

Commercial mushrooms are produced in structures called mushroom houses. Abandoned mines have also been used. Light is not required other than for worker efficiency, because mushrooms can grow in the dark and white types are often discolored by sun. Structures lack windows and are highly sophisticated in terms of environmental control.

FIGURE 17–32 A NRCS Soil Conservationist and a Pennsylvania mushroom farmer discuss the composting process used to create a growing medium for mushrooms. Mushrooms are the largest cash crop in Pennsylvania. (NRCS/USDA photograph.)

Essential features include energy efficiency, ease of cleaning and sterilization, and the ability to carefully control temperature, the ratio of atmospheric gases, humidity, and ventilation. Stacked beds three high can be used for the actual growth. Other systems use trays, which can be manipulated easily by forklifts. Structure sizes vary, but a typical mushroom house might be 65 by 20 by 15 feet high. While alternative, low-input systems can be used such as growth on logs and stumps outdoors, these systems are only commercially feasible for specialty or niche mushrooms not widely produced commercially. For example, button mushrooms are highly commercialized, but oyster and shiitake mushrooms less so.

Directions that follow are only meant to be general. Literature and experts should be consulted for specific details, which can vary for many reasons such as cultivar or type of mushroom production system employed.

Mushrooms are generally produced on compost (not mature). Certain ingredients are common, but the initial mixture is often carefully blended in various ways, and at specific times other ingredients are added (determined by the mushroom to be grown). Nitrogen, phosphorus, potassium, calcium, carbon, oxygen, and water are essential. It is beyond our scope to provide recipes here (see Internet Resource list). Composts are built from ingredients such as horse manure, poultry manure, rice straw, wheat straw, sawdust, paper, banana leaves, distiller grain waste, corncobs, logs, cotton textile waste, cotton seed hulls, coffee pulp, sugar processing wastes, and water hyacinths. Compost preparation is usually by machine and the pile is turned periodically for mixing ingredients and to supply improved aeration. Mushroom production does have a nice, sustainable connection in that organic waste products can become mushroom growing substrates.

The compost is allowed to partially undergo biological oxidation until the temperature reaches 60° to 66°C (140° to 150°F). The material is then layered in the bins 6 to 8 inches deep and is pasteurized using special equipment. Pasteurization requires 60°C (140°F) for 72 hours. Longer times are needed if ammonia is detected. Some production composts might require sterilization at higher temperatures if contaminating/competing microorganisms are present. The temperature must be dropped to 21° to 24°C (70° to 75°F) prior to adding the mushroom spawn (inoculant). The compost at this stage must have a pH of 7.5 to 8.0 and moisture content of 65% to 70%. Adequate ventilation is required throughout this process.

Spawn is procured from reliable suppliers. One broadcasts it over the compost, usually mechanically, at a rate of approximately 1 quart to 30 square feet. The temperature is maintained at 70°F for 1 week and then dropped to 64°F (17.8°C) for another week. During this time, the spawn produces an extensive mycelial mat that gives the compost a cotton-like appearance. During this time, the ratio of atmospheric gases is adjusted to favor mycelial production. Ventilation must also be controlled and water added as needed.

Next a layer of soil (casing) is added (cased). The soil is pasteurized beforehand and has a neutral pH and is damp. One inch is added. The ratio of atmospheric gases is readjusted to favor the production of fruiting bodies (mushrooms). Water is added as needed. Ventilation is controlled. The first mushrooms appear in 2 weeks. Temperatures are usually dropped 13° to 16°C (55° to 61°F) to improve quality. During the next 2 to 3 months, flushes of new mushrooms will appear at various intervals. Water must be added as needed, ventilation controlled to replenish the fresh air and the relative humidity maintained in the 70% to 80% range.

Mushrooms are carefully harvested by hand pulling, hand cutting, or mechanically. The spent compost can be recycled in horticultural landscaping and nursery operations as a source of organic matter. The facility is thoroughly sanitized prior to the next batch. Spent compost should be removed off-site so as to not attract pests that can interfere with mushroom production.

Pest control can be an issue. A common pest is the mushroom fly. Slugs can be another. Competing "weed" fungi can also be a problem. Best control is the use of screening, double doors, and a positive atmospheric pressure to keep flies out. Pools of water that attract insects should also be eliminated. Biological controls should be utilized whenever possible such as for control of the sciarid fly. Sanitation should also be practiced at every stage and every place.

FIGURE 17–33 'Southern Giant Curled' mustard. Mustard crops should not be planted for at least 3 years after other crucifer crops to minimize disease problems. Greens are derived mainly from *Brassica hirta* and *B. juncea* cultivars. Seeds from the *Brassica hirta* cultivar Tilney are used to make the standard American hot dog yellow mustard. Brown and Oriental mustards are *Brassica juncea*. Seed of brown mustard cultivars such as Common Brown, Blaze, and Forge, are used in hot, stone-ground and "French"-style mustards. Oriental mustard cultivars are Lethbridge 22-A and Domo. Black Mustard (*Brassica nigra*) is less common. It can be used for greens or its seed is used for flavoring pickles or salads. Seeding rates for greens and mustard seed are 3 to 4 and 5 to 7 pounds per acre. Harvesters are available for mustard greens. Combines are used for harvesting mustard seed. (Courtesy of National Garden Bureau.)

Mustard and Arugula

Mustard (Figure 17–33) and arugula are hardy, cool seasons annuals grown for their edible leaves used as greens (mustard) and as salad ingredients (arugula and mustard). Some mustard species are grown for their oil seeds used in the preparation of table mustard. Mustard is a cole crop; both arugula and mustard are in the Mustard family (Brassicaceae or Cruciferae). Mustard cultivars used as greens are derived from *Brassica hirta*, *B. juncea*, and *B. rapa*, Perviridis group (greens) and table mustard is derived from *B. nigra*. Arugula or 'Rocket' cultivars are derived from *Eruca vesicaria* subsp. *sativa*.

In the North it is treated as a spring or fall crop, and as a fall, winter, and spring crop in the South. Leaves should be picked early and before heat sets in, or bolting and increased pungency will result. Seeds are sown directly in place (0.5 inch deep) at 2-week intervals to provide successive harvests. Germination is in 4 to 8 days in early spring. Soil pH can be between 6.0 to 7.0. Rows are placed 12 to 24 inches apart and plants in rows are 4 to 6 inches apart. Continuous moisture is essential for good yields of arugular and mustard. Water stress causes tough leaves; drought tolerance is low. Preferred minimal soil moisture is −0.25 Bars or 70% ASM. An inch of water weekly is needed. Harvest is in 35 to 50 days. Leaves can be totally or partially removed. Yields average 7.6 tons/acre and Georgia is the leading producer. Weeds are controllable by cultivation.

Insect and disease problems are minimal. Suggested mustard (greens) cultivars include 'Florida Broad Leaf,' 'Green Wave,' 'Savanna Hybrid,' 'Slobolt,' 'Southern Giant Curled,' and 'Tendergreen.' Arugula cultivars include 'Rocket' and 'Wild Italian Rocket.' Cultivars used in condiment mustard are 'Blaze,' 'Common Brown,' and 'Forge' for French-style mustards and 'Tilney' for yellow mustard.

FIGURE 17–34 New Zealand spinach grows to a height of 1 to 2 feet and spreads 2 to 3 feet across with much branching. (Courtesy of National Garden Bureau.)

Napa Cabbage (*see* Chinese Cabbage)

New Zealand Spinach

New Zealand spinach (*Tetragonia tetragonioides*) is a tender, warm season crop (Figure 17–34) grown for its edible leaves used as greens or in salads. It is a substitute for spinach, since it thrives during the warmer weather when spinach does not. This annual belongs to the New Zealand spinach family (Tetragoniaceae). Commercialization is very limited to specialty producers.

Seed is soaked in water at 49°C (120°F) for 1 or 2 hours to hasten germination. Seeds germinate in 12 to 20 days and are sown in place 1.0 to 1.5 inches deep after all danger of frost is past. Rows are 3 to 4 feet apart and in-row spacing is 12 to 18 inches. Soil pH can be from 6.0 to 7.0. New Zealand spinach has low drought tolerance. Water supply needs to be continuous, as insufficient water results in poor production and tough leaves. One inch of water in 5 days is essential. Preferred minimal soil moisture is −0.25 Bars or 70% ASM. Maturity is 70 days. Successive harvests on each plant are possible if only partial removal of foliage (tips 3 to 4 inches long) is practiced. Insect and disease pests are minimal.

Okra

Okra (Figure 17–35), also known as gumbo, is a very tender, warm season crop. Cultivars are derived from *Abelmoschus esculentus,* which is in the mallow family (Malvaceae). Okra is an annual grown for its edible, immature, ridged seedpod, especially in the South. Okra is used in salads, stews, and soups and is also steamed, boiled, microwaved, batter-fried, and sautéed as a main vegetable. It is a fair source of vitamins A, K, and C, calcium, and iron. Okra is a minor commercial crop with average yields of 3.4 tons/acre. Florida and Texas are the leading producers.

Seeds have a short viability of 2 years. To speed germination, seeds can be soaked in water for 1 to 2 hours at 43.3°C (110°F). Seeds are sown 1 inch deep directly in place after all danger of frost is past. Soil temperature must be above 15.5°C (60°F). Optimal soil temperature range for germination is 21.1° to 35.0°C (70° to 95°F). Plants are spaced 12 to 24 inches apart within rows based on cultivar size and rows are 36 to 48 inches apart. Soil pH should be between 6.0 and 7.0. Shallow cultivation is used. Okra has moderate to high drought tolerance. Too much water can reduce yields. Too little water during flowering can

FIGURE 17–35 Okra has both green (more common) and red cultivars. Seeding rates are 8–12 pounds per acre. For early market, transplants are used. Adequate and constant moisture are required for good growth and high yields. Okra has small spines that cause an allergic reaction in some people. Gloves are suggested for harvesting. (USDA photograph.)

reduce yields and if continued, cause tough pods. Water needs are 1 inch in 2 weeks. Preferred minimal soil moisture is −0.70 Bars or 40% ASM.

Immature pods are ready in 48 to 60 days (4 to 6 days after flowering) and should be picked to ensure a continuing supply. Insect and disease problems are minimal, but nematodes can be a problem. Suggested cultivars include 'Annie Oakley II Hybrid,' 'Baby Bubba Hybrid,' 'Blondy,' 'Burgundy,' 'Cajun Delight,' 'Clemson Spineless,' 'Dwarf Long Pod,' 'Emerald,' and 'North & South Hybrid.'

Onion

The onion (Figure 17–36) is grown for its edible bulb either in the early green top stage with minimal bulb development (green bunching onions or scallions) or later in the dry top stage with maximal bulb enlargement (dry onions). It is a hardy, cool season biennial (monocot) grown as an annual crop. Onion is used cooked as a seasoning addition to many foods, cooked and creamed alone, and can be eaten raw in salads and on sandwiches such as hamburgers. Some is also processed into onion powder, onion rings, and dehydrated flakes. Scallions provide modest amounts of vitamin A and dry onions vitamin C. Onions picked early with small bulbs (pearl onions) are often used in cream sauces and in a drink called a Gibson. California, Oregon, and Washington lead in commercial production. Onion sets are produced by a few specialty growers in Illinois, Oregon, and Colorado. Yields average 20.8 tons/acre with yields as high as 32 tons/acre reported from Idaho.

Onion cultivars are derived from *Allium cepa*, Cepa group, and are in the amaryllis family (Amaryllidaceae). Onion cultivars vary extensively in terms of origin, pungency, skin color, response to photoperiod, shape, and size. Many of the onions purchased in the winter are American or domestic globe onions with their tan/brown (some red) skin and white/yellow flesh, pungent taste, and medium-size globe to oblate shape. These onions store well and are grown either for fresh use in spring and summer and for harvest in late summer and fall as storing onions. About three-quarters of U.S. onions are of this type. Spanish onions are larger, rounder, tan or red, mild and used fresh or processed into onion rings, although they can be stored. Bermuda onions (including Grano-Granex types) are usually also mild and white (although yellow and red forms exist) and do not store well. Their shape is like a toy top. Both Spanish and Bermuda onions are grown as spring and early summer onions. Scallions (essentially spring onions) can be from any type of onion,

FIGURE 17–36 Most commercial onions are direct-seeded followed by seedling transplants. Sets are used mainly in gardens. Onions used for dehydration are usually strains of Southport White Globe such as Dehyso and Dehydrator No. 14. Scallions are usually cultivars of *A. fistulosum* (Hishiko, Ishikura, Japanese Bunching, Kincho, Santa Claus (red), Tokyo Bunching, and Tokyo Long White). (Courtesy of National Garden Bureau.)

but cultivars especially bred for use as scallions are available. Some special onions are also grown such as the Vidalia and Walla Walla. These two are sweet and not storable.

Best conditions for growing onions are good moisture and no extremes of heat or cold, 12.8° to 23.9°C (55° to 75°F). In the North, onions are a spring, summer, and fall crop, and a fall, winter, and spring crop in the South. Onions may be directly seeded in (0.25–1.0 inch deep; deeper range for light, sandy soils) or started from sets planted 1 to 4 inches deep (immature onion bulbs less than 0.75 inch in diameter) or grown from seedling plants. Seeds are cheaper and widely used in California. Sets and transplants are used for early planting where time is a factor and sets mainly in gardens. Soil pH should be from 5.5 to 6.8 with the lower range for muck soils and the upper end for mineral soils.

Seeds or sets can be put out as early as the soil is workable, but seedlings should be delayed until all danger of frost is past. Seeds should be fresh given that seed viability is 1 year. Germination takes 6 to 12 days in the optimal soil temperature range of 10.0° to 35.0°C (50° to 95°F). Seeds and sets are mechanically precision sown to avoid thinning labor costs. Transplants are done by hand; the expense is justified only for early production of high-value onions. Double-row slightly raised beds are very common in commercial production. Rows and in-row spacing vary based on cultivar. Rows can be 3 to 24 inches apart (the lower end for pearl onions and the upper end for large cultivars) with in-row spacing of nearly touching (pearl, scallions, producing onion sets) to 4 inches (large cultivars).

Onions have low drought tolerance. The critical period is during bulb initiation and subsequent bulb expansion. Too little water produces small bulbs. An inch of water weekly is needed. Preferred minimal soil moisture is −0.25 Bars and 70% ASM.

Onion yields are especially reduced by weeds. Herbicides are widely used both before and after planting. Mechanical cultivation must be shallow and limited to between rows. A constant moisture supply is needed because onions are shallow rooted. Irrigation is stopped when the tops start to fall over. Onions may be harvested at pencil size or larger as scallions. Mature dry onions are harvested when the tops fall over. Green bunching onions take 60 to 120 days from seed, while dry onions take 100 to 185 days from seed. Bulbs are lifted and topped either by hand or machine. Bulbs are windrow field cured or cured in off-field curing sheds for 2 to 3 weeks or until outer scales become paper-like and the neck-tops are tight and dry. Some onions such as Spanish onions must be shaded-cured to avoid sunscald. Onions are stored at 0°C (32°F) and 65% relative humidity for up to 8 months. Sprout inhibitors are usually required for stored onions. Scallions have limited storage for a few weeks in polyethylene film packs at 0°C (32°F).

Several insects attack onions. The worst are the onion maggots (*Hylema antigua*) and onion thrips (*Thrips tabaci*). Troublesome diseases include bacterial soft spot (*Erwinia carotovora*), basal rot (*Fusarium oxysporum*), downy mildew (*Peronospora destructor*), neck rot (*Botrytis* sp.), onion smut (*Urocystis cepulae*), pink root rot (*Pyrenochaeta terrestris*), purple blotch (*Alternaria porri*), tip and leaf blight (*Botrytis squamosa*), white rot (*Sclerotium cepivorum*), and yellow dwarf and yellows (viruses).

The time of bulb formation is a function of day length and is influenced somewhat by temperature. As such, cultivars tend to be regionally specific. This condition should be borne in mind when a cultivar is selected. A list of some dry onion cultivars, without regard to region or color (yellow, white, red), follows: 'Brahma,' 'Bravo,' 'Columbia Hybrid,' 'Crystal White Wax,' 'Yellow Ebenezer,' 'First Edition Hybrid,' 'Latin Lover,' 'Mambo,' 'Red Baron Hybrid,' 'Red Grano,' 'Red Hamburger,' 'Redwing Hybrid,' 'Snow White Hybrid,' 'Southport Red Globe,' 'Southport White Globe,' 'Southport Yellow Globe,' 'Stuttgarter,' 'Super Star Hybrid,' 'Sweet Sandwich F_1,' 'Texas Grano,' 'Walla Walla,' 'White Granex Hybrid,' 'White Sweet Spanish,' 'Yellow Granex Hybrid' (Vidalia), and 'Yellow Sweet Spanish Hybrid.' Scallion cultivars include 'Beltsville Bunching,' 'Crimson Forest,' 'Crystal White,' 'Evergreen White Bunching,' 'Gold Princess,' 'He-She-Ko,' 'Ishikuro,' 'Red Baron,' 'Santa Clause,' 'Southport White Bunching,' 'Summer Isle,' and 'White Lisbon Bunching.'

Parsley and Parsley Root

Parsley (*Petroselinum crispum*) is a biennial (see Figure 2–17) that can be treated as an annual. It is a member of the parsley or carrot family (Umbelliferae or Apiaceae). Mature height is 10 to 15 inches. The typical plant with curly, crisped leaves (curly parsley) is the variety *crispum*. The variety *neapolitanum* (Italian parsley) has flat leaves. A third variety (turnip-rooted parsley), *tuberosum*, has tuberous roots. Fresh leaves are used as a garnish, a pesto ingredient, and in soups, stews, salads, and various cooked dishes. Parsley leaves are a very good source of vitamins A, C, and the various B vitamins, as well as calcium, iron, phosphorus, and potassium. The tuberous-rooted type is grown mainly for the root, which is prepared and used like parsnips in soups and stews.

Seeds are very slow to germinate. Seeds are soaked in water prior to sowing in early spring. Germination often takes 21 days without water soaking. Direct seeding (0.5 inch deep) is possible, but transplants are preferable. Seeds or transplants should be placed out in spring after frost danger is past in the North. Spacing is 6 to 8 inches in the row. Rows can be spaced at 12 to 24 inches. An average soil with full sun and continuous moisture is best. Soil can be between pH 5.5 and 7.0. Crops can be partially harvested throughout the growing season. Parsley is also a minor winter greenhouse crop in the North. Parsley is planted as a fall, winter, and spring crop in the South. Once frosts are due, the entire plant can be harvested. Pests are very minimal, including aphids, carrot rust fly (see carrots), and parsley worm (*Papilio polyxenes asterius*). Diseases are not bothersome.

Curly-leaf cultivars include 'Extra Curled Dwarf,' 'Favorit,' 'Forest Green,' 'Green River,' and 'Krausa.' Flat-leafed cultivars include 'Italian Dark Green,' 'Italian Plain Leaf,' and 'Sweet Italian.' Tuberous-rooted cultivars include 'Hamburg.'

Parsnip

The parsnip (see Figure 17–6) is a half-hardy, cool season root crop that can be grown over much of the United States. Cultivars of parsnip are derived from *Pastinaca sativa*, which is in the parsley or carrot family (Umbelliferae or Apiaceae). Roots resemble white carrots. Roots can be steamed, microwaved, or boiled as a main vegetable, served buttered or creamed, or used in soups and stews. Parsnips are modest sources of vitamin C, folate, and potassium. Production is minimal for this very minor crop. States growing commercial parsnips include Pennsylvania, Illinois, California, and New York.

Extremes of hot weather cause poor seed germination and poor-quality roots. In the South the parsnip should be grown such that maturity occurs during the spring or early summer. Seeds should be fresh, as viability is 1 year. Germination is slow at 12 to 20 days. Optimal germination soil temperature is 10.0° to 21.1°C (50° to 70°F). Soil can be between pH 5.5 and 7.0, and it should be free of debris or misshapen roots will be formed. Parsnips are sown 0.5 inch deep directly in place. Rows are 18 inches apart for hand cultivation and 24 to 36 inches for mechanical cultivation. Plants are thinned to 2 to 4 inches apart. Culture and weed control are similar to that of carrots.

Parsnips have high drought tolerance, needing only one inch of water every two weeks. The critical period is during root expansion. Preferred minimal soil moisture is −0.70 Bars or 40% ASM. Maturity takes 105 to 150 days. Parsnips are hand harvested. Frost improves their sweetness. Storage conditions are similar to those for carrots. In many areas, parsnips can be left in the ground during winter and harvested as needed.

Insects and diseases are the same as those for carrots. Suggested cultivars are 'All American,' 'Avonresister,' 'Gladiator F_1 Hybrid,' 'Hollow Crown,' 'Panache F_1 Hybrid,' 'Harris' Model,' and 'Tender and True.'

Peas

The pea (Figure 17–37) is a hardy, cool season annual, vine legume noted for its edible green seeds alone (green or English peas) or combined peas and pods (edible-podded peas). It is a member of the pea or pulse family (Leguminosae or Fabaceae). Edible-podded peas include snow peas (a.k.a. sugar pea or China pea) and snap peas. Cultivars of green peas are

(A)

(B)

(C)

FIGURE 17–37 Pea cultivars shown are: (A) Green pea 'Mr. Big'. (B) Snow pea. (C) Edible-pod snap pea 'Sugar Snap'. Peas found by archaeologists on the Thai-Burmese border have been carbon-dated to 9750 B.C. Peas arrived in North America in 1493 via Columbus, who had them planted on Isabella Island. Peas gained widespread popularity in the United States in the 1600s. (A and C courtesy of All America Selections, B courtesy of National Garden Bureau.)

derived from *Pisum sativum* var. *sativum*, while edible-podded pea cultivars are from *P. sativum macrocarpon*. Minnesota, Washington, and Wisconsin lead in commercial production. Average yield is 1.8 tons/acre. Dry peas are grown mainly in North Dakota and Washington. Average yields are close to 1 ton/acre.

Peas are used alone either raw or cooked and in combination with other vegetables and in salads, soups, stews, and stir-fries. Dried pea seeds are used to make pea soup. The majority of the green pea crop is processed, but gardeners and specialty growers do produce fresh peas. Green peas (no pods) provide good amounts of vitamins A, C, K, folate, and niacin. Edible-podded peas provide very good amounts of vitamin C and good amounts of niacin.

Peas do not tolerate high temperatures and are grown as a spring and fall crop in the North, and as a fall, winter, and spring crop in the South. Seeds have a 3-year viability and an optimal soil temperature range for germination of 4.4° to 23.9°C (40° to 75°F). Minimal soil temperature for germination is 4.4°C (40°F). Seeds are sown directly in place as early as the soil can be worked. An inoculum of *Rhizobium* sp. can be beneficial for yield, if they are not present in the soil. Planting depth is 1 to 2 inches and germination takes 8 to 10 days at soil temperatures above 10.0°C (50°F). Soil pH can be from 5.8 to 6.8.

Cultivars are of either determinate (smaller vine with terminal fruit set, favored commercially) or indeterminate types (preferred by gardeners). The taller cultivars can be grown on brush, trellises, or netting to conserve space. Support is not used in commercial production. Rows for commercial production are spaced 8 inches apart with in-row spacing of 2 inches. Garden spacing is usually on nets (vertical gardening) with 2-inch plant spacing. If rows are used, they are placed 2 to 3 feet apart.

Peas generally have low drought tolerance, especially during flowering and pod fill. Water needs are 1 inch weekly. Dry peas should not be irrigated once pods start to dry. Drought can cause poor pod fill. Preferred minimal soil moisture is −0.70 Bars or 40% ASM.

A few succession plantings at 10-day intervals or planting both early- and late-maturing cultivars can be used to extend the harvest period. Weed competition with peas is minimal given their very early planting dates. High-density commercial plantings also shade out weed growth. Herbicide use is minimal and shallow cultivation is used where allowed by row spacing. Maturity is from 55 to 75 days. Peas for processing are machine harvested; those for fresh use are picked by hand. The best yields are obtained from the spring crop in the North and the winter-spring crop in the South. Green peas should be picked when their pods are well developed and fully green, but before the peas harden or lose their sugar content to starch. Edible-podded peas are harvested when pods reach full length and the seeds are beginning to develop (pods flat). Peas are hydrocooled and moved fast to keep their sugar content high.

Troublesome insects include the pea aphid (*Acyrthosipon pisi*) and pea weevil (*Brushus pisorum*). Diseases include serious problems with root rot (*Fusarium solani* and species of *Aphanomyces, Pythium,* and *Rhizoctonia*) and lesser problems with bacterial blight (*Pseudomonas pisi*), powdery mildew (*Erysiphe polygoni*), wilt (*Fusarium oxysporum*), and several viruses (pea Mosaic, stunt and streak; yellow bean mosaic, enation virus). Some disease resistance can be found in certain cultivars. Rotations and use of disease-free seed are helpful. Viruses are best controlled by eliminating aphids (carriers).

Garden pea cultivars include 'Alderman,' 'Bolero,' 'Dual,' 'Early Alaska,' 'Fortune,' 'Frosty,' 'Garden Sweet,' 'Green Arrow,' 'Hurst Green Shaft,' 'Knight,' 'Lincoln,' 'Little Marvel,' 'Maestro,' 'Misty,' 'Mr. Big,' 'Rondo,' 'Sparkle,' 'Spring,' 'Thomas Laxton,' 'Venus,' and 'Wando.' Edible-podded snap pea cultivars include 'Sugar Ann,' 'Sugar Bon,' 'Sugar Snap,' 'Sugar Sprint,' 'Sugar Star,' 'Super Snappy,' and 'Super Sugar Snap.' Snow pea cultivars include 'Mammoth Melting Sugar,' 'Norli,' 'Oregon Sugar Pod II,' 'Snowbird,' 'Snow Green,' and 'Snow Wind.'

Peppers

Garden peppers (Figure 17–38) are a very tender, frost-sensitive, warm season solanaceous crop in the nightshade family (Solanaceae). Peppers are perennials grown as an annual crop. Cultivars are derived from *Capsicum annuum* var. *annuum*. Peppers are noted for their green (immature) and colorful mature, edible fruits (red, orange, yellow, purple, chocolate), which vary in size and shape. The pungency also varies from mild to very hot. The typical, blocky,

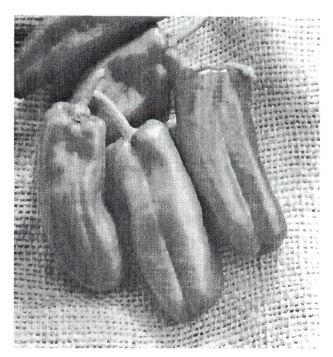

FIGURE 17–38 Pepper F_1 'Giant Marconi.' Peppers should not follow eggplants, potatoes, or tomatoes in rotations. Yields of processing bell peppers are reported as high as 25 tons/acre. Fresh market peppers are often waxed to improve keeping qualities. (Courtesy of All America Selections.)

mild garden pepper is in the Grossum group (bell pepper, green pepper, sweet pepper, and pimiento). The pimiento is a more-heart shaped bell pepper and is generally processed. The Longum group includes pungent, hot peppers (capsicum pepper, cayenne pepper, chile pepper, long pepper). Other hot and ornamental peppers are in the Cerasiforme, Conoides, and Fasciculatum groups. The tabasco pepper is another species of pepper, *C. frutescens.*

Peppers are eaten raw or cooked. Raw forms are used in salads and cooked forms are used alone or in combination with other vegetables and main dishes. Mild forms are popular in salads and Italian cooking. Hot forms are especially popular in Chinese, Mexican, and Thai dishes. Bell forms can be stuffed and longer, tapered ones are good fried and sautéed. Hot peppers add pungency and spice to dishes and salsa. Hot and mild peppers can also be pickled, dried, or made into paprika. Hot peppers are also sources of capsaicin, an alkaloid used medicinally and in protective pepper sprays. Hot peppers vary in hotness from zippy to the break-out-in-sweat-and-tears types. Peppers are excellent sources of vitamin C either green or ripe. Ripe peppers also provide very good amounts of vitamin A.

California, Florida, New Jersey, Georgia, and North Carolina are the leading commercial producers of fresh/processed bell peppers. Bell peppers account for about two-thirds of all pepper production. New Mexico and California are the leading growers of fresh/processed chile peppers. Average yields of bell and chile peppers are 13.5 and 4.8 tons/acre, respectively. California has reported chile pepper yields as high as 11.5 tons/acre.

Peppers must be started indoors in the North, and 6- to 7-week-old transplants are planted when all danger of frost is past. In the South the seeds can be sown outdoors in a seedbed after all danger of frost is past, and the 6-week-old plants can be transplanted to the field. Bell peppers grow well in the North and South, but hot peppers appear to do better in warmer areas with hot, long summers. Seeds have a 3-year viability and should not be sown at soil temperatures below 15.5°C (60°F). The optimal soil temperature range for germination is 18.3° to 35.0°C (65° to 95°F). Seeds are sown 0.5 to 1.0 inches deep and germinate in 6 to 10 days. Soil pH can vary from 5.5 to 6.8. Yields of pepper can increase with the use of black plastic mulch. Seeds or transplants can be machine planted through black plastic mulch. Black plastic mulch is widely used in Florida, but not California. Double rows spaced 1 foot apart with 36 to 40 inches between each set are popular. Plants in rows are 12 inches apart.

Peppers have moderate drought tolerance. An inch of water weekly is best. Critical periods for water stress are transplanting, flowering, and fruit formation. Water stress can result in shriveled peppers, blossom-end rot, small peppers, and reduced yields. Preferred minimal soil moisture is −0.45 Bars or 50% ASM.

Peppers are shallow rooted, so cultivation needs to be very shallow or mulch is used. Some herbicides before and after planting are also used. Peppers in the heavy fruit stage are easily blown over by high winds and staking might be required in some areas.

Blossoms fail to set fruit well if the temperatures drop below 12.8°C (55°F), or the temperatures are high for a prolonged period, or humidity or soil moisture is low, or if excess soil nitrogen is present. Peppers are harvested from transplants in 60 to 70 days and from seed in 100 to 130 days. Generally green peppers can be picked 30 days after flowers set and red at about 50 to 60 days. Green peppers are picked full sized, but before they become ripe (red or yellow); peppers are also utilized ripe. Bell peppers are usually hand harvested, whereas hot peppers or mild peppers to be processed are mechanically harvested. Peppers can be treated with growth regulators to speed ripening. Wax sprays are also used to prevent moisture loss. Peppers can be stored at 7.2° to 10.0°C (45° to 50°F) and 90% relative humidity for 2 to 3 weeks.

Insect pests include the aphid (*Macrosiphum euphorbiae, Myzus persicae*), cucumber beetle (see cucumber), fleabeetle (*Epitrix cucumeris*), cutworms, common stalk borer (*Papaipema nebris*), leafminer (*Liriomyza sativae*), pepper maggot (*Zonosemata electa*), pepper weevil (*Anthonomus eugenii*), and whitefly. Diseases include anthracnose (*Colletotrichum* sp.), bacterial spot (*Xanthomonas campestris* pv. *vesicatoria*), blight (*Sclerotium rolfsii*), damping off (*Pythium* and *Rhizoctonia* sp.), phytophthora blight (*Phytophthora capsici*), and several viruses (alfalfa mosaic virus, cucumber mosaic virus, curly top virus, pepper mottle, potato virus X and Y, tobacco etch, tobacco mosaic virus, and tomato spotted wilt). Viruses are spread by aphids and leafhoppers, which need to be controlled. Physiological disorders that trouble tomatoes, blossom-end rot and sunscald, also are found in peppers (see tomatoes).

Cultivars of the mild or sweet bell-shaped form, eaten green or ripe (red, orange, yellow, purple, brown), include 'Ace Hybrid,' 'Bell King,' 'Bell Tower,' 'Blushing Beauty Hybrid,' 'California Wonder,' 'Chocolate Bell Hybrid,' 'Crispy Hybrid,' 'Early Crisp Hybrid,' 'Early Sensation Hybrid,' 'Golden Giant II Hybrid,' 'Gold Standard Hybrid' (yellow), 'Golden Summer Hybrid,' 'Great Stuff Hybrid,' 'Hungarian Spice Hybrid,' 'New Ace Hybrid,' 'Orange Bell II Hybrid,' 'Pimiento Elite F_1 Hybrid,' 'Purple Beauty,' 'Red Delicious,' 'Skipper,' and 'Valencia Hybrid' (orange). Long, pointed, mild cultivars include 'Big Banana F_1 Hybrid,' 'Bananarama,' 'Cubanelle,' 'Giant Macaroni Hybrid,' 'Gypsy,' 'Key Largo Hybrid,' and 'Sweet Banana.' Hot cultivars are 'Ancho (Poblano),' 'Anaheim Chile,' 'Anaheim TMR,' 'Big Red Hybrid,' 'Big Thai Hybrid,' 'Biker Billy,' 'Caribbean Red,' 'Cherry Bomb F_1 Hybrid,' 'Habanero,' 'Hot Lemon,' 'Hungarian Wax,' 'Jalapa,' 'Jalapeno,' 'Kung Pao Hybrid,' 'Large Chile,' 'Long Red Slim Cayenne,' 'Salsa Delight,' 'Super Cayenne,' 'Tabasco,' 'Tears of Fire Hybrid,' 'Thai Hot,' 'Volcano,' and 'Yellow Cayenne.'

Potato

The white potato (Figure 17–39) is a half-hardy, cool season crop that only has moderate tolerance toward frost. It is noted for its edible, underground, modified stem known as a tuber. Cultivars of the potato are derived from *Solanum tuberosum,* which is in the nightshade family (Solanaceae). The potato is a herbaceous perennial grown as an annual crop. Its main food value is as a carbohydrate source, primarily starch, and potatoes also provide protein, iron, potassium, fiber, and some vitamin C. Idaho, Washington, Oregon, Colorado, and Maine are the leading commercial sources. Yields average 18 tons/acre. The majority of the potato crop is harvested in the fall for storage and use and is grown primarily in the North. Spring, summer, and winter potatoes go into processing and are produced mainly in the South and West.

The dry matter content of the potato determines its use. High dry matter potatoes develop a mealy texture when cooked and are used for baking, potato chips, and french fries and are usually called baking potatoes. Midlevel dry matter potatoes are used for mashed potatoes and boiling, while the lowest dry matter types are used for frying, canning, and in potato salads. Midlevel types are sometimes called general-purpose potatoes because they can also be baked in a pinch.

Because tubers are formed in preference to vegetative growth only when days are on the shorter side with temperatures of 15.5° to 18.3°C (60° to 65°F), potatoes do not suc-

FIGURE 17–39 The Russet Burbank potato accounts for 40% of U.S. potato production. (USDA photograph.) Potato information is available at: http://www.css.orst.edu/potatoes/.

ceed as well in the South as in the North during midsummer. Certain cultivars are therefore recommended for the South.

Potatoes do not come true from seeds, so either seed potatoes, small tubers of 3 ounces or less, or seed pieces (cut up tuber sections) are used. Certified disease-free seed pieces should be utilized. Seed potatoes are preferred, because yields are higher and disease problems fewer, but cost is higher. The tuber is cut into pieces (seed pieces), which have one or two buds ("eyes"). These seed pieces are cured in a cool, 10.0° to 18.3°C (50° to 65°F), dry area for 2 to 3 days until suberization of the cut surface occurs. Soil pH can vary from 5.0 to 7.0. The lower end is used when scab is a problem, because acid conditions inhibit scab.

Potato cultivars of the early type can be planted 10 to 14 days before the last killing frost in the spring. Late cultivars can be planted a few or more weeks later. Seed potatoes or pieces are planted 3 to 4 inches deep in rows spaced about 36 inches apart. In-row spacing depends on several factors such as cultivar, seed potato versus seed piece, and desired final tuber size. It varies from 6 to 16 inches. Soil is hilled along the rows before the plant reaches a height of 10 inches. This prevents exposure of the tubers to light, which would cause greening of the tubers. Solanine, a toxin, is found in the green tissue.

Potatoes require considerable water, and irrigation is needed in the drier parts of the United States. Potatoes have moderate drought tolerance. Water needs are 1 inch weekly. The critical period is after flowering. Yields, tuber size, and shape can be affected negatively by drought. Preferred minimal soil moisture is −0.35 Bars or 70% ASM. Potatoes are good competitors against weeds because of fast, early growth and shading. Often herbicide is used prior to planting. Cultivation is shallow and is used minimally and only before plants start to bloom. Tubers are too easily damaged afterward. Tubers mature in 90 to 120 days depending on cultivar, weather, and other conditions.

Potatoes are harvested after the vines mature and die. Vines are often killed chemically or mechanically to facilitate the harvest. Mechanical harvesting is used. Potato storage is more complicated than for most vegetables. First potatoes are cured for 2 to 3 weeks under good ventilation, high relative humidity, and at 12.8° to 15.5°C (55° to 60°F). Storage relative humidity is 90% to 95%, while temperature varies based on potato use after storage. The range is 3.3° to 4.4°C (38° to 40°F) (seed potatoes or consumer market potatoes), 5.6° to 7.2°C (42° to 45°F) (freezing or dehydration processing), and 10.0° to 12.8°C (50° to 55°F) (chips). For removal to market, potatoes must be conditioned to prevent damage and to restore sugar levels through starch reconversion. Temperatures of 10.0° to 15.5°C (50° to 60°F) for 2 to 3 weeks are used. Sprouting inhibitors are also used on potatoes stored longer than 2 months.

A number of insects attack potatoes. These include aphid (*Macrosiphum euphorbiae, Myzus persicae*), Colorado potato beetle (*Leptinotarsa decemlineata*), fleabeetle (*Epitrix cucumeris*), leafhopper (*Empoasca fabae*), potato tuberworm (*Phthorimaea operculella*), and wireworm. Diseases associated with potatoes include scab (*Streptomyces scabies*), early blight (*Alternaria solani*), late blight (*Phytophthora infestans*), Verticillium wilt, ring rot (*Cornybacterium sepedonicum*), black leg (*Erwinia atroseptica*), and viruses spread by aphids (potato mosaic, rugose mosaic, and leaf roll). Cultivars are available with varying degrees of disease resistance and aphids should be controlled.

Red-skinned cultivars include 'Chieftain,' 'Norland,' 'Red La Soda,' 'Red Pontiac,' and 'Viking.' Gold flesh-colored potatoes include 'Saginaw Gold' and 'Yukon Gold.' White- or cream-skinned cultivars include 'Atlantic,' 'Beltsville,' 'Chipbell,' 'Idaho,' 'Irish Cobbler,' 'Katahdin,' 'Kennebec,' 'Monona,' 'Norchip,' 'Norgold Russet,' 'Russet Burbank' (the most widely used), 'Sebago,' 'Superior,' and 'Wauseon.' A purple/blue cultivar is 'All Blue.' In the South, 'Early Gem,' 'Irish Cobbler,' 'Pungo,' 'Red La Soda,' and 'Red Pontiac' are widely used.

Pumpkin

The pumpkin (Figure 17–40) is a cucurbit or vine crop that is sensitive to both frost and extreme heat. It is noted for its rounded, or at least symmetrical, fruit with orange to pale orange flesh. The lines of distinction between winter squash and pumpkins are not exact, since both pumpkins and winter squash cultivars can be found in four different species of Cucurbita (*C. maxima, C. mixta, C. moschata,* and *C. pepo*). Most of what we consider pumpkins are found in the last two species. The pumpkin is in the gourd family (Cucurbitaceae). Pumpkins are used for pies, pumpkin butter, roasted seeds, and jack-o'-lanterns. Pumpkins are reasonable sources of vitamins A and K and somewhat for niacin. Yields for this minor crop average 12.9 tons/acre and the leading states are Illinois, California, New York, and Pennsylvania.

Because they are frost sensitive, pumpkins cannot be seeded outdoors until all danger of frost is past (soil temperature at least 15.5°C or 60°F). Optimal soil temperature range for germination is 21.1° to 32.2°C (70° to 90°F). Seeds have a 4-year viability and are sown 1 to 2 inches deep. They can also be started indoors and put out as 2-week-old transplants, if care is taken to minimize transplanting shock. Soil pH can range from 5.5 to 7.5. Rows are spaced 5 to 8 feet apart with in-row spacing of 3 to 5 feet. Since pumpkins require much

FIGURE 17–40
Pumpkin F₁ 'Orange Smoothie.' Naked-seeded or hulless varieties are preferred when seeds are to be roasted. Some cultivars are 'Trick or Treat,' 'Triple Treat,' and 'Lady Godiva.' Yield of seed from naked-seeded pumpkins runs from 800 to 1,500 pounds/acre. Yields of whole pumpkins can be as high as 23 tons/acre. (Courtesy of All America Selections.)

space and give minimal returns, they are often grown as a companion crop with corn, because pumpkins can tolerate partial shade. Pumpkins can also be put in after the removal of early crops such as radishes or lettuce.

Pumpkins have moderate drought tolerance. The critical water need is during fruit formation. An inch of water per 14 days is needed to maximize fruit size. Preferred minimal soil moisture is −0.70 Bars and 40% ASM. Cultivation needs to be earlier and close to plants up to the time vines start to run. After that point, cultivation becomes shallow and further out from plants. Black plastic mulch is effective and promotes earlier crops.

Pumpkins are picked when fully matured (80 to 100 days, hard rind and orange color). Harvest is by hand for fresh market and mixed hand or machine harvest for processing pumpkins. Pumpkins are generally not stored.

Insects that trouble pumpkins are those that affect cucumbers and squash. The main ones are the striped cucumber beetle (carrier of bacterial wilt), squash bug, squash vine borer, and aphid. Diseases of concern are leaf blights, powdery mildew, wilt, and mosaic virus (see cucumber and squash).

Pumpkin cultivars include the following: 'Baby Bear,' 'Big Max,' 'Big Moon,' 'Bushkin' (bush form), 'Connecticut Field,' 'Dill's Atlantic Giant,' 'Howden,' 'Howden Biggie,' 'Jack Be Little,' 'Jack-O'-Lantern,' 'Kentucky Field,' 'Lumina' (white skin), 'Munchkin' (small decorative), 'Pankow's Field,' 'Prizewinner Hybrid,' 'Sorcerer F_1 Hybrid,' and 'Triple Treat.'

Numerous cultivars of gourds are also available. Culture and identity (*C. pepo* for most) is similar to pumpkins. Gourds can provide additional sources of income at the time pumpkins are sold. Pumpkins and gourds must be isolated to prevent cross-pollination. Gourds should also be isolated from summer squash.

Radicchio (*see* Endive)

Radish

The radish (Figure 17–41) is a root crop noted for its edible crisp, sharply pungent storage root. It is a hardy, cool season biennial grown as an annual crop that cannot tolerate heat. Cultivars are derived from *Raphanus sativus* in the Mustard family (Brassicaceae or Cruciferae).

(A)

FIGURE 17–41 Typical spring (A) and winter (B) radishes. The Korean dish, Kimchi, is made by pickling winter radishes. No crucifer crops should have been planted in radish fields within at least the last 3 years. (A is USDA photograph and B is courtesy of National Garden Bureau.)

(B)

Spring radishes are fast crops and slower growing, longer maturity radishes are called winter radishes. Spring radishes are a minor commercial crop and winter radishes much less common, except for gardens. Radishes are eaten fresh alone or in salads and have little nutritive value other than fiber and some vitamin C. Florida and California are the leading states. Yields average 4.5 tons/acre, but have been reported as high as 12.5 tons/acre in California.

Radishes are grown from seed during the spring and fall in the North, and in the fall, winter, and spring in the South. Radishes can be directly sown in place as soon as soil is workable. The minimal soil temperature is 4.4°C (40°F) and the optimal soil temperature range is 7.2° to 32.2° C (45° to 90°F) Germination is in 4 to 8 days. Seeds are sown 0.5 inch deep in rows spaced 6 to 12 inches apart. Radishes in rows can be crowded to 0.75 to 1.0 inch apart. Winter types are left 4 inches apart. Precision planters are used for seeding to eliminate thinning. The soil pH can range from 6.0 to 7.0. Spring radishes mature quickly (24 to 42 days) and remain at optimal quality briefly, so succession sowings at 7- to 10-day intervals are possible. Radishes can also be followed by other succession crops or interplanted with other crops requiring longer to reach maturity such as carrots.

Radishes need continuous moisture for rapid growth and to avoid pithy roots. Drought tolerance is low. Water needs are 1 inch in 5 days. Preferred minimal soil moisture is −0.25 Bars or 70% ASM.

Mechanical harvest is used, except for leaf-topped radishes. Winter radishes take 45 to 60 days. Radishes are hydrocooled and can be stored for 3 to 4 weeks at 0°C (32°F) and 95% relative humidity. Winter radishes under these storage conditions can last 3 to 4 months. The only pest of consequence is the radish maggot (*Hylemya brassicae*), although pests of cabbage can be occasionally troublesome.

Suggested spring radish red, white, and red/white cultivars include 'Champion,' 'Cherriette,' 'Cherry Belle,' 'Cherry Bomb II Hybrid,' 'Crimson Giant,' 'French Breakfast,' 'Fuego,' 'Mirabeau,' 'Red Flame Reggae,' 'Salad Rose,' 'Sparkler,' and 'White Icicle.' Winter radishes include 'All Seasons,' 'April Cross Hybrid,' 'Japanese Daikon,' and 'Summer Cross Hybrid.'

Rhubarb

Rhubarb (Figure 17–42) is a very hardy perennial noted for its reddish, acid-flavored, edible petiole or leafstalk produced in the spring. The leafstalk is the only nonpoisonous part of the rhubarb plant. Cultivars are derived from *Rheum rhabarbarum,* a member of the buckwheat family (Polygonaceae). The best rhubarb is grown where summers are cool and moist and winter freezes prevail. The mean summer and winter temperatures, respectively, should not exceed 23.9°C (75°F) and 4.4°C (40°F). As such, it is not suitable for cultivation in much of the South. Rhubarb is usually stewed and served with lots of sugar or made into preserves and pies alone or with strawberries. It is a good source of vitamin A. Rhubarb is a very minor commercial crop in the Northwest and Michigan.

The crown with its fleshy roots can be divided in the spring. This is the method of choice for propagation, since seedlings do not come true. Divisions are planted to the same depth as from the original plant (3 to 4 inches usually) and spaced 3 to 4 feet apart in all directions. Leafstalks should not be pulled the first year, only sparingly the second year, and more heavily thereafter. Flower stalks should be removed as they appear. Leaves are toxic because they contain oxalic acid. Rhubarb is relatively pest free. Soil pH can be 6.0 to 7.0.

Rhubarb has moderate drought tolerance. The critical period is leaf emergence. An inch of water in 21 days is needed. Water stress causes pithy stems. Preferred minimal soil moisture is −2.00 Bars or 20% ASM.

Leafstalks should be harvested in early spring when tender up to 6 to 8 weeks. Stalks can be pulled by hand or cut just above the soil line. Storage for a few weeks at 0°C (32°F) and 95% relative humidity is possible. Crowns will require periodic division at 7 to 8 years to maintain productivity. Pests include rhubarb curculio (*Lixus concavus*; destroy curled dock that serves as feed) and potato stem borer (*Hydroecia micacea*; destroy quack grass that serves as feed). Phytophthora crown rot can sometimes be a problem. Weeds are controlled by mulch or cultivation.

FIGURE 17–42 Rhubarb can be field grown, hothouse produced, forced for January sales, and micropropagated. When first planted, fertilizer applications run about 70–80, 70–80, and 140–160 lb/acre of N, P_2O_5, and K_2O, respectively. In subsequent years, N rates can be doubled, while leaving P_2O_5 and K_2O the same. Nitrogen applications should be split into one application before growth starts in the spring, another after growth starts, and the last side dressing after harvest. Manure works well as a nitrogen source with rhubarb. (Author photograph.)

Newer cultivars have a pleasing reddish color as opposed to the green of the old cultivar, 'Victoria.' Suggested cultivars include 'Cherry Red,' 'Chipman's Canada Red,' 'Crimson Red,' 'Glaskins Perpetual,' 'MacDonald,' 'Valentine,' and 'Victoria.'

Rutabaga (*see* Turnip)

Salsify

Salsify (*Tragopogon porrifolius*) or oyster plant is a hardy biennial grown as an annual crop noted for its edible, carrot-like white root. It can be grown in most parts of the country, except in the extremes of the North, because it requires a long growing season of 120 days. Salsify is in the Compositae or Sunflower family (Compositae or Asteraceae). The root can be steamed, boiled, or microwaved and served as a main vegetable with an oyster-like flavor or can be used in soups and stews. It supplies modest amounts of calcium and iron.

Seed should be sown in place about 0.5 inch deep as soon as the soil is workable. Rows are 18 inches apart for hand cultivation and 36 inches for mechanical cultivation. Plants are thinned to 2 to 4 inches apart. Soil pH can be between 6.0 and 7.0. Culture is similar to that of parsnip. Insect and disease pests are minimal. Frost sweetens the root. In many regions salsify can be left in the ground for later harvesting. Storage conditions are similar to carrots. Salsify can be interplanted with lettuce, radish, and spinach. The suggested cultivar is 'Sandwich Island Mammoth.'

Shallot

The shallot is one of the perennial onion crops grown as an annual. Cultivars are derived from *Allium cepa,* Aggregatum group, and are in the amaryllis family (Amaryllidaceae). It is noted both for its bulbs, which are small and multiple (cloves), and for its use in the immature stage when the young plants are used as green onions or scallions. The cloves are minced and used to flavor soups, stews, and meats with a milder, aromatic onion flavor.

Young leaves and scallions can be used in salads and sautéed. Shallots are a very minor commercial crop in parts of the South.

Being hardy, shallots are sown in the North as soon as the ground can be worked in the spring. Shallots are planted later in the South as a winter crop. They are planted in place by small cloves or bulblets, as for onion sets. Planting depth is 2 inches, rows are spaced 12 to 24 inches apart, and in-row spacing is 4 inches. Seeds can also be used, being sown 0.5 to 1.0 inch deep. Soil pH can be between 6.0 and 7.0. Shallots need a continuous moisture supply, being shallow-rooted. Maturity is in 120 days from seed and 70 to 80 from sets. Insect and disease pests are those of onions, as is seed information, culture, and harvest.

Cultivars include 'Atlantic,' 'Bonilla,' 'Dutch Yellow,' 'French Red,' 'Holland Red,' 'Matador Hybrid,' 'Pikant,' and 'Prisma Hybrid.'

Snap Beans (*see* Beans)

Soybeans (*see* Beans)

Spinach

Spinach (Figure 17–43) is a hardy, cool season annual crop grown for its edible leaves or greens. Cultivars are derived from *Spinacia oleracea,* which is in the Goosefoot family (Chenopodiaceae). Spinach is a very good source of iron, potassium, calcium, thiamin, folate, riboflavin, niacin, and vitamins A, K, and C. It is served raw in salads and cooked as a green. California leads in fresh production and Texas in processed spinach. Average yields of fresh and processing spinach are 6.9 and 9.4 tons/acre, respectively.

In the North it is a spring and fall crop, and a fall, winter, and spring crop in the South. High temperatures and long photoperiod induce seed formation and reduced leaf quality. Seed has a 4-year viability and can be sown in place 0.5 inch deep as soon as the soil is workable. Optimal soil temperature range for germination is 7.2° to 23.9°C (45° to 75°F). Seed germination is in 6 to 10 days. Soil pH can vary from 6.0 to 7.0. Rows are spaced 5 to

FIGURE 17–43   Spinach cultivar, Long Standing Bloomsdale. For resistance to downy mildew, try cultivars such as Baker, Bossanova, Bolero, Cascade, and Olympia. The variety Coho has resistance to White Rust. Cultivars that show exceptional bolting tolerance include 'Bejo 1369,' 'Coho,' 'Olympia,' 'Skookum,' 'Splendor,' and 'Tyee.' Spinach crops are easily damaged by air pollutants such as ozone. (USDA photograph.)

20 inches apart with in row spacing of 0.6 to 3.0 inches. Closer spacing is used for processing spinach relative to fresh market spinach. Spinach can be sown in succession at weekly intervals to extend the harvest period. Precision mechanical seeding is used to avoid thinning. Uniform moisture is required as spinach has low drought tolerance and is shallow-rooted. Weeds are mostly controlled with shallow cultivation, although some preplant applications of herbicides are used.

Spinach must be harvested prior to flower stalk formation. Maturity is 30 to 55 days. Most spinach is mechanically harvested and promptly cooled with either ice, hydrocooling, or vacuum cooling. Storage is limited to 10 to 14 days at 0°C (32°F) and 95% relative humidity.

Troublesome insects include aphids (*Myzus persicae*), which can be carriers of mosaic virus (spinach blight or yellows), and leaf miners (*Pegomyia hyoscyami*). Fusarium wilt and downy mildew (*Peronospora effusa*) can be problem diseases along with mosaic virus.

Suggested wrinkled leaf (Savoy types, preferable for fresh use) and smooth leaf cultivars include 'Dark Green Bloomsdale,' 'Long Standing Bloomsdale,' 'Mazurka,' 'Melody Hybrid,' 'Nordic IV,' 'Olympia Hybrid,' 'Razzle Dazzle Hybrid,' 'Salad Fresh,' 'Skookum,' 'Steadfast,' 'Teton,' and 'Tyee.'

Squash

Squash (Figure 17–44) is a warm season, frost-sensitive crop grown in most of the United States. Squash is in the gourd family (Cucurbitaceae) and includes a large group of cucurbits, many of which are vine crops; some are bush forms. All are noted for their edible fruit. Summer squashes include those fruits eaten when immature up to when the rind just starts to harden; the shell should be easily dented with a fingernail. Summer squashes (*Cucurbita pepo*) are bush types. Winter squash (*Cucurbita pepo, C. mixta, C. maxima, C. moschata*) are eaten after the fruit is mature (hard shelled) and is usually a vine crop, although some bush types are available. The division between winter squash and pumpkin is not absolute. Winter squash has a flesh that is darker orange, sweeter, less fibrous, and higher in dry matter than pumpkins or summer squash.

Summer squash is usually cooked either by steaming or microwaving and used as a main dish vegetable. It can also be used raw in salads. Modest amounts of vitamins A and C and niacin are present. Summer squash can be yellow and club shaped (yellow straightneck

(A)

(B)

FIGURE 17–44 (A) 'Waltham Butternut winter squash.' (B) Various types of summer squash. Genetic engineering has produced squash with high levels of resistance or immunity to some viruses. One of these "biotech" squashes is a summer squash called Freedom II (Asgrow Seed Co.). This squash shows resistance to zucchini yellow mosaic virus and watermelon mosaic virus 2. Spaghetti winter squash cultivars such as Pasta F_1, Tivoli F_1 (bush spaghetti), and Orangetti (higher in vitamin A than other two) are used as spaghetti substitute. Such squashes have become popular recently. (Courtesy of National Garden Bureau.)

and crookneck types), green and club shaped (zucchini and cocozelle types), and white, yellow, or light green pan shaped (scallop or patty pan types). Winter squash is usually baked or microwaved and served as a main vegetable with melted margarine and brown sugar. It can also be used in pies or stuffed with ground meat, sausage, or cheese fillings. Winter squash is a very good source of vitamins A and C, and has reasonable amounts of niacin. Winter squashes are either small or large. Small ones include acorn, butternut (half a dumbbell shape), buttercup, and turban. Large ones include banana, delicious, and hubbard types. Summer squash yields average 8.0 tons/acre and the leading states are Georgia, California, and Florida. Yields and states for winter squash are similar to those for pumpkins.

Because squash is frost sensitive, it cannot be seeded outdoors until all danger of frost is past (soil temperature at least 15.5°C or 60°F). Optimal soil temperature range for germination is 21.1° to 35.0°C (70° to 95°F). Seeds have a 4-year viability and are sown 1 to 2 inches deep. They can also be started indoors and put out as 2-week-old transplants, if care is taken to minimize transplanting shock. Soil pH can range from 5.5 to 7.5. Rows are spaced 4 to 6 feet apart with in-row spacing of 1.5 to 3.0 feet for summer squash and small vine winter squash. Large winter squash is planted in rows 6 to 8 feet apart with in-row spacing of 3 to 5 feet. Since large winter squash require much space and give minimal returns, they can be grown as a companion crop with corn or be put in after the removal of early crops such as radishes or lettuce.

Cultivation needs to be earlier and close to plants up to the time vines start to run. After that point, cultivation becomes shallow and further out from plants. Bush types can be more closely cultivated. Black plastic mulch is effective and promotes earlier crops.

Water needs to be constant and uniform, especially during fruit development. Summer and winter squash require one inch of water in 5 and 10 days, respectively. Summer and winter squash have low and moderate drought tolerance, respectively. Water stress causes undersized misshapen fruits plus reduced yields. Preferred minimal soil moistures are −0.25 and 70% ASM for summer squash and −0.70 and 40% ASM for winter squash.

Summer squash matures in 40 to 50 days and squashes mature fast after flowering (3 to 8 days). Squashes must be picked constantly or productivity will slow. Winter squash is picked when fully matured (85 to 120 days, hard rind not easily dented by fingernail, orange color). Some are improved in sweetness if touched by a light frost prior to harvest. However, if intended for storage, winter squash must not be exposed to frost. Harvest is by hand for fresh market and mixed hand or machine harvest for processing squash. Summer squash can be stored for 1 to 2 weeks at 5.0° to 10.0°C (41° to 50°F) and 95% relative humidity. Winter squash is cured prior to storage at 26.6° to 29.4°C (80° to 85°F) and 80% relative humidity for 10 days. Storage is at 10.0°C (50°F) and 50% to 70% relative humidity for several months.

Insects that trouble squash are those that affect cucumbers. The main ones are the striped cucumber beetle (carrier of bacterial wilt), squash bug, squash vine borer, and aphid. Diseases of concern are leaf blights, powdery mildew, wilt, and mosaic virus (see cucumber and pumpkin).

Insect pests include the squash borer (*Melittia cucurbitae*), squash bug (*Anasa tristis*), squash beetle (*Epilachna borealis*), striped cucumber beetle, white fly (*Trialeurodes vaporariorum*), melon worm (*Diaphania hyalinata*), pickle worm (*Diaphania nitidalis*), and aphid. Powdery mildew and other diseases associated with cucumber can be troublesome.

Cultivars of the yellow summer squash include 'Butterstick Hybrid,' 'Cougar,' 'Crescent Hybrid,' 'Crookneck Park's PMR Hybrid,' 'Dixie Hybrid,' 'Early Golden Crookneck,' 'Enterprise Hybrid,' 'Golden Girl,' 'Multipik,' 'Pic-n-Pic,' 'Prolific Straightneck,' 'Sunglo Hybrid,' 'Superpik,' and 'Yellow Crookneck.' Cultivars of the green summer squash include 'Ambassador,' 'Black Magic,' 'Burpee Hybrid Zucchini,' 'Eightball Hybrid' (round zucchini), 'Embassy Hybrid,' 'Gold Rush Hybrid' (yellow zucchini), 'Greyzini' 'Jackpot Hybrid,' 'Onyx,' 'Sweet Gourmet' (light green), 'Sweet Zuke,' 'Tigress,' and 'Tristen F_1 Hybrid.'

Cultivars of the patty pan summer squash include 'Benning's Green Tint,' 'Pagoda Gold,' 'Peter Pan,' 'Starship Hybrid,' 'Sunburst Hybrid,' and 'Sunny Delight.' 'Pasta Hybrid,' 'Small Wonder,' and 'Tivoli F_1 Hybrid' (bush form) are cultivars of the novel spaghetti squash that cooks up like spaghetti.

Winter types include 'Ambercup,' 'Autumn Cup,' 'Banana,' 'Blue Hubbard,' 'Buttercup,' 'Cornell's Bush Delicata' (bush form), 'Cream of the Crop,' 'Delicata,' 'Discus Bush Buttercup,' 'Early Butternut Hybrid,' 'Gold Nugget,' 'Green Striped Cushaw,' 'Harris Betternut,' 'Nicklow's Delight F_1 Hybrid,' 'Orange Magic Hybrid,' 'Royal Acorn,' 'Show King Giant,' 'Sugar Hubbard,' 'Sweet Dumpling,' 'Sweet Mama Hybrid,' 'Sweet Meat,' 'Table Ace,' 'Table King,' 'Ultra Butternut,' 'Waltham Butternut,' and 'Zenith.'

Sweet Potato

The sweet potato (Figure 17–45) is a very tender, warm season perennial herbaceous plant grown as an annual root crop noted for its tuberous storage roots. Cultivars are derived from *Ipomoea batatas,* which is in the morning glory family (Convolvulaceae). Since it requires a somewhat long growing season with warm days and nights, it is a crop adapted to the South (zones 8 to 10). Some cultivars can be grown in the milder regions of the North. It should not be confused with the yam (*Dioscorea* sp.), which is mainly confined to the tropics. The term *yam* used in the South is misapplied. Yields average 7.3 tons/acre. North Carolina, Louisiana, California, and Mississippi are the primary commercial producers.

Sweet potatoes are baked, fried, boiled, candied, or used in sweet potato pie. Some (industrial types) are also used as sources of flour and starch. Sweet potatoes have extremely high vitamin A content and reasonable amounts of vitamin C and calcium. Food types can be either dry flesh or moist, soft flesh types. The moister soft types have deep, orange flesh and are used for baking, while the drier ones are boiled or fried and have white, yellow, or light orange flesh and more yellowish skin.

Propagation is by transplants created by the rooting of slips forced from small sweet potatoes (seed roots) saved from the previous year's crop. Seed roots can be subjected to 26.6° to 32.2°C (80° to 90°F) and high humidity for several weeks prior to propagation to cause the development of small shoots to speed up the process and to increase shoot numbers. Seed roots are set 1 to 2 inches deep. Rooting is in moist propagation beds, usually of sand. Sand is added as rooting occurs so that the final cover is 4 to 5 inches. At 23.9° to 26.6°C (75° to 80°F), rooted plants can be produced in 6 weeks. Certified seed stock is preferred for fewer disease problems and assurance of cultivar trueness.

Transplants are set out after all danger of frost is past. Soil pH should be between 5.5 and 6.2. Later plantings can also be obtained through rooting tip cuttings of growing vines, which is practiced only in the lower South. The planting system uses ridges 8 to 10 inches tall. Rows are 32 to 48 inches apart. Plants are spaced 8 to 18 inches apart in the row. Spac-

(A)

(B)

FIGURE 17–45 Sweet potato vines (A) and sweet potatoes (B). 'Beauregard' stores well, and 'Jewel' is resistant to root-knot nematode, but does poorly in cool, wet soil. These two cultivars accounted for about 90% of U.S. production. (USDA photographs.)

ing is based on cultivar and other factors. Cultivation is used until the vines run and shade out weeds. Some herbicide is used also. Sweet potatoes are deep rooted and can tolerate drought, so irrigation is minimal in the South and more extensive in the Southwest and West. Temperatures below 12.8°C (55°F) can be detrimental.

Sweet potatoes can be harvested any time they reach a usable size. Maturity from plants is 95 to 110 days, but maximal amount of harvestable material occurs from 130 to 150 days. Harvest must take place before the vines die or before temperatures fall below 50°F. Hand and machine harvests are used, but great care is required to not bruise or damage the sweet potatoes. Vine removal and digging are a required part of the harvest. Sweet potatoes need curing for 4 to 7 days prior to storage. Conditions are 26.6° to 29.4°C (80° to 85°F) and 85% to 90% relative humidity. Storage for 4 to 7 months is at 12.8° to 15.5°C (55° to 60°F) and the same relative humidity.

Insect and other pests include the leaf beetle (*Typophorus nigritus*), sweet potato weevil (*Cylas formicarius elegantulus*), various tortoise beetles, cutworms, and nematodes. Troublesome diseases include black rot (*Ceratocystis fimbriata*), root rot (*Phymatotrichum omnivorum*), scurf (*Monilochaetes infuscans*), soft rot (*Rhizopus stolonifer*), and stem rot (*Fusarium oxysporum* f. *batatas*). Nematode-resistant cultivars can be used. Storage areas should be sanitized prior to use to prevent storage diseases.

The sweet potato has high drought tolerance. Water needs are 1 inch in 3 weeks. The critical period is tuberous root formation, especially the last 40 days. Water stress can cause small and misshapen tuberous roots. Preferred minimal soil moisture is −2.00 Bars or 20% ASM.

Moist or soft-flesh cultivars are 'Centennial,' 'Georgia Jet,' 'Georgia Red,' 'Gold Rush,' 'Julian,' 'Jewel,' 'Jasper,' 'Porto Rico,' 'Varaman,' and 'Velvet.' Dry-flesh cultivars include 'Orlis' and 'Yellow Jersey.' 'Nugget' and 'Nemagold' are intermediate cultivars. Industrial types are 'Pelican Processor' and 'Whitestar.' A cultivar for more northerly locations is 'Beauregard' (90 days) and 'Georgia Jet' (85 days).

Swiss Chard (*see* Chard)

Taro

Taro (*Colocasia esculenta*) or dasheen (Figure 17–46) is a warm season, perennial crop noted for its edible corms. Its long growing season restricts it to zone 10 in the United

FIGURE 17–46 Taro (dasheen) production information is available at: http://www.extento. hawaii.edu/kbase/reports/ taro_prod.htm. (USDA photograph.)

States. There are two kinds of taro: wetland and upland. Wetland taro is planted in flooded land and requires the longer growing season of the two (8 to 12 months); it is grown in Hawaii and is an ingredient in Hawaiian poi. Upland taro is grown in conventional soil, requires a shorter growing season (6 to 8 months), and can be found in zone 10. 'Trinidad' is one good cultivar of upland taro.

Propagation is by corms or corm sections, which can be planted in place 2 to 3 inches deep or started earlier indoors. Soil pH can be from 6.0 to 7.0. Rows are 42 to 48 inches apart and in-row spacing of 24 inches is needed. Peeling and cooking prior to consumption accomplish destruction of the acrid component, calcium oxalate. It is used as a vegetable or added to soups and stews. Taro is a good potassium and fiber source. It tastes somewhat like a cross between potato and chestnut. Irrigation is needed in the drier, less rainfall low land areas.

Tomato

Tomatoes (Figure 17–47) are warm season, frost-intolerant crops in the nightshade family (Solanaceae) noted for their edible fruit. Cultivars are derived from *Lycopersicon lycopersicum* (formerly *L. esculentum*). They are grown over much of the United States. Fruits vary from small, 1-ounce cherry sizes (var. *cerasiforme*) to large baseball sizes (1 pound or more). Shapes range from plum to pear (var. *pyriforme*) to round, and colors vary from greenish-white through yellow, orange, pink, and red. Florida and California are the leading states in fresh tomato production, while California is the leading state for processing tomatoes. Yields are 13.9 and 35 tons/acre for fresh and processing tomatoes, respectively. Fresh tomato yields as high as 18.5 tons/acre have been reported in Florida. Tomatoes also place first in vegetable crops grown commercially in greenhouses.

Tomatoes are eaten raw in salads, on sandwiches, as tomato juice, in drinks (Bloody Mary), and are used in salsa. Cooked tomatoes can be served as stewed tomatoes, baked and stuffed, or as various types of tomato sauce on pasta. Other products include ketchup and soup. Tomatoes have reasonable levels of vitamin A, some vitamin C, niacin, potassium, phosphorus, and iron. Lycopenes in tomatoes also appear to help prevent cancer.

(A)   (B)

FIGURE 17–47 Tomatoes were considered poisonous prior to 1820 in the United States. In that year Colonel Robert Gibbon Johnson consumed a basket full of tomatoes at the Salem, New Jersey courthouse. The crowd fully expected to see the Colonel die. He didn't and the tomato went on to become the number one vegetable garden plant. (A) Indeterminate tomato plant. (B) Determinate tomato plant. (USDA photographs.)

Tomatoes are available in determinate (tomatoes produced heavily on outside of plant, which tends to be smaller than indeterminate types) and indeterminate types. Determinate types need a little less fertilizer and mature earlier than indeterminate ones. Tomatoes of the determinate or self-pruning type are not suited for staking, but can be grown in tomato cages if desired. Indeterminate types have continually growing branches not terminated by fruit set, so they are staked or trellised if space is a problem.

Tomatoes must be started indoors in the North, and 6- to 7-week-old transplants are planted when all danger of frost is past. In the South the seeds can be sown outdoors in a seedbed after all danger of frost is past, and the 6-week-old plants can be transplanted to the field. Transplants from the South are also produced for Northern markets. Direct seeding is widely used for processing tomatoes and for fresh tomatoes in Florida and parts of California. Precision seeders are often used. Seeds have a 3-year viability and should not be sown at soil temperatures below 12.8°C (55°F). The optimal soil temperature range for germination is 18.3° to 35.0°C (65° to 95°F). Seeds are sown at 0.5 to 1.0 inch deep and germinate in 5 to 12 days. Soil pH can vary from 5.5 to 7.5. Earlier tomatoes are produced with the use of black plastic mulch. Seeds or transplants can be machine planted through black plastic mulch. Spacing for rows varies considerably, depending on culture system, cultivar, and whether for fresh or processing markets. Four feet between rows is one possibility with plants in rows spaced from 9 to 36 inches. Processing tomatoes tend to be spaced closer than fresh market tomatoes. Early, midseason, and late cultivars are planted for an extended harvest.

Constant, uniform moisture is essential and irrigation is frequently used. Water stress leads to blossom-end rot, reduced yields, and smaller fruits. The critical period is during fruit expansion. An inch of water every 5 and 7 days are needed for staked and unstaked tomatoes, respectively. The preferred minimal soil moisture is −0.45 or 50% ASM.

Tomatoes are deep rooted, so cultivation is used for weed control. Some herbicides before and after planting are also used. Mulch is an alternative. Tomatoes can be grown either staked or unstaked. Processing tomatoes are unstaked, as are those picked at the mature green stage. Staked tomatoes are found in gardens and in commercial operations producing vine-ripe tomatoes.

Blossoms fail to set fruit well if the temperatures drop below 12.8°C (55°F), or the temperatures are high for a prolonged period, or humidity or soil moisture is low, or if excess soil nitrogen is present. Tomatoes are harvested from transplants in 50 to 75 days and from seed in 100 to 135 days. The range covers early, midseason, and late cultivars. Generally green tomatoes can be picked 30 days after flowers set and are red at about 42 to 60 days.

Tomatoes are harvested at various stages of ripeness, depending on intended use. For shipping long distances, mature green or breaker stages are preferred. Mature green tomatoes are full sized, light green, and yellow-green at the blossom end. Breaker tomatoes are similar, except the lower one-quarter shows pink coloration. Vine-ripened tomatoes for fresh market are picked at breaker, pink (three-quarter pink), or nearly full ripe stages. Processing tomatoes are harvested at full ripeness. Tomato ripening at the mature green/breaker stage can be hastened with ethylene-based growth regulators. Tomatoes to be processed are mechanically harvested. Fresh market tomatoes are often harvested by machine, but mechanical harvesters are available. Tomatoes at the mature green stage and pink stage can be stored at 13.9° to 16.1°C (57° to 61°F) and 10.0°C (50°F), respectively for 2 to 3 weeks. Ripe tomatoes lose quality if refrigerated. Ripe tomatoes can hold well for several days at room temperature.

Insect pests include the aphid (*Macrosiphum euphorbiae, Myzus persicae*), cabbage loopers, cutworms (species such as *Agrotis ipsilon, Amathes c-nigrum,* and *Peridroma saucia*), flea beetle (*Epitrix cucumeris*), leafhoppers, leafminer (*Liriomyza sativae*), mites, tomato (corn) fruitworm (*Heliothis zea*), tomato hornworm (*Manduca quinquemaculata*), wireworms, and white fly. Diseases include anthracnose (*Glomerella phomoides*), bacterial canker (*Corynebacterium michiganense*), bacterial speck (*Pseudomonas tomato*), bacterial spot (*Xanthomonas vesicatoria*), bacterial wilt (*Pseudomonas solanacearum*), early blight (*Alternaria solani*), Fusarium wilt (*Fusarium oxysporum*), late blight (*Phytophthora infestans*), leaf spot (*Septoria Iycopersici*), root knot (nematodes), *Verticillium* wilt (*Verticillium* sp.), and viruses (cucumber mosaic, spotted wilt, and tobacco mosaic). Viruses are spread by aphids and leafhoppers, which need to be controlled. Spread is also possible from handling both by hands and machines. Some resistance to Verticillium, Fusarium, and nematodes is possible with certain cultivars. Rotations are highly advised. Blossom end rot is a physiological disorder caused by overfertilization by nitrogen and an inadequate water supply from root damage or drought that induces localized depletion of calcium. Sunscald of exposed green tomatoes is also seen. Other disorders are catfacing, growth cracks, puffiness, and green shoulder. These growth and ripening disorders are poorly understood, and cultivars vary considerably in terms of susceptibility.

Cultivars of the small, red cherry types used in salads include 'Gardener's Delight,' 'Jolly Hybrid,' 'Red Cherry,' 'Sugar Snack Hybrid,' 'Super Sweet 100,'[*] 'Sweetie,' 'Sweet Million Hybrid'[*] and 'Tiny Tim.' Grape-type red, salad tomatoes are 'Agriset 8279,' 'Juliet Hybrid'[*] and 'Santa F$_1$ Hybrid.' Small yellow cultivars include 'Banana Legs,' 'Chello,' 'Sun Gold F$_1$ Hybrid,' 'Yellow Canary,' 'Yellow Pear' and 'Yellow Plum.'

Standard large red tomatoes include 'Ace 55,'[*] 'Better Boy Hybrid,'[*] 'Big Beef Hybrid,' 'Big Boy Hybrid,' 'Big Girl Hybrid,' 'Bush Big Boy Hybrid,' 'Celebrity Hybrid,'[*] 'Early Girl Hybrid,' 'Heinz 1350,' 'Jet Star,' 'Manitoba,' 'Northern Exposure,' 'Quick Pick Hybrid,'[*] 'Sub-Arctic Plenty,'[*] 'Super Beefsteak,'[*] 'Supersteak F$_1$ Hybrid,' and 'Sweet Cluster Hybrid'[*] to name only some. Paste-type red plum cultivars include 'Amish Paste,' 'Big Mama,' 'La Rossa Hybrid,'[*] 'Roma,'[*] 'San Marzano,' 'Super Marzano Hybrid,'[*] and 'Viva Italia Hybrid.'[*]

Large pink cultivars include 'Florida Pink,' 'Oxheart,' and 'Pink Girl F$_1$ Hybrid.'[*] Large orange cultivars include 'Jubilee,' 'Persimmon,' and 'Sweet Tangerine Hybrid.' Large yellow cultivars include 'Golden Boy Hybrid' and 'Lemon Boy Hybrid.'[*] A cultivar adapted to cellar storage for 6 to 12 weeks is 'Long Keeper.'

Tomatoes for greenhouse culture are 'Dombito F$_1$ Hybrid,' 'Earlypak 707,' 'Greenhouse 656,' and 'Jumbo F$_1$ Hybrid.' Processing types are 'Heinz 722,' 'Heinz 2653,' 'Roma'[*] (paste), and 'UC 204.'

Turnip and Rutabaga

Turnips and rutabagas (Figure 17–48) are a hardy, cool season root crop noted for their edible storage root. Turnip cultivars are derived from *Brassica rapa*, Rapifera group, and rutabaga cultivars from *Brassica napus*, Napobrassica group. Both are in the Mustard family (Brassicaceae or Cruciferae). Turnip and rutabaga (a.k.a. Swede turnip) are somewhat

(A)   (B)

FIGURE 17–48 Turnips (A) and rutabagas (B). Cultivars suggested for turnip greens are All Top, Seven Top, Shogoin, and Topper. (USDA photographs.)

similar, but rutabaga is a bit larger, more pungent in flavor, and requires a longer growing period. Rutabaga foliage is bluish green, fleshy, and smooth, while turnip foliage is green, thin, and hairy. Outside color and flesh differs somewhat in commercial forms, but not necessarily in all cultivars. Turnip skin color is white to tan and flesh is white, while rutabaga skin is yellowish tan, often (but not always) with a purple top band and yellow flesh.

Turnip leaves are also used for greens, but rutabaga tops are not. Both vegetables are served mashed after boiling or steaming. Roots are good sources for vitamin C and niacin, calcium, iron, and potassium. Turnip greens are very good sources of calcium, phosphorus, iron, potassium, vitamins A and C, and niacin. Greens can be eaten raw, but are are usually steamed, sautéed, or boiled. Both are very minor commercial crops found mostly in Wisconsin, Minnesota, and Washington.

Turnips are grown as spring and fall (better quality) crops in the North, and both turnips and rutabagas as fall, winter, and spring crops in the South. Turnips succeed better in the South than rutabagas. In the North, rutabagas are grown only as a fall crop. Turnips usually do better in the South than rutabaga.

Both are sown directly in place. Seed has a 4-year viability and is planted 0.5 inch deep. Germination is in 4 to 8 days. Rows are spaced 10 to 30 inches apart depending on whether hand or mechanical cultivation is used. In-row spacing can be 4 to 8 inches (lower for turnip, higher for rutabaga). Precision seeders are used to eliminate thinning. Soil pH can be 5.5 to 7.0.

Rutabagas and turnips have moderate drought tolerance. The critical time is during root expansion, where water stress can cause toughness and woodiness. Water needs are 1 inch in 10 and 14 days for turnip and rutabaga, respectively. Preferred minimal soil moisture for both is −0.45 Bars or 50% ASM.

Turnips and rutabagas are subject to the same insects and diseases as cabbage. The most serious pest is the root maggot. Turnip greens are ready in 30 days, turnips in 55 to 65 days, and rutabagas in 85 to 100 days. Best sizes are 3 and 5 inches in diameter for turnips and rutabagas, respectively. Harvest can be mechanical or by hand. Storage (4 to 6 months) conditions are similar to those for carrots. Roots sent directly to market are often waxed.

Suggested turnip cultivars include 'Hakurei,' 'Just Right Hybrid,' 'Purpletop White Globe,' 'Royal Crown Hybrid,' 'Shogoin,' 'Tokyo Cross Hybrid,' and 'White Lady Hybrid.' Suggested rutabaga cultivars include 'American Purple Top' and 'Marian.'

Watercress

Watercress (*Nasturtium officinale*), a member of the Mustard family (Brassicaceae or Cruciferae), is a hardy, salad crop noted for its edible leaves. Watercress soup is also made and the greens can be cooked like spinach. Watercress is naturalized in wet surroundings, such as in shallow springs, brooks, or on river edges. Very moist garden sites are also possible. Seeds are planted 1/8 inch deep and thinned to 4 inches. Germination takes 1 week at 12.8°C (55°F). Soil pH of 7.0 to 8.0 is good. Full sun is needed. It can be started from seeds or cuttings in the early spring, and once established will remain, since it is a perennial. Leaves must be picked prior to flowering to avoid bitterness.

Watermelon

The watermelon (Figure 17–49) is a warm season, frost-sensitive vine crop or cucurbit. It is noted for its edible, water-crispy, sweet reddish-fleshed fruit. Cultivars are derived from *Citrullus lanatus,* a member of the gourd family (Cucurbitaceae). Since it requires a long growing season, it is usually limited to the South. Florida, California, Texas, and Georgia are the leading producers. Average yields are 11.3 tons/acre, although yields as high as 25 tons/acre have been reported in California. Shorter maturing cultivars are available for the North.

Watermelons are seeded directly in place in the warmer areas. In the North, where the growing season is short, 3- to 4-week-old transplants can be used, if transplanting shock

FIGURE 17–49 Air pollution from ozone, sulfur dioxide, and sulfur trioxide can cause injury to watermelon vines. Resistant cultivars are 'Charleston Gray,' 'Prince Charles,' and 'Royal Jubilee.' 'Crimson Sweet,' 'Jubilee,' and 'Sugar Baby' are susceptible. Seedless cultivars have poor germination rates and are usually raised as transplants. It is highly recommended that at least one bee colony per acre be used for good fruit set. (Author photograph.)

is minimal (preferably with peat pots or large plugs). Machines are available for direct seeding and placing of transplants including through plastic mulch. Soil temperature should be at least 15.5°C (60°F) and optimal soil temperature range for germination is 23.9° to 35.0°C (75° to 95°F). Seeds are covered with 0.5 to 1.5 inches of soil. Rows are spaced 6 to 10 feet apart and in-row spacing is 24 to 36 inches. Soil pH can be between 5.5 and 6.8. Sandy soils appear to be best, because they warm faster. Clear or black (slower) plastic mulch can be used in the North to produce a soil temperature better suited for melon culture earlier than would occur naturally. Even better is the use of row tunnels or row covers coupled with plastic mulch. Weeds are highly competitive and plastic and cultivation are best for control.

Watermelons have moderate to high drought resistance. Water stress during fruit expansion can lead to smaller fruits; earlier water stress causes reduced yields. Water needs are 1 inch in 21 days. Preferred minimal soil moisture is −2.00 Bars or 40% ASM.

Determining when watermelons are ripe can be tricky. Maturity is in 70 to 100 days with the smaller cultivars on the lesser side. Novice growers often take sample plugs. A ripe watermelon has a hollow sound when thumped. Lightening of the underbelly color can also be an indicator. Melons are harvested by hand. Some mechanization of melon collection and movement is used in large operations. For shipping, melons should be precooled by hydrocooling, ice packs, or forced-air cooling. Refrigerated shipping is used. Melon storage is usually not practiced.

Insect pests include those that attack cucumbers such as aphids, pickleworm, and striped cucumber beetle. Diseases include those that attack cucumbers (anthracnose, bacterial wilt, downy mildew, cucumber mosaic virus, and powdery mildew), plus *Alternaria* leaf spot (*Alternaria cucumerina*), black rot (*Phoma cucurbitacearum*), *Fusarium* wilt (*Fusarium oxysporium*), gummy stem rot (*Didymella bryoniae*), squash mosaic virus, and watermelon mosaic virus.

Larger, later cultivars include 'Allsweet,' 'Black Diamond,' 'Calsweet,' 'Carolina Cross No. 183,' 'Charleston Gray,' 'Cobb Gem,' 'Crimson Jewel,' 'Crimson Sweet,' 'Fiesta,' 'Georgia Rattlesnake,' and 'Yellow Doll' (yellow flesh). Early-maturing small cultivars (icebox melon) include 'Bush Sugar Baby' (bush form), 'Fordhook Hybrid,' 'Garden Baby' (bush form), 'Golden Crown Hybrid,' 'Jade Star Hybrid,' 'New Queen Hybrid,' 'Sugar Baby,' 'Sugar Doll Hybrid,' and 'Tiger Baby Hybrid.' Seedless cultivars include 'Cotton Candy,' 'Everglade Hybrid,' 'King of Hearts,' 'Nova,' 'Seedless Sugar Baby Hybrid,' 'SweetHeart Hybrid,' and 'Triple Crown Hybrid.'

HERB PRODUCTION BASICS

Herbs are treated here with vegetables for several reasons. Herb production can be similar to vegetable production when herbs are grown as field crops. Herbs can also provide a secondary source of income on small family-owned vegetable farms. Herb production can also be a stand-alone activity to provide a part-time farming income. Some herbs are highly attractive to bees, which can aid in the pollination of nearby vegetable crops.

Many vegetable gardeners also raise herbs as a normal component of their vegetable garden activities. Some herbs are attractive to butterflies, making them good candidates for butterfly gardens. Herbs are usually grown in a small, informal section of the garden, or sometimes in an elaborate formal garden devoted entirely to herbs. Some are decorative and used in borders or in rock gardens. Some are suitable for indoor or greenhouse culture.

Herbs need to be harvested at their peak, often before the plant flowers. Harvest must be fast and gentle and is often done by hand. Cooling might be needed and perhaps even washing for fresh market herbs. Washed herbs need to be dried thoroughly to prevent fungal problems. Careful packaging is required. Fresh herbs need to be moved to market quickly, because keeping qualities are limited.

Some herbs are highly valued for their culinary uses such as basil, oregano, and tarragon. Others have minor culinary use such as fenugreek. Culinary herbs, along with fresh vegetables, can be sold directly to upscale restaurants and at farm stands and through community-supported agriculture. Culinary herbs could be sold in bulk lots to wholesalers.

Some culinary herbs can be processed on site into value-added products such as pesto made from basil, parsley, walnuts, and olive oil. Various herbal vinegars such as tarragon vinegar are another possibility as are olive oils infused with garlic, basil, and oregano. Herbs can also be dried, bottled, and sold on site or as bulk, dry lots sold to wholesalers. Dry herbs can extend the market product at your operation when fresh herbs are gone. Dry fragrant herbs such as lavender can also be made into sachets and pillows. Herb growers might even consider selling potted culinary herbs during the bedding plant season and later for use in homes.

Herbs can also be grown for sales to herb product manufacturers that process herbs into oils, dyes, and herbal remedies. Be aware that these markets are much more difficult to break into. Considerable research is in order before these markets are attempted. Always know your market through research and have it established before committing yourself to herbal production.

Herbs can be grown as commercial row crops, row tunnel crops, and greenhouse crops. The latter two approaches can make herbs available earlier on the market when herb prices are higher. Commercial herbs can also be harvested in the wild (wild-crafting). Herbs grown as field crops are essentially treated like vegetables: level to slightly sloped (south facing) land with good soil, water access, and reasonably free of difficult-to-control weeds. Cultural aspects, given their similarity to vegetables, won't be repeated here. These aspects were covered earlier briefly before the vegetable entries and in more detail in terms of sustainable practices in previous chapters (Chapters 7, 11, 12, 13, and 15). Most herbs grow reasonably well at a soil pH of 6.0 to 7.0 on mineral soils and 5.0 to 6.0 on high organic matter content or muck soils.

Pests and diseases are often similar to those found with vegetable crops. The most common pests are aphids, caterpillars, cutworms, earworms, flea beetles, Japanese beetles, leafhoppers, mites, whiteflies, and wireworms. Damping off diseases during seed propagation can be troublesome, as can various fungal diseases such as leaf spot, mildew, and various wilts. Most herbs are grown organically for a number of reasons. In some cases the motivation is personal conviction. Other reasons include the lack of labeled pesticides for herbs and the market preference plus higher prices for organic herbs. With wild-crafted herbs, pesticide usage is not economically feasible and organic certification is somewhat easier than for cultivated herbs.

HERBS A TO Z

Herb crops can be annuals, biennials, or perennials. Many are aromatic or sweet smelling and are found in the mint (Labiatae), parsley or carrot family (Umbelliferae or Apiaceae), and composite or sunflower family (Compositae or Asteraceae). A discussion of the various herbs follows. Given spatial limitations, not all herbs can be covered here. Only a sampling of major or mainstream herbs are given. For other herbs, consult the Internet Resources and Relevant References lists at the end of the chapter.

A note of caution is in order about herb usage. Care must be exercised with the more unusual wild-crafted herbs to avoid the danger of poisonous look-alikes or use of the wrong plant part, which could be poisonous. The use of herbal remedies in general should not be taken lightly. Although some might have medicinal value such as echinacea, others do not and might even be harmful. Harm could come from toxic or carcinogenic effects, or simply as a result of taking them in place of mainstream medicine that could help. An unexpected case in point is St. John's wort, which appeared to be a workable, herbal remedy for depression, that is, until it was discovered that St. John's wort interacts with a number of prescribed drugs, causing them to be metabolized more rapidly than normally. This action diminishes their effectiveness. Certain herbs have been dropped from food or drink use such as wormwood in absinthe. Pregnant women certainly should avoid most herbal remedies.

As a precautionary note, the taking of any herbal remedy should be discussed with your physician. It is also in your best interest to investigate the remedy as thoroughly as you can. Keep in mind that unlike normal medications, herbal remedies are not subject to FDA oversight and have not undergone supervised clinical trials.

Angelica (*Angelica archangelica*)

Angelica is a hardy biennial to short-lived, hardy perennial herb found in the parsley or carrot family (Umbelliferae or Apiaceae). Mature height is 6 feet. Propagation is by seed (seed needs chilling for 6 to 8 weeks in refrigerated moist media). Germination takes 21 to 30 days. Seeds can be directly planted in spring or used to prepare transplants. Seeds are better sown late summer to early autumn and covered with 1/8 (light required) inch of soil. Seed viability is limited. Plants should be spaced 18 to 30 inches apart. Soil should be moist, but well drained. Exposure should be sunny to partial shade. Hot, dry weather ends its usefulness. The life of the plant can be prolonged by topping to prevent flowering. Leaves are used in cooking fish, stems are candied or used to flavor liqueurs, petioles are candied, and seeds and roots are used to flavor liqueurs. Flavor is juniper-like.

Anise (*Pimpinella anisum*)

Anise (Figure 17–50) is a half-hardy annual that reaches a height of 2 feet. It is found in the parsley or carrot family (Umbelliferae or Apiaceae). Seeds are sown 0.25 inch deep (18 to 24 inches apart) after the danger of frost is past. Germination is in 4 to 8 days. Transplanting is difficult because of its long taproot. Well-drained soil with full sun is best. Seeds are used in breads, pastries, and many foods. Flowers are used to flavor liqueurs. Leaves are also used to flavor foods such as salads and carrots. Anise oil is distilled from the seeds. The oil is used in beverages, candy, and medicine. Flavor is licorice-like.

Basil (*Ocimum basilicum*)

Basil or sweet basil (Figure 17–51) is a tender annual that reaches a height of 18 to 36 inches. It is found in the mint family (Labiatae). Propagation is by seeds in the spring. Soil must be warm and seeds are sown 0.25 inch deep. Basil can be started early indoors and transplants placed after all danger of frost is past. Germination is in 5 to 10 days. Spacing is 6 to 12 inches. Average moist soil, summer heat, and full sun are best. Remove flowers to keep basil producing. Leaves are used to flavor meat, fish, eggs, cheese, olive oil, soups, sauces, sausage, salads, vegetables, vinegar, and tomato products. It is also used to make pesto.

Basil has three commonly planted varieties and numerous cultivars. Basils for pesto and general use tend to be either the species or cultivars of the species. These cultivars include 'Basilico Greco,' 'Genovese,' 'Green Ruffles,' 'Italian Large Leaf,' 'Mammoth,' 'Nufar

FIGURE 17–50   Anise. Pliny, the famous Roman scholar, is believed to have stated, "Anise should be chewed in the morning upon awakening in order to remove all bad odors from the mouth." The seeds and distilled oils are used to flavor a number of popular liqueurs and cordials. (USDA photograph.)

FIGURE 17–51   Basil does best at pH 5.5–6.5. Flowers should be pinched off to keep plants productive. Basil has been cultivated in India for more than 5,000 years. (Courtesy of National Garden Bureau.)

Hybrid' (fusarium resistant), 'Thai Siam Queen,' and 'Valentino.' Purple basil cultivars are 'Dark Opal,' 'Osmin Purple,' 'Purpurascens,' and 'Purple Ruffles.' Dwarf cultivars are 'Dwarf Bush,' 'Fineleaf,' 'Minima,' 'Minette,' and 'Spicy Bush.' Lemon basil cultivars are 'Citriodorum,' 'Lemon Sweet Dani,' and 'Mrs. Burns.' Others are 'Cinnamon' and 'Licorice.'

Bay (*see* Sweet Bay)

Borage (*Borago officinalis*)

Borage is a hardy annual herb that reaches a height of 2 feet. It is a member of the borage family (Boraginaceae). Propagation is by seeds sown 0.5 inch deep in the spring or by cuttings. A well-drained soil and full sun are best. It tolerates dryness well. Plants should be spaced 24 inches apart. The tender, upper leaves have a cucumber-like flavor and are used in salads, teas, iced fruit and wine drinks, and pickles. Borage usually self-sows. Bees love borage.

Burnet (*Poterium sanguisorba*)

Burnet is a hardy, perennial herb in the rose family (Rosaceae) reaching a height of 18 to 30 inches. Propagation is by seeds and division. Seeds germinate in 7 to 14 days. Full sun and a dry to moist, well-drained soil are best. Spacing is 12 inches. Grows well in sandy, dry soils. Leaves are used when young in salads, iced drinks, and vinegar. Flowers are removed to encourage productivity. Flavor is cucumber-like.

Caraway (*Carum carvi*)

Caraway (Figure 17–52) is usually grown as a very hardy biennial. It is found in the parsley or carrot family (Umbelliferae or Apiaceae). Mature height is 24 to 36 inches. Propagation is by seeds sown 1/8 (light required) inch deep in spring through early fall. Germination is in 7 to 14 days. Spacing is 6 to 8 inches. A sunny location and soil on the dry side are favored. It

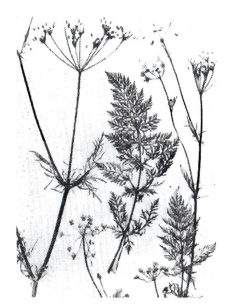

FIGURE 17–52
Caraway does best at pH 6.0–7.5. Seed is usually directly sown, as transplants fare poorly. Roots can be boiled and used like parsnips. Young leaves can be used in soups and salads. (USDA photograph.)

is grown mostly for its seeds, which are used to flavor rye bread and other baked goods, as well as processed foods. It is also used in the cordial called Kummel. One cultivar is 'Arterner.'

Catnip (*Nepeta cataria*)

Catnip is a very hardy, perennial herb in the mint family (Labiatae). Mature height is 18 to 36 inches. Propagation is by seed or transplants in the spring, seeds in the fall, or division in the spring. Seeds are sown 1/8 (light required) inch deep and take 7 to 10 days for germination. An average soil and full sun are favored. Spacing is 18 inches. Growth is vigorous and it can easily become a weed. Its main use is as a tonic for cats, but is also used for tea.

Chamomile (*Matricaria recutita, Chamaemelum nobile*)

Chamomile is a common name for both an annual and perennial herb. One is sometimes called German chamomile (*M. recutita*) and is an annual. The second, often referred to as Roman chamomile, is a perennial (*C. nobile*). The heights of the annual and perennial forms are 24 to 36 and 6 to 12 inches, respectively. Both are found in the composite or sunflower family (Compositae or Asteraceae). Both can be direct sown in the spring or started as transplants. Both should be sown 1/8 (light required) inch deep. Germination for the annual and perennial forms is 10 to 14 and 7 to 10 days, respectively. Spacing is 6 and 12 inches, respectively for German and Roman chamomile. The dried flowers are used to make a soothing, sleep-inducing tea.

Chervil (*Anthriscus cerefolium*)

Chervil is a hardy annual reaching a height of 2 feet. It is a member of the parsley or carrot family (Umbelliferae or Apiaceae). Propagation is by seed sown in early spring or late summer. Seeds are sown 1/8 (light required) inch deep and take 10 to 14 days to germinate. It is treated as a spring or fall crop, because heat is detrimental. Plants are spaced about 6 to 12 inches apart. Culture is simple in most soils and full sun to partial shade. Leaves are ready in 6 to 8 weeks and are somewhat like parsley/anise. They are used as a garnish and to flavor asparagus, butter, egg dishes, potatoes, soup, fish, and meat. One cultivar is 'Brussels Winter.'

Chives (*Allium schoenoprasum*)

Chives (Figure 17–53) are a perennial herb hardy through zone 3. It is a member of the amaryllis family (Amaryllidaceae). Mature height is about 12 inches. Propagation is by seeds, bulbs, or division, usually in early spring. Full sun and average soil are best. Leaves are used

FIGURE 17–53 Chives have been cultivated for at least 5,000 years. Most chives are grown as a garden plant rather than as a commercial crop. (USDA photograph.)

FIGURE 17–54 Purple coneflowers. Individuals with allergies to daisies should not take Echinacea, as it is in the daisy family. Echinacea should also be avoided by those who have autoimmune disorders or are immunocompromised. (Author photograph.)

to flavor salads, omelets, sour cream, soups, vegetables, and many other foods. Flowers can be used to make chive vinegar. Flavor is onion-like. Chives can be grown easily indoors. Garlic chives are a closely related species, *A. tuberosum,* which has slightly broader, flatter leaves and a milder, slightly garlic-like taste. Its culture and culinary uses are similar to common chives. Chives can be started as seeds directly, as transplants, and by division in the spring. Germination is 7 to 14 days. Seeds are planted 0.25 inch deep. Plants are set about 12 inches apart. Sun, moisture, and a rich soil are best. Both will need dividing about every 4 to 5 years. Common chives can also be easily grown potted indoors. Flowers of both are quite attractive.

Cilantro (*see* Coriander)

Coneflowers (*Echinacea* spp.)

Species of *Echinacea* (Figure 17–54) include *E. angustifolia* (narrow-leafed purple coneflower), *E. pallida* (pale purple coneflower), *E. paradoxa* (yellow coneflower), and *E. purpurea* (purple coneflower). The latter is a common perennial plant used for ornamental purposes (see Chapter 16). Heights vary from 18 to 36 inches. The roots, stems, leaves, and flowers of these hardy, perennial plants are a source of the herbal remedy echinacea, which is used to stimulate the immune system. Unlike some herbal remedies, evidence exists that this herbal extract can raise white blood cell levels if taken only as needed at the onset of colds, flu, and sore throat and for a limited period afterward. Seeds are direct sown in the spring or used to produce transplants. Seeds of all, except for purple coneflower, require chilling 3 to 9 weeks (length varies based on species). The purple coneflower does not, but 1 week will enhance germination somewhat. Alternatively, seeds can be fall-sown and stratification omitted. Seeds are sown 0.25 inch deep and germinate in 10 to 21 days in warm conditions. Plants require 18 to 36 inches, depending on species.

Coriander (*Coriandrum sativum*)

Coriander or cilantro is an annual herb that has a mature height of 18 inches. It is a member of the parsley or carrot family (Umbelliferae or Apiaceae). Propagation is by direct seeding or transplants (somewhat difficult) in the spring. Cilantro can self-sow. Seeds are sown at 0.25 to 0.5 inch deep and germinate in 7 to 10 days. Plants should be spaced 4 to 8 inches

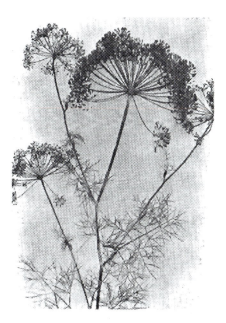

FIGURE 17–55   One to two pounds of seed per acre are sown for fresh market dill production when rows are 2 to 3 feet apart. If planned for use with dill pickle production, sow seeds 1 week earlier than pickling cucumbers. Dill yields run about 5,000 lb/acre for fresh weight. Dry weight is reported to be 1,000–3,000 lb/acre. Too much nitrogen can overproduce leaves and stalk and must be avoided when the crop is grown for seed use. Barnyard manure at 10 tons/acre can be plowed in early in the spring. Fertilizer applications run about 75–100, 80–120 and 80–120, lb/acre of N, P_2O_5, and K_2O, respectively. (USDA photograph.)

apart. An average soil and full sun are best. Fruits and seeds are candied or used as seasoning in poultry dressing, pickles, beverages, baked goods, and processed meats. Ground seeds are also one of the ingredients in curry powder. Leaves are used in salsa, Chinese, Indian, Mexican, Spanish, Thai, and Vietnamese dishes. One cultivar is 'Santo.'

Cumin (*Cuminum cyminum*)

Cumin is a tender annual herb in the parsley or carrot family (Umbelliferae or Apiaceae). Plants are 12 inches or less. Propagation is by direct seeding in the South and transplants in the North, as it requires 4 months to seed harvest. Seeds are shallowly sown in warm media or soil. Plants are thinned to 4 inches. Transplants have taproots, so must be handled carefully. Transplants are set out in warm soil after all danger of frost is past. Moisture must be especially provided during seed formation. Ground seed is an ingredient in curry and chili powder.

Dill (*Anethum graveolens*)

Dill or dill weed (Figure 17–55) is a hardy annual; it reaches a height of 3 feet. It is a member of the parsley or carrot family (Umbelliferae or Apiaceae). Propagation is by seeds sown 0.25 inch deep in the early spring. Germination takes 7 to 14 days. Spacing is 8 to 12 inches. Average to rich, well-drained soil and full sun are best. Dill transplants poorly because of a taproot, but if moved when young and with sufficient soil, success is reasonable. Leaves are called dill weed and used to flavor cottage cheese, new potatoes, potato salad, sauces, soups, vegetables, vinegar, and fish. Seed heads are used in dill pickles, breads, and soups. Hot, dry weather tends to end its usefulness, but with adequate water, it can be extended. Succession sowings can also extend the harvest. Dill self-sows and reappears for many years. A few cultivars are 'Bouquet,' 'Fernleaf,' and 'Superdukat.'

FIGURE 17–56 Fennel is one of the oldest cultivated plants. Roman gladiators ate fennel before battle to make them fiercer, while Roman women consumed fennel to reduce weight gain. Suggested cultivars are 'Bronze,' 'Giant Bronze,' 'Smokey,' and 'Purpureum.' (USDA photograph.)

Fennel (*Foeniculum vulgare* var. *dulce*)

Fennel (Figure 17–56) is a perennial usually treated as an annual. It is a member of the parsley or carrot family (Umbelliferae or Apiaceae). Fennel reaches a height of 24 to 36 inches. Propagation is by direct seeding or transplants in the spring. Seeds are planted 1/8 to 1/4 inch deep. Germination takes 6 to 14 days. An average to dry soil and full sun are best. Plants should be 6 to 12 inches apart. Fennel is grown for its leaves used in salads, cole slaw, and various dressings and to flavor cooked foods such as fish. Flavor is anise-like. Oil is extracted from the seeds.

Fenugreek (*Trigonella foenum-graecum*)

Fenugreek is a hardy annual in the pea or pulse family (Leguminosae or Fabaceae). It has a mature height of 2 feet. Propagation is by direct seeding or transplants in the spring after all danger of frost is past. Seeds are planted 0.5 inch deep. Germination takes 5 to 10 days. An average to dry soil and full sun are best. Plants should be 12 to 18 inches apart. Seeds are used as an ingredient in curry powder. Fresh leaves are cooked as a vegetable curry, dry leaves are used for flavoring in Middle Eastern dishes. Seeds are also sprouted and used in salads and sandwiches. Seeds can also be extracted to produce a maple syrup-like liquid.

Garlic Chives (*see* Chives)

Golden Seal (*Hydrastis canadensis*)

Golden seal is a hardy perennial in the buttercup or crowfoot family (Ranunculaceae). Its mature height is 12 inches. Propagation is by seeds or root divisions. Rich, moist soil with leafmold and shade are best. Golden seal is often wild-crafted, although it can be produced under lath or under trees.

Its main use is medicinal; the golden roots were used in Native American medicine. It was said to strengthen the immune system. It is sometimes blended with echinacea. It does have isoquinoline alkaloids. One of these, berberine, does have antibacterial and amoebicidal activity. Golden seal would appear to be effective when applied to surface tissues to promote healing and prevent infection. No evidence exists that taken internally, it strengthens the immune system. Mixing it with echinacea would appear to be wasteful, unless someday a synergistic promotion effect is discovered through clinical research.

Horehound (*Marrubium vulgare*)

Horehound is a hardy perennial. It is found in the mint family (Labiatae). Mature height is 18 to 24 inches. Propagation is by seeds, cuttings, division, and layering. Seeds are planted 0.25 inch deep in the spring after all danger of frost is past. Transplants can also be used. Germination takes 7 to 21 days. Spacing is 12 to 18 inches. It can become weedy. Leaves are used to flavor candy and cough drops.

Horseradish (*see* Vegetables)

Hyssop (*Hyssopus officinalis*)

Hyssop is a perennial hardy through zone 3. It is found in the mint family (Labiatae). Mature height is about 18 to 24 inches. Propagation is by direct seeding in the spring or transplants. Seeds are planted 0.25 inch deep in the spring after all danger of frost is past. Transplants can also be used. Germination takes 5 to 21 days. Spacing is 12 to 18 inches. A well-drained soil and full sun to partial shade are best. Spacing is 12 inches. Fresh shoot tips have a bitter, mint-like taste and are used to flavor vegetables and soups. Green leaves are used for medicinal tea.

Lavender (*Lavandula*)

English lavender (*L. angustifolia*) and French lavender (*L. dentata*) are shrubby, woody perennials in the mint family (Labiatae). Respective heights are 2 to 3 and 1 to 3 feet. Neither one is hardy, although English lavender can be grown in the lower North. Winter protection might be required. French lavender can be grown in California and other warm areas of the United States. Seeds or cuttings taken in the spring or autumn can be used for both lavenders. Seeds are lightly pressed into soil and take 14 to 28 days to germinate and have poor rates of germination. All danger of frost must be past when seeds or transplants are set outdoors. Spacing is 18 inches. Dry, well-drained soil on the neutral to slightly alkaline side and bright sun are best. English lavender is one source of oil of lavender used in perfumes. Spanish lavender, *L. stoechas,* is the other. Lavender has little culinary use, but is used in sachets, pillows, perfumes, and other scented products. The flowers are the primary sources. Butterflies love lavender. Numerous cultivars of English lavender are available. A few are the popular 'Hidcote' and 'Munstead.'

Lemon Balm (*Melissa officinalis*)

Lemon balm is a hardy perennial found in the mint family (Labiatae). Mature height is 2 feet. Propagation is by seeds, division, or cuttings in the spring. Seeds are planted 1/8 inch deep in the spring after all danger of frost is past. Transplants can also be used. Germination takes 7 to 14 days. Spacing is 18 to 24 inches. An average soil and sun to partial shade are best. Bees love it. The lemon-scented leaves are used in teas, soups, salads, cold drinks, and many cooked foods. A suggested cultivar is 'Quedlinburger Niederliegende.'

Lovage (*Levisticum officinale*)

Lovage is a hardy perennial. It is a member of the parsley or carrot family (Umbelliferae or Apiaceae). Mature height is 6 feet. Propagation is by direct seeding in the spring or fall, transplants, or by division in the spring. Seeds are planted 0.25 inch deep. Germination takes 10 to 14 days. Spacing is about 24 inches. A moist, rich soil with full sun (North) to partial shade (South) is best. Dry or fresh leaves are used to flavor salads, soups, sauces, potato, and poultry dishes. It is used as a celery substitute.

Mint (*Mentha* spp.)

The most popular mints (Figure 17–57) are peppermint (*Mentha* × *piperita*) and spearmint (*Mentha spicata*). Both are found in the mint family (Labiatae). There are also several, lesser used mint species.

Peppermint and spearmint are hardy perennials reaching heights of 2 to 3 feet. Propagation is by seeds, transplants, division, or cuttings. Seeds are planted 1/8 inch deep. Germination takes 7 to 14 days. Spacing is about 12 inches. An average to moist soil and full sun to partial shade are best. Peppermint leaves are used to flavor foods, candy and beverages. Fresh spearmint leaves are used more often to flavor jellies, vinegar, iced drinks (iced tea and mint julep), sauces, soups, candy, confectionery, cooked dishes such as lamb, fruit, and ice cream. Peppermint and spearmint oils, used for flavoring candy, gum, tea, medicines, and other products are derived from the leaves. Like all mints, these plants tend to spread and become invasive.

Oregano (*Origanum vulgare*)

Oregano is found in the mint family (Labiatae). Its nomenclature is somewhat confusing. *Origanum vulgare* has been called both oregano and wild or winter majoram. This latter herb does have an oregano-like quality, but is milder. The stronger, preferred oregano, often called Greek oregano, is the type associated with pizza and Italian cooking. Its identity is difficult, having been said to be *O. heracleoticum, O. hirtum* (a name replaced by *O. heracleoticum*) and, more recently, *O. vulgare* subsp. *hirtum*, which appears to have replaced *O. heracleoticum*. If you want milder types, the best bet is wild majoram (*Origanum vulgare*), and for more pungent oregano, go for types identified as Greek oregano, especially if listed as *O. vulgare* subsp. *hirtum*. It is also listed in catalogues as *O. vulgare hirtum*.

Oregano is a very hardy perennial. Mature height is 18 to 30 inches. Propagation is by seeds, cuttings, layering, and root division in the spring. Seeds are planted 1/8 inch deep. Germination takes 7 to 14 days. Spacing is about 12 inches. Average soil and full sun are

FIGURE 17–57
(A) Peppermint.
(B) Spearmint. Oils from these two mints are used to flavor cough drops, mouthwashes, soap, and toothpaste. Mints are characterized by their square stems and opposite leaves. (USDA photographs.)

(A)

(B)

best. It often self-sows and pops up elsewhere. Flowers should be cut back to encourage more leaves. Fresh and dry leaves are used to flavor salads, soups, stuffing, sausage, sauces (especially tomato-based), cooked dishes, meat, and fish.

Parsley (*see* Vegetables)

Rosemary (*Rosmarinus officinalis*)

Rosemary is a tender, somewhat woody perennial that can, but rarely, reaches a height of 6 feet. It is found in the mint family (Labiatae). Some winter protection might be necessary in the lower North and upper South, or it can be a large container plant that winters indoors. Seeds require stratification for best germination. Propagation is by seeds (slow to germinate and low percentage germination) or cuttings. Seeds are planted 1/8 inch deep. Germination takes 14 to 21 days. Spacing is 24 to 36 inches. A dry, well-drained rich soil and full sun are best. Dried or fresh very pungent leaves are used to flavor breads, pizza, vegetables, meats, fish, potatoes, poultry, soups, sauces, and many other cooked products. Aromatic oil is distilled from the leaves.

Sage (*Salvia officinalis*)

Sage is a somewhat hardy, shrubby woody perennial found in the mint family (Labiatae). Mature height is 16 to 30 inches. Propagation is by seeds, transplants, cuttings, or division. Seeds are planted 1/8 inch deep in the spring. Germination takes 7 to 21 days. Spacing is about 18 to 24 inches. Average soil and full sun to partial shade are best. Fresh and dried leaves are used to flavor cheese, dressing, omelets, pickles, vegetables, and sausage. 'Berggarten' and 'Extrakta' are cultivars.

Sesame (*Sesamum indicum*)

Sesame is a tender annual likely to be only grown from Florida to Texas and the Southwest. It is a member of the pedalium family (Pedaliaceae). Its mature height is 36 inches. Propagation is by direct seeding (0.25 inch deep) and transplants in the spring. The latter are handled carefully because of a taproot. Planting is after all danger of frost is past. Germination is 5 to 7 days. Long, hot summers are required to mature the crop. The toasted seeds are used in breads and other baked goods, both plain and as a carmelized, sugary confection. Sesame seed oil (tahini) is also extracted and used in cooking, salad dressing, and Middle Eastern dishes. It is a minor crop in the United States.

Sorrel (*Rumex acetosa* and *R. scutatus*)

Sorrel or French sorrel is a perennial found in the buckwheat family (Polygonaceae). It reaches a height of 18 inches. Propagation is by seeds and transplants in the spring. Seeds are lightly pressed into soil. Spacing is 12 inches. An average soil with full sun is best. Leaves are ready early. Flower and seed stalks should be cut back to maintain productivity and to prevent volunteer plants, as it can become weedy. Taste is tangy and lemon-like. Leaves are used like spinach as a cooked green and soups or fresh sparingly in salads.

Summer Savory (*Satureja* spp.)

Summer savory (*Satureja hortensis*) is an annual that reaches a height of 12 to 18 inches. Winter savory (*Satureja montana*) is a hardy perennial reaching a height of 6 to 12 inches. Both are found in the mint family (Labiatae). Propagation is by direct seeding or transplants in the spring for both. Winter savory is also propagated by cuttings and division. Seeds of both are sown 1/8 inch deep and both germinate in 7 to 14 days. All frost danger must be past for setting out seeds or transplants of summer savory. Spacing is 8 to 12 inches for both summer and winter savory. An average soil with full sun is best. Leaves are used to flavor fresh and dry beans, peas, lentils, meat, fish, poultry, eggs, vegetables, sausage, sauce, soup, stuffing, and gravy. Winter savory has a stronger, more pungent taste than summer savory. 'Aromata' is a cultivar of summer savory.

Sweet Bay (*Lauris nobilis*)

Sweet bay or laurel is a broad-leaved, tender evergreen tree found in the laurel family (Lauraceae). It reaches a height of 40 to 60 feet. Propagation is by seeds or cuttings. Seeds can take 6 to 12 months to germinate, although under ideal conditions a month is possible. The tree can only take a light degree of frost, so is restricted to the deep South. Other areas must be content with a large tub that can be brought indoors, preferably in a greenhouse. Dry leaves are used to flavor soups, stews, sauces, and various other long-cooking dishes. After the flavor is extracted, the leaf is removed because it is too tough to chew.

Sweet Cicely (*Myrrhis odorata*)

Sweet cicely is a hardy perennial herb that reaches a mature height of 3 feet. It is a member of the parsley or carrot family (Umbelliferae or Apiaceae). Propagation is best by seeds in the fall or division in the spring or fall. Spring sown seeds require considerable stratification, have a very poor germination rate, and take 2 to 3 months. Average soil and partial shade are best. Leaves are anise-like and used to flavor soups and salads. Dried seeds have a licorice taste and are used to flavor foods. Leaves can be harvested throughout the growing season.

Sweet Marjoram (*Origanum majorana*)

Sweet marjoram (Figure 17–58) is a very tender perennial usually treated as an annual. It is found in the mint family (Labiatae). Mature height is 2 feet. Propagation is by seeds or transplants after all danger of frost is past. Division is also possible. Seeds are sown 1/16 inch deep. Germination is in 10 to 14 days. Spacing is 8 inches. Average soil and full sun are best. Leaves are used to flavor tomatoes, omelets, cheese dishes, salads, soups, stuffing, sausage, sauces, cooked dishes, meat, and fish. Taste is somewhat like oregano, but much milder and sweeter.

FIGURE 17–58   Sweet marjoram. Ancient Greeks called marjoram the "Joy of the Mountains." They brewed a marjoram tea for the treatment of asthma, rheumatism, and toothaches. *Majorana hortensis* is a former name for sweet marjoram. It is listed as a host for Pierce's disease, so should not be cropped near other susceptible plants. (USDA photograph.)

Sweet Woodruff (*Galium odoratum*)

Sweet woodruff is a hardy perennial that is found in the Madder family (Rubiaceae). Mature height is 6 to 12 inches. Propagation is by root division in the spring, because seeds need stratification and take a very long time to germinate. Seeds sown in fall fare better. Moist soil and partial shade are best. Spacing is 12 inches. Leaves are used to flavor May wines, liqueurs, and cold drinks. The dry leaves are fragrant. One cultivar is 'Blue Surprise.'

Tarragon (*Artemisia dracunculus* var. *sativa*)

French tarragon is a hardy perennial that is found in the Compositae or sunflower family (Compositae or Asteraceae). Mature height is 2 feet. Propagation is by cuttings or division to avoid the less desirable Russian tarragon. Well-drained soil and full sun are best. Fresh leaves are used in salads and pickles and to make tarragon vinegar. Dry leaves are added to meats, fish, poultry, soups, eggs, cream sauces, dressings, leeks, potatoes, mushrooms, and so on.

Thyme (*Thymus vulgaris*)

Thyme is a hardy, woody, evergreen perennial. Winter protection is needed in colder and windy areas. Mature height is 6 to 15 inches. It is found in the mint family (Labiatae). Propagation is by seeds, transplants, division, and cuttings in the spring. Seeds are lightly pressed into the soil. Germination takes 14 to 21 days. All frost danger must be past before sowing seeds outdoors or placing plants. Spacing is 8 to 12 inches. Well-drained, dry soil and full sun to partial shade are best. Fresh and dry leaves are used to flavor meats, fish, poultry, soup, beans, rice, vegetables, sauces, dressings, pickles, and vinegar. French and Italian dishes use thyme. Thyme is one of the stronger herbs or "a little goes a long way" as with rosemary. Bees like the flowers.

Wild or Winter Marjoram (*see* Oregano)

Winter Savory (*see* Savory)

SUMMARY

Olericulture deals with the production, storage, processing, and marketing of vegetables. What amount of vegetables and herbs are grown in a sustainable manner is unknown. The exact limits of what constitutes sustainable horticulture are not clearly defined. Only one part of sustainable horticulture, organic farming, is clearly defined. According to USDA data, certified organic cropland doubled in the 1992–1997 period. Certified U.S. organic farmers increased to 7,800, 18% more than in 1999. The organic market (all aspects and crops) is now at $7.7 billion.

In 1999 33.74 million American households had a vegetable garden for a total of 35% of the 96.4 million households in America. Retail sales for vegetable-oriented products and services reached $1.21 billion.

During 2000 about 2.10 million acres were harvested for fresh market vegetable and melon production and 1.45 million acres for processed vegetables. California led in vegetable production. Fresh vegetable production value in 2000 totaled $9.33 billion and processed vegetables $1.43 billion.

Greenhouse food crops had a value of $223 million in 1998. Most vegetable seeds sold by firms are produced under contract on the West Coast. Production of vegetable transplants had a wholesale value of about $130 million in 1998.

Careers in vegetable production are numerous. Seed firms need vegetable breeders, propagators, independent growers who contract with seed firms for seed production, and sales personnel. Vegetable production on farms and in greenhouses requires a manager. Processing companies have people in the field that act as a liaison and consultant between independent growers and the processing company. Sales personnel knowledgeable in vegetables sell horticultural machinery and inputs. Financial specialists also deal in vegetable-related activities. There are numerous positions within the vegetable industries. A number of them occur with sales agencies that handle sales of vegetables for growers. These agencies require an overall manager, a storage supervisor, a supervisor for grading and packing, a field person who is a liaison and consultant between the agency and growers, and a buyer. Brokers resell vegetable wholesale lots obtained from growers or sales agencies, and buyers for retail chains obtain lots from growers, agencies, or brokers. There are also promotional agencies on the state or national level for specific vegetables. Extension personnel, professors, and researchers dealing with vegetables are often part of state land-grant universities and the USDA. IPM consultants with vegetable backgrounds can consult and provide services to farmers or work in research or extension.

Soil must be prepared properly for seeds and transplants. Soil fertility, organic matter content, drainage, aeration, and pH must be properly maintained. Organic matter plays a vital role in reducing soil erosion and maintaining soil health, especially with microbiological processes that contribute to nutrient availability from decaying organic matter. Organic matter contributes to the maintenance of beneficial microorganisms that protect plants from soil-borne diseases by competitive and antagonistic processes. Soil pH is critical in that deviations lead to decreased availability of nutrients, thus leading to decreased yield and more chemical fertilizers.

Soil pH should be monitored through at least annual soil testing and limestone or sulfur added as recommended. Many vegetables do well at pH 5.5 to 7.0, while others require 6.0 to 7.0. Organic matter can be maintained through a number of approaches such as crop residues, cover crops, green manure and organic amendments. Manure is an especially attractive option for vegetable growers (if available locally) both in terms of organic matter and nutrients.

The amount of fertilizer to apply to vegetable crops is variable. Several factors contribute to the range. One factor is soil type. Mineral soils need more nitrogen than muck soils and soils with high organic matter. With phosphorus and potassium, the range results from correlation with available phosphorus and potassium. High-density plantings need more nutrients than low-density plantings. For perennial crops, new plots need more nutrients than established plots. A first step in achieving sustainable fertilization is to have soil tests done by labs that have experience with sustainable horticultural needs. Other things to do are to use practices that promote more efficient use of nutrients such as banding and split applications.

Crops are planted at times based on knowledge of dates of local last killing spring frost and first killing fall frost, days needed for maturity, crop responses expected from the variations in weather known to occur during the growing season, and marketing and processing demand. Propagation of many vegetables is by seed, which should be fresh, viable, clean, properly labeled, and insect and disease free. Precision seeding with mechanical seeders can be used to eliminate thinning. Some vegetables are propagated asexually such as asparagus, garlic, horseradish, onion, potato, rhubarb, shallot, sweet potato, and taro. Certain vegetable crops are grown from transplants for reasons of early harvest and for economical use of space. The transplants are produced in a greenhouse or outdoor seedbed at high seeding rates. This leaves some field space free for early crops, which can be followed by transplants. Mechanical transplanting devices are available.

Optimal efficiency can be realized in vegetable production by the use of certain horticultural practices. Keep in mind that increased productivity per unit area increases the loss of nutrients also. Therefore, soil fertility must be maintained if increased productivity is to be realized. Succession planting is one that is widely used commercially to ensure continuous harvests. Another is high-density (intensive) planting.

Water must be supplied by irrigation, if rainfall is insufficient. While uniform soil moisture throughout the growing season is generally required for optimal yields, certain stages are more critical than others. Vegetables subjected to periods of water stress during early development have lower yields and delayed maturity. Water stress during fruiting or maturation produces poorer quality fruits. Requirements also vary somewhat by crop. Crops grown for enlarged parts such as fruits, seeds, bulbs, tubers, tuberous roots and fleshy leaf heads need more of their water budget during enlargement of the harvested structures. Fruiting vegetables also have critical water needs prior to enlargement, from flowering time through fruit set.

Water stress during this period can result in incomplete pollination and less fruit. Crops grown for their leaves need uniform moisture throughout their development. Excessive water at any time can also cause problems such as increased diseases or result in poorer quality fruits or tubers grown for their dry matter, if applied excessively during fruit enlargement.

Weeds must be controlled through cultivation, mulches, or herbicides (last resort). Cultivation depth needs to be matched to the crop. Deeper, more effective cultivation can be used with deep-rooted crops. Shallow cultivation is used with shallow-rooted crops such as lettuce and later in the season with most crops. Closeness of cultivation to plants also changes from closer earlier to further away later. Mulches are very effective and are popular with gardeners. Mulch use with commercial crops is more likely if, besides weed control, higher yields and earlier harvests also result. Preemergent herbicides are frequently used to give young plants a better start and to reduce later weeding problems. Herbicides are also used for difficult to eradicate weeds.

Insects and diseases must be controlled through rotations, resistant cultivars, and with biological controls supplemented with judicious use of pesticides when necessary (or without chemical pesticides in organic production). The best sustainable approach is proactive: sanitation, rotations, and resistant cultivars. Plants with disease or insect problems should not be planted, those with uncontrollable diseases must be destroyed and unused crop residues should be removed. Many hybrid vegetable cultivars are available with resistance to diseases. Rotations are critical to the prevention of soil disease and weed and pest buildups in your fields. Rotations disrupt the feeding and breeding chains of these organisms. Rotations also improve nutrient status and soil health in that different vegetables contribute differing amounts of crop residues and different vegetables remove varying amounts of nutrients. Crops in the same family should not replace one another. Most rotation plans use a 4- to 5-year cycle.

The best preventive practices for pest control are either the use of IPM systems or IPM systems without synthetic pesticides (organic systems). Many commercial biological controls are available against insects, mites, nematodes, and diseases. Examples are parasitoid wasps, predatory insects, predatory mites, predatory fungi, predatory nematodes, and antagonistic fungi. In addition, many botanical, natural product, and environmentally safe (green) pesticides are available.

Determining when to harvest is a critical part of vegetable production. For the home gardener, harvesting usually occurs at peak quality. For commercial operations, harvest is determined by a number of factors. For vegetables destined for processing, harvest can be close to peak maturity, because processing happens shortly after and better quality products result from peak maturity vegetables. For fresh market, harvest can be close to peak maturity or somewhat less, because ripening will continue in many cases. Dead ripe vegetables are generally avoided, because bruising can occur during shipping and decreased quality can result during the time from harvest through sales. In some cases, gassing with ethylene is used prior to harvest to initiate uniform ripening. The determination of the correct picking time can be tricky. Days to maturity from seeding or transplanting are known and can be helpful predictors of maturity. Degree-days or heat units are even more reliable indicators. Other indicators are color (tomatoes and peppers), stem slippage (muskmelons), hollow "thunk" tap noises (watermelons), dry silk and milky kernel juice (sweet corn), rinds easily indented with fingernails (summer squash), or rinds not easily dented with fingernails (winter squash).

Harvesting equipment saves time and money. Plant breeders have aided mechanized harvesting by developing crop varieties with uniform ripening habits, compact growth, less easily bruised fruits, more easily detached vegetables, or an angle of hang better suited for mechanized removal. In addition, growth regulators are used as mechanized harvest aids to promote uniform ripening, color, and abscission.

The oldest harvesting technique is hand harvesting. The home horticulturist uses this approach. Commercial operations also use this technique for easily damaged vegetables intended for fresh market sales. Hand harvest is also employed for many other fresh market vegetables because of nonuniform ripening of various vegetables. Less easily damaged crops such as root crops (and thick-skinned fruits) can be mechanically harvested for fresh markets. Crops intended for processing are mechanically harvested, because damages or blemishes are not of concern. Even hand harvests are facilitated to some degree by aids such as conveyor belts or mechanized undercutting to facilitate hand removal. Other reasons for not using mechanized equipment are that suitable harvesting equipment is not available or is bypassed in favor of pick-your-own operations, or the cost and/or feasibility of the equipment is questionable.

When a horticultural crop is harvested, ongoing metabolic processes continue to produce changes in the picked crop. These changes can enhance or detract from the appeal of the crop. The appeal of the crop is determined by three factors: quality, appearance, and condition. Quality is used in reference to flavor and texture, as determined by taste, smell and mouth-tongue touch. The appearance refers to the visual impression of the crop, whereas the condition refers to the degree of departure from a physiological norm or the degree of bruising or disease.

Postharvest alterations include water loss, the interconversions of starch and sugar, flavor changes, color changes, toughening, vitamin gain and loss, moisture loss, sprouting, rooting, softening, and decay. These processes can improve or detract from the quality, flavor, and texture of the crop. On the whole, postharvest changes produce more detrimental results than beneficial ones. The time between harvesting and marketing or processing should be as short as possible to minimize post harvest alterations.

No one set of conditions will achieve this end, because postharvest alterations are a function of the crop type, and each crop varies as to the optimal air temperature, relative humidity, and concentrations of oxygen and carbon dioxide required for stabilization. Some form of air circulation is usually beneficial in reducing disease problems and removing any released gases that accelerate harmful physiological and biochemical changes. The level of disease pathogens should also be kept to a minimum. The application of postharvest preventative measures will also vary as to when they should be applied. Various means of cooling are used to slow postharvest changes.

The most important aspect of postharvest treatments is in storage facilities designed for long-term crop storage, so that the crop can be sold at a reasonable price over an extended period of time. Storage for the less critical crops, in terms of better postharvest longevity, usually consists of unrefrigerated storage like insulated storage houses, outdoor cellars, and mounds. More reliable storage consists of refrigerated storage with precise control of temperature and humidity. A number of vegetables are held near 32°F (0°C) and 90% to 95% relative humidity. Others need higher temperatures and lower humidity.

Premarketing operations include washing, trimming, waxing, curing, precooling, grading, prepackaging, and shipping. Some steps prevent postharvest damage, others are designed to improve the marketing value of the vegetable. Still others are used to remove unwanted materials. These steps can be done in the field, at specialized facilities, or a combination of both can be used. Field operations reduce transport of and damage to the crop and cost less. The downside is that quality control is less, weather can cause problems, large machines can disrupt field activities and cause soil compaction, and cooling of small lots is less energy efficient than bulk lot cooling.

At the receiving end, vegetables are transported in dry conveyors or running water troughs. The more fragile crops and those needing washing are placed in the latter. Some form of elimination is done to remove rotten or undersized produce and foreign objects. Water is often chlorinated to reduce disease problems. Washing and brushing is especially needed for root crops with adhering soil particles. A similar problem exists with mud splashed leafy greens. Drying by air or giant sponge rollers follows.

Trimming is used to cut back roots or discolored leaves or to cut back excess green tops. Waxing is sometimes used to improve appearance and to slow moisture loss. Some sort of machine or hand sorting is next to remove off-quality materials. Vegetables are also sized mechanically or electronically and manually when necessary. Curing is used to heal cuts or bruises incurred during harvesting of crops.

Precooling is used for rapid removal of heat from recently harvested vegetables. This might be done at the field or later at the packinghouse. Precooling slows the rate of respiration, slows wilting and shriveling by decreasing water loss, and inhibits the growth of decay caused by microorganisms. Major precooling methods include hydrocooling, contact icing, vacuum cooling, and air cooling.

After precooling the crop, it must be kept cool in transit by shipment in refrigerated trucks, railroad cars, and cargo holds in planes. If stored, cold storage is used, and refrigerated display cases are preferred.

Uniformity in quality, size, shape, color, condition, and ripeness is preferred for reasons of consumer appeal, freedom from disease, and ease of handling by automated machinery. Vegetables are often graded according to trade standard grades. The requirements of grading vary depending on factors such as whether the crop is destined for the fresh market or for processing. Grades have been established by federal law and are strongly suggested; however, they are not enforced legally unless the federal grades are used to describe the produce.

At least half of all vegetables are packaged in retail units at some point prior to reaching the retail store. This type of packaging is called prepackaging. Products are placed in bags of transparent plastic film or in trays and cartons covered over by transparent film. Trays and cartons used for the prepackaging unit are usually made of Styrofoam or wood pulp. The master containers for the prepackaged units are usually made of paperboard. Master containers are designed for easy stacking, shipping, and minimal bruising. Packaged units are bulk-packed into larger shipping containers.

Vegetables are shipped a number of ways. Large trucks with refrigerated containers, which sometimes have a controlled atmosphere, are used extensively. Refrigerated railroad cars are also used. Crops are also air freighted by large cargo planes. Cargo boats are another source of transportation. Loading and unloading are often facilitated by the use of wooden pallets and forklifts.

Vegetables destined for fresh consumption can be made available over a longer time by storage. However, crops destined for processing can be preserved by a number of techniques. These include canning and freezing, dehydration, pickling, fermentation, and juicing. Chemical preservatives are often added to maintain quality in terms of taste, texture, and color. Home growers are more likely to only use freezing and sometimes canning. Some vegetables such as salad greens are not processed.

Crops grown for processing usually cost less per unit area of land or per ton. One factor is that, unlike crops grown for the fresh market, appearance is not of major concern. However, size, quality, and uniformity are important. An extended harvest through successive plantings or the use of varieties with different maturation dates is needed to maintain a constant supply. This enables the processing factory to operate at an even flow over a reasonable period of time.

Generally, vegetables can be grouped in the following categories. Salad crops include leafy vegetables that are mainly eaten raw, as in a salad. Arugula, celery, cress, endive, lettuce, parsley, and raddichio are often included in this group. Certainly, other vegetables are eaten raw in salads, but their main use is for other purposes. The edible legumes include lima beans, peas, snap beans, and soybeans. The solanaceous crops include the eggplant, pepper, potato, and tomato. Cucurbits are vine crops and include the cucumber, melon, pumpkin, squash, and watermelon. The cole crops encompass broccoli, brussels sprouts, cabbage, cauliflower, Chinese cabbage, Chinese mustard, kohlrabi, kale, collards, and mustard. The last three are often included in the group of crops called greens, which are the immature leaves and stems of plants prepared by boiling. Chard, New Zealand spinach, spinach, and turnip greens are also greens. The onion group includes the common onion and its close relatives, garlic, leek, and shallot. The starchy root and modified stem crops are grown for their underground edible portions. They include the beet, celeriac, carrot, parsnip, potato, radish, rutabaga, salsify, sweet potato, taro, and turnip. Artichoke, asparagus, and rhubarb are included in the perennial crop group, whereas some vegetables such as horseradish, sweet corn, and okra are classed as miscellaneous vegetables.

Vegetables are a valuable component of the human diet. They contribute vitamins A, C, and K, various B vitamins, and several minerals (calcium, iron, magnesium, phosphorus, and potassium). Vegetables are also valuable sources of fiber and phytochemicals that help prevent and fight various cancers.

Herb production can be similar to vegetable production, when herbs are grown as field crops. Herbs can also provide a secondary source of income on small family-owned vegetable farms. Herb production can also be a stand-alone activity to provide a part-time farming income. Some herbs are highly attractive to bees, which can aid in the pollination of nearby vegetable crops.

Many vegetable gardeners also raise herbs as a normal component of their vegetable garden activities. Some herbs are attractive to butterflies, making them good candidates for butterfly gardens. Herbs are usually grown in a small, informal section of the garden, or sometimes in an elaborate formal garden devoted entirely to herbs. Some are decorative and used in borders or in rock gardens. Some are suitable for indoor or greenhouse culture.

Herbs need to be harvested at their peak, often before the plant flowers. Harvest must be fast and gentle and is often done by hand. Cooling might be needed and perhaps even washing for fresh market herbs. Washed herbs need to be dried thoroughly to prevent fungal problems. Careful packaging is required. Fresh herbs need to be moved fast into sales, as keeping qualities are limited.

Some herbs are highly valued for their culinary usage such as basil, oregano, and tarragon. Others have minor culinary use such as fenugreek. Culinary herbs, along with fresh vegetables, can be sold directly to upscale restaurants and at farm stands and through community-supported agriculture. Culinary herbs could be sold in bulk lots to wholesalers.

Some culinary herbs can be processed on site into value-added products such as pesto. Various herbal vinegars such as tarragon vinegar are another possibility as are olive oils infused with garlic, basil, and oregano. Herbs can also be dried, bottled and sold on site or bulk, dry lots sold to wholesalers. Dry herbs can extend the market product at your operation when fresh herbs are gone.

Herbs can be grown as commercial row crops, row tunnel crops, and greenhouse crops. The latter two approaches can make herbs available earlier on the market when herb prices are higher. Commercial herbs can also be harvested in the wild (wild-crafting). Herbs grown as field crops are essentially treated like vegetables. Level to slightly sloped land with good soil, water access, and soil reasonably free of difficult-to-control weeds is desirable. Cultural aspects are similar to vegetables. Most herbs grow reasonably well at a soil pH of 6.0 to 7.0 on mineral soils and 5.0 to 6.0 on high organic matter content or muck soils. Pests and diseases are often similar to those found with vegetable crops.

Herb crops can be annuals, biennials, or perennials. Many are aromatic or sweet smelling and are found in the mint (Labiatae), parsley or carrot family (Umbelliferae or Apiaceae), and composite or sunflower family (Compositae or Asteraceae).

A note of caution is in order about herb usage. Care must be exercised with the more unusual wild-crafted herbs to avoid the danger of poisonous look-alikes or the use of the wrong plant part, which could be poisonous. The use of herbal remedies in general should not be taken lightly. Keep in mind that unlike normal medications, herbal remedies are not subject to FDA oversight and have not undergone supervised clinical trials.

Some of the herbs grown commercially and in gardens for culinary, medicinal, or perfumery use include angelica, anise, basil, borage, burnet, caraway, catnip, chamomile, chervil, chives, coneflowers, coriander, cumin, dill, fennel, fenugreek, golden seal, horehound, hyssop, lavender, lemon balm, lovage, mint, oregano, rosemary, sage, savory, sesame, sorrel, sweet bay, sweet cicely, sweet marjoram, sweet woodruff, tarragon, and thyme.

EXERCISES

Mastery of the following questions shows that you have successfully understood the material in Chapter 17.

1. Why is the contribution of sustainable and organic vegetable to overall vegetable production poorly understood?
2. What contribution does vegetable gardening make on the economic level?
3. Which vegetable crops are tracked by the federal government?
4. What is the breakdown of acreage devoted to fresh market and processing vegetables? What are the leading states for both categories? What are the leading vegetables for both categories? Which state leads in the production of organic vegetables? Which state produces the largest amount of organic vegetables relative to its total output?

5. In terms of total tonnage, what is the breakdown between fresh market and processing vegetables? Which states lead in each category? Which vegetables lead in each category?

6. In terms of total value, what is the breakdown between fresh market and processing vegetables? Which states lead in each category? Which vegetables lead in each category?

7. What are the economics and production facts of white potatoes, sweet potatoes, herbs, mushrooms, greenhouse crops, vegetable seeds, and vegetable transplants? Are any of these organically produced? To what degree?

8. Which career in olericulture would you pick? Why?

9. From a sustainability viewpoint, organic matter content and pH are critical. Why?

10. Why is a range of fertilizer nutrients recommended for vegetable applications? What does one do to use fertilizers in a sustainable manner?

11. What factors are involved in determining the favorable times for sowing seeds and placing transplants?

12. Why are transplants used for some vegetables?

13. What practices can be used proactively in pest control? Describe each one.

14. What types of preventive controls can be used on vegetables?

15. When is mechanical harvest used? When is hand harvest used instead? What are the advantages and disadvantages of each form?

16. What changes take place after vegetables are harvested? Why is it important to prevent most of these changes? How are the changes prevented?

17. What are premarketing operations? Describe each one and why it is important.

18. Pick your five favorite vegetables and discuss each one in terms of cultural requirements and culinary value. Do the same for three vegetables that you have never consumed, don't like, or eat rarely.

19. What similarity to vegetables do herbs have? What advantage does herb production offer to vegetable producers?

20. Why are most herbs produced organically?

21. Why do you need to approach the use of medicinal herbs cautiously?

22. Pick your five favorite herbs and discuss each one in terms of cultural requirements and culinary value. Do the same for three herbs that you have never consumed, don't like, or use rarely.

INTERNET RESOURCES

(See also Internet Resources list in Chapter 12.)

Commercial vegetable fertilization principles
http://edis.ifas.ufl.edu/BODY_CV009

Commercial vegetable links for several states and topics
http://www.orst.edu/Dept/NWREC/veglink.html

Commercial vegetable production in Florida gateway
http://edis.ifas.ufl.edu/MENU_CV:CV

Commercial vegetable production in New Jersey
http://www.rce.rutgers.edu/pubs/vegetable/general_production_recommendations.pdf

Commercial vegetable production in Northwest gateway
http://www.orst.edu/Dept/NWREC/vegindex.html

Commercial vegetable production in Texas, other states gateway
http://aggie-horticulture.tamu.edu/vegetable/vegetable.html
http://aggie-horticulture.tamu.edu/extension/veghandbook/index.html

Commercial vegetable production in Wisconsin
http://www1.uwex.edu/ces/pubs/pdf/A3422.PDF

Crop production economics search
http://usda.mannlib.cornell.edu/

Fresh herb production and marketing
http://fletcher.ces.state.nc.us/staff/jmdavis/NYCONF.htm

Greenhouse organic herb production
http://www.attra.org/attra-pub/gh-herbhold.html

Greenhouse production Web sites
http://www.attra.org/attra-pub/ghwebRL.html

Greenhouse vegetable list of references
http://www.ces.ncsu.edu/depts/hort/hil/hil-32-a.html

Growing herbs
http://muextension.missouri.edu/xplor/agguides/hort/g06470.htm

IPM and commercial vegetable production, New York
http://www.nysaes.cornell.edu/recommends/

Medicinal herbs: numerous links
http://agebb.missouri.edu/mac/links/linkview2.asp?catnum=132&alpha=H

Mushrooms
http://www.attra.org/attra-pub/mushroom.html

National organic program final rule
http://www.attra.org/attra-pub/nop.html

Organic certification organizations and programs
http://www.attra.org/attra-pub/PDF/organicorgs.pdf

Organic crop production overview
http://www.attra.org/attra-pub/organiccrop.html

Organic/sustainable farming trends
http://www.ers.usda.gov/emphases/harmony/issues/organic/organic.html

Pest management newsletters by state
http://www.pmcenters.org/Public/news.cfm?USDARegion=National%20Site&site=
Producers

Plasticulture for commercial vegetable production
http://www.ces.uga.edu/pubcd/b1108-w.html

Production of medicinal herbs
http://fletcher.ces.state.nc.us/staff/jmdavis/patalk.html

References in commercial vegetable production
http://www.hort.uconn.edu/ipm/veg/htms/vegmanl.PDF

Resources for organic and sustainable vegetable production
http://www.attra.org/attra-pub/PDF/vegetable-guide.pdf

Small farm herb production
http://info.ag.uidaho.edu/Resources/PDFs/CIS1079.pdf

Soil testing labs for sutainable production
http://www.attra.org/attra-pub/soil-lab.html

Sources of information—organic or sustainable vegetable production
http://www.ianr.unl.edu/pubs/horticulture/nf108.htm

Sustainable practices for commercial vegetable production in the South
http://www.cals.ncsu.edu/sustainable/peet/

Vegetable disease information
http://vegetablemdonline.ppath.cornell.edu/

Vegetable garden IPM
http://vegipm.tamu.edu/

Vegetable information, California
http://vric.ucdavis.edu/

Vegetable IPM: California
http://www.ipm.ucdavis.edu/default.html

Vegetable IPM: Midwest
http://www.vegedge.umn.edu/intro/pestfact.htm

Vegetable IPM: Northeast links
http://www.nysaes.cornell.edu/ipmnet/links/veg.html

Vegetable IPM: South
http://ipm.ncsu.edu/TopicResults.cfm?TopSubID=40

Vegetable IPM: Texas
http://horticulture.tamu.edu:7998/vegetable/

Vegetables: Irrigation
http://www.ces.ncsu.edu/depts/hort/hil/hil-33-e.html

Vegetable pest links and IPM
http://www.isis.vt.edu/~fanjun/text/Link_pest14.html

Vegetable production statistics
http://usda.mannlib.cornell.edu/reports/nassr/fruit/pvg-bban/

Vegetable rotations, successions, and intercropping
http://aggie-horticulture.tamu.edu/vegetable/lubbock/vegrote.html

RELEVANT REFERENCES

(See also the Relevant References list in Chapter 12.)

Atha, Antony (2001). *The Ultimate Herb Book: The Definitive Guide to Growing and Using over 200 Herbs.* Collins & Brown, London. 320 pp.

Coleman, Eliot (1995). *The New Organic Grower: A Master's Manual of Tools and Techniques for the Home and Market Gardener.* Chelsea Green Publishing Co., White River Junction, VT. 340 pp.

Corum, Vance, Marcie Rosenzweig, and Eric Gibson (2001). *The New Farmers' Market: Fresh Ideas for Producers, Managers & Communities.* QP Distribution, Winfield, KS. 272 pp.

Dimitri, Carolyn, and Nessa J. Richman (2000). *Organic Food Markets in Transition.* Economic Research Service/USDA and Henry A. Wallace Center for Agricultural & Environmental Policy/Winrock International, Greenbelt, MD. 43 pp.

Duke, James A. (2000). *Handbook of Medicinal Herbs: Herbal Reference Library.* CRC Press, Boca Raton, FL. 696 pp.

Facciola, Stephen (1998). *Cornucopia II: A Source Book Of Edible Plants.* Kampong Publications, Vista, CA 713 pp.

Foster, Getrude B., and Rosemary F. Louden (1980). *Park's Success with Herbs*. George W. Park Seed Company, Greenwood, SC. 192 pp.

Gilbertie, Sal, and Maggie Oster (1998). *Growing Herbs in Containers*. Storey Communications, North Adams, MA. 32 pp.

Grubinger, Vernon P. (1999). *Sustainable Vegetable Production from Start-Up to Market*. Natural Resource, Agriculture and Engineering Service, Cornell University, Ithaca, NY. 280 pp.

Heaton, Donald D. (1997). *A Produce Reference Guide to Fruits and Vegetables from Around the World: Nature's Harvest*. Food Products Press, Binghamton, NY. 244 pp.

Henderson, Elizabeth (1999). *Sharing the Harvest: A Guide to Community-Supported Agriculture*. Chelsea Green Publishing, White River Junction, VT. 254 pp.

Larkcom, Joy (1991). *Oriental Vegetables: The Complete Guide for Garden and Kitchen*. Kodansha International, New York. 232 pp.

Maynard, Donald N., and George J. Hochmuth (1997). *Knott's Handbook for Vegetable Growers*. John Wiley, New York. 582 pp.

NY State Vegetable Growers Association (NYSVGA) (2002). *Proceedings from the 2002 New York State Vegetable Conference*. NYSVGA, Ithaca, NY. 225 pp.

Salunke, D. K., and S. S. Kadan (eds.) (1998). *Handbook of Vegetable Science and Technology: Production, Composition, Storage and Processing*. Marcel Dekker, New York. 721 pp.

Small Farm Center (1998). *Specialty and Minor Crops Handbook* (2nd ed.). University of California Publication 3346, Davis, CA. 184 pp.

Smith, Miranda (1999). *Your Backyard Herb Garden: A Gardener's Guide to Growing, Using, and Enjoying Herbs Organically*. Rodale Press, Emmaus, PA. 160 pp.

Splittstoesser, Walter E. (1990). *Vegetable Growing Handbook: Organic and Traditional Method*. AVI book formerly, now Van Nostrand Reinhold, New York. 362 pp.

Stephens, James M. (1988). *Manual of Minor Vegetables*. Florida Cooperative Extension. Bulletin SP-40. University of Florida, Gainesville. 123 pp.

Sturdivant, Lee, and Tim Blakley (1998). *Medicinal Herbs in the Garden, Field and Marketplace*. San Juan Naturals, Friday Harbor, WA. 375 pp.

Swiader, John M., George W. Ware, and J. P. McCollum (1992). *Producing Vegetable Crops* (4th ed.). Interstate Publishing, Danville, IL. 626 pp.

VIDEOS

Dabney, Seth (1997). *Using Cover Crops in Conservation Production Systems*. USDA-ARS National Sedimentation Lab, Oxford, MS.

Groff, Steve (1997). *No-till Vegetables*. Cedar Meadow Farm, Holtwood, PA.

University of California (1993). *Creative Cover Cropping in Annual Farming Systems*. Division of Agriculture and Natural Resources, University of California, Davis.

Fruits and Nuts

OVERVIEW

Pomology is the branch of horticulture that deals with the production, storage, processing, and marketing of fruit. What percentage of fruit is grown in a sustainable manner is unknown. Current statistics from the USDA's National Agricultural Statistics Service do not differentiate among differing systems of fruit production. Part of the problem is that the exact limits of what constitutes sustainable horticulture have not been clearly defined. Only one part of sustainable horticulture, organic farming, has been clearly described such that a certification process exists. Individual states and federal government standards allow for the tracking of organic crops and products. The acreage devoted to the organic production of fruits and nuts in the United States totaled 49,000 acres in 1997. *Trends in Agriculture 2000,* a survey conducted by the Gallup Organization, indicates that roughly one-fourth of U.S. farmers believe that they are operating as sustainable farmers. About 33% of tree crop producers (tree fruits and nuts) thought their practices were sustainable.

CHAPTER CONTENTS

OBJECTIVES

- To learn about the economics of and the contribution of sustainable and organic operations to commercial fruit/nut production
- To learn about the acreage, production, value, states, and fruits/nuts involved in fresh/processing markets
- To explore possible careers

- To understand the basics of fruit and nut production, harvest, postharvest, and marketing
- To learn all about fruits and nuts: crop and cultivar selection, site selection, site preparation, planting, pruning, orchard floor management, fertilization, irrigation, pollination, fruit thinning, pest management, harvest, storage, and culinary basics

BACKYARD FRUIT PRODUCTION

On the noncommercial level, fruit growing ranks behind ornamentals and vegetables. Two limiting factors are the space requirements for tree, bush, and bramble fruits along with the labor involved with fruit growing such as pruning and protection from birds. Also, the time available to homeowners is limited and the desire to grow things may be exhausted as a result of fulfilling daily work and family commitments. What little time remains is more likely to be devoted to the care and maintenance of ornamental plantings around the well-landscaped home or perhaps to the money-saving and health-related aspects of a vegetable garden. Those who do practice fruit growing (17 million American households) are rewarded with the beauty of certain fruit plantings and the appealing taste of fresh fruits. One only has to look around in Florida, California, Texas, and Arizona for a short time before a backyard with citrus trees is observed. In the New England or Northwest states, one can quickly find a yard with strawberries, raspberries, or blueberries.

COMMERCIAL FRUIT AND NUT PRODUCTION

The federal government tracks over 30 fruits and 6 nuts under the fruits and tree nuts category. The current tracked fruits are apples, apricots, avocados, bananas, blackberries, black raspberries, blueberries, boysenberries, cherries (sweet and tart), cranberries, dates, figs, grapefruits, grapes (and raisins), guavas, kiwifruit, lemons, limes, loganberries, mangos, nectarines, olives, oranges, peaches, pears, papayas, pineapples, plums (and prunes), red raspberries, strawberries, tangelos, tangerines, and temples. Tracked nuts include almonds, hazelnuts, macadamias, pecans, pistachios, and walnuts. Peanuts are tracked separately under crops. Tracking includes consumer trends, acreage, production, and value on the individual crop and state levels as well as the U.S. and export/import levels.

Consumer Trends

Fruits, like vegetables, are a healthy part of the American diet. Fruits supply vitamins A and C plus fiber and carbohydrates. The fiber supplied by fruits is generally more soluble than fibers from vegetables, so it is a complementary fiber source. Fiber of both types is important in cancer prevention. Fruits and vegetables are assigned a high rank on the food pyramid.

In 2000, we consumed 102.9 pounds of fresh fruit per person. At the same time, Americans consumed 8.9 gallons of fruit juice, 16.1 pounds of canned fruits, 3.5 pounds of frozen fruits, 2.5 pounds of dried fruits, and 2.4 pounds of tree nuts. The per capita consumption of peanuts was 5.9 pounds on a kernel weight basis. All citrus fruits combined accounted for 23.8% of our fresh fruit consumption. The leading individual fruit at 28.3% was the banana, followed by apples (17.4%), oranges (11.7%), and grapes (7.5%).

Considerable amounts of fresh and processed fruits and nuts were exported. Large amounts of fruits and nuts were also imported to meet consumption needs year-round and to provide tropical fruits and nuts grown only in a few U.S. locations or not at all. Fresh fruit imports reached 6.9 million tons in 2000, with the major imports being bananas (64.5%), grapes (7.5%), pineapples (5.1%), guavas/mangos (3.7%), plantains (3.4%), limes (2.9%), and apples (2.6%). Exports of fresh fruits in 2000 totaled 2.8 million tons, with the major exports being apples (25.3%), oranges (20.9%), grapefruit (15.6%), grapes (11.5%), pears (6.6%), and peaches (4.6%). Fruit juices were also exported and imported.

Acreage

On the commercial level, fruits are big business. During 2000 about 4.1 million bearing acres existed in fruit and nut production. Of this acreage, citrus accounted for 1.1 million acres, major deciduous fruits for 1.9, miscellaneous noncitrus fruits for 0.3, and tree nuts 0.8. For all bearing fruit acreage combined, the leading fruits in decreasing order were grapes, oranges, apples, peaches, grapefruit, and plums (and prunes). These six fruits accounted for 82.9% of the combined bearing fruit acreage. For all tracked bearing tree nut acreage (pecans not tracked for acreage), the nuts ranked in decreasing order were almonds, walnuts, pistachios, hazelnuts, and macadamias. Peanuts exceeded the combined tree nut acreage by 1.6 times. Acreage for individual fruits and nuts can be found in Tables 18–1 and 18–2.

Certified organic fruit and nut acreage in 1997 was 49,000 acres or 1.2% of the bearing fruit and nut acreage in 1997. The leading certified organic fruit was grapes at 39%, followed by apples, citrus, and tree nuts at 18%, 12%, and 10%, respectively. When viewed on the organic level, California in 1997 led with nearly 66% of the total certified organic U.S. acreage for all fruits and nuts combined. California produced 95.7% of the nation's certified organic grape crop, 72.2% of the organic tree nut crop, and 49.4% of the organic citrus

TABLE 18–1 U.S. Commercial Fresh Fruit Production in 2000

Fruit	Acreage* (1,000 acres)	Production (1,000 tons)	Value ($1,000)	Leading State (% of acreage)
Apples	447.3	5,324	1,373,682	WA (38.1)
Apricots	20.4	99	35,604	CA (93.2)
Avocados	65.2	234	321,210	CA (90.6)
Bananas	1.6	15	10,440	HI (100)
Black raspberries	1.2	1.9	5,687	OR (100)
Blackberries	6.1	22	21,437	OR (100)
Blueberries	40.3[†]	146	220,883	MI (41.4)
Boysenberries	1.7	4.5	5,225	OR (84.8)
Cherries (sweet)	58.4	207	276,710	CA (32.4)
Cherries (tart)	39.9	144	53,949	MI (72.0)
Cranberries	36.6	282	106,827	WI (41.3)
Dates	4.7	15	24,875	CA (100)
Figs	15.0	55	14,465	CA (100)
Grapefruit	153.5	2,762	411,332	FL (74.3)
Grapes	946.5	7,658	3,062,800	CA (87.5)
Guavas	0.7	8.0	2,051	HI (100)
Kiwifruit	5.3	34	13,480	CA (100)
Lemons	62.8	840	298,677	CA (77.2)
Limes	2.8	26	9,728	FL (100)
Loganberries	0.08	0.2	296	OR (100)
Mangos	1.7	2.8**	1,500**	FL (100)
Nectarines	35.5	267	106,266	CA (100)
Olives	36.0	53	32,328	CA (100)
Oranges	812.9	12,997	1,666,100	FL (74.1)
Papayas	1.7	27	16,023	HI (100)
Peaches	155.8	1,300	506,961	CA (43.1)
Pears	66.1	967	255,341	WA (36.9)
Pineapples	20.7	354	101,530	HI (100)
Plums and prunes	128.1	909	261,308	CA (96.9)
Red raspberries	12.4	42.9	33,600	WA (76.7)
Strawberries	47.8	924	1,013,537	CA (56.2)
Tangelos	11.3	99	11,232	FL (100)
Tangerines	40.6	458	108,192	FL (63.9)
Temples	5.8	88	9,173	FL (100)

*Bearing acreage.

[†]Does not include wild-harvested acreage in Maine, but does include wild-harvested blueberries in production and value. If leading state based on production, Maine becomes number 1 with 37.8% of production.

**Estimated.

TABLE 18–2   *U.S. Commercial Nut Production in 2000*

Nut	Acreage* (1,000 acres)	Production (tons)	Value ($1,000)	Leading States (% of acreage)
Almonds	500.0	351,500	710,030	CA (100)
Hazelnuts	28.4	9,628	21,374	OR (98.8)
Macadamias	17.7	5,678	29,500	HI (100)
Peanuts	1,315.5	1,643,800	835,400	GA (41.1)
Pecans	†	46,323	238,768	GA (38.1)**
Pistachios	74.6	57,082	238,140	CA (100)
Walnuts	193.0	102,429	289,190	CA (100)

*Bearing acreage.
†Not reported.
**Based on production.

crop. Florida produced 37.6% of the organic citrus crop. Arizona, California, and Washington were the major organic apple producers in the United States with 35.9%, 21.3%, and 19.3%, respectively. Organic grapes, apples, citrus, and tree nuts accounted for 1.9%, 1.6%, 0.5%, and 0.3% of the total bearing acreage for each fruit, respectively.

Production

Fresh fruit production was 37.2 million tons in 2000. Of this tonnage, citrus accounted for 17.3 million tons, noncitrus fruits for 18.8, and tree nuts 1.1. For all fruit tonnage combined, the leading fruits in decreasing order were oranges, grapes, apples, grapefruit, peaches, and plums (and prunes). These six fruits accounted for 85.8% of the combined fruit tonnage. For all tracked nut production tonnage on a shelled basis, the tree nut ranks in decreasing order were almonds, walnuts, pistachios, pecans, hazelnuts, and macadamias. The production of peanuts exceeded all combined tree nuts by a factor of 2.87. Tonnage for individual fruits and nuts can be found in Tables 18–1 and 18–2.

Value

Fresh fruit value in 2000 totaled $10.4 billion. Of this value, citrus accounted for $2.5 billion and noncitrus for $7.9 billion. For all fruit values combined, the leading fruits in decreasing order were grapes, oranges, apples, strawberries, and peaches. These five fruits accounted for 73.3% of the combined fruit value. The value for combined tree nut crops was $1.5 billion. For all tracked tree and groundnut values, the nut ranks in decreasing order were peanuts, almonds, walnuts, pecans, pistachios, macadamias, and hazelnuts. Values for individual fruits and nuts can be found in Tables 18–1 and 18–2.

CAREERS IN POMOLOGY

Operations involved in fruit-associated production, marketing, and services have need of individuals with expertise in pomology. Production nurseries propagating fruit and nut trees need breeders and propagators knowledgeable in pomology. Fruit growing can be a self-employment business or one can manage tree or bramble fruit production or grape vineyards for an employer. One can be a consultant or provider of fruit integrated pest management (IPM) programs. Wineries and vineyards have various positions for grape specialists. Fruit processors and fruit cooperatives need technical advising staff skilled in fruit knowledge. Marketing careers in fruit wholesaling, distributing, promotion, and retailing are available. Careers in education, extension, and research at land-grant universities and

U.S. government agencies such as the USDA are another opportunity for pomologists. Suppliers of agricultural equipment, chemicals, and financial/technical services can also utilize individuals with degrees in fruit sciences. Inspectors are needed for examining and testing imported fruits. Opportunities in the biotechnology industry for those with fruit knowledge coupled with molecular biology expertise are likely to increase in the near future.

FRUIT AND NUT PRODUCTION BASICS

Basic practices in pomology include the propagation and improvement of the cultivar, site selection, planting and spacing, training and pruning, irrigation, fertilization, weed control, soil management, pollination, thinning, insect and disease control, harvesting, postharvest physiology, and preservation. Most of these practices have been covered in Section Three and only more specific aspects will be touched on here.

Crop and Cultivar Selection

Choices of commercial fruit or nut crops in a given region are heavily determined by environmental constraints. To compete effectively in the marketplace, high yields and minimal cultural problems are essential. Backyard plantings offer more flexibility, because lower yields are not a problem. Home growers can sometimes grow fruit organically or with very minimal pesticides because of their acceptance of fruits with cosmetic blemishes that would normally be culled in commercial production.

For example, chilling requirements dictate that apples are better grown in the Northwest and Northeast instead of Florida. On the other hand, requirements for long, warm growing seasons and mild winters determine that citrus crops are suited to California, Florida, Arizona, and Texas as opposed to New England. Tolerance for very little frost or cold restricts mangos to Florida and pineapples to Hawaii. The need for no late spring frosts with apricots because of early flowering explains the concentration of the apricot industry in California.

Similarly cultural problems play a role. Organic apples are primarily produced in the West as opposed to the East, even though good apples can be produced in either region. The primary factors are serious apple pests (plum curculio and apple maggot) and higher fungal diseases in the East; the latter results from more humid conditions. Fruit rots plague apples in the Southeast. Apple growers in the East can and do produce apples sustainably by the use of IPM methods that involve biological controls and only a few sprays. A recent kaolin particulate, clay-based spray (Surround™ WP Crop Repellent), is likely to extend organic apple production to the East. This product has demonstrated suppression of several eastern apple pests and is approved in organic culture. Organic peaches are more apt to be produced in the West than East, primarily because of the plum curculio. Bramble crops can be grown organically and sustainably in both the East and West because of fewer pest problems.

Cultivars should be selected for high productivity and high-quality fruit. Other considerations include genetic pest and disease resistance, which is a primary factor in organic production and is still a consideration even with IPM systems. Disease resistance is more common; insect resistance is more likely for indirect pests such as aphids or mites. Resistance to pests that eat fruits is generally not available. Keep in mind that resistance is not necessarily immunity. Some damage might still occur, but to a lesser degree, or symptoms might appear with minimal impact on yields. Multiple disease resistance is less likely, so avoidance of all control measures for diseases is unlikely. Cultivar choices should also be selected based on regional considerations, because some cultivars are clearly more adapted to one region than another.

Much of the fruit grown exists as a clone and, as such, is asexually propagated. Grafting and budding are used extensively to improve vigor and reduce size. In some cases rootstocks also offer resistance to certain soil pests and diseases. Topworking is also practiced with fruit trees. This consists of altering the cultivar of an established tree or orchard through grafting or budding of a new cultivar on its trunk or branches. This is a useful tool to update an orchard to a more recent, improved cultivar, to add a pollinating cultivar to a portion of an orchard with a poorly pollinating cultivar, or to alter disease resistance.

Site Selection, Preparation, Planting, and Pruning

Site selection can be important for long-range success. Microclimate and soil conditions of sites are especially important. Spring frosts during or shortly preceding or following bloom are a serious hazard in fruit growing. The role of microclimate in the alleviation of frost hazard could make a site highly desirable. Gentle southern slopes or areas near large bodies of water are often good fruit growing locations. Soil condition is important, since roots of fruit trees must extend 3 feet or more in depth for high productivity. Any soil condition that interferes with root development, such as hardpan, wetness, or high salinity, could reduce yields. Sites with heavy perennial weed growth (Johnson grass, quack grass) are less desirable to organic operations, because successful elimination of these problem weeds usually requires herbicides.

Site preparation is important because it sets the stage for successful development and fruiting of the long-lived fruiting crops planned for the site. Soil testing and subsequent pH adjustments should be done prior to planting. Improvements in soil fertility, organic matter, and even weed suppression can be accomplished with cover crops and green manure (see Chapter 11). Existing vegetation or sod is plowed under, then a cover crop is sown and tilled under. In some instances a number of cover crops are rotated through the area and plowed under prior to planting. For example, both a warm season and winter season cover crop might be used. Choices of these cover crops are based on several factors (see Chapter 11). Alternatively, soil solarization can be used (see Chapter 11) for weed suppression. Cost is high though, so this method is limited in applications, other than for strawberries.

Smaller fruits such as strawberries can be managed in beds with rows, whereas larger fruits such as apples are managed in orchards. Planting systems and spacing are determined by two, sometimes conflicting factors: (1) size, growth vigor, and light requirements and (2) management practices. The aim is for highest productivity with least labor. Trends are toward dwarf trees and management practices such as high-density plantings (Figure 18–1) that are more conducive to mechanized operations. Labor is the largest cost factor in fruit production. Cultivars that require cross-pollination pose special requirements with regard to cultivar mix selection, planting, and spacing.

Many fruit plants need to be pruned for form when young. This type of pruning is called *training*. Generally, trained fruit plants will require maintenance pruning to maintain fruitfulness and ease of other management practices such as pest control or harvesting. The

FIGURE 18–1 Aerial view of apple and pear orchards near Yakima, Washington. Note the high-density planting pattern. (USDA photograph.)

types of training for fruit trees of standard size (modified leader, central leader, and open center) and their maintenance pruning were covered in Chapter 14. Recommended practices for each fruit will be cited in this chapter.

Dwarf and other fruit trees are being increasingly trained in a hedgerow system, which is essentially a continuous tree wall. This practice provides maximal trees per acre and facilitates the use of mechanical equipment for pruning, spraying of growth regulators and pesticides, and harvesting of fruit. This is a consequence of the drive to maximize production and minimize labor. Hedgerow plantings are supported by post or three- and four-wire trellis systems such as those used with grapes or dwarf apples. Hedgerows can be left freestanding on stronger-rooted stocks. Freestanding isolated trees are also used.

Weeds and Orchard Floor Management

Weeds in perennial fruit plantings can be managed in a number of ways. One is clean cultivation (Figure 18–2), an approach that is very effective. Disadvantages include energy usage, increased likelihood of soil erosion (given bare soil), and the loss of organic matter, which causes detrimental changes in soil structure. Winter cover crops could be grown and incorporated into the soil to maintain organic matter and partially offset erosion. Another choice is herbicides. The downside here is costly herbicides and possible groundwater pollution. Both approaches minimize nonfruit plant competition for nutrients, water, and light.

More sustainable options include the use of cover crops and mulches. Permanent cover crops such as sod or legumes prevent soil erosion, but can compete with fruit crops for water and nutrients. Additional fertilizer and water would be needed. However, some organic matter and nutrients are added to the soil by natural decomposition of the sod. Soil structure is improved, and soil moisture fluctuations are less, especially during dry periods. Management systems to minimize this competition are available.

For example, ground covers can be used in aisles and clean cultivation under the plantings (Figure 18–3). Ground covers such as cool season grasses that go dormant during the hot part of the growing season when fruits are forming are another option. Another possibility is ground cover in the aisles and mulch under the plantings (Figure 18–4). Another option is to use legumes instead of grass as the ground cover (Figure 18–5). The legumes can supply nitrogen for the fruit plantings. The clippings from the mowing of the aisle legumes can be used to provide nitrogen-rich mulch for under the plantings, too. Legumes (and grasses) do compete for water, but both also increase water permeability of the soil.

FIGURE 18–2 These strawberries are planted in a matted row system. Note the clean cultivation achieved by machine between the rows. (USDA photograph.)

FIGURE 18–3 Blando brome is used as a cover crop between the tree rows of this orchard. It is mowed, but is allowed to set seed for the following year's cover crop. Since it is an annual grass, dying roots and shoots contribute organic matter to the soil. Soil compaction and loss of organic matter is minimal, as no cultivation is needed for weed control except around the trees. (USDA photograph.)

FIGURE 18–4 This apple orchard is sodded in the rows and the trees are mulched. This system minimizes competition for water and nutrients by keeping the sod away from the trees. It also maintains soil structure and health well and as a bonus, supplies the mulch, too. (USDA photograph.)

Depending on your climate, other options can come into play. A system exists that can provide between-row ground cover, mulch, fertilizer, and even a habitat for beneficial insects. The catch is that this system of orchard ground management only works in areas where winter temperatures do not drop below −17.8°C (0°F) regularly. The ground cover is subterranean clover (*Trifolium subterraneum*). This clover is grown between and under plantings. It dies back during the hot part of the summer, thus eliminating competition. Upon dying, it leaves behind a thick layer of mulch. It is an annual legume, but does reseed itself. Cultivars include 'Mt. Barker' and 'Nangeela,' which are midseason varieties used in areas with low rainfall and for sites that become dry in early summer. 'Tallarook' is a later choice suitable for areas with moisture into July, such as the Oregon coast and lower valley area.

Mixtures of ground covers offer both advantages and disadvantages. Mixtures of cover crops offer a suitable environment for diversity of beneficial insects, leading to fewer orchard pests. Mixtures need to be carefully chosen such that no species support local orchard

FIGURE 18–5 Clover is used as a cover crop in this vineyard in Michigan. Since clover fixes nitrogen, it supplies some nitrogen to the grapevines. The soil structure and health is maintained. Some competition for water between vines and the cover crop is the minor cost. (USDA photograph.)

FIGURE 18–6 Strawberries are grown here using black plastic mulch. These plasticulture systems are used commercially in California, Florida, and North Carolina for strawberries. (Author photograph.)

pests. For example, certain peach and apple viruses use chickweed and dandelions as hosts. Tarnished plant bugs thrive in some winter annual broadleaf weeds in warmer areas. Some legumes also attract tarnished plant bugs and stink bugs. Some plants that support beneficial insects are buckwheat, dwarf sorghum, members of the sunflower family (Compositae or Asteraceae) and members of the Mustard family (Brassicaceae or Cruciferae). Another possibility is to manage adjacent vegetation to enhance beneficial insects (see Chapter 15). This concept is sometimes referred to as farmscaping.

Mulch usage varies from being used only around trees to the drip line or close-in row plantings such as with brambles or grapes. The remaining or aisle space in the orchard, plantation, or vineyard is either in sod or cover cropped. In some cases, the area near the plants plus the entire aisle is mulched such as with strawberries. In most cases the mulch is organic, except for strawberries (Figure 18–6), where it is likely black plastic. If organic, the mulch should be kept 8 to 12 inches from trunks to reduce vole damage and diseases such as crown rot. On the other hand, sawdust mulch on blueberries can be mounded right up to the shoots. Organic mulches can be costly if not on hand or difficult to move around. A possible solution is to mow the ground cover in the aisle periodically with a sickle bar mower and to use the clippings as mulch. Mulch materials can also be grown and harvested from nearby fields or possibly purchased cheaply from local sources.

Mulches do offer a number of compensatory benefits beyond weed suppression: increased water infiltration, slower evaporative loss of soil water, enhanced soil aggregation, and soil temperature moderation (less plant stress). Mulches have been shown to rank first as an orchard floor management system in terms of moisture availability. Another plus is that the mulch releases small amounts of N, P, and K as it decomposes. Some organic mulch also exhibits allelopathic effects against herbaceous weeds, but not against trees.

Newer nonorganic mulches are coming into use. These mulches, woven plastic or geotextiles, are superior to black plastic in that they are permeable and do not encourage matted surface root growth. Trees with matted surface root growth show increased drought susceptibility, increased winter injury, and poorer anchorage. While these geotextiles are expensive initially and do not provide organic matter or nutrients to the soil, their life span is 10 or more years and the labor of continuous remulching is ended. Savings in weed control over the geotextile life are likely to compensate for the initial cost.

If areas around trees or adjacent to row plantings are not mulched, weed control will be required. The same is true for row crops where ground-covered aisles or mulch are not used such as for peanuts or strawberries. For backyard plantings and perhaps small commercial, part-time operations, hand cultivation with a hoe could be sufficient. Larger operations can use mechanical hoes or rotary tillers that can be tractor mounted and set to cultivate shallowly right up to plantings. Other attachments such as scrapers can be used to scrape the soil near plants or to scrape soil further away and mound it around plants.

Other options include possibly organically acceptable herbicides such as various soap-based, fatty acid-based, or potassium salt-based herbicides (see Chapter 15). Such herbicides should be checked against certification requirements. Other nonorganic products might be useful for other sustainable systems in small amounts. All sprays need to be kept away from green tissues. Costs can be high for large operations. Another alternative is to use weeder geese in orchards, strawberries, brambles, and blueberries. Their effectiveness is best against either small or emerging weeds. Bermuda grass and Johnson grass are consumed by geese, as are windfall fruits that might eventually harbor disease organisms. Another possibility is flame weeding. Hand and tractor-mounted flame weeders are available. For close-up plant work, some type of water shielding is needed to protect the crop.

Fertilization

Of all the horticultural crops, fruits are the easiest to fertilize in an organic or sustainable manner. Fruits, unlike other crops, consist mostly of water. Their harvest does not remove nearly as many soil nutrients as do vegetable crops. With fruits, fertility needs are more easily met with nutrients provided from decomposing mulch and efficient cover crop management supplemented with various rock dusts and good pH control. If legumes are part of the cover crop management, supplementary nitrogen needs are minimal. Nuts require considerably more fertilizer, because nuts contain protein and oils.

It is important to keep in mind that the fertility, pH, and organic matter content for soil prior to establishing new plantings in orchards, vineyards, and plantations must be as optimal as possible. This window presents the only time that soil incorporation of compost, fertilizers, limestone or sulfur, cover crops, green manures, and legumes is really possible. Once planted to trees, vines, and bushes along with aisle cover crop systems, soil incorporation becomes unrealistic and surface treatments are the only recourse for adding nutrients. Decaying mulches and soil root residues become the main venue for replenishing organic matter. It is fortunate that fruits do not have high needs for continual nutrient inputs and that cover crop management systems are capable of maintaining organic matter and preventing soil erosion.

Nitrogen can be supplied with manure, compost, and various meals such as blood, cottonseed, or feather meal. Such products should be applied in the early spring and in sufficient amounts to compensate for slow release through decomposition. Ideally, these organic products should be incorporated into the soil to reduce nitrogen loss. Incorporation is not always possible, because root damage can result with surface-rooted fruits. Alternatively, soluble organic fertilizers such as fish emulsion can be used as a foliar spray. Cost can be a detracting feature with these products. Nonorganic, sustainable fruit growers can elect to use conventional fertilizers.

Regardless of what products are used, sustainable applications must be based on data from reliable soil tests and foliar tissue analyses. As mentioned in the previous chapter on vegetables, soil tests are done best by laboratories familiar with the needs of sustainable farmers. Such needs include accounting for nitrogen supplied by legumes and decomposing mulch over time, thus making sustainable recommendations for fertilizer applications. Foliar tissue analyses also supply useful data as they indicate actual nutrient contents. Levels can be related to what amounts of fertilizers (N-P-K and micronutrients) need be applied to keep soil at optimal fertility. A plus is that tissue analyses indicate deficiencies prior to visible plant symptoms and also indicate what the plant actually takes up. Soil tests only show levels of soil nutrients, which might not all be available to the plant because of chemical limitations in a given soil area.

Visual indicators also provide clues to problems. Prompt attention is warranted when such signs are noted. For example, terminal shoot growth can be monitored to give input about nitrogen adequacy. If sufficient nitrogen is present, growth of terminal shoots for nonbearing apples, cherries, and pears should be 12 to 24 inches and for nonbearing peaches, 14 to 24 inches. Figures for these same fruits in the bearing stage are 8 to 12 and 12 to 15 inches, respectively. Older leaves show nitrogen deficiency symptoms first as a uniform pale green to yellow color. Other clues are thin twig growth, early autumn leaf drop, light fruit set, heavy June fruit drop, smaller fruits, and earlier maturity.

Keep in mind that fertilizer amounts suggested here in this chapter and in various other publications are only recommendations. Soils with high levels of fertility and good soil health could require only half as much. Sandy soils could need up to 50% more. Sites in legumes could require less. Heavily sodded orchards would need 20% to 50% more than clean cultivated orchards. Heavy weed growth increases fertilizer needs. Trees after heavy pruning need less fertilizer. Young nonbearing trees need less fertilizer than do mature bearing trees. Needs also change with the densities of trees or plants per acre. To ensure sustainable fertilizer applications, soil tests and foliar tissue analyses are essential.

Irrigation

Fruits are somewhat more drought tolerant than vegetables and herbaceous annuals. However, a constant, moderate soil moisture supply from flowering through harvest is generally needed to ensure good fruit set and good-sized quality fruits. Less irrigation is needed prior to the fruiting stage. Over-watering earlier can slow root development, increase nutrient leaching, increase root and collar rot problems, and reduce the uptake of calcium and phosphorus. Excess vegetative growth can also occur. Deliberate underirrigation in the pre-fruiting stage can be used as a practice (regulated deficit irrigation) to reduce vegetative growth and subsequent pruning, while saving water and power. If increased later at the flowering/fruiting stage, little or no effect will be observed with fruit growth.

While the most sustainable option is trickle or drip irrigation, other forms with up-to-date technology can suffice. Drip systems are best for high-value crops such as many fruits. Irrigation was covered in Chapter 11. Drip irrigation is also adaptable to supplying nutrients to fruits through fertigation. This form of fertilization is highly sustainable in that less nitrogen is needed and less leaching occurs. The one caveat is that system uniformity of at least 80% is required to ensure effective nutrient distribution. Uniformity tests can be rapid and simple and give accurate results (see Internet Resources at the end of the chapter, drip irrigation entry). Causes of nonuniformity include bad system design, emitter plugging, manufacturing variations in emitters, and failure to use or clean filters. Fertilizers are injected (see Internet Resources list, drip irrigation) based on creating pressure differentials (venturi), pressure (positive displacement pump), and pressure gradients (batch tank injector).

Irrigation amount and frequency should be based on tree, shrub or bramble needs at each developmental stage, the water holding capacity of the soil, soil moisture condition, and irrigation efficiency. There is no substitute for local knowledge and frequent measurements of soil moisture. Tensiometers are used to measure soil moisture as soil tension in centibars (0.001 Bars) and are preferred for sandy soils. Soil blocks are best with loams and clays: these measure available soil moisture (ASM) as a percent. ASM is the percentage of soil water between field capacity (-0.1 Bar) and permanent wilting point (-15 Bars). Irrigation is usually started when two-thirds of the available moisture at the 2-, 3-, and 4- foot levels has been used.

Pollination and Thinning

Pollination can be enhanced by the introduction of beehives into the fruit plantings during flowering. Insect control must not be detrimental to bees at this time (see Chapter 15). Hand pollination is not practical because of labor costs. Removal of flowers or young fruit (thinning) is a common practice of fruit growers. Remaining fruit develop larger or more rapidly, and the following year's flowering and fruit set are not affected adversely. Thinning can be done by hand, but mechanical or chemical means are more economical. Mechanical hydraulic or pneumatic limb shakers or growth-regulating substances can be used, as discussed in Chapter 14. Organic fruit growers use nonchemical means. Another possibility is to prune some of the fruiting wood, which is considerably easier than hand thinning.

Pest Management

Pest control is probably the most expensive and time-consuming aspect of fruit growing. Integrated pest control, the judicious use of biological and chemical controls, is being increasingly utilized. Costs and environmental pressures can be reduced with this approach. Organic and IPM approaches were covered in Chapter 15.

One major option in pest management, rotations, is generally not available with most fruit and nut crops. The only possible exceptions are strawberries, peanuts, and perhaps bramble crops. Other efforts must be used to prevent the buildup of pests over time. At the same time, the stability of the fruit growing area offers a good opportunity to establish and maintain populations of beneficial insects and other organisms. One obvious solution is to maintain good plant vigor and health to minimize damage from indirect pests that feed on nonfruit parts of the crop, as well as pathogens.

Both organic and other sustainable fruit growers need to take full advantage of biological controls (see Chapter 15). Release of beneficials is one strategy, but the best long-run economic strategy is to manage nearby natural vegetation and in-site cover crops to provide a favorable habitat for beneficial organisms. These methods will likely provide adequate control for indirect pests, but perhaps not complete control for direct pests that feed on fruit, especially if very little cosmetic or other damage is tolerable.

Additional help using organically acceptable natural or biological pesticides will probably make up the difference for organic growers. Other sustainable growers might opt for "green" or minimally disruptive pesticides. Microbial, botanical, soaps, oils, fungal-derived compounds, green pesticides, and mating disruptors were thoroughly covered in Chapter 15. A new product, kaolin clay, was mentioned earlier in this chapter and shows promise for apples, pears, and possibly brambles and other fruit crops.

Diseases can be controlled to some degree with proactive methods. Keeping fruit trees and other crops (brambles, grapes) properly pruned such that air circulates and sunlight penetrates into the interior canopy of the plant helps dry and reduce moisture conditions favored by bacterial and fungal diseases. Sanitation practices that remove fallen leaves and fruits reduce the overwintering of certain diseases. Good soil drainage is also essential to reduce the potential of various soil-borne pathogens to cause root rot. The use of genetically resistant cultivars is also helpful to limit problems. Finally, good soil health maintained with significant organic matter produces a soil ecosystem full of beneficial microorganisms that either compete with or are antagonistic toward pathogens. One can also use organic fungicides such as copper- or sulfur-based products, biofungicides (competing or antagonistic microbes), or various fungicides (see Chapter 15).

A number of mammals can be especially troublesome in fruit plantings, causing both plant and fruit damage. Deer and birds can be major problems and mice, voles, rabbits, and raccoons to a lesser extent. Control strategies include scare devices, repellents, noise devices, traps, fencing, netting, and tree guards. Poisons are another option, but are not available to organic growers. Most options work reasonably well when pest pressures are low to moderate. Heavy deer pressure requires electrified fencing and heavy bird pressure, netting. These options are expensive, but provide long-term solutions.

HARVESTING FRUIT AND NUT CROPS

Determining when to harvest fruits is critical. For the backyard grower, the choice is clear. Harvesting usually occurs at peak quality, because the fruit is either going to be used quickly, refrigerated only for a short period, or processed soon after harvest. For commercial operations, harvest is determined by a number of factors. For fruits or nuts destined for processing, harvest can be close to peak maturity, as processing happens shortly after and better quality products result from peak maturity as opposed to less ripe stages. For fresh market, harvest is somewhat before peak maturity, because ripening will continue in many cases. Dead ripe fruits are generally avoided, because bruising can occur during shipping and decreased quality can result during the time from harvest through sales. In some cases, gassing with ethylene-producing products is used prior to harvest to initiate uniform ripening as is done with pineapples. The determination of the correct picking time can be tricky. With fruits, degree-days or heat units are reliable indicators. Other indicators are color (apples, bramble fruits, peaches, and strawberries), fragrance (pears), or softening (mangos).

The oldest harvesting technique is hand harvesting (Figure 18–7). The home horticulturist makes extensive use of this approach for fruit and nut crops. Commercial operations also hand harvest when mechanized operations are not available or are bypassed in favor of pick-your-own operations or when the cost and feasibility of the equipment is questionable. At one time commercial operations did not have suitable mechanized

FIGURE 18–7 Hand harvesting blueberries with a rake is a slow process. (USDA photograph.)

harvesting equipment for easily bruised crops such as the bramble fruits, or the equipment was of limited feasibility because of variations in the time of ripening. However, harvesting machinery has been improved even for easily bruised fruit, and plant breeders have aided mechanized harvesting by developing crop varieties with uniform ripening habits, compact growth, less easily bruised fruits, more easily detached fruits, or an angle of hang better suited for mechanized removal. In addition, more and more growth regulators are used as mechanized harvest aids to promote uniform ripening, color, and abscission. Harvesting equipment (Figure 18–8) usually saves time and cuts the harvesting cost considerably.

One type of mechanized harvester consists of motor-driven tree and bush shakers with catching belts and some storage capacity. Possible problems associated with this type of mechanized harvesting are soil compaction, injury to lower fruit-bearing branches (such as spur apples) by falling fruits, and possible damage to the root system. Shaker harvesters have worked fairly well with cherries and most nuts. More recent mechanized harvesters for various fruit trees make use of finger-type pickers, slappers, paddles, and impacter rods (Figure 18–9).

FIGURE 18–8 Machine harvesters for blueberries are designed to minimize fruit bruising of tender blueberries. (USDA photograph now in the National Archives.)

FIGURE 18–9 This machine harvester uses twelve-foot-long nylon rods that rotate and shake foliage as it moves along rows of citrus trees in Parrish, Florida. (USDA photograph.)

These often have decelerators, such as numerous rubber strips, to slow fruit fall and hence minimize damage. Some of these newer machines are designed to be used with dwarf fruit trees or orchards trained to the hedgerow system or trellis system. Other low-growing fruits are harvested with what amounts to mechanized slappers and even mechanical rakes. Mechanized harvesters are constantly being improved, and new ones are being developed. In some instances, these machines can harvest crops with less damage than occurs by hand harvesting.

The mechanization of harvesting varies as to degree and the number of steps involved. For example, many apple orchards are still picked by hand because the size and spacing of large trees necessitates large harvesting machinery that is costly and somewhat difficult to handle. Harvesters designed for compact trees, hedgerows, and trellis systems are more feasible and harvest fruits with minimal bruising. These devices can be expected to increasingly replace hand labor in many orchards. However, the bulk of tree and bush fruits intended for the fresh market are still hand harvested. Some operations may be handled in a partially mechanized manner, for example, the hand picking of fruit like apples followed by mechanized grading and packing. Crops intended for processing are extensively harvested by mechanical means because of the savings in labor dollars and the lesser importance of crop bruises. These operations might be completely mechanized in one location; nuts might be mechanically picked and shelled right in the field.

POSTHARVEST CHANGES OF FRUITS

When a horticultural crop is harvested, it is still a living organism, and life processes continue to produce changes in the picked crop. These changes constitute the science of *postharvest physiology*. The biochemical and physiological changes can enhance or detract from the appeal of the crop. The appeal of the crop is determined by three factors: quality, appearance, and condition. Quality is used in reference to flavor and texture, as determined by taste, smell, and mouth-tongue touch. The appearance refers to the visual impression of the crop, whereas the condition refers to the degree of departure from a physiological norm or the degree of bruising or disease.

Postharvest alterations are numerous and include water loss, the conversions of starch to sugar and sugar to starch, flavor changes, color changes, toughening, vitamin gain and loss, softening, and decay. These processes can improve or detract from the quality, flavor, and texture of the crop. For example, the loss of water by transpiration and evaporation is bad for apples, but certainly useful in producing raisins and prunes from grapes and plums, respectively. Most people are familiar with the loss of quality associated with the rapid enzymatic conversion of sugar to starch that occurs with harvested sweet corn, and the reverse reaction, starch to sugar, that takes place in potatoes. However, the latter reaction results in the improvement of picked bananas and pears, and is one of the reasons that pears and bananas are the few fruit crops that can be picked when immature and ripened to a high-quality product.

On the whole, postharvest changes produce more detrimental results than beneficial ones. If the crops are harvested for subsequent processing, the time between harvesting and processing should be minimal, so postharvest alterations are not a serious problem. However, crops destined for fresh market sales must be handled in a manner that will minimize the detrimental effects and maximize any beneficial effects, if possible.

No one set of conditions will achieve this end, because postharvest alterations are a function of the crop type, and each crop varies as to the optimal air temperature, relative humidity, and concentrations of oxygen and carbon dioxide required for stabilization. Some form of air circulation is usually beneficial in reducing disease problems and removing any released gases that might accelerate harmful physiological and biochemical changes. The level of disease pathogens should also be kept to a minimum.

The application of postharvest preventive measures will also vary as to when they should be applied. For example, if a postharvest change is both rapid and seriously detrimental, some treatment might be required from the moment the fruit is picked. Rapid deterioration would be a problem with strawberries or raspberries, whose postharvest longevity is only a few days, but not with lemons, which have a postharvest longevity of several months. Treatment might be a temporary refrigerated storage area to hold the freshly picked berries until enough is accumulated. Then the berries might have to be shipped to a grading and packing area in a refrigerated truck. The grading and packing area might even need to be temperature controlled, yet the comfort of the workers must be considered. For example, apples are often graded and packed in areas where the temperature is kept between 12.8° to 15.5°C (55° to 60°F), and hot-air blowers or radiant heaters are directed at the feet of the workers to keep them comfortable.

The most important aspect of postharvest treatments is to have good storage facilities designed for long-term crop storage, so that the crop can be sold at a reasonable price over an extended period of time. The best storage conditions slow the life processes but avoid tissue death, which produces gross deterioration and drastic differences in the quality, appearance, and condition of the stored crop.

Although conditions vary, it is possible to make some generalizations. Fruits and nuts intended for processing or storage should be free from mechanical, insect, or disease injuries and at the proper stage of maturity. Reliable storage consists of cold (refrigerated) storage with precise control of temperature and humidity. An improvement on this form of storage is the addition of controlled atmosphere (CA storage) by which the concentration of oxygen and carbon dioxide are kept at an optimal level. The low temperature decreases the rate of respiration, the controlled humidity level slows water losses through transpiration and evaporation, and the controlled atmosphere brings about an overall reduction in the rates of chemical reactions occurring in the stored materials. The overall change in reactions is reflected in the inhibition of respiration and ethylene production. Apples and pears are examples of fruit stored under controlled atmosphere, temperature, and humidity conditions.

Controlled atmosphere conditions are usually 5% oxygen and 1% to 3% carbon dioxide. A compensating factor for the higher cost of controlled atmospheres is the fact that optimal temperatures are often several degrees higher if a controlled atmosphere is present. Not only do savings in refrigeration cost result, but physiological disorders sometimes associated with the lower temperatures are minimized.

Storage temperature and humidity differ somewhat for each fruit, but some generalizations are possible. Most temperate-zone fruits can be held at 0° to 5°C (32° to 41°F). Subtropical and tropical fruits would show chilling injury if held at these temperatures for a prolonged period of time. Bananas, for example, will not tolerate temperatures below 11.7°C (53°F). Relative humidity is usually kept at 85% to 90%. Nuts can be stored at 1.1° to 7.2°C (34° to 45°F) at 65% to 70% relative humidity.

PREMARKETING OPERATIONS

Premarketing operations include washing, waxing, precooling, grading, prepackaging, and shipping. Some steps are part of preventing postharvest damage, others are designed to improve or enhance the marketing value of the fruit. Still others are used to remove unwanted materials. These steps can be done in the field, at specialized facilities, or a combination of both can be used. Each place has advantages and disadvantages. Field operations reduce transport of and damage to the crop and cost less. The downside is that quality control is reduced, the weather can cause problems, large machines can disrupt field activities and cause soil compaction, and cooling of small lots is less energy efficient than bulk lot cooling.

Washing and Waxing

Fruits are often washed or brushed to enhance appearance. Disinfectants are often present in the wash water to eliminate pathogens that affect storage and marketing qualities. Waxing is used to improve appearance and to slow moisture loss, which causes shriveling. Waxing has been employed to varying extents on apples and citrus fruits.

Precooling

Precooling is used for rapid removal of heat from recently harvested fruits. This might be done at the field or at the packinghouse. For some crops cooling can be introduced at the washing step. With other crops cooling might occur after packaging. Precooling permits the grower to harvest near maximal maturity and yet be assured that the crop will have maximal quality when it reaches the consumer. Precooling slows the rate of respiration in the fruit, slows wilting and shriveling by decreasing water loss, and inhibits the growth of decay caused by microorganisms. Precooling is an energy-requiring process; harvesting at night and in the early morning, when the temperature of the crop is lowest, can minimize its cost. Major precooling methods include hydrocooling, contact icing, vacuum cooling, and air cooling. Descriptions of these techniques were covered in the preceding vegetable chapter, Chapter 17. Highly perishable crops such as blackberries, cherries, peaches, raspberries, and strawberries can benefit from precooling. After precooling the crop, it must be kept cool in transit by shipment in refrigerated trucks, railroad cars, and cargo holds in planes. If stored, cold storage is used, and refrigerated display cases are preferred.

Grading

Uniformity in quality, size, shape, color, condition, and ripeness is usually preferred. This is for reasons of consumer appeal, freedom from disease, and ease of handling by automated machinery. To achieve this end, fruits are often graded according to standard grades, which form a basis of trade. The requirements of grading vary depending on factors such as whether the crop is destined for the fresh market or for processing. Grades have been established by federal law and are strongly suggested; however, they are not enforced legally unless the federal grades are used to describe the produce. Government grades include such designations as Fancy or Extra Fancy, No. 1, No. 2, and so on. Fruits and nuts not meeting grading standards are usually processed or employed in products using lesser grades and off-sizes.

Packaging

Many fruits are packaged in retail units at some point prior to reaching the retail store. This type of packaging is called *prepackaging*. Products are placed in paper bags with handles (apples), plastic mesh bags (oranges), or in trays and cartons or covered over by transparent film (blackberries, blueberries, raspberries, and strawberries). Transparent film is usually made of polyethylene, which can be made to have a differential permeability toward oxygen and carbon dioxide. Often a controlled atmosphere can be reached in these containers covered in polyethylene, which consists of less oxygen and more carbon dioxide than is present in normal air. The retention of water vapor is also improved with plastic film, which is usually beneficial for the quality of the fruits. Trays and cartons used for the prepackaging unit are usually made of paperboard, Styrofoam, or wood. The master containers for the prepackaged units are usually made of paperboard. Wooden crates and bushel baskets are seen less and less. Master containers are designed for easy stacking, shipping, and minimal bruising.

Shipping

Fruits are shipped in a number of ways. Large trucks with refrigerated containers, which may or may not have a controlled atmosphere, are used extensively. Refrigerated railroad cars are also used. Crops are also air freighted by large cargo planes. Cargo boats are another source of transportation. Loading and unloading are often facilitated by the use of wooden pallets and forklifts.

Preservation

Fruits and nuts destined for fresh consumption can be made available over a longer time by storage, which was discussed earlier. However, fruits and nuts destined for processing

(A)

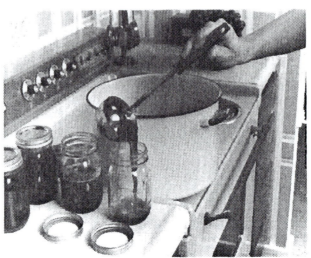

(B)

FIGURE 18–10   (A) Commercial canning of apples. These cans are about to be sealed and heated in a pressure cooker. (B) Pouring of hot jelly into sterilized fruit jars at home. (USDA photographs.)

can be preserved by a number of techniques (Figure 18–10). These methods include wet pack canning (apples, cherries, figs, fruit salads, peaches, pears, plums), vacuum canning (nuts), freezing (blackberries, blueberries, mangos, raspberries, strawberries), dehydration (apples, apricots, bananas, cherries, cranberries, dates, figs, papayas, pineapples, prunes and raisins), plastic bags (shelled nuts), baked goods and candies (various fuits and nuts), sugar concentrates (butters, jellies, jams, preserves, conserves, marmalades, purees, and sauces), juicing (apple, apricot, citrus, cranberry, grape, papaya, peach), and use in oils (coconut, olive, walnut) and brandies/liqueurs/wines (almond, apricot, blackberry, cherry, grape, hazlenut, orange, peach, raspberry, strawberry). All, of course, can be eaten fresh directly or held in storage besides being processed. The banana, for example, is more likely to be eaten fresh than dehydrated, although one can buy dehydrated banana chips.

Crops grown for processing usually cost less per unit area of land or per ton. This is because, unlike crops grown for the fresh market, appearance is not of major concern. However, size, quality, and uniformity are important. An extended harvest through successive plantings or the use of varieties with different maturation dates is needed to maintain a constant supply. This enables the processing factory to operate at an even flow over a reasonable period of time.

FRUIT AND NUT CROPS A TO Z

Horticulturists usually group fruits on the basis of their climatic requirements: temperate, subtropical, and tropical fruits. Temperate-zone fruits are deciduous and tropical fruits are evergreen; subtropical fruits can be either. Temperate fruits include the pome fruits (apple, pear, and quince), stone fruits (apricot, cherry, nectarine, olive, peach, and plum), and the small or berry fruits: brambles (black raspberry, red raspberry, blackberry, dewberry, boysenberry, youngberry, and loganberry), blueberry, cranberry, currant, gooseberry, grape, and strawberry. The brambles, blueberry, currant, and gooseberry are often categorized as bushberries. Subtropical and tropical fruits include avocado, banana, citrus, date, fig, mango, papaya, persimmon, pineapple, and pomegranate. Many other minor fruits have not been mentioned, but some of which will be covered later in the chapter. Fruits are a good source of soluble fiber and are a healthy complement to the less soluble fibers found in vegetables. Both are required for good health. Fruits are also sources of vitamins, minerals, and some cancer-fighting chemicals.

Nut trees have much to offer the horticulturist. Nuts are high in food value, providing protein, oils, minerals, fiber, and vitamins. The tree itself provides shade and is ornamental in the landscape. Nuts can be tropical, subtropical, and temperate. Major commercial nuts produced in the United States include almonds, hazelnuts, macadamias, peanuts, pecans, pistachios, and walnuts. All except for peanuts (row crop) are tree crops. All of these plus some minor nuts can be found in backyard plantings.

Almonds

Almonds are deciduous trees grown for their nuts. They belong to *Prunus dulcis; P. dulcis* var. *amara* is noted for oil of bitter almond, and the variety *P. dulcis* is grown for its edible nut. The flowering ornamental almonds are *P. glandulosa, P. japonica,* and *P. triloba.* Edible commercial cultivars include 'Butte,' 'Carmel,' 'Mission (Texas),' 'Nonpareil' (most widely grown commercial cultivar), and 'Price.' Newer cultivars making their way into commercial orchards include 'Frtiz,' 'Monterey,' 'Padre,' and 'Sonora.' Other cultivars include 'All in One,' 'Ballico,' 'Davey,' 'Eureka,' 'Halls Hardy,' 'IXL,' 'Jordanolo,' 'Kapareil,' 'Merced,' 'Ne Plus Ultra,' 'Peerless,' 'Thompson,' and 'Titan.' Cultivars vary in shell hardness. Soft-shelled cultivars are more important in commercial production.

Hardiness is through zone 7. Actually, hardiness is greater, but flowering is so early that late spring frosts limit the range in terms of nut production. Commercial production is mainly in California, but backyard production is seen in many states. Rainy weather in the spring and summer increases disease problems with blossoms and fruits. Midwinter rains also enhance disease problems.

Propagation is by T budding and grafting (whip and tongue, cleft or bark) of cultivars. Commonly used rootstocks include peach seedling rootstocks ('Lovell' and 'Nemaguard'), peach/almond hybrid rootstocks ('Bright's Hybrid' and 'Hansen'), and plum rootstocks ('Mariana 2624'). One-year-old nursery stock is planted in the late fall to early winter. Trees are placed 21 to 30 feet apart in all directions. Some nuts can be expected by the third or fourth year, and full bearing will be reached in the eighth year.

Foliar tissue analyses supplemented with soil tests are critical for achieving sustainable fertilization with nut trees. Nitrogen needs vary from 60 to 250 pounds per acre, depending on soil levels and cover crop considerations. Fertilizer is usually applied in two to four split applications, starting in early spring. Fertilizer should be watered in, unless it is possible to incorporate it into the soil. Fertigation is an especially good system for nut trees. Irrigation needs in much of California beyond natural rainfall entail 39 or more inches of irrigation water annually. Mature height is about 25 feet.

Almonds are usually trained to the open center or vase shape system. Pruning is light and only practiced every second or third year. Fruit thinning is not practiced. Almond flowers are self-incompatible, so two or more compatible cultivars are necessary, and bees are recommended to ensure cross-pollination. Two to three thousand pounds of unshelled nuts per acre can be expected or about 25 to 40 pounds per tree. Harvesting is done with mechanical shakers. Average unshelled yield in 2000 from California was 1.15 tons/acre.

Troublesome insects and other pests include conserse stink bug (*Euschistus conspersus*), European fruit lecanium (*Parthenolecanium corni*), glassy-winged sharpshooter (*Homalodisca coagulata*), June beetle (*Polyphylla decemlineata*), leaf-footed bug (*Leptoglossus clypealis*), leafrollers (*Archips argyrospila, Choristoneura rosaceana*), mites (*Bryobia rubrioculus, Panonychus ulmi, Tetranychus* spp.), navel orange worm (*Amyelois transitella*), nematodes (*Criconemella xenoplax, Pratylenchus vulnus, Xiphinema americanum*), oriental fruit moth (*Grapholitha molesta*), peach twig borer (*Anarsia lineatella*), San Jose scale (*Quadraspidiotus perniciosus*), shot-hole borer (*Scolytus rugulosus*), tree borers (*Bondia comonana, Euzophera semifuneralis*), the Southern fire ant (*Solenopsis xyloni*), and the Western tent caterpillar (*Malacosoma californicum*).

Viral diseases include ring spot, almond mosaic, and almond calico. Other diseases are anthracnose (*Colletotrichum acutatum*), brown rot (*Monilinia laxa*), cankers (*Ceratocystis fimbriata, Pseudomonas syringae*), Coryneum blight or shot-hole (*Wilsonomyces carpophilus*), crown gall (*Agrobacterium tumefaciens*), crown rot (*Phytophthora* spp.), leaf scab (*Cladosporium carpophilum*), leaf blight (*Seimatosporium lichenicola*), leaf scorch disease (fastidious xylem-limited bacteria, *Xyllela fastidiosa,* carried by glassy-winged sharpshooter), leaf spot (*Alternaria alternata*), oak root fungus (*Armillaria mellea*), rust (*Tranzchelia discolor*), scab (*Cladosporium carpophilum*), and verticillium wilt (*Verticillium dahliae*).

Apples

Apple cultivars (Figure 18–11) are probably derived from *Malus pumila,* or possibly from hybrids of *M. pumila* and *M. sylvestris*. Cultivars number in the hundreds. Some cultivars are early, some are midseason, and others late. Some are best when eaten fresh, others for making cider and others for cooking into pies, cakes, and applesauce. Some are sweet, others tart; some are soft and others crisp. Color varies from green to red and yellow. Apples are good fiber sources and contain small amounts of vitamin C.

Some cultivars of importance are 'Ambrosia,' 'Arkansas Black,' 'Ashmead's Kernal,' 'Baldwin,' 'Beni Oshu,' 'Big Red,' 'Braeburn,' 'Cameo,' 'Campbell Red Delicious,' 'Cortland,' 'Cox Orange Pippin,' 'Crispin,' 'Criterion,' 'Earligold,' 'Egremont Russet,' 'Ellison's Orange,' 'Empire,' 'Fiesta,' 'Fireside,' 'Fuji,' 'Gala,' 'Gold Rush,' 'Gold Star,' 'Gravenstein,' 'Grimes Golden,' 'Granny Smith,' 'Haralson,' 'Hillwell Red Braeburn,' 'Idaho Spur,' 'Jonagold,' 'Jonamac,' 'Jonathan,' 'Jubilee Fuji,' 'Kingston Black,' 'Lodi,' 'Macoun,' 'McIntosh,' 'Melrose,' 'Newtown Pippin,' 'Northern Spy,' 'Novamac,' 'Pacific Gala,' 'Priscilla,' 'Red Cort,' 'Red Delicious,' 'Red Fuji,' 'Red Gravenstein,' 'Red Rome,' 'Red Stayman,' 'Rhode Island Greening,' 'Rome,' 'Royal Gala,' 'Russet,' 'Smokehouse,' 'Spartan,' 'Spitzenberg,' 'Starkspur Dixiered,' 'Sun Fuji,' 'Sunrise,' 'Tompkin's King,' 'Triple Red,' 'Ultrared Jonathan,' 'Yellow Golden Delicious,' 'Yataka Fuji,' 'Yellow Transparent,' 'York Imperial,' 'Washington Spur,' 'Winesap,' and 'Winter Banana.' Many other cultivars are known. Those interested in growing apples should consult local agricultural authorities for the cultivars best suited to their locality. Many older cultivars such as 'Northern Spy' and 'Russet' can be found in collections of older apple cultivars such as those at Sturbridge Village in Massachusetts.

Several cultivars are disease resistant and are useful for organic growers. These cultivars include 'Akane,' 'Aroma,' 'Bramley,' 'Chehalis,' 'Dayton,' 'Enterprise,' 'Freedom,' 'Gold

FIGURE 18–11   Ripe 'Yellow Golden Delicious' apples are shown here. Recently apple juice has been found to contain antioxidants, so the adage an apple a day keeps the doctor away might have some validity. (USDA photograph.)

Star,' 'Greensleeves,' 'Honeycrisp,' 'Jonafree,' 'Liberty,' 'MacFree,' 'Pristine,' 'Rajka,' 'Releika,' 'Resi,' 'Shay,' 'William's Pride,' and 'Wolf River.'

Much of the United States is suitable for the temperate apple, except for Florida and southern extremities of Texas and California. A cold dormancy period of 1,200 to 1,500 hours of 7.2°C (45°F) or lower is needed. Low chill cultivars (400 to 600 hours) such as 'Beverly Hills' can be used to extend production into warmer areas. Winters with temperatures below −28.9°C (−20°F) are not suitable for good apple production. Sites should have good, deep, well-drained agricultural soil. Apples can be grown in orchards or by the homeowner. The latter must realize that some effort and expense are required. Washington, New York, Michigan, and California are the main apple producers.

Propagation of apples is generally by budding or whip grafting of cultivars onto rootstocks. The latter may be used to control the size of the tree. Full-sized trees, which attain a height of 18 to 22 feet, are produced with seedling rootstocks grown from cultivars such as 'Red Delicious' or 'McIntosh.' More recently, a Russian sucker-free rootstock, Antonovka 313, has become available. It produces a full-size tree with excellent winter hardiness and wide soil adaptability.

Other rootstocks are used for size control; these are usually produced by stool beds, semihardwood cuttings, and mound layering of suckers produced by plants with the desired size-controlling rootstock. Mature size for the cultivars ranges from fully dwarfed trees, about 6 feet tall, through semidwarfed trees of 8 to 10 feet and nearly full size. These include rootstock series such as Malling (M), Malling Merton (MM), East Malling/Long Ashton (EMLA), Budagovsky, Geneva, Cornell-Geneva, Ottawa, Poland (P), Mark (formerly Mac-9, Michigan apple clone), and Supporter. All have advantages and disadvantages such as disease resistance or susceptibility and maximal or minimal cold hardiness, in addition to size-controlling characteristics.

Dwarfing can also result with spur apple cultivars. A spur is a mutation whereby a short shoot is produced from internode compression. Spurs grow from 2-year or older wood and produce flowers and fruit and are found in apples and pears. A spur-type cultivar, when grafted to a seedling rootstock, matures into a tree 66% to 75% the size of the nonspur form of the same apple on seedling roots. Dwarfing results from a genetic factor inherent in the spur-type cultivar. If a spur-type cultivar is grafted on a dwarfing rootstock it will have an additive dwarfing effect, that is, the genetic dwarfing effect of the spur cultivar plus the genetic dwarfing effect of the rootstock. Spur-type trees are available minimally among apple cultivars.

Interstem grafting, or the grafting of an interstem onto a rootstock followed by grafting or budding of the desired cultivar onto the interstem, is practiced with apples. Advantages such as earlier ripening, better root growth of a dwarf apple, longer life, or higher productivity can be realized with this type of grafting. However, costs for interstem grafted apples are high, thus limiting their use.

In-row and between-row spacing are influenced by several factors such as rootstock choice, cultivar vigor, soil factors (soil type, fertility, depth, water retention), and production system type. Low-trellis hedgerow systems are popular with their high-density plantings (close spacing) of dwarf and semidwarf rootstocks. Trellis systems produce high yields, but do require considerable time, expense, training, and pruning to establish and maintain the trees on the four-wire trellises. Freestanding trees require more space with their semidwarf to full-size rootstocks that can be semivigorous to vigorous in habit. While these systems give lower yields, labor and money inputs can be lower. Other systems (slender spindle, hoop skirt, and spindle bush) are used for high-density plantings and differ in training techniques from trellis systems. These systems only need support with certain rootstocks. Again yields are high, but considerable training and pruning efforts are needed.

Some examples of in-row spacing as determined primarily by rootstock type and production system are shown in Table 18–3. Ranges are a factor to account for differences in tree vigor and soil type. Between-row spacing is influenced by all of the in-row factors plus the ultimate tree height and to some degree by equipment width. With large equipment and semistandard and standard rootstocks, between-row spacing is the height of the tree plus 8 feet. For a seedling rootstock, this spacing could vary from 24 to 30 feet, depending on the

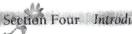

TABLE 18–3   *In-Row Spacing (Feet) for Apples*

Rootstock	Low-Trellis Hedgerow	Slender Spindle, Hoop Skirt, Spindle Bush	Freestanding Central Leader on Semidwarf Rootstocks	Freestanding Central Leader on Semistandard and Standard Rootstocks
EMLA 7			9–12	
EMLA 9	6–7	5–7		
EMLA 26	7–9	7–10	8–10	
EMLA 27		4–5		
EMLA 106			11–14	16–20
EMLA 111			15–18	17–20
M9/106			10–13	
M9/111			10–13	
M2			12–15	
Seedling				18–24

tree vigor and soil type. The 8 feet is a factor to accommodate full-size equipment. With smaller rootstocks and generally smaller sized equipment, this rule is invalid. It wastes too much space and results in lower yield per acre. However, too close spacing shades out the adjacent row and also reduces yields. Rules for determining between-row spacing for smaller rootstocks are based on tree height (either ultimate or pruning controoled). Rules providing the ranges for between-row spacing are:

Desired tree height/0.75
Double the desired tree height and subtract 6
Divide the desired tree height by 2 and add the in-row spacing

As an example, if the rootstock produced 10-foot tall trees in your soil with your cultivar, the above rules yield 13, 14, and 15 feet, so the optimal between-row spacing is 13 to 15 feet. An alternative method is to look at existing plantings that duplicate your system on a sunny day during late afternoon. If no shadow strikes the adjacent row's canopy, trees are too far apart. If the adjacent row has more than 20% of the canopy in shadow, plantings are too close. Ideally, the lower canopy should be shadowed 10% to 20%. More information on spacing can be found in the Internet Resources list and from your local extension office.

Planting is done in early spring before the soil is dry. In mild climates, autumn planting can be preferable. When grafted trees are planted, the bud union must never be planted below the soil surface. Bearing age varies. Dwarf trees on 'EM 9' can begin bearing the second year after planting a 1- or 2-year-old nursery-grown tree. Semidwarf trees require 4 or more years, and standard sizes up to 10 years. In their first year, each tree needs about 0.05 pounds nitrogen and 0.10, 0.15, and 0.20 pounds each additional year. Subsequent needs are 0.30, 0.40, 0.50, and 0.60 pounds/tree for their 6th, 8th, 10th, and 12th year in the orchard. Mature trees need 0.75 pounds each or about 60 pounds/acre. These numbers are suggested for a density of 80 standard-sized trees per acre. As tree densities increase and smaller trees are used, these numbers decrease somewhat.

Pruning should be kept to a minimum the first 3 to 8 years. The main purpose is to train the tree such that it has the desired placement of scaffold branches and proper leader development. Subsequent maintenance pruning is to keep the tree balanced in shape and proportions of old and young wood, such that maximal fruit yield is possible (see Chapter 14). Most pruning is done during late winter before active growth starts in spring. Some summer pruning may be needed with young or dwarf trees to improve shape and remove unwanted growth or with mature trees to remove watersprouts, weak wood, and unproductive wood.

Standard-sized trees are usually trained to the central-leader system. Dwarf trees can be trained that way, also, but the hedgerow system is the choice for commercial orchards. Topworking may be utilized to change cultivars in older apple orchards. Grafts used for this purpose include cleft grafting, bark grafting, and modified kerf grafting. Young trees may be topworked with the tongue or whip graft.

Apple cultivars are self-incompatible to varying degrees, but most cultivars can be pollinated by almost any other single cultivar through bee activity. An exception is Cortland and Early McIntosh, which together need a third cultivar for pollination. A few cultivars are triploids (Gravenstein, Jerseyred, Jonagold, Mutsu, Rhode Island Greening, Stayman, and Winesap) and are not suitable pollinators for the normal diploid apples. These apples require the presence of two other normal cultivars to ensure pollination for all three. Several varieties should be grown to ensure cross-pollination to set the best commercial crop. It is essential to have overlapping flowering periods for the cultivars, which can differ in their time of blooming. Distances between two cultivars should be less than 100 feet. Crab apples are also pollinators. Thinning will be necessary when overcropping occurs and to prevent biennial bearing. Chemical thinners are available for apples. Alternative approaches include hand pruning and mechanical shaking.

About 400 different insect and disease problems occur with apples. Commercial growers use several sprays of various pesticides and the homeowner may need to apply four to eight treatments if unblemished high-quality fruit is desired. Organic apples can be produced, but more easily in the West than the East. Pests and treatments vary from area to area, so a local agricultural authority should be consulted.

A few of the insect pests are aphids, apple maggot (*Rhagoletis pomonella*), apple tree borers (*Chrysobothris femorata, C. mali, Saperda candida*), codling moth (*Cydia pomonella*), curculio (*Conotrachelus nenuphar*), fruitworms (*Amphipyra pyramidoides, Lithophane antennata,* and *Orthosia hibisci*), leafrollers (several species), mites (European red and several other), oriental fruit moth (*Grapholitha molesta*), scale, and tarnished plant bug (*Lygus lineolaris*). Deer, rabbits, voles, and birds can also cause extensive tree and fruit damage.

Scab (*Venturia inaequalis*) is one of the main fungal diseases. Other diseases are bitter rot (*Colletotrichum gloeosporidides*), black rot (*Botryosphaeria obtusa*), blister spot (*Pseudomonas syringae* pv. *papulans*), brooks spot (*Mycosphaerella pomi*), canker (*Cytospora* sp.), crown gall (*Agrobacterium tumefaciens*), crown rot/collar rot (*Phytophthora cactorum*), fire blight (*Erwinia amylovora*), and powdery mildew (*Podosphaera leucotricha*). Bitter pit and cork spot are physiological disorders: both are caused by calcium deficiency. Numerous other diseases also occur (see Internet Resources, apple diseases entry).

Average yields of apples for standard orchards are about 6 bushels per tree. Higher yields of 15 to 20 bushels per tree are possible for standard-sized cultivars with optimal conditions. Yields up to 45 to 60 bushels per tree are possible with high-density plantings of spur semidwarf trees grown in hedgerows. Average yields in the United States in 2000 were 11.9 tons/acre. Apples should be stored under controlled atmosphere. Home storage is best at 0.6°C (33°F) and high humidity.

Apricots

There are at least three varieties of the apricot, a temperate fruit (Figure 18–12). *Prunus armeniaca* var. *armeniaca* is the commercial apricot source. Variety *sibirica* and variety *mandshurica* have greater cold tolerance, but less desirable fruits. Cultivars of commercial or West Coast value include 'Blenheim,' 'Castlebrite,' 'Improved Flaming Gold,' 'Katy,' 'Moorpark,' 'Patterson,' 'Puget Gold,' 'Royal,' 'Smith,' 'Tomcot,' and 'Tilton.' Cultivars with good winter hardiness are 'Brookcot,' 'Debbie's Gold,' 'Goldcot,' 'Harcot,' 'Harglow,' 'Hargrand,' 'Harlayne,' 'Harogem,' 'Harval,' 'Moongold,' 'Sungold,' 'Veecot,' and 'Westcot.' Apricot cultivars are often regional specific, so local authorities should be consulted. An aprium is an apricot/plum cross where the apricot dominates; 'Flavor Delight' is a cultivar. When the plum dominates, it is called a pluot; cultivars include 'Dapple Dandy,' 'Flavor King,' 'Flavor Queen,' and 'Flavor Supreme.'

FIGURE 18–12
Apricots are a good source of vitamins, both in fresh and dry forms. California apricots are available for a short season and total annually about 1.5 million 24-pound boxes. Apricots are very susceptible to soil diseases associated with wet soils, so should be planted in soils with good drainage. (USDA photograph.)

Apricots only require a cold dormancy period of 600 (few cultivars even less) to 1,000 hours below 7.2° C (45°F). Therefore, apricots bloom early and are highly susceptible to late spring frosts, which restrict their culture to areas with fast warming springs free of frosts. California is the primary commercial producer of apricots. Cultivars of smaller size and less sweetness than those of commercial use can be grown by the homeowner where peaches succeed, especially if a favorable microclimate exists.

Apricots are propagated by budding onto apricot seedling rootstocks. For cold climates, Manchurian apricot seedling rootstocks are the best choice. Peach and plum rootstocks are sometimes used to overcome problems, such as those caused by heavy soil conditions. A few dwarf types are available. Apricots have a minimal planting distance of 15 to 20 feet (dwarf cultivars, 10 feet). They may be planted as 1- or 2-year-old trees in early spring in the East or late autumn in California. Apricots start bearing 2 to 4 years after planting. The modified-leader system of training is best, although an open center tree like that used with peaches is satisfactory. Because fruit is borne on spurs that last about 3 years, pruning is aimed at renewing spur growth. Mature height is about 30 feet.

Most apricot cultivars are self-fruitful. Bees are the chief pollinators. Exceptions are 'Earliril,' 'Moongold,' and 'Sungold,' which need another cultivar for cross-pollination. Thinning of flower buds to reduce overbearing is practiced by hand (pruning) or mechanical shakers. Use of chemical thinning sprays is minimal. Fruit thinning is not practiced, since the size increase of the remaining fruit is minimal.

Nitrogen needs should be based on foliar tissue analyses and other nutrients on soil tests. A 1-year-old apricot tree needs roughly 0.07 pounds of nitrogen yearly. For each additional year, another 0.07 pounds should be added until the total in the ninth year is 0.63 pound. Mature trees should be continued at this rate, which is equivalent to 80 pounds of nitrogen per acre.

Various insect and other pests of apricots include branch and twig borer (*Polycaon confertus*), brown mite (*Bryobia rubrioculus*), cribrate weevil (*Otiorhynchus cribricollis*), curculio (*Conotrachelus nenuphar*), European earwig (*Forficula auricularia*), European fruit lecanium (*Parthenolecanium corni*), European red mite (*Panonychus ulmi*), fruit tree leafroller (*Archips argyrospila*), green fruitworms (*Orthosia hibisci, Amphipyra pyramidoides, Xylomyges curialis*), mealy plum aphid (*Hyalopterus pruni*), orange tortrix (*Argyrotaenia citrana*), Pacific flatheaded borer (*Chrysobothris mali*), peachtree borer (*Synanthedon exitiosa*), peach twig borer (*Anarsia lineatella*), redhumped caterpillar (*Schizura concinna*), shothole borer (*Scolytus rugulosus*), spider mites (*Tetranychus pacificus, Tetranychus urticae*), various nematodes (*Criconemella xenoplax, Meloidogyne* sp. *Pratylenchus vulnus, Xiphinema americanum*), and western tussock moth (*Orgyia vetusta*).

Troublesome diseases include bacterial canker (*Pseudomonas syringae*), bacterial leaf spot (*Xanthomonas prunii*), brown rot blossom, and twig blight plus ripe fruit rot (*Monilinia fructicola, M. laxa*), Coryneum blight or shot-hole (*Wilsonomyces carpophilus*), crown gall (*Agrobacterium tumefaciens*), crown rot/collar rot (*Phytophthora* spp.), eutype dieback (*Eutypa lata*), jacket rot complex (*Botrytis cinerea, Sclerotinia sclerotiorum, Monilinia laxa*, and *M. fructicola*), mushroom root rot (*Armillaria mellea*), powdery mildew (*Podosphaera tridactyla, Sphaerotheca pannosa*), and ring spot virus.

Fresh apricots do not last long or ship well and are only available for a short time. Most of the crop is dried, canned, frozen, juiced, or used in brandy. Apricots are good sources of vitamins A and C. Annual yields of 250 pounds per tree are possible. Homeowners might expect 30 to 120 pounds per tree, depending on climate and fertility. Average yields in the United States during 2000 were 4.9 tons/acre. Apricots can be stored for 7 to 14 days at 0°C (32°F) and 90% to 95% relative humidity.

Avocados

The avocado, *Persea americana,* is a tropical fruit produced on an evergreen broad tree reaching a maximal height of 80 feet. Cultivars are divided into three groups that differ in fruit characteristics and climatic requirements. The West Indian and Guatemalan groups are associated with the varieties *americana* and *guatamalensis,* respectively; the Mexican group is found under variety *drymifolia.* The West Indian group is strictly tropical and includes early ripening cultivars such as 'Fuchs,' 'Lula,' 'Pollock,' 'Simmonds,' 'Trapp,' and 'Waldin.' The Guatemalan group is somewhat hardier; common cultivars are 'Anaheim,' 'Gwen' (dwarf), 'Hass,' 'Lyon,' 'Nabal,' 'Pinkerton,' and 'Queen.' The Mexican group is the hardiest and includes cultivars such as 'Duke,' 'Ganter,' 'Mexicola' (hardiest cultivar, good rootstock for more northerly areas), 'Mexicola Grande,' 'Pueblo,' and 'Topa Topa.' Hybrids also exist between the three groups. Important hybrid cultivars are 'Bacon,' 'Booth,' 'Choquette,' 'Creamheart,' 'Fuerte,' 'Hall,' 'Hickson,' 'Murietta Green,' 'Reed,' 'Rincon,' 'Ryan,' 'Simpson,' 'Spinks,' 'Whitsell,' 'Wurtz,' and 'Zutano.'

The West Indian group can only be grown in the warmest parts of Florida, since injury occurs below −2.2°C (28°F). West Indian fruits are the large, round, smooth, glossy green fruits and lowest in oils. The Guatemalan group can withstand temperatures to −3.9°C (25°F) and can be found in much of zone 10 in California and Florida. This group has medium-sized fruits that are pebbly, somewhat pear shaped, and blackish-green. The Mexican group can stand temperatures as low as 6.7°C (20°F) and can be grown in southern California or Florida (zone 9). Mexican avocados are small and have paper-thin skins that are either glossy green or black. Avocados are grown commercially in orchards and by homeowners. California is the primary commercial producer of avocados. Avocados are high-fat fruits like olives, but the fat is highly monosaturated and has been shown to reduce cholesterol in clinical studies.

Propagation is by shield budding and side grafting on seedling avocado or various cultivar rootstocks. This is done in the spring and fall. Shield budding is practiced in California with Mexican group rootstocks and in Florida with West Indian or Guatemalan rootstocks. Mature trees reach a height of 30 feet or more.

Grafted trees are transplanted in the spring or early summer. Spacing is between 20 and 30 feet, depending on desired planting density. Fruit may be expected in as little as 3 years, although heavy fruiting will not occur for 5 to 8 years. Growth habits of the avocado are such that pruning required for training and maintenance will be minimal. Several cultivars should be present to ensure cross-pollination. Within a group, pollen production and pistil receptivity have minimal overlap. Bees are the main pollinators. Fruit thinning is unnecessary.

Soil pH is acceptable from 5.5 to 7.5. Trees are usually fertilized when planted and then after 1 year of growth, four times yearly with a balanced fertilizer. Nitrogen needs per tree per year are year 1, 0.10 lb; year 2, 0.20 lb; year 3, 0.33 lb; year 4, 0.50 lb; and year 5 on, 1.0 to 1.5 lb. Irrigation needs are minimal during the winter rainy season, and during the growing season should not be overdone, because avocados are susceptible to root rot from too much water.

Commercial avocados are harvested when the oil content is at least 8%. Guatemalan types ripen in 12 to 18 months from flowering and Mexican types in 6 to 8 months. Mature trees in orchards can produce from 2 to 6 tons per acre. Average yields in the United States during 2000 were 3.6 tons/acre. One to three bushels can be expected from an individual tree. Storage is possible for 2 to 4 weeks at 85% to 90% relative humidity, but temperatures are based on cold tolerance (4.4° or 12.8°C, 40° or 55°F). Ripe avocados are stored at the colder temperature, and mature green ones at the higher temperature.

Insects that attack the avocado include leafrolling caterpillars (*Amorbia* and *Tortrix* sp.), mites (avocado brown and six-spotted), scales, and thrips. Disease problems include anthracnose (*Glomerella cingulata*), avocado root rot (*Phytophthora cinnamomi,* usually associated with wet soils), canker (*Botryosphaeria ribis*), fruit spot blotch (*Cercospora perseae*), and scab (*Sphaceloma perseae*). One virus is a problem, sun blotch, but can be controlled with virus-free propagating wood and proper sanitizing of grafting and other tools. Rats, snails, and squirrels can be a problem.

Bananas

Bananas are herbaceous perennials. Their height varies from about 5 feet for 'Dwarf Cavendish' to 30 feet for the standard commercial cultivars. Common edible bananas are cultivars of *Musa acuminata.* These cultivars include the diploids, 'Blande' and 'Paka,' and the triploids such as Cavendish types (includes 'Dwarf Cavendish,' 'Giant Cavendish,' 'Lacatan,' and 'Robusta'), 'Grand Nain,' 'Gros Michel,' and 'Valery.' 'Valery' resembles 'Robusta' and might be the same. 'Williams' is the same as 'Giant Cavendish.' Other cultivars are those derived from the hybrid *M. acuminata* × *M. balbisiana* (designated *M.* × *paradisiaca*). Some cultivars are diploid ('Ney Poovan'), triploid ('French Plantain,' 'King,' 'Nadan,' and 'Silk'), or tetraploid ('Tiparoot').

Bananas are tropical fruits. In the tropics, bananas are grown commercially on plantations and as backyard plants. Windy sites are to be avoided. Hawaii is the only U.S. state to commercially produce bananas. Yields in 2000 were 9.4 tons/acre. Bananas are grown by homeowners in some warmer areas of California. Bananas are characterized as sweet and used as dessert fruits. Bananas are excellent sources of potassium and contain some vitamin C. Plantains are starchy and are served as cooked vegetables.

Propagation is by rhizome cuttings with at least two buds (eyes) or by suckers that arise at the base of the pseudostem. 'Dwarf Cavendish' can be set 8 to 10 feet apart. Other cultivars require 14 to 20 feet. In plantations the pseudostem is cut off after fruiting, since it dies after fruiting once. It is possible in the tropics for the next oldest pseudostem to produce fruit in 10 months. Rooted suckers can produce fruit in 12 months to 3 years, depending on climate.

Pruning is mainly to remove spent pseudostems and to allow only one primary stem for each rhizome. Once the main stalk has developed for 6 to 8 months, one sucker is allowed to grow as the next season's producing stalk. Pollination is not required with the majority of the commercial cultivars, which set their fruit parthenocarpically. Diploid cultivars require pollination. Thinning of fruit is not practiced.

Bananas are heavy feeders. A mature tree can require 1 to 2 pounds of an 8-10-8 fertilizer monthly during the growing season. Younger plants need about 25% to 33% of that amount. Soil pH must be 5.5 to 6.5. Bananas need large amounts of water during active growth. Even moisture is best, because drying for long periods or standing water is harmful. Salty soils also cause problems. Mulches can be helpful for reducing water needs and keeping the soil evenly moist.

Insect and disease problems are minimal in the United States. Several insect and disease problems are encountered in the tropics (consult Internet the Resource listings). Disease resistance varies among the cultivars. Root rot is the main problem for U.S. plantings.

Bananas are harvested immature (plump, full sized, and green) and ripened at temperatures above 11.7°C (53°F). Controlled atmospheric storage can extend the time by two more weeks. Bananas can be picked in the mature green stage if shipped under controlled atmospheric conditions. Ethylene can be used to initiate and stimulate ripening. Refrigerated storage of ripened bananas causes brown discoloration and off-flavor. Green bananas can hold for 7 to 28 days at 13.3° to 14.4°C (56° to 58°F) and 90% to 95% relative humidity.

Beechnuts

Beeches have been grown for their edible nuts, but the practice is extremely limited. They are utilized mostly for ornamental aspects. See the American and European beeches in Chapter 16 (Table 16–18) under *Fagus grandifolia* and *F. sylvatica*, respectively.

Blackberries

Blackberries are deciduous bramble fruits derived from various species of *Rubus*, although the exact lineage between some cultivars and species is not clear (Figure 18–13). Unlike raspberries, the fruit and receptacle do not separate, so both are eaten. Blackberries are eaten fresh and used in desserts, pies, juices, jams, and wine/brandy. Vitamin C levels are good. Blackberry cultivars can be broken down into three groups: Eastern erect and thorny, Eastern semi-erect and thorn free, and Pacific Coast trailing (mostly thorny) types, which are best suited to processing.

Erect and nearly erect types exist in the eastern United States; 'Arapaho' (thorn free, genetic exception), 'Apache' (thorn free, genetic exception), 'Cherokee,' 'Cheyenne,' 'Choctaw,' 'Darrow,' 'Ebony King,' 'Kiowa,' 'Navaho' (thorn free, genetic exception), and 'Shawnee' are typical cultivars. Semi-erect Eastern blackberry cultivars include 'Black Satin' (thorn free), 'Chester Thornless,' 'Dirksen Thornless,' 'Hull Thornless,' 'Loch Ness Thornless,' and 'Triple Crown Thornless.' Eastern blackberries can be grown on the Pacific Coast.

Trailing types exist on the Pacific Coast and have a life span of 15 years. Popular cultivars of this last group include 'Black Douglas Thornless,' 'Boysen' (boysenberry), 'Cascade Trailing,' 'Kotata,' 'Logan' (loganberry), 'Marion,' 'Ollala' (Ollalaberry), 'Thornless Evergreen,' 'Waldo,' and 'Young' (youngberry). 'Boysen' and 'Logan' are actually blackberry/raspberry hybrids. These Pacific Coast cultivars generally do not do well in the eastern United States because of cold hardiness limitations. Trailing form blackberries are often referred to as dewberries. Certain loganberry/raspberry hybrids are also called tayberries.

Blackberry culture is limited by cold winters in the North and the Plains States and by heat and drought in the Southwest. Sites should have sufficient moisture, as blackberries are more prone to wilting damage than other fruits because of higher rates of transpiration. Blackberries are grown both commercially and by homeowners. The primary commercial producer of blackberries is Oregon.

Blackberries can be propagated from suckers that arise from the roots, root cuttings, rooting of cane tips, and micropropagation. The latter is preferred for the trailing blackberries. One-year-old plants are placed out in the spring as early as possible. Erect types are

FIGURE 18–13 A trailing blackberry cultivar (A) and an erect cultivar (B) are shown here. Blackberries are adaptable to a wide range of environments. As such culture is easy and these fruits are more easily grown organically than many tree fruits. (USDA photographs.)

(A)   (B)

set about 2 to 3 feet apart for hedgerows and 2 to 4 feet apart for the linear system. In both systems the spacing between rows is 8 to 12 (12 is common) feet apart for tractor use and mechanical harvest. In the hedgerow system, suckers fill in spaces to provide a solid, continuous row in a few years. In the linear system, fruiting canes come from only the crown area. The linear system is more preferred for blackberries. Hills at 3 to 4 feet apart can be used for home gardens. Trailing and semi-erect types are set 4 to 6 feet apart in rows separated by 7 to 10 feet, depending on the vigor of the cultivar. Trellis post and wire systems are usually used and can be adapted to either hedgerow or linear systems.

Blackberries are tolerant of soil pH variation from 4.5 to 7.5, but the optimal pH is 6.0 to 6.5. Erect blackberries require 25 to 30 pounds of nitrogen annually per acre for establishment and 50 to 60 pounds for maintenance. Trailing blackberries need 30 to 35 and 65 to 75 pounds of nitrogen, respectively. Applications can be done once prior to bloom in the spring or split into pre-bloom and immediate postharvest applications. If the cropping system is dryland farming, amounts should be reduced by 10% to 25%. Potassium and phosphorus needs are determined by soil tests. Overfertilization and application of high phosphate fertilizer should be avoided, because too much phosphorus has been implicated in zinc deficiency of brambles. Potassium chloride should be avoided also, because brambles are chloride sensitive. A constant supply of water is needed from early spring through fruit harvest. The top 2 feet of soil need moisture, because that is the rooting depth of blackberries. Blackberries can use 0.20 to 0.33 inch of water each day. Trickle and soaker types of irrigation are best. Overhead irrigation does provide for the option of heat and frost protection.

If left unpruned, heights can vary from 3 to 10 feet, depending on cultivars. First-year canes, primocanes, do not bear fruit; second year, or floricanes, do. Primocanes usually appear while floricanes are fruiting. Trailing types are conveniently trained to two or more wire trellises about 6 feet high. Erect blackberries require summer topping of the tips of new shoots when they are 36 inches high. About 3 to 4 inches is removed, and tips should be pinched out afterward. This method produces a stout plant and good lateral development. If this is not done, erect types should be topped at 5 feet. Laterals can be cut back in late winter to early spring by about 50%. This practice ensures that an excessive part of the crop is not removed, but berries are not too small from overproduction. For trailing types, laterals are cut back to 12 to 18 inches. Floricanes are removed after fruiting. Some spring thinning of canes for all groups is usually desirable such that there is one strong cane every 6 to 8 inches. Rows should not become any wider than 24 inches, because picking becomes difficult. Canes are removed from trailers after harvest, and new canes are trellis trained in early spring before growth starts.

Some cultivars are self-fertile and others are not. Whether all self-fertile cultivars can pollinate themselves is not completely clear. It would appear that the growing of more than one cultivar and encouraging bees would be beneficial. Fruit thinning is unnecessary.

Disease problems include anthracnose (*Elsinoe veneta*), cane and leaf spot (*Septoria rubi*), crown gall (*Agrobacterium tumefaciens*), double blossom (*Cercosporella rubi*), downy mildew (*Peronospora sparsa*), gray mold (*Botrytis cinerea*), orange rust (*Gymnoconia nitens* and *Arthuriomyces peckiana*), purple blotch (*Septocyta ruborum*), spur blight (*Didymella applanata*), and *Verticillium* wilt (*V. dahalie*). Unlike raspberries, balckberries are not usually troubled by viruses.

Troublesome insects include aphids, blackberry gallmaker (*Diastrophus nebulosis*), blackberry leafminer (*Metallus rubi*), blackberry psyllid (*Trioza tripunctata*), cane borers (*Oberea bimaculata* and *Agrilus ruficollis*), cane maggot (*Pegomya rubirora*), consperse stink bug (*Euschistus consperus*), crown borer (*Pennisetia marginata*), Japanese and June beetles (*Popillia japonica* and *Phyllophaga* spp.), red berry mite (*Acalitus essigi*), rose chafer (*Macrodactylus subspinosus*), rose scale (*Aulacaspis*), snowy tree cricket (*Oecanthus fultoni*), and spider mites (*Tetranychus* spp.).

Blackberries are picked when black, slightly sweet, and firm for commercial use. Mechanical harvesters are available and used for large acreage destined for processing. For fresh, on-sight consumption or pick-your-own operations, the best berries are soft and separate easily from the plant. Ripe berries are highly perishable. Precooling and refrigeration are recommended. Storage life is only 2 to 3 days at −0.6° to 0°C (31° to 32°F) and 90% to

95% relative humidity. Yields with average to good care are 2,300 to 5,000 quarts per acre, or up to 15 pounds per plant. Average yields in the United States during 2000 were 3.7 tons/acre. In Oregon yields can be as high as 5 tons/acre.

Blueberries

Blueberries can be grouped into lowbush and highbush species (Figure 18–14). Most cultivars are derived from two highbush species, *Vaccinium corymbosum* and *V. ashei*. The former is more important commercially, and its cultivars include 'Atlantic,' 'Berkeley,' 'Bluechip,' 'Bluecrop,' 'Bluegold,' 'Bluejay,' 'Blueray,' 'Bluetta,' 'Bounty,' 'Briggita,' 'Chandler,' 'Collins,' 'Coville,' 'Croatan,' 'Darrow,' 'Dixi,' 'Duke,' 'Earliblue,' 'Elliott,' 'Herbert,' 'Ivanhoe,' 'Jersey,' 'Lateblue,' 'Legacy,' 'Meader,' 'Nelson,' 'Olympia,' 'Patriot,' 'Pemberton,' 'Rancocas,' 'Rubel,' 'Sierra,' 'Spartan,' 'Sunrise,' and 'Toro.' Cultivars of *V. corymbosum* reach heights of 6 to 12 feet.

Vaccinium ashei is important in the Southeast; it is more tolerant of soil type, heat, and drought, but quality is not as good as *V. corymbosum. Vaccinium ashei* cultivars reach heights of 4 to 8 feet. A few cultivars of *V. ashei* (rabbit-eye blueberry) are 'Bluebell,' 'Brightwell,' 'Centurion,' 'Climax,' 'Delite,' 'Powderblue,' 'Premier,' 'Sunshine Blue,' and 'Tifblue.' A few Southern highbush (*V. ashei* × *carymbosum*) cultivars (better adapted to the South) are 'Bladen,' 'Blue Ridge,' 'Misty,' 'O'Neal,' and 'Reveille.'

Lowbush species, although not cultivated, can be harvested from native stands of *V. angustifolium* and *V. myrtilloides*. Lowbush species vary in height from 1 to 2 feet. Lowbush species are found mostly in New England and maritime Canada. *Vaccinium angustifolium* is especially important in Maine as the basis of their commercial, wild-harvested blueberry industry. Highbush blueberries have been crossed with lowbush blueberries to produce half-high cultivars that are shorter, but have increased cold hardiness and good fruit. These cultivars are 'Northblue,' 'Northcountry,' 'Northland,' 'Northsky,' and 'St. Cloud.'

Vaccinium ovatum (evergreen blueberry) and *V. membranaceum* (mountain blueberry) are blueberries gathered in the wild in the Pacific Northwest. *Vaccinium membranaceum* grows to 4.5 feet and *V. ovatum*, 10 feet or more. Others wild-gathered include *V. deliciousum, V. ovalifolium, V. parvifolium,* and *V. uliginosum*. Wild-gathered Pacific Northwest blueberries are also referred to as huckleberries.

Highbush blueberries grow best in moist acidic soil at pH 4.0 to 5.2 (optimal 4.5 to 4.8), have a cold requirement similar to Elberta peach, cannot withstand temperatures below −28.9°C (−20°F), and require a growing season of at least 160 days. They can be found

FIGURE 18–14
Blueberries are ranked highly as an antioxidant source, contain resveratrol (a cancer inhibitor and cardiovascular health protector) and are as effective as cranberries in promoting urinary tract health. (USDA photographs.)

from Florida to Maine and west to Michigan. Blueberries are grown commercially and by the homeowner. The leading states for the commercial production of blueberries are Maine (wild-gathered), Michigan, New Jersey, and Oregon.

Blueberries are propagated by hardwood or softwood cuttings and rhizome division. Hardwood cuttings are preferred for nursery production of highbush cultivars, whereas rhizome division is used for lowbush types. Softwood cuttings are used for some cultivars that are difficult to root from hardwood cuttings. Some cultivars are starting to be propagated via tissue culture. Two-year-old highbush container plants can be set out in early spring or early to late fall. Plants are placed 2 to 5 feet apart in rows spaced at 8 to 12 feet. Plants start to bear well in their third year and reach full bearing in 6 to 8 years. Hedgerow systems are popular for use with mechanical harvesters. Some trellis systems are also used with mechanical harvesters.

Sulfur is usually needed to keep pH in the appropriate acidic range. Applications are often split between the spring and fall, because the total amounts needed can often impact soil biology negatively. Sulfur has both fungicidal and insecticidal activities. An alternative is the use of peat moss (also a soil acidifier) to provide organic matter and to reduce the need for sulfur. Good levels of organic matter are especially important in blueberry culture, given their symbiotic relation with mycorrhizae. These fungal partners increase the efficiency of water and nutrient uptake, thus reducing the levels of needed inputs.

Blueberry needs for phosphorus and potassium are determined with soil tests. Blueberries generally need more potassium than phosphorus. Generally, mulched blueberries need 100 to 120 pounds of nitrogen per acre. Unmulched plantings need half of that amount. Nitrogen applications are usually split for minimizing losses of nitrogen. One-third is given at bud break, and the next third and last third at 6-week intervals. Organic nitrogen can be applied 1 to 4 weeks earlier to allow time for conversion to available nitrogen. Later applications from mid-July promote lush growth that is sensitive to fall frost damage. Foliar analyses are helpful in monitoring organic applications of nitrogen. Levels of 1.6% or less indicate deficiency and over 2.3%, excess. Leaves from the mid-shoot area of fruiting branches are used for the tests. Fertigation using drip irrigation also works well for blueberries. Organic mulch is used over black plastic mulch, which seems to favor surface rooting and hence greater susceptibility to drought stress and winter damage.

Blueberries are shallow rooted, their roots being found mostly in the upper 18 inches of soil. Drought damage can occur easily. A constant water supply of 1.0 to 1.5 inches weekly is needed. Depending on where blueberries are grown, some supplemental irrigation will be required. The period of fruit expansion is especially critical in terms of water needs. Bud set for next year's crop also occurs during this period and low water can reduce next year's crop. Both overhead sprinklers and drip irrigation setups are used.

Blueberries are borne on wood of the previous season's growth. Pruning is necessary to maintain production of large-sized fruit. Pruning is possible from the time of leaf drop to shortly after blossoms appear. Pruning is generally done during dormancy. Bushy growth at the base of plants is first removed after the third growing season. Thin out bushy or twiggy wood and remove old stems (fourth year or older wood) that have lost vigor in subsequent years. Lower, drooping branches are removed. Many cultivars overbear, so some thinning of fruit buds by pruning is desirable. Exact pruning requirements vary among cultivars, so local sources should be consulted. Because of possible self-sterility and pollen scarcity, best fruit production is obtained with cross-pollination. The best practice is to interplant with compatible cultivars or to alternate different cultivars in blocks of rows (10) and to place bee sources at flowering time.

Insects that attack blueberries include the blueberry maggot (*Rhagoletis mendax*), cranberry fruit worm (*Acrobasis vaccinii*), cranberry weevil (*Anthonomus musculus*), cherry fruit worm (*Grapholitha packardi*), blueberry bud mite (*Acalitus vaccinii*), stem gall wasp (*Hemadas nubilipennis*), stem borer (*Oberea myops*), blueberry thrips (*Frankliniella vaccinii*), scale (several spp.), and leafhopper (*Scaphytopius magdalensis*). Birds can be serious pests as can nematodes, rabbits, and deer.

Troublesome diseases include anthracnose (*Colletotrichum gloeosporioides* and *C. acutatum*), blueberry mosaic (virus), blueberry scorch (virus), blueberry shock (virus), blueberry stunt (a *Phytoplasma* sp. disease spread by leafhopper, *Scaphytopius magdalensis*), botrytis tip blight (*Botrytis cinerea*), canker (*Fusicoccum* sp., *Godronia cassandre*), crown gall (*Agrobacteria tumefaciens*), double spot (*Dothichiza caroliniana*), mummy berry (*Monilinia vaccinii-corymbosi*), red ringspot (virus), root rot (*Phytophthora cinnamomi*), rust (*Pucciniastrum vaccinii*), shoestring disease (virus), stem blight (*Botrysphaeria dothidea*), stem canker (*Physalospora corticis*), phomopsis twig blight (*Phomopsis vaccinii*), and witches'-broom (*Pucciniastrum geoppertianum*). Disease resistance varies with the cultivar.

Blueberries should be picked ripe to near ripe. Blueberries are not as perishable as raspberries. Forced air-cooling of blueberries within 2 hours of harvesting results in about 40% less decay. Storage at −0.6° to 0°C (31° to 32°F) and 90% to 95% relative humidity for 10 to 18 days is possible. A mature plant can yield 6 to 8 pints or, under optimal conditions, 20 pints. With good management yields up to 6,000 pints per acre are possible. Average yields in the United States during 2000 were 2.2 tons/acre. Yields as high as 20 tons/acre from mature established blueberry plantings are reported in Oregon, with averages for such plantings being 7 to 8 tons/acre. Blueberries are eaten fresh, frozen, and used in pancakes, pies, cakes, muffins, sauces, and jellies.

Brazil Nuts

The Brazil nut (*Bertholletia excelsa*) is a tropical evergreen tree found primarily in the Amazon rain forest. Nuts are harvested from natural stands. It is seen occasionally under glass in the United States. Brazil nut culture will not be covered here.

Calamondin Orange (*see* Citrus)

Cashews

The cashew is produced on a tropical evergreen tree (*Anacardium occidentals*). It can be grown in the warmer parts of zone 10, but is seldom seen there. Cashews are produced commercially in India. Their culture will not be covered here.

Cherries

Most cultivated cherries (Figure 18–15) are derived from two species. Sweet cherries are from *Prunus avium* and sour cherries from *P. cerasus*. Duke cherries, hybrids between the preceding two, are *P.* × *effusus*. Lesser cherries include the dwarf or Western sand cherry, *P. besseyi,* and the Nanking cherry, *P. tomentosa*. The former can be used as a dwarfing stock for plum, prune, and peach. Sweet cherry cultivars are mostly red or black skinned; a few are yellow skinned with red blush and some are yellow skinned. Cultivars include 'Angela,' 'Attika,' 'Bing,' 'Black Gold,' 'Black Tartarian,' 'Compact Stella,' 'Early Burlat,' 'Emperor Francis,' 'Glacier,' 'Gold Sweet,' 'Hedelfinger,' 'Kristin' (very winter hardy), 'Lapins,' 'Napoleon,' 'Rainier,' 'Royal Ann,' 'Sam,' 'Schmidt,' 'Somerset,' 'Stella,' 'Sweetheart,' 'Ulster,' 'Van,' 'Venus,' 'Vogue,' 'Windsor,' and 'Yellow Spanish.' Sour cultivars include 'Balaton,' 'Black Beauty,' 'Danube,' 'Early Richmond,' 'English Morello,' 'Meteor,' 'Montmorency,' 'Surefire,' and 'Northstar.' Duke cultivars include 'Late Duke,' 'May Duke,' and 'Royal Duke.'

Growth regions vary for these cultivars. Sweet cherries are the most tender, being comparable to the peach. Sour cherries are hardier. Dormancy requirements are 1,100 to 1,300 hours below 7.2°C (45°F) and 1,200 hours, respectively, for the sweet and sour cultivars. Flowers appear early in the spring and are susceptible to injury from late frosts. The sweet cherry is particularly adapted to the Hudson Valley of New York, the Great Lakes shores, and the Pacific Coast. The sour cherry can be grown in these regions, plus farther north and south. Cherries are grown commercially and to a limited degree by homeowners. The leading states for commercial sweet cherry acreage are California, Washington, Oregon, and Michigan. Michigan is the leader for sour cherries.

Propagation of sweet and sour cultivars is by budding on seedling rootstocks of *P. avium* (Mazzard cherry) and *P. mahaleb* (Mahaleb cherry). Mazzard is the preferred stock, but where problems of extreme temperatures and drought exist, Mahaleb is better. However, Mahaleb is susceptible to *Phytophthora* root rot. 'Morello' stock is used where soil is wet or a semidwarfing effect is desired. The best dwarfing rootstocks for sweet cherry cultivars appear to be the Gisela series that offer size reductions as low as 40% of normal. The cherry cultivar 'Colt' is a semidwarfing rootstock.

Vigorous 1- or 2-year-old nursery trees can be planted in the fall in milder climates and in early spring planting for colder climates. Sweet cherries can be planted from 20 to 36 feet apart in rows set 20 to 36 feet apart. Sour and Duke cherries require 20 by 20 feet. Bearing may start after 5 to 6 years, with full bearing starting at 8 to 9 years for sour cherries, and starting at 8 to 9 years and becoming maximal at 14 years for sweet cherries. Sweet and sour cherries have maximal heights of 60 and 30 feet, respectively.

Pruning to train sweet or Duke cherries is minimal; a spreading habit with several large scaffold branches and height kept to 20 feet is desirable. Sour cherries are trained to either open-center or modified-leader systems; the latter reduces the risk of weak, rotting crotches. Subsequent pruning of both encourages spur development. Fruit thinning of sour cherries is unnecessary, and fruit thinning of sweet cherries by chemical means is not dependable.

Nitrogen needs should be based on foliar tissue analyses and other nutrients on soil tests. A 1-year-old cherry tree needs roughly 0.07 pound of nitrogen yearly. For each additional year, another 0.07 pound should be added until the total in the ninth year is 0.63 pound. Mature trees should be continued at this rate, which is equivalent to 80 pounds of nitrogen per acre.

Sweet cherries are self-unfruitful, except for 'Stella' and 'Sweetheart.' Most cultivars are cross-compatible, but some exceptions exist and sweet cultivar mixtures must be selected with care. For example, the following pairs will not pollinate each other: 'Van'/'Windsor' and 'Emperor Francis'/'Napoleon.' Sour cherries are mostly self-fruitful. Sour cherries can also serve as cross-pollinators, except for 'Meteor' and 'Montmorency,' since bloom overlap with sweet cherries is insufficient. Duke cherries require cross-pollination. Bees are utilized for all to ensure maximal fruit production.

Troublesome diseases include bacterial blight (*Pseudomonas syringae*), black knot (*Dibotryon morbosum*), brown rot (*Monilinia fructicola*), canker (*Cytospora* sp.), Coryneum blight or shot-hole (*Wilsonomyces carpophilus*), crown gall (*Agrobacterium tumefaciens*), crown rot/collar rot (*Phytophthora* spp.), leaf spot (*Blumeriella jaapii*), powdery mildew

(*Podosphaera oxyacanthae*), ring spot virus, rugose mosaic virus, sour cherry yellows virus, Western X (*Phytoplasma* sp.), and wilt (*Verticillium dahliae*). Many other viruses not listed here attack cherry; stock should be purchased as certified virus free.

Insect and other pests are American plum borer (*Euzophera semifuneralis*), black cherry aphid (*Myzas cerasi*), black scale (*Saissetia oleae*), branch and twig borer (*Polycaon confertus*), brown mite (*Bryobia rubrioculus*), cherry maggot (*Rhagoletis cingulata*), cherry slug (*Caliroa cerasi*), cribrate weevil (*Otiorhynchus cribricollis*), European earwig (*Forficula auricularia*), European fruit lecanium (*Parthenolecanium corni*), eyespotted bud moth (*Spilonota ocellana*), fruit tree leafroller (*Archips argyrospila*), green fruitworms (*Orthosia hibisci, Amphipyra pyramidoides, Xylomyges curialis*), leafhopper (*Colladonus montanus, Fieberiella florii*), various nematodes (*Meloidogyne incognita, M. javanica, Pratylenchus penetrans, P. vulnus, Xiphinema americanum*), plum curculio (*Conotrachelus nenuphar*), leafroller (*Archips argyrospilus*), orange tortrix (*Argyrotaenia citrana*), Pacific flatheaded borer (*Chrysobothris mali*), peachtree borer (*Synanthedon exitiosa*), peach twig borer (*Anarsia lineatella*), redhumped caterpillar (*Schizura concinna*), San Jose scale (*Quadraspidiotus perniciosus*), sawfly (*Hoplocampa cookei*), shot-hole borer (*Scolytus rugulosus*), spider mites (*Tetranychus pacificus, Tetranychus urticae*), and Western tussock moth (*Orgyia vetusta*).

In addition, birds are a serious problem, and fruit splitting can occur if heavy rains coincide with ripening.

Cherries can benefit from precooling. Refrigerated shipping is desirable for long-distance shipping. Storage at −1.1° to −0.6°C (30° to 31°F) and 90% to 95% relative humidity for 14 to 21 days is possible. Unripe cherries can be held in controlled atmosphere conditions for 25 days and ripened later. Sweet cherry production averages 3.5 tons per acre, although yields of 5 tons/acre have been reported in Washington and Oregon. Sour cherries yields average 7 tons per acre, but can be as high as 11 to 12. Yields per tree for both are 2 to 3 bushels. Sweet cherries are eaten fresh, in fruit salads, jellies, toppings, baked goods, fruit drinks, and cherry brandy/liqueurs. Sour cherries are used in pies and are canned and frozen.

Chestnuts

The chestnut (Figure 18–16) is found in the genus *Castanea*. There are several species: American chestnut (*C. dentata*), European chestnut (*C. sativa*), Japanese chestnut (*C. crenata*), and Chinese chestnut (*C. mollissima*). There are other species, but they are not important in terms of nut production. In this country the American and European chestnuts

FIGURE 18–16
Chestnuts store poorly and should be purchased fresh and used quickly. Chestnuts from California are likely the freshest as opposed to imports. Much chestnut information is available at: http://www.attra.org/attra-pub/PDF/chestnut.pdf

have essentially been eliminated by the chestnut blight [*Cryphonectria (Endothia) parasitica*]. The Japanese and Chinese chestnuts are somewhat resistant; their cultivars are mostly grown. Cultivars and hybrids of the Chinese chestnut include 'Abundance,' 'Clapper,' 'Crane,' 'Douglas,' 'Dunstan,' 'Eaton,' 'Gellatly #1,' 'Hemming,' 'Kuling,' 'Lyeroka,' 'Manchurian,' 'Meiling,' 'Myoka,' 'Orrin,' 'Ranking,' 'Skioka,' 'Skookum,' 'Sleeping Giant,' and 'Williamette.' Cultivars of the European chestnut include 'Belle Epine,' 'Borra,' 'Caste Del Rio,' 'Montesol,' 'Okei,' 'Schrader,' and 'Silverleaf.' 'Collosal' (European-Japanese hybrid) is grown in California commercially with 'Silverleaf' and 'Okei' as pollinators.

The Chinese chestnut is hardy through most of zone 5, where the temperature does not go below −26.1°C (−15°F). The Japanese chestnut is hardy through zone 6. Cultivars and hybrids of the Chinese chestnut are grown for nuts, since combined nut quality and blight resistance are better than in the Japanese chestnut. Commercial and backyard production of chestnuts is limited compared to the more popular nuts. Most commercial orchards are found in Maryland and Georgia. Flower buds can be damaged by late spring frosts.

Cultivars of Chinese chestnuts are usually propagated by the splice or whip graft. Stocks are usually seedlings of the Chinese chestnut. Rooting of cuttings is possible, but difficult. These cultivars may reach 50 feet in height. Chestnuts are planted in the spring in the North and in the fall or spring in the South. Transplanting is done with care to avoid injury to the taproot system. Spacing can be from 25 to 50 feet, but pruning to keep the tree small will be required at the lesser spacing. Hedgerow systems are spaced at 11 by 22 feet (180 trees/acre). Training is usually open center or central leader. During dry periods, trees need 5 to 8 gallons of water weekly. Grafted trees may bear nuts as early as the second year after being planted. Two or more cultivars should be planted because maximal production is dependent on cross-pollination. Mature height can be 100 to 120 feet.

Foliar tissue analyses supplemented with soil tests are critical for achieving sustainable fertilization with nut trees. Nitrogen needs vary from 40 to 100 pounds per acre, depending on soil levels, tree age, and cover crop considerations. Fertilizer is usually applied in two to four split applications, starting in early spring. Fertilizer should be watered in, unless it is possible to incorporate it into the soil. Fertigation is an especially good system for nut trees. A 1-year-old backyard chestnut tree can use 1 pound of 10-10-10 and 2 pounds in its second year. For each subsequent year of age, add an additional 1 pound up to a maximum of 15 pounds per tree.

Possible insect pests include the Asiatic oak weevil (*Cyrtepistomus castanets*), chestnut moth (*Synanthedon castaneae*), chestnut weevil (*Balanius rectus, Curculio auriger, C. elephas, C. proboscideus*), filbert worm (*Cydia latiferreana*), gall wasp (*Dryocosmus kuriphilus*), and shot-hole borer (*Scolytus rugulosus*). Diseases include crown and root rot (*Phytophthora* spp.) and oak root fungus (*Armillaria mellea*). The chestnut blight is a disease already discussed.

Yields are about 50 to 75 pounds per tree. Nuts are mechanically picked up during droppage periods of 10 to 15 days. Drop can be facilitated with mechanical shakers. Machinery to separate nuts from hulls exists, but is not perfected. Nuts are stored in ventilated cans or polyethylene bags at 0° to 2.2°C (32° to 36°F). Nuts are high in starch and low in protein and oils, unlike most other nuts. Nuts are roasted and used in stuffings, baked goods, Chinese cooking, and as candied nuts.

Citron (*see* Citrus)

Citrus

A number of evergreen citrus fruits are grown commercially and by the homeowner in the subtropic areas (Figure 18–17). Cultivars of the lime, *Citrus aurantifolia,* include 'Bearss' and 'Tahiti.' Seedling trees of the species are named for their point of origin, that is, Key, Mexican, or West Indian lime. These limes are smaller, but more intensely flavored and aromatic. Cultivars of the lemon, *C. limon,* include 'Eureka,' 'Lisbon,' and 'Villafranca,' which are used commercially. Lemon cultivars of value in the home garden or greenhouse include 'Meyer' and 'Ponderosa.' The grapefruit, *C.* × *paradisi* (*C. maxima* × *C. sinensis*), includes

FIGURE 18–17
Chlorophyll fluorescence analysis can be used on Valencia oranges to measure the tree's stress. (USDA photograph.)

white-fleshed cultivars ('Duncan' and 'Marsh') and pink/red-fleshed cultivars ('Flame,' 'Redblush,' 'Ruby,' 'Star Ruby,' and 'Thompson').

The cultivars of the sweet orange, *C. sinensis,* are divided into four horticultural groups: Mediterranean, Spanish, blood, and navel oranges. The sweet orange is the most important commercial citrus fruit. The main cultivars are 'Hamlin,' 'Pineapple,' 'Valencia' (Spanish), and 'Washington' (navel). Others are 'Homasassa,' 'Jaffa' (Mediterranean), 'Lue Gim Gong,' 'Moro,' 'Parson Brown' (good homeowner cultivar), 'Ruby' (blood orange), 'Sanguinelli,' and 'Tarocco.' The last three and 'Moro' are blood oranges (red flesh). The Mandarin oranges, including tangerines and satsumas (*C. reticulata*), are known for tangerine cultivars such as 'China,' 'Clementine,' 'Cleopatra' (also used as a rootstock), 'Dancy,' and 'Ponkan' (larger, low-acid type). 'Owari' is a popular satsuma, a cold hardy mandarin. Cultivars of the tangelo (cross between a grapefruit and mandarin orange), *C.* × *tangelo* (*C.* × *paradisi* × *C. reticulata*), are 'Minneola,' 'Orlando,' 'Sampson,' 'Seminole,' and 'Thornton.' The tangor, *C.* × *nobilis* (*C. reticulata* × *C. sinensis*) is known for its cultivars 'King' and especially 'Temple.'

The sour orange, *C. aurantium,* is not widely cultivated in the United States, but makes an excellent marmalade. Otaheite orange, much used as a tub plant, is a form of *C.* × *limonia* (*C. limon* × *C. reticulata*). The citron, *C. medica,* is grown for its rind, which is candied. The pomelo or shaddock (*C. maxima*) has minimal value except as an ornamental, but when crossed with the sweet orange produced the grapefruit. The calamondin orange, grown as a tub plant, is × *Citrofortunella mitis* (a hybrid genus between the mandarin orange and a kumquat; see kumquats later).

Because of their subtropical requirements, citrus fruits are generally restricted to the southern portion of zone 9 and all of zone 10. This includes Southern California, Florida, Rio Grande Valley in Texas, southwest Arizona, and southern parts of the Gulf states. Frost tolerance varies from almost none to limited as follows in order of approximately increasing hardiness: lime, lemon, grapefruit, sweet orange, tangerine, and calamondin orange.

Citrus cultivars are propagated by budding on to seedling stocks of the same or other species. Seedling stocks are 1 to 2 years old at budding time. Seedling stocks commonly used include sour orange (*C. aurantium*), sweet orange (*C. sinensis*), citrange (× *Citroncirus webberi* cv. 'Troyer,' a hybrid genus from *Citrus sinensis* × *Poncirus trifoliata*), trifoliate orange (*Poncirus trifoliata*), rough lemon (*C. limon* cv. 'Rough'), 'Volkamer' lemon, rangpur lime (*C.* × *limonia*), and Cleopatra tangerine (*C. reticulata* cv. 'Cleopatra'). Sweet orange and rough lemon are suitable for light, sandy soils in warmer areas, sour orange for good-quality soils, and trifoliate orange for heavier, moist soils in colder areas. Height for mature citrus trees varies. For example, the otaheite orange will reach 3 feet, the lemon 25 feet, the citron 16 feet, the sweet orange 40 feet, and the grapefruit 50 feet.

Citrus planting is usually done from March to May in California and in spring through early summer in Florida. Planting distances vary according to the type of citrus. Each of the following has a planting distance range, which depends on cultivar vigor and types of pruning: lime, 15 to 22 feet; lemon, 20 to 30 feet; grapefruit, 24 to 30 feet; tangerine, 20 to 22 feet; sweet orange, 20 to 25 feet; and tangelo, 24 to 30 feet. An all-purpose spacing of 20 to 25 feet between rows with an in-row spacing of 10 to 15 feet is suggested. Trees tend to form continuous hedgerows under these conditions. Tree densities vary from 116 trees per acre for 15- by 25-foot spacing to 218 for 10- by 20-foot spacing. Bearing ages vary from 5 to 10 years for good bearing with budded sweet orange cultivars and 4 to 7 years for lemon, lime, and grapefruit budded cultivars.

Nursery trees are pruned back at planting time to 18 to 24 inches. Pruning for the next year or so is mainly to remove sprouts arising below the bud union. Subsequent pruning is to select the four or five main branches that will form the main framework. Maintenance pruning is minimal, since the growth habit of citrus produces a shaped, symmetrical head. Pruning can be done to reduce size, but fruitfullness is reduced. Thinning of fruit is not necessary, unless the crop is too heavy for the tree, since the size of the remaining fruit is not greatly affected.

Nitrogen needs of citrus trees should be based on foliar tissue analyses and soil tests for other nutrients. Some general guidelines are that a 1-year-old tree requires 0.25 pound of nitrogen, and 0.50, 0.75, and 1.00 pound in years 2, 3, and 4, respectively. The 1-pound rate is continued from year 4 on.

Insects and mites causing the most trouble include aphids (*Aphis* spp.), broad mite (*Polyphagotarsonemus latus*), California orangedog (*Papilio zelicaon*), California red, yellow, purple, and black scale (*Aonidiella aurantii, A. citrina, Lepidosaphes beckii, Saissetia oleae,* respectively), brown soft and citricola scale (*Coccus hesperidum, C. pseudomagnoliarum*), citrus bud mite (*Eriophyes sheldoni*), citrus cutworm [*Egira (Xylomyges) curialis*], citrus flat mite (*Brevipalpus lewisi*), citrus leafminer (*Phyllocnistis citrella*), citrus thrips (*Scirtothrips citri*), cottony cushion scale (*Icerya purchasi*), flower thrips (*Frankliniella bispinosa* and *F. kelliae*), orange tortrix (*Argyrotaenia citrana*), red and rust mites (*Panonychus citri* and *Phyllocoptruta oleivora*), root weevil (*Diaprepes abbreviatus*), six-spotted mite (*Eotetranychus sexmaculatus*), two-spotted mite (*Tetranychus urticae*), Yuma spider mite (*Eotetranychus yumensis*), Western Tussock moth (*Orgyia vetusta*), and whitefly (*Aleurothrixus floccosus, Dialeurodes citri, Parabemisia myricae, Siphoninus phillyreae*). Ants are also troublesome in that they protect scales and mealybugs from introduced predators. A few damage trees by removing leaves (leaf cutting ant, *Atta texana*) and girdling twigs (fire ant).

Troublesome diseases include alternaria brown spot (*Alternaria alternata*), alternaria rot (*Alternaria citri*), anthracnose (*Glomerella cingulata*), armillaria root rot (*Armillaria mellea*), blue mold (*Penicillium italicum*), botrytis rot (*Botrytis cinerea*), brown rot (*Phytophthora citrophthora*), citrus blast (*Pseudomonas syringae*), citrus canker (*Xanthomonas axonopodis*), citrus scab (*Elsinoe fawcettii*), citrus stubborn disease (*Spiroplasma citri*), foot rot (*Phytophthora nicotianae*), exocortis and cachexia (viroids), greasy spot and greasy spot rind blotch (*Mycosphaerella citri*), green mold (*Penicillium digitatum*), melanose (*Diaporthe citri*), phytophthora gummosis (*Phytophthora* spp.), phytophthora root rot (*Phytophthora citrophthora* and *P. parasitica*), phomopsis stem-end rot (*Phomopsis citri*), post-bloom fruit drop (*Colletotrichum acutatum*), septoria spot (*Septoria* spp.), sour rot (*Geotrichum candidum*), stem-end rot (*Lasiodiplodia theobromae*), tristeza disease complex (*Tristeza* virus), and twig blight (*Diplodia natlensis*). Physiological disorders include chilling injury, oil spotting, rind staining, and stem-end rind breakdown. There are numerous other insects and diseases of lesser importance; authorities should be consulted for information on these.

Sweet oranges can be harvested over a long period, because the ripening process is not rapidly followed by deterioration. Oranges can be left on the tree when ripe. Refrigerated storage can keep oranges in good quality for several weeks to months. Lemons and limes can also be picked over a long period and have good storage lives. Grapefruits should not be harvested too long after ripening, and storage life is shorter than for other citrus. Tangerine harvesting should be prompt and storage life is short like that of grapefruit.

Oranges can be stored at 3.3° to 7.8°C (38° to 46°F) up to 90 days; an exception is Valencia oranges at 8.9°C (48°F). Mandarins and tangerines store best at 5.0° to 7.8°C (41° to 46°F) for 2 to 6 weeks, depending on cultivar. Exposure to ethylene is used to degreen oranges, mandarins, and tangerines. Limes are stored for 6 to 8 weeks at 10.0° to 12.8°C (50° to 55°F). Limes should not be exposed to ethylene, because their color becomes yellow. Lemons store well at 12.2° to 13.9°C (54° to 57°F) for up to 6 months. Grapefruits are stored at 12.2°C to 13.9°C (54° to 57°F) for 6 to 8 weeks. Lemons and grapefruits are usually not degreened with ethylene, because the exposure can result in accelerated degradation. All require the same relative humidity in storage, 90% to 95%.

Orange yields during the 2000–2001 season were 5.1 to 5.9 tons/acre in Arizona, 10.5 to 12.9 in California, 14 to 19 in Florida, and 9.1 to 10.6 in Texas. Grapefruit yields during the 2000–2001 season were 4.2 tons/acre in Arizona, 14.1 in California, 17.6 to 19 in Florida, and 14.4 in Texas. Lemon yields during the 2000–2001 season were 10.1 tons/acre in Arizona and 17.4 in California. Lime yields during the same season were 9.1 tons/acre in Florida. Tangelo yields during the 2000–2001 season were 8.7 tons/acre in Florida. Tangerine yields during the 2000–2001 season were 4.4 tons/acre in Arizona, 9.0 in California, and 10.5 in Florida. Temple yields during the 2000–2001 season were 10.2 tons/acre in Florida.

Coconuts

The coconut (Figure 18–18) is the fruit of the tropical palm, *Cocos nucifera*. Few cultivars exist in true form because of cross-pollination and the fact that coconuts are only propagated from seeds. The only maintained cultivars are the popular 'Golden Dwarf Malay' and the less commonly grown 'Green Dwarf Malay' and the 'Orange Dwarf Malay.' Other cultivars include 'Fiji Dwarf' (Niu Leka), 'Jamaica Tall,' 'Malayan Dwarf,' 'Maypan,' and 'Panama Tall.'

Coconuts are grown in the tropics and to some extent in southern Florida, southern California, and Hawaii. Most are grown for ornamental use and the nuts are of secondary value. Commercial operations are essentially nonexistent in the continental United States.

Propagation is from seeds allowed to mature on the tree. The whole fruit is buried about two-thirds in a seedbed. Germination requires 4 to 5 months. One- or two-year-old trees are planted about 18 to 30 feet apart. Coconut palms produce fruits 6 to 10 years from germination and achieve full production in 15 to 20 years. Trees produce for about 80 years. Production is 50 to 200 fruits per tree, varying with cultivar and climate. Fruits mature in 12 months. Trees are not pruned. Both self- and cross-pollination occur. Bees are the main pollinators.

FIGURE 18–18 A Manila dwarf coconut palm is shown here. (USDA photograph.)

Special fertilizers for palms are used and often contain magnesium to correct a common deficiency. A typical one is 8-4-12-4 (N, P, K, and Mg). The N, K, and Mg should be formulated as controlled-release ingredients, because soils in palm areas are prone to leaching. Trace amounts of iron, managanese, zinc, copper, and boron are also helpful. Applications at rates of 1.5 lbs of fertilizer per 100 sq. ft. of canopy area every 3 months or 1 lb/100 sq. ft. every 2 months are best.

Troublesome diseases include bud rot (*Phytophthora palmivora*), lethal yellowing (*Phytoplasma sp.*), and leaf scorch (*Ceratocystis paradoxa*). Aphids, scales, mealybugs, leaf skeletonizer (*Homaledra sabalella*), and palm weevils are insect pests. A good tree may yield 75 coconuts per year, but 20 to 30 is average.

Cranberries

Cranberry cultivars (Figure 18–19) are derived from *Vaccinium macrocarpon*. Cultivars include 'Beaver,' 'Beckwith,' 'Ben Lear,' 'Bergman,' 'Black Veil,' 'Crowley,' 'Early Black,' 'Franklin,' 'Howes,' 'McFarlin,' 'Pilgrim,' 'Searles Jumbo,' 'Stevens,' and 'Wilcox.' 'Stevens' is the leading commercial cultivar followed by 'Early Black' and 'Howes.' A related species, *V. vitis-idaea minus,* produces red cranberry-like berries called lingonberries and includes cultivars called 'Red Pearl,' 'Sanna,' and 'Sussi.'

Requirements for cranberry culture are such that their cultivation is largely restricted to commercial operations in a limited number of areas. An acid bog or swamp (pH 3.2 to 4.5) with nearby water for irrigation and prevention of frost is needed. Flooding is needed in winter (except on the West Coast) to prevent winter desiccation, because the cranberry is shallow rooted. Since bogs or swamps are low, late spring frosts are a threat, which used to be reduced by flooding, but now mostly by sprinkler systems. Flooding is also used in insect control. Leading cranberry areas are found in Massachusetts, New Jersey, Wisconsin, Washington, and Oregon.

Cranberries are propagated by cuttings. The mature plant is under 1 foot in height. Cranberries are planted in spring, and a 12-inch spacing for cuttings is typical. Three years are required before the first harvest. Pruning and fruit thinning are not practiced. Excessive runners can be trimmed after harvest and are generally the source for cuttings because cranberry cultivars are not sold through nurseries. Two exceptions are 'Shaw's Success' and 'Hamilton,' which can be found in nurseries. Bees are placed during flowering to ensure adequate pollination.

FIGURE 18–19
Cranberries. Only about 10% of the cranberry crop in the United States is sold in the fresh fruit form. (USDA photograph.)

Cranberry fertility needs are minimal relative to other fruits. This situation arises from the evolution of cranberries from bog plants. Bogs are naturally nutrient poor. Potassium is hardly ever needed. If soil tests indicate phosphorus needs, the best form is colloidal phosphae, because chemical phosphates are generally locked up by iron or aluminum in acid bogs. A new cranberry plantation, if treated with 500 to 1,500 pounds of colloidal phosphate per acre, is likely not to require phosphate for its life span. New growth is a good indicator of nitrogen needs. If new growth is 4 inches or more, sufficient nitrogen is present. Less growth indicates the need for nitrogen. Compost applications of 5 to 10 tons per acre can be applied when bogs are sanded and will supply enough nitrogen for several years. If chemical nitrogen is applied, it should be in an ammonium form and at a rate of 10 pounds per acre. Actual needs are best based on tissue analyses supplemented with soil tests.

Diseases associated with cranberries include leaf spot (*Glomerella cingulata*), early rot (*Guignardia vaccinii*), and false blossom virus. Troublesome insects include cranberry fruit worm (*Acrobasis vaccinii*), yellow-headed fireworm (*Acteris minuta*), blunt-nosed leafhopper (*Scleroracus vaccinii*, carrier of false blossom), rootworm, (*Rhabdopterus picipes*), gypsy moth, and girdler (*Chrysoteuchia topiaria*).

An acre of cranberries can produce 100 to 200 barrels. Each barrel weighs about 100 pounds. Average yields of cranberries in the United States during 2000 were 7.1 tons/acre. Cranberries can be held at 1.7°C (35°F). Berries are good sources of vitamins A and C. Cranberry juice also has medical value for reducing urinary track infections.

Currants and Gooseberries

Red, pink, and white currants (Figure 18–20) can be *Ribes sativum, R. petraeum,* and *R. rubrum,* with the first being more important. Black currants are *R. nigrum.* Red currant cultivars include 'Cascade,' 'Cherry,' 'Jonkheer van Tets,' 'Fay Perfection,' 'Red Lake,' 'Rovada,' and 'Wilder.' White currants include 'Primus,' 'Weisse aus Juterberg,' 'White Imperial,' and 'White Versailles.' Black currant cultivars include 'Ben Sarek,' 'Ben Lomond,' 'Blacksmith,' 'Boskoop Giant,' 'Crandall Black,' 'Hilltop Baldwin,' 'Kiev Select,' 'Minaj Smyriou,' 'Noir de Bourgogne,' 'Prince Consort,' 'Wellington XXX,' and 'Willoughby.'

(A)   (B)

FIGURE 18–20   Red currant (A) and white currant ('White Imperial')(B). The cultivar, 'New York 68,' from the New York Agricultural Experiment Station (Geneva) is the most resistant of any red currant to gray mold, mildew, or leaf spot. (USDA photographs.)

The American gooseberry is *R. hirtellum* and the European gooseberry is *R. uva-crispa* (*R. grossularia*). American gooseberry cultivars are 'Colossal,' 'Downing,' 'Glenndale,' 'Oregon Champion,' 'Pink,' 'Pixwell,' 'Poorman,' and 'Welcome.' European gooseberry cultivars include 'Achilles,' 'Black Velvet,' 'Careless,' 'Early Sulfur,' 'Hinnonmaki Red,' 'Hinnonmaki Yellow,' 'Invicta,' 'Leepared,' 'Telegraph,' 'Whinham's Industry,' and 'Whitesmith.' Gooseberries can be red, pink, purple, black, yellow, green, and gray-green. Jostaberries are a cross between black currant and gooseberry. 'Captivator' is a hybrid between the American and European gooseberry.

Currants and gooseberries are deciduous shrubs that reach heights of 5 and 3 feet, respectively. Currants and gooseberries make good jams, jellies, sauces, and pies. Currants can be made into wine; the liqueur cassis is made from black currants. Fruits can be eaten fresh, stewed with cream added, or dried. Vitamin C levels are good. *Ribes* sp. are considerably cold hardy and do poorly in warm or dry areas. Their culture is best in the northern or higher altitudes in the United States. The gooseberry is less heat sensitive than the currant. Both are grown commercially (New York and Great Lakes area) and by the homeowner, but their acreage is very limited. *Ribes* spp. can be alternate hosts to white pine blister rust, are legally banned in some areas, and cannot be shipped into some 16 states.

Currants and gooseberries are usually propagated from seed or hardwood cuttings from 1-year-old wood. Difficult gooseberry cultivars can be propagated by mound layering. Bearing from seeds and cuttings is 3 to 5 and 2 years, respectively. Mature heights are 5 and 3 feet, respectively, for currants and gooseberries. Spring planting is practiced in the colder climates and early spring or fall planting in others. Fall planting is desirable because of early blooming. Shrubs are planted in hills set 5 to 6 feet apart in both directions or 4 to 5 feet apart in rows set at 6 to 10 feet. Nurseries usually sell 1- to 2-year-old plants. Both currants and gooseberries can tolerate light shade in the afternoon. Soil pH can be 6.0 to 7.5. Fertilization needs are minimal; split applications of a balanced fertilization are fine. During dry weather, irrigation is needed, especially during fruiting.

Pruning is used to develop a bush form. Most currant and gooseberry cultivars are self-fertile and pollination can be by wind, honeybees, and other insects such as hover flies. Currants produce the best fruit on spurs of 2- and 3-year-old canes, and gooseberries on 1- and 2-year-old canes. Older wood, not being productive, is removed. Thinning of fruit is not practiced.

Insect pests include aphid (*Cryptomyzus ribis*), currant fruit fly (*Epochra canadensis*), currant spanworm (*Itame ribearia*), gall mite, sawworm, spider mite, and San Jose scale. Diseases include powdery mildew (several spp.), anthracnose (*Pseudopeziza ribis*), cane blight (*Botryosphaeria ribis*), and white pine blister rust (*Cronartium ribicola*). Black currants can also get reversion virus from gall mites, but this disease has been seen only in Europe. *Ribes oxycanthoides* cv. Jahns Prairie (Canadian gooseberry cultivar) is resistant to white pine blister rust.

Yields of red currants are around 3 tons per acre and of gooseberries, 300 bushels per acre, or about 3 to 10 and 8 to 10 pounds per bush for each, respectively. Currants can only be stored for a short time under cool, dry conditions. Gooseberries last longer. Both should be utilized reasonably soon after harvesting.

Dates

The subtropical fruit, the date (Figure 18–21), is derived from the dioecious palm *Phoenix dactylifera*. Cultivars are grouped according to fruit texture: soft, semidry, and dry. Soft cultivars grown in the United States include 'Barhee,' 'Halawy,' 'Khadrawy,' 'Khalasa,' 'Medjool,' and 'Saidy.' 'Dayri,' 'Deglet Noor' (most important in California), and Zahidi are semidry cultivars. Dry cultivars are not popular here, but one cultivar is 'Thoory.' Semidry dates are most popular commercially because they handle better after harvest than do soft cultivars. The latter have higher quality and are found in local markets and backyards. Dates are high in sugar and provide potassium and fiber. Dates are eaten fresh. Both fresh and dry dates

FIGURE 18–21 Dates are an ancient fruit crop, dating back to the ancient civilizations found in the Middle East at the Tigris and Euphrates region. (USDA photograph.)

can be used in baked goods and as additions to fruit salads and yogurt. Fresh dates are highly perishable and keep well only in the dry form.

Dates require high temperatures during growing and ripening and minimal rain or humidity during this time. Dates can stand light frosts. This climate is found in the low desert valleys of Southern California (main commercial source), Arizona, Nevada, Texas, and Utah. Other areas of the United States are not suitable for date culture. Dates can tolerate some dryness, but irrigation is needed during growth and fruiting and must be deep when done because dates are deep rooted. Soil needs to be moderately fertile only. High fertility reduces fruiting. An annual application of small amounts of a balanced fertilizer is sufficient. Mature height can vary from 50 to 100 feet.

Dates can be propagated by seeds, but offshoots are utilized for cultivars. Seeds are used to produce male pollinators. Offshoots usually only arise during the first 5 to 10 years of a tree's development. Offshoots need to be in contact with soil to promote rooting at least 1 year prior to removal. Offshoots are removed with a large, broad chisel when 3 to 4 years old in the spring. Lower leaves are removed and offshoots are planted about 30 feet apart in all directions (50/acre). One male is needed for every 20 to 30 females. Dates can tolerate alkalinity and saline soils, but expect some yield reduction. A palm fertilizer (see under coconuts) is applied four times yearly at rates of 5 to 8 pounds for mature palms and 2 to 5 for young date palms. If manure is used, a rate of about 40 pounds per mature palm is suitable.

Pruning is minimal; dead leaves and their spines are removed. Pests are minimal, but bird protection is needed. Ants, beetles, and nematodes can be problems. Diseases are currently minimal.

Pollination is required and usually done by hand, although a machine pollination device can be used. The wasp that pollinates dates in Northern Africa has not been successfully established in the United States. Pollen is gathered from the male cultivar and shaken or blown over the female inflorescence. Another method is to remove some of the upper female flowers and to tie in the male inflorescence and shake vigorously. The first crop comes in 5 to 6 years and full bearing at 10 to 15 years. Trees can bear for up to 50 years. Some fruit thinning is necessary and involves hand labor and ladders. Trees with too many bunches have some removed and some bunch thinning is done.

Dates can be picked individually as they ripen or the whole bunch can be picked when one-third of the fruit is succulent and translucent. Yields per tree average 40 to 60 pounds. Commercial yields per tree can be as high as 300 pounds. Date yields for the United States in 2000 were 3.2 tons/acre. California yields are 4 to 5 tons/acre.

Elderberries

Elderberries are fruits of the common American elder, a deciduous shrub. The species is *Sambucus canadensis,* and a few cultivars are known: 'Adams No. 1,' 'Johns,' 'Nova,' and 'York.' The species *S. caerulea,* blueberry elder, also bears edible fruit, but no recognized cultivars exist. European black elder (*S. nigra*) also bears edible berries; 'Guincho Purple' is a cultivar.

The American elder grows throughout the eastern half of the United States. Moist soil is best. Clumps spread through stolons. Both backyard and commercial plantings exist, and native stands are also used for fruit production. Its status is that of a minor fruit.

Propagation is by seeds, but preferably by division or cuttings for recognized cultivars. Plants may be planted in spring or fall. Spacing should be at least 12 feet because of their spreading nature. Mature height is 12 feet. Fruit may be produced in 3 years. Pruning is minimal except to remove spent or dead wood. Cross-pollination is needed, so two different cultivars should be planted. No fruit thinning is needed. Insect and disease problems are minimal. Berries are picked ripe and consumed, or preserved quickly by canning or drying, or made into wine.

Feijoas

The feijoa or pineapple guava is derived from *Acca (Feijoa) sellowiana.* It is an evergreen shrub that reaches a height of 15 feet. Cultivars include 'Apollo,' 'Choiceana,' 'Coolidge,' 'Edenvale Improved Coolidge,' 'Edenvale Late,' 'Edenvale Supreme,' 'Gemini,' 'Mammoth,' 'Moore,' 'Nazemetz,' 'Pineapple Gem,' 'Trask,' and 'Triumph.'

Its subtropical requirements limit the feijoa to the mild winter areas of the United States. It is grown in Southern California and Florida and to a lesser extent in the southern parts of the Gulf States. It can withstand light frosts, some shade, and slight exposure to salt spray. Commercial and backyard culture is limited, so it is a minor fruit. Fruits are eaten fresh and made into jellies, sauces, chutney, and pies.

Propagation is by young wood tip cuttings, layering (air and soil), and grafting (whip, tongue, or veneer). Seeds are also used, but the former methods are used for cultivars. Spring or fall planting is possible. Soil pH should be 5.5 to 7.0. Plants should be 15 to 18 feet apart. Bearing starts in 3 to 4 years. While drought can be tolerated, fruits drop when soil is too dry. Fertilization needs are not high. A light use of a balanced fertilizer every 2 months is sufficient.

Minimal pruning is needed to thin out after harvest. Plants adapt well to hedge use, but fruit yield is lower. Fruit thinning is unnecessary. Cross-pollination is desirable, so two or more cultivars should be planted. Birds eating the fleshy petals help pollinate the flowers, but bees are the primary pollinators. Nine-year-old plants produce 10,000 pounds of fruit per acre. Insect and disease problems are minimal. Black scale and fruit flies are occasional pests. Ripe fruits fall easily from the shrub. Unripe fruits can be ripened at room temperature with some loss of quality. Keeping qualities are poor; ripe fruits can be held refrigerated for 5 to 7 days.

Figs

Fig form varies from that of a deciduous shrub to a tree with a height from 10 to 50 feet. Shrub shapes are commonly seen in areas where frost damage occurs. Cultivars are derived from *Ficus carica* (Figure 18–22). Figs vary in their pollination requirements, and their qualities determine whether they are consumed fresh, dried, or canned or used in jams. Cultivars include 'Adriatic,' 'Black Mission,' 'Blanche,' 'Brown Turkey,' 'Brunswick,' 'Calimyrna,' 'Celeste,' 'Conadria,' 'Croisic,' 'Desert King,' 'Dottato,' 'Excel,' 'Flanders,' 'Genoa,' 'Judy,' 'Kadota,' 'Lattarula,' 'Len,' 'Magnolia,' 'Mission,' 'Neveralla,' 'Osborn's Prolific,' 'Panachee,' 'Peter's Honey,' 'Petite Negri,' 'Ronde Noire,' 'Tena,' 'Ventura,' 'Verte,' and 'Violette de Bordeaux.'

FIGURE 18–22 Figs are another ancient fruit crop dating back to ancient Greece and Rome and even earlier. References to figs in ancient Sumaria date back to 2900 B.C. (USDA photograph.)

The frost tolerance of the fig is higher than that of most subtropical fruits. Dormant trees are hardy down to −11.1° to −9.4°C (12° to 15°F). Many cultivars can be grown as far north as zone 8 and a hardier few in zone 5 with proper winter protection or movement indoors during the winter. Figs are grown both commercially, mostly in California, and by the homeowner. While growing, figs can be frost damaged at −1.1°C (30°F). The cold dormancy period needed is 200 hours below 7.2°C (45°F), and the growing season should be at least 190 days. Unless the temperatures in the winter are cold enough to produce dormancy, temperatures of wood kill can be considerably higher than the normal −15°C (5°F).

Propagation is by dormant hardwood cuttings from 4 to 12 inches long. These are taken in the winter or very early spring. Both conventional and air layering are also possible, as is whip, crown, or cleft grafting and chip or patch budding. Rooted cuttings are planted in early spring 18 to 25 feet apart. Bearing can start in 2 to 4 years from rooted cuttings, but commercial bearing starts at 7 years.

Figs need regular watering during establishment. Once mature, some drying is possible between watering, but too much dryness causes leaves to yellow and fall off. A deep watering once every 1 to 2 weeks is usually enough. Fertilization needs are low; too much nitrogen causes leaf growth over fruiting. About 0.5 to 1 pound of nitrogen split over three to four applications is best.

Pruning is minimal to extensive, depending on the cultivar. For example, 'Adriatic' is pruned to remove lower spreading branches and to thin out interior branches when they become too thick. Some top pruning is done annually or biennially to induce new vigorous wood. 'Mission' requires only an occasional thinning of branches. 'Kadota' is trained low and flat to facilitate harvesting. 'Calimyrna' requires cutting back of long upright branches, occasional thinning, and heading back of the top.

Pollination requirements vary. Most of the previously listed cultivars can produce fruit by parthenocarpic means, because the flowers are female and require no pollination. Caprifigs are pollinated by the female fig wasp (*Blastophaga grossorum*). When she hatches and emerges from the caprifig, where she has overwintered, she picks up pollen and, in looking for sites to lay eggs, pollinates nearby flowers. Caprifigs are of poor eating quality, but are grown to ensure a supply of wasps. The caprifigs are removed prior to wasp emergence and placed in trees of cultivars requiring pollination (caprification), but which are unsuitable hosts for wasp eggs. The main cultivar requiring this treatment is 'Calimyrna.' The process is also used to some extent with 'Kadota.' Fruit thinning is unnecessary. Figs bear two crops. The first is borne on last season's growth in the spring. This crop is light, often damaged by spring frosts, and is called the breba crop. The main crop is produced on current wood during the fall.

Insects causing problems include the root knot nematode, several fig scales, and several fig mites. Certain beetles (*Mitadulid* and *Carpophilus*) can enter the ripening fruit and cause introduced fungal damage. Troublesome diseases include anthracnose (*Colletotrichum gloeosporioides*), black smut, botrytis, dieback (several spp.), fig canker, internal rot, leaf spot (several spp.), rust (*Cerotelium fici*), and souring. Souring, black smut, and internal rot result from several fungi introduced by the fig wasp. Mosaic virus can be troublesome. Gophers damage roots and birds eat the figs.

'Calimyrna' yields 1.25 to 1.50 tons of dry figs per acre, 'Mission' and 'Adriatic' 2.0 to 2.5 tons per acre, and 'Kadota' up to 5 to 7 tons of fresh figs per acre. Average fig yields in the United States during 2000 were 3.7 tons/acre. Figs for fresh consumption are picked ripe (won't ripen if picked unripe) and held for only 7 to 10 days at 0°C (32°F) and 85% to 90% relative humidity. Dried figs can be held for 6 to 8 months.

Filberts (Hazelnuts)

Filberts (Figure 18–23) are obtained from species of *Corylus*. The European filbert, *C. avellana*, and the giant filbert, *C. maxima*, plus a number of cultivars derived from them, are the main source of nuts. The American hazelnut, *C. americana*, and the beaked hazelnut, *C. cornuta*, are used mainly in breeding programs with the other species to provide increased hardiness and resistance to Eastern filbert blight caused by the fungus *Cryptosporella anomala*. The main commercial cultivar is 'Barcelona,' and 'Daviana' is the pollenizer. Other cultivars include 'Bixby,' 'Buchanan,' 'Butler' (excellent pollinator), 'Clark,' 'Cosford,' 'DuChilly,' 'Halls Giant' (pollinator), 'Italian Red,' 'Lewis,' 'Medium Long,' 'Potomac,' 'Red Lambert,' 'Reed,' 'Royal,' 'Rush,' 'Tonde di Gifone' (pollinator), and 'Winkler.' 'Finger Lakes Filbert Super Hardy' is a very cold hardy hybrid cultivar.

Chilling requirements of filberts are similar to apples. They are about as hardy as peaches, but their pistils are frequently killed by spring frosts because the pistillate flowers open early. This limits commercial production to areas near large bodies of water, mainly Oregon. Filberts are grown by homeowners in other areas, but cultivars should be chosen carefully by the backyard grower because of blight and hardiness problems.

Species of filberts can be raised easily from seeds. Cultivars are propagated mainly by tip layering, since budding is not satisfactory. Two-year-old plants are placed in the fall or spring about 15 feet apart. In the Northwest, filberts are trained to a single trunk with scaffold branches. Basal suckers are removed. In the Northeast, filberts are allowed to sucker and are grown as shrubs. Rejuvenation pruning is minimal, but does increase yields. Cultivars vary in height from 6 to 30 feet.

Trees are monoecious and unfruitful, so a pollinating cultivar is interplanted among the main cultivar(s). Filberts may start bearing in 3 years and reach maximal production between 15 and 25 years of age. Commercial yields are from 2,000 to 3,000 pounds per

FIGURE 18–23
Filberts. Filberts, also called hazelnuts, are used to flavor hazelnut coffee and hazelnut liqueurs. (USDA photograph.)

acre. In the Northeast one tree will produce about 5 to 10 pounds of nuts per year. The average yield in 2000 was 0.85 tons/acre and in 1999, 1.37 tons/acre. These figures are typical of the biennial fluctuations in nut yields.

Troublesome insects include filbert aphid (*Myzocallis coryli*), filbertworm (*Melissopus latiferreanus*), and filbert bud mite (*Phytoptus avellanae*). Diseases are mainly a fungal filbert blight (*Apioporthe anomala*) in the East and a bacterial filbert blight (*Xanthomonas corylina*) in the Northwest.

Gooseberry (*see* Currant)

Grapefruits (*see* Citrus)

Grapes

Grapes are used as fresh (table grapes) and dried fruits (raisins). Many products are processed from grapes: grape juice (fresh and concentrated), grape jam and jelly, wines (table wine, champagne, other sparkling wines,), fortified wines (port, sherry, and other aperitifs), and liquors distilled from wine (armagnac, brandy, cognac, *etc.*). California is the leading commercial grape producer. Grapes are widely grown by homeowners. Grapes and their products contain antioxidants that appear effective at reducing heart disease problems.

Grapes (Figure 18–24), a deciduous vine crop, belong to the genus *Vitis*. Four cultural classes are grown in the United States: American, Old World, French hybrids, and muscadine. American cultivars are derived from northern and northeastern United States species. *Vitis labrusca* is the most important parent species, but others such as *V. rupestris*, *V. aestivalis*, and *V. riparia* are involved in the parentage. The latter, the frost grape, is responsible for improved cold hardiness and shorter times to maturity for some American cultivars. The Old World cultivars are mostly from the European grape *V. vinifera*. French hybrids are crosses between *V. vinifera* and native American species, usually other than *V. lambrusca*. Cultivars of the muscadine grape are derived mostly from the native southeastern U.S. species, *V. rotundifolia*.

Red, blue, or red-purple grape cultivars of the American type used as table, jam and jelly, juice, and wine (fox grape or Concord types) grapes include 'Alden,' 'Bath,' 'Bluebell,' 'Buffalo,' 'Canadice,' 'Catawba,' 'Campbell's Early,' 'Concord,' 'Concord Seedless,' 'Delaware,' 'Erie,' 'Fredonia,' 'Reliance,' 'Steuben,' 'Swenson Red,' 'Valiant,' and 'Venus' (seedless). 'Concord' is the leading cultivar for juice, jelly, and wines. White American cultivars include 'Edelweiss,'

FIGURE 18–24
Grapes. A recent study showed that purple grape juice not only appeared to promote important biological functions like proper blood clotting, but it also appeared to raise antioxidant levels in the body. Subjects drank two cups of purple grape juice daily for 2 weeks. Afterwards, plasma levels of vitamin E rose 13% and total antioxidant capacity increased by 50%. The production of superoxide, a possibly harmful free radical, was also reduced by a third. (USDA photograph.)

'Himrod,' 'Interlaken' (seedless), 'Kay Gray,' 'Niagara,' and 'Ontario.' Old World grape cultivars include those grown as red to pink or white/green table grapes: 'Cardinal,' 'Einset Seedless,' 'Emperor,' 'Flame Tokay,' 'Perlette,' 'Ribier,' 'Thompson Seedless,' and 'Vanessa'
(seedless). Those Old World grape cultivars grown for red wine include 'Cabernet Franc,'
'Cabernet Sauvignon,' 'Carignane,' 'Grenache,' 'Merlot,' 'Pinot Noir,' and 'Zinfandel.' White
cultivars include 'Chardonnay,' 'Chenin Blanc,' 'French Colombard,' 'Gewürtztraminer,' 'Mus-
cat of Alexandria,' 'Pinot Blanc,' 'Sauvignon Blanc,' 'Sémillion,' 'Sylvaner,' and 'White Riesling.'
French hybrids, mainly red and white wine grapes, include 'Aurore,' 'Baco Noir,' 'Cascade,'
'Chancellor,' 'Chelois,' 'Colobel,' 'De Chaunac,' 'Leon Millot,' 'LaCroose,' 'Marechal Foch,'
'Rosette,' 'Rougeon,' 'Seyval Blanc,' 'Ventura,' 'Verdelet,' 'Vidal Blanc,' 'Vignoles,' and 'Vincent.'
Female (pistillate) muscadine cultivars include 'Africa Queen,' 'Black Beauty,' 'Black Fry,' 'Dar-
lene,' 'Early Fry,' 'Farrer,' 'Fry,' 'Higgins,' 'Jumbo,' 'Loomis,' 'Pam,' 'Rosa,' 'Scarlett,' 'Scupper-
nong,' 'Sugar Pop,' 'Sugargate,' 'Summit,' 'Supreme,' 'Sweet Jenny,' and 'Watergate.' Self-fertile
muscadine cultivars include 'Alachua,' 'Carlos,' 'Cowart,' 'Delight,' 'Dixieland,' 'Dixie Red,'
'Fry Seedless,' 'Ison,' 'Janet,' 'Janebell,' 'Late Fry,' 'Magnolia,' 'Nesbitt,' 'Noble,' 'Pineapple,'
'Polyanna,' 'Redgate,' 'Regale,' 'Southland,' 'Sterling,' 'Tara,' and 'Triumph.' These lists are not
complete, but cover most cultivars of importance in the United States.

The cultural requirements of the four groups define their areas of importance within
the United States. American grapes mature in about 150 to 165 days with adequate summer heat. Largest acreage is centered on the Great Lakes. American grapes can be grown
east of the Rocky Mountains and north of the Cotton Belt. Muscadine grapes are suited to
the hot South and are more adapted to humid, hot summers. Temperatures should not go
below −17.8°C (0°F). Old World grapes are less hardy, and early ripening cultivars need at
least 175 days to maturity. Even better is 180 to 200 days. Summers should be cool to hot,
but have low humidity. Old World cultivars are grown mainly in California and in Washington and Oregon to a lesser extent. French hybrids have similar requirements to Old
World grapes, but are a bit more winter hardy and can be grown in areas where American
cultivars grow. Slopes and bodies of water are important in grape culture because of reduction of late spring and early autumn frosts.

Grapes are propagated from seeds, cuttings, layering, grafting, and budding. Layering
is used for grapes that are difficult to root such as the muscadine grape. Leafy cuttings under mist also work for muscadine grapes. Dormant hardwood cuttings are widely used.
These cuttings consist of 12-inch cane prunings with three buds. After removal in the winter, they are callused through storage in moist sand or sawdust and rooted in early spring.
Planting is after the first or second season. Budding and grafting are used with Old World
grapes to overcome nematode and phyloxera problems and to increase vigor. Rootstocks
are derived from American grapes and include 'Freedom,' 'Ramsey,' 'Rupestris St. George,'
'Teleki 5C,' '110 Richter,' and '3309 Couderc.' Seeds are used for breeding and for producing rootstocks. Bench grafting is common, as is field budding. Topgrafting of older cultivars to newer cultivars is done with whip grafting and T budding.

Vines (except muscadine) are planted 6 to 8 feet apart in rows, based on vigor and
training system. Rows are set 8 to 10 feet apart, based on equipment. Spacing of 11 to 12
feet is used with certain training systems such as Geneva double curtain and quadrilateral
systems. Muscadine grapes are more vigorous, so spacing is toward the higher end of these
ranges. Rows are oriented north to south for maximal light interception. Grapes begin bearing 2 to 3 years after planting. Early fruit production in the second year is discouraged by
removal of flowers or young fruit to promote better vine development.

Trellis systems are used as supports in grape vineyards. Several types are available; all
utilize either posts with wire or posts without wire. Choice is made on type of grape, vigor,
planting density, pruning/training system, and regional considerations. When vigor is a
consideration as is canopy density and light interception, the Geneva double curtain,
quadrilateral cordon, and lyre trellises are good choices. High-density plantings also use the
Geneva double curtain. Two-wire trellises are commonly used for Old World and American
hybrids. Local extension specialists should be consulted or see the Internet Resources list
for more on various trellis systems.

Fruit is produced on 1-year-old canes or shoots that have become woody after growth ceased. Shortened (two-bud) canes are referred to as spurs. Wood older than 2 years is unproductive, so annual cane or spur pruning is needed to keep productivity up and to encourage the formation of fruiting wood for the current and following year. Pruning is usually done in early spring after the last hard freeze, but before buds swell. Numerous training systems exist. Examples are the four-cane and six-cane kniffin systems, the umbrella kniffin system, cordon systems (low-wire cordon, bilateral cordon, mid-wire cordon, and high-wire cordon), fan systems, and arbor systems. Cordons are older, horizontal permanent woody stems (arms) from which new fruiting canes or spurs can be produced. Cane and spur systems rely on the main trunk and canes for renewal. Pruning details are beyond the scope of this book (see Internet Resources list and Chapter 14, balanced pruning entries). Power-assisted hand pruners are used, as is machine hedge pruning to facilitate pruning and to reduce labor.

Grapes when mature need about 25 to 50 pounds of nitrogen annually per acre, depending on soil fertility. Young grapes require about 15 to 20 pounds of nitrogen for each field year. Very vigorous or high-intensity double curtain wall trellis systems on poor soils might need as much as 75 to 100 pounds of nitrogen annually per acre. Petiole tissue analyses are helpful for determining exact amounts of nitrogen and other needed nutrients. Soil tests are needed for phosphorus and potassium.

Insect and other pests include berry moth (*Endopiza viteana*), bostrichid beetle (*Scobicia declivis*), branch and twig borer (*Melalqus confertus*), bud beetle (*Glyptoscelis squamulata*), bud mite (*Eriophyes vitis*), click beetle (*Limonius canus*), cutworms (*Peridroma saucia, Amathes c-nigrum, Orthodes rufula*), false chinch bug (*Nysius raphanus*), grapeleaf skeletonizer (*Harrisina brillians*), hoplia (*Hoplia oregona*), leaf-folder (*Desmia funeralis*), leafhoppers (*Erythroneura elegantula, E. variabilis, Platynota stultana*), mealybug (*Planococcus ficus*), orange tortrix (*Argyrotaenia citrana*), Japanese beetle (*Popillia japonica*), mealybug (*Pseudococcus longispinus, P. maritimus, P. viburni*), phylloxera (*Daktulosphaira vitifoliae*), sharpshooters (*Carneocephala fulgida, Draeculacephala minerva, Graphocephala atropunctata, Homalodisca coagulata*), thrips (*Drepanothrips reuteri, Frankliniella occidentalis*), vinegar flies (*Drosophila melanogaster, D. simulans*), and web-spinning spider mites (*Eotetranychus willamettei, tetranychus pacificus, T. urticae*).

Diseases include anthracnose (*Elsinoe ampelina*), armillaria root rot (*Armillaria mellea, A. tabescens*), canker (*Botryodiplodia theobromae*), bitter rot (*Greeneria uvicola, Melanconium fuligineum*), botrytis bunch rot (*Botrytis cinerea*), crown gall (*Agrobacterium tumefaciens, A. rubi*), downy mildew (*Plasmopara viticola*), eutypa dieback (*Eutypa armeniacae*), grape black rot (*Guignardia bidwellii*), leaf spot (*Mycosphaerella personata*), phomopsis cane and leaf spot (*Phomopsis viticola*), Pierce's disease (fastidious xylem-limited bacteria, *Xyllela fastidiosa*, carried by glassy-winged sharpshooter: *Homalodisca coagulata*), and powdery mildew (*Uncinula necator*). Summer bunch rot (sour rot) involves several pathogens: *Aspergillus niger, Alternaria tenuis, Botrytis cinerea, Cladosporium herbarum, Rhizopus arrhizus, Penicillium* sp., and others.

Yields can vary from 7 to 60 pounds per vine, depending on cultivar, training systems, and cultural conditions. Yields per acre vary from 3 to 11 tons. The average grape yield in the United States in 2000 was 8.1 tons/acre. During that same year in California, yields for wine, table, and raisin grapes were 6.9, 8.2, and 9.8 tons/acre, respectively. Some grapes are mechanically harvested. American cultivars can be stored at 0°C (32°F) and 80% to 85% relative humidity for 3 to 8 weeks. Old World grapes can be stored at 0°C (32°F) at 85% relative humidity for 2 to 6 months. Controlled atmosphere conditions will extend storage life.

Guavas

The guava (Figure 18–25) is a tropical evergreen shrubby, small tree fruit of the genus *Psidium*. Mature height is 33 feet, but in California heights rarely exceed 12 feet. The most commonly cultivated species are the common guava, *P. guajava,* and the purple strawberry guava, *P. littorale* var. *longipes. Psidium guajava* is more tender and found only in subtropical Florida, coastal southern California, and Hawaii in the United States. *Psidium littorale*

FIGURE 18–25
Guavas. Vitamin C is found mostly in the skin. Vitamin A is present in the pulp. Much information about guavas is available at: http://www.hort.purdue.edu/newcrop/morton/guava.html#Cultivars. (USDA photograph.)

var *longipes* is more hardy and is seen in southern California and Florida. Cultivars of *P. guajava* include 'Beaumont,' 'Detwiler,' 'Hong Kong Pink,' 'Mexican Cream,' 'Red Indian,' 'Ruby X,' 'Sweet White Indonesian,' 'White Indian,' and 'White Seedless.' Commercial (Hawaii) and backyard culture of the guava is limited, making it a minor fruit. Guavas are high in vitamin C and are eaten fresh or used for juice or processed into jelly and sherbet.

Propagation is possible by seed, but not without variation. Propagation of cultivars is by air layering, root cuttings, patch budding (Forkert method), and side-veneer or approach grafting. Late spring and fall planting are practiced in Florida and California, respectively. The common guava is planted at 15- to 25-foot intervals. The strawberry guava reaches 10 to 25 feet in height and is set at 15 to 20 feet. Good bearing can be expected at 8 to 10 years.

Guavas are somewhat drought tolerant, but for good fruit production, irrigation at regular intervals with deep watering is best. Soil can dry partially between applications. Guavas grow fast and are heavy feeders. Fertilizer should be applied monthly. Mature trees require 0.5 pound of nitrogen annually. Pruning is minimal. Best fruit is borne on 1- to 3-year-old wood. Thinning and heading back of the top can be done at 2- to 3-year intervals. Fruit thinning is not necessary. Self-pollination is possible, but best fruit set is with cross-pollination by bees.

Insects and disease are minimal. In humid areas, anthracnose is a problem. Root rot nematodes can be troublesome. In Florida, guava whitefly, guava moth, and Caribbean fruit flies can be major pests. In California, pests are likely to be mealybugs, scale, thrips, and whiteflies.

Fruits can be picked when ripe for best flavor. Fruits picked earlier can be stored for 2 to 5 weeks at 7.8° to 10°C (46° to 50°F) at 85% to 95% relative humidity and ripened later. Fruit ripening is ethylene dependent. Yields are 200 to 700 pounds of fruit per tree. Guava yields in 2000 for Hawaii were 11.7 tons/acre. Ripe fruit can only be stored briefly under refrigeration.

Hazelnuts (*see* Filberts)

Hickory Nuts

Hickory nuts are the fruits of several species of the genus *Carya*. These include the bitternut, *C. cordiformis;* bitter pecan, *C. aquatica;* mockernut, *C. tomentosa;* the pecan, *C. illinoinensis* (see pecan); the pignut, *C. glabra;* shagbark hickory, *C. ovata;* and the shellbark hickory, *C. laciniosa.* Desirable nuts, in decreasing order, are the pecan, shagbark hickory, and shellbark hickory. The others vary from poor to inedible. The pecan is considered under a separate heading in this chapter. Hickory nuts are not produced commercially, but are grown by noncommercial horticulturists. The shagbark and shellbark hickories are hardy through zones 5 and 6, respectively.

Cultivars should be chosen on the basis of region, since southern cultivars are not apt to mature their nuts in the shorter and less hot growing season of the North. Cultivars are numerous, so only a limited number will be mentioned here. Shagbark hickory cultivars of merit include 'Davis,' 'Fox,' 'Glover,' 'Grainger,' 'Hales,' 'Harold,' 'Kentucky,' 'Kirtland,' 'Porter,' 'Wilcox,' and 'Wilson.' For the shellbark hickory, they are 'Keystone,' 'Nieman,' 'Ross,' 'Stephens,' and 'Weiper.' A number of hybrid cultivars exist, such as between the pecan and shagbark or shellbark (a hican); pecan-shagbark cultivars are 'Burton,' 'Henke,' 'Pixley,' and 'Wapello'; pecan-shellbark cultivars are 'Baress,' 'Bixby,' 'Burlington,' 'Clarksville,' 'Green Bay,' and 'Jay Underwood.'

Species can be propagated from seeds, and cultivars can be propagated by budding or grafting. Seedlings are slow, and great care is required in the latter case. Transplanting must be done with care because of the taproot system. The best approach is to plant the nut in a permanent location in the fall or, if stratified, in the spring. Then the grafting can be done in place. Trees should be placed 60 to 75 feet apart. Because of various degrees of self-unfruitfulness, it is best to plant two or more cultivars. Pruning is only needed to establish lower limb height and to remove dead wood. Seedlings may take 10 to 15 years to bear nuts.

A backyard tree in its first year can use 1 pound of 10-10-10 and 2 pounds in its second year. For each subsequent year of age, add an additional 2 pounds up to a maximum of 50 pounds per tree. Fertilizer should be watered in, unless it is possible to incorporate it into the soil. Yields are about 50 to 75 pounds per tree. Mature height may be 100 to 120 feet.

Troublesome insects include the hickory bark beetle (*Scolytus quadrispinosus*), several species of aphid, hickory shuckworm (*Laspeyresia caryana*), tussok moth (*Halisidota caryae*), case bearer (*Coleophora laticornella*), walnut caterpillar (*Datana integerrima*), webworm (*Hyphan triacunea*), and the painted hickory borer (*Megacyllene caryae*). Diseases include anthracnose (*Gnomonia caryae*) and canker (several spp.).

Huckleberries (*see* Blueberries)

Jujubes

The jujube or Chinese date is *Zizyphus jujuba*. Good cultivars are 'Lang,' 'Li,' 'So,' 'Tanku Vu,' and 'Yu.' It can be grown in zone 9, especially in the hotter, drier, alkaline parts such as the desert valleys in California. It has minimal importance as a commercial or backyard fruit. Hot summers are needed for proper ripening. 'Lang' needs 'Li' as a pollinator. 'Li' is self-fruitful, but bears more and bigger fruit with 'Lang' as a pollinizer.

Cultivars are propagated by grafting or root cuttings. These are set about 25 to 35 feet apart in the spring and can reach a height of 40 feet. Bearing can start as young as 4 years. Pruning is mainly to build a good framework and to remove older, unproductive wood. Fruit is borne on growing shoots of the current year. It appears that maximal fruit production occurs with bees and cross-pollination between two cultivars. Insect and disease problems are minimal.

Juneberries

The juneberry, also called shadbush, serviceberry, saskatoons, and sugarplum, are species of *Amelanchier*. Species from which cultivars are derived are *A. alnifolia* and *A. stolonifera*. Mature heights are 25 and 4 feet, respectively. Cultivars of *A. alnifolia* are 'Altaglow,' 'Forestburg,' 'Indian,' 'Northline,' 'Pembina,' 'Shannon,' and 'Smoky.' 'Success' is a cultivar of *A. stolonifera*.

Juneberries are very winter hardy through zone 5. Commercial production is essentially nonexistent. Some fruits are gathered from the wild and from backyards. It is classed as a minor fruit. Propagation of the species is by seed. Cultivars can be propagated by softwood cuttings, division, suckers, and grafting. Planting is in the spring. Pruning is not necessary other than to remove dead wood and to maintain shape. Cross-pollination by insects is desirable. Insect pests are leaf miner (*Nepticula amelanchierella*), scales, and mites. The fireblight disease (*Erwinia amylovora*) and birds are also troublesome.

Kiwifruits

Kiwifruit cultivars are derived from *Actinidia delicosa,* a woody, dioecious vine that can be 10 to 15 feet wide, 9 to 12 feet high, and as long as 18 to 24 feet. Kiwifruits need a long growing season of at least 240 days and only moderate winters. Kiwifruits are currently commercially grown in California. Another closely related species is *A. chinensis.* Unlike *A. delicosa,* these kiwifruits are fuzz free and yellow inside as opposed to the typical emerald green. Another close species is *A. arguta,* the hardy kiwifruit that only needs a growing season of 150 days and has greater cold tolerance. Fruits are green inside, but are smaller and sweeter than the regular kiwi. Cultivars of this species are very promising and commercial kiwifruit production will likely expand into other areas in the United States.

Female cultivars of *A. delicosa* include the standard New Zealand 'Hayward,' which requires extensive winter chilling. 'Saanichton' has greater cold hardiness. Other New Zealand cultivars include 'Allison,' 'Bruno,' 'Gracie,' and 'Monty.' Cultivars for areas that cannot provide sufficient winter chilling are 'Abbott,' 'Dexter,' 'Elmwood,' 'Tewi,' and 'Vincent.' If early fall freezes pose a problem, the early ripening 'Blake' (6 weeks earlier than 'Hayward') is good. Recently introduced cultivars of *A. chinensis* show promise for areas with low winter chill. Cultivars for the hardy kiwifruit (*A. arguta*) include 'Ananasnaja,' 'Cordifolia,' 'Geneva,' 'Issai,' 'Michigan State,' and 'Red Princess.' Male cultivars (pollinators) for *A. delicosa* are the more commonly used 'Matua' and 'Tomuri.' Other males are 'Chico,' 'CC Early Male,' and 'Fuzzy Male.' These males can be used for all three species, although specific males are available for *A. chinensis* and *A. arguta.* 'Male Arguta' is a pollinator for the hardy kiwi.

Kiwifruits are either grafted or propagated from cuttings. The rootstock is usually the seed-grown species. Cuttings are slightly better in the sense that dieback from an unexpected winter freeze results in regrowth of the same cultivar from the roots. Kiwifruits are grown on trellis systems such as a single wire or T-bar system. Winter pruning is practiced to train shoots along the trellis and to annually replace older canes. In backyards Kiwifruits can grow up a lattice patio cover post and spread across the top.

Soils should have a pH of 5.0 to 6.5. Well-drained soils with good organic matter levels are best. Plants require high nitrogen levels during the first half of the growing season. Split applications are best. Kiwifruit roots are sensitive to fertilizer burn, so organic fertilizers or split applications of chemical fertilizers are best. About 150 pounds per acre of nitrogen are needed. About two-thirds is applied before bloom and the remainder after bloom. For individual plants in the first year, one pound before bloom and 0.75 pound after bloom are good and in subsequent years, these numbers should be 2.0 and 1.0. Phosphorus and potassium are added according to soil test needs. Kiwifruits are also high water consumers. Regular watering, especially when hot, is necessary. Plants that are allowed to undergo drought stress usually become defoliated and sometimes die.

Pest problems are minimal. Boxelder bug (*Leptocoris trivittatus*), leafrolling caterpillars (*Archips argyrospilus, Argyrotaenia citrana, Choristoneura rosaceana,* and *Platynota stultana*), scales (*Aspidiotus nerii, Hemiberlesia lataniae,* and *H. rapax*), snails, deer, gophers, nematodes, and cats can be minor pests. Cats like the smell of young plants and can damage them by rubbing. Diseases include armillaria root rot (*Armillaria mellea*), bacterial blight (*Pseudomonas viridiflava* and *Pseudomonas syringae*), bleeding canker (*Pseudomonas syringae*), crown gall (*Agrobacterium tumefaciens*), fruit rot (*Botrytis cinerea*), and root/crown rot (*Phytophthora* spp.).

Kiwifruits are ripe when seeds are black and sugars reach 6.5 brix on a refractometer. Hard, nearly ripe fruits have excellent storage and shipping qualities. Storage for 28 to 84 days at 0°C (32°F) and 90% to 95% relative humidity is possible. Exposure to ethylene must be avoided. Controlled atmospheric storage can extend the storage time. Kiwifruits have high vitamin C levels. Kiwis are eaten fresh and in pies, ice cream, and jams.

Kumquats

Kumquats are subtropical fruits related to citrus. The nagami kumquat, *Fortunella margarita*, and the marumi kumquat, *F. japonica,* are the main two cultivated. The meiwa kumquat is thought to be a hybrid, possibly between these two species.

Kumquats are grown in the same areas as oranges, and their culture is similiar (see discussion of oranges under citrus entry). Commercial production is almost nil; backyard culture exists. Propagation is by grafting on *Citrus limon* cv. 'Rough' and *Poncirus trifoliata* stocks. Spacing is 15 feet. Mature height is about 10 feet. They are grown more as ornamental than fruit trees. Fruits are used in jellies and marmalades.

Lemon (*see* Citrus)

Ligonberries (*see* Cranberries)

Lime (*see* Citrus)

Litchis

The litchi fruit is borne on a tree that can reach a height of 40 feet. Winter chilling needs are minimal at 100 to 200 hours at 0° to 7.2°C (32° to 45°F). Mature trees are more frost tolerant than young trees. Plantings are found in Hawaii and frost-free areas of coastal California and Florida. Trees appear to be long lived. Cultivars are derived from *Litchi chinensisi* and include 'Amboina,' 'Bengal,' 'Brewster,' 'Groff,' 'Hak Ip,' 'Kwa Luk,' 'Mauritius,' 'No Mai Tsze,' and 'Tai Tsao.'

Propagation is mainly by air layering, because seeds do not come true and grafting is hard, although wedge and bud grafts have been used. Trees are placed 30 to 40 feet apart. Soil pH is best at 5.5 to 7.5. The lower end is better, as the tree is an acid-loving plant. Trees are slow growers, so minimal fertilization is needed when young. Young trees should be pruned to develop a framework best suited for harvesting. As trees mature, heavier fertilization is needed at monthly intervals during the spring and summer. Pruning is minimal and basically to remove fruited out or damaged wood. Even moist soil is best, but avoid waterlogging. Saline conditions are harmful. Insect pests and diseases are minimal. At times aphids, mites, and scales are troublesome. Birds are the biggest problem and netting is likely to be needed.

Fruit should not be picked until ripe. Ripe fruits are pink to strawberry red, depending on cultivars. Interior parts are white. Some cultivars leak juice when the skin is broken, others do not. The latter ones are more desirable. Litchi can sometimes have alternate bearing. Refrigerated fruits can be kept for 30 days. Ripe fruits only last a few days at room temperatures. Fruits can be eaten alone or in fruit salads, dried, canned, and frozen. Fruits contain vitamins B, C, D, and E.

Macadamia Nuts

Two evergreen tree species are grown for the production of the macadamia nut: *Macadamia integrifolia* (smooth-shelled macadamia) and *M. tetraphylla* (rough-shelled macadamia). These trees reach heights of 30 to 40 feet and similar widths. Cultivars of *M. integrifolia* include 'Arcia,' 'Dorado,' 'Faulkner,' 'Ikaika,' 'James' (highest yielding cultivar in California), 'Kakea,' 'Keaau,' 'Keauhou,' 'Kohala,' 'Nuuanu,' 'Pahau,' 'Parkey,' 'Wailua,' and 'Waimanalo.' Cultivars of *M. tetraphylla* are 'Burdick' (used mostly as a rootstock), 'Cate' (wisely used commercial cultivar in California), 'Hall,' and 'Santa Ana.' 'Beaumont' (good for home gardens) and 'Vista' are hybrid cultivars between the two species. Mature trees can tolerate frosts, but frost kills young trees and can kill flower clusters at −2.2°C (28°F).

Commercial production is mostly in Hawaii, but culture is possible in areas producing citrus and avocado. Currently, macadamia nuts are also produced in a limited commercial fashion in warmer, coastal areas of California. The macadamia nut in backyard production is seen mainly in zone 10, primarily in Southern California and somewhat in southern Florida.

Rootstocks are propagated from seeds. Cultivars are mostly grafted by the splice or whip graft in California or in Hawaii; the side-wedge graft and budding is by the patch technique. Trees can also be topworked with the saw-kerf method. Both *M. integrifolia* and *M. tetraphylla* are reciprocally graft compatible. Trees are placed at 25- to 35-foot intervals, usually in early spring. Trees are trained to the central-leader system. Bearing usually takes 6 to 8 years. Bees are important pollinators, although wind pollination also helps. Cultivars vary from completely self-fruitful to almost self-sterile. Planting of two cultivars for cross-pollination appears to be beneficial.

Soil pH should be 5.5 to 6.5. Well-drained, salt-free soils are best. Water needs are similar to those of avocados. Macadamias can withstand drought, but irrigation is needed from the time of nut set through nut filling and midsummer vegetative growth. Split applications of nitrogen two to three times yearly, starting with bud swell, are good. Being slow growers, only modest amounts of nitrogen are needed. Foliar tissue analyses supplemented with soil tests are critical for determining nutrient needs. Organic fertilizers should be considered, because young trees are easily injured with too much nitrogen.

Insect and disease problems appear minimal. Mites, scales, and thrips can be minor pests. Humid conditions can bring about anthracnose, and canker can result from tree wounds.

Ripe nuts fall to the ground. Tree shaking can bring down immature nuts so long pole tree nut harvesters are used for mature nuts high up in the tree. Mechanical harvesters can collect fallen nuts on the ground. Harvested nuts require dehusking and drying by sun or oven. Macadamia nuts can be eaten raw or roasted either salted or unsalted. Yields of nuts per acre range from 1.5 to 2.5 tons per acre or about 150 pounds per tree. Yields are lower in California, being 60 to 70 pounds per tree. In 2000, the yield in Hawaii was 1.4 tons/acre. Nuts store well in cool, dry places, freezers, closed containers in refrigerators, and in vacuum packs.

Mangos

The mango (Figure 18–26), *Mangifera indica,* is a tropical tree fruit that can grow to 65 feet. Because of its tropical nature and frost-free requirement, the mango is only grown in the southern half of zone 10 in Florida and in a very small section of California. Commercial production is currently confined to Florida, but California is penetrating the market. Cultivars found in commercial, greenhouse, and backyard production in the United States include 'Aloha,' 'Brooks,' 'Cambodiana,' 'Carabao,' 'Carrie,' 'Cooper,' 'Costa Rica,' 'Doubikin,' 'Earlygold,' 'Edgehill,' 'Edward,' 'Fascell,' 'Gouveia,' 'Haden,' 'Irwin,' 'Julie,' 'Keitt,' 'Kensington Pride,' 'Kent,' 'MacPherson,' 'Manila,' 'Mulgoba,' 'Ott,' 'Piñta,' 'Pirie,' 'Reliable,' 'Sensation,' 'T1,' 'Thomson,' 'Tommy Atkins,' 'Villaseñtor,' 'Winters,' and 'Zill.' Two races exist: India and Philippines/Southeast Asia. Philippine types from Mexico are the hardiest.

FIGURE 18–26
Mangos are good sources for Vitamins A and C as well as fiber. The mango can be traced back 4,000 years to Southeast Asia. (USDA photograph.)

Seedlings can be a gamble and are usually used as rootstocks. Some seeds are polyembryonic. Grafting can be done on seedlings 4 to 6 weeks old during middle to late summer using a whip graft. In the second year, cultivars are propagated by cleft, tongue or side-veneer grafting, or chip shield budding. Mature trees can be topworked.

Plants are set about 40 feet apart. Grafted trees can commence bearing in 3 to 5 years. Soil pH should be 5.5 to 7.5. Mangos are heavy feeders and respond well to citrus fertilization programs (see citrus entry), but fertilization should stop in midsummer. Young trees are easily damaged by excess fertilizer. Water is needed steadily until fruit is harvested.

Mangos are rarely pruned, other than to encourage new growth for annual bearing. Insect pests include several mites, red-banded thrip (*Selenothrips rubrocinctus*), scale (several spp.), and fruit fly (*Toxotrypana curvicauda*). Gophers can cause root damage. The diseases anthracnose (*Glomerella cingulata*), bacterial spot (*Colletotrichum oleosporides*), and powdery mildew (*Oidium mangiferae*) can be troublesome.

The presence of bees appears to aid fruit set. Fruits mature 100 to 150 days after flowering. Fruits color on ripening and are often yellow to yellow-red. Fruit can be picked at first color and ripened at room temperature, but tree-ripened fruits have better quality. Selective harvesting is needed as the fruit softens and falls at maturity. Keeping qualities of fully ripe fruit are poor. Partially ripe and ripe fruit can be held for 14 to 25 days at 10°C (50°F) and mature green mangos at 12.8°C (55°F) and 90% to 95% relative humidity. Controlled atmospheric storage can increase longevity by a few weeks. Mango ripening responds to ethylene. Yields are about 2 tons/acre in Florida.

Mulberries

Mulberries are borne on deciduous trees of the genus *Morus*. Cultivars of the mulberry specifically planted for fruit in the North are derived from the white mulberry, *M. alba*. These are 'New American,' 'Oscar,' 'Thornburn,' 'Trowbridge,' and 'Wellington' (possibly same as 'New American'). Fruit can be whitish, but more often pinkish to purplish and looks like blackberries. These trees can reach 80 feet in height. Cultivars of the tender black mulberry, *M. nigra,* are grown in the South and include 'Black Persian,' 'Kaester,' and 'Noir de Spain.' Black mulberries are bushy and reach a height of 30 feet. 'Collier' and 'Illinois Everbearing' are hybrid cultivars (*M. alba* × *M. rubra*). The latter is considered one of the best cultivars for the North. Other cultivars for warmer areas include 'Pakistan,' 'Riviera,' 'Shangri-La,' and 'Tehama.' Mulberries make good fresh fruit and are very good in jellies, preserves, and mixed berry pies. Mulberry wine can also be made.

Species are grown from seed, but can take 10 years to bear fruits. Cultivars may be propagated by hardwood or softwood cuttings and by budding. Young trees are planted 25 to 30 feet apart in the spring. Commercial production is nil; most production is in backyards. The dioecious condition may be encountered with some cultivars. Most are wind pollinated. Water needs are modest, but dryness can cause premature fruit drop. An annual application of 10-10-10 fertilizer is sufficient. Pruning is only needed to remove dead wood and to thin out branches.

Insect pests are the mulberry whitefly and San Jose scale. Bacterial blight can be troublesome. Birds can also be annoying during fruit ripening season. Popcorn disease is also observed where the fruit swells and looks like popcorn. Hybrids between the two species (*M. alba* × *M. rubra*) are prone to this problem. Fruits have poor keeping qualities, but can be refrigerated for several days.

Nectarines (*see* Peaches)

Olives

The olive (Figure 18–27) is an evergreen tree of the species *Olea europaea*. The variety *O. europaea* is the typical cultivated common olive. Trees can reach 50 feet in height and 30 feet in spread and can live for 500 years. Cultivars include 'Ascolano,' 'Barouni,' 'Gordal,' 'Manzanillo,' 'Mission,' 'Picholine,' 'Rubra,' and 'Sevillano.' Olives are sold fresh for home

FIGURE 18–27 *Olives originated in Turkey and Syria. References to olives are found in the Bible. Olive oil is rich in the antioxidant vitamin E and very high in monounsaturated fatty acids. These types of fatty acids raise levels of HDL and lower the level of LDL. HDL (high density lipoproteins) help reduce cholesterol. Olive oil is a cornerstone of the Mediterranean diet and is better than most oils for healthy living. (USDA photograph.)*

curing or processed by pickling or pressed for oil. Curing to make table olives is done either with lye (to remove bitter-tasting alkaloid) and water washes and a final salt solution, salt curing, or Greek-style curing. Olives can be cured when unripe (green) or ripe (usually blue/black or black; some are coppery brown). Some cultivars are considerably less bitter than others; these can be directly sun dried and eaten. 'Sevillano' is the largest olive. Because of its low oil content, it is mainly processed through pickling.

Climatic requirements of olives are such that their production is limited to the hot dry areas of the Southwest, especially the hot interior valleys of California and one area near Phoenix, Arizona. Winter chilling for 12 to 15 weeks with night and day temperatures of 1.7° and 15.5°C (35° and 60°F), respectively, is needed to initiate flowering. Temperatures below −11.1°C (12°F) are fatal. Blooming is late enough to miss spring frosts. A long, hot growing season is needed for good fruit development. Bearing is poor in the Southeast, because the late flowering and requirement of a long growing season expose the fruits to fall frost damage there.

Propagation is by rooting leafy cuttings. Grafting or budding is also practiced; the whip graft or side graft is favored. Olive rootstocks are used. Trees are set 30 feet apart in early spring. Hedgerows are used for higher density plantings (15 by 30 feet, 97 trees/acre). Bearing can start 4 to 5 years after planting 2- or 3-year-old nursery stock. Full bearing can start in 12 to 20 years. The olive can survive dry periods, but irrigation is usually needed during the hot, dry summers experienced in inland valleys of California. Lower yields and smaller fruits can result from not enough water. Water requirements are 3 acre-feet of water/year. Homeowners need to provide a monthly, deep watering.

Nitrogen needs should be based on foliar tissue analyses and other nutrients on soil tests. July foliar tissue samples are useful for gauging nutrient needs. Soil pH can be 7.0 to 8.5. A 1-year-old olive tree needs roughly 0.07 pound of nitrogen yearly. For each additional year, another 0.07 pound should be added until the total in the ninth year is 0.63 pound. Mature trees should be continued at this rate, which is equivalent to 80 pounds of nitrogen per acre.

Initial pruning for the modified central-leader system is mainly to select three to five scaffold branches. Subsequent pruning must be carefully done to avoid alternate-year bearing. Annual pruning can be used to maintain height at 20 to 25 feet. Branches selected for removal are done so based on the fact that the olive does not fruit in the same place twice and usually appears on the previous year's growth. Olives can also be espaliered.

The only serious insect is black scale (*Saissetia oleae*). A gall-causing bacteria, *Pseudomonas savastanoz,* and the fungi, *Verticillium* and *Cycloconium oleaginum,* which cause wilt and defoliation, respectively, can be troublesome.

Most cultivars are self-fruitful, but cross-pollination probably increases yields. Bees appear to be effective, but wind is the chief vector for pollination. A few cultivars are not self-fruitful and require other cultivars nearby. Excess fruit set can occur, resulting in smaller fruit and possibly in alternate bearing. A chemical fruit-thinning agent can be used. Olives can be harvested both ripe and unripe. Ripe olives are easily bruised. Yields vary from 4 to 7 tons per acre in California. Olives are usually processed by pickling or for oil.

Orange, Sweet and Sour (*see* Citrus)

Otaheite Orange (*see* Citrus)

Papayas

The papaya (Figure 18–28) is a woody tropical fruit that reaches a height of 10 to 12 feet. It is a fast grower and is short lived. The species is *Carica papaya*. Since it can only withstand a few degrees of frost, it is only cultivated in the warmer parts of zone 10 in Florida and California. Commercial production is confined to Hawaii and backyard production is limited, making it a minor fruit. Cultivars include 'Kamiya,' 'Mexican Red,' 'Mexican Yellow,' 'Solo,' 'Sunrise,' 'Sunset,' 'Vista Solo,' and 'Waimanalo.' Sammler papayas are called Hawaiian types and large ones, Mexican types. The Hawaiian type is the one commonly found in supermarkets. While smaller, the flavor is more intense in the Hawaiian ones. Fruits are eaten raw or made into mixed juices and chutneys. Fruits are rich in vitamin C and contain modest amounts of vitamin A and potassium.

Propagation is by seeds started indoors and placed out in early spring. Semihardwood cuttings are also used. Papayas must be transplanted with great care. Plants are set about 10 to 12 feet apart. Bearing may start with a 1-year-old plant, but they only bear for 3 to 4 years. A sunny, hot area with minimal wind exposure is best. Saline soils, dry soils in hot weather, and wet soils in cold weather are all undesirable. Even moisture is needed during the growing season. Monthly applications of fertilizer are needed. Pruning is nil, except pinching of seedlings or cutting back established plants to encourage multiple trunk formation. Frost is not tolerated.

Insects and disease problems are minimal. Potential pests include fruit flies, fruit spotting bugs, mites, thrips, whiteflies, and nematodes. Possible diseases are anthracnose, mildew, root rot, and a few viruses. Gophers can damage roots. Two cultivars ('Rainbow' and 'SunUp') have been genetically engineered to be resistant to papaya ring spot virus, a problem in Hawaii.

FIGURE 18–28
Papayas. UH Rainbow and UH SunUp are genetically engineered papayas that are resistant to papaya ring spot virus, a serious problem for papaya production in Hawaii. (USDA photograph.)

Some set fruit by parthenocarpic means, others are dioecious requiring a male and female plant, and yet others are self-fruitful. Best fruits appear to arise with dioecious plants pollinated by bees. Fruit is ripe when the skin is yellow-green. Full ripeness is represented by an all-yellow skin. For shipping, pick at one-quarter yellow; for local sales, one-half to three-quarter yellow. Each plant can bear 12 to 30 fruits. Yields in Hawaii vary from 11 to 17 tons/acre. Plants are often replaced after 4 years. Fruits can be stored for 7 to 21 days at 90% to 95% relative humidity and 7.2°C (45°F) for ripe papayas, 10°C (50°F) for partially ripe (one-quarter to one-half yellow) ones, and 12.8°C (55°F) for mature green to one-quarter yellow papayas. While ethylene will improve color, little change in flavor results. Controlled atmospheric conditions can increase longevity by 1 to 2 weeks.

Passion Fruits

Passion fruit or granadilla are species of *Passiflora*. Edible species include purple granadilla (*P. edulis*), sweet granadilla (*P. ligularis*), yellow granadilla (*P. laurifolia*), giant granadilla (*P. quadrangularis*), sweet calabash (*P. maliformis*), and curuba (*P. mollissima*). These are woody tropical vines, and their culture is limited to Hawaii and the southern parts of zone 10 in California and Florida. Commercial production is minimal in Hawaii and essentially nonexistent in the continental United States, but backyard plantings exist. In the United States, cultivars are associated with *P. edulis* (purple fruited) and *P. edulis flavicarpa* (yellow fruited). Purple cultivars include 'Black Knight,' 'Edgehill' (best cultivar in California), 'Frederick,' 'Kahuna,' 'Paul Ecke,' 'Purple Giant,' and 'Red Rover.' Yellow ones are 'Brazilian Gold' and 'Golden Giant.' Many hybrids exist between the purple and yellow cultivars.

Propagation is by seeds and cuttings. Plants are set out in early spring from fall-planted seeds. They are placed 6 to 10 feet apart in rows 10 feet apart. Plants are usually trellised in full sun and can reach heights of 15 to 20 feet in a growing season. Plants can start bearing at 1 year and remain productive for 5 to 7 years. Vines are usually pruned back severely after harvesting the fruit. With yellow cultivars, flowers are self-sterile and must be cross-pollinated; carpenter bees are effective pollinators. Purple types are self-fruitful. Purple types are more frost tolerant than yellow ones. Cold snaps in winter can kill tops, but resprouting occurs. Nematodes and snails can be very troublesome in California.

Because of vigorous growth, split applications of fertilizer (10-5-20) four times during the growing season are suggested. The total yearly application is 10 to 12 pounds per plant. Soil pH can be 6.5 to 7.5. Water needs are high, especially during fruit development. Drought can cause fruit shriveling and dropping. Yields are from 15,000 to 40,000 pounds per acre or 40 pounds at most per vine. Fruits ripen fast from green to purple (or yellow) and should be picked promptly, because they fall to the ground when ripe. At peak ripeness (70 to 80 days after pollination), the outer skin shows signs of wrinkling. Fruits can be stored for 14 to 21 days at 10°C (50°F). Passion fruits are good sources of vitamins A and C plus potassium.

Paw Paws

Paw paw cultivars are derived from *Asimina trioba*. Paw paws are produced on a narrow, deciduous tree that reaches heights of 12 to 20 feet. Prone to root suckering, plantings often become paw paw patches. At least 160 days are required in the growing season and minimally 400 hours of winter chill. Native plants are widespread from New York to Florida and Texas. Growth in parts of northern California and the Pacific Northwest is possible. Currently not in commercial production, it is a good fruit and has commercial potential. Fruit use is mainly from wild plants and backyard plantings. Fruits are eaten fresh and are highly flavored and have a custard-like texture. Fruits can be used in pies, marmalades, and puddings and have good levels of vitamin A and some B vitamins. Known cultivars include 'Davis,' 'Mary Foos Johnson,' 'Mitchell,' 'Overleese,' 'Pennsylvania Golden,' 'Prolific,' 'Rebecca's Gold,' 'Sunflower,' 'Sweet Alice,' 'Taylor,' 'Taytoo,' and 'Wells.'

Paw paws are best propagated by grafting or budding (chip budding works well) onto seedling rootstocks of the species. Seeds require 3 to 4 months of stratification to germinate.

Fruit qualities from seed are too variable. Care is required in transplanting, because roots are easily damaged. Initial plantings need to be shielded from full sun, but established plants do best in full sun. Soil pH can be 5.0 to 7.0. Even moisture is required, but avoid waterlogging when irrigating. An annual application of fertilizer high in potassium is needed. Pruning is minimal and is needed to remove dead or damaged branches and fruited-out wood. Plantings are relatively disease free. An occasional caterpillar, snail, slug, or earwig is found. Cross-pollination is required to set fruit. Bees are not pollinators, only some flies and beetles.

Ripe fruits are soft and pleasant smelling. The green color lightens and can develop black blotches. Ripe fruits keep for only a few days, but if refrigerated, up to 3 weeks.

Peaches and Nectarines

The peach (Figure 18–29), *Prunus persica,* is a temperate-zone fruit. The nectarine is a smooth-skinned variety of the peach, *P. persica* var. *nucipersica.* Both are grown extensively in commercial and backyard production. Life spans are short for peaches, varying from 8 to 20 years. Standard trees reach a height of 24 feet.

Peaches are a little less hardy than apples, so their range is a bit south of the apple. Zone 5 is the limit for peaches. The chilling requirement for peaches is usually 600 to 1,000 hours below 7.2°C (45°F), but values as low as 50 hours and as high as 1,200 hours are known for cultivars adapted to the southern and northern limits, respectively, of the peach growing range. Winter temperatures should not fall below −26.1°C (−15°F). Late spring frosts can damage the fruit buds, and clear hot summers are best for fruit development. Best conditions are found in the South, Midatlantic States, West Coast, and those areas near large bodies of water in the northern states, such as the Great Lakes area.

Cultivars are extremely numerous and vary in region adaptability, time of ripening, and various fruit characteristics such as clingstone, freestone, yellow fleshed, and white fleshed. Some cultivars are better for fresh eating, others for canning, and yet others for freezing. For example, in the extreme north of zone 5, cultivars with the hardiest fruit buds are recommended such as 'Reliance.' At the other end of the peach spectrum in Florida, cultivars with minimal chilling requirements are needed: 'Flordacrest,' 'Flordaglow,' 'Flordaking,' 'Flordaprince,' 'Gulfprince,' 'Tropic Beauty,' 'Tropic Snow,' 'UF Gold,' and 'UF 2000.' More specialized works and local sources should be consulted for which cultivars do best in your region and which are more suitable for fresh or processing market.

FIGURE 18–29
Peaches. The following peach cultivars are resistant to bacterial leaf spot: Ambergem, Belle of Georgia, Biscoe, Candor, Cardinal, Cherryred, Dixired, Early-Red-Free, Mayflower, Redbird, and Southhaven. (USDA photograph.)

For New England, New York, Michigan, the Southeast, California, and the Pacific Northwest, suggested cultivars are 'Autumn Glo,' 'Avalon Pride,' 'Bellaire,' 'Bounty,' 'Carey Mac,' 'Cresthaven,' 'Dixired,' 'Dr. Davis,' 'Early Red,' 'Elberta,' 'Eloise' (white), 'Empress,' 'Encore,' 'Ernie's Choice,' 'Fay Elberta,' 'Fayette,' 'Flamin Fury Jersey 14,' 'Flamin Fury PF # 5B,' 'Flamin Fury PF #15A,' 'Flamin Fury PF #17,' 'Flamin Fury PF #23,' 'Flamin Fury PF #27A,' 'Flavorcrest,' 'Frost,' 'Garnet Beauty,' 'Georgia Belle' (white), 'Glenglo,' Gold Prince,' 'Harcrest,' Harken, 'Harrow Beauty,' 'Harrow Diamond,' 'Harvester,' 'Hesse,' 'Hiley' (white), 'Jefferson,' 'Jerseyglo,' 'Jerseyland,' 'Jersey Queen,' 'J. H. Hale,' 'Jim Dandee,' 'John Boy,' 'June Gold,' 'June Lady,' 'June Prince,' 'Late Sun Haven,' 'Laurol,' 'Loring,' 'Majestic,' 'Mary Jane,' 'Nectar' (white), 'New Haven,' 'O'Henry,' 'Parade,' 'Red Globe, 'Red Haven,' 'Redkist,' 'Redskin,' 'Redtop,' 'Rich Haven,' 'Rio Oso Gem,' 'Rizzi,' 'Ross,' 'Ruston Red,' 'Salem,' 'Sentinel,' 'Sentry,' 'Springcrest,' 'Summer Gold,' 'Sunbrite,' 'Sun Prince,' and 'White Lady' (white). Some dwarf cultivars are 'Eldorado,' 'Pix Zee,' 'Valley Gem,' 'Valley Red,' and 'Valley Sun.'

A few nectarine cultivars are 'Arctiglo,' 'Armking,' 'Easternglo,' 'Fantasia,' 'Flavortop,' 'Hardired,' 'June Glo,' 'Mericrest,' 'Redgold,' 'Roseprincess' (white), 'Summer Beaut,' and 'SunGlo.' Nectarines for Florida are 'Suncoast,' 'Sundollar,' 'Sunmist,' 'Sunraycer,' and 'UF Queen.' 'Necta Zee' is a dwarf nectarine cultivar.

Peach cultivars are propagated by being budded on seedling stocks. 'Lovell' is a common rootstock produced from pits from self-pollinated orchard blocks. 'Halford' is another common rootstock produced from cannery pits derived from cross-pollinated orchards. Areas with nematode problems such as the South use 'Nemaguard.' A newer rootstock that produces a vigorous result with most grafted cultivars and shows better tolerance toward short peach tree life and nematodes is 'Guardian.' Other promising rootstocks are 'Bailey' (good cold hardiness) and 'Tennessee Natural.' Rootstocks used to produce dwarf peaches include 'St. Julien 655/2,' 'St. Julien P53/7,' and 'Citation.' Peach cuttings can also be rooted with moderate success.

Pits are sown in late summer or in spring after stratification. Buds are inserted late on rootstocks after one season of growth; buds remain dormant then until spring. After one season's bud growth, trees are set 18 to 24 feet apart in rows in the spring in the North and in fall or spring in the South. Between-row spacing varies from 12 to 18 feet. Variation results from soil fertility and vigor differences. Bearing can start as early as 3 years, and maximal bearing is from 8 to 12 years.

Soil pH should be maintained at 6.5. Nutrient needs should be based on soil tests and leaf tissue analyses. Leaf tissue analyses are best done during mid-July to mid-August. Wherever possible, apply nutrients at the tree drip line. Nitrogen needs should be based on foliar tissue analyses and other nutrients on soil tests. A 1-year-old peach tree needs roughly 0.07 pound of nitrogen yearly. For each additional year, another 0.07 pound should be added until the total in the ninth year is 0.63 pound. Mature trees should be continued at this rate, which is equivalent to 80 pounds of nitrogen per acre. The same can be said for P_2O_5 and K_2O, when soil tests indicate medium or low nutrient levels. Trees need about 1 inch of water weekly, especially from blossoming through fruit harvest. Weekly deep irrigation is best when water is needed. Wilting is best avoided.

Peach trees are usually trained to the open-center form and, to a lesser extent, the central-leader system. Maintenance pruning is to keep height within control for harvesting, to keep the center open (for open-center form), and to maintain productivity. Peaches are borne on wood produced the previous season. Most cultivars are self-fruitful. Bees are the main pollinators. Fruit thinning is usually necessary, unless a late spring frost has reduced the crop. Larger fruits are produced by thinning either by hand, mechanical, or chemical means.

Troublesome insects are the peach twig borer (*Anarsia lineatela*), peach tree borer (*Sanninoidea exitiosa* and *Synanthedon pictipes*), oriental peach moth (*Grapholitha molesta*), plum curculio (*Conotrachelus nenuphar*), scale (several spp.), aphid (several species), and tarnished plant bug (*Lygus lineolaris*). Deer can cause considerable damage to young orchards and mice and voles can be troublesome to roots and lower trunks on young trees when mulches are used.

Diseases include bacterial leaf spot (*Xanthomonas campestris* pv. *pruni*), brown rot (*Monilinia fructicola* and *M. laxa*), constriction canker (*Phomopsis amygdali*), Coryneum blight or shot-hole (*Wilsonomyces carpophilus*), crown gall (*Agrobacterium tumefaciens*), *Cytospora* canker, little peach (virus), peach leaf curl (*Taphrina deformans*), peach scab (*Cladosporium carpophilum*), peach X disease (*Phytoplasma* organism), phony peach (fastidious xylem-limited bacteria), *Phytophthora* root and collar rot, powdery mildew (*Sphaerotheca pannosa*), rust spot (*Podosphaerea leucotricha*), silver leaf (*Stereum purpureum*), sour rot (*Geotrichum candidum*), wilt (*Verticillium dahliae*), and yellows (virus). Numerous viruses attack peach; stock should be purchased virus-free.

Yields are 3 to 4 bushels per tree. In 2000 U.S. yields for nectarines and peaches, respectively, were 7.5 and 8.3 tons/acre. Yields as high as 19 tons/acre have been reported in California. Peaches destined for storage should be hydrocooled as quickly as possible. Peaches can be held at 0°C (32°F) and 90% to 95% relative humidity for 14 to 28 days. Peaches contain vitamins A and C. Fruits are eaten fresh, processed, and in pies, cakes, cobblers, jellies, sauces, and liquor/brandy.

Peanuts

Peanuts (Figure 18–30) are seeds of an annual, herbaceous legume, *Arachis hypogaea*. Two subspecies are known, *A. hypogaea* and *A. fastigata*. The former subspecies doesn't produce flowers on the main stem, has alternate branching, matures later, requires considerable water, and yields large peanuts. The latter subspecies produces flowers on the main stem, has sequential branching, matures earlier, uses less water, and yields small peanuts. Subspecies *A. hypogaea* contains the two market classes of peanuts known as runner and Virginia types. Subspecies *A. fastigata* contains Spanish and Valencia types. Runner types account for the majority of commercial production.

Runner-type cultivars are 'Andru 93,' 'AgraTech 108,' 'AgraTech 120,' 'Georgia Bold,' 'Georgia Green,' 'Georgia Runner,' 'GK7,' 'Florida MDR 98,' 'Florunner,' 'Marc 1,' 'Okrun,' 'Southern Runner,' 'Tamrun 88,' 'Tamrun 96,' and 'ViruGard.' Runner cultivars with high oleic acid content (longer shelf life for seeds and peanut products) include 'AgraTech 1-1,' 'AgraTech 201,' 'Flavor Runner 458,' 'GK 7 High Oleic,' and 'SunOleic 97.' Virginia cultivars include 'Gregory,' 'Jumbo Virginia,' 'NC 7,' 'NC 12C,' 'VA-98R,' and 'VC-2' (high oleic). Spanish cultivars include 'Spanco' and 'Tamspun 90.' Valencia cultivars include 'McRan,' 'Tennessee Red,' 'Valencia A,' and 'Valencia C.' 'Garroy' and 'Ruby' are Valencia cultivars with short maturation times of 110 to 120 days that might be adaptable to northern regions.

FIGURE 18–30
Peanuts are widely used as snack nuts and in products such as peanut butter, candies, ice cream and Thai cuisine. For some, peanut allergies can be a serious, even deadly problem. (USDA photograph.)

Fruits need a long season with heat to mature. Peanuts are grown commercially in several southern states. Georgia led in peanut production in 2000 (41%), followed by Texas (21%), North Carolina, Alabama, and Florida. Peanuts are grown throughout the South by homeowners. It is possible to grow peanuts in the North, if short season cultivars are used.

Peanuts must be innoculated with *Rhizobium* bacteria, unless sufficient populations of nitrogen-fixing bacteria are present in the soil. Inoculated peanuts are capable of fixing sufficient nitrogen to meet their needs. Peanuts do well with residual fertility from previous crops in rotations. If soils are very low in nitrogen, small amounts of starter nitrogen up to 30 pounds per acre might be beneficial. Soil testing for adequate phosphorus, potassium, calcium, magnesium, and micronutrients is important to maintain aditional nutrient needs. Low levels of calcium cause several production problems. Soil pH is best between 6.0 and 7.0.

Seeds (preferably removed from the pod) are planted when the danger of frost is past. Best results occur when soil temperature is 65°F at a 4-inch depth. Rows are spaced 22 to 30 inches apart, depending on cultivar vigor and field equipment. Seeds are spaced 6 to 8 inches apart in the row and covered with 2 to 3 inches of soil. Rates per acre with 18-inch row spacing are about 100,000, and with 30-inch row spacing, 70,000.

Peanuts grow 12 to 18 inches tall. Flowering will commence 4 to 6 weeks after planting. After pollination the ovary elongates and pushes into the soil. The peanut develops below ground. Water needs vary from 24 to 28 inches during production. During bloom through pod set, 1.5 to 2.0 inches of water are needed weekly. Peanuts are not well suited to conventional cultivation. Cultivation needs to be flat and shallow, essentially scraping, and soil should not be mounded on plants. Certain perennial weeds and some aggressive annual weeds might require judicious use of herbicides.

Mature crops require from 140 to 170 days, depending primarily on temperature. The proper time to harvest is based on field scouting. Best stage for harvesting is when most pods show veining, seed coats show color, and 75% of pods show darkening on the hull's inner surface. Too early or too late harvest results in considerable yield reductions. Harvesting usually requires partial cutting of foliage, followed by machines that combine digging, shaking, and windrowing. Drying in windrows or artificially to 5% to 10% moisture follows. Yields for the United States in 2000 averaged 1.2 tons/acre.

Troublesome insects and other pests include burrowing bug (*Pangaeus* sp.), corn earworm (*Heliothis zea*), corn rootworm (*Diabrotica undecipunctata, D. virgifera*), granulate cutworm (*Feltia subterranea*), grasshoppers, lesser cornstalk borer (*Elasmopalpus lignosellus*), leaf hoppers (*Empoasca fabae*), rednecked peanutworm (*Stegasta bosqueella*), root-knot nematodes (*Meloidogyne arenarai, M. hapla*), armyworms (*Spodoptera exigua, S. frugiperda*), spider mites (*Tetranvchus urticae*), three-cornered alfalfa hopper (*Spissistilus festinus*), thrips (*Frankliniella fusca, F. occidentalis*), velvet-bean caterpillars (*Anticarsia gemmatalis*), whitefringed beetle (*Graphognathus leucoloma*), and white grubs (*Phyllophaga* spp.).

Serious diseases are black hull (*Theilaviopsis basicola*), botrytis blight (*Botrytis cinerea*), charcoal rot (*Marcophomina phaseolina*), black rot or black root rot (*Cylindrocladium parasiticum*), collar rot (*Diplodia arachidis, D. gossypina*), crown rot (*Aspergillus niger*), leaf spot (*Cercospora arachidicola, Cercosporidium personatum*), limb rot (*Rhizoctonia solani*), pod rot (*Pythium* spp., *Rhizoctonia solani*), root rot (*Phymatotrichum omnivorum*), rust (*Puccinia arachidis*), sclerotinia blight (*Sclerotinia minor*), seedling diseases (*Pythium, Rhizoctonia,* and *Fusaruim* spp.), Southern stem blight or white mold (*Sclerotium rolfsii*), verticillium wilt (*Verticillium dahliae*), and web blotch [*Phoma (Didymella) arachidicola*]. Troublesome viruses include tomato spotted wilt and impatiens necrotic spot. Thrips are the usual vectors and should be controlled. Scorched leaves are often the result of an atmospheric pollutant, ozone.

Peanuts must be kept free of aflatoxin, which is harmful to human health (potential carcinogen, liver cancer connection). This compound is produced by *Aspergillus flavus* and *A. parasiticus*. This fungal contaminant is associated with drought-stressed dryland peanuts. Harvested peanuts can also accumulate aflatoxin if drying and curing are poor or in storage from high humidity conditions. Rotations with other crops can reduce this fungal problem, as can irrigation in times of drought.

FIGURE 18–31 Pear cultivars shown are Magness (A), Dawn (B), and Moonglow (C). The Bartlett pear is the leading cultivar in the United States California produces 60% of the Bartlett crop. (USDA photograph.)

Pears

Pears (Figure 18–31) can be divided into three groups: European pear (*Pyrus communis*), Asian pear (*P. pyrifolia, P. ussuriensis*), and Eurasian pear (a hybrid between *P. communis* and *P. pyrifolia, P.* × *lecontei*). The best known cultivars of the latter are 'Kieffer,' 'Leconte,' and 'Spalding.' Cultivars of the European pear are 'Anjou,' 'Aurora,' 'Bartlett,' 'Bosc,' 'Clapp's Favorite,' 'Comice,' 'Conference,' 'Devoe,' 'Douglas,' 'Duchess,' 'El Dorado,' 'Flemish Beauty,' 'Harrow Delight,' 'Harvest Queen,' 'Highland,' 'Honeysweet,' 'Moonglow,' 'Orcas,' 'Red D'Anjou,' 'Red Bartlett,' 'Red Kalle,' 'Rescue,' 'Seckel,' 'Spartlett,' 'Warren,' and 'Winter Nelis.' Commercial pear production is mostly 'Bartlett,' followed by 'Bosc' and 'Anjou.' Cultivars showing resistance to fire blight are 'Harrow Delight,' 'Harvest Queen,' 'Honeysweet,' 'Kieffer,' 'Magness,' 'Maxine,' and 'Moonglow.' Cultivars of the Asian pear are 'Chojuro,' 'Daisui Li,' 'Hosui,' 'Ichiban Nashi,' 'Kikusui,' 'Korean Giant,' 'Kosui,' 'Mishirashu,' 'Niitaka,' 'Nijisseki,' 'Okusankichi,' 'Seigyoko,' 'Seuri,' 'Shinko,' 'Shinseiki,' 'Shin Li,' 'Tsi Li,' 'Yakumo,' 'Ya Li,' 'Yoinashi,' and 'Yongi.' Pears are grown in home orchards and produce acceptable fruit with less attention than apples.

Pears have a more limited range than apples. Their northern range, zone 5, is limited by cold temperatures below −28.9°C (−20°F), and the humidity in the South is limiting. The bacterial disease fireblight, spread by bees and favored by rains, especially limits pear culture. Most of the commercial production centers around the Pacific Coast states of Washington, Oregon, and California. Production occurs in the Great Lakes areas, but is limited. Home plantings occur in these and other areas, but cultivars should be chosen carefully on the basis of local acceptability. About 900 to 1,000 hours below 7.2°C (45°F) are required to break dormancy.

European pear cultivars are widely propagated by budding of cultivars on seedlings of the European pear (*P. communis*), especially 'Bartlett' and 'Winter Nelis.' Seed from canneries processing Bartlett pears is usually used. The one drawback is fireblight susceptibility. *Pyrus calleryana* is used in milder areas and is fireblight resistant. *Pyrus betulaefolia* is also used as a rootstock, but low temperature sensitivity is somewhat less. Dwarf pears may be produced by budding onto a Malting quince 'A' (*Cydonia oblonga* 'Angers') root; an interstem, 'Old Home,' is usually used to overcome incompatibility problems and to increase resistance to fireblight. The quince cultivar 'Provence' is also used for dwarfing purposes. Rootstocks from 'Old Home' × 'Farmingdale' can also be used to produce dwarves and semidwarves to full size. The OH × F333, for example, produces a semidwarf. This rootstock line appears more promising than quince rootstocks. Favored rootstock for cultivars of the Asian pear include *P. betulaefolia, P. calleryana,* and various OH × F clones. All are tolerant of fireblight.

One- or two-year-old nursery stock is planted in the early spring in the North, and in fall or spring in the South. Pears may be planted 8 to 14 feet apart in rows set 12 to 24 feet apart. Dwarf trees may be set 4 to 8 feet apart in rows set 10 feet apart. Pears may commence bearing after 3 years from planting of nursery stock. Pears may live up to 100 years. Mature standard European pears can reach 45 feet in height.

Training is primarily the modified-leader system. Three or four main scaffold limbs are left, except in areas where fireblight is prevalent; six limbs are left in the latter areas in case the blight necessitates some removal. Fruits are borne on spurs that are productive for 7 to 8 years. Maintenance pruning is mainly to remove unproductive spurs and to encourage formation of new ones.

In the first year, each pear tree needs about 0.05 pound nitrogen and 0.10, 0.15, 0.20 pound each additional year. Subsequent needs are 0.30, 0.40, 0.50, and 0.60 pound/tree for their 6th, 8th, 10th, and 12th year in the orchard. Mature trees need 0.75 pound each or about 60 pounds/acre. These numbers are suggested for a density of 80 pear trees per acre. As tree densities increase, these numbers decrease somewhat.

Most pear cultivars are all or partly self-sterile, so two or more cultivars should be interplanted. Bees are recommended for maximal fruit productivity. Unfortunately, they are vectors for the spread of fireblight disease where it is present. Parthenocarpic fruits also occur. Fruit thinning is minimal; it is practiced by hand or chemical and hand treatments when larger fruit is desired. Pears can be thinned with auxin-based fruit thinners.

Troublesome insects include the blister mite (*Eriophyes pyri*), codling moth (*Carpocapsa pomonella*), pear and tarnished plant bug, pear midge (*Contarinia pyrivora*), pear thrip (*Taeniothrips inconsequens*), pear psylla (*Psylla pyricola*, vector of pear decline caused by a *Phytoplasma* sp.), plum curculio, and scale. Rodents can cause extensive girdling damage. Diseases of importance are cankers (*Cytospora* sp. and *Pseudomonas syringae*), fireblight (*Erwinia amylovora*), leaf blight (*Fabraea maculata*), pear decline (phytoplasma), powdery mildew (*Podosphaera leucotricha*), and *Phytopthora* root rot.

Yields of 150 to 240 bushels per acre (4 to 5 bushels per tree) are possible. In 2000 the average pear yield was 14.6 tons/acre in the United States. Pears should be precooled. Storage is at 0°C (32°F) and 90% to 95% relative humidity. With precooling and these temperatures, storage for 2 to 3 months, depending on cultivar, is possible. Controlled atmosphere storage is better, but used very little. Pears offer fiber and modest amounts of vitamins B and C. Pears are eaten fresh, as part of a fresh or canned fruit salad, in baked desserts, and as pear nectar and pear brandy. The best guide to pear ripeness is to apply pressure with the thumb near the stem base. Slight yielding indicates ripeness.

Pecans

Of all the hickory nuts, the pecan, *Carya illinoinensis*, is the one most valued by commercial and backyard nut growers. Native pecan trees tend to have small nuts, thick shells hard to crack, but good flavor. Cultivars are numerous with bigger nuts with thinner shells easier to crack, hence the name *papershells*. Papershells can be planted as cultivar orchards or topworked into native pecan groves. Native groves are currently 20% of pecan production acreage. Native groves are often used in an agroforestry arrangement with cattle or sheep grazing among the trees.

Pecans (Figure 18–32) require a frost-free growing period of 140 to 220 days, depending on the cultivar. This wide range determines the grouping of cultivars into northern and southern (divided into eastern and western southern) categories. Cultivars recommended for your region should be chosen. Flowering is late enough to be uninjured by late spring frosts. Commercial production is limited to the South to as far as New Mexico. Cultivars are grouped as Eastern, Western, or Northern in reference to their placement in the pecan growing range. Eastern cultivars are grown in the humid Southeastern states from Louisiana to Florida and usually have some resistance or tolerance to scab. Western cultivars are adapted to the arid areas of Texas and the Southwest, where scab is not a problem. Northern cultivars mature in shorter growing seasons and are used in Iowa, Illinois, Indiana, Kansas, Kentucky, Missouri, Oklahoma, and Tennessee. While trees can be grown further north, nuts tend to be smaller and are suitable only for backyard production.

FIGURE 18–32 Pecans are high in monounsaturated fats (73%). Pecans were consumed by Native Americans. The pecan is the state tree of Texas. (USDA photograph.)

Some North cultivars include 'Canton,' 'Colby,' 'Fritz,' 'Gibson,' 'Giles,' 'Green River,' 'Hirshi,' 'James,' 'Lucas,' 'Major,' Mullahy,' 'Norton,' 'Pawnee,' 'Peruque,' 'Posey,' 'Starks,' Hardy Giant,' and 'Witte.' Eastern cultivars include 'Caddo,' 'Candy,' 'Chickasaw,' 'Choctaw,' 'Desirable,' 'Elliott,' 'Graking, 'Kiowa,' 'Mahan,' 'Mohawk,' 'Moore,' 'Schley,' 'Shoshoni,' 'Stuart,' and 'Success.' Western cultivars include 'Apache,' 'Cheyenne' (dwarf), 'Grabohls,' 'Ideal,' 'San Saba Improved,' 'Sawnee,' 'Sioux,' 'Tejas,' 'Western Schley,' and 'Wichita.'

Propagation is by patch budding, bark-grafting, or whip grafting of cultivars onto pecan rootstocks. Rootstocks are grown from seeds of 'Riverside,' 'Burkett,' and 'Western' in the West, from 'Curtis,' 'Stuart,' 'Success,' and 'Mahan' in the Southeast, and from 'Giles' in the North. Three- or four-year-old trees are transplanted best in late winter or early spring as the buds begin to swell. Care must be exercised because of the taproot. Trees are placed 80 feet apart, although spacings of 30 to 40 feet can be used if thinning of trees is practiced when crowding occurs, such that the final thinning leaves trees with an 80 by 80 spacing. Bearing can occur by 7 to 10 years. Mature height may be as much as 150 feet. Topworking uses the inlay bark-graft.

Pruning to produce a tree with a single trunk with its lowest branches at 5 to 7 feet is standard practice. Maintenance pruning is minimal. Trees are monoecious and wind pollinated. Trees are self-fertile, but cultivars are often mixed, since cross-pollination may produce more or higher quality nuts. A tree may produce 50 to 100 pounds of nuts per year. Native groves produce about 600 pounds per acre, but can run as high as 1,000. Pecan orchard yields are as high as 2,500 pounds per acre, but usually run 800 to 1,200 pounds per acre. Pecans usually exhibit alternate bearing.

Foliar tissue analyses supplemented with soil tests are critical for achieving sustainable fertilization with pecan trees. Nitrogen needs vary from 60 to 200 pounds per acre, depending on soil levels, tree age, and cover crop considerations. Fertilizer is usually applied in two to four split applications, starting in early spring. Fertilizer should be watered in, unless it is possible to incorporate it into the soil. Fertigation is an especially good system for nut trees. A backyard tree in its first year can use 1 pound of 10-10-10 and 2 pounds in its second year. For each subsequent year of age, add an additional 2 pounds up to a maximum of 50 pounds per tree.

Troublesome insects include several mites, several aphids, shuckworm (*Laspeyresia caryana*), nut casebearer (*Acrobasis juglandis* and *A. caryae*), scales, weevil (*Curculio caryae*), and curculio (*Conotrachelus affinis*). Scab (*Cladosporium effusum*) is a serious disease.

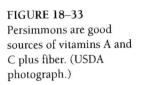

FIGURE 18–33
Persimmons are good sources of vitamins A and C plus fiber. (USDA photograph.)

Persimmons

Persimmons (Figure 18–33) grown for fruit include *Diospyros virginiana* and *D. kaki,* the American and Oriental (or Japanese) persimmon, respectively. Only the Oriental persimmon is grown commercially; both are grown in backyard plantings. American persimmon cultivars are 'Early Golden,' 'Meader,' 'Male' (pollinator), 'Miller,' and 'Ruby.' Cultivars of Oriental persimmons are grouped as astringent and nonastringent. In addition some cultivars are parthenocarpic, but if pollinated, develop seeds (pollination variant varieties). These cultivars have changes in flavor and texture when pollinated.

Astringent cultivars include 'Eureka,' 'Hachiya,' 'Honan Red,' 'Saijo,' 'Tamopan,' 'Tanenashi,' and 'Triumph.' Nonastringent cultivars are 'Fuyu,' 'Gosho,' 'Ichi Ki Kei Jiro,' 'Imoto,' 'Izu,' 'Jiro,' 'Maekawajiro,' 'Okugosho,' and 'Suruga.' Pollination variant varieties that are astringent when seedless include 'Chocolate,' 'Coffee Cake,' 'Gailey,' 'Hyakume,' and 'Maru.' The first two are good cultivars for assuring seed set (and hence nonastringency) with the last two.

Best growing conditions for the Oriental persimmon are found in the Cotton Belt (zone 8). Persimmons have low chilling requirements (<100 hours) and tend to break dormancy during premature warm spells and hence spring frosts can be damaging. Minor commercial production is found in California, Texas, Florida, and southern Georgia. The American persimmon is grown primarily as a backyard plant in the eastern and southeastern states to Texas.

Propagation of cultivars is by cleft grafting, whip grafting, or shield budding on 1- to 2-year-old seedling stocks. The seedling stock in the East and South is *D. virginiana.* In California, both *D. kaki* (most common) and *D. lotus* (date plum) are used. Transplanting is difficult because of the presence of a taproot. Grafted trees are planted in the late fall in warmer areas and in early spring in others. Spacing is 15 to 20 feet between trees. Bearing may commence 2 to 3 years after grafting. Trees tend to be short lived. Mature height is 45 feet for the Oriental type and more for the American species.

Training is to the modified-leader or open-center system. Heading back annually is needed to maintain a manageable size. Thinning of some new growth helps to minimize alternate bearing. Fruit is produced on the previous season's wood. Persimmons can also be pruned to hedge forms or espaliers. Flower types vary considerably; perfect, staminate, and pistillate flowers may even be produced on one tree. Parthenocarpic fruit set also occurs. 'Gailey' is frequently planted as a pollen source. Bees are the primary pollinators. Fruit thinning is usually not needed.

Minimal fertilization is needed. At most 1 pound of 10-10-10 per tree in late winter or early spring is needed. Soil pH can be 6.5 to 7.5. While trees can tolerate drought, fruit size and quality are better with regular watering. Insect and disease problems are minimal. Mealybugs, scale, thrips, and whiteflies can be problems at times. Several animals find the fruit good to eat and gophers damage roots. Avoid overwatering, because root rot can be a problem.

Astringent cultivars have an astringent taste until fully ripe. These cultivars can be allowed to fully ripen on the tree until jelly soft, at which point fruits are no longer astringent. If birds are problems, or shipping is in order, fruits can be picked at the mature hard, colored stage and ripened at room temperature. Nonastringent cultivars can be picked any time at full color even though hard. Flavor does improve somewhat when softer. Hand harvest is used to leave the calyx intact. Care is required, because persimmons bruise easily even though hard. Yields of 50 pounds per tree are seen, although with optimal culture higher yields are possible. Storage of unripe fruit at $-1.1°C$ (30°F) and 85% to 90% relative humidity is possible for 3 to 4 months. Fruits dry well; astringent cultivars at the firm stage lose their astringency during the drying process. Fruits can be eaten or made into juice, wine, jellies, and sorbets or used as flavoring in baked goods. Juice is also sold as grenadine.

Pineapples

The pineapple (Figure 18–34), *Ananas comosus,* is a terrestrial herbaceous bromeliad primarily grown in the tropics. Culture is limited in the continental United States to southern Florida and California. Commercial production is mostly in Hawaii. The main commercial cultivar is 'Smooth Cayenne.' Others are 'Hilo,' 'Kona Sugarloaf,' 'Natal Queen,' 'Pernambuco,' and 'Red Spanish.' Mature height is 3 to 5 feet with a 3- to 4-foot spread.

Propagation is mainly by suckers developed along the stem or slips at the peduncle just below the fruit. The crown (leafy part on top of fruit) can also be rooted, but is much slower to produce fruit. Pineapple slips or suckers are placed 22 inches apart in beds and set 3 to 4 inches deep, usually after they have been removed from harvested fruit and dried for 1 to 4 weeks. Fall-planted slips can set fruit the second summer following. Pineapples do well with black plastic mulch. Cultivation should be shallow. Soil pH should be 4.5 to 6.5. Well-drained soils are essential. Irrigation is essential for good fruit quality, even though pineapples are drought tolerant. During periods of dryness, between 5,000 and 10,000 gallons of water per acre are needed. Pineapples are heavy feeders (up to 400 pounds of nitrogen/acre/year) and

FIGURE 18–34
Pineapple. The pineapple is thought to have originated in Paraguay. The fruit is rich in betacarotene (pre-vitamin A), vitamin C, and B-complex vitamins. (Author photograph.)

fertilizer needs to be applied every 4 months. Pineapples respond well to foliar feeding. Black plastic mulch is widely used in Hawaii with pineapples.

Fruit formation is usually induced chemically for purposes of uniform harvesting. Acetylene gas, calcium carbide solution, α-naphthaleneacetic acid, or β-hydroxyethyl hydrazine have been used. Ethylene gas from an apple in a plastic bag around the plant can be used to cause flowering and fruit formation in individual, homegrown plants. Ethylene in water with an absorbent is also sprayed on pineapple fields. Parthenocarpic fruit formation is possible, because pineapples are self-incompatible. Most commercial pineapples are seedless, because the fields are isolated and consist of the same cultivar. Plantings are usually replaced after 4 years. Yields of pineapples during 2000 in Hawaii were 17 tons/acre.

Insect pests include ants, associated with mealybugs (*Pheidole megacephala, Iridomyrmex humilis, Solenopsis geminata*), mealybugs (*Dysmicoccus brevipes, D. neobrevipes*), mites (*Steneotarsonemus ananas, Dolichotetranychus floridanus*), scales (*Diaspis bromeliae, Melanaspis bromeliae*), symphylids (*Scutigerella sakimurai, Hanseniella unguiculata*), thrips (*Thrips tabaci, Frankliniella occidentalis*), and vectors of yellow spot virus. Nematodes (*Helicotylenchus* spp., *Meloidogyne javanica, Pratylenchus brachyurus, P. elachistus, Rotylenchulus reniformis*) are troublesome.

Wilt is a troublesome disease (possibly viral) spread by mealybugs. Other viruses are yellow spot and possibly mottle. Problem diseases include anthracnose (*Colletotrichum ananas*), black rot (*Ceratocystis paradoxa*), butt rot [*Chalara (Thielaviopsis) paradoxa*], bacterial heart rot (*Erwinia carotovora, E. chrysanthemi*), fruitlet core rot (*Penicillium funiculosum, Fusarium moniliforme* var. *subglutinans*), heart and root rot (*Phytophthora cinnamomi, P. parasitica*), leaf spot (*Curvularia eragrostidis*), pink disease of fruit (*Acetomonas* spp.), and root rot (*Pythium* spp.).

Pineapples are best when picked ripe, but are usually picked about a week before full maturity by commercial growers. Ripe pineapples have a dull, solid sound when snapped on the side by a finger. Another clue is a pleasant fragrance and a yellow coloration. Mature green pineapples can be stored for 14 to 36 days at 10°C to 12.8°C (50° to 55°F) and 85% to 90% relative humidity. Ripe pineapples have a similar storage life and condition, but at a lower temperature of 7.2°C (45°F). Controlled atmospheres can extend storage by a few weeks.

Pistachios

Pistachios (*Pistacia vera*) are grown both commercially and by backyard horticulturists. They are dioecious. Pollinating cultivars are 'Peters' (the standard cultivar) and 'Chico.' Female cultivars include 'Bronte,' 'Buenzle,' 'Ibrahmim,' 'Kerman' (the leading commercial cultivar), 'Minassian,' 'Owhadi,' 'Red Aleppo,' 'Safeed,' 'Sfax,' 'Shasti,' 'Trabonella,' and 'Wahedi' (largest nuts). Hardiness is through zone 9. Long, hot summers with low humidity are favorable. Chilling requirements to break dormancy vary by cultivar from 700 to 1,500 hours. Commercial and amateur production is limited in the United States and primarily restricted to the hot, drier areas in Southern California. Pistachio nuts have good food value with 20% protein and 50% to 55% oil, most of which is unsaturated.

Propagation of cultivars is by shield or T budding and grafting on seedling rootstocks. *Pistacia atlantica, P. integerrima* (*Verticillium* resistant), and *P. terebinthus* are mainly used as rootstocks because of nematode and soil fungi resistance. Container-grown trees have a higher transplanting survival rate than bare-rooted trees because of taproots. Mature trees are 25 to 30 feet tall and have a deciduous, broad, bushy form.

Container-grown trees are set out in any season at 30-foot intervals. Smaller distances can be used, but interplants must be removed when crowding occurs. Trees may begin bearing in 5 to 8 years. Full bearing occurs in the 10th to 20th year. One male cultivar is planted per 10 to 12 female cultivars. Pollination is by wind. Training is minimal and is used to produce a high heading, rather than bushy, spreading tree. The modified-central leader system is the norm. Minimal pruning is needed by the homeowner, but commercial growers prune to shape trees for mechanical harvesting. Pistachios can tolerate considerable dryness. The best regimen is deep, infrequent irrigation.

Soil pH can be 6.5 to 8.0. Fertilizer needs are modest. Foliar tissue analyses supplemented with soil tests are critical for achieving sustainable fertilization with pistachio trees. Nitrogen needs vary from 60 to 150 pounds per acre, depending on soil levels, tree age, and cover crop considerations. Fertilizer is usually applied in two to four split applications, starting in early spring. Fertilizer should be watered in, unless it is possible to incorporate it into the soil. Fertigation is an especially good system for nut trees. A backyard tree in its first year can use 1 pound of 10-10-10 and 2 pounds in its second year. For each subsequent year of age, add an additional 2 pounds up to a maximum of 50 pounds per tree.

Nuts fall easily when mature, so tree shaking by hand or machine works well. Hulls are removed promptly and the naturally split nuts can be boiled in salt water for a short time, dried, packaged, and stored. Yields vary from 15 to 50 pounds of nuts per tree per year. However, production varies because of biennial bearing. Average yields in California during 1999 and 2000 were 0.87 and 1.63 tons/acre, respectively. Mature height is 15 to 30 feet. Pistachios can be held refrigerated for 4 to 6 weeks or many months when frozen.

Disease problems include leaf spot (*Phyllosticta lentisci*), root rot (*Phymatotrichum omnivorum*), oak root fungus (*Armillaria mellea*), and *Verticillium* wilt. The last is the most serious problem in California. Scales, aphids, and mites are minor insect pests. Squirrels, bluejays, and woodpeckers can be a problem.

Plums

Plums (see Figure 9–10), species of *Prunus*, can be divided into four groups. The common or European plum is *P. domestica*. This group is the source of both blue plums eaten fresh and other drier textured blue (prune) plums used to prepare prunes by drying. Cultivars in this group include 'Agen' (largest commercial prune source), 'Arctic,' 'Blue Damson' (damson plum preserves), 'Blufre,' 'Bradshaw,' 'Brooks,' 'Cambridge Gage,' 'Castleton,' 'Coe's Golden,' 'Earliblue,' 'Early Laxton,' 'Emperor,' 'Empress,' 'Ersinger,' 'French' (best for jam), 'Grand Duke,' 'Grand Prize,' 'Green Gage,' 'Herrenhausen,' 'Imperial Epineuse,' 'Mirabelle,' 'Italian Prune,' 'Long John,' 'Mirabelle,' 'Mohawk,' 'Mount Royal,' 'NY # 51,' 'Oullins,' 'Polly,' 'Reine Claude,' 'Richard's Early Italian,' 'Schoolhouse,' 'Seneca,' 'Stanley' (good home garden type), 'Valor,' 'Victoria,' 'Victory,' 'Voyageur,' 'Washington,' and 'Yellow Egg.' Japanese or Oriental plums are usually red or yellow and derive from *P. salicina*; cultivars are 'Beauty,' 'Bryongold,' 'Crimson,' 'Formosa,' 'Fortune,' 'Friar,' 'Oushi-Wase,' 'Ozark Premier,' 'Red Ace,' 'Redheart,' 'Ruby Sweet,' 'Santa Rosa,' 'South Dakota,' and 'Wickson.'

Native American plums are derived from a number of species: wild plum, *P. americana* ('De Soto,' 'Forest Garden,' 'Hawkeye,' and 'Wolf'); sand plum, *P. angustifolia* ('Caddo Chief,' 'Strawberry,' and 'Yellow Transparent'); hortulan plum, *P. hortulana* ('Golden Beauty,' 'Miner,' and 'Wayland'); beach plum, *P. maritima* ('Autumn,' 'Eastham,' 'Hancock,' 'Premier,' 'Raritan,' and 'Stearns'); wild goose plum, *P. munsoniana* ('Newman,' 'Robinson,' and 'Wild Goose'); Canadian plum, *P. nigra* ('Cheney,' 'Hasca,' and 'Oxford'); and the Pacific plum, *P. subcordata* ('Sisson').

The last group is of hybrids between the American and Japanese species; these are represented by cultivars such as 'Delight,' 'Ember,' 'Kaga,' 'Kahinta,' 'La Crescent,' 'Methley,' 'Monitor,' 'Pipestone,' 'Shiro,' 'Sprite,' 'Superior,' 'Toka,' 'Underwood,' and 'Waneta.' The cultivars mentioned are only a partial listing; these and others can be found both in commercial and backyard production. The cultivars used commercially come mainly from *P. domestica* and *P. salicina*.

Because of the wide number of plum species, it is possible to grow plums in just about any part of the United States. Commercial production is centered in California, followed by Oregon, Washington, Michigan, Idaho, Texas, and New York. The European plum is hardy through zone 5, the Japanese plum through zone 8, the American plum as far as zone 2 (*P. nigra*), and Japanese-American hybrids to zones 6 to 7. Local authorities should be consulted for cultivars best suited to your area. Chilling required to break dormancy is 800 to 1,200 hours below 7.2°C (45°F) for the European plum and 700 to 1,000 hours for the Japanese plum.

Propagation is primarily by budding and cuttings to a much lesser extent. Various rootstocks are used. The cherry plum or Myrobalan rootstock (*P. cerasifera*) is widely used in the East and West for European, Japanese, and Japanese-American hybrid plums. Their drawback is they are short lived on sandy or other drought-sensitive soils. Peach rootstocks (Halford and Lovell) are sometimes used for Japanese plums to be grown in light soils or in the South, but tend to be short lived and susceptible to peach problems. The 'Marianna' series of rootstocks works with most plums, but is more sensitive to low winter temperatures. 'St. Julien GF 655.2' has been used for heavy, wet soils. American plum rootstocks are used when increased hardiness is desired. Citation is good for semi-dwarfing.

One-year-old nursery stock is usually planted in early spring. Trees are set 18 feet apart in rows spaced 22 feet apart. Spacing of 8 by 12 feet is possible with cordon-trained trees. Japanese plum may bear fruit in 3 to 5 years, European plums in 5 to 7 years, and American plums in about 8 to 10 years.

Nitrogen needs should be based on foliar tissue analyses and other nutrients on soil tests. A 1-year-old plum tree needs roughly 0.07 pound of nitrogen yearly. For each additional year, another 0.07 pound should be added until the total in the ninth year is 0.63 pound. Mature trees should be continued at this rate, which is equivalent to 80 pounds of nitrogen per acre.

Plums are trained in the modified-leader and open-center systems; choice should be based on growth habits. Spreading types are better suited to the open-center system and upright types to the modified-leader system. Fruit is borne on l-year-old wood and the more vigorous spurs on older wood. Pruning is generally light, with the Japanese plum requiring somewhat more than the European plum. Pruning requirements can vary depending on cultivar. European and Japanese plums can grow to 25 to 30 feet in height. American plums vary from 10 to 30 feet.

European plum cultivars vary from self-fruitful to self-unfruitful. Japanese plums are mostly self-unfruitful. American and hybrid plums vary. Pollination requirements of cultivars must be carefully determined, and cultivars chosen for interplanting must be compatible. Bees are recommended for orchards. Fruit thinning is practiced extensively to increase fruit size and to prevent breakage of the tree. Hand, mechanical, and chemical thinning are practiced.

Troublesome insects include the plum curculio (*Conotrachelus nenuphar*), peach borer (*Sanninoidea exitiosa*), and shot-hole borer (*Scolytus rugulosus*). Diseases include bacterial leaf spot (*Xanthomonas pruni*), black knot (*Dibotryon morbosum*), brown rot (*Monilinia fructicola*), canker (*Cytospora* sp.), crown gall (*Agrobacterium tumefaciens*), plum pockets (*Taphrina pruni*), and ring spot virus.

Yields in the East are 3 to 5 tons per acre or 1 to 1.5 bushels per tree. California yields are 6 to 10 tons per acre. In 2000 in California the average plum and prune yield, respectively, was 5.2 and 8.0 tons/acre. Plums are perishable and can be held at $-1.1°$ to 0°C (30° to 32°F) for 2 to 4 weeks. Precooling is desirable.

Pomegranates

The pomegranate (Figure 18–35), *Punica granatum,* is a deciduous shrub reaching a height of 12 to 16 feet, although in ideal conditions, heights of 20 to 30 feet are possible. Cultivars include 'Balegal,' 'Cloud,' 'Crab,' 'Early Wonderful,' 'Eversweet,' 'Fleshman,' 'Francis,' 'Granada,' 'Green Globe,' 'Home,' 'King,' 'Phoenicia,' 'Sweet,' 'Utah Sweet,' and 'Wonderful' (leading commercial cultivar). Its culture is limited in the United States to the hot desert valleys of California and the Southwest. Good crops require a semiarid condition where heat is present for ripening and water can be supplied to the roots. Commercial and backyard production is limited.

Dormant woody cuttings are used for propagation. Rooted cuttings are placed 12 to 15 feet apart in early spring. Bearing commences in 3 to 5 years. Fruit is borne on spurs found on 2-year-old or older wood. Training involves cutting back at 2 feet and allowing four to

FIGURE 18–35
Pomegranates. Temperatures below 41°F cause chilling injury during pomegranate storage. (USDA photograph.)

five shoots at the 1-foot level to develop. Shoots above and below plus suckers are removed. Pruning of established shrubs is minimal, except to remove suckers and thin dense growth or old fruiting wood.

Pomegranates are drought tolerant, but irrigation is needed for best fruiting. Fertilizer needs are minimal at 2 to 4 ounces of ammonium sulfate per shrub at most. Soil pH can be 6.5 to 7.5. Insects and disease are minimal. Mealybugs, scale, thrips, and whiteflies are minor pests. Leaf and fruit spot are minor problems.

Fruits are ripe when colored and a metallic sound is heard on being tapped. Ripe fruit cracks if left on trees, especially after rain. Long-term storage for 60 to 90 days is at 5°C (41°F) and 90% to 95% relative humidity. Production is about 5 tons/acre. Fruit picked prior to maturity can also be stored for months in a cool, dry place.

Quinces

Fruits of the species *Cydonia oblonga* are those of the common quince. *Cydonia sinensis* is also used for fruit. The Japanese or Oriental quince, *Chaenomeles speciosa,* is also used occasionally for its fruit. Cultivars of the common quince include 'Aromathaya,' 'Champion,' 'Dwarf Orange,' 'Fuller,' 'Havran,' 'Meech,' 'Orange,' 'Pineapple,' and 'Smyrna.' Primary usage is for jelly, although 'Aromathaya' can be eaten fresh in thin slices.

Quinces are comparable to peach in wood hardiness. *Cydonia oblonga* is hardy through zone 5, but *C. sinensis* is hardy only through zone 6. Since it is susceptible to fireblight, this can be a limiting factor. Production is mainly in backyards; commercial production is very limited to small areas in California, Ohio, Pennsylvania, and New York.

Propagation is by rooting hardwood cuttings or by budding of cultivars onto a quince rootstock, usually 'Angers.' Two- or three-year-old nursery stock is set in early spring about 16 feet apart. Bearing may start 2 to 3 years after planting the nursery stock and reach a maximum in 10 years. Mature height is about 20 feet.

Pruning is minimal after training to the open-center system. Fruits are borne on new growth of the current year. Flowers are self-fruitful. Bees are the main pollinators. Some fruit thinning may be needed in some years if the tree overbears. Insect pests include the Oriental fruit moth, quince curculio, aphid, and lace bug. Diseases include the cedar-quince stem rust, black spot, and fireblight.

Yields are about 1 bushel per tree. Quinces can be stored at −1.1° to 0°C (30° to 32°F) for about 2 months.

(A)   (B)

FIGURE 18–36   Raspberry cultivars shown are Heritage, a red raspberry (A) and Jewel, a black raspberry (B). According to legend, the Greek gods went to Mt. Ida in Turkey and brought back wild raspberries. This legend is reflected in the scientific name for red raspberries, *Rubus idaeus*. (USDA photographs.)

Raspberries

Raspberries (Figure 18–36), a bramble fruit, can be divided into three groups. The red raspberry is *Rubus idaeus;* the most common variety in the United States is variety *strigosus*. The black raspberry is *R. occidentalis*. Hybrids between the two preceding groups are *R.* × *neglectus*, the purple raspberry. Cultivars of the red raspberry include 'Algonquin,' 'Amity' (everbearing), 'Autumn Bliss' (everbearing), 'Autumn Cascade' (everbearing), 'Balder' (winter hardy), 'Boyne' (winter hardy), 'Canby,' 'Caroline' (everbearing), 'Chilcotin,' 'Chilliwack,' 'Citadel,' 'Dinkum' (everbearing), 'Dorman Red,' 'Double Delight' (everbearing), 'Encore,' 'Fall Red' (everbearing), 'Festival,' 'Haida,' 'Heritage,' 'Hilton,' 'Killarney' (winter hardy), 'Latham' (winter hardy), 'Lauren,' 'Liberty,' 'Madawaska' (winter hardy), 'Mammoth,' 'Newburgh,' 'Nordic' (winter hardy), 'Nova,' 'Polana' (everbearing), 'Prelude,' 'Qualicum,' 'Red Wing' (everbearing), 'Regency,' 'Reveille,' 'Ruby' (everbearing), 'Sentry' (winter hardy), 'September' (everbearing), 'Southland,' 'Summit' (everbearing), 'Taylor,' 'Titan,' 'Trent,' and 'Tulameen.' Cultivars of the black raspberry include 'Allegany,' 'Allen,' 'Black Hawk,' 'Bristol,' 'Cumberland,' 'Dundee,' 'Early Sweet,' 'Haut,' 'Huron,' 'Jewel,' 'John Robertson' (winter hardy), 'Logan,' 'Lowden,' and 'Mac Black.' Cultivars of the purple raspberry include 'Amethyst,' 'Brandywine,' 'Clyde,' 'Estate,' 'Royalty,' and 'Success' (winter hardy). Yellow cultivars include 'Anne' (everbearing), 'Fall Gold' (everbearing), 'Golden Harvest' (everbearing), 'Goldie' (everbearing), and 'Honey Queen.' 'Boysen' and 'Logan' are hybrid blackberry/raspberry cultivars.

The red raspberry is hardy through zone 3, and the black and purple raspberries are hardy through zone 4. Heat and drought are limiting factors for raspberry culture. As such they are cultivated mostly north of the Mason-Dixon line. Raspberries are produced both in commercial and backyard plantings; the leading commercial areas are Oregon and Washington. Certain cultivars are recommended for various regions; local authorities should be consulted. Raspberries are eaten fresh and used in desserts, pies, juices, jams, and liquors. Vitamin C levels are good.

Tip layering is used for the propagation of black and purple cultivars. This is done in middle to late summer, and roots will have formed by autumn. Red cultivars are increased by removing the suckers that are produced by the crown and roots. Purchased plants should be certified as virus free.

One-year-old plants are set out in the spring. Red raspberries are set 2 to 3 feet apart in rows at 7- to 9-foot intervals. Black raspberries are set 3 to 4 feet apart in rows 8 to 10 feet apart, and purple raspberries have the same spacing within the row, but rows are set 10 feet apart. Suckers arising more than 1 foot on either side of the row should not be allowed to grow. Simple two-wire trellises are often placed to contain and support the drooping plants. Canes are biennial; fruit is produced only once on the second-year cane, except for everbearing types, which set fruit twice in one season (June and August). These are suggested for areas where winter cold kills (if growing season is long enough) or injures summer-fruiting canes (2 years old), since the fall crop is borne on current season canes.

Black and purple raspberries require summer topping; red ones generally do not. When shoots have reached 18 to 24 inches for black cultivars and 30 to 36 inches for purple cultivars, the top 3 to 4 inches is removed. Thereafter, tips should be removed at weekly intervals. All should have fruiting canes removed after harvesting berries (everbearing raspberries later than others) and some thinning out of the weaker canes produced that season. Thinning may be delayed until early spring. In early spring before buds swell, red raspberries should be cut back to 5 to 5.5 feet. Laterals of summer-topped black and purple cultivars should be cut back to 8 and 10 to 18 inches, respectively. Fruit set is improved substantially by bees. Fruit thinning is not practiced. Reds may be productive for 20 years, while purples and blacks last 5 to 10 years.

Raspberries are tolerant of soil pH variation from 4.5 to 7.5, but the optimal pH is 6.0 to 6.5. Summer-bearing purple and black raspberries require 25 to 30 pounds of nitrogen annually per acre for establishment and 50 to 60 pounds for maintenance. Summer-bearing red raspberries need 30 to 35 and 65 to 75 pounds of nitrogen for establishment and maintenance, respectively. The respective establishment and maintenance numbers for everbearing raspberries are 30 to 40 and 95 to 115. Applications can be done once prior to bloom in the spring or split into pre-bloom and immediate postharvest applications. If the cropping system is dryland farming, amounts should be reduced by 10% to 25%. Potassium and phosphorus needs are determined by soil tests. Overfertilization and application of high-phosphate fertilizer should be avoided, because too much phosphorus has been implicated in zinc deficiency of brambles. Potassium chloride should be avoided also, because brambles are chloride sensitive. A constant supply of water is needed from early spring through fruit harvest. The top 2 feet of soil need moisture, because that is the rooting depth of raspberries. Raspberries can use 0.20 to 0.33 inch of water each day. Trickle and soaker types of irrigation are best. Overhead irrigation does provide for the option of heat and frost protection.

Disease problems include anthracnose (*Elsinoe veneta*), cane and leaf spot (*Septoria rubi*), cane blight (*Leptosphaeria coniothyrium*), crown gall (*Agrobacterium tumefaciens*), gray mold or fruit rot (*Botrytis cinerea*), late leaf rust (*Pucciniastrum americanum*), orange rust (*Gymnoconia nitens* and *Arthuriomyces peckiana*), *Phytophthora* root rot, powdery mildew (*Sphaerotheca macularis*), spur blight (*Didymella applanata*), and verticillium wilt (*Verticillium dahaliae*). Mosaic complex virus is troublesome, as is leaf curl, streak, and tomato ring spot. Aphids and nematodes are involved in the spread of these viruses.

Insect pests include aphids (*Aphis rubicola, Amphorophora agathonica*), consperse stink bug (*Euschistus consperus*), crown borers (*Pennisetia marginata*), cane borers (*Oberea bimaculata, Agrilus ruficollis*), gallmaker (*Diastrophus nebulosis*), Japanese beetle (*Popillia japonica*), picnic beetles (*Glischrochilus quadrisignatus, G. fasciatus*), psyllid (*Trioza tripunctata*), raspberry cane maggot (*Pegomya rubivora*), raspberry fruit worm (*Byturus unicolor*), raspberry leafroller (*Olethreutes permundana*), raspberry saw fly (*Monophadnoides geniculatus*), rose scale (*Aulacaspis rosae*), several mites (*Tetranychus* spp.), stalk borer (*Papaiperma nebris*), and tree cricket (*Oecanthus fultoni*).

Yields of purple, black, and red raspberries are, respectively, 4,000 to 5,000; 2,000 to 2,500; and 3,000 to 4,000 quarts per acre. Higher yields are possible with optimal conditions. In 2000, the average yield of black (Oregon) and red raspberries (Washington) was 1.7 and 3.8 tons/acre, respectively. In 2000, the average yield of boysenberries (California) and loganberries (Oregon) was 4.8 and 2.9 tons/acre, respectively. Yields per bush for all vary from 2 to 4 quarts. Raspberries are highly perishable, and precooling is suggested. Storage life is only 2 to 3 days at 0°C (32°F) and 90% to 95% relative humidity.

Saskatoons, Serviceberries (*see* Juneberries)

Strawberries

The cultivated strawberry (Figure 18–37) is a hybrid between *Fragaria chiloensis* and *F. virginiana* and is designated *F. × ananassa* (Figure 18–31). Plants are low-growing herbaceous perennials that produce runners. Chilling requirements are less than 500 hours to break dormancy. Strawberries contain ellagic acid, which acts as an antioxidant, free-radical scavenger, anticarcinogen, and antimutagen. Strawberries also contain high levels of vitamin C. Uses include fresh eating, freezing, jams, jellies, ice cream, pies, and liquors. Commercial production is concentrated primarily in California and somewhat in Florida. Other states include Oregon, Arkansas, Louisiana, Michigan, New Jersey, North Carolina, New York, Ohio, Pennsylvania, Washington, and Wisconsin.

Strawberry cultivars vary considerably in horticultural characteristics, but fall into two groups: June-bearing and everbearing. June-bearing cultivars have one heavy crop in the spring and flowering is dependent on short days and temperatures under 15°C (59°F). Everbearing types produce fruit throughout the growing season, but are of limited value in commercial production. Local authorities should be consulted for cultivars best suited to your area in the United States. The most important cultivars and some of lesser importance at this time are as follows: 'Abundant Sparkle,' 'Aiko,' 'Allstar,' 'Annapolis,' 'Apollo,' 'Atlas,' 'Benton,' 'Camarosa,' 'Cardinal,' Cavendish (very cold hardy), 'Chandler,' 'Douglas,' 'Dover,' 'Dunlap,' 'Earliglow', 'Flordabelle,' 'Fort Laramie' (everbearing), 'Guardian,' 'Honeoye,' 'Jewel,' 'Kent,' 'Midway,' 'Ogallala' (everbearing), 'Ozark Beauty' (everbearing), 'Pajaro,' 'Parker,' 'Primetime,' 'Puget Summer,' 'Redcoat,' 'Seascape' (everbearing), 'Selva' (everbearing), 'Shuksan,' 'Sparkle,' 'Surecrop,' 'Sweet Charlie' (everbearing), 'Tioga,' 'Tribute' (everbearing), 'Tristar' (everbearing), 'Tufts,' 'Veestar,' and 'Winona.'

Propagation is by new plants produced via rooted runners. New plants (certified virus free) are put out in early spring in the North, and usually in the autumn in the South and California. Late frost pocket areas should be avoided to minimize frost damage to the flowers in the spring. Most cultivars are self-fruitful. Information suggests that bees could be used to increase volume of fruit and numbers of perfect fruit.

Spacing depends on the training system (Figure 18–2) used. In the hill system, fall plantings are used. Strawberries are harvested during the following late winter (Deep South) or early spring and then the beds are renovated and replanted. Black plastic mulch is generally used in the hill system. Laying black plastic and drip irrigation tubing at the

FIGURE 18–37
Strawberries are very good sources of vitamin C and fiber. Once picked, berries do not ripen any further. (USDA photograph.)

same time beds are formed and fertilized is done by specialized machines. Plants are set 12 to 14 inches apart in twin rows spaced at 12 to 14 inches, with 3.5 to 4 feet between each set of twin rows. Overhead sprinkling is usually needed the first week to aid establishment after transplanting. Drip irrigation is used after that week. The annual-replacement hill system is used in commercial plantings in California and Florida.

The matted row system (see Figure 18–2) is used mainly in the commercial processing systems and home gardens. In this system plants are set 15 to 30 inches apart in rows, separated by 3.5 to 5 feet. Runners fill in the entire bed, being allowed to root randomly. Runners are not allowed to grow beyond a 15- to 24-inch strip, because cultivation is used for weed control. Fruit set should be avoided the first summer after the plants are set; this is done by removal of flower stems. Fruit production can last 2 to 6 years. Continuous fruit production is usually ensured by putting in new plantings after harvesting over a 2- to 3-year period.

Other approaches are spaced row for cultivars with moderate to weak ability to send out runners, where runners are deliberately placed 4 to 12 inches apart. Specialized growing containers, such as the strawberry pyramid or barrel, are also used in backyards. Runner plants should never be allowed to be closer than 3 to 4 inches; excess growth should be removed. Winter mulch might be needed in very northerly areas, especially if snow cover does not occur.

New plantings of strawberries usually need 35 to 50 pounds of nitrogen per acre. If phosphorus is very low or low (less than 3 ppm), 100 pounds of P_2O_5 should also be added. If potassium is below 50 ppm, 80 pounds of K_2O per acre is added. The soil pH should be 5.0 to 6.5. A second slightly lower application of nitrogen later in the growing cycle after harvest is needed to help aid flower bud formation for the following year. With established plants, additional annual applications of phosphorus and potassium are usually added based on soil tests. Annual applications of nitrogen are also applied with small amounts prior to flowering (if needed) and most or all after harvest. Water needs are high (especially during flowering, fruit set, and fruit growth) and irrigation is usually required. Overhead irrigation is helpful for reducing transplanting shock and for frost protection.

Troublesome insects (and other pests) include aphids (several sp.), cyclamen mite (*Steneotarsonemus pallidus*), nematodes, spider mite, spittle bug (*Philaenus spumarius*), strawberry clipper weevil (*Anthonomus signatus*), strawberry crown borer (*Tyloderma fragariae*), strawberry crown moth (*Synanthedon bibionipennis*), strawberry leafroller (*Ancylis comptana*), strawberry root weevil (*Otiorhyncus ovatus*), strawberry rootworm (*Paria fragariae*), tarnished plant bug (*Lygus lineolaris*), and whitefly (*Trialeurodes* sp.). Aster yellows (*Phyoplasma* sp.) can be serious. Other diseases include gray mold (*Botrytis* sp.), red stele (*Phytophthora fragariae*), leaf spot (several sp.), leaf blight (*Dendrophoma obscurans*), and verticillium wilt. Birds can do considerable damage to strawberries.

Strawberries are harvested by hand. The calyx is left on for fresh market and removed for processing markets. Fresh market strawberries are usually picked when red everywhere except for a white tip. Processing or homegrown strawberries can be picked when completely red, but still firm. Strawberries need to be transported or processed quickly due to their perishable nature. At most strawberries can be held for 5 to 10 days at 0°C (32°F) and 90% to 95% relative humidity. Yields averaged 19.3 tons per acre in 2000, or about 4 quarts per plant.

Walnuts

Walnuts (Figure 18–38) are species of *Juglans*. The Persian (or English) walnut is the only species (*J. regia*) of interest to both commercial and amateur horticulturists. It has both softshell and hardshell cultivars; the former are favored. The following species are mainly of interest to backyard nut growers: black walnut (*J. nigra*), butternut (*J. cinerea*), and Japanese walnut (*J. ailanthifolia*). Cultivars of the English walnut grown on the West Coast include 'Chambers,' 'Concord,' 'Eureka,' 'Franquette,' 'Gustine,' 'Hartley,' 'Lompac,' 'Mayette,' 'Payne,' 'Pioneer,' 'Placentia,' 'Serr,' 'Spurgeon,' 'Tehama,' and 'Vina.' 'Eureka,' 'Franquette,' 'Hartley,' and 'Payne' are the leading commercial cultivars. Those grown in the East are 'Broadview,' 'Colby,' 'Greenhaven,' 'Gratiot,' 'Hansen,' 'Jacobs,' 'Lake,' 'McDermid,'

FIGURE 18–38
Walnuts date back to about 7000 B.C. California produces 98% of the total U.S. commercial crop and accounts for 2/3 of the world's walnut trade. The Franciscan Fathers are thought to have brought walnuts to California from Spain or Mexico. (USDA photograph.)

'McKinster,' 'Metcalfe' (very cold hardy), 'Schafer,' and 'Somers.' Some cultivars of the black walnut are 'Brown Nugget,' 'Elmer Meyers,' 'Emma-K,' 'Football,' 'Huber,' 'Mintle,' 'Michigan,' 'Ohio,' 'Patterson,' 'Rupert,' 'Snyder,' 'Sparks 147,' 'Sparrow,' and 'Thomas.' Butternut cultivars include 'Ayers,' 'Craxeasy,' 'Johnson,' 'Kinneyglen,' 'Love,' 'Thill,' and 'Van Sykcle.' Japanese walnut cultivars include 'Bates,' 'Cardinell,' 'Caruthers,' 'Evers,' 'English,' and 'Wright.' Hybrid cultivars between the butternut and Japanese walnut are 'Corsan,' 'Crietz,' 'Dunoka,' 'Fioka,' and 'Helmick.' These lists are by no means complete. Local authorities should be checked for cultivars best suited to your area.

Persian walnuts are hardy into zone 6 if the proper cultivars are planted (sometimes hardier cultivars are termed Carpathian walnuts). Commercial production is primarily limited to California. Temperatures above 37.7°C (100°F) with low humidity can be harmful; so can late spring frosts. Winter chilling is required to break dormancy. A growing season of at least 150 days is required for most cultivars, with some requiring as much as 260 days. Japanese and black walnuts are hardy through zone 5. The butternut is the hardiest, as it is hardy into zone 4.

Persian walnuts in the West are usually propagated on seedling rootstocks by the whip graft or patch bud. The northern Californian black walnut (*J. hindsii*) is usually the rootstock. 'Paradox,' a hybrid between the Persian and black walnut, is also used. Hardy Persian or Carpathian walnuts are propagated onto black walnut (*J. nigra*) rootstocks by patch or chip budding and by bark grafting. Japanese, black, and butternut walnuts can be propagated by bud grafts. The Japanese walnut cultivars can be grafted on black or butternut seedling rootstocks, butternut cultivars on black walnut, and black walnut cultivars on black walnut seedling rootstocks.

Cultivars of the Persian, black, Japanese, and butternut walnuts are set out as 1-year-old grafts and placed 30 by 30 feet (48 trees/acre). Higher density plantings are set at 24 by 24 feet (76 trees/acre) and hedgerows at 11 by 22 feet (180 trees/acre). Closer distances require the use of specialized pruning systems. Persian walnuts are planted in late winter in the West. In other areas, early spring and late winter are suitable times for any cultivars. Bearing may start in the fifth or sixth year. Mature heights are 70 feet for the Persian walnut, 150 feet for the black walnut, 60 feet for the Japanese walnut, and 90 feet for the butternut. Trees are monoecious and self-fruitful. However, mixed cultivars are often planted because increased yields can result from cross-pollination. The vector is wind.

Persian walnuts are trained to the modified central-leader system. Subsequent pruning consists of thinning top growth to admit light and to stimulate the production of fruiting wood in the central portion. Cultivars of the other species are trained to a single trunk with the lower branches removed at 5 to 7 feet.

Foliar tissue analyses supplemented with soil tests are critical for achieving sustainable fertilization with nut trees. Nitrogen needs vary from 60 to 200 pounds per acre, depending on soil levels, tree age, and cover crop considerations. Fertilizer is usually applied in two to four split applications, starting in early spring. Fertilizer should be watered in, unless it is possible to incorporate it into the soil. Fertigation is an especially good system for nut trees. A backyard tree in its first year can use 1 pound of 10-10-10 and 2 pounds in its second year. For each subsequent year of age, add an additional 2 pounds up to a maximum of 50 pounds per tree.

Well-managed orchards of Persian walnuts may yield 2 to 3 tons of unshelled walnuts per acre per year. In 1999 and 2000 the average yields were 1.48 and 1.24 tons/acre, respectively in California. Homeowners might expect 20 to 35 pounds of nuts per tree per year for the various walnuts discussed here. Walnuts are removed from trees with mechanical shakers, then windrowed, then a mechanical harvester picks them up for cleaning and hulling. Finally walnuts are air dried to 8% moisture for maximal storage quality. Storage is best at 0° to 3.3°C (32° to 38°F) and 55% to 65% relative humidity. Walnuts are good sources of protein, unsaturated fat, several minerals, and vitamins and fiber.

Persian walnuts are troubled by the following insects: codling moth (*Cydia pomonella*), several mites, navel orange worm (*Amyelois transitella*), scales, walnut aphid (*Chromaphis juglandicola*), and walnut husk fly (*Rhagoletis completa*). Diseases include deep bark canker (*Erwinia rubrifaciens*), canker (*Diplodia juglandis*), crown rot (*Phytophthora* spp.), heart rot (*Fomes* sp.), leaf spot (several spp.), root rot (*Armillaria mellea*), and walnut blight (*Xanthomonas juglandis*). Black line, once thought to be delayed grafting incompatibility, is now known to be cherry leafroll virus. For the black walnut, insect pests include black walnut caterpillar (*Datana integerrima*), fall webworm (*Hyphantria cunea*), aphid, and lacebug (*Corythucha juglandis*); anthracnose leaf spot (*Gnomonia leptostyla*) can also be a problem. Butternuts are troubled by canker (*Diplodia juglandis*) and by the butternut curculio (*Conotrachelus juglandis*). All walnuts are susceptible to varying degrees to the witches'-broom or bunch disease, which is suspected to be an insect-transmitted virus.

SUMMARY

Pomology deals with the production, storage, processing and marketing of fruit. The acreage devoted to the organic production of fruits and nuts in the United States totaled 49,000 acres in 1997. About 33% of tree fruit and nut crop producers thought their practices were sustainable.

Backyard fruit growing ranks behind ornamentals and vegetables. Limiting factors are space along with the required labor. The time available to homeowners is another factor. The federal government currently tracks commercial production of over 30 fruits and 6 nuts. Tracking includes consumer trends, acreage, production, and value on the individual crop and state levels as well as the U.S. and export/import levels.

Fruits are a healthy part of the American diet, supplying vitamins A and C plus fiber and carbohydrates. Fruit fiber is generally more soluble than vegetable fiber, so it is a complementary fiber source. Both types of fiber are important in cancer prevention. In 2000, we consumed 102.9 pounds of fresh fruit/person.

Considerable amounts of fresh/processed fruits and nuts are exported. Large amounts are imported to meet out of season needs and to provide tropical fruits and nuts generally not United States grown. Fresh fruit imports reached 6.9 million tons in 2000.

Fruits are big business. During 2000 about 4.1 million bearing acres existed in fruit and nut production. Citrus accounted for 1.1 million acres, major deciduous fruits for 1.9, miscellaneous noncitrus fruits for 0.3 and tree nuts 0.8. Fresh fruit production was 37.2 million tons in 2000. Citrus accounted for 17.3 million tons, noncitrus fruits for 18.8, and tree nuts 1.1. The production of peanuts exceeded all combined tree nuts by a factor of

2.87. Fresh fruit value in 2000 totaled $10.4 billion. The value for combined tree nut crops was $1.5 billion.

Operations involved in fruit-associated production, marketing, and services need individuals with fruit expertise. Production nurseries propagating fruit and nut trees need breeders and propagators. Fruit growing can be a self-employment business or one can manage tree or bramble fruit production or grape vineyards for an employer. One can be a consultant or provider of fruit IPM programs. Wineries and vineyards have various positions for grape specialists. Fruit processors and fruit cooperatives need technical advising staff skilled in fruit knowledge. Marketing careers in fruit wholesaling, distributing, promotion, and retailing are available. Careers in education, extension, and research at land-grant universities and the USDA are another opportunity for pomologists. Suppliers of agricultural equipment, chemicals, and financial/technical services can also utilize individuals with degrees in fruit sciences. Inspectors are needed for examining and testing imported fruits. Opportunities in the biotechnology industry for those with both fruit knowledge and molecular biology expertise are increasing.

Basic practices in pomology include the propagation and improvement of cultivars, site selection, planting and spacing, training and pruning, irrigation, fertilization, weed control, soil management, pollination, thinning, insect and disease control, harvesting, postharvest physiology, and preservation.

Choices of commercial fruit or nut crops are heavily determined by environmental constraints and market competition (high yields and minimal cultural problems are essential).

Chilling requirements dictate that apples grow better in the Northwest and Northeast over Florida. Needs for long, warm growing seasons and mild winters determine that citrus crops are suited to California and Florida as opposed to New England. Tolerance for very little frost restricts mangos to Florida and pineapples to Hawaii. The need for no late spring frosts with apricots because of early flowering explains the Californian concentration of the apricot industry.

Similarly cultural problems play a role. Organic apples are produced in the West instead of the East, even though good apples can be produced in either region. The primary factors are serious apple pests and higher fungal diseases in the East; the latter results from more humid conditions. Apple growers in the East can and do produce apples sustainably by the use of IPM methods. Organic peaches are more apt to be produced in the West than East, primarily because of the plum curculio. Bramble crops are grown organically and sustainably in both the East and West because of fewer pests.

Cultivars are selected for high productivity and high-quality fruit. Other considerations include genetic pest and disease resistance, a factor in organic production. Disease resistance is common; insect resistance is more likely for indirect pests. Resistance to pests that eat fruits is minimal. Multiple disease resistance is uncommon, so avoidance of all control measures is unlikely. Cultivar choices should also be made based on regional considerations.

Most fruits exist as a clone and are asexually propagated. Grafting and budding are used extensively to improve vigor and for size reduction. Rootstocks also offer resistance to certain soil pests and diseases. Topworking is also practiced with fruit trees.

Site selection is important for success. Spring frosts near blooming are a serious hazard in fruit-growing. The role of microclimate in the alleviation of frost hazard can make a site desirable. Gentle southern slopes or areas near large water bodies are often good fruit growing locations. Soil condition is important; roots of fruit trees must extend 3 feet or more in depth for high productivity. Any soil condition that interferes with root development such as hardpan, wetness or high salinity, reduces yields. Sites with heavy perennial weed growth are less desirable to organic operations, because elimination of these weeds usually requires herbicides.

Site preparation sets the stage for successful development and fruiting of long-lived fruiting crops. Soil testing and pH adjustments should be done prior to planting. Improvements in soil fertility, organic matter, and weed suppression can be accomplished with cover crops and green manure. Existing vegetation or sod is plowed under and then a cover crop is sown, then tilled under. In some instances a number of cover crops are rotated through the area and plowed under prior to planting. Alternatively, soil solarization can be used for weed suppression. Cost is high, so this method is limited.

Smaller fruits such as strawberries are managed in beds with rows; larger tree fruits are managed in orchards. Planting and spacing are determined by two, sometimes conflicting, factors: (1) size, growth vigor, and light requirements and (2) management practices. The aim is for highest productivity with least labor. Trends are toward dwarf trees and management practices such as high-density plantings that are more conducive to mechanized operations. Labor is the largest cost factor. Cultivars requiring cross-pollination pose special needs in regards to cultivar mix selection, planting, and spacing.

Many fruit plants need to be pruned for form when young. This type of pruning is called training. Trained fruit plants will require maintenance pruning to maintain fruitfulness and ease of other management practices such as pest control or harvesting. Dwarf and other fruit trees are being increasingly trained in a hedgerow system, essentially a continuous tree wall. This practice provides maximal trees per acre and facilitates the use of mechanical equipment for pruning, spraying of growth regulators and pesticides, and harvesting of fruit. This trend is a consequence of the drive to maximize production and minimize labor.

Weeds in perennial fruit plantings are managed in a number of ways. One is clean cultivation. While effective, disadvantages include energy usage, increased likelihood of soil erosion, and the loss of organic matter, which causes detrimental changes in soil structure. Winter cover crops could be grown and incorporated into the soil to maintain organic matter and partially offset erosion. Another choice is herbicides. The downside here is costly herbicides and possible groundwater pollution. Both approaches minimize nonfruit plant competition for nutrients, water, and light.

More sustainable options include cover crops and mulches. Permanent cover crops such as sod or legumes prevent soil erosion, but can compete with fruit crops for water and nutrients. Some organic matter and nutrients are added to the soil by natural decomposition of the sod. Soil structure is improved and moisture fluctuations are less during dry periods. Management systems to minimize this competition are available. Ground covers can be used in aisles and clean cultivation under the plantings. Ground covers such as cool season grasses that go dormant during hot weather when fruits are forming are another option. Another possibility is ground cover in the aisles and mulch under the plantings. Another option is legumes instead of grass. The legumes supply nitrogen for the fruit plantings. The clippings from the mowing of the aisle legumes can be used to provide nitrogen-rich mulch for under the plantings, too. Legumes (and grasses) do compete for water, but both also increase water permeability of the soil.

Mixtures of ground covers offer both advantages and disadvantages. Mixtures of cover crops offer a suitable environment for diversity of beneficial insects, leading to fewer orchard pests. Mixtures need to be carefully chosen such that no species support local orchard pests. Some plants that support beneficial insects are buckwheat, dwarf sorghum, members of the sunflower family, and members of the Mustard family. Another possibility is to manage adjacent vegetation to enhance beneficial insects.

Mulch is used around trees or row plantings such as with brambles or grapes. The aisle space in the orchard, plantation, or vineyard is either in sod or cover cropped. With some strawberries, the entire area is mulched. The mulch is organic, except for strawberries (black plastic). Organic mulch should be kept 8 to 12 inches from trunks to reduce animal damage and diseases. Organic mulches can be costly if not nearby or difficult to move around. A possible solution is to mow the aisle ground cover with a sickle bar mower and to use the clippings as mulch. Mulch materials can also be grown and harvested from nearby fields or possibly purchased cheaply from local sources.

Mulches offer a number of compensatory benefits beyond weed suppression: increased water infiltration, slower evaporative loss of soil water, enhanced soil aggregation, and soil temperature moderation (less plant stress). Mulches rank first as an orchard floor management system in terms of moisture availability. Another plus is that the mulch releases small amounts of N, P, and K as it decomposes. Some organic mulch also produces allelopathic effects against herbaceous weeds, but not against trees.

Newer nonorganic mulches, woven plastic or geotextiles, exist. These mulches are superior to black plastic in that they are permeable and do not encourage matted surface root growth. Trees with matted surface root growth show increased drought susceptibility, increased winter injury, and poorer anchorage. Geotextiles are expensive initially and do not provide organic matter or nutrients to the soil, but their life span is 10 or more years and the labor of continuous remulching is ended. Savings in weed control over the geotextile life likely compensate for initial cost.

If areas around fruits are not mulched, weed control is required. For backyard plantings and perhaps small commercial, part-time operations, hand cultivation with a hoe could be sufficient. Larger operations can use mechanical hoes or rotary tillers that can be tractor mounted and set to cultivate shallowly right up to plantings. Other attachments such as scrapers can be used to scrape the soil near plants or to scrape soil further away and mound it around plants. Other options include possibly organically acceptable herbicides such as various soap-based, fatty acid-based, or potassium salt-based herbicides. Another alternative is weeder geese in orchards, strawberries, brambles, and blueberries. Another possibility is flame weeding.

Fruits are the easiest crop to fertilize in a sustainable manner. Fruits, unlike other crops, consist mostly of water. Their harvest does not remove nearly as many soil nutrients as vegetable crops. Fertility needs are more easily met with nutrients provided from decomposing mulch and efficient cover crop management supplemented with various rock dusts and good pH control. If legumes are part of the cover crop management, supplementary nitrogen needs are minimal.

The fertility, pH, and organic matter content for new plantings in orchards, vineyards, and plantations should be as optimal as possible. This window presents the only time that soil incorporation of compost, fertilizers, limestone or sulfur, cover crops, green manures, and legumes is possible. Once planted to trees, vines and bushes plus aisle cover crop systems, soil incorporation becomes unrealistic and surface treatments are the only recourse. Decaying mulches and soil root residues become the main venue for replenishing organic matter. It is fortunate that fruits don't have high needs for continual nutrient inputs and that cover crop management systems are capable of maintaining organic matter and preventing soil erosion.

Nitrogen can be supplied with manure, compost, and various meals such as blood, cottonseed, or feather meal. Such products should be applied in the early spring to compensate for slow release through decomposition. Ideally, these organic products should be incorporated into the soil to reduce nitrogen loss. Incorporation is not always possible, because root damage can result with surface-rooted fruits. Alternatively, soluble organic fertilizers such as fish emulsion can be used as a foliar spray. Cost is high with these products. Nonorganic, sustainable fruit growers can use conventional fertilizers.

Regardless of product use, sustainable applications must be based on data from reliable soil tests and foliar tissue analyses. Soil tests are done best by laboratories familiar with the needs of sustainable farmers. Foliar tissue analyses also supply useful data. Visual indicators provide clues to problems. Prompt attention is warranted when such signs are noted. For example, terminal shoot growth can be monitored to give input about nitrogen adequacy. Older leaves show nitrogen deficiency symptoms first as a uniform pale green to yellow color. Other clues are thin twig growth, early autumn leaf drop, light fruit set, heavy June fruit drop, smaller fruits, and earlier maturity.

Fruits are somewhat more drought tolerant than other crops. A constant, moderate soil moisture supply from flowering through harvest is generally needed to ensure good fruit set and good-sized quality fruits. While the most sustainable option is trickle or drip irrigation, other forms with up-to-date technology can suffice. Drip systems are best for high-value crops such as many fruits. Drip irrigation is also adaptable to supplying nutrients to fruits through fertigation. This form of fertilization is highly sustainable in that less nitrogen is needed and less leaching occurs. The one caveat is that system uniformity of at least 80% is required to ensure effective nutrient distribution. Fertilizers are injected based on creating pressure differentials (venturi), pressure (positive displacement pump), and pressure gradients (batch tank injector).

Pollination is enhanced by the introduction of beehives into the fruit plantings during flowering. Insect control must not be detrimental to bees at this time. Hand pollination is not practical because of labor costs. Removal of flowers or young fruit (thinning) is a common practice of fruit growers. Remaining fruit develop larger or more rapidly, and the following year's flowering and fruit set are not affected adversely. Thinning can be done by hand, but mechanical or chemical means are more economical. Mechanical hydraulic or pneumatic limb shakers or growth-regulating substances can be used. Organic fruit growers use nonchemical means. Another possibility is to prune out some of the fruiting wood, which is considerably easier than hand thinning.

Pest control is probably the most expensive and time-consuming aspect of fruit growing. IPM, the judicious use of biological and chemical controls, is increasingly utilized. Costs and environmental pressures can be reduced with this approach. One major option, rotations, is generally not available with most fruit and nut crops. Other efforts must be used to prevent the buildup of pests over time. The stability of the fruit growing area offers a good opportunity to establish and maintain populations of beneficial insects and other organisms. One obvious solution is to maintain good plant vigor and health to minimize damage from indirect pests that feed on nonfruit parts of the crop, as well as pathogens.

Both organic and other sustainable fruit growers need to take full advantage of biological controls. Release of beneficial insects is one strategy, but the best long run economic strategy is to manage nearby natural vegetation and on-site cover crops to provide a favorable habitat for beneficial organisms. These methods will likely provide adequate control for indirect pests, but perhaps not complete control for direct pests that feed on fruit, especially if very little cosmetic or other damage is tolerable. Additional help using organically acceptable natural or biological pesticides will probably make up the difference for organic growers. Other sustainable growers might opt for "green" or minimally disruptive pesticides. Microbial, botanical, soaps, oils, fungal-derived compounds, green pesticides, and mating disruptors are available. A new product, kaolin clay, shows promise for some fruit crops.

Diseases can be controlled proactively. Keeping fruit crops properly pruned such that air circulates and sunlight penetrates into the interior canopy of the plant helps dry and reduce moisture conditions favored by bacterial and fungal diseases. Sanitation practices that remove fallen leaves and fruits reduce the overwintering of diseases. Good soil drainage is essential to reduce the potential of various soil-borne pathogens to cause root rot. The use of genetically resistant cultivars is helpful to limit problems. Finally, good soil health maintained with significant organic matter produces a soil ecosystem full of beneficial microorganisms that either compete with or are antagonistic toward pathogens. One can also use organic fungicides such as copper- or sulfur-based products, biofungicides (competing or antagonistic microbes), or various fungicides.

A number of mammals can be especially troublesome in fruit plantings. Deer and birds can be major problems and mice, voles, rabbits, and raccoons to a lesser extent. Control strategies include scare devices, repellents, poisons, noise devices, traps, fencing, netting, and tree guards.

For the backyard grower, harvesting usually occurs at peak quality, because the fruit is either going to be used quickly, refrigerated, or processed soon after harvest. For commercial operations, harvest is determined by several factors. For processing fruits or nuts, harvest can be close to peak maturity, because processing happens shortly after and better quality products result from peak maturity as opposed to less ripe stages. For fresh market, harvest is somewhat before peak maturity, as ripening will continue in many cases. Dead ripe fruits are avoided. The determination of the correct picking time can be tricky. With fruits, degree-days or heat units are reliable indicators. Other indicators are color, fragrance, or softening.

The home horticulturist hand harvests fruit and nut crops. Commercial operations hand harvest when mechanized operations are not available, or are bypassed in favor of pick-your-own operations, or the cost and feasibility of the equipment is questionable. Harvesting machinery has been improved even for easily bruised fruit, and plant breeders have aided mechanized harvesting by developing crop varieties with uniform ripening habits, compact growth, less easily bruised fruits, more easily detached fruits, or an angle of hang better suited for mechanized removal. Growth regulators are used as mechanized harvest aids to promote uniform ripening, color, and abscission. The mechanization of harvesting varies as to degree and the number of steps involved. The bulk of tree and bush fruits intended for the fresh market are still hand harvested. Some operations may be handled in a partially mechanized manner, for example, the hand picking of fruit like apples followed by mechanized grading and packing. Crops intended for processing are extensively harvested by mechanical means.

A harvested fruit is still a living organism, and life processes continue to produce changes. These biochemical and physiological changes constitute the science of postharvest physiology. The changes can enhance or detract from the appeal of the crop. The appeal of the crop is determined by three factors: quality, appearance, and condition. Postharvest alterations are numerous and include water loss, the conversions of starch to sugar and sugar to starch, flavor changes, color changes, toughening, vitamin gain and loss, softening, and decay. These processes can improve or detract from the quality, flavor, and texture of the crop. On the whole, postharvest changes produce more detrimental results than beneficial ones.

No one set of conditions will achieve this end, because postharvest alterations are a function of the crop type, and each crop varies as to the optimal air temperature, relative humidity, and concentrations of oxygen and carbon dioxide required for stabilization. Some form of air circulation is usually beneficial in reducing disease problems and removing any released gases that might accelerate harmful physiological and biochemical changes. The level of disease pathogens should also be kept to a minimum.

The most important aspect of postharvest treatments is storage facilities designed for long-term crop storage, so that the crop can be sold at a reasonable price over an extended period of time. The best storage conditions slow the life processes but avoid, tissue death, which produces gross deterioration and drastic differences in the quality, appearance, and condition of the stored crop. Reliable storage consists of cold (refrigerated) storage with precise control of temperature and humidity. An improvement upon this form of storage is the addition of controlled atmosphere by which the concentration of O_2 and CO_2 are kept at an optimal level.

Premarketing operations include washing, waxing, precooling, grading, prepackaging, and shipping. These steps can be done in the field, at specialized facilities, or a combination of both can be used. Each place has advantages and disadvantages. Field operations reduce transport of and damage to the crop and cost less. The downside is that quality control is reduced, weather can cause problems, large machines can disrupt field activities and cause soil compaction, and cooling of small lots is less energy efficient than bulk lot cooling.

Fruits are often washed or brushed to enhance appearance. Disinfectants are often present in the wash water to kill pathogens that affect storage and marketing qualities. Waxing is used to improve appearance and to slow moisture loss.

Precooling is used for rapid removal of heat from recently harvested fruits. Precooling permits the grower to harvest at maximal maturity and yet be assured that the crop will have maximal quality later. Precooling slows the rate of respiration in the fruit, slows wilting and shriveling by decreasing water loss, and inhibits the growth of decay caused by microorganisms. Major precooling methods include hydrocooling, contact icing, vacuum cooling, and air cooling.

Uniformity in quality, size, shape, color, condition, and ripeness is usually preferred for reasons of consumer appeal, freedom from disease, and ease of handling by automated machinery. To achieve this end, fruits are often graded according to standard grades, which form a basis of trade. The requirements of grading vary depending on factors such as whether the crop is destined for the fresh market or for processing.

Many fruits are packaged in retail units at some point prior to reaching the retail store. This type of packaging is called prepackaging. Products are placed in paper bags with handles, plastic mesh bags, or in trays and cartons or covered over by transparent film. Trays and cartons used for the prepackaging unit are usually made of paperboard, Styrofoam, or wood. The master containers for the prepackaged units are usually made of paperboard. Wooden crates and bushel baskets are seen less and less. Master containers are designed for easy stacking, shipping, and minimal bruising.

Fruits are then shipped in a number of ways. Large trucks with refrigerated containers, which may or may not have a controlled atmosphere, are used extensively. Refrigerated railroad cars are also used. Crops are also air freighted by large cargo planes. Cargo boats are another source of transportation. Loading and unloading are often facilitated by the use of wooden pallets and forklifts.

Fruits and nuts destined for processing can be preserved by a number of techniques. These methods include wet pack canning, vacuum canning, freezing, dehydration, plastic bags, baked goods, use in candies, sugar concentrates, juices, oils, and brandies/liquors/wines. All, of course, can be eaten fresh directly or held in storage besides being processed.

Crops grown for processing usually cost less per unit area of land or per ton. This is because, unlike crops grown for the fresh market, appearance is not of major concern. However, size, quality, and uniformity are important. An extended harvest through successive plantings or the use of varieties with different maturation dates is needed to maintain a constant supply. This enables the processing factory to operate at an even flow over a reasonable period of time.

Horticulturists group fruits on the basis of climatic requirements: temperate, subtropical, and tropical fruits. Temperate-zone fruits are deciduous and tropical fruits are evergreen; subtropical fruits can be either. Fruits are good sources of soluble fiber and are a healthy complement to the less soluble fibers of vegetables. Both are required for good health. Fruits also provide vitamins, minerals, and some cancer-fighting chemicals.

Nut trees have much to offer, being high in food value, providing both protein and fat. The tree itself provides shade and is ornamental in the landscape. Nuts can be tropical, subtropical, and temperate.

EXERCISES

Mastery of the following questions shows that you have successfully understood the material in Chapter 18.

1. Why is the contribution of sustainable and organic fruit/nut growing to overall fruit and nut production poorly understood?
2. What factors limit backyard fruit and nut growing?
3. Which fruit and nut crops are tracked by the federal government?
4. Why are fruits a healthy component in diets? What current consumer trends are observed with fruit consumption in the United States?
5. What is the breakdown of acreage devoted to fruit and nut production? What are the leading fruits? Nuts? Which state leads in the U.S. production of organic fruits and nuts? What are the leading organic fruits and nuts?
6. In terms of total tonnage, what is the breakdown for fruits and nuts? What are the leading fruits and nuts?
7. In terms of total value, what is the breakdown between fruits and nuts? What are the leading fruits and nuts?
8. Which career in pomology would you pick? Why?
9. What factors influence the choice of crops and cultivars in commercial fruit-growing operations? Why?
10. What factors influence the choice of commercial fruit-growing areas?
11. Why is site preparation important for growing fruit crops? What can be done to improve the site?
12. What two factors determine the planting systems and spacing of fruits? Why have hedgerow systems become increasingly popular?
13. What are the disadvantages of clean cultivation and herbicides for orchard weed control?
14. What sustainable options are available for weed control in orchards? Discuss them.
15. Why are mulches useful in orchards and other fruit plantings? If mulches aren't used, how does one control weeds in proximity to fruit plants?
16. Why are fertility needs more easily met for fruits than for vegetables? Why is it so important to optimize soil fertility and organic matter in areas used for new fruit plantings?
17. What are the best ways of determining soil nutrient needs for fruits and nuts? What are some visual clues that indicate nutrient problems?

18. What period of fruit growing is most critical in terms of water needs? Why is drip irrigation useful for fruit growing?
19. How does one ensure the best pollination with fruits? How does this practice relate to insect control?
20. Why is fruit thinning or flower removal practiced? How are they done?
21. What major option for sustainable pest control is not available to fruit growers? Why? Name some of the sustainable pest and disease control strategies.
22. What is the best time(s) for harvesting fruits? How are fruits harvested? What mechanized help is available?
23. What changes take place after fruits and nuts are harvested? Why is it important to prevent most of these changes? How are the changes prevented?
24. What are premarketing operations? Describe each one and why is it important.
25. Pick your five favorite fruits and discuss each one in terms of cultural requirements and culinary value. Do the same for three fruits that you have never consumed, don't like, or eat rarely.
26. Pick your two favorite nuts and discuss each one in terms of cultural requirements and culinary value.

INTERNET RESOURCES

Apple tree spacing
>http://tfpg.cas.psu.edu/part1/part13h.htm
>http://www.hrt.msu.edu/department/Perry/Apple_Articles/mispacingfinal1.htm

Avocado handbook
>http://ucce.ucdavis.edu/counties/ceventura/Agriculture265/Avacado_Handbook.htm

Characteristics of apple rootstocks and interstem combinations
>http://www.ars-grin.gov/gen/rootstocks.html
>http://tfpg.cas.psu.edu/part1/part13k.htm

Chestnuts
>http://anrcatalog.ucdavis.edu/pdf/8010.pdf

Citrus fruits production statistics
>http://usda.mannlib.cornell.edu/reports/nassr/fruit/pnf-bb/

Citrus handbook
>http://edis.ifas.ufl.edu/MENU_HS:CITH

Citrus links
>http://www.citrusresearch.com/links.html

Citrus publications
>http://fruitsandnuts.ucdavis.edu/citruspb.html

Citrus tree spacing
>http://edis.ifas.ufl.edu/BODY_CH026

Common fruit insect pests
>http://www.extension.umn.edu/distribution/horticulture/DG0574.html
>http://www.nysipm.cornell.edu/factsheets/treefruit/index.html

Drip irrigation for fruits
>http://www.msue.msu.edu/vanburen/e-852.htm#part4

Fertilizing blueberries, raspberries, and strawberries
>http://info.ag.uidaho.edu/Resources/PDFs/CIS0815.pdf

Fertilizing fruit and nut crops

    http://www.utextension.utk.edu/spfiles/sp3072.pdf

    http://www.msue.msu.edu/vanburen/e-852.htm#part3

Fruit and nut information

    http://fruitsandnuts.ucdavis.edu/crops.html

Fruit and nut production statistics

    http://usda.mannlib.cornell.edu/data-sets/specialty/89022

Fruit and nut yearbook

    http://www.ers.usda.gov/publications/fts

Fruits

    http://www.crfg.org/pubs/frtfacts.html

    http://www.hort.cornell.edu/extension/commercial/fruit/index.html

    http://www.rce.rutgers.edu/pubs/treefruitguide/

    http://www.caf.wvu.edu/kearneysville/wvufarm1.html

    http://www.caf.wvu.edu/kearneysville/fruitloop.html

    http://www.nafex.org/

    http://fruitsandnuts.ucdavis.edu

    http://www.umass.edu/fruitadvisor/

    http://www.orst.edu/dept/infonet/

    http://aggie-horticulture.tamu.edu/fruit/fruit.html

    http:/tfpg.cas.psu.edu

    http://www.tfrec.wsu.edu

    http://www.ag.ohio-state.edu/~ohioline/lines/fruit.html

Irrigation basics

    http://eesc.orst.edu/agcomwebfile/edmat/ec1424.pdf

    http://www.ncw.wsu.edu/irighow.htm

Grape trellis and training systems

    http://hgic.clemson.edu/factsheets/HGIC1402.htm

    http://www.nysaes.cornell.edu/hort/faculty/pool/train/trainandstocks.html

    http://cetulare.ucdavis.edu/pubgrape/wg197.htm

    http://www.winemakersemporium.com/Trellis.htm

    http://muextension.missouri.edu/xplor/agguides/hort/g06090.htm

Muscadine production guide

    http://www.smallfruits.org/MuscadineGro/toc.htm

Noncitrus fruits and nuts production statistics

    http://usda.mannlib.cornell.edu/reports/nassr/fruit/pnf-bb/

    http://www.virtualorchard.net

Nut resources

    http://www.icserv.com/nnga/

Organic and low spray apple production

    http://www.attra.org/attra-pub/apple.html

Organic blueberry production

    http://www.attra.org/attra-pub/blueberry.html

Organic culture of bramble fruits

    http://www.attra.org/attra-pub/bramble.html

Organic farming basics

http://www.attra.org/attra-pub/organiccrop.html

Organic grape production

http://www.attra.org/attra-pub/grape.html

http://www.nysaes.cornell.edu/hort/faculty/pool/organicvitwkshp/tabofcontents.html

Organic low-spray peach production

http://www.attra.org/attra-pub/peach.html

Overview of organic fruit production

http://www.attra.org/attra-pub/fruitover.html

http://sustainable.tamu.edu/publications/organicfruit.html

Peanuts

http://lubbock.tamu.edu/peanut/docs/PeanutProdGuide2001.pdf

Pecans

http://aggie-horticulture.tamu.edu/extension/fruit/pecanorchard/pecanorchard.html

Strawberries: organic and IPM options

http://www.attra.org/attra-pub/strawberry.html

RELEVANT REFERENCES

Bowling, Barbara L. (2000). *The Berry Grower's Companion*. Timber Press, Portland, OR. 308 pp.

Caruso, Frank L., and Donald C. Ramsdell (1995). *Compendium of Blueberry and Cranberry Diseases*. APS Press, Saint Paul, MN. 87 pp.

Childers, Norman, F. J. R. Moore, and G. S. Sibbett (1995). *Modern Fruit Science*. Horticultural Publications, Gainesville, FL.

Davidson, Ralph, and William F. Lyon (1987). *Insect Pests of Farm, Garden and Orchard* (8th ed.). John Wiley & Sons, New York. 656 pp.

Ellis, M. A., R. H. Converse, R. N. Williams, and B. Williamson (1991). *Compendium of Blackberry and Raspberry Diseases and Insects*. APS Press, Saint Paul, MN. 122 pp.

Finch, Clarence, and Curtis Sharp (1983). *Cover Crops in California Orchards and Vineyards*. Soil Conservation Service, Davis, CA. 26 pp.

Flint, Mary Loise (1990). *Pests of the Garden and Small Farm* (Division of Agriculture and Natural Resources Publication 3332). University of California, Oakland. 276 pp.

Galletta, Gene, and David Himelrick (eds.) (1990). *Small Fruit Crop Management*. Prentice Hall, Upper Saddle River, NJ. 602 pp.

Heaton, Donald D. (1997). *A Produce Reference Guide to Fruits and Vegetables from Around the World: Nature's Harvest*. Food Products Press, Binghamton, NY. 244 pp.

Jackson, L. K., and F. S. Davies (1999). *Citrus Growing in Florida* (4th ed.). Florida Science Source, Longboat Key, FL. 313 pp.

Jones, A. L., and H. S. Aldwinckle (1990). *Compendium of Apple and Pear Diseases*. APS Press, Saint Paul, MN. 125 pp.

Klein, Maggie Blyth, Claude Sweet, and Richard H. Bond (eds.) (1985). *All About Citrus and Subtropical Fruits*. Meredith Books, Des Moines, IA. 96 pp.

Kokalis-Burelle, N., D. M. Porter, R. Rodríguez-Kábana, D. H. Smith, and P. Subrahmanyam (1997). *Compendium of Peanut Diseases* (2nd ed.). APS Press, Saint Paul, MN. 94 pp.

Lee, Stella (1995). *The Backyard Berry Book: A Hands-on Guide to Growing Berries, Brambles, and Vine Fruit in the Home Garden*. OttoGraphics, Maple City, MI. 284 pp.

Maas, J. L. (1998). *Compendium of Strawberry Diseases* (2nd ed.). APS Press, Saint Paul, MN. 128 pp.

Ogawa, J. M., E. I. Zehr, G. W. Bird, D. F. Ritchie, K. Uriu, and J. K. Uyemoto (1995). *Compendium of Stone Fruit Diseases*. APS Press, Saint Paul, MN. 128 pp.

Otto, Stella (1993). *The Backyard Orchardist: A Complete Guide to Growing Fruit Trees in the Home Garden*. OttoGraphics, Maple City, MI. 248 pp.

Pearson, R. C., and A. C. Goheen (1988). *Compendium of Grape Diseases*. APS Press, Saint Paul, MN. 121 pp.

Ploetz, R. C., G. A. Zentmyer, W. T. Nishijima, K. G. Rohrbach, and H. D. Ohr (1994). *Compendium of Tropical Fruit Diseases*. APS Press, Saint Paul, MN. 118 pp.

Reich, Lee (1991). *Uncommon Fruits Worthy of Mention*. Addison-Wesley, Reading, MA. 273 pp.

Ryugo, Kay (1988). *Fruit Culture: Its Science and Art*. John Wiley & Sons, New York. 344 pp.

Salunke, D. K., and S. S. Kadam (eds.) (1998). *Handbook of Fruit Science and Technology: Production, Composition, Storage and Processing*. Marcel Dekker, New York. 632 pp.

Teviotdale, Beth L., Themis J. Michailides, and Jay W. Pscheidt (2001). *Compendium of Nut Diseases in Temperate Zones*. APS Press, Saint Paul, MN. 100 pp.

Timmer, L. W., S. M. Gamsey, and J. H. Graham (2000). *Compendium of Citrus Diseases* (2nd ed.). APS Press, Saint Paul, MN. 128 pp.

University of California (1999). *Integrated Pest Management for Stone Fruits* (Publication 3389). DANR Communication Services, University of California, Oakland. 264 pp.

Westwood, Melvin Neil (1993). *Temperate-Zone Pomology: Physiology and Culture* (3rd ed.). Timber Press, Portland, OR. 535 pp.

Index